Syntheses in Limnogeology

Series Editors
Antje Schwalb, Institute of Geosystem and Bioindication
Technische Universität Braunschweig
Braunschweig, Germany

Blas L. Valero-Garcés, Instituto Pirenaico de Ecología
Consejo Superior de Investigaciones Científicas (CSIC)
Zaragoza, Spain

The aim of this book series is to focus on syntheses or summaries of modern and/or ancient lake systems worldwide. Individual books will present as much information as is available for a particular lake basin or system of basins to offer readers one distinct reference as a guide to conduct further work in these areas. The books will synthesize the tectonics, basin evolution, paleohydrology, and paleoclimate of these basins and provide unbiased new interpretations or provide information on both sides of controversial issues. In addition, some books in the series will synthesize special topics in limnogeology, such as historical records of pollution in lake sediments and global paleoclimate signatures from lake sediment records.

Robin W. Renaut • Richard Bernhart Owen

The Kenya Rift Lakes: Modern and Ancient

Limnology and Limnogeology of Tropical Lakes in a Continental Rift

Springer

Robin W. Renaut
Department of Geological Sciences
University of Saskatchewan
Saskatoon, Canada

Richard Bernhart Owen
Department of Geography
Hong Kong Baptist University
Kowloon Tong, Hong Kong

ISSN 2211-2731 ISSN 2211-274X (electronic)
Syntheses in Limnogeology
ISBN 978-3-642-25054-5 ISBN 978-3-642-25055-2 (eBook)
https://doi.org/10.1007/978-3-642-25055-2

This Springer imprint is published by the registered company Springer-Verlag GmbH, DE, part of Springer Nature.
The registered company address is: Heidelberger Platz 3, 14197 Berlin, Germany

Paper in this product is recyclable.

Lake Bogoria

For Duncan

Preface

In July 1975, we began a lifetime of interest in the Kenya Rift lakes when we began as graduate students supervised by the late Professor WW (Bill) Bishop at Queen Mary College, University of London.

A year later, having spent several months preparing photogeological maps, we arrived at Embakasi Airport in Nairobi on an East African Airways flight, excited to begin our field research in the Suguta Valley in the northern Kenya Rift. After confusion and delay on arrival from London, we set off with Martin Pickford, and Jimmy and Gill Young (University of Edinburgh, Geography), for a tented camp on the western shore of Lake Baringo where Bill was arranging our onward journey to Suguta by a lorry with two Land Rovers, fuel, food, water, a cook, field assistants, and a nurse. *En route* to Baringo, then a 5–6-hour journey, one Land Rover slid sluggishly into a ditch during heavy rains and several wheel nuts sheared off. A passing lorry loaded with Tusker beer came to the rescue. A few hours later, after dark, our second Land Rover became stuck on its side (>30°) while crossing the gravelly Ndau River near Lake Baringo during a flash flood in torrential rain. That Land Rover became partly filled with water. An empty teapot floated symbolically below the dashboard. To the rescue came that same heroic Tusker lorry and driver, who by then had caught up with us. We eventually arrived at Kampi-Ya-Samaki (midwest shore of Lake Baringo) at about 9.00 pm, wet, tired, and bitten by a myriad of mosquitos and other noisy bugs. No mobile phones back then (1976). Grunting hippos came ashore to feed on grass around our tents during the night but delicately missed the guy ropes. But seeing and hearing Lake Baringo the next morning for the first time, while eating posho and fruit for breakfast, even with yucky, soupy, sugary milky tea, was fantastic. Then, two days later, the Kenya Government revoked our research permit to work at Suguta because the Kenya army was training there, linked to security issues along the Ugandan border. Too late: by then we were already hooked…

So began our interest and geological research in the Kenya Rift and its fantastic lakes. We finally reached Suguta in early 2018. By then, our (1975–1976) photogeological maps were a little out-of-date.

This book is our attempt to summarise the geology, general limnology, environmental setting, and recent history of the modern lakes in the Kenya Rift Valley and Nyanza Rift, and to examine the ancient lake deposits preserved in the Neogene stratigraphic record. We are both geologists by training, so we do not pretend to cover the modern biology of the Kenya Rift lakes from the same perspectives and depth as a biologist or ecologist. Nonetheless, we have published our results in journals that span the interface between geology and biology, reflecting our broad (palaeo)ecological interests, and we discuss and reference much of the relevant biological literature.

We briefly include Lake Natron in northern Tanzania because its history has been intimately associated with neighbouring Lake Magadi, and we mention Chew Bahir in Ethiopia because that lake at times overflowed into Lake Turkana. We have visited most of the Kenyan lakes that we discuss, including many of the smaller lakes and wetlands, and have examined many ancient lacustrine successions that have been described in outcrop or core. We include, but do not always highlight, many unpublished results. We have tried to be inclusive but have been unable to reference all relevant papers on all topics because of access, both online and physical, confidentiality (unpublished government and company reports), and space. For this, we apologise. The relevant literature is growing, now almost exponentially, but this is good. Many new journals are East African with Open Access.

We hope that this volume will provide a useful introduction and general background for those beginning research on the modern Kenya Rift lakes and their precursors, especially graduate students, and for those already familiar with some of the lakes but expanding their areas or breadth of interest. Most of all, we hope that readers will be as fascinated by the Kenya Rift lakes as we have been for … now approaching … almost 50 years. They are special.

San Rafael, Segovia, Spain
Kowloon Tong, Hong Kong
May 2023

Robin W. Renaut
Richard Bernhart Owen

Acknowledgements

Many people have helped us in many different ways over many years.

We owe so much to two Kenyans in particular and their support teams. For more than 30 years, William Kimosop, Senior Warden of the Northern Kenya Rift, has given us invaluable help at Lake Bogoria National Reserve and at Lake Baringo. His knowledge of the region, its history, its wildlife, the lakes, and his enthusiastic support for what we were trying to achieve made our efforts so much easier. Similarly, John Ego, Exploration Manager at the National Oil Corporation of Kenya (and one of Robin's former graduate students), has helped both of us with logistics and field research at Baringo and Magadi. We are extremely grateful to both of them.

The late Jean-Jacques Tiercelin was our close friend and colleague from the late 1970s. Over four decades, we worked together to try to make sense of the sedimentology and history of Lakes Bogoria and Baringo, and later (2013) at Magadi. We miss him deeply as do many others.

We are also deeply indebted to Rick Potts and his Smithsonian Institution field-studies support team at Olorgesailie, and especially Kay Behrensmeyer and Al Deino for their knowledgeable guidance on sedimentation and chronological issues. They provided both comfort and excellent meals in the southern Kenya Rift, as well as wide-ranging and stimulating conversations over many years.

Gail Ashley (Rutgers University) enthusiastically showed us the importance of spring-fed wetlands in Kenya and at Olduvai from many perspectives, including for our hominin ancestors and other mammals. We thank her for putting up with working with the "terrible two" for such a long time (although we think she actually enjoyed most of it despite our reticence to break for lunch).

Rob Crossley, when a PhD student, was one of the pioneering geological mappers in the southern Kenya Rift and worked closely with Bernie when he taught at the University of Malawi. His knowledge of the southern Kenya Rift, and of the Nguruman area in particular, has assisted us in our research efforts in the Magadi region. Martin Pickford similarly played an important role in the early part of our work, both in Kenya and in London. In particular, he guided us on our initial explorations of the Baringo District and the diverse sedimentary sequences exposed in the Tugen Hills.

More recently, Chris Campisano, Andy Cohen, Tim Lowenstein, Annett Junginger, Kennie Leet, Nathan Rabideaux, Emma McNulty, Anthony Mbuthia, and other members of the Hominin Sites and Paleolakes Drilling Project (HSPDP) enabled us to do things we could have only dreamt about 40 years ago. Before HSPDP (2014), our efforts were intense but usually at a local scale. We now acknowledge that such large projects can make an enormous difference in understanding the sedimentology, tectonics, and palaeoclimate of rift lakes. We especially thank Andy for including us and leading HSPDP with such good humour and skill.

Many others helped us in the field, in the laboratory, and with information. We cannot name everyone, but they include Luis Buatois, Joel Casanova, Dan Deocampo, Gijs de Cort, René Dommain, Steve Driese, Michele Goman, Paul Griffiths, Jack Harris, Richard Hay, Andrew Hill, Vicky Hover, Glynn Isaac, Brian Jones, Michael Kimeli, Lothar Krienitz, Claudine Le Mut-Tiercelin, Caroline Le Turdu, Gabriela Mángano, Dan Olago, Steven Rucina, Michael Schagerl, Michael Talbot, Corinne Tarits, Annie Vincens, and Jimmy and Gill Young.

We also thank our former graduate students, who did their thesis research in Kenya, especially Jenni Scott and Veronica Muiruri, from whom we learned so much while and since they did their PhDs, and Ray Lee, Ginette Felske and Britni Brenna.

Peter Achammer demonstrated his helicopter skills at Suguta. We also acknowledge the late Bill French, who helped us to complete our research after Bill Bishop died suddenly.

Bill Bishop made this all possible. He accepted us as graduate students and set our careers (and subsequent lives) in motion. Although Bill died in February 1977, he inspired us (and others) to continue research in Kenya for four more decades, and answer some of the questions he originally posed in 1975–1976. We think that we have answered some … but not all! We also thank Sheila Bishop for her long support.

The funding agencies that helped us at various times include the Natural Environment Research Council and the Royal Society (UK), the Natural Sciences and Engineering Research Council (NSERC Canada), and the Hong Kong Research Council. The HSPDP drilling in 2014 was funded by the International Continental Scientific Drilling Program (ICDP) and National Science Foundation (NSF, USA) to the HSPDP Team. Local Maasai communities and Tata Chemicals Magadi provided field support at Magadi, and the local communities at Lake Bogoria and Lake Baringo gave us support over several decades.

Permission to work in Kenya on various projects was kindly granted by the National Commission for Science, Technology and Innovation (NACOSTI), the Kenya Ministry of Petroleum and Mining, the National Environmental Management Authority of Kenya (NEMA), and their precursors, notably the Office of the President, Government of Kenya.

Lastly, but foremost, we thank our families: Linda, Barbara, Holly, Duncan, Richard, and Christopher for their patience while we were writing this book.

Contents

Part I Background to the Kenya Rift Lakes

1 **Introduction: About This Book** ... 3
1.1 The Kenya Rift Lakes ... 3
References ... 8

2 **Brief History of Geological Research on the Kenya Rift Lakes** ... 11
2.1 Introduction ... 11
2.2 The Early European Explorers ... 12
2.3 The 1920s until the 1940s: Movement on Multiple Fronts ... 15
2.3.1 Expeditions and Limnological Data Collection ... 15
2.3.2 Development and Rejection of Pluvial Theory ... 16
2.4 Development of an Absolute Chronology ... 18
2.5 The 1950s–1970s ... 18
2.5.1 An Era of Geological Mapping ... 18
2.5.2 Palaeoanthropological Research and Lakes ... 21
2.5.3 Other Contributions ... 22
2.6 Applied Research: Geothermal and Petroleum Exploration ... 23
2.7 Palaeoclimate Studies ... 24
2.8 From 2010 Until the 2020s: The Era of Drilling and High-Resolution Studies ... 26
References ... 28

3 **Geology of the Kenya Rift: An Introduction** ... 33
3.1 Tectonic Setting of the East African Rift ... 33
3.2 Major Controls on the East African Rift System ... 36
3.3 Accommodation Zones ... 40
3.4 Kenya Rift Tectonics ... 42
3.5 Kenya Rift Basins ... 45
3.6 Volcanism in the Kenya Rift ... 54
3.7 Geological Evolution of the Kenya Rift Valley ... 57
References ... 63

4 Environmental Background to the Kenya Rift Lakes: An Introduction 77
4.1 Introduction 77
4.2 East African Climate 77
4.3 Hydrology and Hydrogeology 83
4.3.1 Drainage Patterns and Groundwater Flow 83
4.3.2 Types of River System 88
4.4 Weathering and Sediment Production 93
4.4.1 Physical Weathering 94
4.4.2 Chemical Weathering 96
4.4.3 Biological Weathering 98
4.5 Kenya Rift Soils 98
4.6 Aeolian Processes 101
4.7 Kenya Rift Vegetation 103
4.8 Kenya Rift Wildlife 106
4.9 Geothermal Processes and their Impact on Lacustrine Sedimentation 108
4.9.1 Types of Hot Spring in the Kenya Rift 108
4.9.2 Structural Setting of the Hot Springs 112
4.9.3 Hot Spring Water Chemistry 114
4.9.4 Spring Deposits and Hydrothermal Alteration: Records of Hydrothermal Activity in Lake Basins 114
References 122

5 Lake Processes and Sedimentation 129
5.1 The Kenya Rift Lakes and Their Basins 129
5.2 Lacustrine Processes 132
5.3 Aquatic Chemical and Biochemical Processes 136
5.4 Sedimentary Facies 140
5.4.1 Rift Geomorphology and Sedimentation 140
5.4.2 Siliciclastic Coastal Sedimentation 140
5.4.3 Lake Marginal Carbonate Deposition 142
5.4.4 Deltas and Fan-Deltas 144
5.4.5 Offshore Pelagic Settings in Fresh to Brackish Lakes 146
5.4.6 Sedimentation in Highly Saline Deep Lakes 152
5.4.7 Chemical Sedimentation and Evaporites in Shallow Saline Lakes and Ephemeral Playas 152
References 156

Part II The Modern Kenya Rift Lakes

6 Lake Turkana 163
6.1 Introduction 163
6.2 Geological and Geomorphological Setting 169

6.2.1 Lithologies and Stratigraphy of the Turkana Drainage Basin ... 169
6.2.2 Regional Tectonic Framework ... 170
6.3 Climate and Basin Hydrology ... 174
6.4 Peoples, Vegetation and Wildlife ... 177
6.5 Limnology ... 180
6.5.1 Physical Limnology ... 180
6.5.2 Chemical Limnology ... 184
6.6 Biology and Ecology of the Lake ... 186
6.6.1 Phytoplankton ... 186
6.6.2 Zooplankton ... 187
6.6.3 Invertebrates ... 188
6.6.4 Fish ... 189
6.7 Modern Sedimentation ... 190
6.8 Late Quaternary Lacustrine Sediments in the Lake Turkana Basin ... 194
6.8.1 Onshore Spatial Variability and Stratigraphy of the Galana Boi Formation ... 194
6.8.2 Onshore to Offshore Diatom and Chronological Correlations of Holocene Sediments ... 200
6.8.3 Geochemical and Microfossil Stratigraphy of Offshore Cores ... 201
6.8.4 African Humid Period Catchment Expansion, Hydrological Budgets and Sedimentation ... 203
References ... 206

7 Lake Logipi and the Suguta Valley ... 221
7.1 Introduction and History of Research ... 221
7.2 Geological Setting ... 226
7.3 Climate ... 229
7.4 Hydrology of Suguta Valley and Lake Logipi ... 229
7.4.1 Regional Setting and Lake Recharge ... 229
7.4.2 Groundwater Recharge ... 233
7.4.3 Hot Springs in the Suguta Valley ... 234
7.4.4 Physical Limnology of Lake Logipi ... 236
7.4.5 Chemical Limnology ... 238
7.5 Biology and Ecology of the Lakes ... 238
7.6 Sedimentation ... 242
7.7 Late Quaternary Lacustrine Sediments in the Suguta Valley ... 244
References ... 252

8 Lake Baringo ... 257
8.1 Introduction ... 257
8.2 Geological and Geomorphological Setting ... 260
8.2.1 Regional Tectonic Framework ... 260

8.2.2 Lithologies and Stratigraphy within the Baringo Drainage Basin 263
8.3 Climate and Basin Hydrology 264
8.4 Peoples, Vegetation and Wildlife 267
8.5 Limnology 270
8.5.1 Physical Limnology 270
8.5.2 Chemical Limnology 272
8.5.3 Hot Springs 277
8.6 Biology and Ecology of the Lake 278
8.6.1 Phytoplankton and Macrophytes 278
8.6.2 Zooplankton and Benthic Invertebrates 279
8.6.3 Fish 281
8.7 Sedimentation 282
8.8 Late Quaternary Lacustrine Sediments in the Lake Baringo Basin 286
References 291

9 Lake Bogoria 303
9.1 Introduction 303
9.2 Geological and Geomorphological Setting 306
9.3 Climate 310
9.4 Hydrology 310
9.5 Hot Springs and Geothermal Activity 314
9.5.1 Loburu 317
9.5.2 Chemurkeu 317
9.5.3 Ng'wasis – Koibobei – Losaramat (Southern Group) 320
9.5.4 Warm Springs 322
9.5.5 Sublacustrine Hot Springs 322
9.5.6 Life in the Hot Springs and Thermal Wetlands 323
9.5.7 Origin of the Hot Springs 323
9.6 Vegetation and Wildlife 324
9.7 Limnology 325
9.7.1 Physical Limnology 325
9.7.2 Chemical Limnology 327
9.7.3 Evolution of Bogoria Basin Waters 329
9.8 Biology and Ecology of the Lake 331
9.9 Recent Sedimentation in Lake Bogoria 333
9.9.1 Sedimentation Along the Eastern Boundary-Fault Margin (Bogoria-Emsos Escarpment) 334
9.9.2 Sedimentation on Littoral Platforms 334
9.9.3 Sedimentation on Axial Platforms 336
9.9.4 Sedimentation in the Axial Basin 339
9.10 Late Quaternary Sedimentation 341
9.10.1 Stromatolitic Limestones 341
9.10.2 Late Quaternary Sediments of the Sandai Plain 345

9.10.3 Terraces and Incised Valleys 348
9.10.4 Evidence of Lower Lake Levels 348
9.10.5 Sediments Preserved in Lake Cores........................ 350
9.10.6 Former Connections with Lake Baringo 352
References........................ 354

10 Lake Nakuru and Lake Elmenteita........................ 363
10.1 Introduction 363
10.2 Geological Setting 367
10.3 Climate........................ 370
10.4 Hydrology 372
10.4.1 Lake Nakuru........................ 372
10.4.2 Lake Elmenteita 374
10.5 Vegetation and Wildlife 375
10.5.1 Lake Nakuru Basin........................ 375
10.5.2 Lake Elmenteita Basin 376
10.6 Limnology 377
10.6.1 Physical Limnology 377
10.6.2 Chemical Limnology 381
10.7 Biology and Ecology of the Lakes........................ 387
10.7.1 Lake Nakuru........................ 387
10.7.2 Lake Elmenteita 390
10.8 Recent Sedimentation 391
10.8.1 Lake Nakuru........................ 391
10.8.2 Lake Elmenteita 394
10.9 Late Quaternary Lacustrine Sediments in the Nakuru-Elmenteita Basin 395
10.9.1 Early Research – The 'Pluvial Era' 395
10.9.2 Research from the 1950s Until the 1970s 398
10.9.3 Recent Studies (1970s to Present) 399
10.9.4 Archaeological Research 403
References........................ 405

11 Lake Naivasha........................ 417
11.1 Introduction 417
11.2 Geological and Geomorphological Setting 421
11.3 Climate and Basin Hydrology 423
11.4 Peoples, Vegetation and Wildlife........................ 426
11.5 Limnology 428
11.5.1 Physical Limnology........................ 428
11.5.2 Chemical Limnology 430
11.6 Biology and Ecology of the Lake 433
11.6.1 Plant Nutrients and Primary Production 433
11.6.2 Phytoplankton 435
11.6.3 Macrophytes........................ 436
11.6.4 Zooplankton........................ 437

11.6.5 Invertebrates 438
11.6.6 Fish 439
11.7 Sedimentation 440
11.7.1 Sediment Facies and Environments 440
11.7.2 Recent Sediment Stratigraphy 443
11.8 African Humid Period and Holocene Lacustrine Sediments in the Lake Naivasha Basin 445
References 447

12 Lake Magadi and Nasikie Engida 463
12.1 Introduction 463
12.2 Geological and Geomorphological Setting 465
12.3 Climate, Vegetation and Wildlife 468
12.4 Drainage and Hydrology 469
12.4.1 Inflowing Rivers and Lake Water 469
12.4.2 Hot Springs of the Magadi Basin 471
12.5 Hydrochemistry 476
12.5.1 Composition of the Basin Waters and Their Chemical Evolution 476
12.5.2 Origin of the Hot Spring Waters and Circulation 480
12.6 Life in the Lakes 482
12.7 Modern Sedimentation 485
12.7.1 Siliciclastic Sedimentation 485
12.7.2 Alkaline Earth Carbonates 489
12.7.3 Evaporite Precipitation 489
12.7.4 Spring Deposits 494
12.7.5 Siliceous Sediment 494
12.8 Late Pleistocene-Holocene Sedimentation 496
12.8.1 The Evaporite Series 496
12.8.2 The High Magadi Beds (Terminal Pleistocene – Early Holocene) 496
12.8.3 Former Shorelines of the High Magadi Palaeolake 500
12.8.4 Nasikie Engida Beachrock 506
12.8.5 Magadiite and Sodium Silicate Minerals of the High Magadi Beds 506
12.9 Comparisons with Lake Natron, Northern Tanzania 507
12.9.1 Introduction 507
12.9.2 Hydrology and Hydrochemistry 510
12.9.3 Sedimentation and Mineralogy 513
12.9.4 Biology and Ecology 515
12.9.5 Late Quaternary Sedimentation 515
References 516

13 Lake Victoria 527
13.1 Introduction 527
13.2 Geological and Geomorphological Setting 530

13.3 Climate and Basin Hydrology 532
13.4 Peoples, Vegetation and Wildlife 535
13.5 Limnology 536
13.5.1 Physical Limnology 536
13.5.2 Chemical Limnology 540
13.6 Biology and Ecology of the Lake 544
13.6.1 Phytoplankton 544
13.6.2 Macrophytes 549
13.6.3 Zooplankton 550
13.6.4 Invertebrates 552
13.6.5 Fish 552
13.7 Modern Sedimentation 554
13.8 Late Quaternary Lacustrine Sediments in the Lake Victoria Basin 556
References 560

14 The Lesser-Known Lakes and Wetlands of the Kenya Rift 577
14.1 Introduction 577
14.2 Central Island Lakes, Lake Turkana 578
14.3 North Barrier Lakes 581
14.4 South Suguta (Kangirinyang) Maar Lakes 583
14.5 Spring-Fed Lakes and Wetlands, Emuruangogolak to Silali 585
14.6 Spring-Fed Lakes and Wetlands of the Silali-Karosi Region 589
14.7 Lake Kichirtit ('Lake 94') 591
14.8 South Baringo Wetlands 593
14.9 Lake Kamnarok 598
14.10 Lake Solai 600
14.11 Lake Ol' Bolossat 603
14.12 Lake Sonachi, Naivasha 607
14.13 Small and Ephemeral Lakes of the Southern Kenya Rift 611
14.14 Lake Simbi and Other Nyanza (Kavirondo) Wetlands 615
14.14.1 The Yala Swamp Lakes 616
14.14.2 Lake Simbi 616
14.14.3 Nyando (Kusa) Swamp 617
14.15 Other Seasonal and Ephemeral Lakes 618
14.16 Constructed Reservoirs and Ponds 619
References 621

Part III The Ancient Kenya Rift Lakes

15 The Turkana Basin 631
15.1 Introduction 631
15.2 Oligo-Miocene Lacustrine Deposits 633
15.2.1 The Ekitale Basin 633
15.2.2 The Lokichar and North Lokichar Basins 633
15.2.3 Other Oligo-Miocene Sediments in the Turkana Region . 638

15.3 Plio-Pleistocene Lacustrine Deposits. 642
15.3.1 Historical Background 642
15.3.2 Sedimentary Facies 644
15.3.3 Lacustrine Sediments, Faunas and Floras 650
15.3.4 Lake Levels and Sedimentation 658
15.3.5 Plio-Pleistocene Stratigraphy, Palaeolakes and Palaeogeography 660
References. 673

16 The Suguta Basin 693
16.1 Introduction 693
16.2 Mio-Pliocene Sedimentary Formations 699
16.3 Pleistocene Sediments and Lakes 705
16.4 Mio-Pleistocene Palaeogeography. 710
References. 712

17 The Baringo-Bogoria Basin and Adjacent Parts of the Kenya Rift . . 717
17.1 Introduction and Geological Exploration. 717
17.2 Miocene Lacustrine Sedimentary Formations 720
17.2.1 The Kirimun Formation 720
17.2.2 The Kimwarer and Kamego Formations 720
17.2.3 The Tambach Formation 721
17.2.4 The Muruyur Formation. 725
17.2.5 The Ngorora Formation 727
17.2.6 The Ngerngerwa (Ngeringerowa) Formation. 730
17.2.7 The Mpesida Beds 730
17.2.8 The Lukeino Formation 733
17.3 Miocene Lacustrine Palaeogeography and Palaeoenvironments. 734
17.4 Plio-Pleistocene Sediments and Palaeolakes 737
17.4.1 Plio-Pleistocene Sediments and the Toluk Beds 737
17.4.2 The Chemeron and Mabaget Formations. 739
17.4.3 The Kapthurin Formation. 744
17.4.4 Pleistocene Lake Sediments East of Lake Baringo 749
17.4.5 Late Pleistocene Lake Sediments. 750
17.5 Plio-Pleistocene Palaeogeography 754
References. 756

18 The Central Kenya Rift Basins (Nakuru-Elmenteita-Naivasha) 767
18.1 Volcanic Rocks and Geological Background. 767
18.2 Quaternary Shorelines and Pluvial Hypotheses. 770
18.3 Mio-Pleistocene Sediments and Palaeogeography 778
18.3.1 Mio-Pliocene Sediments and Palaeogeography. 778
18.3.2 Lower to Middle Pleistocene Sediments and Palaeogeography 779
18.3.3 Late Pleistocene Sediments and Palaeogeography. 782
References. 787

19 The South Kenya Rift Basins 793
19.1 Introduction 793
19.2 Miocene Lake Sediments of the Lemudong'o Basin 793
19.2.1 General Setting of the Miocene Sediments 793
19.2.2 Sediments, Stratigraphy and Palaeogeography of the Lemudong'o Formation 795
19.3 Pleistocene Sediments of the Munya wa Gicheru Basin 796
19.3.1 Location and History of Research 796
19.3.2 Geochemistry 798
19.3.3 Facies, Lithostratigraphy and Microfossils 801
19.3.4 Palaeoenvironments 803
19.4 Pleistocene Lake Sediments at Olorgesailie 805
19.4.1 Location and History of Research 805
19.4.2 The Olorgesailie Formation (1.2–0.49 Ma) 808
19.4.3 Erosion and the Oltulelei Formation (0.49–0 Ma) 815
19.5 The Koora Basin 821
19.5.1 Geomorphology of the Koora Basin 821
19.5.2 Sedimentation in the Koora Basin 824
19.6 Quaternary Tectonism, Climate and Palaeogeography in the South Kenya Rift 826
References 830

20 The Magadi-Natron Basin 839
20.1 Introduction 839
20.2 Miocene to Early Pleistocene Volcanics and Sediments of the Magadi-Natron Basin 842
20.3 The Oloronga Beds 848
20.3.1 Oloronga Outcrops 848
20.3.2 Lake Magadi Cores and Oloronga Deposition 855
20.4 The Lengorale and Ngare Nyiro (Orkaramation) Lake Beds 861
20.5 The Green Beds 862
20.5.1 General Characteristics and Age 862
20.5.2 The Green Beds Sediments in Core 865
20.5.3 The Green Beds Sediments in Outcrop 866
20.5.4 Geochemistry of the Green Beds Sediments 872
20.5.5 Depositional Environment of the Green Beds 872
20.5.6 Origin of the Green Beds Cherts 874
References 876

21 Lake Victoria Basin 883
21.1 Introduction 883
21.2 Miocene Deposits 884
21.2.1 Miocene Deposits in the Nyanza Rift 884
21.2.2 Miocene Lacustrine Sediments on Rusinga Island 887
21.3 Plio-Pleistocene Lakes of the Nyanza Rift 890
21.3.1 Pliocene to Middle Pleistocene Lacustrine Sediments 890

21.3.2 Middle to Late Pleistocene Origins and Evolution of Lake Victoria 893
21.3.3 Late Quaternary Palaeohydrology and Lake Levels 897
References 898

Part IV Rift Lake Controls, Resources and Environmental Considerations

22 Controls on Lacustrine Sedimentation in Rift Settings 909
22.1 Introduction 909
22.2 Rifting and Volcano-Tectonic Controls on Kenya Rift Sedimentation 910
22.3 Climate Controls on Kenya Rift Sedimentation 914
22.4 Bedrock, Weathering and Diagenesis 920
22.5 Geothermal Influences and Controls 922
22.5.1 Lake Recharge 923
22.5.2 Impact on Lake Water Chemistry 924
22.5.3 Hydrothermal Contributions to the Sedimentary Record of the Lake Basins 925
22.6 Rift Sedimentation Models 926
References 929

23 Economic Aspects of the Kenya Rift Lakes and Their Deposits 935
23.1 Introduction 935
23.2 Mineral Resources 935
23.2.1 Soda Deposits of Lake Magadi 936
23.2.2 Diatomite and Diatomaceous Earth 937
23.2.3 Petroleum Potential of the Kenya Rift Lakes 939
23.2.4 Zeolites 944
23.2.5 Other Lacustrine Mineral Resources 945
23.3 Human-Related Resources 947
References 948

24 Afterthoughts and Perspectives 951
24.1 Drivers of Change in the Contemporary Rift Valley Lakes 951
24.1.1 Pristine Lakes and Catchments? 952
24.1.2 Global Warming and the Rift Lakes: Past, Present and Future 953
24.1.3 Human Impacts on Future Lake Levels 954
24.1.4 Faunal and Floral Stability-Instability of the Lakes and their Catchments 955
24.1.5 Protection Measures and Drivers of Environmental Change 956
24.1.6 Development Issues and their Impacts 956
24.2 Non-Anthropogenic Drivers and the Importance of Time 958
24.2.1 Tectonic, Volcanic and Sedimentological Events and Future Lakes 958

24.2.2 The Climate Versus Tectonic Debate and Change over Geological Time 959
24.2.3 The Development of New Rift Models 959
24.2.4 The Rift Valley Lakes and Hominin-Environment Interactions 960
24.3 A Personal Perspective. 961
References 964

Index 969

About the Authors

Robin W. Renaut is Emeritus Professor in the Department of Geological Sciences at the University of Saskatchewan. He was born in London, UK, in May 1952 and received a BSc degree in Earth Sciences from Bedford College (University of London) in 1974, where he was instructed by Bill Bishop, Basil King, and many PhD students and postdocs (Martin Pickford, Andrew Hill, Greg Chapman, Paul Griffiths) who were then part of the East African Geological Research Unit (EAGRU). He then spent several months at Geosystems, a geoscience information company, before starting PhD research in the autumn of 1975 supervised by the late Bill Bishop at Queen Mary College (now Queen Mary University of London), where he met Bernie Owen, who was also beginning his PhD research. They began fieldwork together in the Kenya Rift in 1976. After Bill died in February 1977, Bill French, an igneous petrologist, led them through to graduation. Robin emigrated to Canada and taught sedimentology, stratigraphy, petroleum geology and related courses at the University of Saskatchewan in Saskatoon for 33 years. During that time, he maintained his interest in limnogeology developed in Kenya and did research in Kenya, Tanzania, Malawi, British Columbia, Bolivia, and Greece. He expanded his early interests in hydrothermal systems acquired at Lake Bogoria and worked on hot spring deposits in New Zealand (with Brian Jones and Michael Rosen), Iceland, western Canada, Bolivia, and Chile. He has co-edited three books: two SEPM Special Publications (saline lakes, rift sedimentation) and one Geological Society of London Special Publication (rift sedimentology). He retired recently, but in name only.

Richard Bernhart Owen is Emeritus Professor in the Department of Geography at Hong Kong Baptist University. Bernie was born in Bangor hospital in North Wales on 28 December 1953, but was raised in St. Helens, near Liverpool. He was awarded a Geology BSc Degree by Sheffield University in 1975 and went straight on to do a PhD at Queen Mary College (now Queen Mary University of London) under Professor Bill Bishop, where he met Robin Renaut as a fellow postgraduate student. Subsequently, he spent several months learning about diatoms and, with Robin, mapping the Suguta Valley (Kenya Rift Valley) using aerial photographs prior to their first field season in the summer of 1976. Unfortunately, permits for the Suguta were refused at the last minute and Bernie ended up reconnoitering diatom-rich sediments in different parts of the Kenya Rift. Bill Bishop, their supervisor, died in February 1977, and Bill French (with no relevant experience) kindly took over and guided them, with Bernie gaining his PhD in 1981. Bernie moved to Malawi as a University of Malawi lecturer and taught a broad range of geology courses, including unfamiliar topics. His research focussed on the recent sediments in Lake Malawi and other lakes in the country. After 8 years, Bernie and family (Barbara and two young children) left Malawi in order to seek another post, which resulted in them moving to Hong Kong in January 1991. He then worked on marine sediments for a number of years, but his African interests led him to return to Kenya in the summer of 2002, with one or two trips per year from then until the present. He also occasionally took part in geological expeditions to other places (Canada, Greece, Iceland, Thailand, Sri Lanka, and New Zealand) to carry out a variety of limnological and/or hot spring studies. Bernie also edited a book on seismicity in eastern Asia and has authored four books on the geology and landscapes of Hong Kong.

Part I
Background to the Kenya Rift Lakes

Chapter 1
Introduction: About This Book

1.1 The Kenya Rift Lakes

Several books and edited volumes have reviewed one or more of the Kenya Rift lakes from a biological or ecological perspective, or both (Beadle 1974, 1981; Johnson and Odada 1996; Lehman 1998; Harper et al. 2003; Awange and Ong'ang'a 2006; Schagerl 2016; El-Sheekh and Elsaied 2023). A few have examined the geology and tectonics of individual or multiple lake basins (Bishop and Clark 1967; Baker et al. 1972; Coppens et al. 1976; Bishop 1978; Frostick et al. 1986; Morley 1999; Renaut and Ashley 2002), but none has approached the Kenyan lakes specifically from a geological (mainly sedimentological and palaeoecological) perspective. Our goal for this book is to provide a general background to the modern Kenya Rift lakes and their precursors back to the Miocene, and to provide a broad framework for future research. We have tried to include most of the important past studies, covering the last 150 years, but do not claim to have included all previous research.

The lakes of the Kenya Rift and their former extensions into Ethiopia (Chew Bahir) and Tanzania (Lake Natron) are highly diverse physically, chemically, and ecologically (Fig. 1.1). Table 1.1 summarises some of the main characteristics of the Kenyan water bodies that illustrate this diversity. Lake Victoria (~169,685 km^2) lies in a tectonic depression between the eastern and western branches of the East African Rift System with the Winam Gulf extending into the Nyanza (Kavirondo) Rift, about 140 km to the west of the main Kenya Rift. The largest lake within the main Kenya Rift is Lake Turkana, which covered 7430 km^2 in December 2020, an area almost 11 times larger than the next nine water bodies combined. These smaller lakes ranged in area between 11 (Solai) and ~200 (Baringo) km^2 in that same month, but all the main lakes vary in area frequently because of changes in annual and interannual precipitation and changing evaporation rates in their respective catchments. Water depths also differ considerably, from more than 120 m (Lake Turkana) to less than 1 m (Lake Magadi). Many shallow lakes and swamps desiccate completely during periods of drought.

R. W. Renaut, R. B. Owen, *The Kenya Rift Lakes: Modern and Ancient*, Syntheses in Limnogeology, https://doi.org/10.1007/978-3-642-25055-2_1

Fig. 1.1 The Kenya Rift lakes, locations and appearance

All the Kenya Rift lakes expanded during the late Pleistocene–Early Holocene, now called the African Humid Period (AHP), when conditions became generally wetter. Some merged or overflowed (e.g. Palaeolake Suguta; Lakes Nakuru and Elmenteita) while others switched from closed hydrology (endorheic) to open-basin lakes (e.g. Turkana, Baringo, Bogoria), only to reverse when the climatic conditions became drier during the later Holocene. Such changes have been common, but tectonics has also strongly influenced, and at times controlled, their evolution (Owen et al. 2018).

All lake waters in the Kenya Rift are alkaline. Modern lakes Naivasha, Baringo and Ol' Bolossat have fresh waters, supporting fish, crocodiles, hippos and associated fauna and flora. In contrast, Lake Turkana and Lake Solai have been brackish during the past few decades. Lakes Logipi, Nakuru and Elmenteita are saline but currently do not precipitate lacustrine salts. Nasikie Engida and Lake Magadi are often highly saline and precipitate evaporites. Lake Bogoria varies between dilute and hypersaline. Lake Natron occasionally precipitates trona, but the salts are generally not preserved. These varied conditions support contrasting phytoplankton and zooplankton, reptiles, mammals and birds. Flamingos, for example, move between

Table 1.1 Major physical and chemical parameters of the Kenya Rift lakes

Lake	Basin type	Area (km^2)	Perimeter (km)	Area/perimeter	Max depth/s (m)	Conductivity – μS cm^{-1}/ or (TDS in ppm)	pH range
Victoria	Inter-rift depression with extension (Winam Gulf) into Nyanza Rift	69,685	3670	19.0	69[11]	97–145[8,9]	6.4–9.8[10]
Turkana	Coalesced border-faulted asymmetric rift grabens. Volcano-dammed to south	7430	730	10.2	109–125[3,12,13]	2763–3190[1,3] (2279–2536)[2]	8.7–9.5[1,7]
Logipi	Axial rift graben. Volcano-dammed to north	51	30	1.7	5	~5000–17,000[14,15]	9.5–9.6[15]
Baringo	Volcanic-dammed rift graben	199	85	2.3	~7	(458–1383)[4]	6.7–8.98[4]
Bogoria	Border-faulted asymmetric graben. Dammed by alluvium to north	42	72	0.6	11.5[18]	Mixolimnion – >50,000[18] Monimolimnion – 90,000[18]	9.3–10.2[18]
Solai	Border faulted rift-marginal graben	11	15	0.7	>8.5[17]	776–2398[16]	8.1–9.0[16]
Ol' Bolossat	Border faulted rift-marginal graben	33	41	0.8	1.83[20]	150–418 (174.9–192.6)[19]	6.7–8.3[19]
Nakuru	Axial rift graben. Volcano-dammed to north	69	36	1.9	4.5[6]	27,600–54,000[6,22]	9.8–10.3[6,21]

(continued)

Table 1.1 (continued)

Lake	Basin type	Area (km^2)	Perimeter (km)	Area/ perimeter	Max depth/s (m)	Conductivity – μS cm^{-1}/ or (TDS in ppm)	pH range
Elmenteita	Axial rift graben. Volcano-dammed to south	22	35	0.6	3.1[22]	39,000[21]	10–10.1[21, 22]
Naivasha	Axial rift graben. Volcano-dammed to south	163	74	2.2	16[24]	74–2480[5, 23, 24]	8.01–8.98[5, 24]
Nasikie Engida	Axial rift graben	10.5	18	0.6	1.6[26]	207,000[26] (30,000–270,000)[26]	9.21–9.9[25]
Magadi	Coalesced axial rift grabens	91	121	0.75	~0–2	(122,000–313,000)[27]	9.65–10.9[27]

Physical data are based on Google Earth and reflect a period of high lake levels in December 2020
1: Beadle (1932); 2: Yuretich and Cerling (1983); 3: Hopson and Hopson (1982); 4: Tiercelin (1981); 5: Mugo (2010); 6: Raini (2007); 7: Olago (1992); 8: Duke (1924); 9: Akiyama et al. (1977); 10: Mwirigi et al. (2005); 11: Johnson and Odada (1996); 12: Yuretich (1979); 13: Källqvist et al. (1988); 14: Junginger and Trauth (2013); 15: Castanier et al. (1993); 16: de Bock et al. (2009); 17: Herrnegger et al. (2021); 18: Renaut and Tiercelin (1994); 19: Karuku and Mugo (2019); 20: NEMA (2012); 21: Oduor and Schagerl (2007); 22: Melack et al. (1982); 23: Njenga (2004); 24: Ndungu et al. (2015); 25: Renaut et al. (2020); 26: de Cort et al. (2019); 27: Jones et al. (1977)

the highly saline lakes seeking suitable food sources, which vary with changing water chemistry as the lakes expand and contract.

Geologically, the modern lakes are the successors of a series of Neogene palaeolakes. Both modern and ancient water bodies are important for many reasons. Their sediments preserve detailed records of past environmental change and also provide models for interpreting ancient lacustrine rift basins and saline, alkaline palaeolakes in other parts of the world. Biologically, the modern lakes are homes for a rich and diverse fauna and flora, as were their predecessors. Most modern lakes are highly productive and home to endemic fish and diverse and abundant birds, as recognised by many areas that have been given Ramsar and UN World Heritage status. They also have palaeoanthropological significance with their ancient deposits associated with important hominin and archaeological sites. Despite their international value, the lakes have been heavily impacted by population growth and related impacts from agriculture, industrial development, dam construction and hydrological change. Consequently, the ecological status of the modern lakes is under threat, and many have already experienced rapid change during the last few decades, although much remains that should be preserved. Many lakes have been unusually high since

2010–11, flooding marginal lands and disrupting livelihoods, and are experiencing physical and ecological changes (Government of Kenya 2021).

In **Part I** of the book, we provide a broad background with general information that is needed to understand subsequent sections. This chapter (1) introduces the rift lakes and structure of this book. Chapter 2 presents a brief history of limnological and sedimentological research undertaken since the nineteenth century. Chapter 3 outlines the geology of the Kenya Rift, placing the modern and ancient lakes in a broader physical context. This includes summaries of the origins of the Kenya Rift and the East African Rift System. The chapter examines the structural and volcanic geology and emphasises the importance of tectonism in the formation of sedimentary basins and the long-term palaeogeographical development of the rift.

Chapter 4 focuses on modern environmental conditions and includes explanations of the climate system and modern weather patterns in the region. In addition, the chapter explores aspects of the river drainage, hydrology and hydrogeology, including the important geothermal systems. The importance of weathering and sediment production is also noted as this partly controls the unusual chemistry of the various lakes. Subsequent portions discuss the nature of soils, vegetation, wildlife and the importance of geothermal springs and meteoric waters.

Chapter 5, the final chapter in Part I, provides a limnological background for understanding the physical, chemical, and biological processes that occur in the Kenya Rift lakes. These include, for example, the importance of physical mixing, stratification, currents, and waves. The chapter also examines the hydrochemistry of lakes in terms of evaporative concentration, solutes and particulate sources, major ions and salinity as well as chemical and biochemical processes. Depositional models are explored, noting controlling factors that influence processes and facies in deltas, littoral zones and pelagic regions.

Part II presents the physical and chemical limnology, ecology, and sedimentation of the modern lakes. Chapters 6 and 7 provide details for the northern Rift lakes (Turkana; Logipi-Suguta). Chapters 8, 9, 10, and 11 examine the major water bodies of the Central Kenya Rift (Baringo; Bogoria; Nakuru; Elmenteita; Naivasha) with details of the southern lakes (Magadi; Nasikie Engida) presented in Chap. 12, including a brief introduction to Lake Natron in Tanzania. Lake Victoria partially lies in the adjacent Nyanza (Kavirondo) Rift of western Kenya and is considered separately in Chap. 13. Chapter 14 presents information on the many smaller and lesser-known lakes (e.g. Solai, Ol' Bolossat, Kwenia, reservoirs).

Each of the main lakes is described in turn using a common format. After a broad introduction, the chapters present the geological and geomorphological setting of the relevant drainage basin, followed by descriptions of the climate and basin hydrology (including springs), vegetation and wildlife. The physical and chemical limnology are then described followed by an examination of the biology and ecology of the individual lakes. Each chapter also explains the dominant sedimentary processes active in the respective lakes and the nature of recent deposition. The chapters conclude with an exploration of the environmental and geological history of the African Humid Period (~15–5 ka) and Holocene precursor lakes from outcrop, core and other evidence, as these have important implications for the modern water bodies. Several chapters mention the archaeological records if relevant to the modern lakes – e.g. lake shoreline sites.

Part III examines the Neogene rift lakes by geographical location. The purpose of this section is to document the known sequences of lacustrine sediments from the oldest pre-Miocene basins through to the Late Pleistocene. Some basins (Turkana-Suguta) were united before the middle to late Quaternary. Others were characterised by the creation and destruction of palaeolakes during the Mio-Pleistocene (Baringo-Bogoria). Chapters 15, 16, 17, and 18 give details for the Turkana, Suguta, Baringo-Bogoria and Central Rift (Nakuru-Elmenteita-Naivasha) basins. The South Kenya Rift and Magadi-Natron basins are described in Chaps. 19 and 20 with the Lake Victoria Basin examined in Chap. 21

As in Part II, a standard format is applied to all palaeolake basins, although the varying information available leads to some departures from this common outline. Chapters begin with an introduction (location, exposure, access) followed by the geological context of the lacustrine deposits and their basin configuration. Descriptions of sediment types (facies), age dating, palaeontology and archaeology follow in stratigraphic order with the oldest deposits described first. Each chapter then presents an interpretation of the respective palaeolake(s) in terms of their type and geological history, and the palaeogeography of the lake basin.

Part IV considers some of the broader implications in terms of controls on sedimentation (Chap. 22) and the importance of rift lakes in terms of resource and economic aspects (Chap. 23). Finally, current environmental issues are covered in Chap. 24 which concludes with a few personal thoughts and perspectives based on more than 40 years of research (1976–2023) in the Kenya Rift by each of the authors.

References

Akiyama T, Kajumulo AA, Olsen S (1977) Seasonal variations of plankton and physicochemical condition in Mwanza Gulf, Lake Victoria. Bull Freshw Fish Res Lab Jpn 27:49–60

Awange JL, Ong'ang'a O (2006) Lake Victoria, ecology, resources, environment. Springer, Berlin

Baker BH, Mohr PA, Williams LAJ (1972) Geology of the eastern rift system of Africa. Geol Soc Am Spec Pap 136

Beadle LC (1932) Scientific results of the Cambridge Expedition to the East African Lakes. 1930-1. 4. The waters of some East African Lakes in relation to their fauna and flora. J Linn Soc Zool 38:157–211

Beadle LC (1974) The inland waters of tropical Africa: an introduction to tropical limnology, 1st edn. Longman, London

Beadle LC (1981) The inland waters of tropical Africa: an introduction to tropical limnology, 2nd edn. Longman, London

Bishop WW (ed) (1978) Geological background to fossil man. Spec Publ Geol Soc Lond 6. Scottish Academic Press, Edinburgh

Bishop WW, Clark JD (eds) (1967) Background to evolution in Africa. University of Chicago Press, Chicago

Castanier S, Bernet-Rollande M, Maurin A, Perthuisot J-P (1993) Effects of microbial activity on the hydrochemistry and sedimentology of Lake Logipi, Kenya. Hydrobiologia 267:99–112

Coppens Y, Clark Howell F, Isaac GL, Leakey REF (eds) (1976) Earliest man and environments in the Lake Rudolf basin: stratigraphy, paleoecology, and evolution. University of Chicago Press, Chicago

De Bock T, Kervyn de Meerendre B, Hess T, Gouder de Beauregard A-C (2009) Ecohydrology of a seasonal wetland in the Rift Valley: ecological characterization of Lake Solai. Afr J Ecol 47:289–298

De Cort G, Mees F, Renaut RW, Sinnesael M, Van der Meeren T, Goderis S, Keppens E, Mbuthia A, Verschuren D (2019) Late-Holocene sedimentation and sodium carbonate deposition in hypersaline, alkaline Nasikie Engida, southern Kenya Rift Valley. J Paleolimnol 62:279–300

Duke HL (1924) Water analyses. Annual report Bacteriology Department, Uganda Protectorate, for year ended 31st Dec. 1923. Entebbe, Uganda

El-Sheekh M, Elsaied HE (eds) (2023) Lakes of Africa: microbial diversity and sustainability. Elsevier, Amsterdam

Frostick LE, Renaut RW, Reid I, Tiercelin J-J (eds) (1986) Sedimentation in the African rifts. Spec Publ Geol Soc Lond 25

Government of Kenya [GoK] (2021) Rising water levels in Kenya's Rift Valley Lakes, Turkwel Gorge Dam and Lake Victoria: a scoping report. GoK, Nairobi

Harper DM, Boar RR, Everard M, Hickley P (eds) (2003) Lake Naivasha, Kenya (Developments in Hydrobiology 168). Kluwer, Dordrecht

Herrnegger M, Stecher G, Schwatke C, Olang L (2021) Hydroclimatic analysis of rising water levels in the Great Rift Valley lakes of Kenya. J Hydrol Reg Stud 36:100857

Hopson AJ, Hopson J (1982) The fishes of Lake Turkana with a description of three new species. In: Hopson AJ (ed) Lake Turkana. A report on the findings of the Lake Turkana Project 1972–1975, vol 1. Overseas Development Administration (ODA), London, pp 283–347

Johnson TC, Odada EO (eds) (1996) The limnology, climatology and paleoclimatology of the East African lakes. Gordon and Breach, Amsterdam

Jones BF, Eugster HP, Rettig SL (1977) Hydrochemistry of the Lake Magadi basin, Kenya. Geochim Cosmochim Acta 41:53–72

Junginger A, Trauth MH (2013) Hydrological constraints of paleo-Lake Suguta in the Northern Kenya Rift during the African Humid Period (15–5 ka BP). Glob Planet Chang 111:174–188

Källqvist T, Lien L, Liti D (1988) Lake Turkana limnological study 1985–1988. Norw Inst Water Res Rep:0–85313

Karuku GN, Mugo EK (2019) Land use effects on lake Ol'bolossat watershed conservation, Nyandarua County. JOJ Wildl Biodivers 1(2):555556

Lehman JT (ed) (1998) Environmental change and response in East African lakes (Monogr Biol 79). Kluwer, Dordrecht

Melack J, Kilham P, Fisher T (1982) Responses of phytoplankton to experimental fertilization with ammonium and phosphate in an African soda lake. Oecologia 52:321–326

Morley CK (ed) (1999) Geoscience of rift systems—evolution of East Africa. AAPG Stud Geol 44

Mugo MJ (2010) Seasonal changes in physico-chemical status and algal biomass of Lake Naivasha, Kenya. MSc thesis, Kenyatta University, Nairobi

Mwirigi PM, Rutagemwa DK, Gikuma-Njuru P, Matovu A, Waya RK, Mwebaza-Ndawula L, Ssenfuma-Nsubuga M, Kinobe J, Abuodha JOZ, Hecky RE (2005) Lake Victoria: the changing lake. In: Mwanuzi FL, Abuodha JOX, Muyodi FJ, Hecky RE (eds) Lake Victoria Environment Management Project (LVEMP) water quality and ecosystem status. South Eastern Kenya University, Kitui, pp 62–80

Ndungu J, Augustijn DCM, Hulscher SJMH, Fulanda B, Kitaka N, Mathooko M (2015) A multivariate analysis of water quality in Lake Naivasha, Kenya. Mar Freshw Res 66:177–110

NEMA (2012) Lake Ol Bolossat integrated management plan, 2008–2013. National Environment Management Authority, Nairobi

Njenga JW (2004) Comparative studies of water chemistry of four tropical lakes in Kenya and India. Asian J Water Environ Pollut 1:87–97

Oduor S, Schagerl M (2007) Temporal trends of ion contents and nutrients in three Kenyan Rift Valley saline-alkaline lakes and their influence on phytoplankton biomass. In: Gulati RD, Lammens E, de Pauw N, van Donk E (eds) Shallow lakes in a changing world. Springer, Dordrecht, pp 59–68

Olago DO (1992) The mineralogy and sedimentology of recent subaqueous sediments in Lake Turkana, Kenya. MSc thesis, University of Nairobi

Owen RB, Renaut RW, Lowenstein TK (2018) Spatial and temporal geochemical variability in lacustrine sedimentation in the East African Rift System: evidence from the Kenya Rift and regional analyses. Sedimentology 65:1697–1730

Raini J (2007) Long-term trends in water quality, water quantity and biodiversity at Lake Nakuru, Kenya. In: Proceedings of the 11th world lakes conference, vol 2, pp 57–62

Renaut RW, Ashley GM (eds) (2002) Sedimentation in continental rifts. SEPM Spec Publ 73

Renaut RW, Tiercelin J-J (1994) Lake Bogoria, Kenya Rift Valley – a sedimentological overview. In: Renaut RW, Last WM (eds) Sedimentology and geochemistry of modern saline lakes. SEPM Sec Publ 50:101–123

Renaut RW, Owen RB, Lowenstein TK, de Cort G, McNulty E, Scott JJ, Mbuthia A (2020) The role of hydrothermal fluids in sedimentation in saline alkaline lakes: evidence from Nasikie Engida, Kenya Rift Valley. Sedimentology 21:1–27

Schagerl M (ed) (2016) Soda lakes of East Africa. Springer, Cham

Tiercelin J-J (1981) Rifts continentaux, tectonique, climates, sediments. Exemples: la sedimentation dans le Nord du Rift Gregory, Kenya, et dans le Rift de l'Afar, Ethiopie, depuis le Miocène. PhD thesis, Univ Aix-Marseille II

Yuretich RF (1979) Modern sediments and sedimentary processes in Lake Rudolf (Lake Turkana) Eastern Rift Valley, Kenya. Sedimentology 26:313–331

Yuretich RF, Cerling TE (1983) Hydrogeochemistry of Lake Turkana, Kenya: mass balance and mineral reactions in an alkaline lake. Geochim Cosmochim Acta 47:1099–1109

Chapter 2
Brief History of Geological Research on the Kenya Rift Lakes

2.1 Introduction

The history of limnological, biological, ecological and palaeoecological studies of the Kenya Rift lakes has been reviewed by Worthington (1996) and Talling (2006) and is discussed in several books (Worthington and Worthington 1933; Beadle 1974, 1981; Hamilton 1982; Payne 1986; Johnson and Odada 1996; Lehman 1998; Talling and Lemoalle 1998; Harper et al. 2003; Odada and Olago 2006; Schagerl 2016; Krienitz 2018). In contrast, the history of geological studies related to the Kenya Rift lakes has been given relatively little attention. The aim of this short chapter is to highlight some of the history of research that has led to our understanding of the geology of the Kenya Rift lakes and their precursors. Historical details about individual lakes are given in the respective chapters.

Homo sapiens and our hominin ancestors have occupied the East African Rift, and areas beyond, for a very long time … our earliest ancestors have been present in Africa for several million years. Historically, the lake shores have been occupied by diverse African people for hundreds of years to millennia. Arab traders and slavers had knowledge of the East African lakes, as well as their resources and peoples, long before nineteenth-century European exploration commenced. Johann Krapf and fellow missionaries, for example, compiled an 1855 map of East Africa based on Arab and local reports that showed a large lake with a narrow southern extension that was named Unyamwezi (Uniamesi) (Krapf and Ravenstein 1860), which may have been a crude combined representation of both Lake Victoria and Lake Tanganyika (Dawson 2008).

R. W. Renaut, R. B. Owen, *The Kenya Rift Lakes: Modern and Ancient*, Syntheses in Limnogeology, https://doi.org/10.1007/978-3-642-25055-2_2

2.2 The Early European Explorers

Europeans explored sub-Saharan Africa for a variety of reasons. Some were involved in missionary work; others were interested in trade, including slavery. Many sought adventure, hunting experiences or new areas for imperial expansion, with the latter group leading the way in the 'scramble for Africa', which eventually resulted in a number of European spheres of influence (Brown 1989). A few early explorers were interested in expanding scientific (mainly geographical) knowledge and clearly, their 'discoveries' made a major impression on them. John Hanning Speke, for example, observed the southern shores of Lake Victoria, west of the Kenya Rift, on 30th July 1858, and returned later to prove it was the source of the Nile, which he found in 1862. On the latter occasion, he reported that "Here at last I stood on the brink of the Nile; most beautiful was the scene, nothing could surpass it! It was the very perfection of the kind of effect aimed at in a highly kept park; with a magnificent stream from 600 to 700 yards wide, dotted with islets and rocks, the former occupied by fishermen's huts, the latter by sterns and crocodiles basking in the sun" (Speke 1864). In contrast with these moments of elation, several explorers died from disease or in clashes with local peoples. Losses were often high among expedition carriers, guards and cooks. Henry Morton Stanley, reported during his 1874–75 expedition to Lake Victoria (Nyanza) that "I mustered the men of the expedition yesterday, and ascertained it to consist of three white men and 166 Wanguana soldiers and carriers, twenty-eight having died since leaving Ituru, thirty days ago. Over one-half of our force has thus been lost by desertions and deaths" (Stanley 1875–76). Despite the hazards, exploration was often followed by trade and the expanding commercial interests of companies and then to formal colonisation.

The earliest geological and limnological observations began in the last decades of the nineteenth century (Kent 1978). Early explorers such as Joseph Thomson (1885; Fig. 2.1a) made general observations in 1883–4 that relate to the environment, wildlife, and geology of some of the lakes. He journeyed along the central Kenya Rift Valley including lakes Naivasha and Baringo plus the Kamasia (Tugen Hills) area, and then to Lake Victoria. At one point Thomson left the rift for Mt. Kenya and subsequently reached its eastern edge again observing "there was the mysterious Lake Baringo, gleaming apparently at our feet, though several thousands of feet below". Having noted his experience with other lakes (Nyassa, Tanganyika) he went on to say "not one of these spectacles approaches in beauty, grandeur and variety of the landscape that now spread out before me on the Lykipia Plateau". At about the same time, Gustav Fischer (1884, 1885), a German naturalist, mapped parts of the Kenya Rift south of Naivasha and parts of northern Tanzania. Ludwig von Höhnel (1894; Fig. 2.1f), with companion Count Teleki (Fig. 2.1e), surveyed much of the African Rift north of Baringo, visiting several lakes and observing high-level lake features. They reached the 'Jade Sea' in March 1888, which he named Lake Rudolf in honour of his friend Prince Rudolf of Austria. The

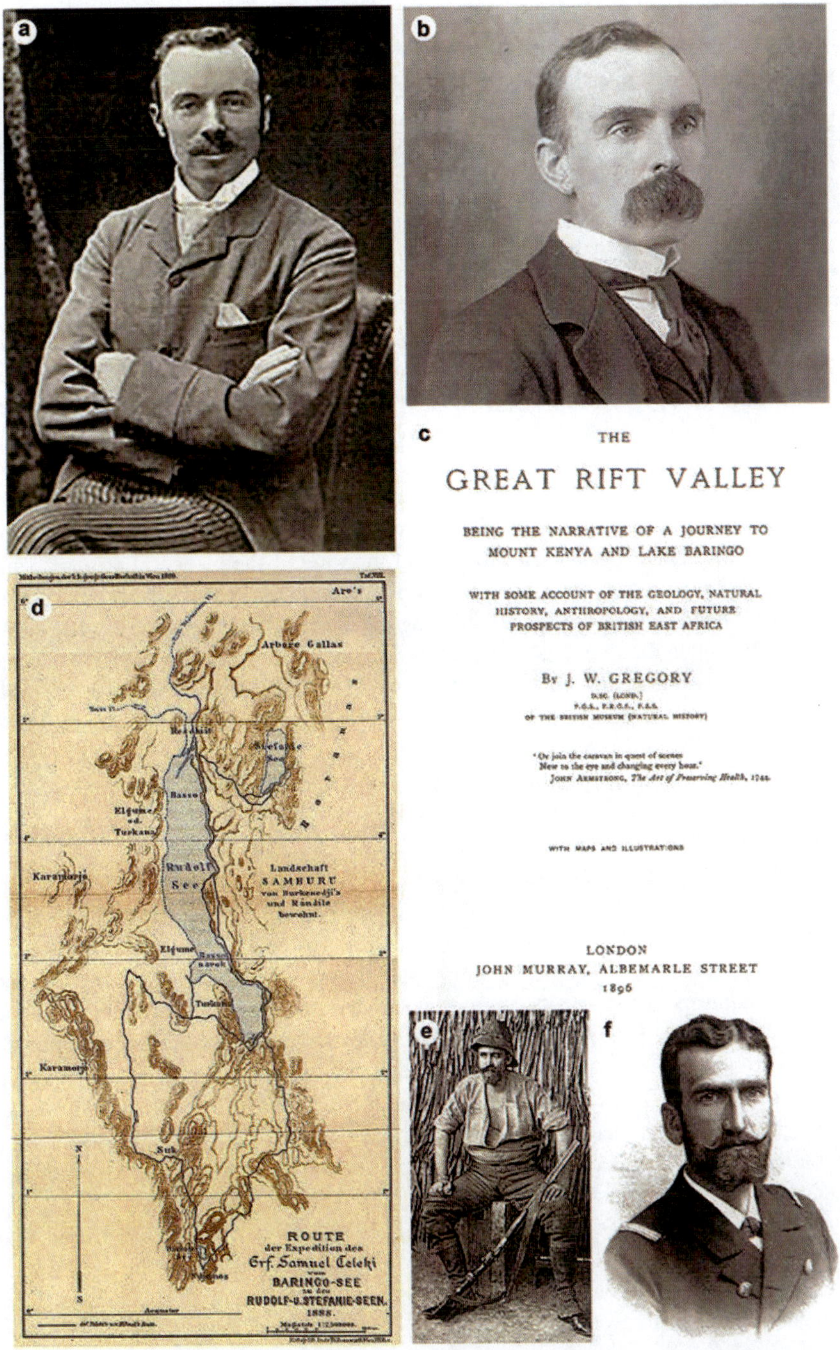

Fig. 2.1 Early European explorers in the Kenya Rift who recorded geological observations. (**a**) Joseph Thomson, who reached Lake Baringo and Lake Victoria. (**b**) JW Gregory, whose name will forever be linked to rift valleys. (**c**) Title page from JW Gregory's (1896) published account of his first expedition to the Kenya Rift. (**d**) Route taken by the Teleki-von Höhnel expedition to lakes Turkana, Baringo and Chew Bahir. (**e**) Count Teleki resting. (**f**) Ludwig von Höhnel. (All images are in the public domain)

lake was renamed Lake Turkana in 1975. They also visited Chew Bahir, a playa lake in the Ethiopian Rift, which he named Lake Stefanie after Princess Stéphanie of Belgium, the Prince's wife. Donaldson Smith (1896) described the eastern shore of Lake Turkana. The 'discovery' of Lake Turkana by Europeans is documented by Brown (1989) and Imperato (1998).

Several geological reports were later published based on observations made and samples collected during some of these early journeys. For example, rocks gathered by Sacci during the 1898 Bottego expedition to Lake Turkana were analysed and described by De Angelis d'Ossat and Millosevich (1900). Suess (1891), who never travelled to Africa, set the scene for future geological studies when he compiled evidence from explorers and suggested that most of the major East African lakes lay within a large north-south fracture system that crossed the continent (von Höhnel 1890; Dawson 2008).

In terms of geology, the most important early researcher was perhaps John W Gregory who first used the term 'rift valley' in 1894. In a series of publications he described the East African rift valleys and interpreted their general structure and origin, recognising that they are produced by extension rather than compression (Gregory 1894a, 1896; Leake 2011). Gregory made observations of the lakes and old lake beds. He made two main expeditions. The goal of the first (1892–3) was to reach Lake Turkana, but this was abandoned. Instead, he visited Lake Bogoria briefly and Lake Baringo, and made observations in the uplifted Kamasia fault block (Tugen Hills). He noted and measured former lake terraces and, after examining old lake sediments, speculated on the existence of an ancient 'Lake Suess' that had occupied much of the rift valley. He returned to Kenya in 1919 and soon afterwards published his second book on the rift (Gregory 1921).

Many early books about the Kenya Rift are little more than monotonous diaries of the exploits of 'brave' hunters, but occasionally they contain information on the lakes and local ecology or a photograph of interest. Some early government publications, however, are very useful. Among others, Harry Johnston, who was part of the colonial administration, made important observations of current relevance. He reported (Johnston 1902), for example, many details about the wildlife and people around Lake Baringo and Lake Bogoria and included early photographs. These showed that Lake Bogoria was then very high and close to overflowing into Lake Baringo, a condition that exists again at the time of writing. John Parkinson (1914) provided the first detailed geological information about the Lake Magadi basin and sediments in the southern Kenya Rift. Edward J Wayland, appointed the first director of the Geological Survey of Uganda in 1919 (Wayland 1920), deduced that drainage into Lake Victoria had reversed, and made observations in the Nyanza (Kavirondo) Rift. Although important sources of information, early journals and books are comparatively few (Fig. 2.2).

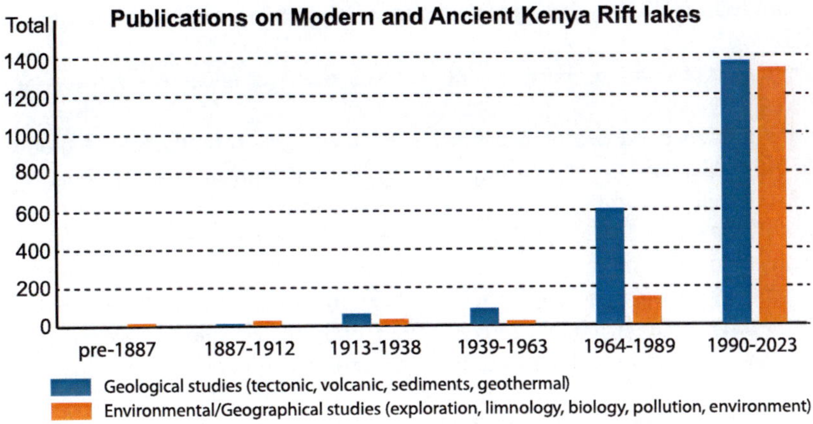

Fig. 2.2 Research publication rates. Graph shows cited publications in this book with slow initial geological, limnological and environmental output rates for modern and ancient lake systems. The first systematic scientific data appeared from the 1920s with the pace of publication increasing from the 1960s with an explosive growth in research during the last three decades

2.3 The 1920s until the 1940s: Movement on Multiple Fronts

2.3.1 Expeditions and Limnological Data Collection

Research activity increased rapidly in the 1920s and several themes emerged. The first multidisciplinary scientific expeditions to the East African lakes began in the late 1920s. Examples include the 1927–28 fisheries and limnology survey of Lake Victoria by Graham and Worthington (Worthington 1929, 1931) and the Percy Sladen Expedition in 1929, which examined the biology of lakes Naivasha, Baringo, Nakuru, Elmenteita, and 'Crater Lake' (Sonachi). Jenkin (1932, 1936) provided useful information about the physical and chemical limnology of the lakes and classified them on salinity and alkalinity. The larger Cambridge Expedition to the East African lakes in 1930–31 was led by E Barton Worthington. Although primarily biological, the resulting papers include general details of the limnology and lake water chemistry (Worthington 1932a; Beadle 1932), and geological investigations by Fuchs (1934).

At Lake Elmenteita, Penelope Jenkin worked with Louis Leakey, who was undertaking archaeological studies at the lake and studying former shorelines in the central Kenya Rift. Presentations at the Royal Geographical Society (RGS) in London in successive years by Leakey (1931: Nakuru, Elmenteita, Naivasha) and Worthington (1932b: Turkana, Baringo) gave details of several Kenya Rift lakes, including photographs.

In some cases, fossils reported during the earliest explorations of the region stimulated subsequent scientific expeditions. This was the case for the 1932–33 Mission

Scientifique de l'Omo, led by Arambourg (1934, 1935), which was prompted by Plio-Pleistocene fossils observed during the 1902–3 Bourg de Bozas French expedition (Harris et al. 2006). Further geological and archaeological research was carried out by pioneering researchers such as Nilsson, Leakey, Murray Hughes, Fuchs, Willis and Dixie. Fuchs, for example, had been a geologist for the mainly biological 1930–31 'Cambridge Expedition', but had been ill with malaria for much of the time. He returned to the Turkana region as leader of the 1934 'Lake Rudolf Rift Valley Expedition', which was sponsored by the RGS and others. His team included a palaeontologist, an archaeologist and a surveyor. The aim of the project was to undertake geological and survey work in the northern part of the Kenya Rift Valley. Results were published by Fuchs et al. (1935) and in discussion by Cox et al. (1935), and in a long paper by Fuchs (1939). The geological history of the lake was the major focus. Details are presented of palaeoshorelines, lake level fluctuations, Pleistocene sediments and fossil molluscs. However, the expedition also demonstrated the risks faced by early researchers when surveyor, WRH Martin and the medical officer and zoologist WS Dyson went missing near the South Island of Lake Turkana (Fuchs 1934). Fuchs noted that "during the search two tins, two oars, and Dyson's hat were found on the west shore of the lake, the latter being nearly 70 miles north of what must've been the scene of the accident".

Systematic science started to develop with the setting up of institutes such as the Department of Mines and Geology in Kenya in 1932 and the East African Fisheries Research Organisation (EAFRO) based in Kampala in 1947 (Talling 2006). However, it was not until the 1960s that there was a major increase in the number of researchers and their publications (Fig. 2.2). In the 1960s, biologists such as Coe, Gaudet and Lind, produced important baseline studies, with others such as Hecky, Kilham, Njuguni and Mavuti continuing their work into the 1970s and 1980s (Talling 2006). Important systematic studies over several years were produced for Lakes Turkana (Hopson and Hopson 1982), Baringo and Naivasha (Källqvist 1987).

Researchers at the time often camped in remote regions that had changed little since the early explorers. Reaching their study areas often entailed long journeys over dirt roads, with vehicles often becoming stuck on muddy surfaces, commonly formed by black cotton soil, or were stuck in or delayed by flooding rivers. The archaeologist Larry Robbins (2006) noted that on a journey down the rift scarp to Lake Turkana there was a sign with a skull and crossbones that stated "private burial ground for reckless drivers". He and others had to take sufficient food, fuel and water for the period of the expedition or would have to embark on long drives to resupply.

2.3.2 *Development and Rejection of Pluvial Theory*

The development of pluvial theory in East Africa is reviewed by Kingston and Hill (2005). Early studies of the lake basins in the East African Rift revealed abundant evidence for wetter periods including high-level shorelines, river terraces and

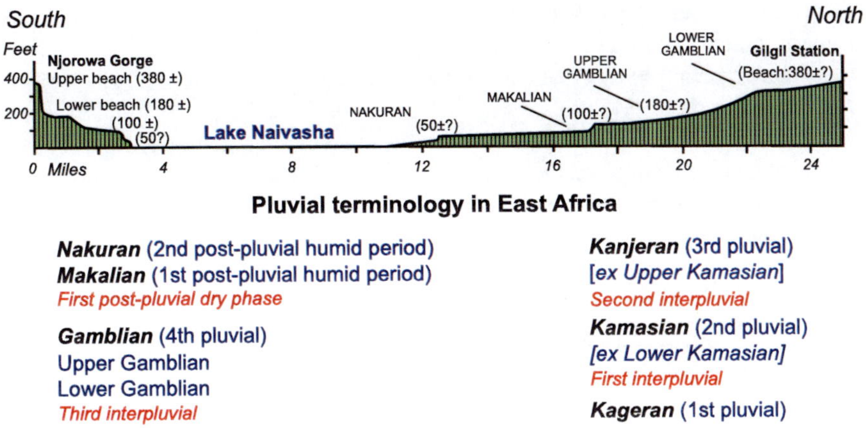

Fig. 2.3 Pluvial 'events' based on Leakey (1931) using Lake Naivasha as an example

sediments, and the sedimentary rock record provided evidence for former lakes that had dried up or become more saline. Gregory (1894b) noticed moraines at low altitudes on Mt. Kenya and speculated that these provide evidence of former higher precipitation and glacial advances, followed by drier conditions and retreat of the alpine glaciers. The next tentative step was to correlate these periods of tropical glacial advance and wetter conditions with the established European glacial periods in the Alps and the intervening drier periods with the interglacials. The wetter periods became known as 'pluvials', and the alternating drier periods became 'interpluvials'.

After a long debate, a four-fold division of climato-stratigraphic nomenclature was agreed (Fig. 2.3) and applied to deposits in the Kenya Rift. Assignments to a pluvial or interpluvial were based on fossils, archaeology and geological evidence. The scheme was widely adopted to provide relative ages for sediments across the region, and even extended across Africa. However, relative dating of archaeological and vertebrate assemblages was an important driver of its adoption in the Kenya Rift.

Usage of the scheme and of the terminology began in the 1920s (Wayland 1929) but was largely abandoned by 1960. The problems with the East African pluvial theory are many (Cooke 1958; Flint 1959; Bishop 1962, 1967) and are summarised by Kingston and Hill (2005), who noted that "It was an attempt to frame human evolution within a series of alternating wet (pluvial) and dry (interpluvial) climatic intervals, primarily as a means of establishing *a relative chronologic framework* [our italics] for fossil and archeological assemblages." The development of radiometric dating made relative dating using pluvial theory largely superfluous. Nonetheless, some phases of wetter and drier climate were real (e.g. the Gamblian) but understanding the causes and significance of climate changes was rarely a priority until much later.

Several studies of the lakes during the 1930s and 1940s used pluvial ideas as a framework for relative dating but made many other observations. Among these, Leakey (1931) reconstructed former levels of lakes Nakuru, Elmenteita and Naivasha using high-level terraces and beaches, and artefacts as indicators of age

(Fig. 2.3). He also provided bathymetric information for Lake Nakuru and presented a tentative correlation of pluvial phases with European ice advances.

A more detailed study of lake level changes and advances in tropical glaciers was made by Nilsson (1931, 1935, 1940). His study examined changes in lake level at lakes Naivasha, Nakuru, Elmenteita, Baringo and Bogoria, including aspects of the archaeology. He recognised the possible influence of tectonics when reconstructing former shorelines. His work on tropical glaciers included those on Mt. Kenya and Mt. Elgon.

2.4 Development of an Absolute Chronology

The development of radiometric dating techniques (K-Ar and ^{14}C in the late 1940s and 1950s, and Ar-Ar in the 1960s) had a dramatic impact on understanding the stratigraphy of Neogene Kenya Rift sediments, and for interpreting the late Quaternary climate history. Much of the stimulus for dating the sediments came from the palaeoanthropological community who wanted to date hominin fossils and stone tools that were increasingly being found in the Kenya Rift (Evernden and Curtis 1965; Bishop and Miller 1972; Bishop 1973). Thus the ages of many lacustrine deposits became known mainly because they were associated with fluvial deposits or paleosols containing fossils or artefacts of interest and were associated with volcanic rocks or tuffs that could be dated. Palaeomagnetism was also employed at an early stage. Indeed the Olduvai normal subchron (1.77–1.95 Ma) was first recognised at Olduvai Gorge near Lake Natron in northern Tanzania.

Radiocarbon dating, and later AMS dating, became common and was used to date former lake levels (e.g. Butzer et al. 1972) and associated occupation sites (e.g. Owen et al. 1982). Molluscs were commonly used to date palaeoshorelines but hardwater effects soon became apparent (Williams and Johnson 1976). Dating lacustrine organic matter in cores became more common though not free from contamination by old inert bicarbonate, especially where lakes are partly fed by hot springs (e.g. Tiercelin and Vincens 1987). Terrestrial plant debris (leaves, twigs, seeds) became the preferred material, and early reported ^{14}C ages often later proved unreliable. Radiometric dating put to sleep much of the pluvial hypothesis but the most recent high lake phase ('Late Gamblian') seems to represent the African Humid Period (broadly 15–5 ka BP).

2.5 The 1950s–1970s

2.5.1 *An Era of Geological Mapping*

Much of the geology of the Kenya Rift was mapped systematically for the first time during the three decades after World War II. Geologists at the Geological Survey of Kenya mapped immense areas using basic field tools. They had no GPS, satellite imagery or computers, and relied upon plane-table surveying, incomplete topographic base maps, and monochrome aerial photographs in difficult terrain. Roads were rough or non-existent, and transport was often unreliable. Notable among those working in the Kenya Rift are J McCall, B Baker, R Dodson, A Thomson, and J Walsh. They mapped the volcanic and sedimentary rocks along the rift and in adjacent regions and established much of the stratigraphic nomenclature in use today (Bishop 1967; McCall et al. 1967). Their contributions provided the framework for our understanding of the geological evolution of the ancient rift lakes and the development of the modern lakes. McCall (1967) mapped >6000 km^2 of the equatorial Kenya Rift including lakes Bogoria, Nakuru and Elmenteita. His emphasis was the volcanic succession and tectonics, but like his colleagues, he recognised that the sedimentary rocks sandwiched between the volcanic rocks preserve an important part of the history of the Kenya Rift. Baker (1958, 1963) similarly mapped and interpreted the large Magadi Basin in the southern Kenya Rift and considered the relationships between tectonics and sedimentation (Baker 1986). Thompson and Dodson (1963) and Thompson (1964) mapped the region around Lake Naivasha. Walsh and Dodson (1969) and Dodson (1963) mapped the vast semi-arid area west of Lake Turkana in reconnaissance style. Dodson's (1963) report and map include Lake Logipi, the Barrier volcanic complex at the south end of Lake Turkana, and northern Suguta Valley.

During the late 1960s and 1970s, a major mapping project was established at Bedford College, University of London, under the direction of Professor Basil C King (Fig. 2.4b) and linked to a group led by Dr. Michael Le Bas at the University of Leicester, with Dr. LAJ (Lawrie) Williams (U Lancaster and Nairobi). The *East African Geological Research Unit* (EAGRU) was set up to map unmapped areas and those examined only in reconnaissance style in the central and northern Kenya Rift, and to provide details in critical regions where the geology and palaeontology merited further investigation. The focus was the region from the junction of the Kenya Rift with the Nyanza (Kavirondo) Rift northwards to southern Suguta, including the Tugen Hills west of Lake Baringo (Fig. 2.4c), but EAGRU also supported work further south at Magadi and in other areas with most of the research undertaken by PhD students. They typically camped in their field areas (Fig. 2.5) and were supported by Kenyans familiar with the study sites and (or) with camp assistants that dealt with food and logistical support. The research led to a series of PhD theses and published geological maps. Analysis of the sedimentary rocks and stratigraphy of the Tugen Hills was led by Professor WW (Bill) Bishop (Fig. 2.4a), then at Bedford College. This region includes many lacustrine sequences (Bishop et al. 1971; Bishop 1978; Chapman et al. 1978; Pickford et al. 2009), and was undertaken when hominin fossils, artefacts and vertebrate fossils were being discovered in sedimentary rocks of Pliocene and Pleistocene age there, at East Turkana, and elsewhere in Kenya, so their importance was raised. Martin Pickford (Fig. 2.5c) and Andrew Hill also played a major role in explaining the geological context of the Tugen Hills discoveries and in their interpretation.

Fig. 2.4 EAGRU leaders and Tugen Hills/East Turkana field sites. (**a**) Professor Bill Bishop, summer 1976. (**b**) Professor Basil King, who was leading an undergraduate field course on Isle of Skye (NW Scotland) in spring 1973. (**c**) The uplifted Tugen Hills fault-block west of Lake Baringo where many lacustrine deposits of Miocene to Pleistocene age are exposed. (**d**) Diatomaceous silts of the Pliocene Chemeron Formation overlain by trachymugearite flow, Tugen Hills. (**e**) Badlands of the Koobi Fora Formation (background) overlain by the Holocene Galana Boi Formation (foreground), NE Lake Turkana. (**f**) Base-camp bandas at Koobi Fora, Lake Turkana, in summer 1976. (**g**) Breakfast at a fly camp east of Koobi Fora. Left-to-right: Bernie Owen, Glynn Isaac, Bill Bishop, Ian Findlater. Bernie Owen is pressing a buzzer to request the next course…

Fig. 2.5 Fieldwork conditions in Kenya. (**a**) PhD students Rob Crossley (second left) and Roger Knight (taking the photo) combined their efforts in an EAGRU-supported camp at the base of the Nguruman Escarpment (west of Magadi) in about 1969. Typically, they were supported by Kenyan teams. Ngua Muli (left) cooked and liaised with local Maasai with his son Julius (far right) who was a general camp assistant. Roger's field assistant stands second from the right. (**b**) Owen and Renaut fly camp, NE Turkana, 1977. Commonly used for short trips from a larger base. (**c**) Martin Pickford (left) and colleague towing a boat through the Molo Swamp (Lake Baringo), 1975. (**d**) Koobi Fora base camp provided better facilities along the remote northeast shores of Lake Turkana

2.5.2 Palaeoanthropological Research and Lakes

Studies in archaeology and human palaeontology expanded rapidly after the 1950s. Many of the most important sites in the Kenya Rift are directly or indirectly linked to lakes. Some occupation sites are present on modern and former shorelines of extant lakes (mainly Late Stone Age and younger); other sites are in fluvial sediments and palaeosols contained within predominantly lacustrine successions, or otherwise associated with lake deposits.

Among the more important localities is Olorgesailie in the southern Kenya Rift, discovered by Gregory (1921) and first studied in detail by Louis Leakey (1952) and Posnansky (1959). The Olorgesailie deposits contain Acheulian and Middle Stone Age artefacts preserved in a Pleistocene succession of diatomaceous lake beds, tuffs and palaeosols. Detailed mapping by Shackleton (1978) and extensive studies by Glynn Isaac (Fig. 2.4g; Isaac 1977 and references therein) set the scene for later work.

Perhaps the most famous studies of this era were those at Koobi Fora at northeast Lake Turkana (Figs. 2.4f and 2.5b, d), where a wealth of archaeological evidence and vertebrates have been found since 1968, including many hominins dating from between 2.1 and 1.3 Ma (Coppens et al. 1976; Harris et al. 2006). The host sediments (Koobi Fora Formation; Fig. 2.4e, f) are interbedded lacustrine and fluvial deposits with multiple tuff beds that have been used for dating. The initial studies on the lake deposits took place in the late 1960s and 1970s.

A third important locality is the Tugen Hills (Fig. 2.4c, d) where archaeological and palaeontological studies have taken place since the early work by EAGRU, Bishop, Pickford and Hill. Pickford (1978; Pickford et al. 2009) recognised that some of those lacustrine deposits formed in saline alkaline lakes. Farrand et al. (1976) studied lacustrine sediments with stone tools on the northern margins of neighbouring Lake Bogoria.

2.5.3 Other Contributions

Studies of the Kenya Rift lakes increased significantly during the 1970s and 1980s, both in number and diversity. French geologists, notably Jean Jacques Tiercelin (Fig. 2.6) and Maurice Taieb, were working around lakes of the central Kenya Rift,

Fig. 2.6 Dr. Jean-Jacques Tiercelin instructing geologists from National Oil Corporation of Kenya at southern Magadi in June 2013

in the Magadi-Natron basin, and in northern Suguta. More Kenyan geologists led the research as the universities in Kenya increased in number, and training in Europe and North America became easier.

The 1960s and 1970s were a period when biological and ecological studies also started to increase (Fig. 2.2). The remote Suguta Valley in the northern Kenya Rift was, for example, the focus of a multidisciplinary study by the *South Turkana Expedition* of the Royal Geographical Society in 1968 and 1969 (Gwynne 1969; Baker and Lovenbury 1971). The geomorphology, geology, and ecology of Suguta valley were studied in reconnaissance style. Geological studies were brief (Rhemtulla 1970). However, the geomorphological and ecological research (e.g. Hemming 1972) provided a useful background for more recent efforts.

Full chemical analyses of lake and inflow waters vary widely from lake to lake. Some lakes (e.g. Nakuru, Elmenteita, Logipi) have few, while others (e.g. Magadi, Bogoria) have >100. Many published analyses include only biologically important components (nutrients, dissolved oxygen, etc.). Talling and Talling (1965) and Kilham (1971) compiled and presented analyses of many Kenya Rift waters. Kilham combined much original and published data to interpret diatoms and fluoride chemistry of the lakes (e.g. Hecky and Kilham 1973; Kilham and Hecky 1973). Their work from this era has provided a wealth of historical data for comparisons with modern waters and their interpretation.

2.6 Applied Research: Geothermal and Petroleum Exploration

The Kenya Rift is rich in geothermal resources. Several power projects are producing energy (Olkaria since 1981) or are in development (Menengai, Akira, Eburru). Much of the initial exploration work was undertaken between the mid-1950s and the 1980s. Geological, geochemical and geophysical work included extensive drilling, notably at Olkaria south of Lake Naivasha, which improved understanding of the regional stratigraphy and geological history, and was accompanied by hydrological and hydrogeological studies that enhanced understanding of groundwater flow and interconnections between the central and southern rift lakes (Allen et al. 1989; Clarke et al. 1990). Such studies are critical when assessing the long-term recharge of the geothermal reservoirs that produce the steam for power generation.

A further major geothermal project was also undertaken in the central and northern Kenya Rift between Lake Baringo and Lake Turkana, as a joint project between the Mines and Geology Department, Kenya, and the British Geological Survey. This included mapping, in various levels of detail, most of the region east of Lake Turkana, with more detailed mapping of the rift between Lake Baringo and Lake

Turkana, a sector that hosts many geothermal prospects. Results of the mapping were published in a series of Kenya Mines-British Geological Survey Reports and maps (Hackman et al. 1990). A separate report by Dunkley et al. (1993) has important data on the northern Kenya Rift lakes including Lake Baringo, the small Suguta lakes and Lake Logipi. Other areas of geothermal interest that have produced stratigraphic and hydrogeological data include the Magadi Basin (Allen et al. 1989) and Homa Bay region in the western Nyanza Rift.

Concurrently with expansion of geothermal exploration, the Kenya Rift and the western branch of the East African rift system became the focus of petroleum exploration. Extensive geophysical surveys were undertaken by international petroleum companies, notably Total, Mobil, Amoco, Exxon, and later Tullow Oil, in liaison with the National Oil Corporation of Kenya (NOCK). These played a critical role in understanding the overall geological structure of continental rifts including the pattern of alternating half-grabens (Rosendahl et al. 1986; Rosendahl 1987; Lambiase 1990; Hendrie et al. 1994; Morley et al. 1999) and the stratigraphic development of basin fills (Lambiase and Bosworth 1995). Surveys included geophysical work at modern Lake Turkana and the Kerio Valley, and in old, buried lake basins to the southwest of the lake (Morley 1999) where lacustrine petroleum has been discovered at south Lokichar.

Although important in Lake Turkana, which was thought to have the highest petroleum potential in the Kenya Rift, the smaller rift lakes were less attractive, partly because of their size and environmental concerns. Nonetheless, some were examined for the clues that they provide for understanding sedimentary and geochemical processes that generate petroleum and produce reservoirs in the volcanic rift environment. For example, Elf (now Total France) supported research in Lake Baringo and Lake Bogoria in the mid-1970s and 1990s that contrasted sedimentation and organic matter evolution in a fresh oxygenated lake with high clastic sedimentation (Baringo) with a stratified saline alkaline lake with high organic productivity and anoxic bottom waters (Bogoria) (Tiercelin and Vincens 1987).

2.7 Palaeoclimate Studies

Reconstructing climate and environmental history by analysing lake sediments has been the primary focus of many projects in the Kenya Rift for several decades. Research has taken place in the modern lakes and wetlands, and in sequences of old sediments and sedimentary rocks, which are commonly well exposed by faulting. Many palaeolake sequences are sandwiched between volcanic rocks or contain tephra, which facilitates dating the rocks. Much of this work is discussed in the succeeding chapters.

Modern lakes and wetlands (Fig. 2.7a, b) have been cored to different depths since the 1960s and 1970s. Among early detailed studies that typify the approach,

Fig. 2.7 Coring lake and wetland sediments using inexpensive methods. (**a**) Recovering soft lake sediments using simple aluminium pipe hammered into soft sediment, Nasikie Engida, June 2013. (**b**) Coring wetland sediments in the Loboi Swamp north of Lake Bogoria, 2002. (**c**) Drilling platform for piston coring built from oil drums and scaffolding pipes, being towed by a Zodiac near Chemurkeu, Lake Bogoria, summer 1996. (**d**) Drilling platform in place near Losaramat, south central-sub-basin Lake Bogoria, summer 1996

Richardson and Richardson (1972) analysed a 28 m piston core from Lake Naivasha and obtained a climate record for the past ~9000 years; Richardson and Dussinger (1986) then studied a core from neighbouring Lake Elmenteita. Tiercelin and Vincens (1987) obtained three long (14.5–18.5 m) cores from Lake Bogoria (Fig. 2.7c, d), and many short cores from Lake Bogoria and Lake Baringo. Johnson and Odada (1996) analysed a piston core from Lake Victoria and showed that the lake desiccated completely in the late Pleistocene. Drill cores were also obtained from Lake Magadi by Magadi Soda Ltd. but with poor recovery. Cores of various lengths and age spans have now been obtained from most of the main rift lakes and some of the smaller (crater) lakes (e.g. Sonachi), which often provide good climate records because most of their sediment originates in a small catchment, and they often have slow sedimentation rates and anoxic bottom waters that inhibit bioturbation.

Multiple climate proxies have been used. Some are biological (e.g. diatoms, pollen), while others are chemical (trace elements, isotopes) or sedimentological/mineralogical (e.g. palaeosols, zeolites, salts). Dating of the cores has depended on the material deposited and its suitability within an individual lake but has included ^{14}C, Ar-Ar, U/Th, palaeomagnetism, and other methods (see Cohen 1993 for a summary). These are supplemented by relative age dating techniques (e.g. correlation of marker beds like volcanic ash), absolute methods (e.g. varve counting), and studies of geomorphological features in the drainage basin such as former shorelines and overflow channels. The wealth of data produced has allowed correlation of climate change indicators between basins and the recognition of wet and dry phases in East Africa. These include the African Humid Period (AHP) placed between ~15,000 and 5000 years ago (DeMenocal et al. 2000), and generally drier phases in the late Pleistocene and in the past few thousand years.

2.8 From 2010 Until the 2020s: The Era of Drilling and High-Resolution Studies

During the last decade, drilling lake sediments became a focus in the eastern branch of the East African Rift, in Kenya, Ethiopia and Tanzania. It was recognised that lake sediments obtained by coring would be a major resource for understanding the history of sedimentation in the Kenya Rift, changes in climate, tectonics and environment, and their implications for human evolution and Neogene palaeontology. Professor Andrew Cohen (U. Arizona) (Fig. 2.8a) led an international project to interpret the palaeoenvironmental histories of lakes in Kenya and Ethiopia, based on multidisciplinary analyses of lake sediments of different ages obtained by drilling near sites of known palaeoanthropological importance (Cohen et al. 2022). The *Hominin Sites and Paleolakes Drilling Project* (*HSPDP*) obtained drill cores from three sites in Kenya (W Turkana, W Baringo and Lake Magadi) and from two sites in Ethiopia (Afar, Chew Bahir) (Fig. 2.8). In parallel, the *Olorgesailie Drilling Project (ODP),* led by Dr. Rick Potts (Smithsonian Institution), drilled a core in the Koora Graben 12 km east of Lake Magadi. Results of these studies are revealing the complex histories of modern and ancient rift lakes, and the interplay between tectonics and climate in producing their record. Climate forcing (astronomical cycles) has been used to explain records of wetter and drier conditions preserved in lake sediments (e.g. Joordens et al. 2011; Lupien et al. 2022). It often works but, in the Kenya Rift, possible tectonic influences cannot be ignored (Bill Bishop, personal communication; Owen et al. 2018; and many others before us).

Fig. 2.8 Coring lake and wetland sediments using expensive methods. HSPDP coring at Lake Magadi in summer 2014. (**a**) Professor Andy Cohen, project leader. (**b**) Portable truck (lorry)-based drill-rig during short break in drilling; mud pit in foreground with white saline trona pan and Nguruman Escarpment in distance. (**c**) Preparing field cores for shipping to sampling laboratory. (**d**) Study and sampling of Magadi cores at LacCore laboratories in Minneapolis. (**e**) Base of one core showing sharp contact between Magadi Trachyte (~1 Ma) and freshwater limestone of the Oloronga Beds. Lack of palaeosol or weathering horizon implies a short interval between trachyte eruption and sedimentation in a then fresh lake

References

Allen DJ, Darling WG, Burgess WG (1989) Geothermics and hydrogeology of the southern part of the Kenya Rift Valley with emphasis on the Magadi-Nakuru area. Brit Geol Surv Res Rep SD/89/1

Arambourg C (1934) Observations sur la bordure nord du lac Rodolphe. CR Acad Sci Paris 197:856–858

Arambourg C (1935) Esquisse géologique de la bordure occidentale du Lac Rodolphe. Mission scientifique de l'Omo. 1932–33. Bull Mus Nat Hist Natur Paris 1

Baker BH (1958) Geology of the Magadi area. Geol Surv Kenya Rep 42

Baker BH (1963) Geology of the area south of Magadi. Geol Surv Kenya Rep 61

Baker BH (1986) Tectonics and volcanism of the southern Kenya Rift Valley and its influence on rift sedimentation. In: Frostick LE, Renaut RW, Reid I, Tiercelin J-J (eds) Sedimentation in the African rifts. Geol Soc Lond Spec Publ 25:45–57

Baker MJ, Lovenbury HT (1971) The South Turkana Expedition scientific papers VII. The 1969 season survey. Geogr J 137:349–360

Beadle LC (1932) Scientific results of the Cambridge Expedition to the East African Lakes. 1930-1.4. The waters of some East African Lakes in relation to their fauna and flora. J Linn Soc Zool 38:157–211

Beadle LC (1974) The inland waters of tropical Africa. An introduction to tropical limnology, 1st edn. Longman, London

Beadle LC (1981) The inland waters of tropical Africa. An introduction to tropical limnology, 2nd edn. Longman, London

Bishop WW (1962) Pleistocene chronology in East Africa. Adv Sci 18:491–494

Bishop WW (1967) Annotated lexicon of Quaternary stratigraphical nomenclature in East Africa. In: Bishop WW, Clark JD (eds) Background to evolution in Africa. University of Chicago Press, Chicago, pp 375–395

Bishop WW (1973) The tempo of human evolution. Nature 244:405–409

Bishop WW (ed) (1978) Geological background to fossil man. Geol Soc Lond Spec Publ 6

Bishop WW, Miller JA (eds) (1972) Calibration of hominoid evolution. Scottish Academic Press

Bishop WW, Chapman GR, Hill A, Miller JA (1971) Succession of Cainozoic vertebrate assemblages from the northern Kenya Rift Valley. Nature 233:389–394

Brown M (1989) Where giants trod: the saga of Kenya's desert lake. Quiller Press, London

Butzer KW, Isaac GL, Richardson JL, Washbourn-Kamau C (1972) Radiocarbon dating of East African lake levels. Science 175:1069–1076

Chapman GR, Lippard SJ, Martyn JE (1978) The stratigraphy and structure of the Kamasia Range, Kenya Rift Valley. J Geol Soc Lond 135:265–281

Clarke MCG, Woodhall DG, Allen D, Darling G (1990) Geological, volcanological and hydrogeological controls on the occurrence of geothermal activity in the area surrounding Lake Naivasha, Kenya. Ministry of Energy, Republic of Kenya

Cohen AS (1993) Paleolimnology: the history and evolution of lake systems. Oxford University Press, Oxford

Cohen AS, Campisano CJ, Arrowsmith JR, Asrat A, Beck CC, Behrensmeyer AK, Deino AL, Feibel CS, Foerster V, Kingston JD, Lamb HF, Lowenstein TK, Lupien RL, Muiruri V, Olago DO, Owen RB, Potts R, Russell JM, Schaebitz F, Stone JR, Trauth MH, Yost CL (2022) Reconstructing the environmental context of human origins in Eastern Africa through scientific drilling. Annu Rev Earth Planet Sci 50:451–476

Cooke HBS (1958) Observations relating to Quaternary environments in east and southern Africa. Trans Geol Soc S Afr 61:1–73

Coppens Y, Clark Howell F, Isaac GL, Leakey REF (eds) (1976) Earliest man and environments in the Lake Rudolf Basin: stratigraphy, paleoecology and evolution. University of Chicago Press, Chicago

Cox P, Wakefield RC, MacInnes DG, Champion AM, Worthington ER, Fuchs VE (1935) The Lake Rudolf Rift Valley Expedition, 1934: discussion. Geogr J 86:137–142

Dawson JB (2008) Discovery of the African rift valleys: early work on the Gregory Rift Valley and volcanoes in Northern Tanzania. Geol Soc Lond Mem 33:3–7

De Angelis d'Ossat G, Millosevich F (1900) Studio geologico sul materiale raccolto da M. Sacchi (seconda spedizione Bottego). Soc Geogr It:1–212

DeMenocal P, Ortiz J, Guilderson T, Adkins J, Sarnthein M, Baker L, Yarusinsky M (2000) Abrupt onset and termination of the African Humid Period: rapid climate responses to gradual insolation forcing. Quat Sci Rev 19:347–336

Dodson RG (1963) Geology of the South Horr area. Rep Kenya Geol Surv 60

Donaldson Smith A (1896) Expedition through Somaliland to Lake Rudolf. Geogr J 8:120–137 and 221–239

Dunkley PN, Smith M, Allen DJ, Darling WG (1993) The geothermal activity and geology of the northern sector of the Kenya Rift Valley. Brit Geol Surv Res Rep, SC/93/1

Evernden JF, Curtis GH (1965) The potassium-argon dating of late Cenozoic rocks in East Africa and Italy. Curr Anthropol 6:342–385

Farrand WR, Redding R, Wolpoff MH, Wright HT (1976) An archaeological investigation of the Loboi Plain, Baringo District, Kenya. Mus Anthropol Univ Mich Tech Rep 4

Fischer GA (1884) Bericht über die im Auftrag der geographischen Gesellschaft in Hamburg unternommene Reise in das Masai-Land. Teil 1. Allgemeiner Bericht. Mit Geogr Ges Hamburg 1883–1884:36–99

Fischer GA (1885) Bericht über die im Auftrag der geographischen Gesellschaft in Hamburg unternommene Reise in das Masai-Land. Wissenschaftliche Sammlungen. Mit Geogr Ges Hamburg 1883–1884:238–279

Flint RF (1959) Pleistocene climates in eastern and southern Africa. Bull Geol Soc Am 70:343–374

Fuchs VE (1934) The geological work of the Cambridge Expedition to the East African Lakes, 1931. Geol Mag 71:97–112 and 145–166

Fuchs VE (1939) The geological history of the Lake Rudolf Basin, Kenya Colony. Phil Trans R Soc Lond Ser B 299:219–274

Fuchs VE, Wakefield RC, Millard JF, MacInnes DG (1935) The Lake Rudolf Rift Valley Expedition, 1934. Geogr J 86:114–137

Gregory JW (1894a) Contributions to the physical geography of British East Africa. Geogr J 4:289–315 and 408–424

Gregory JW (1894b) The glacial geology of Mt. Kenya. Q J Geol Soc Lond 50:515–530

Gregory JW (1896) The Great Rift Valley. Murray, London

Gregory JW (1921) The rift valleys and geology of East Africa. Seeley Service, London

Gwynne M (1969) The South Turkana Expedition scientific papers I. Preliminary report on the 1968 season. Geogr J 135:331–342

Hackman BD, Charsley TJ, Wilkinson AF (1990) The development of the East African Rift system in north-central Kenya. Tectonophysics 184:189–211

Hamilton AC (1982) Environmental history of East Africa. A study of the Quaternary. Academic Press, London

Harper DM, Boar RR, Everard M, Hickley P (eds) (2003) Lake Naivasha, Kenya (Develop Hydrobiol 168). Kluwer, Dordrecht

Harris JM, Leakey MG, Brown FH (2006) A brief history of research at Koobi Fora, northern Kenya. Ethnohistory 53:35–69

Hecky RE, Kilham P (1973) Diatoms in alkaline saline lakes: ecology and geochemical implications. Limnol Oceanogr 18:53–71

Hemming C (1972) The South Turkana Expedition: scientific papers VIII. The ecology of South Turkana: a reconnaissance classification. Geogr J 138:15–40

Hendrie DB, Kusznir NJ, Morley CK, Ebinger CJ (1994) Cenozoic extension in northern Kenya: a quantitative model of rift basin development in the Turkana area. Tectonophysics 236:409–438

Hopson AJ (1982) Lake Turkana. A report on the findings of the Lake Turkana Project 1972–1975, vol 1–6. Overseas Development Administration, London

Imperato PJ (1998) Quest for the Jade Sea: colonial competition around an East African lake. Westview Press, Boulder

Isaac GL (1977) Olorgesailie: archeological studies of a Middle Pleistocene Lake Basin in Kenya. University of Chicago Press, Chicago

Jenkin PM (1932) Reports on the Percy Sladen Expedition to some Rift Valley lakes in Kenya in 1929, I. Introductory account of the biological survey of five freshwater and alkaline lakes. Annal Mag Natur Hist Ser 10:533–553

Jenkin PM (1936) Reports on the Percy Sladen Expedition to some Rift Valley lakes in Kenya in 1929, VII. Summary of the ecological results, with special reference to the alkaline lakes. Annal Mag Natur Hist Ser 18:133–179

Johnson TC, Odada EO (eds) (1996) The limnology, climatology and paleoclimatology of the East African lakes. Gordon and Breach, Amsterdam

Johnston HH (1902) The Uganda protectorate, vol 1, 1st edn. Hutchinson, London

Joordens JCA, Vonhof HB, Feibel CS, Louren LJ, Dupont-Riveted G, van der Lubbe JHJL, Sier MJ, Davies GR, Kroon D (2011) An astronomically-tuned climate framework for hominins in the Turkana Basin. Earth Planet Sci Lett 307:1–8

Källqvist T (1987) Primary production and phytoplankton in Lake Baringo and Lake Naivasha, Kenya. Norwegian Institute for Water Research (NIVA) Report

Kent P (1978) Historical background: early exploration in the East African Rift–the Gregory Rift Valley. In: Bishop WW (ed) Geological background to fossil man. Geol Soc Lond Spec Publ 6:1–4

Kilham P (1971) Biogeochemistry of African lakes and rivers. PhD thesis, Duke University, Durham, NC

Kilham P, Hecky RE (1973) Fluoride: geochemical and ecological significance in East African waters and sediments. Limnol Oceanogr 18:932–945

Kingston JD, Hill A (2005) When it rains it pours: legends and realities of the East African Pluvials. In: Lieberman DE, Smith RJ, Kelley J (eds) Interpreting the past: essays on human, primate and mammal evolution. American School of Prehistoric Research Monograph Series, vol 5. Brill, Boston, pp 189–205

Krapf JL, Ravenstein EG (1860) Travels, researches, and missionary labours, during an eighteen years' residence in Eastern Africa: together with journeys to Jagga, Usambara, Ukambani, Shoa, Abessinia and Khartum, and a coasting voyage from Nombaz to Cape Delgado. Trübner and Company, Paternoster Row, London

Krienitz L (2018) Lesser flamingos: descendants of Phoenix. Springer Nature, Berlin

Lambiase J (1990) A model for tectonic control of lacustrine stratigraphic sequences in continental rift basins. In: Katz BJ (ed) Lacustrine basin exploration: case studies and modern analogs. Am Assoc Petrol Geol Mem 50:265–276

Lambiase JJ, Bosworth W (1995) Structural controls on sedimentation in continental rifts. In: Lambiase JJ (ed) Hydrocarbon habitat in rift basins. Geol Soc Lond Spec Publ 80:117–144

Leake BE (2011) The life and work of Professor J.W. Gregory FRS (1864–1932): geologist, writer and explorer. Geol Soc Lond Mem 34

Leakey LSB (1931) East African lakes. Geogr J 77:497–508

Leakey LSB (1952) The Olorgesailie prehistoric site. In: Leakey LSB, Cole S (eds) Proc Pan Afr Congr Prehist 1947, p 209

Lehman JT (ed) (1998) Environmental change and response in East African lakes (Monogr Biol 79). Kluwer, Dordrecht

Lupien RL, Russell JM, Pearson EJ, Castañeda IS, Asrat A, Foerster V, Lamb HF, Roberts HM, Schäbitz F, Trauth MH, Beck CC, Feibel CS, Cohen AS (2022) Orbital-scale controls on the Pleistocene hydroclimate of eastern Africa. Sci Rep 12:3170

McCall GJH (1967) Geology of the Nakuru–Thomson's Falls–Lake Hannington area. Rep Geol Surv Kenya 78

McCall GJH, Baker BH, Walsh J (1967) Late Tertiary and Quaternary sediments of the Kenya Rift Valley. In: Bishop WW, Clark JD (eds) Background to evolution in Africa. Chicago University Press, Chicago, pp 191–220
Morley CK (ed) (1999) Geoscience of rift systems—evolution of East Africa. AAPG Stud Geol:44
Morley CK, Ngenoh DK, Ego JK (1999) Introduction to the East African rift system. In: Morley CK (ed) Geoscience of rift systems—evolution of East Africa. AAPG Stud Geol 44:1–18
Nilsson E (1931) Quaternary glaciations and pluvial lakes in British East Africa. Geogr Ann 13:249–349
Nilsson E (1935) Traces of ancient changes of climate in East Africa. Preliminary report. Geogr Ann 17:1–21
Nilsson E (1940) Ancient changes of climate in British East Africa and Abyssinia. A study of ancient lakes and glaciers. Geogr Ann 22:1–79
Odada E, Olago D (eds) (2006) The East African Great Lakes: limnology, paleolimnology and biodiversity. Kluwer, Dordrecht
Owen RB, Barthelme J, Renaut RW, Vincens A (1982) Palaeolimnology and archaeology of Holocene deposits north-east of Lake Turkana, Kenya. Nature 298:523–529
Owen RB, Renaut RW, Lowenstein TR (2018) Spatial and temporal geochemical variability in lacustrine sedimentation in the East African Rift System: evidence from the Kenya Rift and regional analyses. Sedimentology 34:1697–1730
Parkinson J (1914) The East African trough in the neighbourhood of the soda lakes. Geogr J 44:33–46
Payne AI (1986) The ecology of tropical lakes and rivers. Wiley, Chichester
Pickford M (1978) Geology, palaeoenvironments and vertebrate faunas of the mid-Miocene Ngorora Formation, Kenya. In: Bishop WW (ed) Geological background to fossil man. Geol Soc Lond Spec Publ 6:237–262
Pickford M, Senut B, Cheboi K (2009) The geology and palaeobiology of the Tugen Hills, Kenya Rift: tectonics, basin formation, volcanics and sedimentation. GeoPal-Kenya. Egerton University, Kenya
Posnansky M (1959) A Hope Fountain site at Olorgesailie, Kenya Colony. S Afr Archaeol Bull 16:83–89
Rhemtulla S (1970) The South Turkana Expedition: scientific papers III. A geological reconnaissance of South Turkana. Geogr J 136:61–73
Richardson JL, Dussinger RA (1986) Paleolimnology of mid-elevation lakes in the Kenya Rift Valley. Hydrobiologia 143:167–174
Richardson JL, Richardson AE (1972) History of an African rift lake and its climatic implications. Ecol Monogr 42:499–534
Robbins LH (2006) Lake Turkana archaeology. Ethnohistory 53:71–93
Rosendahl BR (1987) Architecture of continental rifts with special reference to East Africa. Annu Rev Earth Planet Sci 15:445–503
Rosendahl BR, Reynolds DJ, Lorber PM, Burgess CF, McGill J, Scott D, Lambiase JJ, Dersken SJ (1986) Structural expressions of rifting: lessons from Lake Tanganyika, Africa. In: Frostick LE, Renaut RW, Reid I, Tiercelin J-J (eds) Sedimentation in the African rifts. Geol Soc Lond Spec Publ 25:29–43
Schagerl M (ed) (2016) Soda lakes of East Africa. Springer Nature, Cham
Shackleton RM (1978) Geological map of the Olorgesailie Area, Kenya. In: Bishop WW (ed) Geological background to fossil man. Geol Soc Lond Spec Publ 6, pp 171–172
Speke JH (1864) The discovery of the source of the Nile. Harper and Brothers, New York
Stanley H (1875–76) Letters of Mr. HM Stanley on his journey to Victoria Nyanza, and circumnavigation of the lake. Proc R Geogr Soc Lond 20:134–159
Suess E (1891) Chapter 4: Die Brüche des östlichen Afrika. In: von Höhnel LR, Rosiwal A, Toula F, Suess E (eds) Denkschrifte der Königlich Akademie für Wissenschaften Wien. Math-Naturwissenschaften Klasse 58:555–584

Talling JF (2006) A brief history of the scientific study of tropical African inland waters. Freshw Forum 26:3–37

Talling JF, Lemoalle J (1998) Ecological dynamics of tropical inland waters. Cambridge University Press, Cambridge

Talling JF, Talling IB (1965) The chemical composition of African lake waters. Int Rev Ges Hydrobiol 50:421–463

Thompson AO (1964) The geology of the Kijabe area. Rep Geol Surv Kenya 43

Thompson AO, Dodson RG (1963) The geology of the Naivasha area. Rep Geol Surv Kenya 55

Thomson J (1885) Through Masai land. Sampson Low, Marston, Searle and Rivington, London

Tiercelin J-J, Vincens A (eds) (1987) Le demi-graben de Baringo–Bogoria, Rift Gregory, Kenya: 30,000 ans d'histoire hydrologique et sédimentaire. Bull Centr Rech Explor-Prod Elf-Aquitaine 11:249–540

von Höhnel LR (1890) Ost-Äquatorial Afrika zwischen Pangani und neuentdecken Rudolf-See. Peterm Mitteil Ergänz 99:1–44

von Höhnel LR (1894) The discovery of lakes Rudolf and Stephanie: a narrative of count Samuel Teleki's exploring & hunting expedition in Eastern Equatorial Africa in 1887 & 1888. Two volumes. Longmans, Green and Co.

Walsh J, Dodson RG (1969) Geology of northern Turkana. Geol Surv Kenya Rep 82

Wayland EJ (1920) Annual report of the Geological Survey of Uganda

Wayland EJ (1929) African pluvial periods and prehistoric man. Man 29:118–121

Williams REG, Johnson AS (1976) Birmingham University radiocarbon dates X. Radiocarbon 18:249–267

Worthington EB (1929) Observations on the temperature, hydrogen-ion concentration, and other physical conditions of the Victoria and Albert Nyanzas. Int Rev Ges Hydrobiol 24:328–357

Worthington EB (1931) Vertical movements of freshwater macroplankton. Int Rev Ges Hydrobiol 25:394–436

Worthington EB (1932a) Scientific results of the Cambridge Expedition to the East African Lakes, 1930–31, 1: general introduction and station list. J Linn Soc Zool 38:99–119

Worthington EB (1932b) The lakes of Kenya and Uganda. Geogr J 79:275–293

Worthington EB (1996) Early research on East African lakes: an historical sketch. In: Johnson TC, Odada EO (eds) The limnology, climatology and palaeoclimatology of the East African Lakes. Gordon and Breach, Amsterdam, pp 659–664

Worthington S, Worthington EB (1933) Inland waters of Africa: the result of two expeditions to the Great Lakes of Kenya and Uganda, with accounts of their biology, native tribes and development. Macmillan, London

Chapter 3
Geology of the Kenya Rift: An Introduction

3.1 Tectonic Setting of the East African Rift

Suess (1891) first reported the presence of an East African continental fracture, which Gregory (1894, 1896) called the Great Rift Valley and described in a series of papers (Gregory 1920a, b, 1921, 1923). Other early studies of rift geology included the works of Krenkel (1925), Dixey (1926), and Willis (1936). Subsequently, the various rift segments were referred to as the East African Rift System (EARS) with many authors presenting syntheses of its overall geology (Dixey 1946, 1956; Quennel 1960; McConnell 1967; Baker and McConnell 1970; Baker et al. 1972; Baker and Morgan 1981; Morgan and Baker 1983; Grove 1983; Girdler 1991; Morley 1995, 1999a; Burke 1996; Ebinger et al. 2000; Chorowicz 2005; Ebinger 2005, 2021; Braile et al. 2006; Ebinger and Scholz 2012; Paron et al. 2013; Ring 2014; Woldegabriel et al. 2016; Ring et al. 2018; Purcell 2018). It has also been recognised that the EARS represents only the latest of a varied series of rifts that have included Permian Karoo and Cretaceous to Palaeogene sedimentary basins (Fig. 3.1; Lambiase 1989; Ebinger and Ibrahim 1994; Schlüter 1997; Delvaux 2001; Chorowicz 2005; Macgregor 2015, 2018).

The 3500-km-long EARS extends from the Afar Triple Junction in Ethiopia southwards to Lake Malawi and beyond, with incipient rifts extending also to the Okavango Delta in Botswana (Fig. 3.1b). The Afar Triple Junction also links the EARS with the Red Sea Rift and the Indian Ocean Ridge via the Gulf of Aden (Prodehl and Mechie 1991; Wolfenden et al. 2004; Rooney 2020a, b, c, d). There are two contrasting EARS branches (west and east), which pass around the western and eastern sides of the Archaean Tanzanian Craton (Fig. 3.1b), propagating locally into its margins (Rogers 2006). The eastern branch is associated with widespread igneous activity, whereas only a few volcanic centres are present in the western branch. Although usage varies among authors, the ‘Gregory Rift’ encompasses the Kenya Rift and its southern extension into northern Tanzania. This is distinct from the

R. W. Renaut, R. B. Owen, *The Kenya Rift Lakes: Modern and Ancient*, Syntheses in Limnogeology, https://doi.org/10.1007/978-3-642-25055-2_3

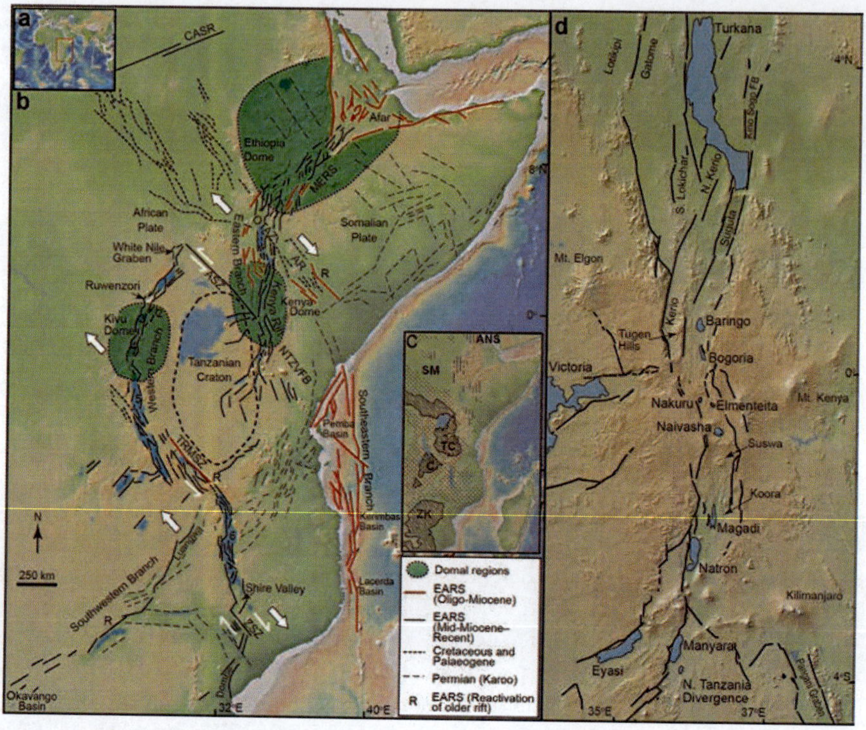

Fig. 3.1 Rift systems in eastern Africa. (**a**) General location. (**b**) The East African Rift System. Base maps constructed from GeoMapApp (https://www.geomapapp.org/). ZSZ Zambesi shear zone, TRMSZ Tanganyika-Rukwa-Malawi shear zone, ASZ Aswa shear zone, NTVFB North Tanzania volcanic and fault belt, OTAZ Omo-Turkana accommodation zone, MERS Main Ethiopian Rift, AR Anza rift, CASR Central African shear zone. White arrows show relative movements. Western rift lakes: Albert/Mobutu (1), George (2), Edward (3), Kivu (4), Tanganyika (5), Malawi (6) and Lake Mweru (7). (**c**) Major cratons – Tanzania (TC), Congo (C) and Zimbabwe-Kalahari (ZK). Sahara megacraton marked SM. ANS (vertical lines) marks the Arabian-Nubian Shield. Stippled pattern represents reworked pre-Neo-Proterozoic crust. Modified from Fritz et al. (2013). (**d**) Kenya Rift faults, topography and lake basins

'Ethiopian Rift' to the northeast. The western branch, characterised by very deep fresh lakes and only local volcanism, has also been called the 'Albertine Rift'.

The eastern EARS also includes two elevated regions, the Ethiopia and Kenya Domes (Fig. 3.1b), which are associated with negative gravity anomalies and reflect tectonism and the accumulation of Neogene volcanic rocks (Fairhead 1986; King 1978; Morley et al. 1999b; Davis and Slack 2002) as well as the balance between uplift and precipitation-induced erosion (Burke and Wilkinson 2016). Xue et al. (2019) also noted that major rivers (Athi, Tana, Ewaso Ngiro North) flowing away from the Kenya Dome and towards the Indian Ocean experienced increased average incision rates after about 4.5 Ma. These domes have been considered the surface expression of mantle plume activity (Furman et al. 2004) and are separated by the Turkana Depression and the early Cretaceous Anza Rift (Fig. 3.1b; Bosworth 1992;

Ebinger and Ibrahim 1994; Bosworth and Morley 1994; Furman et al. 2004; Corti et al. 2019).

The rift varies in age at different locations and was probably initiated in the early Miocene, although earlier Palaeogene igneous activity took place in northern Kenya and Ethiopia (Logatchev et al. 1972; Hendrie et al. 1994; Ebinger et al. 1993; Woldegabriel et al. 2016). Boone et al. (2019), for example, noted pronounced Eocene to Miocene denudational cooling of basement rocks bounding the Lokichar Basin to the southwest of Lake Turkana, suggesting an even earlier onset of rifting. More generally, rifting is considered to have migrated southwards from the Afar, at about 35 Ma, to the southern Kenya Rift by about 5–8 Ma (Fig. 3.2) (Cerling and Powers 1977; Crossley and Knight 1981; Purcell 2018). However, Le Gall et al. (2008) observed that several basins in the north Kenya and Ethiopian rifts formed at about the same time (~35 Ma) as Lower Miocene basins in the Central Gregory Rift (Morley et al. 1992a; Mugisha et al. 1997; Hautot et al. 2000). This led to a suggestion that the rift basins might have initially formed as isolated basins with linkage taking place at a later, more mature stage (Vétel 2005; Vétel and Le Gall 2006).

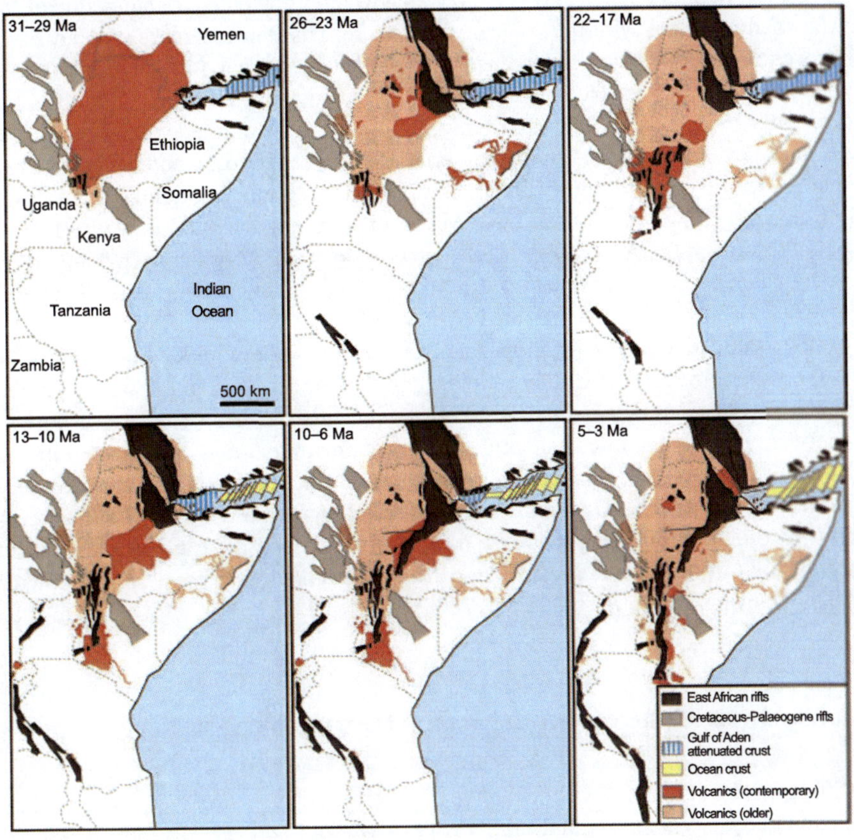

Fig. 3.2 Neogene evolution of the East African Rift System. Modified from Purcell (2018)

Subsequent work on denudation rates in the northern and central Kenya Rift (Acosta 2015) has suggested a three-phased thermal history with rapid uplift of rift shoulders between 65 and 50 Ma and since 15 Ma. Zawacki (2021) used analyses of zircons to provide evidence for Plio-Pleistocene erosion rates in the Turkana and Awash areas and noted consistent and relatively low denudation during this period despite significant tectonic and geomorphic shifts in the landscape. Boone et al. (2017) also examined thermal histories in the Lapur area in northwestern Turkana and observed rapid Early Cretaceous cooling (~120–100 Ma), which they related to denudation. Subsequent reheating between 80 and 20 Ma coincided with the accumulation of the Late Cretaceous-Eocene Lapur Sandstone and eruptions of the latest Eocene-early Miocene Turkana volcanics with faulting beginning around 14 Ma.

Recently, Rooney (2017, 2020a, b, c, d) presented a synthesis of Cenozoic magmatism in East Africa, noting three major pulses of basaltic volcanism in northern Kenya and Ethiopia that gave way to bimodal lavas or silicic volcanism. He recognised an initial basaltic phase (45–34 Ma) located in southern Ethiopia and northern Kenya (Turkana) that was caused by mantle plume activity. This was followed by an 'Oligocene Traps Phase' (~33.9–27 Ma) that produced extensive lavas centred on the NW Ethiopian Plateau and Yemen, which Rooney attributed to interaction of the Afar mantle plume with the African lithosphere. His Phase 3 represents an Early Miocene (~26.9–22 Ma) resurgence in basaltic lavas involving smaller volumes than the earlier phases. He noted uncertainties but suggested that this phase resulted from lithospheric melts and contributions from the Afar plume, coupled with synchronous extension of the continental crust (Stab et al. 2016). Rooney (2020a) further observed that more evolved compositions (flood phonolites in Kenya) subsequently dominated until ~12 Ma with a return to basaltic volcanism at ~12–9 Ma from the Afar to Kenya. Silicic eruptions followed between 9 and 4 Ma with further basaltic activity at ~4–1.6 Ma, the latter giving way to modern patterns of volcanism, which are dominated by central silicic volcanoes. The 20-, 12- and 4-Ma basaltic phases also developed at Turkana, but without major silicic intervals.

The western branch has been considered younger than the eastern branch (Ebinger 1989; Morley et al. 1999b). However, this has been challenged by evidence for rift-related volcanism and lake development that may have been initiated by about 25 Ma in the Lake Rukwa area (Roberts et al. 2010, 2012). A third Tertiary rift lies offshore in the Mozambique Channel (Mougenot et al. 1986) with a young SW-trending rift following reactivated Karoo faults along the Luangwa Valley and terminating in shallow half-grabens of the Okavango Basin (Fig. 3.1b; Chorowicz 2005; Huntsman-Mapila et al. 2005, 2009; Kinabo 2007).

3.2 Major Controls on the East African Rift System

Rifting can be classified as either active or passive with the former characterised by mantle buoyancy and regional uplift, with or without surface volcanism (Ebinger and Scholz 2012). In contrast, the driving forces in passive rifting are derived from distant (far-field) plate-boundary processes. Macdonald et al. (1994) suggested two models for the formation of the EARS, involving either a mantle plume impacting against the base of the lithosphere or extension caused by far-field forces (Saria et al. 2014; Stamps et al. 2020). The latter result in decompression melting of the asthenosphere, which then wells up passively. Evidence for an ascending plume includes gravity anomalies, pre-faulting volcanism, the large volumes of magma erupted in parts of the EARS and the small-scale (<10 km) of lithospheric extension (Ebinger et al. 1989; Hooper 1990; Morley et al. 1992a; Latin et al. 1993; Sippel et al. 2017). Bailey (1992) favoured far-field forces that reactivated ancient crustal weaknesses, with Macdonald et al. (1994) suggesting that rift segmentation was not consistent with a plume below the rift axis. Other researchers have invoked a combination of passive and active rifting to explain the differing observations (Stamps et al. 2010).

Hansen et al. (2012) presented four plume-related models to explain rifting. The first involves small-scale convection due to lithospheric stretching (Fig. 3.3a). Ebinger and Sleep (1998) noted that a single plume might also be channelled to other locations by the basal lithospheric topography with decompression melting taking place where the lithosphere rises. The second Hansen model results from 'edge flow', which occurs at sutures between thinner and thicker parts of the lithosphere (Fig. 3.3b). This is driven by contrasts in heat flow through variable thicknesses of lithosphere and results in small-scale convection without any need for stretching (King and Anderson 1998; King and Ritsema 2000). The Tanzania craton and Mozambique Mobile Belt boundary fit this model, but Nyblade (2002) points out that a lack of magmatism in the western branch, with a similar structural setting, argues against this possibility, as does a lack of volcanism in older African rifts. The third Hansen model infers multiple mantle plumes (Fig. 3.3c) with the fourth suggesting an African superplume (Fig. 3.3d) that transfers thermally heated mantle laterally as well as vertically to overlying rifts.

Several researchers have proposed that the superplume model involves a thermochemical anomaly sourced near the core-mantle boundary below South Africa (Simmons et al. 2007; Bagley and Nyblade 2013; Mulibo and Nyblade 2013; Koptev et al. 2015), which then rises towards East Africa and Arabia, possibly splitting into two or more smaller ascending plumes (George et al. 1998; Davis and Slack 2002; Halldórsson et al. 2014). More recently, this superplume has also been linked to a Large Low Shear Velocity Province (LLSVP), a region of dense material in the deep mantle below the African plate (Rooney 2017). It has also been noted that the superplume hypothesis is consistent with high elevations in eastern and southern Africa (Moucha and Forte 2011), basalt major and trace element data (Furman et al. 2006; Rogers 2006; Furman 2007) and helium isotope evidence (Pik et al. 2006).

Morley (2010) suggested that the stress patterns that determine rift alignments result primarily from external forces related to spreading ridges around Africa and intraplate stresses developed from mantle plumes (Coblentz and Sandiford 1994; Weerarantne et al. 2003). The only major far-field extensional stress is slab pull due to convergence in the north and northeast. In contrast, ridge-push forces, related to

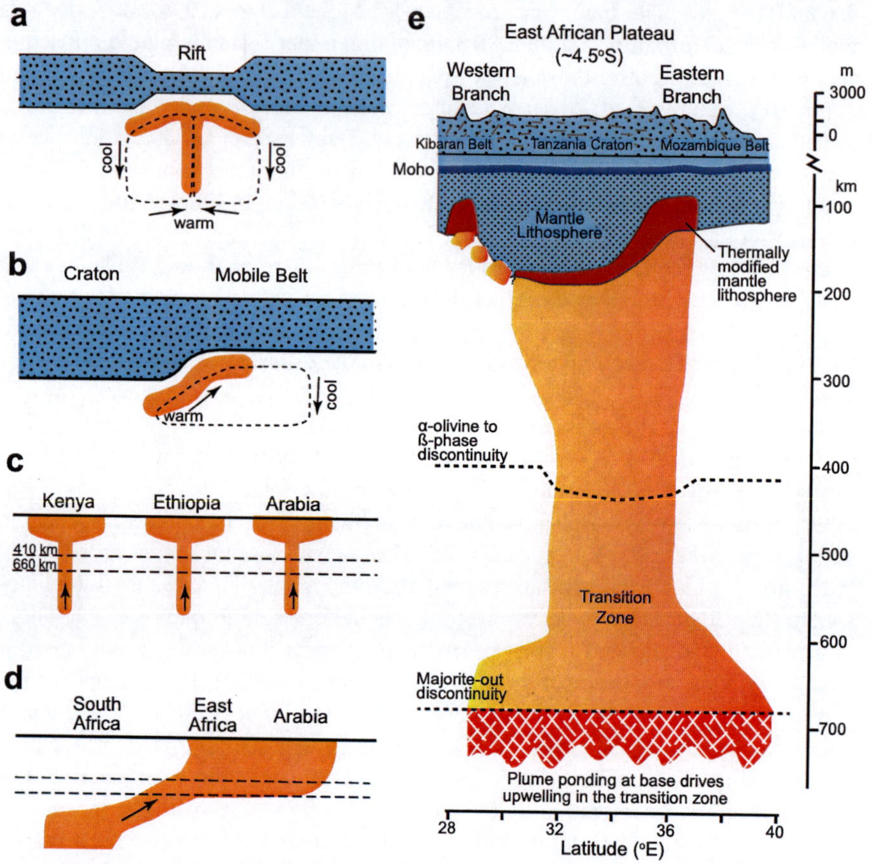

Fig. 3.3 Plume models. After Hansen et al. (2012). (**a**) Small-scale convection due to lithospheric stretching. (**b**) Small-scale convection resulting from edge-flow at craton margin. (**c**) Multiple plume sources. (**d**) Deep core-mantle source below South Africa ascending towards East Africa. Plume may split below the lithosphere. (**e**) Plume ponding at transition zone generates a thermally upwelling mantle below East Africa. After Huerta et al. (2009). Discontinuities at 680 and 430 km represent phase changes

the surrounding ocean ridges, dominate much of Africa causing compression. Consequently, EARS development is largely attributed to asthenospheric upwelling and lateral flow (Demissie 2010; Ebinger and Scholz 2012). Additional variability in stress directions relates to local or regional igneous intrusions, topographic domes, the development of major faults, and variations in basement fabric strength (Morley 2010). Pre-rift structures are thus also important in determining the nature and precise orientation of rifting (Daly et al. 1989; Smith and Mosley 1993; Achauer 1994; Le Turdu et al. 1999).

It has been proposed that there are five major types of pre-rift geological controls that have affected the region (Versfelt and Rosendahl 1989). These include (1) the

50–60-km-thick rigid Archaean cratons (Fig. 3.1c), which consist of granite, gneiss, amphibolite and greenstone belts; (2) the folded Neo-Proterozoic recycled crust or Palaeo- and Meso-Proterozoic mobile (orogenic) belts; (3) transcontinental dislocation zones, mostly within mobile belts, that have been repeatedly reactivated; (4) earlier Permo-Triassic to Cretaceous rifts; and (5) basement foliations (Morley et al. 1999b; Nyblade and Brazier 2002; Fritz et al. 2013).

Both branches of the EARS follow mobile belts around the Tanzanian Craton (Fig. 3.1b). Morley (2010) noted that Tertiary rifts (e.g. Rukwa and Tanganyika) closely followed metamorphic fabrics in Palaeo- and Meso-Proterozoic belts, but with such relationships less clear in the Neo-Proterozoic Mozambique Belt (e.g. the Kenya and Malawi rifts), possibly reflecting gentler foliation dips. In the Rukwa Basin, the steep metamorphic foliations of the NW-trending Ubendian Mobile Belt have repeatedly guided faulting during the formation of a Karoo sedimentary basin, a proto-Rukwa rift and the late Tertiary Rukwa Rift. These alignments follow the Tanganyika-Rukwa-Malawi Shear Zone and contrast strongly with the north-south orientations of the adjacent Malawi and Tanganyika Rifts (Fig. 3.1b; Morley et al. 1992b; Kilembe and Rosendahl 1992; Morley 2010).

The Malawi Rift, for example, follows the major mobile belts in the region with two alternations in half-graben polarity developing at intersections with major transcontinental dislocation zones (TDZ) (Versfelt and Rosendahl 1989). These include the West and East Chamaliro TDZs, which mark a change in polarity between the Livingstone and Usisya half-grabens and the North Chamaliro and Henga Valley TDZs, which coincide with a change between the Usisya and Metangula half-grabens (Fig. 3.4a, b).

Rooney (2020d) discusses the impact of the Tanzania Craton versus Mozambique Mobile Belt settings on Kenya Rift volcanism, whereas Smith and Mosley (1993) emphasise examples of basement control on faults (Fig. 3.4c). The latter authors also report that the lithosphere thins northwards from 35–40 km at the reworked craton margin to 18–20 km below Lake Turkana and the Mozambique Belt. Undersaturated, nephelinitic volcanism has been associated with rifted cratonic areas, whereas alkaline basalt and trachyte activity has developed on thinner crust north of the buried craton (Smith and Mosley 1993; Rooney 2020d). Generally, north-south faults dominate in the Kenya Rift. However, in the central portion over the reworked craton margin additional NW-trending faults and lineaments are common (Fig. 3.4d). Le Turdu et al. (1999), for example, noted a progressive change in this region toward fault orientations of N150° due to interference with the N140° Aswa (Asua) Fault Zone, which marks the southern limit of the reworked craton margin. The rhomb-shaped, Baringo Basin (Lake Baringo and the Loboi Plain), which is >6-km-deep, is defined by north-south faults to the east and west and by the NW-trending Porumbonyanza-Ol Kokwe and Waseges-Marmanet transverse zones to the north and south (Fig. 3.4d). These latter zones, in turn, reflect the major crustal and/or lithospheric discontinuities below (Le Turdu et al. 1999; Le Gall et al. 2000; Hautot et al. 2000).

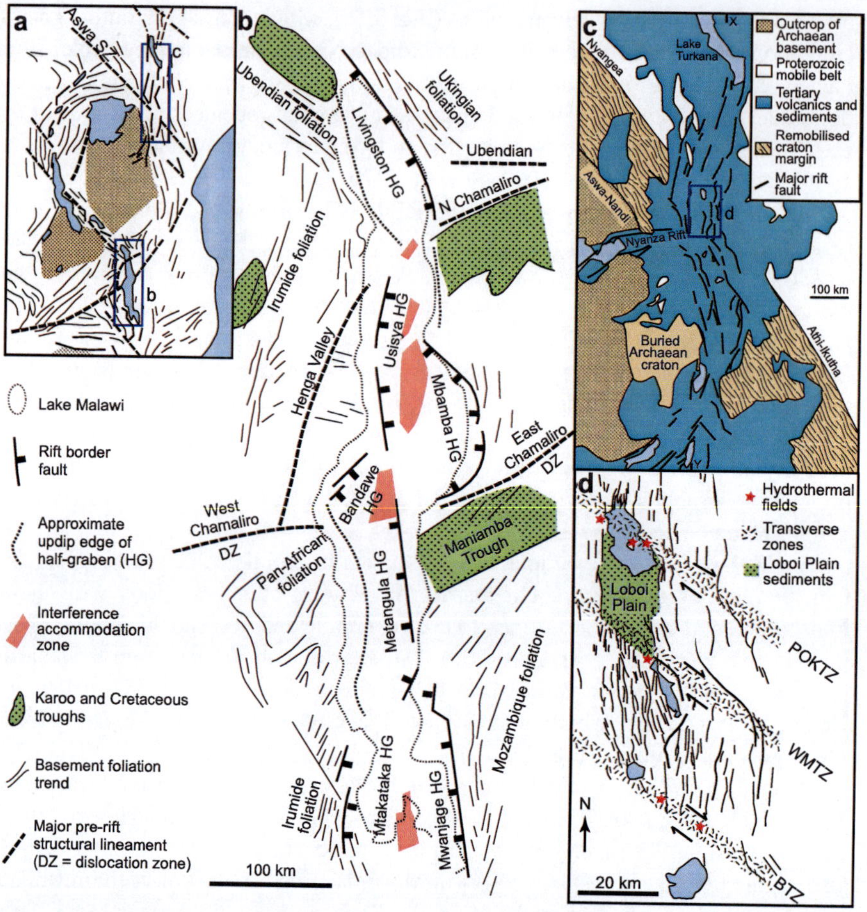

Fig. 3.4 Pre-rift structural controls. (**a**) Tectonic framework with rifts following foliation (thin lines) in mobile belts. (**b**) Detail of major structures near Lake Malawi showing border faults and half-grabens in relation to pre-rift structures. Note fault orientations are similar to basement foliation and the association between accommodation zones and pre-rift dislocation zones. After Versfelt and Rosendahl (1989). (**c**) Kenya Rift Valley. Note change in orientation of faults associated with underlying remobilised craton trends. After Macdonald et al. (2001). (**d**) Structural controls in the Baringo-Bogoria region. Note three major NW-SE substratum heterogeneities associated with fault deflections and basin alignments (e.g. boundary between Loboi Plain sediments and intensely faulted region to the south). POKTZ Porumbonyanza-Ol Kokwe Transverse Zone, WMTZ Waseges-Marmanet Transverse Zone, BTZ Bahati Transverse Zone. After Le Turdu et al. (1999)

3.3 Accommodation Zones

The EARS is comprised of a series of connected basins that form fundamental units (~60–100 km long, ~40–70 km wide) that can be linked in a variety of ways (Bosworth 1985, 1989; Bosworth et al. 1986; Rosendahl et al. 1986; Rosendahl

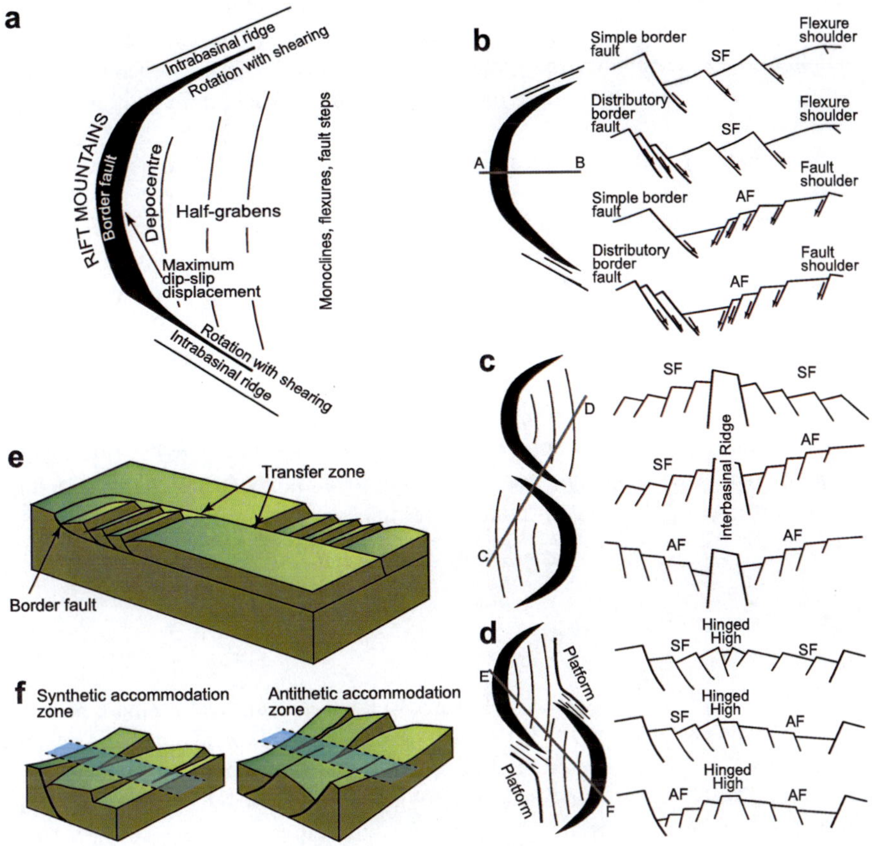

Fig. 3.5 Accommodation zones, half-grabens and border faults. (**a–d**) Modified from Rosendahl (1987); (**e, f**) After Faulds and Varga (1998). (**a**) Schematic border fault and half graben. (**b–d**) Half-graben morphologies. Synthetic faults (SF) and antithetic faults (AF) dip in the same and opposite directions to the border fault, respectively. Note high points in '**c**' and '**d**' correspond to opposing border faults and overlapping half-grabens, respectively. (**e**) Transfer zone with normal faults terminating against strike-slip faults. (**f**) Accommodation zones showing overlapping faults with strain transferred via fault ramps. Synthetic zones involve faults with similar dip polarities. Antithetic zones characterised by opposing dips

1987). Full grabens are present, but commonly the basins are asymmetric with curved border faults that exhibit large throws. These are opposed by monoclines and/or relatively minor antithetic or synthetic faults that form multiple horsts, grabens, tilt blocks and step-faulted platforms (Fig. 3.5). The geometry of these half-grabens also means that the border faults are arcuate in cross-section, flattening with depth (Rosendahl et al. 1986).

Ebinger et al. (2002) noted that the hanging wall of a typical border fault is flexed down with a maximum dip-slip displacement located near the centre of the fault and with slippage decreasing towards the tips, where strike-slip motion may develop

(Fig. 3.5a). They also observed that the rift flanks adjacent to a border fault tend to be uplifted as an isostatic response to exposure of the footwall. The scale and distribution of these buoyancy effects on the rift margin, and rift axis, depend in turn on the geometry of the crustal thinning, the density of the basin infill and lithospheric rigidity.

Maximum subsidence is located adjacent to border faults. Consequently, basin fills tend to be asymmetric in cross section and thickest adjacent to the border fault (Dunkelman et al. 1988). Half-grabens are often separated by accommodation/transfer zone highs with synthetic and antithetic faults forming intrabasin highs that define local barriers and subbasins (Fig. 3.5b–d). Accommodation and transfer zone terminology and definitions have been modified several times (Bosworth 1985; Bosworth et al. 1986; Rosendahl 1987; Morley 1995). In general, both refer to zones where normal (border) faults terminate, and strain is transferred between basins. Faulds and Varga (1998), for example, described 'transfer zones' as discrete fault zones with a large, but not exclusive, component of strike-slip motion (Fig. 3.5a, e). They also observed that accommodation zones are characterised by diffuse areas of overlapping normal faults (Fig. 3.5f).

3.4 Kenya Rift Tectonics

The Kenya Rift extends from Lake Turkana to Lake Magadi (and northern Natron), follows basement trends in the Proterozoic Mozambique Belt, avoids the Archaean Tanzania Craton to the west, and crosses the Kenya Dome (Fig. 3.1b; Baker et al. 1972; Smith 1994; Davis and Slack 2002; Wichura 2010), which is in isostatic equilibrium due to anomalous mantle within the underlying lithosphere (Smith 1994; White et al. 2012). To the south, faults splay out into the Northern Tanzanian Divergence (NTD) (Fig. 3.1d) forming two half-grabens (Eyasi and Manyara) that follow basement trends in the underlying Tanzania Craton (Calais et al. 2008; Albaric et al. 2014; Plasman et al. 2019). A third section of the NTD forms the Pangani Graben to the east of the Masai Block (Chorowicz 2005; Isola et al. 2014).

The main Kenya Rift is characterised by numerous small earthquakes, especially in the area north of Lake Magadi and west of Lake Bogoria (Ibs-von Seht et al. 2001, 2008; Albaric et al. 2009; Kuria 2011; Mulwa et al. 2014; Mulwa and Kimata 2014) but is generally quieter than the western branch of the EARS where larger earthquakes are associated with major border faults (Morley 1999a; Morley et al. 1999b). Earthquakes focussed at depths of 2–7 km in the central Kenya Rift and 4–12 km in the NTD have both been related to failure at an interface with dyke rocks (Mohr 1992; Morley et al. 1999b; Calais et al. 2008). Rifting has propagated southwards with time (Macdonald et al. 1994), perhaps due to the northward movement of the African plate over a mantle plume(s) (Kampunzu and Mohr 1991; Ebinger and Sleep 1998; Nyblade and Brazier 2002; Huerta et al. 2009; Sippel et al.

2017). However, it has also been suggested that the initiation of faulting may have been less systematic with parts of the rift being activated and deactivated at different times, with an additional tendency to migrate eastwards (Baker and Wohlenberg 1971; Morley 1999b).

Changes in fault orientations have been related to variations in lithospheric thickness (Keller et al. 1994) and mechanical heterogeneities, such as Late Proterozoic shear zones, thrusts and brittle fractures (Fig. 3.4c; Smith and Mosley 1993; Hetzel and Strecker 1994; Meju and Sakkas 2007; Koptev et al. 2015; Acosta et al. 2015; Muirhead and Kattenhorn 2018). In the central parts of the Kenya Rift, the crust

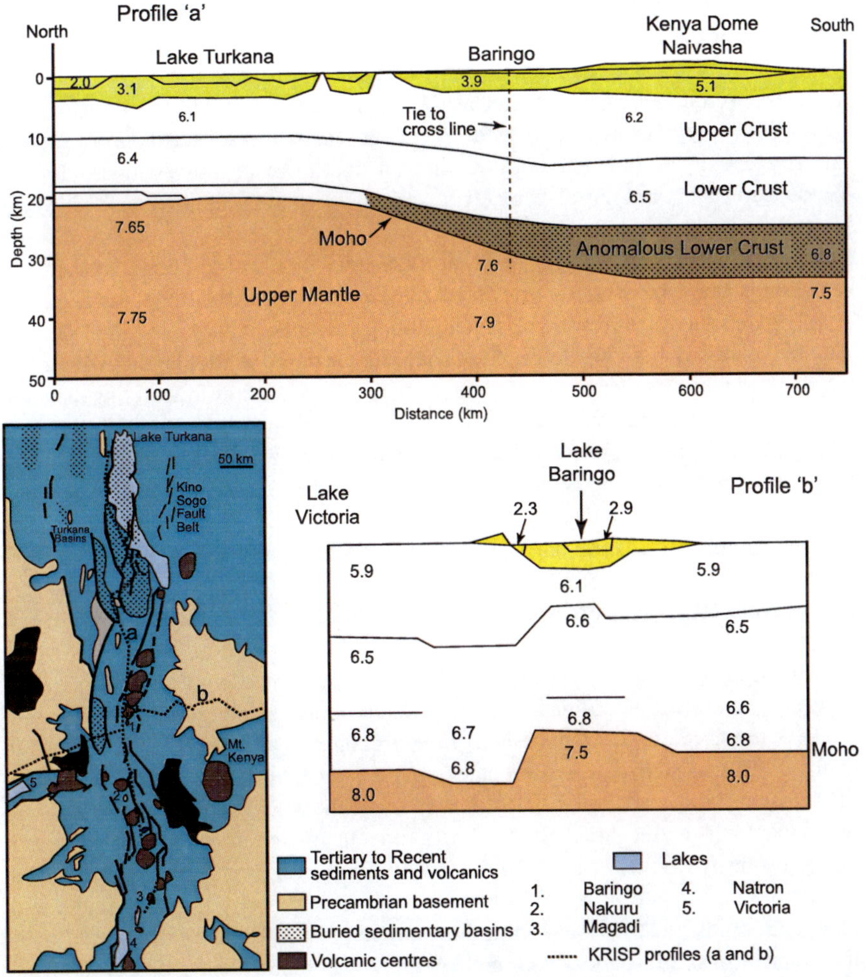

Fig. 3.6 KRISP seismic velocity models for the Kenya Rift. Profile locations shown by dotted lines '**a**' and '**b**' on map. After Morley et al. (1999b), Achauer et al. (1992) and Mariita and Keller (2007). Original data from KRISP (1991). Note thinning of the crust northwards

ranges from 35–40 km thick below the rift flanks to ~30 km below the rift floor (Fig. 3.6) with Emishaw and Abdelsalam (2019) reporting gravity data that suggest depths of 23–28 km below the Turkana Depression. There is also a significant change along the rift axis with the crust thinning from ~35 km below Lake Naivasha to 20 km below Lake Turkana (Henry et al. 1990; KRISP 1991; Mechie et al. 1994; Mooney and Christensen 1994; Khan et al. 1999; Mariita and Keller 2007).

Several studies have indicated that the asthenosphere lies close to the base of the crust with thermal, plume-related, processes driving both igneous activity and faulting (Fig. 3.3e; Williams 1978a; Smith 1994; Huerta et al. 2009). Observations by Macdonald et al. (1994) indicated a steep-sided, low-velocity body that extends to at least 200 km depth, and which might also reflect partial melts (Fairhead 1976; Dallheim et al. 1989; Mechie et al. 1994; Keller et al. 1994). A high-velocity zone at the base of the crust has been used to infer magmatic underplating (Green et al. 1991; Thybo et al. 2000) and that crustal stretching is greater than the present crustal thickness would indicate (Latin et al. 1993; Macdonald 1994; Macdonald et al. 1994). However, Morley et al. (1999b) point out that field evidence does not support this suggestion, noting that both brittle and ductile extension estimates for the Turkana area and the Central Kenya Rift were about 40 km (Hendrie et al. 1994) and 10 km, respectively.

Morley et al. (1999b) suggested that the Kenya Rift is at a more advanced tectonic stage than the western branch of the EARS with early major half-grabens having given way to narrower zones with young volcanic activity and minor faulting above a relatively high brittle-ductile zone. The proposed model begins with initial rifting driven by passive and active mechanisms, characterised by a rising asthenosphere replacing mantle lithosphere and emplacing magma bodies within the crust (Fig. 3.7). Crustal uplift takes place and half-grabens form with major boundary faults developing. During Stage 2 the asthenosphere and the brittle-ductile transition zone both rise as the crust continues to be stretched. Tectonic activity varies between basins with some half-grabens giving way to sag features. Lower-angle boundary faults may develop, and some basins may be abandoned as new ones are created. Subsequently, further extension is associated with a narrowing of the belt of active faulting and igneous activity, including dyke formation (Morley 2020).

Muirhead et al. (2015) noted that during early-stage rifting (<10 Ma) surface-breaching dykes were confined to areas in and around transfer zones, as is the case today in the northern Tanzania Divergence. In contrast, they suggest that dykes in evolved rifts may help to accommodate upper crustal extension along the full length of the basin. Muirhead et al. (2016), in a study of the Magadi and Natron rifts, also noted that fault-related strain changed with time, in agreement with numerical modelling results by Richter et al. (2021). Border faults accommodated extension during early rifting, followed by a shift to strain accommodation by intra-rift faults with magmatic fluids/volatiles weakening lithosphere prior to the development of magmatic intrusions. Subsequently, Shmela et al. (2021) made similar observations for the whole Kenya Rift, noting that strain had been increasingly localised onto larger faults as the system evolved.

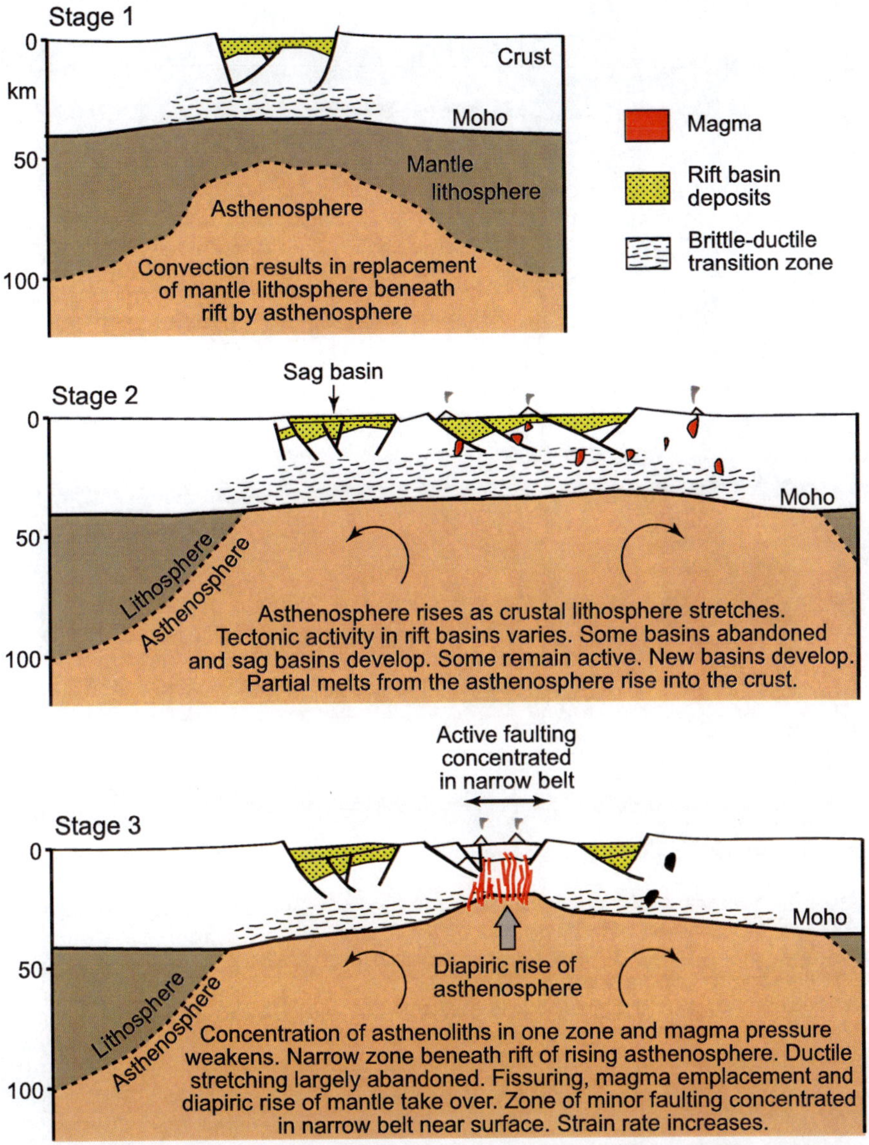

Fig. 3.7 Schematic tectonic model for the Kenya Rift Valley. After Morley (1994)

3.5 Kenya Rift Basins

The Kenya Rift can be subdivided into several sectors with distinctive tectonic features (Bosworth et al. 1986; Morley et al. 1992b; Macdonald et al. 1994). These include the Turkana Rift, the North, Central and South Gregory Rifts and the

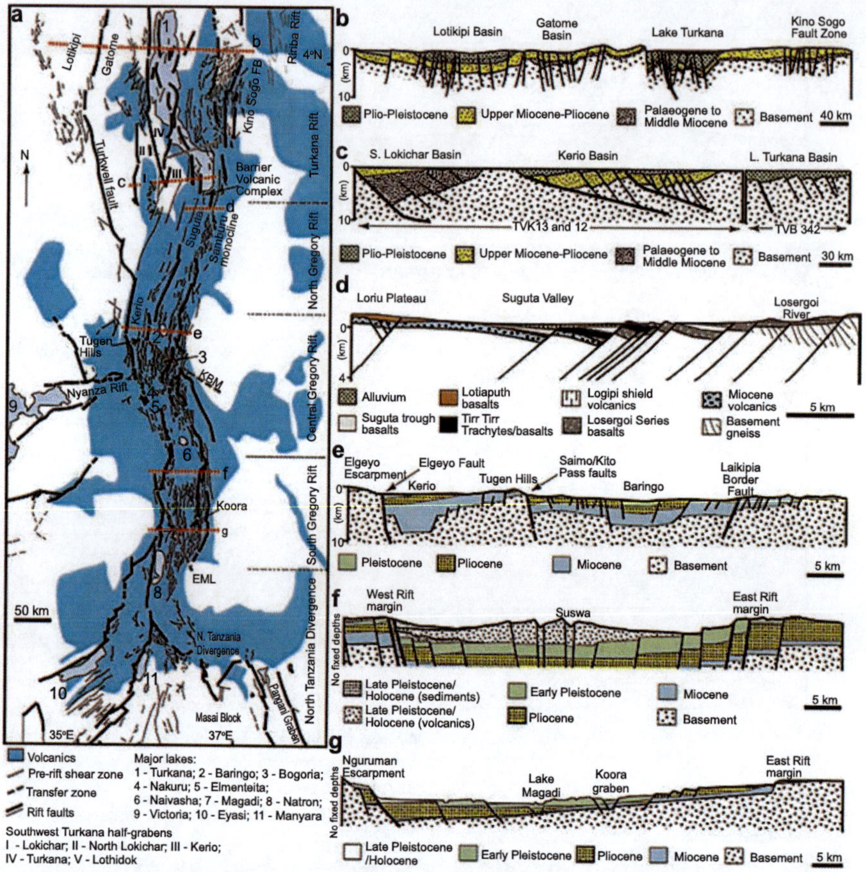

Fig. 3.8 Geological sections from the Kenya Rift. (**a**) Fault patterns. After Baker et al. (1972), with additional faults for Lake Turkana (Le Gall et al. 2005) and the Kino Sogo Fault Belt (Vétel and Le Gall 2005). KBM Kerio-Bogoria-Marmanet transfer zone, EML Engorika-Magadi-Lembolas transfer zone. (**b**, **c**, **e**) based on seismic surveys; (**d**, **f**, **g**) based on geological mapping. (**b**) Lotikipi and Gatome sag basins, Lake Turkana half-graben and Kino Sogo Fault Belt. After Morley et al. (1992b). (**c**) Oligo-Miocene half-grabens of southwest Turkana. After Vétel and Le Gall (2006). (**d**) The Suguta. After Bosworth and Maurin (1993). (**e**) Kerio and Baringo half-grabens. After Hautot et al. (2000). (**f**) South Gregory Rift through Suswa volcano. After Guth and Wood (2014). (**g**) South Gregory Rift through Lake Magadi. After Guth and Wood (2014)

Northern Tanzania Divergence (Fig. 3.8a). The 250-km-wide Turkana Rift lies to the east of the Turkwell (Turkwel) fault, is dominated by NNW-trending faults and overlies Mozambique Mobile Belt rocks. The Lotikipi and Gatome sag basins, to the northwest (Fig. 3.8a, b), may represent Palaeogene thermal subsidence associated with late-stage cooling of the pre-existing, NW-trending, Cretaceous to Palaeogene Anza Rift (Fig. 3.1b; Wescott et al. 1999; Abdelfettah et al. 2016). The two sag basins are filled with Plio-Pleistocene sediments that overlie the Turkana Volcanics (basalt, rhyolite) with upper Eocene basal dates (37–35 Ma) and early to

Middle Miocene ages at the top (Tiercelin et al. 2012a). These may represent eruptions prior to the development of EARS faults (Wescott et al. 1995). Seismic profiles also indicate the presence of a thick sedimentary sequence below the Turkana Volcanics (Wescott et al. 1999; Tiercelin et al. 2012b).

In contrast, Lake Turkana is underlain by six, 25–40 km-long, half-grabens that alternate in polarity and contain 2–4 km of sediment (Dunkelman et al. 1988, 1989; Rosendahl et al. 1992). The border faults have throws of several kilometres and are opposed by synthetic and, less commonly, antithetic faults and monoclines (Fig. 3.8b). The individual half-grabens are linked by accommodation zones characterised by basement highs with one having developed along a Late Mesozoic to Early Tertiary ENE-trending shear zone (Fig. 3.8a; Williamson and Savage 1986; Dunkelman et al. 1988). Holocene extension rates have been estimated at about 3.5–6.7 mm y^{-1} (Melnick et al. 2012). Riedl et al. (2022) also reported that extension today in the inner graben of the North Kenya Rift was a minimum of 1.0–1.6 mm y^{-1} locally with rates of up to 2.0 mm y^{-1}. Contemporary seismic activity is concentrated to the east of Lake Turkana along the Kino Sogo Fault Belt, which cuts Pliocene lavas and connects with the northern Gregory Rift via southern Lake Turkana and the Suguta Valley (Fig. 3.9a–c; Vétel et al. 2004; Vétel and Le Gall 2005, 2006).

To the southwest of lake Turkana are a series of inactive, west-thickening, half-grabens with 5–8 km of sediments and volcanics. They are formed by mostly east-dipping boundary faults that pass down into a zone of ductile shearing at about 10–12 km (Morley et al. 1992a, b, 1999b; Morley 1999c). The oldest are the Late Palaeogene–Middle Miocene Lokichar and Lothidok basins (Fig. 3.8a, c; Hendrie et al. 1994; Morley and Wescott 1999), which ceased to be active by the end of the Middle Miocene. Subsequently, the North Lokichar, Turkana and Kerio half-grabens developed and continued to be active at least through to the Pliocene when faulting stopped, except for Pliocene, and possibly Holocene, tectonic inversion in the Turkana and Kerio basins (Vétel and Le Gall 2005).

The NNE-trending, 100-km-wide, North Gregory Rift is also underlain by mobile belt rocks. This sector includes the low-lying Suguta Valley, which is bordered to the east by a faulted monocline and perhaps a border fault at the base of the Ng'iro Range (Fig. 3.8d; Bosworth and Maurin 1993; Acosta et al. 2015). The less uplifted western margin consists of a gently east-dipping monocline with young, eroded lavas that are only locally faulted (Fig. 3.8d). The Barrier Volcanic Complex separates Lake Turkana from the Suguta (Figs. 3.9a, b and 3.10a, b) and may represent an accommodation zone with an opposing border fault on the western side of southern Lake Turkana. Before ~4 Ma ago faulting took place over a much broader area (Saggerson and Baker 1965; Cerling and Powers 1977) with the early Suguta Rift initiated as a broad basin before faulting became confined to a narrower axial zone (Bosworth and Maurin 1993; Saneyoshi et al. 2006). The Lower and Upper Suguta valleys are separated by Emuruangogolak (Fig. 3.9c) with other Quaternary volcanoes puncturing the rift floor between the Upper Suguta and Lake Baringo. The latter include Silali (Fig. 3.10c), Paka (Fig. 3.10d) and Karosi (Fig. 3.9g), which started to form about 460 ka, 390 ka and 500 ka ago, respectively (Lichoro et al.

Fig. 3.9 Structural settings in the northern Kenya Rift. (**a**) South Basin, Lake Turkana, viewed from the Barrier Volcanic Complex. Wargess and Nabuyatom (Naboiyaton) tuff cones on shoreline with South Island on the horizon. Note recent dark basaltic lava on the Barrier. (**b**) Suguta Valley from the Barrier looking south to Naperito (Cathedral Rock) in Lake Logipi (~276 m). (**c**) Northward view of the Suguta Valley from Emuruangogolak volcano. Note recent basalt flows and north-south faulting. (**d**) Dissected lavas on the Ribkwo Plateau, western rift margin, from Kapedo (Upper Suguta Valley). (**e**) The Tugen Hills tilt block rises to >2900 m west of Lake Baringo (~970 m); view towards the Saimo Fault scarp with recent faults in Baringo Trachyte near the shoreline. (**f**) Eastern rift margin with faults stepping down from the Laikipia Escarpment (>1400 m). Karau volcanic cone to left. (**g**) North-south trending horsts and grabens cutting into Karosi volcano, north of Lake Baringo. (**h**) Active faulting along the Baringo Trachyte Fault, west of Lake Baringo

Fig. 3.10 Volcanic landforms in the Kenya Rift, ordered from north to south. (**a**) Nabuyatom (Naboiyaton) tuff cone on south shore of Lake Turkana. (**b**) Naperito (Cathedral Rock) in Lake Logipi with dark basalts and basanites erupted southeastwards to the viewpoint. Barrier volcanic complex in background. (**c**) Silali crater (~7 km wide) composed mainly of basalts and trachytes. (**d**) Paka lava shield with tuff cones, pumice tuffs and agglomerates. (**e**) Ol Kokwe Island, Lake Baringo. Young basalt scoria cones capping a trachytic volcano. (**f**) Menengai Crater (8 km across) north of Nakuru. Caldera filled with post-caldera lava flows. (Photo taken July 2006 before geothermal power developments). (**g**) Longonot volcano, south of Lake Naivasha. Composed of peralkaline trachyte lavas and pyroclastics. (**h**) Suswa volcano from Longonot. Comprised of sodalite-bearing phonolite lavas and smaller volumes of pyroclastics and featuring a summit caldera within an older caldera

2019). Ol Kokwe lies within Lake Baringo and includes trachyte and basaltic scoria that are variously dated at 364–92 ka (Fig. 3.10e). Gravity lows at Paka and Karosi have been interpreted as low-density bodies, <2 km deep, that may represent unaltered near-surface pyroclastics or deeper tuffs hydrothermally altered to smectite (Lichoro et al. 2019; Mibei et al. 2021).

The northern part of the Central Gregory Rift is about 100 km wide with the crust thinning from ~36 km beyond the rift shoulders to ~30 km below the rift floor (Maguire et al. 1994). A large westerly-dipping tilt block forms the Tugen Hills (Fig. 3.9e), which separates the Kerio Valley from the Baringo and Bogoria basins (Chapman et al. 1978; Williams and Chapman 1986; Tiercelin and Vincens 1987; Morley et al. 1992b; Mugisha et al. 1997; Le Turdu 1998; Hautot et al. 2000). Fault orientations include both northerly and northwesterly components (Figs. 3.4c, d and 3.8a, e), reflecting deep basement controls (King 1978; Smith and Mosley 1993; Hautot et al. 2000). A deep half graben lies below the Kerio Valley with the Elgeyo border fault to the west. The Baringo Graben is delineated by the Saimo/Kito Pass boundary faults to the west (Fig. 3.9e) and a series of synthetic and antithetic faults that step down from the Laikipia Escarpment to the east (Figs. 3.8a, e and 3.9f) with the axial rift cut by numerous north-south faults that cut through Karosi volcano and partially define the margins of Lake Baringo (Fig. 3.9g, h).

Early geophysical research in the Baringo Basin (Rooney and Hutton 1977; Swain et al. 1981) indicated up to 1 km of sediment below Lake Baringo. Subsequently, Hautot et al. (2000) reported 1500 m of sediments and volcanics overlying a resistive layer at 4–5 km depth, which they correlated with mid–Miocene flood phonolites. They also noted a Palaeogene basin below these lavas, which led them to challenge earlier suggestions of gradual eastward migration of rifting from the Kerio Valley towards the Laikipia Escarpment (Bosworth 1985; Morley 1994; Mugisha et al. 1997). The Bogoria half-graben lies south of Lake Baringo (Fig. 3.8a) and contains sediments that thicken to the east where they abut against the Bogoria fault (Fig. 3.11a, b). To the west and south a region of intensely grid-faulted trachyphonolite forms a series of easterly dipping tilt-blocks (Fig. 3.11c; Griffiths and Gibson 1980; Bahat 1984; Le Turdu et al. 1995, 1999) that lie south of the NW–SE trending Kerio-Bogoria-Marmanet (KBM) Transfer Zone (Fig. 3.8a; Smith and Mosley 1993). Fault orientations change to a northeasterly direction across the Emsos Fault (Fig. 3.11b, south of Lake Bogoria, in the Lake Solai area (Fig. 3.11d).

Further south, the Central Gregory Rift is characterised by a major NW–SE trending graben, ~50–70 km wide, which extends from the equator to Suswa Volcano (Fig. 3.8a, f; McCall and Bristow 1965; Smith and Mosley 1993). This complex part of the rift shows the greatest difference between flank elevations and the subvolcanic rift floor. The Sattima Border Fault (Fig. 3.11e) marks the eastern rift margin with the Bahati, Naivasha and South Kinangop fault zones forming a series of step-faults down to the rift floor. Monoclinal flexures constitute the western boundary, which coincides with the Mau Escarpment (Fig. 3.11f; Baker et al. 1988). This sector also lies at the centre of the Kenya Dome and intersects with the Nyanza (Kavirondo) Rift to the west (Baker et al. 1972; Bechtel et al. 1987). Several large caldera volcanoes are present, including Menengai, Longonot and Suswa (Fig. 3.10f–h) along with volcanic fields at Olkaria and Eburru. Crossley (1980)

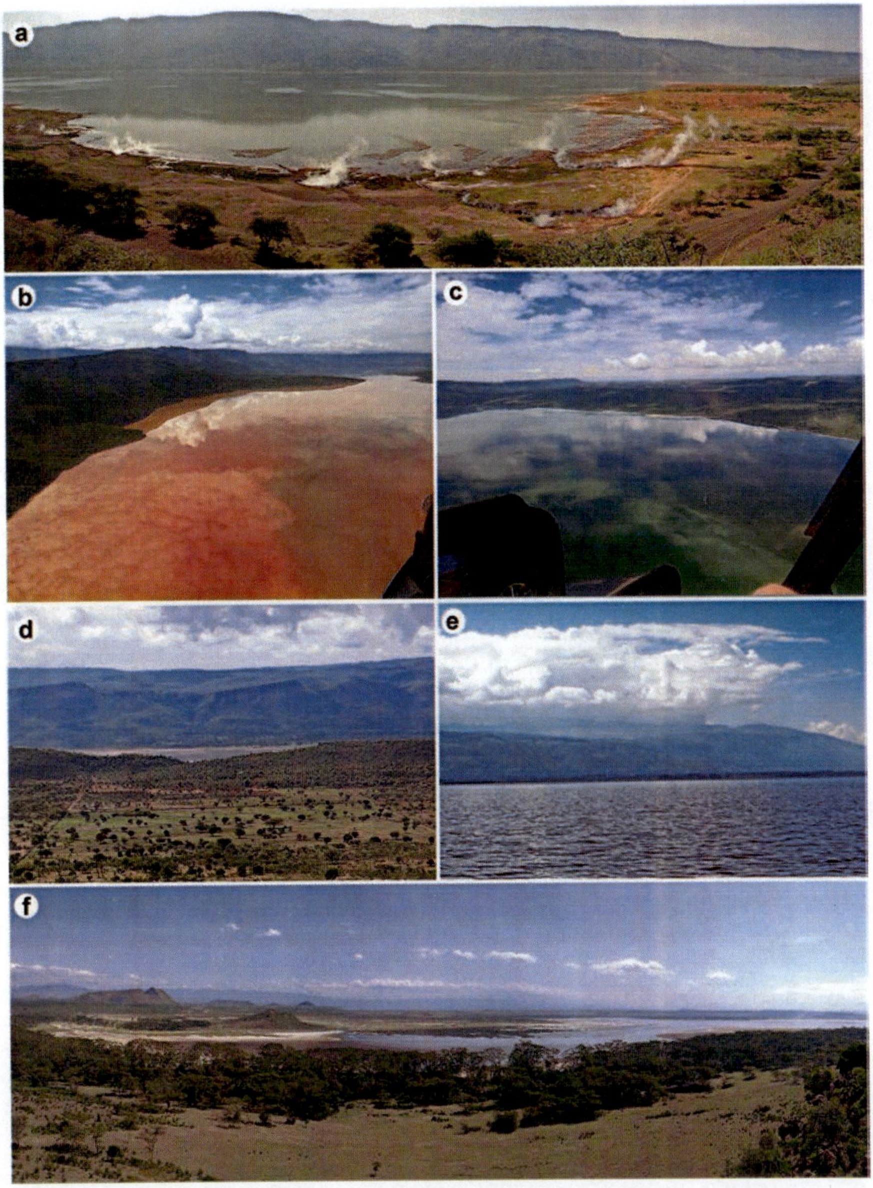

Fig. 3.11 Major fault systems in the central-northern Kenya Rift. (**a**) Bogoria Fault east of Lake Bogoria, forming a ~800-m-high escarpment, with hot springs on the Loburu Delta following fault alignments. (**b**) View south from the turbid waters of the North Sub-basin of Lake Bogoria showing the NNW-trending Lake Bogoria Fault Escarpment (left) with the Waseges-Marmanet Fault at extreme left. Lake Bogoria terminates where the Bogoria Fault curves into the NE-SW-trending Emsos Fault (distant background). (**c**) View from the less turbid Central Basin of Lake Bogoria to the southwest showing tilt blocks dipping eastwards to the Bogoria fault (left). (**d**) NNE-trending, eastward tilted, horsts and grabens southeast of the Emsos Fault. Lake Solai in centre. (**e**) Sattima Fault east of Lake Ol' Bolossat (~2328 m) forms part of the eastern rift margin. Aberdare Mountains (>3100 m) lie on the upthrow side in the background. (**f**) View west over Lake Elmenteita (~1776 m) to the Mau Escarpment (left background; >3000 m), which forms the western rift margin. Eburru volcano at extreme left background, with young cinder cones on the rift floor

suggested that curvatures in the rift margin topography may be evidence of a large Pliocene (?) tectonomagmatic caldera that encompasses the younger Longonot and Suswa volcanoes.

The Nyanza (Kavirondo) Rift, WSW of Lake Bogoria, is developed at a near-right angle to the Central Gregory Rift, trending E-W to NE-SW (Fig. 3.8a). Early reports suggested that it was a full graben structure (Saggerson 1952; McCall 1967), but monoclinal flexures were subsequently also recognised (Jones and Lippard 1979; Mboya 1983). The Nyanza Rift is separated from the Gregory Rift by a series of major central volcanoes, and it was once thought that they intersect (Shackleton 1951; Pulfrey 1960; Walsh 1969; Baker and Wohlenberg 1971; Baker et al. 1971). Pickford (1982), however, reported topographical and structural evidence to suggest that the Nyanza Rift loses definition ~50 km west of the Kenya Rift (Mboya 1983).

The South Gregory Rift extends from Suswa to Lakes Magadi and Natron and is characterised by a return to north-south trending faults (Fig. 3.8a) that have developed mainly through the last million years (Owen et al. 2018, 2019). These form dense networks of horsts, grabens and tilt blocks (Fig. 3.12). Smith and Mosley (1993) suggested that they are related to the NW-trending Engorika-Magadi-Lembolas (EML) transfer zone, which is also apparent from aeromagnetic and gravity data (Fig. 3.8a; Wohlenberg and Bhatt 1972). Hot springs and small volcanic cones locally follow these trends. The EML has also been related to the Aswa-Nandi shear zone within the Proterozoic Basement (Le Turdu et al. 1999), with the KBM transfer zone of the Central Gregory Rift having been correlated with Nyangea-Athi-Ikutha shear zone to the north (Fig. 3.4c).

Gravity data indicate no half-graben polarity reversals in the South Gregory Rift (Simiyu and Keller 2001). At the latitude of Lake Magadi, the rift is formed by a 3.5 km deep asymmetric basin (Richter et al. 2021) that is deepest near the base of the Nguruman border fault, to the west, with step-faults defining its eastern margin (Fig. 3.8g). The crust below the Nguruman Escarpment is thinner (<35 km) than under the eastern rift margin (38–44 km) (Khan and Swain 1978; Khan et al. 1986, 1989, 1999). Simiyu and Keller (1998) studied the upper crustal structure near Lake Magadi and concluded that Neo-Proterozoic tectonic transport had added Mozambique Belt rocks to the top of the Archaean Tanzania Craton. A rift axial-high gravity anomaly has also been reported, which may indicate either a basement horst structure or gabbroic intrusion (Darracott et al. 1972; Simiyu and Keller 1998).

Several Plio-Pleistocene central volcanoes are present south of Suswa that are now extinct, including Ol Esakut (Esayeiti) (~6.7–3.6 Ma), Olorgesailie (2.7–2.2 Ma), Ol Doinyo Nyokie (~0.7 Ma), Lenderut (2.7–2.5 Ma) and Shombole (~2 Ma) (Fig. 3.13a–e; Baker 1975; Le Roex et al. 2001). The active Oldoinyo Lengai volcano lies south of Lake Natron, in Tanzania (Fig. 3.13f). Flood lavas are widespread across the southern rift and are composed of Pleistocene trachyte and trachyphonolite that erupted from numerous fissures, obscuring many older rift structures.

Fig. 3.12 Faulted landscapes, southern Kenya Rift. (**a**) southern Koora Graben (Siriata section). (**b**) northern Koora Graben (Koora Plain) with a normal fault slicing through Ol Doinyo Nyokie volcano. Magadi-Mombasa railway line (for soda shipping) in foreground. (**c**) Northern Lake Magadi Graben, looking north. (**d**) Dissected horst block west of Shanamu and north of Lake Magadi. (**e**) View north over Nasikie Engida lake (~614 m), with a major fault to left and westerly dipping tilt-blocks and grabens in distance. (**f**) Major fault cutting trachyte above Nasikie Engida, looking south

Fig. 3.13 Extinct Plio-Pleistocene volcanoes, southern Kenya Rift. (**a**) Ol Esayeiti (Esakut) with eastern rift margin in the right background. (**b**) Mount Olorgesailie (>1700 m). (**c**) Ol Doinyo Nyokie (>1000 m), northeast of Lake Magadi. (**d**) Lenderut (>1200 m), south of Lake Magadi. (**e**) Shombole (>1500 m) on the Kenya-Tanzania border, south of Lake Magadi. (**f**) Ol Doinyo Lengai (2951 m), south of Lake Natron, Tanzania. Currently an active volcano

3.6 Volcanism in the Kenya Rift

Sodic alkaline volcanism is widespread in the Kenya Rift with abundant intermediate and silicic rocks present (Fig. 3.14a; Baker 1987). The volcanics have been divided into strongly alkaline (nephelinite, basanite, alkali basalt, tephrite, phonolite) and weakly alkaline (olivine basalt, trachyte, trachyphonolite, alkali rhyolite)

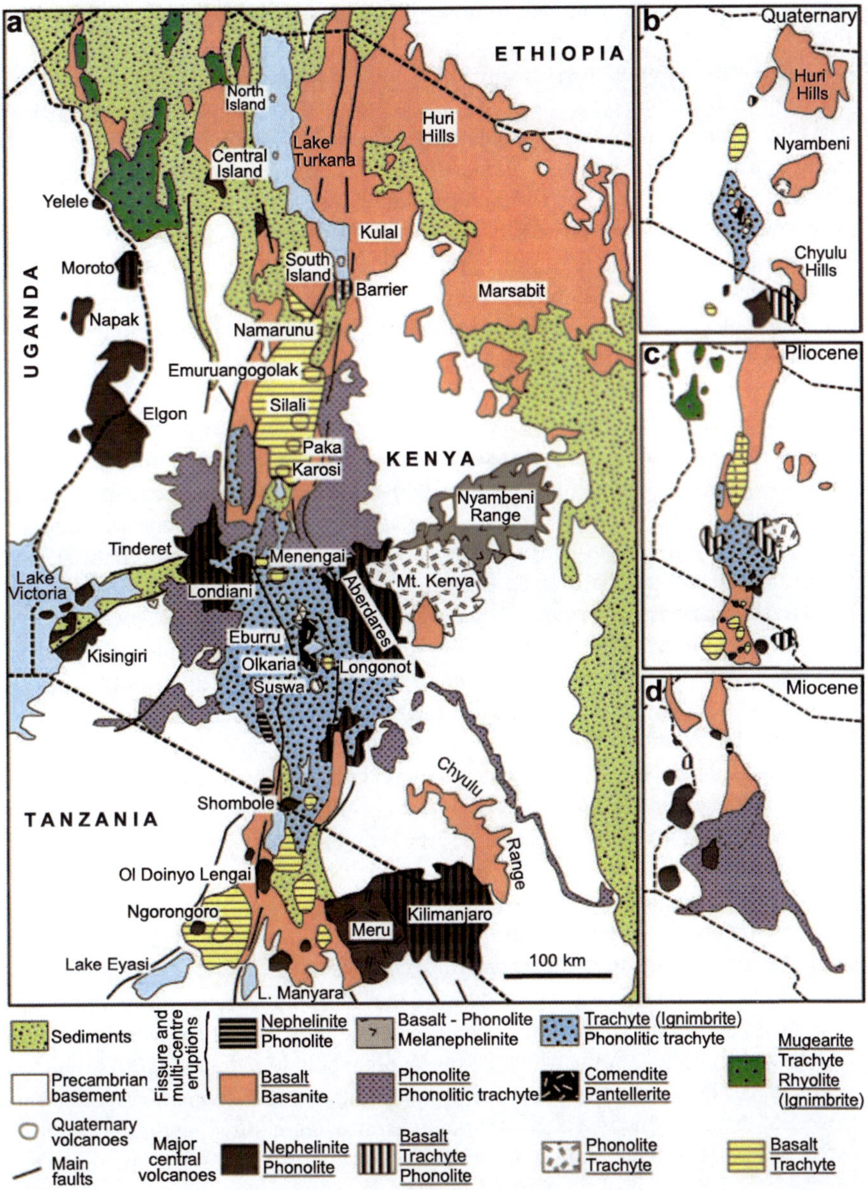

Fig. 3.14 Volcanic associations in the Kenya Rift. Modified from Williams (1970). (**a**) Distribution of volcanic associations. (**b–d**) Spatial and temporal distribution of major volcanics (underlined lithologies in key)

suites, which broadly dominated during the Miocene and post-Miocene, respectively (Fig. 3.14b–d; Baker et al. 1978; Baker 1987). Williams (1969, 1970, 1972) distinguished ten common associations that were characteristic of different volcanic settings. Associations typical of fissure and multi-centre eruptions include basalt-basanite; basalt-phonolite-melanephelinite; phonolite-phonolitic trachyte; quartz trachyte-phonolitic trachyte; comendite-pantellerite and mugearite-trachyte-rhyolite. Williams also noted that nephelinite-phonolite associations were common at central volcanoes and, more rarely, could be related to fissure eruptions. Basalt-trachyte-phonolite, phonolite-trachyte and basalt-trachyte associations were reported only from major central volcanoes.

Latin et al. (1993) estimated the total volume of Oligocene–Recent magma in the Kenya Rift, including underplated material, at 924,000 km^3, which Macdonald (2002) observed was too large to have been produced solely within anhydrous lithosphere. Guth (2013, 2016) calculated eruptive volumes from available seismic data and estimated a total of ~310,000 km^3. Macdonald (2002) suggested that the melt zone is supplied by upwelling mantle with elevated temperatures. Mantle velocity data have been used to suggest that melting starts at about 65 km beneath the northern part of the rift and 45–50 km below Lake Naivasha (Macdonald et al. 1994). Swain (1992) and others inferred dyke intrusion at depth along the rift axis near Lake Baringo and Lake Bogoria. Trace element data from the southern Kenya Rift have been used to infer that melts and/or fluids from an upwelling plume first metasomatised the overlying lithosphere, which was then melted at depths of 75–90 km (le Roex et al. 2001). Other researchers have related the variety of volcanics to the ascent of a lherzolitic asthenospheric wedge, with partial melting at shallow depths producing tholeiitic basalt, with transitional and alkaline magmas forming at intermediate and deep levels (Mohr 1992; Hay et al. 1995).

The varied magmas also appear to reflect location with respect to the underlying lithosphere (Macdonald et al. 2001; Macdonald 2002; Rooney 2020d). Basalt from northern Kenya is the most MgO-rich, reflecting minimal crustal interaction during fractionation and the relatively thin crust (~20 km) (Rogers 2006; Furman 2007). Olivine-poor nephelinite and carbonatite (Baker 1963; Peterson 1989) tends to be restricted to the thick lithospheric areas of the Tanzanian Craton and its margins, whereas olivine-rich nephelinitic lava and alkali olivine basalt are associated with the thinner lithosphere of the Mozambique Mobile belt. Macdonald (2002) reported that mafic rift lavas tend to be less silica-undersaturated than mafic lavas to the east or west of the rift, which he attributed to differences in melting depths.

Several parts of the southern and central Kenya Rift (Fig. 3.14a) are characterised by eruptions of mixed suites of basalt and trachyte/rhyolite/comendite from mostly Quaternary central volcanoes (McCall and Bristow 1965; Johnson 1969; Saggerson 1970; Sutherland 1974; Baker 1975; Webb and Weaver 1975; Baker et al. 1977, 1988; Weaver 1977; Leat et al. 1984; Williams and Macdonald 1984; Macdonald et al. 1995; Clément et al. 2003; Ren et al. 2006; Ridolfi et al. 2006). Macdonald (2002, 2012) and earlier researchers observed that these associations probably reflected fractional crystallisation of basaltic parent magmas, sometimes involving minor assimilation of crustal rocks (Rogers 2006). It has also been noted that the silica content of the parent rock is reflected in daughter products. Mildly alkaline basalt is associated with mildly nepheline or quartz-normative trachyte, whereas nephelinite and basanitic magma produce phonolitic lava (Sceal and Weaver 1971; Weaver et al. 1972; Scott and Bailey 1984; Baker 1987; Kabeto et al. 2001; Macdonald and Baginski 2009; White et al. 2012). In some cases, magma mixing may also be important in determining eruption products (Macdonald et al. 2008).

3.7 Geological Evolution of the Kenya Rift Valley

The Kenya Rift was preceded by Late Carboniferous to Middle Jurassic Karoo faulted troughs that formed within the fragmenting Gondwana super-continent (King 1978). Subsequently, Palaeogene rifts, such as the Anza Rift, developed with contrasting NW–SE orientations (Figs. 3.1b and 3.2). A widespread irregular erosion surface, with areas of deep weathering, laterite and ferricrete, was formed on pre-existing basement rocks. The watershed between the Atlantic and Indian oceans formerly lay to the west of the modern Gregory Rift, but this changed with regional uplifting that would eventually form the Kenya Dome, centred on Lake Naivasha (Saggerson and Baker 1965; King and Chapman 1972; King 1978; Shackleton 1978). Volcanism has also tended to shift eastwards during rifting. Miocene activity took place to the west and within the developing rift (Figs. 3.14d and 3.15). This was followed by late Miocene fissure phonolite eruptions in the Central Gregory Rift. Pliocene activity developed mainly within the rift, whereas Quaternary volcanism occurred in the rift and to the east (Williams 1978b).

Rooney (2020b, c) recognised six major volcanic phases in the Kenya Rift: (1) the Samburu Phase (20–16 Ma), involving mainly basaltic volcanism; (2) the flood Phonolite Phase (16–12 Ma) with phonolites erupting over a wider area to the south of the Samburu Basalts; (3) the Mid-Miocene Resurgence (12–9 Ma) when basalt lavas again became common to the north and phonolites continued to extrude to the south; (4) the Early Rift Development Phase (9–4 Ma), which involved a hiatus at Turkana, but with widespread activity to the south, variously involving phonolites, basalts and explosive eruptions; (5) the Stratoid Phase (4–0.5 Ma), which was characterised by renewed basaltic eruptions at Turkana with varied activity to the south and graben formation; and (6) the Axial Phase (0.5–0 Ma), involving the further development of narrower axial grabens within the broader rift system. Magmatism was focussed along the rift axis with eruptions of fissure basalt/trachyte, central silicic volcanoes and parasitic basaltic cones.

Details of these six eruption phases, local tectonic history and the development of sedimentary basins vary between different segments of the Kenya Rift. For example, the fluvial Lapur Sandstone Formation was laid down during the earlier Palaeogene in northwest Kenya near the modern Turkana Rift (Fig. 3.16a; Wescott et al. 1999; Tiercelin and Lezzar 2002; Thuo 2011; Tiercelin et al. 2012b; Abdelfettah et al. 2016). During the late Eocene and into the Oligocene, the area was characterised by a large sag basin with only minor extensional faulting (Morley et al. 1992b; Morley 1994). Igneous activity started about 36 Ma ago with eruptions of

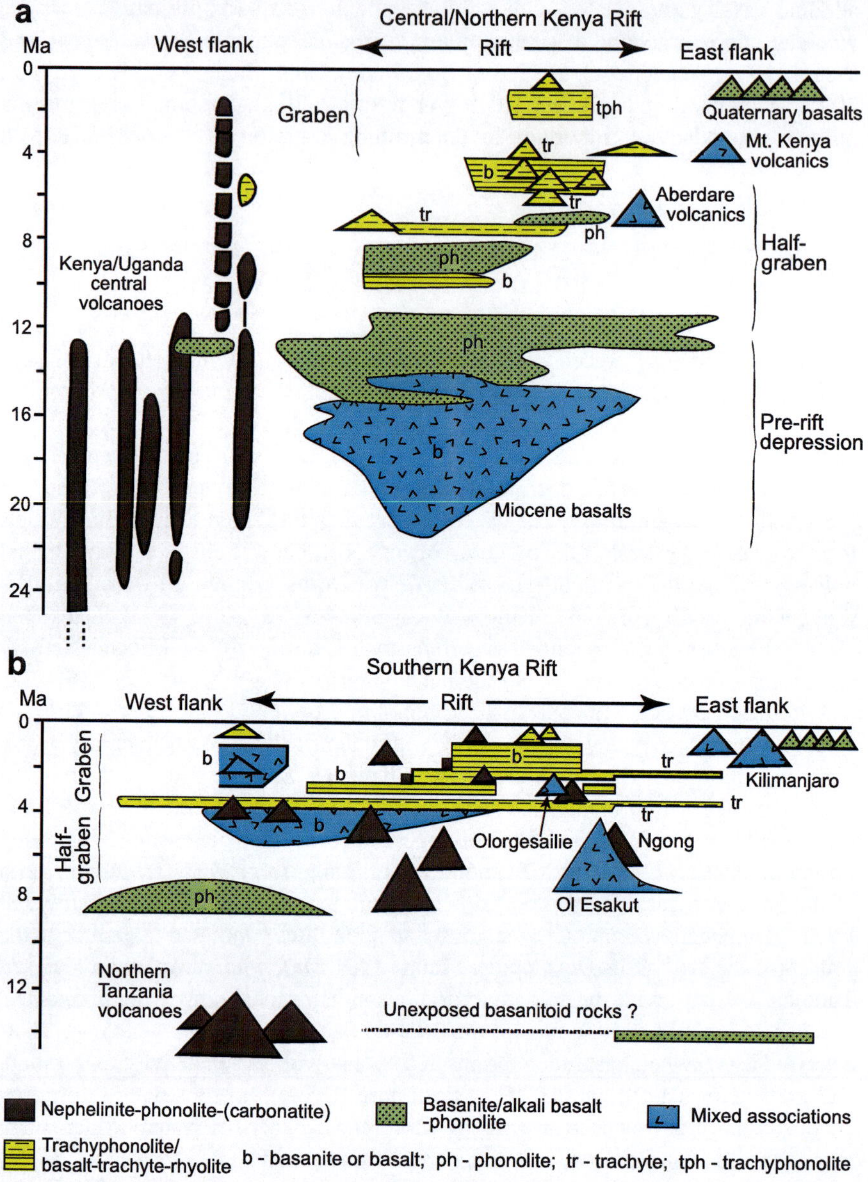

Fig. 3.15 Summary of the temporal and spatial ranges of major Tertiary volcanic associations. After Baker (1987)

voluminous tholeiitic basalt and minor rhyolite (Fig. 3.16b; Walsh and Dodson 1969; Morley et al. 1992a; McDougall and Brown 2008). Volcanism was dominated by rhyolite with less common basalt between ~27 and 23 Ma ago and with alkaline basalt, commonly interbedded with sediments, dominating further south from 19 to

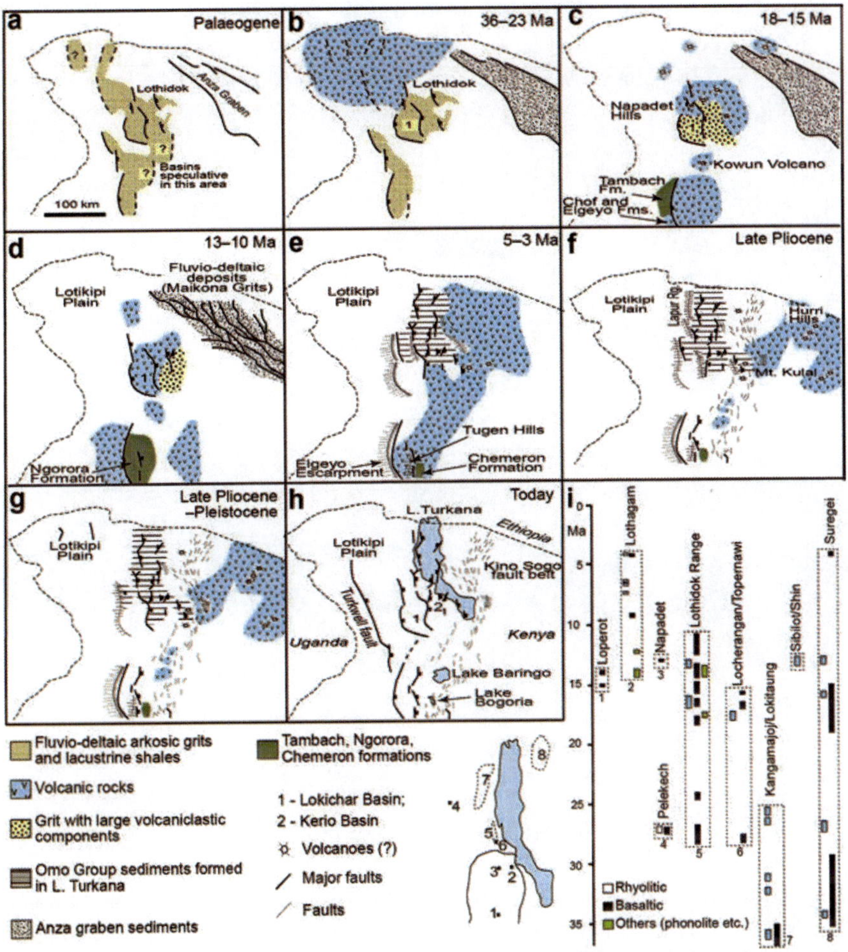

Fig. 3.16 Evolution of the northern Gregory and Turkana Rifts. (**a**–**h**) Palaeogene to Recent changes in basins and areas of volcanism. Note eastward shift in rifting with time. After Morley et al. (1999a) with dates in '**b**' modified after McDougall and Brown (2008). (**i**) Temporal and spatial volcanic stratigraphy, Turkana Rift. After McDougall and Brown (2008)

15 Ma (Zanettin et al. 1983; McDougall and Brown 2008). To the south, the Samburu Basalts were erupted in the North Gregory Rift between about 23 and 18.5 Ma (Baker et al. 1971; Shackleton 1978), prior to a period of erosion and eruption of the Elgeyo Basalts (~15 Ma) on the western side of the present rift.

Large nephelinite-phonolite central volcanoes (Figs. 3.14a, d and 3.15) started to erupt from about 22 Ma (Elgon), 19–22 Ma (Kisingiri), >20 Ma (Tinderet) and around 19 Ma (Napak) along a north-south trending belt some 100 km to the west of the modern rift (King 1949; Williams 1969, 1970; Baker et al. 1971; Shackleton 1978). Kisingiri and Tinderet lie at opposite ends of the WSW-trending Nyanza Rift and activity at these volcanoes may reflect the first stages in the development of the Nyanza (Kavirondo) Rift, although relief at the time was probably subdued (Shackleton 1978).

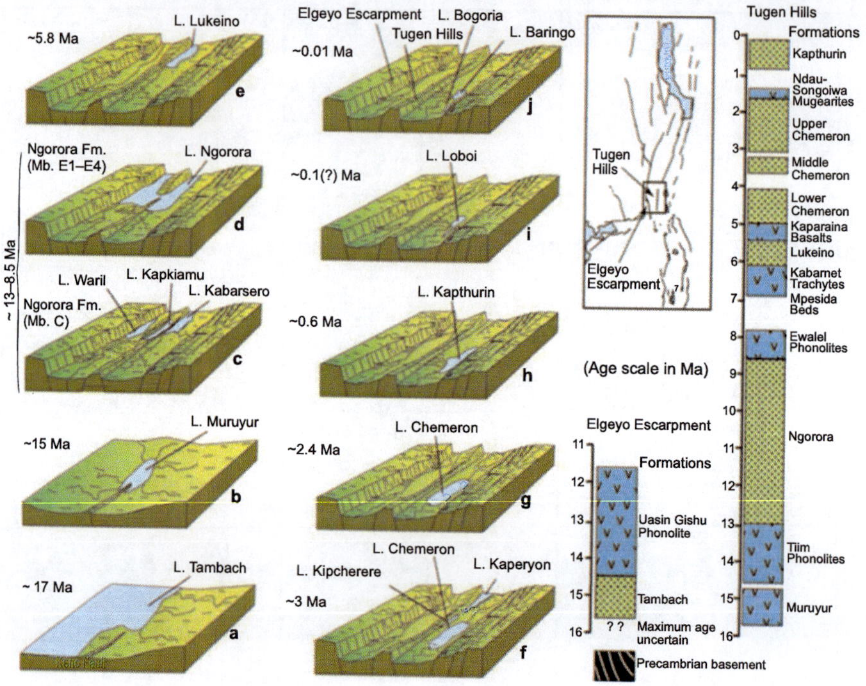

Fig. 3.17 Miocene to Recent rift evolution in the Baringo-Bogoria region. Major sedimentary and volcanic formations to the right. Modified from Tiercelin and Vincens (1987) and Renaut et al. (1999)

Several half-grabens formed to the southwest of modern Lake Turkana with the oldest, the Lokichar Basin ('1' in Fig. 3.16b–d), dating back to the Palaeogene–Middle Miocene. These were filled by up to 7 km of fluvial and lacustrine sediments and volcanics. The Lokichar basin and the Lothidok half-graben were abandoned at the end of the Middle Miocene with new basins developing. During the Upper Miocene or Early Pliocene, extensive parts of the Turkana Rift were covered by ignimbrite, rhyolite, mugearite and trachyte. Middle to Upper Miocene faulting and tilting of the Tugen Hills fault block is demonstrated in the Baringo area by sudden thickness changes in lake and fluvial sediments (e.g. the Muruyur Formation; Fig. 3.17b) and volcanics across faults (Bishop and Pickford 1975; Shackleton 1978; Behrensmeyer et al. 2002). The Samburu monocline, to the northeast (Fig. 3.8a), probably developed in the Late Miocene to Early Pliocene with a monoclinal flexure and associated faults forming on the north side of the Nyanza Rift around 14–10 Ma (Shackleton 1978).

Very large volumes (2500 km^3, up to 700 m thick) of flood phonolite were erupted during the later Miocene (~13.5–11 Ma) in the Central Gregory Rift, near the apex of the Kenya Dome (Figs. 3.14a, d and 3.15a; McCall 1967; Walsh 1969; Baker et al. 1972; Lippard and Truckle 1978). These flood lavas overtopped the subdued contemporary rift, burying earlier volcanics and lake sediments, including

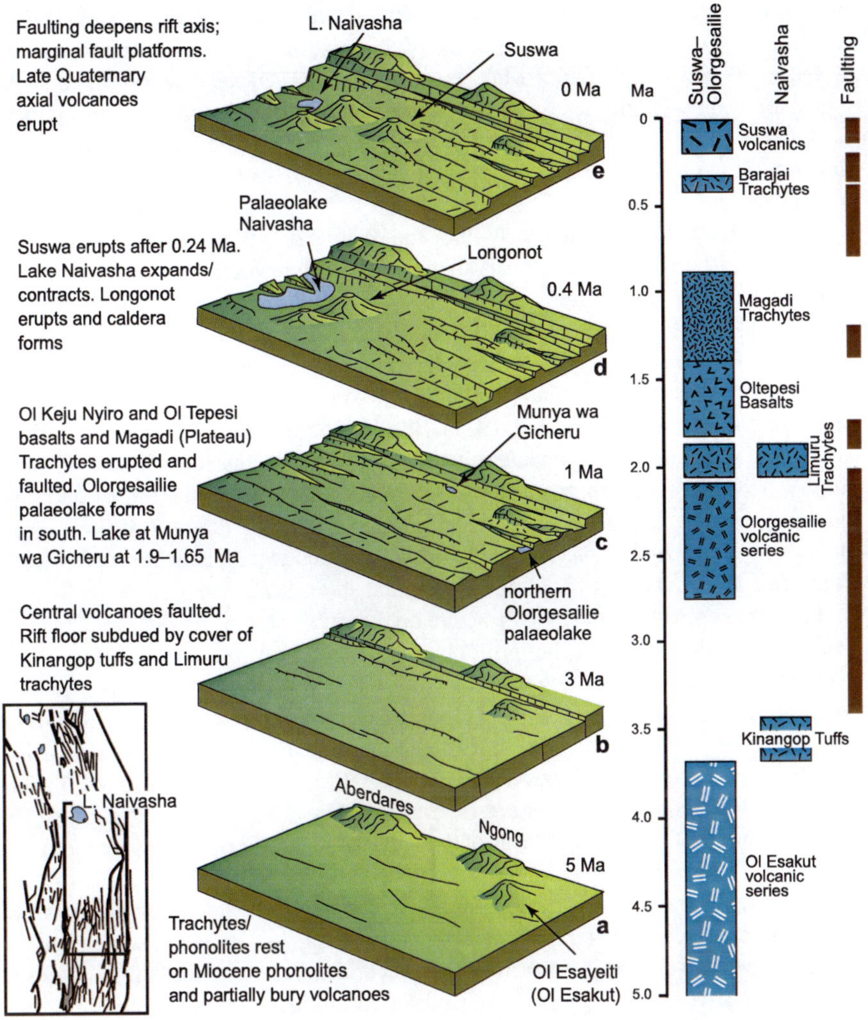

Fig. 3.18 Rift evolution in the eastern South Gregory Rift (Naivasha to Olorgesailie). Major volcanic units and faulting to the right. Based on data from Baker and Mitchell (1976) and Baker et al. (1988)

the Tambach Formation (Figs. 3.16c and 3.17a) and then continued to spread in all directions reaching distances of up to 300 km to the southeast (Fig. 3.14d). King and Chapman (1972) noted that an important period of faulting followed these eruptions.

In the Central Gregory Rift, faulting and volcanism repeatedly created and destroyed accommodation space from the Miocene to the present, with sediments accumulating in a series of basins that tended to shift eastwards with time (e.g. the Miocene Tambach and Ngorora formations; the Plio-Pleistocene Chemeron Formation; Fig. 3.16a–f). Extensive eruptions of trachytes, trachytic tuffs and phonolites took place around 7–5 Ma ago in the Naivasha-Nakuru-Baringo area of the Central Gregory Rift (Fig. 3.15a), giving rise to two main formations: the Kabarnet Trachytes (~7.5–6.2 Ma; Deino et al. 2002) in the Tugen Hills and the Thomson's Falls Phonolites (~6.5 Ma) to the south at Bahati and near the southern Laikipia (Baker et al. 1971). In the Southern Gregory Rift, the Ol Esakut (Esayeiti) and Ngong central volcanoes were active at about 6.7–3.6? Ma (Figs. 3.15b and 3.18a; Baker et al. 1971).

A period of Late Miocene–Early Pliocene basaltic eruptions developed in several parts of the rift. In the South Turkana area, trachyte shield volcanoes formed between 6 and 2 Ma, with the largest (50 km across) being Ribkwo (McClenaghan 1971; Webb 1971; Webb and Weaver 1975). In the Tugen Hills, the Kaparaina Basalt lavas (5.7–5.1 Ma; Deino et al. 2006) extruded over the Kabarnet Trachyte lavas. The predominantly basaltic Aberdare Range (Nyandarua) reached an eruptive maximum around 6.5–5 Ma ago (Baker et al. 1971). Basalts also erupted in the Turkana depression to the north, including the laterally extensive Gombe Basalts with activity ceasing by 3.94 Ma (Haileab et al. 2004). The latter lavas interfinger with lake and fluvial sediments below and near modern Lake Turkana. These Plio-Pleistocene fluvial and lacustrine deposits are up to 4 km thick and were laid down in a series of connected half-grabens (Fig. 3.16e; Dunkelman et al. 1988, 1989).

Pliocene central volcanoes were active on the rift floor and shoulders at Tinderet, Londiani and Kilombe (nephelinite, phonolite, basanite) near the junction of the Nyanza (Kavirondo) and Gregory Rifts, as well as at the Aberdare Range (phonolite, trachyte, basalt, mugearite) and to the east of the rift at Mt. Kenya (phonolite, trachyte) (Fig. 3.14a; Baker et al. 1971). Subsequently, a major period of Plio-Pleistocene faulting cut through and uplifted the Ngong Hills and Ol Esakut (Esayeiti) volcanoes on the eastern rift margin (Fig. 3.18b, c). In contrast, to the west, the five-million-year-old lavas of the Kirikiti Basalts Formation were restricted to the rift by the rising Nguruman Escarpment (Baker et al. 1971, 1988; Shackleton 1978). In the Naivasha area, earlier Mio-Pliocene phonolite and trachyte lavas had filled an east-facing half-graben, but volcanism changed its character producing large-scale trachytic and ignimbritic eruptions of the Kinangop Tuffs (3.7–3.4 Ma), which were subsequently faulted and uplifted to produce the Bahati Platform, east of Lake Naivasha (Strecker et al. 1990; Leat 1991; Roessner and Strecker 1997; Spiegel et al. 2007).

During the Quaternary, eruptions were dominated by basalt to the east of the rift (e.g. Chyulu Range, Huri Hills, Nyambeni; Fig. 3.14b), with basalt, trachyte and phonolite confined mainly to the Gregory Rift. The Mt. Olorgesailie volcanics were erupted in the South Gregory Rift around 2.8–2.2 Ma with the nephelinite-phonolite Shombole volcano also erupting near the Kenya-Tanzania border at about this time. Flood lavas (Gesumeti and Limuru Trachytes) filled much of the South Gregory Rift between 2.1 and 1.7 Ma with a few flows escaping eastwards near the Ngong Hills (Baker et al. 1971, 1988; Baker and Mitchell 1976). Basaltic activity developed south of Suswa prior to widespread eruptions of the Magadi Trachytes (~1.3–0.9 Ma).

North of Nakuru, local eruptions of mugearitic and trachytic lavas developed at around 2–2.5 Ma (Griffiths 1977; Chapman and Brook 1978), which after ~1.5 Ma, gave way to extrusion of the extensive alkali Kwaibus Basalts and Hannington Trachyphonolites near Lake Bogoria, with basalt eruptions taking place after ~0.5 Ma (Griffiths and Gibson 1980; Baker et al. 1971). Intermittent Pleistocene faulting continued to deepen the Central and South Gregory rifts through the Quaternary, with activity becoming focussed on a narrow axial zone and dense networks of Middle to Late Pleistocene grid faults.

Further north, between Lake Baringo and the Suguta Valley, eruptions after about 500 ka produced the Karosi, Paka and Silali volcanoes (Fig. 3.14a) with caldera collapse at Silali and Paka developing at about 64 ka and 10 ka, respectively (Fig. 3.10c, d; Dunkley et al. 1993; Lichoro et al. 2019). Silali includes the second largest caldera in the Kenya Rift, which formed due to withdrawal of magma beneath the volcano (Dunkley et al. 1993). Between 2006 and 2010, subsidence rates were measured at 2 cm y^{-1} (Biggs et al. 2016), which reflects tectonic extension and the generation of sufficient space to allow magma drainage from below Silali (Lichoro et al. 2019).

The chain of Quaternary volcanoes continues northwards at regular intervals through the Suguta Valley, forming Emuruangogolak, Namarunu and the Barrier volcanic complex (Fig. 3.14a). In the Turkana Rift, faulting shifted east to the Kino Sogo Fault Belt (Fig. 3.16h; Vétel and Le Gall 2005). A series of major Late Pleistocene trachytic and basaltic central volcanoes and calderas developed along the narrowed rift axis (Fig. 3.14a). Volcanism was generally at a low ebb in the Turkana Rift, except for eruptions from North, Central and South Islands and the Barrier region between the Suguta and Lake Turkana (Karson and Curtis 1994; Muirhead et al. 2022; Rooney et al. 2022).

Quaternary fluvial and lacustrine sedimentary basins were focussed along the rift axis with rift marginal uplift leading to increased sediment inputs. Several major depositional basins lie below Lakes Turkana and Baringo, and the Kerio Rift (Mugisha et al. 1997; Hautot et al. 2000; Dunkelman et al. 1989). Lake Magadi is underlain by about 200 m of Quaternary lake and minor fluvial sediment that rest on the Pleistocene Magadi Trachyte Formation, with the neighbouring Koora Graben containing >160 m of sediment (Muiruri 2017). Other Pleistocene basins are present at Lake Bogoria and in the Naivasha-Elmenteita-Nakuru area, with outcrops of Pleistocene deposits also present at Munya wa Gicheru and Olorgesailie.

References

Abdelfettah Y, Tiercelin J-J, Tarits P, Hautot S, Maia M, Thuo P (2016) Subsurface structure and stratigraphy of the northwest end of the Turkana Basin, Northern Kenya Rift, as revealed by magnetotellurics and gravity joint inversion. J Afr Earth Sci 119:120–138

Achauer U, KRISP Teleseismic Working Group (1994) New ideas on the Kenya Rift based on the inversion of the combined dataset of the 1985 and 1989/90 seismic tomography experiments. Tectonophysics 236:305–329

Achauer U, Maguire PKH, Mechie J, Green WV, KRISP Working Group (1992) Some remarks on the structure and geodynamics of the Kenya Rift. Tectonophysics 213:257–268

Acosta VT (2015) Denudation processes in a tectonically active rift on different time scales: new insights from thermochronology and CRN dating in the Kenya Rift. PhD thesis, Universität Potsdam

Acosta VT, Bande A, Sobel ER, Parra M, Schildgen TF, Stuart F, Strecker MR (2015) Cenozoic extension in the Kenya Rift from low-temperature thermochronology: links to diachronous spatiotemporal evolution of rifting in East Africa. Tectonics 34:2367–2386

Albaric J, Déverchère J, Petit C, Perrot J, Le Gall B (2009) Crustal rheology and depth distribution of earthquakes: insights from the central and southern East African Rift System. Tectonophysics 468:28–41

Albaric J, Déverchère J, Perrot J, Jakovlev A, Deschamps A (2014) Deep crustal earthquakes in North Tanzania, East Africa: interplay between tectonic and magmatic processes in an incipient rift. Geochem Geophys Geosyst 15:374–394

Bagley B, Nyblade AA (2013) Seismic anisotropy in eastern Africa, mantle flow, and the African superplume. Geophys Res Lett 40:1500–1505

Bahat D (1984) Fracture interaction in the Gregory Rift, East Africa. Tectonophysics 104:47–65

Bailey DK (1992) Episodic alkaline igneous activity across Africa: implications for the causes of continental break-up. In: Storey BC, Alabaster R, Pankhurst RJ (eds) Magmatism and the causes of continental break-up. Geol Soc Lond Spec Publ 68:91–98

Baker BH (1963) Geology of the area south of Magadi. Geol Surv Kenya Rep 42

Baker BH (1975) Geology and geochemistry of the Ol Doinyo Nyokie trachyte ignimbrite vent complex, south Kenya Rift Valley. Bull Volcanol 39:420–440

Baker BH (1987) Outline of the petrology of the Kenya Rift alkaline province. In: Fitton JG, Upton BGJ (eds) Alkaline igneous rocks. Geol Soc Lond Spec Publ 27:293–311

Baker BH, McConnell RB (1970) The structural pattern of the Afro-Arabian Rift System in relation to plate tectonics (and discussion). Proc R Soc Lond Ser A 267:383–391

Baker BH, Mitchell JG (1976) Volcanic stratigraphy and geochronology of the Kedong-Olorgesailie area and the evolution of the South Kenya Rift Valley. J Geol Soc Lond 132:467

Baker BH, Morgan P (1981) Continental rifting: progress and outlook. Eos 62:585–586

Baker B, Wohlenberg J (1971) Structure and evolution of the Kenya Rift Valley. Nature 229:538–542

Baker BH, Williams LAJ, Miller JA, Fitch FJ (1971) Sequence and geochronology of the Kenya Rift volcanics. Tectonophysics 11:191–215

Baker BH, Mohr PA, Williams LAJ (1972) Geology of the eastern rift system of Africa. Geol Soc Am Spec Pap 136:1–67

Baker BH, Goles GG, Leeman WP, Lindstrom MM (1977) Geochemistry and petrogenesis of a basalt-benmoreite-trachyte suite from the southern part of the Gregory Rift, Kenya. Contrib Mineral Petrol 64:303–332

Baker BH, Crossley R, Goles GG (1978) Tectonic and magmatic evolution of the southern part of the Kenya Rift Valley. In: Neumann ER, Ramberg B (eds) Petrology and geochemistry of continental rifts. Reidel, Dordrecht, pp 29–50

Baker BH, Mitchell JG, Williams LAJ (1988) Stratigraphy, geochronology and volcano-tectonic evolution of the Kedong-Naivasha-Kinangop region, Gregory Rift Valley, Kenya. J Geol Soc Lond 145:107–116

Bechtel TD, Forsyth DW, Swain CJ (1987) Mechanisms of isostatic compensation in the vicinity of the East African Rift, Kenya. Geophys J R Astron Soc 90:445–465

Behrensmeyer AK, Deino AL, Hill A, Kingston JD, Saunders JJ (2002) Geology and geochronology of the middle Miocene Kipsaramon site complex, Muruyur Beds, Tugen Hills, Kenya. J Hum Evol 42:11–38

Biggs J, Robertson E, Cashman KV (2016) The lateral extent of volcanic interactions during unrest and eruption. Nat Geosci 9:308–311

Bishop WW, Pickford M (1975) Geology, fauna and palaeoenvironments of the Ngorora Formation, Kenya Rift Valley. Nature 254:185–192

Boone SC, Seiler C, Kohn BP, Gleadow AJW, Foster DA, Chung L (2017) Influence of rift superposition on lithospheric response to East African Rift system extension: Lapur Range, Turkana, Kenya. Tectonics 37:182–207

Boone SC, Kohn BP, Gleadow AJW, Morley CK, Seiler C, Foster DA (2019) Birth of the East African Rift System: nucleation of magmatism and strain in the Turkana Depression. Geology 47:886–890

Bosworth W (1985) Geometry of propagating continental rifts. Nature 316:625–627

Bosworth W (1989) Basin and range style tectonics in East Africa. J Afr Earth Sci 8:191–201

Bosworth W (1992) Mesozoic and early Tertiary rift tectonics in East Africa. Tectonophysics 209:115–137

Bosworth W, Maurin A (1993) Structure, geochronology and tectonic significance of the northern Suguta Valley (Gregory Rift), Kenya. J Geol Soc Lond 150:751–762

Bosworth W, Morley CK (1994) Structural and stratigraphic evolution of the Anza rift, Kenya. Tectonophysics 236:93–115

Bosworth W, Lambiase J, Keisler R (1986) A new look at Gregory's Rift: the structural style of continental rifting. EOS Trans Am Geophys Union 67:577–580

Braile LW, Keller GR, Wendlandt RF, Morgan P, Khan MA (2006) The East African Rift System. In: Olsen KH (ed) Continental rifts: evolution, structure, tectonics (Developments in Geotectonics 25). Elsevier, Amsterdam, pp 213–231

Burke KC (1996) The African plate. S Afr J Geol 99:339–409

Burke K, Wilkinson MJ (2016) Landscape evolution in Africa during the Cenozoic and Quaternary—the legacy and limitations of Lester C. King. Can J Earth Sci 53:1089–1102

Calais E, d'Oreye N, Albaric J, Deschamps A, Delvaux D, Déverchère J, Ebinger C, Ferdinand RW, Kervyn F, Macheyeki AS, Oyen A, Perrot J, Saria E, Smets B, Stamps DS, Wauthier C (2008) Strain accommodation by slow slip and dyking in a youthful continental rift, East Africa. Nature 456:783–787

Cerling TE, Powers DW (1977) Paleorifting between the Gregory and Ethiopian Rifts. Geology 5:441–444

Chapman GR, Brook M (1978) Chronostratigraphy of the Baringo Basin, Kenya. In: Bishop WW (ed) Geological background to fossil man. Geol Soc Lond Spec Publ 6:207–223

Chapman GR, Lippard SJ, Martyn JE (1978) The stratigraphy and structure of the Kamasia Range, Kenya Rift Valley. J Geol Soc Lond 135:265–281

Chorowicz J (2005) The East African rift system. J Afr Earth Sci 43:379–410

Clément J-P, Caroff M, Hémond C, Tiercelin J-J, Bollinger C, Guillou H, Cotton J (2003) Pleistocene magmatism in a lithospheric transition area: petrogenesis of alkaline and peralkaline lavas from the Baringo-Bogoria Basin, central Kenya Rift. Can J Earth Sci 40:1239–1257

Coblentz DD, Sandiford M (1994) Tectonic stresses in the African plate: constraints on the ambient lithospheric stress state. Geology 22:831–834

Corti G, Cioni R, Franceschini Z, Sani F, Scaillet S, Molin P, Isola I, Mazzarini F, Brune S, Keir D, Erbello A, Muluneh A, Illsley-Kemp F, Glerum A (2019) Aborted propagation of the Ethiopian rift caused by linkage with the Kenyan rift. Nat Commun 10:1309

Crossley R (1980) Structure and volcanism in the S Kenya Rift. In: Proc conf geodynamics of the Afro-Arabian Rift System, Rome 1979. Atti dei Covegni Lincei 47:89–98

Crossley R, Knight RM (1981) Volcanism in the western part of the rift valley in Southern Kenya. Bull Volcanol 44:117–128

Dallheim HA, Davis PM, Achauaer U (1989) Deep velocity structure beneath the east African Rift and wave-form analysis using teleseismic events. J Afr Earth Sci 8:461–470

Daly MC, Chorowicz J, Fairhead JD (1989) Rift basin evolution in Africa: the influence of reactivated steep basement shear zones. In: Cooper MA, Williams GD (eds) Inversion tectonics. Geol Soc Lond Spec Publ 44:309–334

Darracott BW, Fairhead JD, Girdler RW (1972) Gravity and magnetic surveys in northern Tanzania and southern Kenya. Tectonophysics 15:131–141

Davis PM, Slack PD (2002) The uppermost mantle beneath the Kenya dome and relation to melting, rifting and uplift in East Africa. Geophys Res Lett 29:1–3

Deino AL, Tauxe L, Monaghan M, Hill A (2002) $^{40}Ar/^{39}Ar$ geochronology and paleomagnetic stratigraphy of the Lukeino and lower Chemeron Formations at Tabarin and Kapcheberek, Tugen Hills, Kenya. J Hum Evol 42:117–140

Deino A, Kingston J, Glen J, Edgar RK, Hill A (2006) Precessional forcing of lacustrine sedimentation in the late Cenozoic Chemeron Basin, Central Kenya Rift, and calibration of the Gauss/Matuyama boundary. Earth Planet Sci Lett 247:41–60

Delvaux D (2001) Karoo rifting in western Tanzania: precursor of Gondwana breakup? In: Weiss RH (ed) Contributions to geology and paleontology of Gondwana in honour of Prof. Dr. Helmut Wopfner. Geologisches Institut, Universität zu Köln, Germany, pp 111–125

Demissie G (2010) Mantle influence, rifting and magmatism in the East African Rift System (EARS): a regional view of the controls on hydrothermal activity. In: Proceedings of the world geothermal congress, Bali, pp 1–12

Dixey F (1926) The Nyasaland section of the Great Rift Valley. Geogr J 68:117–140

Dixey F (1946) Erosion and tectonics in the East African rift system. Q J Geol Soc Lond 102:339–388

Dixey F (1956) The East African rift system. Bull Suppl Colon Geol Min Res 1:1–71

Dunkelman TJ, Karson JA, Rosendahl BR (1988) Structural style of the Turkana rift, Kenya. Geology 16:258–261

Dunkelman TJ, Rosendahl BR, Karson JA (1989) Structure and stratigraphy of the Turkana rift from seismic reflection data. J Afr Earth Sci 8:489–510

Dunkley PN, Smith M, Allen DJ, Darling WG (1993) The geothermal activity and geology of the northern sector of the Kenya Rift Valley. Brit Geol Surv Res Rep SC/93/1

Ebinger CJ (1989) Tectonic development of the western branch of the East African rift system. Geol Soc Am Bull 101:885–903

Ebinger C (2005) Continental break-up: the East African perspective. Astron Geophys 46:2.16–2.21

Ebinger CJ (2021) Recipe for rifting: flavors of East Africa. In: Alderton D, Elias SA (eds) Encyclopedia of geology, vol 4, 2nd edn. Academic Press, London, pp 271–283

Ebinger CJ, Ibrahim A (1994) Multiple episodes of rifting in Central and East Africa: a re-evaluation of gravity data. Geol Rundsch 83:689–702

Ebinger C, Scholz CA (2012) Continental rift basins: the East African perspective. In: Busby C, Azor A (eds) Tectonics of sedimentary basins: recent advances. Wiley-Blackwell, Chichester, pp 185–208

Ebinger CJ, Sleep NH (1998) Cenozoic magmatism in central and East Africa resulting from impact of one large plume. Nature 395:788–791

Ebinger CJ, Deino AL, Drake RE, Tesha AL (1989) Chronology of volcanism and rift basin propagation: Rungwe volcanic province, East Africa. J Geophys Res 94:15785–15803

Ebinger CJ, Yemane T, Woldegabriel G, Aronson JL, Walter RC (1993) Late Eocene-Recent volcanism and faulting in the southern main Ethiopian Rift. J Geol Soc Lond 150:99–108

Ebinger CJ, Yemane T, Harding DJ, Tesfaye S, Kelley S, Rex DC (2000) Rift deflection, migration, and propagation: linkage of the Ethiopian and Eastern rifts, Africa. Geol Soc Am Bull 112:163–176

Ebinger C, Petit C, Burov E (2002) Causes and consequences of lithospheric extension: the ups and downs of continental rifts. In: Renaut RW, Ashley GM (eds) Sedimentation in continental rifts. SEPM Spec Publ 73:11–23

Emishaw L, Abdelsalam MG (2019) Development of Late Jurassic-Early Paleogene and Neogene-Quaternary rifts within the Turkana Depression, East Africa from satellite gravity data. Tectonics 38:2358–2377

Fairhead JD (1976) The structure of the lithosphere beneath the Eastern Rift, East Africa, deduced from gravity studies. Tectonophysics 30:269–298

Fairhead JD (1986) Geophysical controls on sedimentation within the African Rift Systems. In: Frostick LE, Renaut RW, Reid I, Tiercelin J-J (eds) Sedimentation in the African rifts. Geol Soc Lond Spec Publ 25:19–27

Faulds JE, Varga RJ (1998) The role of accommodation zones and transfer zones in the regional segmentation of extended terranes. In: Faulds JE, Varga RJ (eds) Accommodation zones and transfer zones; the regional segmentation of the Basin and Range Province. Geol Soc Am Spec Publ 323:1–45

Fritz H, Abdelsalam M, Ali KA, Bingen B, Collins AS, Fowler AR, Ghebreab W, Hauzenberger CA, Johnson PR, Kusky TM, Macey P, Muhongo S, Stern RJ, Viola G (2013) Orogen styles in the East African Orogen: a review of the Neoproterozoic to Cambrian tectonic evolution. J Afr Earth Sci 86:65–106

Furman T (2007) Geochemistry of East African Rift basalts: an overview. J Afr Earth Sci 48:147–160

Furman T, Bryce JG, Karson J, Iotti A (2004) East African Rift System (EARS) plume structure: insights from Quaternary mafic lavas of Turkana, Kenya. J Petrol 45:1069–1088

Furman T, Kaleta KM, Bryce JG, Hanan BB (2006) Tertiary mafic lavas of Turkana, Kenya: constraints of East African plume structure and occurrence of high-μ volcanism in Africa. J Petrol 47:1221–1244

George R, Rogers N, Kelley S (1998) Earliest magmatism in Ethiopia: evidence for two mantle plumes in one flood basalt province. Geology 26:923–926

Girdler RW (1991) The Afro-Arabian rift system: an overview. Tectonophysics 197:139–153

Green WV, Achauer U, Meyer RP (1991) A three-dimensional seismic image of the crust and upper mantle beneath the Kenya Rift. Nature 354:199–203

Gregory JW (1894) Contributions to the physical geography of British East Africa. Geogr J 4:289–315

Gregory JW (1896) The Great Rift Valley. John Murray, London

Gregory JW (1920a) The African Rift Valley. Nature 104:518–551

Gregory JW (1920b) The African Rift Valleys. Geogr J 56:13–41

Gregory JW (1921) The rift valleys and geology of East Africa. Seeley Service, London

Gregory JW (1923) The structure of the Great Rift Valley. Nature 112:514–516

Griffiths PS (1977) The geology of the area around Lake Hannington and the Perkerra River, Rift Valley Province, Kenya. PhD thesis, University of London

Griffiths PS, Gibson IL (1980) The geology and petrology of the Hannington Trachyphonolite formation, Kenya Rift Valley. Lithos 13:43–53

Grove AT (1983) Geomorphology of the African Rift System. In: Frostick LE, Renaut RW, Reid I, Tiercelin J-J (eds) Sedimentation in the African rifts. Geol Soc Lond Spec Publ 25:9–18

Guth A (2013) Spatial and temporal evolution of the volcanics and sediments of the Kenya Rift. PhD thesis, Michigan Technological University, Houghton

Guth A (2016) Volcanic volumes associated with the Kenya Rift: recognition and correction of preservation biases. In: Wright TJ, Ayele A, Ferguson DJ, Kidane T, Vye-Brown C (eds) Magmatic rifting and active volcanism. Geol Soc Lond Spec Publ 420:31–42

Guth A, Wood J (2014) Geological map of the Southern Kenya Rift (Sheets 1–6). Geol Soc Am Digi Map Chart Ser 16. https://doi.org/10.1130/2014.DMCH016

Haileab B, Brown F, McDougall I, Gathogo P (2004) Gombe Group basalts and initiation of Pliocene deposition in the Turkana depression, northern Kenya and Southern Ethiopia. Geol Mag 141:41–53

Halldórsson SA, Hilton DR, Scarsi P, Abebe T, Hopp J (2014) A common mantle plume source beneath the entire East African Rift System revealed by coupled helium-neon systematics. Geophys Res Lett 41:2304–2311

Hansen SE, Nyblade AA, Benoit MH (2012) Mantle structure beneath Africa and Arabia from adaptively parameterized P-wave tomography: implications for the origin of Cenozoic Afro-Arabian tectonism. Earth Planet Sci Lett 319-320:23–34

Hautot S, Tarits P, Whaler K, Le Gall B, Tiercelin J-J, Le Turdu C (2000) Deep structure of the Baringo Rift Basin (Central Kenya) from three-dimensional magnetotelluric imaging: implications for rift evolution. J Geophys Res 105:23493–23518

Hay DE, Wendlandt RF, Keller GR (1995) Origin of Kenya Rift Plateau-type flood phonolites: integrated petrologic and geophysical constraints on the evolution of the crust and upper mantle beneath the Kenya Rift. J Geophys Res 100:10549–10557

Hendrie DB, Kusznir NJ, Morley CK, Ebinger CJ (1994) Cenozoic extension in northern Kenya: a quantitative model of rift basin development in the Turkana area. Tectonophysics 236:409–438

Henry WJ, Mechie J, Maguire P (1990) A seismic investigation of the Kenya Rift Valley. Geophys J Int 100:107–130

Hetzel R, Strecker MR (1994) Late Mozambique Belt structures in western Kenya and their influence on the evolution of the Cenozoic Kenya Rift. J Struct Geol 16:189–201

Hooper PR (1990) The timing of crustal extension and the eruption of continental flood basalts. Nature 345:246249

Huerta AD, Nyblade AA, Reusch AM (2009) Mantle transition zone structure beneath Kenya and Tanzania: more evidence for a deep-seated thermal upwelling in the mantle. Geophys J Int 177:1249–1255

Huntsman-Mapila P, Kampunzu AB, Vink B, Ringrose S (2005) Cryptic indicators of provenance from the geochemistry of the Okavango Delta sediments, Botswana. Sediment Geol 174:123–148

Huntsman-Mapila P, Tiercelin J-J, Benoit M, Ringrose S, Diskin S, Cotton J, Hémond C (2009) Sediment geochemistry and tectonic setting: application of discrimination diagrams to early stages of intracontinental rift evolution, with examples from the Okavango and Southern Tanganyika rift basins. J Afr Earth Sci 53:33–44

Ibs-von Seht M, Blumenstein S, Wagner R, Hollnack D, Wohlenberg J (2001) Seismicity, seismotectonics and crustal structure of the southern Kenya Rift—new data from the Lake Magadi area. Geophys J Int 146:439–453

Ibs-von Seht M, Plenefisch T, Klinge K (2008) Earthquake swarms in continental rifts – a comparison of selected cases in America, Africa and Europe. Tectonophysics 452:66–77

Isola I, Mazzarini F, Bonini M, Corti G (2014) Spatial variability of volcanic features in early-stage rift settings: the case of the Tanzania Divergence, East African rift system. Terra Nova 26:461–468

Johnson RW (1969) Volcanic geology of Mount Suswa, Kenya. Phil Trans R Soc Ser A 265:383–412

Jones WB, Lippard SJ (1979) New age determinations and the geology of the Kenya Rift-Kavirondo Rift junction, W Kenya. J Geol Soc Lond 136:693–704

Kabeto K, Sawada Y, Iizumi S, Wakatsuki T (2001) Mantle sources and magma-crust interactions in volcanic rocks from the northern Kenya Rift: geochemical evidence. Lithos 56:111–139

Kampunzu AB, Mohr P (1991) Magmatic evolution and petrogenesis in the East African rift system. In: Kampunzu AB, Lubala RT (eds) Magmatism in extensional structural settings. Springer, Berlin, pp 85–136

Karson JA, Curtis PC (1994) Quaternary volcanic centres of the Turkana Rift, Kenya. J Afr Earth Sci 18:15–35

Keller GR, Mechie J, Braile LW, Mooney WD (1994) Seismic structure of the uppermost mantle beneath the Kenya Rift. Tectonophysics 236:201–216

Khan MA, Swain CJ (1978) Geophysical investigations and the rift valley geology of Kenya. In: Bishop WW (ed) Geological background to fossil man. Geol Soc Lond Spec Publ 6:71–83

Khan A, Maguire P, Henry B, Higham M (1986) KRISP 85—an international seismic investigation of the Kenya Rift. Geol Today 2:139–144

Khan MA, Maguire PKH, Henry W, Higham M, Prodehl C, Mechie J, Keller GR, Patel J (1989) A crustal seismic refraction line along the axis of the S. Kenya Rift. J Afr Earth Sci 8:455–460

Khan MA, Mechie J, Birt C, Byrne G, Gaciri S, Jacob B, Keller GR, Maguire PKH, Novak O, Nyambok IO, Patel JP, Prodehl C, Riaroh D, Simiyu S, Thybo (1999) The lithospheric structure of the Kenya Rift as revealed by wide-angle seismic measurements. In: Niocaill CM, Ryan CD (eds) Continental tectonics. Geol Soc Lond Spec Publ 164:257–269

Kilembe EA, Rosendahl BR (1992) Structure and stratigraphy of the Rukwa rift. Tectonophysics 209:143–158

Kinabo BD (2007) Incipient continental rifting: insights from the Okavango Rift Zone, northwestern Botswana. PhD thesis, University of Missouri-Rolla

King BC (1949) Napak area of southern Karamoja, Uganda. Geol Surv Uganda Mem 5

King BC (1978) Structural and volcanic evolution of the Gregory Rift Valley. In: Bishop WW (ed) Geological background to fossil man. Geol Soc Lond Spec Publ 6:29–54

King SD, Anderson DL (1998) Edge-driven convection. Earth Planet Sci Lett 160:289–296

King BC, Chapman GC (1972) Volcanism of the Kenya Rift Valley. Phil Trans R Soc Lond Ser A 271:185–208

King SD, Ritsema J (2000) African hot-spot volcanism: small-scale convection in the upper mantle beneath cratons. Science 290:1137–1140

Koptev A, Calais E, Burov E, Leroy S, Gerya T (2015) Dual continental rift systems generated by plume-lithosphere interaction. Nat Geosci 8:388–392

Krenkel E (1925) Geologie der Erde, Geologie Afrikas. Gebruder Borntraeger, Berlin

KRISP Working Group (1991) The Kenya Rift: pure shear extension above a mantle plume. Nature 345:223–227

Kuria ZN (2011) Seismotectonics of active faults: Magadi fault system, southern Kenya Rift. PhD thesis, University of Twente, Netherlands

Lambiase JJ (1989) The framework of African rifting during the Phanerozoic. J Afr Earth Sci 8:183–190

Latin D, Norry MJ, Tarzey RJE (1993) Magmatism in the Gregory Rift, East Africa: evidence for melt generation by a plume. J Petrol 34:1007–1027

Le Gall B, Tiercelin J-J, Richert J-P, Gente P, Sturchio NC, Stead D, Le Turdu C (2000) A morphotectonic study of an extensional fault zone in a magma-rich rift: the Baringo Trachyte Fault System, central Kenya Rift. Tectonophysics 320:87–106

Le Gall B, Vétel W, Morley CK (2005) Inversion tectonics during continental rifting: the Turkana Cenozoic rifted zone, northern Kenya. Tectonics 24:1–17

Le Gall B, Nonnotte P, Rolet J, Benoit M, Guillou H, Mousseau-Nonnotte M, Albaric J, Deverchère J (2008) Rift propagation at craton margin. Distribution of faulting and volcanism in the North Tanzanian Divergence (East Africa) during Neogene times. Tectonophysics 448:1–19

Le Roex AP, Späth A, Zartman RE (2001) Lithospheric thickness beneath the southern Kenya Rift: implications from basalt geochemistry. Contrib Mineral Petrol 142:89–106

Le Turdu C (1998) Modéles tectono-sédimentaires 3-D des bassins en extension: Exemples du Rift du Kenya (Baringo-Bogoria et Magadi), du Rift Ethiopien (Ziway-Shala) et du Fossé Nord-Tanganyika. PhD thesis, Université de Bretagne Occidentale, Brest

Le Turdu C, Tiercelin J-J, Coussement C, Rolet J, Renaut RW, Richert J-P, Xavier J-P, Coquelet D (1995) Rift basin structure and depositional patterns interpreted using a 3D remote sensing approach: the Baringo–Bogoria basins, central Kenya Rift, East Africa. Bull Centr Rech Explor-Prod Elf-Aquitaine 19:1–37

Le Turdu C, Tiercelin J-J, Richert J-P, Rolet J, Xavier J-P, Renaut RW, Lezzard K, Coussement C (1999) Influence of pre-existing oblique discontinuities on the geometry and evolution of extensional fault patterns: evidence from the Kenya Rift using SPOT imagery. In: Morley CK (ed) Geoscience of rift systems – evolution of East Africa. Am Assoc Petrol Geol Stud Geol 44:173–191

Leat PT (1991) Volcanological development of the Nakuru area of the Kenya Rift valley. J Afr Earth Sci 13:483–498

Leat PT, Macdonald R, Smith RL (1984) Geochemical evolution of the Menengai caldera volcano, Kenya. J Geophys Res 89:8571–8592

Lichoro CM, Árnason K, Cumming W (2019) Joint interpretation of gravity and resistivity data from the Northern Kenya volcanic rift zone: structural and geothermal significance. Geothermics 77:139–150

Lippard SJ, Truckle PH (1978) Spatial and temporal variations in basalt geochemistry in the northern Kenya Rift. In: Neumann ER, Ramberg IB (eds) Petrology and geochemistry of continental rifts. Reidel, Dordrecht, pp 123–132

Logatchev N, Beloussov VV, Milankovsky EE (1972) East African rift development. Tectonophysics 15:71–81

Macdonald R (1994) Petrological evidence regarding the evolution of the Kenya Rift Valley. Tectonophysics 236:373–390

Macdonald R (2002) Magmatism of the Kenya Rift Valley: a review. Earth Env Sci Trans R Soc Edinb 93:239–253

Macdonald R (2012) Evolution of peralkaline silicic complexes: lessons from the extrusive rocks. Lithos 152:11–22

Macdonald R, Baginski B (2009) The central Kenya peralkaline province: a unique assemblage of magmatic systems. Mineral Mag 73:1–16

Macdonald R, Williams LAJ, Gass IG (1994) Tectonomagmatic evolution of the Kenya Rift Valley: some geological perspectives. J Geol Soc Lond 151:879–888

Macdonald R, Davies GR, Upton BGJ, Dunkley PN, Smith M, Leat PT (1995) Petrogenesis of Silali volcano, Gregory Rift, Kenya. J Geol Soc Lond 152:703–720

Macdonald R, Rogers NW, Fitton JG, Black S, Smith M (2001) Plume-lithosphere interactions in the generation of the basalts of the Kenya Rift, East Africa. J Petrol 42:877–900

Macdonald R, Belkin HE, Fitton JG, Rogers NW, Nejbert K, Tindle AG, Marshall AS (2008) The roles of fractional crystallization, magma mixing, crystal mush remobilization and volatile-melt interactions in the genesis of a young basalt-peralkaline rhyolite suite, the greater Olkaria Volcanic Complex, Kenya Rift Valley. J Petrol 49:1515–1547

Macgregor D (2015) History of the development of the East African Rift System: a series of interpreted maps through time. J Afr Earth Sci 101:1–21

Macgregor D (2018) History of the development of Permian–Cretaceous rifts in East Africa: a series of interpreted maps through time. Pet Geosci 24:8–20

Maguire PKH, Swain CJ, Masotti R, Khan MA (1994) A crustal and uppermost mantle cross sectional model of the Kenya Rift derived from seismic and gravity data. Tectonophysics 236:217–249

Mariita NO, Keller GR (2007) An integrated geophysical study of the northern Kenya Rift. J Afr Earth Sci 48:80–94

Mboya B (1983) The genesis and tectonics of the N.E. Nyanza rift valley, Kenya. J Afr Earth Sci 1:315–320

McCall GJH (1967) Geology of the Nakuru–Thomson's Fall-Lake Hannington area. Geol Surv Kenya Rep 78

McCall G, Bristow CM (1965) An introductory account of Suswa volcano, Kenya. Bull Volcanol 28:333–367

McClenaghan MP (1971) Geology of the Ribkwo area, Baringo District, Kenya. PhD thesis, University of London

McConnell RB (1967) The East African rift system. Nature 215:578–581

McDougall I, Brown FH (2008) Timing of volcanism and evolution of the northern Kenya Rift. Geol Mag 146:34–47

Mechie J, Keller GR, Prodehl C, Gaciri S, Braile LW, Mooney WD, Gajewski D, Sandmeier KJ (1994) Crustal structure beneath the Kenya Rift from axial profile data. Tectonophysics 236:179–200

Meju MA, Sakkas V (2007) Heterogeneous crust and upper mantle across southern Kenya and the relationship to surface deformation as inferred from magnetotelluric imaging. J Geophys Res 112:B04103

Melnick D, Garcin Y, Quinteros J, Strecker MR, Olago D, Tiercelin J-J (2012) Steady rifting in northern Kenya inferred from deformed Holocene lake shorelines of the Suguta and Turkana basins. Earth Planet Sci Lett 331–332:335–346

Mibei G, Harðarson BS, Franzson H, Bali E, Geirsson H, Guðfinnsson GH (2021) Eruptive history and volcano-tectonic evolution of Paka volcanic complex in the northern Kenya Rift: insights into the geothermal heat source. J Afr Earth Sci 173:103951

Mohr P (1992) Nature of the crust beneath magmatically active continental rifts. Tectonophysics 213:269–284

Mooney WD, Christensen NI (1994) Composition of the crust beneath the Kenya Rift. Tectonophysics 236:391–408

Morgan P, Baker BH (1983) Introduction—processes of continental rifting. Tectonophysics 94:1–10

Morley CK (1994) Interaction of deep and shallow processes in the evolution of the Kenya Rift. Tectonophysics 236:81–91

Morley CK (1995) Developments in the structural geology of rifts over the last decade and their impact on hydrocarbon exploration. In: Lambiase JJ (ed) Hydrocarbon habitat in rift basins. Geol Soc Lond Spec Publ 80:1–32

Morley CK (ed) (1999a) Geoscience of rift systems–evolution of East Africa. Am Assoc Petrol Geol Stud Geol 44

Morley CK (1999b) Boundary fault angle, with particular reference to the Lokichar Fault, Turkana Region, Kenya. In: Morley CK (ed) Geoscience of rift systems–evolution of East Africa. Am Assoc Petrol Geol Stud Geol 44:115–129

Morley CK (1999c) Patterns of displacement along large normal faults—implications for basin evolution and fault propagation, based on examples from East Africa. Am Assoc Petrol Stud Geol Bull 83:613–634

Morley CK (2010) Stress re-orientation along zones of weak fabrics in rifts: an explanation for pure extension in "oblique" rift segments? Earth Planet Sci Lett 297:667–673

Morley CK (2020) Early syn-rift igneous dike patterns, northern Kenya Rift (Turkana, Kenya): implications for local and regional stresses, tectonics, and magma-structure interactions. Geosphere 16:890–918

Morley CK, Wescott WA (1999) Sedimentary environments and geometry of sedimentary bodies determined from subsurface studies in East Africa. In: Morley CK (Ed) Geoscience of rift systems–evolution of East Africa. Am Assoc Petrol Geol Stud Geol 44:211–231

Morley CK, Cunningham SM, Harper RM, Wescott WA (1992a) Geology and geophysics of the Rukwa Rift, East Africa. Tectonics 11:69–81

Morley CK, Wescott WA, Stone DM, Harper RM, Wigger ST, Karanja FM (1992b) Tectonic evolution of the northern Kenyan Rift. J Geol Soc Lond 149:333–348

Morley CK, Karanja FM, Wescott WA, Stone DM, Harper RM, Wigger ST, Day RA (1999a) Geology and geophysics of the western Turkana basins, Kenya. In: Morley CK (ed) Geoscience of rift systems—evolution of East Africa. Am Assoc Petrol Geol Stud Stud Geol 44:19–54

Morley CK, Ngenoh DK, Ego JK (1999b) Introduction to the East African rift system. In: Morley CK (ed) Geoscience of rift systems–evolution of East Africa. Am Assoc Petrol Geol Stud Geol 44:1–18

Moucha R, Forte AM (2011) Changes in African topography drive by mantle convection. Nat Geosci 4:707–712

Mougenot D, Recq M, Virlogeux P, Lepvrier C (1986) Seaward extension of the East-African Rift. Nature 321:599–603

Mugisha F, Ebinger CJ, Strecker M, Pope D (1997) Two-stage rifting in the Kenya Rift: implications for half-graben models. Tectonophysics 278:63–81

Muirhead JD, Kattenhorn SA (2018) Activation of preexisting transverse structures in an evolving magmatic rift in East Africa. J Struct Geol 106:1–18

Muirhead JD, Kattenhorn SA, Le Corvec N (2015) Varying styles of magmatic strain accommodation across the East African Rift. Geochem Geophys Geosyst 16:2775–2795

Muirhead JD, Kattenhorn SA, Lee H, Mana S, Turrin BD, Fischer TP, Kianji G, Dindi E, Stamps DS (2016) Evolution of upper crustal faulting assisted by magmatic volatile release during early-stage continental rift development in the East African Rift. Geosphere 12:1670–1700
Muirhead JD, Scholz CA, Rooney TO (2022) Transition to magma-driven rifting in the South Turkana Basin, Kenya: part 1. J Geol Soc Lond 179:159
Muiruri VM (2017) Quaternary diatom and palynomorph stratigraphies and palaeoenvironments of the Koora Graben and Lake Magadi Basin, Kenya Rift Valley. PhD thesis, Hong Kong Baptist University
Mulibo GD, Nyblade AA (2013) Mantle transition zone thinning beneath eastern Africa: evidence for a whole-mantle superplume structure. Geophys Res Lett 40:3562–3566
Mulwa JK, Kimata F (2014) Tectonic structures across the East African Rift based on the source parameters of the 20 May 1990 M7.2 Sudan earthquake. Nat Hazards 73:493–506
Mulwa JK, Kimata F, Suzuki S, Kuria ZN (2014) The seismicity in Kenya (East Africa) for the period 1906–2010: a review. J Afr Earth Sci 89:72–78
Nyblade AA (2002) Crust and upper mantle structure in East Africa: implications for the origin of Cenozoic rifting and volcanism and the formation of magmatic rifted margins. In: Menzies MA, Klemperer SL, Ebinger CJ, Baker J (eds) Volcanic rifted margins. Geol Soc Am Spec Pap 362:15–26
Nyblade AA, Brazier RA (2002) Precambrian lithospheric controls on the development of the East African rift system. Geology 30:755–758
Owen RB, Renaut RW, Lowenstein TK (2018) Spatial and temporal geochemical variability in lacustrine sedimentation in the East African Rift System: evidence from the Kenya Rift and regional analyses. Sedimentology 65:1697–1730
Owen RB, Renaut RW, Muiruri VM, Rabideaux NM, Lowenstein TK, McNulty EP, Leet K, Deocampo D, Luo S, Deino AL, Cohen A, Sier MJ, Campisano C, Shen CC, Billingsley A, Mbuthia A, Stockhecke M (2019) Quaternary history of the Lake Magadi basin, southern Kenya Rift: tectonic and climatic controls. Palaeogeogr Palaeoecol Palaeoclimatol 518:97–118
Paron P, Olago DO, Omuto CT (2013) Kenya: a natural outlook: geo-environmental resources and hazards (Developments in earth surface processes 16). Elsevier, Amsterdam
Peterson TD (1989) Peralkaline nephelinites. I. Comparative petrology of Shombole and Oldoinyo L'engai, East Africa. Contrib Mineral Petrol 101:458–478
Pickford M (1982) The tectonics, volcanics and sediments of the Nyanza Rift Valley, Kenya. Z Geomorph N F 42:1–33
Pik R, Marty B, Hilton DR (2006) How many mantle plumes in Africa? The geochemical point of view. Chem Geol 226:100–114
Plasman M, Hautot S, Tarits P, Gautier S, Tiberi C, Le Gall B, Mtelela K, Gama R (2019) Lithospheric structure of a transitional magmatic to amagmatic continental rift system—insights from magnetotelluric and local tomography studies in the North Tanzanian Divergence, East African Rift. Geosciences 9:462–430
Prodehl C, Mechie J (1991) Crustal thinning in relationship to the evolution of the Afro-Arabian rift system: a review of seismic-refraction data. Tectonophysics 198:311–327
Pulfrey W (1960) Shape of the sub-Miocene erosion bevel in Kenya. Geol Surv Kenya Bull 3
Purcell PG (2018) Re-imagining and re-imaging the development of the East African Rift. Pet Geosci 24:21–40
Quennel AM (1960) The rift system and the East African swell. Q J Geol Soc Lond 116:78–86
Ren M, Omenda P, Anthony EY, White JC, Macdonald R, Bailey DK (2006) Application of the QUILF thermobarometer to the peralkaline trachytes and pantellerites of the Eburru volcanic complex, East African Rift, Kenya. Lithos 91:109–124
Renaut RW, Ego J, Tiercelin J-J, Le Turdu C, Owen B (1999) Saline, alkaline palaeolakes of the Tugen Hills-Kerio Valley region, Kenya Rift valley. In: Andrews P, Banham P (eds) Late Cenozoic environments and hominid evolution: a tribute to the late Bill Bishop. Geological Society of London, London, pp 41–58

Richter MJEA, Brune S, Riedl S, Glerum A, Neuharth D, Strecker MR (2021) Controls on asymmetric rift dynamics: numerical modeling of strain localization and fault evolution in the Kenya Rift. Tectonics 40:e2020TC006553

Ridolfi F, Renzulli A, Macdonald R, Upton B (2006) Peralkaline syenite autoliths from Kilombe volcano, Kenya Rift Valley: evidence for subvolcanic interaction with carbonatitic fluids. Lithos 91:373–392

Riedl S, Melnick D, Njue L, Sudo M, Strecker MR (2022) Mid-Pleistocene to Recent crustal extension in the inner graben of the Northern Kenya Rift. Geochem Geophys Geosyst 23:e2021GC010123

Ring U (2014) The East African rift system. Austrian J Earth Sci 107:132–146

Ring U, Albrecht C, Schrenk F (2018) The East African rift system: tectonics, climate and biodiversity. In: Hoorn C, Perrigo A, Antonelli A (eds) Mountains, climate and biodiversity. Wiley-Blackwell, pp 391–412

Roberts EM, O'Connor PM, Stevens NJ, Stevens NJ, Gottfried MD, Jinnah ZA, Ngasala S, Choh AM, Armstrong RA (2010) Sedimentology and depositional environments of the Red Sandstone Group, Rukwa Rift Basin, southwestern Tanzania: new insight into Cretaceous and Paleogene terrestrial ecosystems and tectonics in sub-equatorial Africa. J Afr Earth Sci 57:179–212

Roberts EM, Stevens NJ, O'Connor PM, Dirks PHGM, Gottfried MD, Clyde WC, Armstrong RA, Kemp AIS, Hemming S (2012) Initiation of the western branch of the East African Rift coeval with the eastern branch. Nat Geosci 5:289–294

Roessner S, Strecker MR (1997) Late Cenozoic tectonics and denudation in the Central Kenya Rift: quantification of long-term denudation rates. Tectonophysics 278:83–94

Rogers NW (2006) Basaltic magmatism and the geodynamics of the East African Rift System. In: Yirgu G, Ebinger CJ, Maguire PKH (eds). The Afar Volcanic Province within the East African Rift System. Geol Soc Lond Spec Publ 259:77–93

Rooney TO (2017) The Cenozoic magmatism of East-Africa: part I — flood basalts and pulsed magmatism. Lithos 286–287:264–301

Rooney TO (2020a) The Cenozoic magmatism of East Africa: part II – rifting of the mobile belt. Lithos 360–361:105291

Rooney TO (2020b) The Cenozoic magmatism of East Africa: part III – rifting of the craton. Lithos 360–361:105390

Rooney TO (2020c) The Cenozoic magmatism of East Africa: part IV – the terminal stages of rifting preserved in the northern East African Rift System. Lithos 360–361:105381

Rooney TO (2020d) The Cenozoic magmatism of East Africa: part V – magma sources and processes in the East African Rift. Lithos 360–361:105296

Rooney D, Hutton VRS (1977) A magnetotelluric and magnetovariational study of the Gregory Rift Valley, Kenya. Geophys J R Astron Soc 51:91–119

Rooney TO, Wallace PJ, Muirhead JD, Chiasera B, Steiner RA, Girard G, Karson JA (2022) Transition to magma-driven rifting in the South Turkana Basin, Kenya: part 2. J Geol Soc Lond 179:160

Rosendahl BR (1987) Architecture of continental rifts with special reference to East Africa. Annu Rev Earth Planet Sci 15:445–503

Rosendahl BR, Reynolds DJ, Lorber P, Scott D, McGill J, Lambiase J, Derksen SJ (1986) Structural expressions of rifting: lessons from Lake Tanganyika. In: Frostick LE, Renaut RW, Reid I, Tiercelin J-J (eds) Sedimentation in the African rifts. Geol Soc Lond Spec Publ 25:29–44

Rosendahl BR, Kilembe E, Kaczmarick K (1992) Comparison of the Tanganyika, Malawi, Rukwa and Turkana rift zones from analyses of seismic reflection data. Tectonophysics 213:235–256

Saggerson EP (1952) Geology of the Kisumu District. Geol Surv Kenya Rep 21

Saggerson E (1970) The structural control and genesis of alkaline rocks in central Kenya. Bull Volcanol 34:38–76

Saggerson EP, Baker BH (1965) Post-Jurassic erosion-surfaces in eastern Kenya and their deformation in relation to rift structure. Q J Geol Soc Lond 121:51–72

Saneyoshi M, Nakayama K, Sakai T, Sawada Y, Ishida H (2006) Half graben filling processes in the early phase of continental rifting: the Miocene Namurungule Formation of the Kenya Rift. Sediment Geol 186:111–131
Saria E, Calais E, Delvaux D, Hartnady CJH, Stamps DS (2014) Present-day kinematics of the East African Rift. J Geophys Res Solid Earth 119:3584–3600
Sceal JSC, Weaver SD (1971) Trace element data bearing on the origin of salic rocks from the Quaternary volcano Paka, Gregory Rift, Kenya. Earth Planet Sci Lett 12:327–331
Schlüter T (1997) Geology of East Africa. Gebrüder Borntraeger, Berlin
Scott SC, Bailey DK (1984) Coeruption of contrasting magmas and temporal variations in magma chemistry at Longonot volcano, central Kenya. Bull Volcanol 47:849–873
Shackleton RM (1951) Contributions to the geology of the Kavirondo Rift Valley. Q J Geol Soc Lond 106:345–392
Shackleton RM (1978) Structural development of the East African Rift System. In: Bishop WW (ed) Geological background to fossil man. Geol Soc Lond Spec Publ 6:19–28
Shmela AK, Paton DA, Collier RE, Bell RE (2021) Normal fault growth in continental rifting: insights from changes in displacement and length fault populations due to increasing extension in the Central Kenya Rift. Tectonophysics 814:228964
Simiyu SM, Keller GR (1998) Upper crustal structure in the vicinity of Lake Magadi in the Kenya Rift Valley. J Afr Earth Sci 27:359–371
Simiyu SM, Keller GR (2001) An integrated geophysical analysis of the upper crust of the southern Kenya Rift. Geophys J Int 147:543–561
Simmons NA, Forte AM, Grand SP (2007) Thermochemical structure and dynamics of the African superplume. Geophys Res Lett 34:L02301–L02305
Sippel J, Meeßen C, Cacace M, Mechie J, Fishwick S, Heine C, Scheck-Wenderoth M, Strecker MR (2017) The Kenya Rift revisited: insights into lithospheric strength through data-driven 3-D gravity and thermal modelling. Solid Earth 8:45–81
Smith M (1994) Stratigraphic and structural constraints on mechanisms of active rifting in the Gregory Rift, Kenya. Tectonophysics 236:3–22
Smith M, Mosley P (1993) Crustal heterogeneity and basement influence on the development of the Kenya Rift, East Africa. Tectonics 12:591–606
Spiegel C, Kohn B, Belton D, Gleadow A (2007) Morphotectonic evolution of the central Kenya Rift flanks: implications for late Cenozoic environmental change in East Africa. Geology 35:427–430
Stab M, Bellahsen N, Pik R, Quidelleur X, Ayalew D, Leroy S (2016) Modes of rifting in magma-rich settings: tectono-magmatic evolution of Central Afar. Tectonics 35:2–38
Stamps DS, Flesch LM, Calais E (2010) Lithospheric buoyancy forces in Africa from a thin sheet approach. Int J Earth Sci 99:1525–1533
Stamps DS, Kreemer C, Fernandes R, Rajaonarison TA, Rambolamanana G (2020) Redefining East African Rift System kinematics. Geology 49:150–155
Strecker MR, Blisniuk PM, Eisbacher GH (1990) Rotation of extension direction in the central Kenya Rift. Geology 18:299–302
Suess E (1891) Die Brüche des östlichen Africa. In: Beitrage zur Geologischen Kenntnis des östlichen Africa. Denkschriften Kaiserlichen Akademie der Wissenschaftliche Klasse, pp 555–556
Sutherland D (1974) Petrography and mineralogy of the peralkaline silicic rocks. Bull Volcanol 38:517–547
Swain CJ (1992) The Kenya Rift axial gravity high: a re-interpretation. Tectonophysics 204:59–70
Swain CJ, Khan MA, Wilton TJ, Maguire PHK, Griffiths DH (1981) Seismic and gravity surveys in the Lake Baringo-Tugen Hills area, Kenya Rift Valley. J Geol Soc Lond 138:93–102
Thuo P (2011) Stratigraphic, petrographic and diagenetic evaluation of Cretaceous/Paleogene potential reservoir sandstones of western Turkana, Kenya. Implications on the petroleum potential of northwestern Kenya. PhD thesis, Université de Bretagne Occidentale, Brest
Thybo H, Maguire PKH, Birt C, Perchuc (2000) Seismic reflectivity and magmatic underplating beneath the Kenya Rift. Geophys Res Lett 27:2745–2748

Tiercelin J-J, Lezzar KE (2002) A 300 million years history of rift lakes in Central and East Africa: an updated broad review. In: Odada EO, Olago DO (eds) The East African Great Lakes: limnology, palaeolimnology and biodiversity. Springer, Dordrecht, pp 3–60

Tiercelin JJ, Vincens A (eds) (1987) The Baringo-Bogoria half-graben, Gregory Rift, Kenya, 30,000 years of hydrological and sedimentary history. Bull Ctr Rech-Prod Elf-Aquitaine 11:249–540

Tiercelin J-J, Nalpas T, Thuo P, Potdevin J-L (2012a) Hydrocarbon prospectivity in Mesozoic and Early–Middle Cenozoic Rift Basins of Central and Northern Kenya, Eastern Africa. In: Gao D (ed) Tectonics and sedimentation: implications for petroleum systems. Am Assoc Petrol Geol Mem 100:179–207

Tiercelin J-J, Potdevin J-L, Thuo PK, Abdelfettah Y, Schuster M, Bourquin S, Bellon H, Clément JP, Guillou H, Nalpas T, Ruffet G (2012b) Stratigraphy, sedimentology and diagenetic evolution of the Lapur Sandstone in northern Kenya: implications for oil exploration of the Meso-Cenozoic Turkana depression. J Afr Earth Sci 71–72:43–79

Versfelt J, Rosendahl BR (1989) Relationships between pre-rift structure and rift architecture in lakes Tanganyika and Malawi: East Africa. Nature 337:354–357

Vétel W (2005) Dynamic of intra-continental extension in magmatic rift context: Turkana Rift (Northern Kenya) from Eocene to Present. PhD thesis, Université de Bretagne Occidentale, Brest

Vétel W, Le Gall B (2005) Geometry and growth of an inner rift fault pattern: the Kino Sogo Fault Belt, Turkana Rift (North Kenya). J Struct Geol 27:2204–2222

Vétel W, Le Gall B (2006) Dynamics of prolonged continental extension in magmatic rifts: the Turkana Rift case study (North Kenya). In: Yirgu G, Ebinger CJ, Maguire PK (eds) The structure and evolution of the East African Rift System in the Afar Volcanic Province. Geol Soc Lond Spec Publ 259:211–235

Vétel W, Le Gall B, Johnson T (2004) Recent tectonics in the Turkana Rift (North Kenya): an integrated approach from drainage network, satellite imagery and reflection seismic analyses. Basin Res 16:165–181

Walsh J (1969) Geology of the Eldama Ravine–Kabarnet area. Geol Surv Kenya Rep 83

Walsh J, Dodson RG (1969) Geology of Northern Turkana. Geol Surv Kenya Rep 82

Weaver SD (1977) The Quaternary caldera volcano Emuruangogolak, Kenya Rift, and the petrology of a bimodal ferrobasalt-pantelleritic trachyte association. Bull Volcanol 40:209–230

Weaver SD, Sceal JSC, Gibson IL (1972) Trace-element data relevant to the origin of trachytic and pantelleritic lavas in the East African Rift System. Contrib Mineral Petrol 36:181–194

Webb PK (1971) The geology of the Tiati Hills, Rift Valley Province, Kenya. PhD thesis, University of London

Webb PK, Weaver SD (1975) Trachyte shield volcanoes: a new volcanic form from South Turkana, Kenya. Bull Volcanol 39:294–312

Weerarantne DS, Forsyth DW, Fischer KM, Nyblade AA (2003) Evidence for an upper mantle plume beneath the Tanzanian Craton from Rayleigh wave tomography. J Geophys Res Solid Earth 108(B9):2427

Wescott WA, Stone DM, Wigger ST (1995) Geological and geophysical reconnaissance of the Lotikipi plain of northwestern Kenya and its relationship to the northern Kenya Rift. J Afr Earth Sci 21:241–251

Wescott WA, Wigger ST, Stone DM, Morley CK (1999) Geology and geophysics of the Lotikipi plain. In: Morley CK (ed) Geoscience of rift systems—evolution of East Africa. Am Assoc Petrol Geol Stud Geol 44:55–65

White JC, Espejel-García VV, Anthony EY, Omenda P (2012) Open system evolution of peralkaline trachyte and phonolite from the Suswa volcano, Kenya Rift. Lithos 152:84–104

Wichura H (2010) Topographic evolution of the East African Plateau: a combined study on lava-flow modeling and paleo-topography. PhD thesis, Universität Potsdam

Williams LAJ (1969) Volcanic associations in the Gregory Rift Valley, East Africa. Nature 224:61–64

Williams LAJ (1970) The volcanics of the Gregory Rift Valley, East Africa. Bull Volcanol 34:439–465

Williams LAJ (1972) The Kenya Rift volcanics: a note on volumes and chemical composition. Tectonophysics 15:83–96

Williams LAJ (1978a) The volcanological development of the Kenya Rift. In: Neumann ER, Ramberg IB (eds) Petrology and geochemistry of continental rifts. Reidel, Dordrecht, pp 101–121

Williams LAJ (1978b) Character of Quaternary volcanism in the Gregory Rift Valley. In: Bishop WW (ed) Geological background to fossil man. Geol Soc Lond Spec Publ 6:55–69

Williams LAJ, Chapman GR (1986) Relationships between major structures, salic volcanism and sedimentation in the Kenya Rift from the equator northwards to Lake Turkana. In: Frostick LE, Renaut RW, Reid I, Tiercelin J-J (eds) Sedimentation in the African rifts. Geol Soc Spec Publ 25:59–74

Williams LAJ, Macdonald R (1984) Late Quaternary caldera volcanoes of the Kenya Rift valley. J Geophys Res 89:8553–8570

Williamson PG, Savage RJG (1986) Early rift sedimentation in the Turkana basin, northern Kenya. In: Frostick, LE, Renaut, RW, Reid, I, Tiercelin, J-J (eds) Sedimentation in the African rifts. Geol Soc Lond Spec Publ 25:267–283

Willis B (1936) East African plateaus and rift valleys. Carnegie Inst Wash Publ 470

Wohlenberg J, Bhatt NV (1972) Report on aeromagnetic surveys of two areas in the Kenya Rift Valley. Tectonophysics 15:143–147

Woldegabriel G, Olago D, Dindi E, Ower M (2016) Genesis of the East African Rift System. In: Schagerl M (ed) Soda lakes of East Africa. Springer, Cham, pp 25–60

Wolfenden E, Ebinger C, Yirgu G et al (2004) Evolution of the northern Main Ethiopian rift: birth of a triple junction. Earth Planet Sci Lett 224:213–228

Xue L, Gani ND, Abdelsalam MG (2019) Drainage incision, tectonic uplift, magmatic activity, and paleo-environmental changes in the Kenya Rift, East African Rift System: a morpho-tectonic analysis. Geomorphology 345:106839

Zanettin B, Justin-Visentin E, Bellieni G, Piccirillo EM, Rita F (1983) Le volcanisme du bassin du Nord Turkana (Kenya): âge succession et évolution structurale. Bull Ctr Rech Explor-Prod Elf Aquitaine 7:249–255

Zawacki EE (2021) Plio-Pleistocene sediment provenance and erosion rates along the East African Rift System. PhD thesis, Arizona State University, Phoenix

Chapter 4
Environmental Background to the Kenya Rift Lakes: An Introduction

4.1 Introduction

With their equatorial setting, modern Kenya Rift lakes and their Neogene precursors have shared a similar general setting through many periods of climatic and tectonic change, which have, in turn, had major effects on lacustrine processes, limnology, ecology and sedimentation in the Kenya Rift. In this chapter, we summarise some of the background information (climate, hydrology, weathering, soils, vegetation, wildlife) to provide a regional context for understanding the modern and ancient environments of the Kenya Rift Valley. This is intended to be only a brief introduction rather than a comprehensive review. The final section considers geothermal processes, which play an important, but often underappreciated role in lacustrine sedimentation and ecology in the Kenya Rift.

4.2 East African Climate

Although modern global warming and climate change have been closely linked to anthropogenic controls, climate variability over longer geological time scales is a natural phenomenon. In polar regions, change is most easily reflected in the advance and shrinkage of ice caps. In montane and more temperate regions glaciers have advanced and retreated many times during the last few million years. In tropical latitudes, Quaternary climate fluctuations led to the expansion and contraction of mountain glaciers (e.g. Mt Kenya, Kilimanjaro, Rwenzori: Osmaston 2004), but is more widely recorded by inferred variations in precipitation, temperature and evaporation that changed vegetation and hydrological patterns across the African continent and controlled many lake transgressions and regressions (Fig. 4.1; Adams and Faure 1997). Chevalier et al. (2020), for example, noted a strong similarity between temperature variability in subtropical Africa and global ice volume and CO_2

R. W. Renaut, R. B. Owen, *The Kenya Rift Lakes: Modern and Ancient*, Syntheses in Limnogeology, https://doi.org/10.1007/978-3-642-25055-2_4

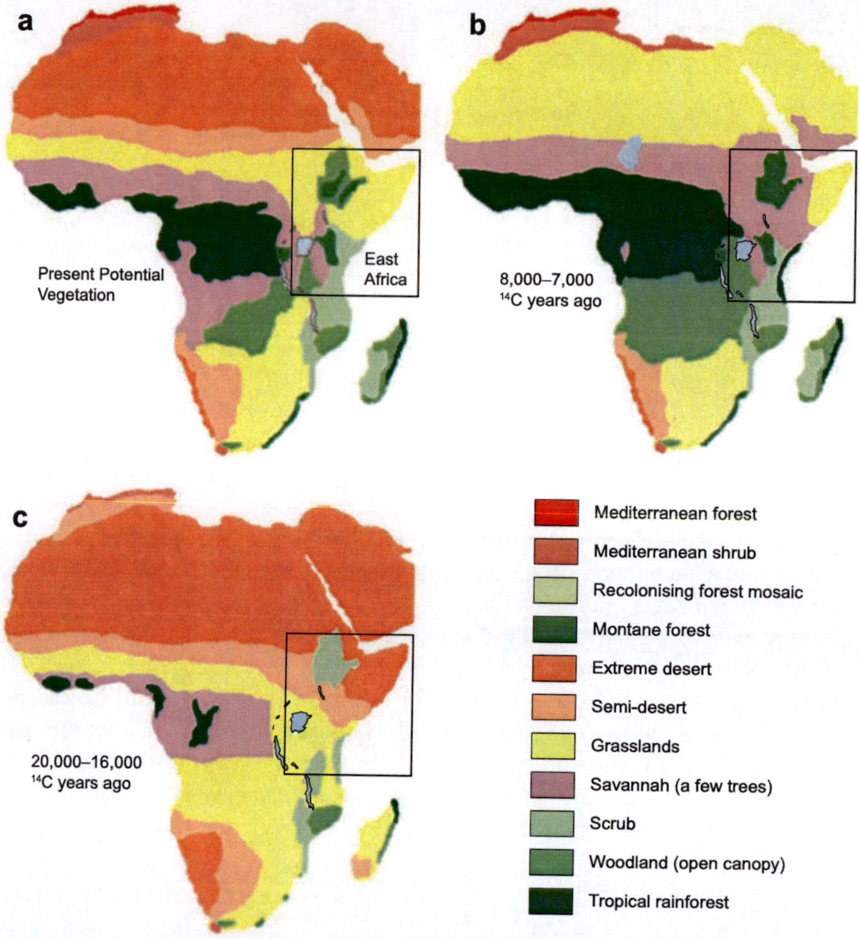

Fig. 4.1 Late Quaternary vegetation during contrasting climate regimes. After Adams and Faure (1997) and http://www.esd.ornl.gov/projects/qen/adams1.htm. Maps plotted using QGIS software. (**a**) Present Potential Vegetation. Note contrast between tropical rainforests in West and Central Africa and savanna in East Africa. (**b**) Early Holocene (African Humid Period). Tropical rainforest maximum expansion. (**c**) Last Glacial Maximum. Increased aridity and cooler conditions compared to today. Sahara Desert extends further south, compressing other vegetation zones

concentrations through the last 790 ka, and suggested that the region followed global climate trends at these time scales, with a range of ~4 °C between glacial minima and interglacial maxima. The most recent large-scale example of an environmental shift is represented by the African Humid Period (~15–5 ka) when many lakes in East Africa expanded in area and depth and became fresher (Butzer et al. 1972; Street and Grove 1979; Foerster et al. 2012; Junginger and Trauth 2013; Junginger et al. 2014).

Modern savannas and semi-deserts of East Africa are supported by a relatively dry climate compared to the wetter equatorial areas to the west that sustain a dense forest cover (Fig. 4.1a). In part, this reflects a rain-shadow effect that reduces moisture from the Atlantic Ocean, linked to uplift of the western branch of the East African Rift system. Today, East Africa has a bimodal rainfall distribution with rainfall maxima during the long 'Masika' (March to May) and short 'Vuli' (October to November) rains, separated by drier periods (Spawls and Matthews 2012; Camberlin 2018; Marchant 2021). Although the cause has been debated (e.g. Nicholson 2017), this precipitation pattern has been widely attributed to shifts in the Intertropical Convergence Zone (ITCZ) and movements of the Congo Air Boundary (CAB) (Verschuren et al. 2009; Junginger et al. 2014; Lashkari et al. 2017), modulated by topography. Figure 4.2a–d shows the average monthly precipitation in the region for four contrasting months with higher rainfall related to the location of the ITCZ and CAB. Mean monthly temperatures and wind speeds also vary seasonally across East Africa (Fig. 4.2e–l).

The climatic variability results partly from contrasts in solar radiation (Fig. 4.2m, n), which are influenced by day length, changing sun angles, latitude, cloud development and altitude (Camberlin 2018). Very-low insolation characterises much of the Ethiopian Highlands in association with shading effects of the main rains. Atmospheric water vapour (Fig. 4.2o) is also important, changing with altitude, contrasting air masses and distance from the ocean. Figure 4.2p shows the predicted variability in precipitation seasonality (O'Donnell and Ignizio 2012), with higher percentages representing areas characterised by greater rainfall variability.

The ITCZ is a low-pressure zone where air rises between the equatorial Hadley cells with consequent expansion, cooling and cloud formation (Fig. 4.3a). The upper air moves southwards in the southerly Hadley Cell and northwards north of the ITCZ before descending to form high-pressure regions associated with aridity and deserts. Importantly, the ITCZ and areas of ascending air move north and south between about 15°N and 15°S, following the overhead sun, but with a 4–6-week time-lag (Junginger and Trauth 2013). This means that precipitation associated with cloud formation moves across Kenya twice during an annual cycle, producing two rainy seasons (Fig. 4.3b). In contrast, regions lying below the extreme northern (e.g. Ethiopia, Eritrea) and southern (e.g. Malawi, Zambia) limits of the ITCZ experience a single rainy season (Fig. 4.3c, d). Movements of the ITCZ are associated with changes in ground level advection, mainly from the Indian Ocean, generating both the Southeast and Northeast Trade Winds. Although many factors complicate flow patterns, these winds curve clockwise and anticlockwise, respectively, due to Coriolis effects (Fig. 4.3a). Although this simple model has been described many times, Grist and Nicholson (2001) and Nicholson (2018) noted that in most years the ITCZ is only a secondary cell of vertical motion independent of the main region of air ascent. They note more specifically that the main ascent and the rainfall maximum, particularly in West Africa, lies between the mid-tropospheric African easterly jet and the upper-tropospheric tropical easterly jet.

Regional climate patterns are complicated further by the CAB, which marks the confluence between drier and relatively stable Indian Ocean and unstable Atlantic

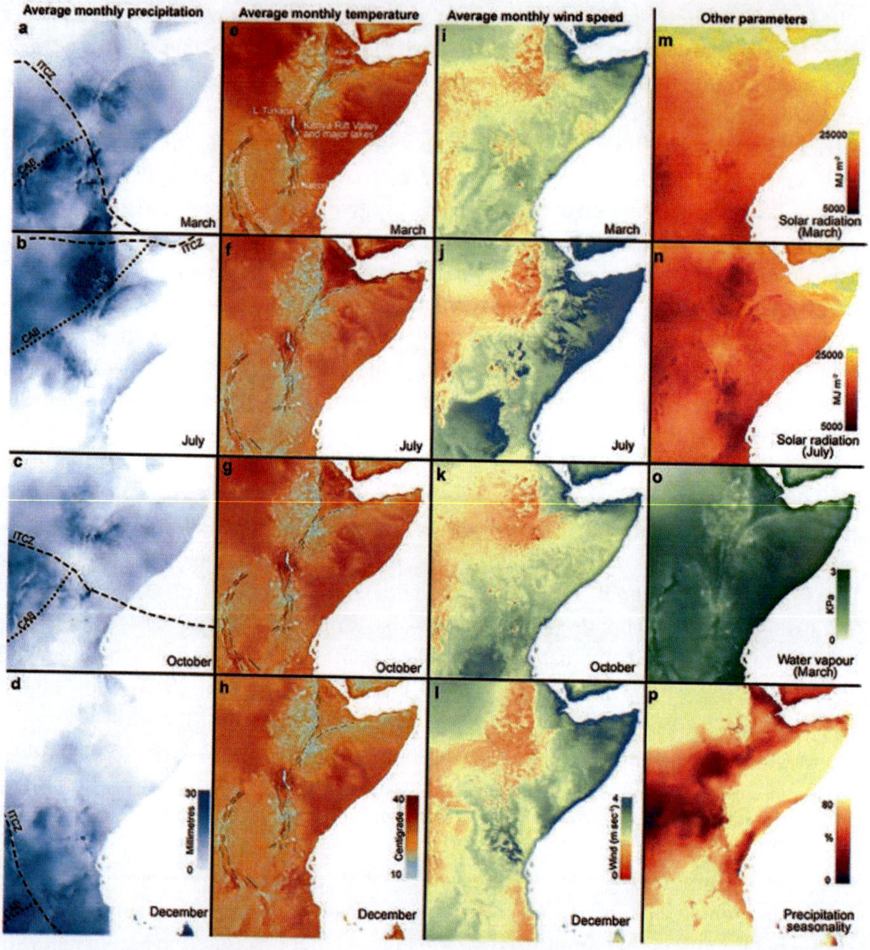

Fig. 4.2 Climate patterns of East Africa. Based on Worldclim data for years 1970–2000 (Fick and Hijmans 2017). Maps plotted using QGIS software. **(a–d)** Average monthly precipitation. Note heavier monthly rainfall takes place close to positions of the Inter Tropical Convergence Zone (ITCZ) and Congo Air Boundary (CAB). **(e–h)** Average monthly temperature. **(i–l)** Monthly wind speeds. Note relatively high velocities along the coast and in the southern Lake Turkana areas during July. **(m, n)** Solar radiation levels in March and July. **(o)** Atmospheric water vapour in March. **(p)** Precipitation seasonality. High percentages indicate greater variability

Ocean air masses. Moisture transfers to the west of the CAB form part of the West African Monsoon (WAM) with considerable recycling related to evapotranspiration from Congo Basin forests (Nicholson 1996; Junginger 2011). The CAB also changes its position seasonally, which results in variations in the timing and location of intrusions of the WAM into western parts of East Africa (Fig. 4.2a–d). Consequently, variations in the position of the CAB can have a significant impact on

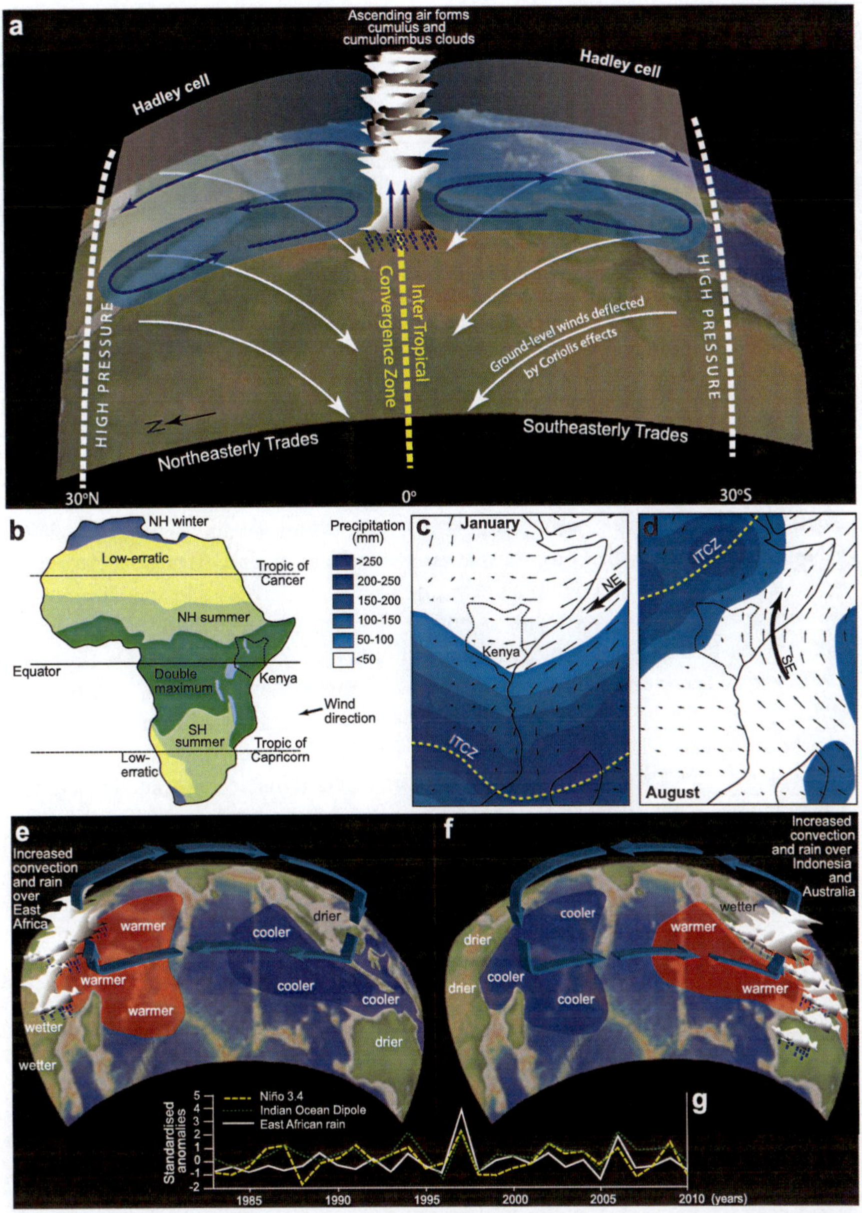

Fig. 4.3 Climate patterns and controls in East Africa. Topography constructed using GeoMapApp. (**a**) Diagrammatic representation of the Hadley circulation over East Africa showing the NE and SE Trade winds directed towards the Inter Tropical Convergence Zone. Curving wind systems reflect Coriolis effects. (**b**) Distribution of seasonal rainfall regimes in Africa. Equatorial areas are characterised by two rainy seasons. (**c, d**) Monthly rainfall and wind patterns with ITCZ at its northerly and southerly extremes. Wind strength proportional to arrow length. (**b–d**) Based on Verschuren et al. (2009). (**e, f**) The Indian Ocean Dipole showing a positive anomaly (**e**) with warm water over the western Indian Ocean and colder waters to the east. Ascending air over the western ocean triggers cloud formation and rain in East Africa. '**f**' shows a negative anomaly with warmer waters to the east and dry conditions in East Africa. (**g**) Standardised IOD, El Niño and East African rainfall anomalies. Note general positive correlations, especially after ~1990. After Wenhaji Ndomeni et al. (2018)

inter-annual precipitation in East Africa. For example, an eastward shift in the CAB can result from increases in the east-west pressure gradient between Africa and India when very low pressure develops over western India (Camberlin 1997, 2018; Camberlin and Philippon 2002; Foerster et al. 2012).

Other teleconnections with East African climate have been reported by several researchers (Nicholson 1996; Saji et al. 1999; Nicholson and Selato 2000; Gitau et al. 2012; Ng et al. 2018; Gebregiorgis et al. 2019). They point out that precipitation variability between years is greatest for the short October to November rains, with anomalies related to changes in 'Indian Ocean Sea Surface Temperatures' (SST) and the 'Indian Ocean Dipole' (IOD), and with more distant correlations reported for the 'El Niño/Southern Oscillation' (ENSO). Wenhaji Ndomeni et al. (2018), for example, reported strong correlations between enhanced East African short rains, a positive 'Dipole Mode Index' (MDI) for the IOD, and El Niño phases in the Pacific Ocean (Fig. 4.3e–g). Notably, the heavy Vuli rains of 1961, 1997 and 2020 took place during strongly positive IOD events (Wainwright et al. 2021). Kaboth-Bahr et al. (2021) report that El Niño years in East Africa are associated with positive precipitation anomalies of up to 60% relative to the annual precipitation in non–El Niño years with opposing dry and humid conditions developing between eastern and western Africa through much of the last 620 ka. De Cort et al. (2021) similarly report anti-phased relationships between eastern and southern Africa during the last 30 ka. The impacts of these anomalies on East African lakes can be profound with the major transgressions of 2020 reaching the highest lake levels since the 1960s. In some cases, enhanced rains lead to sudden change. Lavigne and Ashley (2002), for example, reported rapid hydrological modification when the Sandai River (north Lake Bogoria) migrated during and after enhanced El Niño rains, and Lake Kichirtit formed overnight when the Molo River flooded the adjacent floodplain south of Lake Baringo (Sect. 14.7).

The close links between climate and lake level in East Africa are of vital importance to the region's inhabitants. Rising and high lakes between 2011 and 2022 caused widespread disruption to shoreline populations who had to repeatedly retreat landwards and had major ecological impacts (Onywere et al. 2013; Obando et al. 2016). However, well-documented Holocene and recent variability in East African lake levels also imply that changing climates will induce future regressions and even drought, although the effects of anthropogenic climate change increase the degree of uncertainty across the region. Wainwright et al. (2021), for example, observed that current climate change models predict increased Vuli rains (Dunning et al. 2018) and that IPCC data also indicate increased short rains over East Africa. They note that climate change will likely increase the intensity of storms because warming leads to increases in the saturation vapour pressure (Finney et al. 2020) and that a warming of 1.5 °C above pre-industrial levels may double strongly positive IOD events (Cai et al. 2018).

4.3 Hydrology and Hydrogeology

4.3.1 Drainage Patterns and Groundwater Flow

Drainage in the Kenya Rift is controlled by climate, elevation (topography) and geology, especially faulting. The rift valley, with its elevated shoulders, and the underlying Kenya Dome, exert a major role in controlling water flow directions. Five major drainage basins are recognised in Kenya (Fig. 4.4; World Resources Institute 2007). Three (Athi River, Tana River, Ewaso Ngiro North) drain eastwards

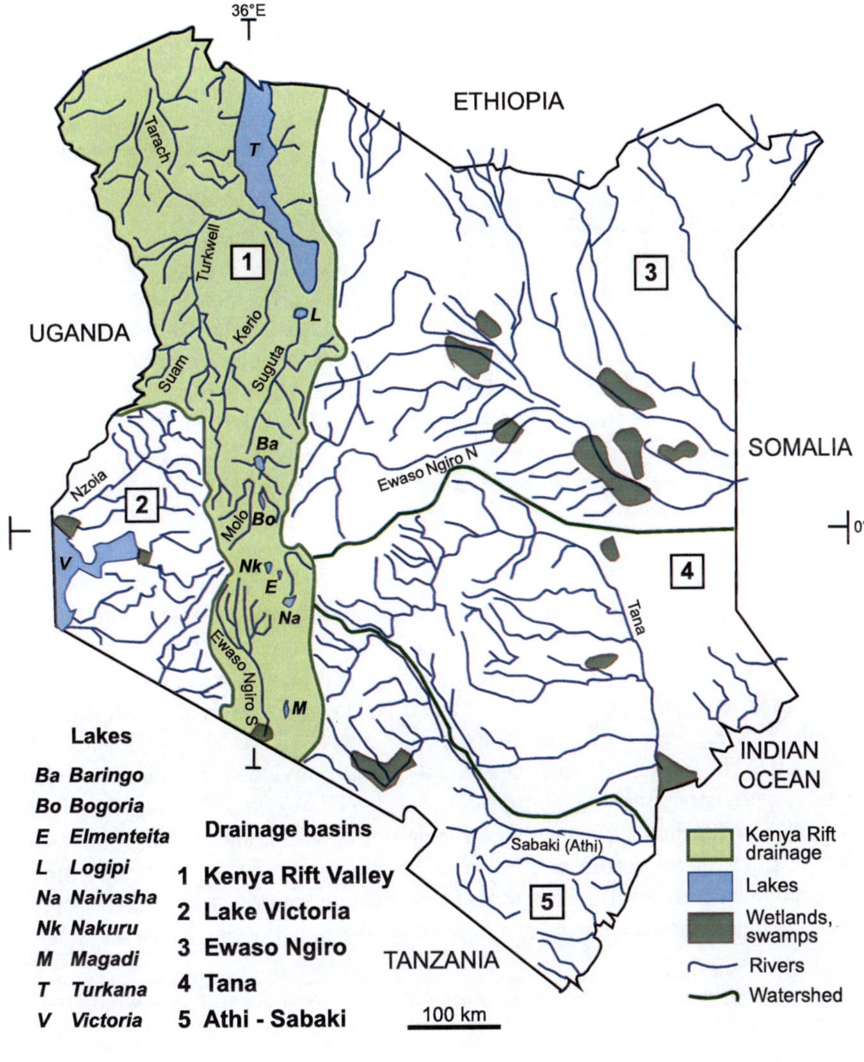

Fig. 4.4 The main drainage basins in Kenya

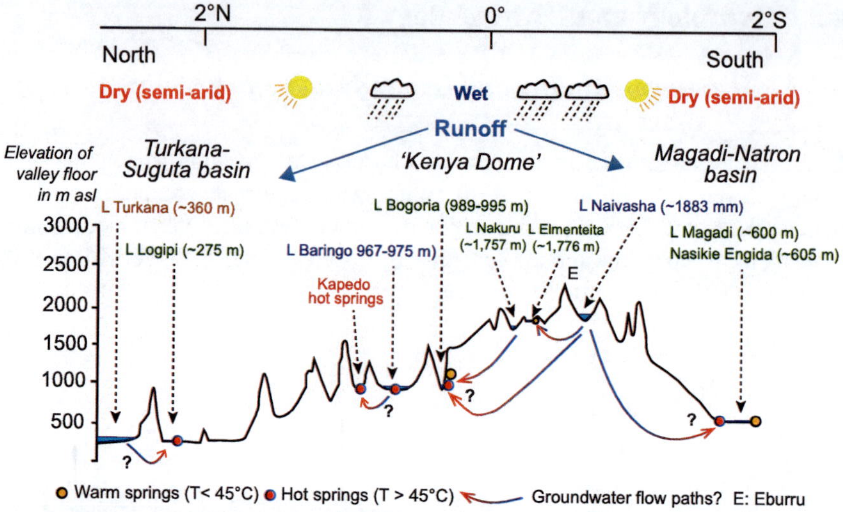

Fig. 4.5 Schematic section along the north-south axis of the Kenya Rift showing inferred groundwater links between lakes. Modified from Becht et al. (2006)

to the Indian Ocean, each with large perennial rivers. The Lake Victoria drainage basin in western Kenya covers the region west of the rift valley as far north as Mt Elgon, along with several perennial rivers that drain the adjacent highlands (Mau, Nandi, western Uasin Gishu). The Kenya Rift Valley is an internally drained basin with almost no outflow to the ocean, which discharges water southwards to Lake Magadi and Lake Natron and northwards to Lake Turkana. It contains many individual sub-basins and lakes with closed or open drainage. Lake Ol' Bolossat, an exception, lies physically within the Kenya Rift but drains externally eastwards to the Ewaso Ngiro North drainage basin.

All lakes in the modern Kenya Rift Valley are fed by direct precipitation (rainfall, hail), surface runoff within their drainage basins (channelled and unchannelled flow), and (or) groundwater. Groundwater derives from runoff that has seeped through soils and surface rocks within the subaerial drainage basin, but some originates from deep sources beyond the surface drainage area. The Kenya Rift floor has its subaerial high-point at Eburru volcano between Lake Naivasha and Lake Elmenteita. This represents the major axial drainage-divide within the Kenya Rift (Fig. 4.5). Eburru massif (~2850 m) is a transverse (E-W) ridge that acts as a watershed for both surface and subsurface rift-drainage. Most runoff and groundwater north of Eburru flows northwards towards Lake Logipi and Lake Turkana. Runoff and groundwater south of Eburru flow southwards towards the Magadi-Natron basin, which acts as the regional sump in the southern part of the Kenya Rift (Becht et al. 2006).

The local hydrology and hydrogeology are complex in detail, as shown by many regional studies (e.g. McCall 1957; Armstrong 2002; Abiye 2009; Kuria 2013). Lake Naivasha and Lake Baringo have fresh water implying that they are not hydrologically closed even though there is no surface outflow. Lake Naivasha water may flow northwards as groundwater to Lake Elmenteita and Lake Nakuru, and perhaps as far as lakes Bogoria and Baringo (Fig. 4.6; Clarke et al. 1990; Becht et al. 2006), emerging in places as springs. There is little surface drainage between Lake Naivasha (~1883 m above sea level) and Lake Magadi (~600 m asl). Water from Lake Naivasha may also move southwards to Lake Magadi as groundwater (Eugster 1970; Armstrong 2002; Becht et al. 2006), but other evidence implies that Lake Magadi is recharged mainly from groundwater from the Ewaso Ngiro South on the western edge of the rift (Eugster 1980; Allen et al. 1989; Clarke et al. 1990; Chap. 12). Surface water on the rift floor north of Lake Baringo flows via the Suguta River into Lake Logipi (~270 m asl), which is the regional sump in the North Kenya Rift. However, much of the water north of Lake Baringo probably moves as groundwater (Fig. 4.6; Dunkley et al. 1993; Sect. 7.4.1). Lake Turkana is fed mainly by the Omo River sourced in the Ethiopian Highlands, but also by the Kerio River which rises in hills surrounding the Kerio Valley west of Lake Baringo (Fig. 4.4). The Kerio Valley lies within the Kenya Rift but is separated from the axial rift lakes (Baringo, Bogoria) by the large Kamasia-Tugen Hills fault-block. The ephemeral Turkwel (Suam) River, which rises on the slopes of Mt Elgon, also flows into Lake Turkana among several non-perennial ephemeral rivers.

The groundwater aquifers in the Kenya Rift Valley are generally well known from boreholes, which are most abundant in the central and southern parts of the rift where the human population is highest, and in the Nyanza Rift (Kuria 2013). The boreholes include those drilled for geothermal exploration, and much is now known about the regional hydrogeology and directions of groundwater flow (Clarke et al. 1990; Dunkley et al. 1993). Flow is generally axial or lateral (Fig. 4.6) from the rift margins.

The aquifers in the Kenya and Nyanza (Kavirondo) rifts are of several types. Most common are:

1. Fractured and faulted volcanic rocks. This includes:
   (i) contractional cooling joints within thick lava flows;
   (ii) brecciated zones, including zones of autobrecciation and faulting;
   (iii) other zones of secondary permeability resulting from fracturing; tectonic joints may have high interconnectivity.
2. Permeable zones at contacts between successive lava flows.
3. Palaeosols, weathering profiles and former land surfaces, especially those with coarse alluvial debris (regosols) overlying or incorporated in a soil profile; these may be metres in thickness; palaeosols can be aquifers but may be aquitards if composed of compacted clay and altered tuff.
4. Lake sediments, including silts and sands, and some diatomaceous deposits.
5. Tuffs, including waterlain tuffs.
6. Palaeovalleys: buried valleys filled with volcanic and sedimentary (typically fluvial) debris.

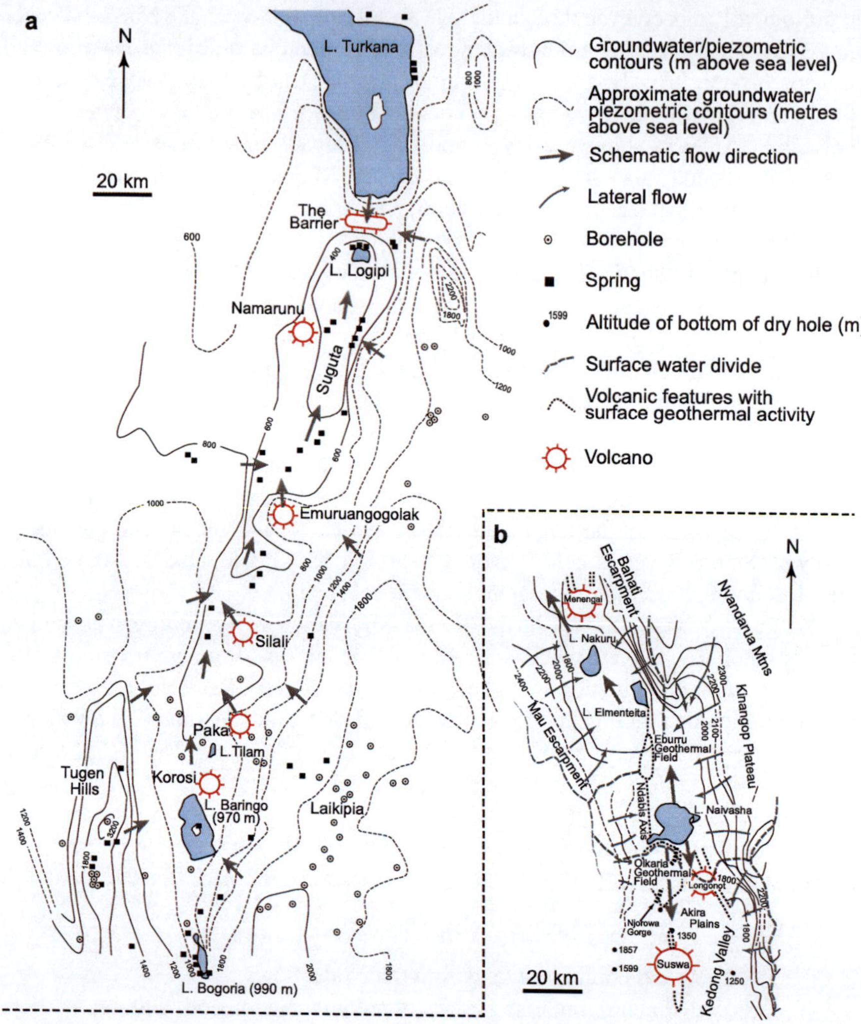

Fig. 4.6 Potentiometric map of the regions between (**a**) Lake Baringo and Lake Turkana, and (**b**) the equatorial central Kenya Rift. Modified from Dunkley et al. (1993), and Clarke et al. (1990)

Aquifers are commonly stacked. Some are interconnected vertically and laterally but others are isolated and separated by impermeable and semi-permeable rocks such as clay that form aquitards and aquicludes (Fig. 4.7). Perched aquifers may allow springs to form in the drainage basin, including the margins of the rift lakes, but others lie metres to tens of metres above lake level. Examples are present around the southeastern margin of Lake Bogoria.

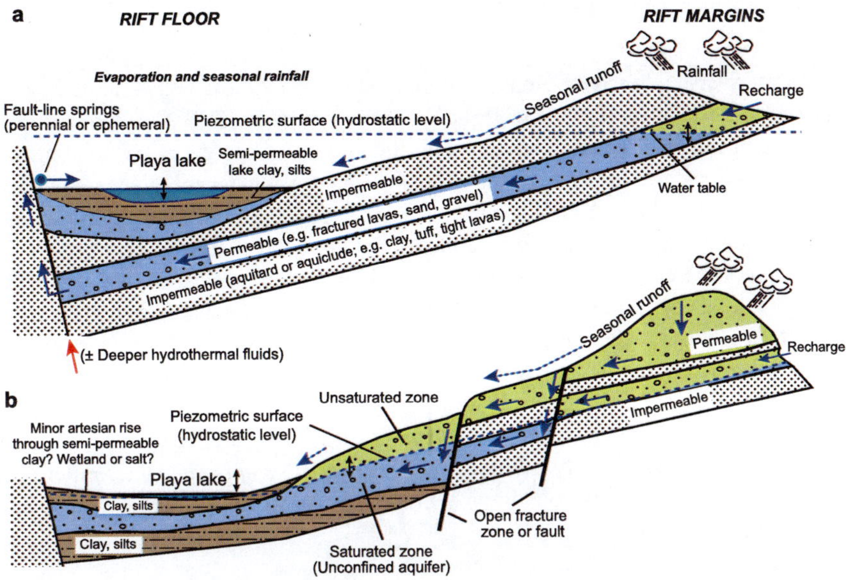

Fig. 4.7 Common settings of aquifers in the Kenya Rift. (**a**) Confined aquifer producing artesian conditions and fault line springs. (**b**) Open, unconfined aquifer typical of shallow playa lakes on the rift floor. The playa lake sometimes represents an outcrop of the water table. Based partly on ideas presented by McCall (1957) and Bristow and Temperley (1964)

Recharge to the Kenya Rift lakes is complex and not constant in time or location (Olaka et al. 2022). As the rift evolves, runoff and groundwater respond and, in turn, can change the hydrology and environment (ecology, limnology) of the rift lakes on a wide range of timescales. Modern studies of the limnology, ecology and chemistry of any rift lake typically record conditions when field data were recorded, but do not always consider the geologically 'recent' hydrological and hydrogeological changes that might have affected the lake basin.

There may also be a time lag between an inferred 'causal event' and its limnological and ecological consequences. Changes in one control often have a knock-on effect on others. Many controls operate on different timescales, and short-term (historical, but also undefined) changes are not always preserved in the geological record or obvious from modern ecological data. Hippos and crocodiles, for example, have recently (observations in 2016–19) lived in Lake Bogoria following unusually heavy rains (Schagerl and Renaut 2016). Their remains and associated biota might not be preserved, especially in core. In contrast, De Cort et al. (2018) have shown that Lake Bogoria was an almost dry saline pan in the last 200 years. Hydrology and hydrogeology are critical factors when interpreting modern and ancient rift lake environments and their biota.

4.3.2 *Types of River System*

Modern rivers that feed the Kenya Rift lakes and Lake Victoria are *perennial* or *non-perennial*. Perennial rivers flow throughout the year (Fig. 4.8). Although they have seasonal peaks in discharge and level during and following the rains, streamflow is maintained except during extreme drought. Most originate in high regions with a humid climate where rainfall usually exceeds evaporation. Groundwater, including springs, may also contribute to perennial flow along the channel.

The non-perennial rivers are more variable and more typical of the large Kenya Rift drainage basin (Fig. 4.9). Their behaviour and the amount and type of sediment that they deliver to the lakes is controlled by climate and geology (e.g. volcanic bedrock or loose sediments). Some rivers are *semi-permanent* and flow throughout most of the year. Others are *ephemeral* and flow less than half of the year. Unlike perennial streams, their channel base usually lies above the local water table, and they are fed predominantly by precipitation. Many are *episodic* (intermittent) and are dry most of the year flowing only briefly (hours to weeks), often with a flash-flood regime.

Fig. 4.8 Perennial rivers flowing into Kenya Rift lakes. (**a**) Endosapia River, Nguruman. (**b**) Ewaso Ngiro South river flowing into Lake Natron, at Olkaramatian bridge. (**c**) Perkerra River at Perkerra Gorge, 1.6 km southwest of Marigat, flowing into Lake Baringo. (**d**) Kerio River flowing north at the outflow (left centre) from Lake Kamnarok

Fig. 4.9 Non-perennial streams. (**a**, **b**) The Suguta River near Namarunu. (**c**) Alluvial fan, north-east Suguta Valley. (**d**) Multichannel braided to anastomising stream, Suguta River south of Lake Logipi. (**e**) Braided Turkwel River, SW Lake Turkana. (**f**) Ephemeral stream, near Koros, South Horr. (**g**) Ephemeral (episodic) stream at Nyogonyek, SW Loboi Plain, Baringo. (**h**) Ephemeral (episodic) stream cut into diatomaceous silts, W Olorgesailie

Each river type can change when the climate changes, across a range of timescales. Individual rivers can also change behaviour along their course from source to lake (talweg). Some rivers are perennial in their upper course but become dry downstream. This is common in rivers that rise in humid rift-marginal highlands that flow into semi-arid parts of the rift floor. In contrast, other rivers are semi-permanent or ephemeral in their upper reaches but become perennial for stretches of their lower course if supplemented by tributaries and shallow groundwater. This distinction is important because rivers may transport sediment and organic remains (e.g. pollen, diatoms, fish, mollusc shells) from a different ecological zone (climatic or vegetation) to a lake. For example, fish in saline Lake Logipi and Lake Bogoria derive from upstream (Suguta River: Chap. 7; Sandai River: Chap. 9). The size of the catchment is important. Most of the major rivers feeding the larger lakes drain large drainage basins (Fig. 4.4).

Channel patterns vary mostly in the ephemeral and normally perennial rivers, and with type of load (Schumm 1985; Miall 2014; Fig. 4.10). A common pattern is for braided patterns (gravel, sand) during the rainy seasons with local overbank

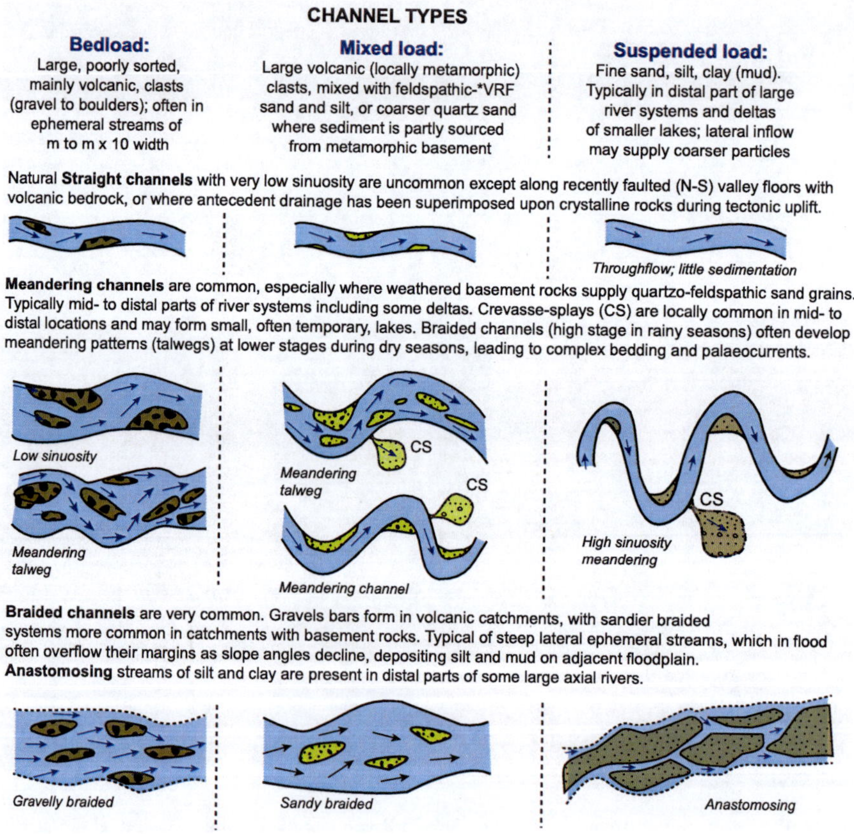

Fig. 4.10 Common river channel patterns in the Kenya Rift

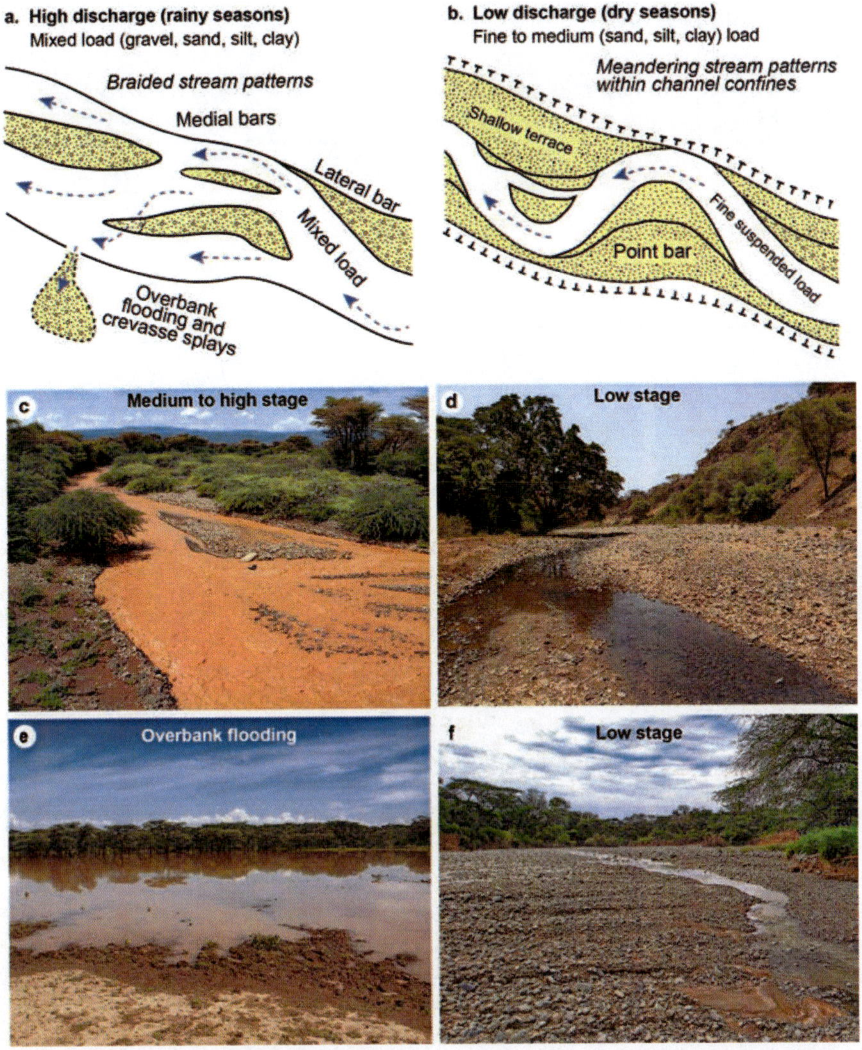

Fig. 4.11 Seasonal changes in channel patterns in the Baringo basin, Kenya Rift Valley. (**a**) Typical channel pattern during flood. (**b**) Typical channel pattern during low-stage flow. (**c**) Perkerra River during rainy season. (**d**) Chemeron River during falling stage. (**e**) Molo River floodplain in flood at Ilosowuani. (**f**) Chemeron River at low stage

flooding. Bedload dominated streams give way to mixed load streams downslope and are commonly braided but with a change to more sinuous (meandering) patterns during low stage (Fig. 4.11). Anastomosing patterns develop where suspended load dominates (e.g. distal parts of Suguta River).

Alluvial fans are common along the axis of the rift where fault scarps abut against the valley floor. They tend to be dominant features in the drier parts of the rift north

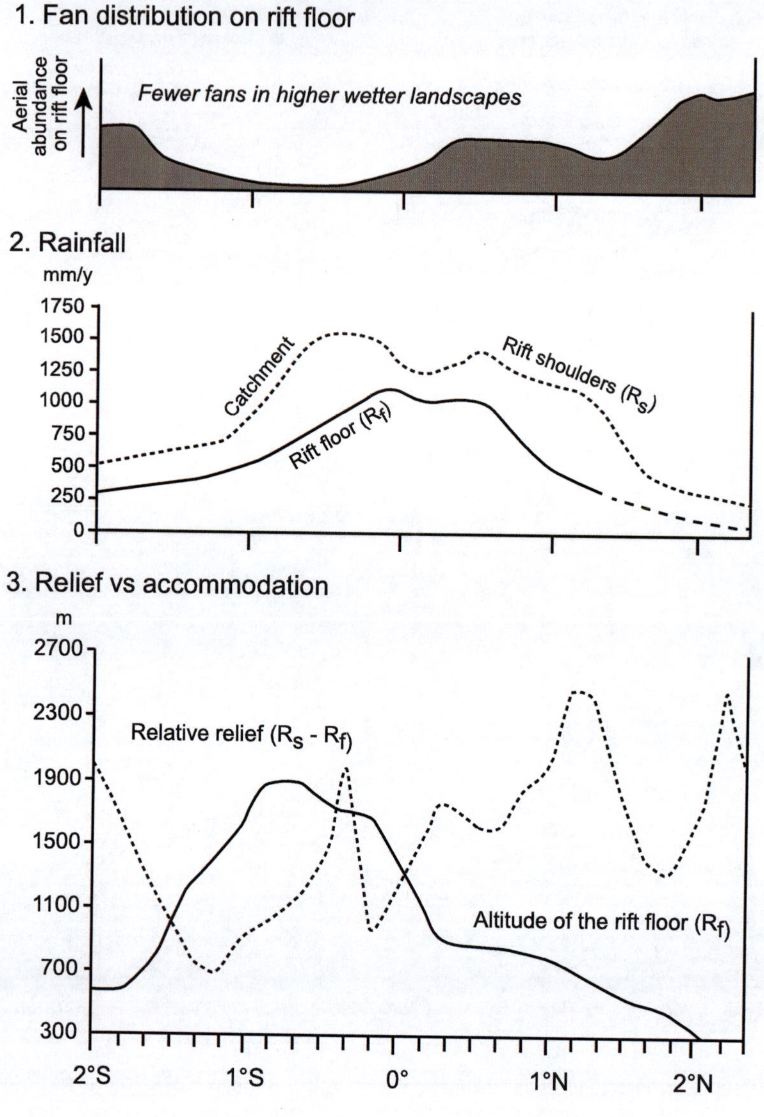

Fig. 4.12 The relationship between alluvial fan development, rainfall and accommodation along the Kenya Rift Valley. Alluvial fans are most prominent in the semi-arid regions north and south of the Kenya Dome

and south of the Kenya Dome between Baringo and Lake Turkana, and in the Magadi-Natron basin. These parts also have the greatest accommodation space between the rift margins and valley floor (Fig. 4.12). Fan deltas, where the distal part of the fan passes directly into a lake are common.

All the rivers are susceptible to changes in base level caused by climatic or tectonic changes, or river capture. Their discharge and channel pattern may respond over different timescales. In closed basins, lake level changes have a major impact on lacustrine sedimentation and ecology. For example, falling lake levels due to climatic change usually lead to channel incision with more confined flow. This may enable the river to transport coarser debris seasonally, but at the same time, the annual discharge may be reduced if a fall in level reflects increasing aridity. In contrast, a fall in level due to tectonic subsidence would not necessarily be accompanied by lower rainfall and lower fluvial discharge. Rising lake level similarly will change channel slope and perhaps lead to lateral channel migration. The flow regime and channel pattern will respond in turn depending on local conditions. It is therefore difficult to predict the consequences of base level changes in terms of fluvial inflow and sediment delivery to any lake. Different lakes may respond differently.

4.4 Weathering and Sediment Production

The chemical composition of the Kenya Rift lake waters, and the physical and chemical compositions of the sediments deposited in the lakes, depend strongly on weathering processes in their drainage basins. In this section, we consider those weathering processes that play an important role in the Kenya Rift. The weathering processes and their solid and aqueous products are influenced mainly by bedrock geology, hydrology and climate.

The dominant lithologies in the drainage basins (surface and subsurface catchments) of the modern and ancient Kenya Rift lakes are volcanic rocks, including associated tephra, and metamorphic basement rocks. Older (pre-Holocene) sedimentary rocks in modern drainage basins and those buried in subsurface sequences or exposed by rift subsidence or marginal uplift originate from the same types of bedrock, but their mineralogy when weathered may differ following diagenesis and chemical sedimentary processes (e.g. evaporites, limestone, chert).

The primary (original) mineralogy of the volcanic rocks varies with magma type during eruption or intrusion (e.g. rhyolitic to trachytic to basaltic to phonolitic to carbonatite), but most rocks exposed at the land surface are dominated in volume by volcanic rocks in which sodic or potassic feldspars and feldspathoids dominate, with mafic and accessory minerals (e.g. zircon, analcime). Most of the exposed volcanic rocks contain little primary quartz. The tephra similarly vary in chemical composition and may be glassy (vitric) or lithic. The tephra are highly reactive and alter readily to clays, zeolites and silica species during weathering and diagenesis.

Volcanic rocks have many unstable minerals because they crystallised at much higher temperatures and pressures and are far from being in equilibrium with earth-surface conditions. Weathering products of trachyte and alkaline volcanics are typically feldspars and mafic minerals. The coarse fraction is commonly composed of sanidine and anorthoclase, both of which have well developed cleavage and are often twinned, and are more prone to physical and chemical weathering than quartz. Consequently, the sand fraction in many fluvial and lacustrine sediments is dominated by K-feldspar grains, some of which were phenocrysts in the parent volcanic rock, and less stable volcanic rock fragments (VRFs), representing aphanitic lavas or groundmass grains from porphyritic volcanic rocks. Basaltic rocks with plagioclase groundmasses often weather more quickly than more alkaline volcanic rocks.

In contrast, older, mainly Precambrian (Mozambiquan Belt: ~1 Ga–500 Ma) basement rocks have a wide range of compositions and mineralogy. They include quartz-bearing gneisses and schists, but also marbles and ultramafic rocks. These differences have an important impact on sedimentation. During and after weathering, drainage basins with quartzose bedrock release abundant detrital quartz grains into the fluvial system that are chemically stable and deliver sand grains to the lakes that are deposited in fluvial systems, deltas and along shorelines, or bypass lake margins and are deposited as turbidites or other mass flows in deeper water offshore.

4.4.1 *Physical Weathering*

Physical weathering processes in the Kenya Rift are related to (1) thermal expansion and contraction related to temperature changes (including the effects of fire); (2) exfoliation induced by pressure release, (3) slaking, and (4) salt weathering. Freeze-thaw weathering plays no role in the Kenya Rift, though is important in volcanic mountains away from the rift margins (e.g. Mt. Kenya: Zeuner 1950).

Pressure release might affect basement rocks in the Turkana basin but is a minor contributor to sediment production. Similarly, possible impacts of thermal expansion and contraction of minerals in heterogeneous rocks might affect rocks in northern (Suguta-Turkana) and southern (Magadi-Natron) Kenya but the diurnal temperature range is too narrow except at high altitudes. Slaking due to repeated wetting (expansion) and drying (contraction) affects weakly lithified shaley sediments in many regions.

Salt weathering encompasses an important group of weathering processes in the Kenya Rift lake basins, especially near saline lakes and in sites where the groundwater is shallow (<50 cm deep) and evaporation is strong. Analogous to freeze-thaw weathering, the pressure exerted by salts crystallising in confined cracks can physically fragment rocks and can expose new surfaces to chemical weathering. In many environments, this is primarily a physical process, but in volcanic environments where the fluids are both saline and alkaline it can be both physical and chemical.

In places, salt weathering is linked to crystallisation of gypsum ($CaSO_4 \cdot 2H_2O$) or anhydrite ($CaSO_4$), and halite (NaCl). Gypsum is present as selenite at and near the land surface in a few areas, notably along the northern margins of Lake Turkana. Sulfate minerals rarely crystallise from evaporating brines in the Kenya Rift because carbonate species usually exceed sulfate except where waters are locally enriched in sulfate from oxidised pyrite or sulfurous geothermal gases.

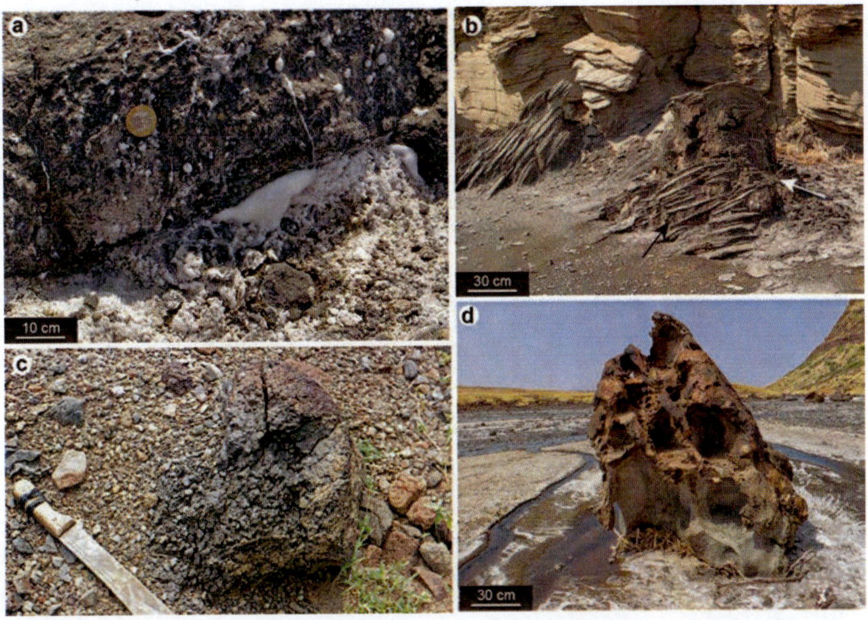

Fig. 4.13 Salt weathering. (**a, b**) Salt (mainly trona) crystallising in joints, fractures and vesicles in Magadi Trachyte on the shoreline of northwestern Nasikie Engida (**a**) and Northeastern Hot Spring group, Magadi (**b**). (**c**) Granular disintegration of phonolite boulder by salt weathering, eastern Sandai Plain (below Siracho Escarpment), Lake Bogoria. Panga (machete) blade is ~45 cm long. (**d**) Chemical corrosion by sodium carbonate. Trachyte boulder with large tafoni, next to hot springs at NW Nasikie Engida

Most salt weathering is linked to saline alkaline water concentrated at or near the land surface (Fig. 4.13). Sodic salts crystallise from waters that periodically flood the littoral zone and evaporate, or, more commonly from shallow pore fluids drawn to the land surface by capillary processes and evaporative pumping, including wicking. These processes affect fractured bedrock and sediments (especially gravel) that lie within a few decimetres of the land surface. The sites where this occurs are often covered by efflorescent salt crusts during the dry seasons, especially along the fluctuating margins of saline lakes.

Salt weathering can also be a chemical weathering process if the fluids are of Na^+–HCO_3^- –CO_3^{2-} composition. Those fluids typically have a high pH, so they not only weather the rocks physically but also chemically. Alkaline pore fluids at and near the land surface often have a pH > 9. Sodic salts may form but at high pH the fluids can also etch and dissolve silicate minerals (Fig. 4.13d). Silica solubility increases with pH above about 9, when monosilicic acid may be accompanied by other species such as $H_3SiO_4^-$ and $H_2SiO_4^{2-}$.

4.4.2 Chemical Weathering

Chemical weathering processes are dominant throughout the Kenya Rift and account for most of the sediment production and the composition of most surface water and much of the groundwater. The intensity of chemical weathering is generally highest in the more humid regions, particularly in the uplands of the rift shoulders, and during more humid periods when vegetation enhances chemical weathering processes (Fig. 4.14; Ivory et al. 2017).

Several processes contribute to the chemical and physical breakdown of the parent rocks at and near the land surface. The most important process is *hydrolysis* in which silicate minerals react with carbonic acid to produce soluble ions, aqueous silica and clays. Atmospheric carbon dioxide dissolves to form carbonic acid, much of which then dissociates into bicarbonate and hydrogen ions:

$$H_2O + CO_2 \leftrightarrow H_2CO_3$$

then

$$H_2CO_3 \leftrightarrow H^+ + HCO_3^-$$

The weak acidity of carbonic acid produced from atmospheric CO_2 is strongly enhanced by soil CO_2 and in places from magmatic (mantle) CO_2 that discharges along the rift floor and at central volcanoes (Lee et al. 2016; Robertson et al. 2016).

Silicate minerals such as olivine (forsterite) dissolve congruently producing soluble ions and aqueous silica with no solid residue (i.e. clay):

$$Mg_2SiO_4 + 4H_2CO_3 = 2Mg^{2+} + 4HCO_3^- + H_4SiO_4$$

In contrast, aluminosilicate minerals such as feldspars react with carbonic acid to produce clay minerals and ions in solution:

$$CaAl_2Si_2O_8 + H_2CO_3 + ½O_2 = Al_2SiO_5(OH)_4 + Ca^{2+} + CO_3^{2-}$$

Aluminium has low solubility so combines with silica to produce silicate minerals or forms Al-oxyhydroxides. In the latter equation, Ca-bearing plagioclase produces kaolinite with calcium and carbonate ions. The weathering products vary with parent material and with intensity. Kaolinite tends to form in more humid areas along the elevated rift margins, whereas smectite and illite are more common in drier areas along the rift floor where the cation status remains higher. The clay minerals deposited in lake sediments can be detrital, transported from weathering profiles and soils, but many are authigenic or diagenetic and form within the water column and pore fluids or from alteration of older minerals and grains.

Oxidation is also important in chemical weathering. Iron in ferromagnesian silicate minerals (e.g. pyroxene, amphibole) dissolves and is converted to ferrous iron, bicarbonate and silicic acid. For example, Fe-olivine (fayalite) is oxidised to iron oxide (e.g. hematite, limonite):

Fig. 4.14 Chemical weathering. (**a**) Red soils and reworked gravel is typical of weathering on the high rift shoulders, Tugen Hills, Baringo. (**b**) Spheroidally weathered trachyte near Kabarnet, Tugen Hills, Baringo. (**c**–**e**) Deeply weathered basalt at Kimalel, 10 km west of Marigat, Baringo. In '**e**' the well-rounded boulders are *in situ* having undergone no transport. (**f**) Weathered trachyte, 1.5 km east of Magadi townsite

$$Fe_2SiO_4 + 4H_2CO_3 \leftrightarrow 2Fe^{2+} + 4HCO_3^- + H_4SiO_4$$

Then

$$2Fe^{2+} + 4HCO_3^- + ½O_2 + 2H_2O \leftrightarrow Fe_2O_3 + 4H_2CO_3$$

Where the bedrock contains sulfides, notably pyrite (FeS_2),

$$2FeS_2 + 7O_2 \leftrightarrow 2Fe^{2+} + H_2SO_4 + 2H^+$$

the sulfide mineral reacts with water and oxygen to form sulfuric acid that etches other rocks. Unless buffered, the pH declines which can mobilise metals (e.g. Pb, Zn, Cu) that are toxic to aquatic organisms.

Dissolution can dissolve carbonates (limestone, marble) and evaporites completely at the surface and in the subsurface from groundwater.

$$CaCO_3 + HCO_3^- + H^+ \leftrightarrow Ca^{2+} + 2HCO_3^-$$

Along the Kenya Rift floor limestones tend to form in lakes or in catchments as spring deposits, calcareous soils and calcretes (Deocampo and Renaut 2016). They become more soluble during humid periods.

4.4.3 Biological Weathering

The activities of plants, microbes and animals can help to break down minerals and rocks in most environments and expose surfaces to physical and chemical weathering, but the evidence is most obvious in the more humid parts of the rift. The roots of higher plants, such as trees and bushes, can physically break up jointed bedrock, sediments and soils, and may release CO_2 into the soil or regolith that after decay may produce carbonic acid. Microorganisms (bacteria, algae, fungi, lichens) are able to weaken and break down rock by using minerals for nutrition (e.g. chelation).

Animal activity (including humans) is also an agent in weathering, but the boundary with erosion is sometimes indistinct. Burrowing, boring and rock abrasion by animals could be classed as biological erosion, but organic acids produced by the decay of the faecal lining of a burrow or guano in a cave or on an island are clearly agents of biological weathering.

4.5 Kenya Rift Soils

Soils throughout Kenya are strongly related to climate, which varies with regional controls and variations in altitude (Marchant 2021). A wide variety of soil types recognised by the Food and Agriculture Organisation (FAO) soil classification

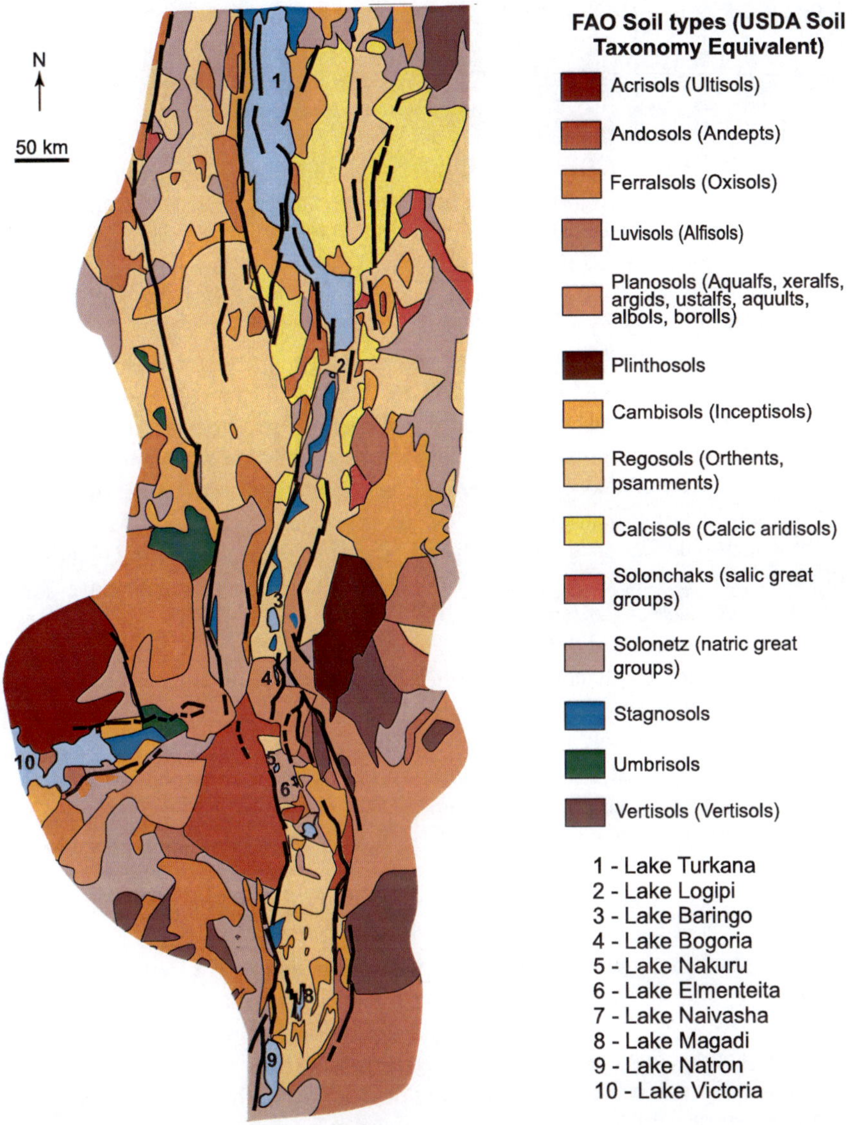

Fig. 4.15 Major soil types in and adjacent to the Kenya Rift Valley. Simplified from Jones et al. (2013)

system are present (Jones et al. 2013). These include regosols, calcisols, cambisols and ferralsols, with acrisols, planosols, luvisols, andosols, plinthosols, umbrisols and vertisols generally located at higher altitudes (Fig. 4.15). Stagnosols, solonchaks and solonetz soils are also present. Average erosion rates for 1990 and 2015 across the rift valley were 6.26 t ha^{-1} y^{-1} and 7.14 t ha^{-1} y^{-1}, respectively, with a total soil loss of 116 Mt $year^{-1}$ in 1990 and 132 Mt $year^{-1}$ in 2015 (Watene et al. 2021).

The greatest soil erosion rates have taken place in areas with increasing cropland with the Lake Nakuru Basin experiencing the highest rate net changes (4.19 t ha^{-1} $year^{-1}$). High rates were also recorded in the Lake Bogoria-Baringo Basin with losses of >10 t ha^{-1} y^{-1}.

The comments in the following paragraphs describe major soil types and are based on Jones et al. (2013). Regosols show weak soil development in unconsolidated fine- to medium-grained materials and are most extensive in the Turkana region. Calcisols include significant calcium carbonate and are typical of low altitudes and high evaporation, also mainly in areas around Lake Turkana. Calcrete may develop locally (Fig. 4.16). Cambisols are immature soils with slight evidence of pedogenesis.

Relatively wet areas north of Lake Victoria with acidic bedrock commonly host acrisols, which are strongly acidic soils with a clay-rich substrate and low capacity to retain nutrients. Ferralsols are strongly weathered highly leached red soils, abundant in residual aluminium and iron, that form in areas of low relief and high rainfall at higher altitudes. Planosols and vertisols occur in areas of low relief and a semi-arid to sub-humid climate with high clay contents leading to poor drainage, typically cracking during dry periods. Planosols are mainly present in the higher parts of the rift and are characterised by a low-permeability layer rich in clay overlain by structureless soil. Vertisols are rich in smectite which readily expands and contracts depending on water availability. Consequently, they display wide cracks during dry periods. They are locally common on the rift valley floor where they create difficult conditions for driving, when wet.

Slightly acidic luvisols are also clay-enriched with a high nutrient-holding capacity and tend to be located at higher altitudes. Andosols are found on volcanic rocks with steep slopes and high precipitation (>1000 mm $year^{-1}$). Plinthosols typically accumulate iron and manganese in the subsoil, which harden when exposed to form a protective cap. Umbrisols are acidic with an organic-rich surface horizon with low nutrient content. Typically, they are found in mountainous areas with high rainfall.

Solonchaks are strongly saline soils located in arid regions where evaporation greatly exceeds precipitation. Solonetz soils are alkaline with a clayey subsoil rich in exchangeable sodium and a columnar structure. Stagnosols may be associated with perched water tables above an impermeable barrier, which causes waterlogging and mobilisation of iron and/or manganese.

Fossil soils (palaeosols) are commonly intercalated with lacustrine and alluvial sediments on the rift floor (Fig. 4.16). Some may be recorded by a simple colour change at a depositional hiatus. However, others have well-developed profiles or are preserved by precipitated minerals. Calcrete is common in the Kenya Rift in Quaternary and Neogene rift sedimentary rocks, though not all calcretes are formed in soil profiles. Some accumulated at the water table, with carbonate precipitated from groundwater (Felske 2016). Root marks, root mats and rhizoliths are commonly present in palaeosols (Owen et al. 2008). Zeolitic palaeosols are common where the surface or groundwater was alkaline, including along lake margins (Hay 1970; Renaut 1993; Mees et al. 2005).

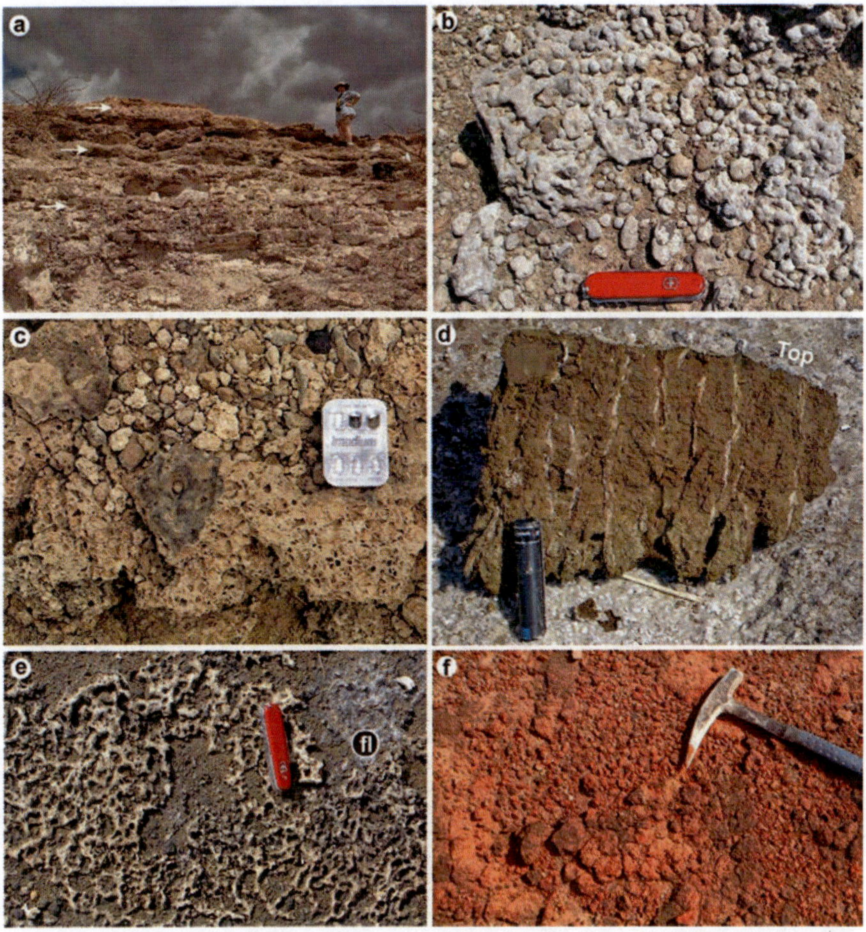

Fig. 4.16 Palaeosols and related rocks. (**a**) Three calcrete layers (arrows) separated by lake sediments, Pleistocene Olorgesailie Formation, western Olorgesailie. (**b**) Nodular calcrete north of Nasikie Engida. (**c**) Honeycomb calcrete, Lainyamok, northwest of Lake Magadi. (**d**) Recent silica (opal-A) rhizoliths on Loburu delta plain, Lake Bogoria. (**e**) Rhizomous horizontal rootcasts composed of silica and fluorite (fl), Loburu delta plain, Lake Bogoria. (**f**) Zeolitic (analcime) palaeosol, Sandai Plain, Lake Bogoria

4.6 Aeolian Processes

Wind plays an important role regionally and locally in the Kenya Rift lakes and in their drainage basins. Wind direction and strength vary seasonally in association with migration of the pressure belts across East Africa (Fig. 4.2). The morphology and orientation of the Kenya Rift itself plays a role by acting as an open funnel that often redirects wind from W-E or E-W flow directions to N-S or S-N flow, affecting water bodies by inducing waves, mixing or changes in stratification (see Chap. 5), and inducing erosion and sediment redistribution. Sustained predominant directions of wind flow, seasonal or less regular changes contribute to the processes and sediment record of rift lake basins (Fig. 4.17).

Fig. 4.17 Aeolian processes. (**a**) Dust devils on mudflats north of Lake Magadi that redistribute much silt, fine sand and trona. (**b**) Dust storm north of Lake Bogoria. Deflated silt is blown from Loboi Plain towards the Tugen Hills. (**c**) Loess banked against the eastern edge of the Magadi horst below Magadi townsite. (**d**) Barchanoid sand dunes on the midwestern edge of Suguta Valley. (**e**) Sand dunes near the southwestern shoreline of Lake Turkana. (**f**) Lake Turkana wind farm. (**g**) Wind driven waves along the southwestern shore of Lake Bogoria

Wind erosion is widespread throughout the Kenya Rift (Ataya 2000; Fenta et al. 2020) but plays a major role in the drier parts both north and south of the equator. Fine sand, silt, clay and efflorescent salts are entrained in dust devils (Fig. 4.17a) and dust storms (Fig. 4.17b) and may be redeposited locally or regionally at the land surface or in lakes. In the arid parts of the northern Kenya Rift, deflation also contributes to the development of stone pavements that mantle the land surface and locally protect underlying sediments (Hemming 1972).

Most sediments in the drier parts of the Kenya Rift likely have an aeolian component, but aeolian sediments are locally important and frequently linked to modern or former lakes. They range from small local accumulations of loess (Fig. 4.17c) to dune fields, notably in the midwestern Suguta Valley (Fig. 4.17d). They are common on the western margins of the rift basins, implying that winds from the east are most effective. Sand dunes are present along both shorelines of Lake Turkana but are well developed along the southern shorelines (Fig. 4.17e). Winds are locally strong and reliable enough on the rift floor (SE Turkana; Fig. 4.17f) and along the rift margins (Ngong Hills) to generate electric power. Nonetheless, the wind farms at Lake Turkana and Ngong Hills lie on the eastern margin of the rift. At Lake Turkana winds tend to increase in strength at night. Wind-driven waves directly impact sedimentation by constructing beaches and producing longshore drift, especially in Lake Turkana and Lake Bogoria (Fig. 4.17g).

The relative importance of aeolian processes varies with location and with changing climate. The higher, wetter, equatorial parts of the Kenya Rift show less evidence of aeolian processes. Wind effects increase during drier periods and decrease during wetter periods when increased vegetation cover and soil moisture reduce the amount of loose sediment that can be entrained.

4.7 Kenya Rift Vegetation

The Kenya Rift ranges between ~275 m at Lake Logipi in the Suguta Valley to >3860 m in the Aberdare Mountains on the highest rift shoulder. Consequently, climates vary considerably as do vegetation ecotypes, which have been much altered by human intervention. The VECEA (Vegetation and Climate change in East Africa) project developed a 'potential natural vegetation' map (Fig. 4.18; Kindt et al. 2013), which estimates the distribution of vegetation that would be in equilibrium with present climatic and edaphic conditions without human interference. This map and associated report are the basis for the following paragraphs.

Figure 4.18 shows the major potential vegetation types in the Kenya and Nyanza (Kavirondo) rifts. Several types of Afromontane forest (rain forest, dry forest, bamboo, undifferentiated) are present along the rift shoulders and junction of these two rift systems, mainly between 1200 and 2500 m altitude in areas where mean annual rainfall is 1250–2500 mm. *Podocarpus* is common with *P. latifolius*, *Prunus*

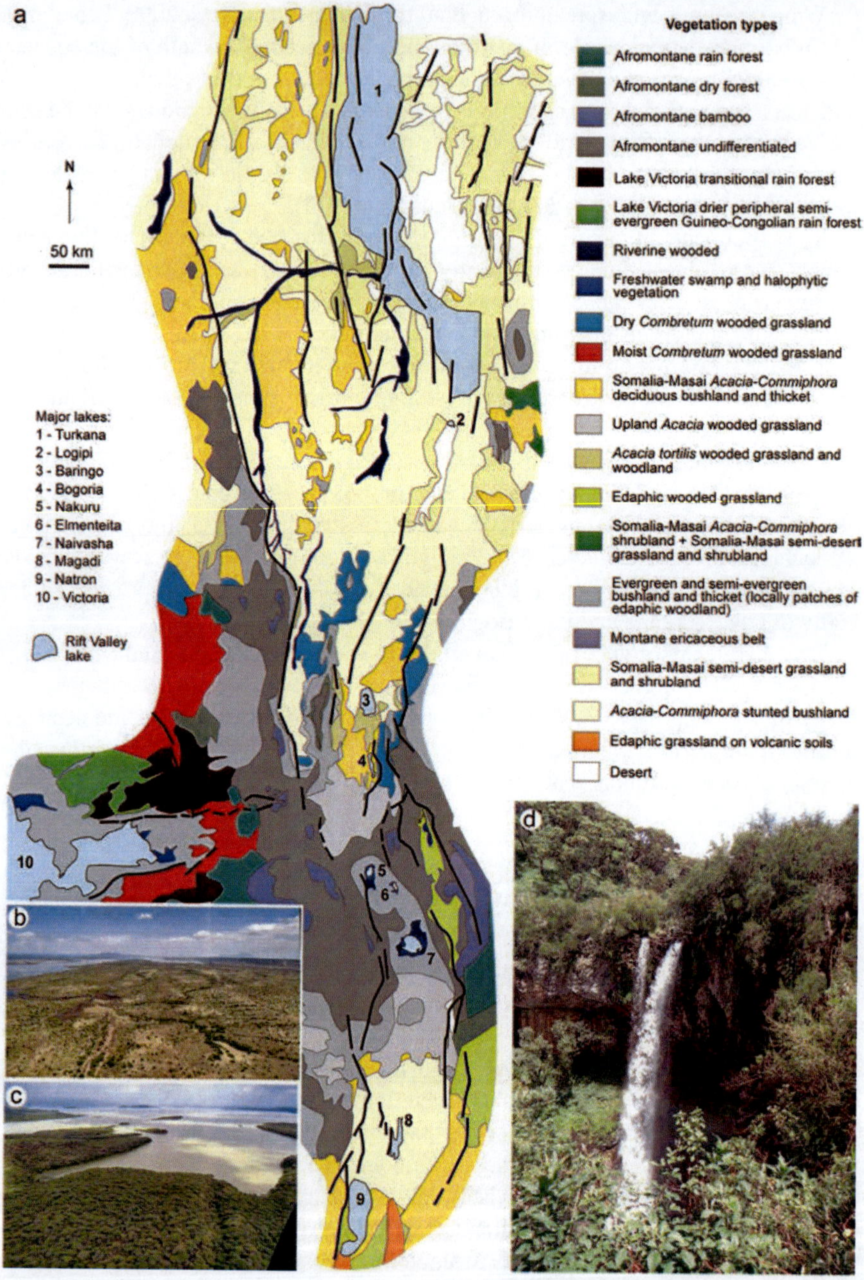

Fig. 4.18 Vegetation map of the Kenya and Nyanza (Kavirondo) rifts. Modified from Kindt et al. (2013). (**a**) Map of potential vegetation in equilibrium with modern climates, excluding human influences. (**b**) *Acacia-Commiphora* stunted bushland, Magadi Basin. (**c**) Somalia-Masai *Acacia-Commiphora* deciduous bushland, Karosi, northern Lake Baringo. (**d**) Afromontane forest on the Aberdare mountain range

africana and *Xymalos monospora* present in both Afromontane rain- and undifferentiated forests. The latter are associated with drier slopes (>850 mm year^{-1}) and single dominant taxa (e.g. *Widdringtonia whytei*, *Hagenia abyssinica*, *Juniperus procera*). Lake Victoria transitional rain forests (1600–1900 m) include characteristic species such as *Alangium chinense*, *Anthonotha pynaertii* and *Apodytes dimidiata*. According to Kindt et al. (2013), drier semi-evergreen Guineo-Congolian rain forests (Fig. 4.18) peripheral to Lake Victoria include *Alstonia boonei*, *Antiaris toxicaria*, *Chrysophyllum albidum* and *Entandrophragma cylindricum*; these forests are a drier variant of semi-evergreen Guineo-Congolian rain forest with reduced floristic diversity.

Somalia-Masai riparian forest occurs along the margins of larger rivers in the drier parts of the northern Kenya Rift. Swamp forests include taxa that are common across tropical Africa with examples along lakes Victoria, Naivasha and Nakuru, including *Anthocleista grandiflora*, *A. schweinfurthii* and *Macaranga monandra*. In shallow standing freshwater and swampy areas *Cyperus papyrus*, *Typha domingensis* and *Phragmites* spp. may be abundant. *Sporobolus spicatus* and *Cyperus laevigatus* often develop on alkaline flats at lakes Bogoria, Nakuru and Elmenteita. Dry *Combretum* wooded grasslands are present east and northwest of Lake Baringo with *C. collinum* and *C. molle* and *Terminalia brownii* common, together with varied *Acacia* (*Vachellia*). Moist *Combretum* wooded grasslands are also present in the Nyanza (Kavirondo) Rift and areas to the north. Somalia-Masai *Acacia-Commiphora* deciduous bushland and thicket are common in arid lands of the northern Kenya rift, especially west of Lake Turkana and Baringo. Smaller areas of *Acacia tortilis* wooded grassland are present in these areas and at intermediate elevations in the southern part of the rift. This vegetation type is replacing evergreen and semi-evergreen bushland, which is mapped as the potential natural vegetation around the Kenyan portion of Lake Victoria (Fig. 4.18).

Edaphic wooded grasslands are characterised by *Acacia* and lie mainly south of Ol' Bolossat and east and south of Lake Natron. Somalia-Masai *Acacia-Commiphora* deciduous bushland and thicket is only present to the northeast of the Suguta Valley, outside the rift. Evergreen and semi-evergreen bushland and thickets are located on drier (850–500 mm year^{-1}) mountain slopes, mainly east of the rift, as well as in the Tugen Hills and near Lake Victoria. The vegetation is varied, but *Acokanthera schimperi*, *Carissa spinarum*, *Euphorbia candelabrum* and *Dodonaea viscosa*, for example, are commonly present. Upland *Acacia* wooded grasslands are found in degraded areas overgrazed by animals. Montane ericaceous woodland occurs only on the Aberdare Range along the eastern rift margins. Drier areas (100–200 mm year^{-1}) in the northern and southern Kenya Rift are characterised by Somalia-Masai semi-desert grasslands with dominant *Centropodia glauca*, *Eragrostis mahrana* and *Panicum turgidum* or *Acacia-Commiphora* stunted bushland. Localised edaphic grasslands are developed on volcanic soils.

4.8 Kenya Rift Wildlife

Many parts of the Kenya Rift once supported abundant and diverse wildlife, but the remaining animals are now largely confined to National Parks, National Reserves and local conservancies (Spawls and Matthews 2012). The northern regions are semi-arid to arid and are consequently associated with species adapted to such conditions with local water resources such as springs, rivers and lake water attracting animals (e.g. hippo and crocodile in Lake Turkana and rivers, with topi, gazelles, lesser kudu, oryx, Burchell's and Grevy's zebra, striped hyaena elsewhere; Fig. 4.19a). Locally, hyaena and lions are present as are rare cheetah. Limited numbers of elephants occupy wetter areas on the rift shoulders and several locations in the Kerio Valley.

Olive and chacma baboons are common in many parts of the central and southern Kenya rift. Lake Baringo supports numerous crocodiles and hippo (Fig. 4.19b, c) and a diverse avifauna with flamingos concentrated at Lake Bogoria National Reserve (Fig. 4.19d), which also supports Grant's gazelle, dik dik, greater kudu, Burchell's zebra and leopard. Lake Nakuru National Park and the Soysambu Conservancy near Lake Elmenteita preserve large herds of buffalo, waterbuck, impala and Burchell's zebra as well as a large range of other fauna, including, for example, warthog, giraffe, white and black rhino, spotted hyena, lion, leopard, rare cheetah, Thompson's and Grant's gazelles, bushbuck, eland, colobus and vervet monkeys and many other species (Fig. 4.19e–h). Buffalo and zebra are also common at Hell's Gate National Park south of Lake Naivasha, which also hosts many hundreds of bird species. Further south, at Lake Magadi, wildlife is dispersed but includes wildebeest, giraffe, zebra, gerenuk, Grant's gazelles, rare lions with small herds of elephants along the Nguruman Escarpment.

Birds are numerous throughout most of the rift (Kennedy 2014; Stevenson et al. 2020) with particular hotspots around freshwater lakes Baringo and Naivasha where hundreds of species are concentrated. Common birds in these areas include martens, ducks, geese, pintails, pigeons, doves, bustards, cuckoos, nightjars, swifts, moorhens, coots, cranes, stilts, avocets, terns, storks, darters, cormorants, pelicans, bitterns, herons, egrets, ibis, spoonbills, eagles, hoopoes, hornbills, kingfishers, rollers, barbets, owls and many other groups. Fish eagles are particularly common at Lake Baringo with superb and splendid starlings in many areas. Ostriches are found in drier locations including areas north of Baringo. Pelicans nest at Lake Elmenteita. Saline alkaline lakes (Logipi, Bogoria, Nakuru, Elmenteita, Magadi, Natron) collectively support millions of lesser flamingos (Krienitz 2018) as well as other birds such as greater flamingos, yellow-billed and marabou storks, spoonbills, teals, pied avocets and stilts.

Fig. 4.19 Kenya Rift wildlife. (**a**) Striped hyena sipping spring water. Anam Naaryangak, Suguta Valley. (**b**) Crocodiles basking on the western shores of Lake Baringo. (**c**) Hippo at Lake Baringo. (**d**) Flamingos wading in saline Lake Bogoria. (**e**) Waterbuck in nearshore woods at Lake Nakuru. (**f**) Buffalo resting on shores of Lake Nakuru. Note flooded trees behind. (**g**) White rhino on shoreline plains, Lake Nakuru. (**h**) Grazing Burchell's zebra, Lake Elmenteita

4.9 Geothermal Processes and their Impact on Lacustrine Sedimentation

Geothermal activity is common in continental rifts (Hochstein 1999; Pirajno 2009). The extensional tectonic regime and crustal thinning, coupled with regionally high geothermal gradients and high heat flow (Crane and O'Connell 1983; Nyblade et al. 1990; Lysak 1992; Wheildon et al. 1994), create favourable conditions for the development of hot springs in regions with adequate groundwater supply. Extensional faulting and fracturing of bedrock, particularly in volcanic sectors of rifts, allow meteoric fluids to infiltrate to depths of up to several kilometres, where they become heated by convection (mainly advection) or conduction. They then rise along permeable pathways to discharge as subaerial and sublacustrine hot springs with a wide range of temperatures and fluid compositions (Fig. 4.20a, b; Table 4.1), sometimes mixing with shallow groundwater *en route* to the surface. Elsewhere steam vents and fumaroles may also be present, especially in high-temperature geothermal areas (including central volcanoes: Fig. 4.20c–e) where the water table is relatively low or shallow boiling occurs.

From a lacustrine perspective, geothermal processes can play a major role in the chemistry, sedimentology and evolution of rift valley lakes and wetlands. Their roles vary in both space and time. Lake Magadi and Nasikie Engida in the southern Kenya Rift receive much of their annual recharge from saline hydrothermal fluids, whereas hot springs are minor contributors at lakes Turkana, Victoria and Naivasha, mainly because of their large size and water volume, the extent of their respective drainage basins, and the regional climate. The role played by hot springs and geothermal gases is nonetheless important in the sedimentology, ecology and history of many Kenyan lakes, as shown in later chapters.

Modern geothermal activity is present along the N-S axis of the Kenya Rift and extends southwards from North Island in Lake Turkana into Lake Natron in northern Tanzania and beyond, including the lateral branch of the Nyanza (Kavirondo) Rift. At the land surface, this is most evident as hot springs, geysers, and fumaroles that discharge steam, CO_2 and other gases. Modern and former activity is recorded by spring deposits of several ages and by hydrothermally altered ground. Sublacustrine hot springs discharge from some lake floors, either from well-defined vents or diffusively. In small rift lakes, hydrothermal inflow can be minor in terms of annual lake recharge but can dominate during drier periods when rainfall and runoff decline and evaporation increases (Sect. 22.5).

4.9.1 *Types of Hot Spring in the Kenya Rift*

There is no single definition for 'hot spring' (Pentecost et al. 2003; Pentecost 2005; Jones and Renaut 2011). The term has been used with reference to local climate (e.g. local non-thermal surface water temperatures or mean air temperature), human

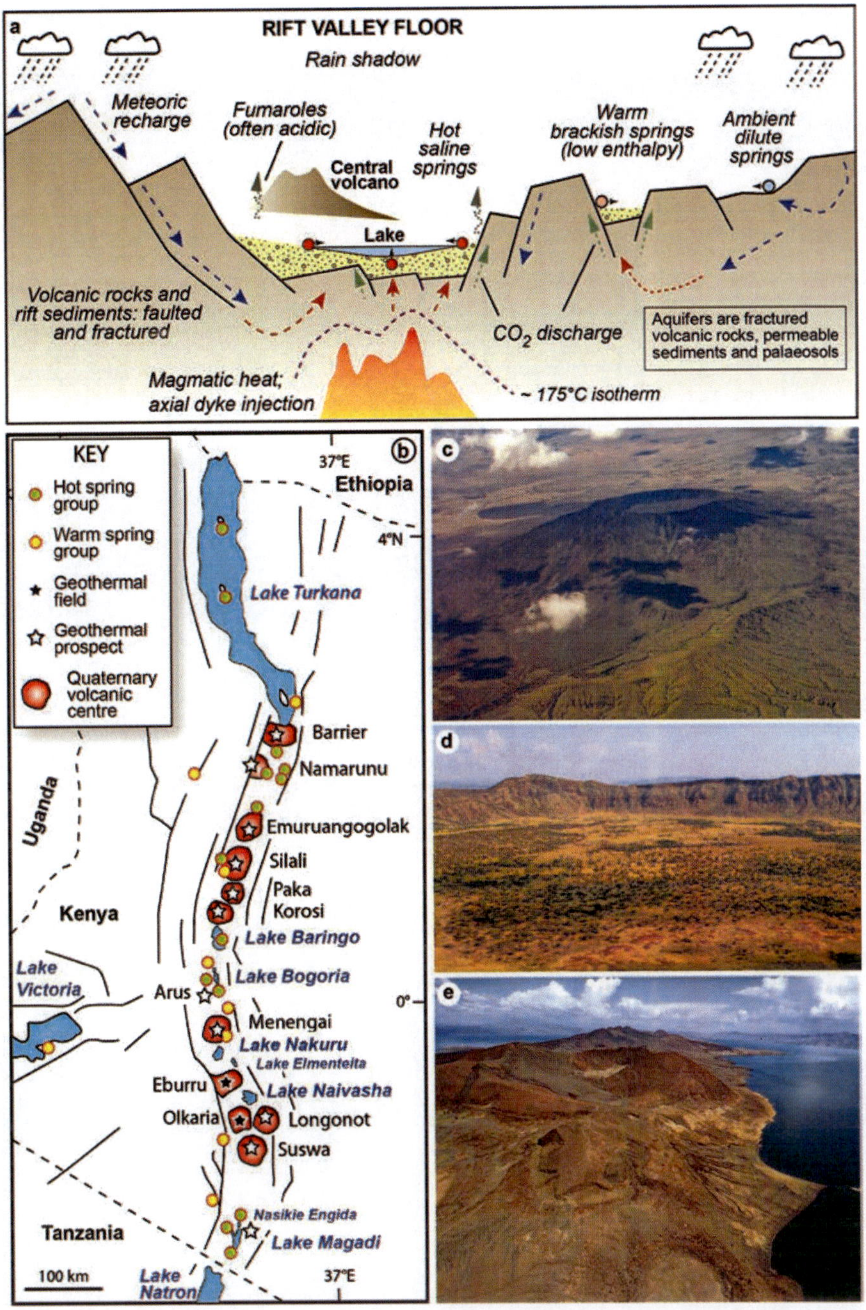

Fig. 4.20 Geothermal contributions to lacustrine environments. (**a**) Simple model for the development of hot springs in the Kenya Rift showing sources of recharge, general flow patterns and sites of fluid discharge. (**b**) Location of geothermal springs and their relationship to central volcanoes. Geothermal prospects are those that have been considered for geothermal power development. (**c**) Longonot volcano. (**d**) Silali volcano. (**e**) South Island volcanic centre, Lake Turkana

Table 4.1 Selected chemical analyses of Kenya Rift hot springs (in mg L^{-1})

Location	Spring reference	Date	Temperature	pH	Na	K	Ca	Mg	HCO_3	CO_3	Cl	SO_4	F	SiO_2	Li	References
Lake Bogoria																
Loburu	KL6	Jul-95	97.5	8.20	1420	15.3	0.85	0.34	2965		235	63.5	63.5	112	0.55	1
Loburu	KL30 geyser	Aug-06	97.0	8.28	1440	42	<0.1	0.34	2850	268	306	63	60	115		2
Chemurkeu	KEN-6	Feb-94	97.0	8.71	1580	35	0.71	0.3	3450		332	60	70	132	0.51	3
Chemurkeu	MM4	Oct-91	98.0	8.75	1600	30	0.5	<0.50	3480		305	58	65	130		3
Ng'wasis	MW1	Jul-95	96.5	8.95	5985	154	0.28	0.09	8560	2210	1525	84	254	179		4
Ng'wasis	NG3	Jul-05	96.0	7.45	36	0.7	0.2	0.02	20		9	44	70	70		4
Koibobei	KOB1	Jul-05	97.9	9.31	4300	140	0.44	0.18	8690		1100	71	175	212		4
Losaramat	LOS-95 geyser	Jul-05	98.0	9.70	6300	120	0.4	0.3	13000		1520	81	220	205		4
Maji Moto		1985-87?	36.5	7.14	113	12.4	6.61	1.12	300		8.9	10.6	4.4		0.06	5
Lake Baringo																
Ol Kokwe	71a	1986	95.8	6.90	22.1	4.5	1.5	0.5	121		1.4	76.1	0.42	82	0.02	6
Ol Kokwe	71b	1988	94.0	9.10	832	35	0.5	<0.7	1960		260	240		186	0.56	6
North Rift																
Kapedo	48a	1986	50.0	8.25	988	22	1.6	0.6	2000		215	80		76	0.06	6
Lorusio	45	1988	82.2	7.45	2120	55.8	3	0.2	5000		300	200	48	33	0.96	6
Suguta																
Logipi	190	1991	61.4	8.30	4170	79.4	0.5	0.1	5160		3420	590	110	40	0.05	6
Elboitong	236	1991	95.0	7.10	6770	129	0.82	<0.1	7940		4400	1070	87	99	0.31	6
Kageinya	188	1989	67.8	9.50	5420	85.5	3	0.3	3350		2280	3700	190	174	0.23	6
Magadi																
Nasikie Engida	NE2-07	Aug-06	82.0	9.01	11200	260	0.6	<0.5	13300	3970	4900	210	165	90		7
Nasikie Engida	NE6-07	Aug-07	76.0	8.72	11500	290	0.5	<0.5	15700	2230	5580	230	165	95		7
NW Lagoon spring	M1009	May-73	42.0	8.98	8060	104			11200	1810	3930	227	69	104		8
SE Lagoon spring	M547	Jul-70	41.8	9.60	12700	127			8820	7030	6540	206	132	46		8

References: 1: Renaut et al. (2013); 2: Renaut et al. (2008); 3: Renaut and Jones (1997); 4: Renaut et al. (2017); 5: Cioni et al. (1992); 6: Dunkley et al. (1993); 7: Renaut et al. (2021); 8: Jones et al. (1977)

body temperature and other biological criteria, or arbitrary temperature divisions. Renaut and Jones (2000), for example, used the term *warm spring* where vent water temperatures are 25–40 °C, *mesothermal spring* for water discharged at 40–75 °C, and *hyperthermal spring* for fluids discharged at >75 °C. *Boiling springs* are those discharged at local boiling point for the altitude. The lack of accepted definitions means that 'hot springs' in one region might be considered only 'warm springs' in another region where hotter springs exist. Many springs in the Kenya Rift discharge at temperatures above those of local surface waters and ambient-temperature springs, exceed the mean annual air temperature, and meet many definitions of 'hot spring'. There is nonetheless often a clear distinction between those associated with relatively high-enthalpy geothermal systems with high temperature and boiling waters, lower enthalpy systems ('warm springs') that discharge at surface temperatures only 5–15° above ambient surface waters, and ambient temperature ('cool') springs many of which are artesian.

Several types of hot spring are present. *Pool springs* flow from pools hosted by sediments or hydrothermally altered bedrock (Fig. 4.21a). *Seepage springs* issue from sediments, typically into thermal wetlands or deltaic-littoral plains, but lack a well-defined vent (Fig. 4.21b), as do *fracture springs*, in which water flows directly from fractured volcanic bedrock (Fig. 4.21c). *Perpetually spouting springs* discharge jets of water constantly from a vent lined by mineral deposits, or directly from sediments or bedrock (Fig. 4.21d). Spouting behaviour can be driven by steam, exsolution of CO_2 or both. *Geysers*, which are rare in the Kenya Rift (Renaut and Owen 2005), discharge a jet of water into the air intermittently on a regular or irregular basis (Fig. 4.21e). Exsolution of magmatic CO_2, which is high in many continental rift springs (Lee et al. 2016), may enhance geyser activity (Renaut et al. 2008). Most geysers discharge at or near local boiling point unless primarily CO_2-charged. Sublacustrine hot springs discharge from vents, or diffusively, on lake floors.

Steam vents and fumaroles release gases in addition to steam such as CO_2 and H_2S. Fumaroles are most common around the margins of craters of central volcanoes in the Kenya Rift (Fig. 4.21f), including those that are active, dormant or recently extinct. Some vents, however, lie along fault lines in areas without recent volcanism such as those at Arus, near Lake Bogoria, Kenya (Lagat 2008; Karingithi and Wambugu 2010). Hot gases are also released subaqueously into lakes. Fumarole vents at recently active volcanic centres tend to be large, vigorous and of high temperature, whereas those in more quiescent areas are often smaller and dispersed. Diffuse release of geothermal gases may affect extensive areas and be identified by local changes or the absence of vegetation cover rather than well-defined vents (Renaut et al. 2017). 'Geothermal grass', a brownish sedge (*Abligaardia lispidula,* also called *Fimbristylis* sp.), is locally common in areas of steaming ground (Dunkley et al. 1993). Areas with abundant fumaroles usually show acidic hydrothermal alteration at the land surface, where hot gases such as H_2S and SO_2 have condensed in shallow groundwaters to produce sulfuric and other acids.

Fig. 4.21 Types of hot springs in the Kenya Rift. All at Lake Bogoria except **b**, which is at Lake Magadi. (**a**) Boiling spring, northern Loburu. (**b**) Seepage spring, SW Lake Magadi. (**c**) Fracture spring emerging from trachyte, Ng'airus. (**d**) Perpetually spouting hot spring (KL8), Loburu. The mound is fossil travertine. (**e**) Geyser (~5 m high), southern Loburu, August 2006. (**f**) Steam from fumarole discharge from trachyte, Koibobei

4.9.2 Structural Setting of the Hot Springs

Most hot springs and fumaroles lie along known or inferred faults, or in zones of fracturing that provide permeable conduits for upward fluid flow. In detail, however, the structural contexts of spring and fumarole vents are variable. Spring vents commonly lie on the floor of topographic depressions, including the faulted margins of lakes or playas, fully subaqueous sites on lake floors, and on the floors of stream channels, many of which lie along faults. Other springs discharge at higher elevations fed by fluids rising along faults and fractures on the lower flanks of volcanoes, and at sites where permeable strata rest upon relatively impermeable aquitards or are sandwiched between impermeable rocks. Changes in the level of groundwater affect the amount and type of discharge. In general, where the groundwater level is high with respect to the land surface, geothermal discharge will be as hot springs; if low, then steam discharge and fumaroles are more likely. Along the rift valley floor and along lake margins hot springs are common. At higher elevations, fumaroles may be present. Hot springs often discharge along fault-controlled shorelines where the hot water, though often saline, is less dense than adjacent lake water.

In their review of the structural setting of hot springs, Curewitz and Karson (1997) observed that high fracture density closely linked to faulting provides most of the permeability for hydrothermal upflow to the surface. They recognised five main structural settings of hot springs, including those in continental rifts (Fig. 4.22). *Fault-tip* line springs are sites of fault termination where open crack networks propagate into non-fractured rocks. *Fault interaction* springs develop where stresses associated with two or more fault-tips interact to form a 'breakdown region' with high fracture density. Fault spacing, distribution and kinematics control the size, shape and fracture density of the breakdown region. *Fault intersection springs* discharge at fault junctions. Curewitz and Karson (1997) separate *locked-intersection junctions* where faults have convergent slip directions from *slipping-intersection junctions*, which have parallel slip vectors and include extensional fault systems and intersection vertical-slip faults. *Fault trace springs* develop in areas where fault planes intersect the earth's surface, and commonly mark the topographic expression on active faults or deep levels of inactive faults exposed by erosion. Springs away from obvious faults are termed *asystematic*.

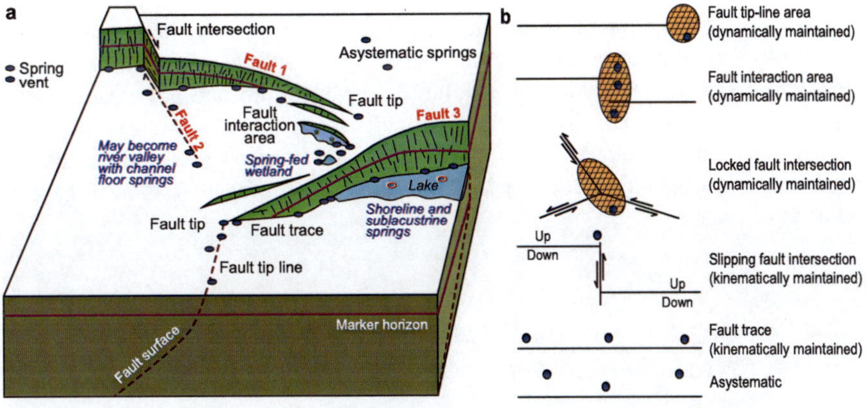

Fig. 4.22 Structural setting of hot springs in continental rifts. Adapted from Curewitz and Karson (1997). (**a**) The various geometric and kinematic characteristics of a fault zone. Fine lines on the fault plane surfaces represent slickensides, whereas the bold line is a marker bed to show displacement across the faults. (**b**) Schematic diagram showing examples of the different structural settings of hydrothermal vents relative to faults. Breakdown regions are shaded

For springs to flow, permeability must be maintained and not be sealed by mineral precipitation, (e.g. $CaCO_3$ or silica). Fracture permeability is formed and sustained by high stresses (mainly tensile) during fault propagation in tip-line areas, fault interaction areas, and locked fault intersections. These breakdown areas, which exist away from the fault itself, were termed *dynamically maintained fracture systems* by Curewitz and Karson (1997). In contrast, in *kinematically maintained fracture systems* slip along existing fault traces or slipping fault intersections can re-open fracture networks along a fault. In their survey, Curewitz and Karson (1997) found that most springs in continental rifts form at fault tips, interaction areas, and along fault traces and that permeability is maintained dynamically more often than kinematically.

4.9.3 Hot Spring Water Chemistry

Most 'very hot' to boiling springs (>70 °C: hyperthermal) in the Kenya Rift are of Na^+–Ca^{2+}–HCO_3^- composition with high concentrations of fluoride and silica, variable chloride and sulfate, and very low calcium and magnesium (Table 4.1). Salinities vary from <1 g L^{-1} to >30 g L^{-1} Total Dissolved Solids. Most waters are alkaline with a pH range of 8.0 to ~10. A few hot springs at Ol Kokwe (Lake Baringo) and Ng'wasis (Lake Bogoria) are dilute with Na^+–SO_4^{2-} composition and neutral pH. Those atypical waters discharge a few metres higher than neighbouring alkaline hot springs near lake shores and represent local condensates of fumarolic steam (Renaut et al. 2017).

The cooler springs (warm and cooler mesothermal springs) fall within two groups, based on salinity and pH. One spring group is of Na^+–Ca^{2+}–HCO_3^- composition with neutral to mildly alkaline pH and a salinity <5 g L^{-1}. Some precipitate $CaCO_3$ deposits. These represent meteoric groundwater that have been heated during subsurface circulation but have not mixed much with deeper, hotter chloride-rich fluids. The second group has similar temperatures, but are more saline with higher pH, and have very low alkaline earths (Ca^{2+}, Mg^{2+}). These represent saline groundwaters that have been heated by geothermal processes (magmatic intrusion, regionally high heat flow). The springs feeding Lake Magadi fall into the latter group.

4.9.4 Spring Deposits and Hydrothermal Alteration: Records of Hydrothermal Activity in Lake Basins

The contribution of geothermal fluids to the modern lakes is generally easy to determine and measure if hot springs are subaerial, but the contribution of sublacustrine springs and gas discharge is less easy to detect, even with isotopes, and the record of hydrothermal contributions is not always preserved as spring deposits. Many springs are present along present and former lake shorelines, but some discharge from vents in the drainage basin that may be several kilometres from the lake into which they flow.

Spring deposits are common in the Kenya Rift though dispersed and few are large (Table 4.2). The two most common groups are composed of calcium carbonate or silica. Hydrothermal alteration is locally present and recorded by a different suite of minerals including sulfates and clays.

Table 4.2 Modern and ancient hot spring deposits of the Kenya Rift

Location	Trav.	Tufa	Silica	Salts	Age	Mineralogy (modern)	Mineralogy (fossil)	References
Lake Bogoria								
Loburu	X		X	X	M, H, P	C, O, F, T	C, A, F	1
Chemurkeu	X		X	X	M, H, P	C, O, F, T, M	C, A, O, F, T, D	2
SW hot springs		X	X	X	M, H, P	O, T, S	C, O, Q, F, Z, P	3
Maji Moto		X			H? P?	C?	C	4
Emsos		X			P?	C?	C	5
Lake Baringo								
Ol Kokwe	X		X	X	M, H, P?	C, O, S	O, Q, P	6
Kapthurin Fm		X			P		C	7
Lake Magadi								
Shoreline springs				X		T, O		8
Oloronga Fm		X			P		C	9
Nasikie Engida			X	X	M	T, O, G?		10
W Magadi	X	X			P		C	11
Olorgesailie-Kedong								
Eremit			X		P		Q	12
SW Olorgesailie		X					C	13
N Kenya Rift								
Lorusio	X		X	X	M, H?	C, O	C	14
West Turkana	X				P		C, Q, D	15
Logipi hot springs				X	M	T, O		16
[a]Central volcanoes			X		H? P		Q	17
The Barrier			X		H? P		Q	18
Losergoi, Suguta	X				?		C?	19

(continued)

Table 4.2 (continued)

Location	Trav.	Tufa	Silica	Salts	Age	Mineralogy (modern)	Mineralogy (fossil)	References
Other sites								
Kanjera		X					C?	20
Lemudong'o		X			Mc		C?	21
Lake Natron W shore				X	M	A, C, T		22

References: 1: Renaut (1982), Casanova (1986, 1994), Jones and Renaut (1995, 1996b), Renaut et al. (1998, 2013); 2: Renaut (1982), Casanova (1986, 1994), Jones and Renaut (1996a, b, 1998), Renaut and Jones (1997); 3: Renaut and Owen (1988), Renaut et al. (2017); 4: Griffiths (1977) and Renaut et al. (2013); 5: Renaut et al. (2013); 6: Dunkley et al. (1993), Renaut et al. (2002a), and Buatois et al. (2017); 7: Tallon (1976) and Roure Johnson et al. (2009); 8: Original observations; 9: Owen et al. (2019); 10: Renaut et al. (2021); 11: Baker (1958), Casanova (1986, 1994), Faure et al. (2002), and Owen et al. (2019); 12: Owen et al. (2014); 13: Lee et al. (2013); 14: Dunkley et al. (1993), Jones and Renaut (1998), Renaut et al. (1999); 15: Walsh and Dodson (1969) and Renaut et al. (2002b); 16: Castanier et al. (1993); original observations; 17: Sceal (1974) and Dunkley et al. (1993); 18: Dunkley et al. (1993); 19: Bosworth and Maurin (1993); 20: Oswald (1914) and Kent (1942); 21: Ambrose et al. (2007); 22: Manega and Bieda (1987)

Trav.: Travertine

Age: M; modern; H: Holocene; P: Pleistocene; Mc: Miocene

Mineralogy: A: aragonite; C: calcite; D: dolomite; F: fluorite; G: Si-gel; M: Mg-silicate; O: opaline silica; P: pyrite; Q: quartz; Z: zeolites; T: Na-salts

[a]Chepchuk, Emuruangogolak, Nakaporon, Cheptomas

Travertine and Tufa

Travertine and tufa are the most common spring deposits in the Kenya Rift and are present at many sites in modern settings and in Neogene rocks. Many are directly associated with the lakes. Calcareous spring deposits, composed of $CaCO_3$ (calcite, less commonly aragonite), precipitate from both ambient-temperature and hot spring fluids, so not all of these deposits are hydrothermal. However, because of the regionally high heat flow, many waters that precipitate alkaline earth carbonates in continental rifts discharge above local ambient air temperatures. Pentecost (2005) recognised two main groups of springs and associated travertines based on the type of fluid that carries the CO_2. *Meteogene travertines* precipitate from waters charged with CO_2 of meteoric origin. These usually are soil waters. Those waters usually discharge as ambient (normal local) temperature springs but include some hot springs ('superambient springs') where the water discharged has been in deeper circulation. Thermogene springs precipitate *thermogene travertines* that are partly charged with meteoric (soil) CO_2, but much of the carbon dioxide derives from thermal processes in or below the earth's crust including mantle-derived and magmatic CO_2, and thermal decarbonation processes that release CO_2 to circulating groundwater (see Chapter 2 in Pentecost 2005 for discussion). Thermogene travertines, therefore, typically form in tectonically active regions and areas of recent volcanism, including continental rifts, where deep sources of CO_2 are more likely.

Emphasis here is given to deposits with evidence for hot (>37 °C) waters being involved in precipitation – the thermogene and superambient spring deposits. The terms travertine and tufa are poorly defined (Jones and Renaut 2010). Some definitions are based on lithology; others are based on inferred genetic mechanisms such as CO_2 degassing (e.g. Pentecost 2005; Brasier 2011; Capezzuoli et al. 2013). The term 'travertine' is commonly used for $CaCO_3$ precipitated from spring water that is banded or laminated, sometimes dense, and usually visibly crystalline. Travertine typically contains few macrovegetal (plant) remains but can include biogenic structures and crystal morphologies that were mediated in part by bacteria, including cyanobacteria. 'Tufa', in contrast, is typically used for more porous, somewhat granular carbonates, many of which contain externally encrusted higher plant and (or) moss remains, although the organic matter has normally decayed to produce mouldic porosity that may be later filled by cement. Tufas can be bedded or banded. Plants fossilised in tufas include rooted and floating aquatic macrophytes that were present in a spring pool or outflow channel, and detrital leaves and wood. This macrovegetal debris derived from local vegetation and included water-adapted plants that were present only because of spring water supply rather than local or regional vegetation. Microbes also mediate calcium carbonate precipitation in tufas. Depending on water temperature, higher organisms such as molluscs and ostracods may also be present. Tufas in dry climates, however, may show little plant debris.

Some spring carbonates show textural and fabric characteristics common to both tufa and travertine or contain interlayered facies of each type. Adding to the confusion, the term tufa has also been used for carbonate mounds and littoral encrustations of several origins that have precipitated subaqueously around the vents of sublacustrine springs and along lake shorelines (mainly USA usage). Some of these tufa mounds have textures similar to rocks that would otherwise be called travertine. This usage demonstrates the problem – lithology or genetic origin?

Travertine commonly forms as mounds around spring vents, terraces on their discharge aprons or in their outflow channels (Fig. 4.23a–d), and a range of small-scale deposits including oncoids, pisoids, lilypads and ledges (Chafetz and Folk 1984; Della Porta et al. 2017). Some mounds develop as linear fissure ridges, where water is discharged along the axis of an open fissure or major fracture with water flowing laterally at ~90° on each side of the fissure axis (Bargar 1978; Renaut et al. 2002a; De Filippis et al. 2013) building up elongate mounds that sometimes form barriers ponding small lakes. Tufas, which tend to be more massive or coarsely bedded, also form mounds and terraces, but many form aggrading deposits on the floors of fluvial channels (*barrage tufas*) or precipitate in groundwater seepage zones on the margins of shallow lakes or wetlands (*paludal tufas*). In general, carbonate rocks with the characteristics of travertine form more commonly at high-temperature thermogene springs, whereas tufas are associated with spring waters that discharge at lower temperatures (meteogene: ambient to ~40 °C), but there are many exceptions, and usage of these terms has been, and remains, inconsistent.

Precipitation mechanisms have long been debated (Pentecost 2005; Jones and Renaut 2010), particularly the respective roles of biotic and abiotic processes. The abiotic process most likely to play a role in $CaCO_3$ precipitation is exsolution of CO_2 linked either to equilibration with respect to atmospheric CO_2 upon surface discharge or to boiling of thermal waters at the surface or in the shallow plumbing system of very high-temperature springs (Fig. 4.23b). Some CO_2 may be of magmatic, of lower crustal or mantle origin. This possible effect is enhanced by the high

Fig. 4.23 Hot spring deposits related to lakes and palaeolakes in the Kenya Rift. (**a**) Calcite travertine terraces (foreground) adjacent to a boiling spring pool (right centre) and spouting hot spring (upper right corner) at spring KL6, Loburu, Lake Bogoria. (**b**) Interbedded dendritic calcite (C) and skeletal aragonite (A) beds at Chemurkeu, Lake Bogoria, that formed just below the contemporary land surface. (**c**) Subaerial Pleistocene terraces of calcite tufa, west of Olorgesailie,

water temperature, which decreases the solubility of calcium carbonate, and agitation due to turbulence that drives off carbon dioxide. Although exsolution of CO_2 is generally thought to induce $CaCO_3$ precipitation, particularly when fluids boil, Shvartsev et al. (2007) suggested that in some springs degassing is a consequence rather than a cause of travertine precipitation. With the warm climates of the rift floor throughout East Africa, hot spring fluids tend to cool downslope from the vents, which slowly increases $CaCO_3$ solubility. Cooling and mixing with other surface and groundwaters (i.e. dilution) limit the area and thickness of the travertine or tufa deposit by decreasing levels of carbonate saturation in the spring outflow waters. Many springs today, however, have low initial concentrations of Ca^{2+} (<2 mg L^{-1}) and rapidly become calcium depleted downstream. In contrast, in arid parts of the rift evaporative concentration can contribute to successive precipitation of calcite or aragonite, opaline silica and sodic salts at sites downflow from the vent (Renaut et al. 1999). Mixing of spring fluids with lake water can also play a role in carbonate precipitation around some sublacustrine hot springs by the common-ion effect, where for example, heated dilute Ca-bearing spring water mixes with saline, alkaline lake water, or Na^+–HCO_3^- spring fluids are in contact with dilute Ca^{2+}-bearing lake water.

Biotic processes that influence carbonate precipitation in hot springs have been linked to photosynthetic uptake of CO_2 by bacteria (including cyanobacteria), fungi, lichens, mosses and higher plants, or to other biological process (e.g. sulfate reduction) that biologically induce nucleation and precipitation of calcium carbonate (Pentecost 2005). Most carbonates with tufa fabrics form at temperatures <45 °C where plants and fauna are more abundant, and microbial processes of biomineralization are frequently reported.

Spring fluid temperature is an important primary control in this context because only microbes are considered potential mediators of $CaCO_3$ nucleation in high-temperature fluids, whereas plants and mosses can potentially play a role at fluid temperatures of <40 °C (cf. Brock 1978). In very hot fluids ($T > 70$ °C at vent), most carbonate precipitation is probably linked to the exsolution of carbon dioxide. In medium to high-enthalpy springs, boiling may lead to exsolution of dissolved CO_2, so that most carbonate precipitates in the feeder conduits to springs rather than at the subaerial vent. Such springs may then discharge Ca-depleted fluids at the surface.

Bacteria (including cyanobacteria) and archaea are common in hot springs of the East African Rift. Thermophilic cyanobacteria and green non-sulfur bacteria (*Chloroflexus* spp.) form mats near spring vents and lie upon substrates where flowing water temperatures are <70 °C; archaea inhabit hotter waters but rarely form thick benthic mats. In many geothermal sites, bacteria are thought to mediate precipitation of calcium carbonate. Chafetz and Folk (1984), for example, suggested

Fig. 4.23 (continued) (**d**) Pleistocene tufa mound that might have formed under a former lake, 4.75 km west of Lake Magadi. (**e**) Modern silica precipitating at the air-water interface in hot spring outflow at Loburu, Lake Bogoria. Cr: opaline silica crusts; Sp: silica spicules. f: fossil (Pleistocene) geyser mound of silica sinter at Eremit, ~12 km north of Ol Tepesi. (**g**) Banded fossil geyserite from the mound shown in '**f**'. (**h**) Seepage spring (v: vent), Northwest Lagoon, Lake Magadi. Mineral deposits are efflorescent sodium carbonate salts and thin opal-A crusts that bacteria mediated calcite crystal growth in hot springs at Yellowstone (Wyoming, USA) and in Italy. Conclusive evidence for bacterial mediation of calcite at T > 70 °C in East African hot springs is unclear. Many sites where bacterial mediation in travertine formation has been suspected are sulfurous and the associated bacteria produce distinctive calcite shrub fabrics (Chafetz and Folk 1984). In contrast, many modern hot springs in the rift with travertine deposits are poor in sulfur and lack characteristic shrub fabrics (Renaut et al. 2013).

The travertine and tufa deposits in the East African Rift precipitate as calcite (or aragonite) from spring fluids with temperatures ranging from ambient to boiling. Most of these fluids are fairly dilute with salinities <5 g L^{-1} TDS. Fossil examples are present throughout the rift and provide evidence for former spring activity. Most are recognized by their fabrics (cf. Brasier 2011), vertical profiles and associated facies, but some paludal (palustrine, wetland) tufas resemble some calcretes.

Siliceous Spring Deposits

Spring deposits composed mainly of silica commonly form at high-enthalpy geothermal sites, including the Taupo Volcanic Zone (New Zealand), Iceland, El Tatio (Chile), Yellowstone (Wyoming, USA), and elsewhere (e.g. Campbell et al. 2001; Fernandez-Turiel et al. 2005; Renaut and Jones 2011). Most silica precipitates at sites where hot spring and geyser fluids with high SiO_2 concentrations undergo rapid cooling and (or) evaporation after surface discharge. Silica precipitates as non-crystalline opal-A ($SiO_2.nH_2O$) or less commonly as chalcedony, but normally alters to more ordered forms of opal (-C or -CT), moganite and quartz during diagenesis, which can be rapid (Jones 2021). Silica sinter typically forms mounds around hot-spring or geyser vents, pool-rim dams around spring margins, and terraces on spring or geyser discharge aprons. Many smaller-scale sinter morphotypes may also develop. The spring outflow discharges into rivers, lakes and thermal wetlands.

Sinters have many textures. Near vents they commonly are dense, banded, laminated and highly ornate, forming the variety termed 'geyserite'. On discharge aprons below the vent, sinters are bedded or laminated, and initially are commonly highly porous though primary porosity is often occluded by cement. Bacterial mats are common on substrates near the vent and on the discharge apron where the water temperature is <75 °C and these may be rapidly silicified by encrustation and (or) replacement to produce stromatolitic textures. Higher plants engulfed by thermal outflow waters are commonly silicified.

To precipitate sinter by cooling alone reservoir fluids need a minimum temperature of ~165 °C (Rimstidt and Cole 1983), which limits most silica precipitation to high-temperature springs. Most sinter formation results from the cooling of thermal fluids that have equilibrated with respect to quartz at depth and then risen rapidly to the surface where they discharge as hot water supersaturated with respect to amorphous silica. Cooling and evaporation of such fluids induce precipitation of opal-A, although some cations (e.g. Fe, Al) can also play a role. Much of the dissolved silica is present as monomeric H_4SiO_4, which polymerises in strongly supersaturated fluids to produce nanospheres and microspheres, forming suspended colloidal silica sols that may aggregate to form gels. At lower saturation levels, aqueous silica precipitates by heterogeneous nucleation on existing substrates, which can be mineral surfaces or reactive biological surfaces.

Sinter abundance depends not only on the silica concentration of the fluids but on their pH and the style of discharge. Neutral and alkaline fluids tend to precipitate much sinter and commonly have extensive microbial mats that preserve a record in the sinter fabrics. In contrast, acidic fluids generally produce much less sinter and have fewer microbial textures. Those present are commonly associated with fungi, diatoms and *Cyanidium*, a unicellular eukaryotic red alga, rather than cyanobacteria. In some acidic areas, including areas around active fumaroles, silica deposits are formed that resemble sinters but have a different origin. Termed 'silica residue' (Rodgers et al. 2002), they are products of hydrothermal alteration of volcanic rocks or volcaniclastic sediments by acidic (pH 2–3) thermal fluids. Unlike sinter, they form *in situ* and for the most part are not precipitated by flowing spring water.

Modern sinter deposits are rare in the Kenya Rift. Silica precipitates as opal-A at modern hot springs as thin crusts on subaerial substrates along pool and outflow channel margins, or at the air-water interface (Fig. 4.23e). It typically forms by evaporation or by wicking through microbial mats (Renaut et al. 1998). In contrast, sublacustrine precipitation of hydrothermal silica has been recorded at Lake Bogoria (Renaut et al. 1988; Renaut et al. 2017) and Lake Baringo (Renaut et al. 2002b; Buatois et al. 2017), where soft and semi-gelatinous silica precipitated on the shallow lake floor when the lake was fresh. Fossil sinter is present at Eremit near Olorgesailie (Fig. 4.23f, g; Owen et al. 2014), and at central volcanoes in the northern Kenya Rift (Sturchio et al. 1993; Dunkley et al. 1993).

Problems of Preservation

Modern and former spring recharge is not always recorded by deposits at spring vents. Modern springs along the Kenya Rift discharge fresh to brackish waters of variable salinity. Some springs discharge hot water that is undersaturated with respect to $CaCO_3$, so spring carbonates do not form. Those waters never become saturated before they flow into lakes, so neither tufa nor travertine form (Fig. 4.23h). In contrast, other springs, generally with lower discharge, become saturated at or near the vent and precipitate $CaCO_3$ as tufa or travertine at the vent or in proximal outflow channels. The subaerial silica crusts (opal-A) are friable and have low preservation potential.

A second problem is that most spring deposits are small and of limited size. Shoreline travertines are typically <100 m^2 in extent though some (e.g. Chemurkeu, Lake Bogoria) extend 250 m along the shoreline and are at least 5 m thick. Subaerial tufa deposits are commonly larger and more extensive, particularly those in valleys fed by springs upstream of lakes (Lee et al. 2013; Renaut et al. 2017). Sublacustrine spring carbonates vary from small chimneys <50 cm high to columns >5 m high (Owen et al. 2019; Fig. 4.23d). Sublacustrine silica sinter extends laterally for several metres but is only decimetres in thickness. The late Quaternary examples from Baringo and Bogoria are still mainly opal-A but should transform to quartz during diagenesis, increasing their preservation potential.

References

Abiye TA (2009) Groundwater dynamics in the East African Rift System. In: Xu Y, Braune E (eds) Sustainable groundwater resources in Africa. CRC Press, London, pp 93–10

Adams JM, Faure H, QEN Members (eds) (1997) Review and atlas of paleovegetation: preliminary land ecosystem maps of the world since the last glacial maximum. Oak Ridge National Laboratory, Oak Ridge

Allen DJ, Darling WG, Burgess WG (1989) Geothermics and hydrogeology of the southern part of the Kenya Rift Valley with emphasis on the Magadi-Nakuru area. Brit Geol Surv Res Rep SD/89/1

Ambrose SH, Nyamai CM, Mathu EM, Williams MAJ (2007) Geology, geochemistry, and stratigraphy of the Lemudong'o Formation, Kenya Rift Valley. Kirtlandia 56:53–67

Armstrong MF (2002) Water balance of the southern Kenya Rift lakes. MSc thesis, International Institute for Geo-information Science and Earth Observation, Enschede, Netherlands

Ataya CO (2000) Wind erosion of volcanic soils. MSc thesis, International Institute for Geo-information Science and Earth Observation, Enschede, Netherlands

Baker BH (1958) Geology of the Magadi area. Geol Surv Kenya Rep 42:1–81

Bargar KE (1978) Geology and thermal history of Mammoth Hot Springs, Yellowstone National Park, Wyoming. US Geol Surv Bull 1444:1–54

Becht R, Mwango F, Muno FA (2006) Groundwater links between Kenyan Rift Valley lakes. In: Odada E, Olago D, Ochola W, Ntiba M, Wandiga S, Gichuki N, Oyieke H (eds) Proc. 11 World Lake Congr., Nairobi, 31 October–4th November 2005, vol 2. Ministr Water Irrig Int Lake Environ Cttee (ILEC), Nairobi, pp 7–14

Bosworth W, Maurin A (1993) Structure, geochronology and tectonic significance of the northern Suguta Valley (Gregory Rift), Kenya. J Geol Soc Lond 150:751–762

Brasier AT (2011) Searching for travertines, calcretes and speleothems in deep time: processes, appearances, predictions and impacts of plants. Earth-Sci Rev 104:213–239

Bristow CM, Temperley BN (1964) Geology and groundwater conditions of central Laikipia. Tech Rep Water Develop Dept Gov Kenya 2

Brock TD (1978) Thermophilic microorganisms and life at high temperatures. Springer, Berlin

Buatois LA, Renaut RW, Scott JJ, Owen RB (2017) An unusual occurrence of the trace fossil *Vagorichnus* preserved in hydrothermal silica at Lake Baringo, Kenya Rift Valley: taphonomic and paleoenvironmental significance. Palaeogeogr Palaeoclimatol Palaeoecol 485:843–853

Butzer KW, Isaac GL, Richardson JL, Washbourn-Kamau C (1972) Radiocarbon dating of East African lake levels. Science 175:1069–1076

Cai W, Wang G, Gan B, Wu L, Santoso A, Lin X, Chen Z, Jia F, Yamagata T (2018) Stabilised frequency of extreme positive Indian Ocean Dipole under 1.5°C warming. Nat Commun 9:1419

Camberlin P (1997) Rainfall anomalies in the source region of the Nile and their connection with the Indian summer monsoon. J Clim 10:1380–1392

Camberlin P (2018) Climate of Eastern Africa. Oxford Research Encyclopaedia, Climate Science. Oxford University Press, pp 1–51

Camberlin P, Philippon N (2002) The East African March-May rainy season: associated atmospheric dynamics and predictability over the 1968–97 period. J Clim 1:1002–1019

Campbell KA, Sannazzaro K, Rodgers KA, Herdianita NR, Browne PRL (2001) Sedimentary facies and mineralogy of the Late Pleistocene Umukuri silica sinter, Taupo Volcanic Zone, New Zealand. J Sediment Res 71:727–746

Capezzuoli E, Gandin A, Pedley M (2013) Decoding tufa and travertine (freshwater carbonates) in the sedimentary record: state of the art. Sedimentology 61:1–21

Casanova J (1986) Les stromatolites continentaux: paléoecologie, paléohydrologie, paléoclimatologie. Application au rift Gregory. Thèse, Université de Marseille II, 2 vols, Marseille

Casanova J (1994) Stromatolites form the East African Rift: a synopsis. In: Bertrand-Sarfati J, Monty C (eds) Phanerozoic stromatolites II. Kluwer, Dordrecht, pp 193–226

Castanier S, Bernet-Rollande M, Maurin A, Perthuisot J-P (1993) Effects of microbial activity on the hydrochemistry and sedimentology of Lake Logipi, Kenya. Hydrobiologia 267:99–112

Chafetz HS, Folk RL (1984) Travertines: depositional morphology and the bacterially constructed constituents. J Sediment Petrol 54:289–316

Chevalier M, Chase BM, Quick LJ, Dupont LM (2020) Temperature change in subtropical southeastern Africa during the past 790,000 yr. Geology 49:71–75

Cioni R, Fanelli G, Guidi M, Kinyariro KJ, Marini L (1992) Lake Bogoria hot springs (Kenya): geochemical features and geothermal implications. J Volcanol Geotherm Res 50:231–246

Clarke MCG, Woodhall DG, Allen D, Darling G (1990) Geological, volcanological and hydrogeological controls on the occurrence of geothermal activity in the area surrounding Lake Naivasha, Kenya. Rep Brit Geol Surv and Ministr Energy Kenya

Crane K, O'Connell S (1983) The distribution and implications of heat flow from the Gregory Rift in Kenya. Tectonophysics 94:253–275

Curewitz D, Karson JA (1997) Structural settings of hydrothermal outflow: fracture permeability maintained by fault propagation and interaction. J Volcanol Geotherm Res 79:149–168

De Cort G, Verschuren D, Ryken R, Wolff C, Renaut RW, Creutz M, Van der Meeren T, Haug G, Olago DO, Mees F (2018) Multi-basin depositional framework for moisture-balance reconstruction during the last 1300 years at Lake Bogoria, central Kenya Rift Valley. Sedimentology 65:1667–1696

De Cort G, Chevalier M, Burrough SL, Chen SY, Harrison SP (2021) An uncertainty-focused database approach to extract spatiotemporal trends from qualitative and discontinuous lake-status histories. Quat Sci Rev 258:106870

De Filippis L, Anzalone E, Billi A, Faccenna C, Poncia PP, Sella P (2013) The origin and growth of a recently-active fissure ridge travertine over a seismic fault, Tivoli, Italy. Geomorphology 195:13–26

Della Porta G, Croci A, Marini M, Sándor K (2017) Depositional architecture, facies character and geochemical signature of the Tivoli travertines (Pleistocene, Acque Albule Basin, Central Italy). Riv Ital Paleontol Stratigr 123:487–540

Deocampo D, Renaut RW (2016) Geochemistry of soda lakes. In: Schagerl M (ed) Soda lakes of East Africa. Springer, Cham, pp 77–93

Dunkley PN, Smith M, Allen DJ, Darling WG (1993) The geothermal activity and geology of the northern sector of the Kenya Rift Valley. Brit Geol Surv Res Rep SC/93/1

Dunning CM, Black E, Allan RP (2018) Later wet seasons with more intense rainfall over Africa under future climate change. J Clim 31:9719–9738

Eugster HP (1970) Chemistry and origin of the brines of Lake Magadi, Kenya. Mineral Soc Am Spec Pap 3:213–235

Eugster HP (1980) Lake Magadi, Kenya and its precursors. In: Nissenbaum A (ed) Hypersaline brines and evaporites. Elsevier, Amsterdam, pp 195–232

Faure H, Walter RC, Grant DR (2002) The coastal oasis: ice age springs on emerged continental shelves. Glob Planet Change 33:47–56

Felske GN (2016) Genesis of calcretes and related carbonate rocks in the southern Kenya Rift. MSc thesis, University of Saskatchewan, Saskatoon

Fenta AA, Tsunekawa A, Haregeweyn N, Poesen J, Tsubo M, Borrelli P, Panagos P, Vanmaercke M, Broeckx J, Yasuda H, Kawai T, Kurosaki Y (2020) Land susceptibility to water and wind erosion risks in the East Africa region. Sci Total Environ 703:135016

Fernandez-Turiel JL, Garcia-Valles M, Gimeno-Torrente D, Saavedra-Alonso J, Martinez-Manent S (2005) The hot spring and geyser sinters of El Tatio, Northern Chile. Sediment Geol 180:125–147

Fick SE, Hijmans RJ (2017) WorldClim 2: new 1-km spatial resolution climate surfaces for global land areas. Int J Climatol 37:4302–4315

Finney DL, Marsham JH, Rowell DP, Kendon EJ, Tucker SO, Stratton RA, Jackson LS (2020) Effects of explicit convection on future projections of mesoscale circulations, rainfall and rainfall extremes over Eastern Africa. J Clim 33:2701–2718

Foerster V, Junginger A, Langkamp O, Gebru T, Asrat A, Umer M, Lamb HF, Wennrich V, Rethemeyer J, Nowaczyk N, Trauth MH, Schaebitz F (2012) Climatic change recorded in the sediments of the Chew Bahir basin, Southern Ethiopia, during the last 45,000 years. Quat Int 274:25–37

Gebregiorgis D, Rayner D, Linderholm HW (2019) Does the IOD independently influence seasonal monsoon patterns in northern Ethiopia? Atmosphere 10:432

Gitau W, Ogallo L, Camberlin P, Okoola R (2012) Spatial coherence and potential predictability assessment of intraseasonal statistics of wet and dry spells over Equatorial Eastern Africa. Int J Climatol 33:2690–2705

Griffiths PW (1977) Geology of the Lake Hannington–Perkerra River area, Kenya Rift Valley. PhD thesis, University of London

Grist JP, Nicholson SE (2001) A study of the dynamic factors influencing the rainfall variability in the West African Sahel. J Clim 14:1337–1359

Hay RL (1970) Silicate reactions in three lithofacies of a semi-arid basin, Olduvai Gorge, Tanzania. Mineral Soc Am Spec Pap 3:237–255

Hemming C (1972) The South Turkana expedition: scientific papers VIII. The ecology of South Turkana: a reconnaissance classification. Geogr J 138:15–40

Hochstein MP (1999) Geothermal systems along the East-Africa Rift. Bull Hydrogéol 17:301–309

Ivory SJ, McGlue MM, Ellis GS, Boehlke A, Lézine AM, Vincens A, Cohen AS (2017) East African weathering dynamics controlled by vegetation-climate feedbacks. Geology 45:823–826

Jones B (2021) Siliceous sinters in thermal spring systems: review of their mineralogy, diagenesis, and fabrics. Sediment Geol 413:105820

Jones B, Renaut RW (1995) Noncrystallographic calcite dendrites from hot-spring deposits at Lake Bogoria, Kenya. J Sediment Res 65:154–169

Jones B, Renaut RW (1996a) Skeletal crystals of calcite and trona from hot spring deposits in Kenya and New Zealand. J Sediment Res 66:265–274

Jones B, Renaut RW (1996b) Morphology and growth of aragonite crystals in hot spring travertines at Lake Bogoria, Kenya Rift Valley. Sedimentology 43:323–340

Jones B, Renaut RW (1998) Origin of platy calcite crystals in hot-spring deposits of the Kenya Rift Valley. J Sediment Res 68:913–926

Jones B, Renaut RW (2010) Calcareous spring deposits in continental settings. In: Alonso-Zarza AM, Tanner L (eds) Carbonates in continental settings (Develop sedimentol 62). Elsevier, Amsterdam, pp 175–223

Jones B, Renaut RW (2011) Hot springs. In: Reitner J, Thiel V (eds) Encyclopedia of geobiology. Springer, Berlin, pp 447–451

Jones BJ, Eugster HP, Rettig SF (1977) Hydrochemistry of the Lake Magadi basin, Kenya. Geochim Cosmochim Acta 41:53–72

Jones A, Breuning-Madsen H, Brossard M, Dampha A, Deckers J, Dewitte O, Gallali T, Hallett S, Jones R, Kilasara M, Le Roux P, Micheli E, Montanarella L, Spaargaren O, Thiombiano L, Van Ranst E, Yemefack M, Zougmor R (2013) Soil atlas of Africa. European Commission, Publications Office of the European Union, Luxembourg

Junginger A (2011) East African climate variability on different time scales: the Suguta Valley in the African-Asian Monsoon Domain. Dr. rer. Nat dissertation, Universität Potsdam

Junginger A, Trauth MH (2013) Hydrological constraints of paleo-Lake Suguta in the Northern Kenya Rift during the African Humid Period (15–5 ka BP). Glob Planet Change 111:174–188

Junginger A, Roller S, Olaka LA, Trauth MH (2014) The effects of solar irradiation changes on the migration of The Congo Air Boundary and water levels of paleo-Lake Suguta, Northern Kenya Rift, during the African Humid Period (15–5 ka BP). Palaeogeogr Palaeoclimatol Palaeoecol 396:1–16

Kaboth-Bahr S, Gosling WD, Vogelsang R, Bahr A, Scerri EM, Asrat A, Cohen AS, Düsing W, Foerster V, Lamb HF, Maslin MA, Roberts HM, Schäbitz F, Trauth MH (2021) Paleo-ENSO influence on African environments and early modern humans. Proc Natl Acad Sci U S A 118:e2018277118

Karingithi C, Wambugu, J (2010) The geochemistry of Arus and Bogoria geothermal prospects. Proc World Geotherm Congr, Bali, Indonesia, 25–29 Apr 2010, 6 p

Kennedy AS (2014) Birds of Kenya's Rift Valley. Princeton University Press, Princeton

Kent PE (1942) The Pleistocene Beds of Kanam and Kanjera, Kavirondo, Kenya. Geol Mag 79:114–132

Kindt R, van Breugel P, Lillesø J-PB, Gachathi F, Omondi WO, Jamnadass R, Graudal L (2013) Potential natural vegetation of Eastern Africa (Ethiopia, Kenya, Malawi, Rwanda, Tanzania, Uganda and Zambia). In: Atlas and tree species composition for Kenya. Forest & Landscape Denmark, vol 8. University of Copenhagen. Forest & Landscape Working Papers

Krienitz L (2018) Lesser Flamingos: descendants of Phoenix. Springer, Cham

Kuria Z (2013) Groundwater distribution and aquifer characteristics in Kenya. In: Paolo Paron P, Olago DO, Omuto CT (eds) Kenya: a natural outlook: geo-environmental resources and hazards (Developments in earth surface processes, vol 16). Elsevier, Amsterdam, pp 83–107

Lagat J (2008) Geology and geothermal resource utilization options in the Arus-Lake Bogoria Prospect, Northern Kenya Rift. Proc ARGeo 2 Conference, Entebbe, Uganda, 24–26 Nov 2008, 6 p

Lashkari H, Mohammadi Z, Keikhosravi G (2017) Annual fluctuations and displacements of Inter Tropical Convergence Zone (ITCZ) within the range of Atlantic Ocean-India. Open J Ecol 7:12–33

Lavigne M, Ashley GM (2002) Climatology and rainfall patterns: Lake Bogoria National Reserve (1976–2001). Unpubl Report, Dept Geol Sci, Rutgers University, Piscataway, New Jersey

Lee RKL, Owen RB, Renaut RW, Behrensmeyer AK, Potts R, Sharpe WD (2013) Facies, geochemistry and diatoms of the Late Pleistocene Olorgesailie tufas, southern Kenya Rift. Palaeogeogr Palaeoclimatol Palaeoecol 374:197–217

Lee H, Muirhead JD, Fischer TP, Ebinger CJ, Kattenhorn SA, Sharp ZD, Kianji G (2016) Massive and prolonged deep carbon emissions associated with continental rifting. Nat Geosci 9:145–149

Lysak SV (1992) Heat flow variations in continental rifts. Tectonophysics 208:309–323

Manega PC, Bieda S (1987) Modern sediments of Lake Natron, Tanzania. Sci Géol Bull 40:83–96

Marchant R (2021) East Africa's human environment interactions, historical perspectives for a sustainable future. Palgrave Macmillan, Cham

McCall GJH (1957) Geology and groundwater conditions in the Nakuru area. Tech Rep Ministr Works (Hydraul Br) Kenya 3

Mees F, Stoops G, Van Ranst E, Paepe R, Van Overloop E (2005) The nature of zeolite occurrences in deposits of the Olduvai Basin, northern Tanzania. Clay Clay Miner 53:659–673

Miall A (2014) Fluvial depositional systems. Springer, Cham

Ng B, Cai W, Cowan T, Bi D (2018) Influence of internal climate variability on Indian Ocean Dipole properties. Sci Rep 8:1–8

Nicholson SE (1996) A review of climate dynamics and climate variability in eastern Africa. In: Johnson TC, Odada EO (eds) The limnology, climatology and paleoclimatology of the East African lakes. Gordon and Breach, Amsterdam, pp 25–56

Nicholson SE (2017) Climate and climatic variability of rainfall over eastern Africa. Rev Geophys 55:590–635

Nicholson SE (2018) The ITCZ and the seasonal cycle over equatorial Africa. Bull Am Meteorol Soc 99:337–348

Nicholson SE, Selato JC (2000) The influence of La Nina on African rainfall. Int J Climatol 20:1761–1776

Nyblade AA, Pollack HN, Jones DL, Podmore F, Mushayandebvu M (1990) Terrestrial heat flow in east and southern Africa. J Geophys Res 95:17371–17384

O'Donnell MS, Ignizio DA (2012) Bioclimatic predictors for supporting ecological applications in the conterminous United States. U.S. Geol Surv Data Ser 691

Obando JA, Onywere S, Shisanya C, Ndubi A, Masiga D, Irura Z, Mariita N, Maragia H (2016) Impact of short-term flooding on livelihoods in the Kenya Rift Valley lakes. In: Meadows ME, Lin J-C (eds) Geomorphology and society, advances in geographical and environmental sciences. Springer, Japan, pp 193–215

Olaka LA, Kasemann SA, Sültenfuß J, Wilke FDH, Olago DO, Mulch A, Musolff A (2022) Tectonic control of groundwater recharge and flow in faulted volcanic aquifers. Water Resour Res 58:e2022WR032016

Onywere SM, Shisanya CA, Obando JA, Ndubi AO, Masiga D, Irura Z, Mariita N, Maragia HO (2013) Geospatial extent of 2011–2013 flooding from the Eastern African Rift Valley lakes in Kenya and its implication on the ecosystems. Papers, Kenya Soda Lakes Workshop, Kenya Wildlife Service Training Institute, Naivasha, Kenya

Osmaston H (2004) Quaternary glaciations in the East African mountains. In: Ehlers J, Gibbard PL (eds) Quaternary glaciations extent and chronology, Part C: Part III: South America, Asia, Africa, Australasia, Antarctica (Developments in Quaternary science, vol 2). Elsevier, Amsterdam, pp 139–150

Oswald F (1914) The Miocene Beds of Victoria Nyanza and the geology of the country between the Lake and the Kisii Highlands. Q J Geol Soc Lond 70:128–198

Owen RA, Owen RB, Renaut RW, Scott JJ, Jones B, Ashley G (2008) Mineralogy and origin of rhizoliths on the margins of saline, alkaline Lake Bogoria, Kenya Rift Valley. Sediment Geol 203:143–163

Owen RB, Renaut RW, Behrensmeyer AK, Potts R (2014) Quaternary geochemical stratigraphy of the Kedong-Olorgesailie section of the southern Kenya Rift Valley. Palaeogeogr Palaeoclimatol Palaeoecol 396:194–212

Owen RB, Renaut RW, Muiruri VM, Rabideaux NM, Lowenstein TK, McNulty EP, Leet K, Deocampo D, Luo S, Deino AL, Cohen A, Sier MJ, Campisano C, Shen C-C, Billingsley A, Mbuthia A, Stockhecke M (2019) Quaternary history of the Lake Magadi Basin, southern Kenya Rift: tectonic and climatic controls. Palaeogeogr Palaeoclimatol Palaeoecol 518:97–118

Pentecost A (2005) Travertine. Springer, Berlin

Pentecost A, Jones B, Renaut RW (2003) What is a hot spring? Can J Earth Sci 40:1443–1446

Pirajno F (2009) Hydrothermal mineral deposits, principles and fundamental concepts for the exploration geologist. Springer, Berlin

Renaut RW (1982) Late Quaternary geology of the Lake Bogoria fault-trough, Kenya Rift Valley. PhD thesis, University of London

Renaut RW (1993) Zeolitic diagenesis of late Quaternary fluviolacustrine sediments and associated calcrete formation in the Lake Bogoria Basin, Kenya Rift Valley. Sedimentology 40:271–301

Renaut RW, Jones B (1997) Controls of aragonite and calcite precipitation in hot spring travertines at Chemurkeu, Lake Bogoria, Kenya. Can J Earth Sci 34:801–818

Renaut RW, Jones B (2000) Microbial precipitates around continental hot springs and geysers. In: Riding R, Awramik S (eds) Microbial sediments. Springer, Berlin, pp 187–195

Renaut RW, Jones B (2011) Sinter. In: Reitner J, Thiel V (eds) Encyclopedia of geobiology. Springer, Berlin, pp 808–813

Renaut RW, Owen RB (1988) Opaline cherts associated with sublacustrine hydrothermal springs at Lake Bogoria, Kenya Rift Valley. Geology 16:699–702

Renaut RW, Owen RB (2005) The geysers of Lake Bogoria, Kenya Rift Valley, Africa. GOSA Trans 9:4–18

Renaut RW, Jones B, Tiercelin J-J (1998) Rapid *in situ* silicification of microbes at Loburu hot springs, Lake Bogoria, Kenya Rift Valley. Sedimentology 45:1083–1103

Renaut RW, Jones B, Le Turdu C (1999) Calcite lilypads and ledges at Lorusio Hot Springs, Kenya Rift Valley: travertine precipitation at the air-water interface. Can J Earth Sci 36:649–666

Renaut RW, Jones B, Tiercelin J-J, Tarits C (2002a) Sublacustrine precipitation of hydrothermal silica in rift lakes: evidence from Lake Baringo, central Kenya Rift Valley. Sediment Geol 148:235–257

Renaut RW, Morley CK, Jones B (2002b) Fossil hot-spring travertine in the Turkana Basin, northern Kenya: structure, facies, and genesis. In: Renaut RW, Ashley GM (eds) Sedimentation in continental rifts. SEPM Spec Publ vol 73, Tulsa, pp 123–141

Renaut RW, Owen RB, Ego J (2008) Recent changes in geyser activity at Loburu, Lake Bogoria, Kenya Rift Valley. GOSA Trans 10:4–14

Renaut RW, Owen RB, Jones B, Tiercelin J-J, Tarits C, Ego JK, Konhauser KO (2013) Impact of lake-level changes on the formation of thermogene travertine in continental rifts: evidence from Lake Bogoria, Kenya Rift Valley. Sedimentology 60:428–468

Renaut RW, Owen RB, Ego J (2017) Geothermal activity and hydrothermal mineral deposits at southern Lake Bogoria, Kenya Rift Valley: impact of lake level changes. J Afr Earth Sci 129:623–646

Renaut RW, Owen RB, Lowenstein TK, De Cort G, McNulty EP, Scott JJ, Mbuthia A (2021) The role of hydrothermal fluids in sedimentation in saline alkaline lakes: evidence from Nasikie Engida, Kenya Rift Valley. Sedimentology 68:108–134

Rimstidt JD, Cole DR (1983) Geothermal mineralization: I, The mechanism of formation of the Beowawe, Nevada, siliceous sinter deposit. Am J Sci 283:861–875

Robertson E, Biggs J, Edmonds M, Clor L, Fischer TP, Vye-Brown C, Kianji G, Koros W, Kandie R (2016) Diffuse degassing at Longonot volcano, Kenya: implications for CO_2 flux in continental rifts. J Volcanol Geotherm Res 327:208–222

Rodgers KA, Cook KL, Browne PRL, Campbell KA (2002) The mineralogy, texture and significance of silica derived from alteration by steam condensate in three New Zealand geothermal fields. Clay Miner 37:299–322

Roure Johnson C, Ashley GM, De Wet CB, Dvoretsky R, Park L, Hover VC, Owen RB, McBrearty S (2009) Tufa as a record of perennial fresh water in a semi-arid rift basin, Kapthurin Formation, Kenya. Sedimentology 56:1115–1137

Saji NH, Goswami BN, Vinayachandran PN, Yamagata T (1999) A dipole mode in the Indian Ocean. Nature 401:360–363

Sceal J (1974) The geology of Paka Volcano and the country to the east, Baringo District, Kenya. PhD thesis, University of London

Schagerl M, Renaut RW (2016) Dipping into the soda lakes of East Africa. In: Schagerl M (ed) Soda lakes of East Africa. Springer, Cham, pp 3–24

Schumm SA (1985) Patterns of alluvial rivers. Annu Rev Earth Planet Sci 13:5–27

Shvartsev SL, Lepokurova OE, Kopylove YG (2007) Geochemical mechanisms of travertine formation from fresh waters in southern Siberia. Russ Geol Geophys 48:659–667

Spawls S, Matthews G (2012) Kenya: a natural history. T and AD Poyser, London

Stevenson T, Fanshaw J, Gale J, Small B (2020) Birds of East Africa, 2nd edn. Bloomsbury, London

Street FA, Grove AT (1979) Global maps of lake-level fluctuations since 30,000 yr B.P. Quat Res 12:83–118

Sturchio NC, Dunkley PN, Smith M (1993) Climate-driven variations in geothermal activity in the northern Kenya Rift Valley. Nature 362:233–234

Tallon PW (1976) The stratigraphy, palaeoenvironments and geomorphology of the Pleistocene Kapthurin Formation, Kenya. PhD thesis, University of London

Verschuren D, Sinninghe Damsté JS, Moernaut J, Kristen I, Blaauw M, Fagot M, Haug GH, van Geel B, de Batist M, Barker P, Vuille M, Conley DJ, Olago DO, Milne I, Plessen B, Eggermont H, Wolff C, Hurrell E, Ossebaar J, Lyaruu A, van der Plicht J, Cumming BF, Brauer A, Rucina SM, Russell JM, Keppens E, Hus J, Bradley RS, Leng M, Mingram J, Nowaczyk NR (2009) Half-precessional dynamics of monsoon rainfall near the East African Equator. Nature 462:637–641

Wainwright CM, Finney DL, Kilavi M, Black E, Marsham JH (2021) Extreme rainfall in East Africa, October 2019–January 2020 and context under future climate change. Weather 76:26–31

Walsh J, Dodson RG (1969) Geology of northern Turkana. Rep Geol Surv Kenya 82:1–38

Watene G, Yu L, Nie Y, Zhu J, Ngigi T, Nambajimana J, Kenduiywo B (2021) Water erosion risk assessment in the Kenya Great Rift Valley region. Sustainability 13:844. https://doi.org/10.3390/su13020844

Wenhaji Ndomeni C, Cattani E, Merino A, Levizzani V (2018) An observational study of the variability of East African rainfall with respect to sea surface temperature and soil moisture. Q J R Meteorol Soc 144:384–404

Wheildon J, Morgan P, Williamson KH, Evans TR, Swanberg CA (1994) Heat flow in the Kenya rift zone. Tectonophysics 236:131–149

World Resources Institute (2007) Nature's benefits in Kenya: an atlas of ecosystems and human well-being. World Resources Institute, Nairobi/Washington

Zeuner FE (1950) Frost soils on Mount Kenya and the relation of frost soils to aeolian deposits. J Soil Sci 1:20–30

Chapter 5
Lake Processes and Sedimentation

5.1 The Kenya Rift Lakes and Their Basins

The lakes of the East African Rift System (EARS) (Fig. 5.1a) are highly diverse. The lakes in the Kenya Rift range from fresh (Naivasha, Baringo) to hypersaline (Magadi, Nakuru, Elmenteita, Logipi) and from the very shallow Nasikie Engida (~2 m deep; 7 × 2 km) to the very large Lake Turkana (110 m deep; 250 × 45 km). Their sediments are similarly varied, including siliciclastics, carbonates, cherts, evaporites, diatomaceous deposits, and organic-rich facies. The lakes contrast strongly with the mainly freshwater bodies of the western branch of the EARS where rainfall is sufficient to fill large subsiding basins such as Lake Tanganyika (1470 m; 630 × 70 km) with sedimentary piles that are up to several kilometres thick (Yuretich 1982).

All the major EARS lakes, both past and present, fill tectonic depressions or volcanic dammed basins, with several smaller floodplain lakes (Kamnarok, Kichirtit) and crater lakes (Sonachi, Simbi, Kangirinyang) also present. Sedimentation reflects tectonic and hydrological controls as well as climatic and biological processes that operate over a variety of time scales (Fig. 5.1b). Sediment inputs and dispersal reflect the varied internal dynamics of a lake (Cohen 2003; Renaut and Gierlowski-Kordesch 2010), including the development of wind-, wave- and gravity-driven currents (density currents, overflows) that redistribute sediment. Wind mixing, waves, and the formation of seiches also impact the development and depth of stratification. Drainage may be open or closed with major impacts on water chemistry and clastic, chemical and organic deposition. The evaporation/precipitation ratio on the rift floor, together with inflow from distal locations and from groundwater, influences the volume of water and sediment entering a lake. Rain delivers solutes such as ocean-derived Cl^- aerosols directly to the lakes with associated winds delivering clastic (silt and finer) particles. Climate, light penetration and nutrient delivery also impact the biology of lakes and the accumulation of autochthonous (phytoplankton

R. W. Renaut, R. B. Owen, *The Kenya Rift Lakes: Modern and Ancient*, Syntheses in Limnogeology, https://doi.org/10.1007/978-3-642-25055-2_5

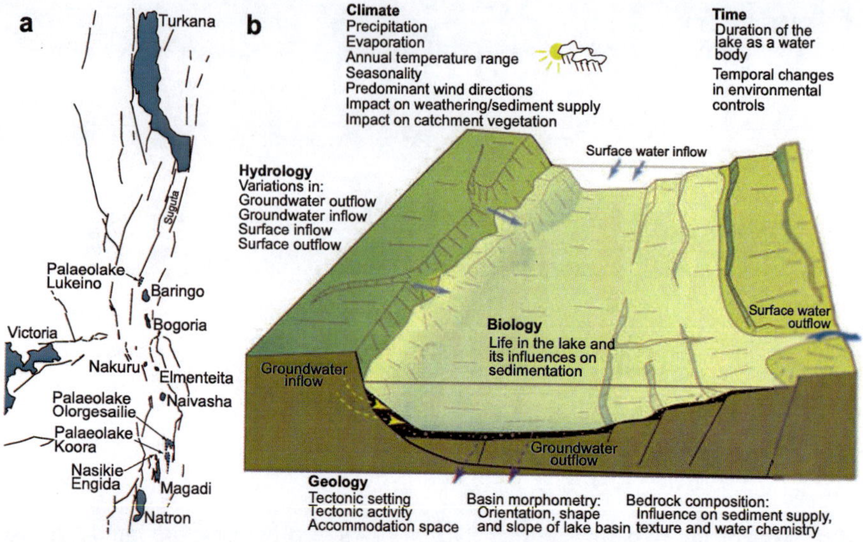

Fig. 5.1 Major controls on rift valley lacustrine sedimentation. Revised from Renaut and Gierlowski-Kordesch (2010). (**a**) Location of major lakes and palaeolakes. (**b**) Major controls on deposition

including diatoms, infauna) and allochthonous (terrestrial plant debris, pollen, spores) components to lake sediments.

Carroll and Bohacs (1999) and Bohacs et al. (2000, 2003) distinguished *overfilled*, *underfilled* and *balanced fill* lakes in several tectonic settings, noting that these were controlled by an equilibrium between sediment + water supply and changes in accommodation space resulting from basin subsidence. Large, overfilled lakes are absent in the modern Kenya Rift, but many lakes rose to their outlet levels and formed fresh or slightly brackish palaeolakes during the African Humid Period (AHP: ~15–5 ka).

Overfilled conditions develop when water and sediment supply exceed the available accommodation space, leading to overflow into a fluvial drainage network and broadly static base levels with the development of progradational parasequences (Fig. 5.2a, d). Erosion occurs during temporary regressions with mixed fluvial and lacustrine facies dominating. These facies include offshore pelagic mud, turbidites, nearshore deltaic sand and silt, littoral sand and coquinas, as well as fluvial sands and conglomerates (Cohen 2003; Renaut and Gierlowski-Kordesch 2010). Evaporation and plankton blooms that remove CO_2 during photosynthesis can lead to minute calcite crystals forming in the water column that settle under gravity to produce micritic mud or mixed carbonate-siliciclastic mud. Mixed terrigenous and planktonic materials that settle in the deeper anoxic zones of stratified lakes produce organic-rich mud containing several percent organic matter (TOC: Total Organic Carbon) that potentially can generate petroleum after shallow burial and heating (Talbot 1988).

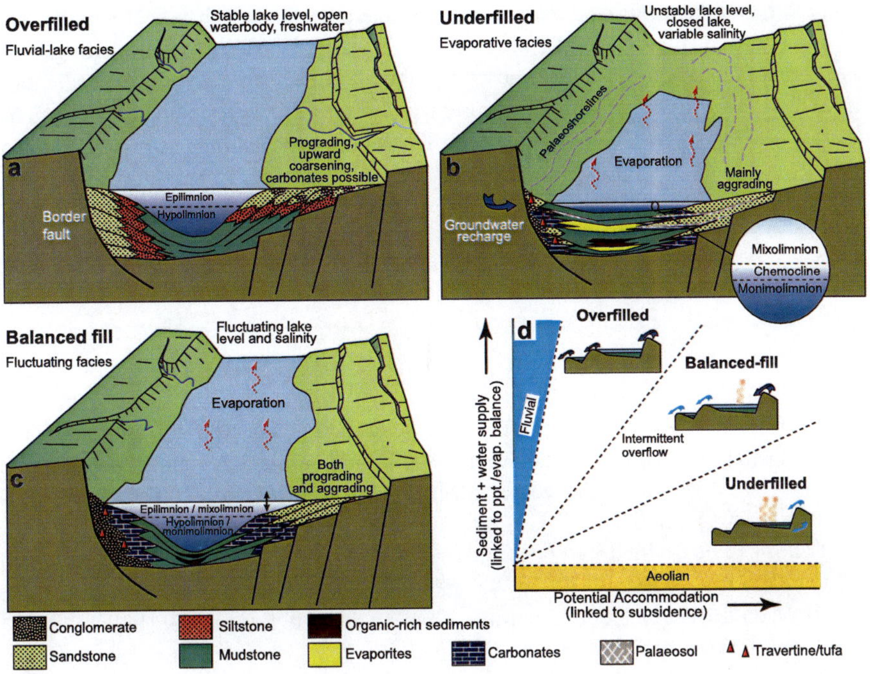

Fig. 5.2 Lake basin classification. Modified from Bohacs et al. (2000). (**a**) Schematic model of an overfilled lake with lacustrine sediments and prograding fluvial/delta deposits. (**b**) Schematic model of an underfilled basin with aggradational sedimentation and evaporative facies at the basin centre. (**c**) Schematic model of a balanced fill basin showing both progradation and aggradation. (**d**) Controlling factors on basin type include sediment + water supply and the development of accommodation space produced by faulting or volcanic damming

Underfilled lakes develop when water and sediment inputs are low compared with the creation of accommodation space (Fig. 5.2b, d) and characterise many modern Kenya Rift lakes. River inflow is commonly ephemeral, and springs, both ambient and hot, may contribute much of the annual recharge. Such water bodies range from seasonally flooded saline playas to perennial saline lakes. These underfilled lakes are characterised by hydrologically-closed surface and subsurface drainage with frequent shoreline changes in response to varying climate-controlled water inflow. Overall, aggradation typically dominates, producing thin parasequences of evaporative facies. During low stands, deposition is dominated by evaporites, mudflat sediments and aeolianites, which may also be overprinted by palaeosols with common desiccation structures. High stands lead to thicker deposits including laminites, micritic mud, evaporites, siliceous sediments, oncolites, stromatolites and siliciclastic shoreline deposits. Organic productivity is commonly very high, but the preservation of organic matter may be low due to frequent oxidation and erosion. Alluvial fans often form adjacent to steep rift scarps.

Balanced fill basins form with intermediate water and sediment supply rates and moderate accommodation (Fig. 5.2c, d). Both progradational and aggradational geometries develop in response to changing lake levels, which reflect variable climates through time. Drainage may be open or closed, leading to alternating fresh and saline lakes on variable timescales. Thermal and (or) chemical stratification may develop with the development of anoxic bottom waters, even in shallow (<2 m) settings. Many lithologies accumulate including, for example, conglomerates, sands, silts, micritic or organic-rich mud, carbonates and evaporites.

5.2 Lacustrine Processes

Our focus is on physical and chemical processes in tropical rift lakes. Biological aspects are discussed at length in several volumes on limnology including Hutchinson (1975a, b, c), Wetzel (2001), Cole and Weihe (2016), and Kalff and Downing (2016). Limnological studies specific to tropical Africa include Beadle (1974, 1981), Johnson and Odada (1996), Odada and Olago (2002), Talling and Lemoalle (2011), Schagerl (2016), and El-Sheek and Elsaied (2023).

Lakes can be divided into distinct zones that reflect depth, light penetration, and water chemistry (Beadle 1981; Wetzel 2001; Renaut and Gierlowski-Kordesch 2010; McGlue 2021). Light penetrates down to the lake floor in the shallow littoral and sublittoral zones around lake margins (Fig. 5.3). However, the depth of light penetration varies with contrasts in suspended sediment (turbidity), plankton, dissolved solids, and the incident angles of solar radiation (Cohen 2003; Melack and MacIntyre 2016). Hennesey et al. (2020) noted that turbid rift lakes (e.g. Lake Baringo) were at least partly controlled by pH and salinity, and tended to develop where the former factor was moderate and the latter was low, which inhibited flocculation of clay particles. Tarits et al. (2006) showed that clay flocculation was linked to higher salinity.

The physical setting also controls light levels within rift lakes with gently sloping fault ramps or platforms extending the areal extent of the littoral zone and with

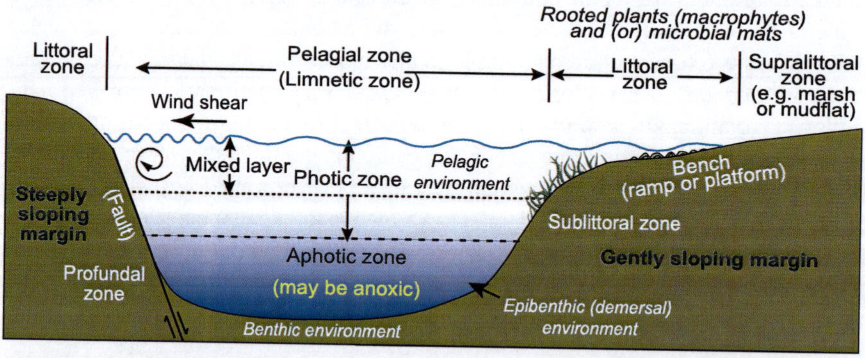

Fig. 5.3 Lacustrine environmental zones. Modified from Cohen (2003) and Renaut and Gierlowski-Kordesch (2010)

faults defining steeper slopes that lead into the deeper profundal zone, which lies below the limits of light penetration. In contrast, the pelagic zone lies in the upper water column and is characterised by sufficient light to allow photosynthesis (Fig. 5.3).

Rift morphology may also determine major changes in processes and sedimentation over time, especially in 'amplifier lakes'. These are found in basins with steep faulted margins (Trauth et al. 2010; Olaka et al. 2010) that separate high rainfall areas on mountainous rift shoulders from low-lying rift floor segments where evaporation rates exceed precipitation (Fig. 5.4). Under these conditions major lake-level fluctuations take place as climate changes. For example, the arid Suguta Valley

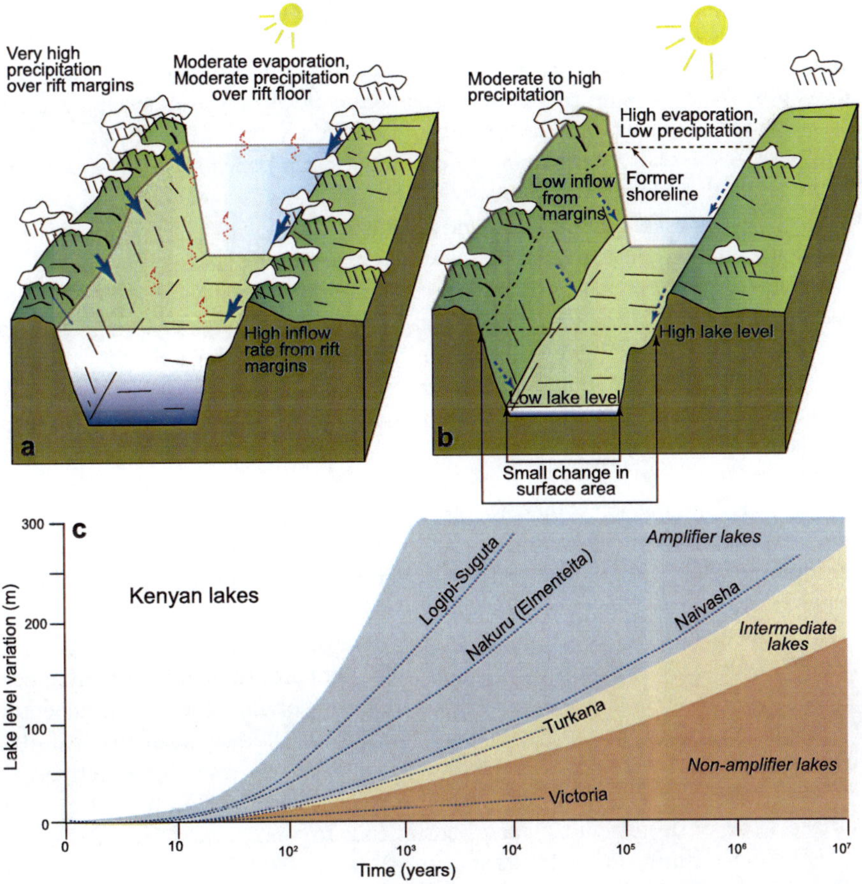

Fig. 5.4 Development of amplifier lakes. These show major variations in lake level and are found in steep-sided basins with evaporation > precipitation on the rift floor and with high inflows from wetter adjacent rift mountain shoulders. (**a**) High-lake stage. (**b**) Low-lake stage. (**c**) Models of lake level variability for contrasting Kenya Rift lakes plotted against differing time scales. Simplified from Trauth et al. (2010)

is today occupied by the very shallow Lake Logipi, but during the AHP was filled by a ~300-m-deep water body.

Sedimentation processes in lakes are partly driven by winds, density contrasts, variations in air pressure and by the inertial inflow of rivers, with tidal effects being minor (Csanady 1978; Sly 1978; Lerman et al. 1995; Håkanson and Jansson 2002; Bengtsson 2022). Coriolis effects may be discernible in the largest lakes but are absent in most Kenya Rift lakes, except Lake Victoria and possibly Lake Turkana. Wind-driven waves redistribute nutrients and gases in the water column and thus impact biological sedimentation. They also directly control particle movement and the accumulation of sediments in coastal areas where depths are shallower than half the wavelength (the wave base). Winds may also drive fine suspended sediments derived from rivers as low-density overflows, which then settle to the lake floor as pelagic deposits. In contrast, lake currents are generally incapable of transporting coarse bedload, except on steeper margins where river-fed or earthquake-induced density (turbidity) currents form turbulent underflows of sediment and water that move down steeper marginal slopes. Denser debris flows may also develop that can transport very large particles, depositing unstratified layers of debris (Cohen 2003). All these processes redistribute sediment in the basin and may lead to sediment focusing, with sedimentation rates increasing in various parts of a lake while remaining low elsewhere (Horppila and Niemisto 2008).

Solar radiation and geothermal sources provide heat to lakes. Depending on water clarity and light penetration, the former directly affects the upper few tens of metres while the latter may introduce heat from adjacent marginal springs and via the lake floor (Cohen 2003; Boehrer and Shultze 2008). Solar heating is more important in most lakes, but geothermal heating can be significant in rift settings, especially in underfilled basins. Solar heating results in the development of a stratified lake (Fig. 5.5a) with an upper warmer (less dense) surface layer termed the epilimnion that overlies a lower, cooler and denser hypolimnion (Beadle 1981; Wetzel 2001). Temperatures increase rapidly upwards within the metalimnion (thermocline), which lies between the epilimnion and hypolimnion. Wind mixing may also transport solar heat to greater depths, assuming the energy is sufficient to overcome density contrasts in stratified waters (Cohen 2003).

In temperate lakes, temperature-controlled density changes may lead to lake mixing as surface water cools once (monomictic) or twice (dimictic) or more times (polymictic) each year. Some tropical lakes are warm polymictic and mix frequently (e.g. northern Lake Turkana: Yuretich and Cerling 1983). Many shallow equatorial lakes may stratify weakly during daytime heating but mix when the water undergoes radiative cooling later in the day (diel cycles: Melack 2009). In contrast, the surface water of deep tropical water bodies tends to remain warm for most of an annual cycle, resulting in stable stratification (Beadle 1981), which is also favoured by the steep-sided basin morphology of rift lakes. If permanently stratified, these waters are referred to as meromictic (Hakala 2004; Stewart et al. 2009; Hall and Northcote 2012). Lake Tanganyika, for example, has a permanent thermocline, whereas the very windy and shallower Lake Turkana is often thoroughly wind mixed with only temporary thermoclines developing (Fig. 5.5a). Stratification

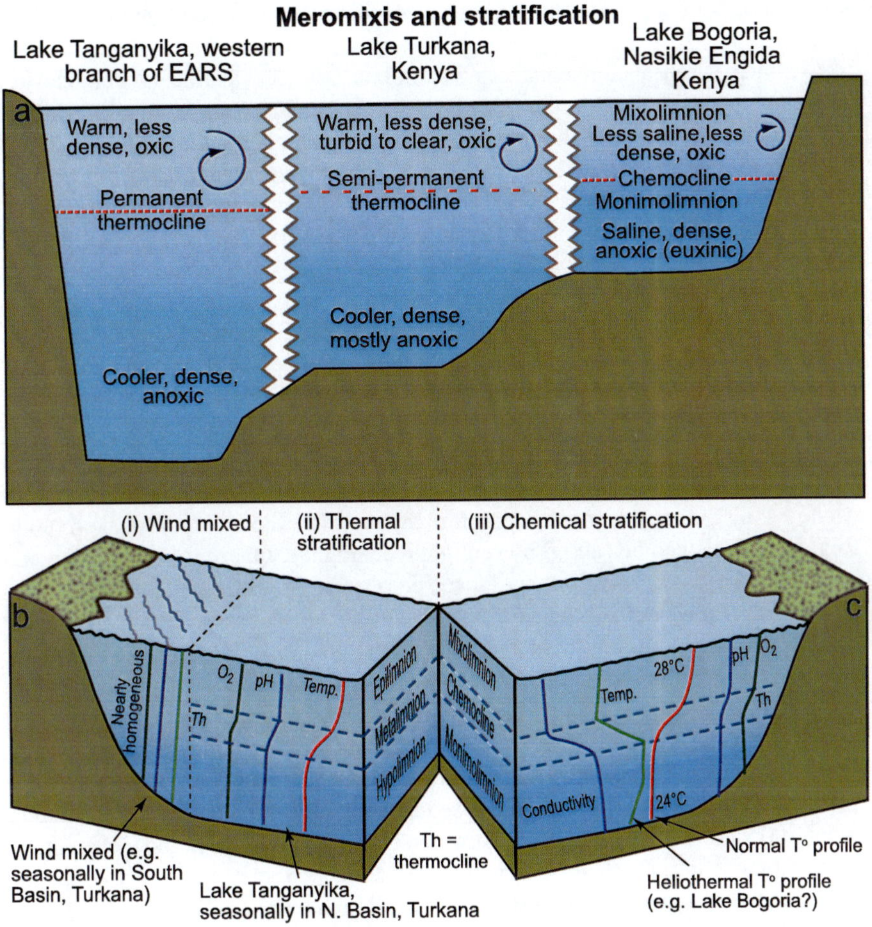

Fig. 5.5 Lacustrine stratification. Modified from Renaut and Gierlowski-Kordesch (2010). (**a**) Comparison of thermal and chemical stratification terminology and the development of thermoclines and chemoclines in contrasting lakes. Meromictic lakes show long-term stability in stratification. (**b**) Wind-mixed water-column profiles (left) compared with thermal stratification. (**c**) Chemical stratification. Note the contrast in heliothermal lake temperature profile with heat trapped in the upper monimolimnion

isolates bottom waters from the atmosphere, which then become dysoxic or anoxic following bacterial oxygen depletion (Talbot 1988; Renaut and Gierlowski-Kordesch 2010). In turn, this often leads to the preservation of organic matter and lamination in bottom sediments due to the absence of burrowing infauna. In some cases, steady axial winds over the larger and deeper lakes tilt the surface and internal boundaries such as thermoclines, producing seiches or standing waves (Cohen 2003). This causes surface water to pile up at the downwind end of elongate rift lakes with gravity causing a return flow of deep water that rises at the upwind end, bringing nutrients to the surface. Seiches may also cause mixing along internal water-mass boundaries causing a return of nutrients to the epilimnion (Plisnier et al. 1999).

Chemical stratification enhances meromixis with less saline (lower density) upper layers of the mixolimnion separated by a chemocline from more saline (higher density) bottom waters of the monimolimnion. The higher salinity and density of the latter may be caused by the dissolution of pre-existing evaporites or inflow from saline springs (shoreline or sublacustrine) (Renaut and Tiercelin 1994). Alternatively, evaporative concentration during dry intervals may increase salinity and density with subsequent freshwater inflow lying on top of the pre-existing saline water. Further evaporation may then increase the mixolimnion density, which will eventually facilitate mixing of the water column (Renaut and Gierlowski-Kordesch 2010).

Lake Bogoria in the central Kenya Rift represents a good example of a meromictic water body with anoxic bottom waters (Tiercelin and Vincens 1987; Renaut and Tiercelin 1994; De Cort et al. 2018). Lake water bottom temperature surveys have shown patches of hot water in its deeper parts. This is most likely due to geothermal seepage, but there remains the possibility of a heliothermal contribution with temperatures in the upper monimolimnion higher than those of the mixolimnion (Fig. 5.5c). This can be caused by sunlight reaching the upper monimolimnion and heating up the water, which cannot escape because of the dense highly saline conditions. Sonnenfeld and Hudec (1980) associated such lakes with NaCl-rich and sulfate-rich waters, noting that a minimum density contrast between the two water layers of 15 g L^{-1} is required. Although a carbonate lake, the density contrasts in Bogoria often exceed this value (Tiercelin et al. 1980) so perhaps this might occasionally be an active process, but this is speculation. If evaporation is extreme, euxinic conditions may develop with combined anoxia, abundant sulfide, and elevated salinity, as appears to have been the case intermittently in Palaeolake Magadi after about 700 ka (Owen et al. 2019; Deocampo et al. 2022). Even shallow chemically stratified saline lakes such as Nasikie Engida (<2 m deep) may develop stratification and bottom water anoxia (De Cort et al. 2019; Renaut et al. 2021).

5.3 Aquatic Chemical and Biochemical Processes

In most freshwater lakes and ponds water leaves the basin through a surface outlet. Such lakes are present where annual recharge from precipitation, runoff and groundwater is balanced or exceeded by losses through outflow and evaporation (Fig. 5.6a) (Wetzel 2001; Renaut and Gierlowski-Kordesch 2010). Two large Kenya Rift lakes, Baringo and Naivasha, have no surface outlet but maintain fresh water. Evidence from several sources implies that these lakes must have significant subsurface outflow of water and solutes through faults, fractures, and permeable sediments or volcanic rocks (Fig. 5.6b; Chaps. 8 and 11). Such lakes are topographically closed (no surface outflow) but hydrologically open (water nonetheless escapes). Today, most Kenya Rift lakes lie in hydrologically closed basins in which evaporation exceeds inflow and direct precipitation on the lakes, resulting in the formation of

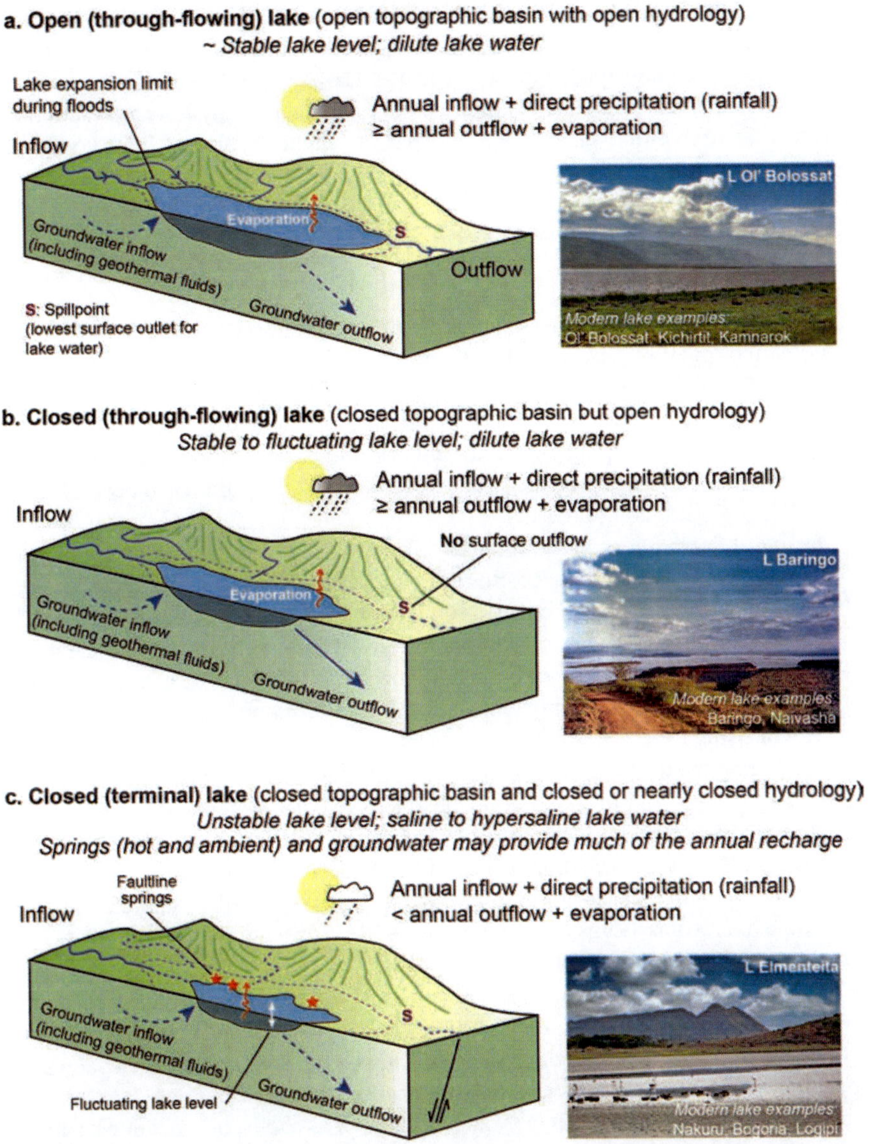

Fig. 5.6 Hydrological conditions in open and closed (through-flowing and terminal) lakes

brackish to hypersaline water bodies of sodium carbonate composition (Fig. 5.6c). Topographic closure is easily demonstrated but hydrological closure is more difficult to confirm.

Saline 'soda lakes' are dominated by Na^+ and carbonate species (HCO_3^-, CO_3^{2-}). They have high pH (>9), mainly reflecting silicate hydrolysis of volcanic rocks (alkali basalt, trachyte, phonolite, related tephra and rare local carbonatites) that form the bedrock in most basins (Eugster and Hardie 1978; Kilham 1990; Deocampo and Renaut 2016; Schagerl and Renaut 2016; Lameck et al. 2023). In silicate hydrolysis, silicate minerals react with carbonic acid (H_2CO_3) produced mainly from rain and organic matter decay in soils (Sect. 4.4.2). An idealised hydrolysis reaction for weathering of Na-rich plagioclase feldspar, a common mineral in the volcanic rocks, is:

$$\underset{Na\text{–feldspar}}{2\mathrm{NaAlSi_3O_{8(s)}}} + 2\mathrm{H}^+_{(aq)} + 9\mathrm{H_2O_{(l)}} \rightarrow \underset{\text{Kaolinite}}{Al_2Si_2O_5(OH)_{4(s)}} + 2Na^+_{(aq)} + \underset{\text{Silicic acid}}{4\mathrm{H}_4SiO_{4(aq)}}$$

This reaction leads to the consumption of H^+ from solution and releases alkali cations, which together raise the pH. Bicarbonate (HCO_3^-) is released as H_2CO_3 is consumed, which increases alkalinity. New ('neoformed') clay minerals, zeolites and ferric oxyhydroxides form at the weathering site, with dissolved Na^+, K^+, Ca^{2+} and SiO_2 (as H_4SiO_4) transported lakewards by dilute runoff and groundwaters. The clay minerals or zeolites produced in the reactions vary with fluid and bedrock composition, and weathering intensity.

Spring inflows are minimal in some lakes (e.g. Turkana) but are an important contributor in others (e.g. Logipi, Bogoria, Nasikie Engida, Magadi) (Dunkley et al. 1993). Meteoric (ambient to warm: <45 °C) springs are generally dilute Na^+–HCO_3^- or Na^+–Ca^{2+}–HCO_3^- waters. In contrast, hotter (~60–97 °C) springs are saline and silica enriched, with CO_2 and sometimes sulfur gases present. The hot fluids interact with subsurface bedrock adding other solutes (cf. Pecoraino et al. 2015). The initial acidity of rising CO_2-rich fluids is often neutralised by reactions with bedrock, akin to weathering, resulting in Na^+ and HCO_3^--rich fluids that discharge at the surface.

Local geology commonly controls the anions in basin waters with sulfate resulting from oxidation of sulfides (e.g. pyrite) in parent rocks, marine aerosols, or from volcanogenic sources including gases. Chloride derives from marine aerosols carried from the Indian Ocean by monsoon winds, supplemented by halite dissolution and fluid inclusions (Eugster and Hardie 1978). Subsequently, these solutes are fractionated by a variety of processes.

Calcium and much of the magnesium are usually the first cations lost from inflow waters through precipitation of calcite, aragonite and dolomite in soils (carbonate concretions, rhizoliths), in sediments (as cement, beachrock, carbonate coatings on pebbles, calcrete), at spring vents (as tufa or travertine) and in dilute lakes (e.g. as fine calcite crystals forming in the water column, or as shells, microbialites, and fish bone) that later evaporate to become saline (Eugster and Hardie 1978; Deocampo and Jones 2014; Deocampo and Renaut 2016). Magnesium is often removed as Mg-silicates in hot springs (Renaut et al. 2013) and in alkaline lakes (Jones and Weir 1983; Deocampo 2005; Chase et al. 2021).

As inflow waters evaporate, they become more saline, and can follow one of several compositional pathways and pass through a series of critical chemical divides (Fig. 5.7). After calcite precipitation, the water is then relatively enriched in Ca^{2+} or alkalinity unless they were equimolar in the inflow. This represents a major chemical divide. If Ca^{2+} exceeds alkalinity in dilute inflow after calcite begins to

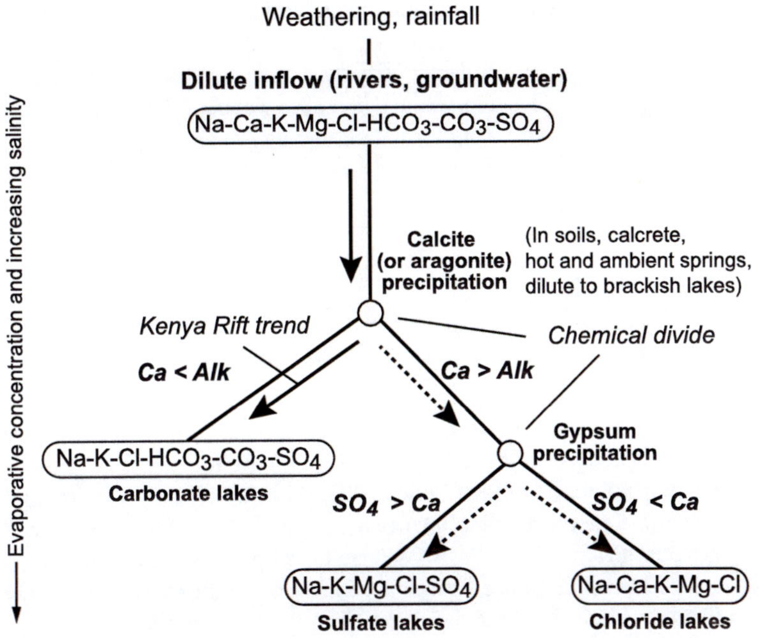

Fig. 5.7 Chemical evolution of Kenya Rift lakes. Dilute inflows evolve towards a series of chemical divides. Kenya surface and groundwater lose Ca^{2+} to soils, calcretes and tufa and gradually shift towards the left of the diagram, eventually producing Na–K–Cl–HCO_3–CO_3–SO_4 lake waters

precipitate, gypsum will next start to precipitate from the saline, carbonate-depleted water, leading to a second chemical divide, after which an excess of SO_4^{2-} over Ca^{2+} leads to the development of a sulfate lake. Alternatively, if there is an excess of Ca^{2+} over SO_4^{2-} after gypsum forms a chloride lake will result (Eugster and Hardie 1978; Eugster and Jones 1979).

In the Kenya Rift lakes, however, carbonate ions (HCO_3^- and CO_3^{2-}) exceed Ca^{2+} and Mg^{2+} after early alkaline-earth precipitation. Consequently, the inflowing waters follow a different path (Fig. 5.7) that depletes any remaining calcium and magnesium by alkaline earth carbonate precipitation, resulting in the formation of Na-rich alkaline water (Na^+–K^+–Cl^-–HCO_3^-–CO_3^{2-}–SO_4^{2-}) or a 'sodium carbonate lake' (soda lake). The Kenya lake brines have substantial Cl^-, but SO_4^{2-} is often lost after bacterial reduction, releasing H_2S and producing pyrite or its precursor if iron is available (Eugster and Hardie 1978; Renaut and Gierlowski-Kordesch 2010). Further evaporation may then lead to precipitation of trona ($Na_2CO_3 \cdot NaHCO_3 \cdot 2H_2O$) and nahcolite ($NaHCO_3$), especially where magmatic (mantle) CO_2 also enters the lake system via faults. After extreme evaporation, halite (NaCl) may eventually precipitate as a late-stage product. In the absence of Ca^{2+}, fluoride may increase to high concentrations in lake waters and may react with existing carbonate minerals (calcite, aragonite) to produce fluorite (CaF_2) or form villiaumite (NaF) in concentrated brines. Silica is very soluble in highly alkaline brines ($pH > 9$) and may precipitate as opaline silica (opal-A) or sodium silicate minerals if the pH falls abruptly or the water evaporates. During their evolution, saline lake brines may interact with lake sediments and form types of cement and early diagenetic minerals such as clays and zeolites, changing the solutes in the lake and pore waters. Potassium is often lost to sorption on active surfaces or the formation of clay minerals but behaves conservatively in some brines (Eugster and Jones 1979).

5.4 Sedimentary Facies

5.4.1 *Rift Geomorphology and Sedimentation*

Rifts are often characterised by alternating half-grabens with a major border fault escarpment that is opposed by a gentler shoaling margin comprised of grabens, tilt blocks and monoclines (Fig. 5.8a; Cohen 2003). The different tectonic settings in rifts generate contrasting geomorphic environments that are, in turn, associated with a variety of facies (Tiercelin 1990). Border fault escarpments shed coarse clastic sediments into rift basins, resulting in stacked sequences of alluvial fans along with more distal offshore turbidites. These are common in the larger rift lakes but are less frequent in the smaller water bodies of the Kenya Rift. The opposing shoaling margins are associated with gentler gradients and may develop broad littoral areas. Both siliciclastic- and carbonate-dominated sediments may accumulate in such settings. The gentler slopes associated with these areas result in major lateral changes as lake levels fluctuate, which can also result in dramatic shifts in the focus of sedimentation.

Rift-axial lake margins are often the site of river inflows and accumulate fluvial and deltaic sediments, such as in the Omo Valley to the north of Lake Turkana. In overfilled lakes, they can also be sites of outflow with the potential for channel incision downstream. Accommodation zones represent areas where the polarity of border faults switches direction (Fig. 5.8a) and mark the separation between adjacent rift basins. These are also areas where inflow to the rift and its lakes may develop. Deeper parts of rift basins, close to border faults, reflect variable subsidence and are associated with pelagic and organic deposition during periods of high lake levels, whereas lowstand conditions favour evaporites and fluvial sediments.

5.4.2 *Siliciclastic Coastal Sedimentation*

The margins of the modern Kenya Rift lakes vary from low-energy mudflats underlain by shallow dilute or saline waters to high-energy beaches with sand and gravel shorelines. Most of the larger lakes also have stretches where faulted bedrock forms steep shorelines with little or no clastic-sediment accumulation. The development of siliciclastic shorelines follows two main pathways. Basalt, trachyte and phonolite

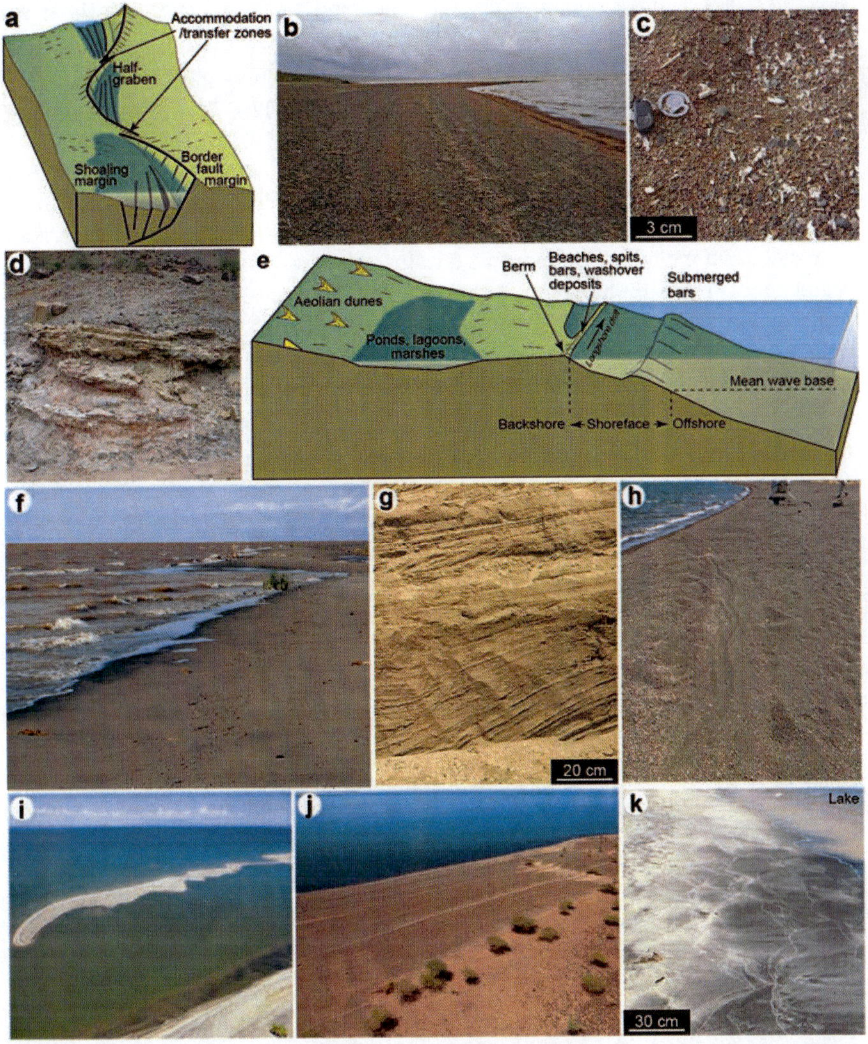

Fig. 5.8 Siliciclastic coastal sedimentation. Lakes dominated by volcanic catchments tend to produce silt and fine sand or granule/pebble clasts whereas metamorphic basins produce sandier shorelines. (**a**) General model of opposing rift valley half-grabens, showing contrasting shoaling and border fault margins, which control sediment inputs to rift lakes. (**b**, **c**) Pebble and granule strandlines, along the shoaling margin of Lake Bogoria. Elongate white particles in '**c**' are opaline rootcasts. (**d**) Fossil fanglomerates below a steep border fault, southeastern Lake Bogoria. Outcrop is about 1.5 m high. (**e**) Terminology and major environments along sandy shorelines, Lake Turkana. (**f**) Modern sandy beach, NW Turkana. (**g**) Planar and cross-bedded beach sand, NE Turkana. (**h**) Granular beach near volcanic hinterland; crocodile trail and helicopter (top) for scale, S Turkana. (**i**) Large spit along western shoreline of Turkana. (**j**) Recessional pebble beaches, SW Turkana. (**k**) Heavy mineral sands sorted by swash and backwash on a sandy beach, W Turkana

generally lack sand-size quartz crystals and tend to fragment into pebble and granule-size clasts or break down further to silt-size mineral detritus. The weathered sand fraction is typically volcanic rock fragments (lithic grains or VRFs) or K-feldspar grains that are less stable than quartz. Consequently, in areas dominated by volcanic rocks (most of the southern and central Kenya Rift) shore zones are characteristically composed of silt and mud, such as at Lakes Nakuru, Magadi and Logipi, but with granule/pebble/breccia shorelines (Lake Bogoria; Fig 5.8b–d) if wave energy is higher (Renaut and Owen 1991). In contrast, the Lake Turkana drainage basin includes metamorphic terrains that supply sand-size quartz grains to segments of its shoreline with silt dominating in the northern Omo Delta region. Consequently, swash-generated cross-bedded and planar stratified sands are common wherever wind and fetch allow significant wave development (Fig. 5.8e–k; Schuster and Nutz 2018). With fluctuating lake levels, sandy palaeoshorelines may be present both above and below the contemporary lake shoreline but can be subject to reworking.

Planar laminated or massive silts may develop in ponds, marshes or shoreline lagoons. Interbedded ripple cross-laminated washover sands may also be present in the latter settings. Sediments may be structureless due to bioturbation or pedogenesis during phases of desiccation. Varied molluscs and ostracod faunas may be preserved as well as *in situ* and reworked plant and root systems (rhizoliths where lithified). Benthic diatoms and other microfossils can be present. Beaches, spits and offshore bars are characterised by silt to sand with sheetlike to bar-shaped morphologies. Low angle, planar, ripple and trough cross-lamination/stratification are variously present. High-energy shorefaces are associated with ripple, hummocky and planar laminated sand, graded beds, gravel and coquinas.

Offshore contourites (planar and ripple cross-laminated sands) may develop in large lakes in response to wind-induced currents. Hummocky cross-stratified sands accumulate in response to storm activity. Both contourites and storm deposits also tend to be interbedded with structureless and laminated clay and mud recording lower energy intervals.

5.4.3 Lake Marginal Carbonate Deposition

Limestone is a minor lithology in the drainage basins of the Kenya Rift lakes, so most Ca^{2+} originates from weathering of volcanic and metamorphic rocks. Extensive Ca-Mg-carbonate deposits generally do not form. Nonetheless, carbonate does accumulate in dilute lakes in a variety of forms and in spring settings. For example, coquinas (e.g. with *Melanoides tuberculata* and *Corbicula africana*) are present along modern and AHP (African Humid Period: ~15–5 ka) shorelines, as are localised *Etheria* bioherms and biostromes (Fig. 5.9a), commonly near palaeoriver inflows (Feibel 2011). Gastropods, molluscs and ostracod shells accumulate offshore and in shallow ponds and lagoons. Oncoids often developed on shallow platforms during Plio-Pleistocene and Holocene higher lake-level stages at Lake

Fig. 5.9 Lake marginal carbonates. Examples of several types of carbonate accumulations in the Kenya Rift. (**a**) Early Holocene *Etheria* biostrome resting on lacustrine silt. NE Lake Turkana. Hat and panga (machete) for scale (bottom left). (**b**) Ovoid Holocene oncolites (~15 cm long) with molluscs as nuclei, NE Lake Turkana. (**c**, **d**) Lacustrine stromatolites and stromatolite-coated branch, W Lake Natron. (**e**) Bedded tufa formed in spring pool giving way upwards to stromatolitic tufa. Section ~4 m high. Orange notebook is 18 cm long. (**f**) Laminated tufa on tree branches (open spaces). (**g**) Travertine at Chemurkeu, Lake Bogoria, records former hot spring activity. Left inset is calcite. Right inset is aragonite from different parts of the mound. (**h**, **i**) Inactive spring travertine at Loburu, Lake Bogoria. Pool-rim dam in '**h**' = 8 m in diameter; outflow terraces in '**i**' are about 1 m wide in foreground

Turkana (Fig. 5.9b; Hargrave et al. 2014), with stromatolitic crusts forming on rocks and plant debris around the margins of several lakes including Lake Baringo, and around Bogoria and Natron during high-level stands (Fig 5.9c, d; Casanova 1994). Both oncoids and stromatolites form by bioinduced precipitation of calcite by bacteria and algae, usually in shallow water.

Modern springs precipitate little carbonate because their waters contain little to no calcium. Pleistocene bedded tufa is associated with several small spring-fed palaeoponds, marshes and shallow lakes located at Olorgesailie in the southern Kenya Rift (Fig. 5.9e; Lee et al. 2013) where thrombolites are also present and stromatolitic tufa/travertine encrusts reeds, twigs and larger branches (Fig. 5.9f). Calcite and aragonite precipitation induced by rapid degassing of CO_2 is present at Chemurkeu and Loburu on the margins of Lake Bogoria (Fig. 5.9g, i), where hot spring activity and the formation of travertine has been linked to changes in lake level (Jones and Renaut 1996; Renaut et al. 2013). Calcite is also present locally as a cement in beachrock at Nasikie Engida and Lake Turkana, and as rhizoliths in many lake-marginal settings (Cohen 1982; Owen et al. 2008).

5.4.4 *Deltas and Fan-Deltas*

Deltas deliver siliciclastic bedload (sand, gravel), suspended sediments (silt, clay, organics) and solutes to all major Kenya Rift lakes, even in underfilled basins (e.g. Lake Logipi) during wetter intervals (Butzer 1970, 1971; Frostick and Reid 1986). The nature of delta sedimentation varies considerably (Fig. 5.10), depending on the relative densities of inflow (water + sediment) and lake water (salinity + suspended sediment) (Renaut and Tiercelin 1994; Renaut and Gierlowski-Kordesch 2010).

If inflow and lake density are similar a homopycnal condition is attained and sediment is deposited rapidly, forming sand and gravel bars in and near the river mouth (Fig. 5.11a). This is common in fresher lakes. Delta foresets accumulate as sediments prograde into the lake, occasionally collapsing on steeper slopes and being transported into deeper water as grainflows, turbidity currents or debris flows. These latter hyperpycnal currents also develop if the river inflow and sediment load is denser than that of the lake water, resulting in a gravity-driven bottom flux of sediment that runs downslope below the lake water column, eventually coming to rest as upward-fining turbidites (Fig. 5.11b). These processes are common in the dilute waters of the large western rift lakes. However, many Kenya lakes are saline and usually denser than the inflow, which results in a hypopycnal overflow or interflow (at stratified lake boundaries). The suspended sediments can be carried as plumes for large distances into relatively deep lakes such as Lake Turkana or Lake Bogoria (Fig. 5.11c, d), often with sharp fronts between river and lake water. Clay minerals may flocculate near the source delta even in shallow saline lakes. Silt and clay-sized particles are carried into deeper areas where pelagic settling takes place. Well-laminated profundal sediments develop where flooding is seasonal.

Fig. 5.10 Rift valley deltas. (**a**, **b**) Channel sands and silts with well-vegetated floodplains on the Turkwell delta, W Turkana. (**c**, **d**) The Kerio delta with muddy inflow to Lake Turkana. (**e**) Ephemeral fan-delta entering SW Lake Turkana with gravel beach along shoreline. (**f**) Malewa River delta, Lake Naivasha. (**g**) Sandai River delta, Lake Bogoria, showing hypopycnal overflow of muddy water into the saline alkaline lake. (**h**) Fan delta with ephemeral flow into southeast Lake Magadi. Efflorescent trona encrusts the shoreline with patches of High Magadi Beds (Late Pleistocene high lake-level deposits) at apex of the delta. (**i**) Delta at northeastern Nasikie Engida with efflorescent trona crusts on saline mudflats and along the shoreline

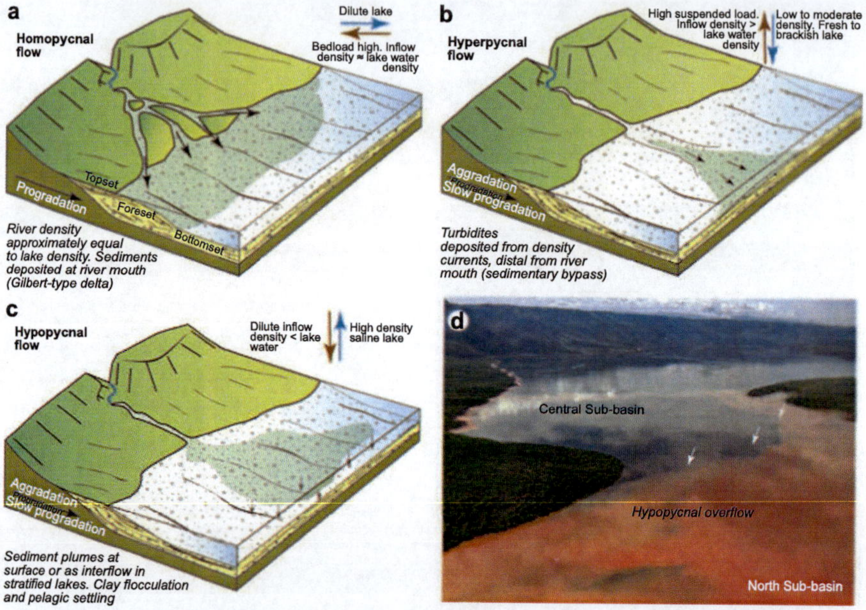

Fig. 5.11 Deltaic sedimentation. Contrasts in relative densities of inflow and lake water determine which types of flow usually develop. (**a**) Homopycnal flow, when inflow density is similar to lake water density, which results in sedimentation near the river mouth. (**b**) Hyperpycnal flow with inflow (mud + water) denser than the lake water, which generates bottom currents, turbidity flow and turbidites. (**c**) Hypopycnal flow results when less dense river water (+ mud) enters a lake with higher density (usually saline) water. (**d**) Hypopycnal overflow plume from Sandai River delta into northern Lake Bogoria. Note sharp boundary with saline lake water (white arrows)

Fan-deltas develop when alluvial fans directly enter a lake. This leads to progradation in shallow waters, but in deeper lakes sediment is often transported as underflows along eroded sublacustrine channels with lake floor levees, composed of massive mud, forming in deeper water with gentler gradients. These density currents eventually decelerate and deposit graded turbidites that may rest upon an eroded surface. In shallow playa basins, unconfined streams form sheetflows close to their lakes, depositing sands with ripples, planar beds and trough cross-beds, and massive to graded sand and clayey silt. These give way lakewards to massive subaqueous clay and silt. On steep alluvial fans, coarse gravel may be delivered to the fan-delta front and form subaqueous debris flows that deposit matrix-supported paraconglomerate.

5.4.5 *Offshore Pelagic Settings in Fresh to Brackish Lakes*

Coarse sediments may be transported to deep offshore locations by underflows (turbidity currents) to produce upward-fining turbidites (Fig. 5.12a). This is common in the large lakes of the western branch of the EARS, but turbidites are scarce in the

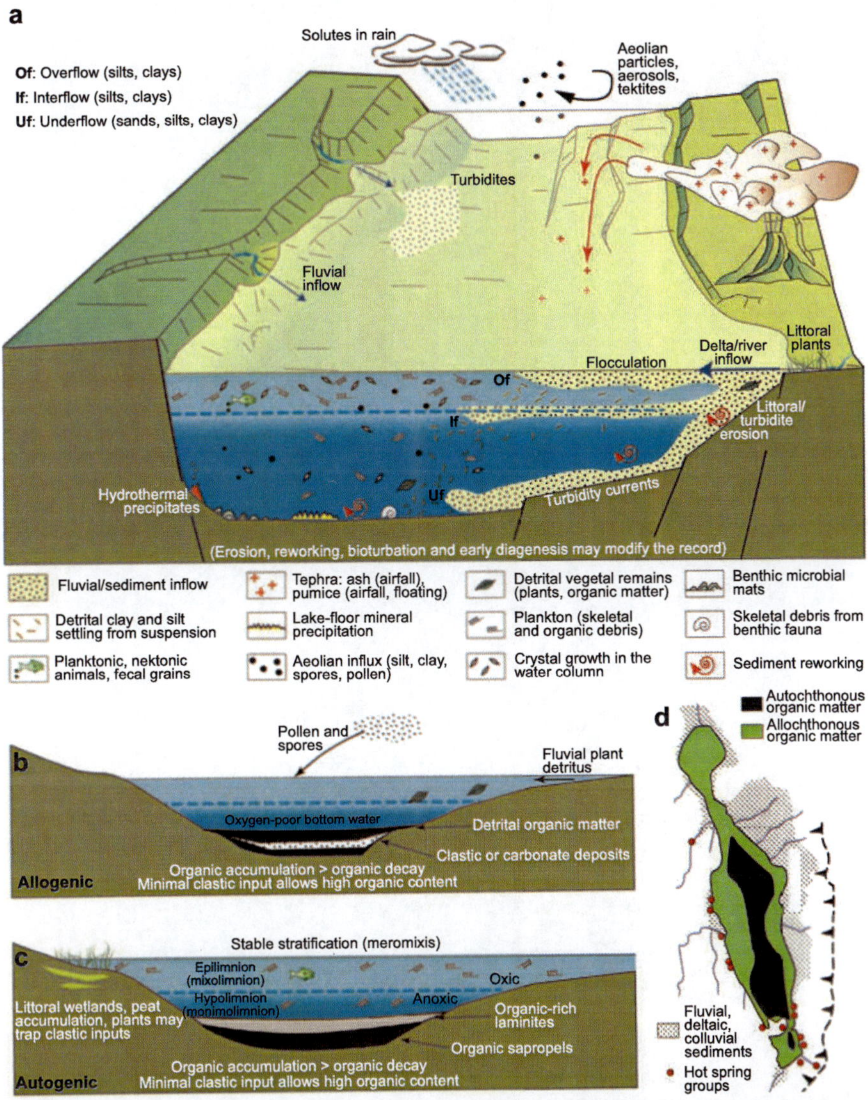

Fig. 5.12 Offshore lacustrine sedimentation. Modified from Renaut and Gierlowski-Kordesch (2010). (**a**) Sediment sources in offshore lake settings. Note that the development of overflow vs. underflow depends on density contrasts between river inputs and lake water. (**b**, **c**) Allogenic and autogenic organic offshore sedimentation in saline Lake Bogoria. (**d**) Distribution of organic sediments in Lake Bogoria

eastern branch, even in the relatively large Lake Turkana, due to gentle lake-floor slopes and density contrasts that do not favour underflows (Cohen 1989). Bouma-type upward-fining sequences have been recorded in Lake Bogoria (Tiercelin and Renaut 1987). More commonly, deposition is dominated by pelagic and

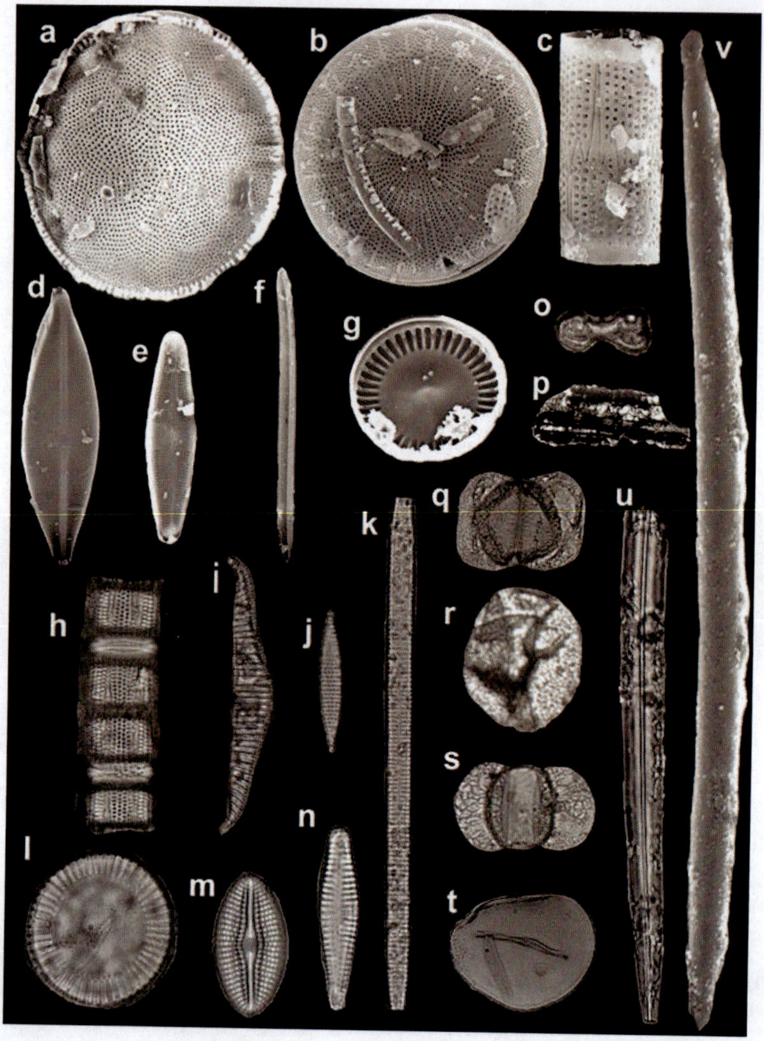

Fig. 5.13 Autochthonous and allochthonous microfossils in Kenya lake sediments. (**a–g**) SEM images of common diatoms (*Thalassiosira, Stephanodiscus, Aulacoseira, Craticula, Caloneis, Nitzschia* spp., in order). (**h–n**) Optical microscope images of common diatoms (*Aulacoseira, Rhopalodia, Nitzschia, Cyclotella* [*Stephanocyclus*], *Diploneis, Gomphonema, Ulnaria* spp.). (**o**) SEM image of a phytolith. (**p**) Optical image of plant fragment. (**q–t**) Optical images of pollen (*Pinus*, cf. *Cordia*, *Podocarpus*, Poaceae). (**u**, **v**) Optical and SEM images of sponge spicules

hemipelagic sediments that settle slowly from hypopycnal overflows and interflows, together with remains of phytoplankton (Fig. 5.13a) and zooplankton, or from rippled mud and silt transported by weak bottom currents (Renaut and Gierlowski-Kordesch 2010). Chemical precipitates are present in some perennial freshwater lakes. These include Fe- and Mn-nodules (e.g. Lake Malawi – a western branch rift lake; Owen et al. 1996) and phosphates (e.g. Miocene Palaeolake Lukeino, Fig. 5.1a; Dericquebourg et al. 2019) with origins that are generally attributed to changing Eh and pH. Authigenic minerals may also precipitate directly on the lake floor or result from early diagenesis in bottom sediments.

Allochthonous sediments include siliciclastic grains (clay, silt, tephra), and rare carbonate debris (tufa, travertine, rhizoliths, soil carbonates). Organic detritus (pollen, spores, leaves, twigs) may also be derived from fluvial and airborne sources (Figs. 5.12a–c, and 5.13o–t). Autochthonous sediments are commonly comprised of siliceous (diatoms, sponge spicules; Fig. 5.13a–n, u, v), carbonate (ostracods, molluscs) and organic microfossils (zooplankton, phytoplankton) as well as the hard and soft parts of fish. In general, organic particles are destroyed in shallow lakes and the littoral zones of larger water bodies due to their well-oxygenated bottom conditions. However, they do accumulate in lakes with high productivity, high preservation potential (anoxia), and low dilution by clastic materials (Cohen 2003). In some lakes (e.g. Bogoria), allochthonous and autochthonous organics are spatially separated with the former present as a ring of deposits around the latter (Fig. 5.12b, d).

Fine calcite crystals may form in the water column, sometimes depositing layers or laminae of micrite, or mixing with detrital mud (Kelts and Hsü 1978; Gierlowski-Kordesch 2010). This can be induced abiotically by evapoconcentration and warming, which reduce the solubility of $CaCO_3$ and CO_2, and sometimes by mixing of waters. Phytoplankton blooms may change pH in rift lakes and trigger calcite precipitation, which appears to have happened in brackish Lake Turkana through the Late Holocene (Halfman et al. 1989, 1994). Calcite, Mg-calcite and less common dolomite also formed intermittently in former deep-water settings in Pleistocene Palaeolake Koora in the southern Kenya Rift (Fig. 5.1a), perhaps partly triggered by diatom growth (Muiruri et al. 2021b).

Diatomaceous silt/clay and pure/impure diatomites have accumulated at the bottom of many Kenya Rift lakes (Figs. 5.14 and 5.15a–c; Owen 2002), although diatom frustules may be dissolved post-mortem in high alkalinity waters with $pH > 9.5$. Offshore non-laminated sediments may reflect current-induced erosion, re-suspension and mixing or bioturbation in permanent or temporarily oxygenated bottom waters. In contrast, lamination and bedding will generally be preserved where anoxia prevents infauna from developing and where water depth is sufficient to prevent wave-induced turbulence at the lake floor (Fig. 5.15d, e). In meromictic lakes the density contrasts are sometimes able to prevent infauna and wind mixing even in relatively shallow lakes, thus maintaining lamination. These may be mm-scale or less and reflect blooms in diatoms, algae and cyanobacteria. Coarser bedding may be associated with the presence of ostracods and other lacustrine organisms that alternate with layers that are dominated by clay and silt, often derived from fluvial sources.

Lamination and bedding in many lakes reflect seasonal or periodic changes. Annual flooding by the Omo River introduces clastics into Lake Turkana's northern basin, producing prominent annual light and dark clay-silt couplets ranging from a few mm to ~1 cm thick (Cohen 1984; Muchane 1996). Pleistocene diatomaceous silt and diatomite in the Suguta Valley reflect seasonal or periodic changes in

Fig. 5.14 Laminated and bedded offshore sediments in Kenya Rift palaeolakes. (**a**) Middle Pleistocene laminated diatomite with a fish bone bed at the base, Rupa, W Suguta Valley. Black scale bar = ~2 cm. (**b**) Bedded and laminated Holocene diatomaceous silt and clay, Baragoi, SE Suguta Valley. (**c**) Diatomaceous silt and diatomite laminae, Miocene Lukeino Formation, Tugen Hills, W Baringo. (**d**) Cyclic diatomite-clay couplets, Early Pleistocene Munya wa Gicheru Formation sediments, SE of Suswa volcano. Thicker (darker) clay units are present approximately every 10 cycles. (**e**) 'Paper shales' in the Miocene Lukeino Formation, Tugen Hills, consisting of mainly *Aulacoseira* diatom frustules with *Synedra* and *Nitzschia* spp., and clay

palaeolake diatom taxa and fluvial clay inputs (Fig. 5.14a, b), as do sediments in the Miocene Lukeino Formation where *Aulacoseira*, *Synedra* or *Nitzschia* spp. alternate with fluvial-sourced clay (Fig. 5.14c, e). Such variations in taxa commonly reflect

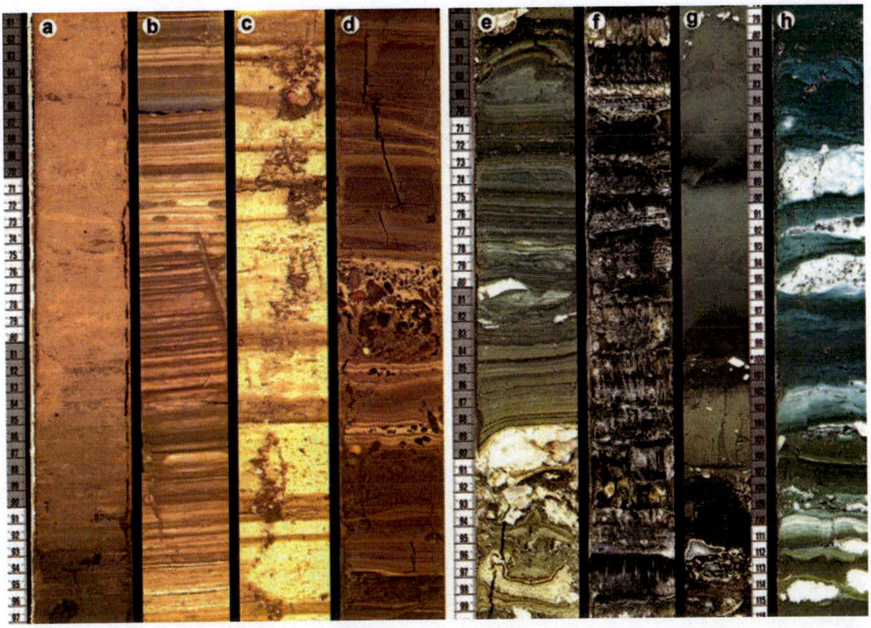

Fig. 5.15 Bedded and laminated lacustrine sediments from the Pleistocene Koora and Magadi palaeolakes, south Kenya Rift. (**a**–**d**) Palaeolake Koora cores courtesy of the Olorgesailie Drilling Project (ODP). (**e**–**h**) Palaeolake Magadi core sediments courtesy of the Hominin Sites and Paleolakes Drilling Project (HSPDP). (**a**) Weakly bedded to massive impure diatomite. (**b**) Laminated diatomite-clay couplets dominated by *Aulacoseira* with rare *Cyclotella* (*Stephanocyclus*). (**c**) Laminae in '**b**' thicken upwards giving way to diatomite-clay beds about 1 m higher in the sequence. Sharp base to darker clay horizons with more diffuse upper boundary implies sharp onset of fluvial influx to the palaeolake and gradual subsequent change. (**d**) Laminated and bedded zeolitic clay, silt and pumice. Zeolites imply high alkalinity, which is supported by a lack of diatoms. (**e**) Zeolitic laminated silt- and clay-sized sediments with pyrite crystals. White to grey chert indicates a saline, alkaline palaeolake with high silica content at the depositional site. (**f**) Bedded trona crystals and darker mud. (**g**) Bedded zeolitic mud with scattered chert near base. (**h**) Cyclical white chert passing into siliceous (darker) silt/clay or zeolitic beds

changing nutrients (silica, phosphate), turbidity, salinity or the strength of wind mixing in the water column. Pleistocene sediments at Munya wa Gicheru also show regular clay-diatomite layers but with thicker diatomaceous clay units every ten couplets or so, suggesting climatic cyclicity on a variety of scales (Fig. 5.14d). In comparison, diatomite laminae in modern Lake Malawi (western branch of the EARS) reflect *Aulacoseira* blooms during the dry windy season with diatomaceous clay accumulating following fluvial-induced overflows during the annual rains (Pilskaln and Johnson 1991). Variability in laminae thickness reflects sedimentation rates that may change within a lake over time. For example, part of the succession at Pleistocene Palaeolake Koora in the southern Kenya Rift gradually changed from thin clay-diatomite laminae to thicker ~5-cm-scale bedding while maintaining patterns of diatom deposition in each horizon (Figs. 5.1a and 5.15b, c).

5.4.6 *Sedimentation in Highly Saline Deep Lakes*

The modern lakes of the Kenya Rift lie mostly in closed basins and are consequently characterised by brackish to saline waters that lead to the formation of different facies to those in freshwater lakes (Fig. 5.6). Relatively deep saline water bodies, such as Lake Bogoria, are typically meromictic and accumulate a variety of laminated to bedded siliciclastic and (or) biogenic (e.g. diatoms, planktonic cyanobacteria, zooplankton) muds below well-developed chemoclines (Figs. 5.5a and 5.12b–d). The development of anoxia in the monimolimnion favours organic matter preservation. An additional key factor is that these closed lakes fluctuate in height with climate change and therefore preserve sediments laid down under both arid and wetter conditions. At Bogoria, this has resulted in basinal associations of massive and laminated organic oozes, laminated pyritic clay, clay and highly saline facies indicators such as trona and nahcolite (Renaut and Tiercelin 1994; De Cort et al. 2018).

Pleistocene palaeolakes in the Magadi basin also developed euxinic conditions that led to the accumulation of laminated and massive organic-rich mud with high concentrations of redox-sensitive trace elements such as Mo, As and V. Sulfate reduction has also resulted in abundant pyrite in parts of the Magadi sediment succession (Fig. 5.15e; Surdam and Eugster 1976; Owen et al. 2019; Deocampo et al. 2022). Furthermore, secondary mineral changes can be important in hypersaline water bodies with much of the initial volcaniclastic sequence at Lake Magadi having been altered to a range of early diagenetic zeolites (e.g. erionite, analcime, clinoptilolite, phillipsite) depending on the location with respect to evaporating lake water and inflowing fresher ground and surface water (Surdam and Eugster 1976; Eugster 1986).

5.4.7 *Chemical Sedimentation and Evaporites in Shallow Saline Lakes and Ephemeral Playas*

Shallow saline lakes develop when evaporation exceeds precipitation and ephemeral stream inflow from the wetter rift shoulders is insufficient to raise lake levels to their outlets (Nakuru, Elmenteita). Such lakes may be maintained by meteoric and (or) hydrothermal spring inflow, which may be dilute or saline. Figure 5.16a shows a schematic model of sedimentation in a (semi-)arid underfilled rift basin, which includes a central ephemeral salt pan or saline lake, bordered by saline and dry mudflats, an alluvial sandflat and alluvial fans (cf. Hardie et al. 1978; Smoot and Lowenstein 1991). In general, alternating mud and evaporites accumulate in saline pans, whereas perennial shallow saline lakes tend to deposit laminites of mud-silt and evaporites, including crystal cumulates (Renaut and Gierlowski-Kordesch 2010).

The margins of the rift basins are commonly fault scarps below which alluvial fans may develop, as for example in the northern Suguta Valley. Channel sediments on fans include fluvial orthoconglomerate and paraconglomerate deposited from

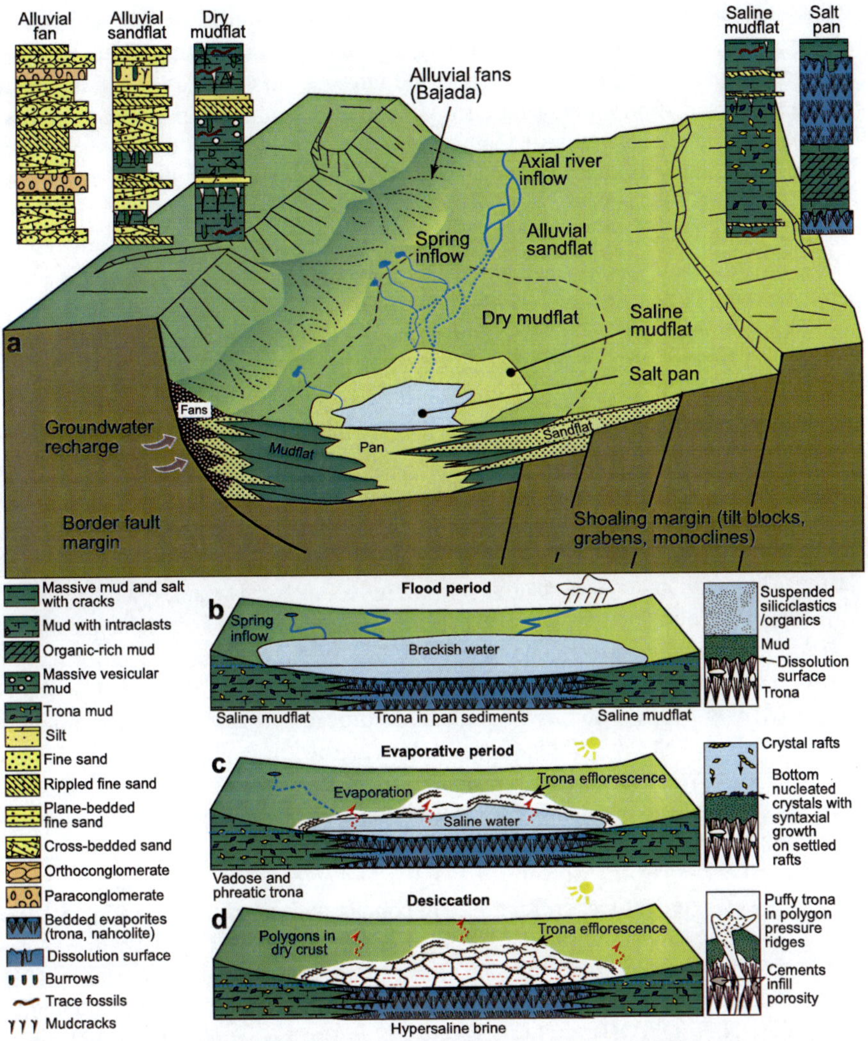

Fig. 5.16 Shallow saline lake to ephemeral playa model. (**a**) General model of an underfilled basin with a shallow ephemeral saline lake surrounded by saline mudflats, dry mudflats and alluvial sandflats with marginal coalescing alluvial fans. Sedimentary logs provide schematic facies for each depositional setting. (**b–d**) Saline playa model based on Lowenstein and Hardie (1985) with minor revision by Renaut and Gierlowski-Kordesch (2010). (**b**) Flooding causes transgression with fresh waters becoming saline following dissolution of trona. Mud deposition on dissolution surface. (**c**) Evaporation decreases volume and increases salinity leading to trona precipitation as floating rafts which grow then sink. Salinity increases further and bottom nucleation of trona starts, producing laterally extensive trona beds. (**d**) Lake dries leaving hypersaline brine groundwater. Polygonal pressure ridges form at surface. Saline cements fill porosity

debris flows. Intermittent flooding of fan surfaces produces shallow ephemeral channels and sheetflow that deposit clay, silt and fine sand with sheet- or lobe-shaped geometries. Alluvial sandflats at the fan toes pass into dry mudflats where decelerating sheetflows produce graded silt and clay laminae with rippled to cross-bedded sand in shallow incised channels. Mud, efflorescent salt crusts, mud intraclasts, mudcracks, burrows and plant roots are variously present with reworking caused by deflation, desiccation and flooding (Fig. 5.16a; Smoot and Lowenstein 1991; Renaut and Gierlowski-Kordesch 2010). Not all rift basins are bordered by alluvial fans. In some basins (e.g. Nasikie Engida, Nakuru) the saline lake abuts against young fault scarps with a very narrow (locally <1 m) sediment fringe.

Saline mudflats (Fig. 5.16a) may surround the lakes or develop only at the axial margins. They typically have very shallow saline groundwaters that precipitate intrasedimentary trona or gaylussite and efflorescent salt crusts at the surface. Salts grow after capillary evaporation, evaporative pumping, and precipitation from evaporating surface water after a playa is flooded. Sheetfloods periodically deposit graded fine sand, silt and mud on the surface. The resulting mud-salt deposits tend to be poorly bedded with a dark colour when wet, which changes to greyish green or reddish reflecting the oxidation state of iron (Renaut and Gierlowski-Kordesch 2010). Thin aeolian deposits may be present reflecting local winds and dust devils.

Trona is the dominant evaporite at ephemeral Lake Magadi and is also present at Nasikie Engida where ascending magmatic CO_2 likely contributes to its formation (Owen et al. 2019; De Cort et al. 2019; Renaut et al. 2021). Polygonal pressure ridges develop in the trona as crystallisation proceeds (Fig. 5.17a). The mineral forms fan-shaped clusters of acicular crystals as it grows in shallow brine (Fig. 5.17c) that dry to a white colour or are orange to purple where archaea are present (Fig. 5.17b). Seasonal flooding also deposits mud and dissolves the trona forming dissolution surfaces and alternating trona and mud beds on a variety of scales (Figs. 5.15f and 5.16b–d; McNulty 2017). Discoid calcium carbonate oncoids form on mudflats where freshwater floods are in contact with alkaline brine (Fig. 5.17f), but most Ca^{2+} is lost to soils and sediments prior to floods entering the saline lakes. Gaylussite ($Na_2Ca_2 \cdot 5H_2O$) crystals also form where dilute Ca-bearing water mixes with alkaline brine. Other unusual Pleistocene chemical deposits at Magadi include magadiite [$NaSi_7O_{13}(OH)_3 \cdot 4(H_2O)$] (Fig. 5.17g; Buatois et al. 2020) and chert (SiO_2) (Figs. 5.15g, h and 5.17h; Behr and Röhricht 2000) with many of the Pleistocene lake sediments altered to zeolites (Surdam and Eugster 1976; Eugster 1986).

Some shallow saline lakes are perennial and charged primarily by groundwater and springs. These include several crater lakes in the Suguta Valley and Nasikie Engida. Nasikie Engida is a permanent shallow saline lake that is maintained by hot spring inflow (>80%) (Renaut et al. 2021) with minor clastics introduced by ephemeral deltas. Consequently, lacustrine sediments in spring-fed lakes are mainly autochthonous consisting of evaporites and organic remains (microbial mats, phytoplankton) mixed with aeolian silt. Offshore sediments at Nasikie Engida are dominated by bedded or laminated feldspathic, erionitic mud that alternates with stubby prismatic crystals of nahcolite (Fig. 5.17d; De Cort et al. 2019), which also forms a

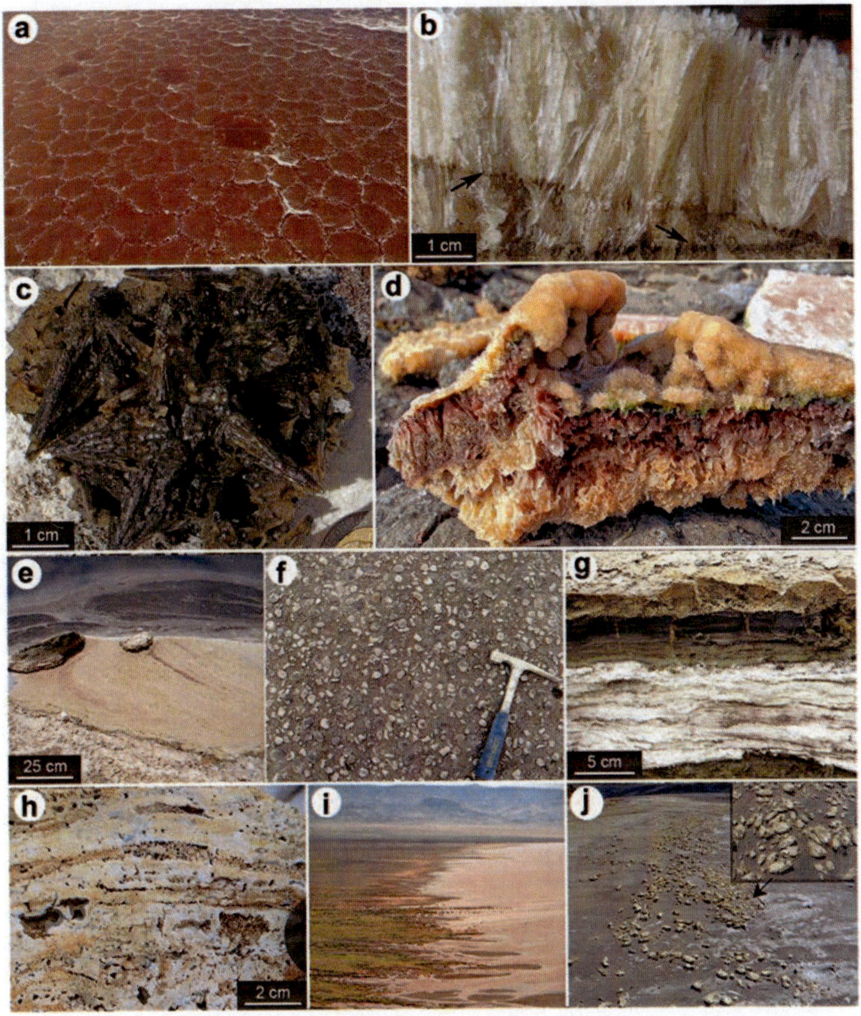

Fig. 5.17 Ephemeral playa sediments. (**a**) Trona polygons (~10 m across) forming in Lake Magadi. (**b**) Desiccated trona with thin detrital layers (arrows). (**c**) Fan-shaped trona crystallising in bottom sediments of Nasikie Engida. (**d**) Nahcolite (stubby crystals at base), Nasikie Engida. These are overlain by muddy flood sediments with acicular trona crystals at the top formed during the start of evaporative phase. (**e**) Organic wind-blown scum concentrating along the shoreline, Nasikie Engida. (**f**) Carbonate oncoids on margins of Lake Magadi form where dilute Ca-bearing water enters the highly alkaline lake. (**g**) Early Holocene (white) magadiite overlain by laminated clay with abundant fish fossils. Lake Magadi. (**h**) Pleistocene bedded chert, Lake Magadi. (**i**) Lake Logipi (off photo to right) saline mudflat and alluvial sandflat. (**j**) Lake Logipi fish-strandline formed during lake regression. Inset shows detail. Fish skeletons are up to 8 cm long

10-cm-thick solid crust across much of the lake floor. The nahcolite also reflects very high magmatic CO_2 inputs to the lake from adjacent and (or) underlying faults (Renaut et al. 2021). Planktonic cyanobacteria are also important sediment components, as are diatoms that are preserved in the highly alkaline lake due to high silica concentrations in the water (Muiruri et al. 2021a). Wind concentrates floating organic scum along shorelines (Fig. 5.17e) with offshore sediments containing up to 25% organic matter (De Cort et al. 2019).

Another playa system is located in the northern Kenya Rift and is centred on Lake Logipi (Fig. 5.17i). This saline lake is fed by springs, ephemeral marginal streams and, intermittently, by the Suguta River. Laterally extensive transgressions and regressions are common on the gently sloping basin floor. Fish (tilapia, catfish) survive in the water during wetter periods but die as it shrinks and increases in salinity, leaving fish strandlines (Fig. 5.17j). The Logipi lacustrine sediments contain little trona and are dominated by black, organic-poor, zeolitic mud (Castanier et al. 1993).

References

Beadle LC (1974) The inland waters of tropical Africa. An introduction to tropical limnology, 1st edn. Longman, London

Beadle LC (1981) The inland waters of tropical Africa. An introduction to tropical limnology, 2nd edn. Longman, London

Behr H, Röhricht C (2000) Record of seismotectonic events in siliceous cyanobacterial sediments (Magadi cherts), Lake Magadi, Kenya. Int J Earth Sci 89:268–283

Bengtsson L (2022) Physical processes in lakes. MDPI, Basel

Boehrer B, Shultze M (2008) Stratification of lakes. Rev Geophys RG2005 46:1–27

Bohacs KM, Carroll AR, Neal JE, Mankiewicz PJ (2000) Lake-basin type, source potential, and hydrocarbon character: an integrated sequence-stratigraphic-geochemical framework. In: Gierlowski-Kordesch EH, Kelts KR (eds) Lake basins through space and time. AAPG Stud Geol 46:3–33

Bohacs KM, Carroll AR, Neal JE (2003) Lessons from large lake systems-thresholds, nonlinearity, and strange attractors. In: Chan MA, Archer AW (eds) Extreme depositional environments: mega end members in geologic time. Geol Soc Am Spec Pap 370:5–90

Buatois LA, Renaut RW, Owen RB, Behrensmeyer AK, Scott JJ (2020) Animal bioturbation preserved in Pleistocene magadiite at Lake Magadi, Kenya Rift Valley, and its implications for the depositional environment of bedded magadiite. Sci Rep 10:67943

Butzer KW (1970) Contemporary depositional environments of the Omo Delta. Nature 226:425–430

Butzer KW (1971) Recent history of an Ethiopian delta: the Omo River and the level of Lake Rudolf (Univ Chicago Res Pap 136). University of Chicago Press, Chicago

Carroll AR, Bohacs KM (1999) Stratigraphic classification of ancient lakes: balancing tectonic and climatic controls. Geology 27:99–102

Casanova J (1994) Stromatolites from the East African Rift: a synopsis. In: Bertrand-Sarfati JJ, Monty C (eds) Phanerozoic stromatolites II. Springer, New York, pp 193–226

Castanier S, Bernet-Rollande M, Maurin A, Perthuisot J (1993) Effects of microbial activity on the hydrochemistry and sedimentology of Lake Logipi, Kenya. Hydrobiologia 267:99–112

Chase JE, Arizaleta ML, Tutolo BM (2021) A series of data-driven hypotheses for inferring biogeochemical conditions in alkaline lakes and their deposits based on the behavior of Mg and SiO_2. Fortschr Mineral 11:106

Cohen AS (1982) Paleoenvironments of root casts from the Koobi Fora Formation, Kenya. J Sediment Res 52:401–414

Cohen AS (1984) Effect of zoobenthic standing crop on laminae preservation in tropical lake sediment, Lake Turkana, East Africa. J Paleontol 58:499–510

Cohen AS (1989) Facies relationships and sedimentation in large rift lakes and implications for hydrocarbon exploration: examples from Lakes Turkana and Tanganyika. Palaeogeogr Palaeoclimatol Palaeoecol 70:65–80

Cohen AS (2003) Paleolimnology: the history and evolution of lake systems. Oxford University Press, Oxford

Cole GA, Weihe PE (2016) Textbook of limnology, 5th edn. Waveland Press, Long Grove

Csanady GT (1978) Water circulation and dispersal mechanisms. In: Lerman A (ed) Lakes: chemistry geology physics. Springer, New York, pp 21–64

De Cort G, Verschuren D, Ryken E, Wolff C, Renaut RW, Creutz M, van der Meeren T, Haug G, Olago DO, Mees F (2018) Multi-basin depositional framework for moisture-balance reconstruction during the last 1300 years at Lake Bogoria, central Kenya Rift Valley. Sedimentology 65:1667–1696

De Cort G, Mees F, Renaut RW, Sinnesael M, Van der Meeren T, Goderis S, Keppens E, Mbuthia A, Verschuren D (2019) Late-Holocene sedimentation and sodium carbonate deposition in hypersaline, alkaline Nasikie Engida, southern Kenya Rift Valley. J Paleolimnol 132:279–300

Deocampo DM (2005) Evaporative evolution of surface waters and the role of aqueous CO_2 in magnesium silicate precipitation: Lake Eyasi and Ngorongoro Crater, northern Tanzania. S Afr J Geol 108:493–504

Deocampo DM, Jones BF (2014) Geochemistry of saline lakes. In: Holland HD, Turekian KK (eds) Treatise on geochemistry, vol 7. Elsevier, Oxford, pp 437–469

Deocampo DM, Renaut RW (2016) Geochemistry of African soda lakes. In: Schagerl M (ed) Soda lakes of East Africa. Springer, Cham, pp 77–96

Deocampo D, Owen RB, Lowenstein TK, Renaut RW, Rabideaux NM, Billingsley A, Cohen A, Deino AL, Sier MJ, Luo S, Shen CC, Gebregiorgis D, Campisano C, Mbuthia A (2022) Orbital control of Pleistocene euxinia in Lake Magadi, Kenya. Geology 50:42–47

Dericquebourg P, Person A, Ségalen L, Pickford M, Senut B, Fagel N (2019) Bone diagenesis and origin of calcium phosphate nodules from a hominid site in the Lukeino Formation (Tugen Hills, Kenya). Palaeogeogr Palaeoclimatol Palaeoecol 536:109377

Dunkley PN, Smith M, Allen DJ, Darling WG (1993) The geothermal activity and geology of the northern sector of the Kenya Rift Valley. Brit Geol Surv Res Rep SC/93/1

El-Sheekh M, Elsaied HE (eds) (2023) Lakes of Africa: microbial diversity and sustainability. Elsevier, Amsterdam

Eugster H (1986) Lake Magadi, Kenya: a model for rift valley hydrochemistry and sedimentation? In: Frostick LE, Renaut RW, Reid I, Tiercelin J-J (eds) Sedimentation in the African rifts. Geol Soc Lond Spec Publ 25, London, pp 177–189

Eugster HP, Hardie LA (1978) Saline lakes. In: Lerman A (ed) Lakes – chemistry, geology, physics. Springer, New York, pp 237–293

Eugster HP, Jones BF (1979) Behavior of major solutes during closed-basin brine evolution. Am J Sci 279:609–631

Feibel CS (2011) A geological history of the Turkana Basin. Evol Anthropol 20:206–216

Frostick LE, Reid I (1986) Evolution and sedimentary character of lake deltas fed by ephemeral rivers in the Turkana basin, northern Kenya. In: Frostick LE, Renaut RW, Reid I, Tiercelin J-J (eds) Sedimentation in the African rifts. Geol Soc Lond Spec Publ 25: 113–125

Gierlowski-Kordesch E (2010) Lacustrine carbonates. In: Alonso-Zarza AM, Tanner LH (eds) Carbonates in continental settings: facies, environments, and processes (Developments in sedimentology 61). Elsevier, Amsterdam, pp 1–101

Hakala A (2004) Meromixis as part of lake evolution – observations and a revised classification of true meromictic lakes in Finland. Boreal Environ Res 9:37–53

Håkanson L, Jansson M (2002) Principles of lake sedimentology, 2nd edn. Blackburn Press, Caldwell

Halfman JD, Johnson TC, Showers WJ, Lister GS (1989) Authigenic low-Mg calcite in Lake Turkana, Kenya. J Afr Earth Sci 8:533–540

Halfman JD, Johnson TC, Finney BP (1994) New AMS dates, stratigraphic correlations and decadal climatic cycles for the past 4 ka at Lake Turkana, Kenya. Palaeogeogr Palaeoclimatol Palaeoecol 111:83–98

Hall KJ, Northcote TG (2012) Meromictic lakes. In: Bengtsson L, Herschy RW, Fairbridge RW (eds) Encyclopedia of lakes and reservoirs. Springer, Dordrecht, pp 519–524

Hardie LA, Smoot JP, Eugster HP (1978) Saline lakes and their deposits: a sedimentological approach. In: Matter A, Tucker ME (eds) Modern and ancient lake sediments. Int Assoc Sedimentol Spec Publ 2, pp 7–41

Hargrave JE, Hicks MK, Scholz CA (2014) Lacustrine carbonates from Lake Turkana, Kenya: a depositional model of carbonates in an extensional basin. J Sediment Res 84:224–237

Hennesey A, Yasindi AW, Zinabu G-M, Taylor WD (2020) Effects of pH and salinity on the sedimentation of clay particles in turbid lakes of the rift valley of Ethiopia and Kenya. Ethiop J Biol Sci 19:117–130

Horppila J, Niemisto J (2008) Horizontal and vertical variations in sedimentation and resuspension rates in a stratifying lake – effects of internal seiches. Sedimentology 55:1135–1144

Hutchinson GE (1975a) A treatise on limnology, volume 1: geography, physics and chemistry, part 1. Geography and physics of lakes. John Wiley & Sons, New York

Hutchinson GE (1975b) A treatise on limnology, volume II. Introduction to lake biology and the limnoplankton. John Wiley & Sons, New York

Hutchinson GE (1975c) A treatise on limnology, volume III. Limnological botany. John Wiley & Sons, New York

Johnson TC, Odada EO (eds) (1996) The limnology, climatology and paleoclimatology of the East African lakes. Gordon and Breach, Amsterdam

Jones B, Renaut RW (1996) Morphology and growth of aragonite crystals in hot-spring travertines at Lake Bogoria, Kenya Rift Valley. Sedimentology 43:323–340

Jones BF, Weir AH (1983) Clay minerals of Lake Abert, an alkaline, saline lake. Clays Clay Miner 31:161–172

Kalff J, Downing J (2016) Limnology: inland water ecosystems. Bibliogenica, Duluth

Kelts K, Hsü KJ (1978) Freshwater carbonate sedimentation. In: Lerman A (ed) Lakes: chemistry, geology, physics. Springer, New York, pp 294–323

Kilham P (1990) Mechanisms controlling the chemical composition of lakes and rivers: data from Africa. Limnol Oceanogr 35:80–83

Lameck AS, Skutai J, Boros E (2023) Review of chemical properties of inland soda and saline waters in East Africa (rift valley region). J Hydrol Reg Stud 46:101323

Lee RKL, Owen RB, Renaut RW, Behrensmeyer AK, Potts R, Sharp WD (2013) Facies, geochemistry and diatoms of late Pleistocene Olorgesailie tufas, southern Kenya Rift. Palaeogeogr Palaeoclimatol Palaeoecol 374:197–217

Lerman A, Imboden DM, Gat JR (eds) (1995) Physics and chemistry of lakes. Springer, New York

Lowenstein TK, Hardie LA (1985) Criteria for the recognition of salt-pan evaporites. Sedimentology 32:627–644

McGlue MM (2021) Lacustrine processes and sediments. In: Encyclopedia of geology, 2nd edn. Elsevier, Amsterdam, pp 876–893

McNulty E (2017) Lake Magadi and the soda cycle: a study of the modern sodium carbonates and of Late Pleistocene and Holocene core sediments. MSc thesis, Binghamton University, Binghamton

Melack JM (2009) Diel variability and community metabolism in African soda lakes. Nat Resour Environ Issues 15:28

Melack JM, MacIntyre S (2016) Morphometry and physical processes of East African soda lakes. In: Schagerl M (ed) Soda lakes of East Africa. Springer, Cham, pp 61–76

Muchane MW (1996) Comparison of the isotope record in micrite, Lake Turkana, with the historical weather record over the last century. In: Johnson TC, Odada EO (eds) The limnology, climatology and paleoclimatology of the East African Lakes. Gordon and Breach, Amsterdam, pp 431–441

Muiruri VM, Owen RB, de Cort G, McNulty E, Rabideaux N, Lowenstein T, Renaut RW, Leet K, Deocampo D, Cohen AS, Campisano CJ, Billingsley A, Mbuthia A (2021a) Middle Pleistocene to recent diatom floras and stratigraphy of the Magadi Basin, southern Kenya Rift. J Paleolimnol 65:315–333

Muiruri V, Owen RB, Potts R, Deino AL, Behrensmeyer AK, Riedl S, Rabideaux N, Beverly EJ, Renaut RW, Moerman JW, Deocampo D, Tyler Faith J, Noren A, Cohen AS, Shannon KB, Dommain R (2021b) Quaternary diatoms and palaeoenvironments of the Koora Plain, southern Kenya Rift. Quat Sci Rev 267:107106

Odada O, Olago SO (2002) The East African Great Lakes: limnology, palaeolimnology and biodiversity (Advances in global change research 12). Kluwer, Dordrecht

Olaka LA, Trauth MH, Olago DO (2010) The sensitivity of East African Rift lakes to climate fluctuations. J Paleolimnol 44:629–644

Owen RB (2002) Sedimentological characteristics and origins of diatomaceous deposits in the East African Rift System. In: Renaut RW, Ashley GM (eds) Sedimentation in continental rifts. SEPM Spec Publ 73, pp 233–246

Owen RB, Renaut RW, Williams TM (1996) Characteristics and origins of ferromanganese nodules in Lake Malawi. In: Johnson TC, Odada E (eds) Limnology climatology and paleoclimatology of the East African Lakes. Gordon and Breach, Amsterdam, pp 461–474

Owen RA, Owen RB, Renaut RW, Scott JJ, Jones B, Ashley GM (2008) Mineralogy and origin of rhizoliths on the margins of saline, alkaline Lake Bogoria, Kenya Rift Valley. Sediment Geol 2013:143–163

Owen RB, Renaut RW, Muiruri VM, Rabideaux NM, Lowenstein TK, McNulty EP, Leet K, Deocampo D, Luo S, Deino AL, Cohen A, Sier MJ, Campisano C, Shen C-C, Billingsley A, Mbuthia A, Stockhecke M (2019) Quaternary history of the Lake Magadi Basin, southern Kenya Rift: tectonic and climatic controls. Palaeogeogr Palaeoclimatol Palaeoecol 518:97–118

Pecoraino G, D'Alessandro W, Inguaggiato S (2015) The other side of the coin: geochemistry of alkaline lakes in volcanic areas. In: Rouwet D, Christenson B, Tassi F, Vandemeulebrouck J (eds) Volcanic lakes. Springer, Berlin, pp 219–237

Pilskaln CH, Johnson TC (1991) Seasonal signals in Lake Malawi sediments. Limnol Oceanogr 36:544–557

Plisnier P-D, Chitamwebwa D, Mwape L, Tshibangu K, Langenberg V, Coenen E (1999) Limnological annual cycle inferred from physical–chemical fluctuations at three stations of Lake Tanganyika. Hydrobiologia 407:45–58

Renaut RW, Gierlowski-Kordesch EH (2010) Lakes. In: James NP, Dalrymple (eds) Facies models 4. Geological Association of Canada, St. John's, pp 541–575

Renaut RW, Owen RB (1991) Shore-zone sedimentation and facies in a closed rift lake: the Holocene beach deposits of Lake Bogoria, Kenya. In: Anadon P, Cabrera L, Kelts K (eds) Lacustrine facies analysis (IAS Spec Publ 13). Blackwell, Oxford, pp 175–195

Renaut RW, Tiercelin J-J (1994) Lake Bogoria, Kenya Rift Valley—a sedimentological overview. In: Renaut RW, Last WM (eds) Sedimentology and geochemistry of modern and ancient saline lakes. SEPM Spec Publ 50, pp 101–123

Renaut RW, Owen RB, Jones B, Tiercelin J-J, Tarits C, Ego JK, Konhauser KO (2013) Impact of lake-level changes on the formation of thermogene travertine in continental rifts: evidence from Lake Bogoria, Kenya Rift Valley. Sedimentology 60:428–468

Renaut RW, Owen RB, Lowenstein T, De Cort G, McNulty E, Scott JJ, Mbuthia A (2021) The role of hydrothermal fluids in sedimentation in saline alkaline lakes: evidence from Nasikie Engida, Kenya Rift Valley. Sedimentology 68:108–134

Schagerl M (ed) (2016) Soda lakes of East Africa. Springer, Cham
Schagerl M, Renaut RW (2016) Dipping into the soda lakes of East Africa. In: Schagerl M (ed) Soda lakes of East Africa. Springer, Cham, pp 3–24
Schuster M, Nutz A (2018) Lacustrine wave-dominated clastic shorelines: modern to ancient littoral landforms and deposits from the Lake Turkana Basin (East African Rift System, Kenya). J Paleolimnol 59:221–243
Sly PG (1978) Sedimentary processes in lakes. In: Lerman A (ed) Lakes – chemistry geology physics. Springer, New York, pp 65–90
Smoot JP, Lowenstein TK (1991) Depositional environments of non-marine evaporites. In: Melvin JL (ed) Evaporites, petroleum and mineral resources. Elsevier, Amsterdam, pp 189–347
Sonnenfeld P, Hudec PP (1980) Heliothermal lakes. In: Nissenbaum A (ed) Hypersaline brines and evaporitic environments. Elsevier, Amsterdam, pp 93–100
Stewart KM, Walker KF, Likens GE (2009) Meromictic lakes. In: Likens GE (ed) Encyclopedia of inland waters, 1st edn. Academic, Amsterdam, pp 589–602
Surdam R, Eugster H (1976) Mineral reactions in the sedimentary deposits of the Lake Magadi region, Kenya. Geol Soc Am Bull 87:1739–1752
Talbot MR (1988) The origins of lacustrine oil source rocks: evidence from the lakes of tropical Africa. In: Fleet AJ, Kelts K, Talbot MR (eds) Lacustrine petroleum source rocks. Geol Soc Lond Spec Publ 40, London, pp 29–43
Talling JE, Lemoalle J (2011) Ecological dynamics of tropical inland water. Cambridge University Press, Cambridge
Tarits C, Renaut RW, Tiercelin JJ, Le Hérissé A, Cotten J, Cabon JY (2006) Geochemical evidence of hydrothermal recharge in Lake Baringo, central Kenya Rift Valley. Hydrol Process 20:2027–2055
Tiercelin J-J (1990) Rift-basin sedimentation: responses to climate, tectonism and volcanism. Examples of the East African Rift. J Afr Earth Sci 10:283–305
Tiercelin J-J, Renaut RW (1987) Cônes alluviaux et deltas éventail. In: Tiercelin J-J, Vincens A (eds) Le demi–graben de Baringo–Bogoria, Rift Gregory, Kenya: 30,000 ans d'histoire hydrologique et sédimentaire. Bull Centr Rech Explor-Prod Elf-Aquitaine 11, pp 312–320
Tiercelin J-J, Le Fournier J, Herbin JP, Richert JP (1980) Continental rifts: modern sedimentation, tectonic and volcanic control. Example from the Bogoria-Baringo graben, Gregory Rift, Kenya. In: Geodynamic evolution of the Afro Arabic Rift System. Acad Naz dei Lincei, Rome, pp 143–164
Tiercelin J-J, Vincens A (eds) (1987) Le demi–graben de Baringo–Bogoria, Rift Gregory, Kenya: 30,000 ans d'histoire hydrologique et sédimentaire. Bull Central Rech Explor-Prod Elf-Aquitaine 11:249–540
Trauth MH, Maslin MA, Deino AL, Junginger A, Lesoloyia M, Odada EO, Olago DO, Olaka LA, Strecker MR, Tiedemann R (2010) Human evolution in a variable environment: the amplifier lakes of Eastern Africa. Quat Sci Rev 29:2981–2988
Wetzel RG (2001) Limnology: lake and river ecosystems, 3rd edn. Academic Press, London
Yuretich RF (1982) Possible influences upon lake development in the East African Rift valleys. J Geol 90:329–337
Yuretich RF, Cerling TE (1983) Hydrogeochemistry of Lake Turkana, Kenya: mass balance and mineral reactions in an alkaline lake. Geochim Cosmochim Acta 47:1099–1109

Part II
The Modern Kenya Rift Lakes

Chapter 6
Lake Turkana

6.1 Introduction

Saline, alkaline Lake Turkana lies in a semi-desert region between the Kenya and Ethiopian Rifts (Figs. 6.1 and 6.2a, b). The north-south trending lake lies ~360 m above sea level and is ~250 km long by 15–30 km wide. Its surface area of ~7000 km^2 (Garcin et al. 2012) makes it much larger than all other Kenyan lakes combined. Bathymetric maps show two major sub-basins. The North Basin is >80 m deep, with the smaller South Basin >100 m deep. Water depths have varied considerably through the later Quaternary with the maximum Holocene shoreline (Fig. 6.2b) standing ~100 m above recent water levels. The drainage basin (~148,000 km^2) is closed with the lake fed mainly by the Omo River (~90%), which drains the Ethiopian Highlands to the north (Fig. 6.2b). Most remaining fluvial inflow comes from the Turkwell (Turkwel) and Kerio rivers, which originate in the Kenya Highlands, south of the lake.

Extensive outcrops of Eocene to Recent volcanic rocks are present in the drainage basin, with Precambrian or Early Palaeozoic basement rocks located mainly in the southwest (Fig. 6.2c). Cretaceous sedimentary rocks and extensive outcrops of Neogene sediments are also present (Fig. 6.2b). These include the Plio-Pleistocene Omo Group, famous for its *Australopithecus* and *Homo* fossils, and its Holocene sediments. Three volcanic centres lie on the lake axis, forming the North, Central and South islands, with a fourth volcanic complex comprising the 'Barrier' (Fig. 6.2b), which separates Lake Turkana from the Suguta Valley to the south.

Its remote location made Lake Turkana the last major water body to be 'discovered' by European explorers. Before their arrival in the late nineteenth century, the area was home to pastoralist nomadic people including the Turkana, Borana, Gabbra, Laikipiak Masai, Rendille and Samburu (Fig. 6.3a), and smaller groups of El Molo fishermen and the Dasenech who farmed along the Omo River floodplains (Brown 1989). The last two decades of the nineteenth century were a time of great difficulty for all inhabitants of the region because of droughts and disease,

R. W. Renaut, R. B. Owen, *The Kenya Rift Lakes: Modern and Ancient*,
Syntheses in Limnogeology, https://doi.org/10.1007/978-3-642-25055-2_6

Fig. 6.1 Lake Turkana islands, crater lakes and volcanoes. (**a**) North Island from Nachakui. (**b**) North Island from Koobi Fora. Note the high-lake-level erosional bench. (**c**) Central Island from the south. (**d**) South Island from Loiyangalani. (**e**) Crocodile Lake, Central Island. (**f**) Flamingo Lake, Central Island. (**g**) Tilapia Lake, Central Island. (**h**) The southern shore of Lake Turkana is formed by the 'Barrier Complex'. Nabuyatom Cone and lake at centre with South Island in the left background. Note also the north-south aligned faults and historic mugearite lava flow from the Barrier (Teleki's cone) in the foreground

Fig. 6.2 (continued) bathymetry. After Halfman et al. (1994) and locations referred to in the text. '**a**' and '**b**' constructed from GeoMapApp (http://www.geomapapp.org). (**c**) Geology of the drainage basin. After (Feibel 2011) and Olago and Odada (1996). (**d**) Distribution of Pliocene Gombe Group and younger basalts. After Haileab et al. (2004)

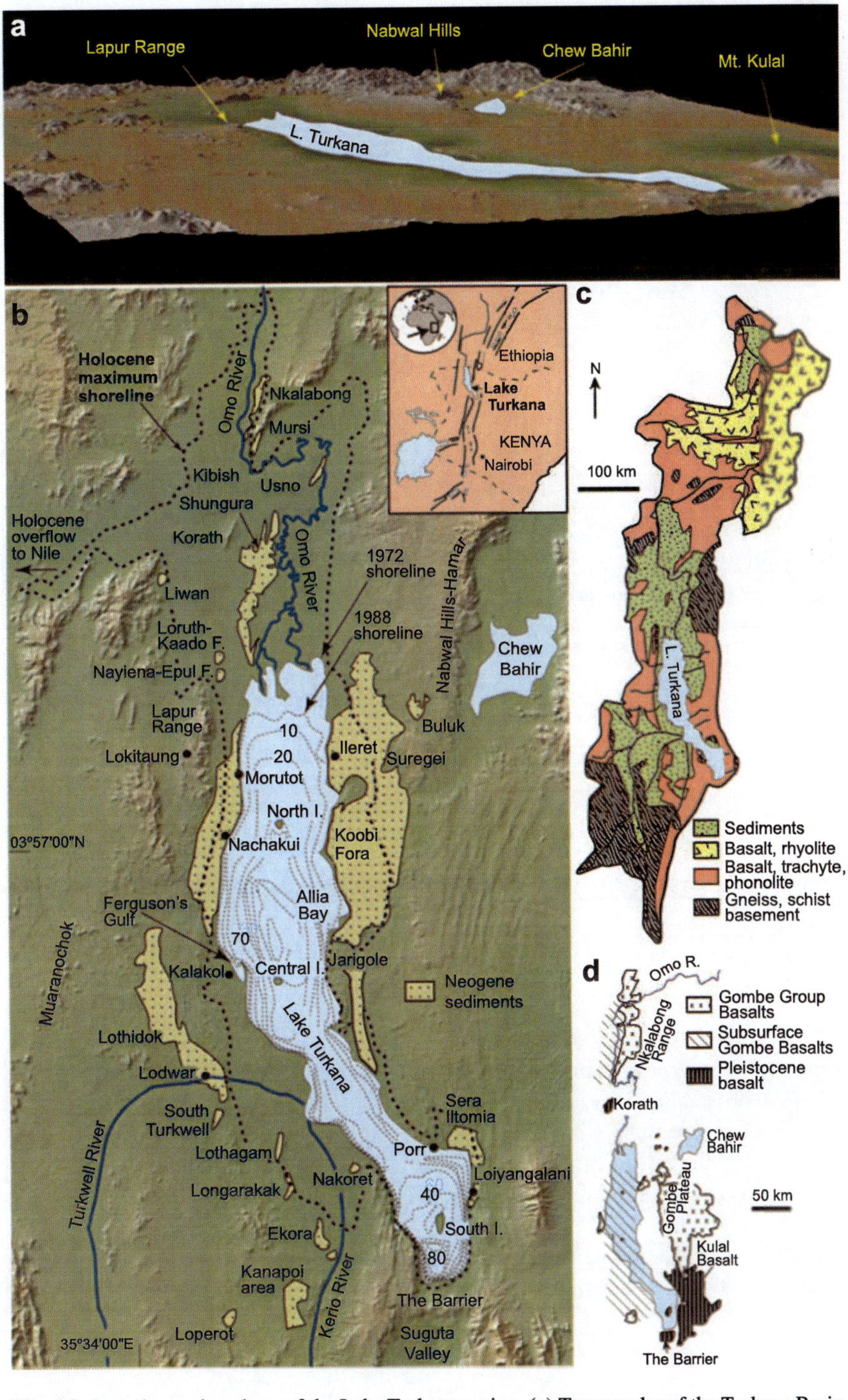

Fig. 6.2 Location and geology of the Lake Turkana region. (**a**) Topography of the Turkana Basin. (**b**) Neogene sediment outcrops, including the Plio-Pleistocene Omo Group. Modified from (Feibel 2011) and Van Bocxlaer (2020) and Holocene maximum shoreline. After Garcin et al. (2012). Lake

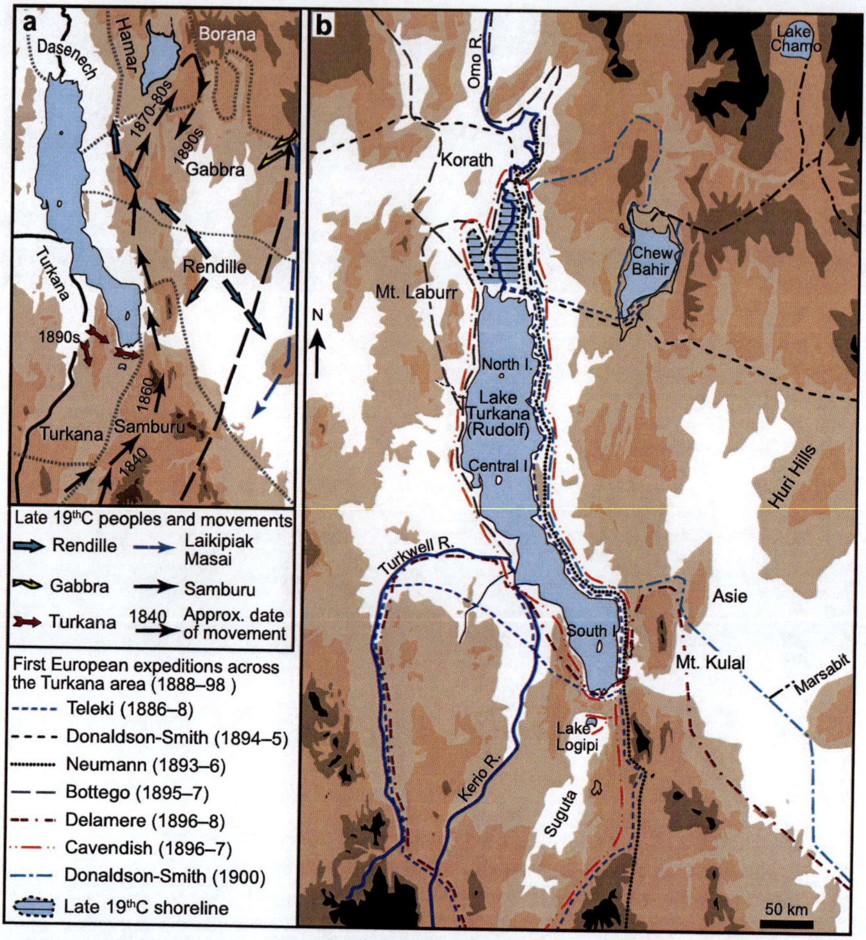

Fig. 6.3 Topography, early explorers and peoples of the Turkana region. Routes modified from Brown (1989). (**a**) Nineteenth-century peoples and migrations in the Turkana region. (**b**) Topography and the first European explorer journeys across the Turkana region

especially rinderpest (cattle plague), which decimated cattle herds and forced changes in traditional ways of life. Imperial pressures increased during the last decade of the century with the expanding influence of the Abyssinian emperor Menelik and the arrival of the first European explorers (Brown 1989).

Count Samuel Teleki de Szék and Ludwig von Höhnel reached the lake on 6th March 1888 (von Höhnel 1894, 1938; Gregory 1901). Von Höhnel et al. (1891) later published geological data from the expedition and a sketch map of the shoreline. Gregory (1894) had referred to the water body as Basso Narok ('Black Lake' in Samburu) but Count Teleki named it Lake Rudolf after the crown prince of

Austria-Hungary. In 1975 Jomo Kenyatta, the president of a newly independent Kenya, officially renamed it Lake Turkana. The lake is also known by the Turkana as Anam Ka'alakol or sea of many fish (Mwakikagile 2010) and it has been informally called the Jade Sea by Europeans (e.g. Hillaby 1964), a reference to its green colour, which reflects microbial blooms. Wellby (1900) referred to it as Lake Gallop.

Over the next decade, many geographical explorers and hunters visited its shores (Fig. 6.3b). Arthur Donaldson-Smith, reached the northeast corner on the 14th July 1895. He had hoped to find a link between the Omo River and Lake Turkana but mistakenly concluded that the Nianam (flowing into Lake Turkana) and Omo rivers were different water courses (Donaldson-Smith 1896). He was followed by Arthur Neumann, an elephant hunter, who reached the lake on his second attempt, in December 1895 (Neumann 1898). Captain Bottego arrived at the Omo River in July 1896 and followed its course south to Lake Turkana (Vannutelli and Citerni 1899). Maurizio Sacchi, a geologist, noted lacustrine sediments along the river valley but died during the expedition, so his work was published posthumously by Angelis d'Ossat and Millosovich (1900). The British explorer Lord Delamere left Berbera for Lake Turkana in December 1896 arriving at its southern shores before following the lake to the Turkwell River and then continuing to Lake Baringo (Sharf 2005). H Andrew and HSH Cavendish reached the mouth of the Omo River on 22 March 1897, having passed Lake Stefanie (Chew Bahir) several days earlier (Cavendish 1898). They separated on 26 March and travelled south along both the eastern and western shorelines, before meeting again at the southern end of the lake.

A decade after Count Teleki first visited Lake Turkana, Herbert Austen approached from the southwest, partly following the Turkwell River and first saw the lake in August 1898 (Austin 1899). His expedition was one of several French and British traverses across the Nile region in the "Scramble for Africa" (Barber 2006; Maxon 2009). Austen had been instructed to "prevent foreign powers establishing themselves in the basin of the Nile between the river and Abyssinia" but was told to avoid conflict with the Abyssinians (Barber 2006). James Harrison, with Butter, Powell-Cotton and Whitehouse, took a different route along the Ethiopian Rift, starting at Zeilla on the Gulf of Aden in November 1899 (Harrison 1901).

Over the next decade, after several military expeditions, British administration became established over the Turkana region. Uganda had struggled with accepting responsibility for the areas west of Lake Turkana, which eventually came under the influence of British East Africa from May 1919 (Barber 2006). Harris et al. (2006) noted that surveying and (or) military missions were sent to Lake Turkana during the early twentieth century, including those of Bulatovich (1900), Maud (1904), Brooke (1905), Gwynn (1911), and Stigand (1910). Early reports about the area were later drafted by Hoey (1911), Archer (1913), Holland (1926), and Hodson (1929).

After the initial period of exploration, military expeditions, and questions over administration had ended, a more intensive phase of scientific exploration began with the 'Cambridge Expedition to the East African Lakes' in 1930–1931, led by the ecologist and administrator, Edgar Barton Worthington. They focused on the limnology and biology of Lake Turkana and other rift lakes (Beadle 1932; Parker 1932; Worthington 1932, 1936; Rich 1933; Worthington and Ricardo 1936). Vertebrate fossils from the Omo Valley were first reported by Brümpt, a member of the Bourge de Bozas expedition of 1902–1903. Those fossils stimulated palaeontological interest (Haug 1912; Joleaud 1920; Boulenger 1920) and led to the 'Mission Scientifique de l'Omo' in 1932–33, organised by Camille Arambourg and René Gabriel Jeannel (Arambourg 1935, 1943, 1947; Arambourg et al. 1969; Coppens et al. 1976; Harris et al. 2006). Although focused on the sedimentary outcrops and fossils, the expedition also examined igneous rock, described by Jérémine (1935).

Vivian Fuchs explored the geology of the Turkana basin for the Cambridge expedition and followed up his studies by leading the 1934 'Lake Rudolf Valley Expedition', with the aim of carrying out geological surveys, and collecting plant and animal specimens (Fuchs 1934, 1935, 1939). Physiographical and igneous research in the southern Lake Turkana basin was carried out by Arthur Champion (Champion 1935, 1937; Champion and Smith 1937; Smith 1938) with additional geological and biological studies taking place during the Sagan-Omo expedition of 1939 (Zavattari 1941). Geological studies continued with the work of Frank Dixey (1948) who carried out a survey of northern Turkana and discovered Miocene vertebrates in south Turkana (Dixey 1945).

After a break, geological research resumed during the late 1950s and early 1960s. Whitworth (1960, 1965) described Pleistocene to Holocene sediments west of Lake Turkana. Robbins (1967, 1972, 1975, 2006) examined the archaeology of Late Pleistocene and Holocene sediments at Lothagam (Fig. 6.2b) and reported the existence of older fossils (Leakey and Harris 2003). Patterson then led Harvard scientists in an examination of Mio-Pliocene vertebrates at Kanapoi (Fig. 6.2b) and Lothagam (e.g. Maglio 1973; Hooijer and Patterson 1972) and discovered fragments of a hominid and a hominin (Patterson et al. 1970; Patterson and Howells 1967). Maglio and Behrensmeyer carried out additional mapping and geological investigations in 1968 (Behrensmeyer 1976).

Geological mapping south of Lodwar (Fig. 6.2b) was carried out during the 1960s (Joubert 1966; Fairburn and Mathieson 1970; Dodson 1971). Walsh and Dodson (1969) mapped much of the area to the north. During that period, Reilly et al. (1966) dated lavas north of Lothidok (Fig. 6.2b). Additional mapping near South Horr included the southeast of Lake Turkana (Dodson 1963). That area was subsequently studied again, together with the Suguta Valley, as part of the 'South Turkana Expedition' of the Royal Geographical Society (Caukwell 1971).

In later decades, research advanced on many fronts, stimulated mainly by academic interest in hominin fossils. Scientific exploration of the region was also encouraged by a search for potential economic benefits that the lake might yield in terms of fisheries. More recently, exploration for hydrocarbons has stimulated environmental impact studies (e.g. Earthview Geoconsultants 2011) and led to seismic investigations that have revealed previously unknown geology below Lake Turkana.

6.2 Geological and Geomorphological Setting

6.2.1 *Lithologies and Stratigraphy of the Turkana Drainage Basin*

Extensive areas of basement gneisses and schists are present to the southwest and in other smaller exposures such as in the Nabwall-Hamar area, Muruanachok Hills and the Lapur (Lapurr, Labur) Range (Fig. 6.2). Sediments and sedimentary rocks are common in many areas. Dinosaur-bearing Cretaceous sandstone, conglomerate and minor mudstone rest on basement rocks in the Lapur Range to the northwest of Lake Turkana and crop out at Sera Iltomia and in the Muruanachok Hills (Fig. 6.2b). The oldest volcanic rocks are of Late Eocene to Middle Miocene age and are comprised of 3.5 km of basalt and less common rhyolite. These form the 'Turkana Volcanics', which rest on the Late Cretaceous to Late Eocene Lapur Sandstone Formation, northwest of Lake Turkana (Thuo 2009; Tiercelin et al. 2012). To the northeast, in the Nabwall Hills, the Asille Group also includes Oligo-Miocene basalts and rhyolites that rest directly on metamorphic basement (Walsh and Dodson 1969; McDougall and Watkins 2006; McDougall and Brown 2009) with Oligocene volcanics cropping out in many other parts of the basin (Brown and McDougall 2011).

Early Miocene strata are dominated by volcanic rocks (Brown and McDougall 2011) with sediments also present in the Bakate Formation (at Buluk and in the Nabwall Hills; Fig. 6.2b) and areas north of Porr and at Lothidok (Fig. 6.2b; Boschetto et al. 1992). Rhyolitic ash-flow tuff, rhyolite, phonolite and basalt dominate Middle Miocene strata both to the west and east of the lake (McDougall and Watkins 1988, 2006; Morley et al. 1992; McDougall and Brown 2009; Brown and McDougall 2011) with Late Miocene volcanics and sediments cropping out at Lothagam (Fig. 6.2b; Powers 1980; McDougall and Feibel 1999; Feibel 2003; Brown and McDougall 2011).

The Plio-Pleistocene Omo Group crops out in many areas, especially in the Omo Valley and inland along the northeast and northwest shores, as well as in scattered outcrops to the south (Fig. 6.2b). The Pliocene Omo Group stratigraphy is documented by (1) the Mursi, Nkalabong and Usno formations in the Omo Valley; (2) the Lower Shungura (Omo Valley) and Lower Koobi Fora (NE Lake Turkana) formations; (3) the Lower Nachakui Formation (NW Lake Turkana); and (4) the Kanapoi Formation to the southwest of the lake (Harris and Leakey 2003; McDougall and Brown 2008; Brown and McDougall 2011). Considerable effort has been devoted to developing a chronology, supported by the geochemical correlation of tephra in the Omo Group in order to understand the rich mammalian record and important ancestral human fossils (e.g. Cerling et al. 1979; Cerling and Brown 1982; Brown et al. 1985; Pickford et al. 1991; deMenocal and Brown 1999; Woldegabriel et al. 2005). The Pliocene Mursi (Omo Valley) and Gombe (East Turkana) Basalts appear to be correlative and have been referred to the Gombe Group with an age of about 4 Ma. These basalts are present across the entire basin south of the Ethiopian Highlands (Fig. 6.2d; Haileab et al. 2004) and may also correlate with prominent subsurface reflectors below Lake Turkana (the Turkana Reflector) and within a series of buried half-grabens to the southwest (the Orange horizon) (Dunkelman et al. 1988, 1989; Morley et al. 1999a). In addition, Davidson (1983) suggested that the Mursi Basalt extends below the Omo Valley Plain (Fig. 6.2d). Other Pliocene basalts, such as the Kankam basalt near Loiyangalani in the southeast, have also been reported from local outcrops (Gathogo et al. 2008).

Pleistocene Omo Group sediments (Upper Shungura, Upper Koobi Fora and Upper Nachukui formations) are widespread, except for the periods between about 1–0.8 and 0.7–0.2 Ma (Brown and McDougall 2011). Much of the present topography (Fig. 6.3b) was determined by Pleistocene faulting and by the eruption of several large central volcanoes such as Mt. Kulal at 2.4–2 Ma (Gathogo et al. 2008), Asie at 2.7–0.4 Ma (Charsley 1987) and the Huri Hills, perhaps around 3.5–2.6 and at 1.6–0.8 Ma (Brotzu et al. 1984). In addition, eruption of the Lenderit Basalt in the southeast of the basin may have blocked the drainage of a proto-Omo River, leading to renewed lacustrine sedimentation at ~2.02 Ma (Bruhn et al. 2011). The Barrier complex to the south of Lake Turkana was formed in several phases after 1.37 Ma (Bloomer et al. 1989; Curtis 1991; Dunkley et al. 1993; Karson and Curtis 1994). Three other basalt-trachyte volcanoes occur along the axis of Lake Turkana, forming its three islands (Fig. 6.1a–d), with a further centre dominated by basanite and tephrite at Korath to the north (Fig. 6.2b; Brown and Carmichael 1969, 1971; Bloomer et al. 1989). The youngest Quaternary sediments are deltaic clay, silt and sand that comprise Members 1–4 of the Kibish Formation in the Omo Valley (Brown and Fuller 2008; Brown and McDougall 2011). Member 4 is of Holocene age and equivalent to the Galana Boi Formation, which is widely distributed around Lake Turkana at heights of up to ~80 m above the 1977 lake surface (Owen et al. 1982).

6.2.2 Regional Tectonic Framework

The East African Rift System (EARS) crosses the Ethiopian and Kenya Plateaus (Gregory 1920, 1923), which rise to heights of >1000 m (Fig. 6.4) and which were formed in response to mantle plume activity that dates to the Eocene–Oligocene (Ebinger and Sleep 1998; Furman et al. 2006; Pik et al. 2008). The Turkana Depression separates these two domal areas and forms part of an NW–SE trending region of low ground (Fig. 6.4; Corti et al. 2019). To the southeast, the Turkana Depression is aligned with the Cretaceous–Tertiary Anza Graben where Neogene reactivation of faults has taken place (Ebinger and Ibrahim 1994; Bosworth and Morley 1994; Furman et al. 2004; Boone et al. 2018). However, gravity data suggests that this ancient rift does not cross the trend of the more recent Tertiary rift that hosts Lake Turkana (Morley et al. 1992).

The Ethiopian and Kenya rifts are linked by a 200-km-wide zone of north–south faulting that crosses the Turkana Depression (e.g. Baker and Wohlenberg 1971; Fitch and Vondra 1976; Cerling and Powers 1977; Karson and Curtis 1989; Chorowicz 2005; Mariita and Keller 2007). In this region, the crust is about 20 km

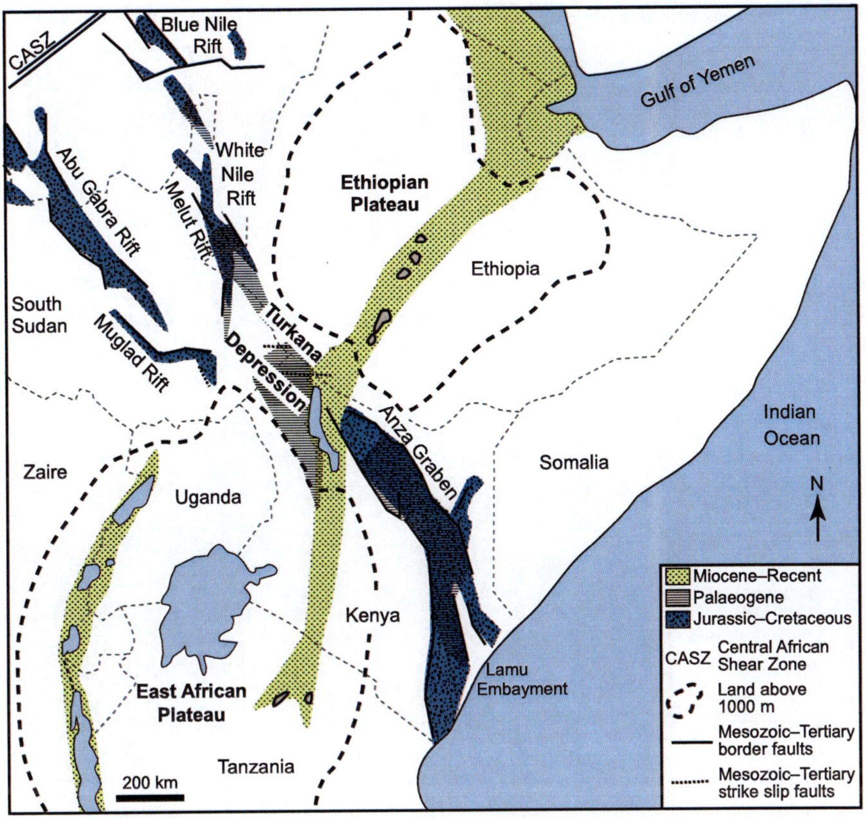

Fig. 6.4 Jurassic to Recent rifting in East Africa. Note the contrasting orientations of the Miocene to Recent East African Rift System and the NW–SE trending Anza and Sudan rifts. After Ebinger et al. (2000)

thick compared with 35 km below the central Kenya Rift (Macdonald et al. 1994; Mutegi 2014). The extension has been estimated at about 35–40 km (Hendrie et al. 1994; Morley et al. 1992), although Le Gall et al. (2005) have suggested smaller values of 20–25 km. Modelling indicates rates of 4.3–3.1 mm y^{-1} (Stamps et al. 2008; Saria et al. 2014) with Knappe et al. (2020) suggesting that extension is confined to a narrow region, not distributed across a broad area. Riedl et al. (2022) indicated current rates of 1.0–1.6 mm y^{-1}, rising locally to 2.0 mm y^{-1}, in the inner graben of the North Kenya Rift. Morley (2020) notes that early to Middle Miocene dykes (typically <2 m wide) accommodated <2% of regional extension at that time.

Rifting within the Turkana Depression developed at different times in different places. The first volcanism (~33 Ma) was centred on the 4-km-thick Lotikipi and 6-km-deep Gatome basins (Fig. 6.5), which lie at the southeast end of the Abu Gabra and Muglad Rifts of South Sudan (Fig. 6.4; Wescott et al. 1999). The Lotikipi Basin is filled with Cretaceous or Palaeogene (?) sediments, which may correlate

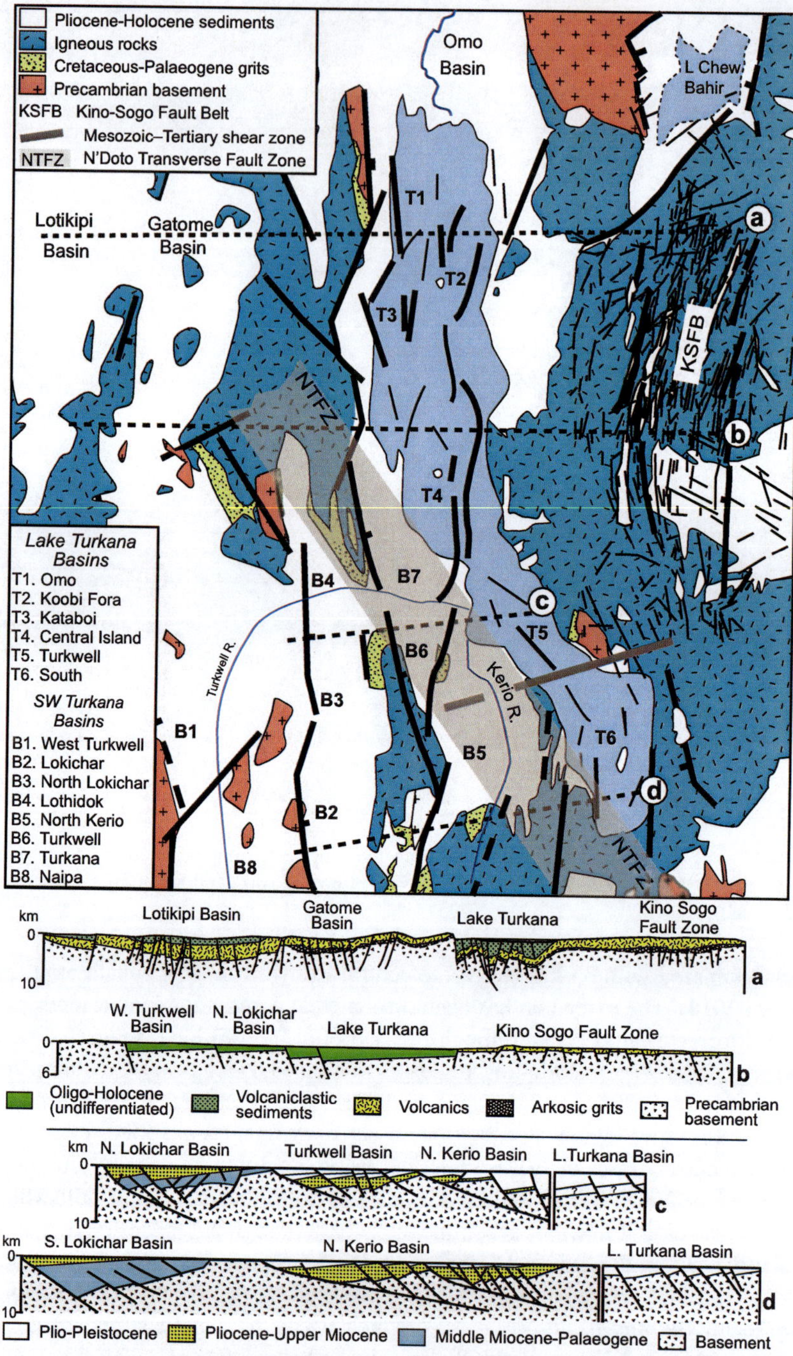

Fig. 6.5 Tectonic setting and geological sections across Lake Turkana. Mapped faults after Morley et al. (1999a) and Vétel and Le Gall (2005). Section '**a**' after Morley et al. (1992); Section '**b**' after Vétel et al. (2004); Sections '**c**' and '**d**' after Le Gall et al. (2005)

with the Lapur Sandstone Formation, and are overlain by Oligocene–Miocene basalt (2.1–2.5 km thick) and rhyolite (>1 km thick) of the Turkana Volcanics. The sequence is capped by >900 m of Pliocene to recent sediments. Wescott et al. (1999) suggest that the volcanics were probably extruded towards the end of an early phase of rifting and during the initial stages of thermal subsidence.

Surface evidence for rifting to the southwest of Lake Turkana is unclear, although drainage anomalies that reflect faulting have been reported (Vétel et al. 2004; Brown and Gathogo 2005). However, seismic studies have revealed eight major Palaeogene to Pliocene, N–S trending, half-grabens (B1–B8 in Fig. 6.5) with up to 7 km of volcanic and fluvio-lacustrine fill (Morley et al. 1992, 1999a, b; Morley 1999; Tiercelin et al. 2004; Abdelfettah et al. 2016). After about 3.7 Ma, many of these basins were affected by a 40 × 100 km elongated zone of tectonic inversion, which overlies the N'Doto Transverse Fault Zone (NTFZ) (Fig. 6.5).

Rifting below Lake Turkana was poorly understood until the seismic surveys of Project PROBE (Duke University) and the work of Dunkelman (1986) and Dunkelman et al. (1988, 1989). Previously, Khan and Swain (1978) had used gravity data to show the existence of a thick sedimentary sequence below the lake. Several researchers (e.g. Baker et al. 1972; Mohr and Wood 1976) had considered that the zone of active rifting passed through the South Basin of Lake Turkana, northwards through the Kino-Sogo Fault Belt (KSFB, Fig. 6.5; Vétel et al. 2005; Vétel and Le Gall 2005, 2006; Vétel 2008) and on to Lake Chew Bahir in southern Ethiopia. However, Project PROBE demonstrated the presence of six 25–40 km long half-grabens below Lake Turkana (T1–T6 in Fig. 6.5). Border faults along these basins were shown to have displacements of several kilometres with monoclines, flexures and step faults forming the opposing side of most basins (Dunkelman et al. 1988, 1989). Holocene extension has been estimated at about 7.5–9.6 mm y^{-1} (Melnick et al. 2012). The thickest sediment sequences were documented close to the border faults, and it was shown that they tended to become thicker in the northern basins (up to 4 km). An accommodation zone between basins B5 and B6, SW of Lake Turkana, appears to follow a Late Mesozoic–Early Tertiary ENE-trending dextral shear zone (Fig. 6.5; Williamson and Savage 1986). Dunkelman et al. (1988) also observed that the Barrier volcanic complex to the south, the three volcanic islands in the lake, a buried volcanic centre, and the volcanic Korath Range to the north, all lie at regular intervals of about 51 km. Consequently, they argued that the active rift passes through Lake Turkana rather than the KSFB.

Furman et al. (2006) also observed that Miocene to Recent fractures were initiated around 25 Ma and suggested that faulting had gradually migrated eastwards from the Lotikipi area towards the KSFB (Fig. 6.5), which is the youngest rifting domain in the Turkana Depression. The latter zone of faulting is comprised of numerous horsts and grabens within a 40-km-long arcuate zone that cuts through Mio–Pliocene lavas that overlie an earlier asymmetric fault block some 150 km long.

6.3 Climate and Basin Hydrology

The Lake Turkana drainage basin (146,000 km^2) (Johnson and Malala 2009) includes high rainfall sectors of the Ethiopian Highlands to the north and arid parts of northwest Kenya. The lake lies in a closed basin and is the largest in the Kenya Rift (250 long; 30 km wide) with a surface area variously reported at 6750 km^2 (Källqvist et al. 1988), 7500 km^2 (Olago and Odada 1996), 8860 km^2 (Spigel and Coulter 1996) and 7300 km^2 (Olaka 2011), reflecting changing lake levels over a 20-m range.

Lake level through the last ~150 years (Fig. 6.6a) has been determined mainly by rains that fall during the biannual passage of the East and West African Monsoon, which in turn reflects sea surface temperatures in the West Indian and East Atlantic oceans (Bloszies and Forman 2015). The greatest shoreline variability occurs along the gently shelving northern coast where the Omo River enters the lake (Fig. 6.3b). The maximum water depth ranges between 109 and 125 m (Yuretich 1979; Hopson 1982; Källqvist et al. 1988) depending on precise survey locations and (or) fluctuating surface height. Yuretich and Cerling (1983) estimated a volume of 245 km^3 with the deepest waters present in the southern of its two major basins (Fig. 6.2b). In recent decades, water depths have averaged about 35 m (Ferguson and Harbott 1982), but this value was considerably greater during the early Holocene when the lake stood up to 100 m higher than the present surface of ~360 m above sea level (Garcin et al. 2012). Shoreline deposits and wave-cut benches (Fig. 6.7) are widespread at heights that reach up to 450–455 m, just below the Early Holocene outlet to the Nile River at 457–460 m (Butzer et al. 1972; Owen et al. 1982; Harvey and Grove 1982; Nyamweru 1989; Garcin et al. 2012).

Mean annual rainfall is erratic and varies across the basin and from year to year (Källqvist et al. 1988; Liti et al. 1991). Butzer (1971) noted average rains of 360 mm y^{-1} for the Omo Delta. Ferguson and Harbott (1982) report an average of 154.9 mm y^{-1} at Lodwar, while Avery and Tebbs (2018) observed that Lodwar is characterised by considerable interannual variations (Fig. 6.6b, c). Precipitation around most of the lake is reported at <250 mm y^{-1}, with the highest probability of rainfall between March–May and November–December (Fig. 6.6b). Rainfall over the lake has been estimated from satellite data, but a lack of ground-truthing makes estimates uncertain (Fig. 6.6d; Avery and Tebbs 2018).

Surface elevation is controlled by the balance between evaporation and fluvial inflow, mainly from the Omo River to the north, which supplies 80–90% of all overland water inputs (Ferguson and Harbott 1982; Yuretich and Cerling 1983). The Omo drains the volcanic Ethiopian Highlands where rainfall (800–1200 mm y^{-1}) is concentrated in a single July to August season (Johnson and Malala 2009). Most of the remaining inflow is derived from the Turkwell and Kerio Rivers, which drain metamorphic and volcanic highlands to the southwest where two rainy seasons prevail (around April and September) (Nicholson 2014, 2015). Groundwater and many smaller streams to the east and west of the lake supply additional minor inputs.

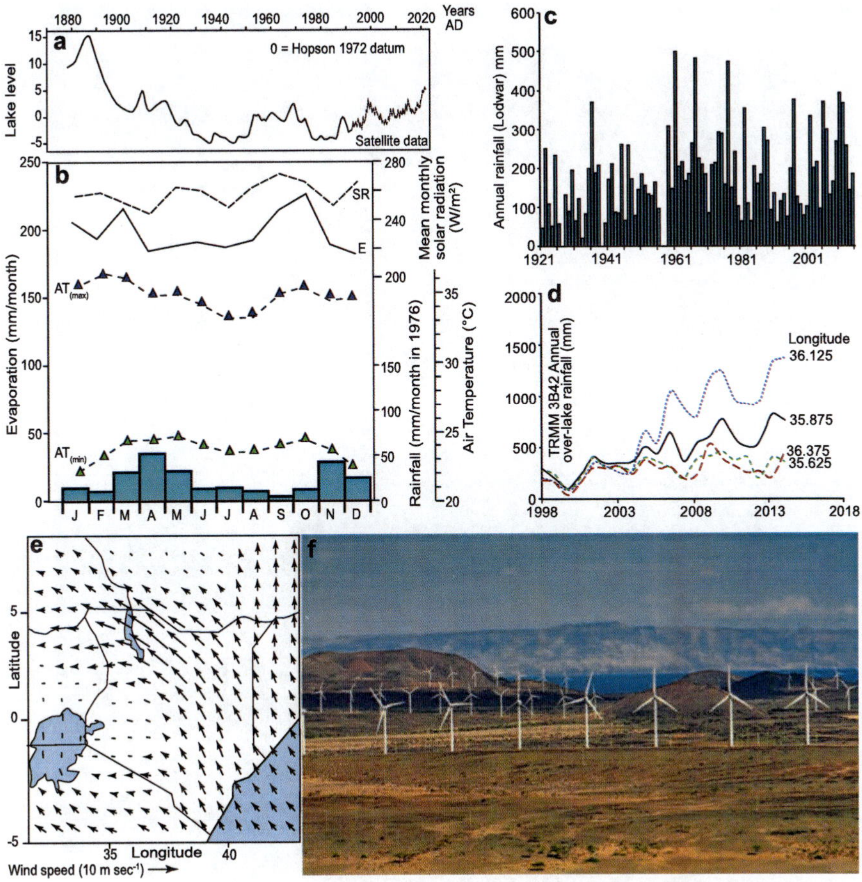

Fig. 6.6 Historical lake levels and climate. (**a**) Lake level fluctuations after Butzer (1971), Avery (2010) and TOPEX/POSEIDON/Jason-1 and Jason-2/OSTM altimetry. (**b**) Climate data modified from Spigel and Coulter (1996) with rainfall representing averages for Ileret and Lodwar (locations in Fig. 6.2b). E Evaporation, SR Solar Radiation; histogram shows rainfall. (**c**) Inter-annual rainfall variability at Lodwar (1921–2015). (**d**) Overlake rainfall estimates based on TRMM 3B42 satellite transects (1998–2005). Data currently lacks ground-truthing and are of uncertain accuracy. ((**c**–**d**) based on Avery and Tebbs (2018)). (**e**) The Turkana Jet blows at speed of 30 m s^{-1} between the Ethiopian and Kenya Highlands. (**f**) Wind power development SE of Loiyangalani takes advantage of locally high wind velocity. The site covers about 160 km^2 and produces >300 MW of power to the Kenya grid via Suswa Transmission Substation, with considerable potential for expansion. Southern Lake Turkana in the background

Spigel and Coulter (1996) report that rivers and storm runoff add a total of 17.5 km^3 of water annually in addition to 3.2 km^3 derived from direct precipitation onto the lake surface (360 mm y^{-1}), which is balanced by a loss of 21 km^3 from evaporation (2340 mm y^{-1}). The net effect of the balance between inflow and evaporation is seasonal variability in lake level of about 1–1.5 m with levels peaking in October to November and being lowest from June to July (Källqvist et al. 1988).

Fig. 6.7 Early Holocene high lake levels. White arrows mark Early Holocene high lake stand. (**a**) Holocene beach ridges on South Island (midwestern shoreline). (**b**) Prominent maximum lake level erosional bench, SE Lake Turkana, north of Loiyangalani

Historical data and gauge records indicate changes in surface height ranging by ~20 m since the 1880s (Fig. 6.6a). Lake levels have been relatively high through much of the last decade, resulting in a 10.4% expansion in the lake area between 2010 and 2020 (Government of Kenya and UNDP 2021).

Daily insolation is similar throughout the year due to proximity to the equator and low cloud cover (Fig. 6.6b). A semi-permanent low-level jet (Fig. 6.6e) with common speeds of 30 m s^{-1} flows from the southeast to the northwest along the

Turkana Depression and between the Ethiopian and East African Highlands (Hartman 2018). The strongest winds blow at night and through the early morning and may suppress rainfall. Vertically, the jet appears to be comprised of two lower (~0.5–2 km) and upper (2–3 km) zones (Nicholson 2015). Evaporation is driven by the strong diurnal winds and by moderately high air temperatures (~19.5–39.9 °C; mean = 29.26 °C; Fig. 6.6b), which vary little through the year — for example, by only 5.86 °C at Longech on Ferguson's Gulf (Fig. 6.2b; Ferguson and Harbott 1982). Vannutelli and Citerni (1899) reported that strong southeasterly winds dominate along northeast coasts, with Beadle (1932) and Butzer (1971) making similar observations. Data for March 1973 to September 1975 at Longech indicate 10-day average speeds of 7.1–19.8 km h^{-1}, with a maximum of 35 km h^{-1}. The same data show that daily winds gradually increase overnight and form strong morning southeasterlies before a calm afternoon period (Fergusson and Harbott 1982). The predictable strong winds have also enabled the development of wind power turbines to the southeast near Loiyangalani (Fig. 6.6f; Achiba 2019).

Major changes to the hydrology of Lake Turkana may follow from the construction of the Gibe Dams in Ethiopia and the potential development of large-scale sugar and cotton irrigation systems near the Omo River, which supplies >80% of the lake's water (Avery 2010, 2012, 2014; Velpuri and Senay 2012; Velpuri et al. 2012; Gownaris et al. 2015, 2016; Carr 2017; Obiero et al. 2022). Lake level fell during reservoir filling and could drop further with future development. Regulation of river flow in subsequent years is likely to alter the current natural pulses in the Omo flow regime and, in turn, Lake Turkana's ecology and fisheries (Tebbs et al. 2015; Hodbod et al. 2019). Avery and Tebbs (2018) modelled lake-level responses to dam construction and irrigation (Fig. 6.8) and noted that the lake dropped 2 m during 2015–16 in response to filling of the Gibe III reservoir. They also suggested that large-scale irrigation could reduce Omo flows by as much as 50%, leading to falls of >15 m, causing an environmental disaster similar in scale to that of the Aral Sea in Asia. Consequently, the Lake Turkana National Parks World Heritage Site was placed on a danger list in 2019.

6.4 Peoples, Vegetation and Wildlife

The population density around Lake Turkana is very low, reflecting its semi-desert character. The largest town is Lodwar to the west of the lake with a population of around 80,000 (2019). The western side of the lake is inhabited today by the Turkana who herd goats, camels, sheep and donkeys (Kaijage and Nyagah 2010) with some of the Turkana also fishing, partly in response to the environmental deterioration of their pastoral lands (Carr 2017). The agro-pastoralist Dasenech live to the east and north and have also taken up some fishing. The pastoralist Gabbra, Rendille, and Samburu are located east of the lake (Avery 2010, 2012; Carr 2017) with the El Molo, who focus on fishing, present to the southeast near Loiyangalani (Vallerini 2019).

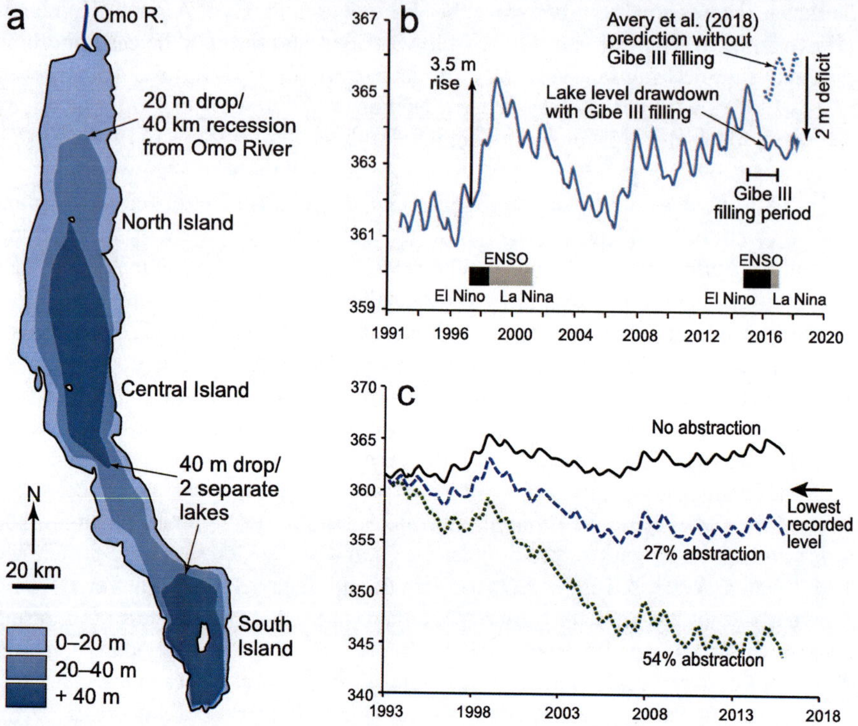

Fig. 6.8 Potential lake-level impacts from dam construction and irrigation in the Omo Valley. Based on Avery and Tebbs (2018). (**a**) Shoreline regression impacts resulting from 20 m and 40 m falls in lake level. (**b**) Impact of dams on Lake Turkana water levels. Note also predicted effects of ENSO events and the 2-m reduction related to filling of Gibe III reservoir. (**c**) Effects of different irrigation abstraction-rates from the Omo on Lake Turkana level

Vincens (1982) described vegetation in the Lake Turkana catchment (Fig. 6.9). Semi-desert shrubs and trees are common with *Acacia* (*Vachellia*), *Commiphora*, *Salvadora persica*, Capparidaceae and Tiliaceae dominating around Lake Turkana (Vincens et al. 2006; Mbaluka and Brown 2016). Larger *Acacia* occur along ephemeral streams with *Euphorbia* present locally. Grasses are common but discontinuous. Xerophyte succulents tend to be found on volcanic outcrops (Walsh and Dodson 1969). Former sandy strandlines are occupied by halophytic *Suaeda monoïca* (Chenopodiaceae). Doum palms are present along larger sand rivers and the lake shoreline. Evergreen and semi-evergreen bush develops between 800 and 1800 m elevation along with *Acacia*, Combretaceae, *Dodonaea viscosa* and *Euclea* (Vincens et al. 2006). Riverine forest is well developed along the permanent Omo River together with Combretaceae, *Ziziphus*, *Ficus* and *Trichilia* (Bonnefille 1971; Mohammed et al. 1995). In the Omo valley, *Acacia*, *Combretum* and *Terminalia* are found at heights of about 800–1500 m, with forests and scrublands of *Podocarpus*, *Juniperus* and *Olea* or *Aningeria*/*Albizzia*. Moorlands are present above 3300 m (White 1983; Friis 1992; Mohammed et al. 1995). To the southwest, semi-tropical

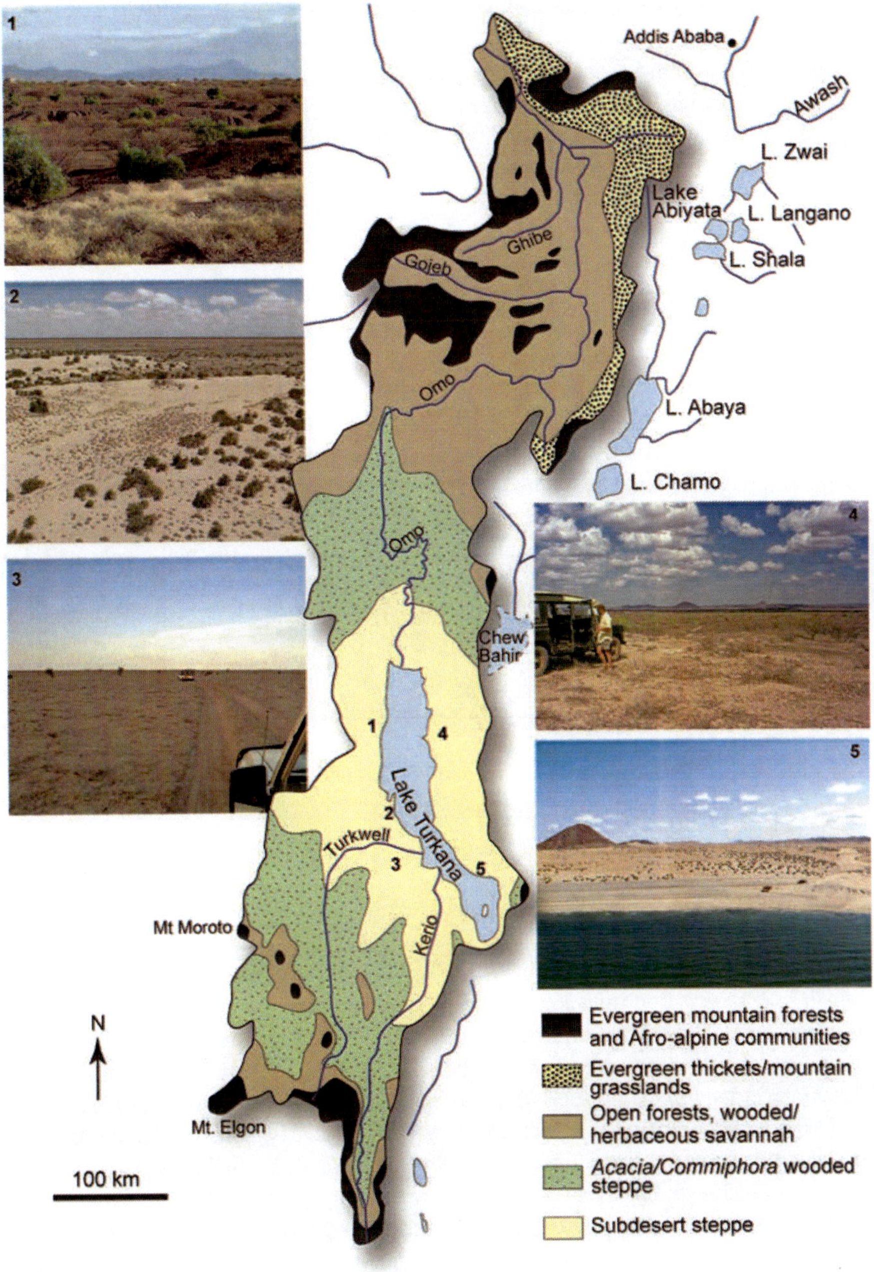

Fig. 6.9 Vegetation map of the Lake Turkana catchment. After Vincens (1982). Bold numbers on photographs match numbered locations on the map

vegetation occurs on high slopes with areas such as the floor of the Kerio Valley covered by dry thorn bush.

Extensive wetland ecosystems are generally absent, except in the far north where the Omo Delta has shifted south in response to falling lake level and sediment progradation (Haack 2007; Ayalew 2009; Kariuki 2011; Ojwang et al. 2016). Most shores lack aquatic macrophytes or papyrus (Beadle 1981). The most common emergent plants are *Sporobolus spicatus* and *Paspalidium geminatum*, which are confined to low-energy shallow water and damp soils. *Potamogeton* also provide shelter for small fish in shallow areas.

Turkana has three National Parks. Sibiloi, gazetted in 1973, is the largest and is located to the northeast of Lake Turkana where it protects a landscape dominated by grass and scattered bush. Other protected areas include the South and Central Island National Parks, which were designated in 1983 and 1985 with all three recognised as a World Heritage Site in 1997. Mt. Kulal was recognised as a Biosphere Reserve in 1978. The formerly diverse fauna at Sibiloi once included reticulated giraffe and black rhino, which are now probably absent. Many animals have disappeared over the last three decades (Cabeza et al. 2016), but wildlife that once formed part of the ecosystem and may still be present include Grevy's and Burchell's zebra, Beisa oryx, topi, greater kudu, gerenuk, Grant's gazelle, dik dik, warthogs, striped hyaena, silver-backed jackal, wildcat, baboon, cheetah, lion, leopard, crocodile, hippopotamus, and over 350 bird species, including 42 dominant water birds (Fasola and Canova 1993). The Central and North islands are home to flamingos, pelicans and other aquatic avifauna as well as hippo and crocodile (Graham 1968; Martins 2012).

6.5 Limnology

6.5.1 Physical Limnology

The surface water temperatures of Lake Turkana varied between 27 and 30 °C during 1973–75, with bottom water generally between 24.5 and 26.5 °C (Ferguson and Harbott 1982). During the same period, north-south temperature profiles (Fig. 6.10a–c) show that cooler water tended to be present to the south. A weak thermal stratification developed in the northern of the two main lake basins between October 1973 and March 1975, but stratification was very weak or nearly isothermal at these times in the southern basin. Stratification was very weak in all areas during August 1974. Figure 6.11 shows data from Site H1 (Ferguson and Harbott 1982), just north of Central Island, where nearly isothermal conditions prevailed for much of the year, but with weak stratification developing during April–August 1973 and April–May 1974. Similar results for Site H1 were documented again in April to May 1987 when a weak thermocline was observed between depths of 10–40 m (Källqvist et al. 1988). Cohen (1984) reported that weak stratification tends to develop daily, in the late afternoon when winds are reduced, only to break down again during the night

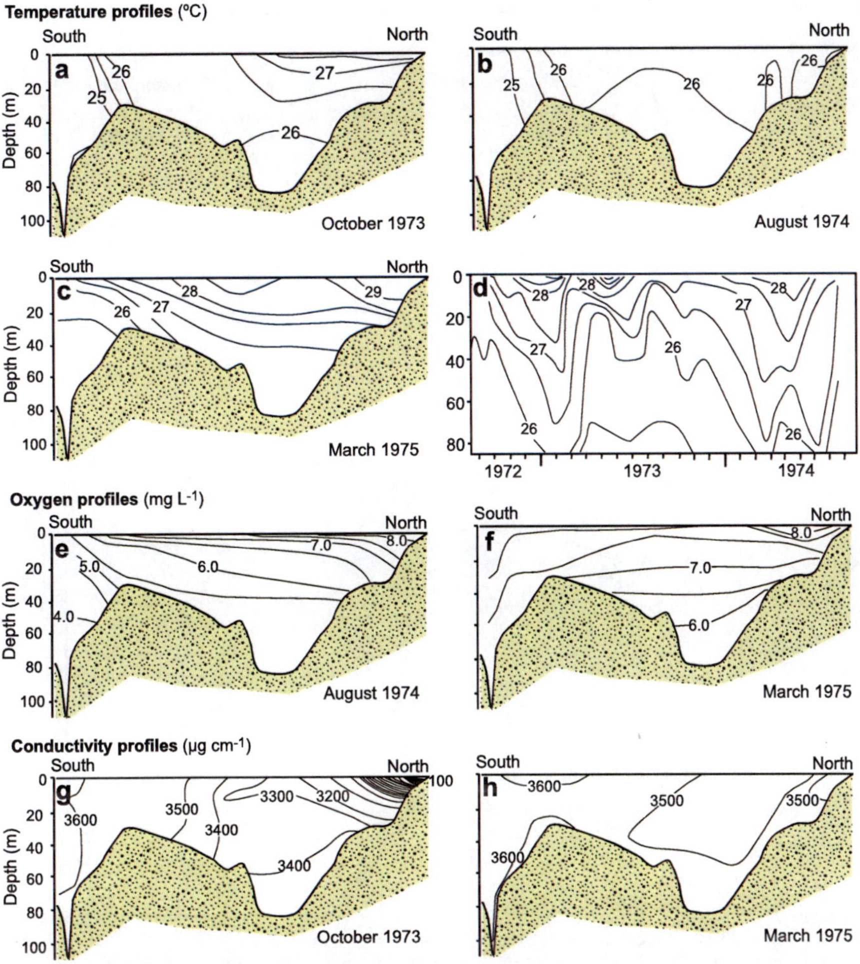

Fig. 6.10 North-south temperature, oxygen, conductivity profiles and seasonal isotherms for Lake Turkana. Based on Ferguson and Harbott (1982). Shaded area is lake floor. Note seasonal development of weak thermoclines, well-oxygenated bottom waters and seasonally strong conductivity gradient formed during Omo River floods. 'd' shows isotherms at station H1, located in Fig. 6.11

as wind speeds increase. He pointed out that thermoclines formed during these daily cycles commonly intercept the water surface with no true epilimnion developing, except in locally sheltered areas such as Ferguson's Gulf.

The influence of wind is also reflected in dissolved oxygen levels (Fig. 6.10e, f), which were relatively high at all depths during 1973–1975, with surface waters containing around 7.5–8 mg L^{-1} O_2 and bottom waters ranging between 4–5.5 mg L^{-1} O_2 (Ferguson and Harbott 1982). In May 1988, a time of relatively strong thermocline development, dissolved oxygen concentrations fell to 0.2 mg L^{-1} O_2 (Källqvist et al. 1988). The generally homogeneous nature of the lake's temperature and

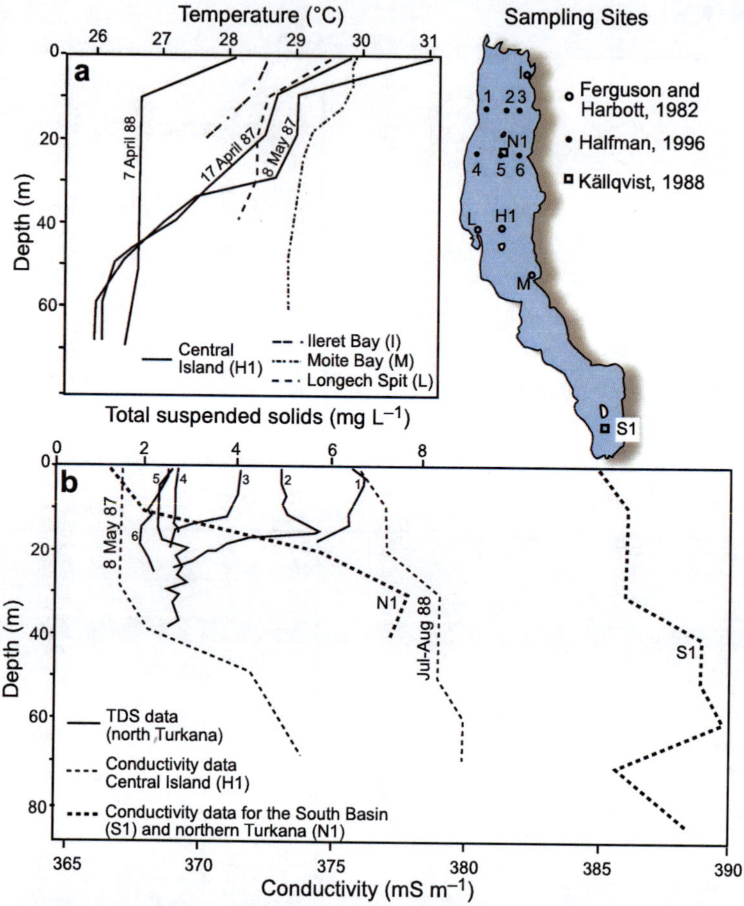

Fig. 6.11 Vertical temperature suspended sediment and conductivity profiles in Lake Turkana. (**a**) Temperature data after Fergusson and Harbott (1982) showing the development of a seasonal thermocline. (**b**) TDS and conductivity data after Halfman (1996). Increased suspended solids in surface layers reflect hypopycnal flow from the Omo River. In contrast, conductivity values show little vertical variability in the South Basin, which reflects fresher water overflows in the North Basin

oxygen profiles reflects its low depth/fetch ratio, which allows thorough mixing to the lake floor in most areas. Wind mixing is probably effective to depths of >30 m in the northern part of Lake Turkana due to long fetch distances, with mixing likely to extend down to >10 m in the south (Cohen 1984).

Upwelling has been reported to the southeast, near Loiyangalani, and along other parts of the eastern lakeshore (Fig. 6.12b; Yuretich 1979). However, across most of the lake, southeasterly winds drive water currents in the upper 6–7 m to the northwest with subsurface counter currents developing (Ferguson and Harbott 1982). A counter-clockwise gyre is formed from May–June to November in the northern part of the lake when currents are reversed by River Omo flood waters, which form a

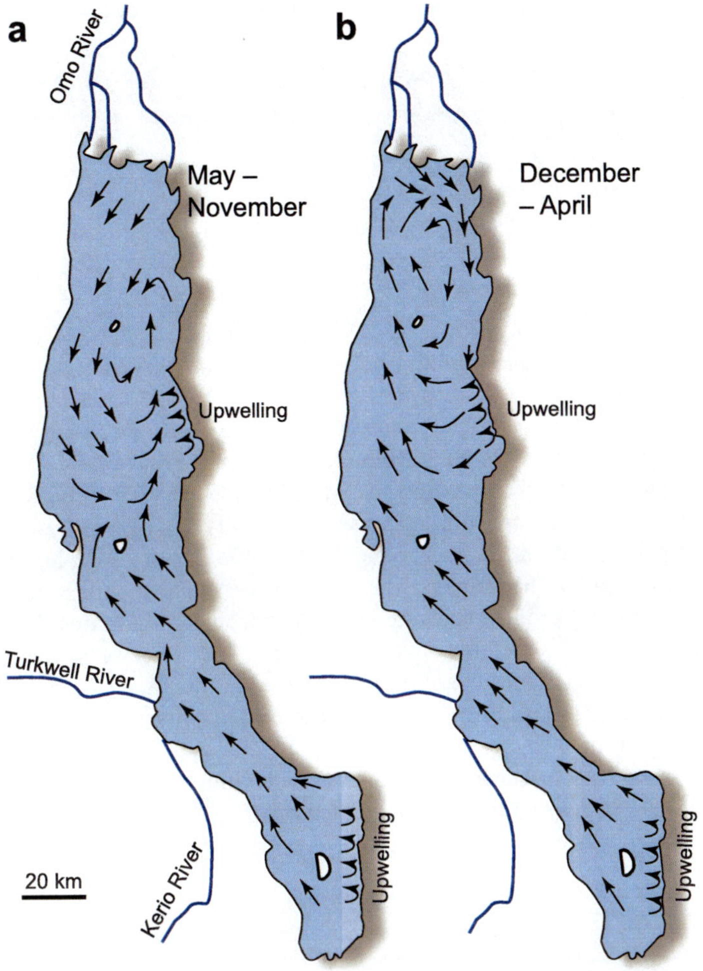

Fig. 6.12 Surface currents in Lake Turkana. After Yuretich and Cerling (1983). (**a**) Surface currents during the May–November Omo flood season. Wind-driven surface flow dominates in the southern lake. Omo overflows move down the west side of the lake as part of a larger circulation. (**b**) Current patterns on the western side of the lake reverse during December–April when the Omo floods decline and southeasterly winds drive surface lake water to the northwest

hypopycnal southward flow along the western side of the lake (Fig. 6.12a). These annual events also have significant impacts on light penetration. For example, in the early 1970s, and 25 km from the Omo River mouth, the depth of the euphotic zone (where 99% of surface light is lost) ranged seasonally from 0.5–3.7 m with depths deepening 230 km to the south to 10.25–13.0 m (Ferguson and Harbott 1982). Källqvist et al. (1988) also reported Secchi depth data along the lake and pointed out how values north of Central Island ranged from 3.0 m in March to 0.4 m in August 1988. They also estimated that the depth of the euphotic zone in the central sector of the lake varied seasonally from about 2.5–12 m with an average of 6 m. Primary production was 0.08–0.68 mg C L^{-1} h^{-1} at light saturation, generally peaking at depths of 1–2 m and ceasing below about 5–6 m.

6.5.2 *Chemical Limnology*

Lake Turkana waters are saline (~2500 mg L^{-1}) and alkaline (pH ~9.2) with Na^+ (664–940 mg L^{-1}), HCO_3^- (19–25 meq L^{-1}) and Cl^- (363–544 mg L^{-1}) forming the dominant ions (Table 6.1; Fish 1953; Talling and Talling 1965; Cerling 1979; Hopson 1982; Yuretich and Cerling 1983; Källqvist et al. 1988; Halfman et al. 1989; Johnson and Malala 2009). Total Dissolved Solids (TDS) and conductivity are high and often spatially and vertically uniform. However, conductivity gradients may develop in response to seasonal fluvial inflow (Källqvist et al. 1988).

The marked conductivity contrast between Omo River water (80 μS cm^{-1}) and the lake (3400–3700 μS cm^{-1}) has allowed conductivity to be used to trace the influence of the Omo inflow (Ferguson and Harbott 1982). For example, in October 1973, near the end of the flood season, the river had a marked effect on the North Basin with vertical profiles showing strong conductivity variations (Fig. 6.10g), whereas the South Basin varied only slightly with depth. In contrast, during March 1975 (Fig. 6.10h) the influence of the Omo River was minimal with conductivities near its delta of 3200 μS cm^{-1} and with little conductivity variation vertically or horizontally through the rest of the lake.

In August 1988, suspended particulates reached high levels (32 mg L^{-1}) near North Island with lower values in the central (2.2–3.2 mg L^{-1}) and southern

Table 6.1 Selected chemical analyses of Lake Turkana waters

Location (Ref.)	Cond. μS cm^{-1}	TDS	pH	Alk	Na	K	Ca	Mg	Cl	SO_4	SiO_2	PO_4
Central Basin (1)	2860	–	9.5	21.7	770	23.0	5.0	4.0	429	56	4.2	1.64
North Basin (2)	3190	–	9.5	23	–	–	2.45	2.4	245	–	72	–
North Basin (3)	3140	–	–	22.65	820	24.3	3.93	2.68	454	–	–	–
North Basin (4)	–	2279	9.2	19.4	678	14.4	5.9	3.1	363	32	–	–
Central Island (4)	–	2432	9.2	19.1	772	17.8	5.9	2.2	429	33	–	–
South Basin (4)	–	2536	–	–	794	19.1	3.5	2.4	520	38	19.5	–
North Basin (5)	2763.1	–	8.7	29.18	823	1.05	5.67	2.28	243		5.76	1.44
Omo River (3)	–	–	7.1	–	17	2.8	11.0	5.6	3.6	–	–	
Omo R./Lake (3)	1156	–	–	10.85	462	12.5	4.68	2.15	265	–	–	–
Kerio R. (4)	–	–	7.8	2.84	29.6	5.8	26.6	5.7	15.0	8.3	–	–

Sources: 1: Beadle (1932); 2: Fish (1953); 3: Hopson (1982); 4: Yuretich and Cerling (1983); 5: Olago (1992)

Major ions, silica and Total Dissolved Solids (TDS) given in mg L^{-1}

Cond. conductivity; *Alk* Total alkalinity in meq L^{-1}

(1.4–1.9 mg L^{-1}) sectors of the lake (Källqvist et al. 1988). Most of the particulates were inorganic, ranging from 88% in the north to 54% in the south with the organic fractions being largely autochthonous. In January 1992, TDS data for several sites north of North Island (Halfman 1996) showed little variation (2.0–6.8 mg L^{-1}) in the upper 20–40 m, with profiles displaying either a slight decrease in turbidity with depth or homogeneous conditions (Fig. 6.11b).

The overall concentration of salts is not as high as expected for a lake that has been closed and evaporating for several thousand years (Beadle 1932). Concentrations of Ca^{2+} (3.5–5.9 mg L^{-1}) and Mg^{2+} (1.9–2.8 mg L^{-1}) are particularly low with K^+ (17.4–19.2 mg L^{-1} in the deeper lake) showing smaller losses (Yuretich and Cerling 1983). In contrast, Cl^- and Na^+ have behaved as relatively conservative ions. These features have been attributed to the loss of ions to the sediments and the burial of solutes in pore waters (Yuretich and Cerling 1983; Cerling 1986). For example, high alkalinity and pH cause the waters to be saturated with respect to calcium and magnesium carbonates, resulting in calcite precipitation in open water (Yuretich 1979). Ca^{2+} is also removed biochemically via ostracods, molluscs and gastropods. Abundant crystalline montmorillonite (smectite) is present in lake sediments, but largely absent in suspended sediment, suggesting that Mg-rich montmorillonite has been formed from amorphous fluvial material following Mg^{2+} uptake in the alkaline lake environment (Yuretich 1979).

Both K^+ and Na^+ are also removed from the lake water but at a slower rate. Illite and K-feldspar are common in the lake sediments reflecting potassium loss through K-feldspar authigenesis and partial regradation of illite (Yuretich 1979). Sodium loss likely results from ion exchange involving clay. In addition, low sulfate levels may reflect removal by reduction to sulfide within the sediments (Yuretich and Cerling 1983; Yan et al. 2002), with methanogenesis having been reported at slightly greater depths within the lake floor deposits (Cerling 1996).

Minor ions, such as Cu, Zn, Co, Pb, Hg, Ni, Cd and Cr are low in concentration and stable in different parts of the lake (Källqvist et al. 1988). In contrast, Al, Fe and Mn are more variable and correlated with turbidity data, suggesting a link with suspended loads from the Omo. The plant nutrient phosphorus ranged from 2.2 to 2.4 mg L^{-1} in the Källqvist study, mostly in the readily available phosphate form, although distinctly lower concentrations of 0.33 mg L^{-1} have also been reported (Hopson 1982). Total nitrogen (0.5–2.1 mg L^{-1}) was high according to Källqvist et al. (1988) but mostly occurred in forms not readily available to algae, with NO_3-N almost entirely <100 μg N L^{-1} and NH_4-N < 40 μg N L^{-1}, with the lowest concentrations occurring in surface layers during periods of stratification. Based on field and experimental data, Källqvist et al. (1988) concluded that nitrogen was the major limiting nutrient for most algal groups with silica (generally <20 mg L^{-1}; Table 6.1) a limiting nutrient for diatoms. Silica loss has been attributed to diatom sedimentation, especially in the south, and the Mg-silicate reactions noted above (Yuretich and Cerling 1983).

6.6 Biology and Ecology of the Lake

6.6.1 Phytoplankton

Källqvist et al. (1988) and Harbott (1982) provide comparative data for Turkana phytoplankton with Olilo et al. (2023) describing microbial ecology. In the 1980s, the cyanobacterium *Microcystis aeruginosa* and the green alga *Botryococcus braunii* were dominant (Källqvist et al. 1988). Both occurred together, but the former tended to bloom during influxes of suspended river sediment. In contrast, *B. braunii* was more common when water was more transparent and lake level was falling. Mean values for biomass integrated over the entire water column showed that algal concentrations (biomass per unit volume) were at a maximum (>10 $mm^3 L^{-1}$) in northern parts of the lake around October–March and that these high values appeared to be linked to the Omo floods (Harbott 1982). In contrast, much lower values (0.25 $mm^3 L^{-1}$) were documented for the same season in the South Basin.

In the early 1970s, open-water phytoplankton showed low diversity, but with distinct seasonal biovolume percentage changes (Fig. 6.13a; Harbott 1982). In November–December 1972, *M. aeruginosa* constituted >93% of the assemblage in northern Lake Turkana, with *Phormidium mucicola* and *Spirulina laxissima* also present. During the same period, *Surirella* dominated in the central sector near the Turkwell River and co-dominated with *B. braunii* to the north of South Island. In contrast, a variety of diatoms (*Surirella engleri, S. birostrata, Rhopalodia gracilis, R. gibberula, Gomphocymbella beccari* and *Nitzschia palea*) co-dominated with *Microcystis* spp. south of South Island. Many of these diatoms are usually found in littoral settings and their presence may reflect wind-induced transportation (Hopson 1982). As the influence of the Omo flood waned into March 1973 algal concentrations declined in the north to about 1.0 $mm^3 L^{-1}$ with *M. aeruginosa* gradually replaced by *B. braunii* southwards and with *Surirella* spp. dominant in the South Basin. The floras changed again in June–July 1973 with lakewide increases in *Planctonema lauterbornii* (17–30%). Other taxa include *Oocystis gigas* and *Sphaerocystis schroeterii* (Källqvist et al. 1988).

Floral diversity was greater in the early 1970s in the littoral zone with 110 species reported (Harbott 1982). Chrysophyta were the largest littoral group and were comprised solely of diatoms, including *Coscinodiscus* [*Thalassiosira*] *rudolfii*, *Cyclotella meneghiniana* (*Stephanocyclus meneghinianus*), *Caloneis aequatorialis, Nitzschia granulata, N. palea, Rhopalodia gibberula, R. gracilis, R. vermicularis*, *Anomoeoneis sphaerophora*, *Cocconeis placentula* and *Cymbella* [*Encyonema*] *muelleri*. Cyanophyta were also very common and were comprised of, for example, *Spirulina platensis*, *S. laxissima*, *Anabaena circinalis* and *Anabaenopsis arnoldii* and *Microcystis* spp. Pyrrophyta were represented only by *Gymnodium aeruginosum*.

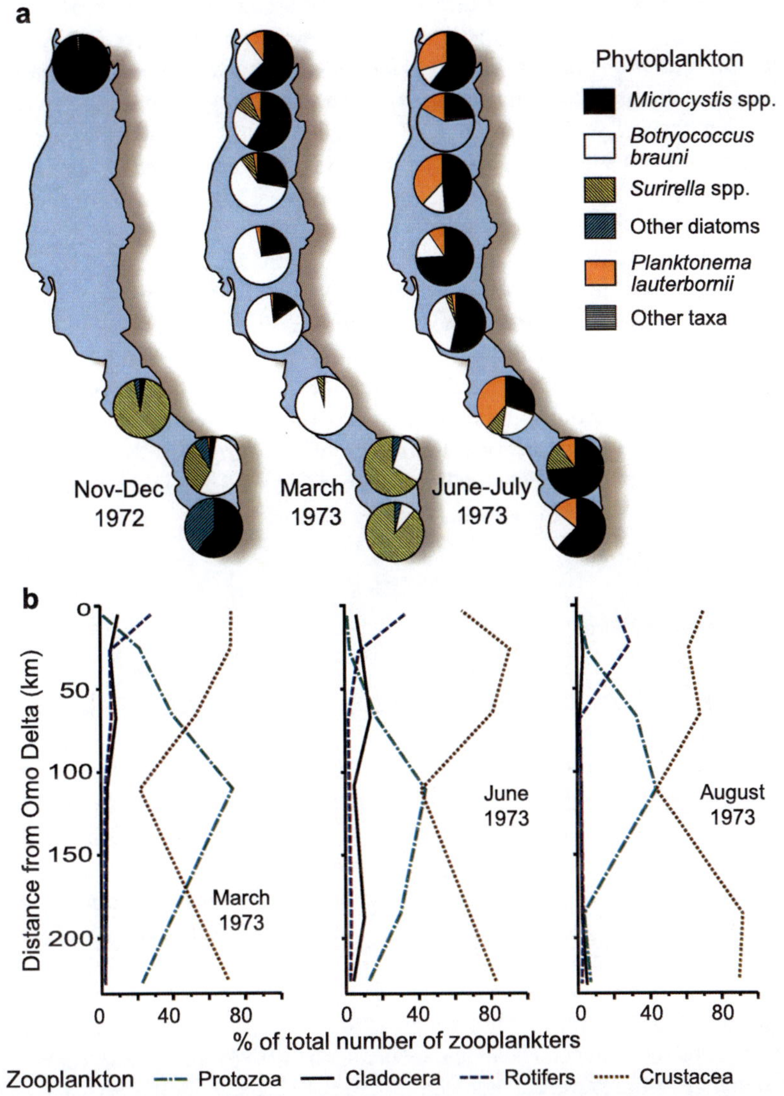

Fig. 6.13 Plankton of Lake Turkana. (**a**) Seasonal variability in phytoplankton at different sampling sites in Lake Turkana. Modified from Harbott (1982). (**b**) Variability in zooplankton with distance from the Omo River delta for different periods of the year. Modified after Ferguson (1982)

6.6.2 *Zooplankton*

During the early 1970s, common zooplankton (Fig. 6.13b) included the protozoa *Raphidiophrys* cf. *pallida*, *Vaginicola* sp. and *Vorticella* spp., as well as the rotifer *Brachionus plicatilis* (Ferguson 1982). Other taxa included *Felinia terminalis*,

Hexartha jenkinae and the crustaceans *Tropodiaptomus banforanus*, *Mesocyclops leuckarti* and *Thermocyclops hyalinus* together with the Cladocera *Diaphanosoma excisum*, *Hyalodaphnia barbata*, *Ceriodaphnia rigaudi*, *Moina brachiata* and *M. micrura*. Johnson and Malala (2009) noted that *T. banforanus* and *M. leuckarti* were later revised to *T. turkanae* and *M. ogunnus* and reported the presence of *M. salinus*. Other studies (Källqvist et al. 1988) have also noted *T. banforanus* as dominating (60%) the crustacean fauna together with common *M. leuckarti* and *T. hyalinus* (combined 30%) and cladocerans (8–10%).

Källqvist et al. (1988) noted that zooplankton abundance peaked twice yearly in March and July–August/September. Ferguson (1982) observed that *M. leuckarti* was mainly carnivorous and that there was little evidence for grazing of the dominant phytoplankton (*M. aeruginosa*, *P. lauterborni*) by the common zooplankton during the 1970s. He further noted high densities of *T. banforanus* in areas of low phytoplankton production and suggested that particulate detritus was used as a food source. It has also been observed that much of the primary carbon production is subject to bacterial decomposition before becoming available to the zooplankton in a pelagic detritus-based food chain (Kolding 1992).

6.6.3 Invertebrates

Among the molluscs, the usually benthic *Cleopatra pirothii* and *Gabiella rosea* appear to be present in deeper water with *Ceratophallus natalensis* and *Bellamya unicolor* found in shallow settings (Beadle 1981). Cohen (1984, 1986) noted that the benthos is impoverished, but includes sponges, a bryozoan, gastropods (8 species), bivalves (3), ostracods (17), insects (17+) and several hydracarines and annelids. More recently, Herren (2013) reported nematodes, rotifers, copepods, ostracods, annelids, mites, chironomids and other non-chironomid insects from hot springs and shallow water sites on deltas and beaches in Lake Turkana, as well as from Flamingo Lake on Central Island (Fig. 6.1f).

Beck (2015) identified seven ostracod genera in sub-recent deposits from Ferguson's Gulf. Littoral, soft bottom areas (<2 m deep) were characterised mainly by insects and the ostracod *Hemicyprus kliei* (Cohen 1986). In contrast, rocky gravel and shingle substrates were dominated by insect larvae and molluscs (e.g. *Gabiella rosea*, *Ceratophallus natalensis*) that grazed on epilithic algae. A third setting includes profundal (>5 m deep) muddy bottoms, which were associated with four chironomid species, stunted gastropods (*Melanoides tuberculata*, *Cleopatra bulimoides*, *Gabbiella rosea*, *Gyraulus* sp.) and ostracods with *H. kliei* more common in shallow settings and *Hemicyprus intermedia* and *Sclerocyprus* cf. *clavularis* dominant in profundal environments.

Two species of detritus-feeding prawns (*Macrobrachium niloticum* and *Caridina nilotica*) were commonly found on and above the lake floor, showing vertical migratory behaviour and tending to rise through the water column at night (McLeod 1982).

6.6.4 Fish

Fish diversity is low with 48 species reported for the 1970s, including 12 restricted to the Omo Delta and 10 that are endemic (Hopson and Hopson 1982). This is a relatively low level of endemism compared with other large African lakes (Victoria, Tanganyika, Malawi) and combined with the dominant Nilotic character of the other taxa suggests former hydrological links to the Nile River (Johnson and Malala 2009; Goodier et al. 2011).

Copley (1941) and Seegers et al. (2003) listed the fish living in Lake Turkana. Four major fish communities were distinguished by Hopson and Hopson (1982) (Fig. 6.14). The first was present in nearshore littoral (0–4 m) communities characterised by *Oreochromis niloticus* (*Tilapia*) and *Clarias lazera* (sharptooth catfish). The second occurred in inshore demersal (4–10 to 15 m) areas with large detritivorous and benthic feeding taxa such as *Labeo horie*, *Citharinus citharus* and *Distichodus niloticus*, with *Bagrus docmac* over rocky substrates. The third community was present in deep-water offshore demersal settings within 3–4 m of the lakebed, and included the predator *Bagrus bayad*, together with small endemic cichlids and barbs (Kolding 1992). A fourth pelagic community was present in offshore waters above the demersal zone, in three sublayers (Fig. 6.14). The surface layer mainly included characins, the tigerfish *Hydrocynus forskalii* and the zooplankton-feeding *Alestes baremose* (Kolding 1992). A 'midwater scattering layer' (Muška et al. 2012) was identified on echo-soundings between 5 and 30 m and was characterised by *Alestes minutus* and *A. ferox*, which formed the main biomass of Lake Turkana. Other taxa in this layer included dwarf perch (*Lates*

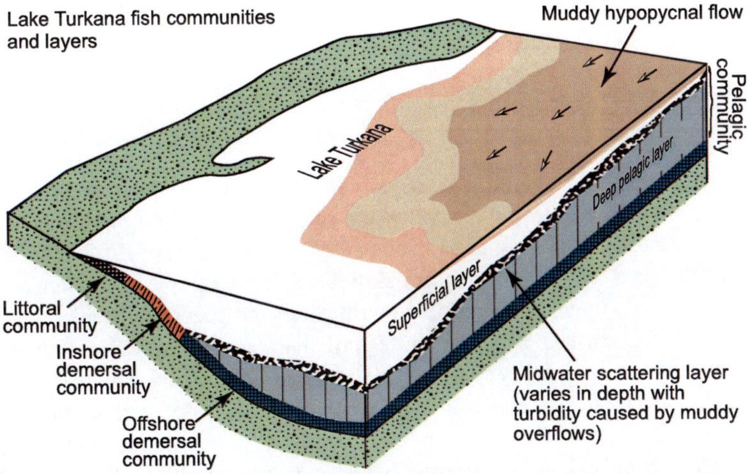

Fig. 6.14 Fish communities in Lake Turkana. Modified after Hopson and Hopson (1982). Three bottom-dwelling communities are recognised with three pelagic layers distinguished. The midwater scattering layer, recognisable on echo soundings in the 1970s, was characterised by abundant *Alestes minutus* and *A. ferox*, which were found at 5–30 m depth, depending on the turbidity of the overlying water

longispinus) and a small catfish (*Schilbe uranoscopus*) (Kolding 1992). Larger fish were not abundant between the scattering layer and the demersal zone, but *Engraulicypris stellae* was widespread and adult prawns (*M. niloticum* and *C. nilotica*) were common. Kolding (1993a) noted that community structure changed between the early 1970s and 1990s in open waters with, for example, slower-growing predators (*Lates* spp.) proliferating in the late 1980s.

Traditional fishing has locally made use of reed fences and cast nets made from doum palm, with a colonial fisheries officer introducing fishing nets from 1955–1960 (Yongo et al. 2010). A commercial fishery concession was started at Kalakol, near Ferguson's Gulf in 1946 (Kolding 1989, 1995) and expanded along the western shores in the 1960s, stimulated by a 1961–62 influx of refugees fleeing a drought. At that time, there was considerable optimism about expanding the commercial fishery with estimates of potential yields of up to 225,000 tons for the whole lake (Kolding 1992). This was scaled back to a maximum of about 30,000 tons following the Lake Turkana Fisheries Research Project (Hopson 1982). Subsequent catches were relatively poor with a target of only 15,000 tons set for 1988. Fish catches appear to correlate well with lake level, but can fall or rise rapidly, especially in relation to conditions at Ferguson's Gulf, which is the main fishery, and which yielded 16,000 tons in 1976 (Harbott and Ogari 1982; Kolding 1993a, b) but periodically dries out. Future declines in water level have been predicted due to dam construction on the Omo River (Avery 2010, 2012), which will likely have major impacts on the fishery. Predictions have been made that *L. niloticus*, *O. niloticus*, and *Synodontis schall* may increase at the expense of species such as *L. horie*, *H. forskalli* and *A. baremose* (Gownaris et al. 2015). Although sediment and water analyses suggest no pollution threats to the fishery, Otachi et al. (2015) have reported that *H. forskalli* contained elevated Li, Zn and Cd in its muscle tissues.

6.7 Modern Sedimentation

Most sand is confined to littoral areas with distributions influenced by local and regional sediment supplies, wind and wave energy, and geomorphological setting (Fig. 6.15) (Reid and Frostick 1985; Cohen et al. 1986; Halfman 1987; Frostick and Reid 1989; Zen et al. 2023). Western shorelines, for example, experience higher wave energies and tend to be associated with silty sand beaches or rocky headlands and gravel. In contrast, the eastern shores are characterised by volcanic headlands and protected bays where silty mud and less commonly sand accumulate (Fig. 6.15a–c). Relatively organic-rich clay and silt are rare and associated with deltas.

Lake Turkana has been categorised as a "wind-driven waterbody" with processes dominated by wind-related circulation (Nutz and Schuster 2016). These produce a variety of landforms including beach ridges, spits, washover fans and associated wind-driven coastal dunes as well as beachface and shoreface deposits (Figs. 6.7a and 6.15d–f; Schuster and Nutz 2017). Locally, these features protect low-energy settings including lagoons with coastal swamps associated with delta settings such

Fig. 6.15 Shoreline types, Lake Turkana. (**a**) Vegetated lagoons, north of Loiyangalani, SE Turkana. (**b**) Sandy, gently-shelving coast, SE Turkana. (**c**) East-west trending strandline near Mt. Porr (background), SE Turkana. (**d, e**) Sandy shorelines near Kalakol, W Turkana. Note swash and backwash sorting of heavy minerals in '**d**' (Chap. 23). Coastal lagoon in (**e**), (**f**) Sandy spit, W Turkana, reflecting northerly longshore drift. (**g, h**) Vegetated distributary channels of the Turkwell Delta, W Turkana. (**i**) Gravel plain showing recessional shoreline, SW Turkana. (**j**) Steep rocky coastline adjacent to Lariu Mountains, SW Turkana

as the Turkwell and Kerio Rivers (Fig. 6.15g, h). Although the Omo Delta has a classic bird-foot morphology that reflects fluvial dominance, most deltas along the western shoreline are wave-dominated with prominent beach ridges. Nutz et al.

(2016) noted a step-wise pattern of regressive beach ridges and spits associated with the Turkwell Delta during the latter stages of the African Humid Period (AHP). They related these geomorphological features to five periods of gradual lake regression separated by episodes of relatively fast lake-level fall and inferred a relationship with reduced rainfall caused by solar activity minima that displaced the Congo Air Boundary westwards, away from the Turkana Basin.

Modelling by Zăinescu et al. (2023) implies that most sediment deposition occurs in the first 10 km offshore from the Omo River mouth and at a depth of <10 m when influenced mainly by inertial flow from the Omo and wind. However, when waves are included in their model, increased bed shear-stresses re-suspend most of the fine sediments, which are then deposited further offshore and in deeper water (>30 m water depth) in the first 30 km offshore. Their study supports the view that wind-driven waves might often be the main agent that transports sediment away from river mouths into deeper areas in large lakes, rather than river-plume (hypopycnal) or gravity-driven (hyperpycnal) transport, which is common in small, and often saline, lakes.

Turbidites are nonetheless common in many East African rift lakes but are generally absent in Lake Turkana, which has been attributed to low lakebed gradients (Cohen 1989). Siliciclastic particles form the bulk of the offshore sediment, except in the South Basin where biogenic and authigenic components are more common (Yuretich 1976, 1979; Olago 1992; Olago and Odada 2000). Yost et al. (2021) noted that about 20% of phytoliths in sediments between Koobi Fora and North Island were sourced from the Ethiopian Highlands. Organic contents are very low (mean <1%) reflecting degradation in the well-mixed, oxygen-rich waters and high detrital inputs. This contrasts strongly with other large rift valley lakes in East Africa, where biogenic deposition is common (Cohen 1989). Laminated (0.5–2 mm), grey to grey-green silty clay, sometimes with microlamination (~1 mm), is present in most cores with the lighter laminae tending to be finer grained. South Basin deposits also contain homogeneous horizons (up to 2 cm) (Yuretich 1979). Lamination has been linked to sediment plumes introduced by the Omo River, which gradually settle and form a soupy flocculent layer at the sediment-water interface (Yuretich 1979). The excellent lamina preservation, despite an oxygenated lake floor, may reflect a paucity of benthic infauna caused by a lack of edible organic detritus and also by a rapid drop in oxygen concentrations in lake floor sediments (Cohen 1984). Halfman and Johnson (1988) suggested that each light and dark sediment couplet reflected an annual succession. Subsequently, Halfman et al. (1994), based on chronological data, proposed that each lamina represented a single year. But controversy continued with the work of Muchane (1996) who supported the earlier view that laminae couplets represented one full year of deposition.

Early estimates of sedimentation rates, based on Ca^{2+} inputs and removal, suggested overall values of about 0.5–1 mm y^{-1} (Yuretich 1979), which compared with rates of 0.2 g cm^{-1} y^{-1} from sediment traps near North Island (Ferguson and Harbott 1982). Subsequently, ^{210}Pb techniques were used to indicate rates of 0.81 and 0.69 cm y^{-1} for the North Basin and 1.38, 0.93 and 2.03 cm y^{-1} for the southern and central lake (Barton et al. 1987). Declination and inclination magnetic data were used to infer values of 1.2 and 1.47 cm y^{-1} for the eastern North Basin and the central sector of the lake (Barton et al. 1987; Barton and Torgersen 1988). Laminae counts, assuming each layer represents 1 year, indicated rates of 1.5 cm y^{-1} for the central sector (Barton et al. 1987) and of 0.2–4 cm y^{-1} for other sites (Halfman 1987). AMS core dating from multiple sites indicated 0.29–0.83 cm y^{-1} (Halfman et al. 1994). Muchane (1996) used a variety of techniques to infer rates of >1.5 cm y^{-1} (carbonate ^{14}C), 2.8 cm y^{-1} (^{210}Pb) and 2.08 cm y^{-1} (laminae counts).

In terms of composition, the North Basin is dominated by diatom-poor siliciclastic sediments derived principally from the Omo River, with smaller inputs from the Turkwell and Kerio Rivers (Yuretich 1979, 1986). In contrast, the South Basin contains coarser-grained sediments due to the higher percentages of biogenic materials, comprising mainly ostracods and diatom frustules. These patterns reflect distance from the main fluvial inputs to the lake, with the South Basin relatively starved of clastic detritus.

Five major mineralogical provinces have been recognised based on clay and silt compositions (Fig. 6.16a; Yuretich 1979). The Omo Delta is characterised by low smectite/kaolinite (S/K) ratios (<1) with higher ratios in the North Basin (>2), the offshore eastern lake (S >> K), the Turkwell-Kerio Basin (>3, plus abundant illite) and the South Basin (~2–2.5). Increase in smectite has been attributed to the crystallisation of an amorphous phase originally delivered by the Omo River (Yuretich 1979).

Silt fractions also show significant variability. The Omo Delta is dominated by quartz, feldspar and magnetite reflecting fluvial inputs from the Ethiopian Highlands. In the North Basin, calcite becomes an important constituent (~30%) with quartz increasing relative to feldspar and with dark minerals forming only 5–7% of the silt (Yuretich 1979). Sand grains near the Omo and throughout the profundal regions are mainly quartz, feldspar and blue-green amphibole that indicate a metamorphic origin and possible aeolian delivery by the strong southeasterly winds. Sediments near the Turkwell and Kerio deltas contain more fine-grained quartz, a lower quartz/feldspar ratio in the silt fraction, and sand with a greater compositional variety, which reflects large areas of exposed metamorphic rocks in the Turkwell-Kerio catchment. The South Basin receives a smaller clastic flux and is characterised by an increase in diatom silica and euhedral, semi-acicular calcite (up to 28%) (Yuretich 1986; Halfman 1987; Halfman and Johnson 1988; Ng'ang'a et al. 1998). Stable isotopes indicate precipitation of low-Mg calcite from surface water (Halfman et al. 1989, 1994). Ostracods also contribute carbonate, especially in sand (Cohen 1984). Diatoms are dominated by *Thalassiosira* and *Nitzschia* (Halfman 1987; Halfman et al. 1992). Volcanic ash and fish bone are also relatively more common in the South Basin sediments.

Maps of sediment geochemistry for Lake Turkana document lake-wide trends (Fig. 6.16; Yuretich 1979, 1986). A southward decrease in Fe_2O_3, TiO_2, Mn and Zn (Fig. 6.16b–e) reflects inputs from the Omo River. The percentages of authigenic and biogenic indicators (CaO and MgO) increase southwards due to lower inputs of siliciclastic materials (Fig. 6.16g, h). The Na_2O and K_2O concentrations reflect other sediment inputs, especially from the Turkwell and Kerio Rivers (Fig. 6.16i, j). Other elements show no significant relationship with the three major rivers and probably reflect local sources.

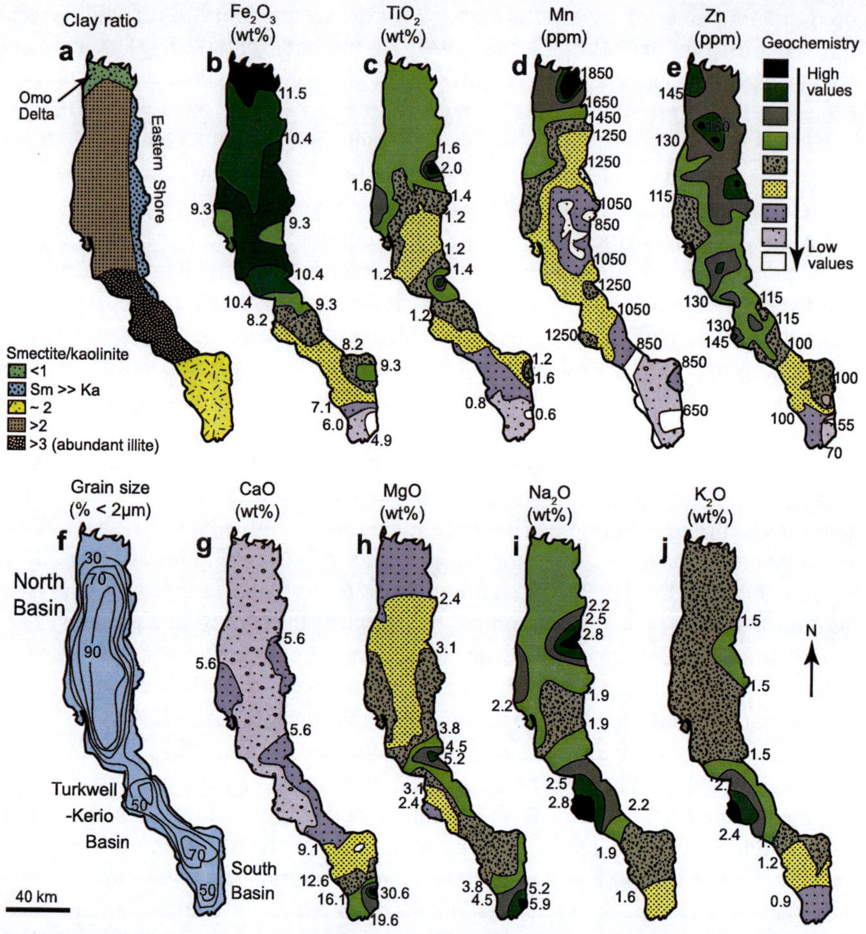

Fig. 6.16 Spatial geochemical variability in surficial sediments. Modified from Yuretich and Cerling (1986). (**a**) Sediment provinces based on clay ratios. (**b–e**) Major and trace elements that tend to decrease southwards. (**f**) Percent fine clay (<2 μm) in sediments. (**g–j**) Major elements that increase in abundance southwards

6.8 Late Quaternary Lacustrine Sediments in the Lake Turkana Basin

6.8.1 Onshore Spatial Variability and Stratigraphy of the Galana Boi Formation

High lake levels during the Late Pleistocene to Early Holocene African Humid Period (AHP) resulted in deposition of laminated diatomaceous silt and bedded to massive sand in the Galana Boi Formation, which is present across most of the Turkana Basin (Fig. 6.17a, b) (Raynolds 1972; Owen 1981; Owen et al. 1982, 2018;

Fig. 6.17 Galana Boi Formation (AHP) sediments, NE Lake Turkana. (**a**) Diatomaceous lacustrine silt and fine sand. (**b**) Eroded Galana Boi Formation, Koobi Fora. (**c**) Extensive spread of unusually large (up to 0.5 m) oncoids near Ileret. (**d–f**) *Etheria elliptica* capping diatomaceous silt and clay, east of Koobi Fora. (**g**) Cross-section through an oncoid showing microbially induced calcite laminae nucleated on *Etheria* shell. (**h**) 'Cauliflower' stromatolites from the Galana Boi Formation. (**i**) Impure diatomite. (**j**) Early Holocene fish-bone harpoon. (**k**) Lacustrine silt with fish vertebrae. (**l**) Cross-bedded beach sand

Owen and Renaut 1986; Martin 2014; Beck 2015). Locally, accumulations of stromatolites and large oncoids mark former shallow nearshore areas (Fig. 6.17c, g, h; Vondra et al. 1971; Johnson 1974; Owen and Renaut 1986) with riverine freshwater *Etheria elliptica* forming well-cemented oyster reefs and biostromes (Fig. 6.17d–f). Other parts of the formation contain impure diatomite, fish bone harpoons in littoral sand, fish bone in lake silt and cross-bedded shoreline sands (Fig. 6.17i–l).

Equivalent high-lake-level silt in the Omo River region forms Member 4 of the Kibish Formation (Fig. 6.18a), which has been divided into two parts using a mid-sequence disconformity (Butzer and Thurber 1969). High Holocene lake levels are documented by the sediments in Members 4a (9500–7500 y BP; 428–459 m asl)

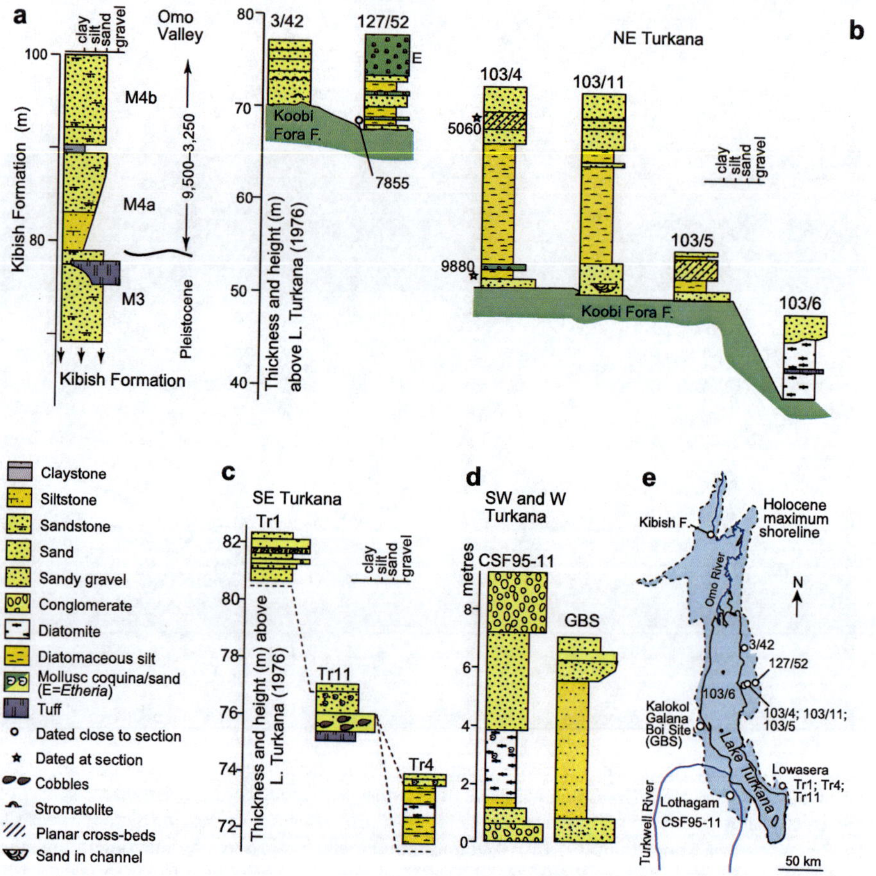

Fig. 6.18 Holocene sediments in the Turkana Basin. (**a**) Upper Kibish Formation. After McDougall et al. (2005). Member 4 is Holocene and broadly equivalent to the Galana Boi Formation. (**b**) The predominantly lacustrine Holocene Galana Boi Formation in the type area, NE Lake Turkana. After Owen et al. (1982) and Owen and Renaut (1986). (**c**) The Galana Boi Formation at Lowasera (Phillipson 1977b) near the Holocene maximum shoreline. (**d**) Sections CSF95-11 and GBS recorded by Feibel (2003) and Bloszies et al. (2015), which include lacustrine silt, diatomite and sand at Lothagam and Kalokol. (**e**) Location of sections in '**a**–**d**'

and 4b (5750–3250 y BP; 444–450 m asl), with one low level of 378 m asl inferred for 6600 y BP (Butzer 1980). Owen et al. (1982) referred to these two high-level periods as Phases 1 and 2 (Fig. 6.19) but raised doubts about the major drop to 378 m due to a lack of an equivalent erosion surface in the Galana Boi Formation

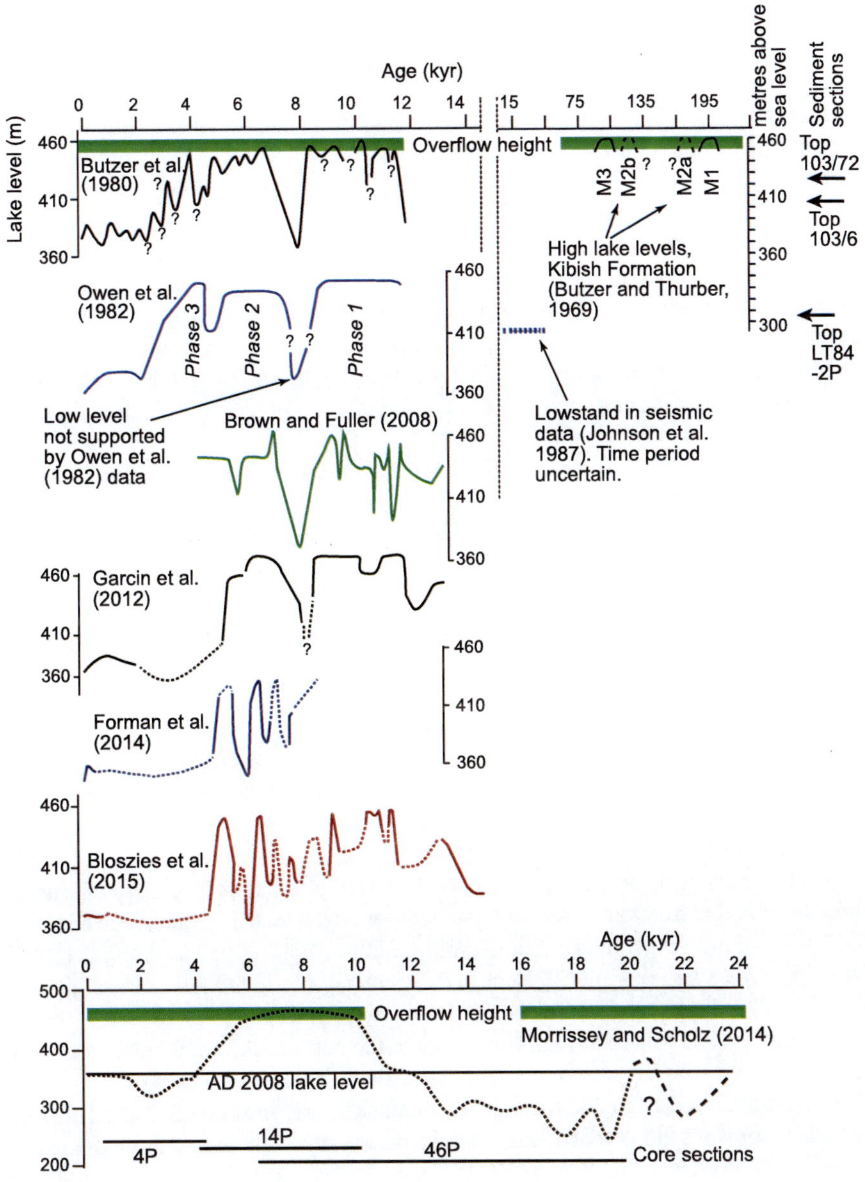

Fig. 6.19 Late Quaternary lake-level changes. Butzer (1980) and Owen et al. (1982) are based on uncalibrated dates; other authors use calibrated dates (see text). Sources are given in the diagram

northeast of Lake Turkana (Fig. 6.17a, b). They used uncalibrated dates and noted that their data did not support a rise to the overflow height at 3250 y BP (Phase 3) as suggested by Butzer (1980). However, they did report an undated diatomite (Section 103/6, Figs. 6.17i, 6.18b, and 6.20) at ~410–415 m elevation that could be related to a Phase 3 transgression. Younger Galana Boi sediments indicate that the lake surface remained below 410 m asl after about 3000 y BP (Owen and Renaut 1986).

Pollen spectra from the Early Holocene Galana Boi Formation northeast of Lake Turkana differ from modern floras (Owen et al. 1982), with montane forest and thicket taxa implying fluvial transport from the Ethiopian Highlands combined with forest expansion. *Graminae* and *Compositae* dominate local elements with few arboreal forms present and a low percentage of *Chenopodiaceae/Amaranthaceae*, which indicates a lack of saline soils. Middle Holocene spectra consist of taxa typical of lake margins in semi-deserts. Regional elements are rare (only *Juniperus* and *Myrica*) as are local arboreal pollen. *Graminae* are very abundant (77–99%) with *Typha* and *Pteridophytes* forming a small component. *Tribulus* is present locally, which may reflect disturbance around human encampments. A core taken offshore from the Koobi Fora shoreline documents the period 2500–600 y BP when *Chenopodiacea/Amaranthaceae* and *Graminae* percentages were relatively high (Mohammed et al. 1995).

During the last 3000 years conditions in the Turkana Basin have been relatively dry with evaporation exceeding precipitation. These semi-arid to arid conditions have, in turn, favoured the accumulation of highly soluble nitratine ($NaNO_3$) in parts of the Galana Boi Formation. Owen and Renaut (1990) reported nitratine near Koobi Fora, and close to the surface of outcrops, where it constitutes up to 7% of the sediment by weight. Its origins are uncertain but may include microbial or photochemical fixation, leaching of plant remains or local volcanic sources.

The Holocene dates noted above have since been recalibrated and combined with new evidence from southern Lake Turkana (Fig. 6.19; Garcin et al. 2012). Consequently, Phase 1 was reassigned to the period 11,500–7800 and Phase 2 was dated at 6500–4200 Cal y BP. Studies in the Mt. Porr area (SE Turkana; Figs. 6.18c and 6.19) also raised questions about parts of the earlier dating and inferred lake levels, especially for Phase 2, during which several additional oscillations in height were documented (Forman et al. 2014). Bloszies et al. (2015) integrated earlier lake-level reconstructions with new data from western lake Turkana, near Kalokol, and inferred nine centennial- to millennial-scale fluctuations between about 15,000 and 4500 years ago (Fig. 6.19). Beck (2015) noted three Late Pleistocene high levels (13,870, 13,040, 11,740 Cal y BP). Subsequently, Beck et al. (2019a, b), based on data from Kalakol and Kabua Gorge, noted the onset of the AHP by 13,900 y BP and a depositional hiatus that corresponded with the Younger Dryas. Additional lake level information is available from three composite cores in southern Lake Turkana, which record a 24,000-year history based on organic and sedimentological data (Fig. 6.19; Morrissey and Scholz 2014). This lake level stratigraphy has confirmed that the lake was high and close to its outlet between ~11,000 and 4000 y BP, during the AHP. Liu et al. (2021) also addressed questions concerning the AHP using rock

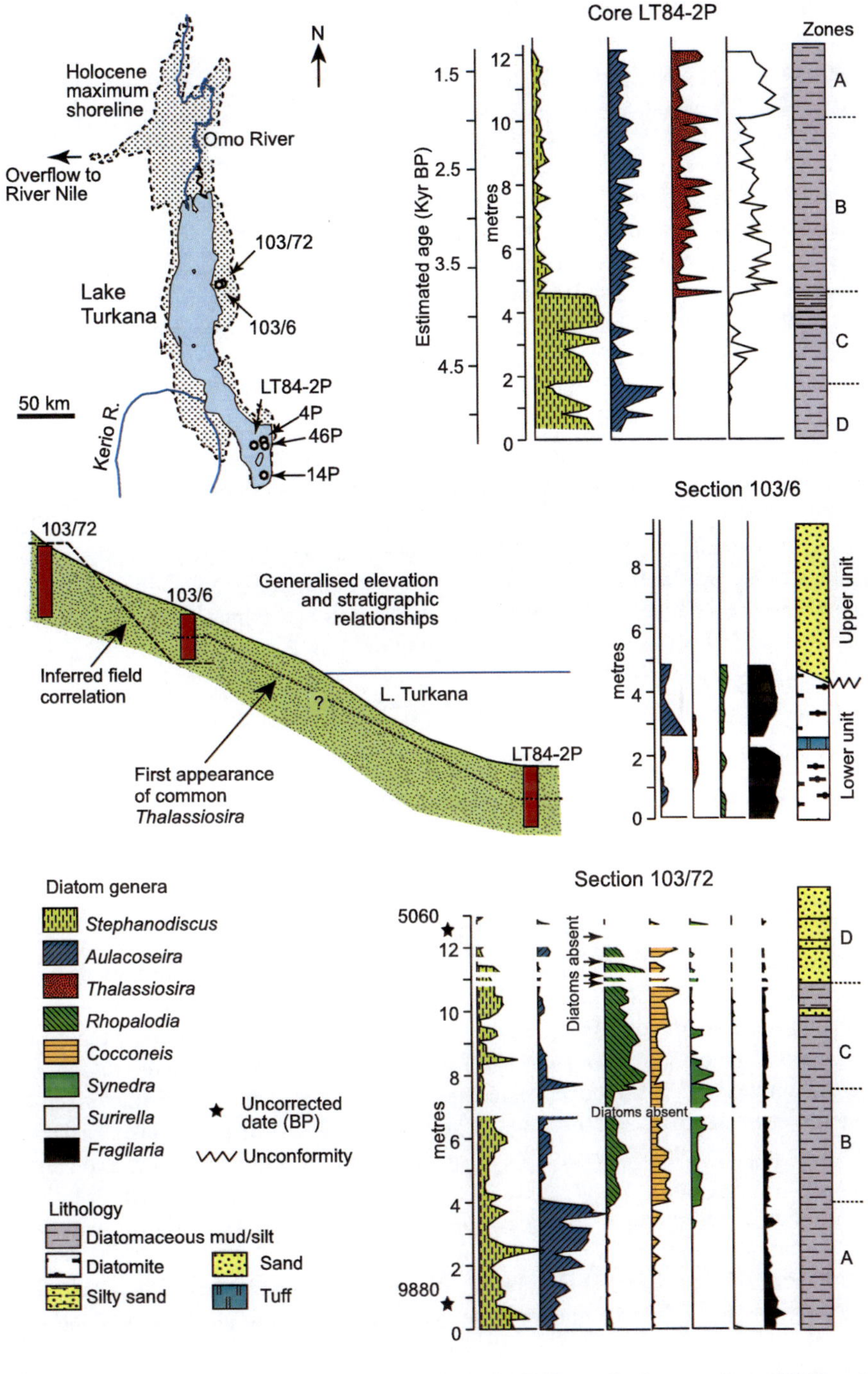

Fig. 6.20 Composite Holocene diatom record for Lake Turkana. Onshore sections 103/72 and 103/6 after Owen et al. (1982) and Owen and Renaut (1986). Core LT84-2P after Halfman et al. (1992). The composite succession shows declining water levels and increasing salinity. Sketch shows a tentative correlation between the three diatom sequences using the first appearance of *Thalassiosira* diatoms as a potential marker

varnish records at Lothagam. They related changes in Mn and Ba to moisture and reported three major wet phases between 11.5 and 8.5 ka, two additional phases between 8 and 5 ka, a single moderate wetter period between 5 and 4.3 ka followed by five minor wet intervals during the late Holocene.

Some of the first studies of Holocene sediments to the southeast (Fig. 6.18c, e; Phillipson 1977b) and southwest (Fig. 6.18d; Robbins 1967, 1968, 1972, 2006; Feibel 2003) of Lake Turkana demonstrated the archaeological significance of lacustrine and lake marginal deposits in these areas (Lahr et al. 2016; Phillipson 1976, 1977a). Subsequent research examined other regions and the relationship between archaeology and lake level (e.g. Barthelme 1985; Beyin 2011; Hildebrand et al. 2011, 2018; Wright et al. 2015; Wright 2019), as well as mammal faunas and Neolithic burials in areas such as Lothagam (Feibel 2003). Correlations have been pointed out, for example, between Early to Middle Holocene high lake beaches and fishing camps associated with fish bone harpoons (Fig. 6.17j) (Wright and Forman 2011; Prendergast and Beyin 2018; Beyin et al. 2017; Goldstein et al. 2017, 2022; Smith 2018), with a shift to domesticated animals and fishing activities after ~4700–4000 y BP (Barthelme 1981; Owen et al. 1982; Marchant and Lane 2014; Ashley et al. 2017; Chritz et al. 2019; Goldstein 2019). OSL dating and the archaeology of the Galana Boi Formation in the northeast have also indicated relationships between the development of subsistence strategies and climatic deterioration that led to Late Holocene falling lake levels (Ashley et al. 2011).

6.8.2 Onshore to Offshore Diatom and Chronological Correlations of Holocene Sediments

Fossil diatoms from Lake Turkana cores reflect variations in lake level and water chemistry. This can be demonstrated in a composite series of onshore (Owen et al. 1982; Owen and Renaut 1986) and offshore (Halfman et al. 1992) sediments. The dated onshore record extends between 9880 and 5060 y BP (uncalibrated dates) in Section 103/72 (Figs. 6.17a and 6.20) from the Galana Boi Formation. Owen (1981) recognised four major zones. His basal Zone A is dominated by planktonic taxa (*Aulacoseira* and *Stephanodiscus* spp.) and indicates a deep, fresh lake. *Aulacoseira* are common but with reduced percentages in Zone B, which is characterised by increases in epiphytic and littoral taxa, signifying a shallowing fresh lake. Elevation data indicate that the water surface must have been higher than ~425 m and part of the shallowing could reflect sediment accumulation. Zone C is characterised by a decline in *Synedra* and an increase in *R. vermicularis* with *Aulacoseira nyassensis* important in several horizons, indicating further lake shallowing, but with lake level still above ~425 m. Zone D lies above a disconformity and is characterised by fluctuating diatom abundance and variable taxa, which suggest nearshore to emergent conditions. Section 103/6 (Figs. 6.17i and 6.20) is undated, but crops out at a lower altitude (~410–415 m) and is younger than Section 103/72. It also contains the first appearance of *Thalassiosira rudolfi* in the section's lower unit, which is otherwise dominated by *Fragilaria* and *Aulacoseira* taxa.

Core LT84-2P, near South Island (Fig. 6.20) overpenetrated, but includes sediments 5450–1070 years old (BP, uncalibrated) (Halfman et al. 1992). The oldest sediments accumulated in a fresh lake (Zone D; 5450–4850 y BP) dominated by *Aulacoseira nyassensis* and *Stephanodiscus* spp., suggesting high lake levels, fresher water and seasonal mixing of the water column. Zone C (4850–3900 y BP) is dominated by *Aulacoseira* spp. with less common *Surirella engleri* and *Stephanodiscus*, indicating relatively high lake levels and seasonal mixing, but with possibly shallower episodes where *Surirella* increase. *Thalassiosira rudolfi* appears in Zone B (3900–1900 y BP) along with a major reduction in *Aulacoseira* and increases in *Surirella*, documenting further reductions in lake level and increasing salinity. Interestingly, and despite the considerable distances between the onshore and offshore sediments, the dates for Zone B and the appearance of *T. rudolfi* in LT84-2P would be consistent with the first appearance of *Thalassiosira* and evidence for lower lake levels in Section 103/6 from northeast Turkana. Halfman et al. (1992) interpreted increased percentages of *Surirella* in Zone A (1900–1070 y BP) as reflecting further lake regression with diatom dissolution suggesting increased pH.

6.8.3 Geochemical and Microfossil Stratigraphy of Offshore Cores

Some of the earliest studies showed that chloride in pore waters increases downwards in cores from the South Basin but decreases in the North Basin (Fig. 6.21d), suggesting that the two basins were once separate (Cerling 1996). This may, for example, reflect Late Pleistocene very low levels (−60 to more than −100 m; Fig. 6.19; Johnson et al. 1987; Morrissey and Scholz 2014) when an isolated saline lake with minimal fluvial inputs would have developed in the South Basin. A large increase in Mg downcore has also been reported for LT84-2P in the South Basin (Fig. 6.21e), especially below 5.5 m, which may reflect diffusion from concentrated water or perhaps the release of Mg^{2+} from Mg-rich smectite in sediments that accumulated in a small very alkaline water body before 10,000 y BP (Yuretich and Ervin 2002). According to this model, an open, freshwater lake developed between about 10,000 and 4000 years ago in which detrital Fe-smectite was deposited and then there was a return to deposition of Mg-smectite in response to falling levels and rising alkalinities.

Studies of other cores collected mainly in the 1980s by Project PROBE focused on carbonate content, stable isotopes, lamination thickness, radiocarbon dates, magnetic susceptibility and time series analyses (Halfman 1987; Halfman and Johnson 1988; Halfman et al. 1989, 1994; Halfman and Hearty 1990). Johnson et al. (1991), for example, related stable isotopes in authigenic calcite to lake level change, with revisions later reported by Johnson (1996) and Muchane (1994, 1996). Other

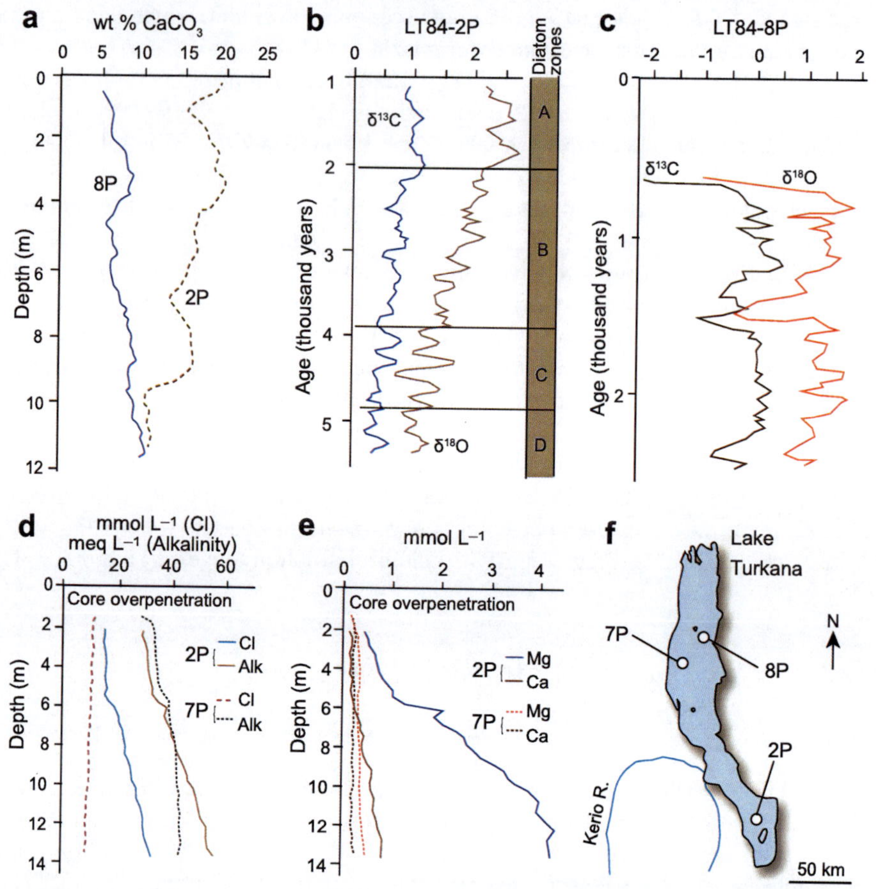

Fig. 6.21 Piston-core geochemical stratigraphies. (**a**) Calcium carbonate data for Cores 2P and 8P showing opposing trends in the North and South Basins. Modified after Johnson (1996). (**b**, **c**) Isotope stratigraphy of authigenic carbonate. After Johnson (1996). Diatom zones (A–D) from Halfman et al. (1992). (**d**, **e**) Selected examples of pore fluid chemistry. After Yuretich and Ervin (2002). (**f**) Core locations

research documented increases in carbonate (ostracods and micrite) in Core LT84-2P (Fig 6.21a) with a sudden change at 9.5 m core depth (4320 Cal y BP) linked to basin closure. The timing of the inferred outlet loss also appears to coincide with a shift from *Aulacoseira* and *Stephanodiscus* to *Thalassiosira* and *Surirella* in the same core (Fig. 6.20; Halfman and Johnson 1988; Halfman et al. 1992; Johnson 1996).

Oxygen isotopes show a complex record (Ricketts and Johnson 1996; Ricketts and Anderson 1998). Studies of the South Basin core LT84-2P (Fig. 6.21b) show a prolonged shift in $\delta^{18}O$ micrite profiles towards heavier values between 3900 and 2200 ^{14}C y BP (4320–2320 Cal y BP) (Johnson and Malala 2009). Halfman et al. (1992) suggested that this steady increase in $\delta^{18}O$ may reflect a gradual shift in the isotopic composition of rain over the Ethiopian Highlands. $\delta^{18}O$ data for core LT84-8P (Fig. 6.21c) from the North Basin suggests that the lake was low and dropping between 2500 and 1600 y BP with a major rise at ~1600–1400 y BP, followed by falling levels to ~700 y BP and another rise before 600 y BP (Johnson 1996). The $\delta^{13}C$ data broadly parallel the $\delta^{18}O$ record (Fig. 6.21b, c), as with many closed basin lakes, with small deviations attributed to the impact of primary productivity (Johnson 1996). Differences have been reported for stable isotope values from ostracods versus authigenic calcite (Ng'ang'a et al. 1998), but with a strong covariance, especially prior to about 4000 y BP when the lake was hydrologically open. Beck et al. (2021) also examined $\delta^{18}O$, $\delta^{13}C$, major elements and ostracods in sediments from Ferguson's Gulf along the western shoreline and inferred higher lake levels during the Little Ice Age followed by increased aridification, higher alkalinities and increasingly frequent desiccation at their study site.

Berke et al. (2012) used TEX_{86} core data from the North Basin of Lake Turkana to infer a thermal maximum at about 5000 y BP involving 3.5 °C of warming after about 5700 y BP. In contrast, a TEX_{86} study spanning 14,000 to 400 y BP from the South Basin suggested that temperatures ranged from 24.3 to 28.6 °C, with a gradual decrease from the Early to Late Holocene, but with many abrupt temperature shifts that were attributed to changes in evaporation and cloud cover (Morrissey 2014; Morrissey et al. 2018).

Strontium isotopes from molluscs in palaeoshorelines of known heights and ostracods from cores show a long-term trend in $^{87}Sr/^{86}Sr$ values across the termination of the AHP that suggests a slow change in hydroclimate that was aligned with the summer insolation curve (van der Lubbe et al. 2017). In addition, the data indicate that the Omo catchment contributed less Sr to the palaeolake during the AHP than today with increased inputs derived from the Turkwell and Kerio rivers to the southwest.

6.8.4 *African Humid Period Catchment Expansion, Hydrological Budgets and Sedimentation*

Expansion of Lake Turkana during the Late Pleistocene to Middle Holocene AHP (Fig. 6.22a) reflects changes in the evaporation/precipitation balance, the addition of new catchment areas following overflow from neighbouring lake basins and the possible effects of an additional rainy season in parts of the palaeocatchment due to changes in the position of the Intertropical Convergence Zone (Fischer et al. 2020). Many lakes expanded at this time. For example, the modern Chalbi Desert (east of Lake Turkana; Fig. 6.22b) was formerly occupied by a lake, although this did not overflow into the Turkana Basin (Nyamweru 1986; Abell and Nyamweru 1988; Nyamweru and Bowman 1989). Similarly, lakes in the southern Ethiopian rift filled their basins to their respective outlet heights, resulting in a series of interconnected water bodies (Fig. 6.22a). This was the case for Lake Chew Bahir (Fig. 6.22c, d), which received additional water inflow from Lake Chamo and Lake Abaya,

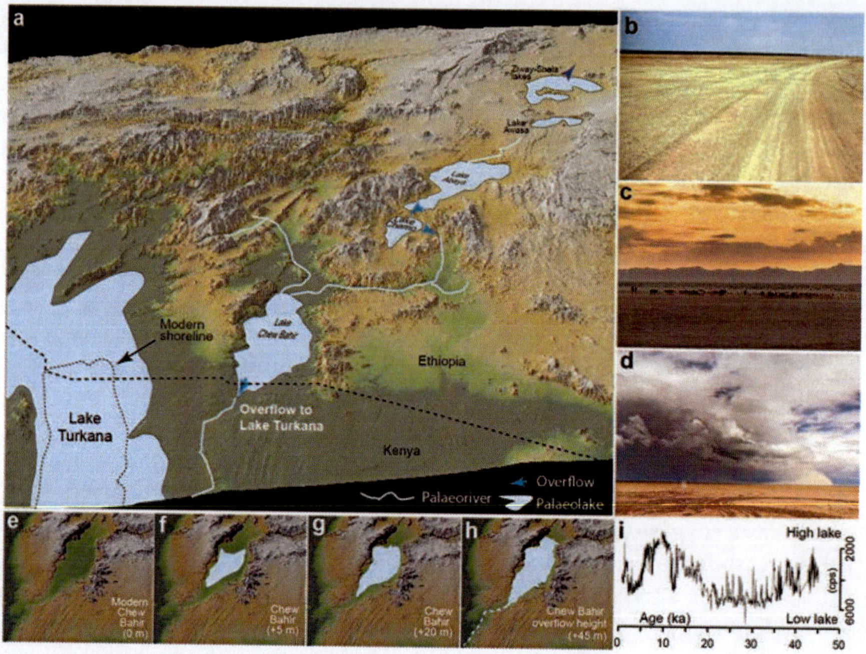

Fig. 6.22 Catchment expansion northeast of Lake Turkana during the African Humid Period (~15–5 ka). The lakes expanded due to an increase in precipitation/evaporation ratios and additional inflows from upstream lakes that reached their respective outlets. **(a)** Palaeogeography maps and lake level curve based on Fischer et al. (2020). **(b–d)** Desiccated AHP palaeolake basins. **(b)** The arid Chalbi Desert in 1977. **(c, d)** Two views of the modern Chew Bahir's dry valley floor, which is subjected to short-lived flooding events. Photographs courtesy of Annett Junginger. **(e–h)** The Chew Bahir depression showing extent of lakes of differing depths. Overflow to Lake Turkana takes place at +45 m in **'g'**. **(i)** Chew Bahir core (XRF) data showing a potassium stratigraphy. From Foerster et al. (2012), which reflects wetter and drier periods in the basin. The relatively wet record for the African Humid Period (~15–5 Ma) is clearly evident

resulting in expansion (Fig. 6.22e–h) and overflow into northeast Lake Turkana (Grove et al. 1975; Forman et al. 2014; Wright 2019).

Foerster (2014) and Foerster et al. (2012, 2015, 2018) provide detailed analyses of the history of Chew Bahir for the last 45 ka noting variations in potassium (Gebregiorgis et al. 2021) in cores collected by the Hominin Sites and Paleolakes Drilling Project (HSPDP). Their K-data was used as a proxy for wetter and drier conditions (Fig. 6.22i), which Schaebitz et al. (2021) subsequently expanded to a record for the last 200 kyr. They noted a well-defined arid phase at 12,800–11,600 y BP and the possible use of highland refugia by humans through dry intervals (Wright and Foreman 2016; Foerster et al. 2016). Trauth et al. (2015) also noted several wetter and drier phases, including drought episodes. Subsequently, Trauth et al. (2018, 2019) suggested that the most dramatic changes at Chew Bahir, through the last 240 ka, took place during ~15.7–15.46 ka and ~4.875–3.885 ka in association with the onset and termination of the AHP. Fischer et al. (2020) also modelled past hydrological conditions, including overflow events, and observed that there must have been a 20–30% increase in precipitation throughout the southern

Ethiopian Rift in order to fill the Chew Bahir paleolake to its overflow. Additional palaeovegetation modelling, using combined lake level data, Chew Bahir phytoliths and Ethiopian pollen records, suggest a Last Glacial Maximum reduction of 15–20% in moisture, compared with today (Fischer et al. 2021). At that time, the region was characterised by open landscapes and scarce forest refugia. In contrast, during the AHP there was a 25–40% increase in moisture above modern levels, which was associated with expanded forests. Krueger et al. (2021) also noted that these climate changes were reflected in sedimentary DNA data from the same HSPDP cores.

During the AHP, additional catchment expansion took place to the south of Lake Turkana, where the presently arid Suguta Valley (Fig. 6.23a) was filled to its outlet, which involved a dramatic ~300 m rise in lake level from the modern shallow and

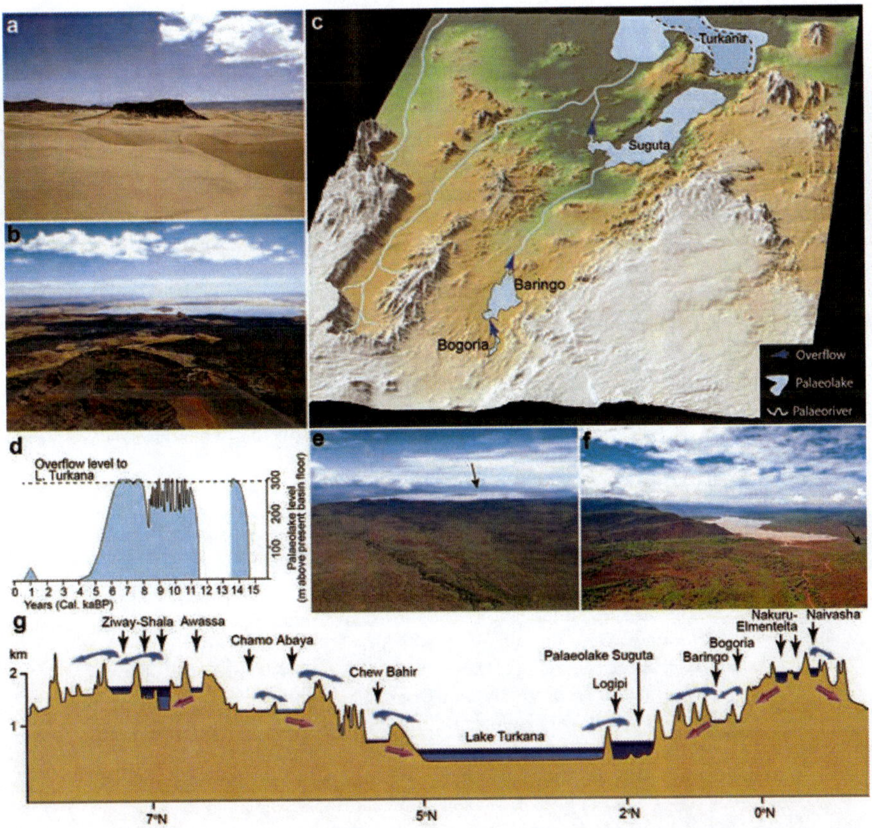

Fig. 6.23 Catchment expansion south of Lake Turkana. (**c**, **d** and **g** after Junginger et al. (2014)). (**a**) Contemporary active dune fields near Rupa (midwestern Suguta). (**b**) Saline Lake Logipi viewed from the Barrier with the arid Suguta Valley to the south. (**c**) Maximum lake expansion and periodic overflows south of Lake Turkana during the African Humid Period. (**d**) Lake levels in the Suguta Valley. (**e**) Lake Baringo with location of possible overflow arrowed (see Chaps. 8 and 9 for discussion). View to northwest. (**f**) Lake Bogoria with location of possible overflow to Lake Baringo arrowed. View to south. (**g**) AHP connections between the Ethiopian and Kenya Rift lakes. Based on Junginger and Trauth (2013, and older sources)

highly saline Lake Logipi (Fig. 6.23b). The Suguta River is today the major water source and was supplemented by overflows from Lakes Baringo and Bogoria further to the south (Fig. 6.23c–f) (Junginger et al. 2014). Figure 6.23g summarises the complex interconnections for both the Ethiopian and Kenya rift lakes that flowed into Lake Turkana during the AHP. Lake Turkana is the largest of the Kenya rift lakes and is often described as a desert lake. Although seemingly a simple waterbody its origins and evolution clearly reflect a complex history controlled by rift tectonics, changes in volcanic activity, climate change and consequent limnological and ecological responses to these primary environmental drivers.

References

Abdelfettah Y, Tiercelin J-J, Tarits P, Hautot S, Maia M, Thuo P (2016) Subsurface structure and stratigraphy of the northwest end of the Turkana Basin, Northern Kenya Rift, as revealed by magnetotellurics and gravity joint inversion. J Afr Earth Sci 119:120–138

Abell PI, Nyamweru CK (1988) Paleoenvironments in the Chalbi Basin of Kenya. Chem Geol 72:283–291

Achiba GA (2019) Navigating contested winds: development visions and anti-politics of wind energy in northern Kenya. Land 8(1):7

Angelis d'Ossat G, Millosovich F (1900) Studio geologico sul materiale raccolto de M. Sacchi. Società Geografica Italiana, Rome

Arambourg C (1935) Mission Scientifique de l'Omo (1932–1933): Géologie-anthropologie-paléontologie. Publ Mus Natl Hist Nat Paris 1(1):1–59

Arambourg C (1943) Contribution à l'étude géologique et paleontologique du Bassin du lac Rodolphe et de Basse Vallée de l'Omo. Publ Mus Natl Hist Nat Paris 1(2):157–230

Arambourg C (1947) Contribution à l'étude géologique et paleontologique du Bassin du lac Rodolphe et de la Basse Vallée de l'Omo. Mission Scientifique de l'Omo (1932–1933). Publ Mus Natl Hist Nat Paris, pp 231–562

Arambourg C, Chavaillon J, Coppens Y (1969) Résultats de la nouvelle Mission de l'Omo (1968). C R Acad Sci, Paris D-268:759–762

Archer GF (1913) Recent exploration and survey in the north of British East Africa. Geogr J 42:421–430

Ashley GM, Ndiema EK, Spencer JQG, Harris JWK, Kiura PW (2011) Paleoenvironmental context of archaeological sites, implications for subsistence strategies under Holocene climate change, northern Kenya. Geoarchaeology 26:809–837

Ashley GM, Ndiema EK, Spencer JQG, Harris JWK, Kiura PW, Dibble L, Du A, Lordan PT (2017) Paleoenvironmental reconstruction of Dongodien, Lake Turkana, Kenya and OSL dating of site occupation during Late Holocene climate change. Afr Archaeol Rev 34:345–362

Austin H (1899) Journeys to the north of Uganda. I Lake Rudolf. Geogr J 14:148–152

Avery ST (2010) Hydrological impacts of Ethiopia's Omo Basin on Kenya's Lake Turkana water levels and fisheries. Final Report African Development Bank, vols I and II

Avery ST (2012) Lake Turkana and the Lower Omo: hydrological impacts of major dam and irrigation developments. University Oxford, African Studies Centre Vol II-Annexes

Avery ST (2014) What future for Lake Turkana and its wildlife. SWARA (Mag E Afr Wildlife Soc). January–March, pp 24–30

Avery ST, Tebbs EJ (2018) Lake Turkana, major Omo River developments, associated hydrological cycle change and consequent lake physical and ecological change. J Great Lakes Res 44:1164–1182

Ayalew L (2009) Analyzing the effects of historical and recent floods on channel pattern and the environment in the Lower Omo Basin of Ethiopia using satellite images and GIS. Environ Geol 58:1713–1726

Baker BH, Wohlenberg J (1971) Structure and evolution of the Kenya Rift Valley. Nature 229:538–542

Baker BH, Mohr PA, Williams LAJ (1972) Geology of the eastern rift system of Africa. Geol Soc Am Spec Pap 136

Barber J (2006) The moving frontier of British imperialism in the Lake Rudolf Region: 1890–1919. Ethnohistory 53:143–172

Barthelme JW (1981) Late Pleistocene-Holocene prehistory northeast of Lake Turkana, Kenya. PhD thesis, University of California, Berkeley

Barthelme JW (1985) Fisher-hunters and Neolithic pastoralists in eastern Turkana, Kenya (Cambridge monographs in anthropology 13). Brit Archaeol Rep, Int Ser 254, Oxford

Barton CE, Torgersen T (1988) Palaeomagnetic and ^{210}Pb estimates of sedimentation in Lake Turkana, East Africa. Palaeogeogr Palaeoclimatol Palaeoecol 68:53–59

Barton CE, Solomon DK, Bowman JR, Cerling TE, Sayer MD (1987) Chloride budgets in transient lakes: Baringo, Naivasha, and Turkana. Limnol Oceanogr 32:745–751

Beadle LC (1932) Scientific results of the Cambridge Expedition to the East African Lakes, 1930–1. 4. The waters of some East African Lakes in relation to their fauna and flora. J Linn Soc Zool 38:157–211

Beadle LC (1981) The inland waters of tropical Africa, 2nd edn. Longman, London

Beck CC (2015) The terrestrial climate record from the Turkana Basin, Kenya: a multi-proxy approach. PhD thesis, Rutgers University, Piscataway, New Jersey

Beck CC, Feibel CS, Wright JD, Mortlock RA (2019a) Onset of the African Humid Period by 13.9 kyr BP at Kabua Gorge, Turkana Basin, Kenya. Holocene 29:1011–1019

Beck CC, Allen MM, Feibel CS, Beverly EJ, Stone JR, Wegter B, Wilson CL (2019b) Living in a swampy paradise: paleoenvironmental reconstruction of an African Humid Period lacustrine margin, West Turkana, Kenya. J Afr Earth Sci 154:20–34

Beck CC, Feibel CS, Mortlock RA, Quinn RL, Wright JD (2021) Little Ice Age to modern lake-level fluctuations from Ferguson's Gulf, Lake Turkana, Kenya, based on sedimentology and ostracod assemblages. Quat Res 101:129–142

Behrensmeyer AK (1976) Lothagam Hill, Kanapoi, and Ekora: a general summary of stratigraphy and faunas. In: Coppens Y, Howell FC, Isaac GL, Leakey REF (eds) Earliest man and environments in the Lake Rudolf Basin: stratigraphy, paleoecology and evolution. University of Chicago Press, Chicago, pp 163–170

Berke MA, Johnson TC, Werne JP, Schouten S, Damsté JSS (2012) A mid-Holocene thermal maximum at the end of the African Humid Period. Earth Planet Sci Lett 351–352:95–104

Beyin A (2011) Recent archaeological survey and excavation around the Greater Kalokol Area, west side of Lake Turkana: preliminary findings. Nyame Akuma 75:40–50

Beyin A, Prendergast ME, Grillo KM, Wang H (2017) New radiocarbon dates for terminal Pleistocene and early Holocene settlements in West Turkana, northern Kenya. Quat Sci Rev 168:208–215

Bloomer SH, Curtis PC, Karson JA (1989) Geochemical variation of Quaternary basaltic volcanics in the Turkana Rift, northern Kenya. J Afr Earth Sci 8:511–532

Bloszies C, Forman SL (2015) Potential relation between equatorial sea surface temperatures and historic water level variability for Lake Turkana, Kenya. J Hydrol 520:489–501

Bloszies C, Forman SL, Wright DK (2015) Water level history for Lake Turkana, Kenya in the past 15,000 years and a variable transition from the African Humid Period to Holocene aridity. Glob Planet Change 132:64–76

Bonnefille R (1971) Atlas des pollens d'Ethiopie: pollens actuels de la basse vallée de l'Omo (Ethiopie); recoltes botaniques 1968. Adansonia Ser 2(11):463–518

Boone SC, Kohn BP, Gleadow AJW, Morley CK, Seiler C, Foster DA, Chung L (2018) Tectonothermal evolution of a long-lived segment of the East African Rift System: thermochronological insights from the North Lokichar Basin, Turkana, Kenya. Tectonophysics 744:23–46

Boschetto HB, Brown FH, McDougall I (1992) Stratigraphy of the Lothidok Range, northern Kenya, and K/Ar ages of its Miocene primates. J Hum Evol 22:4771
Bosworth W, Morley CK (1994) Structural and stratigraphic evolution of the Anza rift, Kenya. Tectonophysics 236:93–115
Boulenger GA (1920) Sur le gavial fossile de l'Omo. CR Acad Sci Paris 170:913–914
Brooke JW (1905) A journey west and north of Lake Rudolf. Geogr J 25:525–531
Brotzu P, Morbidelli L, Nicoletti M, Picirillo EM, Traversa G (1984) Miocene to Quaternary volcanism in eastern Kenya: sequence and geochronology. Tectonophysics 101:75–86
Brown M (1989) Where giants stood: the saga of Kenya's desert lake. Qiller Press, London
Brown FH, Carmichael I (1969) Quaternary volcanoes of the Lake Rudolf region: I. The basanite-tephrite series of the Korath Range. Lithos 2:239–260
Brown FH, Carmichael I (1971) Quaternary volcanoes of the Lake Rudolf region: II. The lavas of North Island, South Island and the Barrier. Lithos 4:305–323
Brown FH, Fuller CR (2008) Stratigraphy and tephra of the Kibish Formation, southwestern Ethiopia. J Hum Evol 55:366–403
Brown FH, Gathogo P (2005) A comment on "Recent tectonics in the Turkana Rift (North Kenya): an integrated approach from drainage network, satellite imagery and reflection seismic analyses." Basin Res 17:329–335
Brown FH, McDougall I (2011) Geochronology of the Turkana depression of northern Kenya and southern Ethiopia. Evol Anthropol 20:217–227
Brown FH, McDougall I, Davies T, Maier R (1985) An integrated Plio-Pleistocene chronology for the Turkana Basin. In: Delson E (ed) Ancestors: the hard evidence. Liss, New York, pp 82–90
Bruhn RL, Brown FH, Gathogo PN, Haileab B (2011) Pliocene volcano-tectonics and paleogeography of the Turkana Basin, Kenya and Ethiopia. J Afr Earth Sci 59:295–312
Bulatovich AK (1900) Dall'Abissinia al Lago Rodolfo per Il Caffa. Boll Soc Geogr Ital Ser 4(1):121–142
Butzer KW (1971) Recent history of an Ethiopian delta: the Omo River and the level of Lake Rudolf. Univ Chicago Dept Geogr Res Pap 136
Butzer KW (1980) The Holocene lake plain of north Rudolf, East Africa. Phys Geogr 1:42–58
Butzer KW, Thurber DL (1969) Some late Cenozoic sedimentary formations of the lower Omo Valley. Nature 222:1138–1143
Butzer KW, Isaac GL, Richardson JL, Washbourn-Kamau C (1972) Radiocarbon dating of East African lake levels. Science 175:1069–1076
Cabeza M, Fernandez-Llamazares A, Fraixedas S, López-Baucells A, Burgas D, Rocha R, Cabeza R, Ajiko A (2016) Winds of hope for Sibiloi National Park. SWARA (Mag E Afr Wildlife Soc) October–December, pp 33–37
Carr CJ (2017) River Basin development and human rights in Eastern Africa – a policy crossroads. Springer, Cham
Caukwell R (1971) The South Turkana Expedition: scientific papers VI. Field survey in South Turkana. Geogr J 137:157–164
Cavendish HSH (1898) Through Somaliland and around and south of Lake Rudolf. Geogr J 11:372–393
Cerling TE (1979) Paleochemistry of Plio-Pleistocene Lake Turkana, Kenya. Palaeogeogr Palaeoclimatol Palaeoecol 27:247–285
Cerling TE (1986) A mass-balance approach to basin sedimentation: constraints on the recent history of the Turkana basin. Palaeogeogr Palaeoclimatol Palaeoecol 54:63–86
Cerling TE (1996) Pore water chemistry of an alkaline lake: Lake Turkana, Kenya. In: Johnson TC, Odada EO (eds) The limnology, climatology and paleoclimatology of the East African lakes. Gordon and Breach, Amsterdam, pp 225–240
Cerling TE, Brown FH (1982) Tuffaceous marker horizons in the Koobi Fora region and the Lower Omo Valley. Nature 299:216–221
Cerling TE, Powers SD (1977) Paleorifting between the Gregory and Ethiopian rifts. Geology 5:441–444

Cerling TE, Brown FH, Cerling BW, Curtis GH, Drake RE (1979) Preliminary correlations between the Koobi Fora and Shungura Formations, East Africa. Nature 279:118–121
Champion AM (1935) Teleki's Volcano and the lava fields at the southern end of Lake Rudolf. Geogr J 85:322–341
Champion AM (1937) Physiography of the region to the west and southwest of Lake Rudolf. Geogr J 89:97–118
Champion AM, Smith WC (1937) The volcanic region around the southern end of Lake Rudolf, Kenya Colony. Z Vulcanol 17:163–172
Charsley TJ (1987) Geology of the North Horr area. Rep Mines Geol Dept Kenya 110
Chorowicz J (2005) The East African rift system. J Afr Earth Sci 43:379–410
Chritz KL, Cerling TE, Freeman KH, Hildebrand EA, Janzen A, Prendergast ME (2019) Climate, ecology, and the spread of herding in eastern Africa. Quat Sci Rev 204:119–132
Cohen A (1984) Effect of zoobenthic standing crop on laminae preservation in tropical lake sediment, Lake Turkana, East Africa. J Paleontol 58:499–510
Cohen A (1986) Distribution and faunal associations of benthic invertebrates at Lake Turkana, Kenya. Hydrobiologia 141:179–197
Cohen A (1989) Facies relationships and sedimentation in large rift lakes and implications for hydrocarbon exploration: examples from Lakes Turkana and Tanganyika. Palaeogeogr Palaeoecol Palaeoclimatol 70:65–80
Cohen A, Ferguson D, Gram P, Hubler S, Sims K (1986) The distribution of coarse-grained sediments in modern Lake Turkana, Kenya: implications for clastic sedimentation models of rift lakes. In: Frostick LE, Renaut RW, Reid I, Tiercelin J-J (eds) Sedimentation in the African rifts. Geol Soc Lond Spec Publ 25:127–139
Copley H (1941) A short account on the freshwater fishes of Kenya. J E Afr Nat Hist Soc 16:1–24
Coppens Y, Howell FC, Isaac GL, Leakey REF (eds) (1976) Earliest man and environments in the Lake Rudolf Basin: stratigraphy, paleoecology and evolution. University of Chicago Press, Chicago
Corti G, Cioni R, Franceschini Z, Sani F, Scaillet S, Molin P, Isola I, Mazzarini F, Brune S, Keir D, Erbello A, Muluneh A, Illsley-Kemp F, Glerum A (2019) Aborted propagation of the Ethiopian rift caused by linkage with the Kenyan rift. Nat Commun 10:1309
Curtis PC (1991) Competing magmatic processes at a continental rift: petrology and geochemistry of Quaternary volcanoes in the Turkana Rift Zone, East Africa. PhD thesis, University of North Carolina, Chapel Hill
Davidson A (1983) The Omo River Project: reconnaissance geology and geochemistry of parts of Ilubabor, Kefa, Gema Gofa and Sidamo, Ethiopia. In: Provisional Military Government of Socialist Ethiopia, Ministry of Mines and Energy, Ethiopian Institute of Geological Surveys. Addis Ababa
deMenocal PB, Brown FH (1999) Pliocene tephra correlations between East African hominid localities, the Gulf of Aden, and the Arabian Sea. In: Agusti J, Rook L, Andrews PA (eds) Hominid evolution and climatic change in Europe, vol 1. Cambridge University Press, Cambridge, pp 23–54
Dixey F (1945) Miocene sediments in South Turkana. J E Afr Nat Hist Soc 18:13–14
Dixey F (1948) Geology of Northern Kenya. Geol Surv Kenya Rep 15
Dodson RG (1963) Geology of the South Horr area. Geol Surv Kenya Rep 60
Dodson RG (1971) Geology of the area south of Lodwar. Geol Surv Kenya Rep 87
Donaldson-Smith A (1896) Expedition through Somaliland to Lake Rudolf. Geogr J 8:221–239
Dunkelman T (1986) The structural and stratigraphic evolution of Lake Turkana, Kenya, as deduced from a multichannel seismic survey. MSc thesis, Duke University, Durham, North Carolina
Dunkelman TJ, Karson J, Rosendahl B (1988) Structural style of the Turkana rift, Kenya. Geology 16:258–261
Dunkelman TJ, Rosendahl B, Karson J (1989) Structure and stratigraphy of the Turkana rift from seismic reflection data. J Afr Earth Sci 8:489–510

Dunkley PN, Smith M, Allen DJ, Darling WG (1993) The geothermal activity and geology of the northern sector of the Kenya Rift Valley. Br Geol Surv Res Rep SC/93/1

Earthview Geoconsultants Ltd (2011) Environmental impact assessment project report for proposed oil and gas seismic project in Block 12a: parts of Turkana East, East Pokot, Marakwet West, Baringo Central, Marigat, Keiyo South, Baringo North, Mogotio and Samburu Districts by Tullow Kenya. Tullow Oil, London

Ebinger CJ, Ibrahim A (1994) Multiple episodes of rifting in Central and East Africa: a re-evaluation of gravity data. Geol Rundsch 83:689–702

Ebinger CJ, Sleep NH (1998) Cenozoic magmatism throughout East Africa resulting from impact of a single plume. Nature 395:788–791

Ebinger CJ, Yemane T, Harding DJ, Tesfaye S, Kelley S, Rex DC (2000) Rift deflection, migration, and propagation: linkage of the Ethiopian and Eastern rifts, Africa. Geol Soc Am Bull 112:163–176

Fairburn WA, Mathieson FJ (1970) Geology of the Loiya–Lorugumu area. Geol Surv Kenya Rep 85

Fasola M, Canova L (1993) Diel activity of resident and immigrant waterbirds at Lake Turkana, Kenya. Ibis 135:442–450

Feibel CS (2003) Stratigraphy and depositional history of the Lothagam sequence. In: Leakey MG, Harris JM (eds) Lothagam: the dawn of humanity in eastern Africa. Columbia University Press, New York, pp 17–29

Feibel CS (2011) A geological history of the Turkana Basin. Evol Anthropol 20:206–216

Ferguson AJD (1982) Studies of the zooplankton of Lake Turkana. In: Hopson AJ (ed) Lake Turkana. A report on the findings of the Lake Turkana Project 1972–1975, vol 1. Overseas Development Administration, London, pp 163–245

Ferguson AJD, Harbott BJ (1982) Geographical, physical, and chemical aspects of Lake Turkana. In: Hopson AJ (ed) Lake Turkana. A report on the findings of the Lake Turkana Project 1972–1975, vol 1. Overseas Development Administration, London, pp 1–110

Fischer ML, Markowska M, Bachofer F, Foerster VE, Asrat A, Zielhofer C, Trauth MH, Junginger A (2020) Determining the pace and magnitude of lake level changes in southern Ethiopia over the last 20,000 years using lake balance modeling and SEBAL. Front Earth Sci 8(197):1–21

Fischer ML, Bachofer F, Yost CL, Bludau IJE, Scheperes C, Foerster VE, Lamb H, Schäbitz F, Asrat A, Trauth MH, Junginger A (2021) A phytolith supported biosphere-hydrosphere predictive model for southern Ethiopia: insights into paleoenvironmental changes and human landscape preferences since the Last Glacial Maximum. Geosciences 11:418

Fish GR (1953) Chemical analyses of Lake Rudolf. E Afr Fish Res Org Rep

Feibel CS (2011) A geological history of the Turkana Basin. Evol Anthropol 20:206–216

Fitch FJ, Vondra CF (1976) Tectonic and stratigraphic framework. In: Coppens Y, Howell FC, Isaac GL, Leakey REF (eds) Earliest man and environments in the Lake Rudolf Basin: stratigraphy, paleoecology and evolution. University of Chicago Press, Chicago, pp 8–11

Foerster VE (2014) Late Quaternary climate variability in the source region of *Homo sapiens*. Dry-wet cycles in Chew Bahir, southern Ethiopia. PhD thesis, Universität Köln

Foerster V, Junginger A, Langkamp O, Gebru T, Asrat A, Umer M, Lamb HF, Wennrich V, Rethemeyer J, Nowaczyk N, Trauth MH, Schaebitz F (2012) Climatic change recorded in the sediments of the Chew Bahir basin, southern Ethiopia, during the last 45,000 years. Quat Int 274:25–37

Foerster V, Vogelsang R, Junginger A, Asrat A, Lamb HF, Schaebitz F, Trauth MH (2015) Environmental change and human occupation of southern Ethiopia and northern Kenya during the last 20,000 years. Quat Sci Rev 129:333–340

Foerster V, Vogelsang R, Junginger A, Asrat A, Lamb HF, Schaebitz F, Trauth MH (2016) Reply to the comment on 'Environmental change and human occupation of southern Ethiopia and northern Kenya during the last 20,000 years' (Quat Sci Rev 129:333–340). Quat Sci Rev 141:130–133

Foerster V, Deocampo DM, Asrat A, Günter C, Junginger A, Krämer KH, Stroncik NA, Trauth MH (2018) Towards an understanding of climate proxy formation in the Chew Bahir basin, southern Ethiopian Rift. Palaeogeogr Palaeoclimatol Palaeoecol 501:111–123

Forman SL, Wright DK, Bloszies C (2014) Variations in water level for Lake Turkana in the past 8500 years near Mt. Porr, Kenya and the transition from the African Humid Period to Holocene aridity. Quat Sci Rev 97:84–101

Friis I (1992) Forests and forest trees of northeast tropical Africa. HMSO, London

Frostick LE, Reid I (1989) Is structure the main control of river drainage and sedimentation in rifts? J Afr Earth Sci 8:165–182

Fuchs VE (1934) The geological work of the Cambridge Expedition to the East African Lakes, 1930–1931. Geol Mag 71:97–112

Fuchs VE (1935) The Lake Rudolf rift valley expedition, 1934. Geogr J 86:114–137

Fuchs VE (1939) The geological history of the Lake Rudolf basin, Kenya Colony. Phil Trans R Soc Lond Ser B 229:219–274

Furman T, Bryce JG, Karson J, Iotti A (2004) East African Rift System (EARS) plume structure: insights from Quaternary mafic lavas of Turkana, Kenya. J Petrol 45:1069–1088

Furman T, Kaleta KM, Bryce JG, Hanan BB (2006) Tertiary mafic lavas of Turkana, Kenya: constraints of East African plume structure and occurrence of high-μ volcanism in Africa. J Petrol 47:1221–1244

Garcin Y, Melnick D, Strecker MR, Olago D, Tiercelin J-J (2012) East African mid-Holocene wet-dry transition recorded in palaeo-shorelines of Lake Turkana, northern Kenya Rift. Earth Plan Sci Lett 331–332:322–334

Gathogo PN, Brown FH, McDougall I (2008) Stratigraphy of the Koobi Fora Formation (Pliocene and Pleistocene) in the Loiyangalani region of northern Kenya. J Afr Earth Sci 51:277–297

Gebregiorgis D, Deocampo DM, Foerster V, Longstaffe FJ, Delaney JS, Schaebitz F, Junginger A, Markowska M, Opitz S, Trauth MH, Lamb HK, Asrat A (2021) Modern sedimentation and authigenic mineral formation in the Chew Bahir Basin, Southern Ethiopia: implications for interpretation of Late Quaternary paleoclimate records. Front Earth Sci 9:607695

Goldstein ST (2019) Lithic technological strategies of the earliest herders at Lake Turkana, northern Kenya. Antiquity 93:1495–1514

Goldstein ST, Hildebrand E, Storozum M, Sawchuk E, Lewis J, Ngugi C, Robbins LH (2017) New archaeological investigations at the Lothagam harpoon site at Lake Turkana. Antiquity 91:143–145

Goldstein ST, Shipton C, Miller JM, Ndiema E, Boivin N, Petraglia M (2022) Hunter-gatherer technological organization and responses to Holocene climate change in coastal, lakeshore, and grassland ecologies of eastern Africa. Quat Sci Rev 280:107390

Goodier SAM, Cotterill FPD, O'Ryan C, Skelton PH, de Wit MJ (2011) Cryptic diversity of African Tigerfish (Genus *Hydrocynus*) reveals palaeogeographic signatures of linked Neogene geotectonic events. PLoS One 6:e28775

Government of Kenya [GoK] and UNDP (2021) Rising water levels in Kenya's rift valley lakes, Turkwel Gorge Dam and Lake Victoria: a scoping report. Government of Kenya, Nairobi

Gownaris NJ, Pikitch EK, Ojwang WO, Michener R, Kaufman L (2015) Predicting species' vulnerability in a massively perturbed system: the fishes of Lake Turkana, Kenya. PLoS One 10:e0127027–e0127024

Gownaris NJ, Pikitch EK, Aller JY, Kaufman LS, Kolding J, Lwiza KM, Obiero KO, Ojwang WO, Malala JO, Rountos KJ (2016) Fisheries and water level fluctuations in the world's largest desert lake. Ecohydrology 10:e1769

Graham A (1968) The Lake Rudolf crocodile (*Crocodylus niloticus* Laurenti) population. MSc thesis, University of East Africa

Gregory JW (1894) Contributions to the physical geography of British East Africa. Geogr J 4:289–315

Gregory JW (1901) The foundation of British East Africa. Temple House, London

Gregory JW (1920) The African Rift Valley. Nature 104:518–519

Gregory JW (1923) The structure of the Great Rift Valley. Nature 112:514–516

Grove AT, Street AF, Goudie AS (1975) Former lake levels and climatic change in the rift valley of southern Ethiopia. Geogr J 141:177–194

Gwynn CW (1911) A journey in southern Abyssinia. Geogr J 38:113–139

Haack B (2007) Environmental auditing – monitoring wetland changes with remote sensing: an East African example. Environ Manag 20:411–419

Haileab B, Brown FH, McDougall I, Gathogo PN (2004) Gombe Group basalts and initiation of Pliocene deposition in the Turkana depression, northern Kenya and southern Ethiopia. Geol Mag 141:41–53

Halfman JD (1987) High-resolution sedimentology and paleoclimatology of Lake Turkana, Kenya. PhD thesis, Duke University, Durham, North Carolina

Halfman JD (1996) CTD-transmissometer profiles from Lakes Malawi and Turkana. In: Johnson TC, Odada EO (eds) The limnology, climatology and paleoclimatology of the East African lakes. Gordon and Breach, Amsterdam, pp 169–182

Halfman JD, Hearty PJ (1990) Cyclical sedimentation in Lake Turkana, Kenya. In: Katz B (ed) Lacustrine basin exploration: case studies and modern analogs. Am Assoc Petrol Geol Mem 50:187–196

Halfman JD, Johnson TC (1988) High-resolution record of cyclic climatic-change during the past 4 ka from Lake Turkana, Kenya. Geology 16:496–500

Halfman JD, Johnson TC, Showers WJ, Lister GS (1989) Authigenic low-Mg calcite in Lake Turkana, Kenya. J Afr Earth Sci 8:533–540

Halfman JD, Jacobson DF, Cannella CM, Haberyan KA, Finney BP (1992) Fossil diatoms and the mid to late Holocene paleolimnology of Lake Turkana, Kenya: a reconnaissance study. J Paleolimnol 7:23–35

Halfman JD, Johnson TC, Finney BP (1994) New AMS dates, stratigraphic correlations and decadal climatic cycles for the past 4 ka at Lake Turkana, Kenya. Palaeogeogr Palaeoclimatol Palaeoecol 111:83–98

Harbott BJ (1982) Studies on algal dynamics and primary productivity in Lake Turkana. In: Hopson AJ (ed) Lake Turkana. A report on the findings of the Lake Turkana Project 1972–1975, vol 1. Overseas Development Administration, London, pp 111–161

Harbott BJ, Ogari JTN (1982) The biology of the larger cichlid fish of Lake Turkana. In: Hopson AJ (ed) Lake Turkana: a report of the findings of the Lake Turkana Project, 1972–1975, vol 5. Overseas Development Administration, London, pp 1331–1355

Harris JM, Leakey MG (2003) Geology and vertebrate paleontology of the Early Pliocene site of Kanapoi, northern Kenya. Contrib Sci Natur Hist Mus Los Angeles 498:1–134

Harris JM, Leakey MG, Brown FH (2006) A brief history of research at Koobi Fora, northern Kenya. Ethnohistory 53:35–69

Harrison J (1901) A journey from Zeila to Lake Rudolf. Geogr J 18:258–275

Hartman AT (2018) An analysis of the effects of temperatures and circulations on the strength of the low-level jet in the Turkana Channel in East Africa. Theor Appl Climatol 132:1003–1017

Harvey CPD, Grove AT (1982) A prehistoric source of the Nile. Geogr J 148:327–336

Haug E (1912) Traité de géologie. Pt. 2. Les périodes géologiques. Armand Colin, Paris

Hendrie DB, Kusznir NJ, Morley CK, Ebinger CJ (1994) Cenozoic extension in northern Kenya: a quantitative model of rift basin development in the Turkana region. Tectonophysics 236:409–438

Herren GL (2013) Thriving in a desert lake: investigations into the meiobenthos of Lake Turkana, Kenya. MSc thesis, Universität Bielefeld

Hildebrand EA, Shea JJ, Grillo KM (2011) Four middle Holocene pillar sites in West Turkana, Kenya. J Field Archaeol 36:181–200

Hildebrand EA, Grillo KM, Sawchuk EA, Pfeiffer SK, Conyers LB, Goldstein ST, Hill AC, Janzen A, Klehm CE, Helper M, Kiura P, Ndiema E, Ngigi C, Shea JJ, Wang H (2018) A monumental cemetery built by eastern Africa's first herders near Lake Turkana, Kenya. Proc Natl Acad Sci 115:8942–8947

Hillaby J (1964) Journey to the Jade Sea. Constable, London

Hodbod J, Stevenson EGJ, Akall G, Akuja T, Angele I, Bedasso EA, Buffavand L, Derbyshire S, Eulenberger I, Gownaris N, Kamski B, Kurewa A, Lokuruka M, Mulugeta M, Okenwa D, Rodgers C, Tebbs E (2019) Social-ecological change in the Omo-Turkana basin: a synthesis of current developments. Ambio 48:1099–1115

Hodson A (1929) Journeys from Maji, southwest Abyssinia. Geogr J 73:401–428

Hoey AC (1911) Lake Rudolf. J E Afr Nat Hist Soc 2:47–51

Holland WP (1926) Volcanic action north of Rudolf in 1918. Geogr J 68:488–499

Hooijer DA, Patterson B (1972) Rhinoceroses from the Pliocene of northwestern Kenya. Bull Mus Comp Zool Harv University 144:1–26

Hopson AJ (ed) (1982) Lake Turkana. A report on the findings of the Lake Turkana Project 1972–1975, vol 6. Overseas Development Administration, London

Hopson AJ, Hopson J (1982) The fishes of Lake Turkana with a description of three new species. In: Hopson AJ (ed) Lake Turkana. A report on the findings of the Lake Turkana Project 1972–1975, vol 1. Overseas Development Administration, London, pp 283–347

Jérémine E (1935) Mission scientifique de l'Omo 1932–1933. Tome 1. Géologie, Anthropologie. Fasc. 1, Étude pétrographique des roches de la bordure occidentale du lac Rodolphe et de quelques régions voisines. Mem Mus Natl Hist Nat Paris

Johnson G (1974) Cainozoic lacustrine stromatolites from hominid-bearing sediments east of Lake Rudolf, Kenya. Nature 247:520–523

Johnson TC (1996) Sedimentary processes and signals of past climatic change in large lakes of the East African Rift Valley. In: Johnson TC, Odada EO (eds) The limnology, climatology and paleoclimatology of the East African lakes. Gordon and Breach, Amsterdam, pp 367–412

Johnson TC, Malala JO (2009) Lake Turkana and its link to the Nile. In: Dumont HJ (ed) The Nile: origin, environments, limnology and human use. Springer, Dordrecht, pp 287–304

Johnson TC, Halfman JD, Rosendahl BR, Lister GS (1987) Climatic and tectonic effects on sedimentation in a rift-valley lake: evidence from high-resolution seismic profiles, Lake Turkana, Kenya. Geol Soc Am Bull 98:439–447

Johnson TC, Halfman JD, Showers WJ (1991) Paleoclimate of the last 4000 years at Lake Turkana, Kenya, based on the isotopic composition of authigenic calcite. Palaeogeogr Palaeoecol Palaeoclimatol 85:189–198

Joleaud L (1920) Contribution à l'étude des hippopotames fossils. Bull Soc Géol Fr Ser 4(20):13–26

Joubert P (1966) Geology of the Loperot area. Geol Surv Kenya Rep 74

Junginger A, Trauth MH (2013) Hydrological constraints of paleo-Lake Suguta in the Northern Kenya Rift during the African Humid Period (15–5 ka BP). Glob Planet Change 111:174–188

Junginger A, Roller S, Olaka LA, Trauth MH (2014) The effects of solar irradiation changes on the migration of the Congo Air Boundary and water levels of paleo-Lake Suguta, northern Kenya Rift, during the African Humid Period (15–5 ka BP). Palaeogeogr Palaeoclimatol Palaeoecol 396:1–16

Kaijage AS, Nyagah NM (2010) Final draft report on socio–economic analysis and public consultation of Lake Turkana communities in northern Kenya. Report commissioned by African Development Bank, Tunis

Källqvist T, Lien L, Liti D (1988) Lake Turkana limnological study 1985–1988. Norw Inst Water Res Rep, pp 0–85313

Kariuki P (2011) Kenyan wetlands – an indicator of climate change and human disturbance. African Conservation Centre Report

Karson JA, Curtis PC (1989) Tectonic and magmatic processes in the eastern branch of the East African rift and implications for magmatically active continental rifts. J Afr Earth Sci 8:431–453

Karson JA, Curtis PC (1994) Quaternary volcanic centres of the Turkana Rift, Kenya. J Afr Earth Sci 18:15–35

Khan MA, Swain CJ (1978) Geophysical investigations and the rift valley geology of Kenya. In: Bishop WW (ed) Geological background to fossil man. Geol Soc Lond Spec Publ 6:71–83

Knappe E, Bendick R, Ebinger C, Birhanu Y, Lewi E, Floyd M, King R, Kianji G, Mariita N, Temtime T, Waktola B, Deresse B, Musila M, Kanoti J, Perry M (2020) Accommodation of East African rifting across the Turkana Depression. J Geophys Res Solid Earth 125:11–14

Kolding J (1989) The fish resources of Lake Turkana and their environment (final report of project KEN-043, NORAD, Oslo). PhD thesis, University of Bergen

Kolding J (1992) A summary of Lake Turkana: an ever-changing mixed environment. Mitt Int Ver Theor Angew Limnol 23:25–35

Kolding J (1993a) Trophic interrelationships and community structure at two different periods of Lake Turkana, Kenya: a comparison using the ECOPATH II box model. In: Pauly D, Christensen V (eds) Trophic models of aquatic ecosystems. Proc ICLARM Conf 26, Manila, pp 116–123

Kolding J (1993b) Population dynamics and life-history styles of Nile tilapia, *Oreochromis niloticus*, in Ferguson's Gulf, Lake Turkana, Kenya. Environ Biol Fish 37:25–46

Kolding J (1995) Changes in species composition and abundance of fish populations in Lake Turkana, Kenya. In: Pitcher TJ, Hart PJB (eds) The impact of species changes in African lakes. Chapman and Hall, London, pp 335–363

Krueger J, Foerster V, Trauth MH, Hofreiter M, Tiedemann R (2021) Exploring the past biosphere of Chew Bahir/Southern Ethiopia: cross-species hybridization capture of ancient sedimentary DNA from a deep drill core. Front Earth Sci 9:683010

Lahr MM, Rivera F, Power RK, Mounier A, Copsey B, Crivellaro F, Edung JE, Fernandez JMM, Kiarie C, Lawrence J, Leakey A, Mbua E, Miller H, Muigai A, Mukhongo DM, Van Baelen A, Wood R, Schwenninger JL, Grün R, Achyuthan H, Wilshaw A, Foley RA (2016) Inter-group violence among early Holocene hunter-gatherers of West Turkana, Kenya. Nature 529:394–398

Le Gall B, Vétel W, Morley CK (2005) Inversion tectonics during continental rifting: the Turkana Cenozoic rifted zone, northern Kenya. Tectonics 24:1–17

Leakey MG, Harris JM (eds) (2003) Lothagam: the dawn of humanity in Eastern Africa. Columbia University Press, New York

Liti D, Källqvist T, Lien L (1991) Limnological aspects of Lake Turkana, Kenya. Verh Int Verein Limnol 24:1108–1111

Liu T, Lepre CJ, Hemming SR, Broecker WS (2021) Rock varnish record of the African Humid Period in the Lake Turkana basin of East Africa. The Holocene 31:1239–1249

MacDonald R, Williams LAJ, Gass IG (1994) Tectonomagmatic evolution of the Kenya rift valley: some geological perspectives. J Geol Soc Lond 151:879–888

Maglio VJ (1973) Origin and evolution of the Elephantidae. Trans Am Phil Soc 63:1–149

Marchant R, Lane P (2014) Past perspectives for the future: foundations for sustainable development in East Africa. J Archaeol Sci 51:12–21

Mariita NO, Keller GR (2007) An integrated geophysical study of the northern Kenya rift. J Afr Earth Sci 48:80–94

Martin RL (2014) Late Quaternary sedimentary geology of North Turkwel, Turkana County, Kenya. MSc thesis, Rutgers University, Piscataway, New Jersey

Martins D (2012) Central Island. Kenya's 'lost world' in the middle of Lake Turkana. SWARA (Mag E Afr Wildlife Soc) April–June:1–26

Maud P (1904) Exploration of the southern borderland of Abyssinia. Geogr J 23:552–559

Maxon R (2009) East Africa: an introductory history, 3rd edn. West Virginia University Press, Morgantown

Mbaluka JK, Brown FH (2016) Vegetation of the Koobi Fora region northeast of Lake Turkana, Marsabit County, northern Kenya. J E Afr Nat Hist 105:21–50

McDougall I, Brown FH, Fleagle JG (2005) Stratigraphic placement and age of modern humans from Kibish, Ethiopia. Nature 433:733–736

McDougall I, Brown FH (2008) Geochronology of the pre-KBS Tuff sequence, Omo Group, Turkana Basin. J Geol Soc Lond 165:549–562

McDougall I, Brown FH (2009) Timing of volcanism and evolution of the northern Kenya Rift. Geol Mag 146:3447

McDougall I, Brown FH, Fleagle JG (2005) Stratigraphic placement and age of modern humans from Kibish, Ethiopia. Nature 433:733–736

McDougall I, Feibel CS (1999) Numerical age control for the Miocene-Pliocene succession at Lothagam, a hominoid-bearing sequence in the northern Kenya Rift. J Geol Soc Lond 156:731–745

McDougall I, Watkins RT (1988) Potassium-argon ages of volcanic rocks from northeast of Lake Turkana, northern Kenya. Geol Mag 125:15–23

McDougall I, Watkins R (2006) Geochronology of the Nabwal Hills: a record of earliest magmatism in the northern Kenyan Rift Valley. Geol Mag 143:25–39

McLeod AAQR (1982) The biology of *Lates longispinis* Wort. In: Hopson AJ (ed) Lake Turkana. A report on the findings of the Lake Turkana Project 1972–1975, vol 5. Overseas Development Administration, London, pp 1305–1330

Melnick D, Garcin Y, Quinteros J, Strecker MR, Olago D, Tiercelin J-J (2012) Steady rifting in northern Kenya inferred from deformed Holocene lake shorelines of the Suguta and Turkana basins. Earth Planet Sci Lett 331–332:335–346

Mohammed MU, Bonnefille R, Johnson TC (1995) Pollen and isotopic records in Late Holocene sediments from Lake Turkana, Kenya. Palaeogeogr Palaeoclimatol Palaeoecol 119:371–383

Mohr PA, Wood CA (1976) Volcano spacings and lithospheric attenuations in the eastern rift system of Africa. Earth Planet Sci Lett 33:126–144

Morley CK (1999) Boundary fault angle, with particular reference to the Lokichar Fault, Turkana Region, Kenya. In: Morley CK (ed) Geoscience of rift systems – evolution of East Africa. AAPG Stud Geol 44:115–129

Morley CK (2020) Early syn-rift igneous dike patterns, northern Kenya Rift (Turkana, Kenya): implications for local and regional stresses, tectonics, and magma-structure interactions. Geosphere 16:890–918

Morley CK, Wescott WA, Stone DM, Harper RM, Wigger ST, Karanja FM (1992) Tectonic evolution of the northern Kenyan Rift. J Geol Soc Lond 149:333–348

Morley CK, Karanja FM, Wescott WA, Stone DM, Harper RM, Wigger ST, Day RA (1999a) Geology and geophysics of the Western Turkana Basins, Kenya. In: Morley, CK (ed) Geoscience of rift systems–evolution of East Africa. AAPG Stud Geol 44:19–54

Morley CK, Ngenoh DK, Ego JK (1999b) Introduction to the East African rift system. In: Morley CK (ed) Geoscience of rift systems—evolution of East Africa. AAPG Stud Geol 44:1–18

Morrissey A (2014) Stratigraphic framework and Quaternary paleolimnology of the Lake Turkana Rift, Kenya. PhD thesis, Syracuse University, Syracuse, New York

Morrissey A, Scholz CA (2014) Paleohydrology of Lake Turkana and its influence on the Nile River system. Palaeogr Palaeoclimatol Palaeoecol 403:88–100

Morrissey A, Scholz CA, Russell JM (2018) Late Quaternary TEX_{86} paleotemperatures from the world's largest desert lake, Lake Turkana, Kenya. J Paleolimnol 59:103–117

Muchane MW (1994) Stable isotope analyses of authigenic calcite from Lake Turkana, Kenya: high resolution paleoclimatic implications for the past 5000 years. PhD thesis, Duke University, Durham, North Carolina

Muchane MW (1996) Comparison of the isotope record in micrite, Lake Turkana, with the historical weather record over the last century. In: Johnson TC, Odada EO (eds) The limnology, climatology and paleoclimatology of the East African lakes. Gordon and Breach, Amsterdam, pp 431–441

Muška M, Vašek M, Modrý D, Jirků M, Ojwang WO, Malala JO, Kubečka J (2012) The last snapshot of natural pelagic fish assemblage in Lake Turkana, Kenya: a hydroacoustic study. J Great Lakes Res 38:98–106

Mutegi BK (2014) Crustal velocity model beneath Lodwar station, Kenya using teleseismic p-wave receiver function. MSc thesis, University of Nairobi

Mwakikagile G (2010) Life in Kenya: the land and the people, modern and traditional ways. Continental Press, Dar es Salaam

Neumann AH (1898) Elephant hunting in east equatorial Africa. Rowland Ward, London

Ng'ang'a P, Muchane MW, Johnson TC, Sturgeon K (1998) Comparison of isotopic records in abiogenic and biogenic calcite from Lake Turkana, Kenya. In: Lehman JT (ed) Environmental change and response in East African lakes. Kluwer, Dordrecht, pp 173–190

Nicholson SE (2014) The predictability of rainfall over the Greater Horn of Africa. Part I: prediction of seasonal rainfall. J Hydrometeorol 15:1011–1027

Nicholson SE (2015) The predictability of rainfall over the Greater Horn of Africa. Part II: prediction of monthly rainfall during the long rains. J Hydrometeorol 16:2001–2012

Nutz A, Schuster M (2016) Stepwise drying of Lake Turkana at the end of the African Humid Period: a forced regression modulated by solar activity variations? Solid Earth 7:1609–1618

Nutz A, Schuster M, Ghienne JF, Roquin C, Bouchette F (2016) Wind-driven waterbodies: a new category of lake within an alternative sedimentologically-based lake classification. J Paleolimnol 16:2001–2012

Nyamweru CK (1986) Quaternary environments of the Chalbi basin, Kenya: sedimentary and geomorphological evidence. In: Frostick LE, Renaut RW, Reid I, Tiercelin J-J (eds) Sedimentation in the African rifts. Geol Soc Lond Spec Publ 25:297–310

Nyamweru C (1989) New evidence for the former extent of the Nile drainage system. Geogr J 155:179–188

Nyamweru CK, Bowman D (1989) Climatic changes in the Chalbi desert, north Kenya. J Quat Sci 4:131–139

Obiero K, Wakjira M, Gownaris N, Malala J, Keyombe JL, Ajode MZ, Smith S, Lawrence T, Ogello E, Getahun A, Kolding J (2022) Lake Turkana: status, challenges, and opportunities for collaborative research. J Great Lakes Res. https://doi.org/10.1016/j.jglr.2022.10.007

Ojwang W, Obiero KO, Donde OO, Gownaris NJ, Pikitch EK, Omondi R, Agembe S, Malala J, Avery ST (2016) Lake Turkana: world's largest permanent desert lake (Kenya). In: Finlayson CM, Milton GR, Prentice RC, Davidson NC (eds) The wetland book. Springer, Dordrecht, pp 1361–1380

Olago DO (1992) The mineralogy and sedimentology of recent subaqueous sediments in Lake Turkana, Kenya. MSc thesis, University of Nairobi

Olago DO, Odada EO (1996) Physical and chemical dynamics of Lake Turkana. In: Johnson TC, Odada EO (eds) The limnology, climatology and paleoclimatology of the East African lakes. Gordon and Breach, Amsterdam, pp 413–430

Olago DO, Odada EO (2000) An inter-basinal comparison of the sedimentology of Late Holocene to recent sediments in the Rift Valley, Lake Turkana, Kenya. J Afr Earth Sci 31:237–252

Olaka LA (2011) Hydrology across scales: sensitivity of East African lakes to climate changes. PhD thesis, Universität Potsdam

Olilo CO, Odoli CO, Obiero MO, Malala JO, Bironga CH (2023) Microbial ecology of desert Lakes Baringo and Turkana, Kenya, East Africa. In: El-Sheekh M, Elsaied HE (eds) Lakes of Africa – microbial diversity and sustainability. Elsevier, Amsterdam, pp 247–268

Otachi EO, Plessl C, Körner W, Avenant-Oldewage A, Jirsa F (2015) Trace elements in water, sediments and the elongate Tigerfish *Hydrocynus forskahlii* (Cuvier 1819) from Lake Turkana, Kenya including a comprehensive health risk analysis. Bull Environ Contam Tox 95:286–291

Owen RB (1981) Quaternary diatomaceous sediments and the geological evolution of Lakes Turkana, Baringo and Bogoria, Kenya Rift Valley. PhD thesis, University of London

Owen R, Renaut RW (1986) Sedimentology, stratigraphy and palaeoenvironments of the Holocene Galana Boi Formation, NE Lake Turkana, Kenya. In: Frostick LE, Renaut RW, Reid I, Tiercelin J-J (eds) Sedimentation in the African rifts. Geol Soc Lond Spec Publ 25:311–322

Owen RB, Renaut RW (1990) Nitrate deposition in the Turkana Basin of northern Kenya. J Afr Earth Sci 10:701–706

Owen RB, Barthelme JW, Renaut RW, Vincens A (1982) Palaeolimnology and archaeology of Holocene deposits north-east of Lake Turkana, Kenya. Nature 298:523–529

Owen RB, Renaut RW, Lowenstein TK (2018) Spatial and temporal geochemical variability in lacustrine sedimentation in the East African Rift System: evidence from the Kenya Rift and regional analyses. Sedimentology 65:1697–1730

Parker H (1932) Scientific results of the Cambridge Expedition to the East African Lakes, 1930–1. 5. Reptiles and amphibians. J Linn Soc Lond Zool 38:213–229
Patterson B, Howells WW (1967) Hominid humeral fragment from early Pleistocene of northwestern Kenya. Science 156:64–66
Patterson B, Behrensmeyer AK, Sill WD (1970) Geology of a new Pliocene locality in northwestern Kenya. Nature 256:279–284
Phillipson DW (1976) Fishermen and the beginnings of East African farming: new light from northern Kenya. Kenya Past Present 7:3–9
Phillipson DW (1977a) The later prehistory of Eastern and Southern Africa. Heinemann, London
Phillipson DW (1977b) Lowasera. Azania 12:1–32
Pickford M, Senut B, Poupeau G, Brown FH, Haileab B (1991) Correlation of tephra layers from the Turkana Basin to the Western Rift Valley. C R Acad Sci Paris 313:223–229
Pik R, Marty B, Carignan J, Yirgu G, Ayalew T (2008) Timing of East African Rift development in southern Ethiopia: implication for mantle plume activity and evolution of topography. Geology 36:167–170
Powers DW (1980) Geology of Mio-Pliocene sediments of the lower Kerio River Valley. PhD thesis, Princeton University, Princeton, New Jersey
Prendergast ME, Beyin A (2018) Fishing in a fluctuating landscape: terminal Pleistocene and early Holocene subsistence strategies in the Lake Turkana Basin, Kenya. Quat Int 471:203–218
Raynolds RGH (1972) East Rudolf stratigraphic report: summary of mapping – area 103 Galana Boi Beds. Dartmouth College (New Hampshire) Report 73–2:1–25
Reid I, Frostick LE (1985) Beach orientation, bar morphology and the concentration of metalliferous placer deposits: a case study, Lake Turkana, N Kenya. J Geol Soc Lond 142:837–848
Reilly TA, Mussett AE, Raja PKS, Grasty RL, Walsh J (1966) Age and polarity of the Turkana lavas, north-west Kenya. Nature 210:1145–1146
Rich F (1933) Scientific results of the Cambridge Expedition to the East African Lakes, 1930–1. 7. The algae. J Linn Soc Lond Zool 38:249–275
Ricketts RD, Anderson RF (1998) A direct comparison between the historical record of lake level and the $\delta^{18}O$ signal in carbonate sediments from Lake Turkana, Kenya. Limnol Oceanogr 43:811–822
Ricketts RD, Johnson TC (1996) Climate change in the Turkana basin as deduced from a 4000 year long $\delta^{18}O$ record. Earth Plan Sci Lett 142:7–17
Riedl S, Melnick D, Njue L, Sudo M, Strecker MR (2022) Mid-Pleistocene to recent crustal extension in the inner graben of the northern Kenya Rift. Geochem Geophys Geosystems 23:e2021GC010123
Robbins LH (1967) A recent archaeological discovery in the Turkana District of northern Kenya. Azania 2:1–5
Robbins LH (1968) Lothagam: a Later Stone Age settlement in the Lake Rudolf area of the Kenya Rift System. PhD thesis, University of California, Berkeley
Robbins LH (1972) Archaeology in the Turkana District, Kenya. Science 176:359–366
Robbins LH (1975) Bone artefacts from the Lake Rudolf basin, East Africa. Curr Anthropol 16:632–633
Robbins LH (2006) Lake Turkana archaeology: the Holocene. Ethnohistory 53:71–93
Saria E, Calais E, Stamps DS, Delvaux D, Hartnady CJH (2014) Present-day kinematics of the East African Rift. J Geophys Res Solid Earth 119:3584–3600
Schaebitz F, Asrat A, Lamb H, Cohen A, Foerster V et al (2021) Hydroclimate changes in eastern Africa over the past 200,000 years may have influenced early human dispersal. Commun Earth Environ 2:123
Schuster M, Nutz A (2017) Lacustrine wave-dominated clastic shorelines: modern to ancient littoral landforms and deposits from the Lake Turkana Basin (East African Rift System, Kenya). J Paleolimnol 59:1–24
Seegers L, Vos LD, Okeyo DO (2003) Annotated checklist of the freshwater fishes of Kenya (excluding the lacustrine haplochromines from Lake Victoria). J East Afr Nat Hist 92:11–47

Sharf FA (ed) (2005) Expedition from Uganda to Abyssinia (1898). The diary of Lieutenant RGT Bright. Tsehai Publishers, Los Angeles

Smith WC (1938) Petrographic description of volcanic rocks from Turkana, Kenya Colony, with notes on their field occurrence from the manuscript of Mr. A.W. Champion. Q J Geol Soc Lond 94:507–533

Smith BD (2018) Hunting in yellow waters: an ethnoarchaeological perspective on selective fishing on Lake Turkana. Quat Int 471:241–251

Spigel RH, Coulter GW (1996) Comparison of hydrology and physical limnology of the East African Great Lakes: Tanganyika, Malawi, Victoria, Kivu and Turkana (with reference to some North American Great Lakes). In: Johnson TC, Odada EO (eds) The limnology, climatology and paleoclimatology of the East African lakes. Gordon and Breach, Amsterdam, pp 103–139

Stamps DS, Calais E, Saria E, Hartnady C, Nocquet JM, Ebinger CJ, Fernandes RM (2008) A kinematic model for the East African rift. Geophys Res Lett 35:L05304

Stigand CA (1910) To Abyssinia through an unknown land: an account of a journey through unexplored regions of British East Africa by Lake Rudolf to the Kingdom of Menelek (1910). Lippincott, Philadelphia

Talling JF, Talling IB (1965) The chemical composition of African lake water. Int Rev Hydrobiol 50:421–463

Tebbs E, Avery S, Odermatt D (2015) Lake Turkana – monitoring the "Jade Sea" from space. Retrieved (26 Feb 2023) from: http://www.diversity2.info/products/documents/DEL8/D2Story-Turkana.pdf

Thuo P (2009) Stratigraphic, petrographic and diagenetic evaluation of Cretaceous/Paleogene potential reservoir sandstones of western Turkana, Kenya. Implications on the petroleum potential of northwestern Kenya. PhD thesis, Université de Bretagne Occidentale, Brest

Tiercelin J-J, Potdevin J-L, Morley CK, Talbot MR, Bellon H, Rio A, Le Gall B, Vétel W (2004) Hydrocarbon potential of the Meso-Cenozoic Turkana Depression, northern Kenya. I. Reservoirs: depositional environments, diagenetic characteristics, and source rock-reservoir relationships. Mar Petrol Geol 21:41–62

Tiercelin J-J, Potdevin J-L, Thuo PK, Abdelfettah Y, Schuster M, Bourquin S, Bellon H, Clément J-P, Guillou H, Nalpas T, Ruffet G (2012) Stratigraphy, sedimentology and diagenetic evolution of the Lapur Sandstone in northern Kenya: implications for oil exploration of the Meso-Cenozoic Turkana depression. J Afr Earth Sci 71–72:43–79

Trauth MH, Bergner AGN, Foerster V, Junginger A, Maslin MA, Schaebitz F (2015) Episodes of environmental stability versus instability in Late Cenozoic lake records of Eastern Africa. J Hum Evol 87:21–31

Trauth MH, Foerster V, Junginger A, Asrat A, Lamb HF, Schaebitz F (2018) Abrupt or gradual? Change point analysis of the late Pleistocene–Holocene climate record from Chew Bahir, southern Ethiopia. Quat Res 90:321–330

Trauth MH, Asrat A, Duesing W, Foerster V, Kraemer KH, Marwan N, Maslin MA, Schaebitz F (2019) Classifying past climate change in the Chew Bahir basin, Southern Ethiopia, using recurrence quantification analysis. Clim Dynam 35:115–119

Vallerini L (2019) The origins of the Earth and man. Dipartimento Di Architettura, Universita Degli Studi Firenza

van Bocxlaer B (2020) Paleoecological insights from fossil freshwater mollusks of the Kanapoi Formation (Omo-Turkana Basin, Kenya). J Human Evol 140:102341

van der Lubbe HJL, Krause-Nehring J, Junginger A, Garcin Y, Joordens JCA, Davies GR, Beck C, Feibel CS, Johnson TC, Vonhof HB (2017) Gradual or abrupt? Changes in water source of Lake Turkana (Kenya) during the African Humid Period inferred from Sr isotope ratios. Quat Sci Rev 174:1–12

Vannutelli L, Citerni C (1899) L'Omo: Viaggi di esplorazione nell'Africa Orientale. Ulrico Hoepl, Milano

Velpuri NM, Senay GB (2012) Assessing the potential hydrological impact of the Gibe III Dam on Lake Turkana water level using multi-source satellite data. Hydrol Earth Syst Sci 16:3561–3578

Velpuri NM, Senay GB, Asante KO (2012) A multi-source satellite data approach for modelling Lake Turkana water level: calibration and validation using satellite altimetry data. Hydrol Earth Syst Sci 16:1–18

Vétel W (2008) Dynamique de l'extension intra-continentale de rift magmatique: le rift Turkana (nord Kenya) de l'Eocene a l'actuel. PhD thesis, Université de Bretagne Occidentale, Brest

Vétel W, Le Gall B (2005) Geometry and growth of an inner rift fault pattern: the Kino Sogo Fault Belt, Turkana Rift (North Kenya). J Struct Geol 27:2204–2222

Vétel W, Le Gall B (2006) Dynamics of prolonged continental extension in magmatic rifts: the Turkana Rift case study (North Kenya). In: Yirgu G, Ebinger CJ, Maguire PGH (eds) The Afar Volcanic Province within the East African Rift System. Geol Soc Lond Spec Publ 259:209–233

Vétel W, Le Gall B, Johnson TC (2004) Recent tectonics in the Turkana Rift (North Kenya): an integrated approach from drainage network, satellite imagery and reflection seismic analyses. Basin Res 16:165–181

Vétel W, Le Gall B, Walsh JJ (2005) Geometry and growth of an inner rift pattern: the Kino Sogo Fault Belt, Turkana Rift, north Kenya. J Struct Geol 27:2204–2222

Vincens A (1982) Palynologie, environnments actuels et plio-pléistocènes à l'Est Turkana (Kenya). PhD thesis, Université d'Aix – Marseille II

Vincens A, Tiercelin J-J, Buchet G (2006) New Oligocene-early Miocene microflora from the southwestern Turkana Basin: Palaeoenvironmental implications in the northern Kenya Rift. Palaeogeogr Palaeoclimatol Palaeoecol 239:470–486

von Höhnel LR (1894) The discovery of lakes Rudolf and Stephanie: a narrative of Count Samuel Teleki's exploring & hunting Expedition in Eastern Equatorial Africa in 1887 & 1888 (two volumes). Longmans, Green and Co, London

von Höhnel LR (1938) The Lake Rudolf region: its discovery and subsequent exploration, 1888–1909. Part II. J R Afr Soc 37:206–226

von Höhnel LR, Rosiwal A, Toula F, Suess E (1891) Beiträge zur geologischen Kenntniss des Östlichen Afrika. Denkschr Akad Wiss Wien Math-Naturw Kl 58:447–584

Vondra CF, Johnson GD, Behrensmeyer AK, Bowen BE (1971) Preliminary stratigraphical studies of the East Rudolf basin. Nature 231:245–248

Walsh J, Dodson RG (1969) Geology of northern Turkana. Geol Surv Kenya Rep 82

Wellby MS (1900) King Menelik's dominions and the country between Lake Gallop (Rudolf) and the Nile Valley. Geogr J 16:292–306

Wescott WA, Wigger ST, Stone DM, Morley CK (1999) Geology and geophysics of the Lotikipi plain. In: Morley CK (Ed) Geoscience of rift systems – evolution of East Africa. AAPG Stud Geol 44:55–65

White F (1983) The vegetation of Africa. A descriptive memoir to accompany the Unesco/AETFAT/UNSO vegetation maps of Africa. UNESCO, Paris

Whitworth T (1960) Fossilized human remains from northern Kenya. Nature 185:947–948

Whitworth T (1965) The Pleistocene lake beds of Kabua, northern Kenya. Durham Univ J 57:88–100

Williamson PG, Savage RJG (1986) Early rift sedimentation in the Turkana basin, northern Kenya. In: Frostick, LE, Renaut, RW, Reid, I, Tiercelin, J-J (eds) Sedimentation in the African rifts. Geol Soc Lond Spec Publ 25:267–283

WoldeGabriel G, Hart WK, Katoh S, Beyene Y, Suwa G (2005) Correlation of Plio-Pleistocene tephra in Ethiopian and Kenyan rift basins: temporal calibration of geological features and hominid fossil records. J Volcanol Geotherm Res 147:81–108

Worthington EB (1932) Scientific results of the Cambridge Expedition to the East African Lakes, 1930–1. 2. Fishes other than Cichlidae. J Linn Soc Lond Zool 38:121–134

Worthington EB (1936) Scientific results of the Cambridge Expedition to the East African Lakes, 1930–1. 15. The fish of Lake Rudolf and Lake Baringo. J Linn Soc Lond Zool 39:353–389

Worthington EB, Ricardo C (1936) Scientific results of the Cambridge Expedition to the East African Lakes 1930–1. 17. The vertical distribution and movements of the plankton in lakes Rudolf, Naivasha. Edward and Bunyoni. J Linn Soc Lond Zool 40:33–69

Wright DK (2019) Long-term dynamics of pastoral ecology in northern Kenya: an old model for new resilience. J Anthropol Arch 55:101068
Wright DK, Foreman SL (2016) Comment on "Environmental change and human occupation of southern Ethiopia and northern Kenya during the last 20,000 years. Quat Sci Rev 129: 333–340". Quat Sci Rev 141:126–129
Wright DK, Forman SL (2011) Holocene occupation of the Mount Porr strand plain in southern Lake Turkana, Kenya. Nyame Akuma 76:47–62
Wright DK, Forman SL, Kiura P, Bloszies C, Beyin A (2015) Lakeside view: sociocultural responses to changing water levels of Lake Turkana, Kenya. Afr Arch Rev 32:335–367
Yan J, Hinderer M, Einsele G (2002) Geochemical evolution of closed-basin lakes: general model and application to Lakes Qinghai and Turkana. Sediment Geol 148:105–122
Yongo EO, Abila RO, Lwenya C (2010) Emerging resource use conflicts between Kenyan fishermen, pastoralists and tribesmen of Lake Turkana. Aquat Ecosyst Health Manage 13:28–34
Yost CL, Lupien RL, Beck C, Feibel CS, Archer SR, Cohen AS (2021) Orbital influence on precipitation, fire, and grass community composition from 1.87 to 1.38 Ma in the Turkana Basin, Kenya. Front. Earth Sci 9:568646
Yuretich R (1976) Sedimentology, geochemistry and geological significance of modern sediments in Lake Rudolf (Lake Turkana). Eastern Rift Valley Kenya. PhD thesis, Princeton University, Princeton, New Jersey
Yuretich R (1979) Modern sediments and sedimentary processes in Lake Rudolf (Lake Turkana) eastern rift valley, Kenya. Sedimentology 26:313–331
Yuretich R (1986) Controls on the composition of modern sediments, Lake Turkana, Kenya. In: Frostick LE, Renaut RW, Reid I, Tiercelin J-J (eds) Sedimentation in the African rifts. Geol Soc Lond Spec Publ 25:141–152
Yuretich RF, Cerling TE (1983) Hydrogeochemistry of Lake Turkana, Kenya: mass balance and mineral reactions in an alkaline lake. Geochim Cosmochim Acta 47:1099–1109
Yuretich RF, Ervin CE (2002) Clay minerals as paleoenvironmental indicators in two large lakes of the African rift valleys: Lake Malawi and Lake Turkana. In: Renaut, RW, Ashley, GM (eds) Sedimentation in continental rifts. SEPM Spec Publ 73:221–232
Zăinescu F, van der Vegt H, Storms J, Nutz A, Bozetti G, May J-H, Cohen S, Bouchette F, May SM, Schuster M (2023) The role of wind-wave related processes in redistributing river-derived terrigenous sediments in Lake Turkana: a modelling study. J Great Lakes Res 49:368–386
Zavattari E (1941) Von Dubja-fluss zum Rudolfsee; geographisch-geologische reisen und entdeckungen in südlichen Äthiopien. Mitt Geogr ges Wien 84:86–118
Zen S, Perona P, Medina-Lopez (2023) River delta eco-morphodynamics under changing scenarios: The case of Lake Turkana, Kenya. Geomorphology 428:108640

Chapter 7
Lake Logipi and the Suguta Valley

7.1 Introduction and History of Research

The Suguta Valley (Figs. 7.1 and 7.2), reputedly the hottest sector of the Kenya Rift Valley, lies 110 km north of Lake Baringo and 16 km south of Lake Turkana. It has been called Kenya's "Death Valley". The main Suguta Valley floor ('Lower Suguta') is ~70 km long (NNE-SSW) by 2–12 km wide (Fig. 7.3). No lakes, apart from small spring-fed wetlands (Chap. 14), lie between Lake Baringo (~970 m asl) and the southern end of the Suguta Valley near Emuruangogolak volcano (~340 m asl). The steep-sided Suguta Valley is the most inaccessible, but still the most pristine, subaerial sector of the Kenya Rift Valley. That might change soon. New roads and resources are being built between Lake Baringo and Lake Turkana to support petroleum and geothermal development at southwestern Turkana and at the Barrier, respectively, and will inevitably increase tourism and associated development (Mohamud and Lopeyok 2018). The Suguta Valley is an amazing place, especially from a geological perspective, but is under pressure from groups whose goals are not necessarily to conserve the environment or to benefit the local communities, many of whom are often in conflict because of historical disagreements and scarce resources, notably reliable drinking water. Suguta urgently needs a balanced program for its protection and the long-term conservation of its spectacular natural environment.

The Suguta Valley is separated from Lake Turkana by the Barrier, a transverse (W-E) wall of Pleistocene volcanic rocks that includes some of the youngest volcanic activity in the Kenya Rift. Lake Logipi – a saline alkaline lake at the northern end of the valley – abuts the Barrier along its northern shoreline and is the only permanent lake (Fig. 7.1). Lake Logipi, also called Lukula or Namakat, is the remnant of much larger palaeolakes that occupied the valley during the African Humid Period (AHP; Fig. 7.3), and of earlier Pleistocene lakes (Chap. 16). Ephemeral Lake Alablab, which lies 10–12 km to the south, is a shallow brackish lake with ill-defined margins that occasionally unites with an expanded Lake Logipi to form a

R. W. Renaut, R. B. Owen, *The Kenya Rift Lakes: Modern and Ancient*,
Syntheses in Limnogeology, https://doi.org/10.1007/978-3-642-25055-2_7

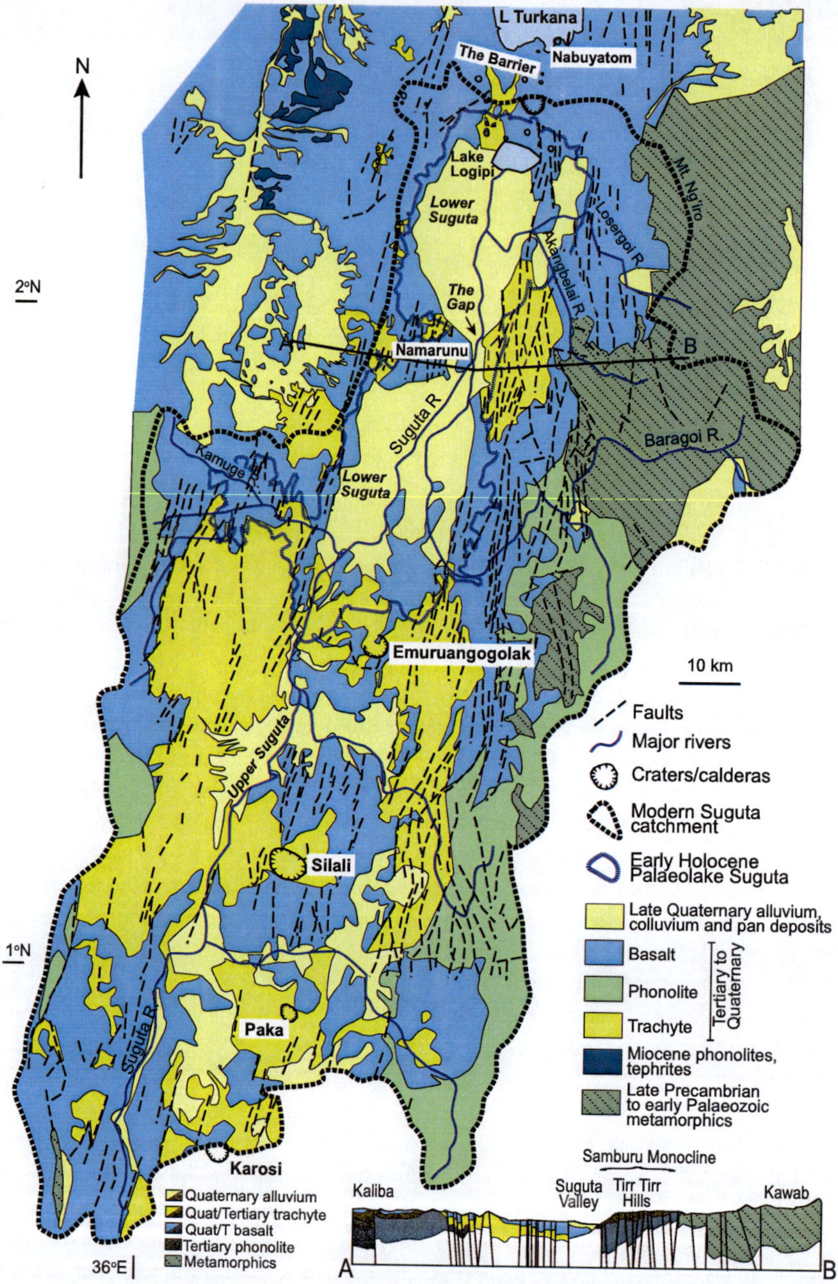

Fig. 7.1 Geological setting of the Suguta Valley. Simplified from Truckle (1977), Key (1987), Ochieng' et al. (1988) and Dunkley et al. (1993)

Fig. 7.2 The Suguta Valley. (**a**) Northern Suguta Valley and Lake Logipi. View south (February 2018) from The Barrier. Andrew's Volcano (Lugugugut) vent in left foreground. (**b**) Central Suguta Valley at The Gap (Kaminanguwadang). Namarunu volcano on left. (**c**) Suguta River in the southern Suguta Valley. Emuruangogolak volcano visible to left of dust storm. (**d**) Lake Logipi and Naperito volcanic cone, June 2019. Compare lake extent with (**a**). (**e**) Cool springs discharging into wetlands at the front of a recent basalt flow (700–900 years old?) (Dunkley et al. 1993), north edge of Emuruangogolak. (**f**) Recent basalt flow, Namarunu, at The Gap. (**g**) Barchanoid dune field south of The Gap, midwestern valley margin

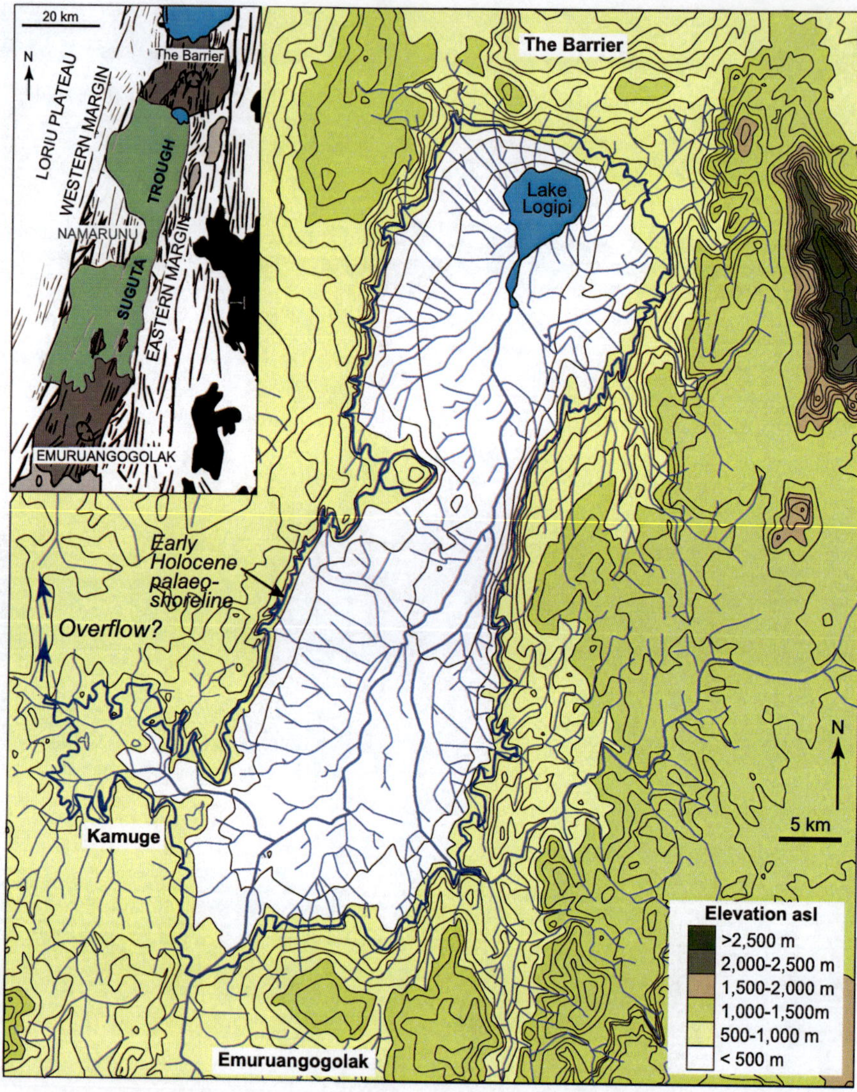

Fig. 7.3 Present topography and drainage in the Suguta Basin with the maximum Early Holocene shoreline and overflow location. Based on maps in Junginger (2011) and Garcin et al. (2009)

single dilute to brackish water body a few metres deep and of variable extent. Both lakes shrink and expand mainly in response to rainfall variations. These lakes provide important resources (water, salt, fish) for the local Turkana people and the animals that inhabit this sparsely populated arid region.

Europeans first visited Suguta and Lake Logipi in 1888, when Count Teleki's expedition traversed the northern part of the valley (Von Höhnel 1894). Geological samples were collected and described (Rosiwal 1891; Toula 1891). Suess (1891) described aspects of the region using information provided by others. Cavendish (1898) briefly visited the Barrier and Lake Logipi in 1897 and noted erroneously that activity at Teleki's volcano, a young volcanic cone on the Barrier, had been extinguished by rising water of Lake Turkana. Welby (1900) visited the region in 1899 but gave few geological insights.

Few of the early explorers commented on the Suguta Valley lakes. The first detailed study of the northern Suguta Valley was made by Arthur Champion (1935, 1937a, b). His focus was the Barrier, but he also observed steam emanating from the "Lokippi Geyser" at Elboitong. The geology of the northern part of the region was first mapped by Dodson (1963) who made general observations of Lake Logipi. Baker (1963) mapped the Baragoi area on the southeastern part of the valley flanks.

In 1968–9, the Royal Geographical Society (London, UK) conducted a multidisciplinary expedition to South Turkana that included the Suguta Valley, and undertook ecological, geomorphological and geological studies (Gwynne 1969; Caukwell 1971; Baker and Lovenbury 1971). Most results were published in a series of papers in the *Geographical Journal*. These papers give useful descriptions of the Suguta environment 50 years ago.

In the late 1960s and 1970s, the geology of northern Kenya Rift (from the equator northwards but focused on the axial parts and western margins) was mapped by the East African Geological Research Unit (EAGRU) based at Bedford College, University of London, directed by Professor Basil King with colleagues from University of Leicester and elsewhere (Sect. 2.5.1). Peter Truckle mapped the southwestern Suguta Valley and its margins. WW (Bill) Bishop led concurrent research on the stratigraphy, sedimentology, palaeontology and geochronology of the northern Kenya Rift, supported by Martin Pickford, Andrew Hill and colleagues.

Bill Bishop accepted us as PhD students in autumn 1975. Our planned role was to study sedimentological and geomorphological processes in the Suguta Valley, and to interpret the late Quaternary history of the region. Bill Bishop died in early 1977 before he could formally publish his reconnaissance results from Suguta. One conference abstract including radiometric dates survives (Bishop 1975). He and Peter Truckle were the first to discuss an expanded palaeolake Suguta during the African Humid Period (AHP: ~15–5 ka) and to recognise that that Suguta palaeolake might have been a contributor to the palaeodrainage of the contemporary River Nile (Truckle 1976, 1977).

Research in the Suguta Valley expanded during the 1980s and 1990s but changed direction. The focus became resource exploitation – petroleum and geothermal energy. The search for lacustrine oil (source and reservoir rocks) in the SW Turkana region contributed much information about its regional structure and geological history (e.g. Bosworth and Maurin 1993; Lambiase and Bosworth 1995; Morley 1999). This project included the only specific study of Lake Logipi and its sediments (Castanier et al. 1993).

The Suguta Valley forms part of a detailed study of the geothermal activity in the northern Kenya Rift undertaken by the British Geological Survey and Kenyan

Ministry of Energy in 1985–1992 (Hackman et al. 1990). The valley and adjacent regions were geologically mapped in detail and associated geochemical and hydrogeological studies were made (Key 1987; Ochieng' et al. 1988; Allen and Darling 1992; Dunkley et al. 1993), followed by specific studies related to geothermal development (e.g. Njau et al. 2020).

More recently the history of the precursor lakes that occupied the Suguta Valley during the Pleistocene and Holocene (African Humid Period: AHP) has become a major focus, undertaken by Annett Junginger (2011 and later papers), Martin Trauth, and their colleagues.

7.2 Geological Setting

The main 'Suguta Valley' (Lower Suguta) is the down-faulted sector of the Kenya Rift between Emuruangogolak volcano and the Barrier, which is floored by young (Pleistocene-Recent) alluvial and lacustrine sediments and young volcanic rocks. The region is seismically active (Pointing et al. 1985). The modern Suguta drainage basin, however, extends NNE from Lake Baringo (Chap. 8) in the south (1.5°N; ~970 m asl) to the Barrier volcanic complex in the north (2.4°N; ~273 m asl), a distance of ~110 km (Figs. 7.1 and 7.4). The faulted succession that forms both flanks of the modern axial rift valley is composed of Pliocene and Quaternary volcanic rocks overlying an older Neogene volcanic succession, interlayered with volcaniclastic sediments and sedimentary rocks of different ages. These rocks record a Quaternary history of faulting and subsidence, eruptions from axial volcanoes, and lacustrine and fluvial sedimentation during periods of changing climate but tending towards aridity during the later Holocene. The older Pleistocene sedimentary rocks, preserved on both flanks of the valley, record the period when the Suguta Valley became separated from Lake Turkana as the Barrier volcanic pile gradually grew, blocked Lake Turkana, and changed regional drainage patterns (Dunkley et al. 1993; Chap. 16).

The northern Kenya Rift at the latitude of Lake Baringo is an asymmetric half-graben bordered to the west by the Saimo fault escarpment and the Tugen Hills (Kamasia) fault-block which pass westwards into the Kerio Valley and the Elgeyo Escarpment border fault (Mugisha et al. 1997; Hautot et al. 2000; Fig. 7.1). To the north, the western rift margin in the Lower Suguta (Fig. 7.3) passes into the Loriu Plateau, which is underlain by Pliocene and Pleistocene volcanic rocks (trachyte, basalt, basanite) and metamorphic basement (Rhemtulla 1970; Truckle 1976, 1977; Dunkley et al. 1993).

The eastern margin of the rift north of Lake Baringo is composed of Neogene volcanic rocks that lie upon Mozambiquan (late Proterozoic-Phanerozoic) metamorphic basement rocks, composed mainly of gneisses, which crop out extensively near Baragoi and South Horr (Baker 1963; Dodson 1963; Key 1987; Ochieng' et al. 1988). High plateaus on the eastern rift flanks, which include Laikipia, pass into the

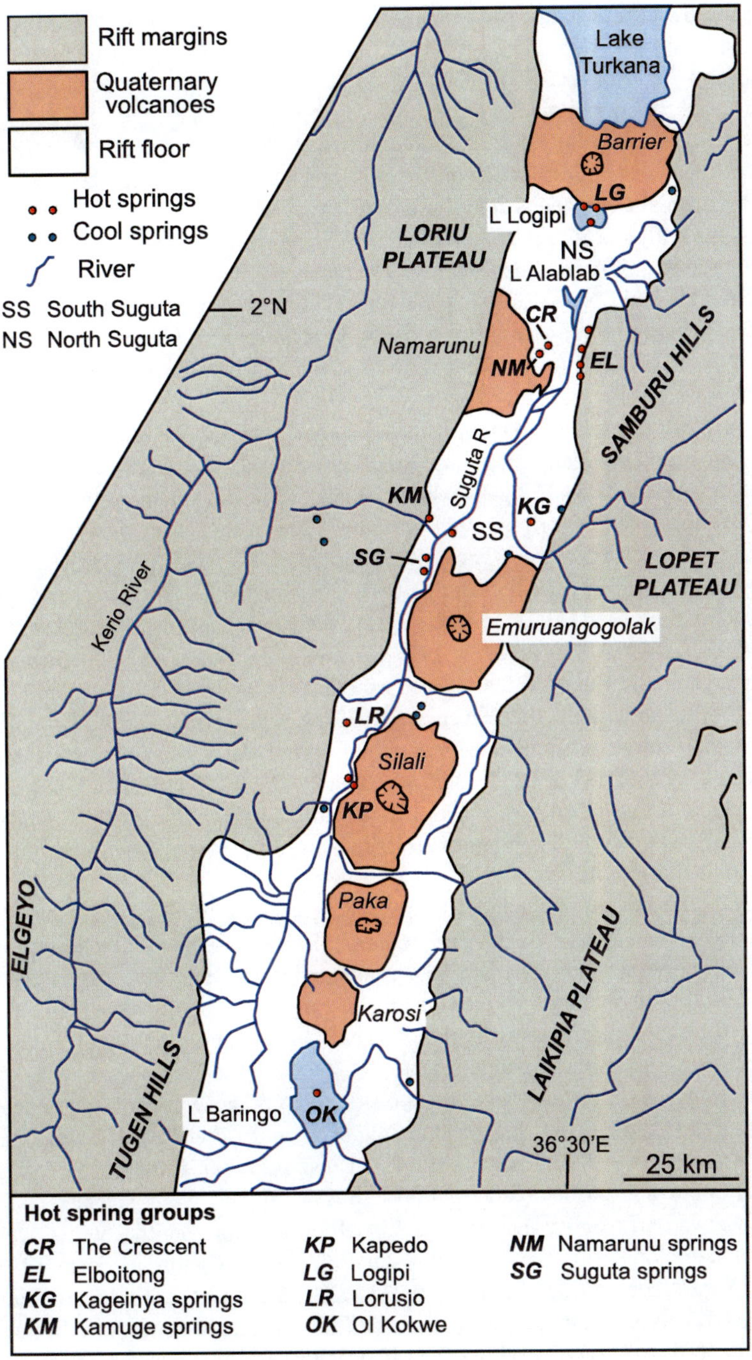

Fig. 7.4 Drainage pattern and springs of the Suguta catchment. (Adapted from published Kenya Geological Survey maps, Dunkley et al. (1993), and original observations. Spring symbols represent groups rather than individual vents)

rift shoulder composed of west-facing, steeply dipping fault-scarps. North of Silali these are replaced by faulted west-facing monoclines with dips of up to 45°, represented by the Samburu Hills. This region contains a succession of volcanic rocks with interbedded sedimentary rocks of Miocene to Recent age that get younger westwards (Koyaguchi 1984; Makinouchi et al. 1984; Sawada et al. 1987; Chap. 16).

The axial inner trough of the northern Kenya Rift is occupied by large low-angle, multivent shield-volcanoes composed of trachytic and basaltic lavas and pyroclastic deposits (Williams et al. 1984; Williams and Chapman 1986; Dunkley et al. 1993; Robertson et al. 2016). Those Quaternary volcanoes (Fig. 7.4) – Karosi, Paka, Silali, Emuruangogolak, Namarunu and the Barrier – are more eroded in the south (e.g. Karosi, which may be extinct) but show fresh lavas and landforms to the north. Minor lakes and spring-fed wetlands are present in this volcanic zone (e.g. Lake Tilam north of Karosi; Chap. 14).

The most important volcanoes in the context of the Suguta Valley lakes are Emuruangogolak, Namarunu and the Barrier. The basalt-trachyte volcano Emuruangogolak (Fig. 16.4g–i) extends as a broad shield spanning the width of the rift, and rising to 1328 m asl, with a higher base in the south than the north and covering an area of 525 km^2. The eastern margin (950 m) is higher than the western margin (550 m) reflecting the tilting of the rift floor from east to west and south to north (Dunkley et al. 1993). Much of the volcano consists of an eroded trachyte shield and caldera (Weaver 1973, 1977). Alluvial plains south of the volcano drain westwards to the Suguta River. The northern margins are the site of some of the youngest volcanic rocks in the Kenya Rift. They include trachyte showing phreatomagmatic features that developed along the former Lake Suguta margins, and basaltic pillow lavas (Truckle 1979; Dunkley et al. 1993). At Kangirinyang northeast of the volcano a linear series of maars, some with lakes, is present (Sect. 14.4). Tuff cones such as Losetom on the northern flanks preserve eroded shorelines of palaeolake Suguta (Sect. 7.7).

Namarunu is a basalt-trachyte complex that forms a hilly terrain on the midwestern side of the Suguta trough (Fig. 16.4d–f). The volcano is a Quaternary centre superimposed on downfaulted Pliocene volcanic rocks. The youngest flows (mid-Holocene?) project eastwards into the Suguta trough, producing 'The Gap'. Truckle (1979) described pillow lavas on the southern part of Namarunu that formed when they erupted into a former Suguta lake.

The transverse Barrier is a composite structure composed of four distinct volcanic centres (Kololenyang, Kakorinya, Likaiu W and Likaiu E) that separates the Suguta Valley from Lake Turkana (Figs. 7.1 and 16.4a–c). It includes two historic spatter cones and associated lava flows. Basalt erupted from Teleki's cone and flowed towards Lake Turkana possibly in 1888 or 1895 (Smith 1897; Cavendish 1898; Champion 1935, 1937a, b; Dodson 1963), and possibly again in 1917 and 1921 (Sigurdsson et al. 2015; Njau et al. 2020). A similar scoria cone (Andrew's volcano, Likaiyu or Lugugugut) discharged fluid basaltic lava flows to the northern shoreline of Lake Logipi at about the same time. Several ash cones on the Barrier, including Namurinyang (Murniau), record high-level palaeoshorelines.

7.3 Climate

Climate data for the Suguta Valley are limited but it is one of the most arid parts of the Kenya Rift Valley (Gwynne 1969; Ojany and Ogenoo 1973; Junginger 2011). The mean annual rainfall on the rift floor is <500 mm y^{-1}, and the mean air temperature is ~30 °C. Air temperatures occasionally exceed 50 °C in the northern Suguta Valley near Lake Logipi. Potential evaporation rates can exceed 4000 mm y^{-1} (Morgan 1971). In common with other parts of the Kenya Rift, rainfall seasons are bimodal with long rains in March–May and short rains in October–November reflecting the seasonal shift in the location of the tropical low-pressure belt (Intertropical Convergence Zone: ITCZ; Nicholson 1996; Camberlin 1997). The southwestern part of the drainage basin is influenced by the humid Congo airstream during August (Junginger 2011). Rainfall increases at higher elevations southwards to Lake Baringo while temperatures decrease.

7.4 Hydrology of Suguta Valley and Lake Logipi

7.4.1 Regional Setting and Lake Recharge

Lake Logipi (~273 m asl) is the 'sump' for axial drainage, for both surface waters and groundwater, in the northern Kenya Rift. Although Lake Turkana lies only 15 km to the north, it is dammed behind The Barrier at ~360 m asl, about 85 m higher than Lake Logipi. Most axial waters in the Kenya Rift north of the rift watershed at Eburru volcano (apex of the 'Kenya Dome' of King 1978) flow towards Lake Logipi (Dunkley et al. 1993; Darling 1996; Becht et al. 2006). The surface drainage (13,000 km^2) feeding the lake is sourced in headwaters north and west of Lake Baringo and in lateral streams feeding the Suguta Valley (Figs. 7.3 and 7.4).

The Suguta River is the main river feeding Lake Logipi and is supplemented by ephemeral streams, hot and ambient springs, unchannelled runoff, and direct rainfall (<500 mm y^{-1}, Fig. 7.5a). The Suguta River rises west of Lake Baringo, ~130 km southwest of Logipi (Fig. 7.4; Chap. 8). It then flows northwards along the western half of the rift floor, crossing a semi-arid volcanic landscape on the western flanks of the Karosi, Paka, Silali and Emuruangogolak central volcanoes before reaching the main Suguta Valley. *En route* to Logipi, the river is supplemented by the ephemeral Kamuge and Baragoi rivers (Fig. 7.5b) and many ephemeral streams, some of which are partly spring-fed (Fig. 7.3).

The Suguta River is often perennial in its upper (near Silali) to mid- (north of Emuruangogolak) courses, but the river dries up in some years. Flow varies with both location and season, and this affects both fluvial and lacustrine sedimentation. The channel patterns and associated sedimentation change markedly from the southern Suguta Valley northwards as elevation and rainfall decline, and evaporation increases. Inflow from lateral streams also plays a role. North of Silali a wide

Fig. 7.5 Inflow to Lake Logipi. (a) Rainstorm falling on Lake Logipi. Losergoi fan on right margin. (b) Gravel and boulder-strewn tributary to Baragoi River feeding SE Suguta Valley. (c) Suguta River near its junction with the Kamuge River, looking north; single channel with submerged sand and gravel bars. (d, e) Suguta River north of Nangarabat (NW Emuruangogolak) is a sandy braided stream (South Saskatchewan type). (f) Braided muddy Suguta River in flood a few kilometers south of The Gap

single channel (locally >3 m deep) is replaced by a shallow (<2 m deep) channel within which mixed sand-gravel braid bars develop that remain within defined margins (Fig. 7.5c, d). Rock outcrops locally create rapids and waterfalls. North of Emuruangogolak, the Suguta River gradually becomes a sandy braided river within a single channel (South Saskatchewan type) that changes to a higher-sinuosity meandering channel that incises its older sand bars during lower stages. That pattern (braided at high discharge but meandering at lower discharge within the same

Fig. 7.6 Inflow to Lake Logipi. (**a**, **b**) Meandering section of Suguta River a few km south of The Gap. Namarunu top left in (**a**). (**c**) Desiccating low-energy channel ~5 km north of Namarunu flowing towards expanded Lake Logipi, July 2018. (**d**) Distal part of the terminal delta south of Lake Logipi. Channels are drying up leaving trona efflorescence, February 2018. (**e**) Northwestern margin of Lake Logipi, 2018. as = active segment (modern fan lobe supplying water and sediment); is = inactive segment (older incised fan-sediments with stone pavement coated by desert varnish); rs = regressional shoreline (late Quaternary lacustrine sediments coated by efflorescent trona); sl = vegetated spring-line where water issues from permeable distal fan sediments. (**f**) Losergoi fan with flow directed towards Lake Logipi

channel confines) is maintained northwards but is gradually replaced by a muddier braided regime as water depth and energy decline (Fig. 7.5e, f). The Suguta River nonetheless maintains high discharge but transforms to a more typical meandering pattern as slope declines. The main meandering channel has vegetated levees (Fig. 7.6a, b) and gradually transforms to braided and anastomosing stream networks during high-stage floods when channel waters overtop their banks. Many anastomosing streams gradually dry up leaving ponded narrow temporary lakes with muddy sediments that dry and become cracked during desiccation or are reworked as intraclasts.

When in flood, the river expands north of 'The Gap' east of Namarunu volcano and becomes a complex network of anastomosing channels that dry up or, in some years, form an ephemeral brackish shallow lake (Lake Alablab; Figs. 7.6c, d, 7.7, and 7.8; Gwynne 1969). That lake in turn flows into Lake Logipi or in wetter years unites with it to form a broad shallow (<5 m deep) expanded Alablab-Logipi lake (e.g. summer 2018). Lake Alablab is recharged by seasonal or longer-term fluvial inflow but often shrinks during dry seasons, sometimes losing its identity as a separate lake or lakes. Lake Logipi, a terminal lake, often shrinks quickly (in months) in the dry season(s) to form a small saline alkaline lake, but it nonetheless usually retains water during dry periods because it is partly recharged by hot springs.

Lateral drainage is represented by ephemeral inflow from the Kamuge and Baragoi rivers, and large alluvial fans along the western and eastern margins (Rubino 2002). Large bajadas lie along the western margin of the valley (Fig. 7.6e). The fans have steep, coarse proximal parts that pass into mid-fan braided channels and extensive distal sandflats and mudflats. The large Losergoi fan east of the lake impinges on Lake Logipi and Lake Alablab (Fig. 7.6f).

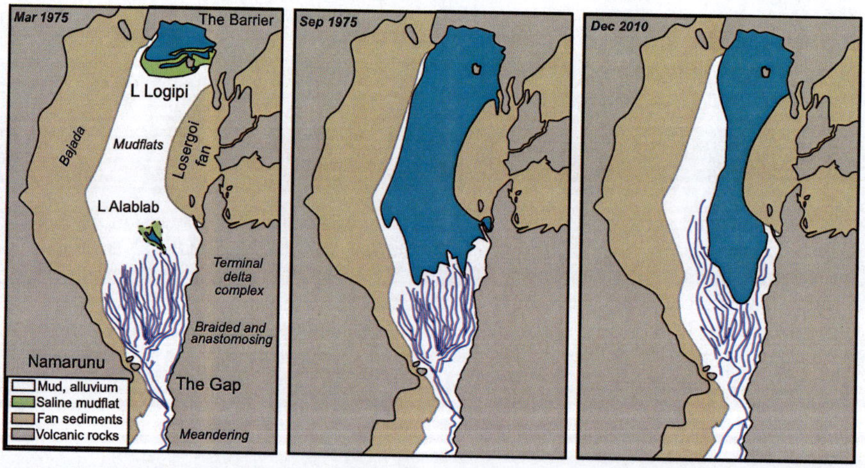

Fig. 7.7 Changes in the level and extent of Lake Logipi and Lake Alablab. Modified from similar figure by Castanier et al. (1993)

Fig. 7.8 Complex river patterns in the terminal delta of the Suguta River, a few km north of Elboitong hot springs and The Gap in general area of Lake Alablab. Contrasting streams are present. Flow from the Elboitong hot springs (white area) and the Suguta River is bordered by vegetated levees. These normally perennial channels transect sluggish muddy braided channels resulting from avulsion from the main river channel and crevasse splays

7.4.2 Groundwater Recharge

The hydrogeology is discussed in detail by Dunkley et al. (1993) who produced a potentiometric map for the region between Lake Baringo and Lake Turkana, using data acquired from boreholes (Fig. 4.6a). They showed that groundwater flows northwards along the rift axis and is supplemented by lateral flow from the sides of the rift, particularly the eastern margin. Central volcanoes modify regional flow systems and provide local recharge; axial dykes may locally divert lateral flow, as do faults which may block the flow or may lead to deeper flow paths or provide conduits for geothermal fluids. They also noted that Lake Baringo is a likely source of groundwater recharge. Lake Turkana also contributes groundwater to Lake Logipi through the Barrier (Dodson 1963; Dunkley et al. 1993; Nunn and Harris 2007; Auko 2013). Champion (1935) reported that water from the Suguta River flows underground near Naperito (Cathedral Rock) but did not find a resurgence.

7.4.3 Hot Springs in the Suguta Valley

Hot springs are common in the drainage basin (Figs. 7.4 and 7.9). Some flow directly into Lake Logipi from vents along the shoreline and have the most influence; others feed the Suguta River and its tributaries, but their waters and their impact are often diluted before they flow into the lake.

Fig. 7.9 Hot springs of northern Suguta Valley. (**a**) Hot spring on the northern shoreline of Lake Logipi. Naperito in distance. (**b**) Hot springs discharging on the northeastern shoreline of Naperito. (**c**) Elboitong. Springs discharge along the base of the fault scarp and landslips below the Tirr Tirr Plateau (right). White efflorescent trona crusts form near the spring vents and along outflow channels. (**d**) Hot spring vents with microbial mats, hosted by horizontally bedded lake? sediments encrusted by trona at Elboitong. (**e**) Line of springs on the northern shore of a small eroded basaltic scoria cone, northeastern Namarunu. (**f**) Outflow from spring in (**e**). View towards northwest

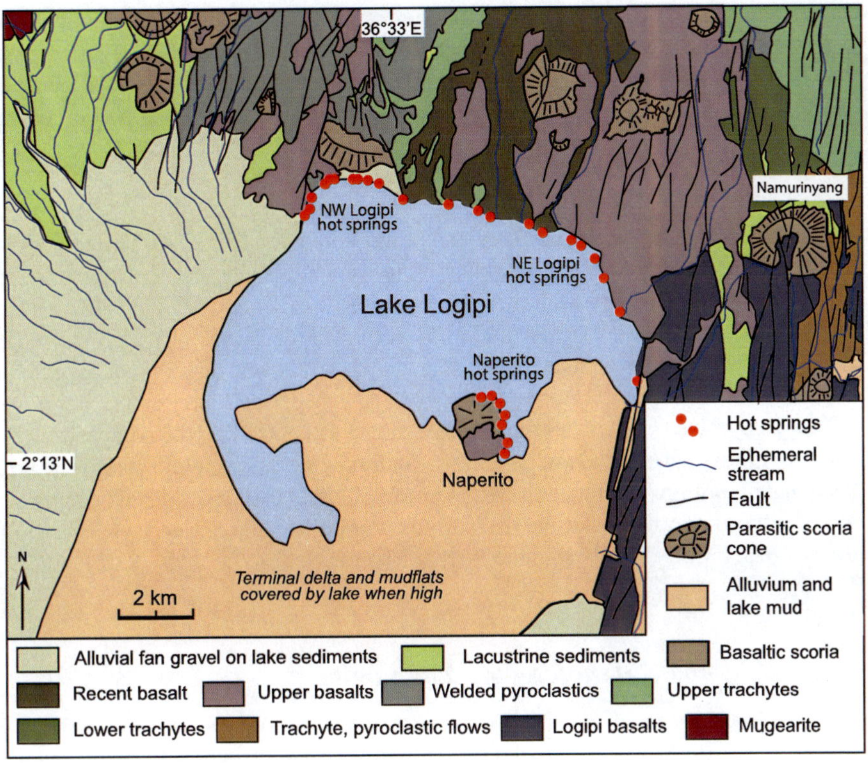

Fig. 7.10 Geology of the region surrounding Lake Logipi. Adapted from Dodson (1963), Dunkley et al. (1993) and original observations

Alkaline hot water discharges quietly into the lake from at least 22 springs along the southern shoreline of The Barrier and the northern and eastern shores of Naperito tuff-cone (Crane and O'Connell 1983; Castanier et al. 1993; Allen and Darling 1992; Dunkley et al. 1993; Figs. 7.9a, b and 7.10). Reported spring water temperatures range from 64.4 to 79.9 °C, with the hottest springs along the northern shoreline of Naperito. Most of the Logipi springs are submerged when the lake level is high, but many continue to flow into the expanded lake. Submerged (<1.5 m depth) circular areas up to 10 m in diameter without benthic microbial mats, seen on the floor of Lake Logipi between Naperito and the northern shoreline from helicopter when lake level is low, are probable sites of hot upwelling fluids. Flamingos avoid them. The Logipi and Naperito springs replenish and maintain Lake Logipi when inflow from fluvial sources, such as the Suguta River, is low. In recent decades this spring recharge might have prevented complete desiccation of Lake Logipi.

About 25 km to the southeast, >50 hot springs discharge along the fault that marks the eastern edge of the valley at Elboitong, at the base of the Tirr Tirr Escarpment (Fig. 7.9c, d). These springs discharge over a distance of 5.5 km at temperatures up to 100.2 °C and are the hottest in Kenya (Dunkley et al. 1993). A few were ebullient with high discharge in the summers of 2018 and 2019 and at least one geyser has been reported (Champion 1935). Their combined perennial outflow contributes to Lake Logipi especially when the Suguta River dries up before reaching the lake (Fig. 7.8). Unlike many other Kenyan hot springs, aridity has precluded the development of geothermal wetlands or local spring-fed lakes.

Other hot spring groups discharge near Namarunu volcano on the midwestern side of the valley, and at the southern end of Suguta Valley (Fig. 7.4). The southernmost hot springs lie south of Emuruangogolak and at Kapedo and Lorusio west of Silali caldera near the headwaters of the Suguta River (Sects. 14.5 and 14.6). These spring waters are up to 120 km from Lake Logipi, but their outflow might contribute to aquifers that flow into the lake. Becht et al. (2006) showed how difficult it can be to assess contributions to lakes from surface and subsurface sources of recharge.

The importance of these springs in influencing or controlling the composition of Lake Logipi is variable. Some are saline and alkaline with low discharge; others have higher or more perennial discharge but are dilute. Most of the distal hot springs are of minor importance but the normally perennial Kapedo hot springs have high discharge and flow directly into the Suguta River (Sect. 14.6).

7.4.4 *Physical Limnology of Lake Logipi*

Little is known about the physical processes in the lake beyond its periodic expansion and deepening, and shrinking and shallowing. Changes in lake area in response to seasonal or longer-term variations in rainfall and the recharge from the Suguta River are well established (Figs. 7.7 and 7.11). The lake is frequently pale brown and turbid during the rainy seasons, reflecting fine suspended detritus from the southern and lateral inflow streams and organic debris from biological (microbial and bird) sources in the lake.

Castanier et al. (1993) estimated the maximum lake depth at 3–5 m during flood, and it is probably <1.5 m during periods of desiccation. They noted that the deepest areas are those floored by black mud that rarely dries out. With its shallow depth and frequent wind mixing any stratification is like to be short-lived. Temperatures taken on the central northern shoreline north of Naperito in February and July 2018 were 36 °C and 32.6 °C respectively.

Energy levels are low even during flooding but high enough for waves to sort littoral fine to medium gravel and skeletal fish remains on Naperito shorelines. Southeastern Lake Turkana is the site of a large windfarm, but the Barrier and Mt. Ng'iro protect Lake Logipi from the strong winds. In contrast, winds from the east have produced a dune field south of Namarunu (Fig. 4.17d).

Fig. 7.11 Lake Logipi. (**a**) Rocky northern shoreline, February 2018, showing fresh basalt flows from Andrew's cone. (**b**) Lake Logipi viewed from the northern shoreline, February 2018. (**c**) Naperito during a dry season, June 2019. Lesser flamingos congregate offshore from northern hot springs. (**d**) High suspended sediment load during high level of July 2018. (**e**) Southern shore of expanded Logipi-Alablab lake, July 2018. (**f**) Trona-encrusted northern shore of Lake Logipi, February 2018. (**g**) Trona encrusted mudflats along the northwest shore, February 2018. (**h**) Submerged stream channels with exposed vegetated levees, northwestern margin of Lake Logipi, July 2018

7.4.5 *Chemical Limnology*

Analyses of river and hot-spring waters have been published (Table 7.1; Dunkley et al. 1993; Auko 2013) but with few of Lake Logipi. Castanier et al. (1993) analysed the water of Lake Logipi during October 1988 (Table 7.1) when the lake was at an 'intermediate' stage (in assumed level), and then again in February 1990 when lake level was lower than in 1988. Lake Logipi water is typically saline (5–30 g L^{-1} TDS) and alkaline (pH: 9.5–10.5), with a Na^+–HCO_3^-–CO_3^{2-}–Cl^- composition. As expected, the salinity increases during drier years when the water volume shrinks and decreases during wetter years when united with Lake Alablab. The Suguta River, which feeds both lakes, is brackish when it floods and recharges both lakes, but Darling (1996) noted that its salinity does not increase significantly with evapo-concentration unlike other perennial rivers (e.g. Ewaso Ngiro S, west of Lake Magadi). In contrast, Gwynne (1969) noted that fluoride concentration increases from 20 ppm F^- near its source at Kapedo to 275 ppm at Lake Alablab.

No chemical analyses have been reported for Lake Alablab, but its waters are likely to be brackish (2–5 g L^{-1} TDS?). Those river waters would be poor in alkaline earths by the time they reach ephemeral Lake Alablab. Pinpointing the sites of inferred Ca^{2+} and Mg^{2+} loss is difficult without evidence, but there are many potential sites: Ca^{2+} and Mg^{2+} could be lost by precipitation of thin micritic cements, coating substrates in ephemeral stream channels and adjacent floodplain sediments, or by bio-induced and abiotic precipitation of carbonates in lakes and adjacent shallow-water sites (Castanier et al. 1993), or on exposed evaporative sites during drawdown of very shallow groundwaters. When delivered to the lakes, the total concentration of alkaline earths is already very low. Ca^{2+} and Mg^{2+} are minor contributors to *modern* Suguta basin hydrochemistry but calcite microbialites preserved at various elevations around the basin record former periods of fresher water. Calcite stromatolites 1–3 cm thick locally encrust bedrock on northern Naperito <2 m above present lake level.

7.5 Biology and Ecology of the Lakes

Few studies have been made of the ecology of Lake Logipi, and ephemeral Lake Alablab seems to have been ignored. Lake Logipi is now recognised as an important site where lesser flamingos and other birds congregate. Castanier et al. (1993) pioneered microbiological work at the lake and recorded a rich microbial population, dominated by cyanobacteria. They also recorded high numbers of aerobic heterotrophic bacteria but fewer anaerobic bacteria, which are mainly sulfate reducers. Microbial mats develop near hot springs and in the cooler parts of their outflow channels (Fig. 7.12), and along the margins of shrinking pools, shorelines and increasingly shallower drainage channels even when the lake is desiccating. Over the last few decades, only the deepest parts have escaped full desiccation. They

Table 7.1 Selected chemical analyses of Suguta basin waters

Sample	Ref no	Date	Source	Temp (°C)	pH	Na	K	Ca	Mg	Cl	HCO_3	DO	NH_4^+	SO_4	F	SiO_2	Salinity
Rivers																	
Suguta 1	206	1989	1			1150	21.6	3.1	1.9	288	2460			124	44	66.3	
Suguta 2	207	1989	1			1030	24.0	3.9	1.9	246	2200			98.5	36	71.7	
Suguta 3	208	1989	1			1104	20.3	3.4	1.7	267	2340			115	90	74.0	
Suguta 4	209	1989	1			1093	22.4	3.8	2.1	273	2380			104	42	64.8	
Suguta trib.	210	1989	1			1340	20.1	1.6	0.3	350	2930			142	47	64.6	
Gerau	211	1989	1		8.7	54	9.0	12.5	3.6	28.5	128			5.0	0.83	26.3	
Amaya	212	1989	1			158	8.7	7.6	2.1	26.0	389			7.2	2.1	44.3	
Naliyo	214	1989	1		8.5	65	10.4	16.5	3.0	16.3	152			15.8	0.8	20.8	
Ambient springs																	
S Nangarabat	198a	1989	1	31.1	9.15	1689	24.9	0.6	0.2	370	3440			213	60	54.1	
N Samburu	193	1989	1		8.85	638	4.7	4.9	1.2	141	1260			140	19	55.6	
Parkati well	243	1991	1		7.5	244	22.8	104	45.7	217	262			306	1.3	121	
Hot springs																	
Crescent Island	CG		2		9.5	4650	194	0.02	0.03				0.04	1150		86.7	
Kamuge	187	1989	1	50.4	7.75	925	6.9	4.2	1.3	160	1930			100	37	82.1	
Namarunu	189	1989	1	66.2	8.80	5500	113	0.5	0.2	3200	5860			814	140	55.6	
Logipi NW	190	1991	1	61.4	8.30	4170	79.4	0.50	0.1	3420	5160			590	110	39.6	
Logipi NE	238	1991	1	69.8	8.85	6200	120	0.48	<0.1	5250	5720			920	82	65.9	
Elboitong N	237	1991	1	91.8	9.0	6690	126	1.4	0.5	4500	8240			1060	87	125.8	
Elboitong S	236	1991	1	95	7.1	6770	129	0.82	<0.1	4400	7940			1070	87	98.8	

(continued)

Table 7.1 (continued)

Lake																	
Lake Logipi	LL		2		10.5	17,500	382	0.63	0.09				0.06	1600		83.8	
Naperito pond	SV		2		9.5	2650	188	0.14	0.09	5				50		140.5	
Lake Logipi	1	1988	2		9.53						2573	12.8	2.44	900			12
Lake Logipi	4	1988	2		9.54						2593	13.1	2.64	750			2
Lake Logipi	18	1988	2		9.53						2625	7.5	2.23	900			10
Lake Logipi	29	1988	2		9.55						2584	7.2	1.39	375			15

More hot spring analyses are given by Auko (2013)
Analyses in mg L^{-1} except Salinity in g L^{-1} and Dissolved Oxygen (DO) in ppm. Reference sources: 1: Dunkley et al. (1993). 2: Castanier et al. (1993)

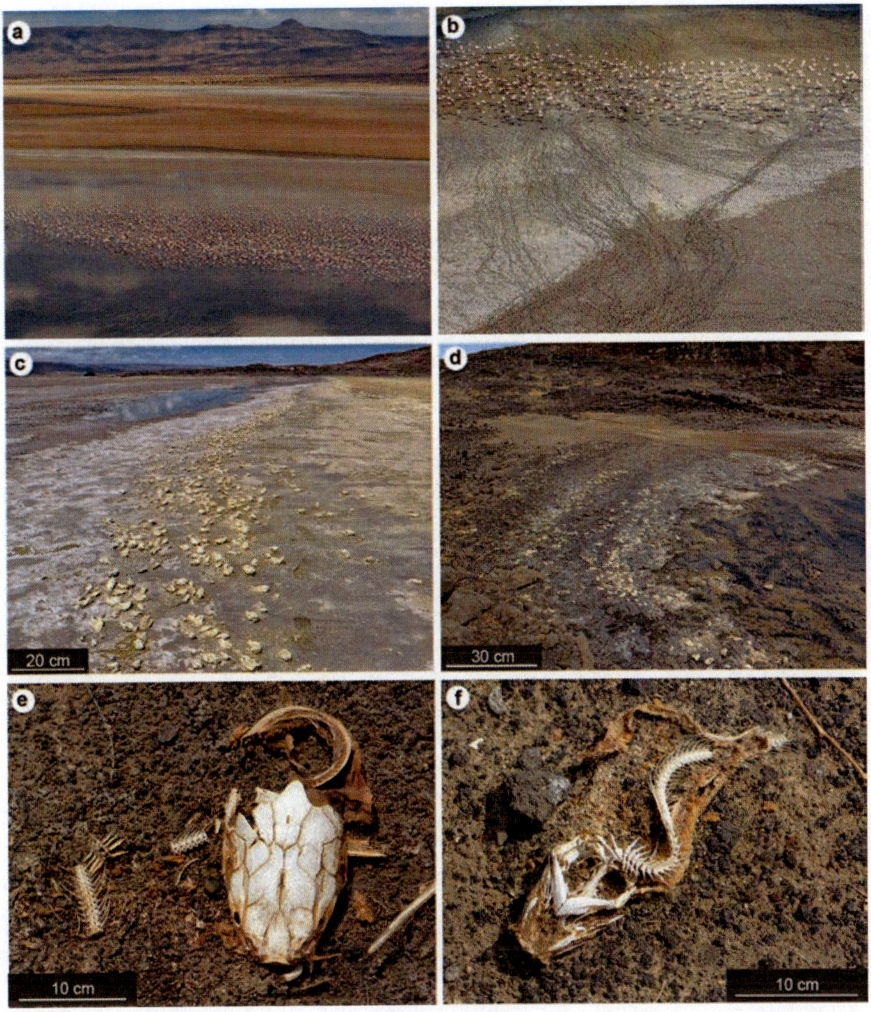

Fig. 7.12 Life in Lake Logipi. (**a**) Flock of lesser flamingos near the northwest shoreline. (**b**) Flamingos taking off from the shallow lake leaving parallel tracks in the mud. (**c**, **d**) Fish skeletons along the regressive shoreline of eastern Naperito. (**e**) Skeleton of catfish (*Clarias* sp.). (**f**) Skeleton of *Tilapia nilotica*

suggested that most of the organic matter produced in the lake is biologically recycled or retained in solution. Little organic matter is preserved in the sediments. Various insects, including beetles, inhabit the shorelines and microbial mats.

A transient population of lesser flamingos (*Phoeniconaias minor*) live at the lake where they feed on microphytobenthos (mainly benthic diatoms: Tuite 1981, 2000). Tebbs et al. (2015) noted that high numbers of flamingos may migrate to Lake Logipi when conditions are unfavourable at the central rift valley soda lakes (Bogoria, Nakuru, Elmenteita). Many flamingos cluster near the northern shoreline where comparatively dilute warm and hot springs flow into the lake (Fig. 7.12a). They avoid areas of hot discharge from the lake floor. When the lake level is low (<1 m deep), they produce distinctive linear tracks in surface sediments when they take off. These traces cluster together forming broad linear parallel patterns or may be focused along clear routes (Fig. 7.12b). During periods of lake expansion and relatively dilute water, fish inhabit the lake including *Tilapia nilotica* (Champion 1935) and catfish (*Clarias*) presumably washed in from the Suguta River. Their skeletons lie along abandoned shorelines as the lake shrinks and becomes more saline (Fig. 7.12c, d), with catfish at the highest levels and more salinity-tolerant *Tilapia* at lower levels.

A variety of animals inhabit the Suguta Valley including grants gazelle, striped hyena, dik dik, baboon, domestic cattle, goats, camels and other mammals (see Watson 1969). Small mammals in south Turkana, mainly rodents and small cats, are described by Coe (1972).

The generally sparse vegetation on the valley floor is mainly limited to river margins but isolated patches and clumps are present on the pediments and mudflats along the margins. Hemming (1972) described riparian vegetation and soils along the Suguta River. Tall grasses such as *Sporobolus consimilis*, which can be 2 m high, are common. *Sueda* (seep weeds) are locally common. Palms (*Hyphaene thebaica*) and *Acacia* (*Vachellia*) are locally common near freshwater sources such as the Suguta River and springs. The vegetation changes with environmental and geomorphological setting, and salinity.

7.6 Sedimentation

Castanier et al. (1993) sampled the sediments of the shallow floor of Lake Logipi and obtained short cores. One (core A: 60 cm long) was studied in detail. They recognised two sediment layers. A layer of black mud underlies an indistinct sediment-water interface. The mud, up to ~1.5 m thick in the deepest parts of the lake, overlies a harder reddish layer. The mud thins southwards until it disappears. The reddish layer, overlain locally by microbial mats, then forms the sediment-water interface.

The black mud contains fish remains (*Tilapia* sp.?). X-ray diffraction analyses implied that zeolites (clinoptilolite?), clay minerals and sodium silicates form the mud, but that carbonates and pyrite are absent (Castanier et al. 1993). The aluminosilicates lack calcium and have variable Mg, implying that some Ca^{2+} and Mg^{2+}, derived from weathering of volcanic rocks in the drainage basin, are removed from the inflowing river- and ground waters as divalent carbonate cements or crusts that accumulate in the catchment. Iron is apparently tied to the silicates and aluminosilicates rather than to sulfides (i.e. pyrite or its precursors), even though H_2S is produced by bacteria in the lake and (or) lake sediments. Low C/N ratios imply that the organic matter in the black mud is mainly microbial, an interpretation supported by the subspherical (bacterial?) morphology of many sedimentary particles when viewed by SEM. High C/N ratios would not have been expected in such an arid setting where higher plants are rare.

The reddish layer has a similar X-ray mineralogy to the black mud, but its colour implies that Fe^{2+} released from silicates and aluminosilicates has been oxidised to Fe-oxides and hydroxides. This was attributed to periods of subaerial exposure when the lake receded. The black mud, therefore, represents the area that is normally covered by lake water.

The sediments of Lake Alablab have not been studied, but are siliciclastic muds and silts transported by the Suguta River during seasonal floods. Riparian vegetation established along the margins of braided channels may survive overbank flooding, so root indicators may be present in the sediments. Mudcracks are common. Efflorescent salt crusts (Na-CO_3 species) and thin microbial mats locally cover the terminal lake plains and overbank areas during desiccation, but evaporites have not formed recently. Salts are more extensive where springs discharge along the base of alluvial fans that impinge on the margins of the expanded lake. They may include alkaline earth carbonates (Fig. 7.6e).

Observations in 2018 and 2019 and examination of satellite imagery show that sedimentation is controlled mainly by processes within the lake when it is small and saline, whereas clastic sedimentation dominates when the lake expands with inflowing clay silt supplied from the Suguta River and local drainage. Unlike Lake Magadi, which lies in a sump at the southern part of the Kenya Rift (Chap. 12), no bedded evaporites are present at Lake Logipi – only efflorescent trona and halite crusts. This implies that the salinity is rarely high enough to attain trona saturation due to dilution by the Suguta River and (or) dilution by underground seepage from Lake Turkana. Darling (2001) noted that the tectonic setting for evaporite formation is more favourable at Lake Magadi where the lake is confined by steep graben walls and more inflow is hydrothermal. He noted that most hot springs at Magadi are more saline and discharge directly into the lake, whereas many at Suguta flow overland before they reach the lake, enabling possibilities for recirculation. Thin buried evaporites are possible at Logipi but unlikely.

Grewe (2016) examined short cores from Lake Logipi. The focus was pore fluid chemistry (especially sulfur) and biomarkers. She found horizons of elemental sulfur that are likely remnants of very low lake levels when anoxygenic photosynthesis or sulfide oxidation occurred at the lake bottoms.

Hot spring deposits, except efflorescent sodium carbonate salts, were not observed along the margins of modern Lake Logipi. Minor calcite and opaline silica might be present in surface crusts. No modern travertine or tufa was observed (2018 and 2019). Silica sinter was reported at Elboitong hot springs (Champion 1935), and fossil travertine was recorded 14 km southeast of the lake (Bosworth and Maurin 1993). A brief search for these deposits in 2019 was unsuccessful. Silica veins reported by Sturchio et al. (1993) were found at the Barrier (SW Kakorinya), but subaerial sinter was not found during our brief visit.

7.7 Late Quaternary Lacustrine Sediments in the Suguta Valley

Unlike the other Kenya Rift lakes, more is known of the Late Quaternary history of the Suguta basin than of the modern lakes. Champion (1937a) reported lacustrine sediments in the Suguta Valley and observed wave-cut terraces along the southern side of the Barrier. He concluded that Lake Turkana and a 'Suguta Lake' formerly united as one, and suggested that growth of the Barrier, assuming upward accumulation of the volcanic pile, eventually divided the palaeolake into separate water bodies. The expanded Suguta Lake must have then shrunk by evaporation because replenishment from the inflow from the Suguta River alone could not maintain a large Suguta Lake. The age of that precursor was unspecified.

Accommodation space in the Lower (northern) Suguta Valley was defined by the development of faulted, basin-dipping, lavas to the east and west ~3–1.8 my ago and by eruptions between 1.3 and 0.5 Ma to the south and north (Weaver 1973, 1977; Truckle 1976, 1977; Dunkley et al. 1993). Topographic isolation from the Turkana Basin was achieved by ~0.2 Ma as eruptions added to the relative height of the Barrier Volcanic Complex (Dunkley et al. 1993). Junginger (2011) described the major volcanic events during the last 50 ka, and these are summarised in Fig. 7.13.

Trauth et al. (2010) and Olaka et al. (2010) noted that lakes in steep-sided rift basins may rise and fall considerably in response to changes in catchment

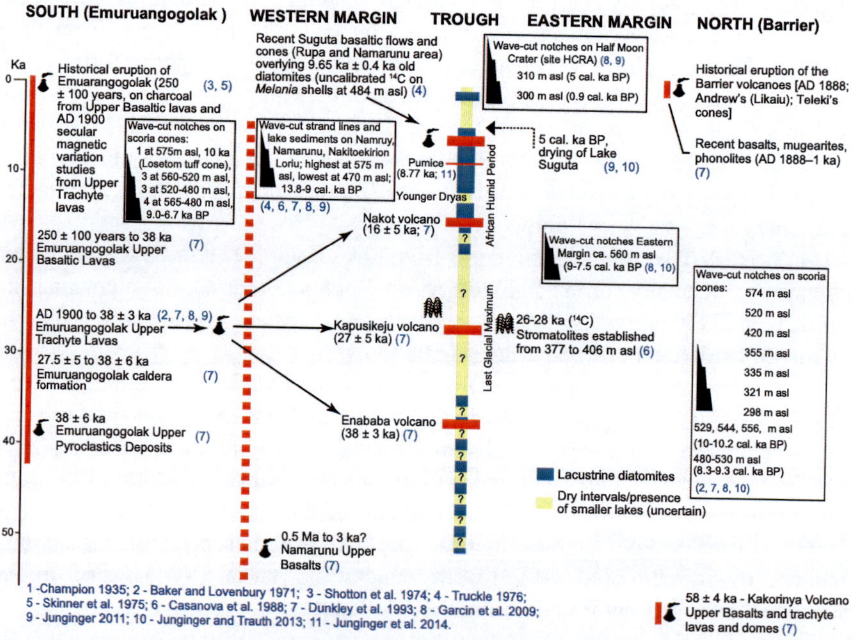

Fig. 7.13 Late Quaternary volcanic stratigraphy and temporal variability in shorelines. Modified from Junginger (2011)

precipitation. They referred to such water bodies as amplifier lakes with the Lower Suguta providing perhaps the best example in the Kenya Rift. Late Quaternary climate change and the steep-sided valley morphology interacted to facilitate a major transgression-regression cycle during the African Humid Period (AHP). At its maximum extent, the palaeolake filled the Suguta basin to a depth of ~300 m and covered an area of ~2150 km^2 (Fig. 7.3; Garcin et al. 2009).

The AHP represents a major wet phase that followed a period of Late Pleistocene aridity that was widespread across East Africa. Owen and Renaut (1980, 1986) noted that many lakes (Turkana, Chew Bahir, Abaya, Chamo, Baringo, Bogoria) and rivers were interconnected during the AHP. In turn, lakes Bogoria and Baringo may have received overflow from an expanded Lake Nakuru via Menengai Crater (Mwaniki et al. 2019; Dommain et al. 2022). Dates for the AHP vary among different studies, but it is generally placed at 15–5 ka (Street-Perrott and Perrott 1990; Garcin et al. 2009; Junginger and Trauth 2013). Research by deMenocal et al. (2000) on cores offshore from Cap Blanc (Mauritania) showed that the start and end of this wetter interval were abrupt and occurred when summer insolation crossed a threshold ~4.2% greater than today. Rapid climate changes in response to gradual increases in insolation imply non-linear feedback processes, such as loss or gain of overflows from adjacent basins, changes in vegetation cover and interacting shifts in soil moisture, albedo, cloud cover or temperature (Junginger and Trauth 2013; van der Lubbe et al. 2017).

Research over the last three decades in Suguta has provided lake-level records for the last 17,000 years that illustrate the scale and timing of climate variability during the AHP. That evidence is based on palaeoshorelines (Fig. 7.14) and lake sediment outcrops (Fig. 7.15). Bishop (1975) reported the results of a geological reconnaissance of the valley including the presence of the "Suguta Beds". Williams and Johnson (1976) provided the first ^{14}C ages of paleoshorelines, dating from the AHP. Truckle (1976) provided descriptions of the Middle Pleistocene and Late Pleistocene to early Holocene lake sediments that comprised the Suguta Beds (Fig. 7.16a). The younger (AHP) sediments were later described by Casanova et al. (1988), Junginger et al. (2014) and Garcin et al. (2017).

Figure 7.16b shows representative sections from different parts of the Suguta palaeolake, based on Junginger et al. (2014). Section BG08 is the lowest in altitude and represents the longest interval (11.3–8.52 ka BP). Those deposits include unconsolidated diatomaceous clay and silt with *Melanoides tuberculata*, a common gastropod in rift palaeolakes. The sequence coarsens upwards into coarse sand and conglomerate deposited in alluvial fans and rivers. Section EL08, in the northeast Suguta, includes fluvial orthoconglomerate overlain by lacustrine silt, and records lacustrine deposition at 11.45–8.87 ka BP, with the uppermost surface representing part of a former lake floor. Section LN08 (SW Suguta), dated between 11.21 and 8.57 ka BP, includes silty diatomite, coquina and pumice layers that accumulated on a lake floor close to the maximum Holocene highstand. Garcin et al. (2017) later described AHP deposits from Baragoi Gorge (Fig. 7.17) and interpreted them as forming in lacustrine, fluvial and deltaic settings. Monty (unpublished confidential report), Casanova et al. (1988) and Casanova (1994) gave details of stromatolites

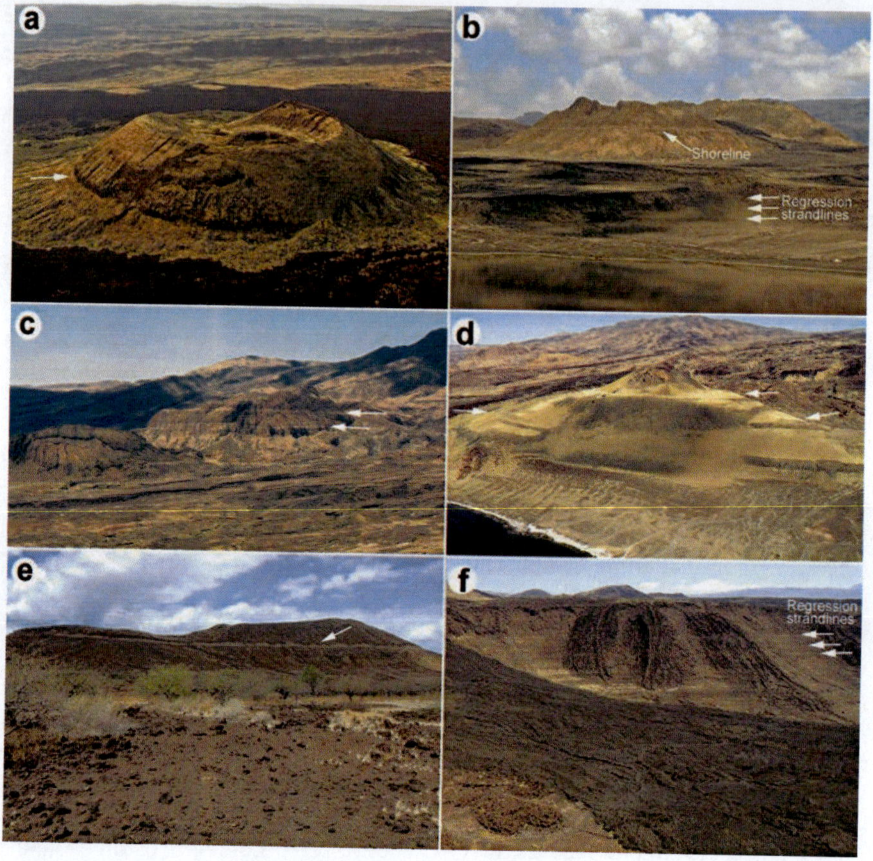

Fig. 7.14 Late Quaternary palaeoshorelines. (**a–e**) Major stillstands cut into pyroclastic cones (white arrows). (**b, f**) Strandlines left during lake regression on volcanic rocks. Note these are partially overlain by more recent lava flows in (**f**)

associated with palaeoshorelines elsewhere in the basin. Calcite stromatolites are present at several levels above the modern shoreline.

Estimates of the former elevation of the palaeolake have ranged from 484 m above sea level (asl) (Casanova 1986; Casanova et al. 1988) to 600 m asl (Truckle 1976). Garcin et al. (2009) carried out a high-resolution survey mainly using biogenic carbonate dates that implied high lake stands at 16.5 to 15 ka BP, reaching at least 535 m asl (Fig. 7.18a) with a maximum highstand at 12.8 ka BP. At that time, the Suguta palaeolake overflowed west and then northwards via the Kamuge Valley (Fig. 7.3), forming a link in a larger drainage system that extended southwards at least as far as Lake Bogoria and its contemporary catchment (Fig. 7.18b). However, reservoir effects due to old carbon resulted in these dates being about 2 ky too old (Junginger et al. 2014). Later deformation of the palaeoshoreline was induced by up to ~10 m of isostatic rebound due to lake regression and fault movements with

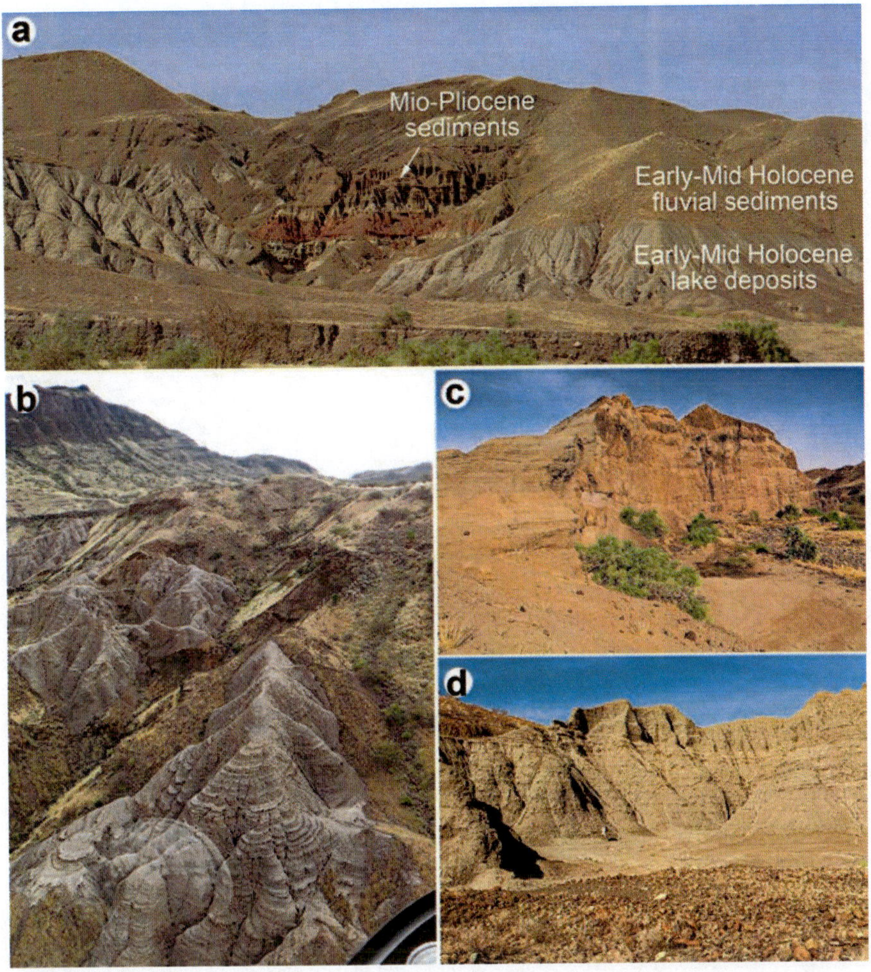

Fig. 7.15 Late Pleistocene to mid-Holocene lake sediments, SE Suguta. (**a**, **b**) Grey lacustrine sediments resting on Mio-Pliocene rocks. (**c**) Holocene diatomaceous silt onlapping Mio-Pliocene rocks left of cliff line. (**d**) 15-m-thick sequence of diatomaceous silt and fine sand deposited during the African Humid Period

throw rates of 6.8–8.5 mm y^{-1} (Melnick et al. 2012). This tectonic activity led to an eastward tilt of palaeoshorelines (Fig. 7.16c), but the average height of 567 m is considered representative and indicates a lake about 300 m deep.

Radiocarbon calibration and reservoir issues are a problem in many basins in the northern Kenya Rift, especially for older lake-level reconstructions, which can lead to misinterpretations and poor correlations between basins. Junginger et al. (2014), for example, applied a 1900-ky correction to their aquatic carbonate dates, which was supported by dates from other types of samples. When this adjustment is applied to earlier data (Garcin et al. 2009), the lake level records agree well

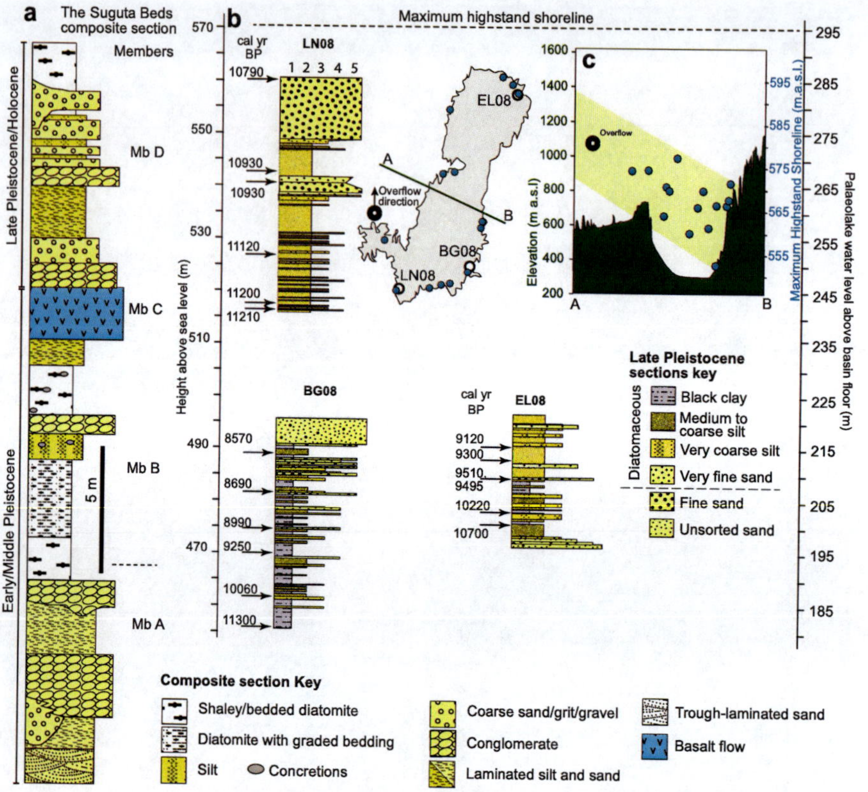

Fig. 7.16 Late Pleistocene-Early Holocene sediment logs and lake levels. (**a**) Composite section of the Suguta Beds (after Truckle 1976, 1977). The full succession includes both Middle Pleistocene and Late Pleistocene to Holocene lake deposits separated by a lava flow (Member C). (**b**) Lake sediments plotted against elevation above sea level and above the basin floor. The map shows the inferred Early Holocene maximum elevation of the former palaeolake. White circles = sediment log locations. Filled circles give positions of preserved maximum shoreline sites based on high-resolution surveys (from Junginger et al. 2014). (**c**) Plots of Early Holocene maximum elevation shorelines. All data plotted on the same (**a**, **b**) profile. Note the eastward tilt of the data set, implying deformation. Modified from Garcin et al. (2009)

(Fig. 7.18a). A fall in lake level of at least 50 m occurred during the Younger Dryas that may have led to desiccation (Fig. 7.18a). Overflow to Lake Turkana would have been lost before being re-established during the Late Pleistocene to Early Holocene, between about 11.8 and 8.5 ka BP, with a dramatic regression of 250 m taking place over the following 1 kyr as drier Middle Holocene conditions developed.

Figure 7.18 also compares former Suguta lake-level changes with data sets from nearby basins (Junginger et al. 2014). All Kenya Rift lakes were high during the early Holocene. Junginger et al. (2014) showed that the Suguta palaeolake was high at 14 ka BP and 11.5–6.5 ka BP, matching height variations at Lake Nakuru to the south (Richardson and Dussinger 1986). Lake Turkana, to the north, was high

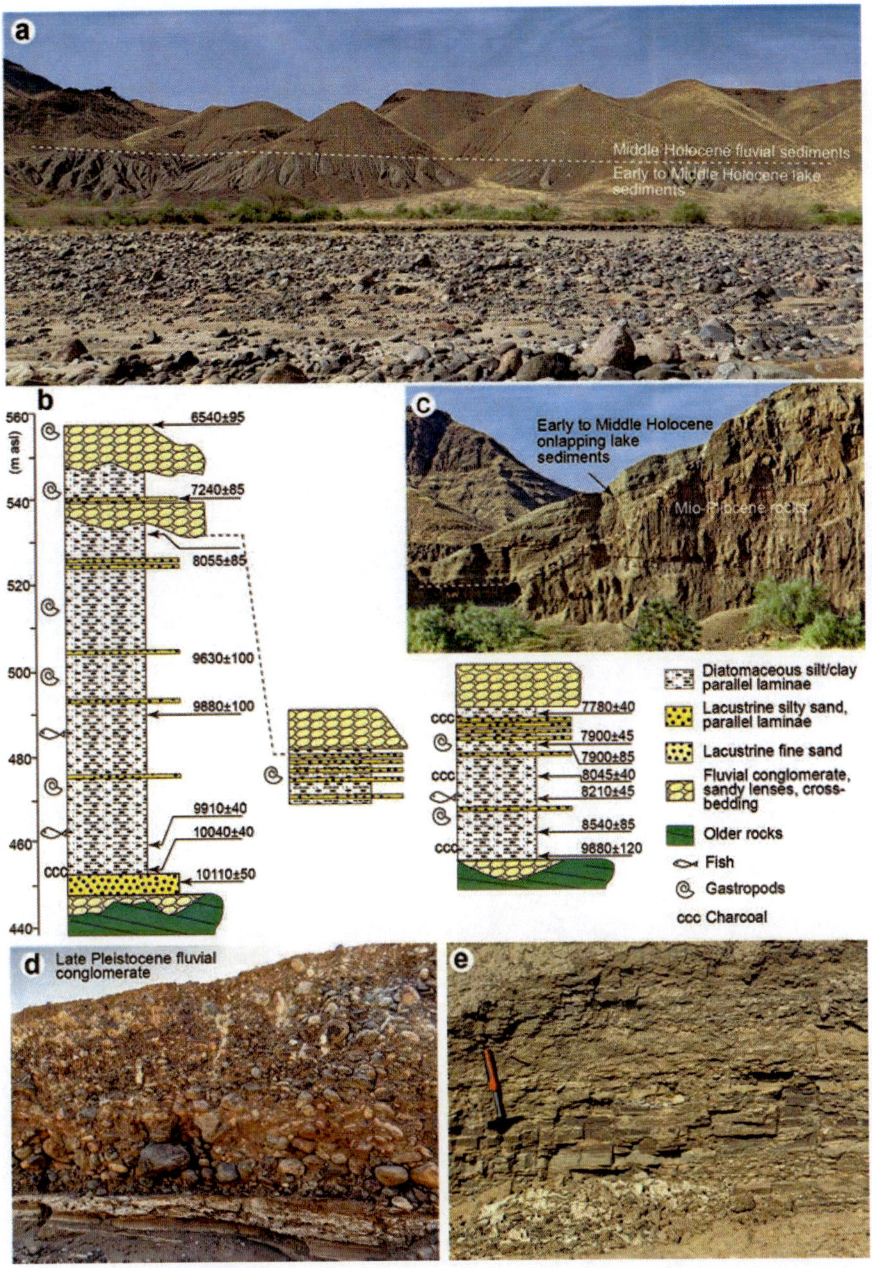

Fig. 7.17 Deltaic and lacustrine sedimentation at Baragoi Gorge, SE Suguta. (**a**) General view of the northern side of Baragoi River entering the Suguta Valley. Middle Holocene fluvial sediments rest on Early to Middle Holocene lake deposits. (**b**) Logged sections based on Garcin et al. (2017). (**c**) Holocene lake sediments onlapping onto Mio-Pleistocene rocks. Section to the right is about 10 m high. (**d**) Late Pleistocene fluvial conglomerates below lake sediments deposited in the African Humid Period. Section is ~2 m high. (**e**) Holocene lacustrine clay overlain by diatomaceous silt. Hammer for scale

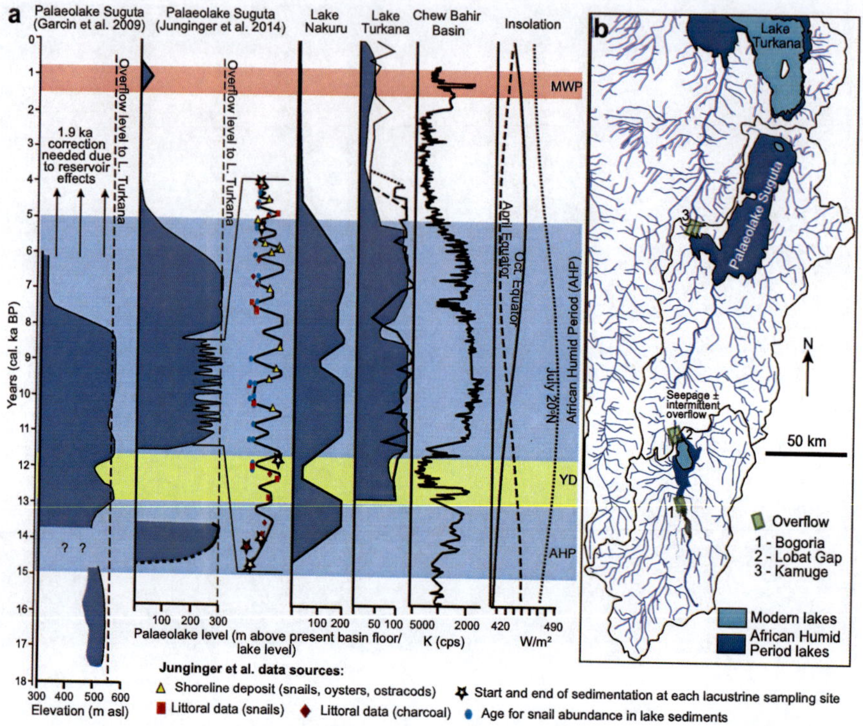

Fig. 7.18 Palaeolake levels in the Suguta and adjacent basins. After Junginger et al. (2014). (**a**) Lake level variability of Palaeolake Suguta and neighbouring basins. Lake Suguta data from Garcin et al. (2009) and Junginger et al. (2014). Lake Turkana shaded curve derived from Garcin et al. (2012), dotted curve from Johnson et al. (1991), and dashed curve from Brown and Fuller (2008). Lake Nakuru curve is from Richardson and Dussinger (1986). The Chew Bahir data are from Foerster et al. (2012) with an inverted scale for K in counts per second. YD = Younger Dryas; MWP = Medieval Warm Period; AHP = African Humid Period. Dated sequences are in calendar years. The discrepancy between the Garcin and Junginger curves disappears when a 1900-year correction for inert carbon is applied to the Garcin data. Junginger's Suguta record is based on several different shoreline indicators as represented by labelled symbols. (**b**) Fluvial connectivity during the Late Pleistocene to Early Holocene. Map based on Garcin et al. (2017)

during this interval and remained so until at least 5–4 ka BP (Johnson et al. 1991; Brown and Fuller 2008; Garcin et al. 2012) with Chew Bahir (lake NE of Lake Turkana) starting to shrink after ~6 ka BP (Foerster et al. 2012).

A lack of modern climate data has hampered attempts to model past hydrological variability and lake levels in the Lower Suguta. Evapotranspiration estimates, for example, have used remote sensing techniques (Borchardt and Trauth 2012), but there is a scarcity of field data. Junginger and Trauth (2013) and Junginger et al. (2014) used modern climate parameters derived from gridded-surface climatological data (New et al. 2002) to develop models of lake level change over differing timescales (Fig. 7.19a, b). With unknown water loss through outflow, their results implied a minimum 26% increase in AHP precipitation to sustain the maximum

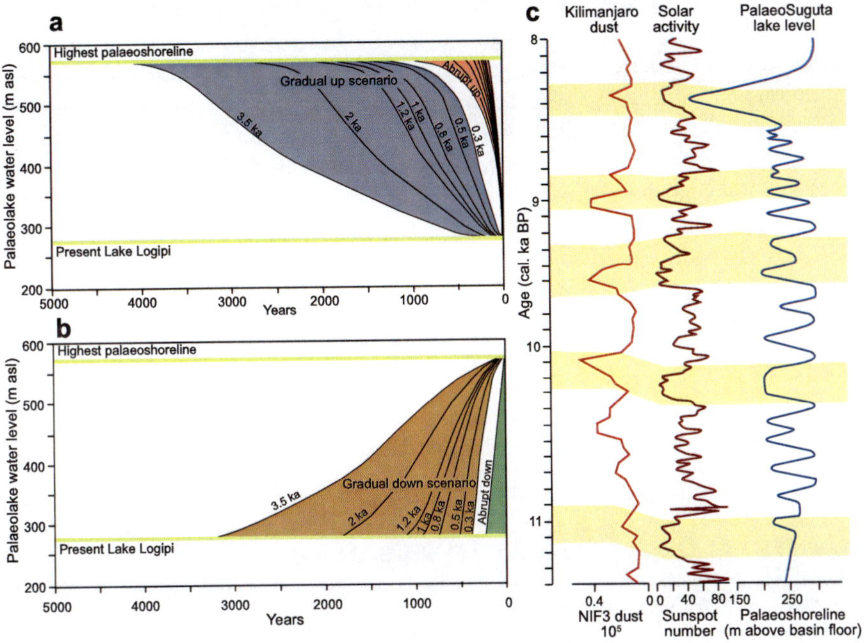

Fig. 7.19 Palaeolake Suguta hydrology and short-term variability. Modified from Junginger and Trauth (2013) and Junginger et al. (2014). (**a**, **b**) Models of lake level change for different transgression and regression scenarios. (**a**) Transgression due to abrupt and gradual precipitation changes for varying periods. (**b**) Regression due to abrupt and gradual precipitation changes. (**c**) Correlation of short-term Palaeolake Suguta lake-level cycles during the AHP with dust records from Kilimanjaro and centennial-scale variability in solar irradiation. Shaded areas show dry intervals during periods of reduced solar activity

lake level. Their modelling results, and data from lake sediments and shorelines, also indicated that smaller-scale lake-level fluctuations were superimposed on the broader climate cycles. They recognised multiple fluctuations during the AHP that ranged up to 100 m over timescales of 100 years or less (Fig. 7.19c). Such rapid variations were possible because the steep-sided Suguta basin lies in a semi-arid area with high evaporation rates and is supplied by water from adjacent elevated regions with high precipitation (Junginger and Trauth 2013). Junginger et al. (2014) also noted that these centennial events correlated well with dust episodes in Mt Kilimanjaro ice cores that implied extreme droughts (Fig. 7.19c). They related these to the weakening of the West African and Indian summer monsoons and observed broader correlations with cooling events in the North Atlantic that were previously attributed to reductions in solar irradiation.

Many researchers have attempted to explain the major Suguta lake-level changes in terms of air masses and orbital forcing. Garcin et al. (2009), for example, concluded that lakes close to the Equator were driven by cycles of spring insolation, whereas more northerly water bodies reflected summer insolation cycles. In contrast, Junginger et al. (2014) inferred that the AHP highstand in the Suguta basin resulted from a northeasterly shift in the Congo Air Boundary, caused by an increased atmospheric pressure gradient between East Africa and India during a northern-hemisphere insolation maximum. They also emphasised the importance of orbitally controlled insolation variations and short-term changes in solar irradiation as controls on lake-level variability at Suguta and in East Africa more generally. The Suguta Valley has been tilted eastwards and northwards (Dunkley et al. 1993) and some late Quaternary sediments have been faulted and deformed (Melnick et al. 2012), so tectonic controls might also have influenced lake levels.

References

Allen DJ, Darling WG (1992) Geothermics and hydrogeology of the Kenya Rift Valley between Lake Baringo and Lake Turkana. British Geological Survey, Research Report SD/92/1

Auko LO (2013) Evaluation of hydrogeochemical facies of the Barrier Volcanic Complex geothermal fluids in developing a conceptual model. MSc thesis, University of Nairobi

Baker BH (1963) Geology of the Baragoi area. Rep Geol Surv Kenya 53:1–74

Baker MJ, Lovenbury HT (1971) The South Turkana Expedition scientific papers VII: the 1969 season survey. Geogr J 137:349–360

Becht R, Mwango F, Muno FA (2006) Groundwater links between Kenyan Rift lakes. In: Odada EO, Olago DO, Ochola W, Ntiba M, Wandiga S, Gichuki N, Oyieke H (eds) Proceedings of the 11th World Lakes conference, Nairobi, Kenya, vol 2. International Lake Environment Committee, Kusatsu-shi, Shiga, pp 7–14

Bishop WW (1975) Geological reconnaissance of the Lower Suguta Valley and its eastern flank, north Kenya Rift Valley. In: Grove AT (ed) Cambridge meeting on desertification, 22–26 September 1975. University of Cambridge, pp 62–63

Borchardt S, Trauth MH (2012) Remotely-sensed evapotranspiration estimates for an improved hydrological modeling of the early Holocene mega-lake Suguta, northern Kenya Rift. Palaeogeogr Palaeoclimatol Palaeoecol 361–362:14–20

Bosworth W, Maurin A (1993) Structure, geochronology and tectonic significance of the northern Suguta Valley (Gregory Rift), Kenya. J Geol Soc Lond 150:751–762

Brown FH, Fuller CR (2008) Stratigraphy and tephra of the Kibish Formation, southwestern Ethiopia. J Hum Evol 55:366–403

Camberlin P (1997) Rainfall anomalies in the source region of the Nile and their connection with the Indian summer monsoon. J Clim 10:1380–1392

Casanova J (1986) Les Stromatolites Continentaux: Paléoécologie, Paléohydrologie, Paléoclimatologie. Application au Rift Gregory. Thèse Docteur d'Etat-Sciences, Université d'Aix Marseille II

Casanova J (1994) Stromatolites from the East African Rift: a synopsis. In: Bertrand-Sarfati J, Monty C (eds) Phanerozoic stromatolites II. Springer, New York, pp 193–226

Casanova J, Hillaire-Marcel C, Page N, Taieb M, Vincens A (1988) Palaeohydrology and Late Quaternary stratigraphy of lacustrine deposits in the Suguta Rift (Kenya). C R Acad Sci Paris Sér II 307:1251–1258

Castanier S, Bernet-Rollande M, Maurin A, Perthuisot J-P (1993) Effects of microbial activity on the hydrochemistry and sedimentology of Lake Logipi, Kenya. Hydrobiologia 267:99–112

Caukwell RA (1971) The South Turkana Expedition scientific papers VI. Field survey in South Turkana. Geogr J 137:157–164

Cavendish H (1898) Through Somaliland and around and south of Lake Rudolf. Geogr J 11:372–393

Champion AM (1935) Teleki's volcano and the lava fields at the southern end of Lake Rudolf. Geogr J 85:323–341

Champion AM (1937a) The volcanic region around the southern end of Lake Rudolf, Kenya Colony. Z Vulkanol 17:163–172

Champion AM (1937b) Physiography of the region to the west and south-west of Lake Rudolf. Geogr J 139:97–118

Coe M (1972) The South Turkana Expedition scientific papers IX. Ecological studies of the small mammals of South Turkana. Geogr J 138:316–338

Crane K, O'Connell S (1983) The distribution and implications of heat flow from the Gregory Rift Valley. Tectonophysics 94:253–275

Darling W (1996) The geochemistry of fluid processes in the eastern branch of the East African Rift system. PhD thesis, Open University, UK

Darling W (2001) Magadi and Suguta: the contrasting hydrogeochemistry of two soda lake areas in the Kenya Rift Valley. In: Cidu R (ed) Proceedings of the 10th international symposium on water-rock interaction (WRI-10), Villasimius, Italy. A.A. Balkema, pp 95–98

deMenocal P, Ortiz J, Guilderson T, Adkins J, Sarnthein M, Baker L, Yarusinsky M (2000) Abrupt onset and termination of the African Humid Period: rapid climate responses to gradual insolation forcing. Quat Sci Rev 19:347–361

Dodson RG (1963) Geology of the South Horr area. Rep Geol Surv Kenya 60:1–54

Dommain R, Riedl S, Olaka LA, deMenocal P, Deino AL, Owen RB, Muiruri V, Müller J, Potts R, Strecker MR (2022) Holocene bidirectional river system along the Kenya Rift and its influence on East African faunal exchange and diversity gradients. Proc Natl Acad Sci 119:e2121388119. https://doi.org/10.1073/pnas.2121388119

Dunkley PN, Smith M, Allen DJ, Darling WG (1993) The geothermal activity and geology of the northern sector of the Kenya Rift Valley. British Geological Survey, Research Report SC/93/1

Foerster V, Junginger A, Langkam O, Gebru T, Asrat A, Umer M, Lamb HF, Wennrich V, Rethemeyer J, Nowaczyk N, Trauth MH, Schaebitz F (2012) Climatic change recorded in the sediments of the Chew Bahir basin, southern Ethiopia, during the last 45,000 years. Quat Int 274:25–37

Garcin Y, Junginger A, Melnick D, Olago DO, Strecker MR, Trauth MH (2009) Late Pleistocene–Holocene rise and collapse of Lake Suguta, northern Kenya Rift. Quat Sci Rev 28:911–925

Garcin Y, Melnick D, Strecker MR, Olago D, Tiercelin J-J (2012) East African mid-Holocene wet–dry transition recorded in palaeo-shorelines of Lake Turkana, northern Kenya Rift. Earth Planet Sci Lett 331–332:322–334

Garcin Y, Schildgen TF, Acosta VT, Melnick D, Guillemoteau J, Willenbring J, Strecker MR (2017) Short-lived increase in erosion during the African Humid Period: evidence from the northern Kenya Rift. Earth Planet Sci Lett 459:58–69

Grewe S (2016) Hydro- and biogeochemical investigations of lake sediments in the Kenyan Rift Valley. PhD thesis, Universität Potsdam

Gwynne MD (1969) The South Turkana Expedition scientific papers I. Preliminary report on the 1968 season. Geogr J 135:331–342

Hackman BD, Charsley TJ, Key RN, Wilkinson AF (1990) The development of the East African Rift system in north-central Kenya. Tectonophysics 184:189–211

Hautot S, Tarits P, Whaler K, Le Gall B, Tiercelin J-J, Le Turdu C (2000) Deep structure of the Baringo Rift Basin (central Kenya) from three-dimensional magnetotelluric imaging: implications for rift evolution. J Geophys Res Solid Earth 105:23493–23518

Hemming C (1972) The South Turkana Expedition: scientific papers VIII. The ecology of South Turkana: a reconnaissance classification. Geogr J 138:15–40

Johnson TC, Halfman JD, Showers WJ (1991) Paleoclimate of the past 4000 years at Lake Turkana, Kenya, based on the isotopic composition of authigenic calcite. Palaeogeogr Palaeoclimatol Palaeoecol 85:189–198

Junginger A (2011) East African climate variability on different time scales: the Suguta Valley in the African-Asian Monsoon Domain. PhD thesis, Universität Potsdam

Junginger A, Trauth MH (2013) Hydrological constraints of paleo-Lake Suguta in the northern Kenya Rift during the African Humid Period (15–5 ka BP). Glob Planet Chang 111:174–188
Junginger A, Roller S, Olaka LA, Trauth MH (2014) The effects of solar irradiation changes on the migration of the Congo Air Boundary and water levels of paleo-Lake Suguta, northern Kenya Rift, during the African Humid Period (15–5 ka BP). Palaeogeogr Palaeoclimatol Palaeoecol 396:1–16
Key RM (1987) Geology of the Maralal area. Rep Geol Surv Kenya 105:1–93
King BC (1978) Structural and volcanic evolution of the Gregory Rift Valley. In: Bishop WW (ed) Geological background to fossil man. Geol Soc Lond Spec Publ 6:29–54
Koyaguchi T (1984) Volcanic rocks in the Samburu Hills, northern Kenya. Afr Stud Monogr Suppl Iss 2:147–179
Lambiase JJ, Bosworth W (1995) Structural controls on sedimentation in continental rifts. In: Lambiase JJ (ed) Hydrocarbon habitat in rift basins. Geol Soc Lond Spec Publ 80:117–144
Makinouchi T, Koyaguchi T, Matsuda T, Mitsushio H, Ishida S (1984) Geology of the Nachola area and the Samburu Hills, west of Baragoi, northern Kenya. Afr Stud Monogr Suppl Iss 2:15–44
Melnick D, Garcin Y, Quinteros J, Strecker MR, Olago D, Tiercelin J-J (2012) Steady rifting in northern Kenya inferred from deformed Holocene lake shorelines of the Suguta and Turkana basins. Earth Planet Sci Lett 331–332:335–346
Mohamud YN, Lopeyok T (2018) The future potential of the Silali Caldera, North Rift, Kenya. In: Proceedings of the 7th African rift geothermal conference (ARGeo C7), Kigali, Rwanda, pp 1–10
Morgan WTW (1971) The South Turkana Expedition scientific papers IV. Land units of the Lokori area. Geogr J 137:14–23
Morley CK (1999) Geoscience of rift systems – evolution of East Africa. AAPG Stud Geol 44:1–18
Mugisha F, Ebinger CJ, Strecker M, Pope D (1997) Two-stage rifting in the Kenya Rift: implications for half-graben models. Tectonophysics 278:61–81
Mwaniki P, Taita T, Sierens T, Triest L (2019) Barriers to genetic connectivity of smooth flatsedge (*Cyperus laevigatus*) among alkaline-saline lakes of Eastern Rift Valley (Kenya). Aquat Bot 155:38–44
New M, Lister D, Hulme M, Makin I (2002) A high-resolution data set of surface climate over global land areas. Clim Res 21:1–25
Nicholson SE (1996) A review of climate dynamics and climate variability in eastern Africa. In: Odada EO, Johnson TC, Whitaker KT (eds) The limnology, climatology and paleoclimatology of the East African lakes. Gordon and Breach, Amsterdam, pp 25–56
Njau K, Kimani F, Wamburu J (2020) Geothermal exploration of the Barrier Volcanic Complex, Kenya. In: Proceedings of the 8th African rift geothermal conference, Nairobi
Nunn JA, Harris NB (2007) Subsurface seepage of seawater across a barrier: a source of water and salt to peripheral salt basins. Geol Soc Am Bull 119:1201–1217
Ochieng' JO, Wilkinson AF, Kagasi J, Kimono S (1988) Geology of the Loiyangalani area. Rep Geol Surv Kenya 107:1–53
Ojany FF, Ogenoo RB (1973) Kenya: a study in physical and human geography. Longman, Nairobi
Olaka LA, Odada EO, Trauth MH, Olago DO (2010) The sensitivity of East African rift lakes to climate fluctuations. J Paleolimnol 44:629–644
Owen RB, Renaut RW (1980) The early Holocene palaeogeography of the northern Kenya Rift. Rech Géol Afr 5:143–146
Owen RB, Renaut RW (1986) Sedimentology, stratigraphy and palaeoenvironments of the Holocene Galana Boi Formation, NE Lake Turkana, Kenya. In: Frostick LE, Renaut RW, Reid I, Tiercelin J-J (eds) Sedimentation in the African rifts. Geol Soc Lond Spec Publ 25:311–322
Pointing AJ, Maguire PKH, Kham MA, Francis DJ, Swain CJ, Shah ER, Griffiths DH (1985) Seismicity of the northern part of the Kenya Rift Valley. J Geodyn 3:23–37
Rhemtulla S (1970) The South Turkana Expedition scientific papers III. A geological reconnaissance of South Turkana. Geogr J 136:61–73

Richardson JL, Dussinger RA (1986) Paleolimnology of mid-elevation lakes in the Kenya Rift Valley. Hydrobiologia 143:167–174

Robertson EAM, Biggs J, Cashman KV, Floyd MA, Vye-Brown C (2016) Influence of regional tectonics and pre-existing structures on the formation of elliptical calderas in the Kenyan Rift. In: Wright TJ, Ayele A, Ferguson DJ, Kidane T, Vye-Brown C (eds) Magmatic rifting and active volcanism. Geol Soc Lond Spec Publ 420:43–67

Rosiwal A (1891) Uber Gesteine aus dem Gebiete zwischen Usambara und dem Stefanie-See. Derkschr K Akad Wiss Wein 58:465–550

Rubino J-L (2002) Influence of lake level fluctuations on Quaternary fluvial and lacustrine sediments in the Suguta Rift (Kenya). In: Abstracts, Conference on dryland rivers: process and product, Aberdeen, 8–9 August 2002

Sawada Y, Tateishi M, Ishida S (1987) Geology of the Neogene System in and around the Samburu Hills, northern Kenya. Afr Stud Monogr Suppl Iss 5:7–26

Shotton FW, Williams REG, Johnson AS (1974) Birmingham University radiocarbon dates VIII. Radiocarbon 16:285–303

Sigurdsson H, Houghton B, McNutt S, Rymer H, Stix J (eds) (2015) The encyclopedia of volcanoes, 2nd edn. Academic Press (Elsevier), London

Skinner NJ, Iles W, Brock A (1975) The secular variation of declination and inclination in Kenya. Earth Planet Sci Lett 25:338–346

Smith AD (1897) Through unknown African countries. Edward Arnold, London

Street-Perrott FA, Perrott RA (1990) Abrupt climate fluctuations in the tropics: the influence of Atlantic Ocean circulation. Nature 343:607–612

Sturchio NC, Dunkley PN, Smith M (1993) Climate-driven variations in geothermal activity in the northern Kenya Rift Valley. Nature 362:233–234

Suess E (1891) Die Brüche des Östlichen Afrika. Derkschr K Akad Wiss Wein 58:555–584

Tebbs EJ, Remedios JJ, Avery ST, Rowland CS, Harper DM (2015) Regional assessment of lake ecological states using Landsat: a classification scheme for alkaline–saline, flamingo lakes in the East African Rift Valley. Int J Appl Earth Obs Geoinf 40:100–108

Toula F (1891) Uber Gesteine aus dem Gebiete zwischen Usambara und dem Rudolf-See, und Begleitworte zu derselben. Derkschr K Akad Wiss Wein 58:551–554

Trauth MH, Maslin MA, Deino AL, Junginger A, Lesoloyia M, Odada EO, Olago DO, Olaka LA, Strecker MR, Tiedemann R (2010) Human evolution in a variable environment: the amplifier lakes of Eastern Africa. Quat Sci Rev 29:2981–2988

Truckle P (1976) Geology and late Cainozoic lake sediments of the Suguta Trough, Kenya. Nature 263:380–383

Truckle P (1977) The geology of the area to the south of Lokori, South Turkana, Kenya. PhD thesis, University of Leicester

Truckle P (1979) Pahoehoe and pillow lavas from the Suguta Trough, South Turkana, Kenya. Geol Mag 116:139–142

Tuite CH (1981) Standing crop densities and distribution of *Spirulina* and benthic diatoms in East African alkaline saline lakes. Freshw Biol 11:345–360

Tuite CH (2000) The distribution and density of Lesser Flamingos in East Africa in relation to food availability and productivity. Waterbirds 23:52–63

van der Lubbe HJL, Krause-Nehring J, Junginger A, Garcin Y, Joordens JCA, Davies GR, Beck C, Feibel CS, Johnson TC, Vonhof HB (2017) Gradual or abrupt? Changes in water source of Lake Turkana (Kenya) during the African Humid Period inferred from Sr isotope ratios. Quat Sci Rev 174:1–12

Von Höhnel L (1894) Discovery by Count Teleki of Lakes Rudolf and Stefanie, volume II. Longmans Green, London

Watson RM (1969) The South Turkana Expedition scientific papers II. A survey of the large mammal population in South Turkana. Geogr J 135:529

Weaver SD (1973) Geology of the Nasaken area, South Turkana, Kenya, and the petrochemistry of the Pliocene volcanic rocks. PhD thesis, University of London

Weaver SD (1977) The Quaternary caldera volcano Emuruangogolak, Kenya Rift, and the petrology of a bimodal ferrobasalt-pantelleritic trachyte association. Bull Volcanol 40:209–230
Welby MS (1900) King Menelek's dominion and the country between Lake Gallop (Rudolf) and the Nile Valley. Geogr J 16:292–306
Williams LAJ, Chapman GR (1986) Relationships between major structures, salic volcanism and sedimentation in the Kenya Rift from the equator northwards to Lake Turkana. In: Frostick LE, Renaut RW, Reid I, Tiercelin J-J (eds) Sedimentation in the African rifts. Geol Soc Lond Spec Publ 25:59–74
Williams REG, Johnson AS (1976) Birmingham University radiocarbon dates X. Radiocarbon 18:249–267
Williams LAJ, MacDonald R, Chapman GR (1984) Late Quaternary caldera volcanoes of the Kenya Rift Valley. J Geophys Res Solid Earth 89:8553–8570

Chapter 8
Lake Baringo

8.1 Introduction

Lake Baringo (Fig. 8.1) lies 60 km north of the equator (0°20′–1°N) and is reached by paved road from Nakuru, ~105 km to the south. The lake has typically been ~21 km long (N-S) by ~13 km wide (E-W) but expanded in area after 2010 in common with most other Kenya Rift lakes (GoK and UNDP 2021). It is the largest freshwater body entirely in Kenya (Johansson and Svensson 2002; Scoon 2018) and lies in the semi-arid, northern part of its 6820 km^2 drainage basin (Fig. 8.2). The lake is located near the eastern margins of the central Kenya Rift Valley and is separated from the Elgeyo Escarpment and the Kerio Valley, 50 km to the west, by the steep 1500-m-high Tugen Hills tilt-block (Fig. 8.2a, b; Tiercelin et al. 1980).

The fresh to mildly alkaline (pH ~8.5) waters of the modern lake are very shallow with a maximum depth of about 5–7 m (Fig. 8.2c) in the northwest and are dominated by muddy siliciclastic sedimentation (Tiercelin and Vincens 1987; Oduor et al. 2003). The lake is topographically closed but hydrologically open, in common with Lake Naivasha. Today there is no surface outlet, but water likely escapes northwards via subsurface drainage through fractured and faulted lavas and permeable sediments (Allen and Darling 1992). Lake level varies in response to seasonal and longer-term climate change (Källqvist 1987), which is driven mainly by shifts in the Inter Tropical Convergence Zone (Johansson and Svensson 2002).

Gregory examined the geology of the region during an 1892–93 expedition to Baringo and Mt. Kenya (Gregory 1896; Leake 2011). Members of EAGRU (East African Geological Research Unit) carried out detailed studies in the 1970s that included mapping, petrological and geochemical analyses, and sedimentological, stratigraphic and palaeontological work (Bishop and Chapman 1970; Bishop et al. 1971; King and Chapman 1972; Andrews and Banham 1999). The structural and tectonic framework of the basin was established, supported by geophysical investigations.

R. W. Renaut, R. B. Owen, *The Kenya Rift Lakes: Modern and Ancient*, Syntheses in Limnogeology, https://doi.org/10.1007/978-3-642-25055-2_8

Fig. 8.1 Lake Baringo viewed from the southeast at the Laikipia Border Fault. Ol Kokwe is the large volcanic island in the centre of the lake

The lake is named from the local word "Mparingo", which translates as 'lake' (Johnston 1902a; Odada et al. 2006). Early European explorers had little understanding of the geography and extent of the East African lakes (Wakefield and Johnston 1870; Daly and Taylor 1875). Consequently, their speculative commentaries on the extent of water bodies such as Baringo were based on, for example, information about caravan routes (Burton 1871). The geologist Joseph Thomson visited Lake Baringo in the early 1880s (Thomson 1887; Brown 1989) and was soon followed by Count Teleki and von Höhnel who stayed at Njemps, near the southwest shores before they left for Lake Turkana on 10th February 1888 (von Höhnel 1894). Other late nineteenth-century European safaris that passed Lake Baringo included the expeditions of Delamere in 1896–97, Cavendish in 1897, Harrison in 1899 and Austin in 1900 (Brown 1989). Johnston (1902a, b) and Powell-Cotton (1904) visited the lake in the early 1900s. Colonial administration was established in the early twentieth century (Brown 1989) at a time when the area became famous for its wildlife and hunting (Chapman 1908).

The lake was one of the first Kenyan lakes to be studied by limnologists, who studied its plankton and fish during the Percy Sladen Expedition in 1929 (Jenkin

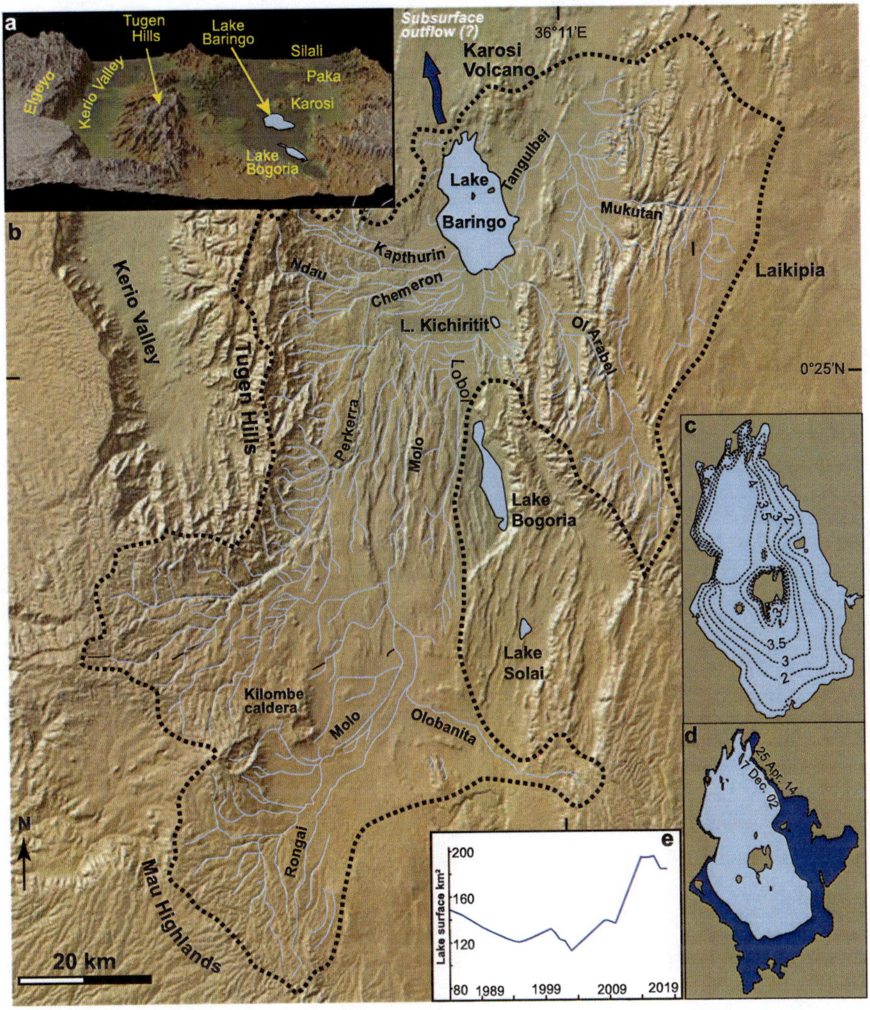

Fig. 8.2 Lake Baringo and its drainage basin. (**a**) Topography of the Lake Baringo region. (**b**) Drainage basin area and major rivers. Base maps constructed from GeoMapApp (http://www.geomapapp.org). (**c**) Lake bathymetry (August 1976). (**d**) Variability in lake surface area between December 2002 and April 2014 based on satellite data. (**e**) Change in surface area. After Kiage and Douglas (2020)

1929, 1932, 1936) and Cambridge Expedition in 1930–1 (Worthington 1932a, b, 1996; Worthington and Ricardo 1936). Changes in lake level were studied in the 1930s (Nilsson 1931). The lake has continued to be a focus of research, especially in the past two decades.

8.2 Geological and Geomorphological Setting

8.2.1 Regional Tectonic Framework

The western boundary of the seismically active (Tongue 1991; Tongue et al. 1994) central Kenya Rift is marked by the Elgeyo Fault, a major listric feature that was active by at least 12–16 Ma (Jones 1988). A large, rotated fault-block constitutes the Tugen (Kamasia) Hills and separates the Kerio Valley, to the west, from the Baringo region to the east (Mugisha et al. 1997). The Tugen Hills tilt-block dips to the west with the major Saimo and Kito Pass Fault scarps facing eastwards, exposing Miocene to Pleistocene sediments and igneous rocks, overlying metamorphic basement (Chapman 1971; Chapman et al. 1978). The Laikipia Boundary Fault marks the eastern margin of the rift (Fig. 8.1) at this latitude.

The western shores are bounded by north-south trending faults that cut into Pleistocene trachyphonolite and trachyte (Fig. 8.3). Volcanic and sedimentary rocks to the east are cut by a series of north-south step faults with the Laikipia Escarpment marking the eastern rift margin (Hackman 1988). Directly north of the lake, Karosi (Korosi) is an extinct Pleistocene-Holocene volcano (Carney 1972; Dunkley et al. 1993; Kipng'ok 2011; Njue 2014, 2015) that provides topographic closure. Late Pleistocene (Loboi Silts) and Holocene fluvio-deltaic sediments form the Loboi Plain to the south, which separates Lake Baringo from Lake Bogoria. The Loboi Silts rest on older sediments and Pleistocene trachyphonolite and basalt that crop out further to the south (Fig. 8.3). Several small fault-controlled trachytic and basaltic islands (Parmalok, Samatian, Longicharo, Lesukut) are present in the central and northern parts of Lake Baringo with the largest being the Late Pleistocene (~175 ka; Clément et al. 2003) volcanic Ol Kokwe Island, which consists of basalt scoria cones on trachyte lavas, with localised hot springs and fumaroles.

The boundary between the upper and lower crust in the Baringo region lies ~10 km below the rift floor with a brittle-ductile transition taking place about 12 km below Lake Bogoria, some 16 km south of Lake Baringo (Mechie et al. 1994). Early estimates inferred a 1-km-thick sedimentary pile below the Loboi Plain (Rooney and Hutton 1977). However, later research suggested that Lake Baringo and the Loboi Plain are the surface expression of a 5-km-deep fault-controlled basin that dates to the Palaeogene (Hautot et al. 2000) and that a 1.5-km-thick sedimentary basin rests on mid-Miocene flood phonolite. The phonolites can be seen elsewhere in outcrop and overlie a basal volcano-sedimentary sequence (~4–8 km thick) of possible Oligo-Miocene age.

Lake Baringo itself is located within the northern part of a rhomb-shaped half-graben controlled by two major tectonic alignments (Dunkley et al. 1993; Grimaud et al. 1994; Le Turdu et al. 1995, 1999). The first is formed by broadly north-south faults that cut an axial graben within the eastern part of the Kenya Rift. Active faulting, volcanism and sedimentation have taken place since the Late Pliocene (Fig. 8.4a–c). The second set of trends consists of northwest-southeast lineaments that follow older structures within the Precambrian basement. These include the

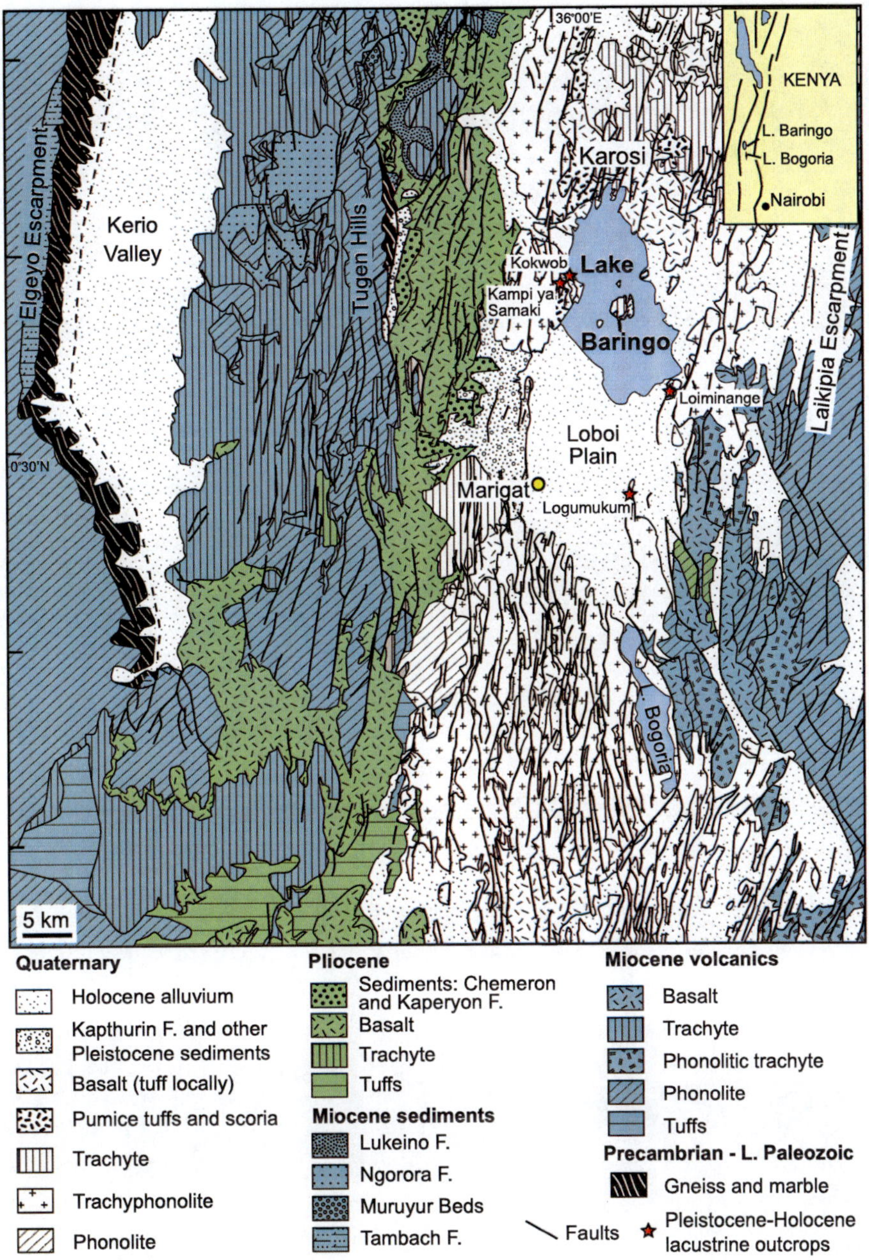

Fig. 8.3 Geological map of the Lake Baringo region based on Chapman et al. (1973) and East African Geological Research Unit Degree Sheet 34, NE Quarter (1973)

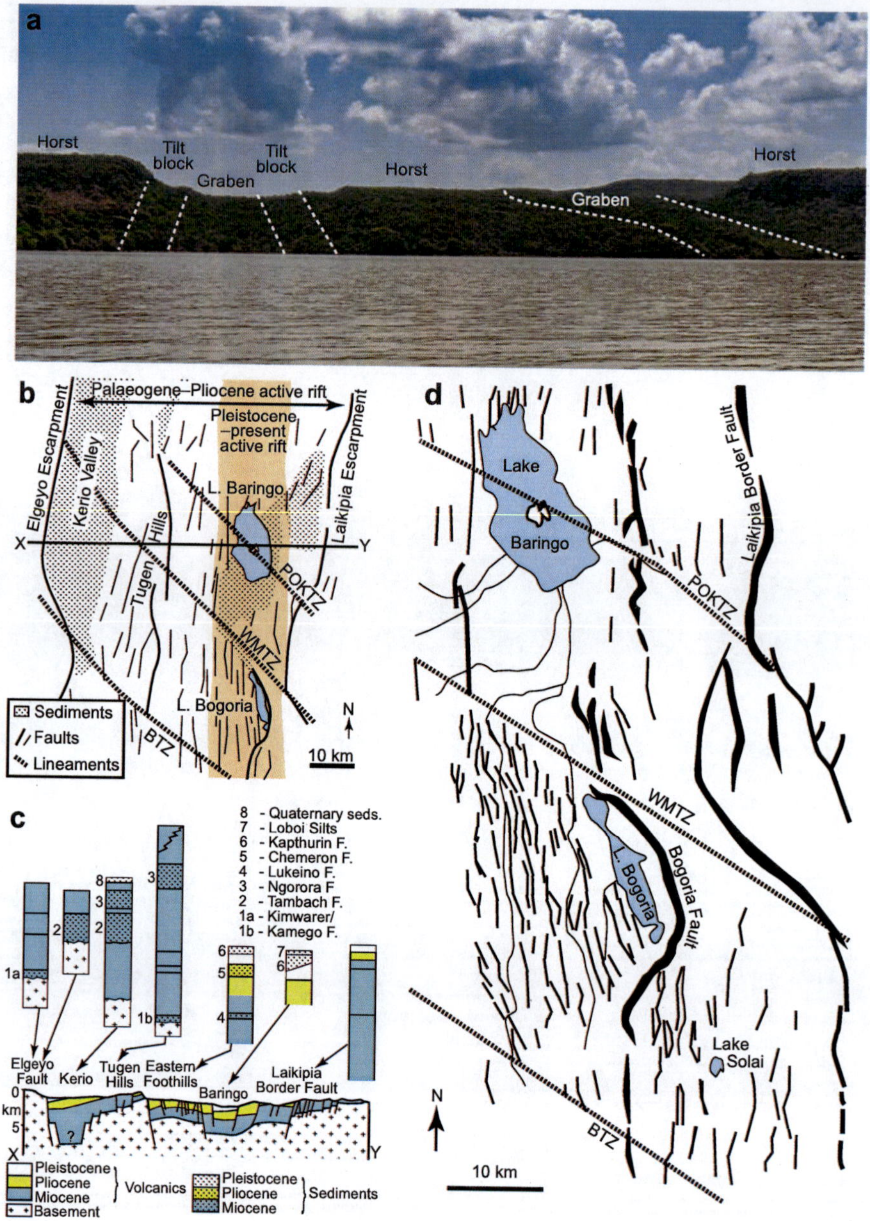

Fig. 8.4 Geological structure of the Baringo region. (**a**) North-south faults (dashed white lines) forming a series of inclined horsts and grabens, Karosi Volcano, north of Lake Baringo. (**b**) Simplified major rift structures in the Central Kenya Rift. Note north-south faulting and intersection with deep northwest-southeast lineaments. POKTZ: Porumbonyanza-Ol Kokwe Transverse Zone; WMTZ: Waseges-Marmanet Transverse Zone; BTZ: Bahati Transverse Zone. Shaded area represents zone of active rifting. (**c**) Summary stratigraphic columns and rift structure between the Elgeyo and Laikipia Escarpments. Section shown in (**b**). (**d**) Detailed fault structure for the Baringo-Bogoria area. Line thickness reflects displacements. (**b**, **c**) After Tarits et al. (2006); (**d**) After Hautot et al. (2000)

Porumbonyanza-Ol Kokwe Transverse Zone (POKTZ), which passes through Lake Baringo, the Waseges-Marmanet Transverse Zone (WMTZ) between Lake Bogoria and the southern Loboi Plain, and the Bahati Transverse Zone (BTZ) further to the south (Fig. 8.4d). Interaction between the two alignments is seen in the basin morphology – the orientation of the southern and northern shorelines, and southern margin of Loboi Plain are controlled by NW-SE lineaments – as is local faulting (Le Gall et al. 2000) and drainage in the basin. Hackman (1988) and Hackman et al. (1990) also recognised a series of subtle W-E lineaments on the southern Loboi Plain ('Marigat lineament') that they related to secondary tensional stress oblique to the rift axis, noting that this direction is parallel to the Nyanza Rift.

8.2.2 *Lithologies and Stratigraphy within the Baringo Drainage Basin*

Lake Baringo is the latest expression of a series of palaeolakes that have shifted eastwards across the Kenya Rift since the Middle Miocene (Bishop et al. 1971; Pickford 1975, 1978; Chapman et al. 1978; Tiercelin 1981; Tiercelin and Vincens 1987; Hill 1995; Renaut et al. 1999, 2000; Owen and Renaut 2000; Deino and Hill 2002; Deino and McBrearty 2002; Kingston et al. 2007). The Tugen Hills comprise a series of Miocene (Ngorora and Lukeino formations), Pliocene (Chemeron Formation) and Pleistocene (Kapthurin Formation, Loboi Silts) palaeolake and fluvio-lacustrine sediments that alternate with sequences of trachyte, phonolite and basalt, and with small outcrops of basement metamorphic rocks present in a few places (Figs. 8.3 and 8.4c). Palaeolake sediments (Tambach Formation) also crop out along the Elgeyo Escarpment, west of the Kerio Valley (see Chap. 17 for details of the precursor lakes).

Late Pleistocene-Holocene lacustrine sediments crop out in small areas north of Kokwob Murren on the mid-west shore and at Logumukum and Loiminange to the south and southeast (Fig. 8.3). Fluvial and fluvio-lacustrine sediments are widespread south of Lake Baringo where they form the Loboi Plain. Soils in the region reflect the geomorphological setting, parent materials and hydrology (Institute of Hydrology 1985). Several Food and Agricultural Organization (FAO) soil types have been described (Sketchley 1978), that include silty loam and clay (fluvisols) on alluvial plains, reddish brown clay (solonchaks) on plateaux, bouldery regosols and luvisols on trachytic and phonolitic uplands with cambisols, ferralsols and lithosols in mountain locations. Several of the lowland fluvisols have been described in detail, especially where they have agricultural potential (Kinyanjui 1982; Engelen 1983; Mwangi 1983; Wararu 1990). Soil erosion resulting from overgrazing and tree felling is a widespread problem, with gullies and badlands developing locally (Schnabel 1990; Oostwoud Wijdenes 1992; Bryan 1994; Bryan and Schnabel 1994; Oostwoud Wijdenes and Bryan 2001; Onyando et al. 2005; Lwenya and Yongo 2010; Boles et al. 2019; Ochuka et al. 2019a; Mwasi 2021; Boitt et al. 2022; Chesire et al. 2022), notably on the southwestern Loboi Plain.

8.3 Climate and Basin Hydrology

Lake Baringo lies in a semi-arid region with rainfall averaging about 450–900 mm y^{-1} (Fig. 8.5a; Kipkorir 2002; Herrnegger et al. 2021), which rises to >1500 mm y^{-1} over neighbouring highlands. The rains reach a peak seasonally around July and November (Fig. 8.5b, f), and reflect movements of the Intertropical Convergence Zone (ITCZ). In addition, rainfall fluctuates over longer (5–7 years) El Niño and La Niña timescales (Fig. 8.5a; Indeje et al. 2000; LaVigne and Ashley 2002; Kiage and Liu 2009a) and correlates with changes in sea surface temperature (SST) in the western Indian Ocean (Clark et al. 2003).

The mean annual temperature at Marigat is about 23–26 °C (Figs. 8.3 and 8.5c) with potential evaporation rates of about 2600 mm y^{-1}. Herrnegger et al. (2021) reported much higher annual evaporation rates (1650–2300 mm) than rainfall. Maximum and minimum mean monthly air temperatures range from 30.0–33.7 °C to 15.9–18.0 °C, with the highest temperatures in February–March and September–October (Källqvist 1987). Average wind speeds are about 113 km day^{-1} (4.7 km h^{-1}), with strong northeasterlies developing in afternoons and weaker southerlies sometimes blowing during mornings (Källqvist 1987).

Most of the water input is derived from the perennial Molo and Perkerra rivers (Fig. 8.2b), which flow from the Mau Highlands in the southwest (Renaut et al. 2000; Chebet et al. 2020). Very heavy El Niño rains in 1994 over the Molo drainage basin resulted in the formation of a new 3 km^2 floodplain lake, Lake Kichirtit (Tarras-Wahlberg et al. 2003; Sect. 14.6). Molo River flow in 1973 was estimated at ~126×10^6 m^3, with Perkerra River flows of about 39×10^6 m^3 (Ojany and Ogendo 1973). However, in very dry years the discharge is greatly diminished, with irrigation to the southwest of the lake also reducing water inputs to Lake Baringo (Renaut et al. 2000). Several seasonal rivers drain the Tugen Hills to the west (e.g. Ndau, Kapthurin, Chemeron) and the Laikipia escarpment to the east (e.g. Ol Arabel, Mukutan, Tangulbei) (Fig. 8.2). Flash floods are frequent in the rainy seasons. There is minimal inflow from Karosi volcano to the north (Fig. 8.6a). Rains falling directly on the lake, especially from March to August, also contribute water, as do small hot springs on northeastern Ol Kokwe Island in the centre of the lake (Renaut et al. 2002). Groundwater to the north of Lake Baringo flows northwards (Dunkley et al. 1993) and is derived from meteoric sources on the rift margins, as well as from Baringo subsurface outflow. No measured groundwater estimates are currently available for groundwater inflow to Lake Baringo (Tarits et al. 2006) although Nyaberi (2010) reports geophysical and borehole details of aquifer flow in the extreme south of the drainage basin. Cherop et al. (2016) report geophysical data for aquifers near Marigat and Ombasa et al. (2022) used GIS studies to identify potential areas with good groundwater potential for borehole-derived water resources.

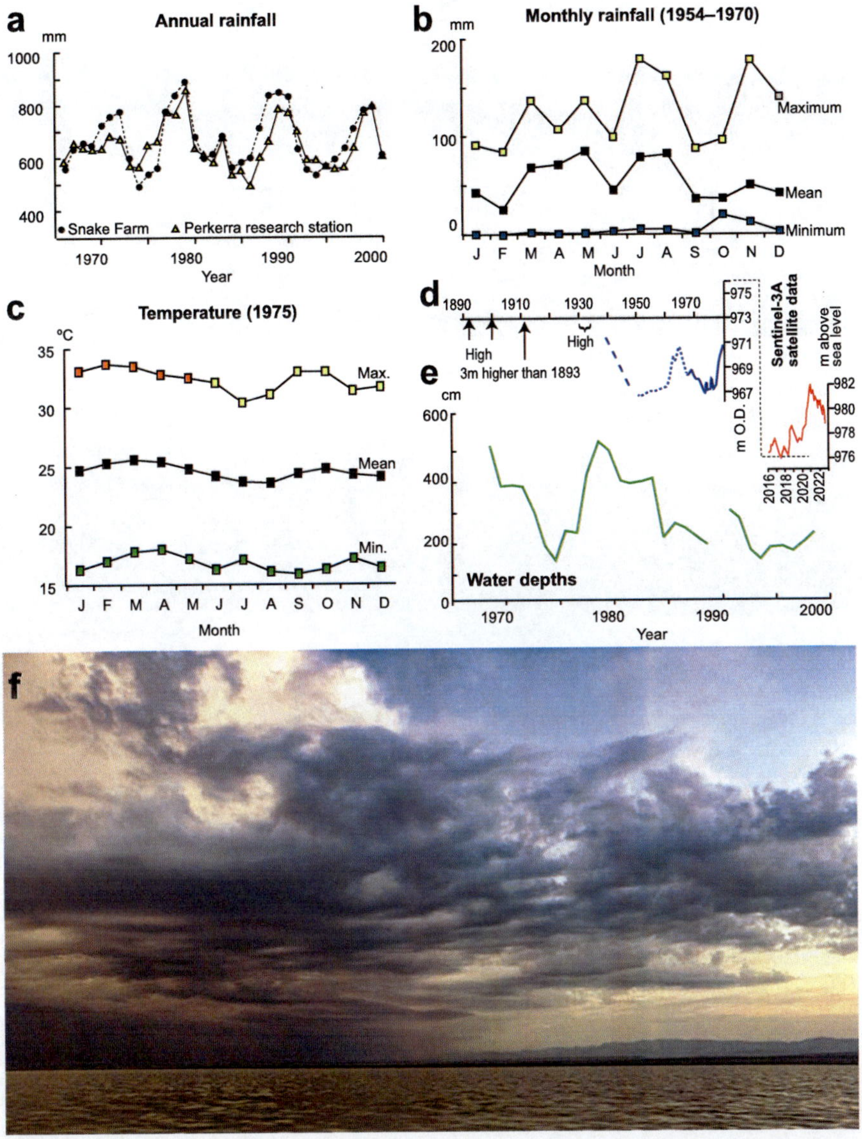

Fig. 8.5 Climate and lake level data for Lake Baringo. (**a**) Three-year running average annual rainfall variability for 1965–2000. Triangles: Perkerra Research Station, Circles: Snake Farm. After Kipkorir (2002). (**b**) Monthly rainfall data for Marigat. After Källqvist (1987). (**c**) Monthly mean temperatures in 1975 for Perkerra Irrigation scheme station. After Mwangi (1983). (**d**) Relative fluctuations in lake level. Long-term variation after Tiercelin and Vincens (1987). Post-2016 based on Sentinal-3A satellite data. (**e**) Water depths from 1970 to 2000. After Hickley et al. (2004). (**f**) Late afternoon clouds over SE Lake Baringo, which commonly build up into thunderstorms (July 2019). Tugen Hills to the right

Fig. 8.6 Lake Baringo. (**a**) General view from northwest of Lake Baringo to Ol Kokwe Island. Karosi volcano forms left background. (**b**) Flooding of hotel facilities at Kampi ya Samaki on mid-western shore, August 2013. (**c**) Coastal inundation along the southern fringe of Lake Baringo, August 2013. (**d**) Shoreline flooding along northern Lake Baringo (July 2018). Note tilted, alternating, horsts and grabens cutting volcanics of Karosi Volcano (background). (**e**) Western shoreline with *Typha* swamps and *Acacia* savannah, backed by the Baringo Trachyte Fault Escarpment and Tugen Hills. (**f**) Subaqueous macrophytes (*Ceratophyllum demersum*) are common in shallow littoral zones

The lake has no surface outflow and consequently shows significant changes in level (Fig. 8.5d, e) that result from variable evaporation-precipitation balances (Ngaira 2006; Walumona et al. 2022). In recent decades, the surface level has ranged between about 965 and 973 m above sea level with the surface area varying from ~108 to 200 km^2 (Fig. 8.2d, e; Tiercelin and Vincens 1987; Renaut et al. 2000, 2002; Odour et al. 2003; Odada et al. 2006; Ouma and Tateishi 2006; Kiage and Douglas 2020). Mean water depth has been variously estimated at ~5.6 m in the 1960s, ~8 m in the late 1970s, ~3 m in 1994, 4.5 m in 1998, 2.65 m in 2003 (Hickley et al. 2004),

8.6 m in 1975 and 2.1 m in 2001 (Johansson and Svensson 2002). In recent years (2013–2021), much of the shoreline has flooded resulting in the lake area increasing by ~50% compared with mean values for 1984–2009 (Herrnegger et al. 2021; GoK and UNDP 2021; Aura et al. 2023).

Kiage and Liu (2009b) used pollen and spore data from cores to infer drought conditions at Lake Baringo between 1750 and 1830 and noted that an absence of *Sporormiella* suggested a lack of large herbivores in the area. Bessems et al. (2008) also reported a severe drought in the first two decades of the nineteenth century, followed by a rapid recovery after about 1826. High lake levels may have occurred in 1893 (Fig. 8.5d; Tiercelin and Vincens 1987), around 1900 (Johnston 1902a, b), and again in 1911 (Gregory 1921) with the lake also reaching high levels in the early 1960s (Grove 1974; Nicholson 1998). The lake stood at about 965–973 m between 1950 and 1979 with a general decline from 1978 through to 2006 (Renaut and Owen 1980; Tarits et al. 2006) that was interrupted by brief highstands, notably in 1997–98. Recent very high levels after 2010 have had considerable impacts on shoreline communities (Fig. 8.6a–d; Onywere et al. 2013; Obando et al. 2016; Muia et al. 2021; Victor et al. 2023).

8.4 Peoples, Vegetation and Wildlife

Early reports about the inhabitants of the Baringo District include the works of Hobley (1906) and Dundas (1910) with Anderson (2002, 2016) summarising how environmental change affected people in the region. Immigration and natural growth resulted in a population increase from 4.4 to 44 persons/km^2 between 1948 and 1999 in the area near Lake Baringo (Johansson and Svensson 2002) with population density continuing to increase to about 72 people/km^2 by 2008 (Olago et al. 2009). According to a 2009 census the population was 555,561. Three groups of local pastoral and agro-pastoral peoples live in the region (Hodder 1977; Thom and Martin 1983; Hartley 1986; Homewood and Lewis 1987; Little 1996; Lewis 1998). The Pokot (~35%), a Kalenjin Nilotic group, are located to the north. The Tugen (~53%), also Kalenjin, occupy hill country to the west (Odada et al. 2006). The Njemps or Il Chamus (~12%), to the south and east of Lake Baringo, are related to the Samburu and speak a variety of Kimaa (Burnett and Rowntree 1990). Archaeological research by Petek (2018) indicated that the basin has been part of the pastoralist cultural sphere for at least 3000 years and may have been a frontier between different cultures, with the Il Chamus settling in the area around the late eighteenth to early nineteenth century and practising limited irrigation from time to time depending on climate changes (Petek-Sargeant and Lane 2021). Fishing is a major resource for local people with the main centre of activity located at Kampi ya Samaki on the mid-western shoreline (Fig. 8.3).

Several of the first Europeans to visit the area described it as one of the most dependable sources of grain (Little 1996; Johansson and Svensson 2002). However, the modern vegetation has been significantly degraded due to pastoral and agricultural activities (Arap-Chellal 1969; Burnett and Rowntree 1990; Snelder and Bryan 1995; Kiage et al. 2007; Lwenya and Yongo 2010; Boles et al. 2019; Mwasi 2021; Chebungei et al. 2022). The drainage basin, for example, lost about 50% of its natural forest cover between 1976 and 2001, decreasing from 829 to 417 km^2 (Odada et al. 2006). The invasive tree *Prosopis juliflora* was introduced in the 1980s to help offset the effects of deforestation, but its spread has since resulted in the loss of agricultural and pastoral land to the south and southwest of Lake Baringo (Mwangi and Swallow 2008; Ng et al. 2017; Castillo 2021).

Loss of wildlife through hunting and competition led to a reduction in grass and an expansion of bush and *Acacia* (Vehrs and Heller 2017; Boles et al. 2019). Lake Bogoria pollen document a decrease in grass from 50% to 18% since 1910 with expansion in woodland, Amaranthaceae and Asteraceae (van der Plas et al. 2019). Today, the area near Lake Baringo is dominated by savannah with common bushland and *Acacia* (*Vachellia*) thickets that include *A. reficiens*, *A. senegal* and *A. mellifera* with *A. tortillis* along water courses (Fig. 8.7; Tiercelin and Vincens 1987). This habitat gives way at higher altitudes to a denser *Acacia* savannah and then a sparse Combretaceae forest with *Combretum molle*, *Terminalia brownii*, *Tarchonanthus camphorates* and *Dodonaea viscosa*. The higher mountains are characterised by coniferous forests that include areas with *Juniperus*, *Podocarpus* and *Olea*. Kiage and Liu (2009b) also reported the presence of *Acacia* and *Commiphora* with little undergrowth near the lake. Swamps are characterised by *Cyperus papyrus*, *Typha domingensis* and *Typha angustifolia*, mainly to the south of the lake (Fig. 8.6e; Muasya et al. 2006; Tarras-Wahlberg et al. 2003), with aquatic macrophytes common (Fig. 8.6f) (Omondi et al. 2016). In addition, a variety of trees are interspersed with flowering plants and grasses at higher elevations where diverse woodlands with *Celtis* spp., *Urticaceae*, *Myrtaceae*, *Croton*, *Holoptelea*, *Prunus*, *Podocarpus*, *Acacia* (*Vachellia*), and *Olea* are also found (Kiage and Liu 2009b).

The region once contained a very abundant and diverse fauna (Thomson 1887; Chapman 1908; Akeley 1931), but most of the large mammals have been lost due to agricultural expansion, habitat degradation and past hunting. Nevertheless, the lake and its surroundings remain an important home for crocodiles, hippopotamus, monitor lizards, snakes and insects (Fig. 8.8a–e) with adjacent trachyte cliffs supporting hyrax. Attempts were made in 2013 to reintroduce giraffes to areas east of the lake where small gazelles are also located. Five species of fish are common and two may have disappeared (Aloo 2002). Increased concentrations of DDE have been reported in fish from Lake Baringo due to the use of agricultural pesticides (Lincer et al. 1981). The region also remains an important location for birds, which are abundant and diverse, with over 500 species reported, including, for example, cormorants, herons, kingfishers, fish eagles and pink-backed pelicans (Fig. 8.8f–k; Hartley 1986; Lincer et al. 1981).

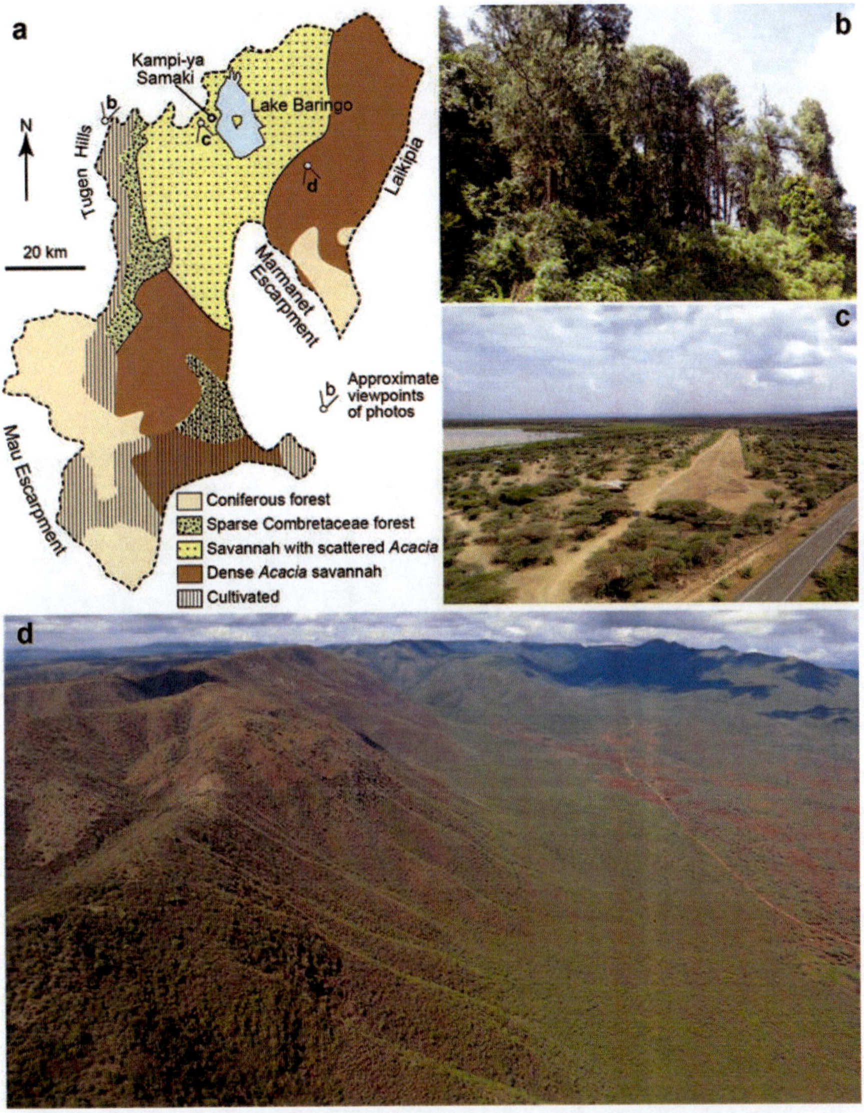

Fig. 8.7 Vegetation in the Baringo District. (**a**) Distribution of major vegetation types in the Baringo Catchment. Based on Tiercelin and Vincens (1987). (**b**) High altitude forest, Tugen Hills. (**c**) Savannah with scattered *Acacia* (*Vachellia*), Kampi ya Samaki Airstrip. (**d**) Dense *Acacia* (*Vachellia*) savannah. Ol Arabel catchment with Laikipia border fault and Marigat to Karandi road

Fig. 8.8 Wildlife in and near Lake Baringo. (**a**) Nile Crocodile. (**b**) Hippopotamus. (**c**) Carpet Viper. (**d**) Scorpion. (**e**) Monitor Lizard. (**f**) Fish Eagle. (**g**) Pied Kingfisher. (**h**) Great Cormorant. (**i**) Grey Heron. (**j**) Little Weaver. (**k**) Pink-backed Pelican

8.5 Limnology

8.5.1 Physical Limnology

The shallow lake floor slopes gently from sites of deltaic input along the southern, southwestern and eastern margins, and is deepest in the northwest quadrant where structural control of bathymetry is evident (Fig. 8.2c). The lake marginal slopes are steep where bordered by volcanic rock, including Ol Kokwe Island.

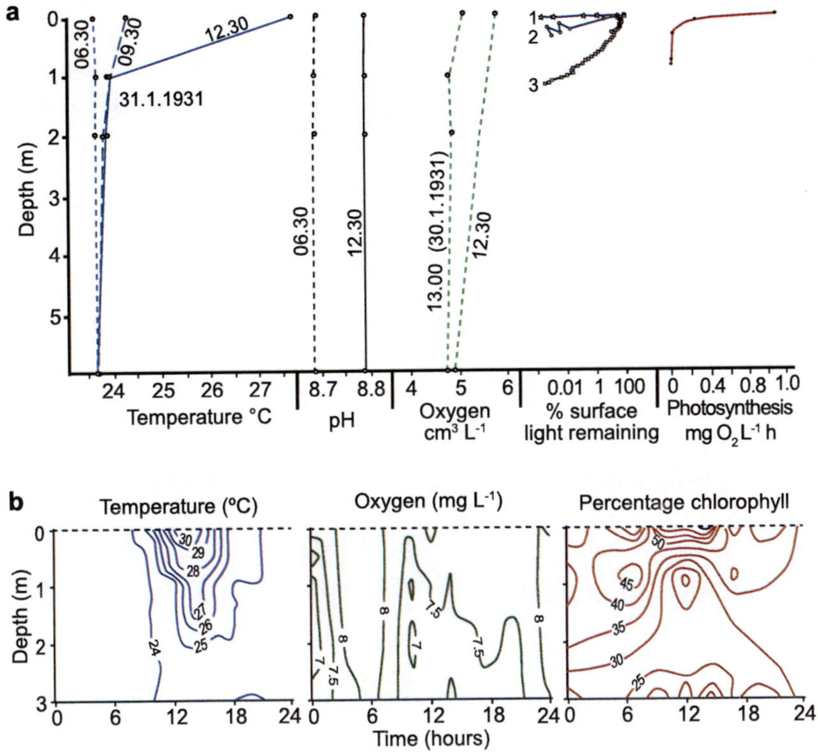

Fig. 8.9 Physical limnology of Lake Baringo. (**a**) Temperature, pH and oxygen records. After Beadle (1932). Light levels for August 2003. After Hickley et al. (2004). Photosynthesis rates after Källqvist (1987). (**b**) Depth-time distributions for temperature, oxygen and chlorophyll. After Patterson and Wilson (1995)

A temporary thermocline develops on calm days at depths of about 0.2–0.5 m with bottom waters up to 4 °C cooler than at the lake surface (Oduor et al. 2003; Källqvist 1987). However, stratification rapidly breaks down to a holomictic condition during the late afternoon (Fig. 8.9a) in response to wind-induced mixing and reduced insolation. Similarly, the lake waters have a generally uniform pH (8.7–8.8) and are well oxygenated down to the lake floor where oxygen concentrations generally exceed 6 mg L^{-1} (Fig. 8.9b; Beadle 1932). The lake is usually calm but small swells often develop during late afternoon.

The lake is very turbid (350–900 NTU [nephelometric turbidity units]; Odada et al. 2006) due to suspended silt and clay, and contrasts strongly in colour with neighbouring Lake Bogoria on satellite imagery. Secchi disc visibilities range from 3.5 to 20 cm (Beadle 1932; Källqvist 1987; Hickley et al. 2004) and the lake has a mean euphotic depth of about 18 cm (Oduor et al. 2003). High siltation rates of ~315–350 tonnes y^{-1} (Aloo 2002) have been attributed to overgrazing, deforestation and consequent soil erosion (Kamar 1992; Snelder 1993; Lwenya and Yongo 2010;

Boitt 2016). Hickley et al. (2004) estimate that almost 90% of the drainage basin is environmentally degraded. The quantity of suspended sediments carried by inflowing rivers can be very high (50–70 g L^{-1}) during peak discharge in the Kapthurin (Katiorin) basin to the southwest of the lake (Sutherland and Bryan 1990). Snelder and Bryan (1995), for example, documented rapid increases in erosion in the Baringo region when vegetation cover fell below 25%, with soil losses of 140 g cm^{-2} taking place for storms of 60-min duration. On longer time scales, estimates of sediment yield suggest erosion of 1.43 million tonnes y^{-1} from the Perkerra catchment alone (Onyando et al. 2005). Degefa et al. (2014) have noted alternating seasonal spikes for sediment influx in short Baringo cores. They interpret the inputs as being derived from rivers and direct run-off with erosion resulting from increased population and livestock pressures, especially since the 1920s. Observations suggest that the recent deeper waters have sometimes reduced the amount of bottom sediment stirred by wind in the northern part of the lake allowing the water to become clearer despite an increased influx of suspended sediment from the south. Similarly, more sediment is trapped at the Molo Delta as the river mouths retreat southwards.

8.5.2 *Chemical Limnology*

Despite its high evaporation rate and a lack of surface outflow Lake Baringo remains fresh. This apparent contradiction has stimulated debate on possible mechanisms that would maintain low salinities, including salt removal to sediments and pore waters, subsurface seepage losses from the basin, and the possible addition of fresh groundwater (Gregory 1921; Barton et al. 1987; Tarits et al. 2006). Gregory (1921, p. 107) famously reported seeing water "pouring freely into the lavas" along the northeastern shore in May 1893. Several researchers have suggested water loss through lake floor sediments and fractures in volcanic rocks with groundwater moving northwards along the generally declining rift floor (Allen and Darling 1992; Dunkley et al. 1993; Darling et al. 1996). Interestingly, in late 2020 there were press reports (Kangogo 2020) of "an outlet formed at Kibenop-Cherogony rocky ranges near Loruk" (NW Baringo) with waters flowing down fractures in the volcanic rocks of the area, much as Gregory had described more than a century earlier from the northeastern shoreline ….

The lake water has a Na^+–Ca^{2+}–HCO_3^- composition with minor Cl^- and SO_4^{2-} (Tarits et al. 2006) that is similar to the dilute river inflows (Table 8.1; Beadle 1932; Jenkin 1936; Talling and Talling 1965; Tiercelin 1981; Tiercelin and Vincens 1987; Onindo and Mwangi 2012). However, water chemistry varies with lake level. For example, there was a threefold increase in Na^+ and Cl^- between 1961 and 2000 in response to reduced water volume (Oduor et al. 2003). Salinity changes annually and with distance northwards from the Molo River, but generally ranges between 0.5 and 1.5 g L^{-1} TDS (Total Dissolved Solids) and averaging 0.6 g L^{-1} TDS (Tarits et al. 2006). However, mean values exceeded 0.8 g L^{-1} TDS in 1976, 1977, 1984, 1990 and 1996 (Allen and Darling 1992; Dunkley et al. 1993; Darling et al. 1996).

Table 8.1 Selected chemical data for Lake Baringo (mg L^{-1})

Sample	Num.	Date	Ref	T (°C)	pH	Na	K	Ca	Mg	Cl	HCO_3	CO_3	SO_4	F	SiO_2	NH_4	NO_3	NO_2	PO_4	Li	TDS
Rivers																					
Kapthurin	BA3-76	Jul-76	1	21.0	7.10	37	4.2	25.0	3.2	31	139	0	4.2	1.2	57	–	–	–	–	–	302
Molo	BA4-76	Jul-76	1	24.0	7.00	49	2.2	16.0	1.1	31	134	0	1.2	0.5	41	–	–	–	–	–	276
Ndau	BA5-76	Jul-76	1	21.0	7.30	56	1.8	33.0	1.9	35	201	0	1.7	0.8	57	–	–	–	–	–	388
Mukutan	BA7-76	Jul-76	1	21.0	7.00	42	5.7	21.0	4.3	31	139	0	2.3	1.6	35	–	–	–	–	–	282
Molo	E62-150	Oct-78	2	–	6.00	57	0.0	18.0	1.0	35	152	0	0.0	0.6	45	14.0	1.0	0.01	0.6	–	324
Perkerra	E62-151	Oct-78	2	–	6.20	57	0.0	20.0	1.0	35	152	0	0.0	0.7	40	16.0	3.0	0.01	0.7	–	325
Perkerra (Bridge)	PKR1	Aug-96	3	19.7	7.77	16	5.1	8.7	1.9	6	71	0	3.3	0.3	24	–	1.5	–	–	–	138
Ol Arabel	OLA	Aug-96	3	26.1	8.71	70	7.7	11.6	3.8	17	205	0	2.8	0.1	23	–	0.0	–	–	–	340
Molo	MOL1	Aug-96	3		7.90	18	9.8	5.6	1.2	10	63	0	4.3	0.2	31	–	3.2	–	–	–	147
Ndau	ND1	Aug-96	3	25.3	8.17	22	5.2	18.6	3.2	9	117	0	3.3	0.4	31	–	1.5	–	–	–	212
E. Bar	EB1	Aug-96	3	29.7	8.91	145	14.6	12.7	3.6	56	330	0	11.9	0.5	54	–	0.0	–	–	–	627
Mukutan	MUK1	Aug-96	3	22.2	8.47	168	8.6	5.8	2.2	31	381	0	19.0	0.4	39	–	12.0	–	–	–	667
Kutwa	KUT	Aug-96	3	26.4	8.78	141	9.5	12.4	2.6	41	339	0	7.5	0.3	30	–	0.6	–	–	–	584
Tigeri	142	1988	4	–	7.65	11	4.9	7.2	1.4	4	483	0	1.9	0.4	31	–	–	–	–	0.01	544
Irtwa	145	1988	4	–	8.60	91	8.2	5.5	2.3	25	500	0	2.5	0.9	40	–	–	–	–	0.01	676
Kabarmel	146	1988	4	–	8.05	45	10.0	16.8	3.9	26	135	0	8.6	0.4	40	–	–	–	–	0.01	287
Wells																					
Kampi-Ya-Samaki	131d	1991	4		8.15	227	18.5	16.0	13.9	47	637	0	11.4		62	–	–	–	–	0.04	1033
Tangulbei	133	1988	4	30.2	7.70	164	5.4	27.0	21.3	27	556	0	11.7	0.9	51	–	–	–	–	0.01	864
Borehole	C616	Jun-71	9		7.25	120		5.8	33.1		492		16		82	–	–	–	–	–	

(continued)

Table 8.1 (continued)

Thermal springs																					
Maji ya Moto	MJM-1-94	Feb-94	5	36.0	6.80	140	11.6	21.0	2.3	37	384	0	10.0	3.9	72	–	–	–	–	–	682
Loboi S.	LOB-91	Nov-91	5	35.0	6.75	138	12.0	8.1	2.5	14	350	0	9.6	3.9	69	–	–	–	–	–	607
Ol Kokwe	OK-1-94	Feb-94	5	95.0	8.77	905	40.0	1.5	0.2	265	1765	55	38.0	20.0	225	–	–	–	–	–	3315
Ol Kokwe	OK-2-94	Feb-94	5	90.0	8.65	860	38.0	2.0	0.1	260	1860	0	35.0	18.0	195	–	–	–	–	–	3268
Ol Kokwe	OK-3-94	Feb-94	5	91.0	8.71	895	37.0	0.9	0.2	260	1950	20	35.0	17.5	230	–	–	–	–	–	3446
Ol Kokwe	OK-1-95	Jun-95	5	96.5	8.95	925	42.0	1.5	0.7	268	1750	65	38.0	16.5	236	–	–	–	–	–	3343
Ol Kokwe	OK-4-95	Jun-95	5	92.5	8.75	800	36.0	5.5	0.7	235	1420	160	30.0	22.0	170	–	–	–	–	0.65	2880
Ol Kokwe	OK1	Aug-96	3	88.0	6.01	64	9.6	9.3	2.62	9	81	0	100	2.5	134	–	7.6	–	–	0.58	420
Lake																					
Baringo N	216	1989	4	–	8.85	192	14.3	11.0	3.5	46	399	0	17.0	7.1	28	–	–	–	–	0.01	718
Baringo S	217	1989	4	–	8.75	170	14.9	11.3	3.5	44	375	0	15.3	6.3	30	–	–	–	–	0.01	670
Baringo	–	Dec-30	6	–		126	15.0	22.0	2.0	36	342	0	40.3	–	–	–	–	–	–	–	583
Baringo	–	May-62	7	–		95	11.0	11.6	1.1	25	271	0	19.7	–	24	–	–	–	–	–	458
Baringo	–	Sep-69	8	–	8.50	128	12.0	13.6	5.0	31	328	0	39.1	5.2	47	–	–	–	–	–	609
Baringo	E62-011	Aug-76	2	23.5	8.98	320	19.0	20.0	4.0	106	671	0	156	–	–	3.2	83.0	0.10	0.5	–	1383
Baringo	E62-044	Aug-76	2	40.7	8.68	322	17.0	22.0	2.0	89	671	0	123	–	–	2.6	77.0	0.10	0.5	–	1326
Baringo	E61-802	Nov-78	2	–	8.20	134	0.0	20.0	2.0	70	275	15	0	2.8	43	0.9	12.0	0.01	0.3	–	575
Baringo	E61-803	28,795	2	–	8	123	0	26	4	70	275	15	0	0.2	45	1.5	30	0.01	0.3	–	590
Baringo	E61-806	Nov-78	2	–	8.20	135	10.0	14.0	5.0	53	336	0	18.0	5.5	40	0.2	4.0	0.01	0.5	–	621
Baringo	E61-807	Nov-78	2	–	8.20	118	10.0	18.0	1.0	53	305	0	0	5.0	35	0.1	5.0	0.01	0.5	–	551
Baringo	E61-810	Nov-78	2	–	7.60	120	10.0	24.0	4.0	53	336	0	0	3.5	35	0.2	12.0	0.01	0.4	–	598
Baringo	E61-811	Nov-78	2	–	7.20	145	10.0	26.0	1.0	53	336	0	5	6.0	23	0.3	12.0	0.01	0.4	–	618
Baringo	E61-813	Nov-78	2	–	8.20	123	10.0	18.0	1.0	53	305	0	0	3.5	30	0.1	10.0	0.01	0.5	–	554
Baringo	E61-814	Nov-78	2	–	8.00	108	0.0	28.0	1.0	53	275	0	10.0	4.5	30	0.4	17.0	0.01	0.5	–	527

Baringo	E61-821	Nov-78	2	–	8.10	125	20.0	26.0	0.0	53	336	0	20.0	4.5	35	1.1	35.0	0.01	0.1	–	656
Baringo	E61-822	Nov-78	2	–	8.10	128	10.0	20.0	5.0	53	336	0	5.0	4.5	30	0.2	12.0	0.01	0.5	–	604
Baringo	E61-825	Nov-78	2	–	8.20	129	10.0	20.0	1.0	53	336	0	0	4.5	40	0.2	12.0	0.01	0.5	–	606
Baringo	E61-826	Nov-78	2	–	8.10	115	20.0	28.0	2.0	53	336	0	5.0	4.5	34	0.5	10.0	0.01	0.4	–	608
Baringo	E62-188	Nov-78	2	25.6	7.00	118	10.0	16.0	4.0	35	336	0	0	5.5	35	0.2	4.0	0.01	0.4	–	564
Baringo	E62-189	Nov-78	2	20.4	6.90	113	10.0	18.0	0.0	53	275	0	5.0	5.5	35	0.4	8.0	0.01	0.4	–	523
Baringo	E62-191	Nov-78	2	24.7	8.90	145	0.0	20.0	2.0	53	244	60	0	1.4	30	0.4	4.0	0.01	0.7	–	560
Baringo	E62-192	Nov-78	2	22.3	6.80	145	0.0	20.0	2.0	53	366	0	0	1.5	33	2.0	1.0	0.01	0.4	–	624
Baringo	E62-195	Nov-78	2	–	6.70	135	10.0	16.0	4.0	53	336	0	10.0	4.5	40	2.0	2.5	0.01	0.3	–	613
Baringo	E62-196	Nov-78	2	–	8.20	123	20.0	20.0	0.0	53	336	0	0	5.0	40	0.1	12.0	0.01	0.5	–	610
Baringo	E62-198	Nov-78	2	–	8.20	135	10.0	16.0	2.0	53	336	0	0	6.0	25	0.2	5.0	0.01	0.4	–	589
Baringo	E62-199	Nov-78	2	–	8.20	137	10.0	16.0	4.0	53	183	90	0	4.0	35	0.1	3.5	0.01	0.4	–	536

1 = Renaut (1982); 2 = Tiercelin (1981); 3 = Tarits et al. (2006); 4 = BGS; 5 = original; 6 = Beadle (1932); 7 = Talling and Talling (1965); 8 = Kilham (1971); 9 = Institute of Hydrology (1985)

All reported TDS values have been less than those reported for thermal springs discharging on Ol Kokwe Island (Dunkley et al. 1993). Ochuka et al. (2019b) noted that TDS values were high at inlets and near islands compared to rivers.

Conductivity values of 980–1200 μS cm^{-1} were recorded in 1976–1977 with alkalinities of 8.7–11.9 meq L^{-1} (Källqvist 1987). Conductivity fell to 520 μS cm^{-1} in 1978 with alkalinity reduced to 4.4 meq L^{-1} due to increased inflow from rivers. Oduor et al. (2003) documented a mean conductivity of 1222 μS cm^{-1} and noted that Baringo waters ranged in pH from 8.0 to 10.5. Other studies have documented lower pH values of 7.97–8.94 for different locations in the lake (Table 8.1; Tarits et al. 2006; Omondi et al. 2014). Total carbonate and bicarbonate alkalinity averages ~10.1 meq L^{-1}, with the carbonate component being small (~1.7 meq L^{-1}) (Oduor et al. 2003).

Oduor et al. (2003) noted mean total phosphorus of 365 μg L^{-1}, low nitrate-N (mean = 169 μg L^{-1}) and low silica (mean = 15 μg L^{-1}) concentrations. Later research (December 2010) documented spatial variability in the surface water chemistry with mean (range) values of dissolved oxygen, pH and electrical conductivity of 6.9 (4.5–8.4) mg L^{-1}, 8.3 (7.8–8.5), and 573 (556–601) μS cm^{-1}, respectively. Data for water transparency and turbidity indicate high values of 33 (28–37) and 43.3 (32.7–54.6) NTU (Ouma and Mwamburi 2014). Hennesey et al. (2020) also reported that turbidity tended to develop in freshwater rift lakes, such as Baringo, with moderate pH and low salinity, which reduced flocculation and particle settling. Mean and range values for plant nutrients in the lake include: total nitrogen (788.4; 278–4486 μg L^{-1}), nitrate-nitrogen (4.5; 2.4–10.0 μg L^{-}), ammonia nitrogen (42.6; 33.8–56.3 μg L^{-1} NH_4-N), total phosphorus (102.9; 20.3–585.3 μg L^{-}) and soluble reactive phosphorus (23.5; 15.2–30.5 μg L^{-}) with silica ranging between 19.7 and 32.7 mg L^{-1}. Walumona et al. (2021) also report data for NO_3^- (9.62–13.1 μg L^{-1}), PO_4^{3-} (18.42–24.1 μg L^{-}), NH_4^+ (20.94–32.47 μg L^{-}) and other chemical data from several sites in Lake Baringo for the period 2008–2020.

A general evaporative concentration factor of about 2–4×, compared with fluvial sources, has been estimated (Renaut et al. 2000), with suggested depletion in Ca^{2+}, HCO_3^- and SiO_2 due to losses through calcite precipitation and deposition of siliceous organisms (diatoms, sponge spicules). In contrast, there is enrichment in HCO_3^-, as well as Cl^-, SO_4^{2-}, F^-, Na^+ and the trace elements B, V, Cr, As, Mo, Ba and U in samples located at some distance from the major Molo and Perkerra river inputs (Tarits et al. 2006). Other elements generally show no enrichment.

Two broad zones have been identified in terms of water composition (Tarits et al. 2006). Southern Lake Baringo is characterised by relatively low suspended sediments (440 mg kg^{-1}) with low (~380 mg L^{-1}) and seasonally variable TDS. Possible etched calcite was documented from these areas. In contrast, suspended sediment values in the northern parts of the lake range up to 820 mg kg^{-1} and show strong evaporation effects with TDS of up to 1400 mg L^{-1}, which is about eight times greater than river inflows. Suspended sediment particles are also smaller (<5 μm) in the north and may include authigenic clay or clay coatings on other grains.

8.5.3 *Hot Springs*

Hot springs are present on the shoreline of Lake Baringo and at several sites in the drainage basin. Their overall impact on lake water composition, however, is minimal because their discharge volume is extremely small compared to that of dilute inflow. Fault-aligned springs are present on the northeastern peninsula of Ol Kokwe Island (Dunkley et al. 1993). Two groups are present at Soro. The first comprises small (<1 m diameter) Na^+–HCO_3^-–Cl springs (81–96 °C) with a salinity of 3 g L^{-1} and pH of 8.7 that discharge at, or slightly above or below, the modern rocky shoreline, depending on lake level (Fig. 8.10a, b; Renaut et al. 2002). Some springs precipitate minor calcite and opaline silica. In places, spring silica has cemented beach gravel. The opaline silica contains diatoms and trace fossils that confirm soft silica precipitation from sub-lacustrine springs (Renaut et al. 2002; Buatois et al. 2017). McCall (1967) suggested that the springs formed from steam-heated (dilute) lake water. However, this does not explain their high chloride content, which implies a deep reservoir fed by groundwater and lake water. Fluid geothermometry studies imply a possible hydrothermal reservoir at 300–900 m depth with temperatures of ~170–200 °C (Glover 1972; Allen and Darling 1992; Dunkley et al. 1993; Ogola et al. 2012). A second group of dilute (<1 g L^{-1}) springs is located a few metres landwards and slightly higher near H_2S-bearing fumaroles and hot steaming ground (Fig. 8.10c–e; Renaut et al. 2002). These may have formed as condensate following

Fig. 8.10 Lake Baringo hot springs and fumaroles at Soro (NE Ol Kokwe Island). (**a, b**) Nearshore springs with cyanophyte growth around margins in (**b**). (**c**) Mud pot. (**d**) Fumarole area with reduced vegetation cover, northern embayment. (**e**) Sulfur deposits at (**d**)

steam separation and are characterised by very low chloride (<10 mg L^{-1}), high sulfate (up to 1500 mg L^{-1}) and a pH of 6–7 (Tarits et al. 2006). These slightly acidic fluids locally alter the volcanic rocks to smectite, kaolinite, illite and alunite and precipitate native sulfur.

Hydrothermal activity has also been documented elsewhere in the Baringo Basin (Ogola et al. 2011). Mpawenayo and Mathooko (2004), for example, described hot springs with temperatures of up to 90 °C on what was the mid-eastern shore at the time of their study. Steaming ground with hydrothermal alteration is also present where the midwestern shoreline intersects the POKTZ trend. Other springs discharge at Releng. In addition, a borehole drilled in 2004 at Chepkoiyo (6.5 km west of Lake Baringo) penetrated shallow hot water (98 °C) that briefly erupted to a height of >20 m (Simiyu et al. 2011) forming an artificial geyser until sealed. Geophysical data from the area suggested that a heat source about 6000 m below sea level (mbsl), with a high resistivity layer present at about 1000 mbsl, which was interpreted as part of a reservoir zone. To the north of the lake, fumaroles have been described from the flanks and summit of Karosi volcano (Dunkley et al. 1993; Ochieng et al. 2012; Ogola et al. 2012; Njue 2015). To the south, warm springs discharge from the channel floor of the Loboi River at Maji Moto (McCall 1967; Griffiths 1977), Arus (McCall 1967; Mulwa et al. 2006; Karingithi and Wambugu 2010) and at Loboi Swamp (Sect. 14.8).

Seurey et al. (2016) noted that basaltic dykes and sills may be possible heat sources for hydrothermal activity in the Baringo area. Groundwater samples from boreholes across the region show correlations between F^- and Mg^{2+}, Ca^{2+} and HCO_3^- ions (Mutonga 2014). High F/Cl ratios have also been used to infer previously unknown subsurface hydrothermal inputs to Lake Baringo (Tarits et al. 2006), with REE data suggesting that the hydrothermal fraction of the lake water ranged between 0.04 and 0.47, much of which probably followed fault systems related to fractures that also guide hot springs at Lake Bogoria to the south.

8.6 Biology and Ecology of the Lake

8.6.1 Phytoplankton and Macrophytes

Submerged macrophytes are generally sparse on the lake floor (Aloo 2002). However, swamps to the south contain *Utricularia* sp., *Ceratophyllum demersum* (Fig. 8.6f) and *Najas minor* (Beadle 1932) with floating *Pistia stratiotes* in wind-sheltered bays (Källqvist 1987). Mats of free-floating water hyacinth have recently appeared on the lake surface.

Chlorophyll-*a* concentrations of 47 and 43 μg L^{-1} have been documented from the open lake surface (Källqvist 1987; Yasindi and Taylor 2006) with values falling to 19 μg L^{-1} at 0.5 m depth. Primary production in the open lake is generally low because of the abundant suspended sediment that produces an aphotic to photic zone ratio of 19 (Schagerl and Oduor 2003). Biomass values range from 1.5 to 8.2 mg L^{-1} (Ballot et al. 2003). Richardson and Richardson (1972) reported a value of 1.6 g O_2 m^{-2} day for photosynthetic oxygen production in 1969, with Källqvist (1987) recording values of <1.0 mg O_2 L^{-1} h^{-1} (Fig. 8.9a).

The turbid lake waters are brown but may develop a greenish tinge that has been attributed to *Microcystis aeruginosa*, which dominates the phytoplankton (Jenkin 1929; Rich 1932; Copley 1948; Tiercelin and Vincens 1987; Aloo 2002; Schagerl and Oduor 2003; Odada et al. 2006; Gichuki et al. 2006; Olilo et al. 2014, 2023) and has been linked to relatively high cyanobacterial toxin levels (microcystin = 0.08–3.25 µg L^{-1}; anatoxin = 0.05–0.2 µg L^{-1}; Kotut et al. 2006). *Microcystis aeruginosa* is also often colonised by *Pseudanabaena* sp. and *Nitzschia* sp. (Ballot et al. 2003). Other taxa present include *Microcystis holsatica* (=*Aphanocapsa holsatica*) *Anabaena circinalis*, *Anabaena flos-aquae* (=*Aphanizomenon flosaquae*), *Anabaenopsis arnoldii*, *Lyngbya autumnale* (=*Phormidium autumnale*), *Phormidium mucicola* (=*Pseudanabaena mucicola*), *Pediastrum simplex* and *P. duplex*, as well as *Botryococcus* spp., *Cryptomonas* sp. and *Euglena* sp.

Diatoms are scarce and dominated by *Aulacoseira granulata* and *Cyclotella* sp. (Källqvist 1987; Ballot et al. 2003; Schagerl and Oduor 2003). Other studies have documented the dominance of planktonic diatoms by *A. granulata* var. *angustissima* (and f. *curvata*) with *Thalassiosira rudolfii* (3.4%), minor *Cyclotella* (*Stephanocyclus*) spp., *Cymbella* sp., *Fragilaria* sp., *Stephanodiscus* sp. and *Synedra* sp. also present (Gasse 1987). Littoral assemblages from the northern lake are dominated by *Achnanthes minutissima* (*Achnanthidium minutissimum*) or *Nitzschia communis* and *Cyclotella* (*Discostella*) *pseudostelligera*. Diatoms in superficial lake mud differ from the modern plankton in containing *Aulacoseira agassizii* and *Stephanodiscus astraea*. Pollen and spores of Pteridophytes are minor in superficial sediments, whereas gymnosperms are common (Podocarpaceae). Spores of lower plants (mushrooms, bryophytes) are also present.

Springs at Soro support mats composed of *Oscillatoria*, *Phormidium*, *Pseudoanabaena*, *Synechococcus* and *Calothrix* and diatoms (*Nitzschia*, *Navicula* and *Anomoeoneis*) (Renaut et al. 2002). Mpawenayo and Mathooko (2004) also reported the presence of diatom floras (*Eunotia*, *Achnanthes*, *Craticula*, *Anomoeoneis*, *Caloneis*, *Gomphonema*, *Navicula*, *Neidium*, *Nitzschia*, *Pinnularia* and *Rhopalodia*) in springs along a former eastern shoreline.

8.6.2 Zooplankton and Benthic Invertebrates

Early research on zooplankton documented the presence of *Diaptomus*, *Cyclops*, *Moina*, *Daphnia*, *Ceriodaphnia* and *Diaphanosoma* (Beadle 1932) with seven species of rotifer (Pejler 1974). Abundant *Daphnia barbata* have been reported together with less common *Daphnia laevis* and *Daphnia pulex* in Baringo sediments (Mergeay et al. 2005, 2006). More recently, Omondi et al. (2011) noted that the zooplankton community includes 31 taxa, with 21 belonging to the Rotifera and only two copepod species (*Thermocyclops consimilis* and *Thermodiaptomus galebi*). Cladocera were represented by eight taxa, with *Diaphanosoma excisum* as the most common taxon. Ciliates are dominated by *Cyclidium* with less abundant *Cristigera* and other species (Yasindi 2001).

Ostracods are widespread in sediments from waters <2 m deep (Fig. 8.11a), but scarce in other samples, possibly due to sampling methods or carbonate dissolution (Tiercelin and Vincens 1987). *Limnocythere africana* was common with other taxa including *Hemicypris* cf. *kliei* and species belonging to *Darwinula*, *Stenocypris*, and *Cypris*. Sponges are also present, including *Dosilia brouni* and a *Radiospongilla* sp. (Tiercelin and Vincens 1987; Vacelet et al. 1991).

The diet of the introduced, benthic-feeding fish *Protopterus aethiopicus* is dominated by molluscs with a mean composition of 94.3% in studies of stomach content (Omondi et al. 2013). However, molluscs have been reported as present only in low abundances in and on the lake floor, except in littoral areas (Aloo 2002).

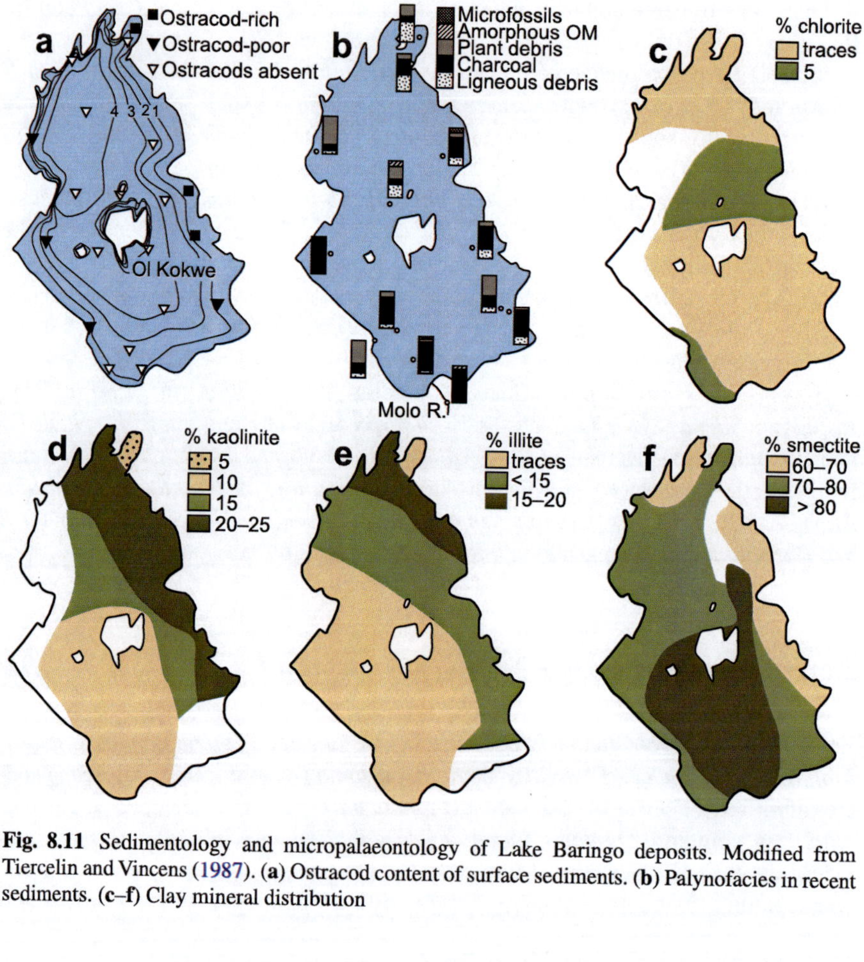

Fig. 8.11 Sedimentology and micropalaeontology of Lake Baringo deposits. Modified from Tiercelin and Vincens (1987). (**a**) Ostracod content of surface sediments. (**b**) Palynofacies in recent sediments. (**c**–**f**) Clay mineral distribution

8.6.3 Fish

Local fishermen still use Ambatch canoes (also called il-kadish; Brown 1991; Fig. 8.12a) as part of their subsistence economy, catching catfish, lungfish, tilapia and *Barbus* (Fig. 8.12b–e). However, recent invasions of water hyacinth may cause future problems for the fishery, which has a history of major changes.

Early fish researchers documented the presence of *Tilapia nilotica* (=*Oreochromis niloticus*), *Clarias mossambicus* (*C. gariepinus*), *Barbus gregorii* (*B. intermedius*) and *Labeo cyclindricus* (Worthington and Ricardo 1936) with Copley (1948) presenting a checklist of fish from Baringo and other freshwater lakes. *Oreochromis*

Fig. 8.12 Lake Baringo fishing. (**a**) An ambatch (Papyrus) canoe paddled mainly by Njemps fishermen (Brown 1991). (**b**) *Tilapia*. (**c**) Catfish. (**d**) Lungfish. (**e**) *Barbus*

niloticus was the basis of a commercial fishery, but by the 1980s the size and numbers of these fish had been reduced substantially, partly due to poor fisheries management (Ssentongo 1974, 1995; Aloo 2002; Ogutu-Ohwayo and Balirwa 2006; Britton and Harper 2008; Nyakeya et al. 2020). Small surviving populations of *O. niloticus* have been reported from the Loboi swamp to the south of Lake Baringo (Nyingi et al. 2009; Ndiwa et al. 2014). Aloo (2002) noted that by the time of their study that *Aplocheilichthyes* sp. and *Barbus lineomaculatus* had disappeared with only five common taxa present, including *O. niloticus*, *C. gariepinus*, *B. intermedius*, *L. cyclindricus* and *Protopterus aethiopicus* (Campbell et al. 2003; Ndiwa et al. 2016), which was introduced from Lake Victoria in June 1975 (Mlewa 2003; Mlewa and Green 2004, 2006; Hickley et al. 2008). The guppy *Poecelia reticulata* was reported as rare. Aloo (2002) indicated that *Barbus* and *Labeo* were near extinction due to lake water quality changes, water abstraction for irrigation (and more recently for geothermal developments) and the damming of rivers (Engelen 1983). However, Britton and Harper (2005) suggested that *L. cyclindricus* numbers may have been underestimated due to overreliance on commercial fishery data.

Fish catches had reached a sufficiently high level for a fish processing plant to be opened in 1957 with the lake yielding about 600–1000 t y^{-1} (Hickley et al. 2004). However, a subsequent decline in catches resulted in the closure of the factory by 1989. A positive correlation between fish catches and water transparency has been reported, which implies that efforts to reduce catchment erosion would benefit the fishery (Hickley et al. 2004). Since the 1990s, the fishery has shifted towards the exploitation of the catfish *C. gariepinus* (Nyamweya et al. 2010) and the introduced lungfish *P. aethiopicus* (Britton et al. 2006; Britton et al. 2009). Efforts to improve fish catches include moratoria, with closures between 1993 and 1994 and 2003–2004, but with total catches remaining only 5 tonnes in 2005 (Britton et al. 2009). Walumona et al. (2022) found a direct link between years of high lake-level and increased fisheries landings and attributed this to increased habitat and breeding grounds.

8.7 Sedimentation

The perennial Molo and Perkerra rivers deliver large quantities of suspended sediment (silt, clay) to Lake Baringo in most years, supplemented by several seasonal rivers with flash-flood discharge that can have a major impact on erosion and sedimentation even within 24 h. In the past 50 years erosion and sedimentation have been accelerated by forest degradation and associated erosion in the drainage basin (Oostwoud Wijdenes and Bryan 2001; Renaut et al. 2000; Lwenya and Yongo 2010; Kareri 2018). Feldspathic (locally quartzose) sand and granules are deposited at some deltas from turbid flows with lava gravel beaches accumulating along rocky volcanic shorelines (e.g. northern Ol Kokwe). However, except along the northern shorelines, most littoral areas are sites where mud and abundant organic detritus accumulate. Extensive lake-marginal marshes to the south are floored by mud and brown silty clay that contains gastropods, ostracods (Fig. 8.11a), *Chara* oogonia, diatoms, sponge spicules and macrophyte debris (Renaut et al. 2000). Several palynofacies have been described (Fig. 8.11b; Tiercelin and Vincens 1987) with the allochthonous organic matter attributed to macrophytes and pollen and spores that have been washed or blown into the lake.

Fine-grained feldspathic mud and finely laminated clay dominate on the lake floor becoming gradually finer with distance northwards from the Molo River (Tiercelin et al. 1980; Tiercelin 1981; Tiercelin and Vincens 1987; Renaut et al. 2000). Smectite (60–80%) dominates in the sediments with kaolinite and illite comprising >15% of deposits in eastern and northern areas (Fig. 8.11c–f; Tiercelin and Vincens 1987). Authigenic clay may also be present in suspended sediments of the more saline northern waters (Tarits et al. 2006).

Massive silty clay and thin silt horizons are present in short cores (Fig. 8.13a, b) with sediments becoming slightly coarser near shorelines and towards deltas (Fig. 8.13c). Gas vesicles are common in core muds. Microbial mats and biofilms, gastropods, ostracods, fish bone, sponge spicules and gemmules, diatoms and allochthonous organic detritus are all present. Small carbonate concretions of possible bacterial origin exist in mud east of Ol Kokwe.

Calcite stromatolites and oncoids with well-preserved cyanobacterial filaments and diatoms are present on volcanic substrates along northern shorelines, including parts of Kokwob Murren (Figs. 8.13c and 8.14a–c) and several of the islands (Renaut et al. 2000). The carbonates are confined to a narrow littoral fringe, or 'stromatolite window' (Renaut et al. 2012). During high lake levels, the waters are dilute and become insufficiently saturated with respect to $CaCO_3$ to form stromatolites. Increased turbidity may also limit the depth of the photic zone, further restricting stromatolite growth. In contrast, calcium depletion takes place during periods of low lake levels as Ca^{2+} is removed by the formation of subsurface cements and calcrete in the drainage basin and through other fractionation processes. This, and the inhibition of many cyanobacteria induced by higher salinity and shallowing water, reduces or restricts stromatolite formation during periods of low lake level. Consequently, the stromatolites develop only within a narrow range of lake levels, forming a fringe along the shore.

Metal concentrations are similar to other rift basin lakes (Table 8.2). Wandiga et al. (1983) provided an early baseline study for the waters and sediments in Lake Baringo and concluded that the lake then lacked any significant indicators of anthropogenic pollution. Ochieng et al. (2007) examined additional elements and reported enrichment in Co, Ni and Cu compared with Lake Naivasha. Mwamburi (2015) focussed on three metals observing that concentrations of Cd < Pb < Zn. Githaiga et al. (2021) undertook a broader study of heavy metals in eight rift lakes.

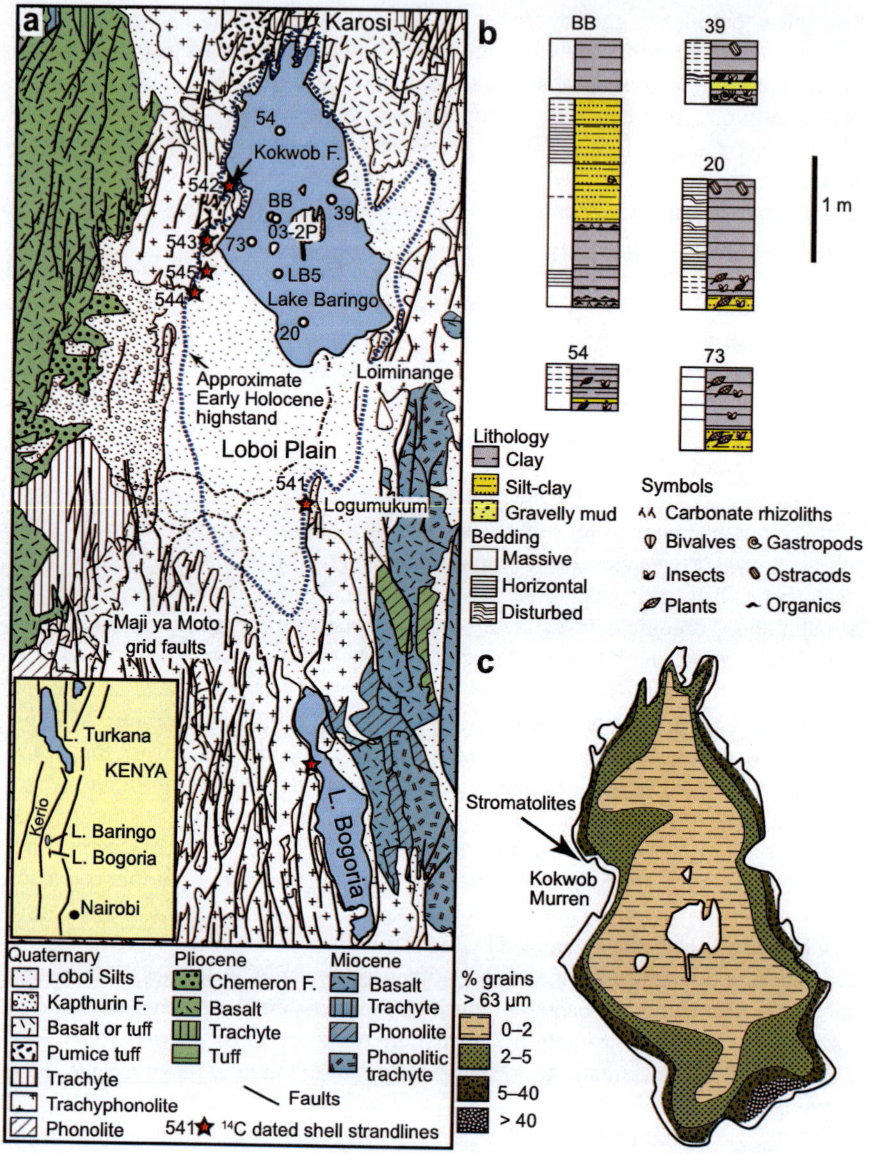

Fig. 8.13 Late Pleistocene–Early Holocene sediments. (**a**) Geology and location of Pleistocene-Holocene high lake sediments. Approximate Early Holocene shoreline marked by a dotted line. (**b**) Shallow cores from Lake Baringo. (**c**) Grain size variation in surficial sediments (after Tiercelin 1981), and location of shoreline stromatolites

Fig. 8.14 Modern and Holocene sediments in and around Lake Baringo. (**a**) Recently formed and lithified calcite stromatolite crusts along the western margin of Lake Baringo, exposed during a period of lowered lake level. (**b**) Sponge encrusting stromatolitic coatings on lava boulder (location in Fig. 8.13c). (**c**) Modern stromatolites forming on lava boulder. (**d**) High-level Early Holocene shells resting on fluvial conglomerate and alluvial silt in river section at Loiminange, SE Baringo. (**e, f**) Shelly (*Melanoides tuberculata*, *Corbicula africana*) Early Holocene strandlines at Logumukum, southeast Baringo

Grewe (2016) described the chemistry of pore waters in three cores from the northern, central and southern parts of Lake Baringo, noting similar trends in all three. Sulfate reduction was comparable to other freshwater lakes (0.5–13 nmol $cm^3 \times d$), with very low NO_3^- levels. She also noted that the total biomarker concentration was the lowest of all lakes in her study (Logipi and Kageinya in the Suguta Valley, Baringo, Bogoria, Naivasha, Oloidien and Sonachi) due to clastic dilution and that stigmasterol was the only biomarker present in the two freshwater lakes (Baringo and Naivasha).

Table 8.2 Metal concentrations in waters and sediments of Lake Baringo

Ref	Ag	Cd	Co	Cr	Cu	Mn	Ni	Pb	Sn	Hg	Zn
Lake waters (μg/L)											
(1)	–	15–16	–	–	6–8	38–40	–	2–18	–	1.0	147–153
(2)	44–185	2–5	5–25.1	125–188	5–20.2	50–280	25–38	25–65.1	300–301	–	40–105
Lake sediment (μg/g)											
(1)	–	1.00	–	–	20.0–21.33	80.0–82.5	–	18.3–18.5	–	0.07–0.09	69.4–72.7
(2)	0.27–0.35	0.57–0.76	0.69–1.38	2.17–4.87	15.12–20.95	942–1464	25.7–39.72	16.57–21.83	23.39–43.41	–	171.5–207.1
(3)	–	1.5–6.5	–	–	–	–	–	15.1–101.1	–	–	59–168

1: Wandiga et al. (1983); 2: Ochieng et al. (2007); 3: Mwamburi (2015)

8.8 Late Quaternary Lacustrine Sediments in the Lake Baringo Basin

Several short cores have been obtained from Lake Baringo (Fig. 8.13a, b; Tiercelin and Vincens 1987), but no long cores have yet provided a full Late Pleistocene-Holocene stratigraphy. Chronological data (^{210}Pb) for a core from the central lake indicate sedimentation rates of ~0.83 to 2.5 cm y^{-1}, depending on ^{226}Ra assumptions (Barton et al. 1987). These rates and data from Cl^- concentrations in pore water have been used to infer a recent closure about 100–200 years ago, although the same data might only indicate changed salinities. Okech et al. (2019) have, for example, provided diatom evidence for fluctuating lake levels and episodes of near desiccation since the late eighteenth century. Core BB (Fig. 8.15a), collected by Barton, represents an ~300-year record with diatom floras indicating three periods of lake contraction and changing salinity (Owen 1981). Zone A, at the top, represents conditions similar to those of today with clay and silt accumulating in a freshwater lake with rare *Aulacoseira granulata*. Zones B and D again indicate fresh conditions with planktonic *A. granulata* and *Stephanodiscus astraea* dominating. In contrast, Zones C and E contain common *Thalassiosira rudolfii*, which indicate elevated salinity and alkalinity, possibly up to 80 meq L^{-1}, which would imply higher salinity and a reduced lake water volume. The basal Zone F represents very shallow and possibly emergent conditions at the core site with epiphytic *Gomphonema parvulum* and *G. lanceolatum* (*G. grunowii*) dominating.

Further data on the sedimentology, magnetic susceptibility, percentage water and organic matter content is available for a second 240-cm-long core (03-2P) recovered close to the BB site (Fig. 8.13a; Bessems et al. 2008). Dating by ^{137}Cs and ^{210}Pb techniques is consistent with a ^{14}C age of 185 ± 30 y BP at 227 cm with the base of the core indicating a desiccation event that has been correlated with a widespread drought in East Africa during the late eighteenth to early nineteenth century

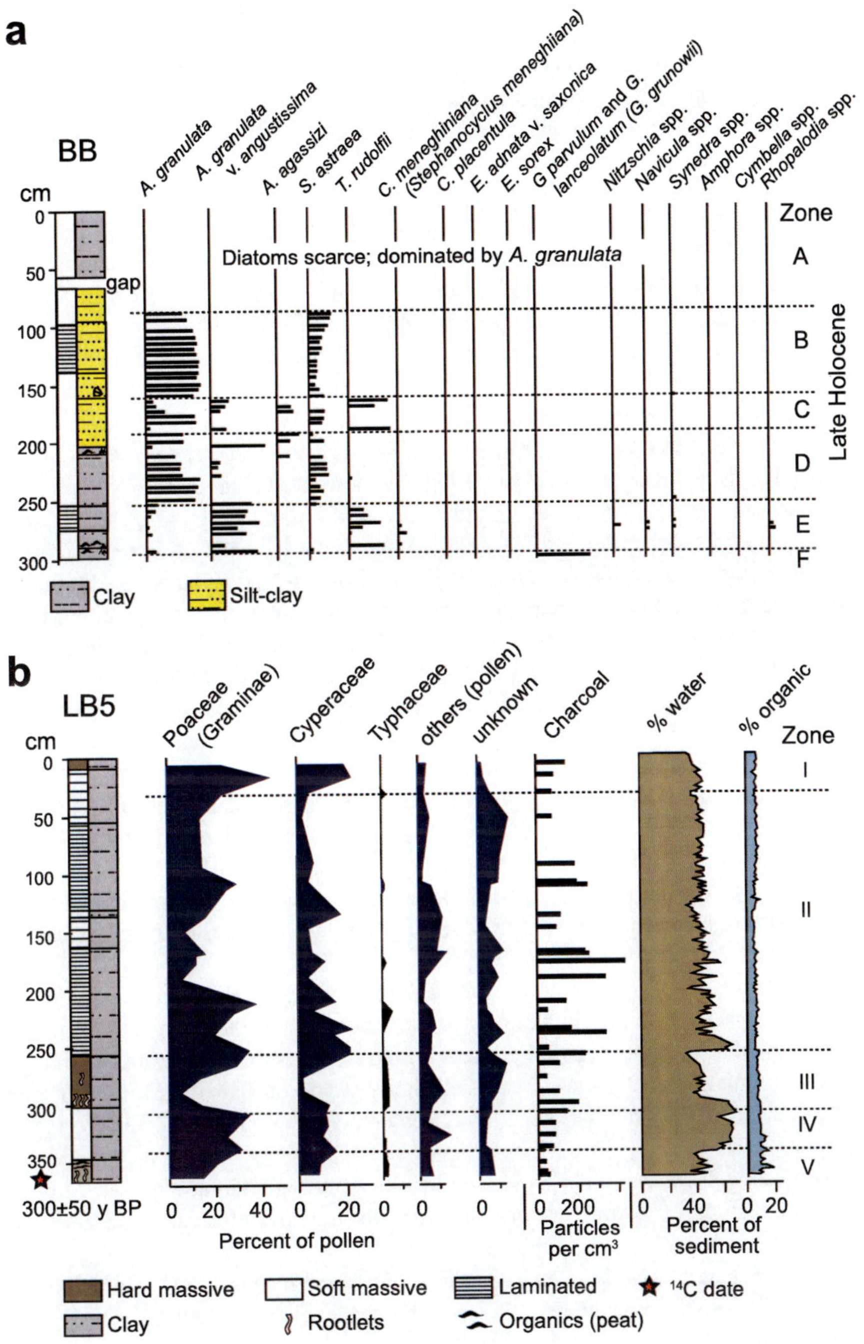

Fig. 8.15 Sediment cores from Lake Baringo. Locations in Fig. 8.13a. (**a**) Core BB with diatom stratigraphy. After Owen (1981). (**b**) Core LB5 with pollen (after Kiage and Liu (2009b) and sedimentological stratigraphies (after Kiage and Liu (2009a)

(Nicholson 1998; Bessems et al. 2008). Dry conditions for this period have also been reported from the Loboi Swamp to the south of Lake Baringo (Ashley et al. 2004, 2013; Driese et al. 2004; Gitau and Verschuren 2016) and at Lake Naivasha in the central Kenya Rift (Verschuren et al. 2000).

A detailed pollen record was obtained for a 350-cm-long core from central Lake Baringo (LB5, Fig. 8.15b; Kiage and Liu 2009a, b) where the basal sediments are radiocarbon dated at 300 ± 50 y BP. Grasses (Graminae) dominate the pollen spectra (20–40%), but fluctuations in other pollen genera indicate variable wet and dry periods. Preservation is poor in the basal Zone V, but it nevertheless contains high percentages of Poaceae, Cyperaceae, Combretaceae, *Podocarpus*, *Olea*, *Tarchonanthus*, *Phyllanthus*, *Acacia* (*Vachellia*) and *Euclea*. *Macaranga*, *Phoenix*, Poaceae, *Acacia*, *Sesbania*, *Euclea* and *Phoenix* increase in Zone IV, with decreases in *Olea*, *Tarchonanthus*, *Phyllanthus* and Combretaceae. *Acacia* (*Vachellia*), *Euclea*, Balanitaceae, Chenopodiaceae/Amaranthaceae, Typhaceae, *Justicia*, and *Phyllanthus* increase through Zone III, with Poaceae and Cyperaceae declining significantly. Cyperaceae vary in Zone II, which shows a gradual increase upwards in the percentages of *Podocarpus*, Cupressaceae, Combretaceae, and *Acacia* (not shown in Fig. 8.15b). *Acacia* (*Vachellia*) pollen is low in the uppermost Zone I, despite abundant *Acacia* in the area today, with Poaceae and Cyperaceae dominant. Pollen, spores, sediments, charcoal content and the water percentage support the evidence for three drier periods in Core LB5, which may have occurred around calendar years 1650, 1720 and possibly 1880 AD (Kiage and Liu 2009a, b).

Middle to Late Holocene environmental records are lacking at Baringo, but Taylor et al. (2005) reported more wooded and humid conditions than the present from the neighbouring Laikipia Plateau to the east, between ~6600 and 1900 y BP with a subsequent change to a fire-modified *Acacia* bushland and grassland at ~1700 y BP. Similarly, Muiruri et al. (2021) documented pollen and charcoal from swamps at Laikipia, noting closed forests between 2268 and 1615 y BP, after which there was a major change to an open woodland/grassland mosaic through to modern times, with this environmental shift reflecting land use changes and an increase in pastoral activity.

In contrast, only a fragmented Late Pleistocene-Early Holocene sedimentary record is preserved on the rift floor by isolated outcrops of the Kokwob Formation to the west of the lake near Kampi ya Samaki, and southeast of the lake at Loiminange and Logumukum (Fig. 8.13a). Dispersed shelly strandlines also mark former lake highstands (Fig. 8.14d–f) (Owen and Renaut 1980; Renaut and Owen 1980; Renaut 1982).

The Kokwob Formation type-section lies in a small graben at Kokwob Murren north of Kampi ya Samaki where the deposits lie disconformably on Baringo Trachyte and volcaniclastic deposits of the Pleistocene Kampi ya Samaki Beds (Owen 1981; Renaut and Owen 1980; Tiercelin and Vincens 1987). The Kokwob deposits (Fig. 8.16a–c) lie at up to 987 m above sea level and include orthoconglomerate, laminated and massive silt, shelly silt and sand, and a sandy coquina dated at

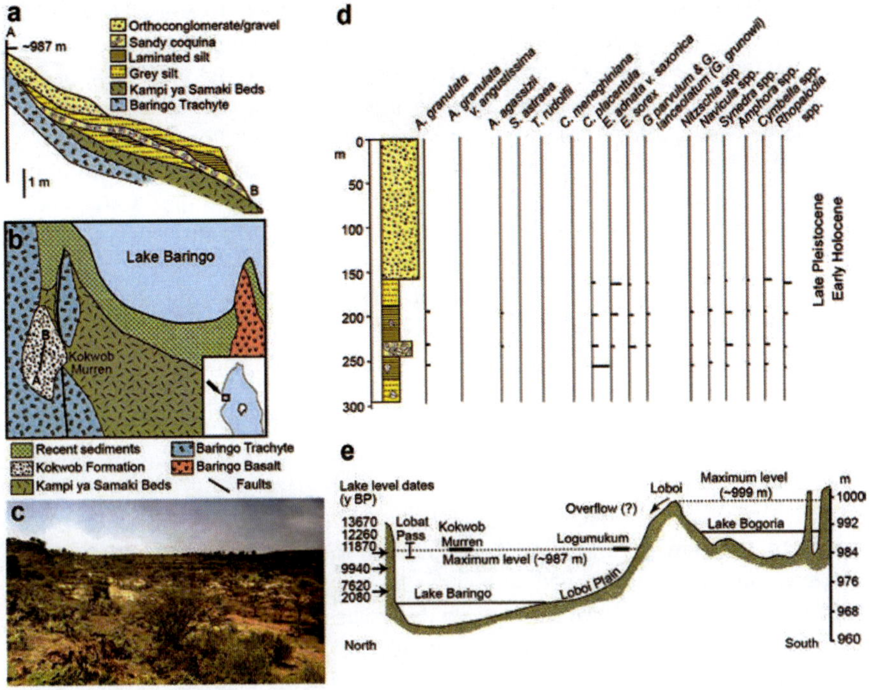

Fig. 8.16 Kokwob Formation sediments and diatoms. (After Owen 1981). (**a**) Section through the Kokwob Formation located in (**b**). (**b**) Geological map of the Kokwob Formation outcrop area, northern Kokwob Murren. (**c**) Kokwob Murren in 1976 viewed from the north. (**d**) Diatom stratigraphy. (**e**) Early Holocene lake levels in the Baringo and Bogoria basins. Contrasting Early Holocene highstands indicate separation of the two lakes

13,850–13,670 years. BP based on uncorrected middle and inner shell fractions, respectively (see below). Those shells are dominated by *Melanoides tuberculata* and *Corbicula africana* together with *Bellamya unicolor*, *Anisus* sp. and *Biomphalaria* sp. *Tilapia* and *Lates niloticus* fish bones are also present. The lowest 1.4 m of the section contains diatoms dominated by *Aulacoseira granulata*, *Cocconeis placentula*, *Epithemia adnata*, *Epithemia sorex* and *Gomphonema parvulum*, indicative of a freshwater lake with abundant epiphytic habitats (Fig. 8.16d).

Middle (MF) and inner (IF) shell fractions from *Melanoides tuberculata* in Kokwob Formation strandlines have been dated at several sites (Williams and Johnson 1976; Young and Renaut 1979), yielding uncalibrated dates of 13,850 (MF) and 13,670 (IF) y BP for Site 542 (Fig. 8.13a) at 15.8 m above the 1973 lake level. A shell from a strandline at Logumukum with a similar elevation (+15.8 m) to the Kokwob Murren locality yielded younger dates of 10,860 and 11,870 y BP (Site 541). Other middle/inner fraction dates have included: 12,600/12,260 (Site 544) at +15.54 m; 10,810/9940 (Site 541) at +10.06 m; and 8460/7620 (Site 543) at

+3.66 m. The contrast in ages for shells at the same height (+15.8 m), the difference between inner and middle shell fractions, and generally older dates than for high levels at other Kenyan lakes imply dating errors related to old inert CO_2, perhaps derived from hot springs (Young and Renaut 1979).

The highest recorded Lake Baringo palaeoshoreline (~987 m; Fig. 8.13a) was about 15 m higher than Lake Baringo in 1973 (then ~972 m?) and ~12 m lower than the maximum known Holocene highstand at Lake Bogoria to the south (~999 m above sea level; Fig. 8.16e), recorded by thin stromatolitic calcite crusts on bedrock at northern Lake Bogoria and by the modern level of the watershed (~1000 m asl) between the two lakes. Those stromatolitic crusts, however, might relate to a later Middle Holocene high lake-level (Vincens et al. 1986) and represent a period after the Sandai fan had aggraded and prograded westwards, helping to separate the Lake Baringo and Lake Bogoria drainage basins. The ^{14}C ages of those stromatolites are insecure because of probable hardwater effects, so the Late Pleistocene-Holocene chronology of lake-level changes in Lake Baringo and Lake Bogoria remains uncertain.

Other potential complications include erosion of higher palaeoshoreline sediments in the Baringo Basin or differential subsidence or faulting between the Baringo and Bogoria basins, although the maximum Late Pleistocene-Early Holocene highstand of Lake Baringo appears to be consistent across that basin (i.e. 987 m at both Kobwob Murren and Logumukum). Another difficulty relates to the amount and timing of groundwater outflow. The Baringo surface outlet at Lobat may not have been the only control on the maximum lake level. Nilsson (1931) discussed possible tectonic influences on former shorelines. Further details of the Baringo-Bogoria linkage question and its complications are discussed in Chap. 9 (Lake Bogoria).

Despite the many uncertainties, the presence of Nilotic fish in Lake Baringo before anthropogenic impacts implies that both Lake Baringo and Lake Bogoria might have been part of a network of lakes and rivers that ultimately fed the Nile via the Suguta and Turkana Basins (Worthington and Ricardo 1936; Owen and Renaut 1980, 1986; Vacelet et al. 1991; Renaut et al. 2000). No clear geomorphological evidence for prolonged overflow at the lowest outlet sill of Lake Baringo, the Lobat Pass (NW Baringo), which lies about 3–7 m above the highest preserved Kokwob strandline has been reported (JAT Young, personal communication; Renaut and Owen 1980). Renaut (1982) noted that the ‘Lobat Pass’ is floored by permeable, pumiceous volcaniclastics (Kampi ya Samaki Beds) which might have acted as a shallow, low-energy ‘soakway’ for waters flowing from an expanded Lake Baringo with minimal suspended sediments and energy for channel incision. Surface drainage (overflow) might have been intermittent, but much of the lake water might have been lost by seepage through underlying fractured volcanic rocks and volcaniclastics at the Lobat Pass and along the northern shoreline. A long core in northwestern Lake Baringo and a shorter core near Loboi village would help to resolve many unanswered questions about the late Pleistocene-early Holocene history of lakes Baringo and Bogoria, and their possible former linkage to Nile drainage.

References

Akeley MLJ (1931) Carl Akeley's Africa. Gollancz, London

Allen DJ, Darling WG (1992) Geothermics and hydrogeology of the Kenya Rift Valley between Lake Baringo and Lake Turkana. British Geological Survey, Research Report SD/92/1

Aloo PA (2002) Effects of climate and human activities on the ecosystem of Lake Baringo, Kenya. In: Odada EO, Olago DO (eds) The East African Great Lakes: limnology, palaeolimnology and biodiversity. Kluwer, Dordrecht, pp 335–348

Anderson DM (2002) Eroding the commons. James Curry, Oxford

Anderson DM (2016) The beginning of time? Evidence for catastrophic drought in Baringo in the early nineteenth century. J East Afr Stud 10:45–66

Andrews P, Banham P (1999) Late Cenozoic environments and hominid evolution: a tribute to Bill Bishop. Geological Society of London, London

Arap-Chellal R (1969) Human problems associated with the Kerio river, Kenya. E Afr Geogr Rev 7:53–60

Ashley GM, Maitima Mworia J, Muasya AM, Owen RB, Driese SG, Hover VC, Renaut RW, Goman MF, Mathai S, Blatt SH (2004) Sedimentation and recent history of a freshwater wetland in a semi-arid environment: Loboi Swamp, Kenya, East Africa. Sedimentology 51:1301–1321

Ashley G, Deocampo D, Kahmann-Robinson J, Driese SG (2013) Groundwater-fed wetland sediments and paleosols: it's all about water table. In: Driese SG, Nordt LC (eds) New frontiers in paleopedology and terrestrial paleoclimatology: paleosols and soil surface analog systems. SEPM Spec Publ 104:7–61

Aura CM, Nyamweya C, Musa S, Ogari Z, Owoko W, Osore M, May L, Njiru JM (2023) The quantification of the extent of flooding on selected major Afrotropical lakes to guide management implications. Front Environ Sci 11:1062289

Ballot A, Pflugmacher S, Wiegand C, Kotut K, Krienitz L (2003) Cyanobacterial toxins in Lake Baringo, Kenya. Limnologica 33:2–9

Barton CE, Solomon DK, Bowman JR, Cerling TE, Sayer MD (1987) Chloride budgets in transient lakes: lakes Baringo, Naivasha, and Turkana. Limnol Oceanogr 32:745–751

Beadle LC (1932) The waters of some East African lakes in relation to their fauna and flora. J Linn Soc Lond Zool 38:157–211

Bessems I, Verschuren D, Russell J, Hus J, Mees F, Cumming B (2008) Palaeolimnological evidence for widespread late 18th century drought across equatorial East Africa. Palaeogeogr Palaeoclimatol Palaeoecol 259:107–120

Bishop WW, Chapman GR (1970) Early Pliocene sediments and fossils from the northern Kenya Rift Valley. Nature 226:914–918

Bishop WW, Chapman GR, Hill A, Miller JA (1971) Succession of Cainozoic vertebrate assemblages from the northern Kenya Rift Valley. Nature 233:389–394

Boitt MK (2016) Impacts of Mau Forest catchment on the Great Rift Valley lakes in Kenya. J Environ Prot Ecol 4:137–145

Boitt M, Gathoni J, Kaelo D, Koech L (2022) GIS-based multi-criteria evaluation to identify areas for soil and water conservation in lower Lake Bogoria landscapes, Baringo County, Kenya. J Geosci Environ Prot 10:64–92

Boles OJC, Shoemaker A, Mustaphi CJC, Petek N, Ekblom A, Lane PJ (2019) Historical ecologies of pastoralist overgrazing in Kenya: long-term perspectives on cause and effect. Hum Ecol 47:419–434

Britton JR, Harper DM (2005) Assessing the true status of the fish species *Labeo cylindricus* (Peters 1868) (Teleostei: Cyprinidae) in Lake Baringo, Kenya. Afr J Aquat Sci 30:203–205

Britton JR, Harper DM (2008) Juvenile growth of two tilapia species in lakes Naivasha and Baringo, Kenya. Ecol Freshw Fish 17:481–488

Britton JR, Ng'eno J, Lugonzo J, Harper DM (2006) Can an introduced, non-indigenous species save the fisheries of Lakes Baringo and Naivasha, Kenya? In: Odada EO, Olago DO, Ochola W, Ntiba M, Wandiga S, Gichuki N, Oyieke H (eds) Proceedings of the 11th World Lakes conference, Nairobi, Kenya, vol 2. International Lake Environment Committee, Kusatsu-shi, Shiga, pp 568–572

Britton JR, Jackson MC, Muchiri M, Tarras-Wahlberg H, Harper DM, Grey J (2009) Status, ecology and conservation of an endemic fish, *Oreochromis niloticus baringoensis*, in Lake Baringo, Kenya. Aquat Conserv 19:487–496

Brown M (1989) Where giants stood: the saga of Kenya's desert lake. Qiller Press, London

Brown J (1991) The ambatch canoes of Lake Baringo. Kenya Past Present 23:35–39

Bryan RB (1994) Soil erosion, land degradation and social transition: geoecological analysis of a semi-arid tropical region, Kenya. Adv Geoecol 27:1–248

Bryan RB, Schnabel S (1994) Estimation of sedimentation rates in the Chemeron reservoir. Adv Geoecol 27:231–248

Buatois LA, Renaut RW, Scott JJ, Owen RB (2017) An unusual occurrence of the trace fossil *Vagorichnus* preserved in hydrothermal silica at Lake Baringo, Kenya Rift Valley: taphonomic and paleoenvironmental significance. Palaeogeogr Palaeoclimatol Palaeoecol 485:843–853

Burnett GW, Rowntree KM (1990) Agriculture, research and tourism in the landscape of Lake Baringo, Kenya. Landsc Urban Plan 19:159–172

Burton RF (1871) On the Ukara, or the Ukerewe Lake of Equatorial Africa. Proc R Geogr Soc Lond 16:129–132

Campbell LM, Osano O, Hecky RE, Dixon DG (2003) Mercury in fish from three rift valley lakes (Turkana, Naivasha and Baringo), Kenya, East Africa. Environ Pollut 125:281–286

Carney JN (1972) The geology of the area east of Lake Baringo, Rift Valley Province, Kenya. PhD thesis, University of London

Castillo ML (2021) Processes and drivers of *Prosopis* invasions in Eastern Africa. PhD thesis, Stellenbosch University, South Africa

Chapman A (1908) On safari: big-game hunting in British East Africa. Edward Arnold, London

Chapman GR (1971) The geological evolution of the northern Kamasia Hills, Baringo District, Kenya. PhD thesis, University of London

Chapman GR, Lippard SJ, Martyn JE, Sinclair-Jones N, King BC (1973) Geological map of the northern Tugen Hills (Degree Sheet 34 North East Quarter). East African Geological Research Unit, Bedford College, University of London

Chapman GR, Lippard SJ, Martyn JE (1978) The stratigraphy and structure of the Kamasia Range, Kenya Rift Valley. J Geol Soc Lond 153:265–281

Chebet EB, Kibet JK, Mbui DM (2020) The assessment of water quality in river Molo water basin, Kenya. Appl Water Sci 10:92

Chebungei M, Jeruto C, Mulungu D, Nobert J (2022) Assessment of biophysical land degradation in Kerio River basin, Kenya. Int J Innov Sci Res Tech 7:1383–1390

Cherop HK, Muchiri L, Githonga GJ, Ambusso WJ (2016) Groundwater investigation and characterisation in Marigat area, Baringo County using electrical resistivity method. J Appl Geol Geophys 4:33–42

Chesire AK, Kollongei JK, Ng'etich W (2022) Sediments yields in Saimo catchment of Tugen Hills in Baringo County, Kenya. Afr Environ Rev 5:15–26

Clark CO, Webster PJ, Cole JE (2003) Interdecadal variability of the relationship between the Indian Ocean zonal mode and East African coastal rainfall anomalies. J Clim 16:548–554

Clément J-P, Caroff M, Hémond C, Tiercelin J-J, Bollinger C, Guillou H, Cotten J (2003) Pleistocene magmatism in a lithospheric transition area: petrogenesis of alkaline and peralkaline lavas from the Baringo–Bogoria Basin, central Kenya Rift. Can J Earth Sci 40:1239–1257

Copley H (1948) Lakes and rivers of Kenya: a short guide to the inland waters and their inhabitants. Highway Press, Nairobi

Daly CP, Taylor B (1875) Remarks on explorations in Central Africa. J Am Geogr Soc 7:296–304

Darling WG, Gizaw B, Arusei MK (1996) Lake-groundwater relationships and fluid-rock interaction in the East African Rift Valley: isotopic evidence. J Afr Earth Sci 22:423–431

Degefa A, Bessems I, Tjallingii R, Cumming B, Verschuren D (2014) Reconstructing 200 years of human-induced soil erosion in the Lake Baringo catchment (Kenya) from the sedimentology and geochemistry of lake sediments. KAOW-ARSOM 2014:P19

Deino AL, Hill A (2002) $^{40}Ar/^{39}Ar$ dating of Chemeron Formation strata encompassing the site of hominid KNM-BC 1, Tugen Hills, Kenya. J Hum Evol 42:141–151

Deino AL, McBrearty S (2002) $^{40}Ar/^{39}Ar$ dating of the Kapthurin Formation, Baringo, Kenya. J Hum Evol 42:185–210

Driese SG, Ashley GM, Li Z-H, Hover VC, Owen RB (2004) Possible Late Holocene equatorial palaeoclimate record based upon soils spanning the Medieval Warm Period and Little Ice Age, Loboi Plain, Kenya. Palaeogeogr Palaeoclimatol Palaeoecol 213:231–250

Dundas KR (1910) Notes on the tribes inhabiting the Baringo District, East Africa Protectorate. J R Anthropol Inst 40:49–72

Dunkley PN, Smith M, Allen DJ, Darling WG (1993) The geothermal activity and geology of the northern sector of the Kenya Rift Valley. British Geological Survey, Research Report SC/93/1

Engelen V (1983) An assessment of the irrigation suitability of the soils of the Amaya, Chemeron, Eldume, and Loiminang areas (Baringo District). Kenya Soil Survey, Site Evaluation Report No. 65

Gasse F (1987) Le lac Baringo. Diatomées. In: Tiercelin JJ, Vincens A (eds) Le demi-graben de Baringo-Bogoria, Rift Gregory, Kenya. Bull Centr Rech Explor-Prod Elf-Aquitaine 11:359–361

Gichuki NN, Oyieke HA, Terer T (2006) Status and root causes of biodiversity loss in the eastern Rift Valley lakes, Kenya. In: Odada EO, Olago DO, Ochola W, Ntiba M, Wandiga S, Gichuki N, Oyieke H (eds) Proceedings of the 11th World Lakes conference, Nairobi, Kenya, vol 2. International Lake Environment Committee, Kusatsu-shi, Shiga, pp 511–517

Gitau PN, Verschuren D (2016) Sedimentological and paleoecological research techniques, with applications to the environmental history of Loboi Swamp, Kenya. J Earth Sci Clim Change 7:1000351

Githaiga KB, Njuguna SM, Gituru RW, Yan X (2021) Water quality assessment, multivariate analysis and human health risks of heavy metals in eight major lakes in Kenya. J Environ Manag 297:113410

Glover RB (1972) Chemical characteristics of water and steam discharges in the Rift Valley of Kenya. Unpublished report, Geothermal Resources Exploration Project, United Nations–East Africa Power and Lighting, Nairobi, Kenya

GoK (Government of Kenya) and UNDP (2021) Rising water levels in Kenya's Rift Valley lakes, Turkwel Gorge Dam and Lake Victoria: a scoping report. Government of Kenya, Nairobi

Gregory JW (1896) The Great Rift Valley. Murray, London

Gregory JW (1921) The rift valleys and geology of East Africa. Seeley Service, London

Grewe S (2016) Hydro- and biogeochemical investigations of lake sediments in the Kenyan Rift Valley. Dr rer Nat thesis, Universität Potsdam

Griffiths PW (1977) Geology of the Lake Hannington – Perkerra River area, Kenya Rift Valley. PhD thesis, University of London

Grimaud P, Richert JP, Rolet J, Tiercelin JJ, Xavier JP, Morley CK, Coussement C, Karanja SW, Renaut RW, Guerin G, Le Turdu C, Michel-Noel G (1994) Fault geometry and extension mechanisms in the central Kenya Rift, East Africa. A 3D remote sensing approach. Bull Centr Rech Explor-Prod Elf-Aquitaine 18:59–92

Grove AT (1974) Desertification in the African environment. Afr Aff 73:137–151

Hackman BD (1988) Geology of the Baringo–Laikipia area. Mines and Geology Department, Nairobi, Kenya, Report No. 104

Hackman BD, Charsley TJ, Key RM, Wilkinson AF (1990) The development of the East African Rift system in north-central Kenya. Tectonophysics 184:189–211

Hartley J (1986) Baringo and Bogoria. Evans Brothers (Kenya) Ltd, Nairobi

Hautot S, Tarits P, Whaler K, Le Gall B, Tiercelin JJ, Le Turdu C (2000) Deep structure of the Baringo Rift Basin (central Kenya) from three-dimensional magnetotelluric imaging: implications for rift evolution. J Geophys Res 105:23493–23518

Hennesey A, Yasindi AW, Zinabu G-M, Taylor WD (2020) Effects of pH and salinity on the sedimentation of clay particles in turbid lakes of the rift valley of Ethiopia and Kenya. Ethiop J Biol Sci 19:117–130

Herrnegger M, Stecher G, Schwatke C, Olang L (2021) Hydroclimatic analysis of rising water levels in the Great Rift Valley lakes of Kenya. J Hydrol Reg Stud 36:100857

Hickley P, Muchiri M, Boar R, Britton R, Adams C, Gichuru N, Harper DM (2004) Habitat degradation and subsequent fishery collapse in Lakes Naivasha and Baringo, Kenya. Int Ecohydrol Hydrobiol 4:503–517

Hickley P, Muchiri M, Britton JR, Boar R (2008) Economic gain versus ecological damage from the introduction of non-native freshwater fish: case studies from Kenya. Open Fish Sci J 1:36–46

Hill A (1995) Faunal and environmental change in the Neogene of East Africa: evidence from the Tugen Hills sequence, Baringo District, Kenya. In: Andrews P, Banham P (eds) Late Cenozoic environments and hominid evolution: a tribute to Bill Bishop. Geological Society of London, London, pp 85–98

Hobley C (1906) Notes on the geography and people of the Baringo District of the East Africa Protectorate. Geogr J 28:471–481

Hodder I (1977) The distribution of material culture items in the Baringo District, western Kenya. Man 12:239–269

Homewood K, Lewis J (1987) Impact of drought on pastoral livestock in Baringo, Kenya 1983-85. J Appl Ecol 24:615–631

Indeje M, Semazzi FH, Ogallo LJ (2000) ENSO signals in East African rainfall seasons. Int J Climatol 20:19–46

Institute of Hydrology, UK (1985) Groundwater conditions in Baringo District, Kenya, pp 1–26

Jenkin PM (1929) Biology of lakes in Kenya. Nature 124:574

Jenkin PM (1932) Reports on the Percy Sladen Expedition to some Rift Valley lakes in Kenya in 1929. I. Introductory account of the biological survey of five freshwater and alkaline lakes. Ann Mag Nat Hist Ser 10:533–553

Jenkin PM (1936) Reports on the Percy Sladen Expedition to some Rift Valley lakes in Kenya in 1929. VII. Summary of the ecological results, with special reference to the alkaline lakes. Ann Mag Nat Hist Ser 18:133–160

Johansson J, Svensson J (2002) Land degradation in the semi-arid catchment of Lake Baringo, Kenya. Deparment of Physical Geography, Göteborgs Universitet, Projketarabete Göteborg B3443

Johnston H (1902a) The Uganda Protectorate, Ruwenzori, and the Semliki Forest. Geogr J 19:1–39

Johnston H (1902b) The Uganda Protectorate, vol 1, 1st edn. Hutchinson, London

Jones WB (1988) Listric growth faults in the Kenya Rift Valley. J Struct Geol 10:661–672

Källqvist T (1987) Primary production and phytoplankton in Lake Baringo and Lake Naivasha, Kenya. Norwegian Institute of Water Research Report, Oslo

Kamar MJ (1992) The effect of mulches on hydrologic and erosional processes, soil moisture and crop yield, in highly crusting tropical semi-arid soils, Baringo, Kenya. PhD thesis, University of Toronto

Kangogo J (2020) The year the Earth unleashed its fury on Baringo residents. The Star, Nairobi, 3 December

Kareri RW (2018). Land use changes and their impacts on wetlands in Loboi Plains, Baringo County, Kenya. PhD thesis, Moi University, Eldoret

Karingithi C, Wambugu J (2010) The geochemistry of Arus and Bogoria geothermal prospects. In: Proceedings of the world geothermal congress, Bali, Indonesia, April

Kiage LM, Douglas P (2020) Linkages between land cover change, lake shrinkage, and sublacustrine influence determined from remote sensing of select Rift Valley Lakes in Kenya. Sci Total Environ 709:136022

Kiage LM, Liu K (2009a) Paleoenvironmental changes in the Lake Baringo Basin, Kenya, East Africa since AD 1650: evidence from the paleorecord. Prof Geogr 61:438–458

Kiage LM, Liu KB (2009b) Palynological evidence of climate change and land degradation in the Lake Baringo area, Kenya, East Africa, since AD 1650. Palaeogeogr Palaeoclimatol Palaeoecol 279:60–72

Kiage LM, Liu KB, Walker ND, Lam N, Huh OK (2007) Recent land-cover/use change associated with land degradation in the Lake Baringo catchment, Kenya, East Africa: evidence from Landsat TM and ETM+. Int J Remote Sens 28:4285–4309

Kilham P (1971) Biogeochemistry of African lakes and rivers. PhD thesis, Duke Univ, Durham, North Carolina

King BC, Chapman GR (1972) Volcanism of the Kenya Rift Valley. Phil Trans R Soc A 271:185–208

Kingston JD, Deino AL, Edgar RK, Hill A (2007) Astronomically forced climate change in the Kenyan Rift Valley 2.7–2.55 Ma: implications for the evolution of early hominin ecosystems. J Hum Evol 20:487–503

Kinyanjui HCK (1982). Detailed soil survey of Sandai irrigation scheme (Baringo District). Kenya Soil Survey, Report No. D19

Kipkorir E (2002) Analysis of rainfall climate on the Njemps Flats, Baringo District, Kenya. J Arid Environ 50:445–458

Kipng'ok J (2011) Geochemical assessment of Korosi-Chepchuk geothermal prospect, Kenya Rift Valley. Geotherm Res Counc Trans 35:637–644

Kotut K, Krienitz L, Ballot A (2006) Toxic cyanobacteria and its toxins in standing waters of Kenya: implications for water resource use. In: Odada EO, Olago DO, Ochola W, Ntiba M, Wandiga S, Gichuki N, Oyieke H (eds) Proceedings of the 11th World Lakes conference, Nairobi, Kenya, vol 2. International Lake Environment Committee, Kusatsu-shi, Shiga, pp 522–528

LaVigne M, Ashley GM (2002) Climatology and rainfall patterns: Lake Bogoria National Reserve (1976–2001). Unpublished report, Rutgers University, Piscataway, New Jersey

Le Gall B, Tiercelin J-J, Richert J-P, Gente P, Sturchio NC, Stead D, Le Turdu C (2000) A morphotectonic study of an extensional fault zone in a magma-rich rift: the Baringo Trachyte fault system, central Kenya Rift. Tectonophysics 320:87–106

Le Turdu C, Tiercelin J-J, Coussement C, Rolet J, Renaut RW, Richert JP, Xavier J-P, Coquelet D (1995) Basin structure and depositional patterns interpreted using a 3D remote sensing approach: the Baringo–Bogoria basins, central Kenya Rift, East Africa. Bull Centr Rech Explor-Prod Elf-Aquitaine 19:1–37

Le Turdu C, Tiercelin J-J, Richert JP, Rolet J, Xavier JP, Renaut RW, Lezzar K, Coussement C (1999) Influence of preexisting oblique discontinuities on the geometry and evolution of extensional fault patterns: evidence from the Kenya Rift using SPOT imagery. In: Morley CK (ed) Geoscience of rift systems – evolution of East Africa. AAPG Stud Geol 44:173–191

Leake BE (2011) The life and work of Professor J.W. Gregory FRS (1864–1932): geologist, writer and explorer. Geological Society of London Memoir 34, Geological Society, London

Lewis MN (1998) A guide to Lake Baringo and Lake Bogoria. Horizon Books, Nairobi

Lincer JL, Zalkind D, Brown LH, Hopcraft J (1981) Organochlorine residues in Kenya's Rift Valley lakes. J Appl Ecol 18:157–171

Little PD (1996) Pastoralism, biodiversity, and the shaping of savanna landscapes in East Africa. Africa 66:37–50

Lwenya C, Yongo E (2010) Human aspects of siltation of Lake Baringo: causes, impacts and interventions. Aquat Ecosyst Health Manag 13:437–441

McCall GJH (1967) Geology of the Nakuru-Thomson's Falls Lake Hannington area. Rep Geol Surv Kenya 78:122

Mechie J, Keller GR, Prodehl C, Gaciri S, Braile LW, Mooney WD, Gajewski D, Sandmeier KJ (1994) Crustal structure beneath the Kenya Rift from axial profile data. Tectonophysics 236:179–200

Mergeay J, Verschuren D, De Meester L (2005) Cryptic invasion and dispersal of an American *Daphnia* in East Africa. Limnol Oceanogr 50:1278–1283

Mergeay J, Declerck S, Verschuren D, Meester LD (2006) *Daphnia* community analysis in shallow Kenyan lakes and ponds using dormant eggs in surface sediments. Freshw Biol 51:399–411

Mlewa C (2003) Biology of the African lungfish *Protopterus aethiopicus* Heckel 1851, and some aspects of its fishery in Lake Baringo, Kenya. PhD thesis, University of Newfoundland, St. John's

Mlewa CM, Green JM (2004) Biology of the marbled lungfish, *Protopterus aethiopicus* Heckel, in Lake Baringo, Kenya. Afr J Ecol 42:338–345

Mlewa CM, Green JM (2006) Translocation of marbled African lungfish, *Protopterus aethiopicus* (Telostei: Protopteridae), and its fishery in Lake Baringo, Kenya. Afr J Aquat Sci 31:131–136

Mpawenayo B, Mathooko J (2004) Diatom assemblages in the hotsprings associated with Lakes Elmenteita and Baringo in Kenya. Afr J Ecol 42:363–367

Muasya AMM, Hover VC, Ashley GM, Owen RB, Goman MF, Kimeli M (2006) Diversity and distribution of macrophytes in a freshwater wetland, Loboi swamp (Rift Valley), Kenya. J East Afr Nat Hist 93:39–47

Mugisha F, Ebinger CJ, Strecker M, Pope D (1997) Two-stage rifting in the Kenya Rift: implications for half-graben models. Tectonophysics 278:63–81

Muia D, Gicheru M, Mutiso J, Mwangi B, Kavinda L, Kairu E (2021) Effects of extreme flooding of Lake Baringo on livelihoods of communities lining around the lake. Adv Appl Sociol 11:404–414

Muiruri V, Marchant R, Rucina SM, Scott L, Lane PJ (2021) Late Holocene environmental change and anthropogenic: ecosystem interaction on the Laikipia Plateau, Kenya. Ambio 51:785–798. https://doi.org/10.1007/s13280-021-01554-6

Mulwa J, Mariita N, Barongo J, Patel J (2006) Preliminary magnetotelluric survey results of Arus-Bogoria geothermal prospect, Kenya. In: Extended abstracts 18th workshop IAGA WG 1.2 on electromagnetic induction in the earth, El Vendrell, Spain, September, S7/E5

Mutonga M (2014) Fluoride ions in groundwater case study; Baringo-Bogoria geothermal area in Kenya. In: Proceedings of the 5th African rift geothermal conference, Arusha, October 2014, pp 1–10

Mwamburi J (2015) Comparative evaluation of the concentrations of lead, cadmium and zinc in surficial sediments from two shallow tectonic freshwater lake basins, Kenya. Afr J Environ Sci Technol 9:531–544

Mwangi S (1983) Soils of the Perkerra Agricultural Research Station – Marigat. Kenya Soil Survey, Report No. D30

Mwangi E, Swallow B (2008) Invasion of *Prosopis juliflora* and local livelihoods: case study from the Lake Baringo area of Kenya. Conserv Soc 6:130–140

Mwasi B (2021) Land use conflicts resolution in a fragile ecosystem using multi-criteria evaluation (MCE) and a GIS-based decision support system (DSS). In: International conference on spatial information for sustainable development, Nairobi, Kenya, October 2021, pp 1–10

Ndiwa TC, Nyingi DW, Agnèse J-F (2014) An important natural genetic resource of *Oreochromis niloticus* (Linnaeus, 1758) threatened by aquaculture activities in Loboi drainage, Kenya. PLoS One 9:e106972

Ndiwa TC, Nyingi DW, Claude J, Agnèse J-F (2016) Morphological variations of wild populations of Nile tilapia (*Oreochromis niloticus*) living in extreme environmental conditions in the Kenyan Rift-Valley. Environ Biol Fish 99:473–485

Ng W-T, Rima P, Einzmann K, Immitzer M, Atzberger C, Eckert S (2017) Assessing the potential of Sentinel-2 and Pléiades data for the detection of *Prosopis* and *Vachellia* spp. in Kenya. Remote Sens 9:74

Ngaira JK (2006) Implications of climate change on the management of Rift Valley lakes in Kenya. The case of Lake Baringo. In: Odada EO, Olago DO, Ochola W, Ntiba M, Wandiga S, Gichuki N, Oyieke H (eds) Proceedings of the 11th World Lakes conference, Nairobi, Kenya, vol 2. International Lake Environment Committee, Kusatsu-shi, Shiga, pp 133–138

Nicholson SE (1998) Historical fluctuations of Lake Victoria and other lakes in the northern rift valley of East Africa. In: Lehman JT (ed) Environmental change and response in East African lakes. Kluwer, Dordrecht, pp 7–35

Nilsson E (1931) Quaternary glaciations and pluvial lakes in British East Africa. Geogr Ann 13:249–349

Njue LM (2014) Stratigraphic model of Korosi geothermal prospect, Kenya. In: Proceedings of the 5th African rift geothermal conference, Arusha, October 2014

Njue L (2015) Geological model of Korosi geothermal prospect, Kenya. In: Proceedings of the world geothermal congress, Melbourne, April 2015

Nyaberi MD (2010) Geophysical characterization of the lithology and structure of the Olobanita well field, lower Lake Baringo Basin, Kenya Rift: implication on groundwater occurrence. MSc thesis, University of Nairobi

Nyakeya K, Chemoiwa E, Nyamora JM, Ogombe CO, Gichana ZM, Frank Onderi Masese FO, Aura CM, Mbaru EK, Nyamweya C, Njiru J, Ondib R, Basweti E (2020) Endemic Lake Baringo *Oreochromis niloticus* fishery on verge of collapse: review of causes and strategies directed to its recovery, conservation and management for sustainable exploitation. Lakes Reserv Res Manag 25:423–428

Nyamweya CS, Mlewa CM, Ngugi CC, Kaunda-Arara B (2010) Validation of daily growth of African catfish *Clarias gariepinus* (Burchell 1822) young-of-the-year from Lake Baringo, Kenya. Lakes Reserv Res Manag 15:341–345

Nyingi D, Vos LD, Aman R, Agnèse JF (2009) Genetic characterization of an unknown and endangered native population of the Nile tilapia *Oreochromis niloticus* (Linnaeus, 1758) (Cichlidae; Teleostei) in the Loboi Swamp (Kenya). Aquaculture 297:57–63

Obando JA, Onywere S, Shisanya C, Ndubi A, Masiga D, Irura Z, Mariita N, Maragia H (2016) Impact of short-term flooding on livelihoods in the Kenya Rift Valley lakes. In: Singh RB (ed) Advances in geographical and environmental sciences. Springer, Tokyo, pp 193–215

Ochieng EZ, Lalah JO, Wandiga SO (2007) Analysis of heavy metals in water and surface sediment in five rift valley lakes in Kenya for assessment of recent increase in anthropogenic activities. Bull Environ Contam Toxicol 79:570–576

Ochieng L, Kipng'ok J, Kanda I, Igunza G, Wanjie C (2012) Geochemistry of fumarole discharges and borehole waters of Korosi-Chepchuk geothermal prospect, Kenya. Geotherm Res Counc Trans 36:17–21

Ochuka M, Ikporukpo C, Mijinyawa Y, Ogendi G (2019a) Land use/land cover dynamics and anthropogenic driving factors in Lake Baringo catchment, Rift Valley, Kenya. Nat Resour 10:367–389

Ochuka M, Ikporukpo CO, Ogendi GM, Mijinyawa Y (2019b) Spatial variability in physico-chemical parameters of water in Lake Baringo Basin, Kenya. Curr World Environ 14:334–457

Odada EO, Onyando JO, Obudho PA (2006) Lake Baringo: addressing threatened biodiversity and livelihoods. Lakes Reserv Res Manag 11:287–299

Oduor SO, Schagerl M, Mathooko JM (2003) On the limnology of Lake Baringo (Kenya): I. Temporal physico-chemical dynamics. Hydrobiologia 506:121–127

Ogola PFA, Davidsdottir B, Fridleifsson IB (2011) Lighting villages at the end of the line with geothermal energy in eastern Baringo lowlands, Kenya – steps towards reaching the millennium development goals (MDGs). Renew Sust Energ Rev 15:4067–4079

Ogola PFA, Davidsdottir B, Fridleifsson IB (2012) Potential contribution of geothermal energy to climate change adaptation: a case study of the arid and semi-arid eastern Baringo lowlands, Kenya. Renew Sust Energ Rev 16:4222–4246

Ogutu-Ohwayo R, Balirwa J (2006) Management challenges of freshwater fisheries in Africa. Lakes Reserv Res Manag 11:215–226

Ojany FF, Ogendo RB (1973) Kenya: a study in physical and human geography. Longman, London

Okech EO, Kitaka N, Omondi S, Verschuren D (2019) Water level fluctuations in Lake Baringo, Kenya, during the 19th and 20th centuries: evidence from lake sediments. Afr J Aquat Sci 44:25–33

Olago D, Opere A, Barongo J (2009) Holocene palaeohydrology, groundwater and climate change in the lake basins of the central Kenya Rift. Hydrol Sci J 54:765–780

Olilo CO, Shivoga WA, Kitetu JJ, Agembe S (2014) Effects of environmental factors on cyanobacteria dynamics in Lake Baringo, Kenya. Lakes Reserv Res Manag 19:240–254

Olilo CO, Odoli CO, Obiero MO, Malala JO, Bironga CH (2023) Microbial ecology of desert Lakes Baringo and Turkana, Kenya, East Africa. In: El-Sheekh M, Elsaied HE (eds) Lakes of Africa. Elsevier, Amsterdam, pp 247–268

Ombasa DT, Kosgei JR, Nyandwaro G, Munishi ES (2022) Multi-criteria approach to assess groundwater potential: a case study of Baringo County, Kenya. Water Pract Technol 17:2199–2223

Omondi R, Yasindi AW, Magana A (2011) Spatial and temporal variations of zooplankton in relation to some environmental factors in Lake Baringo, Kenya. Egerton J Sci Technol 11:29–50

Omondi R, Yasindi AW, Magana A (2013) Food and feeding habits of three main fish species in Lake Baringo, Kenya. J Ecol Nat Environ 5:224–230

Omondi R, Kembenya E, Nyamweya C, Ouma H, Machua SK, Ogari Z (2014) Recent limnological changes and their implication on fisheries in Lake Baringo, Kenya. J Ecol Nat Environ 6:154–163

Omondi R, Ojwang W, Olilo C, Mugo J, Agembe S, Ojuok JE (2016) Lakes Baringo and Naivasha: endorheic freshwater lakes of the Rift Valley (Kenya). In: Finlayson CM, Milton GR, Prentice RC, Davidson NC (eds) The wetland book II. Springer, Dordrecht, pp 1349–1360

Onindo CO, Mwangi IW (2012) Thermodynamic modelling of the equilibria in Lake Baringo, Kenya. J Agric Sci Technol 14:104–113

Onyando JO, Kisoyan P, Chemelil MC (2005) Estimation of potential soil erosion for River Perkerra catchment in Kenya. Water Resour Manag 19:133–143

Onywere SM, Shisanya CA, Obando JA, Ndubi AO, Masiga D, Irura Z, Mariita N, Maragia HO (2013) Geospatial extent of 2011–2013 flooding from the eastern African Rift Valley lakes in Kenya and its implication on the ecosystems. In: International workshop on the soda lakes of Kenya: their current conservation status and management, Naivasha, pp 1–18

Oostwoud Wijdenes DJ (1992) The dynamics of gully-head erosion on a semi-arid piedmont plain, Baringo District, Kenya. PhD thesis, University of Toronto

Oostwoud Wijdenes DJ, Bryan R (2001) Gully-head erosion processes on a semi-arid valley floor in Kenya: a case study into temporal variation and sediment budgeting. Earth Surf Process Landf 26:911–933

Ouma H, Mwamburi J (2014) Spatial variations in nutrients and other physicochemical variables in the topographically closed Lake Baringo freshwater basin (Kenya). Lakes Reserv Res Manag 19:11–23

Ouma YO, Tateishi R (2006) A water index for rapid mapping of shoreline changes of five East African Rift Valley lakes: an empirical analysis using Landsat TM and ETM+ data. Int J Remote Sens 27:3153–3181

Owen RB (1981) Quaternary diatomaceous sediments and the geological evolution of lakes Turkana, Baringo and Bogoria, Kenya Rift Valley. PhD thesis, University of London

Owen RB, Renaut RW (1980) The early Holocene palaeogeography of the northern Kenya Rift Valley. Rech Géol Afr 5:143–146

Owen R, Renaut RW (1986) Sedimentology, stratigraphy and palaeoenvironments of the Holocene Galana Boi Formation, NE Lake Turkana, Kenya. In: Frostick LE, Renaut RW, Reid I, Tiercelin J-J (eds) Sedimentation in the African rifts. Geol Soc Lond Spec Publ 25:311–322

Owen RB, Renaut RW (2000) Miocene and Pliocene diatomaceous lacustrine sediments of the Tugen Hills, Baringo District, Central Kenya Rift. In: Gierlowski-Kordesch EH, Kelts KK (eds) Lake basins through space and time. AAPG Stud Geol 46:399–404

Patterson G, Wilson KK (1995) The influence of the diel climatic cycle on the depth-time distribution of phytoplankton and photosynthesis in a shallow equatorial lake (Lake Baringo, Kenya). Hydrobiologia 304:1–8

Pejler B (1974) On the rotifer plankton of some East African lakes. Hydrobiologia 44:389–396

Petek N (2018) Archaeological perspectives on risk and community resilience in the Baringo Lowlands, Kenya. PhD thesis, University of Uppsala

Petek-Sargeant N, Lane P (2021) Weathering climate change in archaeology: conceptual challenges and an East African case study. Camb Archaeol J 3:437–454

Pickford M (1975) Late Miocene sediments and fossils from the northern Kenya Rift Valley. Nature 256:279–284

Pickford MH (1978) Stratigraphy and mammalian palaeontology of the late-Miocene Lukeino Formation, Kenya. In: Bishop WW (ed) Geological background to fossil man. Geol Soc Lond Spec Publ 6:263–278

Powell-Cotton PHG (1904) A journey through northern Uganda. Geogr J 24:56–65

Renaut RW (1982) Late Quaternary geology of the Lake Bogoria fault-trough, Kenya Rift Valley. PhD thesis, University of London

Renaut RW, Owen RB (1980) Late Quaternary fluvio-lacustrine sedimentation and lake levels in the Baringo Basin, northern Kenya, Rift Valley. Rech Géol Afr 5:130–133

Renaut RW, Ego J, Tiercelin J-J, Le Turdu C, Owen RB (1999) Saline, alkaline paleolakes of the Tugen Hills-Kerio Valley region, Kenya Rift Valley. In: Andrews P, Banham P (eds) Late Cenozoic environment and hominid evolution: a tribute to Bill Bishop. Geological Society, London, pp 41–58

Renaut RW, Tiercelin JJ, Owen RB (2000) Lake Baringo, Kenya Rift Valley, and its Pleistocene precursors. In: Gierlowski-Kordesch E, Kelts K (eds) Lake basins through space and time. AAPG Stud Geol 46:561–568

Renaut RW, Jones B, Tiercelin J-J, Tarits C (2002) Sublacustrine precipitation of hydrothermal silica in rift lakes: evidence from Lake Baringo, central Kenya Rift Valley. Sediment Geol 148:235–257

Renaut RW, Owen RB, Jones B, Tiercelin J-J, Tarits C, Ego JK, Konhauser KO (2012) Impact of lake-level changes on the formation of thermogene travertine in continental rifts: evidence from Lake Bogoria, Kenya Rift Valley. Sedimentology 60:428–468

Rich F (1932) Reports on the Percy Sladen expedition to some Rift Valley lakes in Kenya in 1929. IV. Phytoplankton from the Rift Valley lakes in Kenya. Ann Mag Nat Hist 10:233–262

Richardson JL, Richardson AE (1972) History of an African rift lake and its climatic implications. Ecol Monogr 42:499–534

Rooney D, Hutton V (1977) A magnetotelluric and magnetovariational study of the Gregory Rift Valley, Kenya. Geophys J Roy Astron Soc 51:91–119

Schagerl M, Oduor SO (2003) On the limnology of Lake Baringo (Kenya): II. Pelagic primary production and algal composition of Lake Baringo, Kenya. Hydrobiologia 506–509:297–303

Schnabel S (1990) The application of dendrochronology in soil erosion research in semi-arid tropical areas. Studies in the Baringo District. MSc thesis, University of Toronto

Scoon RN (2018) Geology of national parks of Central/Southern Kenya and Northern Tanzania. Springer, Cham

Seurey P, Githiri G, Ambusso WJ (2016) Application of Euler deconvolution and forward modelling techniques in estimating depth to magnetic structure in Lake Baringo area, in the Kenyan Rift. J Appl Geol Geophys 4:63–72

Simiyu C, Gichira J, Mwakirani R, Wamalwa A, Kangogo D, Noor Y (2011) Resistivity studies of the Lake Baringo Geothermal Prospect, Kenya. In: Proceedings of the Kenya geothermal conference 2011, Nairobi, pp 1–10

Sketchley HR (1978) Reconnaissance soil survey, Lake Baringo-Kerio Valley area, Republic of Kenya. US Department of Agriculture, Soil Conservation Service and Republic of Kenya Ministry of Agriculture

Snelder DJ (1993) Methods of rangeland improvement and revegetation for denuded, semi-arid savanna areas in the Baringo District, Kenya. PhD thesis, University of Toronto

Snelder DJ, Bryan RB (1995) The use of rainfall simulation tests to assess the influence of vegetation density on soil loss on degraded rangelands in the Baringo District, Kenya. Catena 25:105–116

Ssentongo GW (1974) On the fish and fisheries of Lake Baringo. Afr J Trop Hydrobiol Fish 3:95–105

Ssentongo GW (1995) Report on the present fisheries situation of Lake Baringo, Kenya. FAO, Rome

Sutherland R, Bryan R (1990) Flow dynamics and the variability of suspended sediment in a semi-arid tropical stream, Baringo District, Kenya. Geogr Ann 72:23–39

Talling JF, Talling IB (1965) The chemical composition of African lake waters. Int Rev Gesamt Hydrobiol 50:421–463

Tarits C, Renaut RW, Tiercelin JJ, Le Hérissé A, Cotten J, Cabon JY (2006) Geochemical evidence of hydrothermal recharge in Lake Baringo, central Kenya Rift Valley. Hydrol Process 20:2027–2055

Tarras-Wahlberg H, Harper DM, Tarras-Wahlberg N (2003) A first limnological description of Lake Kichiritith, Kenya: a possible reference site for the freshwater lakes of the Gregory Rift Valley. S Afr J Sci 99:494–496

Taylor D, Lane PJ, Muiruri V, Ruttledge A, Gaj McKeever R, Nolan T, Kenny P, Goodhue R (2005) Mid- to late-Holocene vegetation dynamics on the Laikipia Plateau, Kenya. The Holocene 15:837–846

Thom DJ, Martin NL (1983) Ecology and production in Baringo-Kerio Valley, Kenya. Geogr Rev 73:15–29

Thomson J (1887) Through Masai Land. Sampson Low and Co, London

Tiercelin J-J (1981) Rifts continentaux, tectonique, climates, sediments. Exemples: la sedimentation dans le Nord du Rift Gregory, Kenya, et dans le Rift de l'Afar, Ethiopie, depuis le Miocène. PhD thesis, Université d'Aix-Marseille II

Tiercelin JJ, Vincens A (1987) Le demi-graben de Baringo-Bogoria, Rift Gregory, Kenya. Bull Centr Rech Explor-Prod Elf-Aquitaine 11:249–540

Tiercelin J-J, Le Fournier J, Herbin JP, Richert JP (1980) Continental rifts: modern sedimentation, tectonic and volcanic control. Example from the Bogoria-Baringo graben, Gregory Rift, Kenya. In: Geodynamic evolution of the Afro-Arabic rift system. Accademia Nazionale dei Lincei, Rome, pp 143–164

Tongue JA (1991) Seismicity studies in the East African Rift, Kenya. PhD thesis, University of Leicester

Tongue J, Maguire P, Burton P (1994) An earthquake study in the Lake Baringo basin of the central Kenya Rift. Tectonophysics 236:151–164

Vacelet J, Tiercelin J-J, Gasse F (1991) The sponge *Dosilia brouni* (Spongillidae) in Lake Baringo, Gregory Rift, Kenya. Hydrobiologia 211:11–18

van der Plas GW, De Cort G, Petek-Sargeant N, Wuytack T, Colombaroli D, Lane PJ, Verschuren D (2019) Distinct phases of natural landscape dynamics and intensifying human activity in the central Kenya Rift Valley during the past 1300 years. Quat Sci Rev 218:91–106

Vehrs H-P, Heller GR (2017) Fauna, fire, and farming: landscape formation over the past 200 years in pastoral East Pokot, Kenya. Hum Ecol 45:613–625

Verschuren D, Laird KR, Cumming BF (2000) Rainfall and drought in equatorial east Africa during the past 1,100 years. Nature 40:410–414

Victor N, Eric P, Kyeba K (2023) The risk of flooding to architecture and infrastructure amidst a changing climate in Lake Baringo, Kenya. Am J Clim Chang 12:80–99

Vincens A, Casanova J, Tiercelin J-J (1986) Palaeolimnology of Lake Bogoria (Kenya) during the 4500 BP high lacustrine phase. In: Frostick LE, Renaut RW, Reid I, Tiercelin J-J (eds) Sedimentation in the African rifts. Geol Soc Lond Spec Publ 25:323–330

von Höhnel LR (1894) The discovery of lakes Rudolf and Stephanie: a narrative of Count Samuel Teleki's exploring & hunting expedition in eastern equatorial Africa in 1887 & 1888. Longmans, Green & Co. Two volumes

Wakefield T, Johnston K (1870) Routes of native caravans from the coast to the interior of Eastern Africa, chiefly from information given by Sádi Bin Ahédi, a native of a district near Gázi, in Udigo, a little north of Zanzibar. J R Geogr Soc Lond 40:303–339

Walumona JR (2022) Modelling lake level variations, water balance, and fisheries of Lake Baringo, Kenya. PhD thesis, University of Eldoret

Walumona JR, Odoli CO, Raburu P, Amisi FM, Murakaru MJ, Kondowe BN, Kaunda-Arara B (2021) Spatio-temporal variations in selected water quality parameters and trophic status of Lake Baringo, Kenya. Lakes Reserv Res Manag 26:e12367

Walumona JR, Boaz K-A, Ogombe CO, Murakaru JM, Raburu P, Amisi FM, Nyakeya K, Kondowe BN (2022) Effects of lake level changes on water quality and fisheries production of Lake Baringo, Kenya. Ecohydrology 15:e20368

Wandiga O, Molepo M, Alala LN (1983) Concentration of heavy metals in water, sediments and plants of Kenya lakes. Kenya J Sci Technol Ser A 4:89–94

Wararu B (1990) An assessment of the irrigation suitability of the soil and water of the Mukutan, Loiminange and Logumkum sites (Marigat Division, Baringo District). Kenya Soil Survey, Site Evaluation Report No. P83

Williams REG, Johnson AS (1976) Birmingham University radiocarbon dates X. Radiocarbon 18:249–267

Worthington EB (1932a) Scientific results of the Cambridge Expedition to the East African lakes, 1930–31, No. 1: General introduction and station list. J Linn Soc Lond Zool 38:99–119

Worthington EB (1932b) The lakes of Kenya and Uganda. Geogr J 79:275–293

Worthington EB (1996) Early research on East African lakes: an historical sketch. In: Johnson TC, Odada EO (eds) The limnology, climatology and palaeoclimatology of the East African lakes. Gordon and Breach, Amsterdam, pp 659–664

Worthington EB, Ricardo CK (1936) Scientific results of the Cambridge Expedition to the East African lakes, 1930-1. No. 15. The fish of Lake Rudolf and Lake Baringo. J Linn Soc Lond Zool 39:353–389

Yasindi AW (2001) The ecology of planktonic ciliates in tropical lakes of East Africa. PhD thesis, University of Waterloo, Ontario

Yasindi AW, Taylor WD (2006) The abundance, biomass and composition of pelagic ciliates in East African lakes of different salinity and trophy. In: Odada EO, Olago DO, Ochola W, Ntiba M, Wandiga S, Gichuki N, Oyieke H (eds) Proceedings of the 11th World Lakes conference, Nairobi, Kenya, vol 2. International Lake Environment Committee, Kusatsu-shi, Shiga, pp 374–378

Young JAT, Renaut RW (1979) A radiocarbon date from Lake Bogoria, Kenya Rift Valley. Nature 278:243–255

Chapter 9
Lake Bogoria

9.1 Introduction

John Walter Gregory (1864–1932), born in Bow in east London near Queen Mary, University of London, visited Lake Bogoria in 1893 and described the lake as "by far the most beautiful and interesting… " of the Kenya Rift lakes (Gregory 1894, p. 310). Many would agree (Figs. 9.1a and 9.2a, b). In contrast, the American hunter, Carl Akeley (1920) wrote "If the unattractiveness of the place were not sufficient to discourage a disposition to explore the lake, the sickening stench from the green waters must dishearten anyone who has not a definite object in further investigation."

When Gregory visited Lake Bogoria, which lies in a narrow steep-sided half-graben 20 km north of the equator, it was called Lake Losuguta and its shoreline hot springs were named Suva ya Moto (Gregory 1901). Other early lake names were Mbatibat and Makaria (Johnston 1902a). During the colonial period, the lake was renamed Lake Hannington in honour of Bishop James Hannington, an Anglican missionary murdered in Busoga, Uganda in 1885 (Dawson 1890). A local name was reinstated when the lake and its surrounding area were designated the Lake Bogoria National Reserve on 29 November 1973. Bogoria reputedly means 'place of the lost tribe' and the local mythology is that the lake formed when the inhabitants, ancestors of the Endorois community, were swallowed into the ground. The lake is now an international Ramsar site and part of a UNESCO World Heritage site.

We began research at the lake a few years later (1976–7) when there were few visitors. The road north of Nakuru was unpaved and accommodation at Bogoria and Baringo was in its infancy. The Lake Bogoria Hotel was being built. We camped at Loburu hot springs, then named Kiboriit on Kenya Survey maps (Fig. 9.1b). President Jomo Kenyatta was planning a residence overlooking the hot springs, and flagpoles had been cemented in place by summer of 1977 (Fig. 9.1b), but he died (22 August 1978) before its construction.

R. W. Renaut, R. B. Owen, *The Kenya Rift Lakes: Modern and Ancient*, Syntheses in Limnogeology, https://doi.org/10.1007/978-3-642-25055-2_9

Fig. 9.1 (**a**) Lake Bogoria viewed from the north. (Photograph by J-J Tiercelin). Sandai Plain centre left and swampy southern Loboi Plain in the foreground. (**b**) Central Loburu delta in summer 1977; view south from the adjacent scarp

Fig. 9.2 (**a**) South sub-basin, July 2005; Lake Bogoria Escarpment on right side (**b**) Lesser flamingos, North sub-basin. (**c**) Geysers, southern Loburu, July 2005. (**d**) Boiling pool springs aligned along a fault, southern Loburu, August 2006

Over the next decade, Lake Bogoria was becoming an important tourist destination, increasingly famous for its natural beauty, spectacular birdlife, and its hot springs (Fig. 9.2c, d; Tiercelin 1985; Renaut and Tiercelin 1993; North Lewis 1998). We were unable to return until 1991. Lake Bogoria has the most impressive and accessible geothermal activity in the Kenya Rift, with more geysers (at least 20 are known) than anywhere else in Africa (Renaut and Owen 2005; Renaut et al. 2008).

Gregory's opinion (1894, 1896) remains valid, but recent (post-2010) flooding of the shorelines has nonetheless marred, at least temporarily, some of its previous attractions. In the last decade (2010–2022) more than half of the ~200 hot spring vents have been submerged, and dead trees surround much of the western and southern shorelines. Access to the lake and hot springs has become more difficult and many temporary roads have been built.

Lake Bogoria is a perennial, saline alkaline lake with exceptionally high organic productivity dominated by planktonic cyanobacteria that sometimes feed more than a million lesser flamingos. The hot springs help to maintain the lake during drought. Before 2010, the abstraction of inflow water for irrigation, drainage diversions and their potential impact on the lake ecology were a concern. In contrast, heavy rains during and after 2010 raised the lake's surface to its highest level in more than 50 years (since the early 1960s), similar to its state at about 1900 (Johnston 1902b). Before those phases of high lake level, the high salinity of the water limited the biodiversity (except microbial life), but its biomass remained extremely high.

Geological evidence, however, shows that Lake Bogoria has been both fresher and deeper than today with greater biodiversity during the last few millennia. It might have overflowed northwards into Lake Baringo via the swamps on the Loboi Plain. At other times the lake has shrunk in both volume and area, become hypersaline, and precipitated sodium carbonate salts similar to modern Lake Magadi (Tiercelin and Vincens 1987; De Cort et al. 2013, 2018). Recent changes have been rapid and within our lifetimes.

The geology, sedimentology, recent history and hydrothermal activity have been studied in detail. Most geological research took place during the 1970s and 1980s. In addition, the limnology, biology and ecology of the lake have been the focus of many studies during the past two decades (e.g., Harper et al. 2003; Dadheech et al. 2013; Krienitz et al. 2016; Tebbs et al. 2013, 2015; Schagerl 2016). Much is now known of the lake, its ecology and its history, but gaps in understanding remain, including details of the lake hydrochemistry and the chronology of late Quaternary environmental changes.

9.2 Geological and Geomorphological Setting

Lake Bogoria occupies a narrow N-S half-graben in the axial depression of the Kenya Rift (Fig. 9.3), within the greater Baringo-Bogoria half-graben (McCall 1967; Griffiths 1977; Tiercelin and Vincens 1987; Hackman 1988; Grimaud et al. 1994; Le Turdu et al. 1995, 1999). The topography surrounding the lake is strongly asymmetric. The eastern lake margin is a steep, nearly unbroken N-S fault scarp that begins at Ilosowuani, 11 km northeast of the lake (Fig. 9.3a) and continues southwards along the eastern shoreline before changing abruptly to a southwesterly (NE-SW) orientation at the southern limit of the lake (Figs. 9.2a and 9.3a, b). That tectonic change provides topographic closure. This fault scarp, the Bogoria-Emsos Escarpment, rises up to 1700 m above sea level or about 700 m above the lake surface. The dip-slope east of the Bogoria fault-scarp falls steadily to the Waseges River valley before meeting a second major fault escarpment (Laikipia-Marmanet, Fig. 9.3a), which represents the eastern edge of the Kenya Rift at this latitude (Jones 1988). The Laikipia Plateau lies beyond that escarpment.

Bedrock east of the lake is volcanic (lava flows and dykes) with thin interlayers of volcaniclastic and siliciclastic sediments, and palaeosols. The oldest volcanic rocks, exposed in valleys that incise alluvial fans along the foot of the Bogoria-Emsos Escarpment, are the Miocene Samburu Basalts (~20–16 Ma), which are overlain by the Miocene Rumuruti Phonolite series (McCall 1967; Griffiths 1977; Hackman 1988). These are the main parent rocks for sediments delivered to the eastern lake margins and from the Sandai-Waseges River.

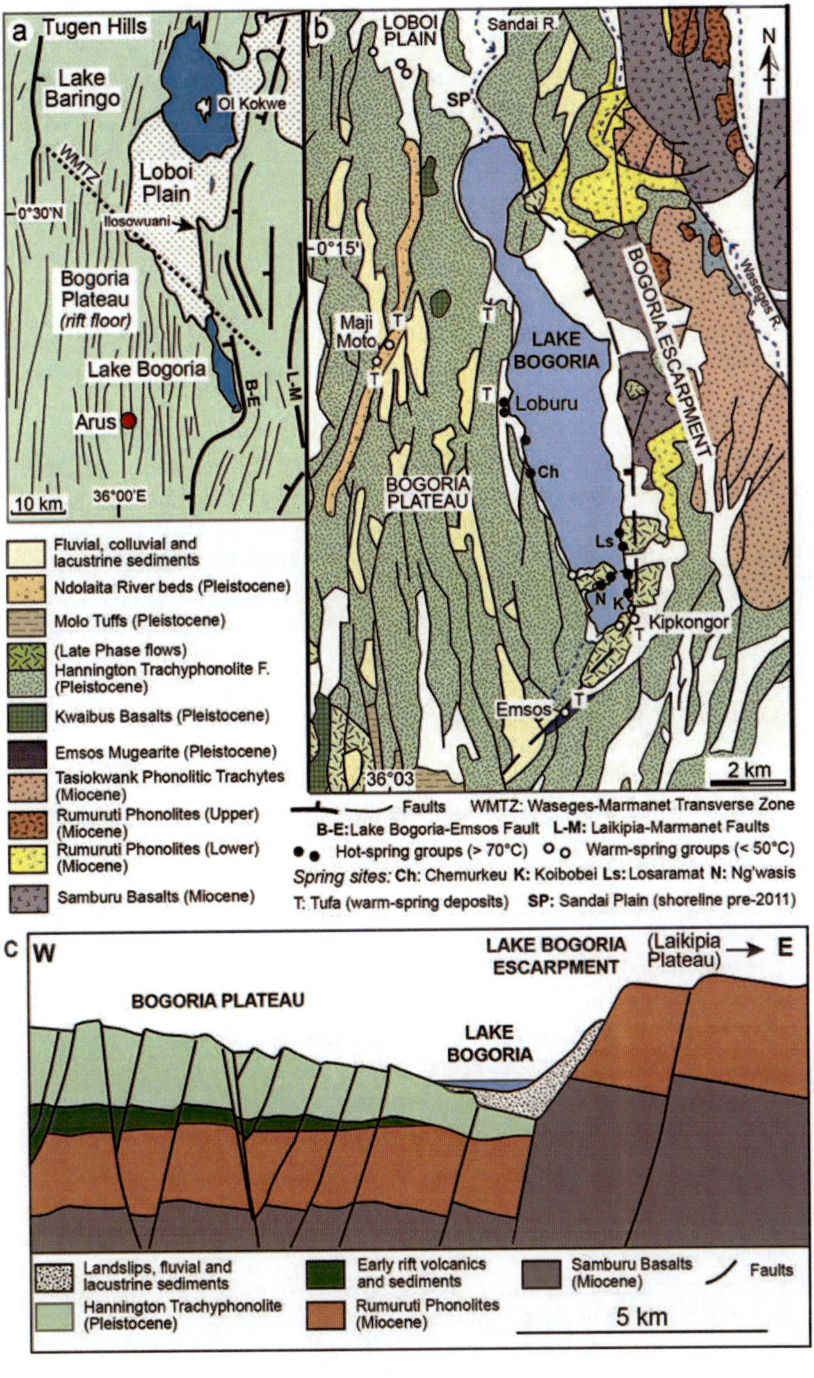

Fig. 9.3 (**a**) Geological setting of Lake Bogoria and Lake Baringo. (**b**) Geological map of Lake Bogoria. Simplified after Griffiths (1977). (**c**) West-to-east cross-section of the Bogoria Basin

The Bogoria Plateau (~1500–1000 m asl), which forms the down-faulted axial rift floor west of Lake Bogoria (Fig. 9.3c), dips northwards and is intensely grid-faulted with many small N-S horsts and grabens, including the larger Molo Graben, with most drainage flowing towards Lake Baringo (Griffiths 1977; Griffiths and Gibson 1980). The exposed volcanic rocks constitute the Pleistocene Hannington Trachyphonolite Formation, dated at ~1–0.3 Ma, and the older Kwaibus basalts (Chapman and Brook 1978; Clément et al. 2003). Two local volcanic centres are present: Kwaibus retains part of its original morphology, while Goitumet ~45 km to the south is preserved as an eroded massif younger than most Hannington lavas and is dated at <0.5 Ma. Griffiths (1977) recognised two phases of eruption of the Hannington Trachyphonolites separated by a Mid-Pleistocene period of faulting. The Lake Bogoria half-graben formed during and after the 'early phase', possibly at ~300 ka? (Griffiths and Gibson 1980, their Fig. 3) and giving a maximum possible age estimate for the lake or its precursor. The 'late phase' flows of the Hannington Trachyphonolites erupted during and after a tectonic phase during which the modern Lake Bogoria depression became defined. They include Ng'wasis, Losaramat, and other small volcanic flows that are now active geothermal sites (Sect. 9.5). Hannington Trachyphonolite that caps older volcanic rocks south, west and east of the lake records volcanic activity that predates the faulting that formed the modern half-graben (Griffiths 1977). Drilling and (or) geophysics will be needed to determine exactly when Lake Bogoria was initiated: the broader basin formed in the late Pleistocene and the oldest dated lake sediments are >20 ka.

North of the narrow watershed near Loboi village, the rift floor expands laterally into the broad alluvial Loboi Plain and Lake Baringo, with the Kamasia (Tugen Hills) fault-block to the west (Fig. 9.3a), and the Laikipia highlands to the east. The Quaternary Karosi (Korosi) volcano lies at the northern end of the Baringo basin.

The lake axis has an NNW-SSE orientation. In the five decades before 2010 the lake was ~17 km long by up to 3.5 km wide. Recent flooding of the Sandai Plain has increased its length to 20.1 km (late 2022) but the width has not increased by much because of the steep margins. The lake has three sub-basins separated by structurally controlled shallow sills where the lake narrows (Fig. 9.4). The bathymetry (Table 9.1) was first measured by Naylor (1972), then by Tiercelin and Vincens (1987, their Fig. 26), and Hickley et al. (2003). Each sub-basin differs in depth, salinity, ecology and sedimentation and they have been isolated from each other in the past few hundred years (De Cort et al. 2017).

The North sub-basin is relatively shallow because it captures much of the clastic sediment from the Sandai River and is separated from the Central sub-basin by the Nyalibuch sill. The sill depth is unclear (Table 9.1), but it does not stop Sandai sediment plumes from reaching the northern Central sub-basin. The latter shows more topography than the North sub-basin. Naylor's (1972) survey revealed several N-S axial 'trenches' in the Central sub-basin based on local depth measurements. Hickley et al. (2003) reported a 'zone of undulating bed'. The origin is unclear but diffuse sublacustrine hydrothermal discharge and (or) recent faulting are possible explanations.

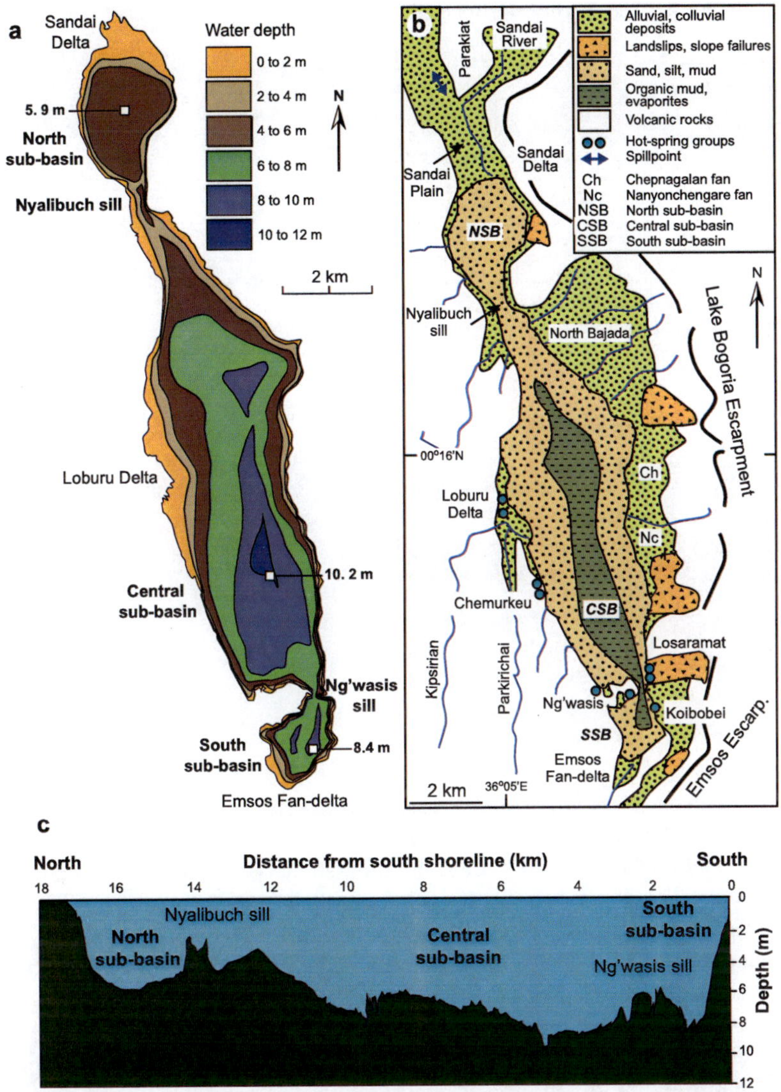

Fig. 9.4 Bathymetry and sedimentary environments of Lake Bogoria. (**a**) Bathymetry. (**b**) Lake sediments. (**c**) Lake floor topography. (**a**) and (**c**) are based on Hickley et al. (2003)

The structural context of the South sub-basin is complex reflecting several regional controls including major faults to the east and south (Griffiths 1977). The transverse Ng'wasis peninsula separates the Central and South sub-basins. The South sub-basin shallows towards the southwest but is deep south of the Ng'wasis sill (Fig. 9.4a, c) where sublacustrine hot springs are known to discharge.

Table 9.1 Recorded depths (m) of Lake Bogoria

Maximum depth (m)	1971 (a)	1977 (b)	1978 (b)	2002 (c)	2014 (d)
North Sub-basin	5–6	6	8.5	5.9	12.1
Nyalibuch sill	4–4.5	~1–2	?	~4	(985.5 m)
Central Sub-basin	9–9.5	9	11.5	10.2	16
Ng'airus sill	~6	5	?	~6	(982.7 m)
South Sub-basin	8–9	8–10	8–10	8.4	15.1
Lake level (m asl)[a]	989.5	990	992.5	980[b]	~996

From (a) Naylor (1972); (b) Tiercelin and Vincens (1987); (c) Hickley et al. (2003); (d) De Cort et al. (2017); quoted sill values expressed as elevation asl rather than depth
[a]As reported in original source
[b]Incompatible with other data

The region is seismically active (Cooke et al. 1989; Young et al. 1991; Zielke and Strecker 2009). A Magnitude 6.9 event in Subukia east of the lake in 1928 led to slope failures, drainage diversions and changes in hydrothermal discharge. The lake reputedly became black following the disturbance of anoxic bottom mud (McCall 1967; Ambraseys 1991).

9.3 Climate

The semi-arid rift valley floor where the lake is located receives about 450–1600 mm y^{-1} rainfall (Onyando et al. 2005; average ~700 mm y^{-1}) in two rainy seasons (April–May [*Masika*] and October–November [*Vuli*]) linked to the passage of the Intertropical Convergence Zone, with an additional July–August peak attributed to westerly airflow (Davies et al. 1985; LaVigne and Ashley 2001). Between December and March, the valley floor is usually dry. Higher rainfall (>1000 mm y^{-1}) is received in the highlands east of the lake (Kiage and Douglas 2020). Herrnegger et al. (2021) reported the mean rainfall across the catchment to be 1036 mm y^{-1} (1981–2020). Longer-term weather patterns include El Niño (wetter than normal) events about every 5–7 years and an occasional La Niña (drier than normal) event (Rowntree 1989; LaVigne and Ashley 2001). Potential evaporation is >2500 mm y^{-1}, so evaporation and transpiration greatly exceed precipitation. Mean annual temperature is 23–25 °C with a range of 14–35 °C. Typical daily maxima and minima are 27–35 °C (max) and 15–18 °C (min), respectively.

9.4 Hydrology

The inflow waters originate from four main sources: rainfall, ambient and warm springs, rivers (Fig. 9.5), and hot springs (discussed separately below). Rain falls on the lake commonly during convectional storms during the late afternoon, some

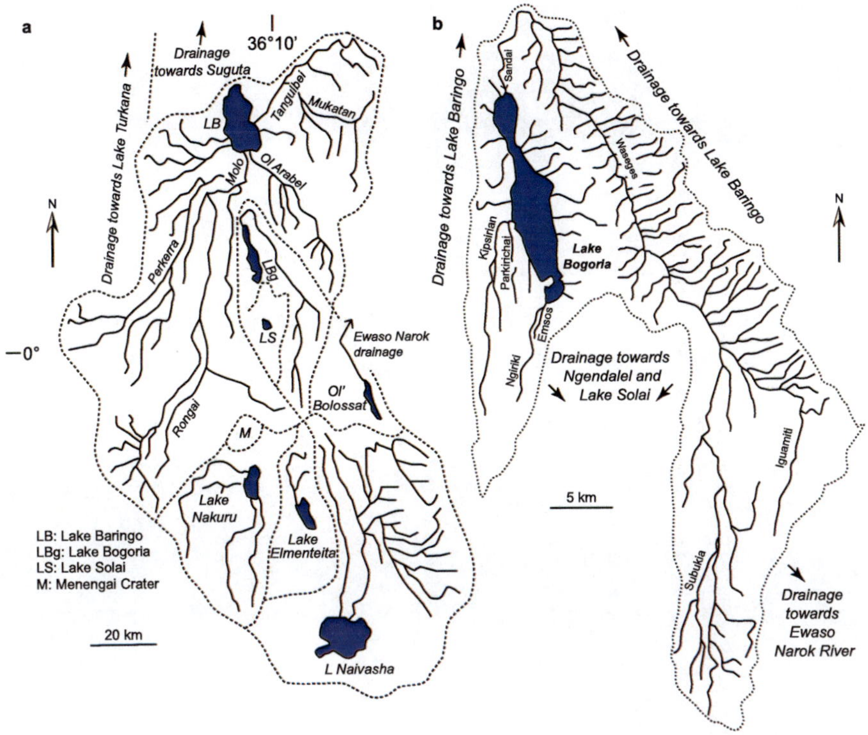

Fig. 9.5 Drainage basin of Lake Bogoria. (**a**) Regional drainage showing context of Lake Bogoria drainage. Based on Tiercelin and Vincens (1987). (**b**) Lake Bogoria drainage basin. From Renaut (1982)

lasting all night (Fig. 9.6a). Ambient and warm (<40 °C) springs discharge at several sites around the lake shore (Fig. 9.6b–d). Ashley et al. (2002) described cool artesian blister wetlands from near the northwest shoreline.

The Sandai River, which is called the Waseges in its upper course, is the main affluent stream. It rises in the Laikipia highlands and flows northwards parallel to the lake before reversing its flow direction and entering the lake along the northern shoreline (Figs. 9.5b and 9.7a, b). The discharge varies seasonally and from year to year. In some years it is perennial; in other years it dries up before reaching the lake at the Sandai delta (Fig. 9.7c, d). Previously, the river flowed along the eastern edge of Sandai Plain, but avulsion in 2008 led to the diversion of the lower course towards the western side of the plain. With the recent flooding, the river mouth is now (2023) located near the Kesubo swamp.

Sandai River inflow is supplemented by several smaller ephemeral streams including the Kipsirian-Parkirichai at Loburu (Fig. 9.7e), and many unnamed streams (Fig. 9.7f). The Emsos River in the southwest corner of the lake is perennial because it is fed by warm springs (Fig. 9.7g). Two small streams (Kipkongor, Kosirin) flow from freshwater springs below the Emsos Escarpment (Fig. 9.6c). The

Fig. 9.6 (**a**) Convectional afternoon thunderstorm over Lake Bogoria. (**b**) Dilute ambient freshwater spring, western shoreline at Nyalibuch. (**c**) Perennial warm dilute spring (Kipkongor) discharges below the Emsos Escarpment and flows into SE corner of the lake at Fig Tree camp site. (**d**) Ng'wasis warm spring

warm spring-fed Loboi River formerly flowed into Lake Baringo but was partly diverted into Lake Bogoria for irrigation (Harper et al. 2003).

Onyando et al. (2005) calculated that 37% of the annual inflow to Lake Bogoria derives from the rivers, ~34% from hot springs, and 29% from direct rainfall. Renaut's (1982) estimates were similar (41.7%, 31.8%, and 26.5%, respectively). The fluids that discharge at the springs originate from the local rift margins and rift floor, but some may derive from deep northward-flowing groundwaters. Becht et al. (2006) proposed that some groundwater may derive from as far south as Lake

Fig. 9.7 River inflow. (a) Northward flowing Sandai-Waseges River before its U-turn at Sandai Gorge. (b) South-flowing Sandai River, Sandai Gorge and its alluvial fan. (c) Sandai River (upstream view) in flood below Siracho Escarpment. (d) Shallow distributary channel, mouth of Sandai delta. (e) Ephemeral Kipsirian-Parkirichai River, proximal Loburu delta. (f) Ephemeral Ng'iniwa stream, northern Loburu. Sp: Small warm (41 °C) seepage spring discharging on the channel floor. (g) Perennial spring-fed Emsos River, proximal Emsos fan-delta

Naivasha, 100 km to the south, and possibly from subsurface discharge from Lake Nakuru and Lake Elmenteita.

McCall (1967) suggested that there is continuous removal of the saline water northwards towards Lake Baringo. Although feasible, there is no evidence to confirm this. Lake Bogoria has frequently produced evaporites and most water loss must be by evaporation.

9.5 Hot Springs and Geothermal Activity

Geothermal activity plays a major role in the hydrochemistry, limnology and sedimentology of Lake Bogoria. In 1977, when the lake level was at ~990 m elevation, Renaut (1982) mapped almost 200 subaerial hot-spring vents around the shoreline. Two main types are recognised – *warm springs*, which are relatively dilute with neutral to alkaline waters containing >2 mg L^{-1} Ca^{2+}, and *hot springs*, which are more saline with few alkaline earths and have higher pH. The terms 'warm' and 'hot' are used as relative terms, but they are distinct groups in temperature and chemistry (Table 9.2). All these springs might be termed 'hot' in other locations.

Most hot springs lie in three groups at Loburu, Chemurkeu, and in three neighbouring sites along the southern shoreline (Fig. 9.4b): Ng'wasis, Koibobei and Losaramat (southern group). Most fluids are of Na^+–HCO_3^- composition, except a few dilute sulfate springs that are condensates (Tables 9.2 and 9.3).

In form, the Bogoria hot springs fall within four categories (Sect. 4.8). *Pool springs* discharge warm to boiling water from pools usually hosted in poorly consolidated sediments. *Seepage springs* flow from vents without a well-defined pool, and seep lakewards quietly from siliciclastic sediments and littoral marsh. *Fracture springs* emerge from joints and fractured volcanic rocks or lithified sedimentary rocks. *Travertine springs* are those where fluids discharge from self-constructed

Table 9.2 General characteristics of the main hot-spring groups at Lake Bogoria

	Number of springs[a]	Range of temperature (°C)	Range of salinity (g L^{-1} TDS)	Range of pH
Loburu	55–60	43–97.5	4–5	7.7–9.5
Chemurkeu	45–50	66–96.5	5–6	8.1–9.1
Southern group				
Ng'wasis	20–24	64–97	1.5–24	7.5–9.4
Koibobei	30–35	80–98.5	1–19	6.8–9.7
Losaramat	15–18	92–98.5	0.8–26	8.8–9.9
Warm springs	10–15	32–45	0.3–0.7	6.6–7.8

From Renaut (1982), Allen et al. (1989), Cioni et al. (1992), and original data

[a]Varies with lake level. Many springs continue to flow when submerged, but some dry up when lake level falls or become fumaroles discharging steam and (mainly) CO_2

Table 9.3 Selected chemical analyses of Lake Bogoria hot-spring waters

Sample	Ref no	Notes	Date	Source	Temp °C	pH	Na	K	Ca	Mg	Cl	HCO_3	CO_3	SO_4	F	SiO_2
Loburu																
Spring KL6			Jul-05	1	97.5	8.2	1420	15.3	0.85	0.34	235	2965		63.5	63.5	112
Spring KL8			Jul-05	1	97	8.55	1350	14.5	1.02	0.43	220	3055		61.5	63	108
hs A	316		1985–7	2	98.9	8.2	1270	26.2	1.6	0.97	220	2810		62.4	60.8	108
B3	B3		1985–9	3	96.4	9.3	1669	37	0.7	0.16	400	2787		52		137
Spring KL30	KL25	geyser	Aug-06	4	98	8.28	1440	42	0.9	<0.1	306	2850	268	63	60	115
Chemurkeu																
KEN-5			Feb-94	5	98.5	9.02	1710	38	0.22	0.05	340	3750		65	75	135
KEN-7			Feb-94	5	91.5	8.34	1590	32.6	1.74	0.55	290	3460		55	68	120
MM4			Oct-91	6	98	8.75	1600	30	0.5	0	305	3480		58	65	130
hs B	210		1985–7	2	98.2	8.85	1630	34	0.58	0.16	323	3540		58	72.2	129
hs B	319		1985–7	2	97.6	8.94	1590	34.7	0.54	0.33	327	3480		58	72.2	127
Ng'wasis																
E Ng'wasis	MW2		Feb-94	7	95.5	9.15	6680	165	0.31	0.11	1650	13,980		85	280	185
S Ng'wasis	MW1		Jul-05	7	96.5	8.95	5985	154	0.28	0.09	1525	8560	2210	82	254	179
N Ng'wasis	NG4		Jul-05	7	96.4	9.0	5680	140	0.2	0.1	1510	12,100		76	250	168
E Ng'wasis	NG3	+5 m[a]	Jul-05	7	96	7.45	36	0.7	0.2	0.02	9	20		44	1	70
Koibobei																
S Kibwu	BH13		Aug-77	7	93	9.7	3995	71	0	0	1070	2575	2908	115	116	210
hs C	323		1985–7	2	97	9.32	4600	141	0.44	0.27	1210	9760		72	190	222
S Kibwu	KOB 1		Jul-05	7	97.9	9.31	4300	140	0.44	0.18	1100	8690		71	175	212
W Kibwu	KOB 2	+3 m[a]	Jul-05	7	94.6	6.8	64	1.5	0.27	0.05	14	28		72	0.6	70
Losaramat																
LOS 1			Feb-94	7	97.5	9.85	5945	119	0.64	0.14	1430	13,900		84.3	193	197
LOS 95 geyser			Jul-05	7	98	9.7	6300	120	0.4	0.03	1520	13,000		81	220	205

(continued)

Table 9.3 (continued)

Sample	Ref no	Notes	Date	Source	Temp °C	pH	Na	K	Ca	Mg	Cl	HCO_3	CO_3	SO_4	F	SiO_2
hs D	324		1985–7	2	95.3	9.5	7360	164	0.8	0.49	1740	15,900		86.5	285	218
hs D	325		1985–7	2	97	9.5	6900	149	0.46	0.28	1490	15,300		72	266	192
Warm springs																
Kipkongor	212		1985–7	2	44.4	7.11	122	14.9	15.2	4.98	8.5	360		15.9	4.6	70
Ng'wasis W	MW14		Feb-94	7	39	7.3	134	15.2	14.3	4.7	12	378		13	4.2	51
Emsos	B4		1985–9	3	34.5	7.6	93	10	8.9	3.8	12	283		20		58

Sources: 1: Renaut et al. (2013); 2: Cioni et al. (1992); 3: Allen et al. (1989); 4: Renaut et al. (2008); 5: Renaut and Jones (1997); 6: Jones and Renaut (1995); 7: Renaut et al. (2017).
All analyses in mg L^{-1}
[a]Elevation of vent above lake level when sampled

travertine deposits, mostly in the form of pool-rim dams and vent mounds, or older exposed and eroded carbonates. Several geysers are active periodically along the shoreline (McCall 1967, 2010; Renaut and Owen 2005; Renaut et al. 2008). Steam vents and fumaroles (with high CO_2 discharge and other gases) are locally common, especially in the southern spring group. Hot springs also discharge from the lake floor, often offshore from the subaerial spring groups.

9.5.1 *Loburu*

The most impressive hot springs in the Kenya Rift are those at Loburu delta on the midwestern shoreline, clustered in two groups (N and S) along fault lines (Figs. 9.1b and 9.8). When not submerged, the northern group consists of 30–35 individual springs that discharge hot and boiling waters from ebullient pools, some >3 m deep, or form smaller (dm to m wide) shallow seepages in deltaic sediments or marsh. A few springs in the northern group flow from (sub)fossil travertine deposits. The two most photogenic springs (KL6 and KL8 using numbering of Renaut 1982 and Renaut et al. 2013) are perpetually spouting springs that are often incorrectly termed geysers. The southern group (~25 springs) are mainly pool- and seepage springs but include several true geysers that are active only when the lake is relatively low (e.g., 2005–6; Renaut et al. 2008, 2013; Fig. 9.2c, d). Vent water temperatures for the main springs at Loburu range from 43 °C to 97.5 °C. The salinity is typically 4–5 g L^{-1} TDS and the pH is 7.7–9.5.

Spring activity is controlled strongly by lake level. The highest springs dry up occasionally when the lake level falls, and activity initially may increase during lake level rise until vents are submerged. The travertine deposits (calcite) have been dated at 32–35 ka using both ^{14}C and U/Th methods (Renaut et al. 2013). This confirms that the springs have been active for thousands of years, implying a mature plumbing system. Nonetheless minor travertine (calcite) and opal-A crusts precipitate from some modern Ca-poor waters (Renaut et al. 1998, 2013). Fluorite is locally present in the travertine.

9.5.2 *Chemurkeu*

At Chemurkeu (Mawe Moto in old literature; Loburu of McCall 1967), which is 2 km south of Loburu, about 45 small hot-spring vents discharge along a steeply sloping rocky shoreline over a N-S distance of ~250 m. The most intense activity is currently (1977–2019) at the northern part of the site (Fig. 9.9). Unlike Loburu, Chemurkeu does not lie along an obvious fault. Most spring vents are fracture springs that discharge from fissile (trachy)phonolite or fossil travertine. A few springs flow from small shallow (<1 m deep) depressions in gravelly sediments.

Fig. 9.8 Loburu hot springs. (**a**) Map of spring locations. From Renaut (1982). (**b**) Northern spring group (NSG) and thermal wetlands. (**c**) Southern spring group (SSG) and delta plain. (**d**) Small geysers, SSG. (**e**) Boiling pool spring, NSG. (**f**) Seeping hot spring, NSG. (**g**) Geyser KL30, SSG. (**h**) Small travertine spring, NSG. For details, see Renaut et al. (2013)

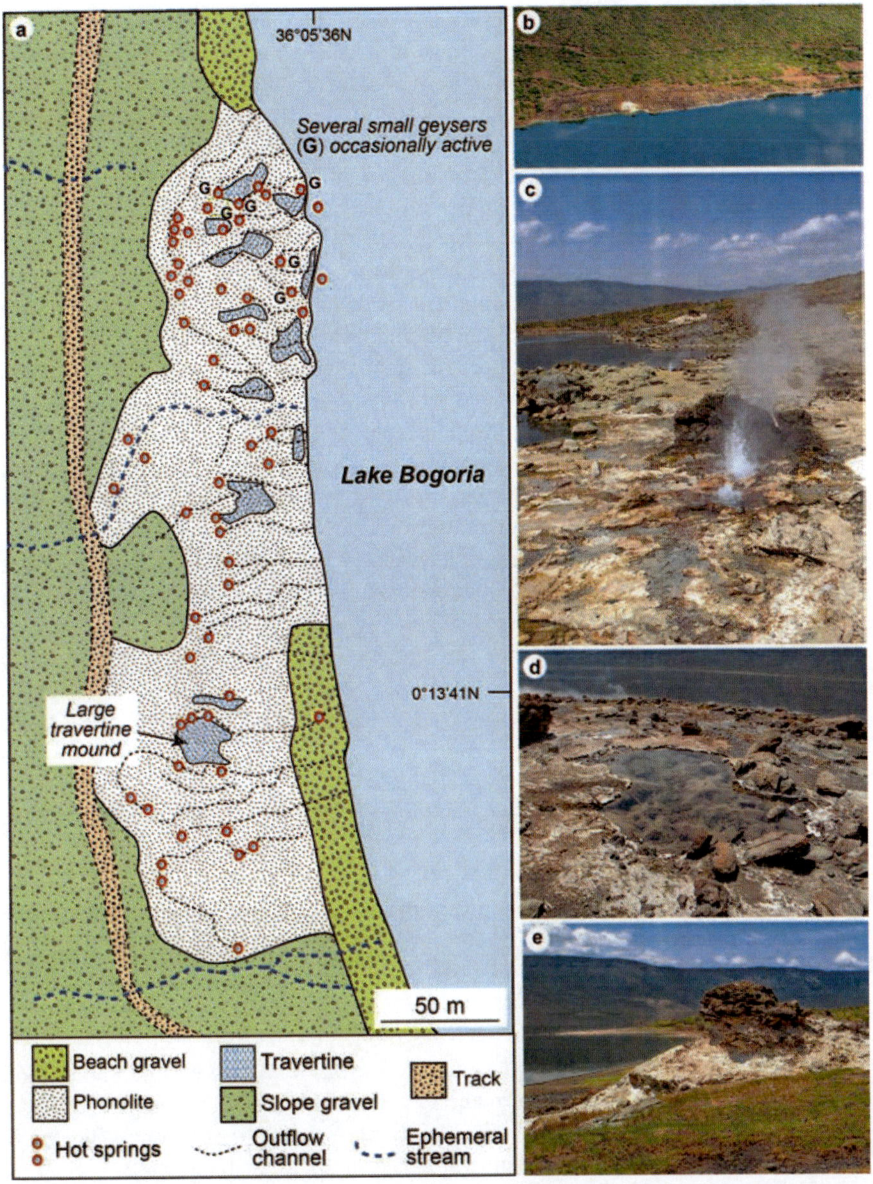

Fig. 9.9 Chemurkeu hot springs. (**a**) Geological map of the spring site. (**b**) View from the air. The white mass is the southern travertine mound. (**c**) Small geyser north central part. Fossil travertine is present at the surface. (**d**) Boiling pool spring hosted by fractured phonolite (a fracture spring), north part of the site. (**e**) Southern (bedded) travertine mound

Several small (<1.5 m high) geysers are sometimes present. Satellite imagery shows at least five offshore springs along a linear N-S trend up to 600 m north of the site.

The springs are slightly more saline (5–6 g L^{-1} TDS) than those at Loburu (Table 9.3) with similar pH. Travertine (mainly fossil) is more abundant than at Loburu, particularly at the southern end where remnants of an eroded mound are present (Fig. 9.9e). The travertine is composed of both calcite and aragonite and formed at the surface and in the shallow subsurface from boiling water undergoing rapid degassing of CO_2 (Jones and Renaut 1995, 1996, 1998; Renaut and Jones 1997). The travertine has been eroded and later encrusted by Holocene lacustrine stromatolites (Sect. 9.10.1), indicating a possible late Pleistocene age. A U/Th age of ~32 ka was obtained for a bulk sample, implying that the travertines formed at about the same time as those at Loburu (Renaut et al. 2013). Minor opal-A and Na-carbonate crusts precipitate around most vents and along outflow channels and line some joints. Rippled pale brown and purple laminated Mg-silicates up to 7 mm thick have precipitated on the proximal discharge apron of the northern springs and geysers.

In common with Loburu, spring activity varies strongly with lake level but not always in a predictable way. Observations over the past 40 years have suggested that the strongest discharge has been when the lake has been at 'intermediate' levels (~990–994 m). When the lake level is higher (e.g., June 2018) activity may decline, perhaps because the higher fluid pressures partly suppress shallow boiling and CO_2 degassing.

9.5.3 Ng'wasis – Koibobei – Losaramat (Southern Group)

Three groups of hot springs lie along the southern shorelines at (1) Ng'wasis (Mwanasis) – a narrow rugged peninsula that projects eastwards from the southwestern shoreline; (2) Koibobei (Kwaipopei, Kwaibeipei) along the southeastern shoreline, and (3) Losaramat, below a steep fractured trachyte cliff, 2 km north of Koibobei (Fig. 9.10a). Renaut et al. (2017) described these spring groups in detail.

The faulted Ng'wasis peninsula separates the Central and South sub-basins, with a narrow gap at Ng'airus, a small peninsula on the opposing eastern shore. Hot springs discharge along the southern and eastern shorelines of Ng'wasis, and along the southern shoreline of Ng'airus (Fig. 9.10b–d). A few springs lie on the northern shoreline of each peninsula near the gap. Lake-floor water temperatures <150 m south of the Ng'wasis sill locally exceed 65 °C indicating lake-floor spring discharge (Naylor 1972). The springs fall into two groups: saline Na^+–HCO_3^- springs that discharge nearly boiling alkaline water from littoral sediments or fractured bedrock; and dilute sulfate-rich springs a few metres higher that discharge condensate. Fossil opaline silica that cements shoreline gravels contains planktonic diatoms that record subaqueous silica precipitation in a former freshwater lake (Renaut and Owen 1988). No travertine is present but Holocene stromatolites coat some eroded siliceous spring deposits.

Fig. 9.10 South Bogoria hot springs. (**a**) Geological setting of the southern spring group. (**b**) Shoreline spring, SE Ng'wasis. (**c**) Hot spring, Ng'wasis. (**d**) Spouting hot spring, Ng'airus. (**e**) Spouting hot spring, northern Koibobei. (**f**) Thermally altered ground, Koibobei. Arrow points to south-facing fumarole cliff shown in (**g**)

Koibobei encompasses the southeastern shoreline from Fig Tree Camp (outflow from Kipkongor stream) to ~200 m north of Kibwu fan-delta (Fig. 9.10a). About 25 hot springs discharge along the shoreline. Most are small except those at Kibwu, which include large boiling springs both onshore and several metres offshore (Fig. 9.10e). East of Kibwu a south-facing cliff face is a site of many fumaroles and mildly acidic hydrothermal alteration with local formation of kaolinite, jarosite and native sulfur (Fig. 9.10f, g). Diatomaceous sublacustrine chert, similar to that at Ng'wasis, is present at south Kibwu hot springs and records a former period of higher lake levels (Renaut et al. 2017).

About 15 hot and boiling springs, some of which are spouting but <1 m high, emerge at the base of a steep cliff at Losaramat, most at or <50 cm above the contemporary shoreline. Minor opaline silica and trona precipitate around the vents. The springs produce a lot of steam visible from across the lake but are minor features on site.

9.5.4 *Warm Springs*

Ca-bearing warm (32–45 °C) springs lie at several sites around the lake (Fig. 9.6). With one exception (NW Ng'wasis, Fig. 9.10a), they discharge away from the recent shoreline. The Ng'wasis spring flows from a 5-m-diameter pool on the southwestern shoreline (Fig. 9.6d). Several dilute springs with fossil plant tufa deposits (Kipkongor, Kosirin, Emsos) discharge along the Emsos fault up to ~290 m above lake level (Fig. 9.6c); a few other springs are scattered around the lake margins. Although near the lake, other warm springs (Loboi, Maji Moto) lie in the Baringo drainage basin (Owen et al. 2004; Chap. 8). A few warm springs have been reported at the base of the alluvial fans of the Central sub-basin (Crane and O'Connell 1983). We have been unable to visit them, and they are now submerged.

9.5.5 *Sublacustrine Hot Springs*

In 2013 all Loburu springs became sublacustrine during a period of exceptionally high lake level. Many of those springs continued to discharge offshore, shown by plumes of hot water breaking the lake surface to form concentric waves disrupting plankton blooms, and by continuous streams of rising gas bubbles (probably CO_2). However, even when lake level is much lower, some springs discharge offshore from the main spring sites along inferred subaqueous extensions of faults or fractures.

A lake-bottom temperature survey of the southern part of the lake (Naylor 1972) showed several 'hot spots' where temperatures were >20 °C above the average. Those areas, in lake water >4 m deep, may be sites where thermal fluids discharge directly from lake floor sediments or fractured volcanic rocks. It is unknown whether these fluids are released from vents or as diffuse hot seepage through lake sediments. Several minerals found in lake cores, such as thermonatrite ($Na_2CO_3{\cdot}H_2O$), are compatible with elevated temperatures of precipitation (Sect. 9.9.4), implying the discharge of shallow thermal fluids from the lake floor.

9.5.6 *Life in the Hot Springs and Thermal Wetlands*

Many hot springs are brightly coloured with microbial mats. Mats are colour zoned according to temperature from green (>60 °C), to shades of orange (40–60 °C), to brown (<40 °C). Most mats develop along the cooler damp margins of spring pools and their outflow channels. Visible mats are absent where water temperatures exceed 65–70 °C in permanently subaqueous sites near the spring vent or in the proximal outflow channel. The microbial flora includes the cyanobacteria *Leptolyngbya*, *Phormidium*, *Spirulina* and *Synechococcus*, and *Chloroflexi* bacteria (Grant and Tindall 1986; Péniguel 1987; Hindák 2001; Krienitz et al. 2003, 2005, 2016; Dadheech et al. 2013). Scott et al. (2019) described bioturbation of the mats around the springs. Renaut et al. (1998) described the silicification of some of the cyanobacteria around the pools and outflow margins attributing much of it to evapotranspiration. Gasse (1987) and Owen et al. (2004, 2008a) recorded the diatom flora in the cooler parts of the springs and thermal wetlands.

9.5.7 *Origin of the Hot Springs*

The geothermal fluids derive mainly from infiltration of rain falling locally on the flanks and floor of the rift. Glover (1972) proposed that the Loburu springs discharge local groundwater that has been heated by steam and has mixed with lake water during its ascent. Karingithi and Wambugu (2008) favoured a similar origin. Cioni et al. (1992) recognised two groups of Na-HCO_3 waters: (1) steam heated waters from a shallow aquifer (100 °C; 200 mg L^{-1} Cl^-) at Loburu and Chemurkeu; and (2) more saline waters from a deeper hotter (170 °C; 660 mg L^{-1} Cl^-) aquifer with high PCO_2. Allen et al. (1989) and Clarke et al. (1990) rejected steam heating of local groundwater and suggested that the fluids originate from a regional shallow groundwater body that has been drawn directly into a 'hot CO_2-rich upflow', which was not clearly defined. These authors proposed that the deeper fluids are contaminated by non-thermal dilute groundwater and lake water during thermal circulation. Hochstein (1999) proposed an origin by advection. In his model, rainwater seeps down faults and fractures to great depth under its own hydrostatic head, where it is heated on contact with hot rocks and rises to emerge on the rift floor. It is a fracture-controlled system (Lagat 2008). Dyke injection along faulted zones may be a magmatic heat source (Naylor 1972; Mwawongo 2006; Omenda 2008; Mulwa and Mariita 2015), but Hochstein (1999) suggested that crustal intrusions are not needed to supply enough heat. The focus of geothermal upflow lies near Arus (Fig. 9.3a), where strong CO_2-rich fumaroles and small hot springs discharge on the lower eastern hillslope in a graben adjacent to the Molo River, 14 km southwest of Lake Bogoria (Mulwa and Mariita 2013; Mwawongo 2017).

9.6 Vegetation and Wildlife

Many higher plant species are found near the lake (Vincens 1986; Harper et al. 2003; Mugo 2007; Wetang'ula et al. 2019). Grasses (Poaceae) include *Dactyloctenium aegyptium*, *Sporobolus ioclados*, *Digitaria velutina*, and *Chloris virgata*. Shrubs include *Grewia tenax*, *G. bicolor* and *Acacia* (*Vachellia*) *mellifera*. Among the trees *Balanites aegyptica*, *Combretum* spp., *Ficus* spp., *Terminalia* spp. and *Acacia tortilis* are common. No macrophytes grow along the shoreline except near hot-spring inflow where salt-tolerant grasses and sedges such as *Cyperus laevigatus*, *Sporobolus spicatus* and *Cynodon dactylon* are often present (Onkware 2000). *Opuntia* sp. (prickly pear cactus) and *Prosopis juliflora* (mathengi) are locally common invasive plants.

Fig. 9.11 Lesser flamingos at Lake Bogoria. (**a**) Flocks along the northwestern shoreline. (**b**) Thick (10 cm) green scum of *Arthrospira fusiformis* (inset) along the northern shoreline of Ng'wasis. (**c**) Modern flamingo nest mound, Loburu delta. (**d**) Subfossil nest mounds, Loburu delta-plain

More than 120 bird species have been recorded at Lake Bogoria of which about 50 are directly associated with the lake (Hartley 1986; North Lewis 1998; Harper et al. 2003; Mugo 2007; Baringo County Government 2020). The lake is also a feeding station for Palaearctic migrants. The most important bird is the lesser flamingo (*Phoeniconaias minor*: Fig. 9.11a) but many other birds use the lake and environs as a permanent or temporary home.

Leslie Brown (1959), who pioneered the field study of East African flamingos, visited “Lake Hannington” many times. He described details of their behaviour at lakes Bogoria, Nakuru and Elmenteita before determining that Lake Natron is their main breeding ground. He estimated that at times up to two million birds inhabited Lake Bogoria, but in some years only a few hundred birds may be present. Many other ecologists and biologists have since studied their behaviour, their migratory patterns between East African lakes, and the factors that influence or control their distribution and abundance. Krienitz (2018) summarises this research and references the important papers.

Scott (2010) and Scott et al. (2012) have shown that flamingos have periodically inhabited Lake Bogoria for thousands of years. ‘Practice nests’ (constructed but unused for breeding) are built in some years (Fig. 9.11c, d). Fossil flamingo nests, and sedimentary structures produced by flamingos are recorded in the late Quaternary sedimentary record of Lake Bogoria and other Kenya Rift sites, including Lake Magadi (Buatois et al. 2016; Scott et al. 2021; Chap. 12).

Although small, the National Reserve provides habitat for many mammals including the rare greater kudu, baboon, vervet monkey, gazelles, impala, dik dik, Burchell’s zebra, leopard, wart hog, jackal, aardvark, serval and others (Mugo 2007; Baringo County Government 2020). Other animals, including buffalo, inhabit the eastern lake margin below the Bogoria Escarpment where visitor access has been restricted.

The post-2010 flooding, which reduced the salinity from >50 g L^{-1} to ~15 g L^{-1} (Krienitz et al. 2016), enabled crocodiles and even hippopotamus to move into the lake from the wetlands and rivers of the Loboi Plain (William Kimosop, personal communication). Even when the lake was hypersaline, freshwater catfish (*Clarias gariepinus*) were washed into the North sub-basin from the Sandai River (Renaut and Owen 1991).

9.7 Limnology

9.7.1 Physical Limnology

Before 2010 Lake Bogoria had a surface area of ~34–36 km^2 (Renaut 1982; Obando et al. 2016) rising to 40.4 km^2 by December 2014 (Obando et al. 2016) and 41.2 km^2 in December 2020 (Herrnegger et al. 2021). Lake level has fluctuated between

~990 ± 1 m (summer 1976: Renaut 1982) and 998 m (Dec 2020). Some early papers quoted the level as 970–975 m, which is incorrect.

Several surveys have shown that the lake stratifies. Tiercelin et al. (1980) reported that the deeper Central and South sub-basins are meromictic with a dense anoxic monimolimnion (salinity: 90 g L^{-1} TDS: total dissolved solids) and an oxic mixolimnion (salinity: 55 g L^{-1} TDS) and with a chemocline at about 6 m depth in October 1977. Harper et al. (2003) similarly recorded less saline surface waters with 60% of the conductivity of bottom waters during an El Niño event in 1978. They also recorded oxygen stratification with an oxycline at 1–2 m depth in the sheltered South sub-basin, but at 5–6 m depth in the more exposed North and Central sub-basins where axial winds are effective. Surface waters were always supersaturated with oxygen. De Cort et al. (2013) found evidence of deep seasonal mixing and concluded that the modern Lake Bogoria is oligomictic: its water column probably does not mix fully every year, but neither is the lake permanently stratified. They noted that stratification is promoted when fresh (lower-density) rain-, river-, and spring- water spreads across the lake surface, hindering mixing with the underlying dense hypersaline brine. However, when the surface layer evaporates and the whole water column has a similar density wind mixing becomes possible (Fig. 9.12; De Cort et al. 2017), but benthic life is still excluded by the low solubility of oxygen in the warm hypersaline water. Lake Bogoria is meromictic most of the time, but deep mixing is occasionally possible when the lake level is low. Even then, however, the high density of the brine inhibits full mixing.

The lake water temperature is typically 23–25 °C at 4 m depth (Harper et al. 2003) rising to 32 °C in the upper metre during early afternoon, forming a temporary shallow thermocline. Bottom water temperatures are sometimes higher in the southern part of the Central sub-basin and in the South sub-basin, locally exceeding 50 °C. This is attributed to sublacustrine geothermal discharge (Naylor 1972; Renaut and Tiercelin 1994; DeCort et al. 2017).

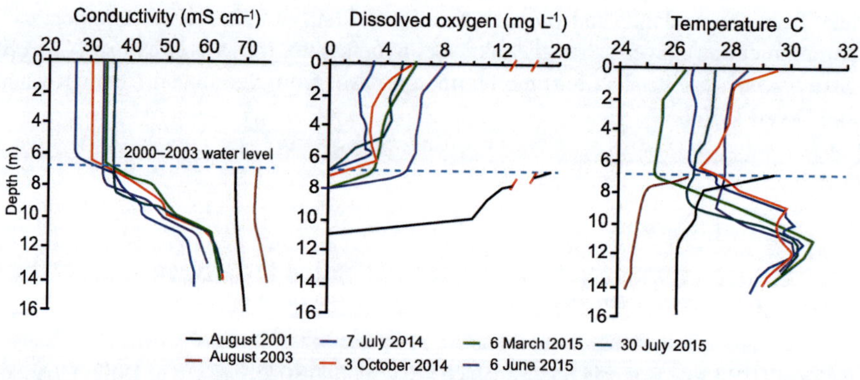

Fig. 9.12 Depth profiles showing the gradients of temperature, dissolved oxygen (DO) and conductivity in the water column of the three sub-basins of Lake Bogoria, recorded at the coring sites in August 2001, August 2003 and once each season between July 2014 and July 2015. From De Cort et al. (2017)

Winds are strong enough at times for waves to construct gravel beaches, but the energy levels are typically low. Surface currents are possible and might be reflected in the evolution of microbial scum lines that move across the lake surface, which tend to be anticlockwise. Small seiches are likely given that the winds are often funnelled along the N-S axis.

9.7.2 *Chemical Limnology*

The hydrochemistry of the Lake Bogoria basin (Table 9.4) is complex and only general relationships are summarised here. The lake is recharged by rain, channelled and unchannelled runoff, warm and hot springs, and groundwater, some of which has been modified by mixing with deeper thermal fluids (Renaut 1982; Renaut et al. 1987; Becht et al. 2006). In addition, CO_2 from deep (mantle) sources may contribute directly to the lake water and affect its alkalinity. Water is lost mainly by evaporation and evapotranspiration, leading to high salinity. McCall (1967) suggested that groundwater is lost by northward seepage towards Lake Baringo based on the difference in elevation, but the evidence is lacking.

The Sandai-Waseges River is dilute (0.2–0.5 g L^{-1}) with Na^+–(Ca^{2+})–HCO_3^- waters typical of most streams in the Kenya Rift. The recorded pH range has been 5.9–8.2 on the Sandai Plain. Ephemeral streams such as the Kipsirian-Parkirichai and many streams feeding the lake along its eastern and western shorelines have a similar composition but are sometimes more saline and alkaline because of evaporation on the rift floor. In most rivers, chloride, potassium and silica are also high but sulfate and magnesium are low. The Emsos, Kipkongor, Kosirin and other perennial streams along the southern shoreline are fed by warm springs but fall within the compositional range of other rivers. The hot spring waters discharging on the lake shoreline vary in composition (Table 9.3) but almost all are of Na^+–HCO_3^- composition, high in F^- and SiO_2, with variable Cl^-, SO_4^{2-} and K^+, but with negligible Ca^{2+} and Mg^{2+} (Renaut 1982; Cioni et al. 1992; Allen et al. 1989; Sect. 9.5).

The lake brine has a Na^+–CO_3^{2-}–HCO_3^-–Cl^- composition (Beadle 1932; Kilham 1971; Renaut et al. 1987; Cioni et al. 1992). K^+ is the subsidiary cation while Ca^{2+} and Mg^{2+} have low but detectable concentrations. After carbonate species (HCO_3^- and CO_3^{2-}), Cl^- is the subsidiary anion. Sulfate concentration is variable and often exceeded by fluoride. Silica concentrations range from <100 to 1210 mg L^{-1}. The pH has ranged from about 9.3 to 10.6, averaging 10.3. The salinity of the mixolimnion varies with location and time. The salinity is usually lowest in the North sub-basin, which receives the greatest river inflow. and increases in the Central and South sub-basins. The salinity of the monimolimnion is highest in the deeper parts of the southern Central sub-basin.

Several studies have considered the trace element composition of the lake water. Kerrich et al. (2002) and Jirsa et al. (2013) provided trace element analyses of lake and spring waters. Other work has examined specific elements and their potential

Table 9.4 Selected chemical analyses of Bogoria river and lake waters

Sample	Ref no	Date	Source	Temp °C	pH	Na	K	Ca	Mg	Cl	HCO_3	CO_3	T-Alk	OH	SO_4	F	SiO_2
Rivers																	
Sandai	SR1-76	Aug-76	1		7.5	70	8	30	9	38	246	0			5.0	0.5	51
Sandai	E62 135	Oct-78	2		5.9	50	tr	16	5	35	122	0			tr	0.1	45
Sandai	SR1-06	Jul-06	3	23.0	7.55	39	7.7	11	3	12	142	0			3.6	1.8	36
W Loburu	KL7-77	Aug-77	1		7.50	91	8	33	3	41	285	0			3	1.3	56
Parkirichai	KL1-91	Oct 91	3	28.4	7.44	57	16.9	18	6	30.1	196	0			6.7	1.8	35
Chepnagalan	E62 148	Oct-78	2		6.7	117	tr	32	4	35	336	0			tr	1.2	40
Lake																	
Surface waters																	
K1		Jan 70	4		10.6	28,400	387	0	2.2	6390			965		216	1060	260
N shore	BW1	Aug 76	1	27.0	10.1	19,550	280	tr	tr	4095			732		186	30	
N subbasin	Station 68	Nov 77	2	25.3	9.4	22,700	432	24	0	5850	0	23,700		1280	222		
N C subbasin	Station 77	Nov 77	2	25.9	9.4	23,500	418	22	0	6030	0	30,000		430	329		
C subbasin	Station 85	Nov 77	2	25.8	9.3	27,800	428	24	0	5850	0	31,500		0	434		
S subbasin	Station 37	Nov 77	2	27.8	9.4	26,500	464	24	0	5500	0	30,000			239		
C subbasin	Station 85	Oct 78	2	33.5	10.2	10,600	195	14	4	3900	4271	8407			68		
S subbasin	Station 37	Oct 78	2	34.0	10.2	10,810	197	17	4	4430	4030	8310			175		
Lake	321	1985–7	5	27.3	10.1	39,300	547	1.82	1.11	7460	85,400				144	1350	378
Lake (median)*		2008–9	6	28.3	10.0	25,860	414	4.2	0.4	5240	34,950	12,700			148	1100	68
Bottom waters																	
N subbasin	Station 68	Nov 77	2	27.2	9.3	41,200	660	28	2	9400	6100	43,500			340		
N C subbasin	Station 77	Nov 77	2	26.7	9.3	39,800	704	36	2	9220	6200	42,000			286		
C subbasin	Station 85	Nov 77	2		10.1	40,900	644	34	2	9040	6100	43,500			549		
S subbasin	Station 37	Oct 78	2		10.1	22,930	547	100	tr	9220	6410	18,900			354	0.1	
C subbasin	Station 85	Oct 78	2	28.9	10.1	26,000	390	14	1	9570	11,900	19,960			300	2.5	

Sources: 1: Renaut (1982); 2: Tiercelin and Vincens (1987); 3: Renaut, unpublished; 4: Kilham (1971); 5: Cioni et al. (1992); 6: Jirsa et al. (2013)
*Median value of several analyses from Jirsa et al. (2013).All analyses in mg L^{-1} except Total alkalinity (T-Alk) in meq L^{-1}

toxicity (e.g., Ochieng et al. 2007; Mercy et al. 2017; Yang et al. 2017; Githaiga et al. 2021). Bettinetti et al. (2011) examined organic toxins (DDT) in the lake water and sediments and found low levels of contamination.

9.7.3 *Evolution of Bogoria Basin Waters*

The evolution of the waters is understood in general terms, but many details are unclear. More than 150 analyses have been made but many remain unpublished. Unfortunately, few analyses have been obtained since the lake began to rise after 2010. Krienitz et al. (2016) found that the salinity had fallen to ~15 g L^{-1} in 2014, but details of the changes in composition during this transition are unreported.

Ion behaviour in the lake is also known but details of many processes are unclear. Eugster and Jones (1979) showed that in most closed-basin waters undergoing evaporation, chloride remains in solution until halite saturation, and can therefore be used as a conservative ion against which the behaviour of other ions and solutes can be compared. On a log-log plot using chloride as a tracer, an ion that is conserved (i.e. remains in solution during evapoconcentration) would lie along a 45° slope. If the water analysis for that ion plots below that slope, it implies that it has been removed or that the chloride concentration has been modified by mixing and does not represent simple evaporation of the original dilute inflow.

Figure 9.13 shows plots of various ions against chloride, and against sodium for some cations. Sodium is usually conservative until evaporites (gaylussite, trona, halite) precipitate. These plots use published analyses by Tiercelin and Vincens (1987) but exclude later analyses. Magnesium was below detection in many samples. The very low silica values reported in Tiercelin and Vincens (1987, Table V) are incorrect reflecting analytical error.

A plot of Na vs Cl in Bogoria waters (Fig. 9.13a) shows conservative behaviour of Na^+ until the most concentrated lake waters are attained when sodium loss is indicated possibly due to sodium carbonate (gaylussite, trona, nahcolite) or sodium silicate precipitation. The plot (Fig. 9.13a) also implies a mixed source of the lake water. Dilute inflow water alone cannot produce the lake water composition unless it has been relatively enriched in sodium through mixing with more sodium-rich fluids (loss of chloride without loss of Na is implausible). The hot spring waters, in contrast, lie along a slope that could produce the ratios measured in the lake waters. A plot of alkalinity vs chloride (Fig. 9.13b) shows that most hot springs have higher alkalinity than would be expected from the evaporation of dilute inflow. This 'excess' alkalinity may reflect an influx of deep CO_2 from geothermal sources or may indicate a source in an aquifer(s) with an alkalinity/Cl ratio higher than modern runoff. At high salinity (high Cl) a loss of alkalinity is implied in common with Na, suggesting some carbonate precipitation, or losses related to biological controls.

Ca^{2+} is removed early in the evolution of waters (Fig. 9.13c) by precipitation of calcite, which represents a critical divide in the evolution of closed-basin waters (Chaps. 4 and 5; Eugster and Jones 1979). Calcite is precipitated today in soils and

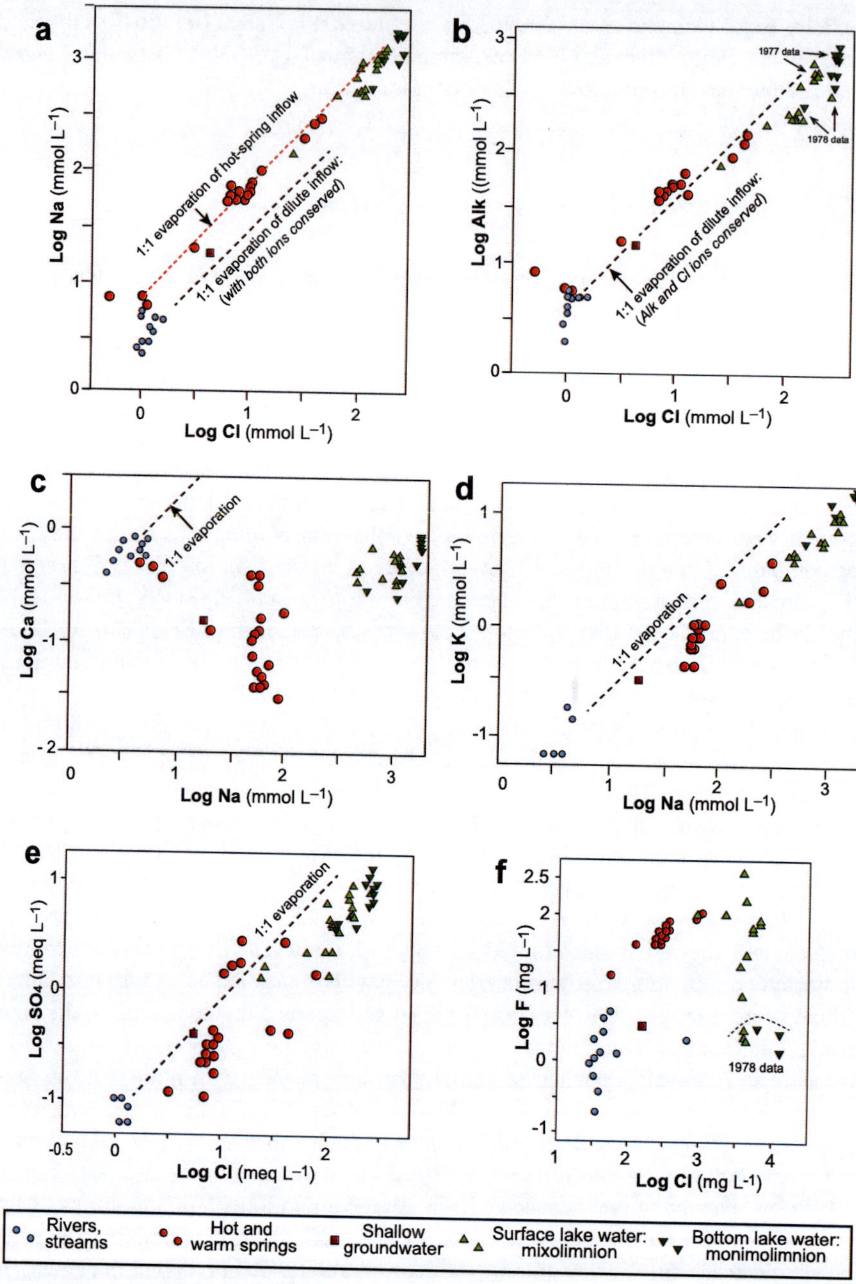

Fig. 9.13 Log-Log plots of Bogoria basin waters to show evolution with increasing salinity. (**a**) Na vs Cl. (**b**) Alkalinity vs Cl. (**c**) Ca vs Na. (**d**) K vs Na. (**e**) SO_4 vs Cl. (**f**) F vs Cl. Analyses are from 1977 and 1978 in Tiercelin (1981), Renaut (1982) and Tiercelin and Vincens (1987))

shallow sediments, often by evaporative concentration, and at hot and warm springs as travertine and tufa (Deocampo and Renaut 2016). In the lake, calcium precipitates in the water column as calcite or gaylussite, or is removed by organisms (molluscs, ostracods) during dilute episodes, or in diagenetic clay minerals.

Potassium for the most part is a conservative ion (Fig. 9.13d). Early removal of K^+ begins with the formation of clay minerals and sorption by volcanic glass in soils. Some is fixed by K-bearing zeolites. In the lake illite, interstratified illite-smectite and zeolites are possible sinks for potassium. Sulfate increases with increasing chloride concentration but the values are widespread (Fig. 9.13e). Some of the southern hot springs, which are more saline, have higher sulfate than Loburu and Chemurkeu. Loss in the lake waters can be attributed to bacterial reduction. Fluoride is naturally high in most Kenya Rift waters (Kilham and Hecky 1973) and is very high around the hot springs (Mutonga 2014). It is depleted in the lake and near the springs by precipitation of fluorite (Renaut 1982) and villiaumite (NaF).

Aqueous silica is high in runoff and shallow groundwaters compared to non-volcanic landscapes. Some groundwater is heated geothermally, which increases aqueous silica solubility, and discharges at the hot springs with concentrations typically from 100 to 115 mg L^{-1} SiO_2, implying that chalcedony controls its solubility (Cioni et al. 1992). Opaline silica precipitates as thin opaline crusts and in open cracks near the springs. Dissolved silica from rivers and hot springs then flows into the lake. At relatively low salinity and during periods of freshening, silica may be removed as diatoms, sponges, and clay minerals. With strong evaporation (salinity increase) and high alkalinity, silica is removed as zeolites such as analcime or as sodium silicates (magadiite, kenyaite, kanemite: Perinet et al. 1982; Icole and Perinet 1984). Abrupt pH reduction at a lacustrine interface such as the chemocline might induce precipitation of magadiite (cf. Eugster 1980), which is present in Holocene sediments, but this mechanism has not been confirmed at Lake Bogoria.

9.8 Biology and Ecology of the Lake

The saline alkaline waters of Lake Bogoria support a flora and fauna with high abundance (biomass) but relatively low diversity compared to other rift lakes. Early biological studies recorded flora and fauna (Johnston 1902b; Beadle 1932; Brown 1959) but detailed work was not undertaken until the 1970s (e.g. Hecky and Kilham 1973; Melack and Kilham 1974; Kilham 1981; Tuite 1981) and later by Lothar Krienitz, Michael Schagerl, David Harper and colleagues (e.g., Harper et al. 2003; Tebbs et al. 2013; Robinson 2015; Ward 2015; Amer 2019). Much of that work is summarised in the collected papers in Schagerl (2016).

A major focus has been on the prokaryotic microbial communities. Grant and Jones (2016) discuss the bacteria and archaea of the lake. The lake supports an extremely high cyanobacterial biomass that has been dominated (often >80%) by the cyanobacterium *Arthrospira fusiformis* (formerly known as *Spirulina platensis*), often present as nearly monospecific blooms that make the water bright green

(Fig. 9.11b). *Arthrospira fusiformis* is the primary food source of lesser flamingos (Harper et al. 2003; Oduor and Schagerl 2007a, b; Krienitz 2018). Other species include *Anabaenopsis arnoldii*, *Anabaenopsis* spp., *Microcystis flosaquae*, *Raphidiopsis* sp., and *Haloleptolyngbya alcalis* (Vareschi 1978; Kaggwa et al. 2013; Krienitz and Schagerl 2016). Other phytoplankton groups in the lake include cryptomonads and green algae (e.g., *Ankistrodesmus* sp., *Crucigenia* sp., and *Monoraphidium minutum*), and euglenoids such as *Euglena* spp. Diatoms include *Nitzschia* sp. and *Navicula* sp.

Krienitz et al. (2016) and Krienitz (2018) illustrated the close relationships between microbial productivity and biomass, species present, salinity and lesser flamingo numbers for the period 2001–2015. Large *Arthrospira fusiformis* was dominant from 2001 to 2005 when the salinity was ~50%, and the lesser flamingo population was >0.5 million. In 2006 lake level fell (Fig. 9.14) and salinity increased. *A. fusiformis* declined and was replaced by *Picocystis salinarum*, a tiny green alga that the flamingos could not ingest. Flamingo numbers dropped to <1000 birds. In 2008 *Arthrospira* recovered, and salinity varied. From 2008 until 2011 salinity varied (26–47%) and other species (*Cyanospira capsulata*) were present. The *Arthrospira* population collapsed by 2013 following heavy rains that began in 2010. Salinity (15.8%) and biomass declined heavily, and *Anabaenopsis abijatae* became dominant.

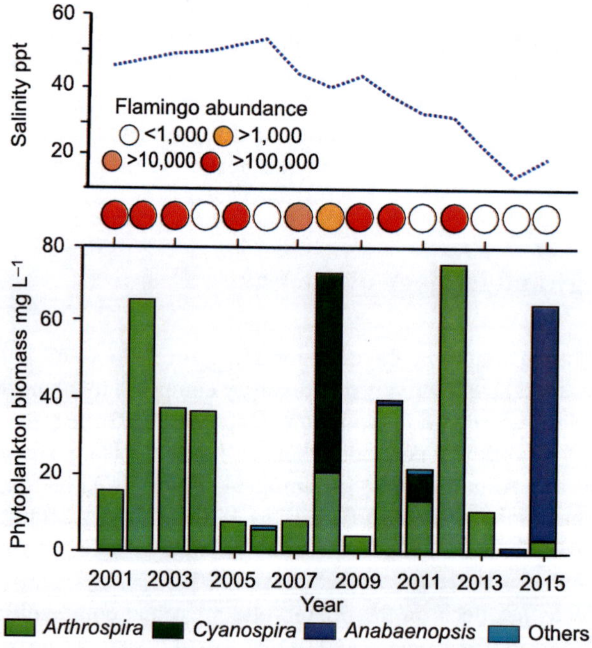

Fig. 9.14 Salinity, phytoplankton biomass and flamingo abundance in Lake Bogoria. After Krienitz et al. (2016)

Many researchers have tried to explain the sudden 'die offs' in the flamingos when numbers decline rapidly, and carcasses litter the shorelines to be consumed by marabou storks and other wildlife. Sometimes more than 10,000 birds die in such events. Many potential causes have been discussed (see Krienitz and Schagerl 2016 for summary) including variations in plankton productivity (Fig. 9.11b), poisoning, infectious diseases, and changes in food quality and abundance (e.g., crashes of *Arthrospira*). Toxins in cyanobacteria, including *Arthrospira*, have been widely discussed (Krienitz et al. 2003; Ballot et al. 2004; Straubinger-Gansberger et al. 2014; Muohi et al. 2018) as have viruses (Amer 2019). A combination of factors may be responsible (Harper et al. 2003; Krienitz 2018). The post-2010 rise in lake level aggravated the situation by diluting the lake water, which had an impact on the plankton composition, but also reduced the open space for landing along the shoreline. Many flamingoes died by being impaled on the thorns of drowned *Acacia* trees when landing or taking off.

Zooplankton are mainly rotifers *Brachionus plicatilis*, *B. dimidiatus* and *Hexathra jenkinae*. The invertebrate community is dominated by the chironomid *Patatendipes* sp. (zoobenthos, a detritivore) which occurs above the anoxic bottom waters at any depth (Harper et al. 2003). Insect communities in the area include Diptera such as tsetse fly *Glossina morsitans* and Odonata (dragon- and damsel flies) larvae, which are adapted to cope with extreme alkaline conditions. Ong'Ondo et al. (2013) also reported many ciliate species such as *Frontonia* spp., *Cyclidium*, *Rimaleptus*, *Halophyra*, and *Acineria*.

Fish and other biota are occasionally washed into the lake from the Sandai River. Their survival in the lake remains unclear. Studies of the changing lake chemistry and ecology during the current period of freshening would provide valuable information for understanding the paleolimnology of Lake Bogoria and other saline, alkaline rift lakes.

9.9 Recent Sedimentation in Lake Bogoria

Modern and late Pleistocene-Holocene sediments of Lake Bogoria have been studied in detail, using evidence from lake deposits exposed around the shoreline (Young and Renaut 1979; Renaut 1982; Casanova 1986; Renaut and Owen 1991) and from cores taken from the lake floor (Tiercelin 1981; Tiercelin et al. 1980, 1981, 1982; Tiercelin and Vincens 1987; Renaut and Tiercelin 1994; De Cort 2016; De Cort et al. 2013, 2017). Hot-spring deposits around the shoreline (Jones and Renaut 1995, 1996, 1998; Renaut and Jones 1997; Renaut et al. 1986, 2013, 2017) and ichnofossils preserved in siliciclastic sediments (Scott 2005; Scott et al. 2007, 2008, 2009, 2010, 2012) have also contributed to understanding sedimentation in the modern and late Quaternary precursor lakes. Most sediments in the Bogoria half-graben are lacustrine because its steep margins limit subaerial sedimentation. Soil erosion in the catchment resulting from land-use changes (Boitt et al. 2022) is having a serious impact on sedimentation.

9.9.1 Sedimentation Along the Eastern Boundary-Fault Margin (Bogoria-Emsos Escarpment)

The Bogoria Escarpment lies along an active normal fault with ~1 km throw, which merges with the transverse (NE-SW) Emsos fault escarpment at its southern end (Figs. 9.2a and 9.4b). Large coalescent alluvial fans (bajadas) and small steep alluvial cones, extend lakewards from the base of the scarp face forming a piedmont zone, broken by landslips and outcrops of Pleistocene volcanic rocks. South of the Sandai Plain the fan toes pass into shallow lake water, so at times the distal margins of the fans and cones alternate between being subaqueous (fan deltaic) and subaerial when lake level changes.

Landslips and rotational slumps have been recorded along the fault margin (Fig. 9.4b), some likely triggered by faulting (Ambraseys 1991; Tiercelin 1990). These poorly exposed deposits include large (>100 m long) slumped blocks of volcanic rock and megabreccias. The megabreccias are chaotic mixtures of large volcanic blocks within an unsorted matrix of gravel, sand and mud in units commonly tens of metres thick, some with erosional bases. Some are interbedded with conglomerates (waterlain and debris flow deposits) and separated by reddish palaeosols that record periods of non-deposition (Renaut and Tiercelin 1994).

Steep alluvial fans issue from re-entrants in the Bogoria Escarpment. Entrenched channels in their upper reaches reveal matrix-supported paraconglomerates deposited by debris flows, especially in the upper half of the fans. Waterlain orthoconglomerates and sands (streamflood deposits), and bedded sands and silts (sheetflood deposits) are common on mid- to distal fan sites (Renaut 1982). Short cores taken offshore from the fans revealed upward-fining cycles composed of massive gravelly sands overlain by stratified silts and fine sands, and capped by black or brown massive to weakly laminated muds, interpreted as turbidites, deposited from low- and high-density flows (Tiercelin 1981; Tiercelin et al. 1982). Offshore paraconglomerates with erosive bases, interbedded with mud, represent debris flows that entered the lake from steep fan-deltas and cones (Renaut et al. 2017, their Fig. 10).

9.9.2 Sedimentation on Littoral Platforms

Depositional surfaces with gentle slopes opposite border faults can be termed *lateral littoral platforms* whereas those at the ends of a lake parallel to the rift axis are *axial littoral platforms*. Sedimentation in both settings is deltaic and usually progradational unless in deep open (overfilled) lakes. At Bogoria, only one major axial littoral platform (Sandai Plain) is present because of the steep transverse structural closure at the southern end of the half-graben. Lateral littoral platforms are represented by the Loburu delta and minor ephemeral-stream deltas. All deltas along the western shoreline are small because most axial drainage (Ol Arabel, Molo rivers) by-passes the lake and flows northwards to Lake Baringo. Littoral (beach) sediments accumulate along steeper parts of the western shoreline between the deltas.

Loburu, the largest (up to ~0.5 km^2) lateral delta consists of two lobes separated by a marshy interdistributary embayment (Figs. 9.1b and 9.15a, b). Each lobe is fed by ephemeral braided streams (the larger Kipsirian-Parkirichai to the south; Ng'iniwa to the north) floored by boulders and poorly developed gravel and sand bars. They are incised in Holocene valley-fills with a proximal delta-plain platform near the lake (Renaut 1982). The Loburu delta plain is composed mainly of planar to gently dipping (but up to 15°) proximal delta-front sand, silt and mud deposited during former high lake levels. These prograding sediments are interbedded with regressive coarser channel and steeply dipping fan-sediments, beach deposits

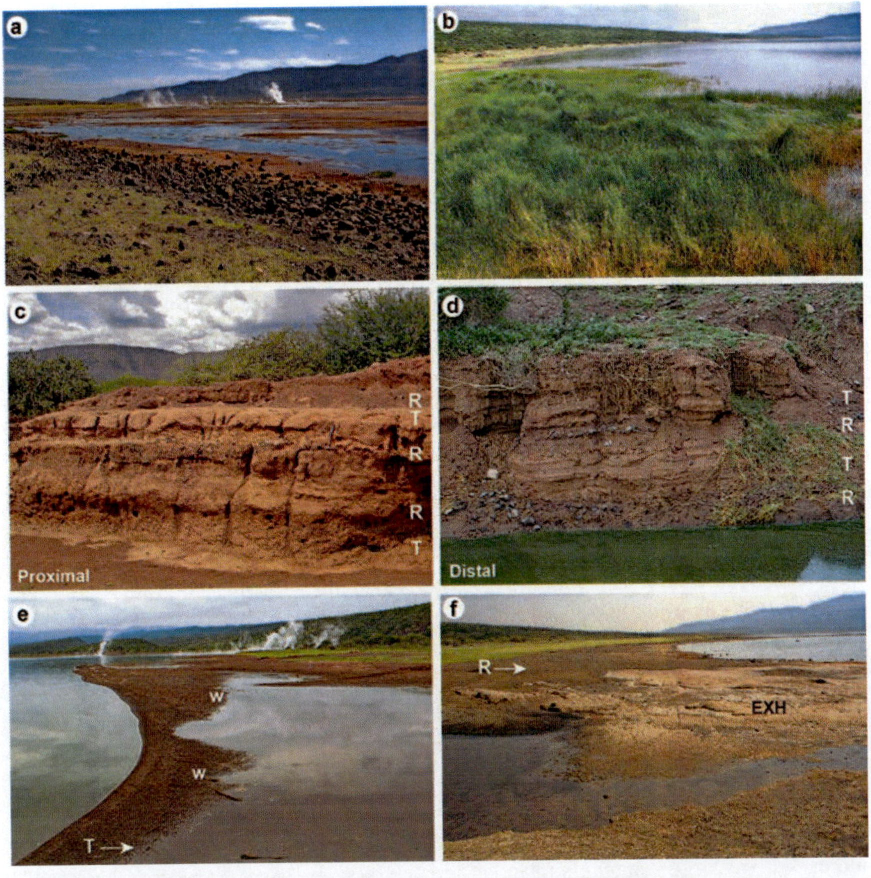

Fig. 9.15 Loburu delta. (**a**) Southern lobe. (**b**) Littoral *Cyperus laevigatus* marsh. (**c, d**) Proximal and distal delta plain sediments exposed in Kipsirian-Parkirichai River channel. The sediments show many small upward-fining cycles with regressive coarse sediments (R) passing upwards into finer transgressive sediments (T). (**e**) Transgressive (T) sandy bar with washover fans (W). (**f**) Regressive (R) beach overlying exhumed (EXH) delta-plain surface

(granules, gravel), and mixed poorly sorted littoral deposits reworked by waves and runoff (Fig. 9.15c, d). The delta stratigraphy is complex in detail reflecting changing lake levels and early cementation of some units, mainly during regressive stages (Renaut 1982). Older deltaic sediments are commonly exhumed during regression and sometimes eroded, producing an irregular substrate. Bioturbation by plant roots, flamingos and insects, efflorescent salt crystallization, and sheetwash often overprint the exhumed sediments (Scott 2005, 2010; Scott et al. 2007, 2009, 2010, 2012). Although small, Lake Bogoria has a dynamic shoreline environment. Beaches, narrow offshore barriers, and spits are common along the western and eastern shorelines (Renaut and Owen 1991; Fig. 9.15e, f).

9.9.3 *Sedimentation on Axial Platforms*

The axial deltas include the northern Sandai delta (1.2–1.5 km^2), which funnels large volumes of water and predominantly fine sediment into the lake that settle in deeper offshore waters, and the small Emsos delta (southwestern corner) which lies in the transition between axial and lateral.

The Sandai-Waseges River (Fig. 9.7a–d) has been the main river feeding Lake Bogoria for millennia. Its delta plain extends from the Siracho Escarpment (Fig. 9.7c) to the eastern edge of the Bogoria Plateau volcanic rocks, and from the contemporary northern lake shoreline almost to the Baringo-Bogoria watershed (~1000 m asl), located near the western edge of the Parakiat volcanic ridge (Figs. 9.4b, 9.16 and 9.17).

The Sandai delta plain (Sandai Plain) has undergone major changes during the past 50 years (Fig. 9.16b–d). In the summers of 1976 and 1977, it was a dry, largely undisturbed surface with shallow (<1 m) ephemeral channels and was partly covered by efflorescent trona and with little vegetation. By 1990 the distal plain had become the first site for tourists to photograph flamingos before driving south to Loburu hot springs. For much of the past decade, however, most of the Sandai Plain has been submerged. The modern (post-2019) delta front has migrated upstream more than 2 km since 2011 and the former delta plain is currently (2023) flooded (Fig. 9.17). Johnston (1902b) provided photographs of the delta plain flooded almost to the watershed at the beginning of the twentieth century, showing that present (2019–2023) high levels have happened before, and that lake level later fell to the same levels as the mid-1970s (when at ~990 m asl).

When exposed, the delta plain (Fig. 9.16c) is a flat to gently undulating surface composed of poorly consolidated silts and sands cut by the Sandai River and shallow ephemeral washes. Vegetation is limited to riparian stands (Harper et al. 2003) along the two larger channels (Sandai and 'Nyoka', west of the plain; Fig. 9.16b). The delta-plain sediments become muddier towards the shoreline. Unchanneled runoff reworks the surface sediments forming ripple cross laminae (Fig. 9.16e). During the dry season, mudcracks commonly develop and efflorescent salts (trona,

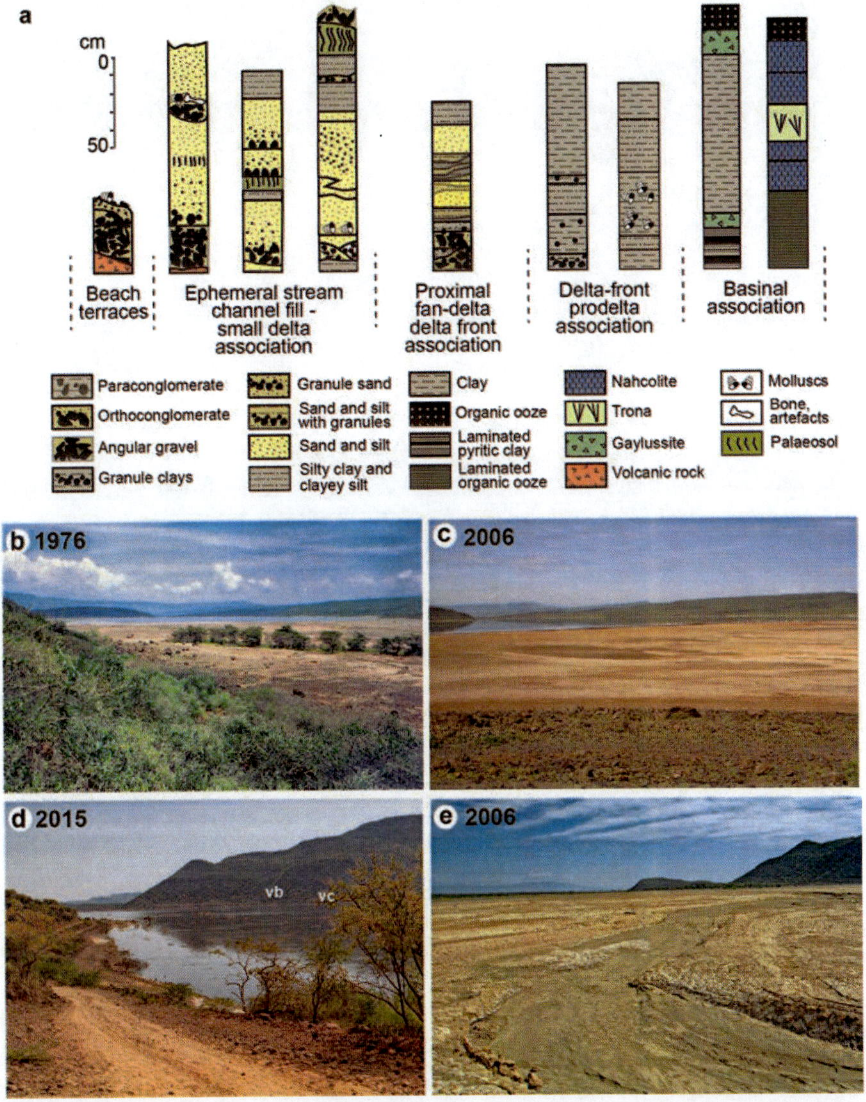

Fig. 9.16 (**a**) Facies associations in the Bogoria basin. (**b–e**) Sandai delta. Delta plain in 1976 (**b**), 2006 (**c**) and 2015 (**d**). (**d**) Ephemeral distributary channel (2005). (**e**) dry mudflat ~1 km from shoreline (1977). In (**d**), va and vb represent approximate locations where the photos in (**b**) and (**c**), respectively, were taken. Flooding after 2010 submerged most of the Sandai Plain

thermonatrite) cover the surface within a few hundred metres of the shoreline, often destroying bedding and lamination.

The modern Sandai delta has undergone many changes during the past five decades in response to changing lake levels, channel migrations (avulsions) and the

Fig. 9.17 Sandai delta in 2019–2020. (**a**) Lake Bogoria showing the extent of the lake in June 2019. The Sandai Plain is submerged and the delta has retreated. The green arrow shows the general location of any overflow at the Baringo-Bogoria watershed. (**b**) Map of the Sandai-delta in 2019. The shoreline has since migrated almost to Loboi village and the watershed. The Sandai Plain is submerged and the delta mouth has migrated upstream to the shrinking Kisibor (Kesubo) Swamp. (**c**, **d**) Mouth of the Sandai River showing main channel (**c**) and crevasse splays (**d**)

impact of drainage diversions and land-use changes in the drainage basin (Mumba 2008; John 2013), some of which have increased erosion and sediment supply but reduced the inflow.

Channel sections and pits show that recent sediments form thin (dm-scale) upward-fining cycles that record periodic rises and falls of the lake (Fig. 9.16a; Renaut 1982; Renaut and Tiercelin 1994). A northward thinning wedge of poorly consolidated sediments lies disconformably upon older Pleistocene deposits ('Loboi Silts') that form the substrate for the plain (Sect. 9.10.2). However, the Loboi Silts are laterally time transgressive. Those south of the Baringo-Bogoria watershed are partly lithified and cemented by zeolites and calcite (Renaut 1993) and are probably older than many of the unconsolidated 'Loboi Silts' to the north.

The small (0.15 km^2) Emsos delta, a fan-shaped bayhead delta at the southwestern corner of the South sub-basin is confined by steep slopes of volcanic rock. It is fed by the ephemeral Ngiriki stream that rises on the Bogoria Plateau and by perennial warm springs that issue below the Emsos Escarpment. The dilute springs have clear water and transport only minor sediment (Fig. 9.7g). The delta is densely vegetated (*Ficus sycamorus, Trichilia emetica*) because of the spring inflow (Vincens, 1986).

9.9.4 Sedimentation in the Axial Basin

The sediments of the axial basin are known from cores obtained in the 1970s (Tiercelin 1981; Carbonel et al. 1983; Tiercelin and Vincens 1987) and during the past decade (De Cort et al. 2013, 2018). Two main zones are recognised: a *peripheral* zone of mainly detrital siliciclastic sediments, and a *central* zone of mainly clay, organic-rich mud and chemical sediments (evaporites, silica) (Figs. 9.4b and 9.16a).

The peripheral zone extends up to several hundred metres from the contemporary shoreline and includes the North sub-basin. The sediments derive from the axial and lateral delta platforms and from the alluvial fans and deltas along the eastern margin. The central zone lies mainly in the southern Central and South sub-basins in waters that were >6 m deep in 1977–8 when coring was undertaken, and when lake level was ~990 m asl.

The sediments of the peripheral zone are mainly feldspathic sand, granules and gravel (mostly volcanic rock fragments) that pass distally into silt, mud and clay in deeper water. Fine sands and silts generally alternate with silty clay laminae, commonly containing allochthonous organic matter from the drainage basin. These appear to be deposited from suspension, often from hypopycnal plumes, some of which are seasonal. Clay flocculation often occurs where fresher delta plumes mix with saline lake water. The Sandai plumes commonly extend to and beyond the Nyalibuch sill and may drift towards the shorelines (Fig. 9.18a, b). Along the eastern shore they mix with fine sediments deposited from turbidity currents. The suspended sediment load has increased overall since the 1970s in response to heavy rains and increased soil erosion (van der Plaas et al. 2019; Boitt et al. 2022). In some dry seasons, however, the Sandai River has dried up south of the Kesubo swamp.

Fig. 9.18 (**a**) Suspended silt and clay from the Sandai delta moving southwards as a hypopycnal plume as far as the Nyalibuch sill, June 2019. (**b**) Plume of suspended fine sediments that has drifted to the western shoreline of the North sub-basin, August 2007. (**c**) Wind-blown microbial scum lines drifting towards the northwestern shoreline of the Central sub-basin, June 2019. (**d**, **e**) Green water rich in *Arthrospira*, South sub-basin. Arrows in d point to sublacustrine hot springs breaking the surface. (**f**) Cores showing interbedding of evaporites (trona) and organic muds in the southern South sub-basin. Trona is present as cumulates and bottom-nucleated fans

The deepest parts of the lake floor of the central zone lie below the chemocline and are composed of black sapropelic mud and ooze, and clay. These are best developed in the southern Central sub-basin and north-central South sub-basin. The muds, oozes and clays are massive or laminated with up to 6% TOC (Total Organic Carbon). Most organic matter is sapropelic derived from filamentous cyanobacteria and *Botryococcus* (Fig. 9.18c–e; Herbin 1979; Tiercelin 1990). Grewe (2016) examined biomarkers in Lake Bogoria sediments, which are well preserved because of anoxicity, and was able to relate certain types to lake-level changes.

The ooze and mud are interbedded with evaporite beds, commonly in cycles 10–40 cm thick (Figs. 9.18f and 9.19a; Tiercelin et al. 1982) which reflect short-term changes in salinity. The evaporite beds consist of trona ($Na_2CO_3 \cdot NaHCO_3 \cdot 2H_2O$), nahcolite ($NaHCO_3$), thermonatrite ($Na_2CO_3 \cdot H_2O$), halite (NaCl) and gaylussite ($Na_2Ca[CO_3]_2 \cdot 5H_2O$). Sodium silicates (magadiite, kanemite) and zeolites are common in some muds (Périnet et al. 1982; Renaut et al. 1986). The trona and nahcolite textures show evidence of cumulate sedimentation and bottom growth, similar to those at Lake Magadi (McNulty 2017).

Figure 9.20 shows the general relationships between climate, lake level and sedimentation. During the humid stages when the lake level is high, the lake waters are fresh and productivity is high. The lake is probably stratified with anoxic bottom waters that allow the preservation of organic matter. At intermediate lake level, the waters are saline. Clay settles from hypopycnal plumes with laminated organic muds forming during drier periods. When the lake is low, saline brines develop in the Central and South sub-basins (see below) surrounded by mudflats. Much of the recharge then derives from saline hot springs at or near the shoreline. Evaporites develop as cumulate crystals and bottom-nucleated crystals. Others form interstitially in brine-saturated mud.

9.10 Late Quaternary Sedimentation

The steep margins east and west of the lake have limited preservation of exposed late Quaternary sediments. The depositional record for this period is best preserved on the Sandai Plain and in cores from the lake floor.

9.10.1 Stromatolitic Limestones

The most obvious indicator of former higher lake levels is stromatolitic limestone (calcite) that encrusts bedrock, gravel, travertine, siltstone-sandstone, and coarse plant debris (tree trunks, twigs and reeds – many *in situ* at Ng'wasis) around the shoreline up to ~997 m elevation (Fig. 9.21a, b, e; Cole 1950; Renaut 1982; Vincens et al. 1986). Most crusts are 0.5–3 cm thick becoming thinner with elevation and to the north. They are grey to pale brown. Locally the stromatolitic crusts are up to

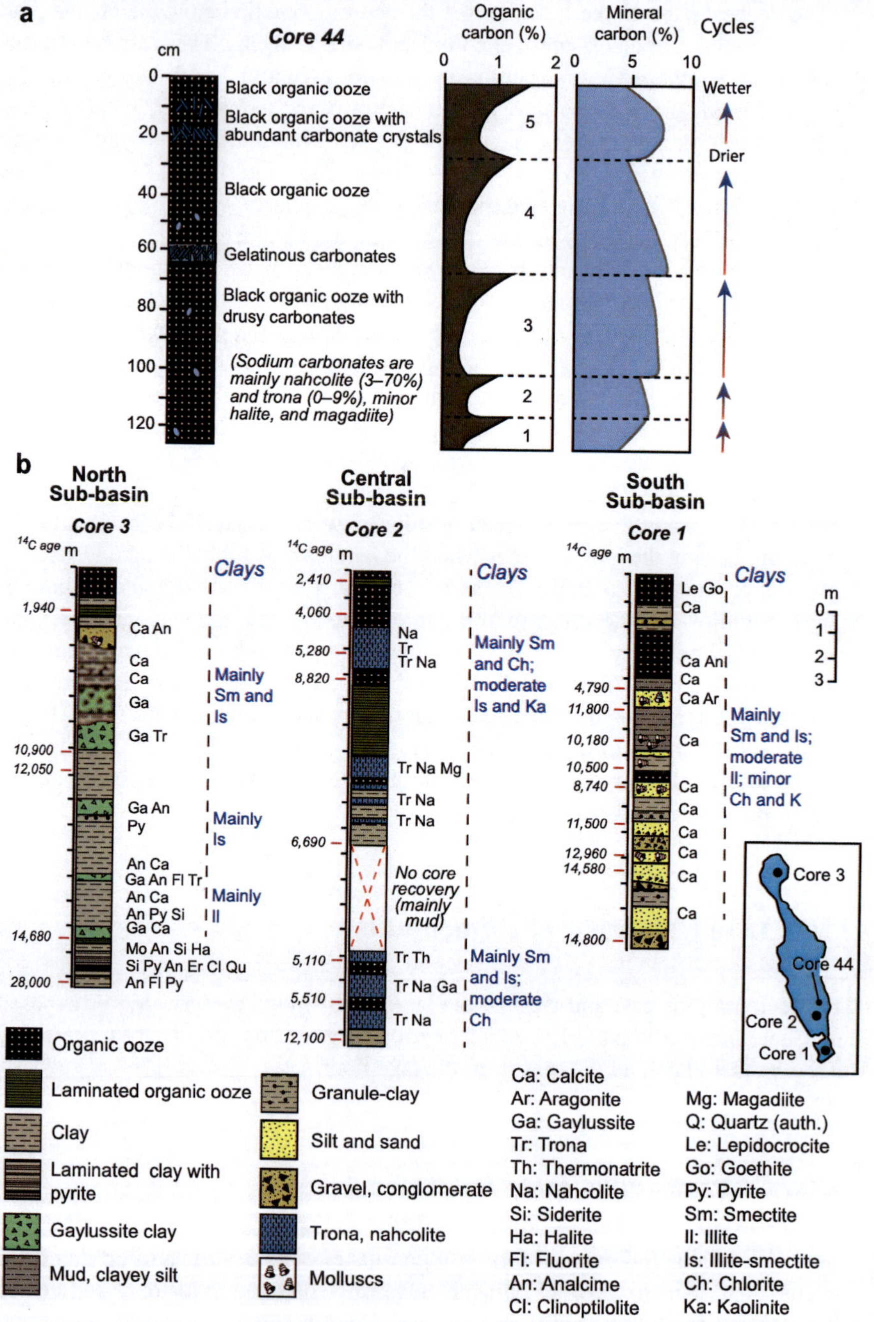

Fig. 9.19 (**a**) Cycles of evaporites and organic mud, Central sub-basin. After Tiercelin (1981). (**b**) Stratigraphy of three drill cores recovered in 1978. From Renaut and Tiercelin (1994)

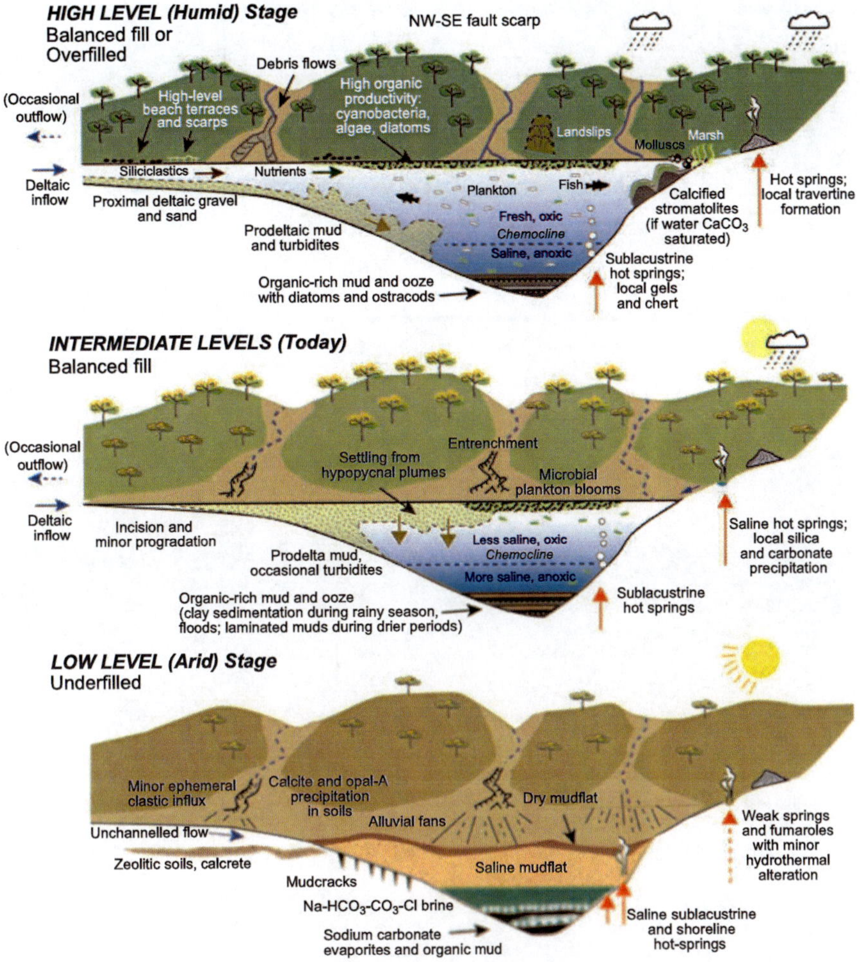

Fig. 9.20 Schematic model showing the main changes in style of sedimentation in Lake Bogoria according to prevailing climate and lake level. Modified from Renaut and Tiercelin (1994)

50 cm thick, particularly in the southern part of the lake, including the northern and southeastern shorelines of Ng'wasis peninsula, where at least two generations have been recognised, separated by discontinuities (Fig. 9.21c; Vincens et al. 1986). The higher (older and thinner) generation was dated at 4140 ± 60 y BP, whereas the younger generation was dated at 3880 ± 60 to 3750 ± 180 y BP. Hardwater effects mask their real age (De Cort et al. 2013) so they are probably younger. Thin (<0.5 cm), soft, malleable calcite crusts were found coating shallow (<5 cm deep) littoral gravel at the northeastern corner of Ng'wasis in 2006 when the lake level was relatively low (Fig. 9.21d). Some recent precipitation of calcite cannot be ruled out.

Fig. 9.21 Carbonates recording former high levels of Lake Bogoria. Stromatolitic limestones showing colour banding coating bedrock at Ng'airus (**a**, **b**) and southern Ng'wasis (**c**). (**d**) Thin microbial coatings on fossil stromatolites, NE Ng'wasis. (**e**) Reed, grass and wood tubules formed when microbial carbonate precipitated around *in situ* plant remains, SW Sandai Plain. (**f**) Oncoid showing concentric laminations. (**g**) Hemispherical stromatolite showing typical laminae. (**h**) Scatter of *Corbicula* shells from Bogoria Silts, SW Sandai Plain

Externally most crusts are composed of mm- to cm-scale laterally-linked hemispheroids (LLH type); internally they consist of microsparitic laminae that commonly contain dark erect or suberect filaments ~2.5 μm in diameter that resemble *Phormidium* in morphology (Fig. 9.21f, g; Renaut 1982; Casanova 1986, 1994; Casanova and Renaut 1987). Also present in the carbonate crusts are small molluscs, rare ostracods, chironomid? burrows, mm- to cm-scale oncoids, pollen grains, and detrital clay and feldspar (Cole 1950; Renaut 1982; Vincens et al. 1986). *In situ* reeds and tree stumps were also encrusted.

The stromatolitic limestones record periods when Lake Bogoria was higher, deeper, and fresher (i.e. with available Ca^{2+}), and probably at or near the contemporary watershed (spillpoint) level. They are absent in Lake Baringo at similar levels, implying that the lakes were not united when the Bogoria stromatolites were forming. Vincens et al. (1986) suggested that the crusts represent two phases of stable lake level separated by a brief regressive phase. The period after 4500 y BP was one of increased aridification, which would favour higher levels of $CaCO_3$ saturation, enabling thicker crusts to form, but as noted by De Cort et al. (2013), the Holocene radiocarbon-based chronology is uncertain.

9.10.2 Late Quaternary Sediments of the Sandai Plain

In an archaeological survey, Farrand et al. (1976) mapped two stratigraphic units on the Sandai Plain, which they termed the 'Loboi Silts' and the 'Bogoria Silts' (Fig. 9.22). The informal term Loboi silts was used by McCall (1967) and McCall et al. (1967) for "Recent sediments of the Lake Baringo Plain", which McCall (1967) reported to be at least 50 feet (15 m) thick but gave no details. Farrand et al. (1976) described the Loboi Silts as massive silty lacustrine beds with local coarse lenses and reddish palaeosols. Renaut (1982) and Renaut and Owen (1987) showed that the 'Loboi Silts' is a laterally time-transgressive unit that extends from Lake Bogoria across the watershed into the Baringo Basin where the silts underlie much of the Loboi Plain and overlain by younger alluvial and lacustrine sediments without a clear unconformity in the subsiding axial parts of the basin. The uppermost Loboi Silts of the Sandai Plain are orange and red and represent a zeolitic palaeosol (Hay 1970; Renaut 1993). This exhumed land surface was preserved by zeolitic (analcime) cementation and calcrete (Fig. 9.23). Similar red zeolitic siltstones and breccias at Ilosowuani, ~8 km south of Lake Baringo, are unconformably overlain by former beaches with molluscs deposited during the African Humid Period (AHP). Though undated, the red zeolitic deposits at Ilosowuani are presumed to be of similar age to those of the Sandai Plain, ~12 km to the south.

The 'Bogoria Silts' of Farrand et al. (1973) disconformably overlie the Loboi Silts and calcrete in the southern half of Sandai Plain (Fig. 9.23). They are stratified unconsolidated pale brown silts and clayey silts. Near their base, they locally contain fine gravel, *Melanoides tuberculata, Corbicula* sp., bird bones and charred wood (Farrand et al. 1976). Calcite rhizoliths, tubules and mollusc shells (mainly

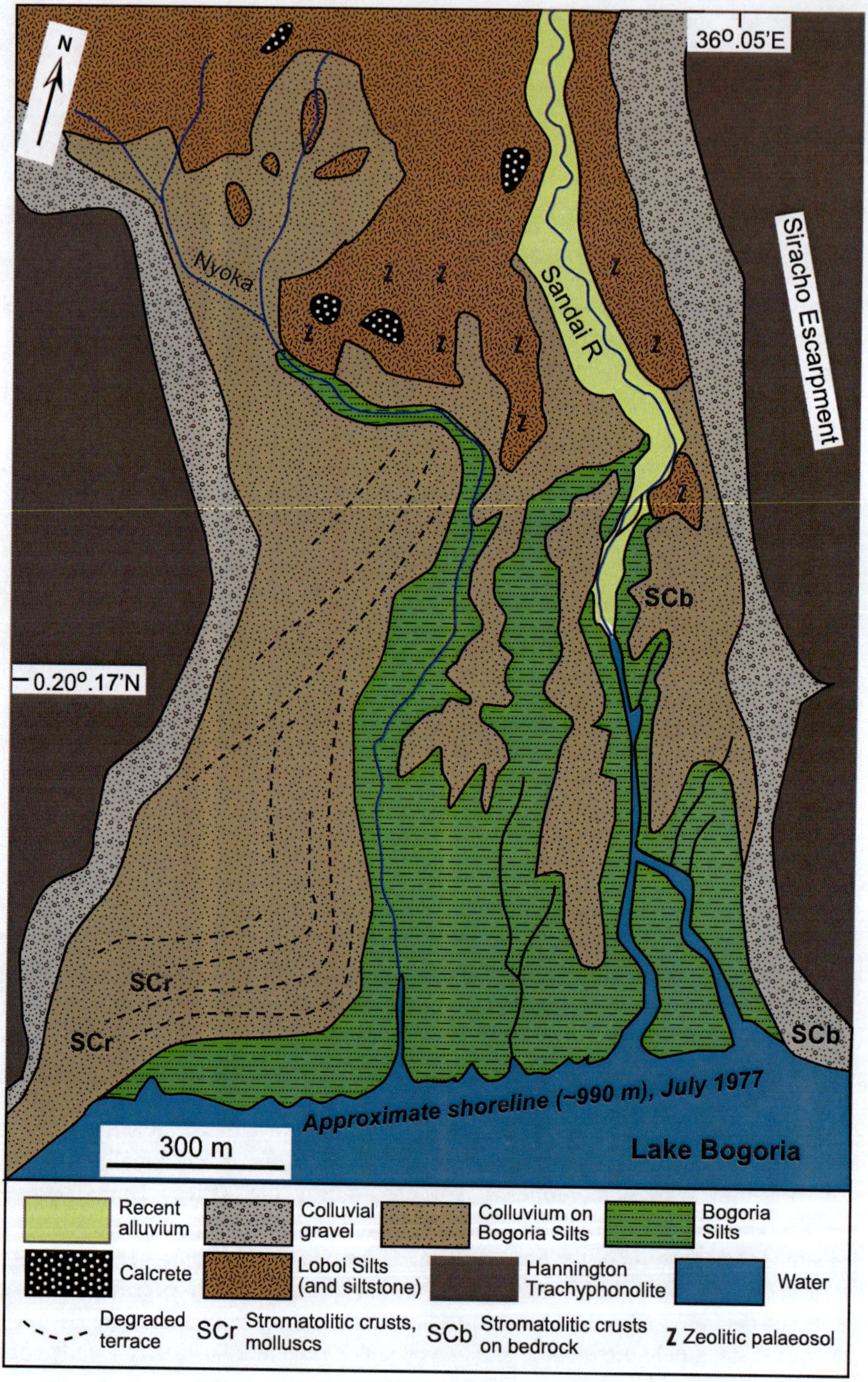

Fig. 9.22 Geological map of the southern Sandai Plain (pre-2011 flooding)

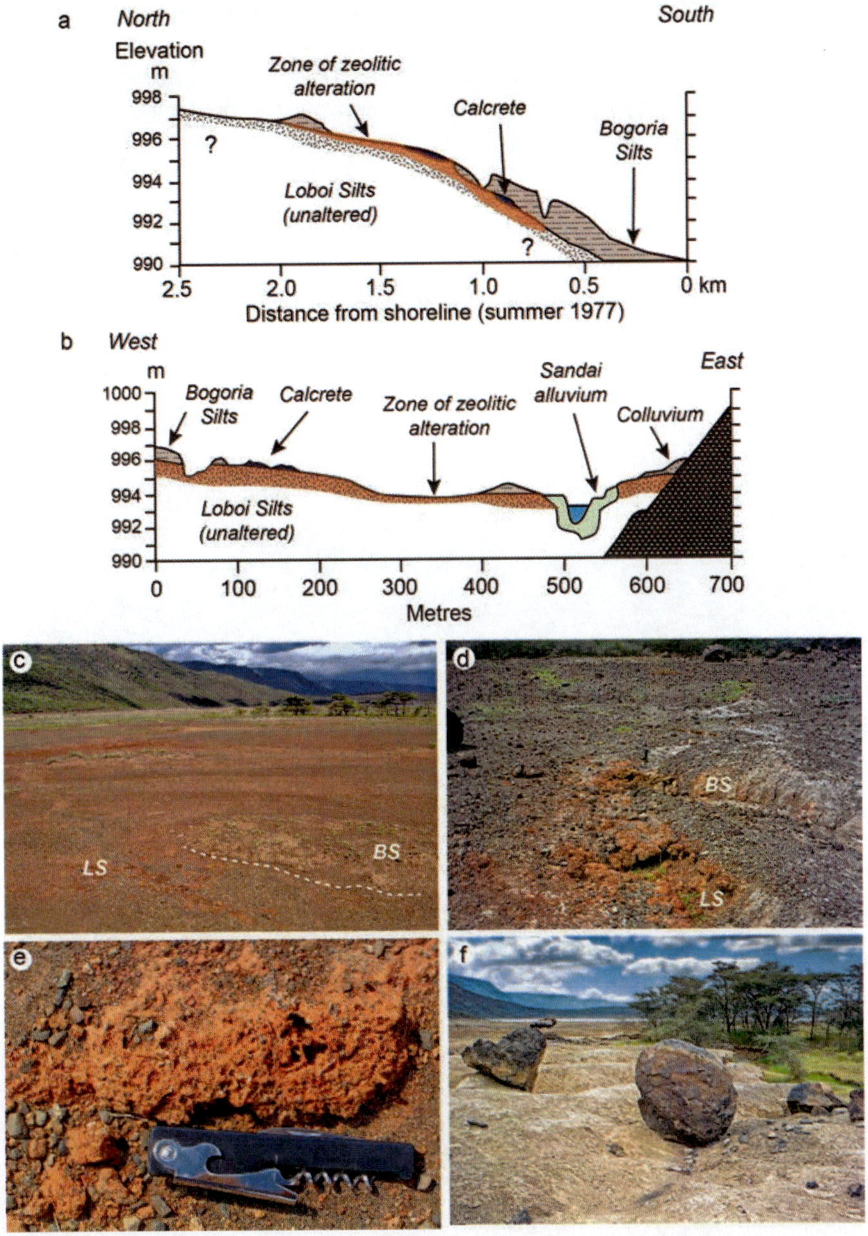

Fig. 9.23 North-south (**a**) and west-east (**b**) cross sections showing relationships among the sediments of the Sandai Plain. (**c**) Exhumed zeolitised land surface (orange) of Loboi Silts (LS) overlain by residual patches of buff Bogoria Silts (BS), south-central Sandai Plain. (Photo taken in 2006) (**d**) Outcrop of Loboi Silts overlain by Bogoria Silt and colluvial gravel, east of Sandai River below Siracho Escarpment. (**e**) Zeolitic rootmat, central Sandai Plain. (**f**) Eroded Bogoria Silts east of Sandai River overlain with tumbled boulders some of which have eroding coatings of stromatolitic limestone. (Photo taken in 1976)

Corbicula) at the SW edge of the plain record a temporary palaeoshoreline (Fig. 9.21e, h) that is undated but lies at the same level as a thin coquina (bioclastic grainstone) 5 km to the south, dated at 10,560 y BP (Young and Renaut 1979). The Bogoria Silts pass upwards into younger delta front and prodelta silts and clays without a well-defined upper contact. Local discontinuities are common. Recent fine gravel, sand and silt deposited by the Sandai River are locally entrenched in older sediments, recording some of the most recent fluctuations of the lake. The most recent sediments will be the deltaic deposits of the last decade. Their relationship to the Bogoria Silts, which have undergone minor lithification (patches of micritic cement), will be revealed when the lake level falls. Recent unconsolidated delta-front silt and mud may be washed into the lake by sheetwash and not be preserved.

The age of the Loboi Silts on the Sandai Plain is uncertain. They are presumed to have formed during the Late Pleistocene (>20 ka?) because the zeolites and calcrete needed time to form and be eroded, and they are overlain disconformably by mollusc-bearing silts dated elsewhere to the African Humid Period (AHP: ~15–5 ka) (Williams and Johnson 1976; Young and Renaut 1979). The associated archaeology supports such an interpretation (Farrand et al. 1976).

9.10.3 Terraces and Incised Valleys

Evidence for former higher lake levels is preserved in siliciclastic sediments around the lake margins in two main forms: littoral terraces and incised valley fills. Littoral terraces are mainly depositional rather than erosional, but both lose definition with increasing elevation above the shoreline. Depositional terraces form and degrade rapidly so are difficult to date. Erosional terraces preserve breaks of slope where the shoreline is composed of sediments but are absent where the lake margin is bedrock. Small, incised valleys fed by lateral ephemeral streams are present along the N–S lake margins. Along the eastern shoreline, many dissect distal fan deposits, some preserving small lateral terraces within incised channels. Along the western shoreline, valley fills are composed of poorly bedded silts, sands and gravel with lenses of darker silts containing rhizoliths (Renaut 1982).

9.10.4 Evidence of Lower Lake Levels

Sedimentological and geomorphological evidence for former lower lake-levels is sparse because most would now be submerged or have been eroded. Low lake levels in June 2006 temporarily exposed fossil travertine offshore from northern Loburu, including small vent mounds and a former subaerial outflow channel (Fig. 9.24a, b). The largest mound has mini-terraces that indicate subaerial formation. The age of the mounds is unknown, but they are probably Late Pleistocene (Renaut et al. 2013).

Fig. 9.24 Exposed evidence for lower lake levels. (**a**) Fossil outflow channel of spring KL6, preserved as travertine. The palaeochannel was visible during low lake-level in summer 2006. Hot-spring waters flowed along the channel into a lake lower than that in 2006. Silica (opal-A) rhizoliths record littoral vegetation predating the spring channel (Owen et al. 2008b). (**b**) Small mounds of travertine that precipitated from two springs (both still active) that were located offshore in 2006. The travertine has evidence of subaerial precipitation (miniterraces), implying formation during a lower lake level (Renaut et al. 2013). (**c**) Large fossil earth-cracks on northern distal margin of southern lobe of Loburu delta in July 1976. The cracks contain authigenic analcime, implying that they formed during a period of low lake level with shallow saline alkaline groundwater. Their age is unknown but analcime often takes millenia to form (Richard Hay, pers comm, 1999)

The exhumed distal delta-plain at Loburu has large (up to 1 m across) fossil mudcracks that could only have formed when lake level was much lower than 990 m (Fig. 9.24c). XRD analysis of a mudcrack fill included analcime, which implies that the associated pore fluids were then saline and alkaline. The zeolitic palaeosols and land surfaces on the Sandai Plain (Fig. 9.23; Renaut 1982, 1993) similarly formed during a relatively dry period of low lake levels.

Lake floor bathymetry (Fig. 9.4a) shows a possible north-south palaeochannel entrenched in the Nyalibuch sill that might represents fluvial incision when the lake shrank and occupied the southern Central sub-basin. Hydrothermal alteration can increase when lake marginal groundwater falls and fluids become more acidic. Evidence is present near the shoreline at Koibobei (Fig. 9.10f; Renaut et al. 2017).

9.10.5 Sediments Preserved in Lake Cores

The late Quaternary history of the lake is known mainly from cores that were obtained in 1977 and 1978. One long core (up to ~19 m) was obtained from each sub-basin (Fig. 9.19b). The cores have been analysed and described in detail, and correlations attempted (Tiercelin and Vincens 1987). A major problem has been obtaining reliable radiometric ages. Hardwater (reservoir) effects account for much of this, including old inert carbon from hydrothermal sources. Cores acquired during the summer of 1996 were not studied and the records were lost.

Core 2 (18.5 m long) from the southern Central sub-basin, taken in water then 10.5 m deep, revealed greenish-black organic-rich clay, mud and oozes alternating with sodium carbonate evaporites (Figs. 9.18f and 9.19b). Mud containing diatoms and ostracods formed when the lake was as high or higher than today and relatively fresh. Organic ooze increases in the upper half of the core. The presence of fine lamination and high organic matter (up to 6% TOC) implies that for periods the deepest parts of the lake were stratified (meromictic) with anoxic bottom water. The evaporites formed when the lake shrank and was restricted to the southern Central sub-basin, bordered by mudflats and often fed mainly by hot springs, analogous to modern Lake Magadi and Nasikie Engida (Chap. 12). The base of the core has a ^{14}C age of 12,100 y BP, but the radiocarbon chronology of the core is clearly unreliable.

Cores 1 (South sub-basin) and 3 (North sub-basin) are mainly deltaic and prodeltaic siliciclastic sediments with complex stratigraphies that record many fluctuations in lake level (Fig. 9.19b). The deltaic sediments contain molluscs, fish bones, sponge spicules and pollen that indicate periods of former fresher waters. Core 1 (14.5 m long) was taken in the prodelta zone of the Emsos delta. The base was dated at 14,800 y BP. The lower 4.5 m consists of fan-deltaic paraconglomerates and granule clays that are overlain by clay, silt and biogenic lithofacies marking a quiet water environment in the delta front favourable for molluscs, fish and ostracods. The upper 4.5 m are clays with thin sands but no molluscs and rare ostracods, recording a more restrictive environment similar to that of today.

Core 3 (16 m long) was taken offshore from the Sandai delta in water 4.7 m deep. The basal muds were dated at 28,000 y BP. The basal 6 m are green to black clays, locally finely laminated, with pyrite, zeolites, gaylussite, siderite and fluorite – indicative of quiet water sedimentation in an anoxic environment. The latter three diagenetic minerals may reflect the reaction of alkaline brines with Ca-bearing minerals or inflow waters. These are succeeded by quiet water clay, silt and sand with periodic development of gaylussite during saline evaporative phases (Renaut et al. 1986). Sands and granules may be thin turbidites. The upper 3 m consist of sand and gravel with mollusc shells capped by clay and organic ooze. An undated disconformity indicates erosion probably linked to a lower level (or southward diversion of the Sandai River?), followed by the onset of conditions similar to the modern North sub-basin.

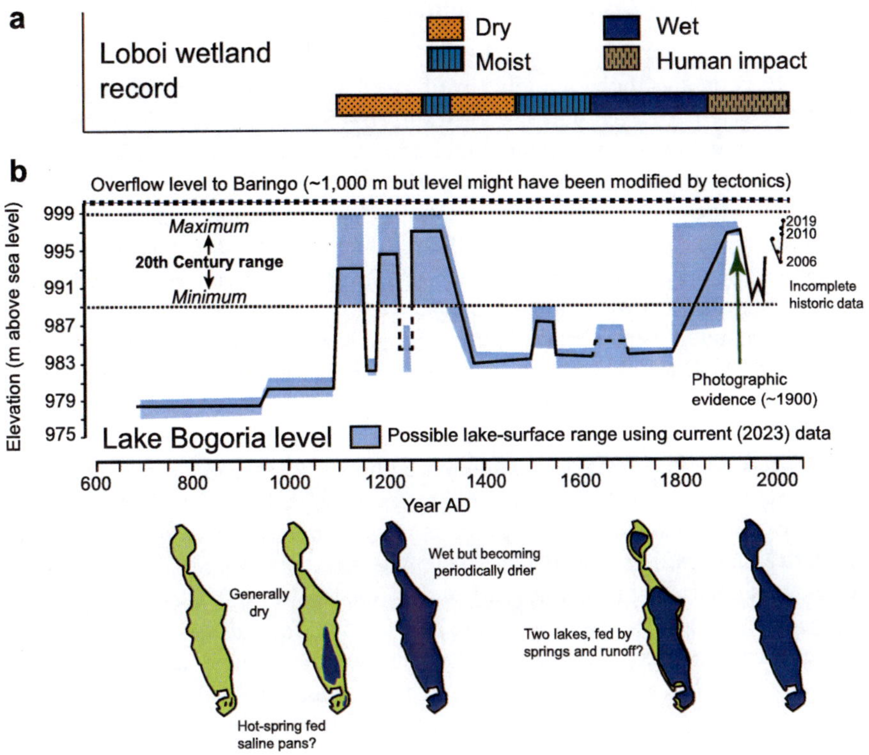

Fig. 9.25 Historical changes in the level of Lake Bogoria based on core evidence (De Cort et al. 2018) compared with a climate record from Loboi Swamp (Goman et al. 2020)

More recently De Cort et al. (2013, 2018) obtained, analysed and dated six short (1.2–3.0 m) cores to determine environmental changes in the lake over the past 1300 years (Fig. 9.25). In summary: the Central and South sub-basins became shallow separate brine pools between ~690 and 950 AD; near complete desiccation in the Central sub-basin led to trona precipitation. From 950 AD to ~1100 AD, slightly higher water levels and evaporative conditions continued and nahcolite precipitated, followed by a high stand between 1100 and 1350 AD. For much of the period between 1350 and 1800 AD, the North sub-basin was separated from the united Central and South sub-basins (i.e., two lakes). The lake has been relatively high for the past two centuries. Terrestrial sediment supply has increased, linked to increased soil erosion and land use changes (van der Plas et al. 2019).

Sanders (2016) and colleagues obtained six cores up to 4 m long primarily for chironomid analysis. Loss-on-ignition analyses were undertaken but radiometric dating was largely unsuccessful.

9.10.6 Former Connections with Lake Baringo

Lakes Bogoria and Baringo might have formed a single lake once or more during the late Quaternary (Nilsson 1931; McCall 1967; Spooner et al. 1969). This is reasonable based on basin morphology, their proximity and present elevations, and the low open topography separating the lakes. A glance at maps and satellite imagery shows that Bogoria could overflow into Lake Baringo and has been close to doing so during the past few years (2019–2023) and at about 1900 (Johnston 1902a, b). Nonetheless, physical evidence for a united Baringo-Bogoria lake during the AHP when other rift lakes were united is inconclusive.

The maximum former level of Lake Baringo based on palaeoshoreline evidence is a raised beach at Kokwob Murren on the mid-western shore of the lake. Its elevation was previously recorded as ~985 m asl (Tiercelin and Vincens 1987, their Fig. 97). In a recently located unpublished report on the locality, Spooner et al. (1969) reported that "The back of the raised beach was found to be 58.4 ft above lake level in August 1969" using levelling instruments. Lake level in August 1969 was 969 m asl as recorded by a gauge at neighbouring Kampi-ya-Samaki, which places the highest raised beach at Kokwob Murren at 986.8 m. Shelly strandline deposits at Ilosowuani (Logumukum) south of the lake were levelled at 987 m using a local datum (Williams and Johnson 1976; Figs. 17.17d, e, 8.14e–f, 8.16e). Google Earth places their height as 988 ± 1 m. Molluscs in those littoral sands at these two sites were dated as terminal Pleistocene to early Holocene (14–10 ka: Williams and Johnson 1976). The validity of those ^{14}C ages is questionable, but they record the AHP. A shell bed along the northwest shoreline of Lake Bogoria, levelled at 992.5 m, was dated at 10,560 y BP (Young and Renaut 1979), 4–5 m higher than the Ilosowuani-Kokwob palaeoshorelines of Lake Baringo but of similar age.

The highest known palaeoshoreline at Bogoria, based on the upper limit of the stromatolitic limestone crusts, is ~997 m, close to the present overflow sill-level of ~1000 m (Renaut and Owen 1991). Those limestones, however, have mid-Holocene (~4500 y BP) ^{14}C dates (Vincens et al. 1986), also of uncertain validity, so caution is needed when making inferences based on radiometric ages and the present elevations. Tectonic movements can displace shorelines and sills as recognised by Gregory (1921) a century ago, Nilsson (1931), and many others.

The spill point (lowest outflow level) for a combined Baringo-Bogoria lake would be the 'Lobat Pass' at the northwestern margin of Lake Baringo near Loruk (Gregory 1921; Nilsson 1931; Renaut and Owen 1980). Its elevation has not been accurately determined but is estimated at 987–991 m using satellite imagery. No obvious geomorphological feature (e.g., a dry channel) records a former overflow in this area. Neither is there clear sedimentary evidence of a late Quaternary Baringo palaeoshoreline higher than ~987 m. The overflow area lies in a narrow gap where Pleistocene lava flows are thinly covered by alluvial sediments. Natural drainage has been modified. Sediments in the northwestern part of Lake Baringo are fine grained (Tiercelin and Vincens 1987; Renaut et al. 2000), implying that any overflow would be of low energy or flow as marsh or seepage, and unlikely to cut a deep channel unless responding to northward tectonic subsidence. The northern end of Lake Ol' Bolossat may be an analogy (Sect. 14.11).

Several factors complicate the interpretation:

1. Lake Baringo levels are controlled not only by the Lobat sill but also by subsurface outflow (Chap. 8). Opening or blocking subsurface channels would affect the level of Lake Baringo or a combined Baringo-Bogoria lake. Faulting, fracturing, and tectonic subsidence could increase lake-floor permeability independently of climatic changes. In contrast, faulting could reduce permeability if fault-plane gouges (clay plugs) or hydrothermal cementation (silica or carbonates) prevented subsurface fluid flow.
2. The NW-SE Waseges-Marmanet Transverse Zone (WMTZ) seems to have acted as an oblique hinge line between the two sub-basins (Le Turdu et al. 1999, their Fig. 14). Old (cemented and zeolitised) Loboi Silts of the Sandai Plain lie south of the tectonic hinge, which coincides with the modern sill between the lakes. A few kilometres north of the watershed, the modern 'Loboi Silts' of the modern Loboi Plain are weakly cemented by calcite in places but with little diagenetic alteration. This implies that the Sandai Plain sedimentary record lies upon a more stable shallow volcanic substrate (Hannington Trachyphonolite) than the sediments north of the watershed.
3. Lake Baringo levels above 987 m might not been recorded by sediments. Beaches often form quickly but can also be eroded quickly. Much of the modern Baringo shoreline consists of marshy low-energy mudflats with limited coarse debris for beach formation. In contrast, the northern shoreline is steep volcanic rock with pockets of sediment preserved only in sheltered areas.

Nilsson (1931) correlated high-level shorelines at Bogoria and Baringo, and used it as evidence to suggest northward tilting of both sub-basins related to tectonics. In contrast, Gregory (1921) had earlier suggested that the Lobat Pass was uplifted with a southward tilt. Northward tilting of the basin is more plausible, but Nilsson assumed that the shorelines that he levelled record the same period(s) of high lake level. Le Turdu et al. (1995, 1999) interpreted the WTMZ as the location of northward (i.e. downward) flexuring of the Hannington Trachyphonolites, implying greater subsidence below the modern Loboi Plain. This is supported by geophysical evidence (Hautot et al. 2000).

The two lakes might have been united as a single lake for periods during the AHP, but more work is needed to confirm the hypothesis. It is difficult to unite the lakes at 987 m without tectonic adjustments. During the AHP it is likely that Lake Bogoria overflowed at times into the Baringo basin through the wetlands north of the present watershed. The lakes might have joined in the late Pleistocene when the Loboi Silts of the Sandai Plain were deposited, then possibly became separated when the Sandai fan grew and prograded westwards (Renaut 1982, 1993). The diversion of the Waseges River from a S-to-N direction to a southward orientation towards Lake Bogoria was influenced by the NW-SE (N130–140°) lineaments that define the orientation of the N and S Baringo shorelines and the WMTZ. Before that diversion the river might have flowed north towards the Loiminange fan.

Development of the Sandai fan would have been a major influence on the hydrology of both lakes. It might have gradually changed Lake Bogoria from an open to closed hydrology. Initially the Sandai River would have flowed west and northwards while the fan was constructed, but eventually flow was diverted southwards as today, perhaps induced by fault movement along the Siracho-Bechot Escarpment or by normal channel switching on the fan.

The timing of such changes, if they happened, is unclear. As well as the different levels of the palaeoshorelines in each basin, Lake Baringo does not have 'Bogoria-type' stromatolitic limestone encrustation around its shoreline. Modern and Holocene stromatolites are present locally at Kampi ya Samaki (Renaut et al. 2013) but at an elevation of ~971–972 m. When stromatolites were forming around the margins of Lake Bogoria (at ~4500 y? BP), the lake was a separate water body but might have overflowed at times into Lake Baringo.

It is also possible that Lake Bogoria received overflow from an expanded Lake Nakuru via Menengai Crater (Washbourn-Kamau 1971; Bergner et al. 2009) during the AHP. Mwaniki et al. (2019) proposed a direct hydrological link based on gene flow evidence. Dommain et al. (2022) later suggested that northward overflow from Menengai crater north of Lake Nakuru reached Lake Baringo, possibly via Lake Bogoria. The proposed route, however, is unclear. In their Fig. 1C, they show a direct link to the South sub-basin. In contrast, in Fig. S3B they show a river connection to Lake Baringo that bypasses Lake Bogoria (via the Molo or Loboi River?).

References

Akeley CE (1920) In brightest Africa. Double Day, Page & Co, Garden City, NY

Allen DJ, Darling WG, Burgess WG (1989) Geothermics and hydrogeology of the southern part of the Kenya Rift Valley with emphasis on the Magadi-Nakuru area. Brit Geol Surv Res Rep SD/89/1

Ambraseys NN (1991) Earthquake hazard in the Kenya Rift: the Subukia earthquake 1928. Geophys J Int 105:253–269

Amer ASE (2019) An environmental study of the factors contributing to the control of *Arthrospira* spp. in East African soda lakes (Lake Bogoria, Kenya). PhD thesis, University of Leicester

Ashley GM, Goman N, Hover VC, Owen RB, Renaut RW, Muasya AM (2002) Artesian blister wetlands, a perennial water resource in the semi-arid rift valley of East Africa. Wetlands 22:686–695

Ballot A, Krienitz L, Kotut K, Wiegand C, Metcalf J, Codd G, Pflugmacher S (2004) Cyanobacteria and cyanobacterial toxins in three alkaline Rift Valley lakes of Kenya--Lakes Bogoria, Nakuru and Elmenteita. J Plankton Res 26:925–935

Baringo County Government (2020) Lake Bogoria National Reserve management plan 2021–2029. Baringo County Government, Kabarnet

Beadle LC (1932) Scientific results of the Cambridge Expedition to the East African Lakes. 1930-1.4. The waters of some East African Lakes in relation to their fauna and flora. J Linn Soc Zool 38:157–211

Becht R, Mwango F, Muno FA (2006) Groundwater links between Kenyan Rift lakes. In: Odada EO, Olago DO, Ochola W, Ntiba M, Wandiga S, Gichuki N, Oyieke H (eds) Proc 11th World Lakes Conf, vol 2. Nairobi, pp 7–14

Bettinetti R, Quadroni S, Crosa G, Harper D, Dickie J, Kyalo M, Mavuti K, Galassi S (2011) A preliminary evaluation of the DDT contamination of sediments in Lakes Natron and Bogoria (Eastern Rift Valley, Africa). Ambio 40:341–350

Bergner AGN, Strecker MR, Trauth MH, Deino A, Gasse F, Blisniuk P, Dühnforth M (2009) Tectonic and climatic control on evolution of rift lakes in the Central Kenya Rift, East Africa. Quat Sci Rev 28:2804–2816

Boitt M, Gathoni J, Kaelo D, Koech L (2022) GIS-based multi-criteria evaluation to identify areas for soil and water conservation in lower Lake Bogoria landscapes, Baringo County, Kenya. J Geosci Environ Prot 10:64–92

Brown LH (1959) The mystery of the flamingos. Country Life, London

Buatois LA, Labandeira CC, Mángano MG, Cohen A, Voigt S (2016) The Mesozoic lacustrine revolution. In: Mángano MG, Buatois LA (eds) The trace-fossil record of major evolutionary events (Topics in Geobiol 40, vol 2). Springer, Dordrecht, pp 179–263

Carbonel P, Grosdider E, Peypouquet JP, Tiercelin J-J (1983) Les ostracodes témoins de l'evolution hydrologique d'un lac de rift: example du lac Bogoria, Rift Gregory, Kenya. Bull Cent Rech Explor Prod Elf Aquitaine 7:301–313

Casanova J (1986) Les stromatolites continentaux: paléoécologie, paléohydrologie, paléoclimatologie. Application au Rift Gregory. Thèse Docteur d'Etat-Sciences, Université d'Aix Marseille II

Casanova J (1994) Stromatolites from the East African Rift: a synopsis. In: Bertrand-Safarti J, Monty C (eds) Phanerozoic stromatolites. Kluwer, Dordrecht, pp 193–226

Casanova J, Renaut RW (1987) Les stromatolites du milieu hydrothermal. In: Tiercelin J-J, Vincens A (eds) Le demi–graben de Baringo–Bogoria, Rift Gregory, Kenya: 30,000 ans d'histoire hydrologique et sédimentaire. Bull Centr Rech Explor-Prod Elf-Aquitaine 11:484–490

Chapman GR, Brook M (1978) Chronostratigraphy of the Baringo Basin, Kenya. In: Bishop WW (ed) Geological background to fossil man. Geol Soc Lond Spec Publ 6:207–223

Cioni R, Fanelli G, Guidi M, Kinyariro JK, Marini L (1992) Lake Bogoria hot springs (Kenya): geochemical features and geothermal implications. J Volcanol Geotherm Res 50:231–246

Clarke MCG, Woodhall DG, Allen D, Darling G (1990) Geological, volcanological and hydrogeological controls on the occurrence of geothermal activity in the area surrounding Lake Naivasha, Kenya. Ministry of Energy, Republic of Kenya

Clément J-P, Caroff M, Hémond C, Tiercelin J-J, Bollinger C, Guillou H, Cotton J (2003) Pleistocene magmatism in a lithospheric transition area: petrogenesis of alkaline and peralkaline lavas from the Baringo-Bogoria Basin, central Kenya Rift. Can J Earth Sci 40:1239–1257

Cole SM (1950) An outline of the geology of Kenya. Pitman, Nairobi

Cooke P, Maguire P, Evans R, Laffoley N (1989) Seismicity near Lake Bogoria in the Kenya Rift Valley. In: Mereu RF, Mueller S, Fountain DM (eds) Properties and processes of Earth's lower crust. AGU Geophys Monogr Ser vol 51, pp 175–177

Crane K, O'Connell S (1983) The distribution and implications of heat flow from the Gregory Rift in Kenya. Tectonophysics 94:253–275

Dadheech PK, Glöckner G, Casper P, Kotut K, Mazzoni CJ, Mbedi S, Krienitz L (2013) Cyanobacterial diversity in the hot spring, pelagic and benthic habitats of a tropical soda lake. FEMS Microbiol Ecol 85:389–401

Davies TD, Vincent CE, Beresford AKC (1985) July–August rainfall in west-central Kenya. J Climatol 5:17–33

Dawson EC (1890) James Hannington - first Bishop of Eastern Equatorial Africa - a history of his life and work 1847–1885. Seeley, London

De Cort G (2016) Studying Africa's hydroclimate history over the last 2000 years. Afrika Focus 29:101–108

De Cort G, Bessems I, Keppens E, Mees F, Cumming B, Verschuren D (2013) Late-Holocene and recent hydroclimatic variability in the central Kenya Rift Valley: the sediment record of hypersaline lakes Bogoria, Nakuru and Elementeita. Palaeogeogr Palaeoclimatol Palaeoecol 388:69–80

De Cort G, Verschuren D, Ryken E, Wolff C, Renaut RW, Creutz M, van der Meeren T, Haug G, Olago DO, Mees F (2018) Multi-basin depositional framework for moisture-balance reconstruction during the last 1300 years at Lake Bogoria, central Kenya Rift Valley. Sedimentology 65:1667–1696

Deocampo DM, Renaut RW (2016) Geochemistry of soda lakes. In: Schagerl M (ed) Soda lakes of East Africa. Springer, Cham, pp 77–93

Dommain R, Riedl S, Olaka LA, deMenocal P, Deino AL, Owen RB, Muiruri V, Müller J, Potts R,Strecker MR (2022) Holocene bidirectional river system along the Kenya Rift and its influence on East African faunal exchange and diversity gradients. Proc Natl Acad Sci 119

Eugster HP (1980) Lake Magadi, Kenya and its precursors. In: Nissenbaum A (ed) Hypersaline brines and evaporites. Elsevier, Amsterdam, pp 195–232

Eugster HP, Jones BF (1979) Behavior of major solutes during closed-basin brine evolution. Am J Sci 279:609–631

Farrand WR, Redding R, Wolpoff MH, Wright HT (1976) An archaeological investigation of the Loboi Plain, Baringo District, Kenya. Mus Anthropol Univ Mich Tech Rep 4

Gasse F (1987) Sédimentation biogène. Diatomées. In: Tiercelin J-J, Vincens A (eds) Le demi–graben de Baringo–Bogoria, Rift Gregory, Kenya: 30,000 ans d'histoire hydrologique et sédimentaire. Bull Centr Rech Explor-Prod Elf-Aquitaine 11:327–335

Githaiga KB, Njuguna SM, Gituru RW, Yan X (2021) Water quality assessment, multivariate analysis and human health risks of heavy metals in eight major lakes in Kenya. J Environ Manag 297:113410

Glover RB (1972) Chemical characteristics of water and steam discharges in the Rift Valley of Kenya. Unpubl Rep UNDP Geothermal Resources Exploration Project, Kenya

Goman MF, Ashley GM, Owen RB, Driese SG, Muasya AM, Hover VC (2020) A high-resolution climate history of geochemical and biological proxies from a tropical freshwater wetland located in the Kenyan Rift Valley. J Afr Earth Sci 162:103703

Grant WD, Jones BE (2016) Bacteria, Archaea and viruses of soda lakes. In: Schagerl M (ed) Soda lakes of East Africa. Springer, Cham, pp 97–147

Grant WD, Tindall BJ (1986) The alkaline saline environment. In: Herbert RA, Codd GA (eds) Microbes in extreme environments. Academic, London, pp 25–54

Gregory JW (1894) Contributions to the physical geography of British East Africa. Geogr J 4:289–315

Gregory JW (1896) The Great Rift Valley. John Murray, London

Gregory JW (1901) The foundation of British East Africa. Horace Marshall, London

Gregory JW (1921) The rift valleys and geology of East Africa. Seeley Service, London

Grewe S (2016) Hydro- and biogeochemical investigations of lake sediments in the Kenyan Rift Valley. PhD thesis, Universität Potsdam

Griffiths PW (1977) Geology of the Lake Hannington–Perkerra River area, Kenya Rift Valley. PhD thesis, University of London

Griffiths PW, Gibson IL (1980) The geology and petrology of the Hannington Trachyphonolite formation, Kenya Rift Valley. Lithos 13:435–453

Grimaud P, Richert J-P, Rolet J, Tiercelin J-J, Xavier J-P, Morley CK, Coussement C, Karanja SW, Renaut RW, Guerin G (1994) Fault geometry and extension mechanisms in the central Kenya Rift, East Africa: a 3-D remote sensing approach. Bull Cent Rech Explor Prod Elf Aquitaine 18:59–92

Hackman BD (1988) Geology of the Baringo–Laikipia area. Rep Mines Geol Dept Kenya 104

Harper DM, Childress RB, Harper MM, Boar RR, Hickley P, Mills SC, Otieno N, Drane T, Vareschi E, Nasirwa O, Mwatha WE, Darlington JPEC, Escuté-Gasulla X (2003) Aquatic biodiversity and saline lakes: Lake Bogoria National Reserve, Kenya. Hydrobiologia 500:259–276

Hartley J (1986) A guide to lakes Baringo and Bogoria. Evans Brothers (Kenya), Nairobi

Hautot S, Tarits P, Whaler K, Le Gall B, Tiercelin J-J, Le Turdu C (2000) Deep structure of the Baringo Rift Basin (central Kenya) from three-dimensional magnetotelluric imaging: implications for rift evolution. J Geophys Res 105:23493–23518

Hay RL (1970) Silicate reactions in three lithofacies of a semi-arid basin, Olduvai Gorge, Tanzania. Mineral Soc Am Spec Pap 3:237–255

Hecky RE, Kilham P (1973) Diatoms in alkaline, saline lakes: ecology and geochemical implications. Limnol Oceanogr 18:53–71

Herbin JP (1979) Sédimentation de rift: géochimie organique des sédiments récents du lac Bogoria. Rapp Inst Fr Pétrol 26881

Herrnegger M, Stecher G, Schwatke C, Olang L (2021) Hydroclimatic analysis of rising water levels in the Great Rift Valley lakes of Kenya. J Hydrol Region Stud 36:100857

Hickley P, Boar R, Mavuti K (2003) Bathymetry of Lake Bogoria, Kenya. J East Afr Nat Hist 92:107–117

Hindák F (2001) Thermal microorganisms from a hot spring on the coast of Lake Bogoria, Kenya. Nov Hedwig Beih 123:77–93

Hochstein MP (1999) Geothermal systems along the East-African Rift. Bull Hydrogéol 17:301–310

Icole M, Perinet G (1984) Les silicates sodiques et les mileuex évaporatiques carbonatés bicarbonatés sodique: une revue. Rev Géol Dyn Géogr Phys 25:167–176

Jirsa F, Gruber M, Stojanovic A, Omondi SO, Mader D, Körner W, Schagerl M (2013) Major and trace element geochemistry of Lake Bogoria and Lake Nakuru, Kenya, during extreme drought. Chem Erde 73:275–282

John V (2013) Water conservation and management in the upper catchment of Lake Bogoria basin. Eur Int J Sci Technol 2:76–84

Johnston H (1902a) The Uganda protectorate, Ruwenzori, and the Semliki Forest. Geogr J 19:1–39

Johnston H (1902b) The Uganda protectorate, vol 1, 1st edn. Hutchinson, London

Jones WB (1988) Listric growth faults in the Kenya Rift Valley. J Struct Geol 10:661–672

Jones B, Renaut RW (1995) Noncrystallographic calcite dendrites from hot-spring deposits at Lake Bogoria, Kenya. J Sediment Res A65:154–169

Jones B, Renaut RW (1996) Morphology and growth of aragonite crystals in hot-spring travertines at Lake Bogoria, Kenya Rift Valley. Sedimentology 43:323–340

Jones B, Renaut RW (1998) Origin of platy calcite crystals in hot-spring deposits in the Kenya Rift Valley. J Sediment Res 68:913–927

Kaggwa MN, Burian A, Odour SO, Schagerl M (2013) Ecomorphological variability of *Arthrospira fusiformis* (Cyanoprokaryota) in African soda lakes. MicrobiologyOpen 2:881–891

Karingithi C, Wambugu J (2008) Geochemical survey case study of Arus and Bogoria geothermal prospects. Papers, short course III on exploration for geothermal resources, UN Univ Geothermal Training Programme, Lake Bogoria-Lake Naivasha, Kenya, 24 Oct–17 Nov

Kerrich R, Renaut RW, Bonli T (2002) Trace element composition of cherts from alkaline lakes in the East African Rift: a probe for ancient counterparts. In: Renaut RW, Ashley GM (eds) Sedimentation in continental rifts. SEPM Spec Publ 73:275–294

Kiage LM, Douglas P (2020) Linkages between land cover change, lake shrinkage, and sublacustrine influence determined from remote sensing of select Rift Valley Lakes in Kenya. Sci Total Environ 709:136022

Kilham P (1971) Biogeochemistry of African lakes and rivers. PhD thesis, Duke Univ, Durham, North Carolina

Kilham P (1981) Pelagic bacteria: extreme abundances in African saline lakes. Naturwissenschaften 68:380–381

Kilham P, Hecky R (1973) Fluoride: geochemical and ecological significance in East African waters and sediments. Limnol Oceanogr 18:932–945

Krienitz L (2018) Lesser flamingos: descendants of phoenix. Springer Nature, Berlin

Krienitz L, Schagerl M (2016) Tiny and tough: microphytes of East African Soda Lakes. In: Schagerl M (ed) Soda lakes of East Africa. Springer, Cham, pp 149–177

Krienitz L, Ballot A, Kotut K, Wiegand C, Pütz S, Metcalf JS, Codd GA, Pflugmacher S (2003) Contribution of hot spring cyanobacteria to the mysterious deaths of Lesser Flamingos at Lake Bogoria, Kenya. FEMS Microbiol Ecol 43:141–148

Krienitz L, Ballot A, Casper P, Kotut K, Wiegand C, Pflugmacher S (2005) Cyanobacteria in hot springs of East Africa and their potential toxicity. Algol Stud 117:297–306

Krienitz L, Krienitz D, Dadheech PK, Hübener T, Kotut K, Luo W, Teubner K, Versfeld WD (2016) Food algae for Lesser Flamingos: a stocktaking. Hydrobiologia 775:21–50

Lagat J (2008) Geology and geothermal resource utilization options in the Arus-Lake Bogoria prospect, Northern Kenya Rift. Proc ARGeo 2 Conference, Entebbe

LaVigne M, Ashley GM (2001) Climatology and rainfall patterns: Lake Bogoria National Reserve (1976–2001). Unpublished report, Dept Geol Sci, Rutgers University, Piscataway, New Jersey

Le Turdu C, Tiercelin J-J, Coussement C, Rolet J, Renaut RW, Richert JP, Xavier J-P, Coquelet D (1995) Basin structure and depositional patterns interpreted using a 3D remote sensing approach: the Baringo–Bogoria basins, central Kenya Rift, East Africa. Bull Cent Rech Explor-Prod Elf-Aquitaine 19:1–37

Le Turdu C, Tiercelin J-J, Richert JP, Rolet J, Xavier JP, Renaut RW, Lezzar K, Coussement C (1999) Influence of preexisting oblique discontinuities on the geometry and evolution of extensional fault patterns: evidence from the Kenya Rift using SPOT imagery. In: Morley CK (ed) Geoscience of rift systems – evolution of East Africa. AAPG Stud Geol 44: 173–191

McCall GJH (1967) Geology of the Nakuru-Thomson's Falls Lake Hannington area. Rep Geol Surv Kenya 78

McCall J (2010) Lake Bogoria, Kenya: hot and warm springs, geysers and Holocene stromatolites. Earth-Sci Rev 103:71–79

McCall GJH, Baker BH, Walsh J (1967) Late Tertiary and Quaternary sediments of the Kenya Rift Valley. In: Bishop WW, Clark JD (eds) Background to evolution in Africa. Chicago University Press, Chicago, pp 191–220

McNulty E (2017) Lake Magadi and the soda lake cycle: a study of the modern sodium carbonates and of Late Pleistocene and Holocene lacustrine core sediments. MSc thesis, Binghamton University, New York

Melack JM, Kilham P (1974) Photosynthetic rates of phytoplankton in East African alkaline, saline lakes. Limnol Oceanogr 19:743–755

Mercy B, Mbaria J, Muchemi G, Gakuya F, Kariuki E (2017) Ecological risk assessment of heavy metals and water bird distribution in rift valley lakes, Kenya. J Ecol Nat Environ 9:30–44

Mugo M (ed) (2007) Lake Bogoria National Reserve, World Ramsar Site (No. 1057), Integrated management plan 2007–2012. County Councils of Baringo, Koibatek and World Wide Fund for Nature, Eastern Africa Regional Programme Office, Nairobi

Mulwa JK, Mariita N (2013) A comparative analysis of gravity and microseismic results from Arus-Bogoria geothermal prospect, Kenya. Schol J Sci Res Essays 2:177–848

Mulwa JK, Mariita N (2015) Dyking processes at Arus-Bogoria geothermal prospect in Kenya revealed using gravity and microseismic data. Proc 37 NZ Geotherm Workshop, Taupo, New Zealand

Mumba M (2008) Adapting to climate change and why it matters for local communities and biodiversity—the case of Lake Bogoria catchment in Kenya. Policy Matters 16:157–162

Muohi AW, Zauke G-P, Brumsack H-J, Mavuti KM (2018) Bioaccumulation of trace elements in *Arthrospira fusiformis* algal blooms in Lakes Bogoria and Sonachi, Kenya: evaluation and verification of toxicokinetic models. Lakes Reserv 23:277–286

Mutonga M (2014) Fluoride ions in groundwater case study; Baringo-Bogoria geothermal area in Kenya. Proc 5 Afr Rift Geotherm Conf, Arusha, Tanzania

Mwaniki P, Taita T, Sierens T, Triest L (2019) Barriers to genetic connectivity of smooth flatsedge (*Cyperus laevigatus*) among alkaline-saline lakes of Eastern Rift Valley (Kenya). Aquat Bot 155:38–44

Mwawongo GM (2006) Kenya's geothermal prospects outside Olkaria: status of exploration and development. UN Univ Geothermal Training Progr, Reykjavík, Rep 4, pp 41–50

Mwawongo GM (2017) Geothermal mapping using temperature measurements. SDG Short Course II on Exploration and Development of Geothermal Resources, UNU-GTP, GDC and KenGen, Lake Bogoria and Lake Naivasha, Kenya, 9–29 Nov 2017

Naylor I (1972) The geology of the Lake Hannington geothermal prospect. Unpublished Report UN Develop Program–East Africa Power and Lighting, Nairobi

Nilsson E (1931) Quaternary glaciations and pluvial lakes in British East Africa. Geogr Ann 13:249–349

North Lewis M (1998) A guide to Lake Baringo and Lake Bogoria. Horizon Books, Nairobi

Obando JA, Onywere S, Shisanya C, Ndubi A, Masiga D, Irura Z, Mariita N, Maragia H (2016) Impact of short-term flooding on livelihoods in the Kenya Rift Valley lakes. In: Meadows ME, Lin JC (eds) Geomorphology and society, advances in geographical and environmental sciences, geomorphology and society. Springer, Japan, pp 193–215

Ochieng E, Lalah J, Wandiga S (2007) Analysis of heavy metals in water and surface sediment in five Rift Valley lakes in Kenya for assessment of recent increase in anthropogenic activities. Bull Environ Contam Toxicol 79:570–576

Oduor S, Schagerl M (2007a) Phytoplankton primary productivity characteristics in response to photosynthetically active radiation in three Kenyan Rift Valley saline–alkaline lakes. J Plankton Res 29:1041–1050

Oduor S, Schagerl M (2007b) Temporal trends of ion contents and nutrients in three Kenyan Rift Valley saline–alkaline lakes and their influence on phytoplankton biomass. Hydrobiologia 584:59–68

Omenda P (2008) Status of geothermal exploration in Kenya and future plans for its development. Papers, short course III on exploration for geothermal resources, UN University Geothermal Training Programme, Lake Bogoria-Lake Naivasha, 24 Oct–17 Nov

Ong'ondo GO, Yasindi AW, Oduor SO, Jost S, Schagerl M, Sonntag B, Boenigk J (2013) Ecology and community structure of ciliated protists in two alkaline–saline Rift Valley lakes in Kenya with special emphasis on *Frontonia*. J Plankton Res 35:759–771

Onkware A (2000) Effect of soil salinity on plant distribution and production at Loburu delta, Lake Bogoria National Reserve, Kenya. Austral Ecol 25:140–149

Onyando JO, Musila F, Awer M (2005) The use of GIS and remote sensing techniques to analyse water balance of Lake Bogoria under limited data conditions. J Civ Eng Res Pract 2:53–65

Owen RB, Renaut RW, Hover VC, Ashley GM, Muasya AM (2004) Swamps, springs and diatoms: wetlands of the semi-arid Bogoria-Baringo Rift, Kenya. Hydrobiologia 518:59–78

Owen RB, Renaut RW, Jones B (2008a) Geothermal diatoms: a comparative study of floras in hot spring systems of Iceland, New Zealand, and Kenya. Hydrobiologia 610:175–192

Owen RA, Owen RB, Renaut RW, Scott JJ, Jones B, Ashley GM (2008b) Mineralogy and origin of rhizoliths on the margins of saline, alkaline Lake Bogoria, Kenya Rift Valley. Sediment Geol 203:143–163

Péniguel G (1987) Algues à test organique. In: Tiercelin J-J, Vincens A (eds) Le demi–graben de Baringo–Bogoria, Rift Gregory, Kenya: 30,000 ans d'histoire hydrologique et sédimentaire. Bull Centr Rech Explor-Prod Elf-Aquitaine 11:327–339

Périnet G, Tiercelin J-J, Barton CE (1982) Présence de kanemite dans les sédiments récents du lac Bogoria, Rift Gregory, Kenya. Bull Minér 105-6:633–639

Renaut RW (1982) Late Quaternary geology of the Lake Bogoria fault trough, Kenya Rift Valley. PhD thesis, University of London

Renaut RW (1993) Zeolitic diagenesis of Late Quaternary fluviolacustrine sediments and associated calcrete formation in the Lake Bogoria Basin, Kenya Rift Valley. Sedimentology 40:271–301

Renaut RW, Jones B (1997) Controls on aragonite and calcite precipitation in hot spring travertines at Chemurkeu, Lake Bogoria, Kenya. Can J Earth Sci 34:801–818

Renaut RW, Owen RB (1980) Late Quaternary fluvio-lacustrine sedimentation and lake levels in the Baringo Basin, northern Kenya Rift Valley. Rech Géol Afr 5:130–133

Renaut RW, Owen RB (1987) Les "Silts de Loboi"; "Les Silts de Bogoria"; Autres sédiments littoraux. In: Tiercelin J-J, Vincens A (eds) Le demi–graben de Baringo–Bogoria, Rift Gregory, Kenya: 30,000 ans d'histoire hydrologique et sédimentaire. Bull Centr Rech Explor-Prod Elf-Aquitaine 11:502–508

Renaut RW, Owen RB (1988) Opaline cherts associated with sublacustrine hydrothermal springs at Lake Bogoria, Kenya Rift Valley. Geology 16:699–702

Renaut RW, Owen RB (1991) Shore-zone sedimentation and facies in a closed rift lake: the Holocene beach deposits of Lake Bogoria, Kenya. In: Anadon P, Cabrera L, Kelts K (eds) Lacustrine facies analysis (IAS Spec Publ 13). Blackwell, Oxford, pp 175–195

Renaut RW, Owen RB (2005) The geysers of Lake Bogoria, Kenya Rift Valley, Africa. GOSA Trans 9:4–18

Renaut RW, Tiercelin J-J (1993) Lake Bogoria, Kenya: soda, hot springs and about a million flamingoes. Geol Today 9:56–61

Renaut RW, Tiercelin J-J (1994) Lake Bogoria, Kenya Rift Valley: a sedimentological overview. In: Renaut RW, Last WM (eds) Sedimentology and geochemistry of modern and ancient saline lakes. SEPM Spec Publ 50:101–123

Renaut RW, Tiercelin J-J, Owen RB (1986) Mineral precipitation and diagenesis in the sediments of the Lake Bogoria basin, Kenya Rift Valley. In: Frostick LE, Renaut RW, Reid I, Tiercelin J-J (eds) Sedimentation in the African Rifts. Geol Soc Lond Spec Publ 25:159–175

Renaut RW, Tiercelin JJ, Owen RB (2000) Lake Baringo, Kenya Rift Valley, and its Pleistocene precursors. In: Gierlowski-Kordesch E, Kelts K (eds) Lake basins through space and time. AAPG Stud Geol 46:561–568

Renaut RW, Tiercelin J-J, Pagé P (1987) Alimentation, hydrologie. In: Tiercelin J-J, Vincens A (eds) Le demi–graben de Baringo–Bogoria, Rift Gregory, Kenya: 30,000 ans d'histoire hydrologique et sédimentaire. Bull Centr Rech Explor-Prod Elf-Aquitaine 11:284–309

Renaut RW, Jones B, Tiercelin J-J (1998) Rapid *in situ* silicification of microbes at Loburu hot springs, Lake Bogoria, Kenya Rift Valley. Sedimentology 45:1083–1103

Renaut RW, Owen RB, Ego JK (2008) Recent changes in geyser activity at Loburu, Lake Bogoria, Kenya Rift Valley. GOSA Trans 10:4–14

Renaut RW, Owen RB, Jones B, Tiercelin J-J, Tarits C, Ego JK, Konhauser KK (2013) Impact of lake level changes on the formation of thermogene travertine in continental rifts: evidence from Lake Bogoria, Kenya Rift Valley. Sedimentology 60:428–468

Renaut RW, Owen RB, Ego JK (2017) Geothermal activity and hydrothermal mineral deposits at southern Lake Bogoria, Kenya Rift Valley: impact of lake level changes. J Afr Earth Sci 129:623–646

Robinson VJ (2015) The ecology of East African soda lakes: implications for lesser flamingo (*Phoeniconaias minor*) feeding behaviour. PhD thesis, University of Leicester

Rowntree KM (1989) Rainfall characteristics, rainfall reliability and the definition of drought: Baringo District, Kenya. S Afr Geogr J 71:74–80

Sanders P (2016) Linkages between saline lakes and their riparian zone over climate change. PhD thesis, University of London

Schagerl M (ed) (2016) Soda lakes of East Africa. Springer, Cham

Scott JJ (2005) Taphonomy of modern and ancient vertebrate traces in the marginal sediments of saline, alkaline and freshwater lakes, Baringo-Bogoria basin, Kenya Rift Valley. MSc thesis, University of Saskatchewan, Saskatoon

Scott JJ (2010) Saline lake ichnology: composition and distribution of Cenozoic traces in the saline, alkaline lakes of the Kenya Rift Valley and Eocene Green River Formation, U.S.A. PhD thesis, University of Saskatchewan, Saskatoon

Scott JJ, Renaut RW, Owen RB, Sarjeant WAS (2007) Biogenic activity, trace formation, and trace taphonomy in the marginal sediments of saline, alkaline Lake Bogoria, Kenya Rift valley. In: Bromley RG, Buatois LA, Mángano G, Genise JF, Melchor RN (eds) Sediment–organism interactions: a multifaceted ichnology. SEPM Spec Publ 88:311–332

Scott JJ, Renaut RW, Owen RB (2008) Preservation and paleoenvironmental significance of a footprinted surface on the Sandai Plain, Lake Bogoria, Kenya Rift Valley. Ichnos 15:208–231

Scott JJ, Renaut RW, Buatois LA, Owen RB (2009) Biogenic structures in exhumed surfaces around saline lakes: an example from Lake Bogoria, Kenya Rift Valley. Palaeogeogr Palaeoclimatol Palaeoecol 272:176–198

Scott JJ, Renaut RW, Owen RB (2010) Taphonomic controls on animal tracks at saline, alkaline Lake Bogoria, Kenya Rift Valley: impact of salt efflorescence and clay mineralogy. J Sediment Res 80:639–665

Scott JJ, Renaut RW, Owen RB (2012) Impacts of flamingos on saline lake margin and shallow lacustrine sediments in the Kenya Rift Valley. Sediment Geol 277–278:32–51

Scott JJ, Buatois LA, Mángano MG, Renaut RW, Owen RB (2019) Bioturbation in matgrounds at Lake Bogoria in the Kenya Rift Valley: implications for interpreting the heterogeneous early Cambrian seafloor. Lethaia. https://doi.org/10.1111/let.12336

Scott JJ, Renaut RW, Buatois LA, Owen RB, McNulty EP, Stockhecke M, Leet K, Lowenstein TK, Mángano GM (2021) Modern and ancient animal traces in the extreme environments of Lake Magadi and Nasikie Engida, Kenya Rift valley. In: Rosen MR, Park-Boush L, Finkelstein DB, Pla Pueyo S (eds) Limnogeology: progress, challenges and opportunities: a tribute to Elizabeth Gierlowski-Kordesch. Springer, New York, pp 19–66

Spooner ETC, Buckland PC, Bishop WW (1969) The Kampi Ya Samaki Beds and Kokwob Beds of the Baringo basin, Kenya. Unpublished Report, Bedford College, University of London

Straubinger-Gansberger N, Gruber M, Kaggwa MN, Lawton L, Oduor SO, Schagerl M (2014) Sudden flamingo deaths in Kenyan Rift valley lakes. Wildlife Biol 20:185–189

Tebbs EJ, Remedios JJ, Harper DM (2013) Remote sensing of chlorophyll-a as a measure of cyanobacterial biomass in Lake Bogoria, a hypertrophic, saline–alkaline, flamingo lake, using Landsat ETM? Remote Sens Environ 135:92–106

Tebbs EJ, Remedios JJ, Avery ST, Rowland CS, Harper DM (2015) Regional assessment of lake ecological states using Landsat: a classification scheme for alkaline-saline flamingo lakes in the East African Rift Valley. Int J Appl Earth Obs Geoinf 40:100–108

Tiercelin J-J (1981) Rifts continentaux: tectonique, climats, sédiments. Exemples: la sedimentation dans le Nord du Rift Gregory, Kenya, et dans le Rift de l'Afar, Ethiopie, depuis le Miocene. PhD thesis, Université d'Aix Marseille II

Tiercelin J-J (1985) Bogoria: Quand la terre c'est entrouverte. L'Univ du Vivant 2:56–69

Tiercelin J-J (1990) Rift-basin sedimentation: responses to climate, tectonism and volcanism. Examples of the East African Rift. J Afr Earth Sci 10:283–305

Tiercelin J-J, Vincens A (eds) (1987) Le demi–graben de Baringo–Bogoria, Rift Gregory, Kenya: 30,000 ans d'histoire hydrologique et sédimentaire. Bull Cent Rech Explor-Prod Elf-Aquitaine 11:249–540

Tiercelin J-J, Le Fournier J, Herbin JP, Richert JP (1980) Continental rifts: modern sedimentation, tectonic and volcanic control. Example from the Bogoria-Baringo graben, Gregory Rift, Kenya. In: Geodynamic evolution of the Afro Arabic Rift system. Acad Naz dei Lincei, Rome, pp 143–164

Tiercelin J-J, Renaut RW, Delibrias G, Le Fournier J, Bieda S (1981) Late Pleistocene and Holocene lake levels in the Lake Bogoria basin, Kenya Rift Valley. Palaeoecol Afr 13:105–120

Tiercelin J-J, Perinet G, Le Fournier J, Bieda S, Robert P (1982) Lacs du rift est africain exemples de transition eaux douces eaux salées: Le lac Bogoria, rift Gregory, Kenya. Mém Soc Géol Fr 144:217–230

Tuite CH (1981) Standing crop densities and distribution of *Spirulina* and benthic diatoms in East African alkaline saline lakes. Freshw Biol 11:345–360

van der Plas GW, De Cort G, Petek-Sargeant N, Wuytack T, Colombaroli D, Lane PJ, Verschuren D (2019) Distinct phases of natural landscape dynamics and intensifying human activity in the central Kenya Rift Valley during the past 1300 years. Quat Sci Rev 218:91–106

Vareschi E (1978) The ecology of Lake Nakuru (Kenya). I. Abundance and feeding of the lesser flamingo. Oecologia 32:11–35

Vincens A (1986) Diagramme polinique d'un sondage Pleistocene supérieur–Holocene du Lac Bogoria (Kenya). Rev Palaeobot Palynol 47:169–142

Vincens A, Casanova J, Tiercelin J-J (1986) Palaeolimnology of Lake Bogoria (Kenya) during the 4500 BP high lacustrine phase. In: Frostick LE, Renaut RW, Reid I, Tiercelin J-J (eds) Sedimentation in the African rifts. Geol Soc Lond Spec Publ 25:323–330

Ward SA (2015) Are changes in the lesser flamingo population a natural consequence of soda lake dynamics? PhD thesis, University of Southampton
Washbourn-Kamau C (1971) Late Quaternary lakes in the Nakuru-Elmenteita basin, Kenya. Geogr J 137:522–535
Wetang'ula GN, Kubo BM, Were JO (2019) Environmental baseline study for geothermal developments: case study of Arus-Bogoria geothermal prospects, Kenya SDG short course IV on exploration and development of geothermal resources (UNU-GTP-KenGen), Lake Naivasha, December 2019, 16 pp
Williams REG, Johnson AS (1976) Birmingham University radiocarbon dates X. Radiocarbon 18:249–267
Yang Y, Wei L, Cui L, Zhang M, Wang J (2017) Profiles and risk assessment of heavy metals in Great Rift Lakes, Kenya. Clean (Weinh) 45:1600825–1600829
Young JAT, Renaut RW (1979) A radiocarbon date from Lake Bogoria, Kenya Rift Valley. Nature 278:243–245
Young PAV, Maguire PKH, Laffoley NDA, Evans JR (1991) Implications of the distribution of seismicity near Lake Bogoria in the Kenya Rift. Geophys J Int 105:665–674
Zielke O, Strecker MR (2009) Recurrence of large earthquakes in magmatic continental rifts: insights from a paleoseismic study along the Laikipia-Marmanet Fault, Subukia Valley, Kenya Rift. Bull Seismol Soc Am 99:61–70

Chapter 10
Lake Nakuru and Lake Elmenteita

10.1 Introduction

Lake Nakuru and Lake Elmenteita are very shallow, saline alkaline lakes only 11 km apart (2021) located in the central Kenya Rift just south of the equator (Figs. 10.1, 10.2a–d and 10.3). The city of Nakuru is ~135 km northwest of Nairobi. Both lakes occupy topographically closed drainage basins composed mostly of Pliocene to Holocene volcanic and volcaniclastic rocks, and Quaternary fluvial and lacustrine sediments. During the African Humid Period (~15–5 ka) they united as a single larger, deeper and fresher lake, but became separated when lake level fell as the climate became drier a few thousand years ago. They share similar hydrochemistry, ecology and sedimentation, and are treated together in this chapter.

In Maasai, *Nakuru* means a "dry and dusty place". Periodically, the lake desiccates and becomes a soda-encrusted muddy playa with efflorescent trona crusts prone to deflation (Fig. 10.2e). Lake Nakuru is perhaps the best known of the Kenya Rift lakes, famous for its flamingos and the wildlife that occupies many habitats surrounding the lake (Fig. 10.2f, g). In 1961, the Kenya Government designated the lake and its environs a conservation area and a National Park in 1978 (188 km^2). It then became a Ramsar site in 1990 and, with Lake Elmenteita and Lake Bogoria, a UNESCO World Heritage Site in 2011. Although "protected", Lake Nakuru National Park and its drainage basin have been under great stress for several decades because of pollution from the expanding city of Nakuru (Odada et al. 2006), which in 2023 was <1 km from its northern shoreline, and from land-use changes in the drainage basin that have severely limited access for faunal migration. It shares common problems with Nairobi National Park. Both national parks risk becoming fenced 'safari parks' similar to those established overseas, where land-based animals become isolated from their former migration routes and gene-flow declines.

Nakuru National Park has three gates: the Main Gate near the northern shore; the Lanet Gate (southeast of Nakuru) which links the park to the Nairobi-Nakuru highway, and the Nderit Gate southeast of the lake. Pressures from the growing local

R. W. Renaut, R. B. Owen, *The Kenya Rift Lakes: Modern and Ancient*,
Syntheses in Limnogeology, https://doi.org/10.1007/978-3-642-25055-2_10

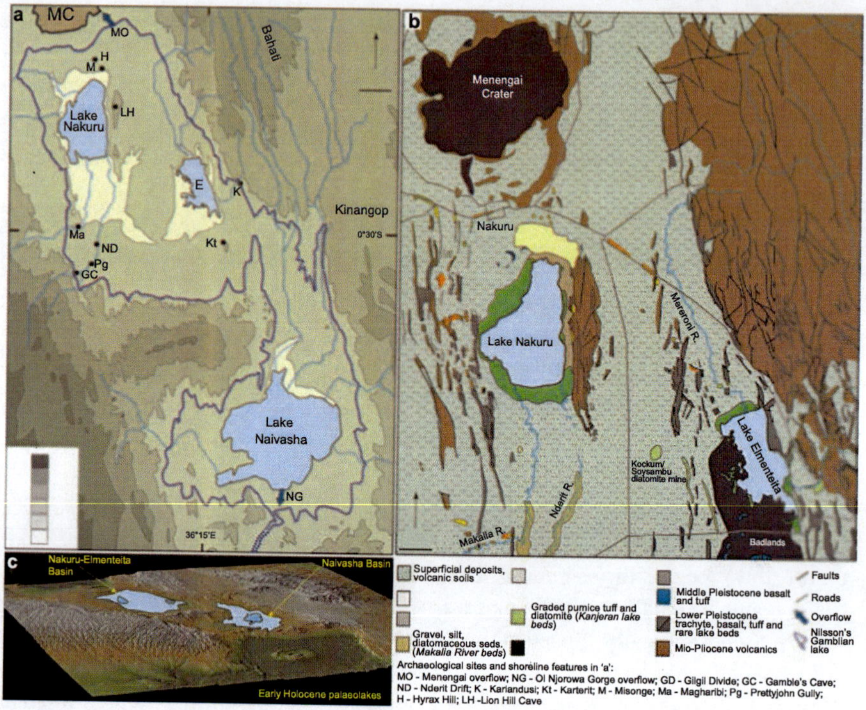

Fig. 10.1 Topography and geology of the Nakuru-Elmenteita Basin. (**a**) Topography of the Central Kenya Rift and highest shoreline suggested by Nilsson (1931). Archaeological sites associated with lake sediments shown on the map. (**b**) Geology map of the Nakuru-Elmenteita Basin, including 'Kanjeran', 'Larmudiac', 'Gamblian' and 'Makalian' sedimentary outcrops. Simplified after the 1966 Kenya Government 'Nakuru Area, Degree Sheet 43, NW Quarter' in McCall (1967). (**c**) 3D representation of the Early Holocene Central Kenya Rift palaeolakes. Based on Butzer et al. (1972) with topography constructed from GeoMapApp (http://www.geomapapp.org)

population (2023 est. >420,000) have been balanced partly by conservation successes over the last 70 years. Lake Nakuru nonetheless remains threatened by climate change, social pressures, and related pollution from several sources.

Lake Elmenteita (Fig. 10.3), which is bordered partly by wetlands and mudflats, is sometimes almost dry. This more secluded lake has a diverse bird population and also periodically attracts hundreds of thousands of lesser flamingos. Unlike other Kenya Rift lakes, most of its shoreline and access routes are on private land (Delamere Estates, Soysambu Conservancy, and the former Ututu Ranch). Until recently (post-2000) tourism was discouraged.

The name Elmenteita (= Elementeita or Elmentaita) originates from another Maasai word '*muteita*', meaning 'place of dust' – a reference to sodium carbonate dust deflated from the peripheral mudflats, which blows across the area during the dry season (January–March) when lake level is typically at its lowest. Both lake names emphasise the potential importance of aeolian components in the recent lake sediments. Lake Elmenteita is ~120 km northwest of Nairobi, and 7 km northwest

Fig. 10.2 General views of Lake Nakuru. (**a**) Lake viewed from north with Nakuru in foreground. (**b**) View from southwestern ridge; Menengai top left, August 2017. (**c**) View south from Baboon Cliff viewpoint, northwestern shore, December 2013. (**d**) Western shoreline and Sirrkon Ridge (right), June 2019. Nderit River in foreground. (**e**) Western shoreline from Sirrkon Ridge showing deflation of trona dust from mudflats, July 2005. (**f**) Lesser flamingos, northern shoreline, July 2005. (**g**) Buffalo, southeastern shoreline, July 2009. (**h**) Dead trees resulting from flooding, southwest shoreline, August 2017

Fig. 10.3 General views of Lake Elmenteita. (**a**) Lake viewed from north with Mereroni delta plain and old lake sediments in foreground, June 2019. (**b**) View across the south part of the lake from the southeastern margin, July 2005. (**c**) Southern shore of Lake Elmenteita showing white saline mudflats with 'Badlands' ash cones in distance and the Mau Escarpment on horizon, July 2005. (**d**) Holocene basalt flow, midwestern shoreline, June 2019. (**e**) Northeastern corner of the lake at Mereroni delta, with Bahati Escarpment in distance, June 2019. Lesser flamingos line the shoreline. Inset: Great white pelicans and offspring on basalt islands off the midwestern shore. (**f**) Sandy Kariandusi delta, southeastern shore, June 2005. Badlands ash cones in middle distance; Eburru massif on left horizon. (**g**) Bioturbated saline mudflat, southeastern shoreline, June 2006

of the town of Gilgil. Access is from the Nairobi-Nakuru highway on its steep eastern side or from an unpaved track to the south that passes through Elmenteita village.

Before Kenya's independence in 1963, most of the drainage basin was part of the 'White Highlands' and under European settler control from 1902 to 1961 (Morgan 1963). Population increase, resettlement and farming after 1963 led to the loss of most upland forest and grassland, which in turn affected the basin hydrology and increased soil erosion, with consequences for recent lake sedimentation (Prasad 1995).

10.2 Geological Setting

The Nakuru-Elmenteita-Naivasha sector is the highest part of the Kenya Rift. The lake elevations range from ~1883 m asl (Lake Naivasha) to ~1776 m asl at Lake Elmenteita to ~1760 m asl at Lake Nakuru, which compare with up to ~3100 m on the adjacent, uplifted rift margins. Despite lake level fluctuations, Lake Elmenteita has typically stood about 16 m higher than Lake Nakuru during the past 50 years.

The geological setting of this part of the rift is complex (Fig. 10.1b; McCall 1967; Strecker et al. 1990; Strecker 1991; Reidl et al. 2020; Conti et al. 2021). The lakes lie near the intersection of the broadly N-S Kenya Rift with the E-W Nyanza (Kavirondo) Rift. The tectonic influence of the Nyanza Rift on the two lakes and their sedimentation, however, is less important than that of the Kenya Rift. The Nyanza Rift fades as a topographic feature ~50 km west of their junction (Jones and Lippard 1979). The intersection of these rifts is located near the apex of the 'Kenya Dome', an area of uplift and (or) accumulation of volcanic rocks (King 1978), which acts as the axial drainage divide of the modern Kenya Rift (Becht et al. 2006). The highest part of the Kenya Rift floor is at Eburru volcanic centre, about 20 km south of Lake Elmenteita.

Both lakes lie within the axial inner graben of the rift, which is bordered by the Mau Fault and Mau Plateau to the west, and by the Bahati and Kinangop faults to the east (Figs. 10.1b and 10.4). The Bahati-Kinangop plateau lies east of those faults and in turn, is bordered by the Sattima fault and the eastern rift shoulder represented by the Aberdare (Nyandarua) range.

Precambrian metamorphic rocks, exposed in the southwestern Mau Plateau distal from the lakes, are overlain by a thick Neogene succession of volcanic rocks. Lower Miocene flood basalts are present in the eastern part of the Kenya Rift in this region. Middle to Upper Miocene phonolite lavas with local basalt covered most of the central Kenya Rift during this time. During the Pliocene, as the axial rift became more defined, the region was covered by a thick succession of tuffs, ignimbrites, trachytes and phonolites. Explosive volcanic activity extended southwards from the Equator. These rocks are exposed in the Mau Plateau and axial rift depression but are absent along the eastern rift margin.

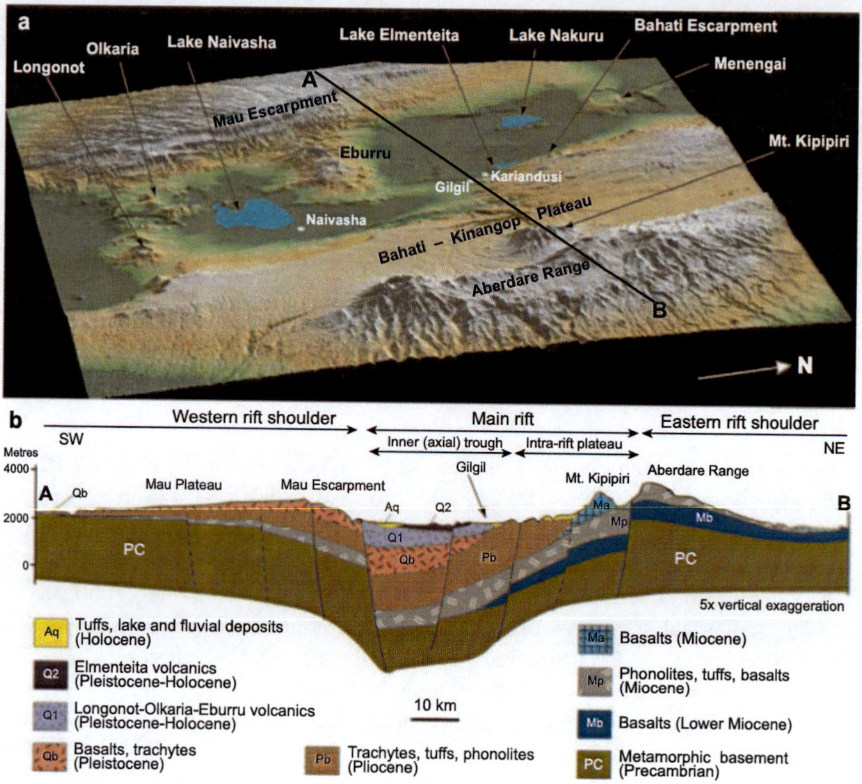

Fig. 10.4 Geological setting of the Nakuru-Elmenteita basin. (**a**) 3D representation of the Central Kenya Rift using topography constructed from GeoMapApp (http://www.geomapapp.org). (**b**) West-to-east geological cross-section of the Kenya Rift at the latitudes of the two lakes. Geology based on Conti et al. (2021) and sources that they quoted

The younger Quaternary volcanic rocks, which have had the most direct influence on the lakes, are restricted to the inner rift trough. Central volcanoes, including Eburru and Longonot, were active south of the lakes, while Menengai, a major caldera volcano, developed to the north (Leat 1984, 1991; Reidl et al. 2020). All these volcanoes produced basalt and trachyte lava flows and thick volcaniclastic deposits.

While these central volcanoes were active, fluvial and lacustrine sediments were being deposited on the subsiding rift floor. Ash layers produced by volcanism (both airfall and reworked) have been interbedded with siliciclastic, biogenic and chemical sediments, and have provided material for radiometric dating. In the Nakuru-Elmenteita region, those sediments include coarse to fine siliciclastic deposits, diatomite and diatomaceous mud, and tuffs, separated by unconformities and disconformities, some present as faulted and erosional contacts and palaeosols. Young volcanic rocks of late Quaternary age are interlayered with the sediments. They include those south of Lake Elmenteita and those associated with Menengai volcano north of Lake Nakuru.

Lake Nakuru lies centrally within the axial graben, ponded to the north by the slopes of Menengai and tightly constrained laterally by faulted volcanic rocks that rise as parallel N-S ridges in a broad plain of mainly Quaternary sediments and volcaniclastic deposits. East of both Lake Nakuru and Lake Elmenteita, the Pliocene Bahati trachytes, ignimbrites and tuffs form faulted uplands; these rocks also crop out locally southwest of Lake Nakuru. They are succeeded by trachytic tuffs, intercalated sediments and welded ignimbrites (Meroronyi Tuffs: ~2 Ma; Leat 1991), and the trachyte, basalt and fluvial sediments of the Mbaruk Basalts and Trachytes (~1 Ma). The latter rocks crop out (1) along the Bahati Fault; (2) between Lake Nakuru and Lake Elmenteita; and (3) west of Lake Nakuru. Sirrkon (Lion) Hill east of Lake Nakuru is an N-S faulted late Pleistocene volcanic centre composed of trachyte and pyroclastic deposits.

Menengai caldera volcano (Fig. 10.5a, b) developed during the Late Quaternary. Conti et al. (2021) recognised three stages:

1. Formation of a shield volcano with low-angle slopes (pre-caldera stage).
2. Collapse of the volcano and formation of a caldera (syn-caldera stage).
3. Eruption of lavas on the caldera floor (post-caldera stage).

Fig. 10.5 Volcanic barriers to Lake Nakuru and Lake Elmenteita. (**a, b**) Margins of the Menengai caldera, filled with post-caldera volcanic flows and local sediments; (**b**) shows local green wetlands around the inner rim and Mau Escarpment. (**c**) Eburru massif south of Lake Elmenteita; ash cones of the Badlands in the foreground. (**d**) faulted ash cone (the 'sleeping warrior'), 3.2 km south of Lake Elmenteita

The volcanic rocks are trachyte and ignimbrite with airfall pyroclastics, some of which are interbedded with contemporary lake sediments in the Nakuru and Elmenteita basins. Two trachytic tuff cones, Honeymoon Hill and Crescent Hill, located near the northwest shoreline of Lake Nakuru, are late vents of the post-caldera phase.

West of Lake Elmenteita, a gently sloping plain at ~1800 m elevation and up to 19 km wide, extends across the rift floor towards the steep Mau Escarpment, which rises to ~3000 m asl (above sea level). The rift valley floor is covered by Holocene and mid-to-late Pleistocene lacustrine, volcaniclastic and alluvial sediments with volcanic and volcaniclastic rocks exposed locally in channel walls. The lacustrine sediments include diatomites that have been mined at Kockum (Soysambu) (Fig. 10.1b; Mboya 1993).

About 16 km south of the lake, the Eburru volcanic centre is a prominent west-east ridge that lies 1000 m above the rift floor (Fig. 10.5c). Eburru is composed of trachyte, pantellerite, tuff and pumice that range in age from 1.2 Ma to <400 y (Clarke et al. 1990; Omenda 1997; Beltran 2003; Kiende and Kandie 2015; Ronoh 2015). Two main eruptive phases are recognised, separated by a period of faulting. The Eburru ridge, which rises to ~2840 m asl, represents the highest elevation and drainage divide on the modern Kenya Rift floor, separating northward- from southward-flowing drainage (Fig. 10.4a). The volcano has extensive fumarolic and steam discharge along N-S fault-lines but few hot springs because of its elevation. Eburru is a site of geothermal power development (Simiyu 2010).

Between Eburru and Lake Elmenteita a series of Holocene basaltic and trachyandesitic lava flows and eroded, late Quaternary faulted ash cones form a region known as the 'Badlands' or 'Elmenteita basaltic field' (Omenda 1997; Figs. 10.1b and 10.3c). These basaltic rocks crop out along the southern and southwestern shorelines of the lake forming a distinctive hummocky topography that locally ponds hypersaline water. The ash cones, the most photogenic geological features of the lake basin (Fig. 10.5d), include subaerial volcaniclastic deposits and hyaloclastites (McCall 1967; Clarke et al. 1990). Nyawire et al. (2018) have suggested that the Badlands is an area of active rift extension (~3 mm y^{-1}) within the main rift that they termed the Otutu Rift. They recognised two segments of the 'rift-in-rift', offset by a transform zone.

10.3 Climate

The climate in the Nakuru-Elmenteita drainage basins (Fig. 10.6a) is semi-arid but the higher elevation of the rift shoulders means that parts of the drainage basin are cooler and wetter. Mean annual temperatures are from ~12 °C in the higher cooler parts of the catchment to the east, rising to ~18 °C on the rift floor (Mwaura and Moore 1991). Maximum and minimum temperatures at Nakuru are ~35 °C and 12 °C, respectively (Koyo 2005). This elevated region is cooler than the northern and southern sectors of the Kenya Rift. Mean annual rainfall in the neighbouring

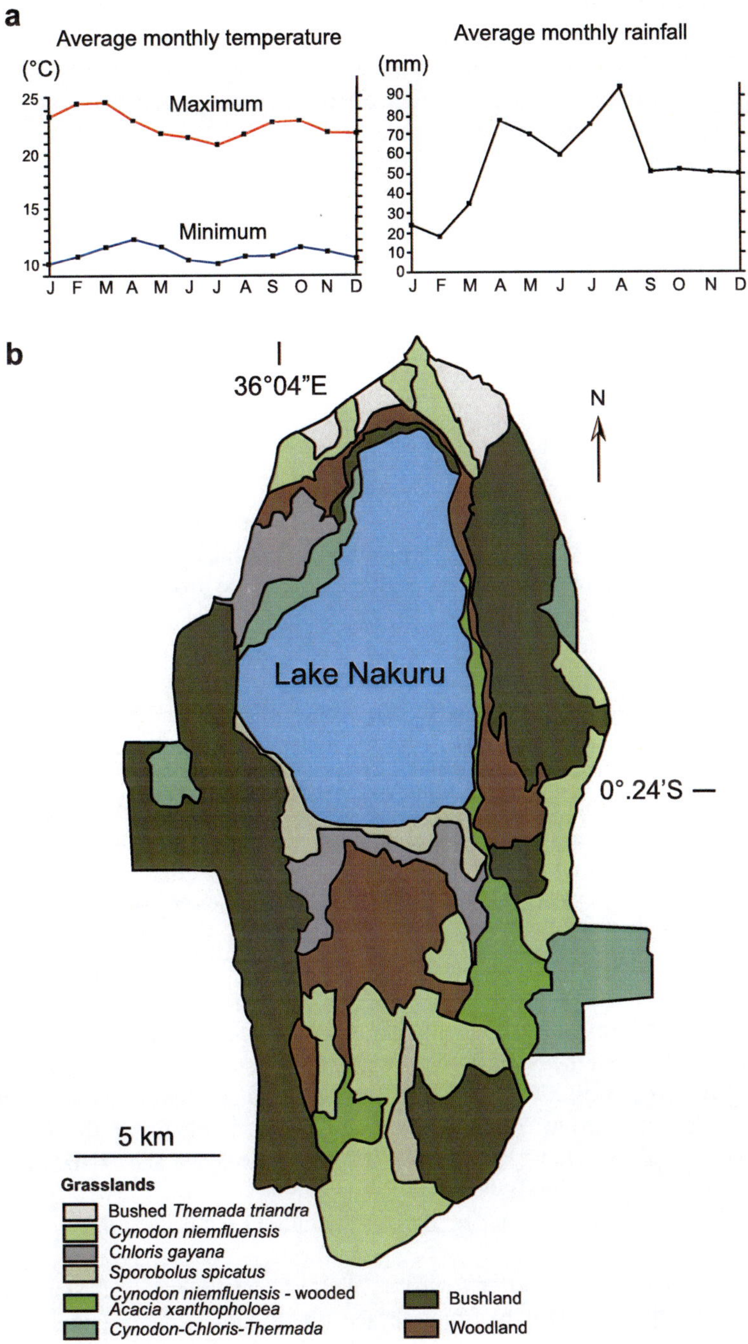

Fig. 10.6 Climate and vegetation at Lake Nakuru. (**a**) Climatic data for Nakuru weather station from Kenya Meteorological Department. (**b**) Vegetation map of Lake Nakuru National Park. Modified from Ng'weno et al. (2009)

uplands is >1000 mm, falling to ~730 mm on the valley floor with short rains from October to November and long rains from April until June (Fig. 10.6). Kiage and Douglas (2020) reported a range from ~600 to 1400 mm y^{-1} for 1980–2014 at Nakuru weather station. Mean annual catchment rainfall was 1013 mm for the period 1981–2020; Lake Elmenteita receives less rainfall (914 mm y^{-1}), particularly in the southern part of the lake basin (Herrnegger et al. 2021). Mean annual evaporation rates have been >1400 mm across both basins in most years. Winds blow throughout the year but are strongest (~10 km h^{-1}) between December and March, and weakest in June and July.

10.4 Hydrology

10.4.1 Lake Nakuru

Lake Nakuru lies at ~1759 m asl (Raini 2006; Herrnegger et al. 2021), but its surface level often changes monthly and from year to year in response to short-term variations in the precipitation/evaporation budget, and from occasional tectonic movements. Its catchment area is ~1475 km^2 (Herrnegger et al. 2021). Lake Nakuru is a playa lake but does not lie in a regional sump, so may not be fully hydrologically closed. In some years tourists flock to see lesser flamingos clustered along the shorelines; in other years (e.g. 2013) they leave disappointed by the meagre flamingo population because the lake is either too fresh or too saline to provide their food (Krienitz 2018).

Lake Nakuru is fed by the perennial Ngosur River to the northeast, and four seasonal rivers: the Njoro, Nderit (Enderit), Makalia and Larmudiac (Lamudhiak) (Fig. 10.7a), all of which originate in the uplands of the eastern Mau Forest (Omondi 2011; Kimaru et al. 2019). The Ngosur River flows southwestwards from the Bahati uplands to the east. Its waters pass underground to feed the water table below Lake Nakuru, some discharging at Baharini springs along the northeastern shoreline (McCall 1967; Shivoga 2001). Using shallow geophysics, Kutukhulu (2010) confirmed that a buried Ngosur river channel feeds an aquifer (Kabatini) northeast of Nakuru. To the west, the Njoro River is an important source of water and sediment. The ephemeral Nderit, Makalia and Larmudiac rivers are minor by comparison. A little inflow comes from two sewage-treatment plants that lie north of the lake. Shallow groundwater movement locally leads to subsidence and related pollution (Ngecu and Nyambok 2000).

Jenkins et al. (2009) concluded that direct rainfall provides the largest inflow to the lake followed by river inflows and groundwater seepage. They noted that the hydrological balance has shifted since the 1970s as land use changed. More water is flowing into the lake rapidly from surface runoff rather than perennially from groundwater. This in turn has led to more frequent fluctuations in level and salinity and a declining water table.

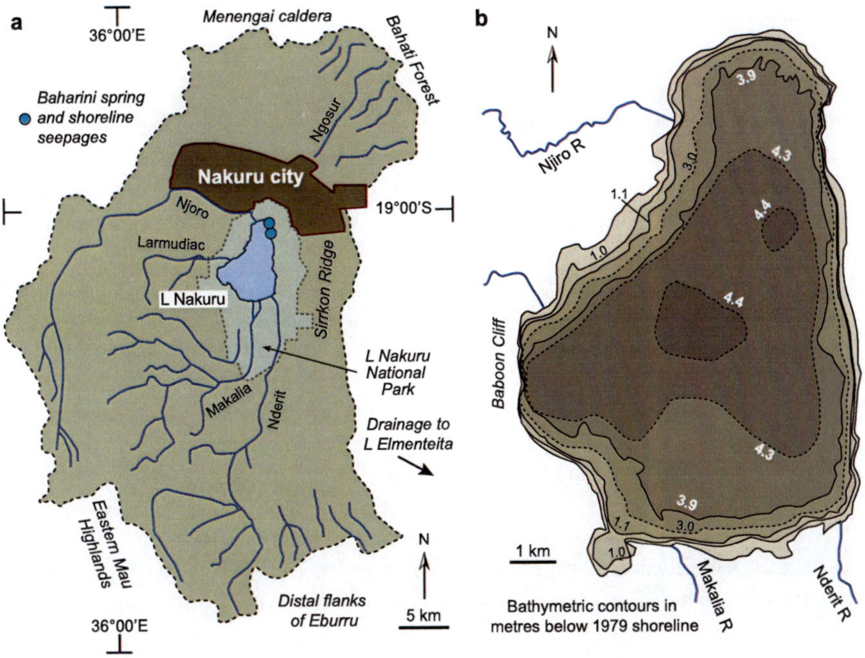

Fig. 10.7 (**a**) Drainage basin of Lake Nakuru; (**b**) Bathymetry of Lake Nakuru. After Vareschi (1982)

Most groundwater feeding Lake Nakuru is from the local rift margins with minor inflow from Menengai. Hydrogeological connections between Lake Naivasha and Lake Elmenteita, and between Lake Elmenteita and Lake Nakuru, have been suggested based on piezometric data (groundwater level and inferred flow directions, usually from well data) and stable isotopic analyses (McCall 1967; Allen et al. 1989; Clarke et al. 1990; Becht and Harper 2002; Muno 2002; Nabide 2002; Ojiambo et al. 2003; Olago et al. 2009; Olaka et al. 2010). Becht et al. (2006) suggested that some Lake Elmenteita water derives from groundwater outflow from Lake Naivasha and that some Elmenteita water may flow towards Lake Nakuru as shallow groundwater. This is reasonable based on their relative groundwater levels, but details of potential aquifers are uncertain because volcanic ashes and diatomaceous sediments can be aquicludes or aquitards if compacted or cemented. The groundwater component feeding Lake Nakuru from the south was estimated at 24 m^3 y^{-1} but the amount of artesian upflow is too small to prevent the lake from drying out (Becht et al. 2006). Allen et al. (1989) raised the possibility of northward groundwater outflow from northwestern Lake Nakuru, based on high chloride in one well, but this is unconfirmed. Geochemical data has shown that Lake Nakuru water likely contributes to the geothermal groundwater reservoirs below Menengai, implying northward leakage (Kanda and Suwai 2013; Igunza and Kanda 2017).

10.4.2 Lake Elmenteita

Lake Elmenteita, which has a surface area that varies from ~15 to 24 km^2 and a drainage basin area of ~550 km^2, lies centrally in its drainage basin. Although superficially simple, the hydrology of this small basin is complex. The lake is recharged by direct rainfall, surface runoff and groundwater seepage, but also by hot springs (Fig. 10.8a). The lake loses water by evaporation and evapotranspiration, with possible shallow groundwater seepage northwards to Lake Nakuru.

Most dilute inflow originates in three rivers and their tributaries that drain the partly wooded Bahati Escarpment. The Mereroni and Mbaruk rivers flow from the Ndunduri-Ngorika Hills to the north, while the Kariandusi River drains the drier Gitare Hills to the east. Minimal surface inflow comes from the dry rift floor to the west and south, where the young volcaniclastic deposits are porous and permeable. The Mereroni is the main affluent river. Flow in all of the rivers, however, decreases before reaching the lake, due partly to evaporation and seepage into the river beds, but increasingly to removal for irrigation upstream and in tributary headwaters. The flow was perennial last century (Leakey 1931a).

Although the river discharge varies greatly, the thermal springs tap a deeper and generally more reliable source. Springs are present along rivers in the drainage

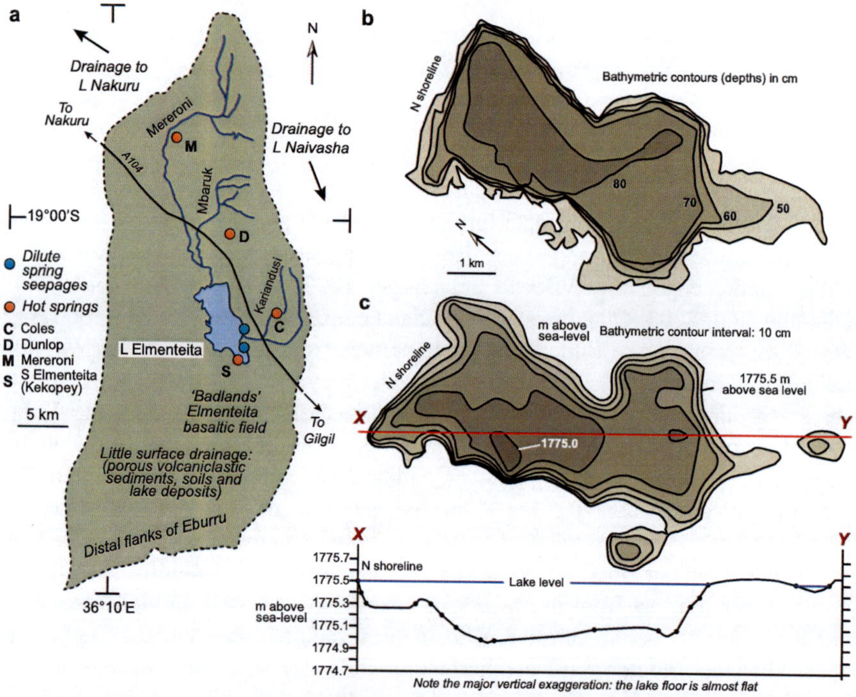

Fig. 10.8 (**a**) Drainage basin of Lake Elmenteita; (**b**) Bathymetry of Lake Elmenteita. After Melack (1988); (**c**) Bathymetry of Lake Elmenteita after Muno (2002) and N-S cross section (X-Y)

basin (Chamuka, Maji Moto) and on the southern lakeshore (McCall 1967; Mwaura 1999). Water flows into the lake from quiet springs along its southern shoreline and from vents located along affluent streams east of the lake. Some springs discharge ambient temperature water; others are 'warm', but none are very hot. Most have remained perennial during recent decades and in some years provide a more reliable source of recharge than the affluent rivers. Springs are therefore important in the lake hydrology and ecology (Leakey 1931a; McCall 1967; Becht et al. 2006), and thus in sedimentation.

About 6–10 alkaline warm springs (T: 35–45 °C) discharge clear water from shallow pools up to a few metres across and partly floored with microbial mats located along the southwest shoreline (Spink 1945; Glover 1972; Kamondo 1988; Mpawenayo and Mathooko 2004; Montcoudiol et al. 2019; Butturini et al. 2020). In wet years many springs are submerged by the lake; in drier years some flow lakewards in low sinuosity channels or dry up. Other hot springs (including Mereroni, Dunlop, Land and Cole springs: McCall 1967) lie in uplands along the foot of the Bahati Escarpment east of the lake. The terminology of the springs in the literature is confusing. Some authors refer to the southern shoreline springs as the Kekopey hot springs; others use the term Lake Elmenteita hot springs. Becht et al. (2006) mention the Maji Moto (Kiswahili for hot or warm water) springs along the faulted southern shoreline, but these do not appear to be the same site.

It has long been suspected that some groundwater recharge into Lake Elmenteita originates from the south, partly because Lake Naivasha (~1883 m asl) remains fresh even though it lacks surface outflow. Both northward and southward underground drainage flow-paths from Lake Naivasha have been proposed using evidence from wells and contouring of piezometric surfaces (Sikes 1936; Clarke et al. 1990). Using stable isotope and water balance data, Becht et al. (2006) estimated that 4.8 million m^3 of Lake Naivasha water flows northwards towards Lake Elmenteita annually. The flow is probably shallow, through faulted volcanic rocks and sediments between Eburru and Gilgil. Some of this water discharges from the springs on the southeastern shoreline, but some of it may flow into the lake diffusively through the lake floor. Other studies support this general interpretation (Yihdego 2005). The groundwater gradient northwest of Lake Elmenteita slopes towards Lake Nakuru (Becht et al. 2006), which is ~16 m lower, but shallow outflow has not been confirmed.

10.5 Vegetation and Wildlife

10.5.1 Lake Nakuru Basin

The vegetation around the lake (Fig. 10.6b) is wooded and bushy grassland with ~550 different plant species (Agembe et al. 2018). Lake Nakuru is bordered by alkaline swamps with sedges and *Typha* marsh along the river and spring inflows.

There are no aquatic macrophytes (Burgis and Mavuti 1987). Marginal grasslands of *Sporobolus spicatus* salt grass pass into dry transitional savanna with grasslands of *Hyparrhenia hirta* and Rhodes grass (*Chloris gayana*) in the lower areas. Higher areas have dry forest with *Acacia* (*Vachellia*) *xanthophloea*, olive *Olea hochstetteri*, and *Euphorbia candelabrum* forest. The bushland is dominated by *Tarchonanthus camphoratus* and *Psiadia arabica*. Were et al. (2013) have documented the rapid modification of natural vegetation resulting from changing land use.

The National Park contains more than 450 bird species, and at least 50 species of mammals including rare white rhinoceros (*Ceratotherium simum cottoni*) and Rothschild's giraffe (*Giraffa camelopardalis rothschildi*), and buffalo, lion, leopard, zebra, eland, waterbuck gazelles, baboon and monkeys. Hongo and Mulaku (2021) discuss the negative impact of flooding on the wildlife population. The lake expanded from 35 km^2 in 2009 to 54 km^2 in 2018 removing much habitat.

10.5.2 Lake Elmenteita Basin

The wooded upland vegetation is variable. Mwaura and Moore (1991) recognised several plant associations: *Olea-Juniperus-Dombeya* upland forest association; *Acacia-Olea* lowland forest association; *Acacia-Thus-Tarchonanthus* woodland association, *Acacia* woodland association; and *Tarchonanthus-Thus-Euclea* bushland association. Near the shoreline, vegetation is sparse bushland except near river mouths where *Acacia xanthophloea* (fever tree) is locally common. Grasslands near the lake are dominated by *Cyanodon dactylon*, *Themeda triandra* and *Chloris gayana* (Koyo 2005), and small shrubs. Soils and old lake beds around the shoreline are saline except near dilute springs, limiting many plants.

Agriculture is the main land use in the northern wetter parts of the basin, especially in Ndunduri, Ngorika and Nyaituga, where maize, potatoes, beans and cabbage are grown. Fodder is produced for local dairy farming. Much of the land is used for herding domestic cattle, sheep and goats or set aside for wildlife at the 200 km^2 Soysambu Conservancy west of the lake. Deforestation in the catchment has led to increased runoff and soil erosion, which have reduced groundwater recharge.

The wildlife around the shoreline includes zebra, Thomson's gazelle, Grant's gazelle, eland, buffalo, giraffe, warthog, colobus monkey, baboon and occasional carnivores. More than 450 bird species have been recorded in the lake basin, including lesser and greater flamingos, and pelicans (Fig. 10.3e). Lesser flamingos have bred at the lake occasionally (Brown 1958).

10.6 Limnology

10.6.1 Physical Limnology

Lake Nakuru

Leakey (1931a) sounded the lake floor in May 1929 and found the lake to be 9 feet 2 inches (2.79 m) deep. Jenkin (1936) reported comparable results. Vareschi (1982) produced the first bathymetric map (Fig. 10.7b), compiled using soundings and aerial photographs. This showed that the deepest part of the lake floor was ~4.4 m below the 1979 shoreline. The lake floor is saucer-shaped with gently sloping margins (0.6–1.1°) and a nearly flat central base (Vareschi 1982).

Climatic variations have led to changes in area, depth and salinity on a range of time-scales with major consequences for the ecology of the lake (see below). Lake level has undergone frequent changes within the past century (Figs. 10.9 and 10.10; Leakey 1931a; Jenkin 1936; Vareschi 1982; Burgis and Morris 1987; Kiage and Douglas 2020; Herrnegger et al. 2021). The lake dried up in 1944–6, July 1961, and in 1997 (Raini 2006). In common with other Kenya Rift lakes, Nakuru increased in level and surface area during the past decade (Onywere et al. 2013; Obando et al. 2016). Its level is related to rainfall and evaporation, but not in a simple way. Vareschi (1982) recognised a time lag of about 3 months between changes in rainfall and changes in lake level, reflecting factors such as the retention of water in the soil and the timing of its delivery to the lake as groundwater and seepage. Although several causes for lake level rise have been proposed (Kiage and Douglas 2020), Herrnegger et al. (2021), using remotely sensed and meteorological data, proposed that changes in lake area and level after 2010 (Fig. 10.10b) can be explained by increased rainfall alone. Others have reached similar conclusions (Multa et al. 2021; Onzere et al. 2022). Nonetheless, other factors, such as changes in land use, might have contributed to lake level variability at Lake Nakuru and other lakes (James 2022). The causes of the post-2010 rise in the levels of the Kenya Rift lakes have not yet been fully resolved.

The equatorial location means that there is little seasonal variation in water temperature. Montcoudiol et al. (2019) reported a range of 19–33 °C. Daily fluctuations in heating and cooling can lead to strong diel cycles of stratification and mixing (Fig. 10.11a). Melack and Kilham (1974) provided an example (Fig. 10.11b) that shows the development of a temporary thermocline around noon, which breaks down as winds increase and surface water cools. Dissolved oxygen and photosynthesis rates are high even when the lake is stratified with the bottom waters remaining oxygenated (Melack and Kilham 1974; Vareschi 1982).

Lake currents are induced by wind and mixing. Small seiches (4.5 cm) have been recorded after storms. With no perennial rivers, inertial inflow effects are minimal. Vareschi (1982) noted that winds typically mix the lake water for up to 7 h daily, breaking down any thermal stratification, but are rarely strong enough to re-suspend bottom sediments. The lake is polymictic (Oduor and Schagerl 2007a, b; Kaggwa et al. 2013).

Fig. 10.9 Changing levels of Lake Nakuru in different years between 1976 and 2018. (**a–f**) View across northern Lake Nakuru from the Baboon Cliff viewpoint, NW lake margin. (All photographs taken in June–July except **d**, taken in December)

Lake Elmenteita

Historically, Lake Elmenteita has been a shallow (<2 m deep) saline playa lake fed by springs, groundwater and ephemeral streams, which dries up periodically to expose trona-encrusted mudflats (Leakey 1931a). Google Earth imagery shows that it was a salt encrusted pan in December 2005 with most remaining surface water in the southern and eastern half near sites of spring inflow. Krienitz (2018, his Fig. 5.7) illustrates the lake when dry in September 2006. In common with Lake Nakuru, it expanded and deepened during rains after 2010.

Lake level and water depth respond to variations in rainfall and evaporation, but also to water abstraction for agriculture in the catchments of the Mereroni and other rivers in the drainage basin, and to land-use changes (Murimi 1989, 1993, 1994). In recent decades, maximum lake depths have varied from <1 m to ~3 m, averaging 1.5 m. Murimi (1994) reported a general decline in lake levels between 1958 and

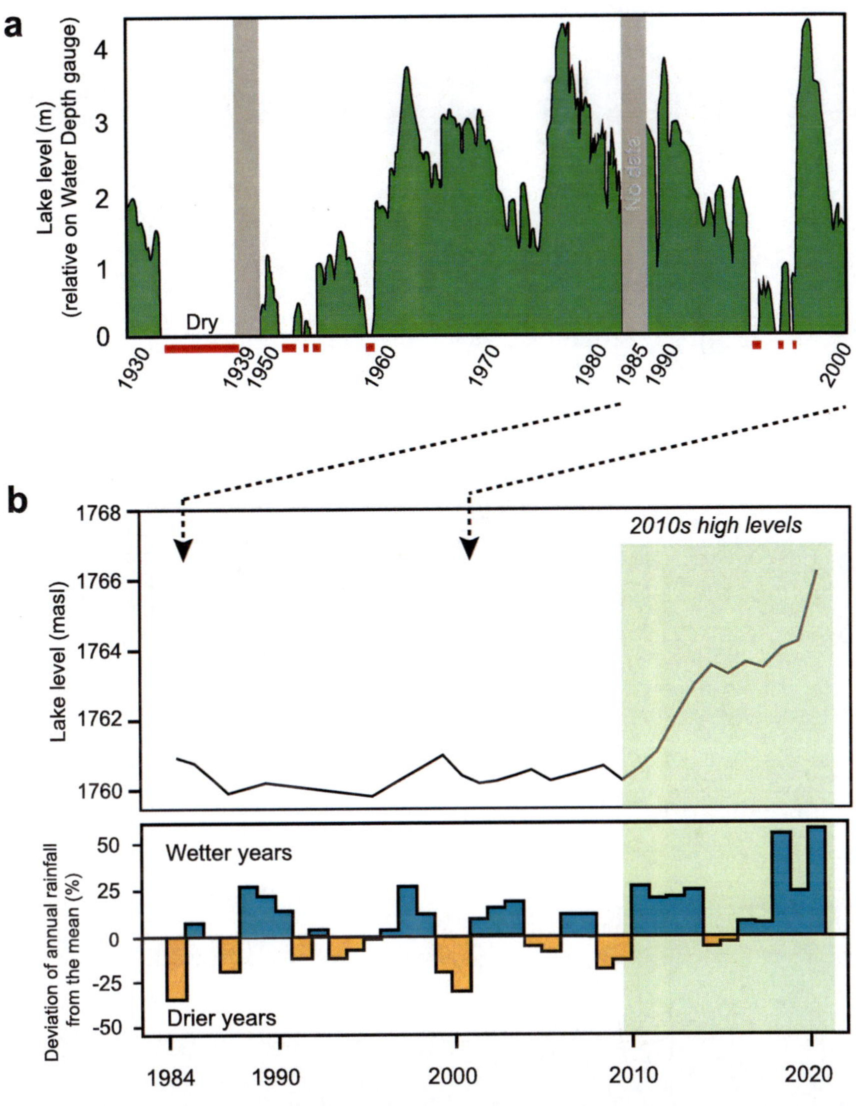

Fig. 10.10 (**a**) Record of lake level change at Lake Nakuru using data from Vareschi (1982) and from Herrnegger et al. (2021). (**b**) Lake level change compared to annual rainfall (percentage deviations from the mean). Data from Herrnegger et al. (2021)

1986 and, after comparing those records with climatic data, found that rainfall and evaporation alone did not explain all observed lake-level trends. He noted that (1) decreasing river inflow was linked to water abstraction upstream, and (2) increases in sediment supply, caused by soil erosion, was an important control on lake level. Mwaura and Moore (1991) also noted that decreasing forest cover linked to expanding human settlement and agriculture was a major cause of increased runoff and soil

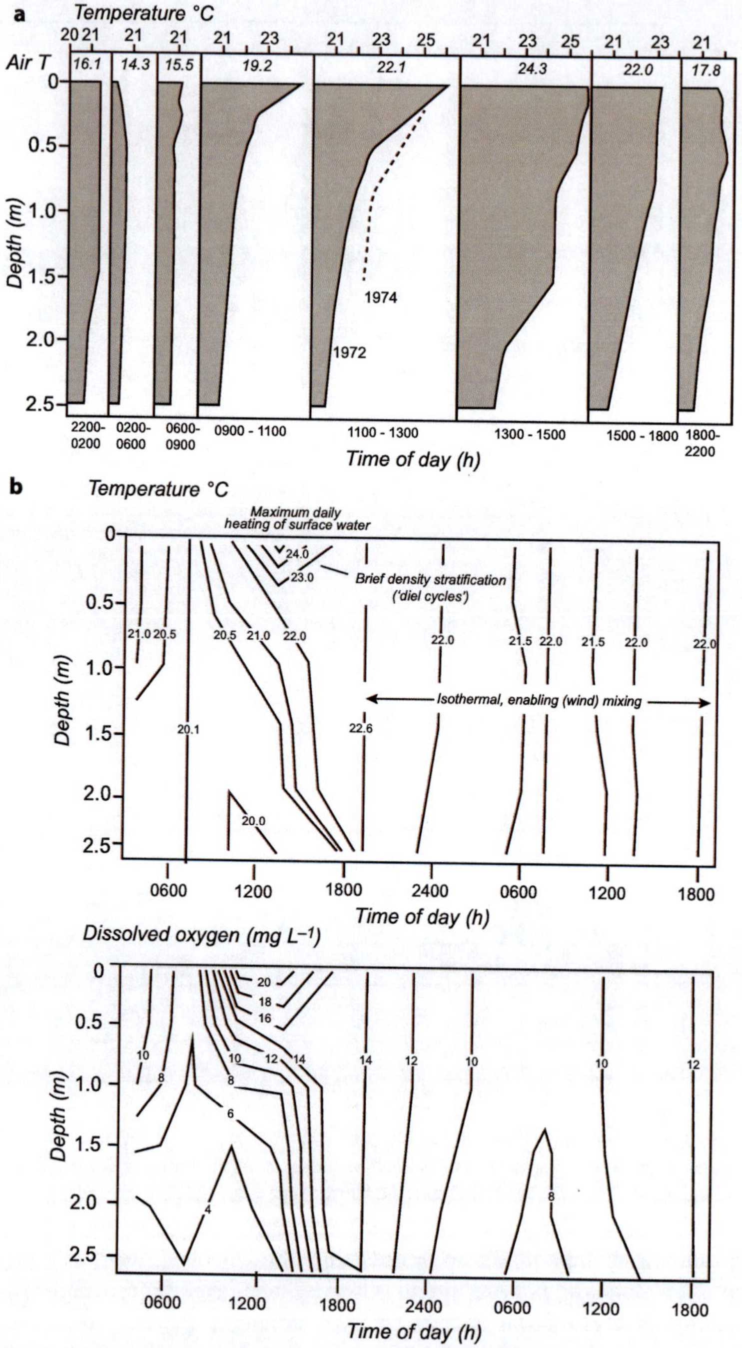

Fig. 10.11 Thermal profiles in Lake Nakuru. (**a**) Daily profile from Vareschi (1982) showing incipient stratification from late morning until early afternoon, followed by cooling and mixing. (**b**) Time vs depth distribution of temperature and dissolved oxygen, 13–14 June 1971. From Melack and Kilham (1974)

erosion. Sediment supply appears to have increased at about the same time that runoff has decreased, which can have a significant impact in shallow lakes. Direct anthropogenic effects, however, are difficult to separate from short-term (decadal) climatic changes. Longer-term records will be needed to assess the consequences of these factors on the recent changes in hydrology and physical processes. The bathymetry (Fig. 10.8b, c; Melack 1988; Muno 2002) shows the deepest part to be near the mid-western shoreline. All lake margins are of low angle, even near the steeper hills east of the lake.

Physical processes in Lake Elmenteita are linked to variations in lake level and are controlled by evaporation and precipitation, and the impact of wind funnelled along the rift axis in both directions. Montcoudiol et al. (2019) reported a temperature range of 23–26 °C The lake has been too shallow to stratify for extended periods and its equatorial location (0.26'S) precludes seasonal variation in water temperature. In common with other soda lakes, however, the lake undergoes strong diel cycles. Diel cycles in temperature and oxygen were documented by Melack (2009), who recorded a lake-water temperature range at 20 cm depth of 16.1–27.4 °C. Higher water temperatures likely exist near sites of hydrothermal inflow. Energy levels are always low. Unidirectional winds with velocities up to ~10 m s^{-1} (Melack 2009) blow water across the southern mudflats but are too low to have much impact on sediment sorting.

10.6.2 Chemical Limnology

Lake Nakuru

Before the post-2010 flooding, Lake Nakuru was a typical saline, alkaline rift lake with sodium carbonate-bicarbonate water and with high chloride and silica, but negligible calcium and magnesium. Its chemistry (salinity, pH) had nonetheless fluctuated for several decades although its ionic composition remained the same. Longer-term changes linked to climate and tectonics had long been recognised (Sect. 10.9; Chap. 18). Most research, however, has focused on ecology rather than hydrochemistry, so the data acquired have emphasised criteria such as dissolved oxygen (DO), conductivity and nutrients (e.g., Peters and MacIntyre 1976; Kulecho and Muhandiki 2006; Okoth et al. 2009), and potential toxins for biota, especially the flamingos, rather than the ionic composition and brine evolution. There are few published (major ion) analyses of Lake Nakuru and Lake Elmenteita basin waters compared to other Kenya Rift lakes (Table 10.1); some are incomplete or have anion/cation ratios (milliequivalents) that do not balance within an acceptable range. Furthermore, the chemistry of the inflow waters is poorly known because most runoff before 2010 was ephemeral and rivers only occasionally flowed into the lake.

Kilham (1971) analysed the main affluent streams (Table 10.1). The Makalia and Nderit rivers have near-neutral pH and Na^+-HCO_3^- composition but little Ca^{2+}. In contrast, the mouths of the Nderit and Njoro rivers are more saline and alkaline (pH >9) having evaporated *en route* to the lake and (or) mixed with shallow

Table 10.1 Selected chemical analyses of Nakuru basin waters

Sample	Ref no.	Notes	Date	Source	T°C	pH	Na	K	Ca	Mg	Cl	HCO_3	CO_3	T-Alk	SO_4	F	SiO_2
Rivers																	
Makalia			Jan-70	1		7.40	21.4	16.3	10.40	2.00	9.8	82.4			18.9	1.22	111.30
Nderit			Sep-69	1		7.30	24.8	12	9.40	2.00	8.6	99.2			8.4	1.46	50.30
Nderit mouth			Dec-69	1		9.40	890	78.0	5.20	0.70	259	1440	636		24.0	32.3	57
Njoro			Dec-69	1		9.25	490	25.3	4.60	1.70	92.6	235.2			66.0	0	78
Makalia	K26		July-16	12	18.4	8.18	15	8.6	5.1	1.3	5.4	57	0		4.7	0.83	
Makalia Falls	K26		Jan-17	12	24.5	7.72	41	36	22	5.2	31	96	0		8.2	1.6	
Makalia Falls	K26		Jun-17	12	25	8.52	34	23	20	3.5	17	30	0		10	1.8	
Springs																	
Artesian spring	25	ppm		2			1460				837	1599	351		342.0	24	
Warm spring	133/1/71	ppm	Aug-71	3		7.25	18.5	12.7	2.75	0.80	6	54			6.5	1.7	58
Warm spring	118/4/1	ppm	Dec-71	3	34	7.60	114	12	1.50	2.40	16.7	258			15.5	4.05	95
Lake																	
Lake Nakuru	65a		Apr-29	4		11.00								296			21.5
Lake Nakuru	65b		May-29	4							1950			265			
Lake Nakuru	65c		Dec-30	5			5550	256	10.00	0	1375			205	253.0		
Lake Nakuru	5		1940s	6			33,855				7160		37,400		1457.0	1400	
Lake Nakuru	65d		Feb-61	7			38,000	1312	<10	<30	13,000			1440	1800.0		730
Hippo Pool			Jul-69	1		9.75	3300	237	0	0.90	1020	3200	4100		62.0	129	208
Lake Nakuru NW			Jan-70	1		9.96	2840	148	0	1.00	834	2124	4116.7		72.0	148	27
Lake Nakuru			Dec-71	3			3350	135	1.75	0.25	1042	1545	2915		180.0	135	185
Lake Nakuru NW	NAK-76		Aug-76	8	23	10.20	26,500	1020	0	0.50	8750			872	1150.0	425	480

Lake Nakuru E	NAK-77		Aug-77	8	22	10.20	20,100	770	0	0.00	6200			698	830.0	385	372
Lake Nakuru	12	ppm	Jan-84	9		10.10	5692	212					12,309				
Lake Nakuru			Jul-85	10		10.00	8028	301	5.7	0.35	2629			306	365		
Lake Nakuru		Mean values	June-02	11		10.35	9498	332.4			2386			327	240.2	331	
Lake Nakuru NW	K25		July-16	12	19.8	9.72	2324	249	17	10	692	2869	563		63	323	
Lake Nakuru S	K27		July-16	12	31.6	9.68	2330	322	nd	13	679	3211	797		63	345	
Lake Nakuru NW	K25		Jan-17	12	23	9.61	2372	127	5.7	16	690	2051	353		77	68	
Lake Nakuru NW	K25		July-17	12	22	9.57	2612	120	1.8	12	743	1603	1636		73	83	
Lake Nakuru S	K27		July-17	12	32.5	9.48	2853	139	0.25	6.2	810	1779	2023		71	106	

Reference sources: 1: Kilham (1971); 2: McCall (1967); 3: Glover (1972); 4: Jenkin (1936); 5: Beadle (1932); 6: (Baker 1958); 7: Talling and Talling (1965); 8: Renaut (1982); 9: Moritsugu (1985); 10: Finlay et al. (1987); 11: Oduor and Schagerl (2007b); 12: Montcoudiol et al. (2019). Other analyses of Nakuru basin waters are found in Njenga (2004), Kanda (2010) and Okech (2012)

Analyses in mg L^{-1} except total alkalinity (T-Alk) given in meq L^{-1}

groundwaters. Springs in the area range from dilute to saline though details of locations are sparse. Groundwater from local wells is dilute (<0.5 g L^{-1} Total Dissolved Solids: TDS), and of Na^+–HCO_3^- type with high silica and low Cl^- and SO_4^{2-} (Clarke et al. 1990; Kanda 2010).

Lake Nakuru has had a salinity ranging from ~3 to >50 g L^{-1} TDS, a conductivity from 9000 to >160,000 µS cm^{-1}, and a pH range of 9.4–11, with pH values of 10.0–10.3 being typical (Talling and Talling 1965; Vareschi 1982; Ballot et al. 2004a; De Cort et al. 2013; Montcoudiol et al. 2019). Surface water temperatures are typically ~25–27 °C but can reach 38.5 °C, and the bottom waters are usually 20–22 °C (Vareschi 1982). Alkaline earths are low (0–17 mg L^{-1} Ca^{2+}; 0–6 mg L^{-1} Mg^{2+}), but recent lake muds contain ~20% $CaCO_3$ (De Cort et al. 2013) implying precipitation in the water column. Aqueous silica is high, as expected from the high pH and local volcanic geology, but there is no record yet of its removal other than by biological sequestration: no chert has been reported from Lake Nakuru (or Lake Elmenteita) although silica removal by authigenic clays or zeolites is possible. Sodic evaporites have not been reported except efflorescent sodic crusts (McCall 1967): the inflow and accommodation space (adequate subsidence), and evaporation have been too low for trona-nahcolite accumulation in both lakes. Any thin trona beds that might have formed would have dissolved.

Lake Elmenteita

Lake Elmenteita waters have a similar composition to those of Lake Nakuru (Table 10.2). Both lakes have common bedrock and climatic setting, and each is partly recharged by rift-floor groundwater. Both lakes lie in topographically closed basins with high evaporation/precipitation ratios. The smaller Lake Elmenteita differs by having active hot springs in the drainage basin and along the lake shore. Both lakes, nonetheless, may be influenced by northward-moving hydrothermal fluids at an unknown depth.

The seasonal rivers have typical dilute (<350 mg L^{-1} TDS), neutral to slightly alkaline (pH 6.9–7.8) Na^+–Ca^{2+}–HCO_3^- waters, some of which are modified by thermal spring discharge into their channels. The Kariandusi River is more saline and is enriched in sulfate and (anthropogenic?) phosphate in some analyses (Adeka et al. 2008). Local dilute groundwaters near the lake include 'unmodified' Na^+–Ca^{2+}–HCO_3^- waters derived from weathering of the igneous bedrock. Clark et al. (1990) ascribed bicarbonate to the dissolution of carbonates, but most can be linked to silicate hydrolysis. They also recognised 'modified' well waters with higher salinity than other wells and discussed the possible mixing of unmodified waters with thermal fluids that evolved along the northern flanks of Eburru.

The hot spring waters are also of Na^+–Ca^{2+}–HCO_3^- composition (McCall 1967; Mwaura 1999; Adeka et al. 2008; Clarke et al. 1990). The upland (rift margin) springs are the most dilute with neutral pH (6.5–7.5) and with calcium values similar to affluent rivers. The southern lakeshore springs and Cole's hot spring are warmer (up to 46 °C), more saline (up to ~3 g L^{-1} TDS) and alkaline (pH 8.9–9.5) with low alkaline earths but high silica (~125 mg L^{-1} SiO_2) and phosphorus (Adeka et al. 2008; Montcoudiol et al. 2019). The high silica reflects its abundance in the

Table 10.2 Selected chemical analyses of Elmenteita basin waters

Sample	Ref.	Units	Date	Source	Temp °C	pH	Na	K	Ca	Mg	Cl	HCO_3	CO_3	T.Alk*	T.Alk	SO_4	F	SiO_2
Rivers																		
Mereroni	39	ppm		1		6.9	17.3				5	32.38		2.5		0	1.5	25
Mereroni			Jul-69	2		6.9	17.9	6.5	4.7	0.9	3.9	62.9				4.4	0.78	48.1
Mereroni			Jul-69	2		7.5	21	11	1.5	1.4	4.9	71.4				4.4	0.78	44
Kariandusi			Sep-69	2		7.3	44.4	10.7	10.1	1.4	1.1	135.5				25.4	7.19	74.9
Mbaruk		ppm	Dec-71	3		7.5	25.7	13	9.6	2.22	17.9	73				13	1.4	49.5
Kariandusi		ppm	Dec-71	3		7.8	58.5	14.5	11.1	1.76	5.5	143				13	9.8	68
Baruku			Jul-16	4	17.5	6.97	22	6.1	7.6	2.0	4.9	96	0			2.2	0.52	
Springs																		
Cole's Spr	6	ppm	1940s	5			1252				461		1245			104	95	
Dunlop	47	ppm		1		6.9			3	0	5	60				10	1.3	90
Mereroni	48	ppm		1		6.7			10	0	6	135				4	1.8	90
Land's	49	ppm		1		6.7			4	0	2	55				4	1	85
Warm spr	3/1	ppm	Jan-72	3	31	6.5	60	19.5	10.3	2.55	5.6	192				13.1	1.75	72
Warm spr	4/4	ppm	Jan-72	3	32	6.5	28.5	12	3.5	0.46	5.1	69				16	2.2	89
Warm spr	4/1	ppm	Dec-71	3	41	7.6	56.7	12.7	9.2	1.6	4.6	154				10.9	8.5	60
Cole's Spr	4/5	ppm	Dec-71	3	45	9.2	1060	42.7	0.27	0.03	490	980	255			94	117	132
Majimoto			Feb-89	6		8.1	44	14	8.8	2.4	8				144	10	2	
Chamuka			Feb-89	6		7.2	13	3.6	4	2.4	13				76	5	1.5	
Warm spr			Feb-89	6		9.5	945	8.1	0.8	0.1	552				1200	85	2	
L Elm spr	41		1985–7	7	43	9.3	1050	35.0	0.46	<0.1	490	1610				82		125
L Elm spr	42		1985–7	7	38	9.1	773	31.0	0.54	0.10	360	1190				68		122
Kekopey	K35P1		Jul-16	4	45	9.31	1036	123	12	4.0	466	1052	130			73	500	
Kekopey	K35P1		Jan-17	4	40	8.98	1075	56	13	0.07	477	1255	73			77	103	
Kekopey	K35P3		Jun 17	4	42	9.17	1107	27	5.9	0.9	485	1269	114			77	134	

(continued)

Table 10.2 (continued)

Sample	Ref.	Units	Date	Source	Temp °C	pH	Na	K	Ca	Mg	Cl	HCO_3	CO_3	T.Alk*	T.Alk	SO_4	F	SiO_2
L Elm spr				8	36.6	9.2	784	29	10.6								79.8	
Lake																		
Lake Elm	56a		May-29	9		10.9					4370			227				0.9
Lake Elm	56b		Dec-52	10					1.1	1.4	4000			147		100	400	
Lake Elm	23	ppm	1940s	5			23271				10460		20100			1317	1627	
Lake Elm	56c		Feb-61	11			9450	381	<10	<30	5200			289		650		295
Lake Elm			Jul-69	2		9.4	3800	284			1970	4162	2328			133	323	177
Lake Elm		ppm	Dec-71	3		9.5	4450	174	1.25	0.16	2280	2500	2217			327	400	159
Lake Elm	K34		Jul-16	4	23.6	9.81	3258	437	11	nd	1857	832	1475			137	1242	
Lake Elm	K35		Jul-16	4	24.9	9.68	3270	450	11	nd	1863	1076	1415			132	1247	
Lake Elm	K38		Jul-16	4	25.8	9.67	3223	457	12	nd	1948	1032	1562			137	1297	
Lake Elm	K34		Jan 17	4	24.8	9.50	3704	205	22	nd	1986	1689	1467			165	288	
Lake Elm	K35		Jan 17	4	24.3	9.49	3597	202	53	nd	1920	1696	1440			149	309	

Reference sources: 1: McCall (1967); 2: Kilham (1971); 3: Glover (1972); 4: Montcoudiol et al. (2019); 5: Baker (1958); 6: Mwaura (1999); 7: Clarke et al. (1990); 8: Butturini et al. (2020); 9: Jenkin (1936); 10: Fish (in Talling and Talling 1965); 11: Talling and Talling (1965). Other analyses are given in Njenga (2004) and Montcoudiol et al. (2019)

All analyses in mg L^{-1} or ppm as indicated (in Units) except T.Alk* in meq L^{-1}, *nd* not detected reactive ash and diatomaceous sediments in the drainage basin and the high pH of the water. Montcoudiol et al. (2019) noted a shift to higher $\delta^{18}O$ isotopes in the Kekopey springs in the dry season and attributed it to a seasonal decrease in meteoric influence and the inclusion of more evaporated lake water in the outflow.

The lake waters usually have a Na^+–HCO_3^-–CO_3^{2-}–Cl composition and, apart from one measurement by Jenkin (1936: pH 10.9), have had a pH range of 9.4–10.0, a typical salinity of ~25–30 g L^{-1} TDS (Oduor and Schagerl 2007a; Njenga 2004) and conductivity of 12,000–40,000 μS cm^{-1} (Hughes and Hughes 1992). The salinity, however, has ranged from <10‰ to ~300‰ when drying out (Krienitz et al. 2013, 2016). The Na/Cl (mol) ratio of most spring and river inflow waters (~4–8) exceeds that of the lake water (~3), indicating that the lake is either fed by groundwater with a lower Na/Cl ratio than surface inflow or that a mechanism for Na^+ depletion is effective. Although efflorescent sodium-carbonate crusts form seasonally around the lake shoreline (see below), the lake sediments do not preserve evidence of evaporites. This implies that groundwater with a lower Na/Cl ratio than surface inflow might also feed the lake. Montcoudiol et al. (2019) found that the lake water recently had higher alkaline earths than Lake Nakuru (10–53 mg L^{-1} Ca^{2+}; 0–25 mg L^{-1} Mg^{2+}). They also noted that Cl^- and F^- account for a higher proportion of the anions compared to Nakuru.

Several studies at Lake Nakuru and Lake Elmenteita have focused on the trace element and biotic controls on flamingo health and their periodic stress and decline, and on heavy metals and other (mainly cyanobacterial) toxins in lake and river waters and sediments (e.g., Greichus et al. 1978; Wandiga et al. 1983; Maskall and Thornton 1989, 1996; Ochieng et al. 2007; Okech 2012; Jirsa et al. 2013; Yang et al. 2017; Davdra 2020; Florence 2021). Some flamingo declines might be related to pollution, but others cannot be attributed to a single cause.

10.7 Biology and Ecology of the Lakes

10.7.1 Lake Nakuru

Lake Nakuru has been a focus of ecological research for almost a century. Much of the interest stems from the global importance of its bird population (especially lesser flamingo), but many studies have focused on the related plankton and their exceptionally high productivity. Early ecological studies were undertaken by the Percy Sladen expedition (1929) led by Penelope Jenkin, and the Cambridge Expedition to the East African Lakes (1930–31) led by Edgar Worthington and supported by Leonard Beadle and colleagues (Worthington 1932a, b; Beadle 1932; Talling 2006). Jenkin (1932, 1936), Florence Rich (1932) and co-workers pioneered ecological studies of the soda lakes in the Kenya Rift. Lind (1965, 1966, 1968) recorded the phytoplankton at Lake Nakuru and Lake Elmenteita.

In the 1970s Ekkehard Vareschi published a series of papers that focused on the microbial ecology and productivity, and the impacts of short-term changes on the lake ecology, and on lesser flamingos and other birds (Vareschi 1978, 1979, 1982; Vareschi and Jacobs 1984, 1985; Vareschi and Vareschi 1984). He provided a framework for detailed ecological research on soda lakes in the East African Rift and elsewhere.

In common with earlier workers (Jenkin 1936), Vareschi and Jacobs (1985) recorded the dominance of the coiled cyanobacterium *Arthrospira fusiformis,* then usually termed *Spirulina platensis*, the main food of the lesser flamingo. *Spirulina* is now recognised as separate genus that inhabits hot springs and sediment (Ballot et al. 2004b; Krienitz 2018). *A. fusiformis* thrives at salinities between 22 and 62‰ and tolerates temperatures of >35 °C (Krienitz and Schagerl 2016). Kaggwa et al. (2013) showed that the morphology and size of *Arthrospira* coils are influenced by their environment.

Other common cyanobacteria include *Anabaenopsis abijatae, A. arnoldii, Synechococcus minutus*, *Synechocystis* sp., *Cyanospira rippkae, Haloleptolyngbya alcalis,* and *Picocystis salinarum* (Milbrink 1977; Dadeech et al. 2012). Other primary producers include cryptomonads (cryptophytes: small biflagellate algae), coccoid green algae ('*Chlorella*'), the alga *Monoraphidium minutum*, *Euglena,* and diatoms (for details see Krienitz et al. 2016; Krienitz 2018; and references therein).

Many have discussed the significance of diatoms at Lake Nakuru and Lake Elmenteita. Hecky and Kilham (1973) found that *Nitzschia frustulum* and *Anomoeoneis sphaerophora* were dominant in most benthic samples with other species reflecting lower salinities. Other common benthic taxa include *Rhopalodia gibberula*, *Craticula elkab* and *Nitzschia sigma* (Melack 1982, 1988; Raini et al. 2010; Schagerl and Oduor 2008), which increase in abundance whenever *Arthrospira* crash. Planktonic diatoms such as *Cyclotella meneghiniana* (*Stephanocyclus meneghinianus*) are present at lower salinities and often in shallow water, with *Thalassiosira rudolfii* at intermediate alkalinity (Hecky and Kilham 1973). Reduced salinity results from lakewide freshening (typically increased rainfall and runoff) or mixing offshore with seasonal river or spring inflow.

Lakes Nakuru and Elmenteita have high surface-area-to-volume ratios and respond rapidly to the physical changes in their environments (Odour and Schagerl 2007a). Such shifts affect the biota, which respond quickly through changes in both composition and biomass (Vareschi 1978). Such changes are often extreme, resulting in major 'crashes' in planktonic biomass (Schagerl et al. 2015; Krienitz and Schagerl 2016, their Table 6.1) and rapid modification in the community structure (Melack 1976; Vareschi et al. 1981). These upheavals in turn affect the avifauna, including lesser flamingos.

The high and fluctuating salinity limits diversity, but productivity is exceptionally high – some of the highest ever recorded (Schagerl and Burian 2016; Fazi et al. 2018). Schagerl and Oduor (2008) showed that nitrate-N, conductivity, phosphorus and light are key variables influencing microbial biomass in both Lake Nakuru and Lake Elmenteita.

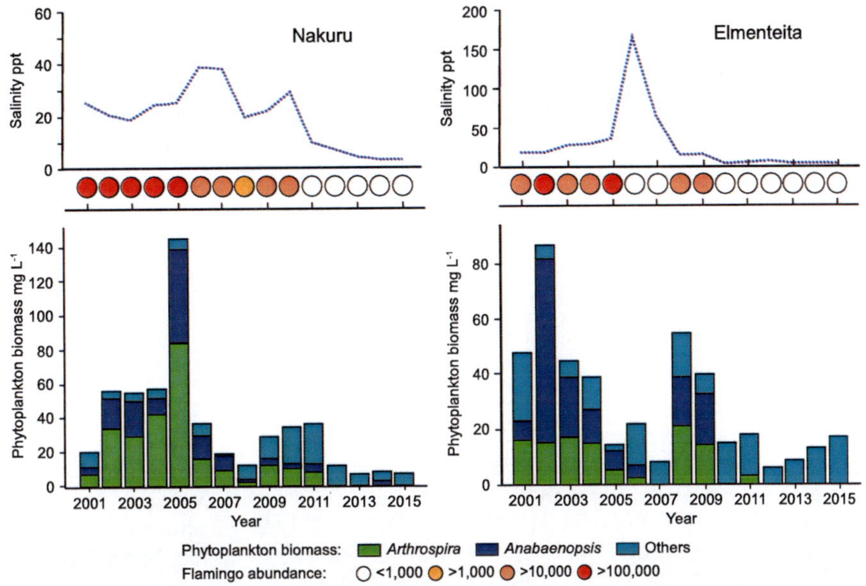

Fig. 10.12 Salinity, phytoplankton biomass, and flamingo abundance in Lake Nakuru and Lake Elmenteita. After Krienitz et al. (2016)

Krienitz et al. (2016) revealed the relationships among microbial flora and biomass, salinity and flamingo abundance between 2001 and 2015 (Fig. 10.12). Between 2001 and 2005, *Arthrospira fusiformis*, *Anabaenopsis abijatae* and *A. arnoldii* dominated the phytoplankton but none had monospecific dominance. The salinity was 19–29‰ and the phytoplankton biomass was 25–135 mg L^{-1}. Diatoms, dominantly *Nitzschia*, were present but in small numbers. Lake level fell in 2007, as did *Arthrospira* biomass (<12 mg L^{-1}), which caused flamingo numbers to decline. In January 2010 the lake shrank, and its salinity increased to 50.7‰. Minute *Picocystis salinarum* then outcompeted the filamentous cyanobacteria and the flamingos departed for lack of food (see also Schagerl et al. 2015).

After the 2010 rains, the water level rose, and the salinity had dropped to <5‰ by January 2015. The phytoplankton biomass declined, light penetration increased in clearer water, and diatom communities became established, including freshwater species. The few remaining lesser flamingos survived by consuming diatoms from the sediment surface.

The zooplankton have low diversity and are dominated by rotifers (*Brachionus dimidiatus*; *B. plicatilis*) and ciliates (Beadle 1932; Jenkin 1936; Nogrady 1983; Finley et al. 1987; Yasindi 2001; Oyoo-Okoth et al. 2011; Burian et al. 2012; Ong'ondo et al. 2013; Yasindi and Taylor 2016) and a corixid (*Micronecta jenkinae*, a water boatman). Other primary consumers include lesser flamingos, the copepod *Lovenula africana*, and midge larvae (*Leptochironomus derib*). Secondary consumers include greater flamingos (*Phoenicopterus roseus*), pelicans (*Pelecanus onocrotalus*; *P. rufescens*), fish eagles (*Haliaeetus vocifer*), and marabou storks (*Leptoptilos *crummeniferus*). Magadi tilapia *Alcolapia grahami* (= *Sarotheradon alcalicus grahami*) were introduced to the lake between 1952 and 1962 to combat mosquito larvae though their status is unknown (Kavembe et al. 2016).

Studies of the lesser flamingos (*Phoeniconaias minor*) and their ecology (Brown 1955; 1959a, b; 1971; Vareschi 1978; Bartholomew and Pennycuick 1973; Tuite 1979, 1981, 2000; Childress et al. 2007; Krienitz and Kotut 2010; Kaggwa et al. 2013; Krienitz et al. 2016; Githaiga 2022) have dominated avifaunal research. Details are summarised in the very readable account by Krienitz (2018). During the last decade, much attention has been given to cyanobacterial toxins as their numbers have declined (e.g. Koenig 2006; Kotut et al. 2006) but the cause of the decline is still being debated.

10.7.2 Lake Elmenteita

Until recently, systematic biological data collection at Lake Elmenteita was rarely a primary focus except for the work by Leslie Brown (1958, 1959a, b) on flamingos and by John Melack, Peter Kilham, Robert Hecky and colleagues on biological productivity and diatoms. Lake Elmenteita has often been a site of sporadic, supplementary, or incidental biological observations, by some *en route* to lakes Nakuru, Bogoria and Baringo. This has been partly because of restricted access to the lake by local land owners. That has gradually changed during the past decade.

In common with Lake Nakuru and Lake Bogoria, life in Lake Elmenteita is often dominated by massive blooms of *Arthrospira fusiformis*, which have often accounted for >90% of the phytoplankton. Other organisms recorded include other cyanobacteria and archaea, diatoms (Hecky and Kilham 1973; Melack 1988; Mwirichia et al. 2010, 2011; Akanga et al. 2015; Tebbs et al. 2015) and zooplankton such as copepods (*Lavenula africana*) and rotifers (*Brachionus spp.*). Chironomid larvae are often common (Mengistou 2016).

The abundant *Arthrospira* support a major, but ephemeral, population of lesser flamingos that visit and stay at the lake when their microbial food is stable and plentiful (Krienitz et al. 2016). The saline alkaline waters limit options for many other aquatic organisms so the biodiversity remains low even when the productivity is exceptionally high or increasing. The lake water surface varies from green, pink or red when the lake is productive, to dark blue or grey when phytoplankton productivity is relatively low or declining. During the past four decades, the lake ecology has also become less predictable because of anthropogenic changes in the catchment.

Melack (1988) examined microbial productivity in the lake during a transition from a high stage (February 1973) to a lower more saline stage (August 1974), linked to increased evaporative concentration. He found that *Spirulina* (*Arthrospira*) and *Anabaenopsis arnoldii* declined rapidly, as did chlorophyll-*a* concentration and phytoplankton photosynthetic rates. The decline led to clearer water that allowed benthic algae to flourish.

Krienitz et al. (2016) studied the lake over a 15-year period (2001–2015) and confirmed the strong links among lake level, salinity, microbial flora and biomass (Fig. 10.12). They showed that the biomass was dominated by *Arthrospira fusiformis* and *Anabaenopsis arnoldii* from 2001 to 2009 except in 2006 when the lake nearly dried out and the lake became hypersaline (>150 g L^{-1} TDS). In 2002 and 2005 the biomass supported >100,000 lesser flamingos. After 2010 the salinity dropped to <10 g L^{-1} TDS and the plankton were replaced by coccoid cyanobacteria and small green algae, both of which are unsuitable food for filter-feeding lesser flamingos.

Mpawenayo and Mathooko (2004) described the diatom flora from the hot springs located on the southern lake shore and found that *Anomoeoneis sphaerophora* and *Rhopalodia gibberula* were dominant. The introduced cichlid fish *Oreochromis alcalicus grahami* occurs in the southern hot springs and has occasionally spread to the main lake (Kavembe et al. 2016; Birdlife International 2023). Salano et al. (2017) investigated the fungal diversity in the hot springs.

In addition to the lesser flamingos, the lake supports a large and diverse bird population including greater flamingo, great white pelicans, little grebes and many palaearctic migrants including the little stint and several types of sandpiper. Pelicans and greater flamingos both have bred at Elmenteita, often competing for space on the basaltic islands off the western shores when they are exposed (Brown et al. 1973).

10.8 Recent Sedimentation

10.8.1 Lake Nakuru

Modern sedimentation in Lake Nakuru is poorly known. Information is based on cores taken over several decades for palaeoenvironmental analyses (Richardson and Dussinger 1986; De Cort et al. 2013; Robakiewicz et al. 2019) and from engineering studies that have focused on the impact of land-use changes on sedimentation and the water balance (Iradukunda and Nyadawa 2021).

Without perennial river inflow, and with the high groundwater recharge, recent lacustrine sediments have been fine grained (fine sand to clay). Our occasional observations (1976–2018), with many before the post-2010 flooding, confirm that most lake margins were low-angled slopes (peripheral mudflats with high water table) cut locally by shallow rills but with low (dm-scale) erosional terraces subparallel to the receding shoreline, some with weak Ca-carbonate cement. The exposed peripheral deposits were vegetated by patchy grasses and reeds (*Sporobolus spicatus, Cyperus laevigatus*: Ng'weno et al. 2009; Kipkemboi 2016), passing into trona-encrusted (<5 mm thick) mudflats near the shoreline when lake level was low. The transition from a grassy sandy mudflat to a saline mudflat was often sharp. Beaches and sorted coarse debris were absent but thin (10–100 cm wide, <5 cm thick) dispersed granule and fine gravel spreads locally mantled littoral flats (mainly NE), deposited by decelerating flow from ephemeral streams and unchannelled sheet-floods. Bioturbation by plants, insects, birds and mammals frequently disturbed the shoreline sediments, destroying primary structures (cf. Scott et al. 2021).

In common with most other shallow lakes in the Kenya Rift, the 'shoreline' at both Lake Nakuru and Lake Elmenteita is a transitional zone, sometimes hundreds of metres in lateral extent rather than a defined line on a map. Unless physically constrained by bedrock, faulting or wave-cut notches, most published shorelines in ephemeral playa lakes in the Kenya Rift record a temporary location during field-work or on an aerial photograph or satellite image.

The depositional environments of the Nakuru basin are relatively simple (Fig. 10.13). Much of the drainage basin is undergoing erosion, accelerated by deforestation and other land-use changes (Ehrensberger and Mbugah 2004; Raini

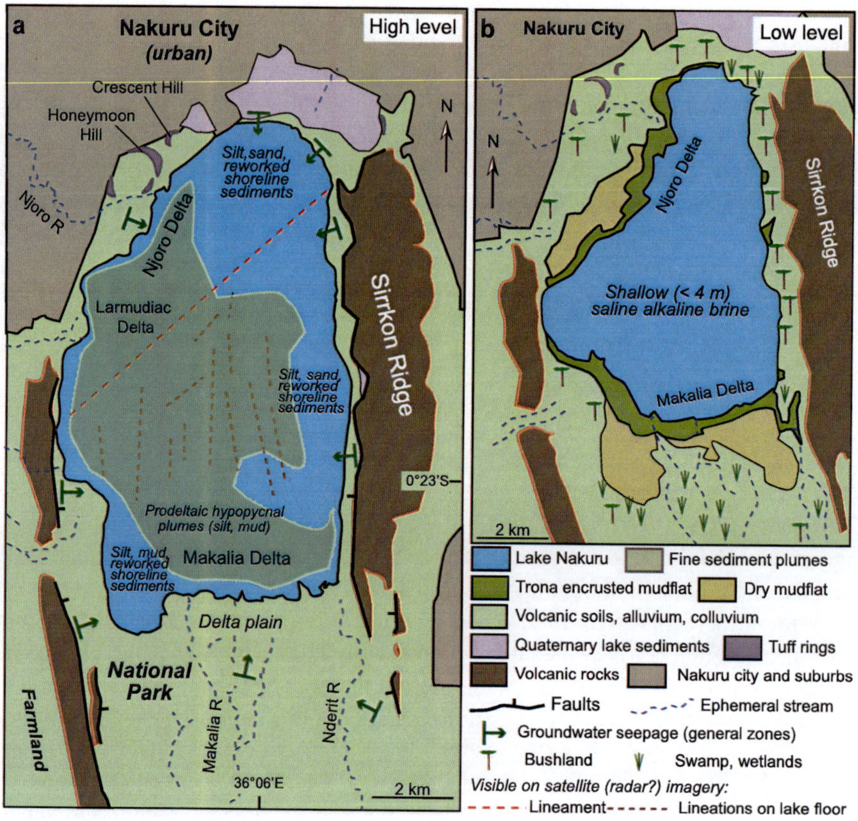

Fig. 10.13 Depositional environments of Lake Nakuru during high and low lake levels. Inferences are based on our occasional observations (1976–2018) and satellite imagery rather than field study. Satellite imagery (radar at some magnifications?) on AppleMaps™reveals irregular N-S and NW-SE lineations on the central lake floor

2009). Fissuring, collapse and underground erosion supply sediment in parts of the basin (Liu et al. 2022). When in flood, the few rivers feeding the lake deliver brown siliciclastic sands, silts and mud (Krienitz 2018, his Fig. 5.23). Maina-Gichaba et al. (2006) found that the Njoro River supplies 70% of the sediment to Lake Nakuru, followed by the Makalia (21%) and Nderit (4%) rivers, sewage effluent (4%), and Baharini springs (1%). These are supplemented by aeolian saline dust (Fig. 10.2e).

Most fluvial load is fine grained when it reaches the lake. Satellite imagery shows that hypopycnal plumes of suspended fines extend at least 1.5 km from the shoreline of the Makalia River when in flood. Those sediments then settle in the shallow water column but are stirred up by winds and reworked across the lake floor, so laminae are rarely preserved. Deltas are poorly defined because of the shallow, low-energy, ephemeral inflow. Nonetheless, Iradukunda et al. (2020), using an echo sounder, found that the thickest sediment fill is in the northwestern quadrant offshore from the Njoro River, and that relatively little sediment has been deposited near the Makalia and Nderit river mouths. Littoral and riparian marshes locally baffle and trap some clastic sediments. De Cort et al. (2013) recorded ~20% $CaCO_3$ in recent sediments, which implies that calcium carbonate precipitates in the water column. Unlike other rift lakes, there are no known carbonate deposits (tufa, travertine, stromatolites) linked to modern or former higher lake levels or hydrothermal activity. Although the water is saline, sodium carbonate salts (trona, nahcolite, gaylussite) do not usually form in the water column, but both minerals and minor halite form efflorescent crusts on cracked muds around the shoreline during dry periods. Thin (cm-scale) ephemeral trona beds, which are highly soluble, might form locally in channels and ponds when the lake dries up, but would dissolve quickly after seasonal rainfall or any rise in lake level. The tectonic and hydrological setting does not favour evaporite preservation in either lake, now or during the past few millenia.

Although most recent sediments are fine grained, Iradukunda and Nyadawa (2021) reported sand (63%) and gravel (16%) in a short (55 cm long) core taken ~1 km offshore of the Njoro River mouth. They noted that sediments in their core are upward fining, implying coarse sediment deposition during a former (undated) phase of lower lake level. In contrast, an offshore core (80 cm depth) taken near the centre of the lake contained 80–87% clay, silt and fine sand, compatible with other cores and satellite observations.

Richardson and Dussinger (1986) described and interpreted (Sect. 10.9.3) a 19.5 m-long piston core obtained from Lake Nakuru in 1969–70. Cohen et al. (1983) had previously examined the ostracod record. The shallowest sediments are dense hard muds that overlie organic-rich muds, which in turn overlie clays (Fig. 10.17c), but few sedimentological details were given. Ash beds are present, some from Menengai. The authors note that sediments are lost by deflation when the lake dries up. De Cort et al. (2013) found evidence of recent (pre-1870) exposure in their short (45 cm) core, with low water and organic matter content from 41 to 45 cm depth, whereas the upper part of the core was grey mud with higher organic matter and carbonate.

10.8.2 Lake Elmenteita

Although the modern lake is saline, the recent lake sediments are predominantly clastic, consisting of mud, fine sand, silt and reworked volcaniclastic deposits with local accumulations of gravelly sand. Detailed studies of the modern sediments have not been made. A 22 m-core recovered in 1969–70 revealed the types of sediment recently accumulating offshore (Cohen et al. 1983; Cohen and Nielsen 1986; Fig. 10.14). They are diatomaceous mud and clay with local sandier layers. Buried,

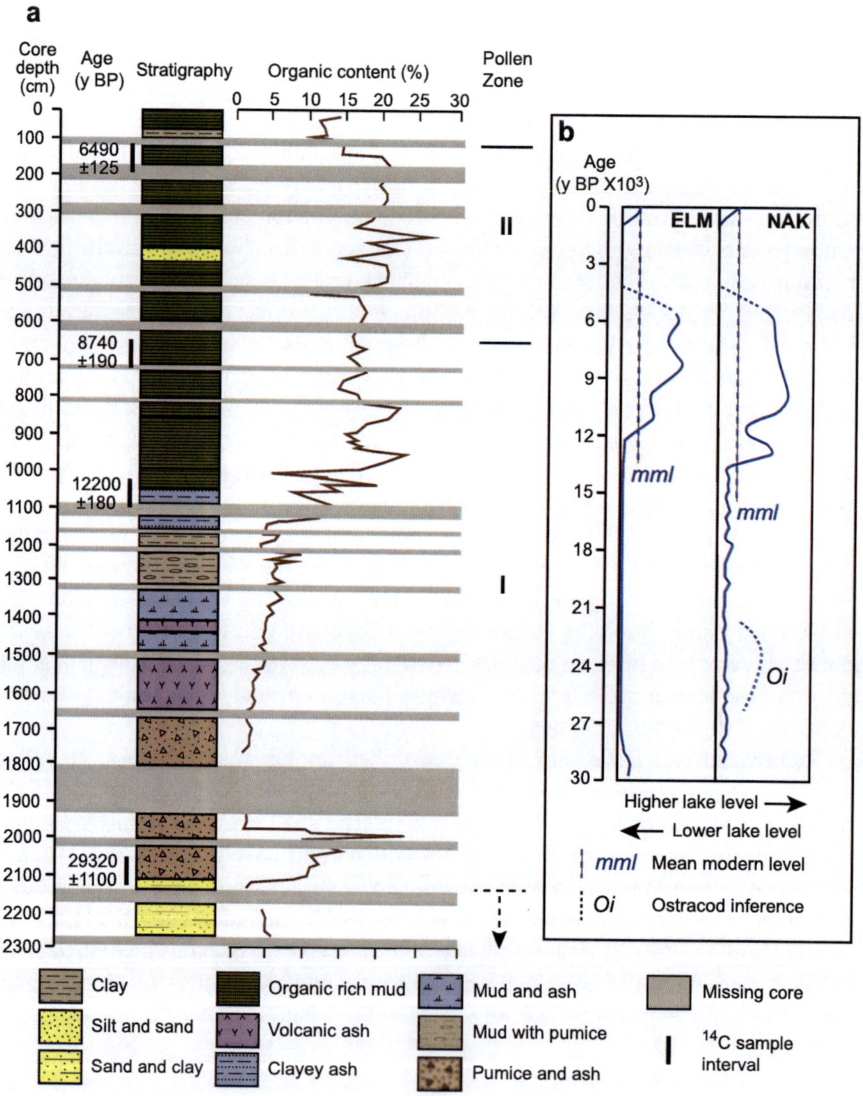

Fig. 10.14 Stratigraphy of the Elmenteita core obtained by Richardson and Dussinger (1986). For diatoms and pollen details, see Fig. 10.17a

deeper older sediments are composed of abundant volcanic ash and pumice mixed with lake mud. The muds are organic-rich at the top of the core, implying low oxidation. Some volcaniclastics are diatomaceous and waterlain. De Cort et al. (2013) recovered a short (40 cm) core composed of uniform olive-brown mud with variable carbonate content. The detrital fine sediments originate mostly from the Mereroni River, ephemeral streams and unchannelled flow, and are mixed with aeolian deposits including reworked Pleistocene diatomaceous silts and ash from local outcrops.

The Kariandusi River, which drains the Bahati-Kinangop uplands, has formed a small lobate delta along the southeast shoreline (Fig. 10.15a–d). The greyish-brown peripheral muds, exposed when lake level is low, include swelling clays (smectites) (Fig. 10.15c). These are locally incised by shallow (<20 cm deep, <2 m wide) ephemeral washes filled with feldspathic sands and fine gravel that locally form thin (<1 cm) spreads across the muddy littoral plains where they leave the confines of the ephemeral channels (Fig. 10.15a, b). The marginal muds are often heavily bioturbated.

In contrast, the Mereroni River has not built an obvious delta because the inflowing sediment is fine grained and littoral energy prevents deposition. The southern shorelines are frequently covered with dense rooted wetlands with *Typha*, *Cyperus laevigatus* and other genera that trap muds from several sources (Fig. 10.15g). The southern part of the western shoreline is a submerged basalt flow (Fig. 10.15f) that locally ponds green saline, alkaline brine. Beaches are absent around Lake Elmenteita except poorly sorted littoral gravels along the northeastern lakeshore where the lake impinges on the steep, faulted volcanic bedrock, and muddy sands along the northern shoreline (Fig. 10.15e).

Modern sedimentation remains predominantly clastic because the salinity is rarely high enough to produce evaporites other than thin efflorescent crusts that form seasonally on the exposed mudflats (Fig. 10.15d) and local thin lenses of trona that accumulate in shallow lake-marginal depressions. The lake water evaporates completely before thick evaporites (trona, nahcolite) can form. Most sodium carbonate-bicarbonate salts redissolve annually. Cerling (1979) reported fluorite in sediments around the lake. Fluoride is high in the lake and in the Kekopey hot springs (Montcoudiol et al. 2019).

10.9 Late Quaternary Lacustrine Sediments in the Nakuru-Elmenteita Basin

10.9.1 Early Research – The 'Pluvial Era'

Late Quaternary sediments and palaeoshorelines are well preserved at several locations in the Nakuru-Elmenteita basin, and their significance has been discussed for more than 90 years. During the late 1920s, Louis Leakey, assisted by geologist J.D. Solomon, and in collaboration with Erik Nilsson, tried to unravel the Quaternary

Fig. 10.15 Shoreline environments of Lake Elmenteita. (**a**) Rippled sandy wash on mid-part of the Kariandusi delta, August 2006. (**b**) Low-angled deltaic beds exposed on the northern bank of a shallow channel, Kariandusi delta, August 2006. (**c**) Mud-cracked swelling clays on peripheral mudflats, Kariandusi delta, August 2006. (**d**) Saline (efflorescent trona) mudflat, southern shore of Kariandusi delta, July 2005. (**e**) Northern shoreline with sandy mudflats. Mereroni River mouth at top left. (**f**) Rugged basalt pavement along the northwest shoreline, July 2019. (**g**) Reedbeds at the site of dilute spring discharge south of Kariandusi delta, August 2006

history of the Nakuru-Elmenteita basin. At that time, Quaternary stratigraphy in East Africa had become dominated by the 'pluvial hypothesis' – the idea that wet periods (= lakes) and dry periods (= land surfaces, palaeosols, fluvial incision and erosion) recorded by sediments near the equator could be correlated with the glacial and interglacial intervals that were becoming recognised at higher latitudes.

Leakey's interest was driven by the implications of climate change for understanding the palaeoanthropological and archaeological records preserved in the Kenya Rift sediments (Leakey 1930). This sector of the rift is rich in Pleistocene and Holocene archaeological sites of different ages that have yielded vertebrate remains, including hominins, and stone tools of several types (Kilombe, Kariandusi, Gamble's Cave and many other sites). Leakey and Solomon (1929), Leakey (1931a, b, c, 1934) and Solomon (1931) established a general stratigraphic scheme based on former shorelines that they related to a series of 'pluvials' (wet periods) and 'interpluvials' (dry phases) (Table 10.3; Coetzee and Van Zinderen Bakker 1989; Kingston and Hill 2005; Sect. 2.3.2).

In 1927–1928, Erik Nilsson, a Swedish geologist under the guidance of Professor Gerard De Geer (who pioneered the study of lacustrine varves), undertook research to determine the Quaternary history of East Africa. His ambitious research combined evidence from both montane glacial landscapes, such as moraines and erosional features, and from 'pluvial' lacustrine sediments and geomorphology. Nilsson (1931, 1935, 1940) mapped and levelled two types of palaeoshoreline around the two lakes and at Lake Naivasha: constructive barrier beaches with berms up to 3 m high and wave-eroded terraces and notches. In places, both were traced for several hundred metres. Using the evidence from palaeoshorelines, he proposed that Lake Nakuru and Lake Elmenteita had united as one lake possibly up to seven times. He suggested that this combined lake was also linked during the first two episodes with a contemporary expanded Lake Naivasha to form a single large rift lake ~80 km long (Fig. 10.1a). The enlarged lake was termed the 'Gamblian Lake' after the Gamble's Caves archaeological site (wave-cut rock-shelters) ~13 km south of Lake Nakuru (Fig. 10.1a; Leakey 1935, 1964; Oakley 1961; Protsch 1978). De Geer (1934) made comparisons between Nilsson's lacustrine varves that formed during seasonal changes in sedimentation and glacial varves in Sweden.

Table 10.3 Major Late Quaternary pluvial units adopted by Leakey and Solomon (1929), Leakey (1931a, b, c, 1934) and Solomon (1931). Oldest pluvial at the bottom

Nakuran *post-pluvial wet phase* *Shoreline at + 145 feet (44.2 m) above Lake Nakuru*
Dry interval
Makalian *post-pluvial wet phase* *Shoreline at + 375 feet (114.3 m)*
Dry interval
Gamblian *pluvial, with three lake stillstands:* *Shoreline at + 510 feet (155.4 m)* *Shoreline at + 600 feet (182.9 m)* *Shoreline at + 750 feet (228.6 m)*

10.9.2 Research from the 1950s Until the 1970s

Based on work at Lake Baringo, Fuchs (1950) proposed a Pleistocene age for the 'Kamasia sediments' west of that lake, which were later divided into the Kamasian (lower Kamasian) and Kanjeran (upper Kamasian) (Leakey 1950, 1955; Bishop 1967). In the Nakuru-Elmenteita Basin, the Kanjeran was considered to include the Larmudiac sediments (Nakuru) and the Kariandusi lake beds (Elmenteita). McCall (1957, 1967) mapped the geology around both lakes Nakuru and Elmenteita, and his 1967 report includes a discussion of the Quaternary deposits. McCall et al. (1967) noted that the deposits did not fit into the accepted Kanjeran-Gamblian-Makalian-Nakuran pluvial stratigraphy of that time and, following Flint (1959) and Bishop (1967), objected to the pluvial nomenclature, which was based primarily on climate, and proposed a revision based on faunal and archaeological relative-age dating (Table 10.4). McCall et al. (1967) reported that the Larmudiac sediments included impure diatomaceous earths capped by a brown tuff or greenish claystone (altered tuff). They inferred that the deposits were younger than the Pleistocene Kariandusi lake beds (Sect. 18.3) because they were coarser, contained less diatomite, and lacked Acheulian hand axes. They also noted that the Larmudiac beds lay close to a shoreline on Menengai volcano but were faulted, unlike the Upper Pleistocene (?) Gamblian sediments, and should therefore be older.

Washbourn (1967) also objected to the pluvial-interpluvial terminology and recommended that it be abandoned. Between 1965 and 1967, she levelled palaeoshorelines across the basin and examined the distribution of exposed lacustrine sediments (Washbourn-Kamau 1970, 1971; Butzer et al. 1972; Kamau 1972). She could not identify Leakey and Solomon's 750 ft (228 m) shoreline and questioned its existence. However, she described good evidence for a wave-cut cliff-line and associated sediments along the southern flanks of Menengai volcano, which implied a lake standing at ~6370 feet (1941.5 m) above sea level (asl). A palaeoshoreline at a similar height was also levelled at Karterit (Fig. 10.1a). Kamau also reported beach pebbles, littoral sands and diatomaceous silts, some associated with the former cliffs, either at or just below, the same elevation. The maximum recorded lake level was ~600 feet (182.9 m) above the contemporary surface of Lake Nakuru. Other

Table 10.4 Stratigraphy of the Central Kenya Rift following the classification of McCall et al. (1967)

	Earth movements	Nakuru basin		Elmenteita basin
Holocene	Fourth faulting episode (minor, very localized, very closely spaced	Evaporites, sediments of present lake		Evaporites, sediments of present lake
		"Nakuran shoreline"		Kariandusi
		Makalia sediments		
Pleistocene		Nderit "Gamblian" sediments		
		"Kanjeran"	Tuff capping Larmudiac sediments	Kariandusi lake beds

former cliff-lines were recorded at 1865 m asl (at Magharibi) and possibly at 1814 m asl (at Misonge, Fig. 10.1a), although the latter was considered "extremely uncertain" because slope breaks also result from faulting (Washbourn-Kamau 1971, p. 528).

10.9.3 Recent Studies (1970s to Present)

Comparatively little research has been carried out on lacustrine sediments independently of archaeological investigations. In earlier decades, such studies involved the Geological Survey of Kenya (McCall 1957, 1967; McCall et al. 1967). Mboya (1993) undertook lithostratigraphic and geochemical analyses of Quaternary sedimentary outcrops (Fig. 10.16a). Subsequently, Bergner (2003), Bergner and Trauth (2004) and Bergner et al. (2003, 2009) investigated Pleistocene sediments at Kariandusi and Soysambu (Elmenteita Basin) and in the Naivasha Basin. Kübler et al. (2016, 2018) explored possible links between rock-controlled soil nutrients and animal migration routes near Kariandusi and across the region.

Dühnforth et al. (2006) described additional sections from the Nakuru-Elmenteita Basin (Fig. 10.16b) and compared the modern and Early Holocene precipitation-evaporation balances needed to maintain a combined Early Holocene lake at elevations of 1940 m asl. Their data suggested a 45% increase in mean annual precipitation, a 9% increase in cloud cover and a 0.5 °C temperature decrease. This compares with previous estimates of increased Early Holocene precipitation of 65% (Washbourn 1967; Butzer et al. 1972) and 30–35% (Hastenrath and Kutzbach 1983). Dühnforth et al. (2006) considered the higher estimates, including their own, to be unrealistic and suggested that Lake Naivasha must have supplied additional groundwater to the Nakuru-Elmenteita Basin, given a lack of evidence for overflow between the two areas.

A different approach to Late Quaternary environmental reconstructions in East Africa began with the advent of piston coring in the late 1960s and 1970s (Kendall 1969; Richardson 1966, 1972; Richardson and Richardson 1972; Richardson et al. 1978). In the Nakuru-Elmenteita Basin, Cohen et al. (1983) and Richardson and Dussinger (1986) described and interpreted cores from Lake Elmenteita (22.5 m) and Lake Nakuru (19.5 m) that had been recovered in 1969–1970 (Richardson 1972). In 2004, two duplicate 17 m-cores were obtained from Lake Nakuru by the 'Lake Naivasha coring project', although full results have not yet been published (Robakiewicz et al. 2019).

Figure 10.17 shows a compilation of coring results at Lake Nakuru and Lake Elmenteita. Basal ash, pumice and mud in the cores contained a sparse diatom flora dominated by saline and shallow-water taxa (Richardson and Dussinger 1986). The fragmentary data shows that two separate saline lakes were present between ~30 and 20 ka that dried out from time to time. From ~20 ka to ~10 ka both lakes were low and highly alkaline (50–150 meq L^{-1}), while neighbouring Lake Naivasha was fresh. Lake Nakuru had a rapid transition to more dilute water and higher lake levels

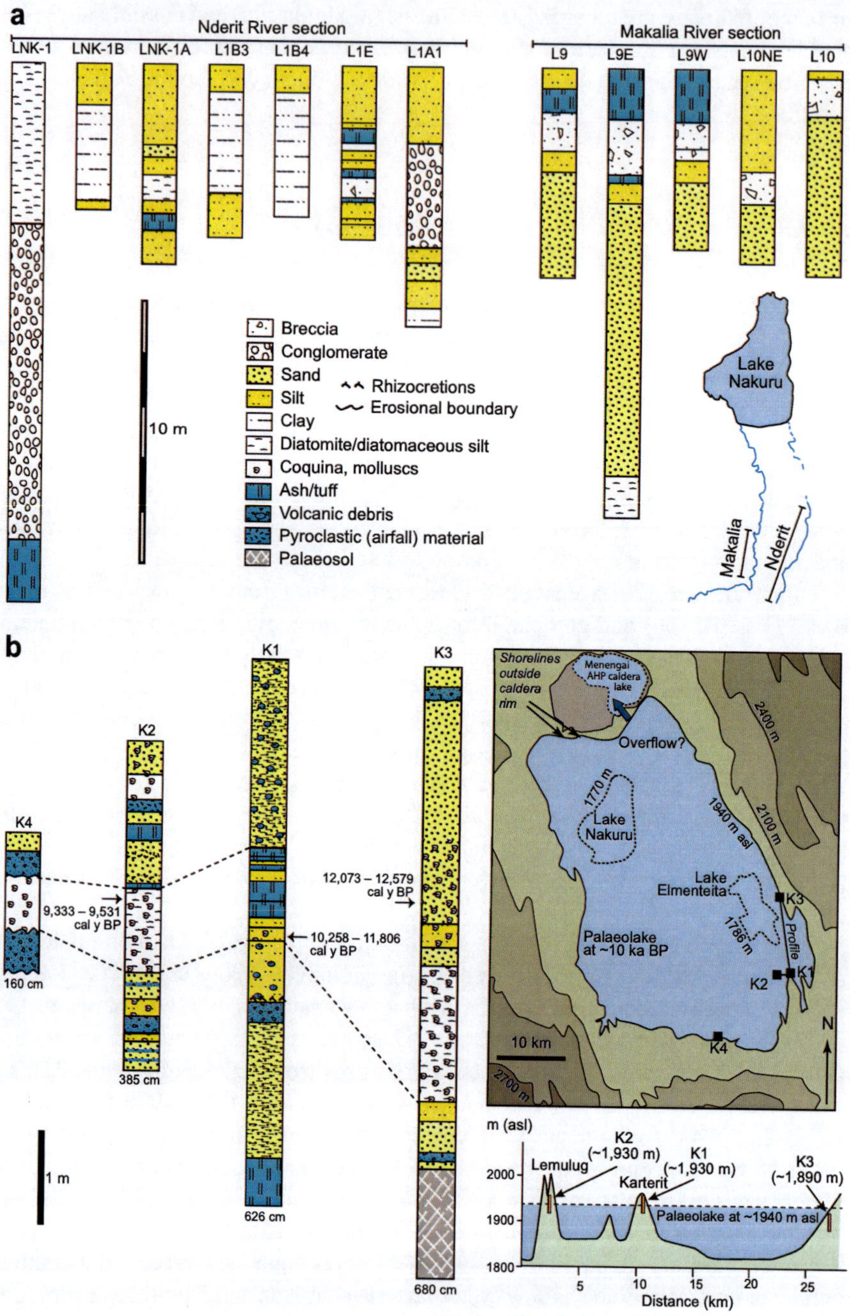

Fig. 10.16 Representative Late Quaternary sediment logs. (**a**) Based on sediment sections along the Nderit and Makalia Rivers, Nakuru Basin. After Mboya (1993). (**b**) Sediment logs and shoreline reconstructions in the Elmenteita Basin based on Dühnforth et al. (2006). Menengai early Holocene caldera lake after Leat (1984)

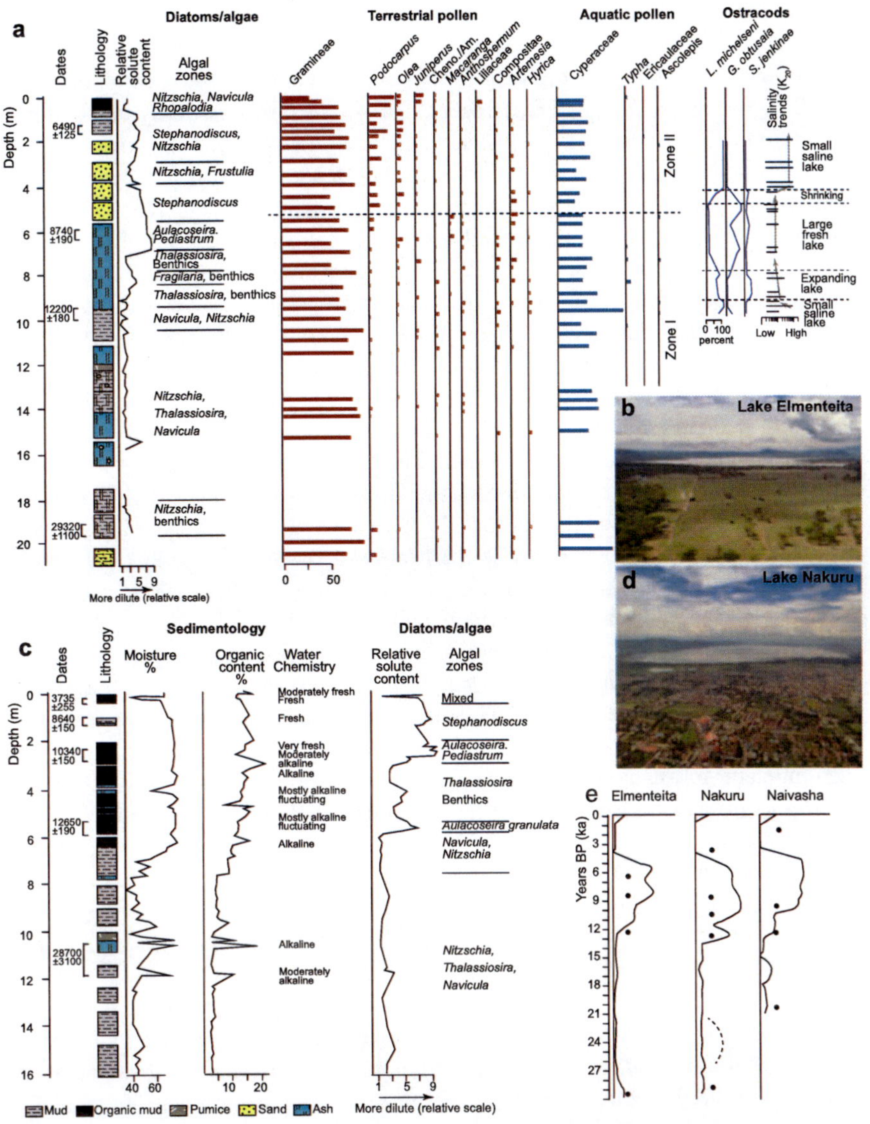

Fig. 10.17 Late Quaternary stratigraphy of the Nakuru-Elmenteita Basin. Diatoms and sediments modified from Richardson and Dussinger (1986). Pollen after Mworia-Maitima (1999). Ostracod data after Cohen and Nielsen (1986). (**a**) Lake Elmenteita sediment and microfossil core stratigraphy. Selected taxa shown; original papers provide more detail. (**b**) Lake Elmenteita. Core from central lake. (**c**) Lake Nakuru sediment and algal core stratigraphy. (**d**) Lake Nakuru. Core from central lake. (**e**) Inferred water level changes plotted against time for Lakes Nakuru, Elmenteita and Naivasha. Note broadly synchronous patterns. Black circles = ^{14}C dates

at 12,800 y BP, before a return to more saline conditions, followed by a major freshening at ~10,500 y BP. In contrast, Lake Elmenteita became fresher at ~12,000 y BP, followed by a return to higher salinities and then a major freshening after about 10,500 y BP. Richardson and Dussinger (1986) divided the last 10 ky into two periods. A major Early Holocene highstand is recorded in both lakes between about 10 and 6 ka with *Aulacoseira* spp. giving way to *Stephanodiscus*, perhaps reflecting stratification and silica-depletion in the epilimnion. An increase in *Nitzschia frustulum* and *Thalassiosira rudolfi* in the Elmenteita core implies a brief phase of more alkaline conditions at about 7500 y BP. Richardson and Dussinger inferred a major regression about 5000 years ago with the lakes separating at, or just before, this period. Both lakes then remained low with fluctuating levels. Figure 10.17e summarises the lake-level fluctuations of the three central Kenya Rift lakes.

Richardson and Dussinger (1986) and Cohen and Nielsen (1986) interpreted the ostracod record of the cores, using evidence from earlier studies of the modern ostracods, including those from the Nakuru-Elmenteita basin (Cohen et al. 1983). Their results showed a similar environmental history to that indicated by diatoms, but with a more dilute and deeper Lake Nakuru at about 25 ka. Vegetation showed a major change at ~8500 y BP (Fig. 10.17a; Mworia-Maitima 1999). The pollen record before this date indicates mainly dry arid open vegetation with minor input from distant forests (mainly *Gramineae* with traces of *Anthospermum*, *Artemesia*, *Myrica*). Subsequently, increased high-altitude forest pollen obscured increases in lowland woody vegetation (mainly *Podocarpus*, *Olea*, *Artemesia*, *Juniperus*). Late Pleistocene to Holocene pollen data has also been reported from marshes in the Mau Highlands west of Elmenteita (Githumbi et al. 2021). The record there indicates that Afromontane forest dominated during the terminal Pleistocene to Holocene with taxa changing from *Apodytes*, *Celtis*, *Dracaena*, *Hagenia* and *Podocarpus* to *Cordia*, *Croton*, *Ficus*, *Juniperus* and *Olea*. In contrast, Late Holocene environments were dominated by more open Afromontane forest with extensive grass and herbs.

De Cort et al. (2013) analysed short cores from Lake Nakuru and Lake Elmenteita to obtain historical records of environmental change. Basal core sections with low organic matter and water content, oxidation and compaction represent a desiccation horizon when the lakes were almost dry. Overlying massive sediments at Nakuru imply a shallower but permanent lacustrine environment with microalgae as the dominant primary producers. The stratigraphic transition occurred during the 1870s when Lake Nakuru was refilled, a change recognised in other Kenya Rift lakes.

A long-standing question has been whether the highest Nakuru-Elmenteita palaeolake overflowed northwards for prolonged periods as an 'overfilled' lake, or intermittently as a 'balanced-filled' lake, using the terminology of Bohacs et al. (2000). The lowest outlet or sill that Washbourn-Kamau (1971) recorded was at 1948 m asl, which is 6 m higher than the maximum shoreline that she levelled (~1942 m asl) at Menengai (Fig. 10.1a). That site lies between the eastern slopes of Menengai and the Bahati Plain (Fig. 10.1a). No clear overflow channel is evident on satellite imagery, but brief high stands are not always recorded by morphology (i.e., from wave or fluvial erosion) or littoral sediments. Evidence based on observations at Lake Bogoria, for example, confirms that high-level beach terraces are ephemeral features that can be destroyed in only a few decades by littoral erosion, slope processes or mass wasting. If cemented by calcite or silica they may survive as beachrock. Leat (1984) later reported lacustrine sediments within northeastern Menengai crater (Fig. 10.16) that provide evidence for an early Holocene intra-caldera lake at 1860 m asl that was present from about 10,300 to 8300 y BP. Reidl et al. (2020) described faulted lacustrine diatomaceous sediments in the crater injected by magmatic fluids and dated to the early AHP (11,600–12,728 cal ka BP). Mwaniki et al. (2019) proposed that an expanded Lake Nakuru overflowed into Lake Bogoria via Menengai, using gene flow evidence. Dommain et al. (2022) also suggested that northward overflow occurred during the Holocene. Further coring during geothermal drilling might resolve finer details of the extent and duration of the Menengai palaeolake and its relation, including timing, to any overflow from an expanded Lake Nakuru-Elmenteita.

10.9.4 Archaeological Research

Archaeological research has driven much of the analysis of former lake levels in the Nakuru-Elmenteita Basin since the early work by Leakey, Solomon, and Nilsson. Most stone-tool assemblages are associated with land surfaces and fluvial sediments. However, these commonly interfinger with lacustrine deposits reflecting shifting palaeoenvironments under climatic, tectonic, and other controls (Owen et al. 2018). Consequently, many sites where stone tools are present also include lake deposits. Dating of the tool assemblages thus, fortuitously, assists with understanding past changes in lake environments.

The early archaeological work in the Nakuru-Elmenteita Basin inferred correlations between pluvials (now discredited) and stone-tool cultural stages (Fig. 10.18; Leakey 1931b). Both relied heavily on a European framework, based upon northern-hemisphere glaciations and stone-tool classifications that have since been modified. The literature can thus be confusing. The oldest industries in the Nakuru-Elmenteita Basin are comprised of Chellean and Acheulian tools associated with fluvial deposits and former land surfaces at Kariandusi (Figs. 10.1a and 10.18). However, the rolled Chellean assemblages show evidence for transport, implying that they predate the deposits in which they are found, whereas Acheulian hand axes are often in excellent condition indicating that they are contemporary with deposition (Gowlett 1980; Gowlett and Crompton 1994; Shipton 2011).

The 'Gamblian' sediments yield Stillbay and 'Kenya Aurignacian' industries (Leakey 1931b; Cole 1954, 1963; Van Baelen et al. 2019). The latter is a good example of changing terminology. It was recognised by Leakey in the 1920–30s and was compared to similar European tool assemblages (abundant backed blades, microliths, burins, common end scrapers) (Fig. 10.18; Wilson 1977; Wilshaw 2016). It was later renamed the East African Aurignacian as new sites were located and was then modified again and termed the 'East African Capsian' (Leakey 1947). Cole

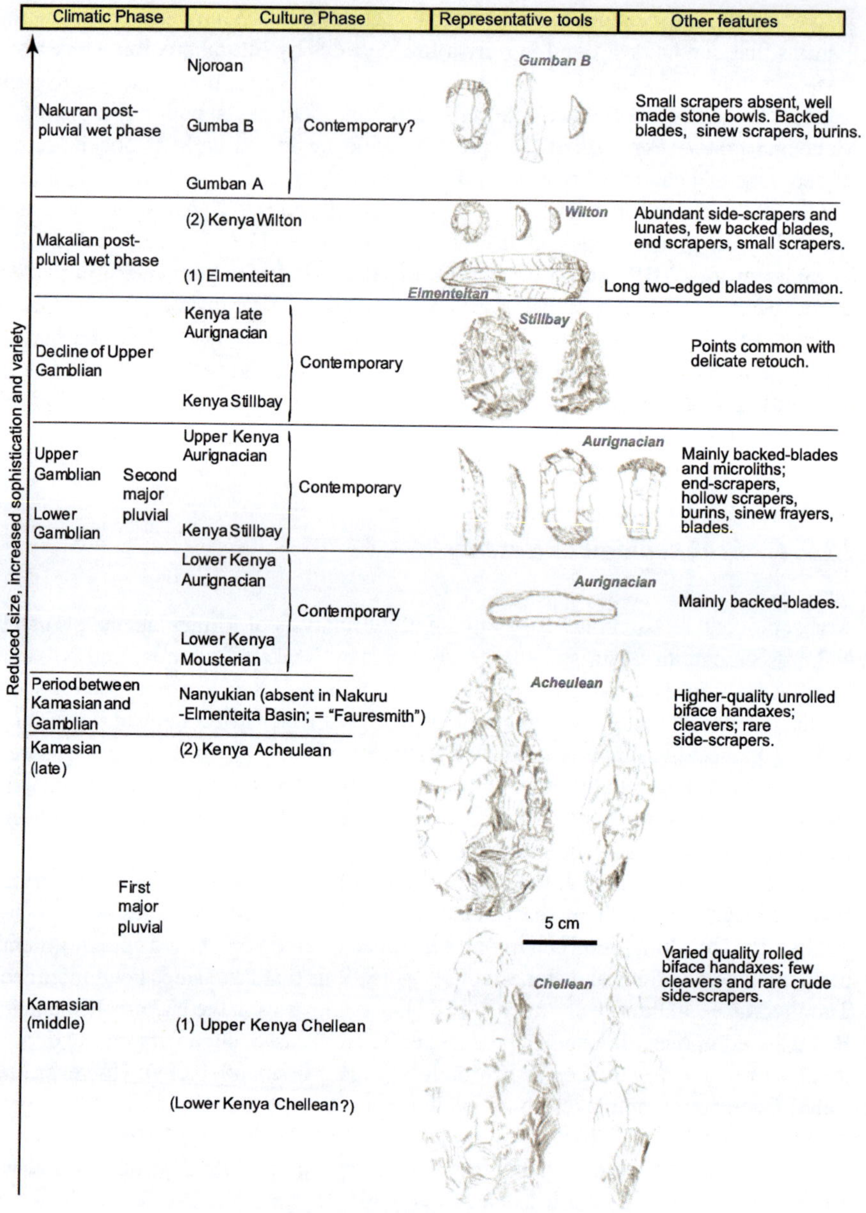

Fig. 10.18 Early classification scheme of stone tool industries in the Nakuru-Elmenteita Basin. After Leakey 1931b. Climate phases (left) have since been abandoned as formal stratigraphic units. Drawings based on figures in Leakey (1931b). Sketches of representative tools are examples only for the cultural phases shown to the left. See text next to sketches for which phase is represented. All tools drawn to the same scale

(1954, 1963) preferred 'Kenya Capsian' and recognised two additional phases (Upper Kenya Capsian D and Neolithic variants). In the 1970s, "part" of the Kenya Capsian was redefined as the Eburran (Ambrose 1984, 1985, 2014), which referred specifically to technologies in the Nakuru-Elmenteita-Naivasha basin dated at 12,710–1110 y BP. Wilshaw (2016) noted that the Eburran, as defined by Ambrose, was related to climate changes that shifted altitudinal vegetation zones (savannah/lowland; bush-woodland/mid-range; montane-forest/highland). He also noted that wet phases led to the expansion of bush and montane forest into lowlands with arid phases leading to the expansion of savannah to higher altitudes. Furthermore, Eburran toolmakers were tied to the savannah-montane zone and shifted their location in response to such zonal movements. Consequently, the revised earlier geographical terminology now incorporated an ecological aspect.

The Mesolithic Elmenteitan period has been dated at about 3000–1300 y BP (Ambrose 2014; Goldstein 2021) and is associated with tools dominated by long two-edged blades, which replaced the backed blades of the Kenya Capsian (Fig. 10.18). A variety of pottery is also present. The Wilton industry has been linked to the 'Makalian' to 'Nakuran' interval and is distinguished by abundant 'thumbnail' scrapers (Cole 1963). Neolithic tools in the Nakuru-Elmenteita Basin include the Gumban and Njoroan with earlier small scrapers absent (Fig. 10.18). Well-made pottery and animal bones are present at many sites (Cole 1963).

Nakuru-Elmenteita archaeological sites associated with beach deposits include the Gamble's caves, Lion Hill Cave and Hyrax Hill (Fig. 10.1a). Gambles Cave II contains beach sediments dated at 8740 ± 190 y BP that are overlain by several occupation levels yielding Upper Kenya Capsian, Kenya Stillbay and Elmenteitan tools (Protsch 1978). Lion Hill Cave also includes beach gravel at about 620 ft (188.9 m), interpreted by Solomon as a stillstand during a lower Gamblian regression. Deposits overlying the beach sediments include Kenya Capsian assemblages and abundant hyrax bones (Leakey 1931b; McCall et al. 1967). Makalian beach deposits located 335 ft (102.1 m) above Lake Nakuru on Hyrax Hill (Leakey et al. 1943; Burkitt 1945) are similarly overlain by an occupation level, but one associated with Gumban B obsidian implements and stone tools (Fig. 10.18) considered to be correlated with the Nakuran wet phase (Table 10.3). Eighteen, mainly female, burials have been reported along with obsidian stone bowls and pottery fragments. Pottery and stone tools have been reported from forty stratified layers at Nderit Drift (Bower and Nelson 1978). Nearly continuous Pastoral Iron Age and Pastoral Neolithic usage has been inferred from the area between 4000 and 250 y BP, but with few remains for the 10,600–4500 y BP interval, when lake levels were high. Kenya Capsian tools were also reported in alluvial sediments dating from 12,000 to 10,600 y BP.

References

Adeka JE, Strobl RO, Becht R (2008) An environmental system analysis of Lake Elmenteita, Kenya, with reference to water quality. In: Sengupta M, Dalwai R (eds) Proceedings of the world lakes conference (Taal 2007), vol 12, pp 1365–1372

Agembe S, Ojwang W, Olilo C, Omondi R, Ongore C (2018) Soda lakes of the Rift Valley (Kenya). In: Finlayson CM, Everard M, Irvine K, McInnes RJ, Middleton BA, van Dam AA, Davidson NC (eds) The wetland book. Springer, Heidelberg, pp 1381–1391

Akanga B, Boga HI, Klenk HP (2015) Studies of sediments, dry mud and water samples from Lake Elmenteita reveal the presence of thaumarchaeota and euryarchaeota. J Biodivers Environ Sci 6:341–355

Allen DJ, Darling WG, Burgess WG (1989) Geothermics and hydrogeology of the southern part of the Kenya Rift Valley with emphasis on the Magadi-Nakuru area. British Geological Survey Research Report SD/89/1

Ambrose SH (1984) Holocene environments and human adaptations in the Central Rift Valley, Kenya. PhD thesis, University of California, Berkeley

Ambrose SH (1985) Excavations at Masai Gorge Rockshelter, Naivasha. Azania 20:29–67

Ambrose SH (2014) The introduction of pastoral adaptations to the highlands of East Africa. In: Clark JD, Brandt SA (eds) From hunters to farmers: the causes and consequences of food production in Africa. University of California Press, Berkeley, pp 1–28

Baker BH (1958) Geology of the Magadi area. Geol Surv Kenya Rep 42

Ballot A, Krienitz L, Kotut K, Wiegand C, Metcalf J, Codd G, Pflugmacher S (2004a) Cyanobacteria and cyanobacterial toxins in three alkaline Rift Valley lakes of Kenya – Lakes Bogoria, Nakuru and Elmenteita. J Plankton Res 26:925–935

Ballot A, Dadheech PK, Krienitz L (2004b) Phylogenetic relationship of *Arthrospira*, *Phormidium* and *Spirulina* strains from Kenyan and Indian waterbodies. Algol Stud 113:37–56

Bartholomew GA, Pennycuick CJ (1973) The flamingo and pelican populations of the Rift Valley lakes in 1968-69. Afr J Ecol 11:189–198

Beadle LC (1932) Scientific results of the Cambridge Expedition to the East African Lakes. 1930-1: 4. The waters of some East African Lakes in relation to their fauna and flora. J Linn Soc Lond Zool 38:157–211

Becht R, Harper DM (2002) Towards an understanding of human impact upon the hydrology of Lake Naivasha, Kenya. Hydrobiologia 488:1–11

Becht R, Mwango F, Muno FA (2006) Groundwater links between Kenyan Rift lakes. In: Odada EO, Olago DO, Ochola W, Ntiba M, Wandiga S, Gichuki N, Oyieke H (eds) Proc 11th World Lakes Conf, Nairobi, pp 7–14

Beltran JMV (2003) The origin of pantellerites and the geology of the Eburru volcanic complex, Kenya Rift, Africa. MSc thesis, University of Texas, El Paso

Bergner A (2003) Lake-level fluctuations and Late Quaternary climate change in the Central Kenya Rift. PhD thesis, Universität Potsdam

Bergner A, Trauth M (2004) Comparison of the hydrological and hydrochemical evolution of Lake Naivasha (Kenya) during three highstands between 175 and 60 kyr BP. Palaeogeogr Palaeoclimatol Palaeoecol 215:17–36

Bergner A, Trauth M, Bookhagen B (2003) Paleoprecipitation estimates for the Lake Naivasha basin (Kenya) during the last 175 ky using a lake-balance model. Glob Planet Change 36:117–136

Bergner AGN, Strecker MR, Trauth MH, Deino A, Gasse F, Blisniuk P, Dühnforth M (2009) Tectonic and climatic control on evolution of rift lakes in the Central Kenya Rift, East Africa. Quat Sci Rev 28:2804–2816

BirdLife International (2023) Important bird areas factsheet: Lake Elmenteita. Downloaded from http://www.birdlife.org on 10/02/2023

Bishop WW (1967) Annotated lexicon of Quaternary stratigraphical nomenclature in East Africa. In: Bishop WW, Clark JD (eds) Background to evolution in Africa. University of Chicago Press, Chicago, pp 375–395

Bohacs KM, Carroll AR, Neal JE, Mankiewicz PJ (2000) Lake-basin type, source potential, and hydrocarbon character: an integrated sequence-stratigraphic–geochemical framework. In: Gierlowski-Kordesch EH, Kelts KR (eds) Lake basins through space and time. AAPG Stud Geol 46:3–34

Bower J, Nelson C (1978) Early pottery and pastoral cultures of the Central Rift Valley, Kenya. Man 13:554–566
Brown LH (1955) The breeding of lesser and greater flamingoes in East Africa. J East Afr Nat Hist Soc 22:159–162
Brown LH (1958) The breeding of the greater flamingo *Phoenicopterus ruber* at Lake Elmenteita, Kenya Colony. Ibis 100:388–420
Brown (1959a) The mystery of the flamingos. Country Life, London
Brown LH (1959b) Flamingoes in Africa. Ostrich Suppl 3:79–83
Brown LH (1971) The flamingoes of Lake Nakuru. New Sci 51:97–101
Brown L, Powell-Cotton D, Hopcraft J (1973) The breeding of the greater flamingo and great white pelican in East Africa. Ibis 115:352–374
Burgis MJ, Mavuti KM (1987) The Gregory Rift. In: Burgis MJ, Symoens JJ (eds) African wetlands and shallow water bodies directory. Travaux et Documents, vol 211. ORSTOM, Paris, pp 331–359
Burgis MJ, Morris P (1987) The natural history of lakes. Cambridge University Press, Cambridge
Burian A, Schagerl M, Yasindi A (2012) Microzooplankton feeding behaviour: grazing on the microbial and the classical food web of African soda lakes. Hydrobiologia 710:61–72
Burkitt MC (1945) Excavations at Hyrax Hill, Nakuru, Kenya Colony. Nature 156:722–723
Butturini A, Amalfitano S, Herzsprung P, Lechtenfeld OJ, Venturi S, Olaka LA, Pacini N, Harper DM, Tassi F, Fazi S (2020) Dissolved organic matter in continental hydro-geothermal systems: insights from two hot springs of the East African Rift Valley. Water 12:3512. https://doi.org/10.3390/w12123512
Butzer KW, Isaac GL, Richardson JL, Washbourn-Kamau C (1972) Radiocarbon dating of East African lake levels. Science 175:1069–1076
Cerling T (1979) Paleochemistry of Plio-Pleistocene Lake Turkana, Kenya. Palaeogeogr Palaeoclimatol Palaeoecol 27:247–285
Childress B, Hughes B, Harper D, van den Bossche W (2007) East African flyway and key site network of the lesser flamingo (*Phoenicopterus minor*) documented through satellite tracking. Ostrich 78:463–446
Clarke MCG, Woodhall DG, Allen D, Darling G (1990) Geological, volcanological and hydrogeological controls on the occurrence of geothermal activity in the area surrounding Lake Naivasha, Kenya. Ministry of Energy, Republic of Kenya
Coetzee JA, van Zinderen Bakker EM (1989) Palaeoclimatology of East Africa during the last glacial maximum: a review of changing theories. In: Mahaney WC (ed) Quaternary and environmental research on East African Mountains. CRC Press, London, pp 189–198
Cohen AS, Nielsen C (1986) Ostracodes as indicators of paleohydrochemistry in lakes: a Late Quaternary example from Lake Elmenteita, Kenya. PALAIOS 1:601–609
Cohen AS, Dussinger R, Richardson J (1983) Lacustrine paleochemical interpretations based on eastern and southern African ostracodes. Palaeogeogr Palaeoclimatol Palaeoecol 43:129–151
Cole SM (1954) The prehistory of East Africa. Am Anthropol 56:1026–1050
Cole SM (1963) The prehistory of East Africa, 2nd edn. McMillan, New York
Conti P, Pistis M, Bernardinetti S, Barbagli A, Zirulia A, Serri L, Colonna T, Guastaldi E, Ghiglieri G (2021) Tectonic setting of the Kenya Rift in the Nakuru area, based on geophysical prospecting. Geosciences 11:80. https://doi.org/10.3390/geosciences11020080
Dadheech PK, Mahmoud H, Kotut K, Krienitz L (2012) *Haloleptolyngbya alcalis* gen. et sp. nov., a filamentous cyanobacterium from the soda Lake Nakuru, Kenya. Hydrobiologia 691:269–283
Davdra AR (2020) Status and causes of deterioration of wetlands in the Eastern Rift Valley, case study of Lake Nakuru – Kenya. MA thesis, University of Nairobi
De Cort G, Bessems I, Keppens K, Mees F, Cumming B, Verschuren D (2013) Late-Holocene and recent hydroclimatic variability in the Central Kenya Rift Valley: the sediment record of hypersaline lakes Bogoria, Nakuru and Elementeita. Palaeogeogr Palaeoclimatol Palaeoecol 388:69–80
De Geer G (1934) Equatorial Palæolithic varves in East Africa. Geogr Ann 16:75–96

Dommain R, Riedl S, Olaka LA, deMenocal P, Deino AL, Owen RB, Muiruri V, Müller J, Potts R, Strecker MR (2022) Holocene bidirectional river system along the Kenya Rift and its influence on East African faunal exchange and diversity gradients. Proc Natl Acad Sci 119. https://doi.org/10.1073/pnas.2121388119

Dühnforth M, Bergner AGN, Trauth MH (2006) Early Holocene water budget of the Nakuru-Elementeita basin, Central Kenya Rift. J Paleolimnol 36:281–294

Ehrensperger A, Mbuguah S (2004) Fostering sustainable urban development in Nakuru, Kenya Rift Valley. Mount Res Dev 24:210–214

Fazi S, Butturini A, Tassi F, Amalfitano S, Venturi S, Vazquez E, Clokie M, Wanjala SW, Pacini N, Harper DM (2018) Biogeochemistry and biodiversity in a network of saline–alkaline lakes: Implications of ecohydrological connectivity in the Kenyan Rift Valley. Ecohydrol Hydrobiol 18:96–106

Finlay BJ, Curds CR, Bamforth SS, Bafort JM (1987) Ciliated protozoa and other microorganisms from two African soda lakes (Lake Nakuru and Lake Simbi, Kenya). Arch Protist 133:81–91

Flint RF (1959) Pleistocene climates in eastern and southern Africa. Geol Soc Am Bull 70:343–374

Florence NM (2021) Levels of selected water quality parameters, heavy metals and pesticide residues in water, sediments and Nile tilapia of Lake Nakuru, Kenya. MSc thesis, Egerton University, Njoro

Fuchs V (1950) Pleistocene events in the Baringo Basin. Geol Mag 87:149–174

Githaiga JM (2022) Impacts of hydro climatic conditions on lesser flamingo, *Phoeniconaias minor,* Geoffroy populations in three Kenyan alkaline lakes. Afr J Phys Sci 7:9–29

Githumbi EN, Mustaphi CJC, Marchant R (2021) Late Pleistocene and Holocene Afromontane vegetation and headwater wetland dynamics within the Eastern Mau Forest, Kenya. J Quat Sci 6:1–16

Glover RB (1972) Chemical characteristics of water and steam discharges in the Rift Valley of Kenya. Unpublished Report, UNDP Geothermal Resources Exploration Project, Kenya

Goldstein ST (2021) Lithic technological organization of the "Elmenteitan" early herders in southern Kenya: implications for mobility, exchange, and climatic resilience. J Anthropol Archaeol 61:101259

Gowlett JAJ (1980) Acheulean sites in the Central Rift Valley, Kenya. Proc Pan-Afr Congr Prehist Quat Stud 8:213–217

Gowlett J, Crompton R (1994) Kariandusi: Acheulean morphology and the question of allometry. Afr Archaeol Rev 12:3–42

Greichus YA, Greichus A, Ammann BD, Hopcraft J (1978) Insecticides, polychlorinated biphenyls and metals in African lake ecosystems. III. Lake Nakuru, Kenya. Bull Environ Contam Toxicol 19:454–461

Hastenrath S, Kutzbach JE (1983) Palaeoclimate estimates from water and energy budgets of East African lakes. Quat Res 19:141–153

Hecky RE, Kilham P (1973) Diatoms in alkaline, saline lakes: ecology and geochemical implications. Limnol Oceanogr 18:53–71

Herrnegger M, Stecher G, Schwatke C, Olang L (2021) Hydroclimatic analysis of rising water levels in the Great Rift Valley lakes of Kenya. J Hydrol Reg Stud 36:100857

Hongo PO, Mulaku GC (2021) Flooding of Lake Nakuru National Park and its effects on the resident wildlife. J Geogr Info Syst 13:660–670

Hughes RH, Hughes JS (1992) A directory of African wetlands. IUCN, Gland and Cambridge, UK/UNEP, Nairobi/WCMC, Cambridge, UK

Ignuza GM, Kanda I (2017) Geochemical evaluation of groundwater flow processes and mixing in Menengai geothermal field, Kenya. Geotherm Res Counc Trans

Iradukunda P, Nyadawa MO (2021) Impact of sedimentation on water seepage capacity in Lake Nakuru, Kenya. Appl Environ Soil Sci 2021:1–10

Iradukunda P, Sang JK, Nyadawa MO, Maina CW (2020) Sedimentation effect on the storage capacity in Lake Nakuru, Kenya. J Sustain Res Eng 5:149–158

James MM (2022) Establishing the relationship between LULC changes, temperature, precipitation and water level rise: case study of Lake Nakuru Basin, Kenya. MSc thesis, University of Nairobi

Jenkin PM (1932) Reports on the Percy Sladen expedition to some Rift Valley lakes in Kenya in 1929, I. Introductory account of the biological survey of five freshwater and alkaline lakes. Ann Mag Nat Hist Ser 9:533–553

Jenkin PM (1936) Reports on the Percy Sladen expedition to some Rift Valley lakes in Kenya in 1929, VII. Summary of the ecological results, with special reference to the alkaline lakes. Ann Mag Nat Hist Ser 18:133–181

Jenkins M, McCord S, Edebe J (2009) Sustaining lake levels in Lake Nakuru, Kenya: development of a water balance model for decision making. Global livestock CRSP research brief 09-05-SUMAWA

Jirsa F, Gruber M, Stojanovic A, Omondi SO, Mader D, Körner W, Schagerl M (2013) Major and trace element geochemistry of Lake Bogoria and Lake Nakuru, Kenya, during extreme drought. Chem Erde 73:275–282

Jones WB, Lippard SJ (1979) New age determinations and the geology of the Kenya Rift-Kavirondo rift junction, W Kenya. J Geol Soc Lond 136:693–704

Kaggwa MN, Burian A, Oduor SO, Schagerl M (2013) Ecomorphological variability of *Arthrospira fusiformis* (Cyanoprokaryota) in African soda lakes. MicrobiologyOpen 2:881–891

Kamau CK (1972) Lake Quaternary lake level fluctuations in the Nakuru-Elmenteita Basin, Kenya. In: Reeves CC (ed) Playa lake symposium, ICASALS Publ 4, Lubbock, Texas,pp 185–192

Kamondo WC (1988) Possible uses of geothermal fluids in Kenya. Geothermics 17:489–501

Kanda IK (2010) Aquifer stratigraphy and hydrogeochemistry of the Lake Nakuru Basin, Central Kenya Rift. MSc thesis, University of Nairobi

Kanda IK, Suwai J (2013) Hydrogeochemistry of shallow and deep water aquifers of Menengai geothermal area, Kenya Rift Valley. Geotherm Resour Counc Trans 37:403–410

Kavembe GD, Meyer A, Wood CM (2016) Fish populations in East African saline lakes. In: Schagerl M (ed) Soda Lakes of East Africa. Springer Nature, Cham, pp 227–257

Kendall R (1969) An ecological history of the Lake Victoria basin. Ecol Monogr 39:121–176

Kiage LM, Douglas P (2020) Linkages between land cover change, lake shrinkage, and sublacustrine influence determined from remote sensing of select Rift Valley lakes in Kenya. Sci Total Environ 709:136022

Kiende R, Kande R (2015) Structural geology of Eburru Volcano and Badlands geothermal prospects in Kenya. Proceedings of the 40th Workshop on Geothermal Reservoir Engineering, Stanford University, Stanford, California. Paper SGP-TR-204

Kilham P (1971) Biogeochemistry of African lakes and rivers. PhD thesis, Duke University, Durham

Kimaru AN, Gathenya JM, Cheruiyot CK (2019) The temporal variability of rainfall and streamflow into Lake Nakuru, Kenya, assessed using SWAT and hydrometeorological indices. Hydrology 6:88. https://doi.org/10.3390/hydrology6040088

King BC (1978) Structural and volcanic evolution of the Gregory Rift Valley. In: Bishop WW (ed) Geological background to fossil man. Geol Soc Lond Spec Publ 6: 29–54

Kingston J, Hill A (2005) When it rains it pours: legends and realities of the East African pluvials. In: Lieberman DE, Smith RJ, Kelley J (eds) Interpreting the past: essays on human, primate, and mammal evolution in honor of David Pilbeam. Brill, Boston, pp 189–205

Kipkemboi J (2016) Vascular plants in Eastern Africa Rift Valley saline wetlands. In: Schagerl M (ed) Soda lakes of East Africa. Springer Nature, Cham, pp 285–293

Koenig R (2006) The pink death: die-offs of the lesser flamingo raise concern. Science 313:1724–1725

Kotut K, Ballot A, Krienitz L (2006) Toxic cyanobacteria and their toxins in standing waters of Kenya: implications for water resource use. J Water Health 4:233–245

Koyo A (2005) Lake Elmenteita. Information sheet on Ramsar wetlands (RIS). Ramsar Bureau, Gland

Krienitz L (2018) Lesser flamingos: descendants of Phoenix. Springer Nature, Berlin

Krienitz L, Kotut K (2010) Fluctuating algal food populations and the occurrence of lesser flamingos (*Phoeniconaias minor*) in three Kenyan Rift Valley lakes. J Phycol 46:1088–1096

Krienitz L, Schagerl M (2016) Tiny and tough: microphytes of East African Soda Lakes. In: Schagerl M (ed) Soda lakes of East Africa. Springer Nature, Cham, pp 149–177

Krienitz L, Dadheech PK, Kotut K (2013) Mass developments of the cyanobacteria *Anabaenopsis* and *Cyanospira* (Nostocales) in the soda lakes of Kenya: ecological and systematic implications. Hydrobiologia 703:79–93

Krienitz L, Krienitz D, Dadheech PK, Hübener T, Kotut K, Luo W, Teubner K, Versfeld W (2016) Food algae for lesser flamingos: a stocktaking. Hydrobiologia 775:21–50

Kübler S, Rucina S, Reynolds S, Owenga P, Bailey G, King GCP (2016) Edaphic and topographic constraints on exploitation of the Central Kenya Rift by large mammals and early hominins. Open Quat 2:199–119

Kübler S, King GCP, Devès MH, Inglis RH, Bailey GN (2018) Tectonic geomorphology and soil edaphics as controls on animal migrations and human dispersal patterns. In: Rasul NMA, Stewart ICF (eds) Geological setting, palaeoenvironment and archaeology of the Red Sea. Springer, Cham, pp 653–673

Kulecho A, Muhandiki V (2006) Water quality trends and input loads to Lake Nakuru. In: Odada EO, Olago DO, Ochola W, Ntiba M, Wandiga S, Gichuki N, Oyieke H (eds) Proc 11 World Lakes Conf, vol 2. Nairobi, pp 529–533

Kutukhulu AW (2010) Geophysical mapping of buried river channels and other shallow structures recharging major aquifers in the Lake Nakuru basin, Kenya Rift: Case study from Kabatini aquifer. MSc thesis, University of Nairobi

Leakey LSB (1930) East Africa past and present. Geogr J 76:494–500

Leakey LSB (1931a) East African lakes. Geogr J 77:497–508

Leakey LSB (1931b) The stone age cultures of Kenya Colony (with appendices by JD Solomon and others). Cambridge University Press, London

Leakey LSB (1931c) Prehistoric man in Kenya. Nature 127:814–815

Leakey LSB (1934) Changes in the physical geography of East Africa in human times. Geogr J 84:296–305

Leakey LSB (1935) The stone age races of Kenya. Oxford University Press, London

Leakey LSB (1947) Capsian or Aurignacian? Which term should be used in Africa? In: Leakey LSB (ed) Proc 1st Panafrican Congr Prehist, Nairobi, 1947. Blackwell, Oxford, pp 205–206

Leakey LSB (1950) The lower limit of the Pleistocene in Africa. Proc 18th Int Geol Congr London, 1948, pp 62–65

Leakey LSB (1955) The climatic sequence of the Pleistocene in Africa. In: Balout L (ed) Act 2 session Congr Panafr Prehist, Alger, 1952, pp 293–294

Leakey LSB (1964) Gamble's Cave, Kenya. Stud Speleol 1:22–25

Leakey LSB, Solomon J (1929) African pluvial periods and prehistoric man. Man 29:213–214

Leakey MD, Leakey LSB, Game PM, Goodwin AJH (1943) Report on the excavations at Hyrax Hill, Nakuru, Kenya Colony, 1937–1938. Trans R Soc S Afr 30:271–409

Leat PT (1984) Geological evolution of the trachytic caldera volcano Menengai, Kenya Rift Valley. J Geol Soc Lond 141:1057–1069

Leat PT (1991) Volcanological development of the Nakuru area of the Kenya rift valley. J Afr Earth Sci 13:483–498

Lind EM (1965) The phytoplankton of some Kenya waters. J East Afr Nat Hist Soc 25:76–91

Lind EM (1966) The phytoplankton of some Kenya lakes and reservoirs. Br Phycol Bull 3:154–155

Lind EM (1968) Notes on the distribution of phytoplankton in some Kenya waters. Br Phycol Bull 3:481–493

Liu Y, Peng J, Jiang F, Lu Q, Zhu F, Xu Q (2022) Model test study on the formation and development of underground erosion ground fissures in the Kenya Rift Valley. J Mount Sci 19:1037–1050

Maina-Gichaba C, Shivoga WA, Enanga EM, Kibichii S, Miller SN (2006) Sediment loading on inland lakes/wetlands: a case study of Lake Nakuru, Kenya. In: Odada EO, Olago DO, Ochola

W, Ntiba M, Wandiga S, Gichuki N, Oyieke H (eds) Proc 11 World Lakes Conf, vol 2. Nairobi pp 389–391
Maskall JE, Thornton I (1989) The mineral status of Lake Nakuru National Park, Kenya: a reconnaissance survey. Afr J Ecol 27:191–200
Maskall JE, Thornton I (1996) The distribution of trace and major elements in Kenyan soil profiles and implications for wildlife nutrition. In: Appleton JD, Fuge R, McCall GJH (eds) Environmental geochemistry and health: with special reference to developing countries. Geol Soc Lond Spec Publ 113:47–62
Mboya BGO (1993) Lithostratigraphic analysis and palaeoenvironmental interpretation of the Late Tertiary and Quaternary sediments in the Nakuru, Elmentaita and Naivasha Basins. PhD thesis, University of Nairobi
McCall GJH (1957) Geology and groundwater conditions in the Nakuru area. Tech Rep Ministr Works (Hydraul Branch) Kenya 3
McCall GJH (1967) Geology of the Nakuru–Thomson's Falls–Lake Hannington area. Rep Geol Surv Kenya 78
McCall GJH, Baker BH, Walsh J (1967) Late Tertiary and Quaternary sediments of the Kenya Rift Valley. In: Bishop WW, Clark JD (eds) Background to evolution in Africa. University of Chicago Press, Chicago, pp 191–220
Melack JM (1976) Limnology and dynamics of phytoplankton in equatorial African lakes. PhD thesis, Duke University, Durham, North Carolina
Melack JM (1982) Photosynthetic activity and respiration in an equatorial African soda lake. Freshw Biol 12:381–400
Melack JM (1988) Primary producer dynamics associated with evaporative concentration in a shallow, equatorial soda Lake (Lake Elmenteita, Kenya). Hydrobiologia 158:1–14
Melack JM (2009) Diel variability and community metabolism in African soda lakes. Natur Resour Environ Iss 15:28
Melack JM, Kilham P (1974) Photosynthetic rates of phytoplankton in East African alkaline, saline lakes. Limnol Oceanogr 19:743–755
Mengistou S (2016) Invertebrates of East African soda lakes. In: Schagerl M (ed) Soda lakes of East Africa. Springer Nature, Cham, pp 205–226
Milbrink G (1977) On the limnology of two alkaline lakes (Nakuru and Naivasha) in the East Rift Valley system in Kenya. Int Rev Ges Hydrobiol Hydrogr 62:1–17
Montcoudiol N, Burnside NM, Györe D, Mariita N, Mutia T, Boyce A (2019) Surface and groundwater hydrochemistry of the Menengai caldera geothermal field and surrounding Nakuru County, Kenya. Energies 12:3131
Morgan WTW (1963) The 'White Highlands' of Kenya. Geogr J 129:140–155
Moritsugu M (1985) A preliminary investigation on chemical quality of inland waters in Kenya. Ber Ohara Inst Landwirtsch Biol Okayama Univ 19:25–34
Mpawenayo B, Mathooko JM (2004) Diatom assemblages in the hotsprings associated with lakes Elmenteita and Baringo in Kenya. Afr J Ecol 42:363–367
Muita R, Gikungu D, Aura S, Njogu A, Ndichu R, Nyinguro P, Kiptum C (2021) Assessment of rising water levels of Rift Valley lakes in Kenya: the role of meteorological factors. Environ Sci Ecol Curr Res 2:1035–1039
Muno FA (2002) Water balance of the southern Kenya Rift Valley lakes. MSc thesis, International Institute for Geo-Information Science and Earth Observations, Enschede, Netherlands
Murimi SK (1989) An investigation of some aspects of recent geomorphological and hydrological changes in the Lake Elmenteita catchment area. MSc thesis, Kenyatta University, Nairobi
Murimi SK (1993) The drying up of the East African Rift Valley lakes in recent times: the case of Lake Elmenteita. J East Afr Res Dev 23:40–62
Murimi SK (1994) Falling water-levels in saline lakes of the central Rift Valley of Kenya: the case of Lake Elmenteita. Int J Salt Lake Res 3:65–74

Mwaniki P, Taita T, Sierens T, Triest L (2019) Barriers to genetic connectivity of smooth flatsedge (*Cyperus laevigatus*) among alkaline-saline lakes of Eastern Rift Valley (Kenya). Aquat Bot 155:38–44

Mwaura F (1999) A spatio-chemical survey of hydrogeothermal springs in Lake Elmenteita, Kenya. Int J Salt Lake Res 8:127–138

Mwaura F, Moore TR (1991) Forest and woodland depletion in the Lake Elmenteita Basin, Kenya. Geoforum 27:17–26

Mwirichia R, Cousin S, Muigai A, Boga HI, Stackebrandt E (2010) Archaeal diversity in the halo-alkaline Lake Elmenteita in Kenya. Curr Microbiol 60:47–52

Mwirichia R, Cousin S, Muigai A, Boga HI, Stackebrandt E (2011) Bacterial diversity in the halo-alkaline Lake Elmenteita, Kenya. Curr Microbiol 62:209–221

Mworia-Maitima J (1999) Late Quaternary pollen record from Lake Elmenteita, Kenya. Palaeoecol Afr 26:103–114

Nabide I (2002) Development of 3-D conceptual hydrogeological model for Lake Naivasha area. MSc thesis, International Institute for Geo-Information Science and Earth Observations, Enschede, Netherlands

Ng'weno CC, Mwasi SM, Kairu JK (2009) Distribution, density and impact of invasive plants in Lake Nakuru National Park, Kenya. Afr J Ecol 48:905–913

Ngecu W, Nyambok I (2000) Ground subsidence and its socio-economic implications on the population: a case study of the Nakuru area in Central Rift Valley, Kenya. Environ Geol 39:567–574

Nilsson E (1931) Quaternary glaciations and pluvial lakes in British East Africa. Geogr Ann 13:249–349

Nilsson E (1935) Traces of ancient changes of climate in East Africa. Preliminary report. Geogr Ann 17:1–21

Nilsson E (1940) Ancient changes of climate in British East Africa and Abyssinia. A study of ancient lakes and glaciers. Geogr Ann 22:1–79

Njenga JW (2004) Comparative studies of water chemistry of four tropical lakes in Kenya and India. Asian J Water Environ Pollut 1:87–97

Nogrady T (1983) Succession of planktonic rotifer populations in some lakes of the Eastern Rift Valley, Kenya. Hydrobiologia 98:45–54

Nyawire E, Barrat J-A, Varet J (2018) Geological, volcanological, petrological and structural studies to evaluate geothermal occurrence within an axial rift segment of the central Kenya Rift Valley. Proc 7th Afr Rift Geotherm Conf, Kigali, 31 Oct–2 Nov 2018

Oakley KP (1961) Bone harpoon from Gamble's Cave, Kenya. Antiqu J 41:86–87

Obando JA, Onywere S, Shisanya C, Ndubi A, Masiga D, Irura Z, Mariita N, Maragia H (2016) Impact of short-term flooding on livelihoods in the Kenya Rift Valley lakes. In: Meadows ME, Lin J-C (eds) Geomorphology and society. Springer, Tokyo, pp 193–215

Ochieng EZ, Lalah JO, Wandiga SO (2007) Analysis of heavy metals in water and surface sediments in five rift valley lakes in Kenya for assessment of recent increase in anthropogenic activities. Bull Environ Contam Toxicol 79:570–576

Odada E, Raini J, Ndetei R (2006) Lake Nakuru: experience and lessons learned brief. Main Report, Global Environment Facility Lake Basin Management Initiative, ILEC-LakeNet, pp 299–319

Oduor S, Schagerl M (2007a) Phytoplankton primary productivity characteristics in response to photosynthetically active radiation in three Kenyan Rift Valley saline alkaline lakes. J Plankton Res 29:1041–1050

Oduor S, Schagerl M (2007b) Temporal trends of ion contents and nutrients in three Kenyan Rift Valley saline-alkaline lakes and their influence on phytoplankton biomass. Hydrobiologia 584:59–68

Ojiambo SB, Lyons WB, Welch KA, Poreda RJ, Karen H, Johannesson KH (2003) Strontium isotopes and rare earth elements as tracers of groundwater–lake water interactions, Lake Naivasha, Kenya. Appl Geochem 18:1789–1805

Okech SO (2012) Anthropogenic influence on surface and ground water quality in Lake Nakuru basin, Central Kenya Rift Valley. MSc thesis, University of Nairobi

Okoth OE, Mucai M, Shivoga WA, Miller SN, Rasowo J, Ngugi CC (2009) Spatial and seasonal variations in phytoplankton community structure in alkaline–saline Lake Nakuru, Kenya. Lakes Reserv 16:35–47

Olago D, Opere A, Barongo J (2009) Holocene palaeohydrology, groundwater and climate change in the lake basins of the Central Kenya rift. Hydrol Sci J 54:765–780

Olaka LA, Odada EO, Trauth MH, Olago DO (2010) The sensitivity of East African rift lakes to climate fluctuations. J Paleolimnol 44:629–644

Omenda PA (1997) The geochemical evolution of Quaternary volcanism in the south-central portion of the Kenya Rift. PhD thesis, University of Texas, El Paso

Omondi AD (2011) Potential effects of changes in climate, land cover and population on the quantity of water resources in Lake Nakuru and Lake Elmenteita areas, Kenya. MSc thesis, University of Nairobi

Ong'ondo GO, Yasindi AW, Oduor SO, Jost S, Schagerl M, Sonntag B, Boenigk J (2013) Ecology and community structure of ciliated protists in two alkaline–saline Rift Valley lakes in Kenya with special emphasis on *Frontonia*. J Plankton Res 35:759–771

Onywere S, Shisanya C, Obando J, Ndubi A, Masiga D, Irura Z, Mariita N, Maragia H (2013) Geospatial extent of 2011-2013 flooding from the Eastern African Rift Valley lakes in Kenya and its implication on the ecosystems. Papers, Kenya Soda lakes workshop. Kenya Wildlife Service Training Institute, Naivasha

Onzere M, Biamah E, Paul O (2022) Analysis of the impacts of climate change on rising water levels of Lake Nakuru. J Eng Agric Environ 8:11–16

Owen RB, Renaut RW, Lowenstein TK (2018) Spatial and temporal geochemical variability in lacustrine sedimentation in the East African Rift System: evidence from the Kenya Rift and regional analyses. Sedimentology 65:1697–1730

Oyoo-Okoth E, Muchiri M, Ngugi CC, Njenga E, Ngure V, Orina PS, Chemoiwa EC, Wanjohi BK (2011) Zooplankton partitioning in a tropical alkaline–saline endorheic Lake Nakuru, Kenya: spatial and temporal trends in relation to the environment. Lakes Reserv 16:35–47

Peters R, MacIntyre S (1976) Orthophosphate turnover in East African lakes. Oecologia 25:313–319

Prasad H (1995) Monitoring the impact of land cover changes on some geomorphological processes in the catchment of Lake Elmenteita by use of remote sensing techniques. In: Sing RB, Haigh MJ (eds) Proc 3 int symp headwater control, New Delhi. Balkema, Rotterdam, pp 389–402

Protsch R (1978) The chronological position of Gamble's Cave II and Bromhead's site (Elmenteita) of the Rift Valley, Kenya. J Hum Evol 7:101–109

Raini J (2006) Long-term trends in water quality, water quantity and biodiversity at Lake Nakuru, Kenya. In: Odada EO, Olago DO, Ochola W, Ntiba M, Wandiga S, Gichuki N, Oyieke H (eds) Proc 11 World Lakes Conf, vol 2. Nairobi, pp 57–62

Raini JA (2009) Impact of land use changes on water resources and biodiversity of Lake Nakuru catchment basin, Kenya. Afr J Ecol Suppl 1(47):39–45

Raini J, Mpawenayo B, Munyeki I (2010) Algal composition, distribution and succession at Lake Nakuru, Kenya. In: Odada EO, Olago DO, Ochola W, Ntiba M, Wandiga S, Gichuki N, Oyieke H (eds) Proc 11 World Lakes Conf, vol 2. Nairobi, pp 1–7

Renaut RW (1982) Late Quaternary geology of the Lake Bogoria fault-trough, Kenya Rift Valley. PhD thesis, University of London

Rich F (1932) Reports on the Percy Sladen expedition to some rift valley lakes in Kenya in 1929. IV. Phytoplankton from some Rift Valley lakes in Kenya. Ann Mag Nat Hist Ser 10:233–262

Richardson JL (1966) Changes in level of Lake Naivasha, Kenya, during postglacial times. Nature 209:290–291

Richardson JL (1972) Palaeolimnological records from rift lakes in Central Kenya. In: Van Zinderen Bakker EM (ed) Palaeoecology of Africa, the surrounding islands and Antarctica, vol 6. Balkema, Cape Town, pp 131–136

Richardson JL, Dussinger R (1986) Paleolimnology of mid-elevation lakes in the Kenya Rift Valley. Hydrobiologia 143:167–174

Richardson JL, Richardson AE (1972) History of an African rift lake and its climatic implications. Ecol Monogr 42:499–534

Richardson JL, Harvey TJ, Holdship SA (1978) Diatoms in the history of shallow East African lakes. Polsk Arch Hydrobiol 25:341–353

Riedl S, Melnick D, Mibei GK, Njue L, Strecker MR (2020) Continental rifting at magmatic centres: structural implications from the Late Quaternary Menengai Caldera, central Kenya Rift. J Geol Soc Lond 177:153–169

Robakiewicz E, Bergner AGN, Mingram J, Trauth MH, Junginger A (2019) 25,000 years of moisture variability at Lake Nakuru, Central Kenya Rift. Geophys Res Abs EGU General Assembly 21, EGU2019-11229

Ronoh I (2015) Evolution and geology of Eburru-Badlands geothermal prospect—Central Kenyan Rift. Goeotherm Resour Counc Trans 39:247–254

Salano AS, Makonde HM, Kasili RW, Wangai LN, Nawiri MP, Boga HI (2017) Diversity and distribution of fungal communities within the hot springs of soda lakes in the Kenyan rift valley. Afr J Micobiol Res 11:764–775

Schagerl M, Burian A (2016) The ecology of African soda lakes: driven by variable and extreme conditions. In: Schagerl M (ed) Soda lakes of East Africa. Springer Nature, Cham, pp 295–320

Schagerl M, Oduor SO (2008) Phytoplankton community relationship to environmental variables in three Kenyan Rift Valley saline-alkaline lakes. Mar Freshw Res 59:125–136

Schagerl M, Burian A, Gruber-Dorninger M, Oduor SO, Kaggwa MN (2015) Algal communities of Kenyan soda lakes with a special focus on *Arthrospira fusiformis*. Fottea 15:245–257

Scott JJ, Renaut RW, Buatois LA, Owen RB, McNulty EP, Stockhecke M, Leet K, Lowenstein TK, Mángano MG (2021) Modern and ancient animal traces in the extreme environments of Lake Magadi and Nasikie Engida, Kenya Rift Valley. In: Rosen MR, Finkelstein DB, Park Boush L, Pla-Pueyo S (eds) Limnogeology: progress, challenges and opportunities. Springer, Cham, pp 19–66

Shipton C (2011) Taphonomy and behaviour at the Acheulean site of Kariandusi, Kenya. Afr Archaeol Res 28:141–155

Shivoga W (2001) The influence of hydrology on the structure of invertebrate communities in two streams flowing into Lake Nakuru, Kenya. Hydrobiologia 458:121–130

Sikes HL (1936) Notes on the hydrology of Lake Naivasha. J East Afr Uganda Nat Hist Soc 13:73–89

Simiyu SM (2010) Status of geothermal exploration in Kenya and future plans for its development. In: Proc World Geotherm Congr, Bali, 25–29 April 2010

Solomon JD (1931) The geology of the implementiferous deposits in the Nakuru and Naivasha basins and the surrounding area in Kenya Colony. Appendix A in leakey LSB, the stone age cultures of Kenya Colony. Cambridge University Press, London, pp 245–257

Spink PC (1945) Thermal activity in the Eastern Rift Valley. Geogr J 105:197–207

Strecker MR (1991) Das Zentrale und Südliche Kenia Rift unter besonderer Berücksichtigung der neotektonischen Entwicklung. Habilitation thesis, Universität Karlsruhe

Strecker MR, Blisniuk PM, Eisbacher GH (1990) Rotation of extension direction in the Central Kenya rift. Geology 18:299–302

Talling JF (2006) A brief history of the scientific study of tropical African waters. Freshw Forum 26:3–37

Talling JF, Talling IB (1965) The chemical composition of African lake waters. Int Rev Ges Hydrobiol Hydrogr 50:421–463

Tebbs EJ, Remedios JJ, Avery ST, Rowland CS, Harper DM (2015) Regional assessment of lake ecological states using Landsat: a classification scheme for alkaline–saline, flamingo lakes in the East African Rift Valley. Int J Appl Earth Obs Geoinform 40:100–108

Tuite CH (1979) Population size, distribution and biomass density of the lesser flamingo in the eastern Rift Valley, 1974-76. J Appl Ecol 16:765–775

Tuite CH (1981) Standing crop densities and distribution of *Spirulina* and benthic diatoms in East African alkaline saline lakes. Freshw Biol 11:345–360

Tuite CH (2000) The distribution and density of lesser flamingo in East Africa, in relation to food availability. Waterbirds 23:52–63

Van Baelen A, Wilshaw A, Griffith P, Noens P, Maillo-Fernandez JM, Foley RA, Lahr MM (2019) Prospect Farm and the Middle and Later Stone Age occupation of Mt. Eburru (Central Rift, Kenya) in an East African context. Afr Archaeol Rev 36:397–417

Vareschi E (1978) The ecology of Lake Nakuru (Kenya) I. Abundance and feeding of the lesser flamingo. Oecologia 32:11–35

Vareschi E (1979) The ecology of Lake Nakuru (Kenya) II. Biomass and spatial distribution of fish (*Tilapia grahami* Boulenger=*Sarotherodon alcalicum grahami* Boulenger). Oecologia 37:321–335

Vareschi E (1982) The ecology of Lake Nakuru (Kenya) III. Abiotic factors and primary production. Oecologia 55:81–101

Vareschi E, Jacobs J (1984) The ecology of Lake Nakuru (Kenya) V. Production and consumption of consumer organisms. Oecologia 61:83–98

Vareschi E, Jacobs J (1985) The ecology of Lake Nakuru VI. Synopsis of production and energy flow. Oecologia 65:412–424

Vareschi E, Vareschi A (1984) The ecology of Lake Nakuru (Kenya) IV. Biomass and distribution of consumer organisms. Oecologia 61:70–82

Vareschi E, Melack JM, Kilham P (1981) Saline waters. In: Symoens JJ, Burgis MJ, Gaudet JJ (eds) The ecology and utilization of African inland waters. UNEP Rep Proc Ser 1, Nairobi, pp 93–102

Wandiga SO, Molepo JM, Alalah LN (1983) Concentration of heavy metals in water, sediments and plants of Kenyan lakes. Kenya J Sci Tech Ser A 4:89–94

Washbourn CK (1967) Lake levels and Quaternary climates in the Eastern Rift Valley of Kenya. Nature 216:672–673

Washbourn-Kamau CK (1970) Late Quaternary chronology of the Nakuru-Elmenteita basin, Kenya. Nature 226:253–254

Washbourn-Kamau CK (1971) Late Quaternary lakes in the Nakuru-Elmenteita basin, Kenya. Geogr J 137:522–535

Were KO, Dick ØB, Singh BR (2013) Remotely sensing the spatial and temporal land cover changes in eastern Mau forest reserve and Lake Nakuru drainage basin, Kenya. Appl Geogr 41:75–86

Wilshaw A (2016) The current status of the Kenya Capsian. Afr Archaeol Rev 33:13–27

Wilson TH (ed) (1977) A survey of the prehistory of eastern Africa. Seventh Pan African Congress on Prehistory and Quaternary Studies. National Museums of Kenya, Nairobi

Worthington EB (1932a) Scientific results of the Cambridge Expedition to the East African Lakes, 1930–31, 1: general introduction and station list. J Linn Soc Zool 38:99–119

Worthington EB (1932b) The lakes of Kenya and Uganda. Geogr J 79:275–293

Yang Y, Wei L, Cui L, Zhang M, Wang J (2017) Profiles and risk assessment of heavy metals in great Rift Lakes, Kenya. Clean-Soil Air Water 45:1600825–1600829

Yasindi AW (2001) The ecology of planktonic ciliates in tropical lakes of East Africa. PhD thesis, University of Waterloo, Ontario

Yasindi AW, Taylor WD (2016) The protozoa of soda lakes in East Africa. In: Schagerl M (ed) Soda lakes of East Africa. Springer Nature, Cham, pp 179–204

Yihdego Y (2005) A three dimensional ground water model of the aquifers around Lake Naivasha area, Kenya. MSc thesis, International Institute for Geo-Information Science and Earth Observations, Enschede

Chapter 11
Lake Naivasha

11.1 Introduction

Lake Naivasha lies 80 km south of the equator and is the second largest freshwater body in Kenya. It is also the highest (~1890 m) of three closed-basin lakes (Fig. 11.1a) with contrasting chemical characteristics. Lakes Nakuru and Elmenteita, to the north, are saline whereas Lake Naivasha is fresh (250–350 μs cm^{-1}; Uku and Mavuti 1994; Ouma and Tateishi 2006) despite a lack of surface outflow. The basins are located within a 50–70 km wide segment of the Central Kenya Rift (Fig. 11.1a–c), an area that is also the highest part of the rift, with the valley floor dropping both to the north and south. The Naivasha drainage basin (3400 km^2; Fig. 11.1c) is dominated by the Malewa River, which accounts for 90% of the water input (Gaudet and Melack 1981) with secondary contributions from the Gilgil and Karati rivers. The watershed passes over Longonot volcano and the Eburru volcanic highlands to the south and northwest, respectively (Fig. 11.1a, b). The Mau Escarpment (>3000 m) forms the western boundary with the Aberdare Mountains (>3960 m) and fault-controlled Kinangop Plateau marking the eastern rift margin.

The lake was designated a Ramsar site in 1995, fluctuates in size (110–170 km^2) and had a volume of about 5 km^3 in 1976 (Källqvist 1987; Harper 2005; Maina et al. 2018). Lake Naivasha is comprised of a main water body (Figs. 11.1d and 11.2a) with a partially separated extension to the southwest called Lake Oloidien (Oloiden) which is 10 m deep with an area of about 5.5 km^2. A semi-isolated embayment to the southeast lies within a former volcanic crater (Crater Lake or Crescent Island Bay; Fig. 11.2b) with a maximum depth of 16 m in 1976 (Källqvist 1987). The relatively small and completely isolated sodic Lake Sonachi (Sect. 14.12) lies 4 km to the west. Many parts of the shoreline are bordered by fringing swamps (~64 km^2), especially in the north (Fig. 11.2c), which fluctuate in size. These are dominated by floating and submerged aquatic plants and extensive stands of *Cyperus* (Fig. 11.1d; Åse et al. 1986; Harper and Mavuti 2004) with high clonal diversity (Triest et al. 2014).

R. W. Renaut, R. B. Owen, *The Kenya Rift Lakes: Modern and Ancient*, Syntheses in Limnogeology, https://doi.org/10.1007/978-3-642-25055-2_11

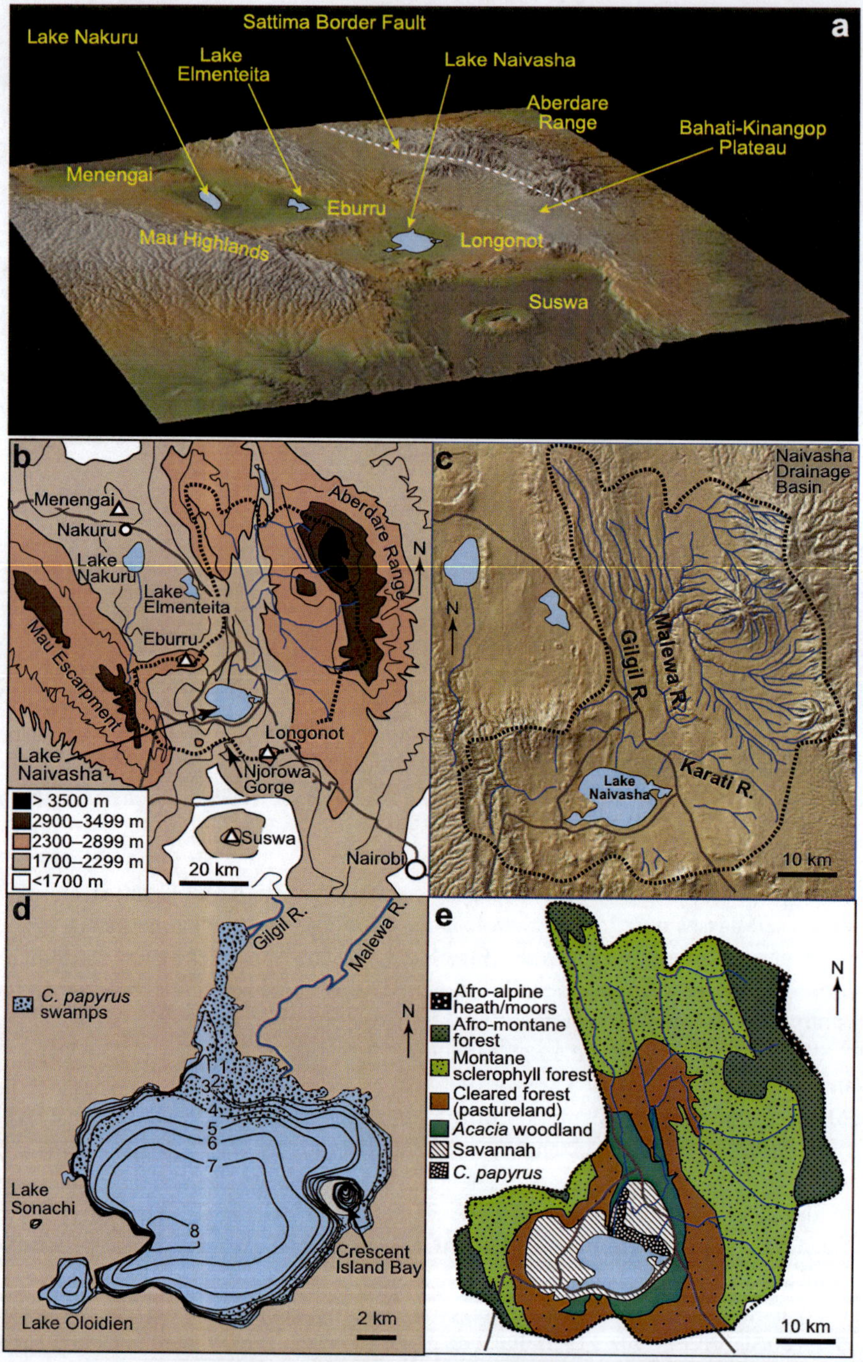

Fig. 11.1 Topography, drainage and vegetation in the Naivasha Basin. (**a**) 3D representation of the Central Kenya Rift. (**b**) Relief map of the Naivasha area and drainage basin. (After Bergner and Trauth 2004). (**c**) Drainage systems and major river inflows. (After Roessner and Strecker 1997). Basemap constructed from GeoMapApp (http://www.geomapapp.org). (**d**) Lake bathymetry and distribution of papyrus swamps. (After Stoof-Leichsenring et al. 2011; Mavuti 1992). (**e**) Vegetation zones in the Naivasha Drainage basin. (After Bergner et al. 2003)

Fig. 11.2 Lake Naivasha and adjacent areas. (**a**) Lake Naivasha from the north. (**b**) Crescent Island, SE shore. (**c**) Malewa River delta on northern shoreline. (**d**) Horticulture along southern shore. (**e**) Ol Njorowa Gorge, carved by channelled overflow from Lake Naivasha

Much of the area around Lake Naivasha and on the Kinangop Plateau is cultivated (Fig. 11.1e) with drier regions dominated by grass that is grazed by cattle and wildlife. Horticulture is important south of the lake where there are extensive greenhouses (Fig. 11.2d). Further south, lies the Njorowa Gorge, a breached outlet of a former expanded Lake Naivasha (Fig. 11.2e). Higher slopes on the Aberdares are forested. Access to the lake is gained via the main Nairobi–Nakuru (C5) and Moi South Lake tarmac roads. A good earth road extends around the lake to the north and west, with access provided by smaller earth roads that service private farms. Geothermal energy is produced at Olkaria to the southwest near Ol Njorowa Gorge with further geothermal energy developments being investigated in the Eburru highlands to the north of Lake Naivasha (Allen et al. 1989; Odongo 1993; Simiyu and Keller 2000; Kubo 2004; Simiyu 2010; Wetang'ula 2010; Kubai and Kandie 2014; Nyakundi et al. 2020).

Wakefield and Johnston (1870) included Lake Naivasha on their maps based on reports from caravan expeditions. The first European traveller to reach the lake was Dr. Gustav Fischer who passed through the Njorowa Gorge in June 1883 (Fischer 1885; Thompson and Dodson 1963). He collected rock and hot-spring water samples, which were later analysed and described by Mügge (1885, 1886). The Scottish geologist Joseph Thomson reached the area a few months later and examined the volcanic rocks, hot springs and lakes (Thomson 1887; Thompson and Dodson 1963). In 1887, Lieutenant von Höhnel (1890) described the topography and volcanic landscape while travelling with the Count Teleki expedition to Lake Turkana. In 1893, Gregory (1896) climbed Longonot volcano before moving northwards to examine craters in the Elmenteita badlands. In the following decades, other travellers and geologists reported their observations on the volcanic rocks in the region (Gibson 1893; Gorges 1900; Walker 1903; Muff 1908; Collie 1912). The rock collections of Gregory were later described by Prior (1903) with Gregory (1921) returning for further studies in 1919 when he developed a Neogene terminology for sedimentary rocks that he thought were of Oligocene or Miocene age. Subsequently, these same rocks were recognised as Pleistocene in age (Thompson and Dodson 1963).

Research through to World War II continued with studies of faulting (Sikes 1926; Willis 1936; Busk 1939), geology (Smith 1931; Bowen 1935, 1937; Shand 1936), soils (Milne 1936), groundwater (Sikes 1934), lake hydrology (Sikes 1936) and hot springs (Spink 1945). However, most efforts went into archaeological research and the stratigraphic context of hominins and their stone tools, which involved detailed work on lacustrine deposits (Leakey 1931a, b, 1934, 1942, 1950, 1952, 1955; Nilsson 1931, 1935, 1938, 1940). Research dealing with mammal fossils was reported by Lönnberg (1933, 1937) and Nilsson (1945). Early studies of modern Lake Naivasha included the work of the 'Cambridge Expedition to the East African Lakes', which examined the water chemistry, plankton, macrophytes and other aspects of the lake environment (Beadle 1932a, b; Worthington 1932; Rich 1933; Worthington and Ricardo 1936).

Interest in the sediments, archaeology and waters of the Naivasha Basin continued into the late colonial era through the work of Cole (1950, 1954) and Cole and Leakey (1952). Shackleton (1955) contributed new information on the timing of faulting in the area. Subsequently, the Kenya Geological Survey published a comprehensive report on the volcanic rocks, sediments and structure (Thompson and Dodson 1963). The post-colonial period brought many advances in the understanding of the basin geology, history and modern ecosystems that are outlined in the following sections.

11.2 Geological and Geomorphological Setting

The Aberdare Range (~4400 m) and the Sattima Border Fault form the eastern boundary of the SE-NW trending Central Kenya Rift with the Mau Escarpment (~3080 m) located 70 km to the west (Figs. 11.1a, b and 11.3). The eastern margin is also marked by the ~25 km wide step-faulted Bahati-Kinangop Plateau at about 2000–2500 m (Thompson and Dodson 1963). Lake Naivasha lies on the rift floor, which ranges between about 1800 and 2000 m in elevation. The late Quaternary volcanoes of the Olkaria Volcanic Complex and Longonot lie directly south of the lake with Eburru volcano lying directly to the north (Woodhall and Clark 1988; Clarke et al. 1990). The youngest lava flows at Longonot date from the nineteenth century.

Mio-Pliocene and younger rocks dominate (Fig. 11.3) and rest on Mozambique metasediments dated at 650–475 Ma. Foliations in the latter trend NW adjacent to the rift axis and appear to control the overlying Neogene fault trends (Shackleton 1986; Williams et al. 1983; Roessner and Strecker 1997). Strecker et al. (1990) recognised a four-fold division of the rocks (Table 11.1). The oldest are the Samburu basalts (about 23–15 Ma), which are unconformably overlain by laterally extensive

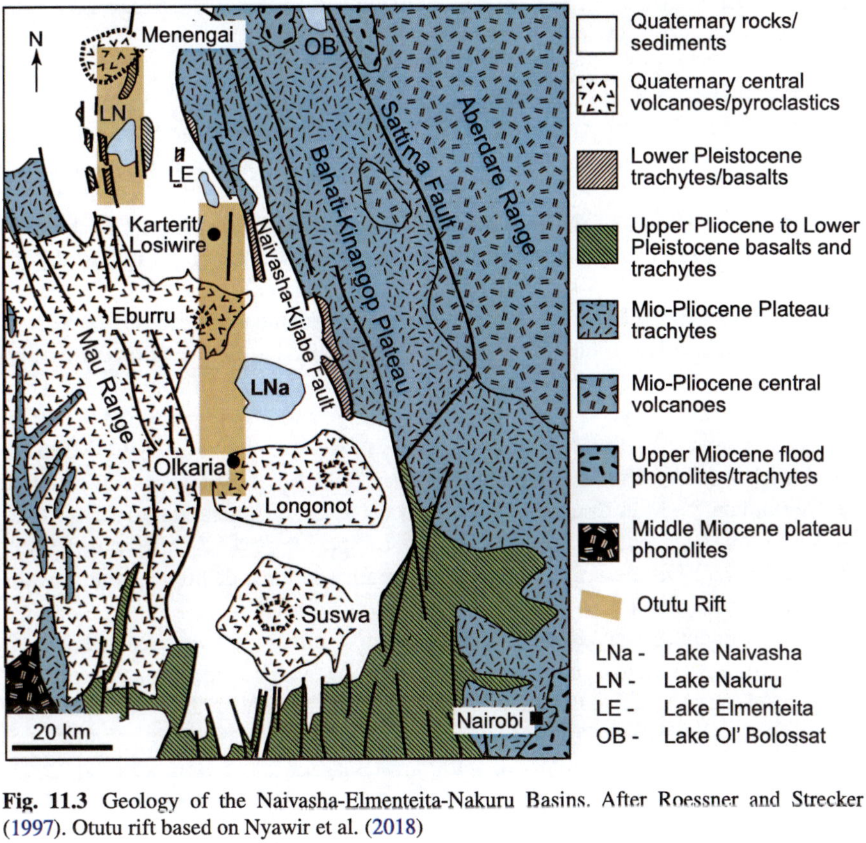

Fig. 11.3 Geology of the Naivasha-Elmenteita-Nakuru Basins. After Roessner and Strecker (1997). Otutu rift based on Nyawir et al. (2018)

Table 11.1 Stratigraphic sequence and evolution of the Central Kenya Rift (including adjacent Lake Bogoria and Suswa areas)

Age (Ma)		Stratigraphy	Structural evolution
>1.9 Ma to Holocene		**Quaternary sediments.** Fluviolacustrine sediments with human artefacts, diatomites, reworked pyroclastics.	NNE to NE normal faults with strong dextral component.
		Upper Pliocene to Quaternary volcanics. Gilgil and Magadi trachytes, Lake Hannington trachyphonolite; Goitumet basalt; pyroclastics, rhyolites, comendites.	N-S to NNE normal faults; reactivation of NNW and NW faults with dextral component.
		Limuru trachytes, Kwaibus and Mbaruk basalt.	Reactivation of faults at western edge of Kinangop Plateau.
3.7 to 2.6		**Pliocene volcanic rocks.** Mainly tuffs (Mau tuff on eastern shoulder; Kinangop tuff to west).	W- to WSW-dipping normal faults with dextral components create Aberdare Escarpment along Sattima Fault; formation of intra-rift Kinangop Plateau.
5.5 to 3.7		**Mio-Pliocene volcanic complexes.** Predominantly basalts (Simbara basalt, Sattima series, Laikipia basalt).	
		Miocene volcanics.	
6.7 to 6.2		Thompson's Falls phonolite.	
14 to 7.6		Nepheline and flood phonolite (Rumuruti, Kericho-Uasin Gishu phonolite; Timboroa assemblage).	ENE-dipping NNW normal faults create central half graben.
23 to 15.1		Samburu basalt.	Minor faulting.

Simplified after Strecker et al. (1990)

phonolite flows (Williams 1970; Baker et al. 1971; Strecker et al. 1990). Between 5.5 and 3.7 Ma, Mio-Pliocene volcanics (Simbara basalt, Sattima Series, Laikipia basalt) were erupted that are now exposed along the eastern rift margin. The third phase of Strecker et al. (1990) gave rise to the Kinangop Tuffs. The upper portion is dated at 3.7–3.3 Ma, but the lower sequence may correlate with part of the Mau Tuffs (6–4.3 Ma) on the western Mau Escarpment (Jones and Lippard 1979). Locally, the Kinangop Tuffs are overlain by basalt and trachyte (2.6 Ma). The basal part of Phase 4 (>1.9 Ma to Holocene; Table 11.1) includes the Limuru Trachytes (>1.9 Ma) (Baker et al. 1988) near Suswa (Fig. 11.3) and basalts (Kwaibus, Mbaruk). Younger trachyte and rhyolite-comendite are exposed on the rift floor in the Lake Naivasha Basin (Fig. 11.4a), forming domes, plugs and lava flows (Griffiths 1977; Jones and Lippard 1979; Griffiths and Gibson 1980; Strecker et al. 1990; Roessner and Strecker 1997). Pyroclastics younger than 0.33 Ma are present near eruptive centres and on the Mau Escarpment with basalts at Lake Elmenteita, to the north of Naivasha, that are younger than 0.7 Ma (Jones and Lippard 1979; Gowlett 1980). Strecker et al. (1990) described the fault kinematics in some detail with their results summarised in Table 11.1.

Lake sediments observed in cores and outcrops show two widespread sequences (Fig. 11.4b) buried below much of the Naivasha Basin. The oldest formed at about 135 ka and represents a major Late Pleistocene lake highstand. An Early Holocene lake expansion is recorded by a second series of buried lake sediments and by sand and gravel strandlines at heights of ~2000 m (Bergner 2003; Bergner et al. 2003; Bergner and Trauth 2004).

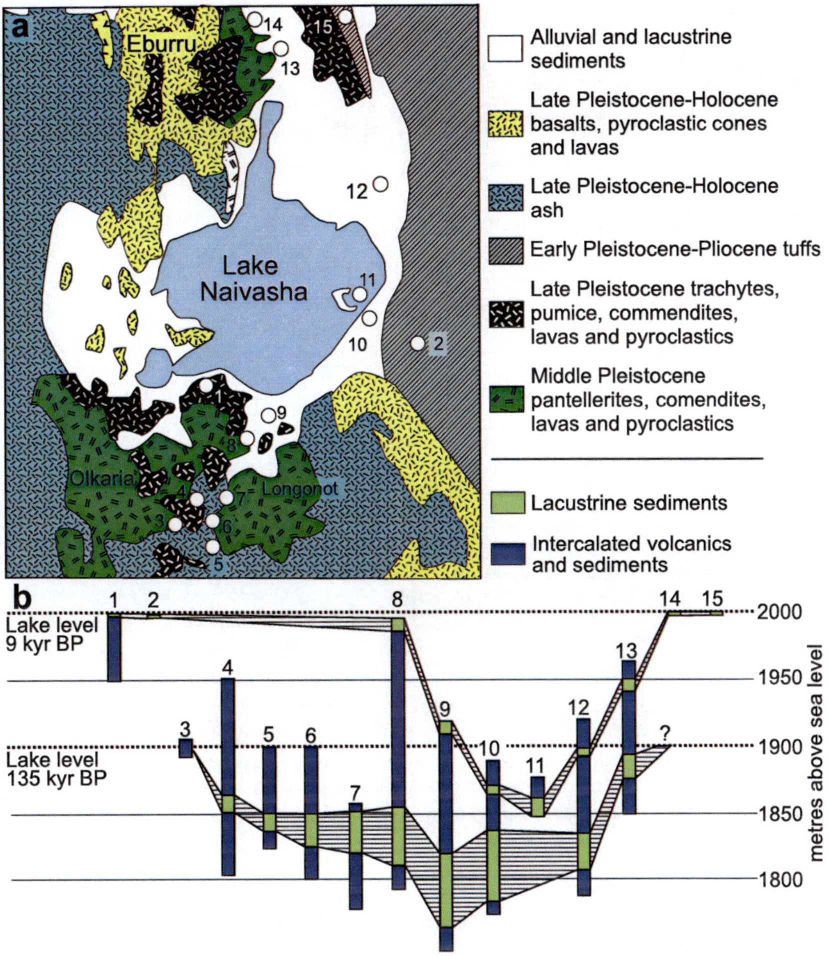

Fig. 11.4 Geology of the Naivasha Basin. After Bergner et al. (2003). (**a**) Lithologies in the Naivasha area. (**b**) Stratigraphic correlation of core and outcrop lake deposits showing two major lake sequences dated to about 135 ka and 9 ka

11.3 Climate and Basin Hydrology

The hottest conditions occur between January and March when humidity is low and daily temperature ranges are at their greatest (Fig. 11.5a). Rainfall varies considerably with altitude. The mountainous Aberdares receive >1500 mm annually, which contrasts with ~610 mm near Lake Naivasha (Richardson and Richardson 1972). Average annual pan evaporation rates of about 1790 mm per year significantly exceed rainfall near the lake, which has a broadly bimodal distribution with variable peaks in April and November (Farah et al. 2004; Kuhn et al. 2012). Winds blow mainly from the southeast or east with speeds of about 0.5–5 m s^{-1} that generally

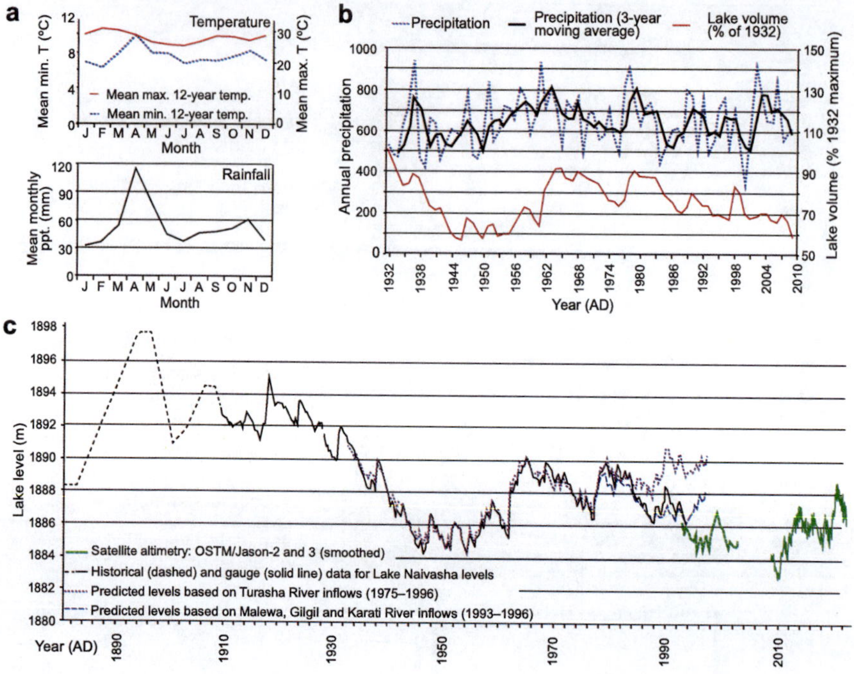

Fig. 11.5 Climate and lake levels. (**a**) Annual temperature and rainfall distributions showing peaks in April and November. After Muthwatta (2004). (**b**) Historical trends in precipitation and Lake Naivasha volume. After Reta (2011). (**c**) Historical lake level trends. Note the comparison with predicted levels based on models that remove the effects of water abstraction for horticulture. Based on Becht and Harper (2002) with subsequent levels based on gauge data and Jason-3 satellite altimetry

increase into the afternoon and which drive currents within the lake (Källqvist 1987; Ndungu et al. 2015b).

Olaka et al. (2010) and Olaka (2011) noted that Lake Naivasha has been particularly sensitive to lake level change through the last 20,000 years due to its basin shape and climate regime. The historical rainfall pattern is very variable with wet periods alternating with prolonged episodes of drought (Fig. 11.5b; Kuhn et al. 2012). The precipitation also shows a 7-year periodicity, reflecting El Niño cycles (Vincent et al. 1979) with stream gauge data indicating significant increases in annual maxima between 1959 and 2008 (Kyambia and Mutua 2014). In the last few decades, rapid population growth and especially the demands for irrigation water for horticultural developments have placed increasing demands on water resources (Muthuwatta 2004; Gitonga 1999; Becht and Nyaoro 2006; Kariuki 2011; de Jong 2011; Mekonnen et al. 2012). Becht and Harper (2002) estimated abstraction rates of 60×10^6 m^3 a^{-1}, which were considered six times higher than safe levels estimated in the 1980s. They also developed lake level models, based on Turasha River inflows, that faithfully reflected historical data up to the time when water abstraction for horticulture began, after which actual lake levels were about 3–4 m lower than predicted by their data (Fig. 11.5c). Statistical comparisons between lake level fluctuations for the periods 1951–1980 and 1981–2000 also suggest significant human impacts on the surface height of Lake Naivasha (Tarafdar and Harper 2008).

Early reports on historical and gauge data indicated that the lake was particularly high in the late eighteenth and early nineteenth centuries (Fig. 11.5c; Sikes 1936) with levels falling between 1915 and 1944. Subsequently, the surface ranged between about 1884 and 1890 m, peaking in the 1960s and in the years 1978–1984, 1997 and 2012–2023. The latest of these peaks resulted in an expansion from 107.7 km^2 in January 2010 to 169.9 km^2 by 2012 (Onywere et al. 2012) with another peak reached in 2020. Herrnegger et al. (2021), for example, report an increase in the lake area of ~15–20% for the years 2015–2021 compared with 1984–2009. The Government of Kenya and UNDP (2021) report that 108.57 km^2 of land were submerged by lake transgression between 2010 and 2020 with significant impacts on riparian vegetation and local communities (Karanja 2022). Drops in lake level have variously been attributed to drier conditions, increased use of groundwater, water used for flower cultivation and unregulated extraction for irrigation from the Malewa River (Awange et al. 2013; Awange and Kyalo Kiema 2013), but in contrast, much recent concern has been driven by high lake levels (Muita et al. 2021).

Several researchers have suggested mechanisms to account for the lake being fresh despite a lack of surface outflow. These include salt loss through a subterranean outlet or deflation, as well as losses by burial of alkaline pore fluids and ion removal by plants (Beadle 1932b; Nilsson 1938; Thompson and Dodson 1963; Richardson and Richardson 1972; Källqvist 1987). Sikes (1934, 1936) first suggested a subsurface outflow rate of 43 million $m^3\ a^{-1}$, with other studies inferring rates of 18–55 million $m^3\ a^{-1}$, of which 50–90% escaped via shallow fractured trachyte aquifers (<500 m deep) to the south (Clarke et al. 1990; Ojiambo 1992; Becht and Harper 2002; Reta 2011).

Other studies have used piezometric and stable isotope data to determine groundwater outflow, which has been traced as far as 30 km to the south. Olaka et al. (2022), for example, pointed out the importance of faults in guiding groundwater flow via four sub-compartments and observed that ^3H-^3He dating of groundwater indicated ages of a few to more than 50 years. Most of the water escapes between Longonot and Olkaria with a smaller proportion moving north between Eburru and Gilgil (Allen et al. 1989; Darling et al. 1990, 1996; Muno 2002; Becht et al. 2006; Ayenew et al. 2007; Ayenew and Becht 2008; Olago et al. 2009; Hogeboom 2013; Hogeboom et al. 2015; van Oel et al. 2013; Yihdego and Becht 2013). Ojiambo et al. (2001) estimated that 40–50% of the Olkaria geothermal reservoir water is recycled from Lake Naivasha with Butturini et al. (2020) reporting heat-modified solid phase-extracted Dissolved Organic Matter in the hot spring water.

11.4 Peoples, Vegetation and Wildlife

Pastoral Maasai formerly occupied the Lake Naivasha area. However, in 1905, they were moved south of the newly built Uganda Railway to make way for European farmers (Wanjiku 2015). The rail station gradually evolved into Naivasha town, which Kenya census data indicates was inhabited by 181,966 people in 2009. The town itself was once a mix of settlers and Maasai, but now includes a diversity of Kenyans, especially from central and western parts of the country. The population in the Naivasha area has expanded rapidly in recent years, averaging 6.6% growth per year up to 2009 (Kuhn et al. 2012). Partly as a result of an influx of people seeking employment in the horticultural industry, the population of the Naivasha Division increased from about 158,679 in 1999 to 376,243 by 2009, with a density of about 124 people km^{-2} (Onywere et al. 2012).

Cyperus papyrus dominate a large swamp along the north shores of Lake Naivasha with lagoons, formed by fringing macrophytes, around other parts of the lake (Fig. 11.1d). The swamps and lagoons are subject to considerable change (Owino and Ryan 2006; Morrison and Harper 2009; Morrison 2013) due to lake level fluctuations and variously contain *Ceratophyllum demersum*, *Utricularia* spp., *Potamogeton* spp., *Najas indica* and *Nymphaea caerulea*. Invasive floating plants such as *Salvinia molesta* and *Eichhornia crassipes* have been growing in bays and in other parts of the lake since the 1970s (Fig. 11.6a–c; Gaudet 1975, 1977a, b, 1979; Källqvist 1987; Åse 1987a, b; Hubble and Harper 2001a; Ndebele-Murisa et al. 2010).

Tropical grasses, bushes and herbaceous plants (e.g. *Rhus natelensis*, *Senecio hadiensis*, *Cynodon dactylon, Urtica massaica*) cover the lower rift valley slopes and floor with *Acacia* (*Vachellia*) *xanthophloea* (fever tree) forming forests near the lake that are periodically inundated by high water levels (Fig. 11.6d). The Kinangop plateau, to the east, was formerly occupied by forests and grasslands, but by 2005 about 30% of the area was covered with maize or vegetables and scattered exotic trees (Mavuti and Harper 2006). Submontane tropical forests grow along upland river systems and on high ground (Gaudet and Melack 1981; Mavuti and Harper 2006).

Lake Naivasha supports both a commercial fishery and wildlife. Birds are especially varied and abundant including, for example, large numbers of cormorants, ducks, geese, rails, coots, stilts, avocet, and other waders as well as herons and birds of prey such as fish eagles (e.g. Watson 1970; Smart 1988a, b; Owino et al. 2001; Childress et al. 2002; Harper et al. 2002a). Birds associated with the lake are most abundant and diverse in areas with submerged or emergent vegetation (Henderson and Harper 1992; Hubble and Harper 2001a). Zebra, antelope, giraffe, hippo and buffalo are found in several areas near the lake and in the Longonot and Ol Njorowa Gorge National Parks to the south.

Large areas of riparian vegetation have been cleared for alfalfa (*Medicago sativa*), cereals, and vegetables (Mutungwa 2011). By the 1990s, former stock-rearing, ranching and sisal cultivation had given way to approximately 100 km^2 of

Fig. 11.6 Lakes Naivasha and Oloidien. (**a–d**) Lake Naivasha; (**e**) Lake Oloidien. (**a**) Floating islands of vegetation are a variable feature of the lake surface, reflecting growth and wind regimes. (**b**) Water hyacinth, black bar = 5 cm. (**c**) Dense floating *Salvinia* mat, black bar = 2 cm. (**d**) Flooding of Lake Naivasha shoreline (2016). (**e**) Lake Oloidien with open waters and forested margin. (Photo courtesy of Annett Junginger)

irrigated horticulture (Johnson et al. 1999; Fayos 2002; Mekonnen et al. 2012) with flowers air freighted to Europe (Becht and Harper 2002; Becht et al. 2005). Other human impacts include, for example, the loss of 49 km^2 of forest on the Kinangop Plateau between 1986 and 2000 (Onywere et al. 2012).

11.5 Limnology

11.5.1 Physical Limnology

Limnological data from several studies are summarised in Table 11.2. Beadle (1932a) provided the earliest (1930 and 1931) detailed descriptions of the physical limnology from Crescent Island Bay (Fig. 11.7a, b). His data indicated broadly uniform temperature profiles with a weak thermocline developing during midday in the upper 0.5 m of the water column. The data also documented a uniform pH with dissolved oxygen declining downwards from about 5.5 to 3.5 ccs L^{-1}, with weak stratification later confirmed by Richardson (1964). Similar limnological data were reported for 1976–1977 for a nearshore site in southern Lake Naivasha (Fig. 11.7a, c; Källqvist 1987) with surface temperatures of 19.5–21.5 °C for profiles that were also characterised by weak to moderate thermoclines. Surface oxygen ranged from 6.6 to 7.4 mg L^{-1} and decreased rapidly towards the lake floor, except on days with minimal temperature change when differences in oxygen levels were also reduced.

Mean spatial surface temperature data for March 2007 to January 2008 (Mugo 2010) ranged from 20.8 to 22.3 °C (Fig. 11.7a, d) with overall average lake temperatures of 21.5 °C (Mar-Apr) to 22.0 °C (Dec-Feb). Average dissolved oxygen levels fluctuated between 5.5 and 6.6 mg O_2 L^{-1} across the lake surface. Mean conductivity and pH ranged from 249 to 278 µS cm^{-1} and 7.7–8.2, respectively, with Secchi depths of about 25–75 cm. Other studies have suggested increased Secchi depths when lake level rises, reaching ~60 cm (Mavuti and Harper 2006) and ~1.26 m

Table 11.2 Limnological data for Lake Naivasha

Parameter	Njuguna (1982)	Kitaka (1991)	Mugo (2010)	Njenga (2004)	Mugo (2010)	Ndungu et al. (2015a)
Secchi depth (cm)	76–227	70–180	67.6 ± 46	–	24–82	–
pH	7.3–8.6		8.6 ± 0.6	8.2–8.9	6.0–8.9	7.97–9.5
Conductivity (µS cm^{-1})	300–347	220–297	320 ± 70	220–2480	204–338	74–392
Temp. °C	19–25	19–24	21.6 ± 1.6	–	18.5–24.8	18.1–29.6
Dissolved O_2 (mg L^{-1})	3.5–7.9		7.2 ± 1.4	5.8–10.4	2.3–9.1	–
Turbidity (NTU)	–	–	–	–	–	2.17–282

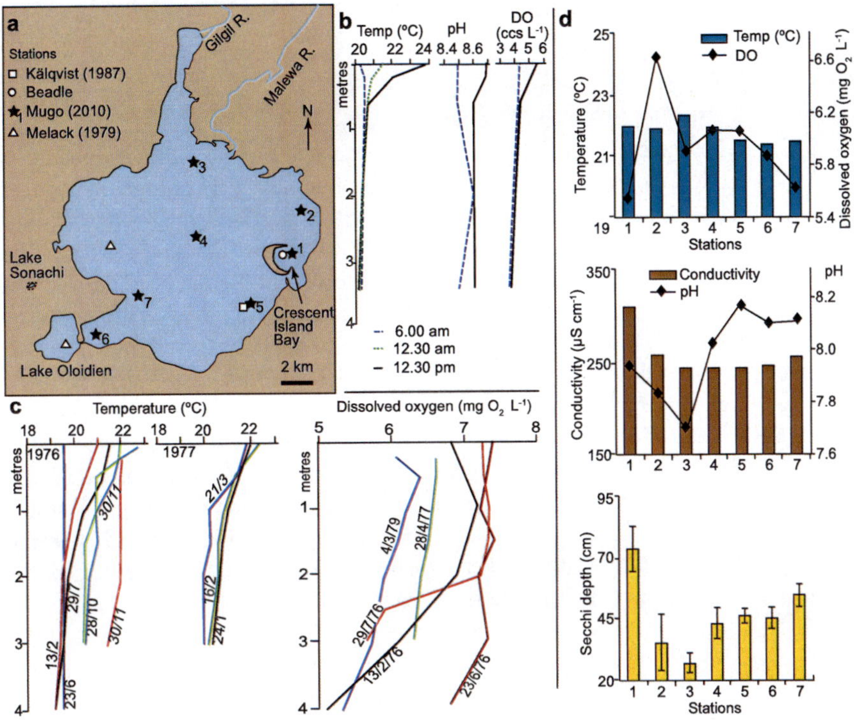

Fig. 11.7 Physical limnology of Lake Naivasha. (**a**) Location of sampling stations. (**b**) Temperature, pH and dissolved oxygen (DO) profiles showing weak thermocline near the lake surface around midday. After Beadle (1932b). (**c**) Temperature and oxygen profiles with weak thermoclines and oxic bottom waters. After Källqvist (1987). (**d**) Spatial variability in temperature, dissolved oxygen, conductivity, and Secchi depths. (After Mugo 2010)

(Ngari et al. 2009). More recent Secchi data for the main lake have ranged between 10 and 75 cm, with Crescent Island Bay values fluctuating from 50 to 120 cm (Ndungu et al. 2013). GIS techniques have also been used to map suspended matter (Ghansah et al. 2016) and the depth of the euphotic zone, which ranged between 0.68 and 2.66 m at the time of sampling (Majozi et al. 2014). In a GIS study, Johnson (2015) reported several trends for the years 2003–2011, noting that Total Suspended Matter (TSM) reached a maximum in November 2006 of 16.61 mg L^{-1} and a low of 4.27 mg L^{-1} in November 2011. Turbidity was highest at 14.16 Formazin Nephelometric Units (FNU) in 2007 and lowest (2.31 FNU) in 2011.

Lake Oloidien (Figs. 11.6e and 11.7a) was connected to Lake Naivasha until 1982 when water level dropped and caused it to become detached (Harper et al. 2011). Due to its variable levels and connections, its waters are much more variable in chemistry than those of Lake Naivasha with conductivity ranging from 660 µS cm^{-1} (1979–80) to 5270 µS cm^{-1} in 2005 (Luo et al. 2016).

11.5.2 Chemical Limnology

Several researchers have examined the chemistry of freshwater Lake Naivasha (Table 11.3; Beadle 1932b; Talling and Talling 1965; Kilham 1971; Gaudet and Melack 1981; Macintyre and Melack 1982; Burgis and Mavuti 1987; Njenga 2004; Mugo 2010; Ndungu et al. 2015a; Keyombe and Waithaka 2017; Montcoudiol et al. 2019), which is dominated by sodium bicarbonate water with relatively low chloride and high fluoride (Kilham and Hecky 1973; Olaka et al. 2016). Mean Total Dissolved Solids (TDS), in mg L^{-1}, vary, being lowest under the northern *Papyrus* swamp (217) with intermediate values reported for Lake Naivasha (329), nearshore lagoons (367) and Crescent Island Bay (394). Higher and more variable values are recorded for Lake Oloidien (831) depending on lake levels and the status of connections with Lake Naivasha (Gaudet and Melack 1981). In 2002, the pH ranged from 8.0 to 8.9 with conductivity between 220 and 2480 μS cm^{-1} (Njenga 2004). Njiru et al. (2015) reported pH values between 6.63 and 9.61 in December 2009 and January 2010 with conductivities from 163 to 453 μS cm^{-1} in October and February 2010. Chloride and bicarbonate accounted for 90% of the anions with sulfate representing only ~1%. Fluoride ranged between 0.1 and 25.1 mg L^{-1}. In general, sodium dominated the cations with important contributions from K^+ and Ca^{2+}.

Most rift valley solutes result from silicate weathering (Kilham and Hecky 1973) with Lake Naivasha receiving about 80% of its dissolved load via the Malewa River (Gaudet and Melack 1981). Loss of dissolved ions through surface outflow does not occur. However, seepage results in a significant loss of solutes (Ojiambo and Lyons 1996) with other mechanisms having also been noted. Hecky and Kilham (1973), for example, observed that Ca^{2+} and Mg^{2+} can be removed through the formation of calcite and (or) dolomite, which appears to be the case in Crescent Island Basin and in Lake Oloidien, both of which are supersaturated with respect to calcite and accumulate calcite in bottom sediments (Gaudet and Melack 1981). Other losses are likely due to incorporation in algae and macrophytes or skeletal parts of molluscs, crayfish and fish. Sulfate is lost below the northern swamps, possibly due to sorption onto small particles, uptake by papyrus and through limited formation and degassing of H_2S (Gaudet 1976, 1979). Silica is also depleted, reaching its lowest concentrations in Lake Oloidien (Gaudet and Melack 1981), with losses resulting from phytolith formation in aquatic plants and the growth of diatoms (Richardson 1968, 1969; Richardson and Richardson 1972; Richardson et al. 1978). Potassium may be lost through uptake by macrophytes, with K^+ and other ions, such as Ca^{2+} and Mg^{2+}, also lost through cation exchange reactions (Cerling 1979).

Other metals tend to be present in relatively low concentrations (Githaiga et al. 2021). Wandiga et al. (1983), for example, reported concentrations (in ppb) for Mn (34–53), Cu (12–14), Zn (134–148), Cd (2–4), and Pb (5–7). Surface sediments from Lake Oloidien contain slightly elevated Pb (42 ppm) with lower values for other heavy metals, including Co (11), Cu (1.8), Zn (16) and Cr (0.2) with Cd below detection limits (Mercy et al. 2017). Similarly, Yang et al. (2017) noted low concentrations of Cr, Ni Cu, As and Hg. Studies of the modern lake have also detected

Table 11.3 Lakes Sonachi and Naivasha water chemistry (mg L^{-1})

	Lake Sonachi		Lake Naivasha										
	Macintyre and Melack (1982)		Gaudet and Melack (1981)				Njenga (2004)	Talling and Talling (1965)	Ochieng et al. (2007)	Ndungu et al. (2015a, b)	Montcoudiol et al. (2019)		
Ions	Depth 0.05 m	Depth 4.5–6 m	Mean lake value	Mean lagoon value	Mean Crescent Island basin	Mean Lake Oloidien	Main lake	Main lake	Main lake	Mid-lake	Crescent Island Basin (K22)	South littoral (K24)	Lake Oloidien littoral (K29)
Na^+	2150–3483	3725–4950	40	43	52	125	101–256	41–45	–	–	22	23	628
K^+	253–470	452–600	20	26	30	32	15–146	21.6–22.6	–	–	16	15	117
Ca^{2+}	3	2–3	21	23	17	9	2.8–9.4	15.2–21.9	–	16.0–43.2	20	19	13
Mg^{2+}	2–3	1–2	6.4	6.9	7.5	6.9	2.0–10.1	6.1–7.7	–	0–7.6	4.7	4.4	10
HCO_3^-	2426–3815	4129–6346	192	14	15	43	–	–	–	–	–	–	–
CO_3^{2-}	1477–2324	1899–2918	10.6	216	231	496	–	–	–	–	–	–	–
Cl^-	274–413	351–502	14	17	17	32	69–797	14.4–16	–	4–32	8.3	8.1	85
SO_4^{2-}	46–69	58–84	6.2	4.3	4.8	7.1	1.6–9.5	8.4–12.1	–	0–4	1.8	1.7	1.3
F^-	83–125	106–152	1.5	1.7	1.5	8	0.1–25.1	–	–	–	1	0.98	85
NO_3^-	–	–	–	–	–	–	–	–	–	–	0.13	–	–
Br^-	–	–	–	–	–	–	–	–	–	–	0.054	0.071	2.6

(continued)

Table 11.3 (continued)

SiO_2	108–128	134–174	34	30	36	44	–	31.5–32.5	–	–	–	–	–
Mn^{2+}	–	–	–	–	–	–	–	–	–	0.044–0.361	–	–	–
Cd^{2+}	–	–	–	–	–	–	–	–	0.008	–	–	–	–
Co^{3+}	–	–	–	–	–	–	–	–	0.019	–	–	–	–
Cr^{3+}	–	–	–	–	–	–	–	–	0.025	–	–	–	–
Cu^{3+}	–	–	–	–	–	–	–	–	0.005	–	–	–	–
Pb^{4+}	–	–	–	–	–	–	–	–	0.042	–	–	–	–
Sn^{4+}	–	–	–	–	–	–	–	–	1.05	–	–	–	–
Zn^{3+}									0.029				

agricultural organophosphates and pesticide residues in water, sediment and fish tissues (Otieno et al. 2015).

11.6 Biology and Ecology of the Lake

11.6.1 Plant Nutrients and Primary Production

Data for major plant nutrients are presented in Table 11.4. Early investigations indicated that total P ranged from 58 to 152 μg L^{-1} with total N values of 411 to 1140 μg L^{-1} (Talling and Talling 1965; Millbrink 1977). Orthophosphate concentrations of 7–10 μg P L^{-1} were recorded in 1973–1974 (Melack 1976), with higher values of 55 μg L^{-1} observed in Crescent Island Bay during 1973 (Millbrink 1977). In 1960–1961, NO_3-N levels were 32–63 μg L^{-1} (Talling and Talling 1965) but fell to only 7 μg L^{-1} by 1973, when NH_4-N and NO_2-N concentrations were 19 and 3 μg L^{-1}, respectively (Millbrink 1977). River inflows in 1997–98 contributed ~1.4 g m^{-2} y^{-1} to the total phosphorus loading in surface lake waters (Kitaka et al. 2002a). More recently, Ndungu et al. (2013) have reported total phosphorus concentrations of 27–410 μg L^{-1}, commenting that the higher values occurred after heavy rains flushed nutrients into the lake and that the water had been briefly hypereutrophic. Silicate has ranged from 18 to 45 μg L^{-1} (Talling and Talling 1965; Melack 1976; Millbrink 1977) with sulfate concentrations of 9.4–15.8 mg L^{-1} (Källqvist 1987).

Spiking experiments with P, N and sulfate indicate that N is the most important limiting nutrient in Lake Naivasha (Källqvist 1987) with soluble reactive phosphorus also limiting (Melack 1996). The minor nutrients silicon, manganese, chloride, iron, sulfate and sodium appear to be non-limiting (Hubble and Harper 2002b). Other controls on species composition and primary production have also been described, including grazer-prey preferences, with several planktonic taxa using motility, spines, organic sheaths or toxicity to avoid predation (Hubble and Harper 2000).

Table 11.4 Major plant nutrients in Lake Naivasha

Parameter	Njuguna (1982)	Malala and Mwamburi (2002)	Mwamburi et al. (2007)	Mugo (2013)	Ndungu et al. (2015a, b)
PO_4–P (μg L^{-1})	–	0.1–2.2	22.2 ± 3.2	3–35	1–79
NO_3–N (μg L^{-1})	–	6 ± 6.8	103 ± 45	14–238	50–800
NH_4–N (μg L^{-1})	–	36.4 ± 31.2	480 ± 158	34–346	10–510
TN (μg l^{-1})	1106–2509	–	4347 ± 607	113–698	–
TP (μg L^{-1})	39.8–116	–	157 ± 12	10–138	27–410
N/P	12–63	–	–	5–11	–

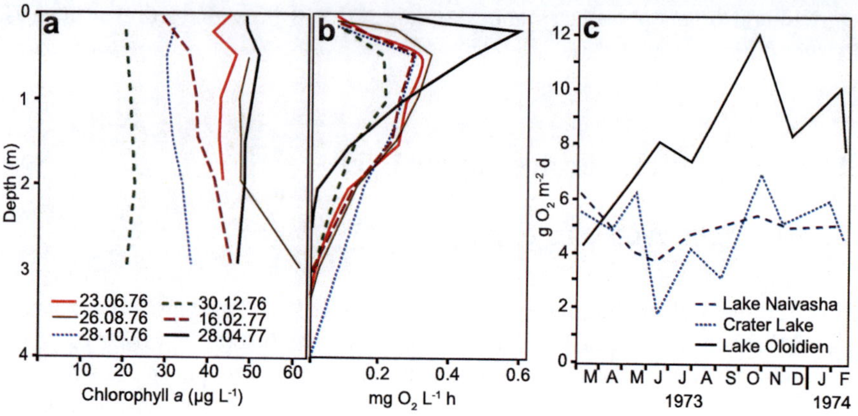

Fig. 11.8 Productivity in Lake Naivasha **a** and **b** after Källqvist (1987). (**a**) Vertical distribution of chlorophyll *a* through 1976–1977. (**b**) Photosynthetic oxygen production with depth. Note the general maxima at about 0.5 to 1 m below the surface. (**c**) Annual variations in photosynthetic rates of plankton in three Naivasha water bodies. After Melack (1979)

In 1976–1977, Chlorophyll-*a* ranged from 21 to 50 µg L^{-1} with increases in April–May 1977 coinciding with a period of greater rainfall (Fig. 11.8a; Källqvist 1987). In contrast, two analyses for the same data set from Crescent Island Bay indicated values of only 7.4 and 25 µg L^{-1} in January 1976 and February 1977. In 1990–1991, median chlorophyll-*a* values were 24.5 µg L^{-1} for both central Lake Naivasha and Lake Oloidien (Uku and Mavuti 1994). By the turn of the millennium, concentrations had increased, reaching 178 µg L^{-1} in 1995 (Mavuti and Harper 2006) and 40–210 µg L^{-1} in 2011 (Ndungu et al. 2013), reflecting increasing eutrophication.

In 1976–1977, maximum photosynthetic oxygen production occurred between 0.5 and 1 m depth (Fig. 11.8b) and gradually decreased down to 3–4 m, with a range of 3.5–7.2 g O_2 m^{-2} day^{-1} (Källqvist 1987). Other studies of the main lake indicate rates of 7.15 g O_2 m^{-2} day^{-1} in 1960 (Richardson 1964) and 3.7–6.2 g O_2 m^{-2} day^{-1} in 1969, with rates of 8.1 g O_2 m^{-2} day^{-1} for Lake Oloidien (Melack 1976, 1979, 1982) and 9.8 g O_2 m^{-2} day^{-1} in Crescent Island Bay (Millbrink 1977). In general, rates tended to be higher in Lake Oloidien than in Lake Naivasha or Crescent Island Bay (Fig. 11.8c).

Photosynthesis reaches a peak about 0.5 m below the surface, depleting nutrients. Below this depth photosynthesis decreases downwards due to light attenuation, ceasing below ~5 m, partly due to particulates and self-shading (Hubble and Harper 2001b). High dissolved organic (16.21 mg L^{-1}) and inorganic (31.97 mg L^{-1}) carbon concentrations have also been linked to high extinction coefficients for a variety of wavelengths of light in Lake Naivasha (Lalah and Wandiga 2006).

Recent studies (2019–20) of trophic state by Yongo et al. (2022) indicated high turbidity, total nitrogen (TN), ammonium-nitrogen (NH_4^+-N), nitrate-N (NO_3-N) and total phosphorus (TP) levels during the wet season, and at the Malewa and

Karati River mouths, while pH, electrical conductivity (EC) and chlorophyll-*a* values were relatively high during the dry season. In contrast, water temperature and dissolved oxygen (DO) concentration did not vary between sampling sites or seasons.

11.6.2 Phytoplankton

Historical sampling has shown considerable variations in phytoplankton composition, both seasonally and from year to year. In 1929, Lake Naivasha had clear water with no major phytoplankton blooms (Rich 1933). Common taxa included *Phacotus*, *Cosmarium*, *Coelosphaerium*, *Synedra* (*Ulnaria*) and *Surirella* with no single dominant taxon (Jenkin 1936). *Botryococcus protuberans* was common in open water. In 1930, Crescent Island Bay and the central portion of Lake Naivasha showed similarities in their phytoplankton assemblages, which were dominated by *Aulacoseira* and a few Chlorophyceae and Cyanophyceae. The southwest bay (Lake Oloidien) phytoplankton included a few *Aulacoseira* and no Chlorophyceae. In contrast, abundant flora and fauna inhabited the northern swamp regions (Beadle 1932a). Several broad-based surveys of lake ecosystems in the Rift Valley during the 1960s also included partial data on Lake Naivasha (Evans 1962; Lind 1965, 1968; Richardson 1964, 1968, 1969).

Human activities have since degraded the lake system and have altered the phytoplankton composition as a result of changing sedimentation rates (Junginger 2011; Stoof-Leichsenring et al. 2011), the introduction of exotic fish species (Gherardi et al. 2011) and pollution-induced eutrophication (Mavuti et al. 1996; Kitaka 1991; Kitaka et al. 2002a, b; Mutungwa 2011; Harper 2006). By 1964 the lake was dominated by diatoms (*Aulacoseira ambigua* and *Synedra ulna* var. *danica*) and green algae (*Staurodesmus convergens* and *Staurastrum pingue*). In 1976–77, the southern part of the lake was dominated by cyanobacteria (*Lyngbya* spp., *Spirulina* sp. and *Aphanocapsa elachista*) and an unidentified filamentous form (Källqvist 1987). Many green algal taxa were present, but low in abundance, except for *Sphaerocystis schroeteri*. By March 1977, diatoms had become scarce, except for blooms of *Synedra acus*, with a *Synedra* sp. dominating again two years later.

Diatoms dominated seasonally prior to moderate eutrophication in the 1980s, but this pattern subsequently shifted towards dominance by cyanobacteria (Hubble and Harper 2001a). A total of 143 taxa were identified in 1979–80 (Kalff and Watson 1986), with 170 reported in 1997–1998 (Hubble and Harper 2001a, 2002a). *Cryptomonas*, *Rhodomonas* and *Ankyra* were abundant in 1982 and *Aulacoseira*, *Microcystis* and filamentous Cyanophyceae were the most common taxa in 1984 (Hubble and Harper 2001a).

Various species were recorded from Lake Naivasha in 1990, including taxa belonging to *Closterium*, *Aulacoseira*, *Synedra*, *Nitzschia*, *Microcystis*, *Scenedesmus*, *Pediastrum*, *Cosmarium* and *Ceratium*, with no clear seasonal succession reported, and with *Microcystis* and *Closterium* dominating in Lake Oloidien (Uku and Mavuti 1994; Hubble and Harper 2001a). By 1998, *Aulacoseira* had become abundant, with reduced light levels and increased nutrients favouring *A. granulata* and *A. italica* over *A. ambigua*. Cocquyt and Verschuren (2023) provide a checklist of diatoms recorded from Lake Naivasha and its satellite lakes.

Ballot et al. (2009) reported data for the 2001–2005 period, noting that Lake Naivasha was dominated by cyanobacteria, Chlorophyceae, and Bacillariophyceae. Coccoid cyanobacteria (*Cyanocatena planctonica*, *Chroococcus limneticus*, *Aphanocapsa* sp.) accounted for 83–98% of the cyanobacterial biomass. Chlorophyceae formed the most abundant group in 2002–2003. Other taxa that temporarily reached a high biomass included the Desmidiaceae (*Gonatozygon* sp.) and Bacillariophyceae with *Microcystis aeruginosa* present in low abundance throughout 2001–2005. Krienitz et al. (2013) reported a progressive increase in toxic cyanobacteria in Lake Naivasha through the early twenty-first century.

Other studies using DNA in sediments genetically distinguished at least two *Aulacoseira* and one *Amphora* spp. (Epp et al. 2011). Lake Oloidien was dominated by cyanobacteria and Chlorophyceae, with coccoid forms most common in June/August 2001. Krienitz et al. (2012) reported an increase in *Anabaenopsis* between 2001 and 2012 as a percentage of total phytoplankton biomass. Increased rainfall resulted in a lake level rise and freshening of the waters between 2010 and 2015 and a change from dominance by *Arthrospira*, with abundant flamingos in Lake Oloidien, to diverse blooms of cyanobacteria and algae. One new fossil taxon, *Staurophora caljonii* (Cocquyt 2004) was reported from sediments in Lake Oloidien (Ballot et al. 2005).

More recently (February to July 2019) Omondi (2019) and Omondi et al. (2020) noted assemblage changes that reflected continued siltation and nutrient loading in the lake, pointing out that this had led to expansions of free-floating and emergent macrophytes and a higher phytoplankton biomass. Phytoplankton at seven of their sampling locations included 124 species belonging to Chlorophyceae (43 species), Bacillariophyceae (38 species), Cyanophyceae (24 species), and smaller numbers of Zygnematophyceae, Euglenophyceae, and Dinophyceae.

11.6.3 Macrophytes

Lake Naivasha has also experienced major changes in macrophyte populations. In 1929, the northeast shores were bounded by wide mudflats with the eastern lake floor largely free of water weeds (Jenkin 1932, 1936). By November 1930, following heavy rains, the lake margins were occupied by dense growths of *Cyperus papyrus* with the offshore waters choked by *Ceratophyllum demersum* and *Potamogeton* (Beadle 1932a, b). At least 25 macrophyte taxa have been reported from the lake (Jones and Muthuri 1985; Gaudet 1977a; Owino and Ryan 2006; Omondi et al. 2016) with *Cyperus papyrus* swamps accumulating nutrients and acting as a nutrient buffer (Gaudet 1977a, 1980) with high annual productivity levels of 5150 g m^{-2} y^{-1} (Muthuri et al. 1989).

During 1987, the swamps were fragmented and reduced in size from 50 to <5 km^2, due to a lake level fall and farming activities that spread onto former lake floor areas (Harper 1991; Harper et al. 1995; Hubble and Harper 2001a). Flooding in the following years resulted in a recovery of the northern swamp (Boar 2006; Morrison 2013), which covered about 12 km^2 by 1992 and again in 2014 during periods of relatively high lake levels. Growth of papyrus thus relates to lake level. Large amounts of carbon (~1.6 kg C m^{-2} y^{-1}) are sequestered as detritus that is preserved by anaerobic conditions below the swamps in wet periods, only to be released again (1.0 kg C m^{-2} y^{-1}) when lake regression exposes plant debris to aerobic conditions (Jones and Humphries 2002). Canopy transpiration rates for the papyrus have also been estimated at about 12.5 mm, similar to other wetland communities (Jones and Muthuri 1984).

In 1995–1996, offshore submerged macrophytes were dominated by *Chara globularis*, *Najas horrida* (*pectinata*), *Nitella hyalina*, *Potamogeton pectinatus*, *P. octandrus* and *P. schweinfurthii*, which expanded from 1984 to 1987 to cover an area of 20 km^2, before a subsequent decline set in (Harper et al. 1995; Harper and Mavuti 1996; Hubble and Harper 2001a). Invasive floating water ferns (*Salvinia molesta*) appeared in Lake Naivasha in the 1960s (Adams et al. 2002) with water hyacinth (*Eichhornia crassipes*) entering the lake in 1988. Windy conditions on the lake may favour the latter over native *Ludwigia stolonifera*, which otherwise outcompetes it under non-limiting nutrient conditions (Njambuya and Triest 2010). The invasive plants have disrupted navigation on the lake and have provided breeding grounds for bilharzia-carrying snails and malarial mosquitoes. They have also altered photosynthesis rates through shading effects (Mironga 2006; Ngari et al. 2009). The general sensitivity of the lake ecosystem was demonstrated, for example, by a breakup of *Eichhornia crassipes* in 2000, which was probably caused by the introduction of *Cyrtobagus eichhorniae* (hyacinth weevil). In turn, this led to a loss of protection for the crayfish *P. clarkii* and subsequent recovery of submerged plants (Harper et al. 2011).

11.6.4 Zooplankton

Lake Naivasha, in 1930–31, showed a dominance of the zooplankton by *Cyclops* and *Moina*, with an increase in rotifers taking place in the latter part of the study period (Beadle 1932b). *Thermocyclops oblongatus*, *Diaphanosoma excisum*, *Brachionus caudatus*, *B. calyciflorus*, *Mesocyclops equatorialis* and *Simocephalus vetulus* were recorded from open waters in 1979 (Mavuti 1990, 1992) with most of the zooplankton species aggregating in the upper 3–4 m in a zone coincident with maximum photosynthesis. The zooplankton also demonstrated a tendency to concentrate at about 2.5 m depth during the development of a weak midday thermocline. *Ceriodaphnia cornuta* has also been reported as a dominant species in both

Lakes Naivasha and Oloidien (Uku and Mavuti 1994). In a two-year study, Yasindi (2001) noted the presence of the ciliates *Dileptus*, *Paradileptus*, *Linostoma*, *Spirostomum*, *Stentor*, *Strombidium*, *Vorticella*, *Halteria* and other taxa with biomass peaking in October 1998.

Fossil ephippial capsules from sediments representing a 200-year record have demonstrated the variable development of eight *Daphnia* taxa through time, with only two species present in the modern lake (Mergeay et al. 2004, 2005, 2006a, b, 2007). The four most common species were *D. barbata*, *D. laevis*, *D. magna* and *D. pulex*, with a further four less common taxa, including *D. lumhotzi*, *D. curvirostris*, *D. longispina* and an indeterminate *Daphnia* species. Species composition has varied with time. For example, core data indicate a first invasion of Lake Naivasha by *D. pulex* around 1630, which then became extinct and was re-established during major lowstands in the early nineteenth century (Mergeay et al. 2006a, b, 2011; Verschuren and Russell 2009).

Sediments spanning about 120 years from Lake Oloidien include chironomids (*Procladius brevipeteolatus*, *Chironomus formosipennis*, *Michrochironomus deribae*, *Cladotanytarsus pseudomancus*) and cladocerans (*Alona diaphana*, *A. rectangula*, *Euryalona orientalis* and *Leydigia propinqua*) with 26 species reported as common (Verschuren et al. 2000b, 2004).

11.6.5 Invertebrates

Major changes in the oligochaetes and molluscs (Clark et al. 1989) in Lake Naivasha and Lake Oloidien have been attributed to variations in land use, changes in macrophytes and the introduction of new fish taxa. Studies by Raburu et al. (2002) indicated that oligochaetes in Lake Naivasha were dominated by *Branchiura sowerbyi* and *Limnodrilus hoffmeisteri* with each producing 7.4 and 0.6 g m^{-2} y^{-1} (dry weight), respectively. These oligochaetes represented 85% of all profundal macroinvertebrate benthic biomass, with the balance comprised of chironomids (8%), other oligochaetes (4%) and microturbellarian worms (3%). Five larval helminth parasites on *Oreochromis leucostictus* and *Tilapia zillii* were described by Aloo (2002) including the nematode, *Contracaecum* sp., the anthocephalan *Polyacanthorhynchus kenyensis*, the trematode *Clinostomum* sp. and two cestodes belonging to the genera *Amirthalingamia* and *Cyclustera.*

Variations in invertebrate composition have been recognised from different parts of the Naivasha aquatic system, mainly reflecting substrate type and water depth. For example, *Chironomus formosipennis* and *Procladius brevipetiolatus* form >10% of the chironomids in Lake Naivasha, whereas the dominants in Crescent Island Bay were, in decreasing order, *Dicrotendipes septemmaculatus*, *C. pseudomancus*, *Psectrocladius viridescens*, *P. brevipetiolatus* and *C. formosipennis* (Eggermont et al. 2007).

Rumes (2010) noted that ostracods were common in Lake Oloidien. Verschuren et al. (2000b) reported up to 58 ostracods in a short core: mainly *Physocypria capensis*, *Zonocypris costata* and *Potamocypris mastiogophora*.

11.6.6 Fish

Fossil *Aplocheilichthys antinorii* may have represented the only common fish in Lake Naivasha prior to twentieth-century fish introductions (Verschuren 1996; Harper et al. 1990; Hubble and Harper 2001a; Hickley et al. 2008) with this species probably becoming extinct by about 1962 (Elder et al. 1971). By 1971, the most common fish was *Oreochromis leucostictus* and related hybrid forms with the introduced cichlid *O. spilurus* having disappeared due to loss of breeding habitat (Siddiqui 1979). The commercial crayfish *Procambarus clarkii* was introduced in 1970 (Oluoch 1990; Harper et al. 2002a, b; Foster and Harper 2006) and became the main food source for larger *Micropterus salmoides* (Hickley et al. 1994; Smart et al. 2002).

Fish observed in Lake Naivasha and its rivers since 1990 include introduced *Oreochromis leucostictus*, *Tilapia zillii*, *Micropterus salmoides*, *Barbus amphigramma* (*paludinosus*) and *Poecilia reticulata* (Muchiri and Hickley 1991; Muchiri et al. 1994; Aloo 1996; Hubble and Harper 2001a; Hickley et al. 2002; Britton and Harper 2008; Gherardi et al. 2011; Aloo et al. 2013) with the latter taxon reported as no longer present (Njiru et al. 2015). The first three species were the basis of an important fishery, which had experienced very variable yields between 21 and 1150 t y^{-1} (Hickley et al. 2002). The fishery started in 1959 but had collapsed by the 1970s, recovering from 1974 to 1988, after which it performed poorly.

Fishing was banned in Lake Naivasha throughout 2001 and in subsequent summer months (Hickley et al. 2004a; Njiru et al. 2008; Kundu et al. 2010). However, a radical change in the fishery occurred after the introduction of the carp *Cyprinus carpio* (including subspecies *specularis* and *coiaceus*) around 1998 (Mutethya and Yongo 2021). This taxon started to appear in catches during 2002 and by May 2005 comprised >90% of commercial catches (Hickley et al. 2004b; Britton et al. 2006, 2007) and 97% by the end of 2009. This was also the main species, followed by *M. salmoides, O. niloticus* and *O. leucostictus*, affected by a fish kill in 2010 that was attributed to eutrophication (Njiru et al. 2015, 2017). The rainbow trout *Oncorhynchus mykiss* has also occasionally been reported in landings by local fishermen (Aloo et al. 2013) as has *Clarias gariepinus* (Yongo et al. 2013).

Many of the ecological changes have involved natural and human-induced lake level change, species introductions, papyrus clearances, the development of intensive cultivation, water abstraction for geothermal power and irrigation, and loss of inflow through damming of rivers (Harper et al. 1990, 1995, 2011; Harper 1991, 1992; Hubble and Harper 2001a; Everard and Harper 2002; Mavuti and Harper 2006; Odongo et al. 2014; Terer et al. 2014). Industrial effluents and urban and rural sewage loads entering the lake have increased as a result of population growth, with agro-chemical pollutants (fertilisers, pesticides) also increasing due to the expansion of a large-scale horticultural industry (Tang 1999; Gitahi et al. 2002; Becht et al. 2005). Ecotoxicological studies are limited, but an examination of Hg in Naivasha fish indicated levels within WHO guidelines (Campbell et al. 2003; Ochieng et al. 2007). Overall, Lake Naivasha has a dynamic history of major ecological changes, reflecting a lake with variable natural controls and increasing human impacts. In recent decades, the multiple environmental problems have, in turn, led to increasing interest in sustainable management of the lake ecosystem (Mathooko 2005; Otiang'a-Owiti and Oswe 2007; Harper et al. 2011).

11.7 Sedimentation

11.7.1 Sediment Facies and Environments

The sediments in Lake Naivasha derive from weathering, erosion and landslides in the Naivasha catchment with sediment transported mainly via the Malewa, Gilgil and Karati rivers (Syrén 1990; Kitaka et al. 2002a, b; Everard et al. 2002a, b; Mwaniki et al. 2017). Maina et al. (2018, 2019) noted variable sedimentation rates between 1996 and 2016, and between 1966 and 2016 (20- and 50-year periods), with the greatest build-up of sediments taking place in the southwest and eastern parts of the lake. In these areas maximum sediment accumulation increased by 0.55 and 1.9 m during the 20- and 50-year intervals with mean annual sedimentation rates of 0.25 and 0.56 m, respectively. The composition of the lacustrine sediments also varies spatially with six depositional sectors having been recognised: a central area, a delta front, as well as the northeast, northwest, southeast and southwest sectors (Tarras-Wahlberg et al. 2002). Organic mud dominates in most locations, with deltaic and prodeltaic massive brown clay and sandy mud prevailing near the Malewa River inflow and in the northeast bay area, near Crescent Island. In the latter sector, stiff grey clay is also present that may record low lake-level exposure (Tarras-Wahlberg et al. 2002).

Geochemically, the deposits are similar to the sediments introduced by rivers, but with increased sulfate, reflecting higher abundances of animals and plants in the lake that concentrate sulfur (Tarras-Wahlberg et al. 2002). The Lake Naivasha and Crescent Island Bay sediments contain 1.6–6.7% carbonates and 1.0–27.9% organic matter (OM), which compares with 0.3–2.5% $CaCO_3$ and 0.6–6.2% OM in the Malewa and Gilgil river sediments (Mwamburi 2013). Trace metal concentrations (Table 11.5) in sediments are generally similar to those in other basins with urban influences, although cadmium concentrations may be rising (Wandiga et al. 1983; Njogu et al. 2011; Mwamburi 2013, 2015). Ochieng et al. (2007) noted higher zinc values (229.6 μg g^{-1}) with trace metals in water samples (Table 11.3) falling within WHO guidelines, except for high Sn values (1050 μg L^{-1}).

Table 11.5 Trace metal concentrations (μg g^{-1}) in Lake Naivasha sediments

Source	Cd	Co	Cr	Cu	Mn	Ni	Pb	Sn	Zn
1	0.73	0.17	1.95	10.33	1118	11.69	25.34	23.39	229.6
2	–	–	0.17–4.67	<5.7	41.7–888.3	–	1.67–16.7	–	9–54
3	2–4	–	–	12–14	34–53	–	5–7	–	134–148
4	1–3	–	–	–	–	–	4.6–67.1	–	93–168
5	–	–	–	1.49–3.22	–	46.71–118.42	26.14–43.94	–	34.86–70.22

1: Ochieng et al. (2007); 2: Mwamburi (2013); 3: Wandiga et al. (1983); 4: Njogu et al. (2011); 5: Mwamburi (2015)

Magnetic susceptibility data indicate that clay, silt, sand and gravel have been introduced from eroding shorelines as well as inflowing streams, with gravel present up to 60 m offshore (Boar and Harper 2002). The papyrus fringes around the lake serve as sediment traps, intercepting both trace metals (Kamau et al. 2008) and clastic inputs before they reach open waters, where sedimentation rates have been estimated at 1–3 cm y^{-1} (Verschuren 1999a; Tarras-Wahlberg et al. 2002) and 2.3 cm y^{-1} (Walker et al. 2022). The movement of floating *Salvinia molesta* may drive floating pumice towards southern shorelines where it tends to accumulate (Åse 1987b).

Short cores (Fig. 11.9a) from Lake Naivasha, Lake Oloidien and Crescent Island Bay have been dated between 1871 and 1993, and are dominated by organic mud and clay (Verschuren 1996, 1999a) with a high water content (>95%) near the lake floor. Inorganic carbon is minimal in Lake Naivasha and Crescent Island Bay but ranges from 0.5% to 4.5% in Lake Oloidien (Verschuren 1999a). Similarly, carbonate in Lake Naivasha and Crescent Island Bay is minimal but increases in the closed basin of Oloidien (7–18%).

Five stratigraphic units occur in most Naivasha short cores (Fig. 11.9b; Verschuren 1999a, b). Unit 0, in Core NS91.1-S for example, represents a drought period (late 1700s until 1820–30) when oxidation and compaction affected the sediments. The overlying Unit 1 has low organic matter (OM) and formed during late nineteenth-century lowstands (Åse et al. 1986). OM increases in Unit 2, reflecting higher lake levels (~1890–1930) when lakes Oloidien and Naivasha were joined. Unit 3 has a lower OM content and was laid down during low to intermediate lake levels. The deposits tend to thicken towards basin centres, which implies sediment focusing. Nevertheless, the stratigraphy in Lake Oloidien and Crescent Island Bay are similar, which has been attributed to minimal disturbance by bottom turbulence due to shallow wave-bases, and to the generally gentle bottom gradients (<4%) that minimise gravity-driven sliding and slumping (Verschuren 1999a, b). In contrast, cores from Lake Naivasha are more variable as a result of wind-driven resuspension of bottom sediments.

Two major types of organic-rich facies have also been identified. In shallow parts of Lake Naivasha, organic-rich beds and peaty mud have formed from the roots and stems of the submerged macrophyte *Ceratophyllum demersum,* supplemented by

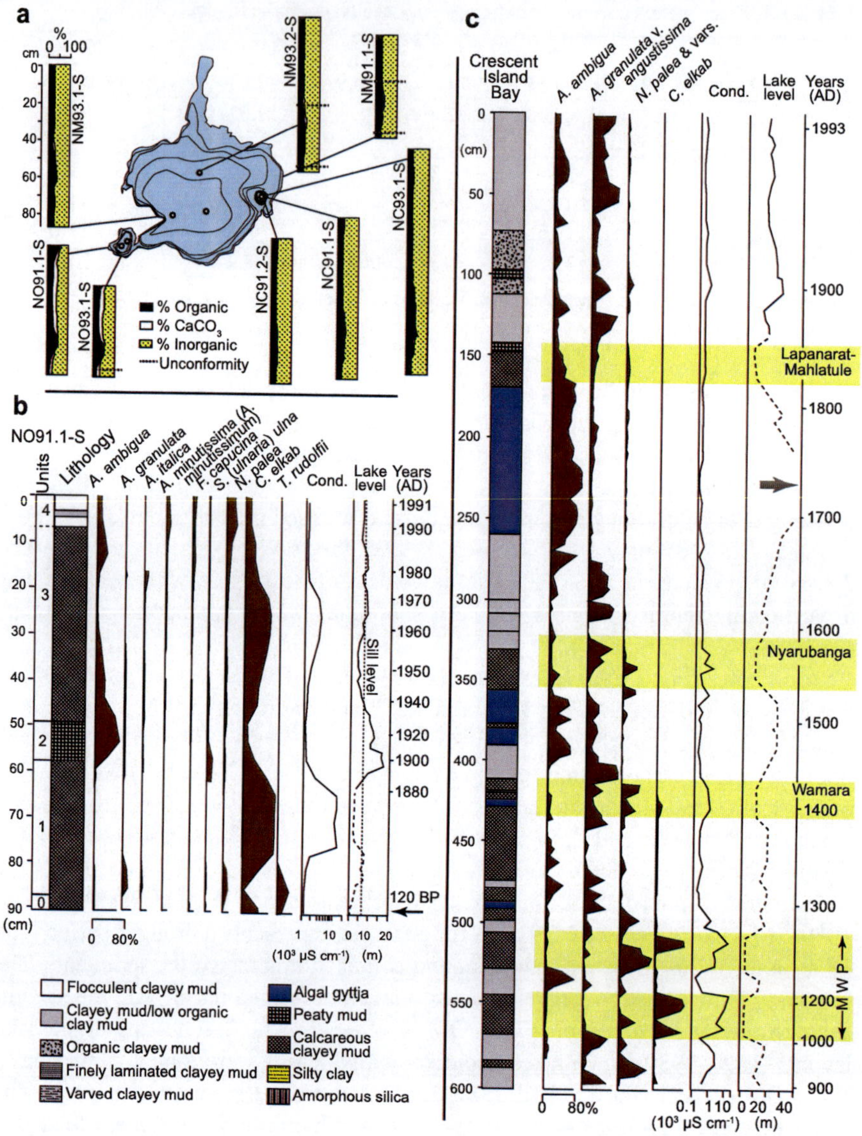

Fig. 11.9 Sediment and diatom stratigraphy in Lake Naivasha. (**a**) Organic and $CaCO_3$ content of Naivasha cores. After Verschuren (1999a). (**b, c**) Stratigraphy of selected diatoms, diatom-inferred conductivity and lake levels. (**b**) Lake Oloidien. After Verschuren et al. (1999b, 2000b). (**c**) Crescent Island Bay. After Verschuren et al. (2000a). Shaded areas reflect drought and low lake levels, named to the right. MWP is the Medieval Warm Period

papyrus debris. In contrast, layers of gelatinous, olive-green gyttja, in places alternating with brown flocculent mud, have accumulated under holomictic hemipelagic deep-water when siliciclastic inputs were more distally located (Verschuren et al. 2000a). Calcareous mud in Lake Oloidien has been related to evaporative concentration of dissolved salts (Verschuren et al. 2000a) in water with a high pH (9.0–9.2), supersaturation with respect to calcite, and low dissolved Ca/Na ratios (Gaudet and Melack 1981; Brierley et al. 1987; Verschuren 2001).

Grewe (2016) noted that Cl^-, F^-, Br^-, SO_4^{2-} and NO_3^- in short-core pore waters fluctuated considerably between different horizons with higher values below about 25 cm where Cl^- is dominant and H_2S content is below detection limits. Sulfate reduction ranged between 3.5 and 30.8 nmol/cm^3 × d. She also noted a close relationship between n-C_{17} alkane, long-chain n-alkane, phytosterol, and carboxylate ester biomarkers and the development of floriculture, the release of fertilizers to lake water, and the growth of algae.

11.7.2 Recent Sediment Stratigraphy

The recent history of Lake Naivasha and its satellite water bodies have been described using a variety of proxies (Marchant et al. 2018). Excellent diatom records indicate changing lake level and salinity (Fig. 11.9b, c) with conductivity reconstructions derived from salt-tolerant (e.g. *Cladotanytarsus pseudomancus*, *Chironomus formosipennis*) and halobiontic (e.g. *Microchironomus deribae*, *Kiefferulus disparilis*) chironomid taxa. Additional palaeoenvironmental studies have documented pigments (e.g. β-carotene, myxoxanthophyll, chlorophyll-*b*), plant debris and aquatic invertebrates (Mergeay et al. 2004, 2005, 2007, 2011; Verschuren et al. 1999b, 2004; Verschuren 2003).

Lake level fluctuations over the last two millennia compare well with reconstructions of past climates across equatorial Africa (Nash et al. 2016). The Crescent Island Crater basin was dry at ~200 AD, but this was followed by relatively high lake levels until ~600 AD with increased rainfall inferred around 900 AD (Verschuren et al. 2000a; Verschuren 2001). Diatom-based lake-level records from Crescent Island Bay have also been correlated with historical evidence for several major droughts in equatorial East Africa (Verschuren 2004). Low lake levels occurred during the Medieval Warm Period at around 1000–1270 A.D. (Fig. 11.9c), with three subsequent dry episodes at 1380–1420, 1560–1620 and 1760–1840 that have been linked to the Wamara, Nyarubanga and Lapanarat-Mahlatule droughts and widespread social disruptions (Verschuren et al. 2000a; Kiage and Liu 2006; Lüning et al. 2018).

Diatom reconstructions for recent (~1820–2000) sediments in central Lake Naivasha show three distinct zones (Stoof-Leichsenring et al. 2011). The lowermost Zone I (Fig. 11.10a) represents the period 1820–1896 and is dominated by littoral and epiphytic taxa with significant percentages of planktonic and littoral planktonic species. Zone II (1896–1930) is transitional with the planktonic littoral diatom

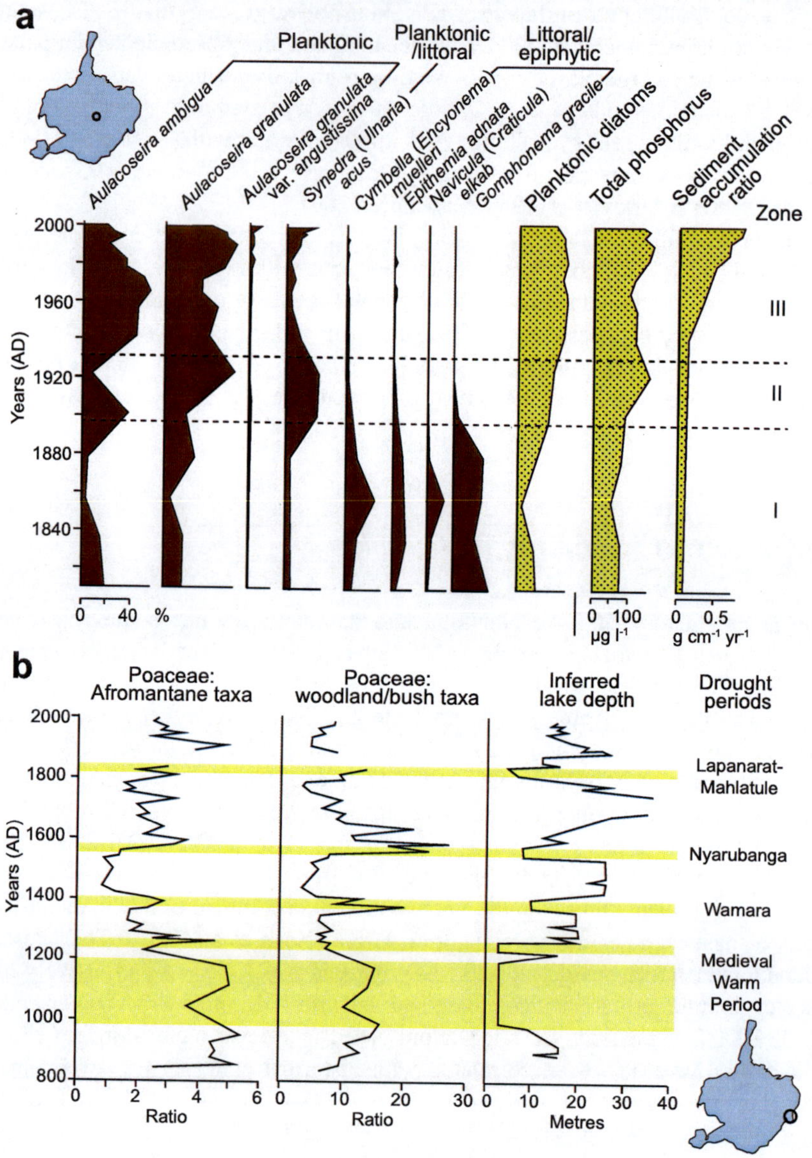

Fig. 11.10 Diatom and pollen stratigraphies for Lake Naivasha and Crescent Island Bay. (**a**) Lake Naivasha diatom record showing three major zones that characterise the last two centuries. After Stoof-Leichsenring et al. 2011). (**b**) Crescent Island Bay pollen stratigraphy and historical drought episodes. After Lamb et al. (2003)

Synedra acus dominant. Zone III (1938–2000) is comprised of planktonic taxa (mainly *A. granulata* and *A. ambigua*) with abundant *Synedra* (*Ulnaria*) *acus*, which dominates in the most recent sediments. Conductivity reconstructions indicate a mean of 300 μS cm^{-1} (peaking at 731 μS cm^{-1}) for Zone I and 150–185 μS cm^{-1} for Zones II and III. Transfer function values, for phosphorus, increase near the top of the core to a maximum of 164 μg L^{-1}, reflecting recent eutrophication. Increased sedimentation rates at the top of the core document recent catchment disturbance.

Reconstructions using pollen (Fig. 11.10b) indicate variations in the ratio of Poaceae (grasses) to woody Afromontane and woodland plants, with peak ratios interpreted as representing times of low rainfall (Lamb et al. 2003). The same data sets also document a decrease in lower montane-forest elements after ~1700, which appears to coincide with human population expansion and the introduction of *Zea mays* (maize) to the area. Herbaceous and exotic tree pollen increased through the twentieth century.

11.8 African Humid Period and Holocene Lacustrine Sediments in the Lake Naivasha Basin

Studies by Butzer et al. (1972) documented a major Early Holocene expansion of Lake Naivasha, which has been cited as an example of an amplifier lake (Trauth et al. 2010). These typically occupy rift settings with steep margins and climates characterised by high precipitation on rift shoulders and low rainfall over the rift floor. Amplifier lakes are very sensitive to climate change and show major fluctuations in water level, with evidence for such changes, in the Naivasha Basin, being preserved by highstand beach gravels and diatomaceous sediments (Washbourn-Kamau 1975).

Richardson (1972), Richardson and Richardson (1972) and Richardson and Dussinger (1986) described the Late Pleistocene and Early Holocene history of the area based on cores from Crescent Island Bay and the central parts of Lake Naivasha (Fig. 11.11). They noted that the Late Pleistocene (20,000–13,000 y BP) was a time of low water levels with fresh conditions (alkalinity 1–15 meq L^{-1}) and that this period was characterised by a variety of planktonic and tychoplanktonic diatoms. The lake briefly became a weedy dilute pond with *Eunotia* spp., *Pinnularia* spp., *Navicula zanonii* and *Cocconeis placentula* var. *euglypta* (Core 2, Fig. 11.11). A short-lived rise in levels took place ~12,500 y BP when the palaeolake had a similar level to the modern lake. However, the Late Pleistocene lake had a different ecology with dominance by *Stephanodiscus rotula* var. *minutula*, *Aulacoseira granulata* and *Thalassiosira rudolfi*. It was also more alkaline and richer in nutrients and suspended clay and silt. During the major Early Holocene expansion (10,000–5500 y BP), Lake Naivasha became fresh and overflowed to the south through the Njorowa Gorge (Fig. 11.2e), which was being gradually incised (Washbourn-Kamau 1975). Diatoms included several *Aulacoseira* spp. and a variety of planktonic taxa (Cores 1 and 2, Fig. 11.11). After about 4145 y BP, the dominant diatoms switched from *A. agassizii* to *A. ambigua* and then *A. distans* and *Eunotia* spp. Lake levels in the

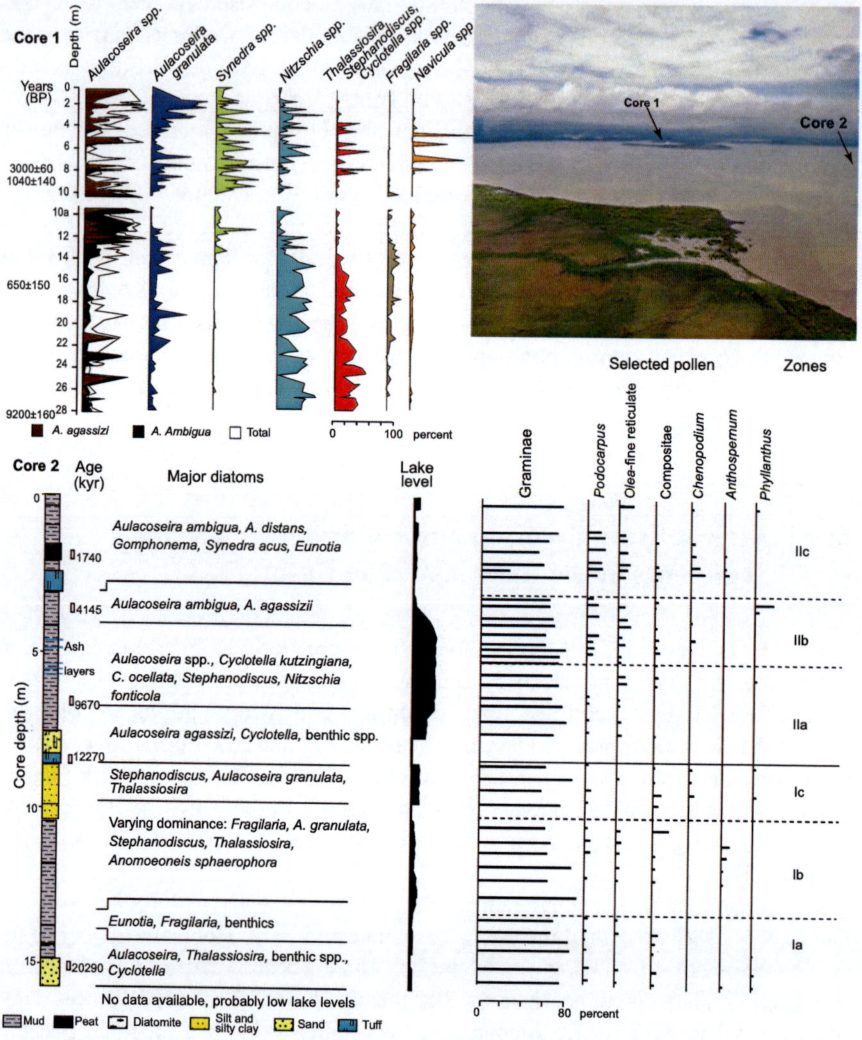

Fig. 11.11 Holocene and Late African Humid Period diatom and pollen stratigraphies for the Naivasha Basin. Core 1 diatom data based on Richardson and Richardson (1972). Core 2 diatom data based on Richardson and Dussinger (1986). Core 2 pollen record after Maitima (1991)

Naivasha palaeolake reached their lowest heights about 3000 years ago. Lake Naivasha broadly assumed its modern state (shallow, fresh, abundant macrophytes) from about 1700 years ago, with varied dominance by *A. ambigua*, *S. acus* and *Gomphonema* taxa.

Maitima (1991) provided data from Core 2 from the main portion of Lake Naivasha in which he recognised six pollen zones and subzones. Gramineae generally dominate, but with *Podocarpus*, *Olea*, *Aeschynomene*, *Cheonopodium*,

Amaranthus, *Liliaceae* and *Rapanea* becoming variously dominant. Maitima (1991) observed that open grasslands dominated before 20,290 y BP and through to about 12,000 y BP, and inferred more arid conditions than today during that period. However, moderate increases in montane and lowland forest pollen implied a wetter climate within the interval 17,000–15,000 y BP. Trees and forests expanded into lower regions from 12,000 to 6500 y BP, with a subsequent decline after 6000 y BP. Shallow aquatic plants became abundant in Lake Naivasha and catchment vegetation became similar to that of more recent times after ~4000 y BP.

Lamb et al. (2003) constructed a detailed pollen record for the last 1110 years from a 625-cm core in Crescent Island Bay. They note that initially low *Podocarpus* percentages increased after about 1280 AD. A second major transition was associated with the first appearance of *Zea mays* (maize) at 1650 AD, together with a peak abundance in *Ricinus communis*, suggesting the onset of cultivation. A third important pollen change took place at ~1920 AD, with an increase in introduced *Pinus* and *Cupressus*.

Beach gravels and barrier beaches associated with the Late Pleistocene/Early Holocene palaeolake suggest that it reached heights of up to 2000 m above sea level (Washbourn-Kamau 1975). That palaeolake would have been about 180 m deep, 685 km^2 in extent with a volume of about 59 km^3. Overflow was likely to the south via the Njorowa Gorge (Kamau 1977; Bergner et al. 2003) although strong differentiation in modern *C. laevigatus* populations suggests a significant "ancestral genetic barrier" between Lake Magadi and rift lakes from Naivasha northwards (Mwaniki et al. 2019). The waters of the Naivasha palaeolake subsequently dropped until the lake surface stabilized at ~5700 y BP (Richardson and Richardson 1972) at a height of about 1950 m, the highest elevation along the floor of the Njorowa Gorge (Washbourn-Kamau 1975). During the Late Holocene, lake levels continued to fall to their historical range of 1884–1895 m.

The Holocene climate-driven changes in Lake Naivasha also impacted the activities and stone technologies of humans. Backed-blades and scrapers, for example, varied in size during earlier wet periods (Goldstein et al. 2022). Hunter-gatherers occupied both highlands and lower-elevation grasslands as rains decreased after about 6 ka (Ambrose 1998; Ambrose 2001, 2002; Goldstein et al. 2022) with differences in tools subsequently developing between these areas. Herders entered the region from the north by ~5 ka with pastoralism developing rapidly with increasing rains and grasslands at about 3.2 ka (Ambrose and Sikes 1991; Prendergast et al. 2019).

References

Adams CS, Boar RR, Hubble DS, Gikungu M, Harper DM, Hickley P, Tarras-Wahlberg N (2002) Dynamics and ecology of exotic tropical species in floating plant mats: Lake Naivasha, Kenya. Hydrobiologia 44:115–122

Allen FJ, Darling WG, Burgess WG (1989) Geothermics and hydrogeology of the southern part of the Kenya Rift Valley with emphasis on the Magadi-Nakuru area. BGS Re Rep SD/89/1

Aloo PA (1996) Anthropogenic impacts on fisheries resources at Lake Naivasha. In: Johnson TC, Odada EO (eds) The limnology, climatology and paleoclimatology of the East African Lakes. Gordon and Breach, Amsterdam, pp 325–335

Aloo PA (2002) A comparative study of helminth parasites from the fish *Tilapia zillii* and *Oreochromis leucostictus* in Lake Naivasha and Oloidien Bay, Kenya. J Helminthol 76:95–104

Aloo PA, Oyugi DO, Morara GN, Owuor MA (2013) Recent changes in fish communities of the equatorial Lake Naivasha, Kenya. Int J Fish Aquac 5:45–54

Ambrose SH (1998) Chronology of the Later Stone Age and food production in East Africa. J Archaeol Sci 25:377–392

Ambrose SH (2001) East African Neolithic. In: Encyclopedia of prehistory. Springer, Boston, pp 97–109

Ambrose SH (2002) Small things remembered: origins of early microlithic industries in Sub-Saharan Africa. Archeol Pap Am Anthropol Assoc 12:9–29

Ambrose SH, Sikes NE (1991) Soil carbon isotope evidence for Holocene habitat change in the Kenya Rift Valley. Science 253:1402–1405

Åse LE (1987a) A note on the water budget of Lake Naivasha, Kenya — especially the role of *Salvinia molesta* Mitch and *Cyperus papyrus* L. Geogr Ann 69A:415–429

Åse LE (1987b) A note on ridges of pumice formed through the influence of floating *Salvinia* in Lake Naivasha, Kenya. Geogr Ann 69A:283–288

Åse LE, Sernbo K, Syren P (1986) Studies of Lake Naivasha, Kenya and its drainage area. Forskn Naturgeogr Instit Stockholm 63

Awange JL, Kyalo Kiema JB (2013) Water resources. In: Environmental geoinformatics: monitoring and management (Environmental science and engineering). Springer, Berlin, pp 341–380

Awange JL, Forootan E, Kusche J, Kiema JBK, Omondi PA, Heck B, Fleming K, Ohanya SO, Gonçalves RM (2013) Understanding the decline of water storage across the Ramsar-Lake Naivasha using satellite-based methods. Adv Water Resour 60:7–23

Ayenew T, Becht R (2008) Comparative assessment of the water balance and hydrology of selected Ethiopian and Kenyan rift lakes. Lakes Reserv 13:81–196

Ayenew T, Becht R, van Lieshout A, Gebreegziabher Y, Legesse D, Onyando J (2007) Hydrodynamics of topographically closed lakes in the Ethio-Kenyan Rift: the case of lakes Awassa and Naivasha. J Spatial Hydrol 7:81–100

Baker BH, Williams LAJ, Miller JA, Fitch FJ (1971) Sequence and geochronology of the Kenya rift volcanics. Tectonophysics 11:191–215

Baker BH, Mitchell JG, Williams LAJ (1988) Stratigraphy, geochronology and volcano-tectonic evolution of the Kedong-Naivasha-Kinangop region, Gregory Rift Valley, Kenya. J Geol Soc Lond 145:107–116

Ballot A, Krienitz L, Kotut K, Wiegand C, Pflugmacher S (2005) Cyanobacteria and cyanobacterial toxins in the alkaline crater lakes Sonachi and Simbi, Kenya. Harmful Algae 4:139–150

Ballot A, Kotut K, Novelo E, Krienitz L (2009) Changes of phytoplankton communities in Lakes Naivasha and Oloidien, examples of degradation and salinization of lakes in the Kenyan Rift Valley. Hydrobiologia 632:359–363

Beadle L (1932a) Scientific results of the Cambridge Expedition to the East African Lakes, 1930–31. Part 3. Observations on the bionomics of some East African swamps. J Lin Soc Lond Zool 28:135–155

Beadle L (1932b) Scientific results of the Cambridge Expedition to the East African Lakes, 1930–31. Part 4. The waters of some East African Lakes in relation to their flora and fauna. J Lin Soc Lond Zool 38:157–211

Becht R, Harper D (2002) Towards an understanding of human impact upon the hydrology of Lake Naivasha, Kenya. Hydrobiologia 488:1–11

Becht R, Nyaoro JR (2006) The influence of groundwater on lake–water management: the Naivasha case. In: Odada EO, Olago DO, Ochola W, Ntiba M, Wandiga S, Gichuki N, Oyieke H (eds) Proc 11th World Lakes Conf, vol 2. Nairobi, pp 384–388

Becht R, Odada EO, Higgins S (2005) Lake Naivasha: experience and lessons learned. In: Managing lakes and basins for sustainable use, a report for lake basin managers and stakeholders. International Lake Environment Committee Foundation (ILEC), Kusatsu, pp 277–298

Becht R, Mwango F, Muno FA (2006) Groundwater links between Kenyan Rift lakes. In: Odada EO, Olago DO, Ochola W, Ntiba M, Wandiga S, Gichuki N, Oyieke H (eds) Proc 11th World Lakes Conf, vol 2. Nairobi, pp 7–14

Bergner AGN (2003) Lake-level fluctuations and Late Quaternary climate change in the Central Kenya Rift. PhD thesis, Universität Potsdam

Bergner AGN, Trauth MH (2004) Comparison of the hydrological and hydrochemical evolution of Lake Naivasha (Kenya) during three highstands between 175 and 60 kyr BP. Palaeogeogr Palaeoclimatol Palaeoecol 215:17–36

Bergner AGN, Trauth MH, Bookhagen B (2003) Paleoprecipitation estimates for the Lake Naivasha basin (Kenya) during the last 175 ky using a lake-balance model. Glob Planet Change 36:117–136

Boar RR (2006) Responses of a fringing *Cyperus papyrus* L. swamp to changes in water level. Aquat Bot 84:85–92

Boar RR, Harper DM (2002) Magnetic susceptibilities of lake sediment and soils on the shoreline of Lake Naivasha, Kenya. Hydrobiologia 488:81–88

Bowen NL (1935) Ferrosilite as a natural mineral. Am J Sci 30:481–494

Bowen NL (1937) Recent high temperature research in silicates and its significance in igneous geology. Am J Sci 32:1–21

Brierley B, Harper DM, Thomas R (1987) Water chemistry and phytoplankton studies at Lake Naivasha: short-term spatial and temporal variations. In: Harper DM (ed) Studies on the Lake Naivasha ecosystem 1982-84. University of Leicester, Leicester, pp 35–72

Britton JR, Harper DM (2008) Juvenile growth of two tilapia species in lakes Naivasha and Baringo, Kenya. Ecol Freshw Fish 17:481–488

Britton JR, Ng'eno JBK, Lugonzo J, Harper DM (2006). Can an introduced, non-indigenous species save the fisheries of Lakes Baringo and Naivasha, Kenya? In: Odada EO, Olago DO, Ochola W, Ntiba M, Wandiga S, Gichuki N, Oyieke H (eds) Proc 11th World Lakes Conf, vol 2. Nairobi, pp 568–572

Britton JR, Boar RR, Grey J, Foster J, Lugonzo J, Harper DM (2007) From introduction to fishery dominance: the initial impacts of the invasive carp *Cyprinus carpio* in Lake Naivasha, Kenya, 1999 to 2006. J Fish Biol 71:239–257

Burgis MJ, Mavuti KM (1987) The Gregory Rift. In: Burgis MJ, Symoens JJ (eds) African wetlands and shallow water bodies. Institut Français de Recherche Scientifique pour le Développement en Coopération, p 211

Busk HG (1939) Block diagram in isometric projection of the Great Rift Valley from Nakuru to Lake Magadi with explanatory note. Q J Geol Soc Lond 95:231–233

Butturini A, Amalfitano S, Herzsprung P, Lechtenfeld OJ, Venturi S, Olaka LA, Pacini N, Harper DM, Tassi F, Fazi S (2020) Dissolved organic matter in continental hydro-geothermal systems: insights from two hot springs of the East African Rift Valley. Water 12:3512–3512

Butzer KW, Isaac GL, Richardson JL, Washbourn-Kamau C (1972) Radiocarbon dating of East African lake levels. Science 175:1069–1076

Campbell LM, Osano O, Hecky RE, Dixon DG (2003) Mercury in fish from three rift valley lakes (Turkana, Naivasha and Baringo), Kenya, East Africa. Environ Pollut 125:281–286

Cerling TE (1979) Paleochemistry of Plio-Pleistocene Lake Turkana, Kenya. Palaeogeogr Palaeoclimatol Palaeoecol 27:247–285

Childress RB, Bennun LE, Harper DM (2002) Population changes in sympatric Great and Long-tailed Cormorants (*Phalacocorax carbo* and *P. africanus*): the effects of niche overlap or environmental change? Hydrobiologia 44:163–170

Clark F, Beeby A, Kirby P (1989) A study of the macro-invertebrates of Lakes Naivasha, Oloidien and Sonachi, Kenya. Rev Hydrobiol Trop 22:21–33

Clarke MCG, Woodhall DG, Allen D, Darling G (1990) Geological, volcanological and hydrogeological controls on the occurrence of geothermal activity in the area surrounding Lake Naivasha, Kenya. Ministry of Energy, Republic of Kenya, Nairobi
Cocquyt C (2004) *Staurophora caljonii* spec. nov. (Bacillariophyceae, Anomoeoneidaceae), a new halophilic diatom species from sub-recent lake deposits in Kenya. Hydrobiologia 511:37–46
Cocquyt C, Verschuren D (2023) Checklist of the diatoms (Bacillariophyta) from Lake Naivasha, Kenya, with some historical notes. PhytoKeys 224:101–174
Cole SM (1950) An outline of the geology of Kenya. Pitman, London
Cole SM (1954) The prehistory of East Africa. Am Anthropol 56:1026–1050
Cole S, Leakey LSB (eds) (1952) Proceedings of the Pan-African Congress on prehistory, 1947. Philosophical Library, New York
Collie GL (1912) Plateaux of British East Africa. Geol Soc Am Bull 13:297–316
Darling WG, Allen DJ, Armannsson H (1990) Indirect detection of outflow from a Rift Valley lake. J Hydrol 113:297–305
Darling WG, Gizaw B, Arusei M (1996) Lake-groundwater relationships and fluid-rock interaction in the East African Rift Valley: isotopic evidence. J Afr Earth Sci 22:423–431
De Jong T (2011) Review on riverwater resource monitoring and allocation planning in the Lake Naivasha Basin, Kenya. BSc thesis, University of Wageningen
Eggermont H, Deyne P, Verschuren D (2007) Spatial variability of chironomid death assemblages in the surface sediments of a fluctuating tropical lake (Lake Naivasha, Kenya). J Paleolimnol 38:309–328
Elder HY, Garrod DT, Whitehead PJP (1971) Natural hybrids of the African cichlid fishes *Tilapia spirulus nigra* and *T. leucosticta*, a case of hybrid introgression. Biol J Linn Soc 3:103–146
Epp LS, Stoof-Leichsenring KR, Trauth MH, Tiedemann R (2011) Molecular profiling of diatom assemblages in tropical lake sediments using taxon-specific PCR and Denaturing High-Performance Liquid Chromatography (PCR-DHPLC). Mol Ecol Resour 11:842–853
Evans JH (1962) Some records and forms of algae in Central East Africa. Hydrobiologia 20:59–86
Everard M, Harper DM (2002) Towards the sustainability of the Lake Naivasha Ramsar site and its catchment. Hydrobiologia 488:191–203
Everard M, Kuria A, Macharia M, Vale JA, Harper DM (2002a) Aspects of the biodiversity of the rivers in the Lake Naivasha catchment. Hydrobiologia 488:43–55
Everard M, Vale JA, Harper DM, Tarras-Wahlberg H (2002b) The physical attributes of the Lake Naivasha catchment rivers. Hydrobiologia 488:13–25
Farah HO, Bastiaanssen WGM, Feddes RA (2004) Evaluation of the temporal variability of the evaporative fraction in a tropical watershed. Int J Appl Earth Obs Geoinf 5:129–140
Fayos CB (2002) Competition over water resources: analysis and mapping of water related conflicts in the catchment of Lake Naivasha (Kenya). MSc thesis, International Institute for Geo-information Science and Earth Observation, Enschede
Fischer GA (1885) Das Masai-Land (Ost-Aequatorial-Africa). Separat-Abdruck aus den Mittheilungen der Geographischen Gesellschaft, Hamburg
Foster J, Harper DM (2006) The alien Louisianan Red Swamp Crayfish *Procambarus clarkii* Girard in Lake Naivasha, Kenya 1999–2003. Freshw Crayfish 15:195–202
Gaudet JJ (1975) Mineral concentration in papyrus in various African swamps. J Ecol 63:483–491
Gaudet JJ (1976) Nutrient relationships in the detritus of a tropical swamp. Arch Hydrobiol 78:231–239
Gaudet JJ (1977a) Natural drawdown on Lake Naivasha, Kenya, and the formation of papyrus swamps. Aquat Bot 3:1–47
Gaudet JJ (1977b) Uptake, accumulation, and loss of nutrients by papyrus in tropical swamps. Ecology 58:415–422
Gaudet JJ (1979) Seasonal changes in nutrients in a tropical swamp: North Swamp, Lake Naivasha, Kenya. J Ecol 67:953–981
Gaudet JJ (1980) Papyrus and the ecology of Lake Naivasha. Nat Geogr Soc Res Rep 1:267–272
Gaudet JJ, Melack JM (1981) Major ion chemistry in a tropical African lake basin. Freshw Biol 11:309–333

Ghansah B, Harper DM, Forkuo EK, Appoh R (2016) Using remote sensing techniques for monitoring ecological changes in lakes: case study of Lake Naivasha. Int J Eng Res Appl 6:1–12

Gherardi F, Britton JR, Mavuti KM, Pacini N, Grey J, Tricarico E, Harper DM (2011) A review of allodiversity in Lake Naivasha, Kenya: developing conservation actions to protect East African lakes from the negative impacts of alien species. Biol Conserv 144:2585–2596

Gibson W (1893) Geological sketch of Central East Africa. Geol Mag 10:561–563

Gitahi SM, Harper DM, Muchiri SM, Tole MP, Ng'ang'a RN (2002) Organochlorine and organophosphorus concentrations in water, sediment, and selected organisms in Lake Naivasha (Kenya). Hydrobiologia 488:123–128

Githaiga KB, Njuguna SM, Gituru RW, Yan X (2021) Water quality assessment, multivariate analysis and human health risks of heavy metals in eight major lakes in Kenya. J Environ Manag 297:113410

Gitonga MS (1999) Study of long term waterbalance of Lake Naivasha, Kenya. MSc thesis, International Institute for Geo-information Science and Earth Observation, Enschede

Goldstein ST, Shipton C, Miller JM, Ndiema E, Boivin N, Petraglia M (2022) Hunter-gatherer technological organization and responses to Holocene climate change in coastal, lakeshore, and grassland ecologies of eastern Africa. Quat Sci Rev 280:107390

Gorges GH (1900) A journey from Lake Naivasha to the Victoria Nyanza. Geogr J 16:78–89

Government of Kenya [GoK] and UNDP (2021) Rising water levels in Kenya's Rift valley Lakes, Turkwel Gorge Dam and Lake Victoria: a scoping report. Government of Kenya, Nairobi

Gowlett JAJ (1980) Acheulean sites in the Central Rift Valley, Kenya. Proc Pan-Afr Congr Prehist Quat Stud 8:213–217

Gregory JW (1896) The Great Rift Valley. John Murray, London

Gregory JW (1921) The rift valleys and geology of East Africa. Seeley Service, London

Grewe S (2016) Hydro- and biogeochemical investigations of lake sediments in the Kenyan Rift Valley. PhD thesis, Universität Potsdam

Griffiths P (1977) The geology of the area around Lake Hannington and the Perkerra River, Rift Valley Province, Kenya. PhD thesis, University of London

Griffiths PS, Gibson IL (1980) The geology and petrology of the Hannington Trachyphonolite formation, Kenya Rift Valley. Lithos 13:43–53

Harper DM (1991) Primary production in Lake Naivasha, Kenya. Verh Int Verein Limnol 24:1112–1116

Harper DM (1992) The ecological relationships of aquatic plants at Lake Naivasha, Kenya. Hydrobiologia 232:65–71

Harper DM (2005) A tale of two lakes. Earthwatch J (Nov):14–16

Harper DM (2006) The sacrifice of Lake Naivasha. Nairobi, Swara Mag (J E Afr Wildl Soc) 29:27–37

Harper D, Mavuti K (1996) Freshwater wetlands and marshes. In: McClanahan TR, Young TP (eds) East African ecosystems and their conservation. Oxford University Press, Oxford, pp 217–239

Harper DM, Mavuti K (2004) Lake Naivasha, Kenya: ecohydrology to guide the management of a tropical protected area. Ecohydrol Hydrobiol 4:287–305

Harper DM, Mavuti KM, Muchiri SM (1990) Ecology and management of Lake Naivasha, Kenya, in relation to climatic change, alien species introductions and agricultural development. Environ Conserv 17:328–335

Harper DM, Adams C, Mavuti K (1995) The aquatic plant communities of the Lake Naivasha wetland, Kenya: pattern, dynamics and conservation. Wetl Ecol Manag 3:111–123

Harper DM, Harper MM, Virani MA, Smart A, Childress RB, Adatia R, Henderson I, Chege B (2002a) Population fluctuations and their causes in the African Fish Eagle (*Haliaeetus vocifer* (Daudin)) at Lake Naivasha, Kenya. Hydrobiologia 488:171–180

Harper DM, Smart AC, Coley S, Schmitz S, Gouder de Beauregard A-C, North R, Adams C, Obade P, Kamau M (2002b) Distribution and abundance of the Louisiana red swamp crayfish *Procambarus clarkii* Girard at Lake Naivasha, Kenya between 1987 and 1999. Hydrobiologia 488:143–151

Harper DM, Morrison EH, Macharia MM, Mavuti KM, Upton C (2011) Lake Naivasha, Kenya: ecology, society and future. Fr Rev 4:89–114

Hecky RE, Kilham P (1973) Diatoms in alkaline, saline lakes: ecology and geochemical implications. Limnol Oceanogr 18:53–71

Henderson IG, Harper DM (1992) Bird distribution and habitat structure on Lake Naivasha, Kenya. Afr J Ecol 30:223–232

Herrnegger M, Stecher G, Schwatke C, Olang L (2021) Hydroclimatic analysis of rising water levels in the Great Rift Valley lakes of Kenya. J Hydrol Reg Stud 36:100857

Hickley P, North E, Muchiri SM, Harper DM (1994) The diet of largemouth bass, *Micropterus salmoides*, in Lake Naivasha, Kenya. J Fish Biol 44:607–619

Hickley P, Bailey R, Harper DM, Kundu R, Muchiri M, North R, Taylor A (2002) The status and future of the Lake Naivasha fishery, Kenya. Hydrobiologia 488:181–190

Hickley P, Muchiri M, Boar R, Britton R, Adams C, Gichuru N, Harper DM (2004a) Habitat degradation and subsequent fishery collapse in Lakes Naivasha and Baringo, Kenya. Int J Ecohydrol Hydrobiol 4:503–517

Hickley P, Muchiri SM, Britton JR, Boar RR (2004b) Discovery of carp (*Cyprinus carpio*) in the already stressed fishery of Lake Naivasha, Kenya. Fish Manag Ecol 11:139–142

Hickley P, Muchiri M, Britton JR, Boar R (2008) Economic gain versus ecological damage from the introduction of non-native freshwater fish: case studies from Kenya. Open Fish Sci J 1:36–46

Hogeboom HJ (2013) On the influence of groundwater abstractions on Lake Naivasha's water level. MSc thesis, University of Twente

Hogeboom RH, van Oel PR, Krol MS, Booij MJ (2015) Modelling the influence of groundwater abstractions on the water level of Lake Naivasha, Kenya under data-scarce conditions. Water Resour Manag 29:4447–4463

Hubble DS, Harper DM (2000) Top-down biological controls on tropical lake productivity. Lakes Reserv 5:187–194

Hubble DS, Harper DM (2001a) What defines a "healthy" lake? Evidence from Lake Naivasha, Kenya. Aquat Ecosyst Health 4:243–250

Hubble DS, Harper DM (2001b) Impact of light regimen and self-shading by algal cells on primary productivity in the water column of a shallow tropical lake (Lake Naivasha, Kenya). Lakes Reserv 6:143–150

Hubble DS, Harper DM (2002a) Phytoplankton community structure and succession in the water column of a shallow tropical lake (Lake Naivasha, Kenya). Hydrobiologia 488:89–98

Hubble DS, Harper DM (2002b) Nutrient control of phytoplankton production in Lake Naivasha, Kenya. Hydrobiologia 488:99–105

Jenkin PM (1932) Reports on the Percy Sladen expedition to some rift valley lakes in Kenya 1929. I. Introductory account of the biological survey of five freshwater and alkaline lakes. Ann Mag Nat Hist 9:533–553

Jenkin PM (1936) Reports on the Percy Sladen expedition to some rift valley lakes in Kenya 1929. VII. Summary of the ecological results. Ann Mag Nat Hist 18:133–181

Johnson VJ (2015) Assessment of water quality trends using remote sensing data: case study of Lake Naivasha in the rift valley of Kenya. MSc thesis, University of Nairobi

Johnson GM, Harper DM, Mavuti KM (1999) Information for the sustainable management of shallow lakes: Lake Naivasha, Kenya. In: Harper DM, Brown AG (eds) The sustainable management of tropical catchments. Wiley, Chichester, pp 259–274

Jones MB, Humphries SW (2002) Impacts of the C4 sedge *Cyperus papyrus* L. on carbon and water fluxes in an African wetland. Hydrobiologia 488:107–113

Jones WB, Lippard SJ (1979) New age determinations and the geology of the Kenya Rift-Kavirondo Rift junction, W Kenya. J Geol Soc Lond 136:693–704

Jones MB, Muthuri FM (1984) The diurnal course of plant water potential, stomatal conductance and transpiration in a papyrus (*Cyperus papyrus* L.) canopy. Oecologia 63:252–255

Jones MB, Muthuri FM (1985) The canopy structure and microclimate of papyrus (*Cyperus papyrus*) swamps. J Ecol 73:481–491

Junginger A (2011) East African climate variability on different time scales: the Suguta Valley in the African-Asian Monsoon. PhD thesis, Universität Potsdam

Kalff J, Watson S (1986) Phytoplankton and its dynamics in two tropical lakes, a tropical and temperate zone comparison. Hydrobiologia 138:161–176

Källqvist T (1987) Primary production and phytoplankton in Lake Baringo and Lake Naivasha, Kenya. Norwegian Institute of Water Research Report

Kamau CK (1977) The Ol Njorowa Gorge, Lake Naivasha basin, Kenya. In: Greer DC (ed) Proc Int Conf Desertic Terminal Lakes. Utah Water Research Laboratory, Logan, Utah, pp 297–307

Kamau JN, Gachanja A, Ngila JC, Kazungu JM, Zhai M (2008) Anthropogenic and seasonal influences on the dynamics of selected heavy metals in Lake Naivasha, Kenya. Lakes Reserv 13:145–154

Karanja JM (2022) Effects of water level rise on riparian areas of Lake Naivasha, Kenya. MSc thesis, Kenyatta University

Kariuki P (2011) Kenyan wetlands – an indicator of climate change and human disturbance. African Conservation Centre Report

Keyombe JL, Waithaka E (2017) Analysis of some aspects of water quality of Lake Naivasha. Int J Chem Life Sci 6:2001–2005

Kiage L, Liu K (2006) Late Quaternary paleoenvironmental changes in East Africa: a review of multiproxy evidence from palynology, lake sediments, and associated records. Prog Phys Geogr 30:633–658

Kilham P (1971) Biogeochemistry of African lakes and rivers. PhD thesis, Duke University, Durham, North Carolina

Kilham P, Hecky RE (1973) Fluoride: geochemical and ecological significance in East African waters and sediments. Limnol Oceanogr 18:932–945

Kitaka N (1991) Phytoplankton productivity in Lake Naivasha. MSc thesis, University of Nairobi

Kitaka N, Harper DM, Mavuti KM (2002a) Phosphorus inputs to Lake Naivasha, Kenya, from its catchment and trophic state of the lake. Hydrobiologia 44:73–80

Kitaka N, Harper DM, Mavuti KM, Pacini N (2002b) Chemical characteristics, with particular reference to phosphorus of the rivers draining into Lake Naivasha, Kenya. Hydrobiologia 44:57–71

Krienitz L, Dadheech PK, Kotut K (2012) Mass developments of the cyanobacteria *Anabaenopsis* and *Cyanospira* (Nostocales) in the soda lakes of Kenya: ecological and systematic implications. Hydrobiologia 703:79–93

Krienitz L, Dadheech PK, Fastner J, Kotut K (2013) The rise of potentially toxin producing cyanobacteria in Lake Naivasha, Great African Rift Valley, Kenya. Harmful Algae 27:42–51

Kubai R, Kandie R (2014) Structural geology of Eburru-Badlands geothermal prospect. Proc 5 African Rift Geothermal Conference, Arusha, Oct 2014

Kubo BM (2004) Environmental management of Olkaria geothermal power project, Kenya. Paper presented at the African Rift Geothermal Conference, Nairobi, Kenya. UN Univ Rep, pp 221–247

Kuhn A, van Oel PR, Meins FM (2012) The Lake Naivasha Hydro-Economic Basin Model (LANA-HEBAMO). Technical documentation. Institute for Food and Resource Economics, Netherlands

Kundu R, Aura CM, Muchiri M, Njiru JM, Ojuok JE (2010) Difficulties of fishing at Lake Naivasha, Kenya: is community participation in management the solution? Lakes Reserv 15:15–23

Kyambia MM, Mutua BM (2014) Detection of trends in extreme streamflow due to climate variability in the Lake Naivasha basin, Kenya. Int J River Basin Manag 13:97–103

Lalah JO, Wandiga SO (2006) Extinction coefficients and dissolved organic carbon content in freshwater in Kenya. Bull Environ Contam Toxicol 77:533–542

Lamb H, Darbyshire I, Verschuren D (2003) Vegetation response to rainfall variation and human impact in central Kenya during the past 1100 years. The Holocene 13:285–292

Leakey LSB (1931a) East African Lakes. Geogr J 77:497–508

Leakey LSB (1931b) The Stone Age Cultures of Kenya Colony. Cambridge Univ Press, Cambridge

Leakey LSB (1934) Changes in the physical geography of East Africa in human times. Geogr J 84:296–305

Leakey LSB (1942) The Naivasha fossil skull and skeleton. J East Afr Nat Hist Soc 16:169–177

Leakey LSB (1950) The lower limit of the Pleistocene in Africa. Proc 18 Int Geol Congr, pp 62–65

Leakey LSB (1952) Capsian or Aurignacian? Which term should be used in Africa. Proceedings First Pan-African Congress on Prehistory, 1947, pp 205–206

Leakey LSB (1955) The climatic sequence of the Pleistocene in Africa. In: Balout L (ed) Act 2 session Congr Panafr Prehist, Alger, 1952, pp 293–294

Lind EM (1965) The phytoplankton of some Kenya waters. J East Afr Nat Hist Soc 25:76–91

Lind EM (1968) Notes on the distribution of phytoplankton in some Kenya waters. Brit Phycol B 3:481–493

Lönnberg E (1933) Description of a fossil buffalo from East Africa. Ark Zool Stockholm 25:1–32

Lönnberg E (1937) On some fossil mammalian remains from East Africa. Ark Zool Stockholm 29:1–23

Lüning S, Gałka M, Danladi IB, Adagunodo TA, Vahrenholt F (2018) Hydroclimate in Africa during the medieval climate anomaly. Palaeogeogr Palaeoclimatol Palaeoecol 495:309–322

Luo W, Li H, Kotut K, Krienitz L (2016) Molecular diversity of plankton in a tropical crater lake switching from hyposaline to subsaline conditions: Lake Oloidien, Kenya. Hydrobiologia 788:1–25

MacIntyre S, Melack J (1982) Meromixis in an equatorial African soda lake. Limnol Oceanogr 27:595–609

Maina CW, Sang JK, Mutua BM, Raude JM (2018) Bathymetric survey of Lake Naivasha and its satellite Lake Oloiden in Kenya; using acoustic profiling system. Lakes Reserv 23:324–332

Maina CW, Sang JK, Raude JM, Mutua BM, Moriasi DN (2019) Sediment distribution and accumulation in Lake Naivasha, Kenya over the past 50 years. Lake Reserv 24:162–172

Maitima JM (1991) Vegetation response to climatic change in Central Rift Valley, Kenya. Quat Res 35:234–245

Majozi NP, Salama MS, Bernard S, Harper DM, Habte MG (2014) Remote sensing of euphotic depth in shallow tropical inland waters of Lake Naivasha using MERIS data. Remote Sens Environ 148:178–189

Marchant R, Richer S, Boles O, Capitani C, Courtney-Mustaphi CJ, Lane P, Prendergast ME, Stump D, De Cort G, Kaplan JO, Phelps L, Kay A, Olago D, Petek N, Platts PJ, Punwong P, Widgren M, Wynne-Jones S, Ferro-Vázquez C, Benard J, Boivin N, Crowther A, Cuní-Sanchez A, Deere NJ, Ekblom A, Farmer J, Finch J, Fuller D, Gaillard-Lemdahl M-J, Gillson L, Githumbi E, Kabora T, Kariuki R, Kinyanjui R, Kyazike E, Lang C, Lejju J, Morrison KD, Muiruri V, Mumbi C, Muthoni R, Muzuka A, Ndiema E, Nzabandora CK, Onjala I, Schrijver AP, Rucina S, Shoemaker A, Thornton-Barnett S, van der Plas G, Watson EE, Williamson D, Wright D (2018) Drivers and trajectories of land cover change in East Africa: human and environmental interactions from 6000 years ago to present. Earth-Sci Rev 178:322–378

Mathooko JM (2005) Application of traditional ecological knowledge in the management and sustainability of fisheries in East Africa: a long-neglected strategy? Hydrobiologia 537:1–6

Mavuti KM (1990) Ecology and role of zooplankton in the fishery of Lake Naivasha. Hydrobiologia 208:131–140

Mavuti KM (1992) Diel vertical distribution of zooplankton in Lake Naivasha, Kenya. Hydrobiologia 232:31–41

Mavuti KM, Harper DM (2006) The ecological state of Lake Naivasha, Kenya, 2005: turning 25 years research into an effective Ramsar monitoring programme. In: Odada EO, Olago DO, Ochola W, Ntiba M, Wandiga S, Gichuki N, Oyieke H (eds) Proc 11th World Lakes Conf, vol 2, Nairobi, pp 30–34

Mavuti KM, Moreau J, Munyandorero J, Plisnier PD (1996) Analysis of trophic relationships in two shallow equatorial lakes, Lake Naivasha (Kenya) and Lake Ihema (Rwanda) using a multispecies trophic model. Hydrobiologia 321:89–100

Mekonnen MM, Hoekstra AY, Becht R (2012) Mitigating the water footprint of export cut flowers from the Lake Naivasha Basin, Kenya. Water Resour Manag 26:3725–3742
Melack JM (1976) Primary productivity and fish yields in tropical lakes. Trans Am Fish Soc 105:575–580
Melack JM (1979) Temporal variability of phytoplankton in tropical lakes. Oecologia 44:1–7
Melack JM (1982) Photosynthetic activity and respiration in an equatorial African soda lake. Freshw Biol 12:381–400
Melack J (1996) Saline and freshwater lakes of the Kenya Rift valley. In: McClanahan TR, Young TP (eds) East African ecosystems and their conservation. Oxford University Press, Oxford, pp 171–190
Mercy B, Mbaria J, Muchemi G, Gakuya F, Kariuki E, Wamiti W (2017) Ecological risk assessment of heavy metals and water bird distribution in rift valley lakes, Kenya. J Ecol Nat Environ 9:30–44
Mergeay J, Verschuren D, van Kerckhoven L, de Meester L (2004) Two hundred years of a diverse *Daphnia* community in Lake Naivasha (Kenya): effects of natural and human-induced environmental changes. Freshw Biol 49:998–1013
Mergeay J, Verschuren D, de Meester L (2005) Cryptic invasion and dispersal of an American *Daphnia* in East Africa. Limnol Oceanogr 50:1278–1283
Mergeay J, Verschuren D, de Meester L (2006a) Invasion of an asexual American water flea clone throughout Africa and rapid displacement of a native sibling species. Proc R Soc B Biol Sci 273:2839–2844
Mergeay J, Declerck S, Verschuren D, de Meester L (2006b) *Daphnia* community analysis in shallow Kenyan lakes and ponds using dormant eggs in surface sediments. Freshw Biol 51:399–411
Mergeay J, Vanoverbeke J, Verschuren D, de Meester L (2007) Extinction, recolonization, and dispersal through time in a planktonic crustacean. Ecology 88:3032–3043
Mergeay J, De Meester L, Eggermont H, Verschuren D (2011) Priority effects and species sorting in a long paleoecological record of repeated community assembly through time. Ecology 92:2267–2275
Millbrink G (1977) On the limnology of two alkaline lakes (Nakuru and Naivasha) in the East African Valley system in Kenya. Int Rev Hydrobiol 62:1–17
Milne G (1936) A provisional soil map of East Africa. Mem E Afr Agric Res Station. Amani, Tanganyika Territory
Mironga JM (2006) Environmental implications of Water Hyacinth infestation in Lake Naivasha, Kenya. In: Odada EO, Olago DO, Ochola W, Ntiba M, Wandiga S, Gichuki N, Oyieke H (eds) Proc 11th World Lakes Conf, vol 2, Nairobi, pp 580–592
Montcoudiol N, Burnside NM, Györe D, Mariita N, Mutia T, Boyce A (2019) Surface and groundwater hydrochemistry of the Menengai Caldera geothermal field and surrounding Nakuru County, Kenya. Energies 12:3131
Morrison EHJ (2013) Ecological restoration of papyrus wetlands at Lake Naivasha, Kenya: social and ecological considerations. PhD thesis, University of Leicester
Morrison EHJ, Harper DM (2009) Ecohydrological principles to underpin the restoration of *Cyperus papyrus* at Lake Naivasha, Kenya. Ecohydrol Hydrobiol 9:83–97
Muchiri SM, Hickley P (1991) The fishery of Lake Naivasha, Kenya. In: Cowx IG (ed) Catch effort sampling strategies. Blackwell, Oxford, pp 382–392
Muchiri SM, Hickley P, Harper DM, North E (1994) The potential for enhancing the fishery of Lake Naivasha, Kenya. In: Cowx IG (ed) Rehabilitation of freshwater fisheries. Blackwell, Oxford, pp 348–357
Muff HB (1908) Report relating to the geology of the East African Protectorate. Misc Colon Rep 45, HMSO, London
Mügge O (1885) Untersuchung der von Dr. G. A. Fischer gesammelten Gesteine des MassaLandes. Mitt Geogr Ges Hamburg 5:238–264
Mügge O (1886) Ueber einige Gesteine des Massailandes. N Jahrb Mineral Geol Palaont Beil-Bd 4:576–609

Mugo MJ (2010) Seasonal changes in physico–chemical status and algal biomass of Lake Naivasha, Kenya. MSc thesis, Kenyatta University, Nairobi

Muita R, Gikundu D, Aura S, Njogu A, Ndichu R, Nyinguru P, Kiptum C (2021) Assessment of rising water levels of rift valley lakes in Kenya: the role of meteorological factors. Environ Sci Ecol Curr Res 2:1035

Muno FA (2002) Water balance of the southern Kenya Rift Valley lakes. MSc thesis, International Institute for Geo-information Science and Earth Observation, Enschede

Mutethya E, Yongo E (2021) A comprehensive review of invasion and ecological impacts of introduced common carp (*Cyprinus carpio*) in Lake Naivasha, Kenya. Lakes Reserv 26:e12386

Muthuri FM, Jones MB, Imbamba SK (1989) Primary productivity of papyrus (*Cyperus papyrus*) in a tropical swamp; Lake Naivasha, Kenya. Biomass 18:1–14

Muthuwatta LP (2004) Long term rainfall-runoff-lake level modelling of the Lake Naivasha basin, Kenya. MSc thesis, International Institute for Geo-information Science and Earth Observation, Enschede

Mutungwa DM (2011) Effects of climate change and human activity on lake levels in the Kenyan rift (case study Naivasha). BSc thesis, University of Nairobi

Mwamburi J (2013) Comparative spatial metal concentrations and partitioning in bottom sediments of two tropical freshwater lake basins, Kenya. Lakes Reserv 18:329–355

Mwamburi J (2015) Comparative evaluation of the concentrations of lead, cadmium and zinc in surficial sediments from two shallow tectonic freshwater lake basins, Kenya. Afr J Environ Sci Technol 9:531–544

Mwaniki MW, Kuria DN, Boitt MK, Ngigi TG (2017) Image enhancements of Landsat 8 (OLI) and SAR data for preliminary landslide identification and mapping applied to the central region of Kenya. Geomorphology 282:162–175

Mwaniki P, Taita T, Sierens T, Triest L (2019) Barriers to genetic connectivity of smooth flatsedge (*Cyperus laevigatus*) among alkaline-saline lakes of Eastern Rift Valley (Kenya). Aquat Bot 155:38–44

Nash DJ, De Cort G, Chase BM, Verschuren D, Nicholson SE, Shanahan TM, Asrat A, Lézine A-M, Grab SW (2016) African hydroclimatic variability during the last 2000 years. Quat Sci Rev 154:1–22

Ndebele-Murisa MR, Musil CF, Raitt L (2010) A review of phytoplankton dynamics in tropical African lakes. S Afr J Sci 106:13–18

Ndungu J, Augustijn DCM, Hulscher SJMH, Kitaka N, Mathooko J (2013) Spatio-temporal variations in the trophic status of Lake Naivasha, Kenya. Lakes Reserv 18:317–328

Ndungu J, Augustijn DCM, Hulscher SJMH, Fulanda B, Kitaka N, Mathooko M (2015a) A multivariate analysis of water quality in Lake Naivasha, Kenya. Mar Freshw Res 66:177–110

Ndungu JN, Chen W, Augustijn DCM, Hulscher SJMH (2015b) Analysis of the driving force of hydrodynamics in Lake Naivasha, Kenya. Open J Mod Hydrol 5:95–104

Ngari A, Kinyamario J, Ntiba M, Mavuti K (2009) Factors affecting abundance and distribution of submerged and floating macrophytes in Lake Naivasha, Kenya. Afr J Ecol 47:32–39

Nilsson E (1931) Quaternary glaciations and pluvial lakes in British East Africa. Geogr Ann 13:249–348

Nilsson E (1935) Traces of ancient changes of climate in East Africa. Geogr Ann 17:1–21

Nilsson E (1938) Pluvial lakes in East Africa. Geol Fören Stockholm Förh 60:423–433

Nilsson E (1940) Ancient changes of climate in British East Africa and Abyssinia. Geogr Ann 22:1–79

Nilsson E (1945) Omett fynd av fossil buffel i Ostafrika. Geol Fören Stockholm Förh 67:241–248

Njambuya J, Triest L (2010) Comparative performance of invasive alien *Eichhornia crassipes* and native *Ludwigia stolonifera* under non-limiting nutrient conditions in Lake Naivasha, Kenya. Hydrobiologia 656:221–231

Njenga JW (2004) Comparative studies of water chemistry of four tropical lakes in Kenya and India. Asian J Water Environ Poll 1:87–97

Njiru M, Ojuok JE, Morara G, Mugo J (2008) Does seasonal closure have effect on fishery?: the case of common carp, *Cyprinus carpio* L in Lake Naivasha, Kenya. Proc 12 World Lake Conf, Taal (2007), pp 137–140

Njiru J, Morara G, Waithaka E, Mugo J (2015) Fish kills in Lake Naivasha, Kenya: what was the probable cause? Int J Fish Aquat Stud 3:179–184

Njiru J, Waithaka E, Aloo PA (2017) An overview of the current status of Lake Naivasha fishery: challenges and management strategies. Open Fish Sci J 10:1–11

Njogu PM, Keriko JM, Wanjau RN, Kitetu JJ (2011) Distribution of heavy metals in various lake matrices; water, soil, fish and sediments: a case study of the Lake Naivasha Basin, Kenya. J Agric Sci Tech 13:91–106

Njuguna SG (1982) Nutrient productivity relationships in tropical Naivasha basin lakes, Kenya. PhD thesis. University of Nairobi, Kenya

Nyakundi ER, Githiri J, K'Orowe M (2020) Depth estimation of geothermal heat structures by Euler deconvolution of gravity data at Eburru area, Kenya. J Geosci Environ Prot 8:148–158

Nyawir E, Barrat JA, Varet J (2018) Geological, volcanological, petrological and structural studies to evaluate geothermal occurrence within an axial rift segment of the central Kenya Rift Valley. Proc 7th Afr Rift Geotherm Conf, Kigali, Rwanda

Ochieng EZ, Lalah JO, Wandiga SO (2007) Analysis of heavy metals in water and surface sediment in five rift valley lakes in Kenya for assessment of recent increase in anthropogenic activities. Bull Environ Contam Toxicol 79:570–576

Odongo MEO (1993) A geological review of Olkaria geothermal reservoir based on structure. Proc 15 N Z Geothermal Workshop, pp 169–173

Odongo VO, Mulatu DW, Muthoni FK, van Oel PR, Meins FM, van der Tol C, Skidmore AK, Groen TA, Becht R, Onyando JO, van der Veen A (2014) Coupling socio-economic factors and eco-hydrological processes using a cascade-modeling approach. J Hydrol 518:49–59

Ojiambo BS (1992) Hydrogeologic, hydrogeochemical and stable isotopic study of possible interactions between Lake Naivasha, shallow subsurface and Olkaria geothermal waters, Central rift valley, Kenya. MSc thesis, University of Nevada, Reno

Ojiambo BS, Lyons WB (1996) Residence times of major ions in Lake Naivasha, Kenya, and their relationship to lake hydrology. In: Johnson TC, Odada EO (eds) The limnology, climatology and paleoclimatology of the East African Lakes. Gordon and Breach, Amsterdam, pp 267–278

Ojiambo BS, Poreda RJ, Lyons WB (2001) Groundwater/surface water interactions in Lake Naivasha, Kenya, using $\delta^{18}O$, δD and $^{3}H/^{3}He$ age-dating. Groundwater 39:526–533

Olago D, Opere A, Barongo J (2009) Holocene palaeohydrology, groundwater and climate change in the lake basins of the Central Kenya Rift. Hydrol Sci J 54:765–780

Olaka LA (2011) Hydrology across scales: sensitivity of East African lakes to climate changes. Doctoral thesis, Universität Potsdam

Olaka LA, Trauth MH, Olago DO (2010) The sensitivity of East African rift lakes to climate fluctuations. J Paleolimnol 44:629–644

Olaka LA, Wilke FDH, Olago DO, Mulch A, Musolff A (2016) Groundwater fluoride enrichment in an active rift setting: Central Kenya Rift case study. Sci Total Environ 545-546:641–653

Olaka LA, Kasemann SA, Sültenfuß J, Wilke FDH, Olago DO, Mulch A, Musolff A (2022) Tectonic control of groundwater recharge and flow in faulted volcanic aquifers. Water Resour Res 58:e2022WR032016

Oluoch AO (1990) Breeding biology of the Louisiana red swamp crayfish *Procambarus clarkii* Girard in Lake Naivasha, Kenya. Hydrobiologia 208:85–92

Omondi AO (2019) Diversity, distribution and biomass on phytoplankton in Lake Naivasha, Kenya. MSc thesis, Kisii University, Kenya

Omondi R, Ojwang W, Olilo C, Mugo J, Agembe S, Ojuok JE (2016) Lakes Baringo and Naivasha: endorheic freshwater lakes of the Rift Valley (Kenya). In: Finlayson CM, Milton GR, Prentice RC, Davidson NC (eds) The wetland book. Springer, The Netherlands, pp 1349–1360

Omondi AO, Ogendi GM, Onchieku MJ, Oduor SO, Omondi R, Omweno JO (2020) Phytoplankton community structure and ecology in Lake Naivasha, Kenya. Int J Fish Aquat Stud 8:478–483

Onywere SM, Mironga JM, Simiyu I (2012) Use of remote sensing data in evaluating the extent of anthropogenic activities and their impact on Lake Naivasha, Kenya. Environ Eng J 5:9–18
Otiang'a-Owiti GE, Oswe IA (2007) Human impact on lake ecosystems: the case of Lake Naivasha, Kenya. Afr J Aquat Sci 32:79–88
Otieno P, Owuor PO, Lalah JO, Pfister G, Schramm K-W (2015) Monitoring the occurrence and distribution of selected organophosphates and carbamate pesticide residues in the ecosystem of Lake Naivasha, Kenya. Toxicol Environ Chem 97:51–61
Ouma Y, Tateishi R (2006) A water index for rapid mapping of shoreline changes of five East African Rift Valley lakes: an empirical analysis using Landsat TM and ETM+ data. Int J Remote Sens 27:3153–3181
Owino AO, Ryan PG (2006) Recent papyrus swamp habitat loss and conservation implications in western Kenya. Wetl Ecol Manag 15:1–12
Owino A, Oyugi J, Nasirwa O, Bennun L (2001) Patterns of variation in waterbird numbers on four Rift Valley lakes in Kenya, 1991–1999. Hydrobiologia 458:45–53
Prendergast ME, Lipson M, Sawchuk EA, Olalde I, Ogola CA, Rohland N, Sirak KA, Adamski N, Bernardos R, Broomandkhoshbacht N, Callan K, Culleton BJ, Eccles L, Harper TK, Lawson AM, Mah M, Oppenheimer J, Stewardson K, Zalzala F, Ambrose SH, Ayodo G, Gates HL, Gidna AO, Katongo M, Kwekason A, Mabulla AZP, Mudenda GS, Ndiema EK, Nelson C, Robertshaw P, Kennet DJ, Manthi FK, Reich D (2019) Ancient DNA reveals a multistep spread of the first herders into sub-Saharan Africa. Science 365:eaaw6275
Prior GT (1903) Contributions to the petrology of British East Africa. Mineral Mag 13:228–263
Raburu P, Mavuti KM, Harper DM, Clark FL (2002) Population structure and secondary productivity of *Limnodrilus hoffmeisteri* (Claparede) and *Brachiura sowerbyi* Beddard in the profundal zone of Lake Naivasha, Kenya. Hydrobiologia 44:153–161
Reta GL (2011) Groundwater and lake water balance of Lake Naivasha using 3-D transient groundwater model. MSc thesis, International Institute for Geo-information Science and Earth Observation, Enschede
Rich F (1933) Scientific results of the Cambridge Expedition to the East African Lakes, 1930–1. 7. The Algae. J Lin Soc Lond Zool 38:249–275
Richardson JL (1964) Plankton and fossil plankton studies in certain East African Lakes. Verh Int Ver Limnol 15:993–999
Richardson JL (1968) Diatoms and lake typology in East and Central Africa. Int Rev Hydrobiol 53:299–338
Richardson JL (1969) Characteristic planktonic diatoms of the lakes of tropical Africa. (Addendum to: Diatoms and lake typology in East and Central Africa). Int Rev Hydrobiol 54:175–176
Richardson JL (1972) Palaeolimnological records from rift lakes in central Kenya. Palaeoecol Afr 6:131–136
Richardson J, Dussinger R (1986) Paleolimnology of mid-elevation lakes in the Kenya Rift Valley. Hydrobiologia 143:167–174
Richardson JL, Richardson AE (1972) History of an African rift lake and its climatic implications. Ecol Monogr 42:499–534
Richardson JL, Harvey TJ, Holdship SA (1978) Diatoms in the history of shallow East African lakes. Pol Arch Hydrobiol 25:341–353
Roessner S, Strecker MR (1997) Late Cenozoic tectonics and denudation in the Central Kenya Rift: quantification of long-term denudation rates. Tectonophysics 278:83–94
Rumes B (2010) Regional diversity, ecology and palaeoecology of aquatic invertebrate communities in East African lakes. PhD thesis, University of Ghent, Belgium
Shackleton RM (1955) Pleistocene movements in the Gregory Rift Valley. Geol Rundsch 43:257–263
Shackleton RM (1986) Precambrian collision tectonics in Africa. In: Coward MP, Ries AC (eds) Collision tectonics, Geol Soc Lond Spec Publ 19, London, pp 329–349
Shand SJ (1936) Rift valley impressions. Geol Mag 73:307–312

Siddiqui A (1979) Changes in fish species composition in Lake Naivasha, Kenya. Hydrobiologia 64:131–138

Sikes HL (1926) The structure of the eastern flank of the Rift Valley near Nairobi. Geogr J 68:385–402

Sikes HL (1934) The underground water resources of Kenya Colony. Crown Agents, London

Sikes HL (1936) Notes on the hydrology of Lake Naivasha. J East Afr Uganda Nat Hist Soc 13:73–89

Simiyu S (2010) Status of geothermal exploration in Kenya and future plans for its development. Proc World Geothermal Congr, Bali, April 2010

Simiyu S, Keller G (2000) Seismic monitoring of the Olkaria geothermal area, Kenya Rift valley. J Volcanol Geotherm Res 95:197–208

Smart AC (1988a) The population density and distribution of the Fish Eagle *Haliaeetus vocifer* on Lakes Naivasha and Oloidien. In: Lake Naivasha vertebrate survey 1987. University of Leicester, Leicester, pp 83–94

Smart AC (1988b) Population census of Lake Naivasha Hippopotamus *Hippopotamus amphibius*. In: Lake Naivasha vertebrate survey 1987. University of Leicester, Leicester, pp 65–82

Smart AC, Harper DM, Malaisse F, Schmitz S, Coley S, Gouder de Beauregard A-C (2002) Feeding of the exotic Louisiana red swamp crayfish, *Procambarus clarkii* (Crustacea, Decapoda), in an African tropical lake: Lake Naivasha, Kenya. Hydrobiologia 44:129–142

Smith W (1931) A classification of some rhyolites, trachytes and phonolites from part of Kenya Colony, with a note on some associated basaltic rocks. Q J Geol Soc Lond 87:212–258

Spink PC (1945) Thermal activity in the Eastern Rift Valley. Geogr J 105:197–207

Stoof-Leichsenring KR, Junginger A, Olaka LA, Tiedemann R, Trauth MH (2011) Environmental variability in Lake Naivasha, Kenya, over the last two centuries. J Paleolimnol 45:353–367

Strecker MR, Blisniuk PM, Eisbacher GH (1990) Rotation of extension direction in the central Kenya Rift. Geology 18:299–302

Syrén P (1990) Estimated transport of the suspended load of the River Malewa, Kenya, between 1931-1959. Geogr Ann Ser A. Phys Geogr 72:285–299

Talling JF, Talling IB (1965) The chemical composition of African lake waters. Int Rev Hydrobiol 50:421–463

Tang ZX (1999) Water quality assessment and pesticide fate modelling in the Lake Naivasha area, Kenya. Water Resources Survey. MSc thesis, International Institute for Aerospace Survey and Earth Sciences (ITC), Enschede

Tarafdar S, Harper D (2008) Anti-persistence in levels of Lake Naivasha: assessing effect of human intervention through time-series analysis. Physica A 387:296–302

Tarras-Wahlberg H, Everard M, Harper DM (2002) Geochemical and physical characteristics of river and lake sediments at Naivasha, Kenya. Hydrobiologia 488:27–41

Terer T, Muasya AM, Higgins S, Gaudet JJ, Triest L (2014) Importance of seedling recruitment for regeneration and maintaining genetic diversity of *Cyperus papyrus* during drawdown in Lake Naivasha, Kenya. Aquat Bot 116:93–102

Thompson AC, Dodson RG (1963) Geology of the Naivasha area. Geol Surv Kenya Rep 43

Thomson J (1887) Through Masai Land. Sampson Low, Marsten, Gilbert and Rivington, London

Trauth MH, Maslin MA, Deino AL, Junginger A, Lesoloyia M, Odada EO, Olago DO, Olaka LA, Strecker MR, Tiedemann R (2010) Human evolution in a variable environment: the amplifier lakes of Eastern Africa. Quat Sci Rev 29:2981–2988

Triest L, Sierens T, Terer T (2014) Diversity and fine-scale spatial genetic structure of *Cyperus papyrus* populations in Lake Naivasha (Kenya) using microsatellite markers. Hydrobiologia 737:131–144

Uku J, Mavuti K (1994) Comparative limnology, species diversity and biomass relationship of zooplankton and phytoplankton in five freshwater lakes in Kenya. Hydrobiologia 272:251–258

van Oel PR, Mulatu DW, Odongo VO, Meins FM, Hogeboom RJ, Becht R, Stein A, Onyando JO, van der Veen A (2013) The effects of groundwater and surface water use on total water

availability and implications for water management: the case of Lake Naivasha, Kenya. Water Resour Manag 27:3477–3492
Verschuren D (1996) Comparative paleolimnology in a system of four shallow tropical lake basins. In: Johnson TC, Odada EO (eds) The limnology, climatology and paleoclimatology of the East African Lakes. Gordon and Breach, Amsterdam, pp 559–572
Verschuren D (1999a) Sedimentation controls on the preservation and time resolution of climate-proxy records from shallow fluctuating lakes. Quat Sci Rev 18:821–837
Verschuren D (1999b) Influence of depth and mixing regime on sedimentation in a small, fluctuating tropical soda lake. Limnol Oceanogr 44:1103–1113
Verschuren D (2001) Reconstructing fluctuations of a shallow East African lake during the past 1800 yrs from sediment stratigraphy in a submerged crater basin. J Paleolimnol 25:297–311
Verschuren D (2003) Lake-based climate reconstruction in Africa: progress and challenges. Hydrobiologia 500:315–330
Verschuren D (2004) Decadal and century-scale climate variability in tropical Africa during the past 2000 years. In: Battarbee RW, Gasse F, Stickley CE (eds) Past climate variability through Europe and Africa (Dev Paleoenviron Res 6). Springer, Dordrecht, pp 139–158
Verschuren D, Russell J (2009) Paleolimnology of African lakes: beyond the exploration phase. J Geophys Res 113:112–116
Verschuren D, Tibby J, Leavitt PR, Roberts CN (1999b) The environmental history of a climate-sensitive lake in the former 'White Highlands' of central Kenya. Ambio 28:494–501
Verschuren D, Laird K, Cumming B (2000a) Rainfall and drought in equatorial East Africa during the past 1,100 years. Nature 403:410–414
Verschuren D, Tibby J, Sabbe K, Roberts N (2000b) Effects of depth, salinity, and substrate on the invertebrate community of a fluctuating tropical lake. Ecol 81:164–182
Verschuren D, Cumming BF, Laird KR (2004) Quantitative reconstruction of past salinity variations in African lakes: assessment of chironomid-based inference models (Insecta: Diptera) in space and time. Can J Fish Aquat Sci 61:986–998
Vincent CE, Davies TD, Beresford UC (1979) Recent changes in the level of Lake Naivasha, Kenya, as an indicator of equatorial westerlies over East Africa. Clim Chang 2:175–191
von Höhnel L (1890) Ostäquatorial-Afrika Zwischen Pangani und dem neuentdecken Rudolf-See: Ergebnisse der Graf S. Telekischen Expedition, 1887–1888 (Ergänzungsheft 99 zu Petermanns Mittheilungen). Justus Perthes, Gotha
Wakefield T, Johnston K (1870) Routes of native caravans from the coast to the interior of Eastern Africa, chiefly from information given by Sádi Bin Ahédi, a native of a district near Gázi, in Udigo, a little north of Zanzibar. J R Geogr Soc Lond 40:303–339
Walker EE (1903) Reports on the geology of the East African Protectorate. Colonial Office, London
Walker D, Shutler JD, Morrison EHJ, Harper DM, Hoedjes JCB, Laing CG (2022) Quantifying water storage within the north of Lake Naivasha using sonar remote sensing and Landsat satellite data. Ecohydrol Hydrobiol 22:12–20
Wandiga O, Molepo M, Alala LN (1983) Concentration of heavy metals in water, sediments and plants of Kenya lakes. Kenya J Sci Tech Ser A 4:89–94
Wanjiku CR (2015) Origin and development of Naivasha town, 1895–2013. MA thesis, Egerton University, Njoro
Washbourn-Kamau CK (1975) Late Quaternary shorelines of Lake Naivasha, Kenya. Azania 10:77–92
Watson RM (1970) The diet of duck and coot on Lake Naivasha. Afr J Ecol 8:131–144
Wetang'ula G (2010) Public participation in environmental and socioeconomic considerations for proposed 2.5 MW pilot Eburru Geothermal Power Project, Kenya. Proc World Geotherm Congr, Bali, April 2010
Williams LAJ (1970) The volcanics of the Gregory Rift Valley, East Africa. Bull Volcanol 34:439–465

Williams LAJ, Macdonald R, Leat PT (1983) Magmatic and structural evolution of the central part of the Kenya Rift. In: Proc region seminar geotherm energy in Eastern and Southern Africa. UNESCO, Nairobi, pp 61–67

Willis B (1936) East African Plateaus and Rift Valleys. Carnegie Institute, Washington

Woodhall DG, Clarke MCG (1988) Geological map of Longonot Volcano, the Greater Olkaria and Eburru volcanic complexes and adjacent areas. Government of Kenya, Ministry of Energy, Geothermal Section

Worthington E (1932) Scientific results of the Cambridge Expedition to the East African Lakes, 1930-1. General introduction and station list. J Lin Soc Zool 38:99–119

Worthington E, Ricardo C (1936) Scientific results of the Cambridge Expedition to the East African Lakes 1931. No 17. The vertical distribution and movements of the plankton in Lakes Rudolf, Naivasha. Edward and Bunyoni. J Lin Soc Lond Zool 40:33–69

Yang Y, Wei L, Cui L, Zhang M, Wang J (2017) Profiles and risk assessment of heavy metals in Great Rift Lakes, Kenya. Clean-Soil Air Water 45:1600825

Yasindi AW (2001) The ecology of planktonic ciliates in tropical lakes of East Africa. PhD thesis, University of Waterloo, Ontario

Yihdego Y, Becht R (2013) Simulation of lake-aquifer interaction at Lake Naivasha, Kenya using a three-dimensional flow model with the high conductivity technique and a DEM with bathymetry. J Hydrol 503:111–122

Yongo EO, Morara GN, Ojuok JE, Nyamweya C, Ojwang WO, Maasai DM, Wasike C (2013) Socio-economic aspects of fisheries management in Lake Naivasha. Afr J Trop Hydrobiol Fish 13:27–32

Yongo EO, Agembe SW, Manyala JO, Mutethya E (2022) Assessment of the current trophic state and water quality of Lake Naivasha, Kenya using multivariate techniques. Lakes Reserv 27:e12422

Chapter 12
Lake Magadi and Nasikie Engida

12.1 Introduction

Lake Magadi (Figs. 12.1 and 12.2) is the most saline of the main Kenya Rift lakes. Its topographic basin lies in a semi-arid region at the southern end of the Kenya Rift Valley and passes southwestwards across a low watershed into the Lake Natron basin in Tanzania. Both lakes were united during the Late Pleistocene. The appearance of Lake Magadi, which occupies a narrow axial trough surrounded by grid-faulted Pleistocene volcanic rocks, often varies dramatically throughout the year. During the dry season, a white to grey salt crust covers most of the lake surface (Fig. 12.1c, e) with several small bodies of shallow open water along the lake margins where saline hot springs discharge. In contrast, in the rainy season, most of the lake surface is covered by a thin layer of saline alkaline brine with clear reflections of extinct volcanoes lying to the south and east (Fig. 12.1d, f). The salts are predominantly trona [$Na_3(CO_3)(HCO_3)\cdot 2H_2O$], which has been mined to produce soda ash (sodium carbonate) since 1911. The soda plant, which lies on the mid-eastern shoreline (Fig. 12.1b), has been the focus of the Magadi community and a major employer for the local Maasai for more than a century (Hill 1964; Hughes 2008). Access is by road from Nairobi (~90 km). Roads west of the lake lead to Olkaramatian and the Nguruman Escarpment. To the south, a track extends from southern Magadi to the delta of the Ewaso Ngiro River (South) on the northwestern shore of Lake Natron.

Lake Magadi and its small cousin to the northwest, Nasikie Engida (Little Magadi or Engaramai), are remarkable lakes, both biologically and geologically. These extreme alkaline lakes are famous for their cichlid fish (*Alcolapia grahami*) which inhabit the hot spring-fed waters and adjacent lagoons, and for the microbial communities that often colour them bright red, in common with Lake Natron. Lake Magadi is well known to geologists for its sodium carbonate evaporites, the abundant cherts present in its Pleistocene sediments, and its diagenetic mineral suites rich in zeolites. Lake Magadi is often considered the modern 'type-example' for hypersaline alkaline lakes.

R. W. Renaut, R. B. Owen, *The Kenya Rift Lakes: Modern and Ancient*,
Syntheses in Limnogeology, https://doi.org/10.1007/978-3-642-25055-2_12

Fig. 12.1 Modern environment of Lake Magadi. (**a**) Northern basin when the lake is relatively dry. Nguruman Escarpment in distance. (**b**) North-central lake when wet; soda plant and Magadi town in distance. (**c**) Northern basin with trona pan; Mt Olorgesailie and Ol Doinyo Nyokie volcanoes (Plio-Pleistocene) on horizon. (**d**) View northwards from southeastern shoreline with most of the lake covered by brine. (**e, f**) View south from causeway of Lake Magadi when a dry saline pan, and when wet and covered with brine

The lake basin has been studied for more than 100 years. Early work by German scientists, notably Gustav Fischer, is summarised by Gregory (1921, p. 179), but this was focused west of the lake near the Nguruman Escarpment. The first reports about the lake itself were by Smith (1907), Parkinson (1914), Gregory (1921) and Coates (as Anonymous 1923). Parkinson made observations on the structural geology and lake sediments and speculated on the origin of the alkaline waters. J.A. Stevens, in an unpublished report of Magadi Soda Company (Stevens 1932), analysed the Magadi spring waters and discussed their origin.

The first detailed studies were by Baker (1958, 1963, 1986) who mapped the region for the Geological Survey of Kenya, focussing on the volcanic and sedimentary rocks, regional structure and tectonics, and the origin of the basin brines. During the 1960s and 1970s, Hans Eugster, a Swiss-American geologist, undertook detailed studies of the hydrochemistry, stratigraphy and mineralogy of the basin, including the origin of the evaporites and chert and its precursors (summarised in Eugster 1980, 1986). Study of the late Quaternary history of the basin began in the 1970s with the dating of palaeoshorelines and analyses of short cores (Butzer et al. 1972) and continued during the 1980s, with parallel studies in the Lake Natron basin (see collected papers in Sci Géol Bull v 40, 1987). Biological studies of Magadi fish and microbes were undertaken throughout this period and have continued to the present (see papers in Schagerl 2016). In June 2014, Lake Magadi was drilled to its volcanic basement as part of the *Hominin Sites and Paleolakes Drilling Project* (HSPDP). Those cores have provided a wealth of new detail about the history and evolution of the lake, and the processes that formed its sediments and their unusual mineralogy (Campisano et al. 2017).

12.2 Geological and Geomorphological Setting

Lake Magadi, with an elevation of ~600 m, occupies the axial depression of the southernmost part of the Kenya Rift, ~25 km north of Lake Natron, which also lies at ~600 m. The local landscape is dominated by young fault scarps exposing volcanic rocks (mainly trachyte), with extinct Pleistocene volcanoes, mostly to the south and northeast (Figs. 12.2 and 12.3).

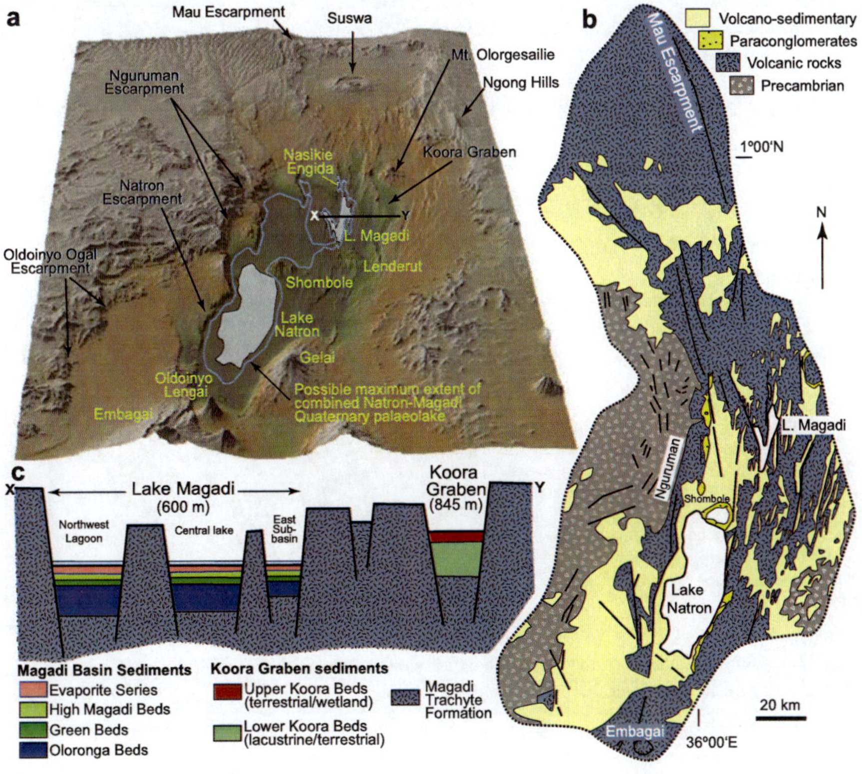

Fig. 12.2 Location and geological setting of the Magadi Basin. (**a**) Topography of the Magadi-Natron Basin. Note potential maximum extent of a combined Natron-Magadi palaeolake. Drafted using GeomapApp. (**b**) Simplified geology of Magadi-Natron catchment. Modified after Vincens and Casanova (1987). (**c**) Schematic east-west section through Lake Magadi and the Koora Graben

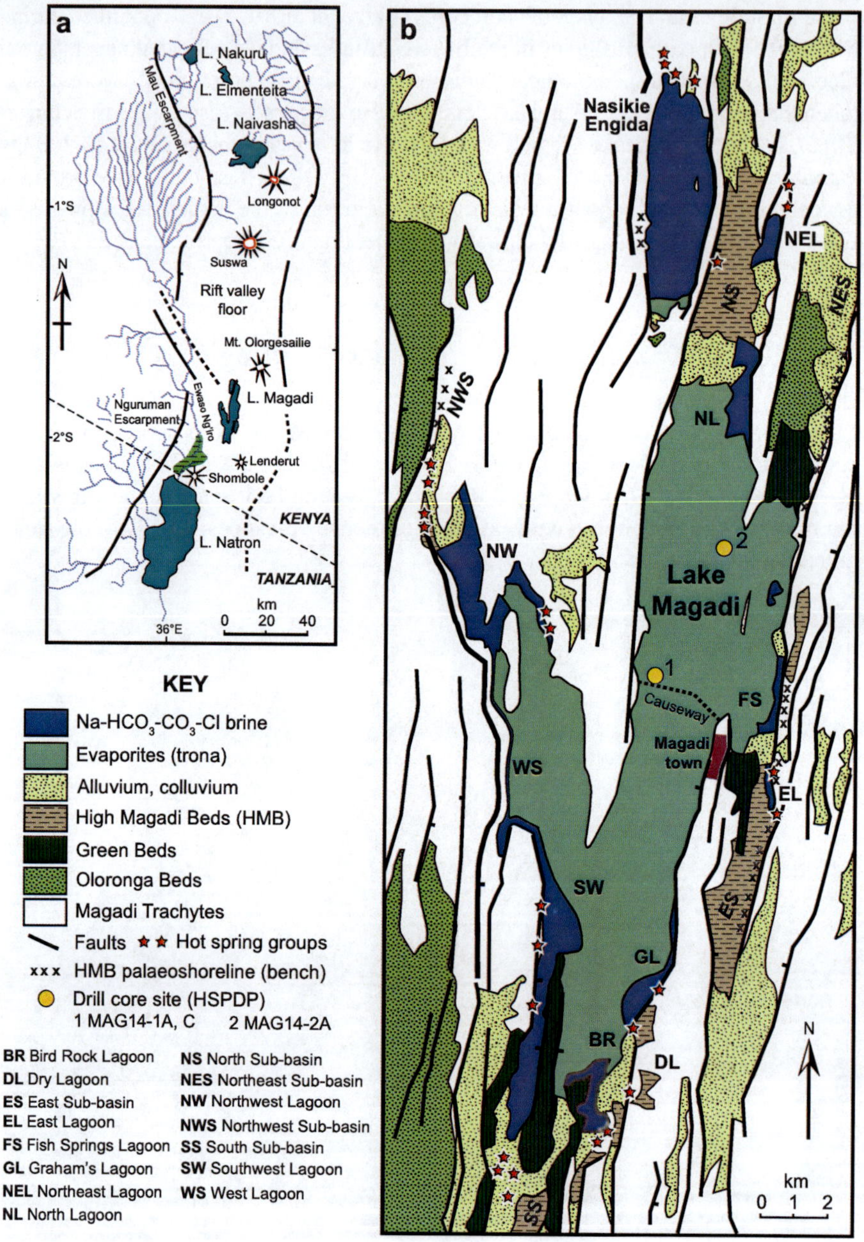

Fig. 12.3 Geological map of the Lake Magadi area. (**a**) Regional context. (**b**) Local geology and Quaternary sediments. Compiled from multiple sources

The rift in the southern sector is ~70 km wide. About 30 km west of Lake Magadi the Nguruman Escarpment, which rises to 2300 m, represents a major N-S border fault. East of the lake, the land rises in a series of stepped fault blocks to the eastern rift margin. The rift floor has an elevation between ~600 and 750 m and is intensely grid faulted with small parallel faults oriented mainly N-S. Small grabens between the faults are commonly floored by soils, alluvial and fluvial sediments, and local swamps and ephemeral lakes, some of which hold water for several years after major floods (Sect. 14.13). Pleistocene lake sediments are present in some of the grabens, including the Olorgesailie and Oltulelei formations at Olorgesailie and deposits in the Koora Graben, recording lakes that predate modern Lake Magadi and Lake Natron. Older pre-Pleistocene lake beds are preserved in outcrops in the Nguruman Escarpment (Chap. 20).

The southern part of the Kenya Rift was first mapped by the Geological Survey of Kenya (Baker 1958, 1963; Saggerson 1966; Wright 1967; Randel 1970). The history of the southern Kenya Rift has been one of faulting with intermittent phases of volcanic activity (Baker and Mitchell 1976; Crossley 1979, 1980; Crossley and Knight 1981; Baker 1986; Le Gall et al. 2008; Guth and Wood 2014a, b; Guth 2015) and with lakes periodically developing in the resultant troughs, sometimes ponded by volcanoes and lava flows. Rifting began initially with faulting and volcanism along the Nguruman Escarpment at ~7 Ma, before the eruption of Lengitoto Trachyte along the western rift margin (Crossley 1979, 1980). Volcanism east of Magadi began at 6.7 Ma with the eruption of the Ol Eyaseti volcanics (Baker and Mitchell 1976) followed by the Kirikiti lavas between 5.1 and 2.5 Ma along the rift centre and western margin (Crossley 1979). Volcanism then began in the Natron region followed by faulting at ~3 Ma (Le Gall et al. 2008). East of the Magadi trough the Ngong-Turoka fault was active between 3.3 and 2.3 Ma (Baker and Mitchell 1976). Several composite central volcanoes erupted during this period including Olorgesailie (2.7–2.2 Ma), Lenderut (2.7–2.5 Ma) and Ngong (2.26–2.5 Ma) (Baker and Mitchell 1976). Widespread basin-filling eruptions after 2 Ma include the Limuru Trachytes (2.05 Ma) and Singaraini basalts (1.96 Ma) with local eruptions from Shombole volcano south of Lake Magadi.

During the final phase, voluminous volcanism occurred in the axial trough. The Magadi Trachytes (1.4–0.95 Ma) were erupted along the Magadi axis, with the Gelai volcanics (1.23–0.95 Ma) to the south. These lavas flooded and flattened the pre-existing topography of the rift floor (Baker and Mitchell 1976). South of Magadi, the location of the border fault migrated from the Oldoinyo Ogol escarpment to the Natron border fault between 1.2 and 1.0 Ma (Foster et al. 1997). For the past million years, volcanism has been quiet near Magadi, but the mid-Pleistocene (~0.7 Ma) Ol Doinyo Nyokie trachyte-ignimbrite volcano, which also produced obsidian flows, erupted east of Lake Magadi (Fairhead et al. 1972; Baker 1975), while Ol Doinyo Lengai has been intermittently active south of Lake Natron (Nyamweru 1988; Dawson 2008). The flood lavas were then broken by intra-rift normal faulting parallel to the rift margins (Baker 1958; Crossley 1979; Le Gall et al. 2008; Muirhead et al. 2016). Although N-S faulting dominates, subvolcanic basement trends are evident and have played a role in compartmentalising sub-basins (Le Turdu et al. 1999) and influencing drainage. Notable are the NW-SE trends (N150–160°) that separate the Northwest Lagoon from the N-S axial basin, and the Kisamis-Nasikie Engida transverse trend (N135°) that controls the location of the northern hot springs.

Sedimentation occurred during and following Pleistocene faulting and volcanism in the southern Kenya Rift. After eruption of the Magadi Trachytes, the Oloronga Beds (~1–0.3 Ma) were deposited in a large lake that at times extended southwards into the Natron basin and northwards beyond Lake Magadi and Nasikie Engida (Fig. 12.2a). Its sediments include fluvial sandstone, and lacustrine tuffaceous silt, carbonates and chert (Sect. 20.3). It is commonly capped by a thick calcrete marking a break in sedimentation. The succeeding Green Beds (~191–40 Ma) are exposed only in the axial part of the basin and were followed by a phase of faulting. They consist of fine lacustrine sediments and abundant chert, much of which was physically intruded into younger deposits (Sect. 20.5).

The region is tectonically and seismically active (Kipkwony et al. 2022). Earthquakes occur frequently, particularly in the area directly north of Lake Magadi and Nasikie Engida (Birt et al. 1997; Ibs-Von Seht et al. 2001, 2008; Atmaoui and Hollnack 2003; Najoks et al. 2008; Sequar 2009; Kuria et al. 2010; Mulwa et al. 2014; Onguso 2019). Other epicentres lie south of the lake and along the western border fault (Weinstein et al. 2017). Depth to the basement in the Magadi-Natron basin has been estimated at ~3.5 km (Birt et al. 1997; Godfrey 2018).

12.3 Climate, Vegetation and Wildlife

The climate across the Magadi-Natron drainage basin is semi-arid to subhumid becoming driest in the lowest axial parts where the modern lakes exist. Annual rainfall is low on the valley floor (~425 mm), reflecting rainshadow effects, and rises to >1000 mm in the highest western parts of the drainage basin. Monsoon winds over East Africa are seasonally dry or wet depending on the movements of the equatorial Inter Tropical Convergence Zone (ITCZ) or tropical low-pressure belt. Northward movement brings winds from the Indian Ocean producing the 'long rains' between March and May. A long dry season is then followed by the 'short rains' in November as the ITCZ moves south.

Temperatures are closely linked to altitude. The Magadi area has a monthly mean maximum temperature of 34.8 °C and a mean minimum of 23.1 °C (Vincens and Casanova 1987). At Narok (1890 m) in the headwaters of the Magadi-Natron basin the respective values are 24.5 °C and 8.5 °C. There is little seasonal variation (3–4 °C), with January to March the warmest months. Potential mean annual evaporation across the basin exceeds mean annual rainfall except at high altitudes in western Mau to the northwest. Local annual evaporation at Magadi exceeds rainfall by 500–2400 mm (Damnati and Taieb 1995).

The modern vegetation of the Magadi-Natron region has been described by Vincens and Casanova (1987), based on White (1983). The low-lying area

(600–900 m elevation) surrounding both lakes is semi-desert grassland and shrubland, consisting of ephemeral grasses (50–90% cover) and widely spaced trees 3–5 m high, dominated by *Commiphora* sp. and *Acacia* (*Vachellia*) *reficiens*. Dense *Acacia-Commiphora* bushland lies between 900 and 1500 m. Also present are members of families Capparidaceae, Tilliaceae and Ballanitiaceae, succulents and perennials. The drier lower slopes of the Nguruman Escarpment have mosaics of semi-evergreen bushland and *Acacia* (*Vachellia*) wooded grassland. Afromontane forest lies at >2000 m. Cyperaceae swamps lie locally on the deltaic margins of the two main lakes.

Wildlife around lakes Magadi and Natron has declined significantly in numbers and diversity since our first visit in 1976, although the development of Maasai conservancies to the west may assist recovery. Gazelles, impala, zebra, giraffe, gerenuk, vervet monkey, baboon and wildebeest are common (2005–22), with occasional lion and hyena. Details of an aerial survey in 2013 are given by Kenana et al. (2013). Domestic goats consume much of the sparse vegetation.

12.4 Drainage and Hydrology

12.4.1 Inflowing Rivers and Lake Water

The Magadi-Natron drainage basin is ~23,000 km^2 (Vincens and Casanova 1987; Figs. 12.3a and 12.4). No major streams flow into Lake Magadi today. Recharge is mainly by hot springs, ephemeral streams and direct rainfall. The only large permanent river in the Magadi-Natron area is the Ewaso Ngiro (Uaso Ngiro) South, which rises in the Mau Highlands in western Kenya, and then flows southwards via the Shombole (Shompole) Swamp and into ephemeral wetlands forming the northern shoreline of Lake Natron. It is supplemented by a few large, often perennial streams (e.g. Oloibortoto, Endosapia) that drain the steep Nguruman Escarpment and Loita Hills. Other perennial streams flow into Lake Natron along its western and southern shores. Hot and warm springs issue along the faulted shorelines of Lake Magadi, Nasikie Engida and Lake Natron. At Lake Magadi, they flow into shallow (<1 m deep) ponded areas between the spring outflow and evaporites, clastic sediment or volcanic shoreline, that Baker (1958) called 'lagoons', which pre-concentrate lake brine by evaporation. Lagoon waters are directly connected to the lake brine when it floods the lake surface or form a moat that passes laterally into permeable trona as shallow groundwater when the saline pan surface is exposed.

Lake Magadi covers an area of ~90 km^2. In most years this changes in response to seasonal rains. It typically floods to a depth of a few decimetres to a metre but most of the lake surface dries out by summer to a form a crust with groundwater at a depth of <1 m. The lagoons usually have surface waters in all seasons. Nasikie Engida (~10 km^2), in contrast, is a perennial saline alkaline lake (Fig. 12.5a–c) because of its hot spring inflow and small size. It shrinks with evaporation and

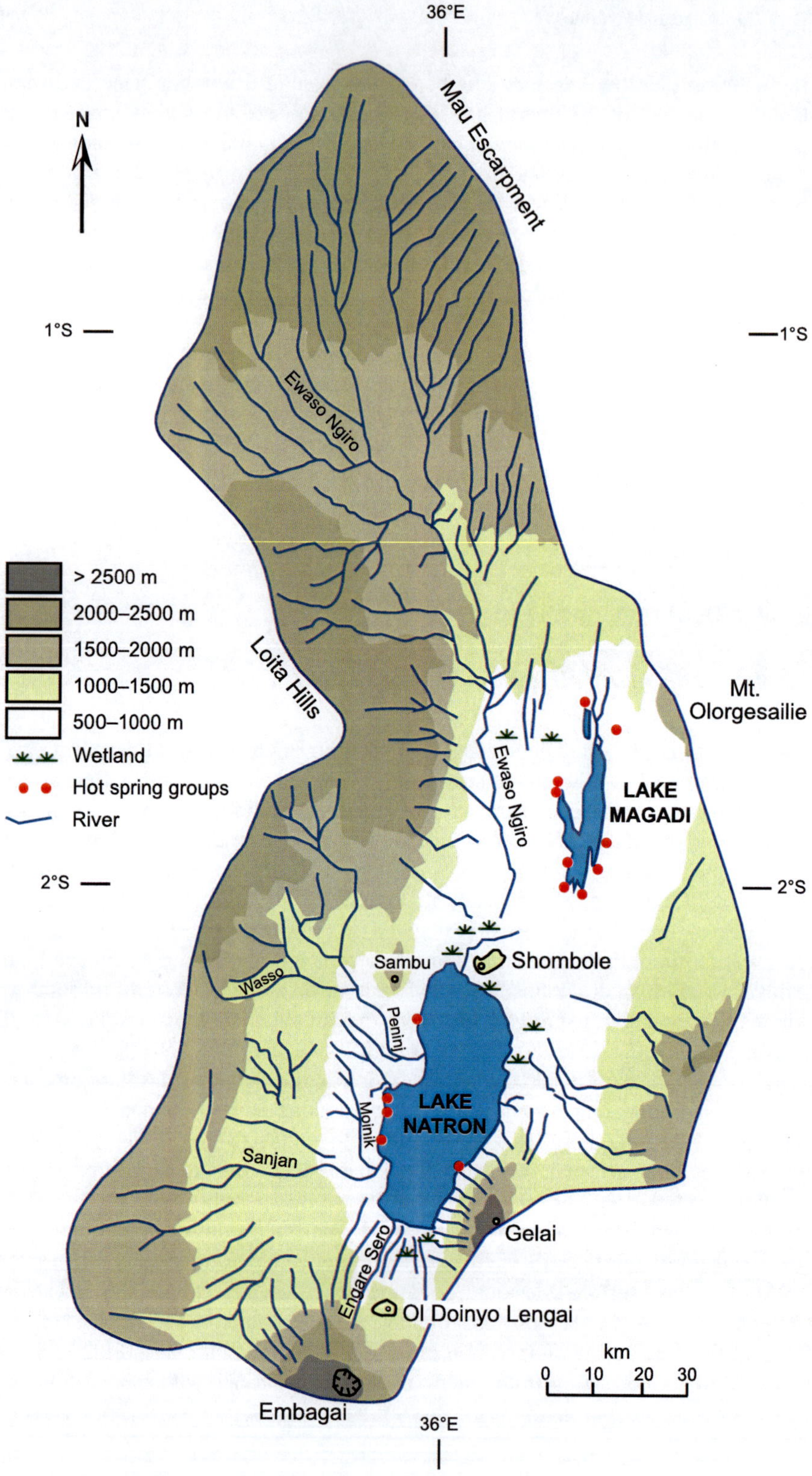

Fig. 12.4 Drainage basin of Lake Magadi and Lake Natron

precipitates evaporites but maintains surface water up to 1.6 m deep throughout the year. Although shallow, its waters are poorly mixed and anoxic at depth (De Cort et al. 2019).

The Magadi saline pan was frequently flooded during the regionally high levels after 2010 (GoK 2021) but with only limited areal expansion mainly on the northern and southern margins. The surface water level is ~600 m. In July 2018 the central part of the motorable causeway that crosses the lake was submerged preventing direct access between the western and eastern shorelines. Steep faulted margins reduced the amount of shoreline flooding in Lake Magadi and Nasikie Engida.

12.4.2 Hot Springs of the Magadi Basin

Alkaline hot springs are a major source of inflow to Magadi waters. The hottest and most dilute springs lie to the north while cooler, more saline springs lie to the south (Baker 1958; Burgess 1986; Allen et al. 1989). The hot springs at Nasikie Engida are hotter (75–86 °C) than all those feeding Lake Magadi (34–45 °C; Table 12.1).

Most Magadi hot springs lie within seven main groups (Fig. 12.6a). The springs at northern Nasikie Engida flow from two sites near the northern shoreline. The ~20 hot springs of the *western Nasikie subgroup* (Fig. 12.6a, b) flow lakewards along a N-S fault below the Olkeju Ole Ntaga escarpment. They discharge from alluvial gravel, colluvium and fractures in the trachyte bedrock. Springs of the *eastern Nasikie subgroup* flow southwards across deltaic mudflats. Most are seepage springs (Sect. 4.9.1; Fig. 12.6c) but include small pool-springs a few dm deep. Small springs also flow along fault lines between those two subgroups including a small vigorous pool-spring in gravel. Discharge from these sites, estimated at >18,000 L min^{-1} (Anonymous 1923), is the highest in the Magadi basin.

At Lake Magadi itself, the *northeastern* group discharges into the adjacent Northeast Lagoon from the faulted margins of a small graben. This group comprises small seepage springs with low discharge and a few shallow pool springs (Fig. 12.6d). Most discharge into mudflats along the base of faulted trachyte along the eastern side of a small graben. The *northwestern group* flow quietly into the Northwest Lagoons at the base of fault scarps (W and E). Most are quiet seepage springs with low water and gas discharge (Fig. 12.7a, b). Several springs of the *southwestern group* discharge into the Southwest Lagoon (Fig. 12.7c, d). The northerly springs (N in Fig. 12.6a) flow along the low faulted western margin of the lagoon; southern springs (S) discharge northwards from mudflats south of the lagoon. Some vents are drowned during high lake level; others dry up during the dry seasons. The *eastern group* discharges from several seepage springs at the northeastern margin of the Eastern Lagoon (Figs. 12.6a and 12.7e, f), and some larger pool springs at its south-eastern margin. The *southeastern group* comprises scattered small shoreline springs near the margins of Bird Rock Lagoon and Graham's Lagoon (Fig. 12.7g). Some discharge subaqueously from loose sandy sediments, with others from fractured trachyte.

Fig. 12.5 Nasikie Engida. (**a**) View south (July 2019) showing the narrow horst that separates the lake from Lake Magadi, and the northeastern (NE) and northwestern (NW) subgroups of hot springs. (**b**) View north from southern shoreline showing trona pan (June 2011). (**c**) The Olekeju Ole Ntaga bounding fault-scarp and mouth of the hot river that drains the NW subgroup of hot springs. (**d**) Siliceous gels covered by microbial mats, northeastern shoreline. (**e**) Black siliceous gel from northeastern shoreline. (**f**) *Alcolapia* swimming along northern lake shore

Table 12.1 General characteristics of the main hot-spring groups at Lake Magadi and Nasikie Engida

Spring group	Number of springs[a]	Temperature range (°C)	Salinity range (g L^{-1} TDS)	pH range
Nasikie Engida				
Western subgroup	~20	43–86	25–29	9.01–9.44
Eastern subgroup	~15	76–85	26–28	8.72–9.50
Lake Magadi				
Northeastern group	7–10	65–69	26–29	8.88–9.50
Northwestern group	15–20	36–45	11–42	8.85–9.72
Southwestern group	10–15	41–46	37–45	9.30–10.0
Eastern group[b]	5–7	31–35	16–33	9.70–10.18
Southeastern group	10–15	37–41	18–22	9.45–9.89

Data from Baker (1958), Jones et al. (1977), Crane (1981), Allen et al. (1989), Denson et al. (2010), Kambura et al. (2016a), Renaut et al. (2021), and original measurements
[a]Approximate number of vents. Total varies with lake level (some are submerged, others dry up or undergo evaporation; most are small with vents 0.2–1 m diameter)
[b]Includes Fish Springs Lagoon springs

The spring locations are controlled by tectonics. They lie along N-S faults, but some groups are influenced by transverse NW-SE tectonic lineaments that probably represent subvolcanic basement trends (Le Turdu et al. 1999). Infrared surveys have confirmed extensive hot ground near many hot spring vents (Crane 1981; Crane and O'Connell 1983).

Although the springs have been largely stable in discharge and composition since the 1950s, Crane (1981) used unpublished reports from Magadi Soda Company to show that activity has changed since the early 1900s. She noted that geysers and boiling springs were active in the southern part of the lake in the 1940s and 1950s. Temperatures in several Magadi spring groups changed by up to 7 °C, increasing in both eastern and western spring groups. She then related the changes in spring activity to possible causes including rainfall fluctuations and seismic activity. Rainfall clearly influences the southern springs. The springs at Nasikie Engida reached maximum temperature (an increase of 23 °C) after an eruption of Ol Doinyo Lengai volcano in Tanzania, 90 km south of Lake Magadi.

Hot springs have been active at Lake Magadi for several hundred thousand years. Lacustrine sediments of the Oloronga Beds near the southwestern spring group and several outcrops to the west of Lake Magadi contain many small travertine chimneys that were formed by hot springs that were contemporary with the Oloronga palaeolake (Sect. 20.3). Tall travertine chimneys west of Lake Magadi have been dated as late Pleistocene (~14.4–10.9 ka: Hillaire-Marcel et al. 1986) but that age is difficult to reconcile with their present elevation unless they have been tectonically uplifted (Owen et al. 2019; Sect. 12.8.3).

The origins of the hot spring fluids are discussed below. Most researchers have concluded that the fluids derive from the Ewaso Ngiro River and local river waters

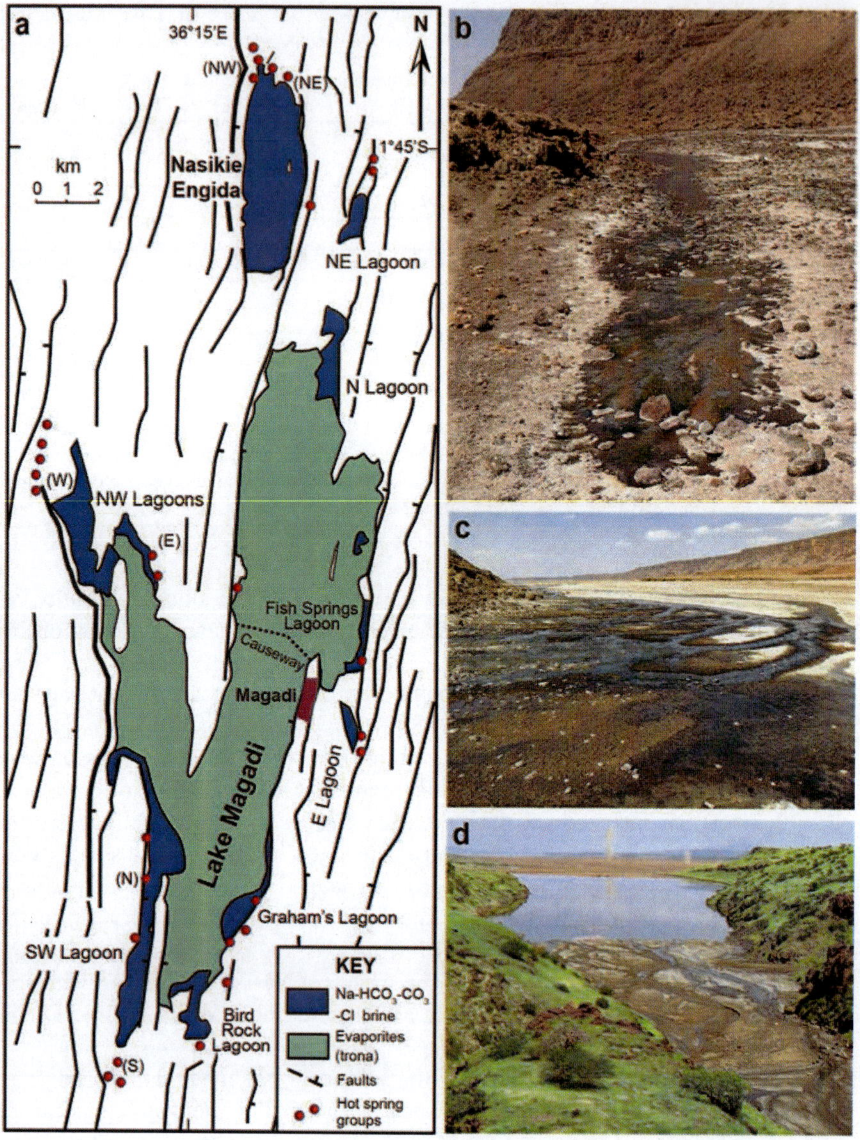

Fig. 12.6 Hot spring groups of Nasikie Engida and northern Magadi. (**a**) Map showing the distribution of the hot springs at Lake Magadi and Nasikie Engida. (**b**) Hot spring (82.5 °C) discharging from gravel, Nasikie Engida (NW) subgroup. (**c**) Seepage springs along fault line, Nasikie Engida (NE) subgroup. (**d**) Hot spring site, NE Lagoon, Lake Magadi

that have passed through channel alluvium and faults to feed groundwater reservoirs before being geothermally heated and rising to the surface as hot and warm springs along fault lines.

Fig. 12.7 Hot spring groups of the Magadi Basin. Locations in Fig. 12.6a. (**a**) Faultline seepage springs, western shore of NW Lagoon. (**b**) Small spring outflow, NW Lagoon (W). Arrow top right points to small vent. (**c**) Shoreline spring, SW Lagoon (N). (**d**) The SW Lagoon (S) spring group, looking south. Field of view at base of photograph is ~500 m. (**e, f**) Hot spring vents, E Lagoon. (**g**) Hot spring vent, Bird Rock Lagoon

12.5 Hydrochemistry

12.5.1 Composition of the Basin Waters and Their Chemical Evolution

The origin of the sodium carbonate brines at Lake Magadi and Nasikie Engida has long been a focus of research. Early work was partly speculative (Anonymous 1923; Gregory 1921). Stevens (1932) began systematic research on the water chemistry. Baker (1958, 1963) made many observations of the springs, lagoons and brine, but the first detailed studies were made by Hans Eugster and colleagues (Eugster 1970; Jones et al. 1977; Eugster and Jones 1979), who provided the most comprehensive overview of lake water evolution, based on the behaviour of major and some minor ions. Although a few stable isotopic analyses had been reported previously (O'Neil and Hay 1973), the first specific studies were undertaken by Hillaire-Marcel (1987) and Hillaire-Marcel and Casanova (1987), and later by Allen et al. (1989). Further water analyses were provided by Glover (1972), Darling (2001), Muiruri et al. (2021a), Renaut et al. (2021, for Nasikie Engida), and Getenet et al. (2020, 2022, 2023).

Inflowing river waters (Table 12.2) are dilute and of Na^+–HCO_3^-–Cl^- composition, typical of most rivers in the Kenya Rift. They vary in pH from 6.9 to 8.2 and in salinity from ~60 to 400 mg L^{-1}, increasing downstream. Shallow groundwaters are more saline (up to 2 g L^{-1} Total Dissolved Solids: TDS) with pH 8.0–9.3 and retain alkaline earths. The hot and warm springs are saline and alkaline, of Na^+–HCO_3^-–Cl^- composition, and vary in salinity from 10 to 40 g L^{-1} TDS and pH from 9.2 to 10.2. They have negligible Ca^{2+} and Mg^{2+} but have high silica and fluoride.

The lake brines vary seasonally in composition and from year to year (Table 12.3). The surface brines have Na^+–CO_3^{2-}–HCO_3^-–Cl^- composition and are rich in F^- and SiO_2. Surface brines in contact with trona have salinities of 100–300 g kg^{-1} TDS or higher and a pH of 10.0–11.2. The underlying interstitial brines in contact with Pleistocene sediments are variable but are typically half as concentrated as the surface brines and four times as concentrated as the most saline warm spring, with a pH range of ~9.85–10.45 (Eugster 1980).

Nasikie Engida waters are similar in composition to Lake Magadi, but the closed lake is perennial because it is fed mainly by hot springs. Renaut et al. (2021) showed that its shallow (<1.5 m) water increases in salinity from 30 g L^{-1} along the northern shoreline where hot springs enter the lake to >270 g L^{-1} at its southern end, a distance of 7 km, where evaporites often form.

The following summary of water evolution and its changing composition is based on Eugster (1980, 1986). He recognised that the Na:Cl ratio was nearly constant between inflow and concentrated brine and so was able to use chloride as a tracer to measure the gain and loss of other ions (plus silica) in fluids from the most dilute to the most saline. Dilute inflow consists of Na^+–HCO_3^-–Cl^- waters that initially have approximately equimolar amounts of Na^+ and Ca^{2+}. During weathering Na^+ is released more rapidly than Ca^{2+}. As the solute load increases through evaporative concentration, the Na:C1 ratio of the dilute inflow increases from 0.025 to 1, indicating input of Na^+ and HCO_3^- through silicate weathering reactions. At this point, the chloride content of the water has increased only from 0.1 to 0.25 mmol kg^{-1}

Table 12.2 Selected chemical analyses of Lake Magadi inflow waters

Sample	Ref no.	Date	Ref	Temp °C	pH	Na	K	Ca	Mg	Cl	HCO_3	CO_3	SO_4	F	SiO_2
Dilute inflow															
Ewaso Ngiro R		Jul-70	1	16.5	7.32	15	7.1	5.9	1.2	1.2	55		6.2	1.3	39
Ewaso Ngiro R		May-73	1	24	7.62	63	17	12	2.3	2.3	194		14	3.6	38
Oloibortoto R		Jul-70	1	16	7.93	9	2.3	6.5	3.7	3.7	48		2.4	0.2	20
Er.dosapia R		May-73	1	19	6.9	5.4	3.3	5.2	3.0	3.0	36		6.7	0.2	21
Narosura R		Jul-68	1	18	7.6	26	12	16	6.9	6.9	110		26	1	25
Shallow groundwater															
Ndopa wells	M1020	May-73	1	24	8.0	310	33	24	3.7	178	615		9	3.1	64
Ngejepa well	M217	Jan-68	1		9.3	1000	28	1.9	0.9	375	1000		25	11	60
Hot and warm springs															
Nasikie Eng W	NE106	Aug-06	2	82	9.13	10,500	280	<1	0	5620	11,500	3320	220	155	93
Nasikie Eng E	NE607	Aug-07	2	76	8.72	11,500	290	0.5	<0.5	5580	15,700	2230	230	165	95
NE hot spr gp	M686	Jan-71	1	68	8.95	11,300	169			5480	14,800	3070	185	152	99
NE hot spr gp	11	1985–6	3	65–7		11,100	157	0.6	<0.4	5850	20,900		134	130	83.4
NW hot spr gp	M1009	May-73	1	42	8.98	8060	104			1550	11,200	1590	122	69	104
NW hot spr gp	9	1985–6	3			9640	112	0.7	<0.4	4900	16,700		204	80	106
SW hot spr gp	M588	Jul-70	1	44.2	9.60	13,200	190			6390	8640	7120	231	141	79
SW hot spr gp	18	1985–6	3	45		12,300	115	0.6	<0.4	6450	23,400		189	130	68
SE hot spr gp	M1008	May-73	1	37	9.66	3870	49			1550	4080	1590	122	50	55
SE hot spr gp	M1107	May-73	1	42.5	9.84	12,500	121			6240	8740	7270	196	126	41
SE hot spr gp	13	1985–6	3			12,200	100	0.6	<0.4	6400	23,100		160		60
SE hot spr gp	16	1985–6	3			12,700	109	0.7	<0.4	6450	23,800		169	140	54.8
E hot spr gp	15	1985–6	3	31–35	9.7	7000	75.4	0.6	<0.4	3550	13,000		52.4	70	35.8

Sources (ref no): 1: Jones et al. (1977); 2: Renaut et al. (2021); 3: Allan et al. (1989). Further analyses are available in Burgess (1986), Tole (1990), Kambura et al. (2016a, b), and Muiruri et al. (2021a, b)

Analyses are in mg L^{-1} except those from Jones et al. (1977) which are quoted in mg kg^{-1}

Table 12.3 Selected chemical analyses of Lake Magadi and Nasikie Engida brines

Sample	Ref no.	Date	Ref	Temp	pH	Na	K	Ca	Mg	Cl	HCO_3	CO_3	SO_4	F	SiO_2
Surface brine – Lake Magadi															
North	M1029	May-73	1	30	9.5	88,100	1020			36,900	16,000	78,200	729	950	261
NW Lagoon	M664	Jun-70	1	25	9.8	78,500	1010			37,800	13,300	67,000	882	759	502
Central[a]	M520	Jun-70	1	30	10.77	111,000	2090			87,700	1790	71,300	2160	1980	1100
W central	M536	May-73	1	40	10.9	124,000	1790			74,100	0	93,000	1890	1560	1040
NW Lagoon	LM2	Aug-08	2	23.2	9.61	33,100	630	4	<1	17,400	20,000	20,300	550	320	306
Central[a]	LM3	Aug-08	2	33.4	10.5	168,000	2980	<1	<1	122,000	<1	120,000	1900	2100	1660
Central[a]	LM4	Aug-08	2	34.8	10.81	157,000	4300	<1	<1	144,000	244	99,500	2100	1840	1470
E Lagoon	LM10	Aug-08	2	28.2	9.96	10,000	275	2	<0.1	5360	4200	6820	120	100	81
Subsurface brine – Lake Magadi															
J10: 31.7 m		Jan-71	1		9.85	69,400	1170			42,100	6160	49,300	1060	807	1540
K: 297 m		Jan-71	1		10.0	40,500				21,600	11,000	29,200		452	288
Nasikie Engida															
North	NE2–06	Aug-06	3	31	9.24	13,900	370	1	0	7090	13,200	5300	270	116	193
South	NE5–06	Aug-06	3	35	9.62	107,000	3100	2	0	61,800	16,100	80,300	1400	1580	585

Sources (ref no): 1: Jones et al. (1977); 2: Muiruri et al. (2021a); 3: Renaut et al. (2021)
Analyses are in mg L^{-1} except those from Jones et al. (1977) which are quoted in mg kg^{-1}
[a]Sampled from Causeway

representing a concentration factor of 2.5. Inflow waters then infiltrate and move underground until they discharge at the surface as hot springs. Dilute streams have an average molar (Ca + Mg):HCO_3 ratio of 0.28. The four-fold dominance of bicarbonate over alkaline earths ensures that Ca^{2+} and Mg^{2+} are depleted early during evaporative concentration, with calcite precipitation acting as the chemical divide (Eugster and Jones 1979). Ca^{2+} and Mg^{2+} are precipitated in soils and shallow sediments often by evaporation (Surdam and Eugster 1976; Deocampo and Renaut 2016; Felske 2016). Calcium is also removed as calcite in oncoids and in efflorescent crusts as gaylussite. Bicarbonate is thus enriched, as are Na^+ and Cl^-. Assuming that Cl^- is a conservative solute, Eugster and Jones (1979) recorded gradual depletion in total carbon species with respect to chloride from dilute inflow to hot springs and then to lake brines. Carbonate alkalinity (HCO_3^- + $2CO_3^{2-}$) is almost constant (Eugster and Jones 1979), which implies a mechanism for CO_2 release, of the type:

$$2HCO_3^- \rightarrow CO_3^{2-} + CO_2 + H_2O$$

During this process, the charge balance (alkalinity) is preserved, and the PCO_2 is controlled by the atmosphere. This degassing of CO_2 contributes to the large pH increase observed during the evaporative concentration of the water, with photosynthesis also an important mechanism in the springs and lakes. When trona [$Na_3(CO_3)(HCO_3)\cdot 2H_2O$] precipitates in saturated brines, HCO_3^- and CO_3^{2-} are removed on an equimolar basis. Carbonate is enriched, bicarbonate is depleted, and the pH rises. Residual brines become depleted in HCO_3^- and undersaturated with respect to trona and can then only continue to precipitate trona through the addition of CO_2. Eugster (1980) proposed that CO_2 from organic matter decay would be the probable source. However, CO_2 also derives from magmatic sources that discharge along faults (Lee et al. 2016, 2017; Muirhead et al. 2016). This geothermal CO_2 also enables nahcolite ($NaHCO_3$) to crystallise in Nasikie Engida brine (De Cort et al. 2019; Renaut et al. 2021; Getenet et al. 2022). Trona and nahcolite precipitation also depletes the brine in Na^+ and relatively enriches it in Cl^-. However, the brines do not usually reach halite saturation (Kaufman et al. 1990) except in artificial evaporation pools near the soda plant (Fig. 12.1b). F^-, B, Br^- and PO_4^{3-} are also enriched in residual brines.

Potassium is lost between inflowing streams and springs indicating a loss during subsurface transport. The high K^+ loss is attributed to removal by exchange on reactive surfaces. These are mainly trachytic glass and silicate gels because clays are rare in modern settings (Herrick 1972; Surdam and Eugster 1976; Nikonova 2016). K^+ is also lost through the formation of zeolites (erionite, phillipsite; Rabideaux 2018). In contrast, there is no further loss of K^+ relative to Na^+ between springs and concentrated brine. Late-stage enrichment in K^+ relative to Na^+ occurs because of trona or nahcolite precipitation. Sulfate is strongly depleted in Magadi brines (Eugster 1970; Jones et al. 1977). Eugster (1970) noted that the deepest borehole brines are most impoverished in SO_4^{2-}. Bacterial sulfate reduction is likely the most effective removal mechanism over the whole concentration range. Pyrite is common in the black muds in Magadi cores (Owen et al. 2019). Wynn et al. (2018) reported that the pyrite is isotopically very heavy in terms of ^{34}S.

Silica remains at ~50 ppm between inflow and the springs while chloride increases 500-fold. Its solubility is controlled by that of trachytic glass and silicate gels. Recent removal is attributed to the precipitation of opaline silica cements and mm-scale crusts, which are common. As the pH rises, silica remains in the brine and is concentrated at the same rate as chloride because above pH 9 silica dissociates into $H_3SiO_4^-$ and $H_2SiO_4^{2-}$ in addition to H_4SiO_4. Silica has also been removed by precipitation of sodium silicates (see Sect. 12.7.5) and diatoms.

Fluoride is strongly enriched in the Magadi brines. Ca^{2+} is generally unavailable to form fluorite (CaF) because it has already been removed as $CaCO_3$. Villiaumite, NaF, is common in the trona deposits (Baker 1958; Nielsen 1999; Getenet et al. 2022) and, with detrital silt, is the principal impurity during trona extraction. Where calcite is present, such as in the Oloronga Beds and in equivalent drill-core sediments older than 370 ka (Chap. 20), fluorite may form by interaction with the interstitial brines (Surdam and Eugster 1976).

12.5.2 Origin of the Hot Spring Waters and Circulation

Parkinson (1914) suggested that the spring water is juvenile, containing magmatic CO_2, and that the fluid reacts with silicates in the volcanic rocks through which it passes to produce aqueous sodium carbonate and silica. The latter reactions are valid but stable isotope analyses showed that the water is meteoric (Allen et al. 1989). Stevens (1932) noted that the carbonate/chloride ratio in spring waters is constant implying a common source reservoir at depth. Guest and Stevens (1951) considered the rates of spring discharge versus lake evaporation and inferred that evaporated lake brine must sink and mix with rising spring water. Baker (1958) proposed that groundwater leaches salt from bedrock, which then becomes concentrated by repeated evaporation in the catchment and in the lake(s), and is then re-leached, accumulating as saline groundwater below the pan. Some recycled brine would then be heated and discharged from warm springs. Glover (1972) suggested that the springs originate from local steam-heated groundwater.

Eugster and colleagues provided the most detailed model for the evolution of the spring and lake fluids (Eugster 1970, 1980, 1986; Eugster and Jones 1979; Jones et al. 1967, 1977) using evaporative concentration and various solute fractionation mechanisms to explain the development and composition of the brine. The Eugster-Jones model (Fig. 12.8) was based mainly on the behaviour of major ions (and aqueous silica) between dilute inflow waters and the most concentrated brines. They proposed that inflow from rivers descending to the rift valley floor and local ephemeral runoff feeds local, dilute shallow groundwater bodies. They also envisaged a hot saline groundwater body within the fractures of the trachyte bedrock below the valley floor, based on the constancy of spring temperatures and compositions over 40 years although Crane (1981) showed that temperatures do change at some springs). The hottest spring (86 °C) north of Nasikie Engida, which has a similar Na/Cl ratio to the dilute inflow, was assumed to represent water that evolved from

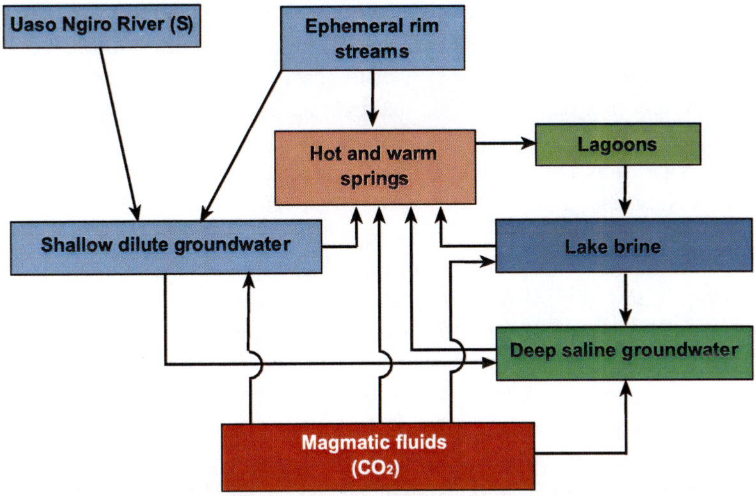

Fig. 12.8 Schematic circulation system for Magadi basin. Revised from Eugster (1980) and later sources

the original dilute recharge. To get the different spring temperatures, compositions and salinities, they proposed three inputs that are mixed in different proportions for each spring: (1) shallow groundwater (dilute cold), (2) deep groundwater (saline hot), and (3) recirculated surface brine (cold, hypersaline, very high pH) (Eugster 1970; Jones et al. 1977). Eugster emphasised the importance of surface and capillary evaporation and fractional dissolution of efflorescent crusts as mechanisms to increase the solute load. Precipitation of efflorescent crusts with their later fractional dissolution by runoff strongly influences brine evolution. Na and Cl in crusts are contained only in very soluble minerals, so no fractionation occurs between Na and Cl. Alkaline earths (Ca, Mg), in contrast, are removed as carbonate minerals (mainly calcite) with silica lost as opaline cement.

Hillaire-Marcel (1987) and Hillaire-Marcel and Casanova (1987) added stable isotopic analyses (δ^2H, $\delta^{18}O$) of the rain, springs and lake to the model to try to identify the types of basinal fluids. They showed that deep thermal waters are depleted isotopically with respect to modern recharge and suggested that this reflects recharge at ~10,000 y BP during the African Humid Period (AHP) when Lake Magadi was high and relatively fresh.

Allen et al. (1989) concluded from stable isotopic results that the hottest springs (N Nasikie Engida) represent unaltered meteoric water that has not mixed much with other types of water. The stable isotopic composition of the spring water is nearly identical to the lower Ewaso Ngiro River (S), 25 km to the west. Reservoir recharge would be by slow percolation through the floodplain alluvium. They noted that the Cl^- concentration of the springs is inversely related to their temperature, implying mixing of a cooler saline fluid with a hotter more dilute fluid. Similar relationships are found among other components. Baker (1986), Hay (1986) and earlier workers considered the possible influence on brine composition of natrocarbonatite ash from Oldoinyo Lengai volcano 90 km to the south. Although ash was a probable contributor, the composition of the basin waters is clearly related to the chemical weathering of the volcanic bedrock and ash by silicate hydrolysis (Sect. 4.4.2; Eugster 1980).

Lake Magadi lies in a regional sump. It has been speculated that groundwater from Lake Naivasha (a dilute lake at 1883 m elevation 110 km to the north; Chap. 11) may reach Lake Magadi and discharge at the springs (Eugster 1970; Becht et al. 2006) but the geochemical (stable isotopic) evidence does not support this (Allen et al. 1989; Darling et al. 1990). The heat source is magmatic – probably dyke injections along the rift axis (Komolafe 2010; Githiri et al. 2011; Atuya et al. 2014). Githiri et al. (2012a, b) showed that the heat flow at Magadi ranges from 69.9 to 111.5 °C km^{-1}, with the highest values north of Lake Magadi and near Nasikie Engida. This agrees with earlier workers, including Allen et al. (1989), who thought the locus of geothermal upflow might lie near the hot springs west or northwest of the lakes. Reservoir temperature estimates using standard chemical geothermometers lie in the 100–150 °C range (Allen et al. 1989; McNitt et al. 1989). The overall influence of mantle CO_2 on basinal fluid evolution is still to be assessed.

12.6 Life in the Lakes

Lake Magadi and Nasikie Engida are best known for their special fish, but they are also home to birdlife, including flamingos, and their diverse array of microorganisms including alkaliphilic and haloalkaliphilic bacteria and archaea (Fig. 12.9a–c).

Considerable attention has been given to the single species of fish that inhabits the warm springs and lagoons at Lake Magadi (Coe 1966; Tichy and Seegers 1999; Wilson et al. 2004). The fish, the endemic tilapia *Alcolapia grahami,* has also been called *Oreochromis alcalicus grahami*, *Sarotherodon alcalicus grahami* or *Tilapia grahami* (Fig. 12.9e, f). The ecology and biology of the fish has been reviewed by Kavembe et al. (2016). The fish live in the warm spring pools and adjacent lagoons of Lake Magadi. They are absent in the very hot springs (>80 °C) of northern Nasikie Engida but are present in cooler (25–40 °C) water offshore (Fig. 12.5f). The fish live in extreme conditions: pH ~10, alkalinity of 300 mmol L^{-1}, temperature up to 42 °C, high UV radiation, daytime hyperoxia with night-time hypoxia, high salinity and intense predation by birds (Kavembe et al. 2016). They feed mainly on cyanobacteria (*Arthrospira* sp.), *Dunaliella* spp. (green alga that gives the lake its frequent pink colour because of β-carotene), dipteran larvae and copepods.

The fish in the four main lagoons are physically separated by trona evaporites; the Nasikie Engida fish are separated from Lake Magadi by a horst. Coe (1966, 1967) suggested that mixing of the different populations might occur occasionally during floods, but genetic studies imply that there is little or no gene flow among modern populations (Kavembe et al. 2014). Three different tilapia species inhabit Lake Natron. Seegers et al. (1999) suggested that a common haplotype might have

Fig. 12.9 Life in the Magadi lakes. (**a**) Green microbial bloom (*Arthrospira*?) at the northern shore of Nasikie Engida. (**b**) Red-pink bloom of haloalkaliphilic archaea (?) in evaporation pans 750 m east of soda plant. (**c**) Warm seepage spring (~35 °C), Northwest Lagoon (W), floored by microbial mat. (**d**) Ephydrid flies feeding on microbial mat, Northwest Lagoon (W). (**e**) Young *Alcolapia* in hot spring pool feeding Graham's Lagoon. (**f**) Adult *Alcolapia* with nests, Northwest Lagoon (E). (**g**) Lesser flamingos, Northwest Lagoon (W). (**h**) Warm spring with sedges (*Cyperus laevigatus*), eastern shoreline of Northwest Lagoon (E). Inset shows small vent. Scale bar = 20 cm

been present in the ancestral fish population of Pleistocene Lake Oloronga (Sect. 20.3) which eventually gave rise to different populations in Lake Magadi and Lake Natron after that large palaeolake separated into different water bodies.

Many birds inhabit Lake Magadi and its environs. These include pelicans and numerous wading birds such as yellow-billed storks, spoonbills, egrets and herons, which are common in the spring-fed shallow water lagoons, especially the SW and SE lagoons, the NW lagoon, and northern Nasikie Engida. Both lesser and greater flamingos are often present (Fig. 12.9g). Typically, a few thousand lesser flamingos are present (Bartholomew and Pennycuick 1973; Krienitz et al. 2016). Although nesting has not taken place recently, Brown and Root (1971) described a breeding colony of more than a million pairs in 1962 when Lake Natron was very wet following floods in 1961. However, few survived when the lake dried up, thus enabling predation. A few birds bred again in 1978 (Tebbs et al. 2013). Subfossil and fossil flamingo nest mounds are present in the southern part of East Lagoon (Buatois et al. 2016; Scott et al. 2021).

Lake Magadi has no zooplankton or other aquatic invertebrates (Mengistou 2016). The invertebrates around Lake Magadi are dominated by Diptera. Ephydrids and other flies are common near hot and warm springs and their exuviae are redistributed along some lake margins (Fig. 12.9d). Chironomids are locally common. Predatory tiger beetles, earwigs and spiders are also present (Scott et al. 2021).

The microbiota of Lake Magadi has been a focus of many studies because the lake hosts so many extremophiles (e.g. Zhilina and Zavarzin 1994; Grant et al. 1999; Davis 2012; van Zyl et al. 2016; Mwirichia and Orwa 2023). Grant and Jones (2016) reviewed the bacteria, archaea and viruses of soda lakes in general including those at Lake Magadi. The colonies (cyanobacteria and algae) are dominated by *Arthrospira* and *Synechococcus*. *Picocystis salinarum*, a chlorophyte, was also recorded (Krienitz et al. 2012; Krienitz and Schagerl 2016). Krienitz (2018) noted that the saline mudflats are locally filled by shallow red brine and that springs produce a mosaic of habitats. Wet areas are commonly covered by cyanobacterial mats or layers of benthic diatoms (Muiruri et al. 2021a). Planktonic microphytes occur only where the water is ~20 cm deep. As salinity changes the microbiota respond, often rapidly. Krienitz (2018), for example, noted that a southern lagoon fluctuated in salinity from 30 to 81 ppt during a multiyear study and that planktonic cyanobacterial diversity (including *Aphanothece* sp. and *Synechocystis*) increased at higher salinity. Shoreline fungi have been studied by Bondarenko et al. (2018) with fungi and spores also recorded in Magadi drill cores (Muiruri 2017). Kambura et al. (2016a, b) and Mulango et al. (2020) have recorded the bacteria, archaea, and the fungi of the Magadi and Nasikie Engida hot springs. Higher plants are common near hydrothermal outflow (Fig. 12.9h).

12.7 Modern Sedimentation

12.7.1 Siliciclastic Sedimentation

No perennial rivers flow into Lake Magadi or Nasikie Engida today although large braided streams fed Magadi during the Pleistocene (Sect. 20.3). Ephemeral streams feed both lakes (Fig. 12.10a). The largest of these are axial, flowing seasonally from both north and south into the Magadi graben, and from the north into Nasikie Engida. The axial streams deposit gravel in their upper courses, but mainly feldspathic (anorthoclase) and lithic sand, silt and mud in their lower courses (Fig. 12.10d–f). Mudflats north and northwest of Lake Magadi, and north of Nasikie Engida, are partly fed by perennial hot-spring outflow that supplies water but relatively little sediment (Figs. 12.5c and 12.10b, c). In places, their channels coalesce to form anastomosing networks. Most lateral streams are minor with small catchments.

Axial deltas are poorly developed as morphological features but muddy hypopycnal plumes may extend offshore across lake brine with mud settling offshore during floods. In contrast, lateral deltas and fan deltas (e.g. W Nasikie Engida, SE and NE Lake Magadi) commonly develop radial channel networks with coarser sediments unless structurally confined (e.g. Renaut et al. 2021; Figs. 12.10e and 12.11). All are small features (<1 km radius) and are frequently submerged during the rainy seasons, sometimes completely (e.g. W Nasikie Engida). The lake margins in deltaic settings are bordered by dry (proximal) and saline (distal) mudflats that commonly have mudcracks and intraclasts (Fig. 12.10g) and are often bioturbated (Fig. 12.10f; Scott et al. 2021). Fan-deltas commonly prograde across older mudflats. Many mudflats are old exposed lacustrine sediments (High Magadi Beds, Sect. 12.8.2; Eugster 1986). The saline mudflats around the shoreline frequently are covered by efflorescent trona crusts with exposed microbial mats and crusts near the shoreline (Fig. 12.12). Aeolian sediments are locally present, particularly along the western margin of the East Sub-basin, where loess drapes the sediments along the slopes below the Magadi Horst. Windblown silt is redistributed across the basin by dust devils and occasional dust storms that deflate dry mudflats (Fig. 4.17a, c). Deforestation and cultivation of the Mau uplands north of the lake (mainly Narok) have increased soil erosion and siltation in Lake Magadi over the last two decades (Kiama et al. 2019).

Modern fine clastic sediments are rich in erionite (detrital and authigenic) and also contain patchy coatings of calcite, gaylussite, opaline silica and fluorite (Surdam and Eugster 1976; Renaut et al. 2021) that are likely evaporative precipitates. The clay fraction is composed mainly of smectite and illite, but clay minerals are rare in the lake sediments (Surdam and Eugster 1976). The sand and silt fractions are mainly feldspar, lithic grains (including volcanic glass), and particles (e.g. chert) reworked from older sediments.

Fig. 12.10 Dry mudflats. (**a**) Ephemeral stream channel that flows into mudflats south of Lake Magadi. People (arrow) for scale. (**b**) Hot-spring fed anastomosing streams crossing mudflats north of Lake Magadi. View from col at southeastern Nasikie Engida. (**c**) Straight hot-spring-fed channel cutting mudflats in **b**. (**d**) Dry mudflats west of Bird Rock Lagoon. (**e**) Deltaic mudflats, Bird Rock Lagoon, SE Lake Magadi. (**f**) Soft mudflats, North Sub-basin. (**g**) Dry cracked mudflat, southern East Sub-basin. (**h**) Flat oncoids, southern margin of Bird Rock Lagoon

Fig. 12.11 Fan-deltas. (**a**) Bayhead fan-delta Northeast Lagoon. hs hot springs. (**b–d**) Fan delta feeding Graham's Lagoon. hs hot springs, HMB High Magadi Beds. (**e**) Small fan-delta southwest Nasikie Engida. (**f, g**) Gravelly fan-delta plain, midwestern shore, Nasikie Engida

Fig. 12.12 Saline mudflats. (**a**) Dry mudflat to saline mudflat transition, fan delta west of Nasikie Engida. (**b**) Black reduced zone below efflorescent crust in **a**. (**c**) Saline mudflat NE Nasikie Engida. (**d**) Saline mudflats northern Northwest Lagoon. (**e**) Saline mudflat on gravel colluvium, Graham's Lagoon. (**f**) Small pressure ridges in efflorescent crust. (**g, h**) Efflorescent salt on crinkly microbial mats north of Bird Rock Lagoon

12.7.2 Alkaline Earth Carbonates

Ca^{2+} and Mg^{2+} have low concentrations (<2 mg L^{-1}) in the spring and lake waters but the dilute inflow waters, including unchannelled runoff, contain both ions (Jones et al. 1977). Calcite is commonly present in exposed lake sediments and in fractured trachyte where it forms by evaporation of dilute fluids. Eugster (1980) described and illustrated discoid pisoids (pisolites) from the mudflats south of Lake Magadi and attributed their formation to the precipitation of calcite on chert fragments where Ca-bearing inflow waters mixed with alkaline brine. Casanova and Tiercelin (1982) reinterpreted them as flat oncoids (Fig. 12.10h), recognising microbial filaments in the carbonate laminae. Travertine and tufa are absent because the modern hot-spring waters contain very little Ca^{2+} and Mg^{2+}.

12.7.3 Evaporite Precipitation

The sodium carbonate evaporites are the most distinctive recent sediments of Lake Magadi and Nasikie Engida, forming a seasonally exposed saline pan at Magadi (Fig. 12.13a–d) and sometimes at the southern part of Nasikie Engida (Fig. 12.14a, b). They are also common as ephemeral efflorescent crusts (Fig. 12.12). Trona dominates, but nahcolite, halite and thermonatrite are sometimes present (Fig. 12.13a, b), with a few mixed salts (e.g., gaylussite) and minor sulfate minerals (Walter 1922; Eugster 1980; Kodikara et al. 2016). Baker (1958) first described the trona deposits in detail and termed the uppermost trona present in cores 'The Evaporite Series'. In a sequence of papers, Eugster (1971, 1980, 1986) described and interpreted their origin from a geochemical and mineralogical perspective. Trona is present at the surface and in core >60 m below the surface, commonly interbedded with black mud and (or) chert in beds, cm-scale bands, laminae, and as interstitial crystals in mud. The oldest trona in core is dated at ~105 ka (Owen et al. 2019). The trona typically forms a hard crust of porous interlocking crystals that radiate upwards. The deposits are banded on a 2–5 cm scale. Tightly-spaced clear crystals radiate upwards in bunches (shrubs) that make light bands, which are separated from each other by thinner bands of trona with dark material consisting of aeolian dust, or mud and microbial debris (Fig. 12.13e–g).

Formation of the trona crusts has been described by Eugster (1980), McNulty (2017) and Renaut et al. (2021). During the rainy season (February–April) Lake Magadi is flooded by shallow brine derived from dilute to brackish inflow and rain upon the lake surface. The uppermost existing crust may be partly dissolved creating a discontinuity and secondary porosity. Fine debris washed in settles on the corroded substrate forming the 'dirt bands'. The brine then begins to evaporate producing crystal rafts (Fig. 12.14c) at the air-water interface and bottom-nucleated crystals that grow upwards as radiating fans and vertical crystals, which interlock and coalesce laterally to produce a bed of trona (Figs. 12.14d–f and 12.15b–d).

Fig. 12.13 Na-CO_3 evaporites. (**a**) Acicular trona needles. (**b**) Nahcolite from floor of Nasikie Engida. (**c**) Dry saline pan of northern Lake Magadi. (**d**) Graham's Lagoon moat separating trona pan from Magadi Horst. (**e**) Fans of bottom-nucleated trona showing syntaxial overgrowth though muddy layer deposited during flooding. (**f, g**) Successive trona beds showing radial growth and dirt bands resulting from flooding and dust deposition. (**g**) shows enlarged view of area highlighted in (**f**)

Crystals and plates from rafts sink in the shallowing brine (Fig. 12.15e) and may be redistributed by wind forming barchanoid bedforms of imbricate crystals (Figs. 12.15f and 12.16a, b) and crystal beaches along the shoreline (Fig. 12.16c, d). The brine becomes shallower, its density increases with evaporation, and the brine sinks and eventually withdraws below the saline pan surface. Syntaxial overgrowths and cement then commonly occlude primary and secondary porosity. Eventually, the surface desiccates fully. When exposed during the dry seasons, the white, grey

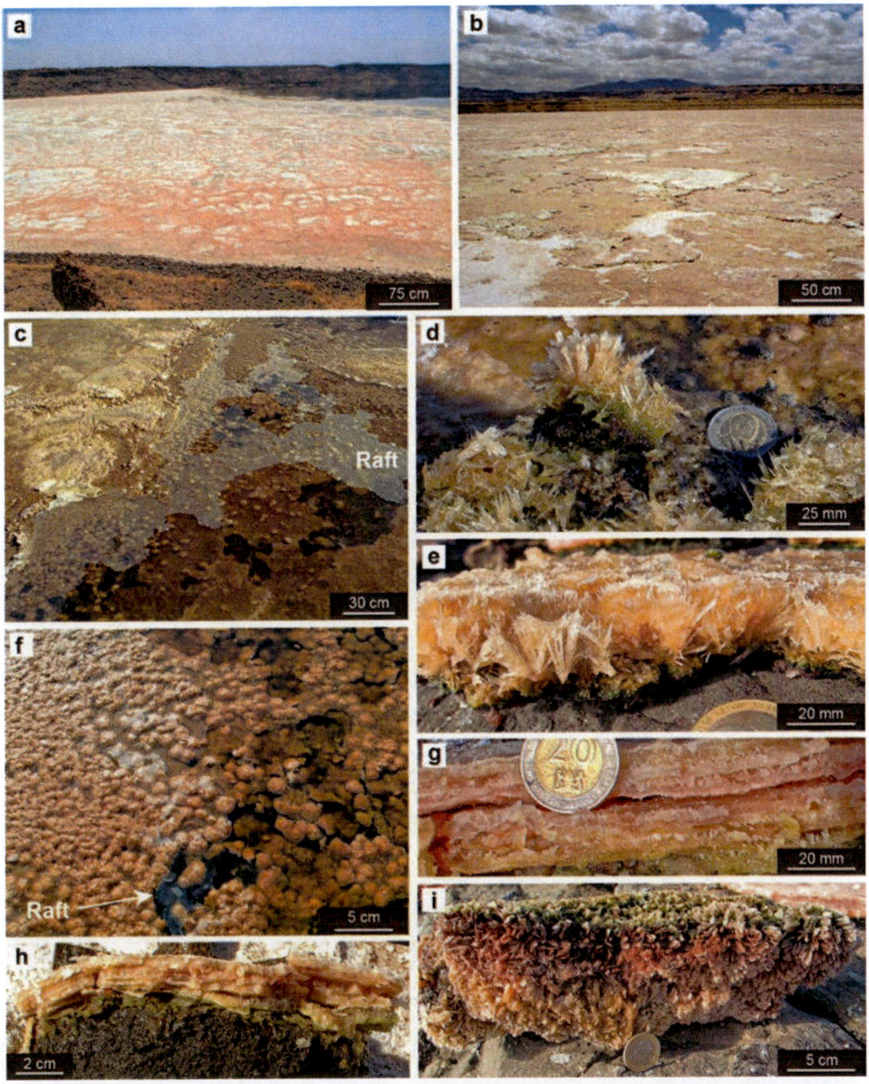

Fig. 12.14 Modern evaporites, Nasikie Engida. (**a**) View from SE shoreline of trona crust, July 2009. (**b**) Trona crusts, SW shoreline, July 2009. (**c**) Trona raft and sunken rafts near SE shoreline. (**d**) Bottom nucleated trona crystal shrubs upon green microbial layer. (**e, f**) Coalescent trona shrubs forming crystal bed. (**g**) Sunken stacked crystal rafts thickened by overgrowth salts on upper and lower surfaces. (**h**) Thin trona crystal beds lying on muddy trona, near SE shoreline. (**i**) Nahcolite crystal beds near mid-eastern shoreline

or pink salt crust receives dust, is periodically flooded by runoff (Fig. 12.16e) and may be broken by metre-scale expansion polygons bordered by pressure ridges a few dm high (Figs. 12.14a, b, 12.15a, b and 12.16f, g).

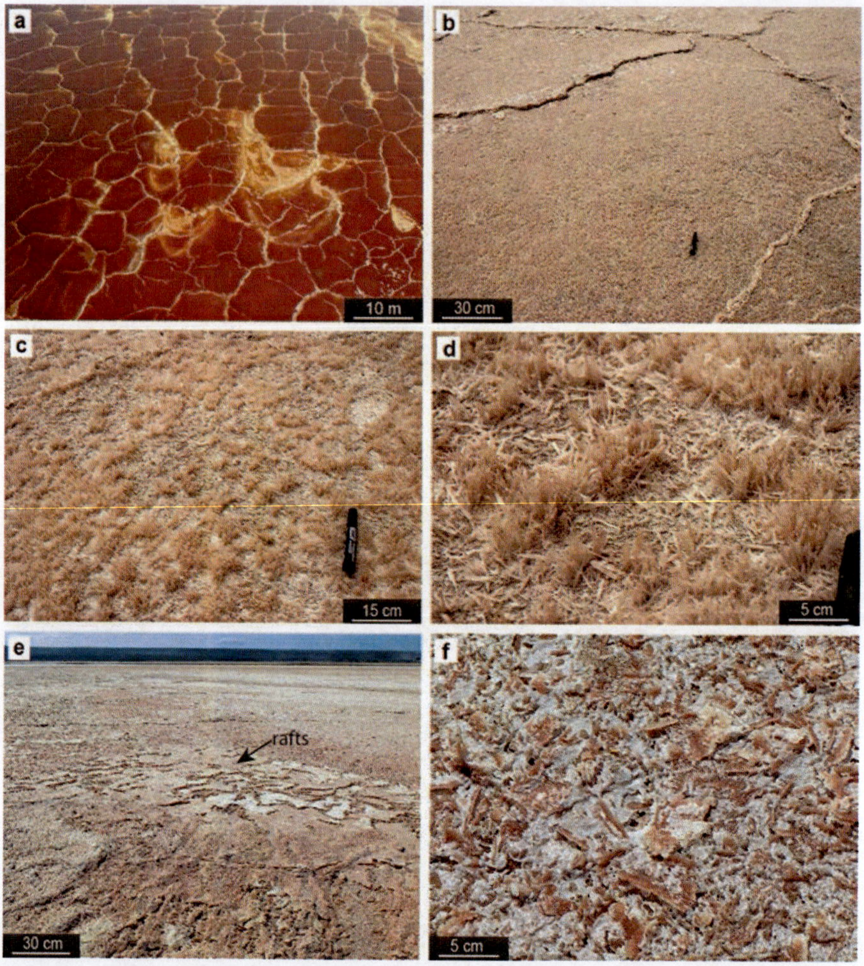

Fig. 12.15 Trona evaporites, Lake Magadi saline pan. (**a**) Polygonal pressure ridges south of the causeway (aerial view from helicopter). (**b–d**) Shallow dry pool with trona shrubs and crystal debris. (**e**) Sunken rafts near shoreline. (**f**) Raft debris with thin cements

Eugster (1966, 1971, 1980), Bradley and Eugster (1969) and Monnin and Schott (1984) discussed the evolution of sodium carbonate fluids and showed that they are strongly dependent on PCO_2. Trona is the dominant carbonate at the surface and its stability field straddles the surface and atmospheric PCO_2. CO_2 from bacterial decay of organic matter maintains trona precipitation when bicarbonate becomes depleted. Nahcolite is stable when the PCO_2 is higher than the atmosphere (Eugster 1966; Bradley and Eugster 1969; Jagniecki et al. 2015). Nahcolite found in cores was originally attributed to CO_2 produced biogenically by bacteria in anaerobic

Fig. 12.16 Common features of the Magadi saline pan. (**a, b**) Salt barchans composed of wind-driven salt-crystal rafts. (**c**) Regressive salt beaches composed of drifted raft debris. (**d**) Imbricated raft debris reworked near shoreline. (**e**) Muddy channels produced by runoff across trona pan. (**f**) Pressure ridge on pan surface. (**g**) Pressure ridge network that filled with brine which dried to produce bottom nucleated trona

sediments. While that may explain some nahcolite, it is now known that CO_2 released from mantle sources (Lee et al. 2016, 2017; Muirhead et al. 2016), along rift faults, is a factor in both trona and nahcolite genesis. Nahcolite is present on the floor of Nasikie Engida as a bottom-nucleated crust of clear crystals with pockets of probable CO_2 gas below that can be related to a high flux of CO_2 from local geothermal sources (De Cort et al. 2019; Renaut et al. 2021).

Getenet et al. (2020, 2023) have determined the evolution of the Magadi brines and hot spring inflow waters and examined the sequence of mineral precipitation using geochemical modelling and evaporation experiments. They confirmed that pH and the CO_3^{2-}/HCO_3^- ratios determine the order and amount of mineral precipitation, and the resulting mineral paragenesis in the East African soda lakes. Precipitation of Na carbonates and bicarbonates increases the pH (hyperalkalinity) of the brines. Their modelling shows that the starting composition of the brine is critical in controlling the Na-carbonate mineralogy. In Lake Magadi, Lake Natron and Nasikie Engida, carbonate alkalinity is dominated by CO_3^{2-}, but in the Nasikie Engida hot springs, which have lower pH than the lake brines, HCO_3^- is significantly higher than CO_3^{2-}. The lower pH makes enough bicarbonate available for nahcolite precipitation in northern Nasikie Engida and across much of the lake except near the southern shoreline where the salinity and pH increase (Renaut et al. 2021) and trona becomes dominant. At Lake Magadi and Lake Natron, nahcolite should precipitate briefly during evaporation but redissolve after further evaporation and increasing pH favour trona precipitation. Thermonatrite and halite form in efflorescent crusts and small shallow pools in which intense rapid evaporation limits the dissolution of atmospheric CO_2.

12.7.4 *Spring Deposits*

The modern springs precipitate only efflorescent salt crusts (mainly trona, nahcolite, with lesser halite and thermonatrite) on subaerial sites and upon exposed microbial mats around the vent margins and along the edges of outflow channels. Thicker salts may form in very shallow ponds as the channel outflow dries up. Some crusts at Nasikie Engida also contain minor opaline silica (opal-A) that may be attributed to the cooling and evaporation of the silica-rich waters, but silica sinter is absent.

12.7.5 *Siliceous Sediment*

The modern brines have some of the highest concentrations of silica measured in natural waters (>1000 ppm SiO_2: Jones et al. 1967; Garcia-Ruiz 1998) so it is unsurprising that the Magadi and Natron basin sediments are rich in chert and sodium silicate minerals such as magadiite [$NaSi_7O_{13}(OH)_3 \cdot 3H_2O$] and kenyaite [$NaSi_{11}O_{20.5}(OH)_3 \cdot H_2O$]. The fate of aqueous silica in the modern lakes is unclear.

Silica is present as magadiite and kenyaite in the late Pleistocene High Magadi Beds (Sect. 12.8.2) where some has already transformed to quartzose chert, and in older Pleistocene sediments, where some is also present as diagenetic makatite [$Na_2Si_4O_8(OH)_2 \cdot 4(H_2O)$] and kanemite [($NaHSi_2O_5 \cdot 3(H_2O)$)] (Eugster 1967, 1969; Sheppard et al. 1970). However, recent magadiite is not obvious in the modern lake. More common is opaline silica (opal-A), which is present in evaporative efflorescent crusts, mm-scale coatings on mineral grains and upon exposed rock surfaces, and as shallow pore-filling cements. Modern diatoms are rare (Eugster and Chou 1973) often dissolving in the high-pH water. Barker (1990), Barker et al. (1990) and Muiruri et al. (2021a) reported the diatom assemblages in the modern springs and lagoons of Lake Magadi and Nasikie Engida, but also noted well-preserved diatoms in parts of the Pleistocene and Holocene sedimentary sequence, which they attributed to high concentrations of silica in palaeolake waters.

Fritz et al. (1987) examined silica behaviour in neighbouring Lake Natron. They concluded that sodium silicate precipitation is enhanced by falling pH in a strong alkaline solution, with a decline in pH that forces a rapid decrease in silica dissociation which provides excess dissolved silica. Gueddari (1984) found magadiite near a sodic shoreline spring with $pH < 10$ but not in the salt crust at Lake Natron. Renaut et al. (2017) similarly found magadiite at modern hot springs at southern Lake Bogoria. Cooling may contribute to magadiite precipitation because of its decreasing solubility with temperature (Maglione 1976). Modelling by Getenet et al. (2023) showed that the silica-rich lake brines should precipitate silica during the initial stages of evaporation at relatively low pH values compared to final brines, but that silica likely redissolves when the pH increases with the precipitation of trona or nahcolite. This redissolution does not happen in the less alkaline waters near the inflowing hot springs. Silica accumulation in stratigraphic sequences, therefore, requires either low evaporation levels or a relatively lower pH of the brines. This is supported by the presence of chert in the Pleistocene Oloronga Beds and Green Beds (Owen et al. 2019; Buatois et al. 2020; Chap. 20), where it is associated with sediments containing calcite and trace fossils.

Silica also precipitates as gels (Fig. 12.5d, e). Gels were originally found along the northern shoreline of Nasikie Engida (Eugster and Jones 1968; Eugster 1986) near the hot springs, but were later found elsewhere in the basin, including lake cores (Herrick 1972; Surdam and Eugster 1976; Leet et al. 2021; Leet 2022). Some cherts in cores were interpreted to have had gel precursors (Behr and Röhricht 2000; Behr 2002; Leet et al. 2021). Modern gels at Nasikie Engida were found to be of Si-Na-Al composition. Others are primarily Si + Na with relatively low Al. Eugster and Jones (1968) attributed gel formation to the alteration of trachyte by highly alkaline waters and reprecipitation as a gel either along the shoreline or during diagenesis. Renaut et al. (2021) showed that gels at Nasikie Engida might have formed by evaporative concentration of the lake brine and might form during the evaporation of shallow pore fluids. Silica is also lost during early diagenetic reactions, particularly the alteration of the abundant volcanic glass to zeolites (Hay 1968; Herrick 1972; Surdam and Eugster 1976; Rabideaux 2018).

12.8 Late Pleistocene-Holocene Sedimentation

12.8.1 The Evaporite Series

The Evaporite Series (Baker 1958, 1963) is the youngest stratigraphic unit in the Magadi basin and represents the Holocene evaporites that underlie the surface of the saline pan in the axial part of the basin (Zone G6d in Core Mag14-2A: Owen et al. 2019). The evaporites are interbedded with anoxic black, brown, and olive-green zeolitic mud, clay and silt. However, evaporites first appear at 65 m depth in Core Mag14-2A and are dated at 105 ka. Butzer et al. (1972) reported ^{14}C ages for trona of 10,100 and 4600 y BP; Behr (2002) reported an age of 9120 y BP for trona at Nasikie Engida. The validity of these ages is questionable given that they lie within the African Humid Period (AHP: ~15–5 ka).

12.8.2 The High Magadi Beds (Terminal Pleistocene – Early Holocene)

The High Magadi Beds (HMB) underlie trona and interbedded mud of the Evaporite Series in core and are preserved in scattered outcrops around the lake margins. They consist of lacustrine, volcaniclastic, colluvial and alluvial deposits (Figs. 12.17a–f and 12.18). They contain fossils indicative of fresh to mildly saline conditions in a few horizons, including fish and diatoms that record an expanded, but fluctuating, deeper lake than today and a wetter environment (Temperley and Thompson 1950; Temperley 1951; White 1953). In outcrop, the HMB are up to ~6 m thick. Surdam and Eugster (1976) suggested that they are up to 80 m thick in core but Owen et al. (2018, 2019) inferred a thickness of <40 m in the northern half of the lake. The HMB record sedimentation in the axial graben of the Magadi Basin during the AHP followed by progressive aridification that led to trona precipitation (Evaporite Series) which continues today. The HMB were preceded by a generally drier erosional phase and low lake levels, lasting millennia when there was no net sedimentation except in the axial trough. HMB are preserved mainly in the central trough but the palaeolake may have expanded beyond those limits during its maximum stages.

In outcrop, the transition from the underlying Pleistocene Green Beds (~190–40 ka; Sect. 20.5) to the HMB is typically a disconformity (Fig. 12.17g). In outcrop, the resistant Green Beds commonly form the higher land surfaces, especially south, east and northeast of Lake Magadi. In other places, the HMB rest unconformably on the older Pleistocene Oloronga Beds (Sect. 20.3). Eugster (1980) considered a calcrete ('caliche cap') that commonly lies upon the Oloronga Beds to be a stratigraphic marker that separates those sediments from the overlying deposits including the HMB (Fig. 12.17h). Behr (2002), supported by Felske (2016), showed that there are several calcretes of different ages in the Magadi Basin – not just the

Fig. 12.17 High Magadi Beds at Lake Magadi. (All photographs from Dry Lagoon except **g**). (**a**) Proximal deposits. Lower member silts and gravels lie upon Magadi Trachyte and are overlain by volcaniclastic silts of upper member. Red notebook (arrow) is 15 cm long. (**b**) Laminated erionite silts and fine sands of upper member. (**c**) Interbedded silts, sand and gravel of lower member overlain by upper member. Gravel is cross bedded with normal grading. (**d**) Proximal offshore sediments. Lower member sediments are finely bedded mud and fine gravel. Compaction at upper contact. (**e**) Erosional contact between upper and lower member. (**f**) Bioturbated muds of HMB lower member with black fossil fish on bedding planes. Arrows point to burrows with different diameters. (**g**) Green Beds chert mound disconformably draped by buff volcanic silts of upper member, East Sub-basin, ~700 m east of Magadi Soda Club. (**h**) Section at eastern edge of Dry Lagoon showing volcanic silts of upper member lying on pisolitic calcrete that caps zeolitic silts and sands of Oloronga Beds

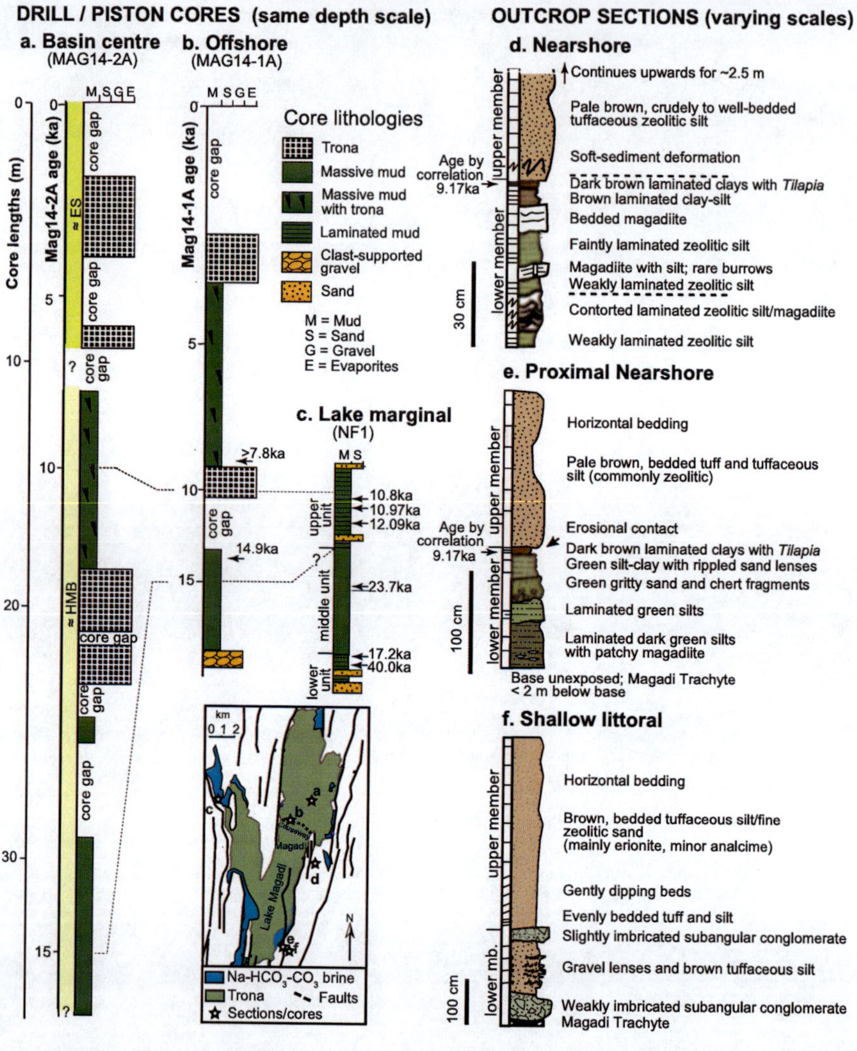

Fig. 12.18 Stratigraphic sections of the High Magadi Beds in core (**a**–**c**) and outcrop (**d**–**f**). Deeper to shallower settings from **a** to **f**. Core lithologies in key; outcrop lithologies described to right

thick example capping the Oloronga Beds. The youngest calcrete formed during the period of deposition of the HMB (Behr 2002).

Figure 12.18 shows the HMB in cores and outcrop arranged according to basin location (deep to shallow water settings). Two informal members are recognised in outcrop (Eugster 1969) (Figs. 12.17a, d and 12.18d–f). The upper member (~4 m thick) consists of pale brown tuffaceous silt(stone)-fine sandstone that shows vague horizontal bedding (1–5 cm units) (Fig. 12.17b) but with dips at shallow angles where draped upon an irregular substrate, ripple cross lamination and rare fish fossils. The volcanic glass forming the tuffaceous silt is largely altered to erionite. The source of the ash is unknown. The volcanic silt is probably mostly waterlain: evidence for subaerial exposure and channelling is generally absent. The lower member consists mainly of organic-rich green to brown mud, some containing magadiite, Magadi-type chert, and tuff. Detrital anorthoclase, amphibole, illite and zeolites are common (Eugster 1969; Surdam and Eugster 1976). Directly below the contact a dark brown mud, locally laminated, contains black *Tilapia nilotica* (Baker 1958) on bedding planes that indicate relatively fresh water (Fig. 12.17f). Plant debris and bioturbation are common near the palaeolake margins (Fig. 12.17f). The contact between the 'upper member' and 'lower member' in outcrop is locally erosive (Fig. 12.17e) implying a break in sedimentation. Flame structures and syndepositional slump features are present at a few sites.

In brine saturated core, the HMB appear different and are not easy to correlate lithologically with outcrop other than by dating of the sediments. Eugster (1969) reported that the upper tuffaceous silt was directly overlain by evaporites in his cores, and this was taken as the top of the HMB. An 8.7 m piston core (NF1; Fig. 12.18c) recovered from a saline mudflat in the Northwest Lagoon was studied in the 1980s (Barker 1990; Barker et al. 1990; Taieb et al. 1991; Vincens et al. 1991; Damnati et al. 1992, 2007; Williamson et al. 1993). The core excluded most of the Holocene record because of erosion or non-deposition at the lake marginal core-site. The core consists of three units separated by hiatuses that record breaks in sedimentation, with thin sandy deposits also present. The upper and lower units are laminated silt and clay ('varves') with magadiite laminae alternating with organic-rich mud (>5% TOC). The middle unit is homogeneous clay and silt (Taieb et al. 1991). The lower unit was dated by U/Th methods at ~40 ka, and the middle unit at ~24 ka (Goetz and Hillaire-Marcel 1992), but with a ^{14}C age of ~17 ka for the lower unit (Vincens et al. 1991). The upper laminated unit records the Late Pleistocene to Holocene transition. A lake highstand between 12 and 11 ka when lakes Magadi and Natron were joined was followed by a fall between 11 and 10.7 ka when the level was below the 635 m sill between the lakes. The lakes then separated before rising and uniting again at ~10 ka. The dry interval represents the Younger Dryas event (Roberts et al. 1993). The lake remained stratified during this lower stand event (Damnati and Taieb 1995).

Sediments that are equivalents of the HMB in core HSPDP-MAG14-2A, collected by the Hominin Sites and Paleolakes Drilling Project, include detrital siliciclastic grains, primary and reworked ash and pumice, detrital and authigenic clay minerals and zeolites, magadiite, chert, and organic matter (pollen, spores, charcoal, diffuse organic matter, microbial matter). Pollen are dominated by Poaceae, but with a diverse flora of other taxa and increases in aquatic taxa such as Cyperaceae, *Typha*, *Hydrocotyle* and *Potamogeton* (Muiruri et al. 2021b). Charcoal reaches a peak abundance during deposition of the HMB, compared with the rest of the one-million-year core record, which likely reflects increased plant growth during the relatively wet African Humid Period (Muiruri et al. 2021b). The pollen record from Core NF1 (Vincens et al. 1991) also provided evidence for conditions wetter than today for the periods 12–10 ka and at ~17 ka. Diatoms are rare to absent in core sediments younger than ~16 ka in MAG14-2A and in HMB outcrops (Muiruri et al. 2021a). However, diatoms in older sediments (66.5–18 m depth; 120–16 ka), contain well-preserved freshwater *Aulacoseira granulata* and saline genera (*Thalassiosira* and *Cyclotella* [*Stephanocyclus*]). The mixed flora could indicate freshwater lakes that became saline following evaporation or stratified lakes with a fresh mixolimnion caused by periodic flooding. Owen et al. (2019) divided the upper parts of MAG14-2A into geochemical zones with their Zone G6 (~105–0 ka), poorly preserved in outcrops, representing a small hypersaline lake or trona pan that was confined to the axial graben. Subzone G6c (~28–5 ka), however, includes black organic-rich mud and trona rich in zeolites (analcime, erionite, mordenite, phillipsite) that partly correlate with the HMB outcrops (Fig. 12.18d–f).

12.8.3 Former Shorelines of the High Magadi Palaeolake

Former shorelines in rift lakes can be preserved as clastic sediments (beaches and littoral terraces), beachrock, erosional benches, corrosional etch zones, minerals precipitated in littoral groundwater seepage or fluid mixing zones, or microbialites. In the Magadi basin, steep faulted volcanic margins limit where palaeoshoreline indicators form and can be preserved, and favour encrustation of the bedrock by stromatolites or erosional benches over the development of beaches (Fig. 12.19a–c).

Guest (1953) placed a palaeoshoreline for a united Magadi-Natron palaeolake at ~3000 ft (~914 m) although its age was unknown. The maximum lake level during the period when the HMB were deposited is unclear. Casanova (1986, 1987) and Casanova and Hillaire-Marcel (1987) reported a maximum level of 656 m above sea level (asl) based on the elevation of calcite stromatolites around the inferred shoreline of a united fresh Magadi-Natron palaeolake (Fig. 12.2a). They recognised three generations of stromatolites. The older generations were dated using the $^{230}Th/^{234}U$ method at >240 ka and ~135 ka, respectively (Hillaire-Marcel et al. 1986). The third generation was dated using ^{14}C at ~12–10 ka and thus formed during the AHP.

The stromatolites developed in shallow water near the former shoreline. They are often assumed to represent the maximum lake level, but they record precipitation within the 'stromatolite window' – a vertical range in which conditions are favourable for microbial carbonate precipitation. The contemporary lake might have been higher or lower at times but with water undersaturated with respect to $CaCO_3$ (Renaut et al. 2013). At that level (656 m), the lake would have flooded the lower Ewaso Ngiro floodplain, lapped against the lower slopes of the Nguruman Escarpment southwest of Lake Magadi, and united with an expanded Lake Natron

Fig. 12.19 (continued) crust on trachyte (East Sub-basin). (**d**) Magadiite beds interbedded with mud, lower member, HMB, in northern East Sub-basin (see Buatois et al. 2020 for details). (**e**) Magadiite bed, HMB, in northern East Sub-basin. (**f**) Nodular magadiite in erionite silts, HMB, at Graham's Lagoon. (**g**) Magadi-type chert showing characteristic reticulation, HMB, at Graham's Lagoon. (**h**) Deformed magadiite, northern East Sub-basin. (**i**) Thin Magadi-type chert, Oloronga Beds, 500 m south of SW Lagoon (S) springs, Lake Magadi

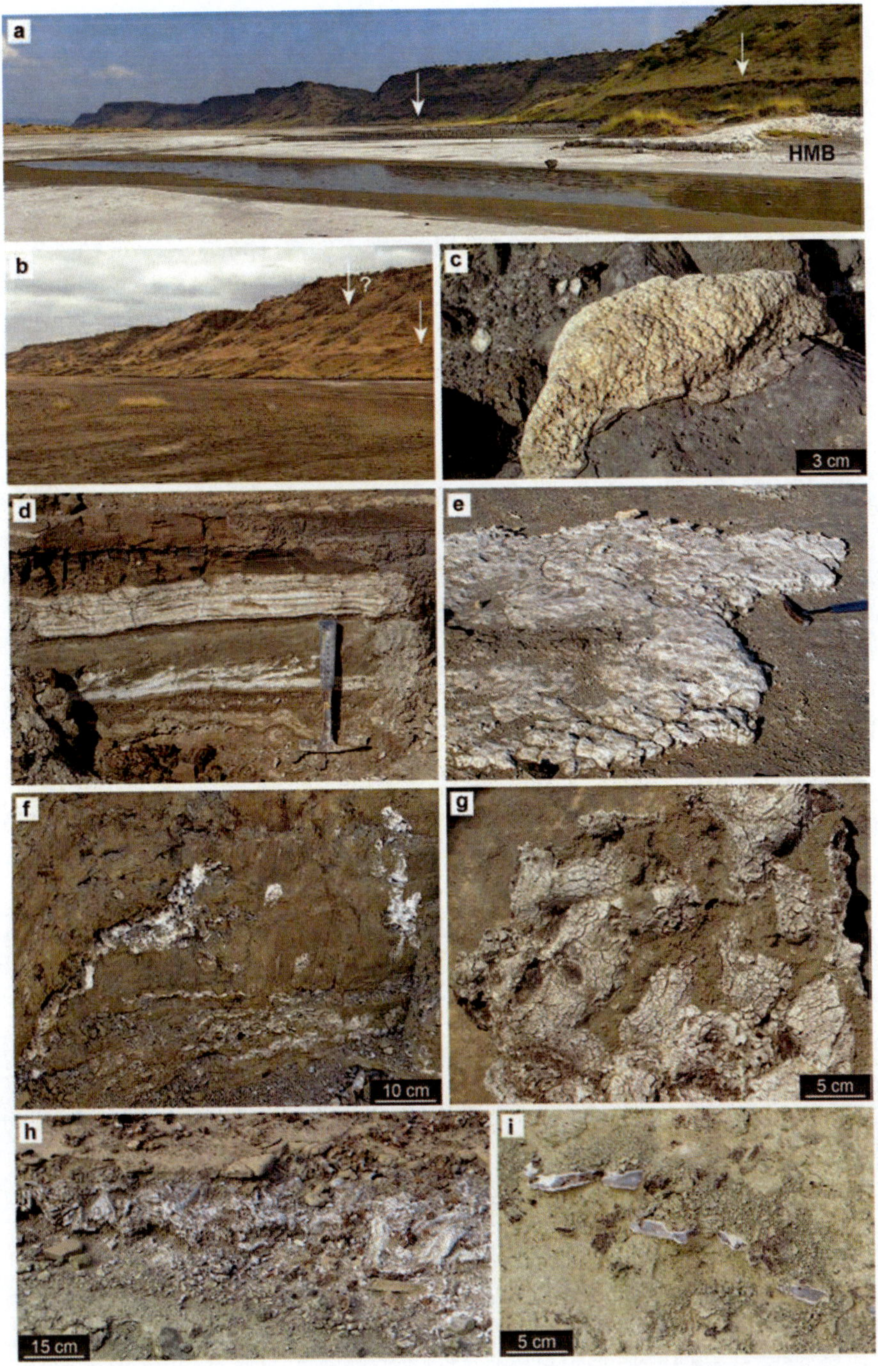

Fig. 12.19 High Magadi Beds features. (**a**) Palaeoshoreline along the western margin of the Northwest Lagoon (Northwest Sub-basin). (**b**) Paleoshoreline (+13 m) Northeast Sub-basin. A break of slope is present at a higher level, but it is not known if this is a shoreline. (**c**) Stromatolitic

to form a single lake. Lake level fluctuated within that window twice during the Pleistocene (at ~135 ka and ~12–10 ka: Hillaire-Marcel et al. 1986). Hillaire-Marcel and Casanova (1987) suggested that this common level was controlled by the water table in adjacent basaltic rocks but did not consider the level of lake water saturation with respect to $CaCO_3$. They reported that their 656 m AHP shoreline was uplifted by ~20 m to ~680 m along the Nguruman fault escarpment ~21 km west of the southern shoreline of Lake Magadi, and also ~3.5–4 km north of Northwest Lagoon.

Although evident at Lake Natron, a palaeoshoreline at 656 m is less obvious near Lake Magadi. That elevation is >50 m above the present level of Lake Magadi (~600 m). The palaeogeographic map for the 656 m lake (Casanova and Hillaire-Marcel 1987) shows that it lies along the eastern edge of the East Sub-basin where a well-defined palaeoshoreline is present and, although broken in places, extends several km N-S. That shoreline, however, is located at ~40 feet (12–13 m) above the modern lake, placing its elevation at ~612–615 m (Baker 1958). A break of slope (bench) along that margin is locally present between 612 and 620 m elevation, but no clear evidence for a lake shore at ~656 m has been reported in that area. That '613 m' palaeoshoreline east of Magadi has stromatolitic calcite coatings on pebbles and plant debris (White 1953; Fig. 12.19c). It is also present along the northwestern margins of the Northwest Lagoon and midwestern shoreline of Nasikie Engida at about the same elevation where stromatolites are also present and encrust bedrock.

Baker (1958) suggested that the HMB were undisturbed by tectonics, but in a footnote, Eugster (1969, p. 4) reported evidence of westward tilting of the Magadi floor and ~10 m of down-faulting of the HMB along the western edge of the Northwest Lagoon. He suggested that tilting may explain why HMB outcrops are rarer west of the lake than to the east.

The explanation is unclear, but the elevation (656 m) reported for the Late Pleistocene-Early Holocene palaeoshoreline is difficult to reconcile with the evidence in many places. Behr (2002) noted that "Sediments, shorelines, relics of lagoons and inlets of these large extensive paleolakes could not be demonstrated." Several dated samples at 656 m come from northwest of Lake Magadi. Patches of light sediment at ~655–660 m are present in grabens east and west of Lake Magadi but many are alluvial and are undated. Some may be local accumulations. Breaks of slope at that elevation are present northeast of the lake but some appear to be fault related.

The well-defined palaeoshoreline at ~13 m above the present level of Lake Magadi is preserved at many locations around the lake (Fig. 12.19b). Behr (2002, his Fig. 2) suggested that this shoreline predated a relatively dry phase (~18–13 ka) before the very high phase (+30–60 m AHP lake) of Casanova (1987). From the evidence, it is difficult to confirm the timing and ages of the HMB palaeoshorelines. The lower (+13 m) shoreline and its stromatolites, which are not discussed by Casanova and Hillaire-Marcel (1987), might represent a later temporary highstand after the two lakes become separated.

Casanova and Hillaire-Marcel (1987) also reported ^{14}C ages of ~14.4–10.9 ka for a large chimney-like tufa mound ('pipe'), about 8 m high, located 4.62 km west of Lake Magadi (Chap. 20, Fig. 20.8f, g). The base of the mound lies at ~713 m, >50 m above the 656 m stromatolites. They interpreted the mound as forming on the palaeolake margin or subaqueously (Casanova and Hillaire-Marcel 1987, their Fig. 9). If the AHP radiocarbon ages are correct, then the tufa mound formed in a lake >50 m higher than the 656 m palaeoshoreline. Growth subaqueously in a dilute lake higher than the stromatolite window might have been feasible if the spring fluids became saturated with respect to $CaCO_3$ upon mixing with lake bottom waters and degassing CO_2. However, no other evidence for an early Holocene lake at >700 m has been reported. Alternatively, the horst upon which the mound formed has been uplifted >40 m during the Holocene, or the mound is older than its apparent radiometric ages – possibly of Oloronga age when tufa commonly formed in the basin (Sect. 20.3). Several toppled mounds lie a few hundred metres north of the main mound.

Trough cross-bedded tuffaceous sandstones at 1.7 km east of Graham's Lagoon are overlain unconformably by upper member silts(stones) of the HMB. Owen et al. (2019) suggested that they are part of the HMB, but from their location and palaeochannel orientation they could be part of the Pleistocene Oloronga Beds.

The HMB are present at Nasikie Engida. They crop out at the SE corner of the lake (Figs. 12.20 and 12.21), at the fan deltas west of the lake, and in the two main valleys north of the lake. The SE section (Fig. 12.20b, e, f) reveals tuffaceous silts with *Tilapia*, siltstone, Magadi-type chert, and sand lenses, locally on laminated calcrete (Fig. 12.21a–d). Greenish zeolitic silts and gravels up to ~5 m above lake level are exposed in terrace deposits of fan deltas of the western shoreline. Some gravel is encrusted by thin calcite stromatolites at ~618–620 m elevation. The HMB north of the lake disconformably overlie calcrete and Oloronga Beds (Fig. 12.21g, h) and are mainly brown zeolitic silts up to 3 m thick similar to those of the upper member.

Lake Magadi and Nasikie Engida were probably united during the high stages of the African Humid Period, but the timing of linkages is unclear. The present spillpoint between the two sub basins is a col (small dry valley) at the southeastern corner of the Nasikie Sub-basin (Fig. 12.20a–c). The spillpoint lies at ~618 m elevation, ~4 m above Nasikie Engida, and ~6 m above the N arm of Lake Magadi. The palaeoshoreline on the western fan-deltas is at about the same elevation (±2 m) as the col. The timing of any linkages is unclear, as is the direction of water flow. The *Tilapia* in the Nasikie HMB outcrop and the presence of stromatolites confirm that waters in Nasikie Engida were dilute when at maximum levels. The Magadi-type chert may be evidence of evaporation after Nasikie Engida water was separated from Lake Magadi. Patches of probable HMB are present ~1 km south of Nasikie Engida implying expansion and possible temporary linkage to a high Lake Magadi to the south. It is not known which lake rose first to the spillpoint, if they formed a single lake, or how long they were united. Although Nasikie Engida is smaller and might fill quickly, its drainage basin is very small. The recent (Late Holocene) history of Nasikie Engida was examined by De Cort et al. (2019), who analysed a 4.6 m core from near the northern shoreline. They showed that the lake has been present for at least 2850 years sustained in part by hydrothermal inflow.

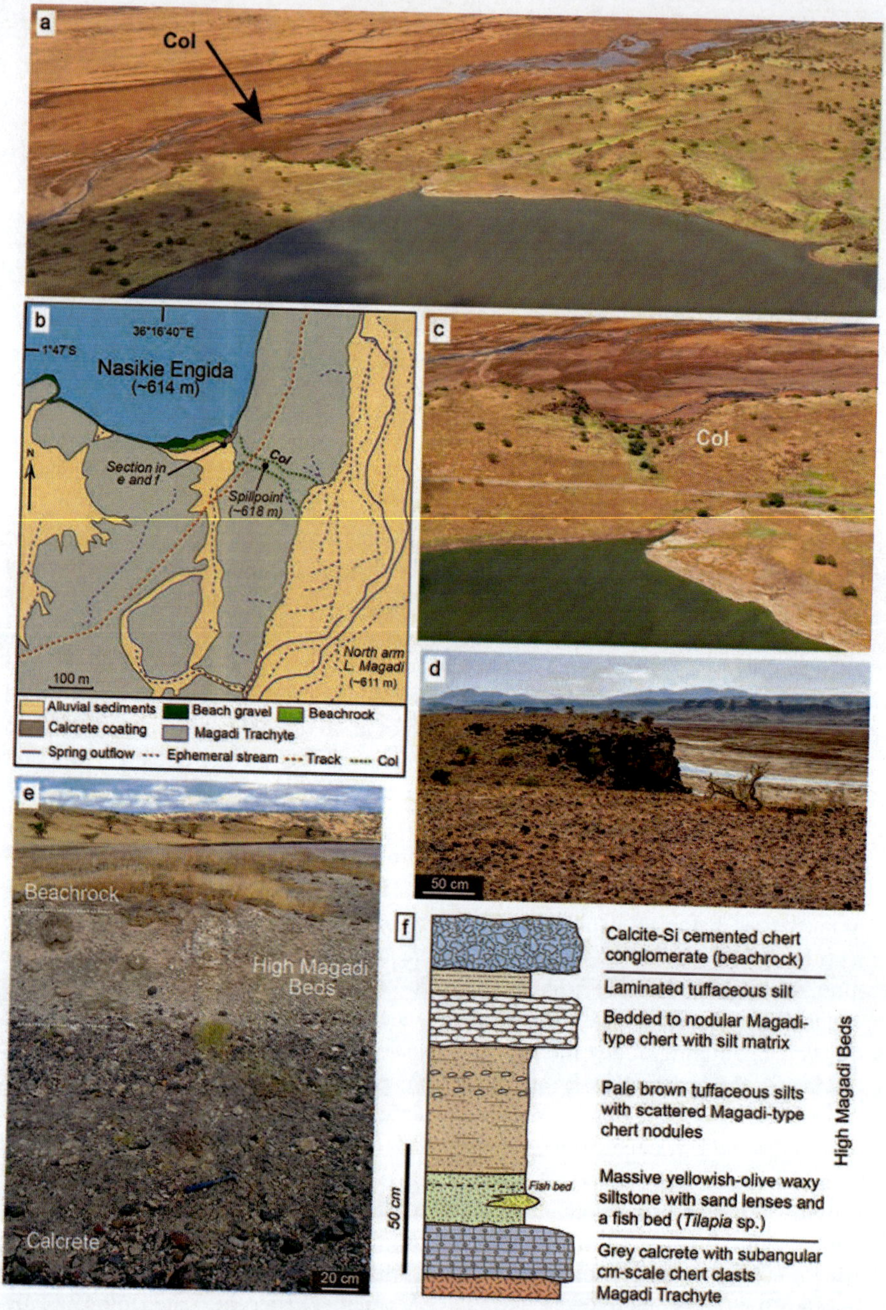

Fig. 12.20 High Magadi Beds (HMB) at Nasikie Engida. (**a–c**) Outcrop locality and col (possible overflow sill) at SE corner of Nasikie Engida. (**d**) Col summit with mudflats of Lake Magadi north arm. (**e, f**) HMB Section at SE corner of the lake at site of overflow

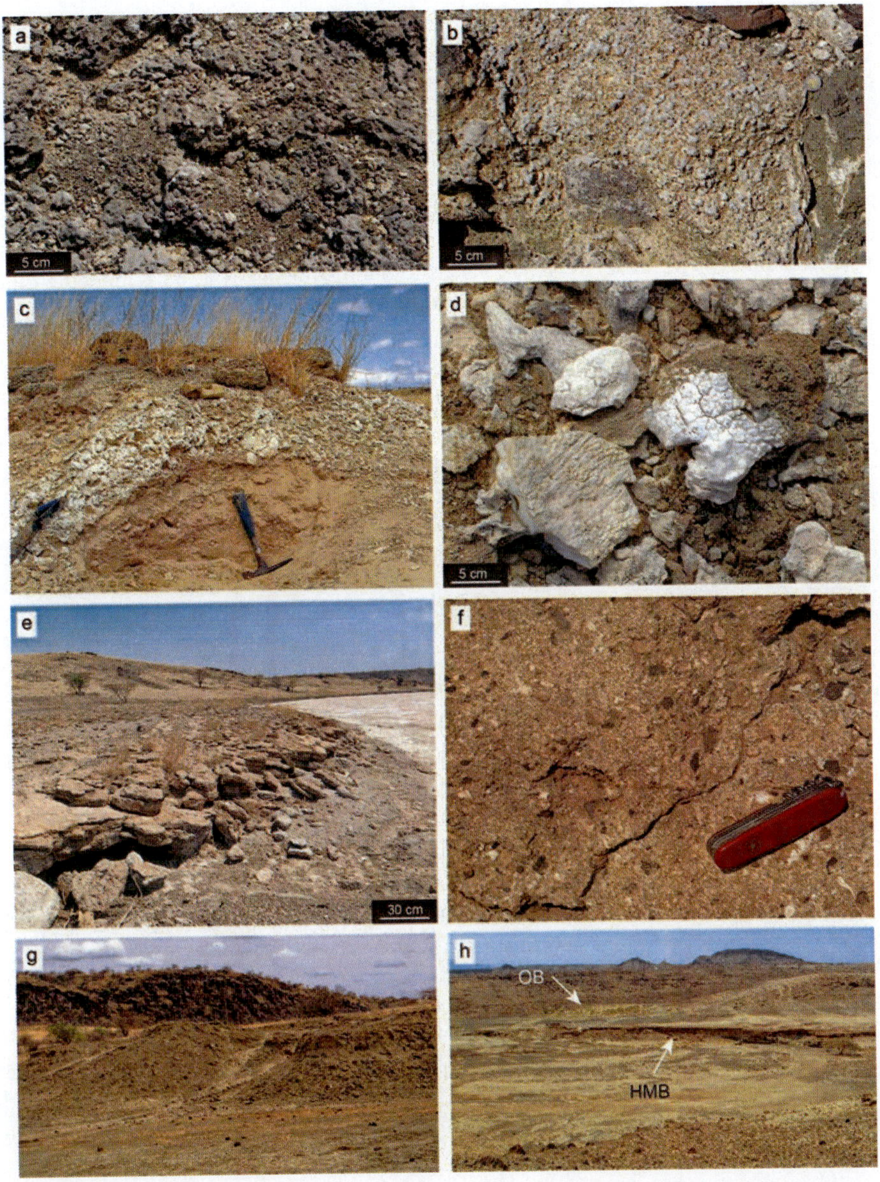

Fig. 12.21 High Magadi Beds (HMB) at Nasikie Engida. (**a**, **b**) Thin (1 cm) laminated calcrete? coating trachyte below High Magadi Beds, SE shoreline. (**c**, **d**) Magadi-type chert in HMB section, SE corner of the lake (Fig. 12.20e, f). (**e**, **f**) Beachrock that overlies chert in **c**. The rock contains angular chert and trachyte clasts. (**g**) Tuffaceous silts of HMB (upper member?) in eastern graben, ~400 m N of northern shoreline. (**h**) HMB, eastern graben, 1.3 km N of northern shoreline. The flat lying HMB lie disconformably upon deformed Oloronga Beds (OB), locally separated by calcrete

During the AHP the expanded Lake Magadi was fed by the Ewaso Ngiro River (S) and by several rim streams to the west, together with hot springs, most of which were probably submerged. Axial streams from the north and south supplemented the inflow. For periods, the lake also probably received inflow from the east including overflow from former Lake Koora in the adjacent N-S graben (Baker 1958, 1986; Crossley 1976; Marsden 1979; Roberts and Barker 1993; Dommain et al. 2022). That palaeolake overflowed through an older Pleistocene overflow channel and fed the southeastern basin of Lake Magadi at Karamai, constructing an alluvial fan (Chap. 19; Fig. 19.13). Rock varnish coating former waterfalls linked to the overflow implies that they have been dry for a long time.

Barthelme et al. (2003, 2009) have outlined the results of excavations of an archaeological site (Olkena) west of Lake Magadi that include the HMB. They reported ^{14}C ages of 10,640–9940 y BP for shell and tooth fragments and found Late Stone Age microlithic tools and a rich vertebrate fauna. Olkena was interpreted to be a Late Pleistocene occupation site in a low-energy channel system draining into palaeolake Magadi during a cool dry phase that preceded the rise of the HMB lake.

12.8.4 *Nasikie Engida Beachrock*

A small ephemeral stream has deposited a gravel alluvial fan at the southeast shoreline of Nasikie Engida. That stream was captured and now flows eastwards to Lake Magadi (Fig. 12.20a–c). The fan conglomerate contains subangular trachyte and Magadi-type chert clasts and is cemented by calcite and opaline silica (Felske 2016). The gravel shows steep southward-dipping foresets that are back-berm washover deposits of a moderately high-energy beach (much higher energy than along the shoreline today). This deposit is interpreted as a beachrock (Fig. 12.21e, f). It lies upon trachyte, a thin grey calcrete layer on the trachyte, and the HMB (Fig. 12.20e, f). Laminar calcite coats many grains. The silica has precipitated in the most saline part of the lake presumably after the lake surface fell below the col level. The carbonate cement may be associated with more dilute waters derived from overflow from Lake Magadi or from groundwater. No hot springs are present in this location, though warm springs are present on the Magadi side of the Nasikie Engida horst a few hundred metres away.

12.8.5 *Magadiite and Sodium Silicate Minerals of the High Magadi Beds*

Eugster (1967) first reported magadiite from Lake Magadi although Baker (1958) had previously described “soft white colloidal silica beds”. Much earlier, Evans (in Discussion in Parkinson 1914) suggested that sodium silicate would precipitate where sodium carbonate solutions leach silica from bedrock and precipitate “siliceous encrustations” on encountering “carbonic acid of the atmosphere”.

Magadiite and related kenyaite are hydrous sodium silicate minerals that are present in many outcrops of the lower ‘member’ of the High Magadi Beds and in cores from the lake floor (Eugster 1967, 1969; Hay 1968; Surdam and Eugster 1976; Behr 2002; Owen et al. 2019). Magadiite forms irregular patches, nodules, thin (<3 mm) discontinuous laminae, and laterally extensive beds (<15 cm thick) interlayered with zeolitic silts and dark green to black muds, some containing microbial organic matter (Fig. 12.19d–f, h). Kenyaite forms cm-scale concretions in some magadiite horizons in outcrop.

Some magadiite forms interstitially in sediments where the pH of pore fluids falls. Eugster (1969, 1980) proposed that bedded magadiite precipitated at the interface of a stratified lake where dilute inflow water came in contact with, but did not mix with, dense sodium carbonate brine. Buatois et al. (2020) reported burrows in bedded magadiite near Magadi townsite and showed that it commonly caps cm to dm-scale cycles. They argued that those cycles are upward shallowing and that the magadiite formed by evaporative concentration in shallow water rather than at the chemocline of a stratified lake with anoxic bottom waters.

Magadiite is soft and puttylike when immersed in brine but hardens irreversibly when it dries, transforming into quartz (Schubel and Simonson 1990; Röhricht 1998). Eugster (1969) excavated quartzose chert in outcrop and showed that it passed laterally into soft magadiite, confirming a genetic link. Magadi-type chert commonly shows soft sediment deformation features and reticulate crack networks attributed to dewatering when it is changed from magadiite to quartz (Fig. 12.19g–i). The transformation involves loss of sodium, water and volume, and has been attributed to the leaching of magadiite, dehydration, or spontaneous recrystallization upon sorption of silica by clays. Intermediate kenyaite, opaline silica and moganite may form during the magadiite-chert transformation (Hay 1968; Eugster 1969; Icole and Perinet 1984).

12.9 Comparisons with Lake Natron, Northern Tanzania

12.9.1 Introduction

Although not a ‘Kenya Rift Lake’ *sensu stricto*, Lake Natron (Figs. 12.22 and 12.23) was united with Lake Magadi during the Pleistocene and Early Holocene. The climate setting, vegetation and wildlife are similar to those of Lake Magadi. A full description is inappropriate here, but some topics merit discussion. Early geological observations were made by Uhlig and Jaeger (1942), Spink and Stevens (1946), Guest and Stevens (1951), Guest (1953), Hay (1968) and Eugster (1970). A compilation of papers on the lake and its geology is given in Sciences Géologique Bulletin v 40 (1987). General geological introductions to the region are given by Guest et al. (1961), Dawson (2008, 2010) and Scoon (2018, 2021) with more detailed

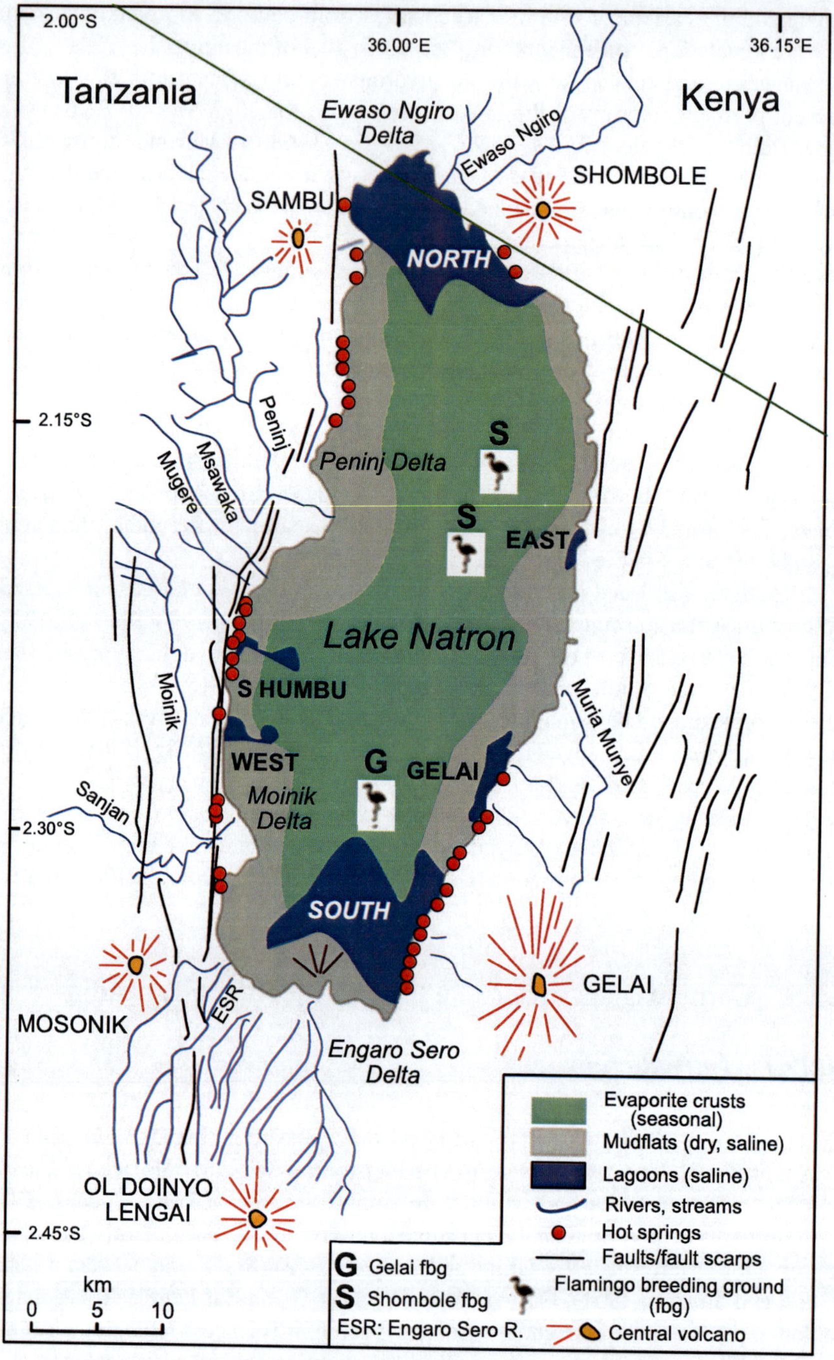

Fig. 12.22 General setting of Lake Natron. Compiled from data in Page and Simon (1987), Tebbs et al. (2013), and Maswi and Mshiu (2020)

Fig. 12.23 Lake Natron. (**a**) View north from SW shoreline. Shombole volcano on the horizon. (**b**) Warm spring discharging onto mudflats, SW shoreline. (**c**) Hot seepage spring showing microbial mats and thin subaerial crusts of trona and opaline silica. (**d**) Ol Doinyo Lengai volcano with trona crusts of saline pan in foreground. (**e**) Trona crusts showing polygonal pressure ridges. (**f**) Trona crystal rafts that settled on bottom-nucleated salts when small pool between ridges dried up. (Photos **d**, **e** and **f** taken in south-central part of lake). (**g**) Terrace recording higher lake level and fresher waters. (**h**) Calcite stromatolites recording former higher lake level

discussions of its context (North Tanzania Divergence Zone) in Saggerson (1966), Baker et al. (1972), Foster et al. (1997), Muirhead et al. (2016), Le Gall et al. (2021) and Reiss et al. (2022).

12.9.2 Hydrology and Hydrochemistry

In common with Lake Magadi, Lake Natron is recharged by rainfall, rivers, cool and hot springs, with potential for artesian groundwater influx from adjacent uplands, as indicated by blister mounds (cf. Ashley et al. 2002) along the western margins. Lake Natron lies in a closed topographic basin and occupies the same regional hydrological sump as Lake Magadi, but the surface level of Natron is taken to be ~600 m asl, similar to Magadi (e.g. Vincens and Casanova 1987). Sherrod et al. (2013), however, quote the elevation of Lake Natron as 597 m. Unlike Lake Magadi, Lake Natron is fed by a large river and is not known to have thick buried evaporites. Water analyses (Table 12.4) are given by Guest and Stevens (1951), Hay (1966), Eugster (1970), Kilham (1971), Jones et al. (1977), Gueddari (1984) and Fritz et al. (1987).

The Ewaso Ngiro River (South), which is perennial in some years, drains ~45% of the catchment, nearly all of it in Kenya, and feeds the lake from the north (Figs. 12.4, 12.22 and 12.24a–c). The river contributes ~27% of the annual inflow, compared with ~25% from hot and warm springs, 9% from ephemeral streams, and ~39% from direct rainfall (~425 m y^{-1}) (Tebbs et al. 2013). Other streams drain the rift margins and volcanic terrain to the south, including the braided and anastomosing Engare Sero, Peninj and Moinik, which are also often perennial. The rivers have dilute (0.15–2.85 g L^{-1}) alkaline (pH 7.9–9.3) waters but many have increased their salinity with evaporation before they reach the lake (Table 12.4). Eugster (1970) and Gueddari (1984) noted that rivers draining basaltic terrain (e.g. Peninj) are enriched in alkaline earths and bicarbonate, whereas rivers draining nephelinitic terrain in the south have higher sulfate and Na/Cl ratios.

Lake Natron (Figs. 12.23a and 12.24d) is ~75 km long and 22–35 km wide and is partly bordered by shallow hot-spring-fed lagoons similar to those at Lake Magadi. Six have been recognised (Tebbs et al. 2013; Fig. 12.22). The largest, the North and South lagoons, are partly fed by rivers. More than 25 springs with temperatures of 32–52 °C discharge from scree and colluvium next to the saline pan as seeps and small springs but with a few larger springs (Fig. 12.23b, c; Guest and Stevens 1951; Gueddari 1984; Hochstein 1999; Hochstein et al. 2000; Maswi and Mshiu 2020; Abdallah et al. 2022). The spring waters are of Na^+–HCO_3^- composition with a salinity range of 3–40 g L^{-1} (Table 12.4), and a pH range of 8.0–10.7. Gueddari (1984) recognised two groups. Springs of the western group that discharge along the shoreline have temperatures of 30–50 °C and salinity of ~25 g L^{-1}. They contain little Ca^{2+} and Mg^{2+}. Those of the eastern (Gelai) group are less saline (<10 g L^{-1}) and cooler, with pH < 9.2. Spring waters are enriched in fluoride and the silica increases with temperature. Abdalla et al. (2022) showed from stable isotope analyses that recharge derives from local groundwater, the rift flanks and rainfall.

Table 12.4 Selected chemical analyses of Lake Natron basin waters

Sample	Ref no.	Source	Date	Type	Temp	pH	Na	K	Ca	Mg	Cl	HCO_3	CO_3	SO_4	F	SiO_2
Ewaso Ngiro R	18	E70	1950	R			84.7				36		68	15	1.8	
Eastern River	9	E70	1950	R	40.5		2950	53			1190		2700	118	35	67
Peninj R	39	E70	1950	R			6.4						3.4		3.2	
Engare Sero R	961	E70	1950	R			800				127		686	365	14	
NW shore	17	E70	1950	S	38.5		27,200	360			15,100		21,820	526	323	390
W shore	30	E70	1950	S	48		5900	90			2530		5290	227	70	
SW shore	37	E70	1950	S	33		979				375		868	113	14	30
NE shore	43	E70	1950	S												
SE lake	14	E70	1950	L			138,100				90,670		99,070	3720	1312	
Central lake	33	E70	1950	L			134,900				97,650		89,440	3040	1312	
NW lake	24	E70	1950	L			136,000				95,220		92,840	3380	1221	
N C lake		H64	7/64	L		9.70	62,700	1080			18,300	11,500	58,100	4420	966	150
Pit S shoreline		H64	7/64	G		9.85	4420	287			1160	2700	2780	937	165	127
Peninj R		K71	1/70	R		7.2	13.3	4.3	18.6	5.8	7.2	101.3	0	12	0.42	28.9
Lake Natron		K71	12/69	L		9.80	3600	263	0.6	0	1830	2526	3457	173	76	64
Lake Natron		K71	12/69	L		9.81	9080	217	0	0	4110	4857	10,946	278	213	43
Lake Natron		K71	6/69	L		9.22	8700	145	0	0	3580	11,307	6133	300	137	35.4
Lake Natron		K71	6/69	L		9.26	9000	150	0	0	3570	11,533	6421	325	141	34

Sources: E70: Eugster (1970), which are recalculated from Guest and Stevens (1951) in ppm; H64 from Hay (1966) in ppm; K71: Kilham (1971) in mg/l. More analyses are given in Jones et al. (1977), Gueddari (1984) and Fritz et al. (1987)

R River, *S* Hot springs, *L* Lake, *G* Shallow groundwater, *R* river water, *S* spring water, *L* lake water, *G* Groundwater

Fig. 12.24 Northern Lake Natron and the Ewaso Ngiro inflow. (**a**) Ewaso Ngiro incised into older alluvium. (**b**) View northwards over the Ewaso Ngiro floodplain, bounded to west by the Nguruman Escarpment. (**c**) The river enters the Shombole Swamp before reaching Shombole and its delta into Lake Natron. (**d**) View south from the Kenya-Tanzania border (white border marker post). The northern lagoon of the lake is in the distance (see expanded inset). Photo taken during a regressive phase (July 2022); high lake/lagoon shoreline marked by dashed white line. (**e**) High Magadi Beds outcrop (small section cut into the flat plain in inset), north of Shombole and close to the former connection between Lake Magadi and Lake Natron

Eugster (1970) concluded that the hydrogeology of Lake Natron is more complex than that of Lake Magadi, with three reservoirs, each having different Cl^- and SO_4^{2-}-concentrations, feeding the hot springs.

The lake varies in depth from 0 to 4 m (Bettinetti et al. 2011), and in area from 81 to 804 km^2 (Tebbs et al. 2013) occasionally drying almost completely (Page and Simon 1987; Clarisse et al. 2019). Some lagoons may be deeper than 4 m. The south-central area often becomes an exposed saline pan (Fig. 12.23d–f) during periods of drought (June–October) with residual patches of muddy brine, but the marginal lagoons usually retain some saline water because of local inflow from springs and streams. The salinity and microbiota of the lagoons can be quite different from each other.

Lake water salinity varies annually sometimes reaching ~35%, increasing by evaporation in the dry seasons, but can vary from year to year and place to place. Schagerl (2016 Appendix) for example, reports a salinity of 12.3%. The recorded pH range is from 9 to 11.5. Surface water temperatures are typically 20–35 °C in the open lake but sometimes exceed 55 °C. Any stratification is likely to be brief given the shallow water depth though might be possible in the deeper lagoons. The waters in the lagoons, surface brines and shallow interstitial brines are dominated by Na^+, CO_3^{2-} and Cl^- with $CO_3^{2-} + HCO_3^- > Cl^- >> SO_4^{2-}$. Fluoride is exceptionally high (60–150 mol L^{-1}; Jones et al. 1977; Gueddari 1984). Clarisse et al. (2019) have shown that substantial amounts of atmospheric ammonia (NH_3) are released from the saline mudflats during lake recession, likely from decaying organic matter. The chemical evolution of the waters is broadly similar to that of Magadi but the Na:Cl ratio is not constant from spring group to spring group making its use as a tracer less reliable.

12.9.3 Sedimentation and Mineralogy

Lake Natron is a large shallow sodium-carbonate lake bordered by hot spring-fed lagoons and extensive saline mudflats (Fig. 12.24d). The central area is commonly exposed with a trona crust that expands in area during the dry seasons and periods when the lake level falls through evaporation. The depositional environments are broadly like those at Lake Magadi except that the lake receives fluvial inflow and clastic sediments from the Ewaso Ngiro River (S), which is perennial in some years, and the saline pan is less well developed with only a thin evaporite crust.

The Ewaso Ngiro (S) River delivers water and fine muddy sediment to northern Lake Natron (Manega and Bieda 1987), the Shombole Swamp having filtered most of the remaining coarser fractions (Fig. 12.24a–c). Other streams (Peninj, Moinik, Engare Sero), which are also perennial in some years, deliver mud and silt with coarser sediments on the fans upstream. Mudflats are commonly cut by channelled flow from the hot springs (Yanda and Madulu 2005).

Short (1–2 m) cores from across the lake contain interbedded mud and trona. The mud contains scattered crystals of trona, gaylussite, pirssonite [$Na_2Ca(CO_3)_2{\cdot}2H_2O$], and authigenic analcime (Hay 1966, 1968). Manega and Bieda (1987) examined the sediments in rivers, outcrops and in short lake cores, and recognised two main zones of sedimentation: (1) a northern zone with positive Eh conditions, precipitation of Mn^{2+} and Fe^{2+} oxides, domination by zeolites, and the mineralogical influence of basement sourcing; and (2) a southern zone with negative Eh conditions and geochemical regime favourable for fluorite formation and accumulation of organic-rich mud.

The soda pan lies mainly in the central and northeast parts of the lake where the trona is up to 1.5 m thick. The pan shows similar features to Lake Magadi (Fig. 12.23d–f) with polygonal pressure ridges and barchanoid salt accumulations, broken in places by resurgent brine erupting through the salt crust. The crust is partly dissolved during the rainy seasons when mud from stream inflow and aeolian dust settle on the surface with microbial organic matter.

Deocampo (2001, 2002) examined sedimentary and geochemical processes on the marginal mudflats of several Tanzanian lakes including Natron. Frequent wetting and drying, associated mineral precipitation (early cementation) and bioturbation are important. Luitkus-Pierce et al. (2016) described the Lake Natron Footprint Tuff, which preserves human footprints on the alluvial mudflats of the lower Engare Sero floodplain ~2 km from the southern shoreline. Originally its age was estimated at ~120 ka, but later work (Balashova et al. 2016) implies a Holocene age, with Ol Doinyo Lengai the likely source of the ash.

Samples of 'magadi' (salt used locally as a food additive) from the saline mudflats of the southern shoreline consist of trona, halite, thermonatrite and kogarkoite ($Na_2SO_4.NaF$) (Nielsen 1999; Darragi et al. 1983). In common with Lake Magadi, opaline silica and Na-CO_3 crusts precipitate around many hot springs (Hewson et al. 2020). Some hot springs at Ndahane on the southeastern shoreline have precipitated carbonates (Abdallah et al. 2022).

As noted by Eugster (1970), Lake Natron differs from Lake Magadi in several respects. Lake Magadi lies in a sump that receives inflow from the adjacent mountains via runoff and groundwater. Strong evaporation at Magadi produces hypersaline brines that precipitate thick trona beds, implying initially high accommodation space or, more likely, continuing subsidence in the axial trough during sedimentation. Lake Natron, in contrast, is a much larger lake in area and depth. Unlike Lake Magadi, it receives inflow from a major, often perennial, river. Its groundwater reservoir is inferred to be saline but poorly mixed (Eugster 1970). The sedimentary and geochemical processes and mineralogy are essentially the same as Magadi, but its sedimentary record is still largely unknown. A significant difference is the proximity of recent volcanism with several young volcanoes near the lake including Ol Doinyo Lengai (Sherrod et al. 2013), which periodically contributes sodic ash to the lake (although ash from large eruptions would also reach Lake Magadi given favourable winds). Modern Lake Natron may be an analogue for the less hypersaline parts of the Lake Magadi record.

12.9.4 Biology and Ecology

Lake Natron, a Ramsar Site, has been recognised as the main breeding site for the lesser flamingo (*Phoeniconaias minor*) since the pioneering work of Brown (1959). More than 75% of *all* lesser flamingos born in recent decades hatched at Lake Natron (Childress et al. 2008; Clemson et al. 2011), with more than 570,000 breeding pairs recorded in 1957 (Brown and Root 1971). Several studies have shown that their breeding patterns and abundance vary with changing lake environment and that they typically are irregular (Krienitz 2018). Tebbs et al. (2013) showed how several conditions have to be 'just right' for successful breeding, including (1) lake level that is high enough to eliminate predation but low enough to prevent flooding of nests—a very delicate balance, (2) local sources of mud for building nest mounds on marginal mudflats or on the trona pan itself ('salt islands'), and (3) adequate food. Ideally, those conditions also need stability for 65–70 days and are favoured between October and December (short rains) but may occur at any time. Fluctuations in lake level, water chemistry and food sources respond to changes in rainfall and evaporation. Tebbs et al. (2013) showed breeding success is favoured by intermediate and receding lake levels.

Microbiological studies at Lake Natron are surprisingly few compared to other soda lakes (Grant and Jones 2016). The lake is highly productive with microorganisms that thrive in its waters which have abundant CO_2, nutrients (P), and high temperatures and light intensities (Robinson 2015; Nonga et al. 2017; Mgimwa et al. 2021). *Arthrospira fusiformis* is an important food source for the lesser flamingos, but other cyanobacteria can be more important (Krienitz et al. 2016). The lesser flamingos also feed on benthic (epipelic) diatoms (*Anomoeneis sphaerophora, Nitzschia* spp., *Navicula* spp.) in the open lagoons where waters are more dilute (Tuite 1981; Robinson 2015; Harper et al. 2016; Krienitz et al. 2016). Haloalkaliphilic archaea bloom under high salinities and give the lake an orange-pink-red colour, especially in its northern half (Clarisse et al. 2019).

Several cichlid fish have been reported from the modern lake, including *Alcolapia alcalicus*, *A. ndalalani* and *A. latilabris* (Seegers and Tichy 1999; Tichy and Seegers 1999; Kavembe et al. 2016) and occasional strays from rivers. Seegers et al. (1999) found a shared haplotype in Lake Magadi and Lake Natron tilapia species and proposed that the common haplotype could have been present in the ancestral fish population of Palaeolake Oloronga (Sect. 20.3) and (or) the Late Pleistocene High Magadi palaeolake (Sect. 12.8) when the two lakes were periodically united. Rosner et al. (2003) also discussed the fossil fish from Lake Natron.

12.9.5 Late Quaternary Sedimentation

The Late Quaternary history of Lake Natron is less well known than that of Lake Magadi. High level shorelines and lake terraces (Fig. 12.23g) record times when the two lakes were united (Sect. 12.8.3). An extensive plain to the north is underlain by poorly exposed HMB with a few scattered small outcrops (Fig. 12.24e), but more is known of its Pleistocene history (Sect. 20.2) than its Late Pleistocene to Holocene record.

Stromatolites and stromatolitic crusts provide evidence for former high-level shorelines (Fig. 12.23h) when conditions became favourable for their formation within the contemporary stromatolite window (Renaut et al. 2013). As noted (Sect. 12.8.3), Casanova (1986, 1987) and Casanova and Hillaire-Marcel (1987) recorded three main phases of stromatolite formation. Two levels of shore zone encrustation lie at ~656 m, 50–60 m above the present lake. The youngest generation were dated at ~12–10 ka during the African Humid Period (Hillaire-Marcel et al. 1986). The lake was then deep and generally fresh before shrinking to its present status.

During these high-level phases, fan-deltas extended into the lake and littoral benches were cut into their sediments that record temporary stillstands. Warren (2016) suggested that fan delta sands might have served as aquifers for dilute water, inhibiting the formation of thick trona beds.

Holocene sediments south of the lake contain minor lacustrine silt and clay interbedded with alluvium. Older Lake Natron deposits (Late Pleistocene?) exposed on the basin floor along the eastern edge of the lake at 620–640 m altitude included highly deformed lake sediments interbedded with debris avalanche deposits from Oldoinyo Lengai (Sherrod et al. 2013).

References

Abdallah EA, Kasanzu CH, Kinabo CP, Imai A, Butler MJ (2022) Constraining the origin and age of the thermal and cold water in the Lake Natron Basin, northern Tanzania. Tanz J Sci 48:804–415

Allen DJ, Darling WG, Burgess WG (1989) Geothermics and hydrogeology of the southern part of the Kenya Rift Valley with emphasis on the Magadi-Nakuru area. British Geological Survey research Report SD/89/1

Anonymous (1923) The soda deposits of Lake Magadi in Kenya. Bull Imp Inst 21:431–444

Ashley GM, Goman N, Hover VC, Owen RB, Renaut RW, Muasya AM (2002) Artesian blister wetlands, a perennial water resource in the semi-arid rift valley of East Africa. Wetlands 22:686–695

Atmaoui N, Hollnack D (2003) Neotectonics and extension direction of the Southern Kenya Rift, Lake Magadi area. Tectonophysics 364:71–83

Atuya RG, Githiri JG, Kinyua R (2014) Geothermal mapping using shallow hole temperature measurement: a study of Magadi prospect, Kenya. Int J Curr Trends Res 3:1–7

Baker BH (1958) Geology of the Magadi area. Geol Surv Kenya Rep 42

Baker BH (1963) Geology of the area south of Magadi. Geol Surv Kenya Rep 61

Baker BH (1975) Geology and geochemistry of the Ol Doinyo Nyokie trachyte ignimbrite vent complex, South Kenya Rift Valley. Bull Volcanol 39:420–440

Baker BH (1986) Tectonics and volcanism of the southern Kenya Rift Valley and its influence on rift sedimentation. In: Frostick LE, Renaut RW, Reid I, Tiercelin J-J (eds) Sedimentation in the African rifts. Geol Soc Lond Spec Publ 25:45–57

Baker BH, Mitchell JG (1976) Volcanic stratigraphy and geochronology of the Kedong-Olorgesailie area and the evolution of the South Kenya rift valley. J Geol Soc Lond 132:467–484

Baker BH, Mohr PA, Williams LAJ (1972) Geology of the eastern rift system of Africa. Geological Soc Am Spec Publ vol 36. https://doi.org/10.1130/SPE136-p1

Balashova A, Mattsson HB, Hirt AM, Almqvist BSG (2016) The Lake Natron footprint tuff (northern Tanzania): volcanic source, depositional processes and age constraints from field relations. J Quat Sci 31:526–537

Barker PA (1990) Diatoms as paleolimnological indicators: a reconstruction of Late Quaternary environments in two East African salt lakes. PhD thesis, Loughborough University

Barker PA, Gasse F, Roberts N, Taieb M (1990) Taphonomy and diagenesis in diatom assemblages; a Late Pleistocene palaeoecological study from Lake Magadi, Kenya. Hydrobiologia 214:267–272

Barthelme J, Murimi S, Ngari L, DeLong R, Statler T, Chartier R, Munyazi W (2003) Recent archaeological research in the Lake Magadi Basin, southern Kenya. Nyame Akuma 60:8–13

Barthelme J, Hunt K, Ngari L, Kipintoi WO, Kweyu R, Murimi S (2009) Renewed survey and excavations in the Lake Magadi Basin, southern Kenya. Nyame Akuma 71:54–62

Bartholomew GA, Pennycuick CJ (1973) The flamingo and pelican populations of the Rift Valley lakes in 1968-69. Afr J Ecol 11:189–198

Becht R, Mwango F, Muno FA (2006) Groundwater links between Kenyan Rift Valley lakes. In: Odada EO, Olago DO, Ochola W, Ntiba M, Wandiga S, Gichuki N, Oyieke H (eds) Proc 11th World Lakes Conf, vol 2. Nairobi, pp 7–14

Behr H-J (2002) Magadiite and Magadi-chert: a critical analysis of the silica sediments in the Lake Magadi Basin, Kenya. In: Renaut RW, Ashley GM (eds) Sedimentation in continental rifts. SEPM Spec Publ 73:257–273

Behr HJ, Röhricht C (2000) Record of seismotectonic events in siliceous cyanobacterial sediments (Magadi cherts), Lake Magadi, Kenya. Int J Earth Sci 89:268–283

Bettinetti R, Quadroni S, Crosa G, Harper D, Dickie J, Kyalo M, Mavuti K, Galassi S (2011) A preliminary evaluation of the DDT contamination of sediments in Lakes Natron and Bogoria (Eastern Rift Valley, Africa). Ambio 40:341–350

Birt CS, Maguire PKH, Khan MA, Thybo H, Keller GR, Patel K (1997) The influence of pre-existing structures on the evolution of the southern Kenya Rift Valley – evidence from seismic and gravity studies. Tectonophysics 278:211–242

Bondarenko SA, Georgieva ML, Bilanenko EN (2018) Fungi inhabiting the coastal zone of Lake Magadi. Contemp Probl Ecol 11:439–448

Bradley WH, Eugster HP (1969) Geochemistry and paleolimnology of the trona deposits and associated authigenic minerals of the Green River Formation of Wyoming. US Geol Surv Prof Pap 496-B

Brown LH (1959) The mystery of the flamingos. Country Life, London

Brown LH, Root A (1971) The breeding behaviour of the lesser flamingo *Phoeniconaias minor*. Ibis 113:147–172

Buatois LA, Labandeira CC, Mángano, GM, Cohen A, Voigt S (2016) The Mesozoic lacustrine revolution. In Mángano GM, Buatois LA (eds) The trace-fossil record of major evolutionary events, vol 2: Mesozoic and Cenozoic. Springer, Dordrecht, pp 179–263

Buatois LA, Renaut RW, Owen RB, Behrensmeyer AK, Scott JJ (2020) Animal bioturbation preserved in Pleistocene magadiite at Lake Magadi, Kenya Rift Valley, and its implications for the depositional environment of bedded magadiite. Sci Rep 10:6794

Burgess WG (1986) Kenya Rift Valley Geothermal Project: hydrogeology and geothermics of the Magadi–Longonot sector. Brit Geol Surv Res Rep WD/OS/86/5

Butzer KW, Isaac GL, Richardson JL, Washbourn-Kamau C (1972) Radiocarbon dating of East African lake levels. Science 175:1069–1076

Campisano CJ, Cohen AS, Arrowsmith JR, Asrat A, Behrensmeyer AK, Brown ET, Deino AL, Deocampo DM, Feibel CS, Kingston JD, Lamb HF, Lowenstein TK, Noren A, Olago DO, Owen RB, Pelletier JD, Potts R, Reed KE, Renaut RW, Russell JM, Russell JL, Schäbitz F, Stone JR, Trauth MH, Wynn JG (2017) The Hominin Sites and Paleolakes Drilling Project: acquiring high-resolution paleoclimate records from the East African rift system and their implications for understanding the environmental context of hominin evolution. PaleoAnthropology 2017:1–43. https://doi.org/10.4207/PA.2017.ART104

Casanova J (1986) Les Stromatolites Continentaux: Paléoécologie, Paléohydrologie, Paléoclimatologie. Application au Rift Gregory. Thèse Docteur d'Etat-Sciences, Université d'Aix Marseille II

Casanova J (1987) Stromatolites et hauts niveaux lacustres Pléistocènes du basin Natron-Magadi (Tanzanie-Kenya). Sci Géol Bull 40:135–153

Casanova J, Hillaire-Marcel C (1987) Chronologie et paléohydrologie des hauts niveaux quaternaires du basin Natron-Magadi (Tanzanie-Kenya) d'après la composition isotopique (^{18}O, ^{13}C, ^{14}C, U/Th) des stromatolites littoraux. Sci Géol Bull 40:121–134

Casanova J, Tiercelin J-J (1982) Construction stromatolitiques en milieu carbonate sodique: les oncolites des plaines inondables du lac Magadi (Kenya). C R Acad Sci 295:1139–1144

Childress B, Nagy S, Hughes B (Compilers) (2008) International single species action plan for the conservation of the lesser flamingo (*Phoeniconaias minor*). CMS Tech Ser 18, AEWA Tech Ser 34. Bonn, Germany

Clamsen TEM, Maliti H, Fyumagwa R (2011) Current population status, trend and distribution of lesser flamingo *Phoeniconaias minor* at Lake Natron, Tanzania. Flamingo. Bull IUCN-SSC Wetl Int 18:54–57

Clarisse L, Van Damme M, Gardner W, Coheur P-F, Clerbaux C, Whitburn S, Hadji-Lazaro J, Hurtmans D (2019) Atmospheric ammonia (NH_3) emanations from Lake Natron's saline mudflats. Sci Rep 9:4441. https://doi.org/10.1038/s41598-019-39935-3

Coe MJ (1966) The biology of *Tilapia grahami* Boulenger in Lake Magadi, Kenya. Acta Trop 23:146–177

Coe MJ (1967) Local migration of *Tilapia grahami* Boulenger in Lake Magadi, Kenya in response to diurnal temperature changes in shallow water. Afr J Ecol 5:171–174

Crane K (1981) Thermal variations in the Gregory Rift Valley of southern Kenya (?). Tectonophysics 74:239–262

Crane K, O'Connell S (1983) The distribution and implications of heat flow from the Gregory Rift valley. Tectonophysics 94:253–275

Crossley R (1976) Structure, stratigraphy and volcanism in the Nguruman Escarpment area of the western side of the Kenya Rift Valley. PhD thesis, University of Lancaster

Crossley R (1979) The Cenozoic stratigraphy and structure of the western part of the rift valley in southern Kenya. J Geol Soc Lond 136:393–405

Crossley R (1980) Structure and volcanism in the S Kenya Rift. Att Conveg Lincei 47:89–98

Crossley R, Knight RM (1981) Volcanism in the western part of the Rift Valley in southern Kenya. Bull Volcanol 44:117–128

Damnati B, Taieb M (1995) Solar and ENSO signatures in laminated deposits from Lake Magadi (Kenya) during the Pleistocene/Holocene transition. J Afr Earth Sci 21:373–382

Damnati B, Taieb M, Williamson D (1992) Laminated deposits from Lake Magadi; climatic contrast during the maximum wet period 12,000–10,000 BP. Bull Soc Géol Fr 163:407–414

Damnati B, Icole M, Taieb M (2007) The application of organic carbon and carbonate stratigraphy to the reconstruction of lacustrine paleoenvironments from Lake Magadi, Kenya. In: Runge J (ed) Dynamics of forest ecosystems in Central Africa during the Holocene: past, present, future. Taylor and Francis, London, pp 155–168

Darling WG (2001) Magadi and Suguta: the contrasting hydrogeochemistry of two soda lake areas in the Kenya Rift Valley. In: Cidu R (ed) Proc 10 int symp water-rock interaction, WRI-10, Villasimius, Italy 10–15 July 2001. Balkema, Lisse, pp 95–98

Darling WG, Allen DJ, Armannsson H (1990) Indirect detection of subsurface outflow from a rift valley lake. J Hydrol 113:297–306

Darragi F, Gueddari M, Fritz B (1983) Mise en evidence d'un fluoro-sulfate de sodium, la kogarkoite (Na_3SO_4F), dans les croutes salines du lac Natron en Tanzanie. C R Acad Sci Ser 2(297):141–144

Davis LE (2012) An astrobiological study of an alkaline-saline hydrothermal environment, relevant to understanding the habitability of Mars. PhD thesis, University of London

Dawson JB (2008) The Gregory Rift Valley and Neogene-recent volcanoes of northern Tanzania. Mem Geol Soc Lond 33:3–7

Dawson JB (2010) Aspects of rift valley faulting and volcanicity in North Tanzania: report of a Geologists' Association field meeting in 2008. Proc Geol Assoc 121:342–349

De Cort G, Mees F, Renaut RW, Sinnesael M, Van der Meeren T, Goderis S, Keppens E, Mbuthia A, Verschuren D (2019) Late-Holocene sedimentation and sodium carbonate deposition in hypersaline, alkaline Nasikie Engida, southern Kenya Rift Valley. J Paleolimnol 62:279–300

Deocampo DM (2001) Geochemistry and sedimentology of modern East African wetlands and a Pleistocene paleo-wetland at Olduvai Gorge, Tanzania. PhD thesis, Rutgers University, Piscataway, New Jersey

Deocampo DM (2002) Sedimentary processes and lithofacies in lake-margin groundwater-fed wetlands in East Africa. In: Renaut RW, Ashley GM (eds) Sedimentation in continental rifts. SEPM Spec Publ 73:295–308

Deocampo DM, Renaut RW (2016) Geochemistry of African soda lakes. In: Schagerl M (ed) Soda lakes of East Africa. Springer, Cham, pp 77–96

Denson JD, Mack Ivey D, Leuko S, Rothschild LJ, Hand KL (2010) A survey of the microbial diversity of an alkaline biofilm associated with a unique geological feature within a hot spring of the hypersaline alkaline basin of Lake Magadi, Kenya. Proceedings Astrobiology Science Conference, League City, Texas. Paper 5510

Dommain R, Riedl S, Olaka LA, deMenocal P, Deino AL, Owen RB, Muiruri V, Müller J, Potts R, Strecker MR (2022) Holocene bidirectional river system along the Kenya Rift and its influence on East African faunal exchange and diversity gradients. Proc Natl Acad Sci 119. https://doi.org/10.1073/pnas.2121388119

Eugster HP (1966) Sodium carbonate-bicarbonate minerals as indicators of PCO_2. J Geophys Res 71:3369–3377

Eugster HP (1967) Hydrous sodium silicates from Lake Magadi, Kenya: precursors of bedded chert. Science 157:1177–1180

Eugster HP (1969) Inorganic bedded cherts from the Magadi area, Kenya. Contrib Mineral Petrol 22:1–31

Eugster HP (1970) Chemistry and origin of the brines of Lake Magadi, Kenya. Mineral Soc Am Spec Publ 3:215–235

Eugster HP (1971) Origin and deposition of trona. Rocky Mount Geol 10:49–55

Eugster HP (1980) Lake Magadi, Kenya and its precursors. In: Nissenbaum A (ed) Hypersaline brines and evaporites. Elsevier, Amsterdam, pp 195–232

Eugster HP (1986) Lake Magadi, Kenya: a model for rift valley hydrochemistry and sedimentation. In: Frostick LE, Renaut RW, Reid I, Tiercelin J-J (eds) Sedimentation in the African rifts. Geol Soc Lond Spec Publ 25:177–189

Eugster HP, Chou I-M (1973) The depositional environment of Precambrian bedded iron-formations. Econ Geol 68:1144–1168

Eugster HP, Jones B (1968) Gels composed of sodium-aluminum silicate, Lake Magadi, Kenya. Science 161:160–163

Eugster HP, Jones BF (1979) Behavior of major solutes during closed-basin brine evolution. Am J Sci 279:609–631

Fairhead J, Mitchell J, Williams L (1972) New K/Ar determinations on rift volcanics of S. Kenya and their bearing on age of rift faulting. Nature 238:66–69

Felske GN (2016) Genesis of calcretes and related carbonate rocks in the southern Kenya Rift. MSc thesis, University of Saskatchewan, Saskatoon

Foster A, Ebinger C, Mbede E, Rex D (1997) Tectonic development of the northern Tanzanian sector of the East African Rift System. J Geol Soc Lond 154:689–700

Fritz B, Zins-Pawlas M-P, Gueddari M (1987) Geochemistry of silica-rich brines from Lake Natron (Tanzania). Sci Géol Bull 40:97–110

Garcia-Ruiz JM (1998) Carbonate precipitation into silica-rich environments. Geology 23:843–846

Getenet M, García-Ruiz JM, Verdugo-Escamilla C, Guerra-Tschuschke I (2020) Mineral vesicles and chemical gardens from carbonate-rich alkaline brines of Lake Magadi, Kenya. Crystals 10:467. https://doi.org/10.3390/cryst10060467

Getenet M, García-Ruiz JM, Otálora F, Emmerling F, Al-Sabbagh D, Verdugo-Escamilla C (2022) A comprehensive methodology for monitoring evaporitic mineral precipitation and hydrochemical evolution of saline lakes: the case of Lake Magadi soda brine (East African Rift Valley, Kenya). Cryst Growth Design 22:2307–2317. https://doi.org/10.1021/acs.cgd.1c01391

Getenet M, Otálora F, Emmerling F, Al-Sabbagh D, García-Ruiz JM (2023) Mineral precipitation and hydrochemical evolution through evaporitic processes in soda brines (East African Rift Valley). Chem Geol 616:121222

Githiri JG, Patel JP, Barongo JO, Karanja PK (2011) Application of Euler deconvolution technique in determining depths to magnetic structures in Magadi area, southern Kenya Rift. J Agric Sci Technol 13:142–146

Githiri JG, Patel JP, Barongo JO, Karanja PK (2012a) Spectral analysis of ground magnetic data in Magadi area, Southern Kenya rift. Tanz J Sci 38:1–14

Githiri JG, Patel JP, Barongo JO, Karanja PK (2012b) An investigation of the structure beneath Magadi area in southern Kenya rift using gravimetric data. J Agric Sci Technol 14:42–160

Glover RB (1972) Chemical characteristics of water and steam discharges in the Rift Valley of Kenya. Unpublished Report, UNDP Geothermal Resources Exploration Project, Kenya

Godfrey G (2018) Sediments thickness and structural framework of Lake Natron Basin and their implications for hydrocarbon potentiality. MSc thesis, University of Dodoma

Government of Kenya [GoK] (2021) Rising water levels in Kenya's Rift Valley lakes, Turkwel Gorge Dam and Lake Victoria: a scoping report. GoK, Nairobi

Goetz C, Hillaire-Marcel C (1992) U-series disequilibria in early diagenetic minerals from Lake Magadi sediments, Kenya: dating potential. Geochim Cosmochim Acta 56:1331–1341

Grant WD, Jones BE (2016) Bacteria, archaea and viruses of soda lakes. In: Schagerl M (ed) Soda lakes of East Africa. Springer, Cham, pp 97–148

Grant WD, Jones BE, Kato C, Li L (1999) Novel archaeal phylotypes from an East African alkaline saltern. Extremophiles 3:139–145

Gregory JW (1921) The rift valleys and geology of East Africa. Seeley Service, London

Gueddari M (1984) Géochimie et thermodynamique des évaporites continentales. Etude du lac Natron en Tanzanie et du Chott el Jerid en Tunisie. Sci Géol Mém 76

Guest NJ (1953) Geology and petrology of the Engaruka - Oldonyo L'Engai - Lake Natron area of Northern Tanganyika Territory. PhD thesis, University of Sheffield

Guest NJ, Stevens, JA (1951) Lake Natron, its springs, rivers, and visible saline reserves. Geol Surv Tangan Miner Resour Pamph 58

Guest NJ, James TC, Pickering R, Dawson JB (1961) Angata Salei. Geological Survey of Tanganyika, Quarter Degree Sheet 39

Guth AL (2015) Volcanic volumes associated with the Kenya Rift: recognition and correlation of preservation biases. In: Wright TJ, Ayele A, Ferguson DJ, Kidane T, Vye-Brown C (eds) Magmatic rifting and active volcanism. Geol Soc Lond Spec Publ 420: 43–67

Guth A, Wood J (2014a) Geology of the Magadi area. Geol Soc Am Digital Map Chart Ser DMCH016S3

Guth A, Wood J (2014b) Geology of the area south of Magadi. Geol Soc Am Digital Map Chart Ser DMCH016S5

Harper DM, Tebbs E, Bell O, Robinson VJ (2016) Conservation and management of East Africa soda lakes. In: Schagerl M (ed) Soda lakes of East Africa. Springer, Cham, pp 345–364

Hay RL (1966) Zeolites and zeolitic reactions in sedimentary rocks. Geol Soc Am Spec Pap 85

Hay RL (1968) Chert and its sodium-silicate precursors in sodium-carbonate lakes of East Africa. Contrib Mineral Petrol 17:255–274

Hay RL (1986) Role of tephra in the preservation of fossils in Cenozoic deposits of East Africa. In: Frostick LE, Renaut RW, Reid I, Tiercelin J-J (eds) Sedimentation in the African rifts. Geol Soc Lond Spec Publ 25:339–344

Herrick R (1972) Authigenic minerals in the Pleistocene and recent sediments of Lake Magadi, Kenya. MSc thesis. University of Wyoming, Laramie

Hewson R, Mshiu E, Hecker C, van der Wert H, van Ruitenbeek F, Alkema D, van der Meer F (2020) The application of day and night time ASTER satellite imagery for geothermal and mineral mapping in East Africa. Int J Appl Earth Observ Geoinfo 85:101991

Hill MF (1964) Magadi – the story of Magadi Soda Company. Kynoch Press, Birmingham

Hillaire-Marcel C (1987) Hydrologie isotopique des lacs Magadi (Kenya) et Natron (Tanzanie). Sci Géol Bull 40:111–120

Hillaire-Marcel C, Casanova J (1987) Isotopic hydrology and paleohydrology of the Magadi (Kenya)-Natron (Tanzania) basin during the late Quaternary. Palaeogeogr Palaeoclimatol Palaeoecol 58:155–181

Hillaire-Marcel C, Carro O, Casanova J (1986) ^{14}C and Th/U dating of Pleistocene and Holocene stromatolites from East African palaeolakes. Quat Res 25:312–329

Hochstein MP (1999) Geothermal systems along the East-African Rift. Bull Hydrogéol 17:301–310

Hochstein MP, Temu EP, Moshe CMA (2000) Geothermal resources of Tanzania. Proc World Geotherm Congr Kyushu-Tohoku, Japan, May 28–June 10, 2000, pp 1233–1238

Hughes L (2008) Mining the Maasai reserve: the story of Magadi. J East Afr Stud 2:134–164

Ibs-von Seht M, Blumenstein S, Wagner R, Hollnack D, Wohlenberg J (2001) Seismicity, seismotectonics and crustal structure of the southern Kenya Rift – new data from the Lake Magadi area. Geophys J Int 146:439–453

Ibs-von Seht M, Plenefisch T, Klinge K (2008) Earthquake swarms in continental rifts — a comparison of selected cases in America, Africa and Europe. Tectonophysics 452:66–77

Icole M, Perinet G (1984) Les silicates sodiques et les milieux évaporitiques carbonatés bicarbonates sodiques: une revue. Rev Géol Dynam Géogr Phys 25:167–176

Jagniecki EA, Lowenstein TK, Jenkins DM, Demicco RV (2015) Eocene atmospheric CO_2 from the nahcolite proxy. Geology 43:1075–1078

Jones BJ, Rettig SL, Eugster HP (1967) Silica in alkaline brines. Science 158:1310–1314

Jones BJ, Eugster HP, Rettig SF (1977) Hydrochemistry of the Lake Magadi basin, Kenya. Geochim Cosmochim Acta 41:53–72

Kambura AK, Mwirichia RK, Kasili RW, Karanja EN, Makonde HM, Boga HI (2016a) Bacteria and archaea diversity within the hot springs of Lake Magadi and Little Magadi in Kenya. BMC Microbiol 16:136–148

Kambura AK, Mwirichia RK, Kasili RW, Karanja EN, Makonde HM, Boga HI (2016b) Diversity of fungi in sediments and water sampled from the hot springs of Lake Magadi and Little Magadi in Kenya. Afr J Microbiol Res 10:330–333

Kaufman A, Margaritz MPM, Hillaire-Marcel C, Hollos G, Boaretto E, Taieb M (1990) The ^{36}Cl ages of the brines in the Magadi-Natron Basin, East Africa. Geochim Cosmochim Acta 54:2827–2833

Kavembe GD, Machado-Schiaffino G, Meyer A (2014) Pronounced genetic differentiation of small, isolated and fragmented tilapia populations inhabiting the Magadi Soda lake in Kenya. Hydrobiologia 739:55–71

Kavembe GD, Meyer A, Wood CM (2016) Fish populations in East African saline lakes. In: Schagerl K (ed) Soda lakes of East Africa. Springer, Cham, pp 227–257

Kenana, LM, Bakari, SK, Bitok E, Machoke NM, Hamza K, Mukeka J, Chepkwony RK, Mwiu, SN, (2013) Total aerial count for Amboseli – West Kilimanjaro and Magadi-Natron cross border landscape, April 2013. Technical Report for KWS, Kenya and TAWIRI, Tanzania

Kiama D, Masai F, Ojwala M, Nyarondia R, Macharia D, Juma E, Njino L, Otero H (2019) Mapping land degradation hotspots in the Suswa/Lake Magadi Basin. Draft Report, Directorate of Resource Surveys and Remote Sensing (DRSRS), Ministry of Environment and Forestry, Nairobi

Kilham P (1971) Biogeochemistry of African lakes and rivers. PhD thesis, Duke University Durham, North Carolina

Kipkwony SJ, Barongo JO, Dindi EW, Mulwa JK, Mavonga GT (2022) Seismic hazard assessment in Kenya and its vicinity. In: Meghraoui M, Sundararajan N, Banerjee S, Hinzen K-G, Eshagh M, Roure F, Chaminé HI, Maouche S, Michard A (eds) Advances in geophysics, tectonics and petroleum geosciences. Springer, Cham, pp 57–61

Kodikara GRL, Woldai T, van Ruitenbeek FJA, Shepherd KD, van Hummel GJ (2016) Mapping evaporite minerals and associated sediments in Lake Magadi, Kenya, using hyperspectral hyperion data. J Univ Ruhuna 4:22–27

Komolafe AA (2010) Investigations into the tectonic lineaments and thermal structure of Lake Magadi, southern Kenya Rift using integrated geophysical methods. MSc thesis, International Institute for Geo-information Science and Earth Observation, Enschede

Krienitz L (2018) Lesser flamingos: descendants of Phoenix. Springer Nature, Berlin

Krienitz L, Schagerl M (2016) Tiny and tough: microphytes of East African Soda Lakes. In: Schagerl M (ed) Soda lakes of East Africa. Springer, Cham, pp 149–177

Krienitz L, Bock C, Kotut K, Luo W (2012) *Picocystis salinarum* (Chlorophyta) in saline lakes and hot springs of East Africa. Phycologia 51:22–32

Krienitz L, Krienitz D, Dadheech PK, Hübener T, Kotut K, Luo W, Teubner K, Versfeld W (2016) Food algae for lesser flamingos: a stocktaking. Hydrobiologia 775:21–50

Kuria ZN, Woldai T, van der Meer FD, Barongo JO (2010) Active fault segments as potential earthquake sources: inferences from integrated geophysical mapping of the Magadi fault system, southern Kenya Rift. J Afr Earth Sci 57:345–359

Le Gall B, Nonnotte P, Rolet J, Benoit M, Guillou H, Mousseau-Nonnotte M, Albaric J, Déverchère J (2008) Rift propagation at craton margin: distribution of faulting and volcanism in the North Tanzanian divergence (East Africa) during Neogene times. Tectonophysics 448:1–19

Le Gall B, Remigius G, Koptev A, Chazot G, Boniface N, Loget N, Daoud MA, Tarits P, Plasman M, Hautot S (2021) The anomalously-propagating South Kenya rift in the context of the north Tanzanian Divergence Zone, East Africa. Tectonophysics 814:228968

Le Turdu C, Tiercelin J-J, Richert JP, Rolet J, Xavier JP, Renaut RW, Lezzar KE, Coussement C (1999) Influence of preexisting oblique discontinuities on the geometry and evolution of extensional fault patterns: evidence from the Kenya Rift using SPOT imagery. In: Morley CK (ed) Geoscience of rift systems—evolution of East Africa. AAPG Stud Geol 44:173–191

Lee H, Muirhead JD, Fischer TP, Ebinger CJ, Kattenhorn CA, Sharp ZD, Kianji G (2016) Massive and prolonged deep carbon emissions associated with continental rifting. Nat Geosci 9:145–149

Lee H, Fischer TP, Muirhead JD, Ebinger CJ, Kattenhorn SA, Sharp ZD, Kianji G, Takahata N, Sano Y (2017) Incipient rifting accompanied by the release of subcontinental lithospheric mantle volatiles in the Magadi and Natron basin, East Africa. J Volcanol Geotherm Res 346:118–133

Leet K (2022) Origins of chert: Lake Magadi Basin, Kenya. PhD thesis, Binghamton University, New York

Leet K, Lowenstein TK, Renaut RW, Owen RB, Cohen A (2021) Labyrinth patterns in Magadi (Kenya) cherts: evidence for early formation from siliceous gels. Geology 49:1137–1142

Liutkus-Pierce CM, Zimmer BW, Carmichael SK, McIntosh W, Deino A, Hewitt SM, McGinnis KJ, Hartney T, Brett J, Mana S, Deocampo D, Richmond BG, Hatala K, Harcourt-Smith W, Pobiner B, Metallo A, Rossi V (2016) Radioisotopic age, formation and preservation of Late Pleistocene human footprints at Engare Sero, Tanzania. Palaeogeogr Palaeoclimatol Palaeoecol 463:68–82

Maglione G (1976) Géochimie des évaporites et silicates néoformés en milieu confiné. Trav Doc ORSTOM Paris 60

Manega PC, Bieda S (1987) Modern sediments of Lake Natron, Tanzania. Sci Géol Bull 40:83–96

Marsden M (1979) Origin and evolution of the Pleistocene Olorgesailie Lake Series, Kenya Rift Valley. PhD thesis, McGill University, Québec

Maswi MS, Mshiu E (2020) Delineation of geothermal potential sites using remote sensing satellite data –Lake Natron, northern Tanzania. Proc 8 Afr Rift Geotherm Conf Nairobi

McNitt JR, Klein CW, Koenig JB (1989) Probable subsurface temperature at Lake Magadi, Kenya, as indicated by hot springs chemistry, and the potential for development of geothermal electric power. Natl Geotherm Assoc, Davis, CA, USA Rep 12850

McNulty E (2017) Lake Magadi and the soda lake cycle: a study of the modern sodium carbonates and of Late Pleistocene and Holocene lacustrine core sediments. MS thesis, Binghamton University, New York

Mengistou S (2016) Invertebrates of East African soda lakes. In: Schagerl M (ed) Soda lakes of East Africa. Springer, Cham, pp 205–226

Mgimwa EF, John JR, Lugomela CV (2021) The influence of physico-chemical variables on phytoplankton and lesser flamingo (*Phoeniconaias minor*) abundances in Lake Natron, Tanzania. Afr J Ecol 59:556–675

Monnin C, Schott J (1984) Determination of the solubility products of sodium carbonate minerals and an application to trona deposition in Lake Magadi (Kenya). Geochim Cosmochim Acta 48:571–581

Muirhead JD, Kattenhorn SA, Lee H, Mana S, Turrin BD, Fischer TP, Kianji G, Dinde E, Stamps DS (2016) Evolution of upper crustal faulting assisted by magmatic volatile release during early-stage continental rift development in the East African Rift. Geosphere 12:1670–1700

Muiruri VM (2017) Late Quaternary diatom and palynomorph stratigraphies and palaeoenvironments of the Koora Graben and Lake Magadi Basin, Kenya Rift Valley. PhD thesis, Hong Kong Baptist University

Muiruri VM, Owen RB, de Cort G, McNulty E, Rabideaux N, Lowenstein T, Renaut RW, Leet K, Deocampo D, Cohen AS, Campisano CJ, Billingsley A, Mbuthia A (2021a) Middle Pleistocene to recent diatom floras and stratigraphy of the Magadi Basin, southern Kenya Rift. J Paleolimnol 65:315–333

Muiruri VM, Owen RB, Lowenstein TK, Renaut RW, Marchant R, Rucina SM, Cohen AC, Deino AL, Sier MJ, Luo S, Leet K, Campisano C, Rabideaux NM, Deocampo D, Shen C-C, Mbuthia A, Davis BC, Aldossari W, Wang C (2021b) A million year vegetation history and paleoenvironmental record from the Lake Magadi Basin, Kenya Rift Valley. Palaeogeogr Palaeoclimatol Palaeoecol 567:110247. https://doi.org/10.1016/j.palaeo.2021.110247

Mulango E, Vasili R, Mwirichia R, Kambura AK, Muhonja C (2020) Isolation and characterization of haloalkaliphilic bacteria from the hot springs of Lake Magadi. Afr J Microbiol Res 14:294–302

Mulwa JK, Kimata F, Suzuki S, Kuria ZN (2014) The seismicity in Kenya (East Africa) for the period 1906–2010: a review. J Afr Earth Sci 89:72–78

Mwirichia R, Orwa P (2023) Diversity of extremophiles in African brine lakes. In: El-Sheekh M, Elsaied HE (eds) Lakes of Africa: microbial diversity and sustainability. Elsevier, Amsterdam, pp 269–287

Naujoks M, Jar T, Jentzsch G, Kurz JH, Hofmann Y (2008) Investigations about earthquake swarm areas and processes. In: Tregoning P, Rizos C (eds) Dynamic planet: monitoring and understanding a dynamic planet with geodetic and oceanographic tools (Int Assoc Geodesy Symposia Vol 130). Springer, Berlin, pp 528–535

Nielsen JM (1999) East African magadi (trona): fluoride concentration and mineralogical composition. J Afr Earth Sci 29:423–428

Nikonova EL (2016) Authigenic clay formation and diagenetic reactions, Lake Magadi, Kenya. MSc thesis, Georgia State University, Atlanta

Nonga HE, Mdegela RH, Sandvik M, Lie E, Miles CO, Skaare JU (2017) Cyanobacteria and cyanobacterial toxins in the alkaline-saline lakes Natron and Momela, Tanzania. Proc Tanz Vet Assoc 34:106–118

Nyamweru CK (1988) Activity of Ol Doinyo Lengai volcano, Tanzania, 1983–1987. J Afr Earth Sci 7:603–610

O'Neil JR, Hay RL (1973) $^{18}O/^{16}O$ ratios in cherts associated with the saline lake deposits of East Africa. Earth Planet Sci Lett 19:257–266

Onguso BN (2019) Crustal structure evolution in Magadi area, Kenya, using 3D seismic tomography. MSc thesis, University of Nairobi

Owen RB, Muiruri VM, Lowenstein TK, Renaut RW, Rabideaux N, Luo S, Deino AL, Sier MJ, Dupont-Nivet G, McNulty EP, Leet K, Cohen AS, Campisano C, Deocampo D, Shen C-C,

Billingsley A, Mbuthia A (2018) Progressive aridification in East Africa over the last half million years and implications for human evolution. Proc Natl Acad Sci 115:11174–11179

Owen RB, Renaut RW, Muiruri VM, Rabideaux NM, Lowenstein TK, McNulty EP, Leet K, Deocampo D, Luo S, Deino AL, Cohen A, Sier MJ, Campisano C, Shen C-C, Billingsley A, Mbuthia A, Stockhecke M (2019) Quaternary history of the Lake Magadi Basin, southern Kenya Rift: tectonic and climatic controls. Palaeogeogr Palaeoclimatol Palaeoecol 518:97–118

Page N, Simon B (1987) Téléanalyse diachronique de différents états du lac Natron (Rift Est-Africain, Tanzanie). Mise en évidence de variations saisonnières et pluriannuelles de son remplissage. Sci Géol Bull 40:23–39

Parkinson J (1914) The East African trough in the neighbourhood of the soda lakes. Geogr J 44:33–46

Rabideaux N (2018) Late Quaternary East African environmental change based on mineralogical and geochemical analysis of outcrop and core material from the Southern Kenya Rift. PhD thesis. Georgia State University, Atlanta

Randel RP (1970) Geology of the Suswa area. Kenya Geol Surv Rep 97

Reiss MC, De Siena L, Muirhead JD (2022) The interconnected magmatic plumbing system of the Natron Rift. Geophys Res Lett 49:e2022GL098922

Renaut RW, Owen RB, Jones B, Tiercelin J-J, Tarits C, Ego JK, Konhauser KK (2013) Impact of lake level changes on the formation of thermogene travertine in continental rifts: evidence from Lake Bogoria, Kenya Rift Valley. Sedimentology 60:428–468

Renaut RW, Owen RB, Ego JK (2017) Geothermal activity and hydrothermal mineral deposits at southern Lake Bogoria, Kenya Rift Valley: impact of lake level changes. J Afr Earth Sci 129:623–646

Renaut RW, Owen RB, Lowenstein TK, de Cort G, McNulty E, Scott JJ, Mbuthia A (2021) The role of hydrothermal fluids in sedimentation in saline alkaline lakes: evidence from Nasikie Engida, Kenya Rift Valley. Sedimentology 68:108–134

Roberts N, Barker P (1993) Landscape stability and biogeomorphic response to past and future climatic shifts in intertropical Africa. In: Thomas DSG, Allison RJ (eds) Landscape sensitivity. Wiley, Chichester, pp 65–82

Roberts N, Taieb M, Barker P, Damnati B, Icole M, Williamson D (1993) Timing of the Younger Dryas event in East Africa from lake-level changes. Nature 366:146–148

Robinson VJ (2015) The ecology of East African soda lakes: implications for lesser flamingo (*Phoeniconaias minor*) feeding behaviour. PhD thesis, University of Leicester

Röhricht C (1998) Lithologie und der Chertserien des Magadi Beckens, Lake Magadi, Kenia. PhD thesis, Universität Göttingen

Rösner TM, Dolle A, Tichy H (2003) Fischfossilien aus zwei Giftseen mit extremen Lebensbedingungen, Natron- und Magadisee. Götting Arb Geol Paläontol 5:77–83

Saggerson EP (1966) Geology of the Loita Hills area. Geol Surv Kenya Rep 71

Schagerl M (ed) (2016) Soda lakes of East Africa. Springer, Cham

Schubel KA, Simonson BM (1990) Petrography and diagenesis of cherts from Lake Magadi, Kenya. J Sediment Petrol 60:761–776

Scoon RN (2018) Geology of national parks of central/southern Kenya and northern Tanzania. Springer, Cham

Scoon RN (2021) Lake Natron and the Ngorongoro conservation area, northern Tanzania. In: The Geotraveller. Springer, Cham, pp 117–137

Scott JJ, Renaut RW, Buatois LA, Owen RB, McNulty EP, Stockhecke M, Leet K, Lowenstein TK, Mángano GM (2021) Modern and ancient animal traces in the extreme environments of Lake Magadi and Nasikie Engida, Kenya Rift Valley. In: Rosen MR, Park-Boush L, Finkelstein DB, Pla Pueyo S (eds) Limnogeology: progress, challenges and opportunities: a tribute to Elizabeth Gierlowski-Kordesch. Springer, Cham, pp 19–66

Seegers L, Tichy H (1999) The *Oreochromis alcalicus* flock (Teleostei: Cichlidae) from lakes Natron and Magadi, Tanzania and Kenya, with descriptions of two new species. Ichthyol Explor Freshw 10:97–146

Seegers L, Sonnenberg R, Yamamoto R (1999) Molecular analysis of the *Alcolapia* flock from lakes Natron and Magadi, Tanzania and Kenya (Teleostei: Cichlidae), and implications for their systematics and evolution. Ichthyol Explor Freshw 10:175–199

Sequar GW (2009) Neotectonics of the East African Rift System: new interpretations from conjunctive analysis of field and remotely sensed datasets in the Lake Magadi area, Kenya. MSc thesis, International Institute for Geo-information Science and Earth Observation, Enschede, Netherlands

Sheppard RA, Gude AJ, Hay RL (1970) Makatite, a new hydrous sodium silicate mineral from Lake Magadi, Kenya. Am Mineral 35:358–366

Sherrod DR, Magigita MM, Kwelwa S (2013) Geological map of Oldoinyo Lengai Volcano and surroundings, Arusha Region. United Repub Tanz, USGS Open-File Rep 2013–1306 with accompanying pamphlet

Smith GE (1907) From the Victoria Nyanza to Kilimanjaro (and accompanying map). Geogr J 20:249–269 and 30:77–78

Spink JC, Stevens JA (1946) Notes on the Magadi section of the Eastern Rift Valley. Geogr J 10:236–241

Stevens JA (1932) Lake Magadi and its alkaline springs. Unpublished report to the Magadi Soda Company, Ltd

Surdam RC, Eugster HP (1976) Mineral reactions in the sedimentary deposits of the Lake Magadi region, Kenya. Geol Soc Am Bull 87:1739–1752

Taieb M, Barker P, Bonnefille R, Damnati B, Gasse F, Goetz C, Hillaire-Marcel C, Icole M, Masault M, Roberts N, Vincens A, Williamson D (1991) Histoire paléohydrologique du lac Magadi (Kenya) au Pleistocène superieur. C R Acad Sci II 313:339–346

Tebbs EJ, Remedios JJ, Avery ST, Harper DM (2013) Remote sensing the hydrological variability of Tanzania's Lake Natron, a vital lesser flamingo breeding site under threat. Ecohydrol Hydrobiol 13:148–158

Temperley BN (1951) Some geological and geophysical investigations in the vicinity of Lake Magadi. Unpublished report, Mines Geol Dept, Nairobi (cited by Baker, 1958; not seen in original)

Temperley BN, Thompson (1950) Report on work done on the Lake Magadi problem up to July 1950. Unpublished report, Mines Geol Dept, Nairobi

Tichy H, Seegers L (1999) The *Oreochromis alcalicus* flock (Teleostei: Cichlidae) from lakes Natron and Magadi, Tanzania and Kenya, a model for the evolution of "new" species flocks in historical times? Ichthyol Explor Freshw 10:147–174

Tole MP (1990) Chemical geothermometry and resource potential of low enthalpy geothermal systems in Kenya. Geothermal Resour Counc Trans 14:187–193

Tuite C (1981) Standing crop densities and distribution of *Spirulina* and benthic diatoms in East African alkaline saline lakes. Freshw Biol 11:345–360

Uhlig C, Jaeger F (1942) Die ostafrikanische Bruchstufe und die angrenzende Gebiete zwischen den Seen Magad und Lara ja Mweri sowie dem Westfluss des Meru. Part II Wissen Verruf deutsch Institit f Landeskunde NS 10

van Zyl LJ, Nemavhulani S, Cass J, Arthur Cowan DA, Marla Trindade M (2016) Three novel bacteriophages isolated from the East African Rift Valley soda lakes. Virol J 13:204

Vincens A, Casanova J (1987) Modern background of Natron-Magadi basin (Tanzania-Kenya): physiography, climate, hydrology and vegetation. Sci Géol Bull 40:9–21

Vincens A, Bonnefille R, Buchet G (1991) Étude palynologique du sondage Magadi NF1 (Kenya) implications paléoclimatiques. Geobios 24:549–558

Walter P (1922) Sodium carbonate minerals of the Mogadi lakes, British East Africa. Am Mineral 7:86–88

Warren J (2016) Evaporites, 2nd edn. Springer, Cham

Weinstein A, Oliva SJ, Ebinger CJ, Rocker S, Tiberi C, Aman M, Lambert C, Witkin E, Albaric J, Gautier S, Peyrat S, Muirhead JD, Muzuka ANN, Mulibo G, Kianji G, Ferdinand-Wambura R,

Msabi M (2017) Fault-magma interactions during early continental rifting: seismicity of the Magadi-Natron-Manyara basins, Africa. Geochem Geophys Geosyst 18:3662–3686
White TH (1953) Some speculations on the sudden occurrence of floods in the history of Lake Magadi. J East Afr Nat Hist Soc 22:69–71
White F (1983) The vegetation of Africa, a descriptive memoir to accompany the UNESCO/AETFAT/UNSO vegetation map of Africa. UNESCO Nat Resour Res 20:1–356
Williamson D, Taieb M, Damnati B, Icole M, Thouveny N (1993) Equatorial extension of the Younger Dryas event: rock magnetic evidence from Lake Magadi (Kenya). Glob Planet Change 7:235–242
Wilson PJ, Wood CM, Walsh PJ, Bergman AN, Bergman HL, Laurent P, White BN (2004) Discordance between genetic structure and morphological, ecological, and physiological adaptation in Lake Magadi tilapia. Physiol Biochem Zool 77:537–555
Wright JB (1967) Geology of the Narok area. Rep Geol Surv Kenya 80
Wynn JG, Lowenstein TK, Renaut RW, Owen RB, McNulty EP (2018) Superheavy pyrite formation in hypersaline lakes of East Africa: a paleo salinity proxy for Lake Magadi HSPDP cores. Geol Soc Am Abstr Progr 50:184–183. https://doi.org/10.1130/abs/2018AM-323852
Yanda PZ, Madulu NF (2005) Water resource management and biodiversity conservation in the Eastern Rift Valley lakes, northern Tanzania. Phys Chem Earth 30:717–725
Zhilina TN, Zavarzin GA (1994) Alkaliphilic anaerobic community at pH 10. Curr Microbiol 29:109–112

Chapter 13
Lake Victoria

13.1 Introduction

The northeastern extension of Lake Victoria, the Winam or Nyanza Gulf (Fig. 13.1), floods part of the east-west trending Nyanza Rift (Fig. 13.2a, b), which is comprised of a complex series of east-west trending faults and monoclines that are oriented perpendicular to the main Kenya Rift. The lake is accessed from Kenya via Kisumu on its northeastern shore (Fig. 13.2a) and is the largest water body in Africa by surface area (68,800 km^2), containing the third largest volume (2760 km^3) of water in Africa (Awange and Ong'ang'a 2006). It is also the second largest lake in the world by surface area (Danley et al. 2012; Awange et al. 2019), occupying a drainage basin of 258,000 km^2 at an elevation of ~1134 m above sea level. Despite its great size, it is relatively shallow with a mean depth of only 40 m and a maximum of ~69 m (Johnson et al. 1996). Its irregular and often swampy 3500-km-long shoreline includes ~760 km within Kenya and reflects the partial flooding of former river valleys (Awange and Ong'ang'a 2006). The outlet lies to the north near Jinja and is a major source of the River Nile with important inflows including the Kagera and Katonga rivers to the west (Fig. 13.2a). Major towns along its shores include Kampala, Entebbe and Jinja in Uganda, Kisumu in Kenya, and Mwanza, Musoma and Bukoba in Tanzania.

The first known map has been attributed to three missionaries (Krapff, Rebmann and Erhardt) who labelled a large inland lake as the "Uniamesi Sea" (Moorehead 1960; Crul 1995). Subsequently, the Royal Geographical Society in London supported an expedition by Richard Burton and John Speke to explore the region. Speke reached southeastern Lake Victoria in 1858 and named it after the reigning monarch in the United Kingdom. He returned in 1862, with James Grant, and explored the Nile outlet (Speke 1863). Wakefield and Johnston (1870) presented a map and descriptions based on reports from caravan routes to the interior. Subsequently, Henry Stanley circumnavigated the lake in 1875 (Stanley 1875, 1877, 1889; Grant 1876) with Joseph Thompson reaching its northeastern shores in 1884

R. W. Renaut, R. B. Owen, *The Kenya Rift Lakes: Modern and Ancient*, Syntheses in Limnogeology, https://doi.org/10.1007/978-3-642-25055-2_13

Fig. 13.1 Lake Victoria. (**a**) Winam (Nyanza) Gulf from Kisumu. (**b**) Southern shoreline of Winam Gulf

(Thomson 1884). Isaak Eduard Schnitzer (Emin Pasha), along with Stanley, reached the area in 1889. Arab traders had known of the lake before the European explorations and called it Lake Ukerewe. Local peoples have several other names, including Nyancha/Nyanja (the Basuba of Kenya), Nam Lolwe or "endless waters" (the Luos, Kenya), Nalubale (the Baganda, Uganda) and Sukuma (the Sukuma of Tanzania) (Awange and Ong'ang'a 2006).

The population in the catchment expanded from 4.6 to 27.7 million between 1932 and 1995 (Verschuren et al. 2002) and it now supports >30 million people in Uganda, Tanzania, Rwanda, Burundi and Kenya (Fig. 13.2a). Several Bantu peoples occupy the basin, having developed local kingdoms in the fourteenth century, such

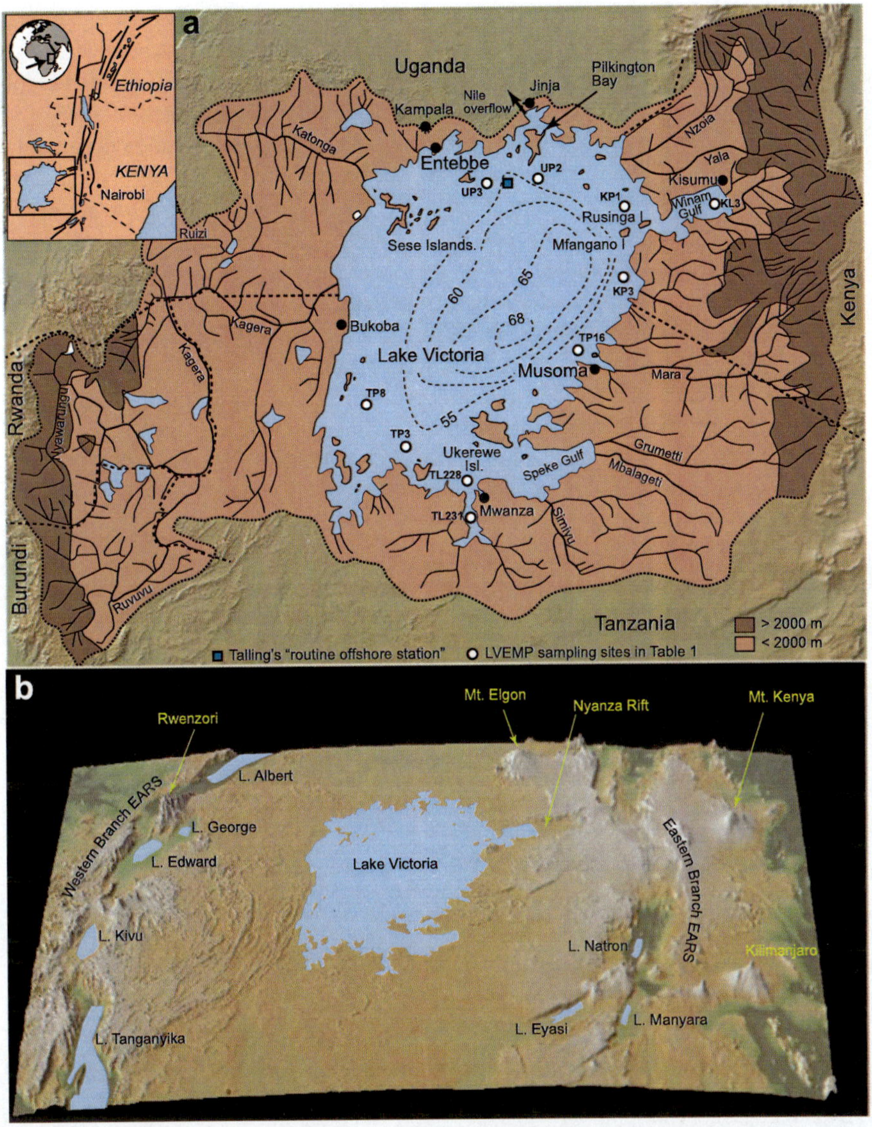

Fig. 13.2 Topography and drainage in the Lake Victoria catchment. Maps constructed from GeoMapApp (http://www.geomapapp.org). Lake bathymetry. After Johnson et al. (2000). (**a**) Catchment drainage and locations. (**b**) Topography and relationship to the East African Rift System (EARS)

as the Baganda, Banyoro, Toro, Ankole and Basoga (Awange and Ong'ang'a 2006). Colonial rule was formalised in 1885 when Tanganyika, Rwanda and Burundi came under German control. After WWI Tanganyika became part of the British Empire, with Burundi and Rwanda controlled by Belgium. In 1887, most of modern Kenya was placed under the administration of the British East Africa Company and was subsequently transferred to direct British rule in 1895. Tanganyika and Zanzibar were unified and gained independence as Tanzania in 1961 with independence following for Uganda, Rwanda and Burundi in 1962 and for Kenya in 1963.

In recent decades, the lake has experienced major ecological and environmental transformations related to population growth, introductions of exotic species and increased pollutants (Kudhongania et al. 1996; Calamari et al. 2006; Downing et al. 2014) driven by land use changes in the catchment with the conversion of forests, woodlands, grasslands, and wetlands to farmlands or settlements (Mugo et al. 2020). Research prior to the 1960s was infrequent and localised. However, several early limnologists (Graham 1929; Worthington 1930; Fish 1957; Talling and Talling 1965; Talling 1966, 2006) carried out important baseline studies. The subsequent environmental changes include reduced biodiversity, eutrophication, deoxygenation of bottom water, alteration in phytoplankton composition and the expansion of invasive species such as water hyacinth and Nile Perch (Mwanuzi et al. 2005; Sitoki et al. 2010; Otieno et al. 2022; Njagi et al. 2022). The literature dealing with the lake is extensive with many articles having been compiled by Crul et al. (1995) who focussed on its fisheries.

The damage caused by these varied environmental pressures led Kenya, Uganda and Tanzania to set up the Lake Victoria Environmental Management Project phase I (LVEMP I) in 1997 as a five-year project that aimed to improve natural ecosystems. Eventually, this led to further cooperation between the three countries on lake management and sustainable development that continues today (Mgaya and Mahongo 2017).

13.2 Geological and Geomorphological Setting

Lake Victoria fills a tectonic sag on the Tanzania Craton, which lies between the eastern and western branches of the East African Rift System (EARS) (Fig. 13.3a). The lake is surrounded mainly by Precambrian metamorphic basement with localised outcrops of Karoo-age sediments. The Nyanza (Kavirondo) Rift, to the northeast of the lake, is dominated by alkali volcanics with smaller outcrops of Tertiary and Recent sediments (Fig. 13.3b; Kent 1944; King et al. 1972; Jones and Lippard 1979; Johnson et al. 2000). The major soils on these lithologies include oxisols (red and yellow) as well as acrisols and vertisols (black clay) (Okalebo et al. 2005; Guya 2013).

The Kenya Rift generally trends north-south, but the Nyanza Rift is bounded by east-west and southwest-northeast trending monoclines, antithetic, normal and oblique faults (Pickford 1982; Mboya 1983) that may follow relatively weak Proterozoic mobile belts (Nyblade et al. 1996). The distinctive elongate shape of Speke Gulf, to the southeast of Lake Victoria (Fig. 13.3c), may be controlled by a late Precambrian graben or pull-apart basin located between two shear zones in the basement rocks. These are the Suguti and Lwashi lineaments, which experienced

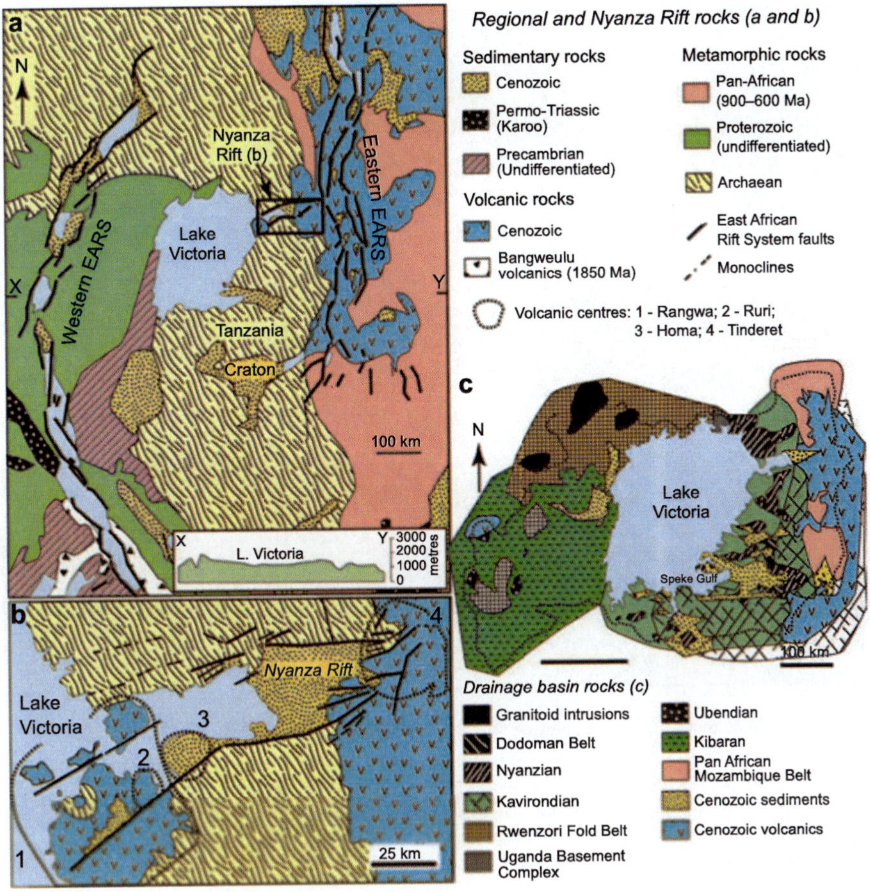

Fig. 13.3 Regional geology of the Lake Victoria drainage basin. (**a**) Geology, major East African Rift System faults and topographic cross section. Modified from Ebinger (1989). (**b**) Nyanza (Kavirondo) Rift sediments, major volcanics and structures. After Pickford (1982). (**c**) Drainage basin geology. Modified from Mulibo and Nyblade (2013)

strike-slip motions during the Precambrian and which may have been partially reactivated in the Cenozoic (Daracott 1974; Rach and Rosendahl 1989).

Early researchers suggested that widespread erosion led to the formation of a series of planation surfaces, including, for example, the pre-Miocene or Mesozoic Buganda and early to mid-Tertiary Kyoga surfaces (Wayland 1933; Ruhe 1954; Bishop and Trendall 1967; Doornkamp 1968; Temple and Doornkamp 1970). The region that is now occupied by Lake Victoria was formerly crossed by westward flowing rivers (Fig. 13.4; Temple 1966) with a proto-Lake Victoria created by blockage as a result of uplift along the margins of the western branch of the EARS (Wayland 1930, 1931; Groves 1934; Doornkamp and Temple 1966; Temple, 1971; Ebinger 1989; Johnson et al. 1996, 2000; Goudie 2005; Salzburger et al. 2014). The lake may have initially developed from two smaller water bodies along the Katonga

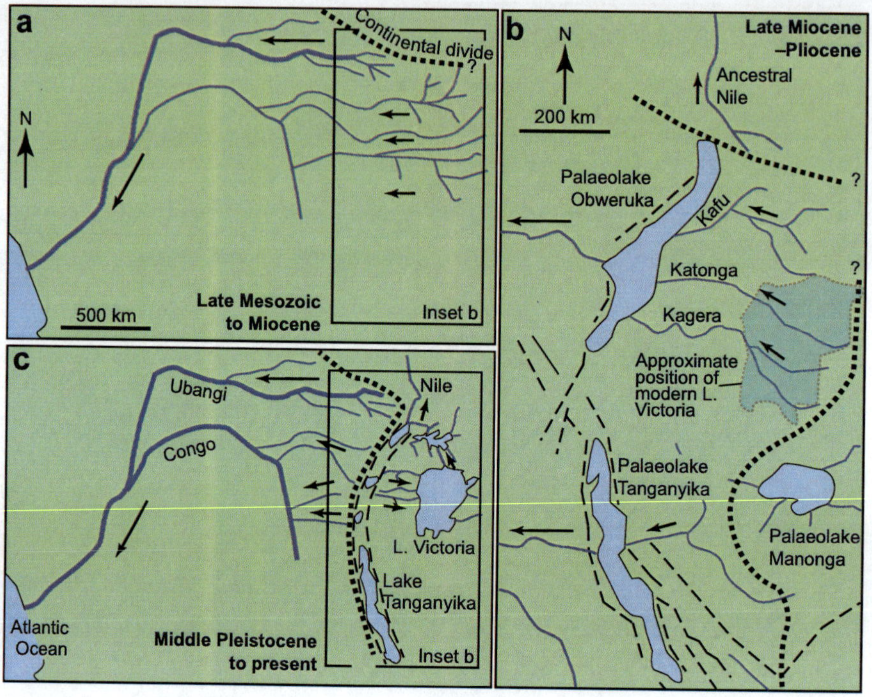

Fig. 13.4 Palaeogeographical evolution of the Victoria Basin. After Talbot and Williams (2009). (**a**) Late Mesozoic to Miocene with westward flowing rivers to the Atlantic. (**b**) Late Miocene–Pliocene times. A large lake Obweruka intercepts westward-flowing rivers and eventually develops into Lakes Albert and Edward. An east-west continental divide develops to the east and a north-south divide forms in the north. Fish in Palaeolake Manonga show Nilotic affinities. (**c**) Location of modern drainage divide and palaeolakes

and Kagera Rivers, which previously fed into an "Obweruka" palaeolake to the northwest (Fig. 13.4b; Bishop 1969; van Damme and Pickford 2003; Talbot and Williams 2009). Palaeolake Manonga lay to the south of modern Lake Victoria and drained to the east (Fig. 13.4b; Stewart 1997). Further rift uplift resulted in flooding and the expansion of the lake with Late Pleistocene outflow draining northwards to Lake Kyoga, Lake Albert and eventually the Nile (Fig. 13.4c).

13.3 Climate and Basin Hydrology

Several aspects of the climate of Lake Victoria are shown in Fig. 13.5a. Air temperatures, for example, vary little through the year, averaging 19–25 °C with daily ranges of 15–30 °C (Awange and Ong'ang'a 2006). Surface water temperatures are also nearly constant (~24 °C). Rainfall is controlled by local topography and by the north-south movements of the Inter Tropical Convergence Zone (ITCZ) and Congo

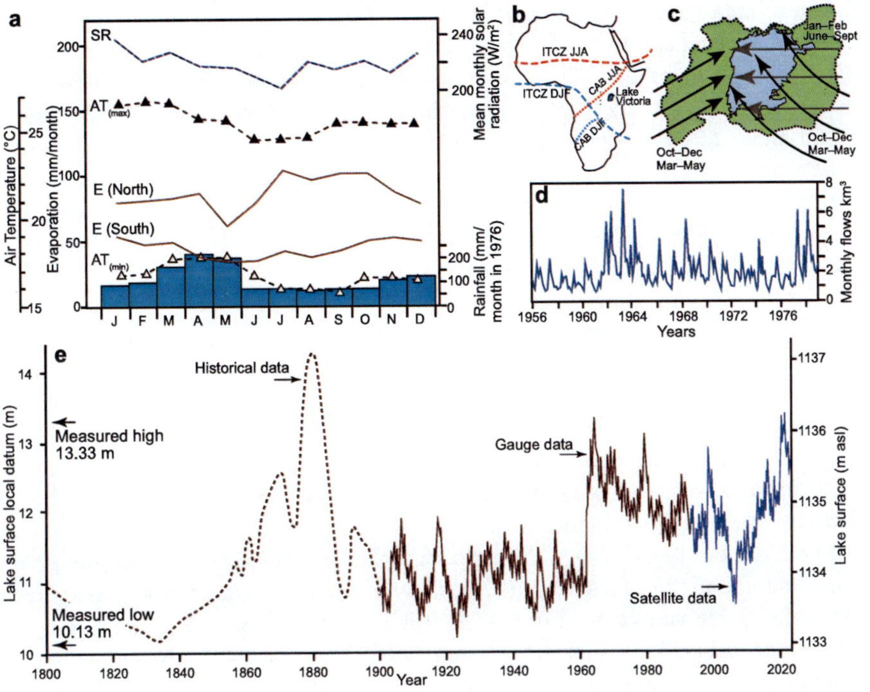

Fig. 13.5 Climate, river and lake level variability. (**a**) Climate pattern for 1976. After Spigel and Coulter (1996). (**b**) Seasonal positions of the InterTropical Convergence Zone (ITCZ) and Congo Air Boundary (CAB). JJA = June, July, August; DJF = December, January, February. (**c**) Wind regimes. (After Okonga et al. 2005). (**d**) Total fluvial inputs from 1958 to 1979 After Sutcliffe and Parks (1999). (**e**) Historical and recorded Lake Victoria water levels. Historical data after Nicholson (1998). Gauge data from Tungaraza et al. (2012). Satellite data based on USDA information for TOPEX, POSEIDON, Jason-1/2, OSTM altimetry

Air Boundary (CAB) (Fig. 13.5b), which control the wind regime (Nicholson 1996; Lehman 2009; Danley et al. 2012; Awange et al. 2013; Awange and Kyalo Kiema 2013; Loiselle et al. 2014). Dry winds blow from east to west in January–February and June–September (Fig. 13.5c) and pick up moisture as they cross the lake with resulting rains developing in western catchments (Okonga et al. 2005). Southeasterly and Congo westerly winds blow between October–December and March–May with the former dominant. Precipitation averages about 1300 mm annually, ranging from 2000 mm in the savannah highlands to 1000 mm near the lakeshore. Nocturnal land breezes converge at the lake centre resulting in upward air movements and thunderstorms (Flohn and Fraedrich 1966), which tend to be displaced to the northwest by low-elevation southeasterly winds (Lehman 2009).

More than 80% of the water falls directly on the lake surface with much of this precipitation being recycled by evaporation, which accounts for up to 90% of the water loss (Baker 1974; Crul 1995; Piper et al. 1986; Nicholson and Yin 2002; Anyah et al. 2006). Seka et al. (2022) also noted that overall water storage in the lake had increased by 0.85 mm y^{-1} between 2002 and 2020. Minor spring activity and spring water chemistry data have been reported in a few locations near Winam Gulf (Henderson et al. 1987). The dry southeasterly winds lead to higher evaporation to the east (1.8–2.3 m y^{-1}) than to the west (1.2–1.8 m y^{-1}) and higher rainfall in the west (2.2 m y^{-1} near Bukoba), with the lowest values occurring in the southeast near Mwanza and Musoma (0.9–1.2 m y^{-1}) (MacIntyre et al. 2014). The diurnal variation in land/lake breezes and cloud cover, as well as increased cloudiness in high rainfall years, also control evaporation (Yin and Nicholson 1998, 2000, 2002; Nicholson and Yin 2002; Lehman 2009).

Early hydrological investigations include the work of Hobbs (1910) who used gauge records to infer earth movements, and Hurst and Phillips (1938) who examined the water balance and fluctuations in height of the water surface. The main inflow is the Kagera River, which drains 60,000 km^2, including upland areas with relatively high rainfall. However, the Kagera flow is reduced somewhat before it reaches Lake Victoria due to evaporation from wetlands and lakes. Nevertheless, it still accounts for about one-third of stream inflow to Lake Victoria (Sutcliffe and Parks 1999) with other important inputs derived from northeastern rivers (Tate et al. 2004; Kitheka et al. 2021). Total inflow is relatively small (18%) compared with direct rainfall to the lake surface (82%) (Kayombo and Jorgensen 2006), but due to its considerable variability (Fig. 13.5d) still exerts a control on lake level and volume (Piper et al. 1986; Sene and Plinston 1994; Tong et al. 2016). Nevertheless, the overall low water contributions from rivers and the large lake-to-catchment-area ratio result in precipitation, interception and runoff being less important than direct rainfall and evaporation over Lake Victoria (Beverly et al. 2020).

Discharge via the Nile, near Jinja (Fig. 13.2a), accounts for about 24% of the water loss and has been controlled since the opening of the Nalubaale (Owens Falls) Dam in 1959 (Kull 2006). The Nile outflow averaged 1046.2 $m^3 s^{-1}$ in the second half of the twentieth century, rising to about 1200 $m^3 s^{-1}$ between 2001 and 2004 (Okonga et al. 2006; Lehman 2009). The flow increase has been partly attributed, at least in part, to excessive power generation at the Nalubaale Dam, and its extension at Kiira, which involved departures from the "agreed curve" between Uganda and Egypt — an international scheme designed to control Lake Victoria water usage (Piper et al. 1986; Kull 2006; Okonga et al. 2006; Lehman 2009; Bastawesy et al. 2012).

In addition to changes related to hydropower generation, lake level responds to seasonal variability in rainfall, which is heaviest around May–June (Tungaraza et al. 2012) and may also reflect sunspot cycles (Yousef et al. 2000; Stager et al. 2005, 2007). Historical data (Fig. 13.5e; Ravenstein 1901; Nicholson 1998, 2002; Nicholson et al. 2000; Nicholson and Yin 2002) indicate long-term variations in the water surface of about 4 m with the lake tending to rise during El Niño years (Birkett et al. 1999; Conway 2002; Mistry and Conway 2003; Awange et al. 2019), which in turn affects hydrodynamic conditions and phytoplankton growth (Cózar et al. 2012). Van Rijssel et al. (2016a) also noted that periods with reduced wind speed coincided with lower lake oxygen levels and decreases in both water temperature and transparency, which, in turn, led to enlarged gill surface areas in selected cichlid fish. Lake level generally ranged between ~1133 and 1134.5 m above sea level (asl) between 1800 and 1960 but with a major rise to 1137 m asl around 1880. The water level rose suddenly in the early 1960s to >1136 m and then declined, with brief transgressive episodes. The lake reached about 1135 m in 2014 and the lake level rose again to >1136 m by 2020, causing widespread flooding and loss of agricultural land and homes (Government of Kenya and UNDP 2021). Over geological timescales, the lake has fluctuated over a much greater range with complete desiccation having occurred several times during the Late Pleistocene (Johnson et al. 1996, 2000; Talbot and Lærdal 2000; Stager et al. 2002, 2011).

Several attempts have been made to model the hydrology of Lake Victoria in relation to lake level fluctuations. Vanderkelen et al. (2018a, b) made detailed calculations for the Lake Victoria water balance and noted that precipitation was the main driver of lake level variability. Broecker et al. (1998) and Milly (1999) had earlier estimated that surface area would be reduced to <10% of the modern lake area with average annual precipitation of 0.75–1.01 m y^{-1}. Beverly et al. (2020) also modelled past lake level change and extrapolated the possible impacts of future climate projections on lake levels, suggesting that Lake Victoria could potentially lose its outlet to the White Nile within 10 years of their study, although the following decade was mostly associated with transgression. They also inferred that with continued lake regression Kenya would lose its access to the lake in <400 years.

13.4 Peoples, Vegetation and Wildlife

Archaeological evidence shows that 'Kansyore' pottery-making fisher-foragers occupied riverine and lakeshore settings in the Victoria Basin between 8 and 1 ka (Goldstein et al. 2022). Today, several different peoples occupy the region. The northeastern areas in Kenya include the Luo, Samia and Suba. The Buganda and Busoga dominate in the Uganda catchment (Awange and Ong'ang'a 2006) whereas Tanzanian regions are inhabited mainly by the Sukuma, along with the Kerewe, Jita, Kara, and Zinza. The Jita, Luo, Kuria, Zanaki, and Ruli live to the southeast with the Haya, Hangaza, Nyambo, Subi and Sukuma occupying locations to the west and southwest.

The drainage basin is comprised of a mosaic of five major vegetation zones (Trapnell and Langdale-Brown 1972; White 1983; Ssemmanda and Vincens 2002). These include the humid Afromontane zone and, with increasing aridity, the Guineo-Congolian, Zambesian, Sudanian and Somali-Masai vegetation types. The latter form the most extensive areas of natural vegetation and are variously dominated by *Acacia* (*Vachellia*), *Albizzia*, *Butyrospermum* (*Vitellaria*) or *Combretum*. Grasses are very diverse and include, for example, *Cymbopogon*, *Hyparrhenia* and *Loudetia* (Kendall 1969). Miombo woodlands (*Brachystegia*, *Isoberlinia*) are common in the relatively dry southwest.

Cyperus papyrus and *Vossia cuspidata* swamps are common along several rivers with swamp grass areas composed of *Miscanthidium violaceum* near Lake Victoria (Sutcliffe and Parks 1999; Kansiime et al. 2007). Swamps also provide an important filtering role in controlling the transfer of matter from the catchment to Lake Victoria (Cózar et al. 2004). *Sphagnum* is present in dilute lakes. *Podocarpus* is present locally, with forests present in elevated areas. In recent decades, human impacts have resulted in deforestation and soil erosion (Masese et al. 2012; Peterson et al. 2018; Humphrey et al. 2022) with consequent eutrophication (Descy and Sarmento 2008; Twesigye et al. 2011) as well as the extensive growth of water hyacinth (*Eichornia crassipes*) (Otieno et al. 2022).

About 215 bird species have been recorded with Chelonia, Sauria, Crocodilia and a wide variety of snakes also common (Lake Victoria Basin Commission 2011). Other animals living close to parts of the lake include sitatunga, hippo and a variety of primates (Spawls and Mathews 2012). The drainage basin contains major wildlife sanctuaries such as the Serengeti and Mara National Parks to the southeast (Fig. 13.6). Several smaller reserves, such as the Ruma National Park and Kisumu Impala Sanctuary in Kenya, provide a range of habitats for many species (e.g. ostrich, hyena, leopard, jackal, baboon, buffalo, roan, impala, reedbuck, topi and hartebeest).

13.5 Limnology

13.5.1 Physical Limnology

The first data on lake temperatures were recorded during fisheries surveys in 1927–1928 (Graham 1929; Worthington 1930) with many later investigations reporting details of temperature and oxygen stratification, water column mixing and nutrient loadings (e.g. Fish 1957; Talling 1957, 1966; Newell 1960; Akiyama et al. 1977; Ochumba and Kibaara 1989; Bugenyi and Magumba 1990, 1996; Hecky and Bugenyi 1992; Hecky 1993; Gophen et al. 1993; Hecky et al. 1994, 1996; Spigel and Coulter 1996; Ochumba 1996; Gophen 2015a). Surface water temperatures increased by ~1.2 °C between 1927 and 2009 with a greater increase (~1.57 °C) at depths below 50 m (Sitoki et al. 2010). Talling (1966) calculated an annual heat budget of 9000–11,000 cal cm^{-2} for the lake. MacIntyre et al. (2014) also noted a greater latent heat flux on the eastern side of the basin than to the west due to lower humidity and cloud cover, as well as the generally higher wind velocities. Horizontal water movements are controlled by surface winds and by temperature differences between water masses, which result in density-driven circulation with Coriolis forces being minimal due to the equatorial location (Ssebuggwawo et al. 2005).

Talling (1957, 1964, 1966) reported that the lake was generally monomictic in 1960–61, with three stratification phases at his routine sampling station (Fig. 13.2a). Surface warming during Phase 1 (September–December) resulted in gradual

Fig. 13.6 Examples of diverse wildlife in Lake Victoria catchments. (**a**, **b**) Grumeruti River catchment (Serengeti National Park, Tanzania), (**c**–**f**) Mara River catchment (Mara Triangle and adjacent areas, Kenya). (**a**) Thompson's gazelle. (**b**) Cheetah. (**c**) Zebra crossing Mara River. (**d**) Wildebeest and zebra crossing Mara River. (**e**) Elephant with Ololooloo Escarpment in background. (**f**) Lioness

temperature gradients that extended to the lake floor with stratification weakening through the period (Fig. 13.7a). A well-defined mixed layer formed during Phase 2 (January–May), which extended down to a deep thermocline with surface water reaching a maximum temperature (>25.8 °C) in March. During Phase 3 (June–August) temperature gradients were minimal, and the lake became nearly

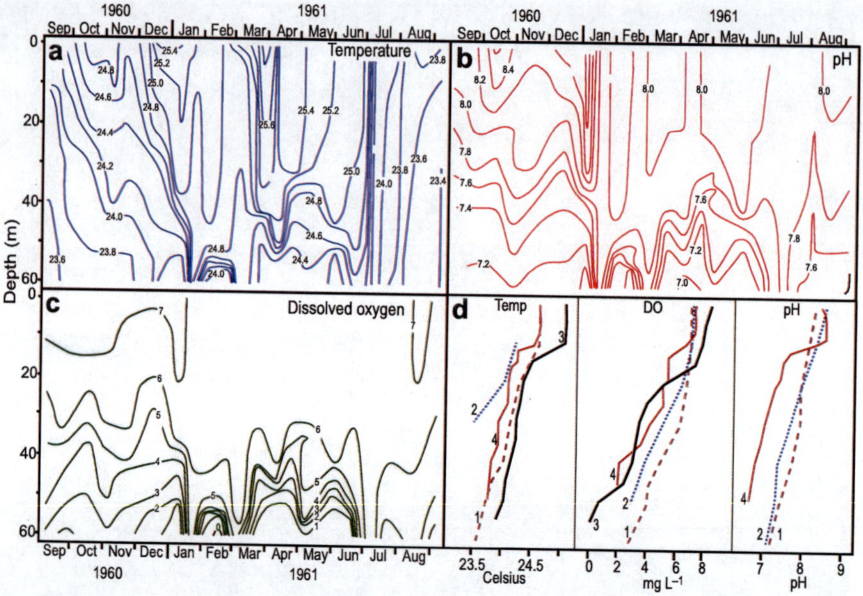

Fig. 13.7 Temporal variability in selected limnological parameters, northern Lake Victoria. (**a–c**) Time-depth profiles. After Talling (1966); (**d**) After Hecky et al. (1994) and Mwirigi et al. (2005)). (**a**) Temperature (°C). (**b**) pH. (**c**) Dissolved oxygen. (**d**) Temperature, dissolved oxygen and pH profiles for different time periods. 1 = Talling, Nov 1, 1960; 2 = Talling, Oct 13, 1960; 3 = Hecky, Nov 1, 1990; 4 = Mwanuzi, Oct 28, 2000

isothermal in June to July, the coolest and windiest time of the year. Ssebuggwawo et al. (2005) noted that this pattern was widespread in many areas, but with pelagic stations in the west and southwest recording higher water temperatures and with a thermocline present between February and May. Nearly complete mixing characterised the rest of the year in his study.

Mixing of the water column is induced by dominant southeasterly winds and by cyclonic storms (Kitaka 1972; MacIntyre 2013). The prevailing winds drive surface waters northward with a southerly return flow at depth (Graham 1929; Fish 1957; Newell 1960; Crul 1995; Song et al. 2002, 2004). Cool littoral water formed during the night may generate gravity flows that sink to reinforce stratification in deeper regions (Talling 1966; Cozar et al. 2012; MacIntyre 2013), with data for the 1920s indicating the presence of a broad lens-shaped cold (<23.1 °C) water mass at greater depths (Graham 1929). Fish (1957) thought a seiche might explain some of his temperature observations, but Newell (1960) and Talling (1966) later disputed this. Tilting of the temperature contours is evident in more recent north-south transects across Lake Victoria (Fig. 13.8a), which show warmer water to the north. Oxygen and conductivity isopleths also dip down to the north, particularly in datasets for May 1995 that may reflect the onset of a southeasterly monsoon (MacIntyre et al. 2014).

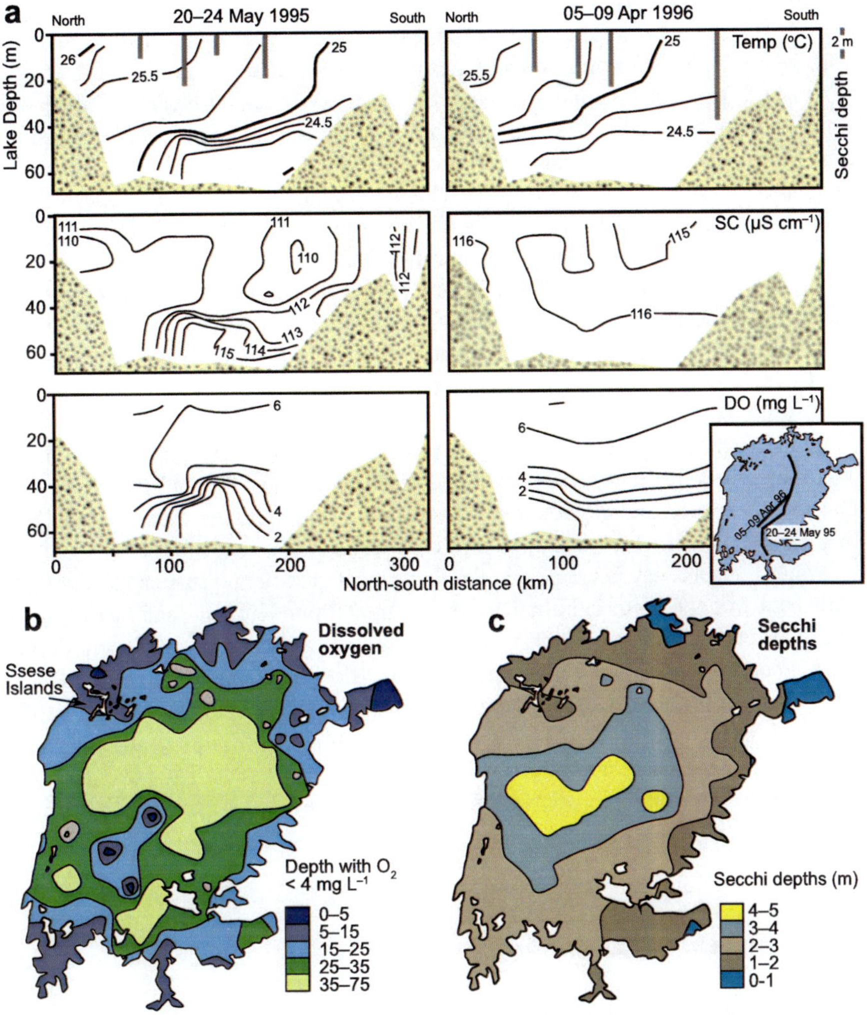

Fig. 13.8 Physical limnology of Lake Victoria. (**a**) North-south transects for May 1995 and April 1996 showing temperature, specific conductance (SC) and dissolved oxygen (mg L^{-1}). Note the northward inclined 25 °C contour plus the 112 µS cm^{-1} conductivity and 6 mg isopleths. May 1995 data collected during onset of northeast monsoon. Grey bars represent Secchi depths. After MacIntyre et al. (2014). (**b**, **c**) Spatial variation in dissolved oxygen and Secchi depths. After Ssebuggwawo et al. (2005)

The precise stratification patterns probably vary between years in relation to the duration and strength of the winds, which appear to reflect ENSO cycles (MacIntyre et al. 2014) and with winds being more variable in nearshore areas (Ochumba 1996). Relatively shallow (<10 m) locations, such as the Winam (Nyanza) Gulf and Pilkington Bay (Fig. 13.2a), develop only slight diurnal thermal stratification

(Worthington 1930; Talling 1966). Due to sheltering effects from winds, deeper bays (>30 m) may develop stratification earlier in the year than offshore regions, with layering also breaking down earlier (Crul 1995). In a few confined locations, such as the Rusinga Channel, between the open lake and the Nyanza Gulf, barotropic forcing results in local currents that may exceed 20 cm s^{-1}, with a flushing time for the gulf of 1–1.5 years (Antenucci et al. 2006; Romero et al. 2006).

In the 1960s and earlier, seasonal changes in dissolved oxygen reflected the annual stratification cycle (Fig. 13.7a, c). Partially deoxygenated water below the main thermocline disappeared when stratification broke down leading to a uniform vertical oxygen profile (Fish 1957; Talling 1966). By the 1990s, deep (>45 m) offshore water generally contained less oxygen with anoxia developing over extensive areas during increasingly stable lake stratification (Hecky et al. 1994; Lehman et al. 1998). Increased photosynthesis also led to surface water becoming saturated or supersaturated with oxygen throughout the year (Ochumba and Kibaara 1989; Muggide 1993; Crul 1995). Locations protected from wind mixing in northern Lake Victoria, especially to the northwest of the Ssese Islands (Fig. 13.8b), tended to develop especially low oxygen levels near the surface, with open parts of the lake being more oxygen-rich (Ssebuggwawo et al. 2005).

Water transparency is variable and reflects algal growth and suspended sediment with much of Lake Victoria's algal flora being light limited by self-shading (Muggide 1993; Ssebuggwawo et al. 2005). In the early twenty-first century, Secchi depths were greater offshore and during the Southeast Monsoon (Fig. 13.8a, c) with average depths at different stations ranging from about 0.3–2.3 m — half the depths recorded during the 1960s (MacIntyre et al. 2014). Sitoki et al. (2010) suggested water clarity increased in offshore areas during the late twentieth and early twenty-first centuries, noting offshore Secchi visibilities of ~3.5–4.0 m, but with minimal changes in littoral areas.

13.5.2 *Chemical Limnology*

Table 13.1 summarises water chemistry data from several sources. Offshore research, mainly in the 1920s, indicated that Ca^{2+} in open water ranged from 5–15 mg L^{-1}, with data for other ions including Na^{+} (10–13.2 mg L^{-1}), K^{+} (3.7–4.2), Mg^{2+} (2.3–8), Cl^{-} (3.9–6), SO_4^{2-} (< 0.8–0.96) and SiO_2 (3–9) (Duke 1924; Tottenham's 1926 data in Talling and Talling 1965; Graham 1929; Fish 1952). Results for $HCO_3^{-} + CO_3^{2-}$ in the same studies ranged from 0.25–1.1 meq L^{-1}. Early chemical analyses from the Winam Gulf indicated Ca^{2+} concentrations of 9.25 mg L^{-1}, with Mg^{2+}, Cl^{-} and SO_4^{2-} values of 1.59, 7.0 and 1.84 mg L^{-1}, respectively (Graham 1929; Worthington 1930). Similar concentrations were again documented from the Winam Gulf in the 1950s (8.6 mg L^{-1} for Ca^{2+}; 6.8 mg L^{-1} for Cl^{-}; 20 mg L^{-1} for SiO_2) (Fish, in Talling and Talling 1965). Studies in the 1960s documented much lower silica (5.9 mg L^{-1}) (Talling and Talling 1965).

Table 13.1 Selected chemical data for Lake Victoria water

Location (date/ref.)	pH	Conductivity (μmho)	TDS	Sum cations (meq L^{-1})	Sum anions (meq L^{-1})	Alkalinity (meq L^{-1})	Na	K	Ca	Mg	Cl	SO_4	SiO_2	Temp °C	TP (μg L^{-1})	TN (μg L^{-1})	PO_4-P (μg L^{-1})	NO_3-N (μg L^{-1})	EC (μS cm^{-1})
L Victoria (4.1923/1)	–	–	76	–		0.97	–	–	15	8	5.0	<0.5	–	–	–	–	–	–	–
L Victoria (11.1927/2)	8.5	–	–	–	1.16	1.003	–	–	7.02	2.3	5.0	0.96	–	–	–	–	–	–	–
L Victoria (9.1952/3)	–	98	–	–	–	0.25	–	–	11.8	–	–	–	8	–	–	–	–	–	–
L Victoria (3.1961/4)	8.0	97	–	1.04	1.08	0.92	10.4	3.8	5.6	2.7	3.9	2.3	4.2	–	47	–	–	–	–
Kavirondo (12.1960/4)	8.2	143	–	–	–	1.39	–	–	8.85	3.06	–	–	4.3	–	–	–	–	–	–
Kavirondo (1960–61/5)	8.75	145	–	–	–	–	–	–	–	–	–	–	2.75	–	67	–	–	–	–
Tanz. River (1973–74/5)	6.9–9.0	–	–	–	–	–	–	–	–	–	–	–	0.6–2.8	22–26	–	–	0–122	0–18	–
Kenya pelagic (1986/6)	7.42–9.4	–	–	–	–	–	–	–	–	–	–	–	–	–	–	–	–	–	88.1–102
Kenya pelagic (1990/7)	–	–	–	–	–	–	–	–	–	–	–	–	–	23.6	–	–	15.7	–	–
UP2 (1990/9)	–	–	–	–	–	–	–	–	–	–	–	–	0.1–2.5	–	43	336–448	12–43	11–84	–
Kenya pelagic (1989–92/8)	–	–	–	–	–	–	–	–	–	–	–	–	–	–	70–103	440–1160	4-73	10–30	–
Kenya pelagic (2000–2005/9)	6.1–9.8	–	–	–	–	–	–	–	–	–	–	–	–	20.9-27.7	63–92	–	3–7	–	98–104

LVEMP comparative data for 1960–61 vs 2000–2005. Ref. 9

(continued)

Table 13.1 (continued)

Location (date/ref.)		pH	Conductivity (μmho)	TDS	Sum cations (meq L^{-1})	Sum anions (meq L^{-1})	Alkalinity (meq L^{-1})	Na	K	Ca	Mg	Cl	SO_4	SiO_2	Temp °C	TP (μg L^{-1})	TN (μg L^{-1})	PO_4-P (μg L^{-1})	NO_3-N (μg L^{-1})	EC (μS cm^{-1})
KL3	2000–2005	8.2–9.0	–	–	–	–	–	–	–	–	–	–	–	2.66–7.14	25.5–26.9	20–113	892–3090	5–163	3-11	159–165
	1960–61	8.75	–	–	–	–	–	–	–	–	–	–	–	2.75	27.5	67	29	–	–	145
KP1	2000–2005	6.7–8.9	–	–	–	–	–	–	–	–	–	–	–	0.04–0.145	24.5–26.8	63–92	410–860	3–7	1–182	94
	1960–61	–	–	–	–	–	–	–	–	–	–	–	–	2.66	25.9	49	–	–	–	–
KP3	2000–2005	6.4–8.3	–	–	–	–	–	–	–	–	–	–	–	0.37–3.63	24.5–26.8	75–92	–	6–7	7–9	94–96
	1960–61	–	–	–	–	–	–	–	–	–	–	–	–	2.33	25.9	26	–	–	–	–
TL228	2000–2005	7.5–9.3	–	–	–	–	–	–	–	–	–	–	–	0.83–0.93	24.2–26.7	102–107	–	13-54	13–119	95.8–97
	1960–61	–	–	–	–	–	–	–	–	–	–	–	–	2.15	–	53	–	–	–	–
TP3	2000–2005	7.0–7.6	–	–	–	–	–	–	–	–	–	–	–	0.63–0.99	24.8	77–141	–	18–33	32–64	82.1–82.5
	1960–61	–	–	–	–	–	–	–	–	–	–	–	–	1.96	25.05	33	–	–	–	96
TP8	2000–2005	6.8–7.9	–	–	–	–	–	–	–	–	–	–	–	0.22–0.32	26–25.6	82-126	–	20–31	5–75	89.9–90.7
	1960–61	–	–	–	–	–	–	–	–	–	–	–	–	2.00	24.9	33	–	–	–	–

TP16	2000–2005	6.4–6.9	–	–	–	–	–	–	–	–	–	–	–	0.68–1.13	25.1–25.2	87–107	–	27–39	5–79	91.1–91.9
	1960–61	8.5	–	–	–	–	–	–	–	–	–	–	–	2.05	25.4	47	–	–	–	95
TL231	2000–2005	6.9–9.0	–	–	–	–	–	–	–	–	–	–	–	0.6–2.8	22.0–26.0	102–107	–	–	0–18	–
UP2	2000–2005	6.8–9.8	–	–	–	–	–	–	–	–	–	–	–	0.14–9.20	25.9–26.0	50–160	25–55	23–60	11.9–50	92.5–92.8
	1960–61	–	–	–	–	–	–	–	–	–	–	–	–	4.05	25.1	43	–	5–15	18	–
UP3	1960–61	8.05	–	–	–	–	–	–	–	–	–	–	–	4	25.1	45	–	–	6	–

LVEMP sampling sites are located in Fig. 13.2a

1. Duke (1924); 2. Graham (1929); 3. Fish (1953); 4. Talling and Talling (1965); 5. Akiyama et al. (1977); 6. Ochumba and Kibaara (1989); 7. Mavuti and Litterick (1991). 8. Gophen et al. (1995); 9. Mwirigi et al. (2005). *LVEMP* Lake Victoria Environmental Management Project

Lake Victoria is very dilute compared with other large lakes in East Africa. Its water also has a relatively short residence time and is dominated by rainfall inputs (Hecky and Bugenyi 1992). In the early 1960s, Talling and Talling (1965) observed very little offshore variability with depth or season for several major ions, including Ca^{2+} (mean = 5.6 mg L^{-1}), Mg^{2+} (2.6), Na^+ (10.4), K^+ (3.8), Cl^- (3.9) and SO_4^{2-} (2.3), with HCO_3^- + CO_3^{2-} alkalinity and conductivity being uniform at about 9.2 meq L^{-1} and 97 μmho, respectively. Temporal variations in pH during the early 1960s reflected thermal stratification and changes in dissolved CO_2 due to photosynthetic removal in the upper 20 m and accumulation in bottom water following organic decomposition (Fig. 13.7b; Talling 1966). The pH of the upper water column generally ranged from 8.0–8.6, with values falling to ~7.0 near the lake floor during periods of deep thermal stratification.

Hecky and Bugenyi (1992) observed that sulfate concentrations were among the lowest of any large lake worldwide and may limit phytoplankton growth. They also reported data for major ions for the years 1961 and 1988, documenting a general increase in all elemental concentrations: Na^+ (0.34 to 0.45 mM), K^+ (0.09 to 0.097 mM), Ca^{2+} (0.12 to 0.14 mM), Mg^{2+} (0.09 to 0.11 mM), Si (7.1 to 69.8 mM), Cl^- (0.093 to 0.11 mM), SO_4^{2-} (0.003 to 0.024 mM).

Chemical analyses for 2000–2005 (Mwirigi et al. 2005) indicated that littoral stations varied in water chemistry seasonally and that they tended to be characterised by slightly higher pH (mean = 8.08–8.53), turbidity (6.36–22.96 NTU) and electrical conductivity (109.69–123.29 μS cm^{-1}) than pelagic stations (mean pH = 7.26–8.43; turbidity = 2.25–8.88 NTU; EC = 91.53–97.72 μS cm^{-1}), which likely reflects the relative influence of runoff in nearshore areas and rainfall in open water. Data for 2000–2009 (Sitoki et al. 2010) show that surface conductivities, in water <5 m deep, ranged between 97.2 and 107 mS cm^{-1} in February–March, with slightly lower values (92.9–104 mS cm^{-1}) in August–September.

13.6 Biology and Ecology of the Lake

13.6.1 Phytoplankton

Nutrients and Primary Production

Data from the 1960s for NO_3-N, PO_4-P and silica indicate that these nutrients were more abundant near the lake floor and depleted near the surface (Fig. 13.9a–c; Talling 1966), except during mixing periods when distributions were uniform and after which diatom blooms developed. In subsequent decades, diatom growth reduced silica to 3 mg L^{-1} or less, especially in inshore areas (Crul 1995; Branstrator et al. 1996). NO_3-N was low all year and typically <10 μg L^{-1} in the upper lake water with PO_4-P mostly >7 μg L^{-1} and with total phosphorus between 30–50 μg L^{-1}. Sulfate has been and remains exceptionally low (~0.1 mg L^{-1}) due to uptake by phytoplankton. Although there has been speculation that it might limit growth, most

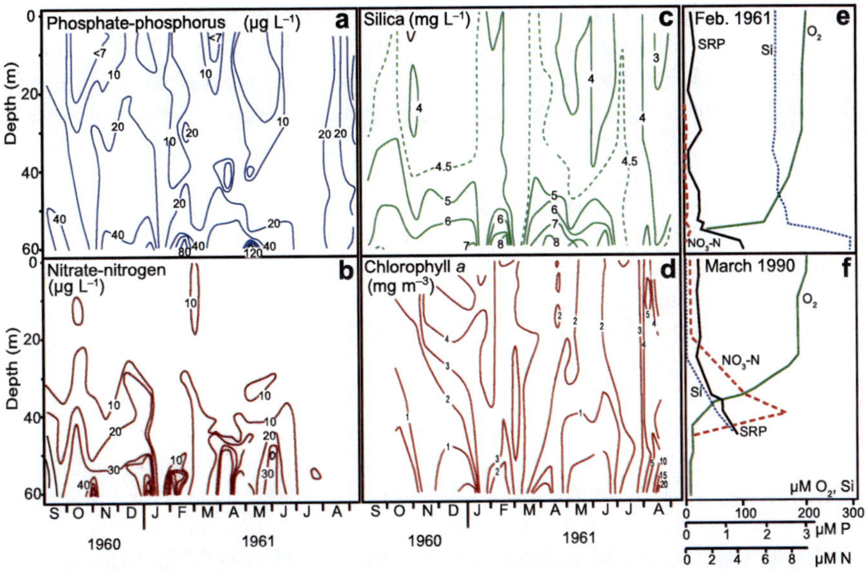

Fig. 13.9 Nutrient variability in Lake Victoria. (**a–d**) Depth-time changes in major nutrients in the early 1960s. After Talling (1966). Note intensification and breakdown of vertical profiles through annual cycles. (**e, f**) Vertical variations in nutrients for 1961 and 1990, northern Lake Victoria. After Bootsma and Hecky (1993). Note increased SRP and decreased NO_3-N in anoxic bottom layers

research suggests that it is not a limiting nutrient (Beauchamp 1953; Fish 1957; Crul 1995). Studies of $\delta^{13}C$ and $\delta^{15}N$ in particulate organic matter indicate sufficient CO_2 from the atmosphere and decomposition to sustain photosynthesis and denitrification processes that operate in the lake (Ramlal et al. 2001; Ramlal 2002). Iron and manganese are concentrated in deeper anoxic water (Crul 1995). The nutrient flux is derived from several sources, including an important wind-blown dust component deposited during rains, as well as from fluvial sources, nitrogen fixation and the recycling of bottom water (Bootsma and Hecky 1993).

Photosynthesis in surface water has increased significantly in recent decades with algal concentrations up to five times greater than in the 1950s and early 1960s (Talling 1965; Lehman and Branstrator 1993; Muggide 1993; Johnson et al. 2000; Mgaya and Mahongo 2017). Experimental studies suggest nitrogen limitation is an important control, with phosphorus not a limiting nutrient (Lehman and Branstrator 1994). Phytoplankton primary productivity yielded 8–50 g O_2 m^{-2} d^{-1} between 2000 and 2005, with an average of 18.3 g O_2 m^{-2} d^{-1}, which was about double the rates recorded in the 1960s (Talling 1966; Mugidde 1993; Gikuma-Njuru et al. 2005). Increased primary productivity and nutrient changes probably relate to population pressures that have led to increased tillage and soil erosion, overgrazing and deforestation (Machiwa 2003; Odada et al. 2004, 2009a, b; Juma et al. 2014). Other causes may include fertilizer usage and human and industrial waste from urban areas and shoreline settlements (Bootsma and Hecky 1993; Lipiatou et al. 1996; Mwanuzi et al. 2005; Henry and Kishimba 2003, 2006; Musa et al. 2011). Much of the increased phosphorus in the lake may also partly reflect weathering of apatite derived from disturbed soils and local carbonatite outcrops (Holtzman and Lehman 1998; Guya 2013).

Additional phosphorus and nitrogen loads are delivered via the atmosphere (Tamatamah et al. 2005; Hecky et al. 2006) with estimates suggesting 39,978 and 167,650 tons of TP and TN being deposited annually between 2000 and 2005 (Mnyanga et al. 2005). Bakayoko et al. (2021) reported four major nitrogen contributions to direct precipitation: (1) crustal and biomass sources (28%), (2) Indian Ocean air masses (14%), (3) volatile organic carbon from biomass burning and vegetation (15%), and (4) livestock and fertilizers, biomass burning and neighbouring agricultural fires (39%). They estimated that atmospheric deposition of nitrogen was ~15 kg ha^{-1} y^{-1}.

Consequently, the open lake water is now mesotrophic to eutrophic with hypertrophic conditions developing near shorelines (Hecky 1993; Bugenyi and Magumba 1996; Muli 1996; Guildford and Hecky 2000; Lung'ayia et al. 2000, 2001; Verschuren et al. 2002; Kishe 2004; Gikuma-Njuru and Hecky 2005; Mwirigi et al. 2005). Based on diatoms in piston cores, Stager et al. (2009) suggested that the onset of eutrophication varied with location, beginning at northern coastal sites in the mid-1970s to mid-1980s, taking place in offshore locations from the 1960s to 1970s and from the 1940s to early 1950s along the Kenyan shoreline.

Estimates of biogenic-Si in short cores imply an initially stable accumulation of diatom silica leading to concentrations of ~5% in sediment between 1820 and 1930, which increased to 8% by 1940 and 12% by 1990. These changes have also been correlated with anthropogenic eutrophication (Johnson et al. 2000; Verschuren et al. 1998, 2002; Shayo and Limbu 2018), as well as increases in Ca, Ti and P in northern nearshore cores (Holtzman and Lehman 1998). Eventually, uptake by diatoms and loss to the sediments resulted in the depletion of dissolved Si and a change to Si-limited diatom growth, which is typical of the modern lake. Lake data for 2000–2005 indicate that dissolved silica ranged between 0.525–0.902 mg L^{-1}, reaching its highest concentration in the northeast and southwest (Mwirigi et al. 2005). Declines in silica since the 1960s have been relatively small in bays and gulfs where silica is introduced by rivers. However, open areas have been reported with 3–20-fold declines (Talling 1966; Hecky et al. 1996; Lehman et al. 1998; Gikuma-Njuru et al. 2005; Mwirigi et al. 2005).

Total phosphorus has increased from 20–47 to 78–140 μg L^{-1}, with total nitrogen, chlorophyll and organic matter higher in littoral settings (Mwirigi et al. 2005). Between 2000 and 2005, NO_3-N and PO_4-P concentrations in the upper water layers ranged between 16.2 and 87.9 μg L^{-1} and 39.6–92 μg L^{-1}, respectively, with both being highest in the northeast (Gikuma-Njuru et al. 2005). Vertical differences in concentrations occur for most nutrients (Fig. 13.9e, f; Hecky et al. 1996). In the open lake, denitrifying bacteria in the anoxic deeper layers reduce nitrate concentrations, with denitrification processes in major wetlands also leading to higher N limitation in adjacent coastal settings (Cózar et al. 2007). Under the low redox conditions of anoxic depths, Fe^{2+} is released along with sorbed Soluble Reactive Phosphorus (SRP), which has an affinity for iron oxides, thus increasing SRP concentrations in bottom water (Fig. 13.9f).

Chlorophyll-*a* concentrations were about 3.0–4.6 μg L^{-1} in the 1960s in all parts of the lake. From about 1980, chlorophyll-*a* values started to rise to averages of up to 40 μg L^{-1} and 60 μg L^{-1} in offshore and inshore waters, respectively, with a maximum value of 650 μg/L (Sitoki et al. 2010). Data for offshore stations showed a drop after 2000 to values just above those of the 1960s. Gikuma-Njuru et al. (2005), for example, record a range of 3.6–11.7 μg L^{-1} with Silsbe et al. (2006) reporting values of 55–70 μg L^{-1} for several Ugandan bays.

Phytoplankton Composition

Studies of short cores from central Lake Victoria have demonstrated that the pelagic-diatom community was broadly stable between ~1820 and 1940, with *Cyclostephanos* (~80%) and *Aulacoseira* (~15%) co-dominant and *Nitzschia acicularis* absent (Verschuren et al. 2002). All these taxa increased through to the early 1960s, with *N. acicularis* forming up to 50% of fossil assemblages in the late 1970s to early 1980s — a considerably greater percentage than has been reported from long cores with sediments dating back 12,400 years (Stager and Johnson 2000).

Early phytoplankton reports from the contemporary lake were brief but include the research of Stuhlman in 1892 (Schmidle 1902), West (1907), Ostenfeld (1908, 1909), Virieux (1913) and Hustedt (1922). Michael Graham carried out systematic biological studies during a 1927–1928 fisheries survey (Graham 1929) with Worthington (1930, 1931) focussing on zooplankton and limnology. In the 1960s, diatoms dominated the phytoplankton during periods of mixing with cyanobacteria prevailing in the epilimnion when stratification developed (Talling 1966, 1987). The main diatoms at that time included *Aulacoseira* (mostly *A. nyassensis* var. *victoriae*) with less common *A. agassizii*, *Surirella nyassae*, *Stephanodiscus astraea*, *Nitzschia acicularis*, *Cymatopleura* and *Cyclotella* spp. (Talling 1966; Cocquyt et al. 1993; Crul 1995). Cyanophyta were dominated by *Anabaena flos-aquae*, *Anabaenopsis tanganyikae*, *Lyngbya circumcreta* and *Microcystis wesenbergii*. Chlorphytes were common and included *Staurastrum*, *Scenedesmus*, *Pediastrum* and *Coelastrum*. The dinoflagellate *Ceratium brachyceros* was observed but varied considerably in abundance (Lung'ayia et al. 2000; Kling et al. 2001).

Diatom biomass remained high in the 1990s, especially in pelagic areas, with *Nitzschia* forming the dominant genus (Hecky 1993; Muggide 1993; Stager 1998; Kling et al. 2001) and with chlorophyte algae having been reduced (Lehman et al. 1998). Chlorophyll-*a* concentrations (3.6–11.7 μg L^{-1}) were highest in littoral regions (Gikuma-Njuru et al. 2005). By the early twenty-first century, diatoms had been largely replaced by cyanobacteria in terms of biomass, accounting for >50% of the phytoplankton (Gikuma-Njuru et al. 2005). A total of 103 phytoplankton species were recorded in the more restricted Winam (Nyanza) Gulf (Fig. 13.10a), including 35 cyanobacteria and 34 diatoms plus 33 Chlorophyceae and 1 Dinophyceae taxa (Lung'ayia et al. 2000). Subsequently, the Lake Victoria Basin Commission (2011) recorded up to 31 cyanophyte, 26 chlorophyte and 111 diatom

Fig. 13.10 Winam Gulf. (a) View towards Homa Mountain from Kisumu. (b) Extensive growth of water hyacinth choking littoral areas in the southern Winam Gulf

taxa. Simiyu and Kurmayer (2022) and Simiyu et al. (2022) also reported a shift from *Aulacoseira*-dominance among diatoms to *Nitzschia*-dominance due to increased eutrophication.

The many changes in Lake Victoria reflect increased phosphorus loads and increased rates of N-fixation and denitrification (Lehman and Branstrator 1994; Lehman et al. 1998). Microbial productivity may also be light limited due to self-shading, especially in littoral areas (Muggide 1993). Mwanuzi et al. (2005) and

Gharib et al. (2023) described the modern plankton of the lake, with the former authors noting that low TN:TP ratios (<20) favour the nitrogen-fixing cyanobacteria, some of which have been associated with toxin production (Krienitz et al. 2002; Sekadende et al. 2005; Haande 2008; Okello and Kurmayer 2011; Poste et al. 2013; Miruka et al. 2021). Cyanobacterial genera reported from the lake include *Microcystis*, *Anabaena*, *Anabaenopsis*, *Aphanocapsa*, *Chroococcus*, *Coelosphaerium*, *Gomphosphaeria*, *Leptolyngbya*, *Cylindrospermopsis*, *Planktolyngbya*, *Merismopedia* and *Spirulina* (Talling 1987; Komarek and Kling 1991; Branstrator et al. 1996; Lung'ayia et al. 2000; Haande et al. 2011; Aura et al. 2023). Lehman et al. (1998) suggested that the changes since the 1960s have been determined by climate-induced variations in mixing depth, pointing out that diatoms in open water were formerly favoured by deep mixing that kept them in suspension and provided bottom water nutrients. During the last few decades, warming of surface lake water has increased stratification and reduced nutrient availability, with denitrification contributing to N-limitation, thus favouring cyanobacteria. In contrast, mixing in nearshore areas reaches the lake floor and inhibits denitrification, resulting in even greater algal concentrations.

13.6.2 Macrophytes

Cyperus papyrus is the most common macrophyte in adjacent swamps and within Lake Victoria (Lind and Visser 1962; Crul 1995; Thenya 2006; Owino and Ryan 2007; Kiwango and Wolanski 2008) with this aquatic plant showing greater genetic diversity in the lake than in isolated wetlands (Terer et al. 2015). Other common macrophytes include *Typha* spp., *Miscanthus*, *Phragmites*, *Echinochloa*, *Vossia*, *Limnophyton*, *Loudetia*, *Dissotis* and *Leersia* (Lake Victoria Basin Commission 2011). Floating plants include *Pistia*, *Eichhornia*, *Lemna*, *Azolla*, *Spirodella*, *Wolfia*, *Nymphaea* and *Salvinia*.

The water hyacinth (*Eichhornia crassipes*) has been present in low numbers on the lake surface since the 1960s but it experienced an explosive expansion from 1989 (Twongo 1996; Sitoki et al. 2010), when it covered over 200 km^2 and choked many nearshore bays (Fig. 13.10b). Fusilli et al. (2013) used GIS techniques to suggest a link between heavy rainfall, increased nutrient delivery from the catchment and expansion in the area covered by water hyacinth in the Winam Gulf. Large expansive mats, especially in littoral bays, limit light penetration and reduce the oxygen level (Bugenyi and Magumba 1996; Mwanuzi et al. 2005; Kateregga and Sterner 2007). Several researchers have also explored the impact of increased rainfall and lake level on hyacinth growth during El Niño years (Kitunda 2003; Williams et al. 2005; Wilson et al. 2007; Kiage and Obuoyo 2011). Williams et al. (2007), for example, suggested that the 1998 El Niño event, together with the impact of *Neochetina* weevils (introduced in 1995), contributed to a lakewide reduction in water hyacinth. In addition to its impacts on fisheries (Willoughby et al. 1996),

Feikin et al. (2010) noted that expansion of water hyacinth may be correlated with increased cholera in the region. More recently, Otieno et al. (2022) have described its life cycle and interactions with weevils, fish and lake nutrients.

13.6.3 Zooplankton

The zooplankton have varied little in overall composition since records began, with copepods, cladocerans and rotifers dominating in various parts of the lake (Gikuma-Njuru et al. 2005) since the early studies of Worthington (1931), MacDonald (1956) and Rzoska (1957). The zooplankton are well dispersed through the water column in shallow areas but tend to be more concentrated in upper and middle levels at offshore stations where phytoplankton food sources are more readily available (Fig. 13.11a, b). Zooplankton in Lake Victoria are also relatively inefficient grazers of the cyanobacteria that dominate in the modern lake (Lehman and Branstrator 1993; Gikuma-Njuru et al. 2005).

Cyclopoid copepods are the most abundant group with calanoid forms increasing in percentage towards littoral regions (Mwebaza-Ndawula 1994, 1998; Mwebaza-Ndawula et al. 2004). The major copopod genera include *Thermocyclops* (e.g. *T. emini*, *T. neglectus*, *T. oblongatus*, *T. incisus*, *T. hyalinus*), *Tropocyclops* (e.g. *T. confinis*, *T. tenellus*), *Eucyclops*, *Mesocyclops*, *Microcyclops*, *Diaptomus*, *Thermodiaptomus* and *Tropodiaptomus* (Sars 1909; Rzoska 1957; Mwebaza-Ndawula 1994; Branstrator et al. 1996; Lehman 1996; Jonna and Lehman 2002; Mwebaza-Ndawula et al. 2004; Semyalo et al. 2009; Lake Victoria Basin Commission 2011; Ngupula 2013). Cladocerans include *Daphnia longispina*, *D. lumholtzi*, *D. magna*, *Diaphanosoma excisum*, *Bosmina longirostris* and *Ceriodaphnia dubia*, *Ceriodaphnia cornuta/rigaudi*, *Hyalodaphnia barbata*, *Moina macrura*, which are variously present in both littoral and offshore areas.

In contrast, rotifers tend to be more common in littoral zones and are represented by a wide range of genera, including *Brachionus* (e.g. *B. angularis*, *B. caudatus*, *B. calyciflorus*, *B. patulus*), *Filinia* (e.g. *F. longiseta*), *Keratella* (e.g. *K. cochlearis*, *K. quadrata*, *K. tropica*), *Asplancha*, *Lecane*, *Eiphanes*, *Euclanis*, *Trichocerca*, *Polyartha*, *Pompholex*, *Platyias*, *Testudinella* and *Macrochaetus* (Waya 2004; Waya and Mwambungu 2004; Lake Victoria Basin Commission 2011; Ngupula 2013). They are common, for example, in the Napolean Gulf in northern Lake Victoria (Mavuti and Litterick 1991) where >40 zooplankton taxa have been recorded. This partly reflects a high degree of eutrophication in the area (Kiggundu et al. 2012) with rotifers dominating the species composition, but with copepods accounting for a greater biomass (Fig. 13.11c). $\delta^{13}C$ and $\delta^{15}N$ isotopes suggest that copepods and cladocera in Lake Victoria form part of a food chain leading to the fish *Rastrineobola argentea* and to lake flies (*Chaoborus*) (Branstrator et al. 2003).

Temporal changes have also been reported for specific taxa such as the cladocerans *Simocephalus vetullus* and *Bosmina longirostris*, which gave way to *Daphnia*

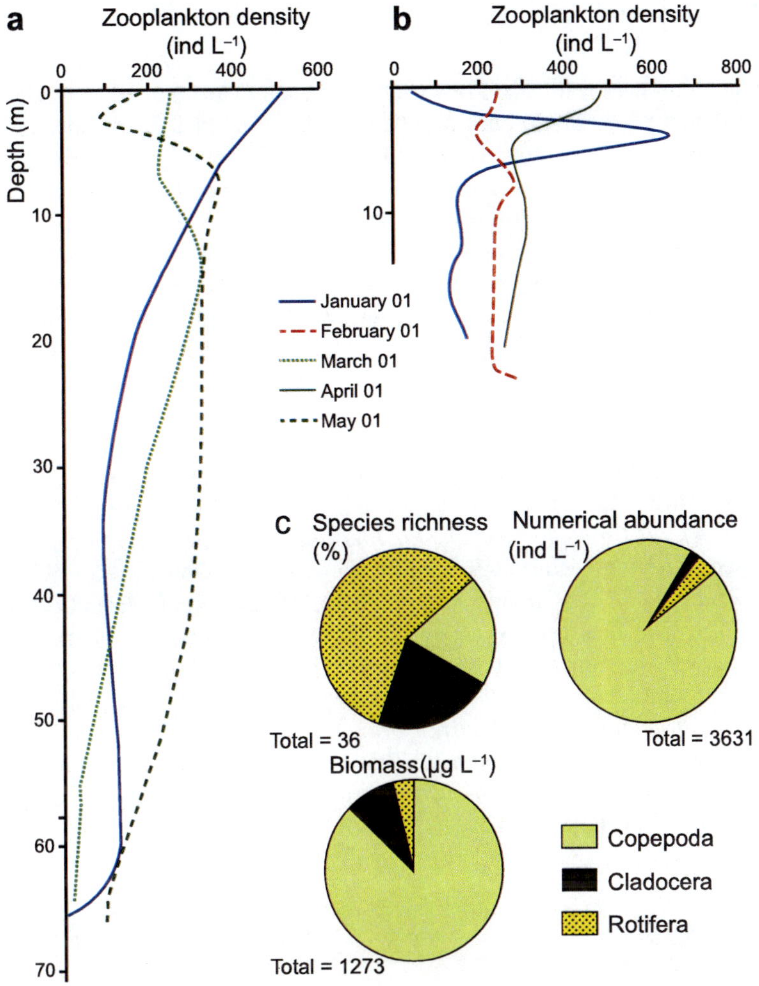

Fig. 13.11 Zooplankton populations in Lake Victoria. (**a**, **b**) Vertical zooplankton density profiles for pelagic (**a**) and littoral (**b**) sites in Lake Victoria. After Gikuma-Njuru et al. (2005). (**c**) Relative species richness, abundance and biomass for relatively shallow waters (<20 m) in Napolean Gulf, northern Lake Victoria. After Kiggundu et al. (2012)

barbata in the Nyanza (Kavirondo) Gulf (Rzoska 1957; Bridgeman 2001; Gikuma-Njuru et al. 2005). In the late twentieth century, Cladocera were reportedly reduced from >30 to 6% of assemblages with calanoid copepods yielding dominance to small-bodied cyclopoid taxa (Mwebaza-Ndawula 1994, 1998; Gikuma-Njuru et al. 2005; Kiggundu et al. 2012; Ngupula 2013). Ciliates have been less studied, but are reportedly dominated by *Strombidium*, *Halteria*, *Strobilidium*, *Vorticella* and *Cyclidium* (Yasindi 2001).

13.6.4 Invertebrates

Benthic biomass in Murchison Bay (northern Lake Victoria) totalled 146,000 tons (148,343 tonnes) over 200 km^2 in the late 1980s. This included bivalves (68,940 tons; 70,046 tonnes), gastropods (30,840 tons; 31,335 tonnes), insects (25,760 tons; 26,173 tonnes) and annelids (20,460 tons; 20,788 tonnes) (Okedi 1990). The freshwater prawn *Caridina nilotica* is common in most of the lake and feeds on particulates and benthos in bottom waters. It also represents an important food source for Nile perch. The prawns survive in low-oxygen settings that probably provide a partial refuge from fish predation (Branstrator et al. 1996, 2003; Mbahinzireki et al. 1998; Mnaya et al. 2006). Stable isotope studies of *C. nilotica* indicate that it has a wide range of $\delta^{13}C$ values along with low $\delta^{15}N$, which is consistent with its appearance in nearly all lake habitats and widespread presence in the stomachs of most fish species (Campbell et al. 2003a).

Other invertebrates include chironomid and chaoborid (*Chaoborus anomalus*, *C. edulis*) insect larvae and insect nymphs (*Diptera*, *Trichoptera*, *Hemiptera*, *Ephemeroptera*, *Coleoptera*, *Odonata*, *Plecoptera*), several of which may have increased following declines in haplochromine fish (MacDonald 1956; Witte et al. 1992; Bridgeman 2001; Gikuma-Njuru et al. 2005; Lake Victoria Basin Commission 2011). The lake flies usually emerge in response to lunar periodicities (Corbet 1958). A variety of molluscs are also present (e.g. *Melanoides tuberculata, Bellamya unicolor, Bulinus globosus, Pila ovata, Cleopatra guillemi, Biomphalaria choanomphala, Caelatura hauttecouri, Corbicula fluminalis, Sphaerium victoriae, Pisidium* and *Mutela* spp.) along with *Potamon* crabs, ostracods, and oligochaetes such as *Branchiura sowerbyii, Alma emini* and *Tubifex* spp. (Mothersill et al. 1980; Okedi 1990; Muli and Mavuti 2001).

13.6.5 Fish

The fish fauna was formerly characterised by diverse haplochromine cichlids (Kaufman and Ochumba 1993; Kaufman et al. 1996; Mgaya and Mahongo 2017) with fewer non-cichlids (Corbet 1961). The fauna must have experienced a period of explosive Late Pleistocene and Holocene speciation and adaptive radiation (Fryer 1997, 2001, 2004; Meyer et al. 1996; Seehausen et al. 1997a, b, 2003; Seehausen 2000, 2002; Elmer et al. 2009; Danley et al. 2012; Salzburger et al. 2014) because sediment records suggest the lake dried out completely before 12,400 y BP (Johnson 1996). During this dry phase, fish would have only been able to survive in small ponds and (or) river systems (Mwanja et al. 2001).

In 1957, the Kenya and Uganda fisheries were dominated by *Oreochromis esculentus* and *O. variabilis*. *Labeo victorianus*, *Bagrus docmak* and *O. esculentus* were dominant in Tanzanian fish catches with *Mormyrus kannume*, *Protopterus aethiopycus* and *Clarius gariepinus* important locally (Kudhongania et al. 1992, 1996;

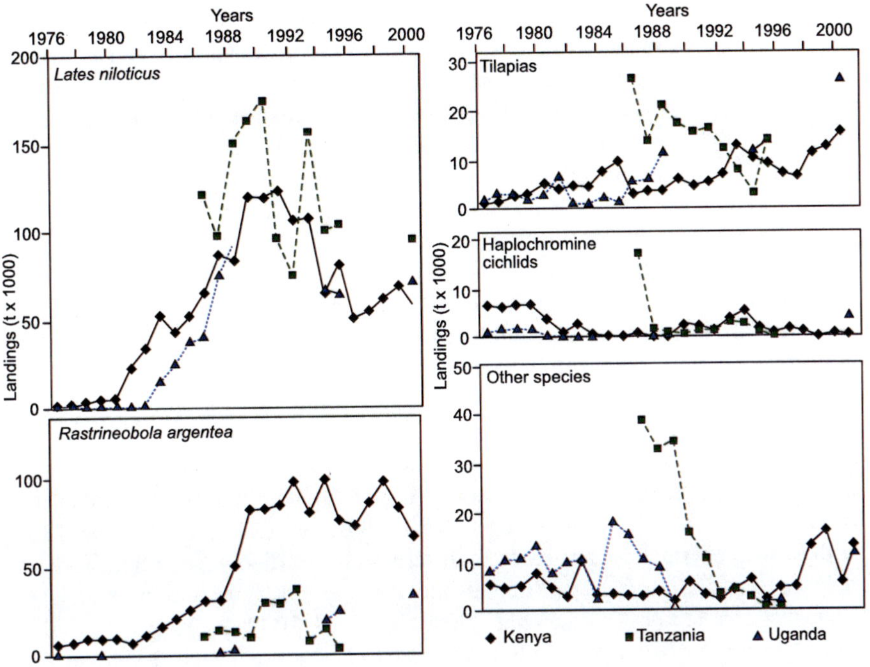

Fig. 13.12 Changes in fish catches for major species and groups in Lake Victoria. After Cowx et al. (2003)

Katunzi 1996; Balirwa et al. 2003; Kolding et al. 2014a). Declines in catches set in from the 1970s with haplochromines becoming important as a source for newly opened fish-meal processing factories (van der Knaap et al. 2010), but with *B. docmak* continuing to be widely fished in Ugandan waters. Over the next few decades, three species became dominant (Fig. 13.12; Cowx et al. 2003), including the introduced Nile perch (*Lates niloticus*), Nile tilapia (*Oreochromis niloticus*) and the native fish *Rastrineobola argentea*.

The major changes in the fish population reflect a variety of factors (Ochumba et al. 1994; Lehman 1998; Njiru et al. 2011; Nyamweya et al. 2020) including climate change and its effects on stratification within the lake, the impacts of eutrophication on the food chain (van Rijssel et al. 2016b), overfishing and the effects of deliberate introductions of exotic predator species (*Oreochromis niloticus*, *O. leucostictus*, *Tilapia zillii*, *Tilapia rendalli*). Nile perch were introduced in the early 1950s, but numbers remained low until a rapid expansion from the mid-1980s with this fish now constituting the top trophic predator (Fryer 1960; Harrison 1991; Goldschmidt et al. 1993; Gophen et al. 1995; Campbell 2001; Campbell et al. 2003a; Pringle 2005; Downing et al. 2013; van Zwieten et al. 2016). This explosive expansion contributed to a collapse in the formerly abundant (>80% of the fish biomass) and diverse (>500 taxa) haplochromine cichlids, which had included many endemic forms (Kudhongania and Cordone 1974; Ligtvoet and Witte 1991;

Ogutu-Ohwayo 1990, 1992; Witte et al. 1992; Kaufman and Ochumba 1993; Johnson et al. 1998; Bootsma and Hecky 2003; Witte et al. 2007a; Sitoki et al. 2010). At about the same time, the endemic *Rastrineobola argentea* increased (Gophen et al. 1995; Gophen 2015b). However, the changes in the fish fauna also coincided with a period of decreased wind speed and warmer temperatures that resulted in strengthened lake stratification, which makes it difficult to directly attribute the altered ecosystem to fish introductions alone (Kolding et al. 2014b).

In the early 1990s, haplochromines started to recover to pre-collapse levels, forming >50% of the total fish biomass (Seehausen et al. 1997a, b; Tumwebaze et al. 2007; Wanink 1999; Witte et al. 2000, 2007b; Kayanda et al. 2009; Downing et al. 2014). In addition, the rapid expansion of water hyacinth seems to have encouraged some recovery in numbers of Nile tilapia, lungfish (*Protopterus aethiopicus*) and mud fish (*Clarias gariepinus*) (Njiru et al. 2005). Total catches stood at about 0.1 million tons per annum before 1980, consisting mainly of haplochromines and other native species. Nile perch accounted for 70% of the 0.5 million ton catch in 1989 and represented 25% of a one million ton catch in 2007 (Reynolds et al. 1995; Sitoki et al. 2010). Peak Nile perch catches occurred around 1990 (Fig. 13.12) with subsequent declines attributed to the effects of overfishing, eutrophication and reduced oxygen levels (e.g. Simonit and Perrings 2005; Mkumbo et al. 2007; Chapman et al. 2008; Kolding et al. 2008; Njiru et al. 2008a, b; Yongo et al. 2017). More generally, natural fish kills have increased along with the development of more extensive cyanobacterial blooms (Ochumba and Kibaara 1989; Ochumba 1990; Olokotum et al. 2020). Other pressures on fish include decreased light intensity due to greater algal growth and increased siltation from inflowing rivers, which may disrupt mating among species reliant on visual clues (Seehausen et al. 1997a, b; Sitoki et al. 2010).

13.7 Modern Sedimentation

The earliest documented survey of lake floor mud reported extensive diatomaceous greenish-black mud and blue or green clay with sand occurring locally (Graham 1929). Bottom facies near Entebbe were mapped using echo-sounding (Gee 1966) with a semi-flocculent bottom layer observed and reported as ranging in thickness from 1–3 m (Gee and Gilbert 1968; Cordone and Kudhongania 1971; Verschuren et al. 2002). Subsequently, high-resolution echo-sounding data, collected by Project PROBE, showed up to 8 m of Late Pleistocene and Holocene mud on a desiccated surface in deeper (>50 m) parts of the lake (Scholz et al. 1990), with surface waves locally preventing the accumulation of mud or eroding the lake floor in shallower areas, especially in the western third of Lake Victoria (Johnson et al. 2000). In areas of net deposition, dry weight sedimentation rates changed from about 57 to 90 g m^{-2} y^{-1} following eutrophication of the lake (Adams and Ochola 2002). Other estimates of offshore deposition suggest overall recent accumulation rates of 0.2–0.5 mm y^{-1} (IDEAL 1990) to 2.3 mm y^{-1} (Verschuren et al. 2002).

Mothersill (1976) described nearshore pebbly sand in northwest Lake Victoria that was dominated by quartz and minor K-feldspar, reflecting the granitic and metamorphic lithologies in the drainage basin. Offshore, dark yellowish-brown mud was mainly composed of quartz, K-feldspar, plagioclase, kaolinite, illite and minor vermiculite, with the kaolinite/illite ratio decreasing away from Murchison Bay in the north and the Katonga River to the west. Vivianite-rich nodules, up to 1.5 mm in diameter, have also been reported from offshore grab samples in northwestern parts of the lake in a 1–3 cm thick layer, about 1 cm or more below the sediment-water interface (Mothersill 1996a). Carbon and nitrogen contents, in the northwest, ranged from 0.43–20.06% and 0.01–2.73%, respectively (Mothersill 1976).

Chemical analyses of dried mud from northern Lake Victoria indicated elemental ranges of 24–64% (silica), 4–25% (carbon), 1.5–4.8% (iron), 0.6–2.2% (nitrogen), 0.5–2.0% (sulfur), 0.04% (phosphate), 0.3% (calcium), 0.07% (magnesium), 0.05% (potassium) and 0.03% (sodium), with the latter rising to 0.45% in Pilkington Bay (Crul 1995). Compositionally, sediments from Pilkington Bay were dominated by cyanobacteria and bacteria, with detritus including diatoms and plant fragments (Newell 1958; Crul 1995). Hesse (1956) suggested that sulfur in lake floor mud was combined with organic matter, but Bloomfield et al. (1970) later noted the presence of pyrite in heavy fractions from Pilkington Bay that accounted for most of the sulfur in their samples.

Studies of Mara River wetlands to the east of Lake Victoria indicate broadly stable conditions from 2000–1000 years ago, but with rapid sedimentation changes over the last 250 years and particularly from the 1950s, coincident with increasing mercury concentrations (Dutton et al. 2019). Heavy metal pollution varies with location in Lake Victoria and nearby water bodies, with some areas characterised by minimal enrichment (Ochieng et al. 2006, 2008; Mgaya and Mahongo 2017) and other sites showing moderate to high levels of contamination. Increased Cu, Pb and Zn have, for example, been reported from wetlands south of Lake Victoria, near Nungwe Bay, where enrichment has been attributed to local mining (Mutakyahwa et al. 2009). Increased heavy metals in wetlands elsewhere have been attributed to aerial deposition with sources derived mainly from cars (Nyangababo et al. 2005). In other locations, a variety of potential industrial sources (e.g. battery factories, smelters, tanneries, pharmaceutical industries, soap and cosmetics facilities) have resulted in increased Cu, Pb, Ni, Co, Cd, Cr and Hg in both lake (Muwanga and Barifaijo 2006; Ongeri et al. 2008, 2010; Mwamburi 2013) and river (Mwamburi 2003) sediments. Total Hg, for example, in short cores, collected in the late 1990s from Napolean Gulf and from a nearby offshore site, ranged from 78–458 ng g^{-1} with the moderately high concentrations being explained in terms of post-World War II industrial development (Campbell 2001; Campbell et al. 2003b).

Sewage inputs to the lake have increased with pathogenic human adenoviruses reported near Homa Bay town in Kenya (Opere et al. 2021). A few studies report Polycyclic Aromatic Hydrocarbon (PAH) concentrations in sediments. Close to Kisumu (Winam Gulf, Kenya), PAH in nearshore deposits and water samples ranged from 0.4 to 31.95 μg g^{-1} dry weight and 3.32–55.8 μg L^{-1}, respectively (Kwach et al. 2009). Other examples of pollutants include diazinon and malathion with high levels reported in lake sediments due to agricultural activities (Musa et al. 2011). Nantaba et al. (2020, 2021) reported the presence of organic micropollutants and pharmaceutical products, with the highest concentrations reported from Murchison Bay.

13.8 Late Quaternary Lacustrine Sediments in the Lake Victoria Basin

Offshore seismic profiles show 60 m of later Quaternary deposits, which form five sedimentary packages separated by erosion surfaces (Fig. 13.13; Johnson et al. 2000). The uppermost 'Unit A' was deposited during the Late Pleistocene–Holocene (Scholz et al. 1998; Johnson 1996, Johnson et al. 2000). Palaeosols indicate desiccation that generally took place at ~18–14 ka, possibly with two drier events at 17 ka and 15 ka (Stager 1984b; Johnson et al. 2000; Stager and Johnson 2008; Beverly et al. 2020). Discontinuities and accumulations of shells have also been used to infer repeated lake desiccation during this time interval (Johnson 1996; Stager 1984a). Estimates of the length of time that the lake was dry range from about 500 years (Kendall 1969) to several thousand years (Johnson 1996). Stager et al. (2002) reported a significant reduction in lake size between 18–17 ka and 15.9–14.2 ka. The diatom *Thalassiosira rudolfi* in samples close to the desiccation hiatuses indicate saline conditions when the lake was much reduced, with saline taxa disappearing at around 14–13 ka as the lake rose and became fresher (Stager 1984b; Stager and Johnson 2000, 2008; Talbot et al. 2000), which Broecker et al. (1998) suggested could be linked to abrupt global climatic changes at the onset of

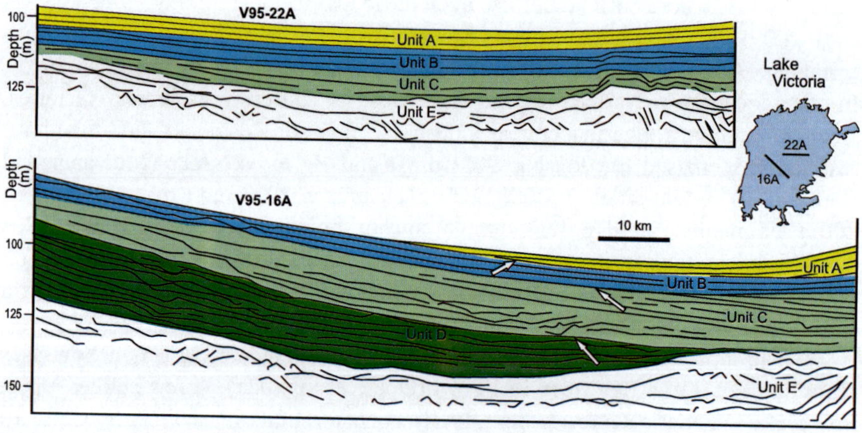

Fig. 13.13 Airgun profile sketches from Lake Victoria showing five major seismo-stratigraphic Units (A–E) . Map shows profile track-lines. Derived from seismic profiles in Johnson et al. (2000)

the Bølling-Allerød warm interval. Beverly et al. (2020) attributed these particular changes to migration of the Congo Air Boundary.

Magnetic susceptibility profiles from several cores show minimal variability through the Holocene, except for high values on and just above a Late Pleistocene palaeosol (Mothersill 1996b; Ngobi et al. 1998; Johnson et al. 2000). Lund et al. (2020) also discuss palaeomagnetic secular variation data from northern Lake Victoria for two cores (V95-1P and V95-7P) and point out that their detailed record permits correlation with Lake Malawi to the south. Carbonate concentration is high and water content low (Fig. 13.14a) in sediments dated between 14 and 12 ka in Core P2 from Pilkington Bay in northern Lake Victoria (Kendall 1969), indicating lowstand conditions. Sand in Core P2 (13.5–14 ka) suggests that the lake stood at elevations below the coring site. Diatom floras change from initial dominance by *Stephanodiscus* to abundant *Aulacoseira* and *Nitzschia* after about 12 ka, indicating fresher conditions (Kendall 1969). These data are supported by increased percentages of moist forest pollen and diatoms in Core IBIS 1 from the Damba Channel (Fig. 13.14b) (Stager 1984a; Talbot and Livingstone 1989; Johnson 1996; Stager and Mayeski 1997). Beuning et al. (1997) used oxygen isotopes to infer that overflow started about 8 ka (7200 ^{14}C y BP) ago, but this was revised to about 13 ka (Talbot et al. 2000; Beuning et al. 2002; Johnson 2002). Studies of TEX_{86}, $\delta^{13}C$ and δD (Core V95-P1) indicate coherence between temperature and rainfall through the Holocene, with a warm, wet interval peaking at ~10–9 ka, followed by gradual cooling and drying (Berke et al. 2012).

Diatoms in several cores have been used to characterise later Quaternary lacustrine palaeoenvironments, lake level and regional climate change (Stager et al. 1986, 1997, 2002, 2003, 2005, 2011). Figure 13.14c, for example, shows a diatom stratigraphy for a 7-m-long core (VP95-2P) recovered from waters 67-m-deep from central Lake Victoria (Stager and Johnson 2000). A palaeosol at the base formed before 12,400 years ago with benthic diatoms and saline taxa such as *Thalassiosira rudolfi* (Zone E) indicating relatively shallow-water and high conductivity. Saline species continue through Zone D, but decline as *Stephanodiscus astraea*, *S. hantzschii*, *Aulacoseira granulata* and *Nitzschia fonticola* increase, indicating fresher, possibly stratified conditions, but still with higher salinities than those of today. High biogenic silica and declines in *Nitzschia* spp. in Zone C (Early Holocene) reflect higher productivity and greater mixing. Diatoms declined during the Middle Holocene, possibly due to reduced Si-availability caused by the initiation of overflow around 7200 y BP or because of increased stability in the water column (Johnson et al. 1998; Stager and Johnson 2000). Increases in *Aulacoseira nyassensis* and needle-like *Nitzschia* in Zone B1, at the expense of *A. granulata*, have similarly been explained in terms of reduced wind-driven turbulence. Reductions in offshore diatom accumulation after 8300–5800 y BP and after 3500 y BP have also been used to infer declines in the duration and (or) intensity of mixing with increasing aridity and deforestation suggested for the period post-dating 1400 y BP.

Geochemical signals in Core V95-2P and other cores have shown that flooding of the basal palaeosol led to the development of abundant shallow-water macrophytes and the accumulation of sediments with high Total Organic Carbon (TOC),

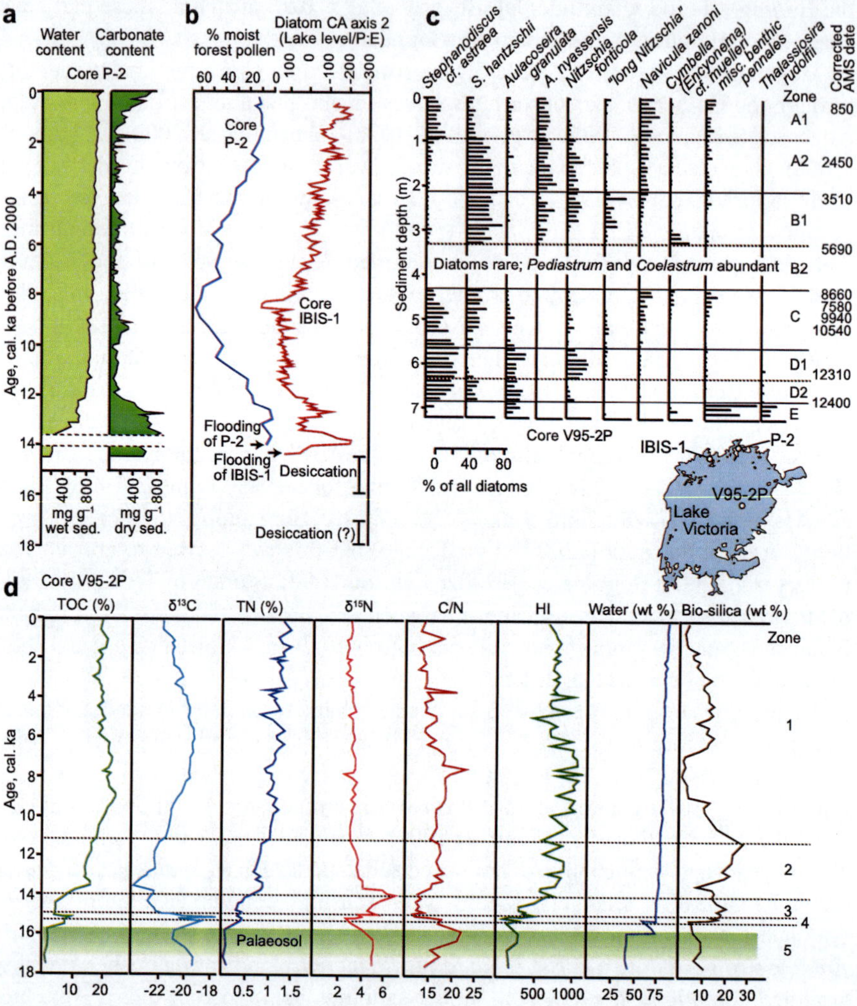

Fig. 13.14 Late Pleistocene and Holocene stratigraphy of Lake Victoria from selected cores. (**a**) Carbonate and water content in Core P-2, northern Lake Victoria. After Kendall (1969). Note basal dewatering and high carbonate content associated with a major desiccation event. (**b**) Holocene moist forest pollen and diatom-based data for Cores P-2 and IBIS 1. Diatom Correspondence Axis 2 represents either lake level height or the precipitation/evaporation balance. Modified from Stager and Johnson (2008). (**c**) Diatom stratigraphy of Core V95-2P from central Lake Victoria. Note increased percentage of the saline taxon *T. rudolfi* near the base indicating high conductivity. Modified from Stager and Johnson (2000). (**d**) Geochemical stratigraphy of Core V95-2P. After Talbot and Lærdal (2000)

C/N ratios and plant debris (Fig. 13.14d; Talbot and Livingstone 1989; Talbot and Lærdal 2000; Talbot et al. 2006) with low algal matter biomarker signals (Cockerton et al. 2015). The subsequent filling of Lake Victoria is marked in V95-2P by increases in TOC and the Rock-Eval Hydrogen Index (HI), reflecting increases in phytoplankton content and bottom preservation of organic matter. Changes in Total Nitrogen (TN), $\delta^{15}N$ and $\delta^{13}C$ reflect variations in biological productivity with much of the Holocene characterised by stability (Talbot 2002). Low and stable $\delta^{15}N$ suggests that the Holocene nitrogen budget may have been dominated by atmospheric N-fixation. Lejju (2012) examined a short core from Napolean Gulf (northern Lake Victoria) and observed stability in TOC, TN, C/N ratios and Total Sulfur (TS) from 4700–1830 y BP, followed by slight increases in TOC, decreases in C/N and variable TS until 1150 y BP. From 1150–640 y BP, he notes major fluctuations in all proxies with subsequent declines in TS and C/N ratios.

Biogenic silica profiles reflect diatom abundance and diatom productivity. Silica concentrations increased rapidly during the Late Pleistocene transgression, reaching a maximum around 10,000 years ago (11.2–9.8 ka), which has been attributed to flooding of pre-existing grasslands that caused phytolith dissolution, the release of silica and consequent diatom blooms (Cockerton et al. 2015). At its highest level, the lake stood 18 m higher than today (Kent 1942; Temple 1964a, b, 1969) with outflow via the Nile leading to loss of silica and river incision that eventually lowered lake level to modern heights (Johnson et al. 1998; Johnson 2002). Cockerton et al. (2015) suggested that reduced dissolved silica since 5500 y BP reflected increasingly dry climates that led to more open vegetation, reinforced by the development of agriculture over the last 3–4 thousand years. Biogenic silica started to rise again in the 1930s in parallel with population and agricultural expansion, but with subsequent silica declines reflected by cyanobacteria replacement of diatoms (Johnson 2002).

Pollen from northern Lake Victoria show a change from grass-dominated landscapes to semi-deciduous forest about 14,000 years ago (Kendall 1969; Cockerton et al. 2015). Subsequently, rising moisture in the Early Holocene resulted in an expansion of evergreen forests, but with an increase in grass during the Younger Dryas. Semi-deciduous forest increased after about 6800 y BP due to increasingly dry seasonal conditions. From 4100–3000 y BP, further increases in aridity, and possibly the development of agriculture, led to a rise in open vegetation (Kendall 1969; Ssemmanda and Vincens 2002). Githumbi et al. (2021) also present data from highland swamps northeast of Lake Victoria, on the Mau Escarpment, that suggest dominance by Afromontane forests during the latest Pleistocene to Holocene. Taxa included *Apodytes*, *Celtis*, *Dracaena*, *Hagenia* and *Podocarpus*, which gave way to forests with *Cordia*, *Croton*, *Ficus*, *Juniperus* and *Olea*. Subsequently, more open vegetation developed with Afromontane forest, grass and herbs. Njagi et al. (2021), for example, report pollen data from Kapsabet Swamp, 50 km northeast of Kisumu, that indicate a progressive change from a forested landscape to an open woodland during the last 3000 years. Forest decline on the Mau Escarpment has continued into modern times with a decline of about 25% in forest cover over the last four decades (Jebiwott et al. 2021).

Analysis of isotopic data from macrofauna south of Winam Gulf shows dominance by C4 grazers with very few C3 browsers, suggesting a grassy setting in that area (Chritz et al. 2015, 2019). In addition, research by Lejju (2012), using phytoliths in cores from western Lake Victoria, indicates that forest decline was significant after 1090 y BP. The substantial floral variability, at least in part, reflects climate change, which has been related to shifts in the east-west trending ITCZ and the north-south aligned Congo Air Boundary (Verschuren and Russell 2009).

References

Adams DD, Ochola SO (2002) A review of sediment gas cycling in lakes with reference to Lake Victoria and sediment gas measurements in Lake Tanganyika. In: Odada EO, Olago DO (eds) The East African Great Lakes: limnology, palaeolimnology and biodiversity. Kluwer, Dordrecht, pp 277–305

Akiyama T, Kajumulo AA, Olsen S (1977) Seasonal variations of plankton and physicochemical condition in Mwanza Gulf, Lake Victoria. Bull Freshw Fish Res Lab Jap 27:49–60

Antenucci JP, Alexander R, Attwater G, Ewing T, Feaver S, Imberger J, Khisa P, Njuguna H, Romero JR, Shimizu K (2006) The physical limnology of Winam Gulf and Rusinga Channel of Lake Victoria during Apr.-May and Aug. of 2005. In: Odada EO, Olago DO, Ochola W, Ntiba M, Wandiga S, Gichuki N, Oyieke H (eds) Proc 11th World Lakes Conf, vol 2, Nairobi, pp 1–6

Anyah RO, Semazzi FHM, Xie L (2006) Simulated physical mechanisms associated with climate variability over Lake Victoria basin in East Africa. Mon Weath Rev 134:3588–3609

Aura CM, Odoli CO, Musa S (2023) Microbial dynamics within Lake Victoria Basin in relation to human activities. In: El-Sheekh M, Elsaied HE (eds) Lakes of Africa. Elsevier, Amsterdam, pp 455–473

Awange JL, Kyalo Kiema JB (2013) Water resources. In: Awange JL, Kyalo Kiema JB (eds) Environmental geoinformatics: environmental science and engineering. Springer, Berlin, pp 341–380

Awange JL, Ong'ang'a O (2006) Lake Victoria: ecology, resources, environment. Springer, Berlin

Awange JL, Anyah R, Agola FE, Omondi P (2013) Potential impacts of climate and environmental change on the stored water of Lake Victoria Basin and economic implications. Water Resour Res 49:8160–8173

Awange JL, Saleem A, Sukhadiya RM, Ouma YO, Kexiang H (2019) Physical dynamics of Lake Victoria over the past 34 years (1984-2018): is the lake dying? Sci Total Environ 658:199–218

Bakayoko A, Galy-Lacaux C, Yobou V, Hickman JE, Roux F, Gardrat E, Julien F, Delon C (2021) Dominant contribution of nitrogen compounds in precipitation chemistry in the Lake Victoria catchment (East Africa). Environ Res Lett 16:045013

Baker SJK (1974) Background to the study of drought in East Africa. Afr Affairs 73:170–177

Balirwa J, Chapman CA, Chapman LJ, Cowx IG, Geheb K, Kaufman L, Lowe-McConnell RH, Seehausen O, Wanink JH, Welcomme RL, Witte F (2003) Biodiversity and fishery sustainability in the Lake Victoria Basin: an unexpected marriage? Bioscience 53:703–715

Bastawesy ME, Gabr S, White K (2012) Hydrology and geomorphology of the Upper White Nile lakes and their relevance for water resources management in the Nile basin. Hydrol Process 27:196–205

Beauchamp RSA (1953) Sulphates in African inland waters. Nature 171:769–771

Berke MA, Johnson TC, Werne JP, Grice K, Schouten S, Damsté JSS (2012) Molecular records of climate variability and vegetation response since the Late Pleistocene in the Lake Victoria basin, East Africa. Quat Sci Rev 55:59–74

Beuning KRM, Kelts K, Ito E, Johnson TC (1997) Paleohydrology of Lake Victoria, East Africa, inferred from $^{18}O/^{16}O$ ratios in sediment cellulose. Geology 25:1083–1086

Beuning KRM, Kelts K, Russell J, Wolfe BB (2002) Reassessment of Lake Victoria–Upper Nile River paleohydrology from oxygen isotope records of lake-sediment cellulose. Geology 30:559–562

Beverly EJ, White JD, Peppe DJ, Faith JT, Blegen N, Tryon CA (2020) Rapid Pleistocene desiccation and the future of Africa's Lake Victoria. Earth Planet Sci Lett 530:115883

Birkett C, Murtugudde R, Allan T (1999) Indian Ocean climate event brings floods to East Africa's lakes and Sudd marsh. Geophys Res Lett 26:1031–1034

Bishop WW (1969) Pleistocene stratigraphy in Uganda. Geol Surv Uganda Mem 10:1–128

Bishop WW, Trendall AF (1967) Erosion-surfaces, tectonics and volcanic activity. Quat J Geol Soc Lond 122:385–420

Bloomfield C, Brown G, Catt JA (1970) The distribution of sulphur in the mud of Lake Victoria. Plant Soil 33:479–481

Bootsma HA, Hecky RE (1993) Conservation of the African Great Lakes: a limnological perspective. Conserv Biol 7:644–656

Bootsma HA, Hecky RE (2003) A comparative introduction to the biology and limnology of the African Great Lakes. J Great Lakes Res 29:3–18

Branstrator DK, Mwebaza-Ndawula L, Lehman JT (1996) Zooplankton dynamics in Lake Victoria. In: Johnson TC, Odada E (eds) Limnology, climatology and paleolimnology of the East African lakes. Gordon and Breach, Amsterdam, pp 337–355

Branstrator DK, Mwebaza-Ndawula L, Montoya JP (2003) Resource–consumer relationships in Lake Victoria, East Africa. Hydrobiologia 493:27–34

Bridgeman TB (2001) The ecology and paleolimnology of food web changes in Lake Victoria, East Africa. PhD thesis, University of Michigan, Ann Arbor

Broecker WS, Peteet D, Hajdas I, Lin J, Clark E (1998) Antiphasing between rainfall in Africa's Rift Valley and North America's Great Basin. Quat Res 50:12–20

Bugenyi F, Magumba KM (1990) The physico-chemistry of the northern waters of Lake Victoria: prevailing activities on the Lake Victoria basin with particular reference to the fisheries of the lake. Workshop on "Prevailing activities on the Lake Victoria Basin with particular reference to the fisheries", Mwanza, Tanzania (8–10 March), pp 43–55

Bugenyi F, Magumba KM (1996) The present physicochemical ecology of Lake Victoria, Uganda. In: Johnson TC, Odada EO (eds) The limnology, climatology and paleoclimatology of the East African Lakes. Gordon and Breach, Amsterdam, pp 141–154

Calamari D, Akech M, Ochumba P (2006) Pollution of Winam Gulf, Lake Victoria, Kenya: a case study for preliminary risk assessment. Lakes Reserv 1:89–106

Campbell LM (2001) Mercury in Lake Victoria (East Africa): another emerging issue for a beleaguered lake? PhD thesis, University of Waterloo, Ontario

Campbell LM, Hecky RE, Sylvester B, Wandera SB (2003a) Stable isotope analyses of food web structure and fish diet in Napoleon and Winam Gulfs, Lake Victoria, East Africa. J Great Lakes Res 29:243–257

Campbell LM, Hecky RE, Muggide R, Dixon DG, Ramlal PS (2003b) Variation and distribution of total mercury in water, sediment and soil from northern Lake Victoria, East Africa. Biogeochemistry 65:195–211

Chapman LJ, Chapman CA, Kaufman L, Witte F, Balirwa J (2008) Biodiversity conservation in African inland waters: lessons of the Lake Victoria region. Verh Int Verein Limnol 30:16–34

Chritz KL, Marshall FB, Zagal ME, Kirera F, Cerling TE (2015) Environments and trypanosomiasis risks for early herders in the later Holocene of the Lake Victoria basin, Kenya. Proc Natl Acad Sci 112:3674–3679

Chritz KL, Cerling TE, Freeman KH, Hildebrand EA, Janzen A, Prendergast ME (2019) Climate, ecology, and the spread of herding in eastern Africa. Quat Sci Rev 204:119–132

Cockerton HE, Street-Perrott FA, Barker PA, Leng MJ, Sloane HJ, Ficken KJ (2015) Orbital forcing of glacial/interglacial variations in chemical weathering and silicon cycling within the upper White Nile basin, East Africa: stable-isotope and biomarker evidence from lakes Victoria and Edward. Quat Sci Rev 130:57–71

Cocquyt C, Vyverman W, Compère P (1993) A check-list of the algal flora of the East African Great Lakes (Malawi, Tanganyika and Victoria). Scripta Botan Belg 8:1–54

Conway D (2002) Extreme rainfall events and lake level changes in East Africa: recent events and historical precedents. In: Odada EO, Olago DO (eds) The East African Great Lakes: limnology, Palaeolimnology and biodiversity. Kluwer, Dordrecht, pp 63–92

Conway D (2003) Remote forcing of East African rainfall and relationships with fluctuations in levels of Lake Victoria. Int J Climatol 23:67–89

Corbet PS (1958) Lunar periodicity of aquatic insects in Lake Victoria. Nature 182:330–331

Corbet PS (1961) The food of non-cichlid fishes in the Lake Victoria basin, with remarks on their evolution and adaptation to lacustrine conditions. Proc Zool Soc Lond 136:1–101

Cordone AJ, Kudhongania AW (1971) Identification of organisms in Lake Victoria responsible for echosounder traces. EAFFRO Annu Rep 1970:28–41

Cowx IG, van der Knaap M, Muhoozi LI, Othina A (2003) Improving fishery catch statistics for Lake Victoria. Aquat Ecosys Health Manag 6:299–310

Cózar A, Bracchini L, Dattilo A, Loiselle S, Azza N (2004) Characterization of the Ugandan inshore waters of Lake Victoria based on temperature-conductivity diagrams. Water Resour Res 40:1–7

Cózar A, Bergamino N, Mazzuoli S, Azza N, Bracchini L, Dattilo AM, Loiselle SA (2007) Relationships between wetland ecotones and inshore water quality in the Ugandan coast of Lake Victoria. Wetl Ecol Manag 15:499–507

Cózar A, Bruno M, Bergamino N, Úbeda B, Bracchini L, Dattilo AM, Loiselle SA (2012) Basin-scale control on the phytoplankton biomass in Lake Victoria, Africa. PLoS One 7:e29962

Crul RCM (1995) Limnology and hydrology of Lake Victoria, UNESCO

Crul R, Silvestre GT, Postma DJ, van Oijen MJP, Acere TO, Bongers G (1995) A bibliography of Lake Victoria: East Africa; technical documents in hydrology, International Hydrological Programme, UNESCO

van Damme D, Pickford MHL (2003) The late Cenozoic Thiaridae Mollusca, Gastropoda, Cerithioidea of the Albertine Rift Valley Uganda-Congo and their bearing on the origin and evolution of the Tanganyikan thalassoid malacofauna. Hydrobiologia 498:1–83

Danley PD, Husemann M, Ding B, DiPietro LM, Beverly EJ, Peppe DJ (2012) The impact of the geologic history and paleoclimate on the diversification of East African cichlids. Int J Evol Biol 2012:574851

Daracott BW (1974) The structure of the Speke Gulf, Tanzania, and its relation to the East African Rift System. Tectonophysics 23:155–175

Descy J-P, Sarmento H (2008) Microorganisms of the East African Great Lakes and their response to environmental changes. Fr Rev 1:59–73

Doornkamp JC (1968) The geomorphology of southwest Uganda. E Afr Geogr Rev 6:75–81

Doornkamp JC, Temple PH (1966) Surface drainage and tectonic instability in part of southern Uganda. Geogr J 132:238–251

Downing AS, Galic N, Goudswaard KPC, van Nes EH, Scheffer M, Witte F, Mooij WM (2013) Was *Lates* late? A null model for the Nile Perch boom in Lake Victoria. PLoS One 8:e76847

Downing AS, van Nes EH, Balirwa JS, Beuving J, Bwathondi POJ, Chapman LJ, Cornelissen IJ, Cowx IG, Goudswaard K, Hecky RE, Janse JH, Janssen A, Kaufman L, Kishe-Machumu MA, Kolding J, Ligtvoet W, Mbabazi D, Medard M, Mkumbo OC, Mlaponi E, Munyaho AT, Nagelkerke LAJ, Ogutu-Ohwayo R, Ojwang WO, Peter HK, Schindler D, Seehausen O, Sharpe D, Silsbe GM, Sitoki L, Tumwebaze R, Tweddle D, Van de Wolfshaar KE, Van Dijk H, Van Donk E, Van Rijssel JC, Van Zwieten PA, Wanink JH, Witte F, Mooij WM (2014) Coupled human and natural system dynamics as key to the sustainability of Lake Victoria's ecosystem services. Ecol Soc 19:31

Duke HL (1924) Water analyses. Annual Report Bacteriology Dept, Uganda Protectorate, for year ended 31st Dec 1923. Entebbe, Uganda

Dutton CL, Subalusky AL, Hill TD, Aleman JC, Rosi EJ, Onyango KB, Kanuni K, Cousins JA, Staver AC, Post DM (2019) A 2000-year sediment record reveals rapidly changing sedimentation and land use since the 1960s in the upper Mara-Serengeti ecosystem. Sci Total Environ 664:148–160

Ebinger C (1989) Tectonic development of the western branch of the East African rift system. Geol Soc Am Bull 101:885–903

Elmer KR, Reggio C, Wirth T, Verheyen E, Salzburger W, Meyer A (2009) Pleistocene desiccation in East Africa bottlenecked but did not extirpate the adaptive radiation of Lake Victoria haplochromine cichlid fishes. Proc Natl Acad Sci 106:13404–13409

Feikin DR, Tabu CW, Gichuki J (2010) Does water hyacinth on east African lakes promote cholera outbreaks? Am J Trop Med Hyg 83:370–373

Fish GR (1952) Appendix A. Annual report for 1951. East African Fisheries Research Organisation, Jinja

Fish GR (1953) Appendix A. Chemical analyses. Annual report for 1952. East African Fisheries Research Organisation, Jinja

Fish GR (1957) A seiche movement and its effects on the hydrology of Lake Victoria. Colonial Office Fishery Publ (HMSO), London, p 10

Flohn H, Fraedrich K (1966) Tagesperiodische Zirkulation und Niederschlagsverteilung am Victoria-See (Ostafrika). Meteorol Rundsch 19:157–165

Fryer G (1960) Concerning the proposed introduction of Nile perch into Lake Victoria. East Afr J Agri 25:267–270

Fryer G (1997) Biological implications of a suggested Late Pleistocene desiccation of Lake Victoria. Hydrobiologia 354:177–182

Fryer G (2001) On the age and origin of the species flock of haplochromine cichlid fishes of Lake Victoria. Proc R Soc Ser B 268:1147–1152

Fryer G (2004) Speciation rates in lakes and the enigma of Lake Victoria. Hydrobiologia 519:167–183

Fusilli L, Collins MO, Laneve G, Palombo A, Pignatti S, Santini F (2013) Assessment of the abnormal growth of floating macrophytes in Winam Gulf (Kenya) by using MODIS imagery time series. Int J Appl Earth Obs 20:33–41

Gee JM (1966) Some preliminary observations on the distribution and behaviour of fish in Lake Victoria as shown by an echosounder. EAFFRO Ann Rep 1965:66–68

Gee JM, Gilbert MP (1968) The establishment of a commercial fishery for *Haplochromis* in the Uganda waters of Lake Victoria, Part II 1968. Jinja, Uganda. EAFFRO Occ Pap 8

Gharib SM, Khairy HM, Abdel-Aal EI, El-Sheekh M (2023) Phytoplankton ecosystem dynamics and seasonal changes in African Lakes: the case of Lake Victoria. In: El-Sheekh M, Elsaied HE (eds) Lakes of Africa: microbial diversity and sustainability. Elsevier, Amsterdam, pp 325–350

Gikuma-Njuru P, Hecky RE (2005) Nutrient concentrations in Nyanza Gulf, Lake Victoria, Kenya: light limits algal demand and abundance. Hydrobiologia 534:131–140

Gikuma-Njuru P, Rutagemwa DK, Muggida R, Hecky RE, Mwebaza-Ndawula L, Mwirigi PM, Abuodha JOZ, Waya RK, Matovu A, Kinobe J (2005) Eutrophication of the Lake Victoria ecosystem. In: Mwanuzi FL, Abuodha JOX, Muyodi FJ, Hecky RE (eds) Lake Victoria environment management project (LVEMP) water quality and ecosystem status. South Eastern Kenya University, Kitui, pp 81–102

Githumbi EN, Mustaphi CJC, Marchant R (2021) Late Pleistocene and Holocene Afromontane vegetation and headwater wetland dynamics within the Eastern Mau Forest, Kenya. J Quat Sci 36:239–254

GoK [Government of Kenya] and UNDP (2021) Rising water levels in Kenya's Rift Valley lakes, Turkwel Gorge Dam and Lake Victoria: a scoping report. Government of Kenya, Nairobi

Goldschmidt T, Witte F, Wanink J (1993) Cascading effects of the introduced Nile perch on the detritivorous/phytoplanktivorous species in the sublittoral areas of Lake Victoria. Conserv Biol 7:686–700

Goldstein ST, Shipton C, Miller JM, Ndiema E, Boivin N, Petraglia M (2022) Hunter-gatherer technological organization and responses to Holocene climate change in coastal, lakeshore, and grassland ecologies of eastern Africa. Quat Sci Rev 280:107390

Gophen M (2015a) Ecological devastation in Lake Victoria: part A: thermal structure and anoxia. Open J Ecol 5:287–298

Gophen M (2015b) Ecological devastation in Lake Victoria: part B: plankton and fish communities. Open J Ecol 5:315–325

Gophen M, Pollingher U, Ochumba P (1993) Feeding habits of tilapias and the limnology of Lake Victoria. Final Report (1989–1991). Kinneret Limnological Laboratory T6/93

Gophen M, Ochumba PBO, Kaufman LS (1995) Some aspects of perturbation in the structure and biodiversity of the ecosystem of Lake Victoria (East Africa). Aquat Liv Resour 8:27–41

Goudie A (2005) The drainage of Africa since the Cretaceous. Geomorphology 67:437–456

Graham M (1929) The Victoria Nyanza and its fisheries. A report on the fishing survey of Lake Victoria 1927–1928 and appendices. Crown Agents for the Colonies, London

Grant JA (1876) On Mr. H.M. Stanley's exploration of the Victoria Nyanza. Proc R Geogr Soc Lond 20:34–54

Groves A (1934) The physiography of Uganda: the evolution of the Great Lakes and the Victoria Nile drainage system. J R Afr Soc 33:59–69

Guildford SJ, Hecky RE (2000) Total nitrogen, total phosphorus, and nutrient limitation in lakes and oceans: is there a common relationship? Limnol Oceanogr 45:1213–1223

Guya FJ (2013) Bioavailability of particle-associated nutrients as affected by internal regeneration processes in the Nyanza Gulf region of Lake Victoria. Lakes Reserv 18:129–143

Haande S (2008) On the ecology, toxicology, and phylogeny of cyanobacteria in Murchison Bay of Lake Victoria, Uganda. PhD thesis, University of Bergen

Haande S, Rohrlack T, Semyalo RP, Brettum P, Edvardsen B, Lyche-Solheim A, Sørensen K, Larsson P (2011) Phytoplankton dynamics and cyanobacterial dominance in Murchison Bay of Lake Victoria (Uganda) in relation to environmental conditions. Limnologica 41:20–29

Harrison K (1991) The taxonomy of East African Nile perch, *Lates* spp. (Perciformes, Centropomidae). J Fish Biol 38:175–186

Hecky RE (1993) The eutrophication of Lake Victoria. Verh Int Verein Limnol 25:39–48

Hecky RE, Bugenyi FWB (1992) Hydrology and chemistry of the African Great Lakes and water quality issues: problems and solutions. In: Lowe-McConnell RH, Crul RCM, Roest FC (eds) Symposium on resource use and conservation of the African Great Lakes, Bujumbura 1989 (Mitteilungen der IVL 23). E. Schweizerbart, Stuttgart, pp 45–54

Hecky RE, Bugenyi F, Ochumba P, Talling J, Mugidde R, Gophen M, Kaufman L (1994) Deoxygenation of the deep water of Lake Victoria, East Africa. Limnol Oceanogr 39:1476–1481

Hecky RE, Bootsma HA, Mugidde RM, Bugenyi F (1996) Phosphorus pumps, nitrogen sinks, and silicon drains: plumbing nutrients in the African Great Lakes. In: Johnson TC, Odada EO (eds) The limnology, climatology and paleoclimatology of the East African Lakes. Gordon and Breach, Amsterdam, pp 205–224

Hecky RE, Bootsma HA, Odada EO (2006) African lake management initiatives: the global connection. Lakes Reserv 11:203–213

Henderson P, Pickford M, Williams CT (1987) A geochemical study of rocks and spring waters at Kanam and Kanjera, Kenya, and the implications concerning element mobility and uptake. J Afr Earth Sci 6:221–227

Henry L, Kishimba MA (2003) Levels of pesticide residues in water, soil and sediments from southern Lake Victoria and its basin. Tanz J Sci 29:77–90

Henry L, Kishimba MA (2006) Pesticide residues in Nile tilapia (*Oreochromis niloticus*) and Nile perch (*Lates niloticus*) from southern Lake Victoria, Tanzania. Environ Pollut 140:348–354

Hesse PR (1956) Sulphate deficiency of Lake Victoria. Nature 177:389–390

Hobbs WH (1910) Earth movements at Lake Victoria in Central East Africa. Science 31:306–307

Holtzman J, Lehman JT (1998) Role of apatite weathering in the eutrophication of Lake Victoria. In: Lehman JT (ed) Environmental change and response in East African lakes. Springer, Dordrecht, pp 89–98

Humphrey OS, Osano O, Aura CM, Marriott AL, Dowel SM, Blake WH, Watts MJ (2022) Evaluating spatio-temporal soil erosion dynamics in the Winam Gulf catchment, Kenya for enhanced decision making in the land-lake interface. Sci Total Environ 815:151975

Hurst HE, Phillips P (1938) The hydrology of the Lake Plateau and Bahr el Jebel. The Nile Basin, vol 5. Government Press, Cairo

Hustedt F (1922) Bacillariales. In: Schröder B (ed) Zellpflanzen Ostafrikas. Hedwigia 63:117–173

IDEAL (1990) An international decade for the East African Lakes. Workshop report 1 on the palaeoclimatology of African Rift Lakes, Bern, Switzerland, March 29–31, 1990. Technical Report Duke University Marine Laboratory

Jebiwott A, Ogendi GM, Agbeja BO, Alo AA, Kibet R (2021) Mapping the trends of forest cover change and associated drivers in Mau Forest, Kenya. Rem Sens Appl Soc Environ 23:100586

Johnson TC (1996) Sedimentary processes and signals of past climatic change in the large lakes of the East African Rift Valley. In: Johnson TC, Odada EO (eds) The limnology, climatology and paleoclimatology of the East African Lakes. Gordon and Breach, Amsterdam, pp 367–412

Johnson TC (2002) Biogenic silica profiles in the sediments of large tropical lakes: examples from East Africa. In: Renaut RW, Ashley GM (eds) Sedimentation in continental rifts. SEPM, Tulsa, SEPM Spec Publ 73:247–255

Johnson TC, Scholz CA, Talbot MR, Kelts K, Ricketts RD, Ngobi G, Beuning K, Ssemanda I, McGill JW (1996) Late Pleistocene desiccation of Lake Victoria and rapid evolution of cichlid fishes. Science 273:1091–1093

Johnson TC, Chan Y, Beuning KRM, Kelts K, Ngobi G, Verschuren D (1998) Biogenic silica profiles in Holocene cores from Lake Victoria: implications for lake level history and initiation of the Victoria Nile. In: Lehman JT (ed) Environmental change and response in East African lakes. Springer, Dordrecht, pp 75–88

Johnson TC, Kelts K, Odada E (2000) The Holocene history of Lake Victoria. Ambio 29:2–11

Jones WB, Lippard SJ (1979) New age determinations and the geology of the Kenya Rift-Kavirondo Rift junction, W Kenya. J Geol Soc Lond 136:693–704

Jonna R, Lehman JT (2002) Invasion of Lake Victoria by the large bodied herbivorous cladoceran *Daphnia magna*. In: Odada EO, Olago DO (eds) The East African Great Lakes: limnology, palaeolimnology and biodiversity. Kluwer, Dordrecht, pp 321–333

Juma DW, Wang H, Li F (2014) Impacts of population growth and economic development on water quality of a lake: case study of Lake Victoria Kenya water. Environ Sci Pollut Res 21:5737–5746

Kansiime F, Saunders MJ, Loiselle SA (2007) Functioning and dynamics of wetland vegetation of Lake Victoria: an overview. Wetl Ecol Manag 15:443–451

Kateregga E, Sterner T (2007) Indicators for an invasive species: water hyacinths in Lake Victoria. Ecol Indic 7:362–370

Katunzi E (1996) A review of Lake Victoria fisheries with recommendations for management and conservation. In: Johnson TC, Odada EO (eds) The limnology, climatology and paleoclimatology of the East African Lakes. Gordon and Breach, Amsterdam, pp 357–366

Kaufman L, Ochumba P (1993) Evolutionary and conservation biology of cichlid fishes as revealed by faunal remnants in northern Lake Victoria. Conserv Biol 7:719–730

Kaufman L, Chapman LJ, Chapman CA (1996) The Great Lakes. In: McClanahan Y, Young TP, McClanahan T (eds) East African ecosystems and their conservation. Oxford University Press, Oxford, pp 191–216

Kayanda R, Taabu AM, Tumwebaze R, Muhoozi L, Jembe T, Mlaponi E, Nzungi P (2009) Status of the major commercial fish stocks and proposed species-specific management plans for Lake Victoria. Afr J Trop Hydrobiol Fish 21:15–21

Kayombo S, Jorgensen SE (2006) Lake Victoria. In: Jorgensen SE (ed) Experiences and lessons learned. Royal Danish University of Pharmaceutical Sciences, Copenhagen, pp 431–446

Kendall R (1969) An ecological history of the Lake Victoria basin. Ecol Monogr 39:121–176

Kent PE (1942) The country around the Kavirondo Gulf of Victoria Nyanza. Geogr J 100:22–31

Kent PE (1944) The Miocene beds of Kavirondo, Kenya. Q J Geol Soc Lond 100:85–118

Kiage LM, Obuoyo J (2011) The potential link between El Nino and water hyacinth blooms in Winam Gulf of Lake Victoria, East Africa: evidence from satellite imagery. Water Resour Manag 25:3931–3945

Kiggundu V, Mwebaza-Ndawula L, Makanga B, Nachuha S (2012) Variations in zooplankton community structure and water quality conditions in three habitat types in northern Lake Victoria. Lakes Reserv 17:83–95

King BC, Le Bas MJ, Sutherland DS (1972) The history of the alkaline volcanoes and intrusive complexes of eastern Uganda and western Kenya. J Geol Soc Lond 128:173–205

Kishe MA (2004) Physical and chemical characteristics of water in selected locations in Lake Victoria, Tanzania. Tanz J Sci 30:65–72

Kitaka GEB (1972) An instance of cyclonic upwelling in the southern offshore waters of Lake Victoria. Afr J Trop Hydrobiol Fish 1:85–92

Kitheka JU, Okoyo S, Mboya N (2021) Hydology and climate impacts on streamflow and sediment yield in the Nyando River Basin, Kenya. In: Diop S, Scheren P (eds) Climate change and water resources in Africa. Springer, Cham, pp 219–238

Kitunda J (2003) The spread of water hyacinth and other aquatic weeds in the African Great Lakes region: an environmental history of Lake Victoria up to 1999. PhD thesis, University of Wisconsin, Madison

Kiwango YA, Wolanski E (2008) Papyrus wetlands, nutrients balance, fisheries collapse, food security, and Lake Victoria level decline in 2000–2006. Wetl Ecol Manag 16:89–96

Kling HJ, Muggide R, Hecky RE (2001) Recent changes in the phytoplankton community of Lake Victoria in response to eutrophication. In: Munawar M, Hecky RE (eds) The Great Lakes of the World (GLOW): food-web, health and integrity. Buckhuys Publishers, Kerkwerve, pp 47–65

van der Knaap M, Ntiba MJ, Cowx IG (2010) Key elements of fisheries management on Lake Victoria. Aquat Ecosyst Health Manag 5:245–254

Kolding J, van Zwieten P, Mkumbo O, Silsbe G, Hecky R (2008) Are the Lake Victoria fisheries threatened by exploitation or eutrophication? In: Bianchi G, Skjodal HR (eds) The ecosystem approach to fisheries. CAB International, London, pp 309–354

Kolding J, Medard M, Mkumbo O, van Zwieten PAM (2014a) Status, trends and management of the Lake Victoria fisheries. In: Welcomme RL, Valbo-Jørgensen J, Halls AS (eds) Inland fisheries evolution and management – case studies from four continents, FAO fisheries and aquaculture technical paper, vol 579. Food and Agriculture Organization of the United Nations, Rome, pp 1–19

Kolding J, van Zwieten P, Manyala J, Okedi J, Mgaya YD, Orach-Meza F (2014b) Regional synthesis report on fisheries research and management, Lake Victoria Environmental Management Program

Komarek J, Kling H (1991) Variation in six planktonic cyanophyte genera in Lake Victoria (East Africa). Archiv Hydrobiol Suppl 62:21–45

Krienitz L, Ballot A, Wiegand C, Kotut K, Codd GA, Pflugmacher S (2002) Cyanotoxin-producing bloom of *Anabaena flos-aquae*, *Anabaena discoidea* and *Microcystis aeruginosa* (Cyanobacteria) in Nyanza Gulf of Lake Victoria, Kenya. J Appl Bot 76:179–183

Kudhongania AW, Cordone AJ (1974) Batho-spatial distribution pattern and biomass estimate of the major demersal fishes in Lake Victoria. Afr J Trop Hydrobiol Fish 3:15–31

Kudhongania AW, Twongo T, Ogutu-Ohwayo G (1992) Impact of the Nile perch on the fisheries of Lakes Victoria and Kyoga. Hydrobiologia 232:1–10

Kudhongania AW, Ocenodongo DL, Okaronon JO (1996) Anthropogenic perturbations on the Lake Victoria ecosystem. In: Johnson TC, Odada EO (eds) The limnology, climatology and paleoclimatology of the East African Lakes. Gordon and Breach, Amsterdam, pp 625–632

Kull D (2006) Connections between recent water level drops in Lake Victoria, dam operations and drought. International Rivers Report, Oakland

Kwach BO, Lalah JO, Shem WO (2009) Spatial and seasonal variations in concentrations of polycyclic aromatic hydrocarbons in water and sediment of Kisumu City Bay of Winam Gulf, Lake Victoria-Kenya. Bull Environ Contam Tox 83:734–741

Lake Victoria Basin Commission (2011) A study on aquatic biodiversity in the Lake Victoria Basin, Kenya. ACTS Press, African Centre for Technology Studies, Nairobi

Lehman JT (1996) Pelagic food webs of the East African Great Lakes. In: Johnson TC, Odada EO (eds) The limnology, climatology and paleoclimatology of the East African lakes. Gordon and Breach, Amsterdam, pp 281–301

Lehman JT (1998) Role of climate in the modern condition of Lake Victoria. Theor Appl Climatol 61:29–37
Lehman JT (2009) Lake Victoria. In: Dumont HJ (ed) The Nile: origin, environments, limnology and human use. Springer, Dordrecht, pp 215–241
Lehman JT, Branstrator DK (1993) Effects of nutrients and grazing on the phytoplankton of Lake Victoria. Verh Int Verein Limnol 25:850–855
Lehman JT, Branstrator DK (1994) Nutrient dynamics and turnover rates of phosphate and sulfate in Lake Victoria, East Africa. Limnol Oceanogr 39:227–233
Lehman JT, Muggide R, Lehman DA (1998) Lake Victoria plankton ecology: mixing depth and climate-driven control of lake condition. In: Lehman JT (ed) Environmental change and response in East African Lakes. Springer, Dordrecht, pp 99–116
Lejju JB (2012) The influence of climate change and human-induced environmental degradation on Lake Victoria. Organisation for Social Science Research in Eastern and Southern Africa, Addis Ababa
Ligtvoet W, Witte F (1991) Perturbation through predator introduction effect on the food web and fish yield in Lake Victoria (East Africa). In: Ligtvoet W, Witte F (eds) Proceedings of the symposium on perturbation and recovery of ecosystems, Sienna, 1990. Ellis Horwood, Hemel Hempstead, pp 263–268
Lind E, Visser S (1962) A study of a swamp at the north end of Lake Victoria. J Ecol 50:599–613
Lipiatou E, Hecky RE, Eisenreich SJ, Lockhart L, Muir D, Wilkinson P (1996) Recent ecosystem changes in Lake Victoria reflected in sedimentary natural and anthropogenic organic compounds. In: Johnson TC, Odada EO (eds) The limnology, climatology and paleoclimatology of the East African Lakes. Gordon and Breach, Amsterdam, pp 523–541
Loiselle S, Cózar A, Adgo E, Ballatore T, Chavula G, Descy JP, Harper DM, Kansiime F, Kimirei I, Langenberg V, Ma R, Sarmento H, Odada E (2014) Decadal trends and common dynamics of the bio-optical and thermal characteristics of the African Great Lakes. PLoS One 9:e93656
Lund S, Platzman E, Johnson TC (2020) Paleomagnetic secular variation records from Holocene sediments of Lake Victoria (0.5°S, 33.3°E). The Holocene 31:1651–1661
Lung'ayia HBO, M'harzi A, Tackx M, Gichuki J, Symoens JJ (2000) Phytoplankton community structure and environment in the Kenyan waters of Lake Victoria. Freshw Biol 43:529–543
Lung'ayia HBO, Sitoki L, Kenyanya M (2001) The nutrient enrichment of Lake Victoria (Kenyan waters). Hydrobiologia 458:75–82
MacDonald WW (1956) Observations on the biology of chaoborids and chironomids in Lake Victoria and on the feeding habits of the elephant snout fish (*Mormyrus kanume* Forsk). J Anim Ecol 25:36–53
Machiwa P (2003) Water quality management and sustainability: the experience of Lake Victoria Environmental Management Project (LVEMP)—Tanzania. Phys Chem Earth 28:1111–1115
MacIntyre S (2013) Climatic variability, mixing dynamics, and ecological consequences in the African Great Lakes. In: Goldman CR, Kumagai M, Robarts RD (eds) Climatic change and global warming of inland waters. Wiley-Blackwell, Chichester, pp 311–336
MacIntyre S, Romero JR, Silsbe GM, Emery BM (2014) Stratification and horizontal exchange in Lake Victoria, East Africa. Limnol Oceanogr 59:1805–1838
Masese FO, Raburu PO, Mwasi BN, Etiégni L (2012) Effects of deforestation on water resources: integrating science and community perspectives in the Sondu-Miriu River Basin, Kenya. In: Oteng-Amoako AA (ed) New advances and contributions to forestry research. Intech Open, Rijeka, pp 3–18
Mavuti KM, Litterick MR (1991) Composition, distribution and ecological role of zooplankton community in Lake Victoria, Kenya waters. Verh Int Ver Limnol 24:1117–1122
Mbahinzireki GB, Lehman JT, Ochieng H (1998) *Caridina nilotica*: spatial distribution and egg production in Lake Victoria, Uganda. In: Lehman JT (ed) Environmental change and response in East African Lakes. Springer, Dordrecht, pp 117–124
Mboya B (1983) The genesis and tectonics of the NE Nyanza Rift Valley, Kenya. J Afr Earth Sci 1:315–320

Meyer A, Montero CM, Spreinat A (1996) Molecular phylogenetic inferences about the evolutionary history of East African cichlid fish radiations. In: Johnson TC, Odada EO (eds) The limnology, climatology and paleoclimatology of the East African Lakes. Gordon and Breach, Amsterdam, pp 303–323

Mgaya YD, Mahongo SB (2017) Lake Victoria fisheries resources: research and management in Tanzania. Springer, Cham

Milly PCD (1999) Comment on "Antiphasing between rainfall in Africa's Rift Valley and North America's Great Basin". Quat Res 51:104–107

Miruka JB, Getabu A, Sitoki L, James O, Mwamburi J, George O, Chrisphine N, Odoli C (2021) Water quality, phytoplankton composition and microcystin concentrations in Kisumu Bay (Kenya) of Lake Victoria after a prolonged water hyacinth infestation period. Lakes Reserv Res Manage 26:e12380

Mkumbo O, Nsinda CP, Ezekiel CN, Cowx IG, Aeron M (2007) Towards sustainable exploitation of Nile perch consequential to regulated fisheries in Lake Victoria. Aquat Ecosyst Health Manag 10:449–457

Mnaya B, Wolanski E, Kiwango Y (2006) Papyrus wetlands a lunar-modulated refuge for aquatic fauna. Wetl Ecol Manag 14:359–363

Mnyanga VP, Osio JO, Mwebembezi L, Myanza OI, Opango PO, Majaliwa JGM, Oleko CO, Okwerede L, Gor SA, Byamukama D, Mathayo AA, Masongo FO, Kanyesigye C, Kitamirike JM, Semalulu O, Hecky RE (2005) Nutrient loading. In: Mwanuzi FL, Abuodha JOX, Muyodi FJ, Hecky RE (eds) Lake Victoria Environment Management Project (LVEMP) water quality and ecosystem status. South Eastern Kenya University, Kitui, pp 103–132

Moorehead A (1960) The White Nile. Hamish Hamilton, London

Mothersill JS (1976) The mineralogy and geochemistry of the sediments of northwestern Lake Victoria. Sedimentology 23:553–565

Mothersill JS (1996a) The formation of vivianite-rich nodules in Lake Victoria. In: Johnson TC, Odada EO (eds) The limnology, climatology and paleoclimatology of the East African Lakes. Gordon and Breach, Amsterdam, pp 543–548

Mothersill JS (1996b) Paleomagnetic results from lakes Victoria and Albert, Uganda. Stud Geophys Geod 40:25–35

Mothersill JS, Freitag R, Barnes B (1980) Benthic macroinvertebrates of northwestern Lake Victoria, East Africa: abundance, distribution intraphyletic relationships between taxa and selected concentrations in lake-bottom sediments. Hydrobiologia 74:215–224

Muggide R (1993) The increase in phytoplankton primary productivity and biomass in Lake Victoria (Uganda). Verh Int Verein Limnol 25:846–849

Mugo R, Waswa R, Nyaga JW, Ndubi A, Adams EC, Flores-Anderson AI (2020) Quantifying land use land cover changes in the Lake Victoria Basin using satellite remote sensing: the trends and drivers between 1985 and 2014. Rem Sens 12:2829

Muli J (1996) Environmental problems of Lake Victoria (East Africa): what the international community can do. Lakes Reserv 2:47–53

Muli J, Mavuti K (2001) The benthic macrofauna community of Kenyan waters of Lake Victoria. Hydrobiologia 458:83–90

Mulibo GD, Nyblade AA (2013) The P and S wave velocity structure of the mantle beneath eastern Africa and the African superplume anomaly. Geochem Geophys Geosyst 14:2696–2715

Musa S, Gichuki JW, Raburu PO, Aura CM (2011) Risk assessment for organochlorines and organophosphates pesticide residues in water and sediments from lower Nyando/Sondu-Miriu river within Lake Victoria Basin, Kenya. Lakes Reserv 16:273–280

Mutakyahwa M, Ikingura JR, Mtui G (2009) Monitoring of heavy metal loading into the wetlands south of Lake Victoria basin, northern Tanzania. Tanz J Sci 35:17–36

Muwanga A, Barifaijo E (2006) Impact of industrial activities on heavy metal loading and their physico-chemical effects on wetlands of Lake Victoria basin (Uganda). Afr J Sci Technol 7:51–63

Mwamburi J (2003) Variations in trace elements in bottom sediments of major rivers in Lake Victoria's basin, Kenya. Lakes Reserv 8:5–13
Mwamburi J (2013) Comparative spatial metal concentrations and partitioning in bottom sediments of two tropical freshwater lake basins, Kenya. Lakes Reserv 18:329–355
Mwanja WW, Armoudlian AS, Wandera SB, Kaufman L, Wu L, Booton GC, Fuerst PA (2001) The bounty of minor lakes: the role of small satellite water bodies in evolution and conservation of fishes in the Lake Victoria region, East Africa. Hydrobiologia 458:55–62
Mwanuzi FL, Abuodha JOZ, Muyodi FJ, Hecky RE (eds) (2005) Lake Victoria Environment Management Project (LVEMP) water quality and ecosystem status. LVEMP, Nairobi
Mwebaza-Ndawula L (1994) Changes in relative abundance of zooplankton in northern Lake Victoria, East Africa. Hydrobiologia 272:259–264
Mwebaza-Ndawula L (1998) Distribution and abundance of zooplankton and *Rastrineobola argentea* (Pisces: Cyprinidae) and their trophic interactions in northern Lake Victoria, East, Africa. PhD thesis, University of Vienna
Mwebaza-Ndawula L, Kiggundu V, Pabire Ghandhi W (2004) The status and significance of invertebrate communities. In: Balirwa JS, Mugidde R, Ogutu-Ohwayo R (eds) Challenges for management of the fisheries resources, biodiversity and environment of Lake Victoria. Fisheries Resources Research Institute, Jinja, pp 153–171
Mwirigi PM, Rutagemwa DK, Gikuma-Njuru P, Matovu A, Waya RK, Mwebaza-Ndawula L, Ssenfuma-Nsubuga M, Kinobe J, Abuodha JOZ, Hecky RE (2005) Lake Victoria: the changing lake. In: Mwanuzi FL, Abuodha JOZ, Muyodi FJ, Hecky RE (eds) Lake Victoria environment management project (LVEMP) water quality and ecosystem status. South Eastern Kenya University, Kitui, pp 62–80
Nantaba F, Wasswa J, Kylin H, Palm WU, Bouwman H, Kümmerer K (2020) Occurrence, distribution, and ecotoxicological risk assessment of selected pharmaceutical compounds in water from Lake Victoria, Uganda. Chemosphere 239:124642
Nantaba F, Palm WU, Wasswa J, Bouwman H, Kylin H, Kümmerer K (2021) Temporal dynamics and ecotoxicological risk assessment of personal care products, phthalate ester plasticizers, and organophosphorus flame retardants in water from Lake Victoria, Uganda. Chemosphere 262:127716
Newell BJ (1958) Additional hydrological data from Lake Victoria. EAFFRO Ann Rep 1956(57):29–31
Newell BJ (1960) The hydrology of Lake Victoria. Hydrobiologia 15:363–383
Ngobi GN, Kelts K, Johnson TC, Silheid PA (1998) Environmental magnetism of Late Pleistocene/Holocene sequences from Lake Victoria, East Africa. In: Lehman JT (ed) Environmental change and response in East African lakes. Springer, Dordrecht, pp 59–74
Ngupula GW (2013) How does increased eutrophication and pollution in the Lake Victoria waters impact zooplankton? J Environ Ecol 4:151–164
Nicholson SE (1996) A review of climate dynamics and climate variability in eastern Africa. In: Johnson TC, Odada EO (eds) The limnology, climatology and paleoclimatology of the East African lakes. Gordon and Breach, Amsterdam, pp 25–56
Nicholson SE (1998) Historical fluctuations of Lake Victoria and other lakes in the northern Rift Valley of East Africa. In: Lehman JT (ed) Environmental change and response in East African lakes. Springer, Dordrecht, pp 7–35
Nicholson SE (2002) Historical fluctuations of Lake Victoria and other lakes in the northern Kenya Rift Valley of East Africa. In: Odada EO, Olago DO (eds) The East African Great Lakes: limnology, Palaeolimnology and biodiversity. Kluwer, Dordrecht, pp 7–35
Nicholson SE, Yin X (2002) Mesoscale patterns of rainfall, cloudiness and evaporation over the Great Lakes of East Africa. In: Odada EO, Olago DO (eds) The East African Great Lakes: limnology, Palaeolimnology and biodiversity. Kluwer, Dordrecht, pp 93–119
Nicholson SE, Yin X, Ba M (2000) Use of a water balance model to interpret historical fluctuations of Lake Victoria. Hydrol Sci J 45:75–96

Njagi DM, Routh J, Olago D, Gaynatha K (2021) A multi-proxy reconstruction of the late Holocene climate evolution in the Kapsabet Swamp, Kenya (East Africa). Palaeogeogr Palaeoecol Palaeoclimatol 574:110475

Njagi DM, Routh J, Odhiambo M, Luo C, Basapuram LG, Olago D, Klump V, Stager C (2022) A century of human-induced environmental changes and the combined roles of nutrients and land use in Lake Victoria catchment on eutrophication. Sci Tot Environ 835:55425

Njiru M, Waithaka E, Muchiri M, van Knaap M, Cowx I (2005) Exotic introductions to the fishery of Lake Victoria: what are the management options? Lakes Reserv 10:147–155

Njiru M, Getabu A, Jembe T, Ngugi C, Owili M, van der Knaap M (2008a) Management of the Nile tilapia (*Oreochromis niloticus* (L.)) fishery in the Kenyan portion of Lake Victoria, in light of changes in its life history and ecology. Lakes Reserv 13:117–124

Njiru M, Kazungu J, Ngugi C, Gichuki J, Muhoozi L (2008b) An overview of the current status of Lake Victoria fishery: opportunities, challenges and management strategies. Lakes Reserv 13:1–12

Njiru M, Nyamweya C, Gichuki J, Mugidde R, Mkumbo OC, Witte F (2011) Increase in anoxia in Lake Victoria and its effects on the fishery. Anoxia:99–128

Nyamweya CS, Natugonza V, Taabu-Munyaho A, Aura CM, Njiru JM, Ongore C, Mangeni-Sande R, Kashindye BB, Odoli CO, Ogari Z, Kayanda R (2020) A century of drastic change: human-induced changes of Lake Victoria fisheries and ecology. Fish Res 230:105564

Nyangababo JT, Henry L, Omutange E (2005) Heavy metal contamination in plants, sediments, and air precipitation of Katonga, Simiyu, and Nyando wetlands of Lake Victoria Basin, East Africa. Bull Environ Contamin Tox 75:189–196

Nyblade AAC, Birt CA, Langston CA, Owens TJ, Last RJ (1996) Seismic experiment reveals rifting of craton in Tanzania. Eos 77:517–521

Ochieng EZ, Lalah JO, Wandiga SO (2006) Heavy metals in water and surface sediments in Winam Gulf of Lake Victoria, Kenya. Bull Environ Contam Tox 77:459–468

Ochieng EZ, Lalah JO, Wandiga SO (2008) Water quality and trace metal distribution in a pristine lake in the lake basin in Kenya. Bull Environ Contam Tox 80:362–368

Ochumba P (1990) Massive fish kills within the Nyanza Gulf of Lake Victoria, Kenya. Hydrobiologia 208:93–99

Ochumba P (1996) Measurement of water currents, temperature, dissolved oxygen and winds on the Kenyan Lake Victoria. In: Johnson TC, Odada EO (eds) The limnology, climatology and paleoclimatology of the East African Lakes. Gordon and Breach, Amsterdam, pp 155–168

Ochumba P, Kibaara D (1989) Observations on blue-green algal blooms in the open waters of Lake Victoria. Kenya Afr J Ecol 27:23–34

Ochumba P, Goshen M, Pollinger U (1994) Ecological changes in Lake Victoria after the invasion of Nile perch (*Lates Niloticus*): the catchment, waters quality and fisheries management. In: Okemwa E, Wakwabi EO, Getabu A (eds) Proceedings of the second EEC regional seminar on recent trends of research on Lake Victoria fisheries. ICIPE Science Press, Nairobi, pp 29–39

Odada E, Olago DO, Kulindwa K, Ntiba M, Wandiga S (2004) Mitigation of environmental problems in Lake Victoria, East Africa: causal chain and policy options analyses. Ambio 33:13–23

Odada EO, Ochola WO, Olago DO (2009a) Drivers of ecosystem change and their impacts on human well-being in Lake Victoria basin. Afr J Ecol 47:46–54

Odada EO, Ochola WO, Olago DO (2009b) Understanding future ecosystem changes in Lake Victoria basin using participatory local scenarios. Afr J Ecol 47:147–153

Ogutu-Ohwayo R (1990) The decline of the native fishes of Lake Victoria and Kyoga (East Africa) and the impact of introduced species, especially the Nile perch, *Lates niloticus* and the Nile tilapia, *Oreochromis niloticus*. Environ Biol Fish 27:81–86

Ogutu-Ohwayo R (1992) The purpose, costs and benefits of fish introductions: with specific reference to the Great Lakes of Africa. Proc Int Ass Theor Appl Limnol 23:34–44

Okalebo JR, Othieno CO, Maritim HK, Iruria DM, Kipsat MJ, Kisinyo PO, Kimenye L, Woomer PL, Mukhwana EJ, Batiano A, Adipala E, Njoroge RK, Thuita M, Nekesa AO, Ruto EC (2005)

Management of soil fertility in western Kenya: experience working with smallholder farmers. Afr Crop Sci Conf Proc 7:1465–1473

Okedi JY (1990) Observations on the benthos of Murchison Bay, Lake Victoria, East Africa. Afr J Ecol 28:111–121

Okello L, Kurmayer R (2011) Seasonal development of cyanobacteria and microcystin production in Ugandan freshwater lakes. Lakes Reserv 16:123–135

Okonga JR, Sewagudde SM, Mngodo RJ, Sangale FD, Mwanuzi FL, Hecky RE (2005) Water balance. In: Mwanuzi FL, Abuodha JOX, Muyodi FJ, Hecky RE (eds) Lake Victoria environment management project (LVEMP) water quality and ecosystem status. South Eastern Kenya University, Kitui, pp 25–41

Okonga JR, Sewagudde SM, Mngodo RJ, Sangale FD, Mwanuzi FL, Hecky RE (2006) Water balance for Lake Victoria. In: Odada EO, Olago DO, Ochola W, Ntiba M, Wandiga S, Gichuki N, Oyieke H (eds) Proc 11th World Lakes Conf, vol 2, Nairobi, pp 47–56

Olokotum M, Mitroi V, Troussellier M, Semyalo R, Bernard C, Montuelle B, Okello W, Quiblier C, Humbert J-F (2020) A review of the socioecological causes and consequences of cyanobacterial blooms in Lake Victoria. Harmful Alg 96:101829

Ongeri DMK, Lalah JO, Wandiga SO, Schramm KW, Michalke B (2008) Levels of toxic metals in multisectoral samples from Winam Gulf of Lake Victoria. Bull Environ Contam Tox 82:64–69

Ongeri DMK, Lalah JO, Wandiga SO (2010) Speciation of Cd, Cu, Zn, Fe and Pb in sediments sampled from Lake Victoria basin, Kenya. Environmentalist 30:254–259

Opere WM, John M, Ombori O (2021) Molecular detection of human enteric adenoviruses in water samples collected from Lake Victoria waters along Homa Bay Town, Homa Bay County, Kenya. Food Environ Virol 13:32–43

Ostenfeld CH (1908) Phytoplankton aus dem Victoria Nyanza. Sammelausbeute von A. Bogert, 1904–1905. Bot Jb 41:330–350

Ostenfeld CH (1909) Notes on the phytoplankton of Victoria Nyanza, East Africa. Bull Mus Comp Zool Harv Univ 52:171–181

Otieno D, Nyaboke H, Nyamweya CS, Odoli CO, Aura CM, Outa NO (2022) Water hyacinth (*Eichhornia crassipes*) infestation cycle and interactions with nutrients and aquatic biota in Winam Gulf (Kenya). Lake Victoria Lakes Reserv 27:e12391

Owino A, Ryan P (2007) Recent papyrus swamp habitat loss and conservation implications in western Kenya. Wetl Ecol Manag 15:1–12

Peterson RB, Kapiyo RA, Campbell EM, Nyabua PO (2018) Gully rehabilitation trusts: fighting soil erosion through community participation in western Kenya. J Rural Stud 58:67–81

Pickford MHL (1982) The tectonics, volcanics and sediments of the Nyanza Rift Valley, Kenya. Z Geomorph NF 42:1–33

Piper BS, Plinston DT, Sutcliffe JV (1986) The water balance of Lake Victoria. Hydrol Sci J 31:25–37

Poste AE, Hecky RE, Guildford SJ (2013) Phosphorus enrichment and carbon depletion contribute to high *Microcystis* biomass and microcystin concentrations in Ugandan lakes. Limnol Oceanogr 58:1075–1088

Pringle R (2005) The origins of the Nile perch in Lake Victoria. Bioscience 55:780–787

Rach NM, Rosendahl BR (1989) Tectonic controls on the Speke Gulf. J Afr Earth Sci 8:471–488

Ramlal PS (2002) Sources, transport and sinks of organic matter in Lake Malawi and Lake Victoria, East Africa. PhD thesis, University of Waterloo, Ontario

Ramlal PS, Kling GW, Ndawula LM, Hecky RE, Kling HJ (2001) Diurnal fluctuations on PCO_2, DIC, oxygen and nutrients at inshore sites in Lake Victoria, Uganda. In: Munawar M, Hecky RE (eds) The Great Lakes of the World (GLOW): food-web, health and integrity. Michigan State University Press, East Lansing, pp 67–82

Ravenstein E (1901) The lake-level of the Victoria Nyanza. Geogr J 18:403–406

Reynolds JE, Gréboval DF, Mannini P (1995) Thirty years on: the development of the Nile perch fishery in Lake Victoria. In: Pitcher TJ, Hart PJB (eds) The impact of species changes in African Lakes (Chapman and Hall Fish and fisheries series 18). Springer, Dordrecht, pp 181–214 van Rijssel JC, Hecky RE, Kishe-Machumu MA, Meijer SE, Pols J, Tienderen KM, Ververs JD, Wanink JH, Witte F (2016a) Climatic variability in combination with eutrophication drives adaptive responses in the gills of Lake Victoria cichlids. Oecologia 182:1187–1201

van Rijssel JC, Hecky RE, Kishe-Machumu MA, Witte F (2016b) Changing ecology of Lake Victoria cichlids and their environment: evidence from C^{13} and N^{15} analyses. Hydrobiologia 791:175–191

Romero JR, Ryan R, Atenucci JP, Attwater G, Ewing T, Feaver S, Imberger J, Khisa P, Lam C, Njuguna H, Shimizu K (2006) Management implications of the physical limnological studies of Rusinga Channel and Winam Gulf in Lake Victoria. In: Odada EO, Olago DO, Ochola W, Ntiba M, Wandiga S, Gichuki N, Oyieke H (eds) Proc 11th World Lakes conf, vol 2, Nairobi, pp 63–68

Ruhe RV (1954) Erosion surfaces of central Africa interior high plateaus. Inst Nat L'Etude Agron Congo Belge 59:1–38

Rzoska J (1957) Notes on the crustacean plankton of Lake Victoria. Biol J Linn Soc 168:116–125

Salzburger W, Van Bocxlaer B, Cohen AS (2014) Ecology and evolution of the African Great Lakes and their faunas. Annu Rev Ecol Evol Syst 45:519–545

Sars GO (1909) Zoological results of the third Tanganyika expedition, conducted by Dr. W.A. Cunnington 1904–1905. Report on the Copepoda. Proc Zool Soc Lond 79:31–77

Schmidle W (1902) Beiträge zur Flora von Afrika. XXIV. Das Chlorophyceenplankton des Nyassa und einiger anderer innerafrikanischer Seen. Botan Jb 33:1–33

Scholz CA, Rosendahl BR, Versfelt JW, Rach N (1990) Results of high-resolution echo-sounding of Lake Victoria. J Afr Earth Sci 11:25–32

Scholz CA, Johnson TC, Cattaneo MH, Shana S (1998) Initial results of 1995 IDEAL seismic reflection survey of Lake Victoria, Uganda and Tanzania. In: Lehman JT (ed) Environmental change and response in East African lakes. Springer, Dordrecht, pp 47–57

Seehausen O (2000) Explosive speciation rates and unusual species richness in haplochromine cichlid fishes: effects of sexual selection. Adv Ecol Res 31:237–274

Seehausen O (2002) Patterns in fish radiation are compatible with Pleistocene desiccation of Lake Victoria and 14,600 year history for its cichlid species flock. Proc R Soc Ser B Biol Sci 269:491–497

Seehausen O, van Alphen JJM, Witte F (1997a) Cichlid fish diversity threatened by eutrophication that curbs sexual selection. Science 277:1808–1811

Seehausen O, Witte F, Katunzi EF, Smits J, Bouton N (1997b) Patterns of the remnant cichlid fauna in southern Lake Victoria. Conserv Biol 11:890–904

Seehausen O, Koetsier E, Schneider MV, Chapman LJ, Chapman CA, Knight ME, Turner GF, van Alphen JJ, Bills R (2003) Nuclear markers reveal unexpected genetic variation and a Congolese-Nilotic origin of the Lake Victoria cichlid species flock. Proc R Soc Ser B Biol Sci 270:129–137

Seka AM, Zhang J, Ayele GT, Demeke YG, Han J, Prodhan FA (2022) Spatio-temporal analysis of water storage variation and temporal correlations in the East Africa lake basins. J Hydrol 41:101094

Sekadende BC, Lyimo TJ, Kurmayer R (2005) Microcystin production by cyanobacteria in the Mwanza Gulf (Lake Victoria, Tanzania). Hydrobiologia 543:299–304

Semyalo R, Nattabi JK, Larsson P (2009) Diel vertical migration of zooplankton in a eutrophic bay of Lake Victoria. Hydrobiologia 635:383–394

Sene KJ, Plinston DT (1994) A review and update of the hydrology of Lake Victoria in East Africa. Hydrol Sci J 39:47–63

Shayo S, Limbu SM (2018) Nutrient release from sediments and biological nitrogen fixation: advancing our understanding of eutrophication sources in Lake Victoria, Tanzania. Lakes Reserv 23:312–323

Silsbe GM, Hecky RE, Guildford SJ, Mugidde R (2006) Variability of chlorophyll *a* and photosynthetic parameters in a nutrient-saturated tropical great lake. Limnol Oceanogr 51:2052–2063

Simiyu BM, Kurmayer R (2022) Response of planktonic diatoms to eutrophication in Nyanza Gulf of Lake Victoria, Kenya. Limnologica 93:125958

Simiyu BM, Amukhuma HS, Sitoki L, Okello W, Kurmayer R (2022) Interannual variability of water quality conditions in the Nyanza Gulf of Lake Victoria, Kenya. J Great Lakes Res 48:97–109

Simonit S, Perrings C (2005) Indirect economic indicators in bio-economic models: agricultural price indicators and fish stocks in Lake Victoria. ICES J Mar Sci 62:483–492

Sitoki L, Gichuki J, Ezekiel C, Wanda F, Mkumbo OC, Marshall BE (2010) The environment of Lake Victoria (East Africa): current status and historical changes. Int Rev Hydrobiol 95:209–223

Song Y, Semazzi FH, Xie L (2002) Development of a coupled regional climate simulation model for the Lake Victoria basin. Int J Climatol 24:57–75

Song Y, Semazzi FH, Xie L, Ogallo LJ (2004) A coupled regional climate model for the Lake Victoria basin of East Africa. In: Odada EO, Olago DO (eds) The East African Great Lakes: limnology, Palaeolimnology and biodiversity. Kluwer, Dordrecht, pp 153–186

Spawls S, Mathews G (2012) Kenya: a natural history. Bloomsbury, London

Speke C (1863) The Nile and its sources. Proc R Geogr Soc Lond 7:217–224

Spigel RH, Coulter GW (1996) Comparison of hydrology and physical limnology of the East African Great Lakes: Tanganyika, Malawi, Victoria, Kivu and Turkana (with reference to some North American Great Lakes). In: Johnson TC, Odada EO (eds) The limnology, climatology and paleoclimatology of the East African Lakes. Gordon and Breach, Amsterdam, pp 103–139

Ssebuggwawo V, Kitamirike JM, Khisa P, Njuguna H, Myanza O, Hecky RE, Mwanuzi FL (2005) Hydraulic/hydrodynamic conditions of Lake Victoria. In: Mwanuzi FL, JOX A, Muyodi FJ, Hecky RE (eds) Lake Victoria environment management project (LVEMP) water quality and ecosystem status. South Eastern Kenya University, Kitui, pp 42–61

Ssemmanda I, Vincens A (2002) Vegetation changes and their climatic implications for the Lake Victoria region during the late Holocene. In: Odada EO, Olago DO (eds) The East African Great Lakes: limnology, palaeolimnology and biodiversity. Kluwer, Dordrecht, pp 509–523

Stager JC (1984a) Environmental changes in Lake Cheshi, Zambia, since 40,000 years BP with additional information from Lake Victoria, East Africa. PhD thesis, Duke University, Durham

Stager JC (1984b) The diatom record of Lake Victoria (East Africa): the last 17,000 years. In: Mann DG (ed) Proc 7 Int diatom symposium, 1982. Otto Koeltz Science, Philadelphia, pp 455–476

Stager JC (1998) Ancient analogues for recent environmental changes at Lake Victoria, East Africa. In: Lehman JT (ed) Environmental change and response in East African Lakes. Kluwer, Dordrecht, pp 37–46

Stager JC, Johnson TC (2000) A 12,400 ^{14}C years offshore diatom record from east central Lake Victoria, East Africa. J Paleolimnol 23:373–383

Stager JC, Johnson TC (2008) The late Pleistocene desiccation of Lake Victoria and the origin of its endemic biota. Hydrobiologia 596:5–16

Stager JC, Mayewski PA (1997) Abrupt early to mid-Holocene climatic transition registered at the equator and the poles. Science 276:1834–1836

Stager JC, Reinthal PN, Livingstone DA (1986) A 25,000-year history for Lake Victoria, East Africa, and some comments on its significance for the evolution of cichlid fishes. Freshw Biol 16:15–19

Stager JC, Cumming B, Meeker L (1997) A high resolution 11,400-yr diatom record from Lake Victoria, East Africa. Quat Res 47:81–89

Stager JC, Mayewski PA, Meeker LD (2002) Cooling cycles, Heinrich event 1, and the desiccation of Lake Victoria. Palaeogeogr Palaeoclimatol Palaeoecol 183:169–178

Stager JC, Cumming BF, Meeker LD (2003) A 10,000-year high-resolution diatom record from Pilkington Bay, Lake Victoria, East Africa. Quat Res 59:172–181

Stager J, Ryves D, Cumming B, Meeker L, Beer J (2005) Solar variability and the levels of Lake Victoria, East Africa, during the last millenium. J Paleolimnol 33:243–251

Stager JC, Ruzmaikin A, Conway D, Verburg P, Mason PJ (2007) Sunspots, El Niño, and the levels of Lake Victoria, East Africa. J Geophys Res 112:D15106

Stager JC, Hecky RE, Grzesik D, Cumming BF, Kling H (2009) Diatom evidence for the timing and causes of eutrophication in Lake Victoria, East Africa. Hydrobiologia 636:463–478

Stager JC, Ryves DB, Chase BM, Pausata FS (2011) Catastrophic drought in the Afro-Asian monsoon region during Heinrich event 1. Science 331:1299–1302

Stanley H (1875) Letters of Mr. HM Stanley on his journey to Victoria Nyanza, and circumnavigation of the lake. Proc R Geogr Soc Lond 20:134–159

Stanley H (1877) A geographical sketch of the Nile and Livingstone (Congo) basins. Proc R Geogr Soc Lond 22:382–410

Stanley H (1889) Letter from Mr. HM Stanley, on his journey from the Albert Nyanza to the southern side of Victoria Nyanza. Proc R Geogr Soc Month Rec Geogr 11:720–726

Stewart KM (1997) Fossil fish from the Manonga Valley, Tanzania. In: Harrison T (ed) Neogene palaeontology of the Manonga Valley, Tanzania (topics in geobiology 14). Springer, New York, pp 333–350

Sutcliffe JV, Parks YP (1999) The hydrology of the Nile. Int Assoc Hydrol Sci Spec Publ 5:1–179

Talbot MR (2002) Nitrogen isotopes in palaeolimnology. In: Last WM, Smol JP (eds) Tracking environmental change using lake sediments (Developments in paleoenvironmental research vol 2). Springer, Dordrecht, pp 401–439

Talbot MR, Lærdal T (2000) The Late Pleistocene–Holocene palaeolimnology of Lake Victoria, East Africa, based upon elemental and isotopic analyses of sedimentary organic matter. J Paleolimnol 23:141–164

Talbot MR, Livingstone DA (1989) Hydrogen index and carbon isotopes of lacustrine organic matter as lake level indicators. Palaeogeogr Palaeoclimatol Palaeoecol 70:121–137

Talbot MR, Williams MAJ (2009) Cenozoic evolution of the Nile Basin. In: Dumont HJ (ed) The Nile: origin, environments, limnology and human use. Springer, Dordrecht, pp 37–60

Talbot MR, Williams MAJ, Adamson DA (2000) Strontium isotopic evidence for Late Pleistocene re-establishment of an integrated Nile drainage network. Geology 28:343–346

Talbot MR, Jensen NB, Lærdal T, Filippi ML (2006) Geochemical responses to a major transgression in giant African Lakes. J Paleolimnol 35:467–489

Talling J (1957) Some observations on the stratification of Lake Victoria. Limnol Oceanogr 2:213–221

Talling JF (1964) The annual cycle of stratification and phytoplankton growth in Lake Victoria (East Africa). Int Rev ges Hydrobiol 50:421–463

Talling JF (1965) The photosynthetic activity of phytoplankton in East African lakes. Int Rev Ges Hydrobiol 50:1–32

Talling JF (1966) The annual cycle of stratification and phytoplankton growth in Lake Victoria (East Africa). Int Rev Ges Hydrobiol Hydrogr 51:545–621

Talling JF (1987) The phytoplankton of Lake Victoria (East Africa). Arch Hydrobiol Beih Ergebn Limnol 25:229–256

Talling JF (2006) A brief history of the scientific study of tropical African inland waters. Freshw Forum 26:3–37

Talling JF, Talling IB (1965) The chemical composition of African lake waters. Int Rev Ges Hydrobiol Hydrogr 50:421–463

Tamatamah RA, Hecky RE, Duthie HC (2005) The atmospheric deposition of phosphorus in Lake Victoria (East Africa). Biogeochemistry 73:325–344

Tate E, Sutcliffe J, Conway D, Farquharson F (2004) Water balance of Lake Victoria: update to 2000 and climate change modelling to 2100. Hydrol Sci J 49:563–574

Temple PH (1964a) Raised features along the southern shoreline of Lake Victoria. E Afr Acad Proc 1:13–22

Temple PH (1964b) Evidence of lake-level changes from the northern shoreline of Lake Victoria, Uganda. In: Steel RW, Prothero RM (eds) Geographers and the tropics, Liverpool essays. Longman, London, pp 31–56

Temple PH (1966) Evidence of changes in the level of Lake Victoria and their significance. PhD thesis, University of London

Temple PH (1969) Raised strandlines and shoreline evolution in the area of Lake Nabugabo, Masaka District, Uganda. In: Wright HE (ed) Quaternary geology and climate. Proc VII Cong Int Ass Quat Res, vol 16, National Academy of Sciences, Washington DC, pp 119–129

Temple PH (1971) The lakes of Uganda. In: Ominde SH (ed) Studies in East African geography and development. University of California Press, Oakland, pp 86–98

Temple PH, Doornkamp JC (1970) Influences controlling lacustrine overlap along the northwestern margins of Lake Victoria. Z Geomorphol 14:301–317

Terer T, Muasya AM, Triest L (2015) Strong isolation by distance revealed among *Cyperus papyrus* populations in the Rift Valley lakes, Lake Victoria, and isolated wetlands of Kenya. Aquat Bot 121:57–66

Thenya T (2006) Analysis of macrophyte biomass productivity, utilization and its impact on various eco-types of Yala Swamp, Lake Victoria Basin, Kenya. PhD thesis, University of Nairobi

Thomson J (1884) Through the Masai country to Victoria Nyanza. Proc R Geogr Soc Month Rec Geogr 6:690–712

Tong X, Pan H, Xie H, Xu X, Li F, Chen L, Luo X, Liu S, Chen P, Jin Y (2016) Estimating water volume variations in Lake Victoria over the past 22 years using multi-mission altimetry and remotely sensed images. Remote Sens Environ 187:400–413

Trapnell CG, Langdale-Brown I (1972) Natural vegetation. In: Morgan WTW (ed) East Africa: its peoples and resources. University of Oxford Press, Oxford, pp 127–140

Tumwebaze R, Cowx I, Ridgway S, Getabu A, MacLennan DN (2007) Spatial and temporal changes in the distribution of *Rastrineobola argentea* in Lake Victoria. Aquat Ecosys Health Manag 10:398–406

Tungaraza C, Eliapenda E, Osewe KO, Palapala PM (2012) Long-term climate impact on the Lake Victoria region influences water level fluctuation and resource availability. Int J Environ Sci 2:1717–1732

Twesigye CK, Onywere SM, Getenga ZM, Mwakalila SS, Nakiranda JK (2011) The impact of land use activities on vegetation cover and water quality in the Lake Victoria watershed. Open Environ Eng J 4:66–77

Twongo T (1996) Growing impact of water hyacinth on nearshore environments on lakes Victoria and Kyoga (East Africa). In: Johnson TC, Odada EO (eds) The limnology, climatology and paleoclimatology of the East African Lakes. Gordon and Breach, Amsterdam, pp 633–642

Vanderkelen I, van Lipzig NPM, Thiery W (2018a) Modelling the water balance of Lake Victoria (East Africa) – part 1: observational analysis. Hydrol Earth Syst Sci 22:5509–5525

Vanderkelen I, van Lipzig NPM, Thiery W (2018b) Modelling the water balance of Lake Victoria (East Africa) – part 2: future projections. Hydrol Earth Syst Sci 22:5527–5549

Verschuren D, Russell JM (2009) Paleolimnology of African lakes: beyond the exploration phase. J Geophys Res 113:112–116

Verschuren D, Edgington DN, Kling HJ, Johnson TC (1998) Silica depletion in Lake Victoria: sedimentary signals at offshore stations. J Great Lakes Res 24:118–130

Verschuren D, Johnson TC, Kling HJ, Edgington DN, Leavitt PR, Brown ET, Talbot MR, Hecky RE (2002) History and timing of human impact on Lake Victoria, East Africa. Proc R Soc Ser B Biol Sc 269:289–294

Virieux J (1913) Plancton du lac Victoria Nyanza. Voyage de Ch. Alluaud et R. Jeannel en Afrique orientale (1911–1912). Resultats Scientifiques. Albert Schulz, Paris

Wakefield T, Johnston K (1870) Routes of native caravans from the coast to the interior of Eastern Africa, chiefly from information given by Sádi Bin Ahédi, a native of a district near Gázi, in Udigo, a little north of Zanzibar. J R Geogr Soc Lond 40:303–339

Wanink JH (1999) Prospects for the fishery on the small pelagic *Rastrineobola argentea* in Lake Victoria. Hydrobiologia 407:183–189

Waya R (2004) Diel vertical migration of zooplankton in the Tanzanian waters of Lake Victoria. Tanz J Sci 30:123–134

Waya R, Mwambungu JA (2004) Zooplankton communities of selected stations of Lake Victoria. Tanz J Sci 30:11–20

Wayland E (1930) The African Bulge. Geogr J 75:381–383

Wayland E (1931) The Rift Valley and Uganda waterways. In 'Summary of Progress 1919–1929', Geol Surv Uganda, pp 40–44

Wayland EJ (1933) Peneplains and some other erosional platforms. Geol Surv Uganda Annu Rep:77–79

West GS (1907) Report on the freshwater algae, including phytoplankton, of the third Tanganyika expedition conducted by Dr W.A. Cunnington, 1901–1905. Bot J Linn Soc 38:81–197

White F (1983) The vegetation map of Africa, a descriptive memoir to accompany the UNESCO/ AETFAT/UNSO vegetation map. UNESCO, Paris

Williams AE, Duthie HC, Hecky RE (2005) Water hyacinth in Lake Victoria: why did it vanish so quickly and will it return? Aquat Bot 81:300–314

Williams AE, Hecky RE, Duthie HC (2007) Water hyacinth decline across Lake Victoria—was it caused by climatic perturbation or biological control? A reply. Aquat Bot 87:94–96

Willoughby NG, Watson IG, Twongo T (1996) Preliminary studies on the effects of water hyacinth on the diversity, abundance and ecology of the littoral fish fauna in Lake Victoria Uganda. In: Johnson TC, Odada E (eds) Limnology, climatography and paleolimnology of the East African Lakes. Gordon and Breach, Amsterdam, pp 643–655

Wilson JRU, Ajuonu O, Center TD, Hill MP, Julien MH, Katagira FF, Neuenschwander P, Njoka SW, Ogwang J, Reeder RH, Van T (2007) The decline of water hyacinth on Lake Victoria was due to biological control by *Neochetina* spp. Aquat Bot 87:90–93

Witte F, Goldschmidt T, Wanink J, van Oijen M, Goudswaard K, Witte-Maas E, Bouton M (1992) The destruction of an endemic species flock: quantitative data on the decline of the haplochromine cichlids of Lake Victoria. Environ Biol Fish 34:1–28

Witte F, Msuku BS, Wanink JH, Seehausen O, Katunzi EFB, Goudswaard PC, Goldschmidt T (2000) Recovery of cichlid species in Lake Victoria: an examination of factors leading to differential extinction. Rev Fish Biol Fisher 10:233–241

Witte F, Wanink J, Kishe-Machumu M (2007a) Species distinction and the biodiversity crisis in Lake Victoria. Trans Am Fish Soc 136:1146–1159

Witte F, Wanink JH, Kishe-Machumu M, Mkumbo OC, Goudswaard PC, Seehausen O (2007b) Differential decline and recovery of haplochromine trophic groups in the Mwanza Gulf of Lake Victoria. Aquat Ecosys Health Manag 10:416–433

Worthington EB (1930) Observation on the temperature, hydrogen-ion concentration and other physical conditions in the Victoria and Albert Nyanzas. Int Rev Hydrobiol Hydrogr 24:328–357

Worthington EB (1931) Vertical movements of freshwater macroplankton. Int Rev Hydrobiol 25:394–436

Yasindi AW (2001) The ecology of planktonic ciliates in tropical lakes of East Africa. PhD thesis, University of Waterloo, Ontario

Yin X, Nicholson SE (1998) The water balance of Lake Victoria. Hydrol Sci J 43:789–811

Yin XG, Nicholson SE (2000) On the diurnal cycle of cloudiness over Lake Victoria and its influence on evaporation from the lake. Hydrol Sci J 45:407–424

Yin XG, Nicholson SE (2002) Interpreting annual rainfall from the levels of Lake Victoria. J Hydrometeorol 3:406–416

Yongo E, Outa N, Kito K, Matsushita Y (2017) Some aspects of the biology of Nile perch, *Lates niloticus*, in the open waters of Lake Victoria, Kenya. Lakes Res 389:272–276

Yousef SM, Amer M, Aty MA (2000) The sharp rise of Lake Victoria, a positive indicator to solar Wolf-Gleissberg cycles turning points. In: International conference on environmental hazards mitigation. Cairo University, Cairo, pp 222–234

van Zwieten PAM, Kolding J, Plank MJ, Hecky RE, Bridgeman TB, MacIntyre S, Seehausen O, Silsbe GM, Giacomini H (2016) The Nile perch invasion in Lake Victoria: cause or consequence of the haplochromine decline? Can J Fish Aquat Sci 73:622–643

Chapter 14
The Lesser-Known Lakes and Wetlands of the Kenya Rift

14.1 Introduction

Wetland environments are transitional between aquatic and terrestrial settings where the water table lies either at or just below the land surface, or in areas where the land surface is occasionally submerged by shallow water to form small lakes that may be permanent or intermittent (Roggeri 1995). The Ramsar Convention defines wetlands as "areas of marsh, fen, peatland or water, whether natural or artificial, permanent or temporary, with water that is static or flowing, fresh, brackish or salt . . ." (Dugan 1990).

In addition to the large lakes of the Kenya Rift already discussed, many small and shallow lakes and wetlands lie within the axial rift depression; others are perched in structural basins, such as back-tilted fault blocks, above the main rift floor. Some small lakes and wetlands are fed by rivers or groundwater and are normally perennial. Others are ephemeral and rely upon seasonal runoff, a high water table or both sources, but become dry seasonally and (or) during drought, or when the water table falls for other reasons (e.g. neotectonics or anthropogenic causes). Such small lakes and wetlands may be fresh or saline depending on the source and composition of inflow, the presence or absence of hydrological closure, and the local precipitation-to-evaporation ratio. Some wetlands are natural, whereas others have resulted from, or been modified by, anthropogenic changes to the natural drainage. Crafter et al. (1992) and MEMR (2012) provide useful overviews of the wetlands in Kenya including those in the Kenya Rift Valley and Nyanza (Kavirondo) Rift.

Gichuki et al. (2001) estimated that small lakes and wetlands cover ~2.5% of Kenya's land surface but noted their decline because of the increased use of surface and ground water for irrigation. Where present, small lakes and wetlands provide refugia for wildlife and livestock and stop-over sites for migrating birds. Many pastoral communities depend on them for water, particularly in the drier regions of the rift. The types of wetlands and small lakes vary with their physical location in the landscape (Table 14.1).

R. W. Renaut, R. B. Owen, *The Kenya Rift Lakes: Modern and Ancient*, Syntheses in Limnogeology, https://doi.org/10.1007/978-3-642-25055-2_14

Table 14.1 Common landscape settings in which wetlands develop in the Kenya Rift Valley

Landscape setting	Typical topography and type of wetlands	Kenya Rift examples
Low lying areas prone to periodic flooding by runoff	River valleys and alluvial floodplains	Ewaso Ngiro River (lower course)
Gently sloping areas with groundwater discharge or seepage	Seepage slopes and springs	Discharge zones in toes of alluvial fans or above aquicludes: NW Suguta Valley Emuruangogolak springs
Structural depressions bordered by uplands	Lakes, ponds, pools, and seasonally waterlogged ground – with open or closed drainage	Lake Solai Lake Ol' Bolossat
Playas and other plains without drainage outlets	Seasonal playa lakes and closed depressions fed by runoff and (or) groundwater	Lake Kwenia Lake Logipi Lake Alablab
Flat and sloping areas adjacent to lakes and rivers	Fringing marshes and seepage wetlands, delta plains	Kesubo Swamp Lake Kichirtit
Hot springs located along fault lines	Thermal wetlands fed mainly by hot springs, especially in ephemeral lake settings	Loburu, Lake Bogoria Lorusio near Kapedo

Modified from Nyamweru (1992), Gichuki et al. (2001) and Tiner (2017). Many wetlands develop through multiple causes

14.2 Central Island Lakes, Lake Turkana

Three young volcanic islands lie along the axial part of Lake Turkana (Fig. 6.3b). North Island has no lakes. South Island, the largest (~40 km^2) of the group, has a few small saline lagoons along its southeastern shoreline and on a small island (Enwoiti) 100 m east of the main island. In contrast, Central Island (9 km^2), which rises up to 170 m above the lake surface, has three lakes that occupy former tuff cones and tuff rings (Figs. 14.1 and 14.2) (Karson and Curtis 1994; Dunkley et al. 1993). Rouwet et al. (2021) suggested that they formed by phreatomagmatic (maar) eruptions, but most pyroclastic deposits are from airfall and the lavas appear to be subaerial without subaqueous indicators such as pillow lavas (Karson and Curtis 1994).

Central Island is composed mainly of bedded lapilli tuffs, ashes, breccias and small lava flows. Those volcanic rocks are mainly basalts, basanites and phonolitic tephrites (Karson and Curtis 1994; Bloomer et al. 1989; Wilkinson 1988). Seismic reflection data confirm that other volcanic structures lie below the lake bed near the island (Dunkelman et al. 1989). Though young, the age range of the volcanic rocks and volcaniclastic sediments is unknown. Karson and Curtis (1994) suggested that the exposed rocks are of Upper Pleistocene and Holocene age. Wave-cut palaeo-shoreline features (+370–377 m) record former higher lake levels.

Worthington (1932) was the first to illustrate, map and describe the lakes. He labelled the three crater lakes, A, B and C. Those lakes each have different

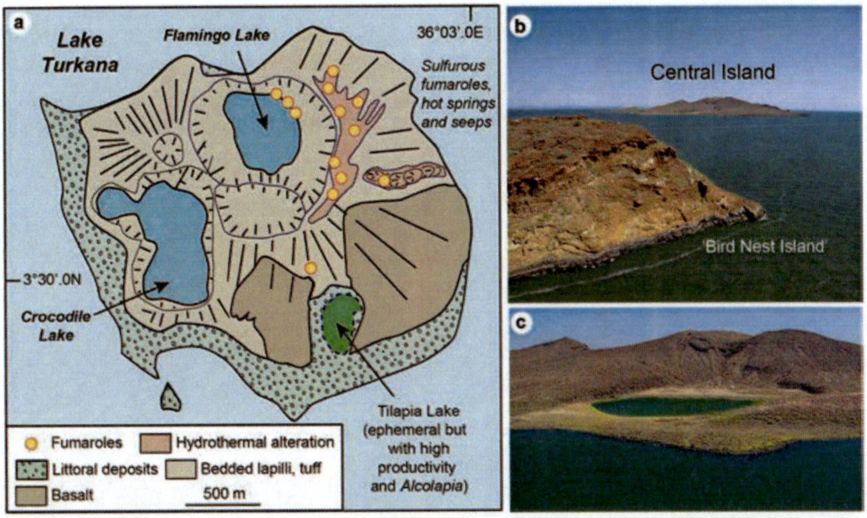

Fig. 14.1 Central Island, Lake Turkana. (**a**) Simplified geological map, based on Dunkley et al. (1993) and original observations. (**b**) Central Island and outlying 'Bird's Nest Island' to the south. (**c**) Tilapia Lake

characteristics and fauna and were renamed Crocodile Lake (A), Flamingo Lake (B) and Tilapia Lake (C) (Karson and Curtis, 1994). Crocodile Lake (0.33 km^2) on the western margin fills a depression formed of three craters and is separated from Lake Turkana by a narrow (~170 m wide) ridge up to ~6 m high. Flamingo Lake (0.16 km^2) occupies another composite crater and is separated from Lake Turkana by a steep ridge ~150 m wide. Alkaline hot springs (up to 70.5 °C: Dunkley et al. 1993) discharge along its northeastern shoreline. Tilapia Lake (0.06 km^2) occupies a single crater (Karson and Curtis 1994) that opens southwards towards Lake Turkana across a ridge about 2 m above the main lake. The Central Island lakes lie at approximately the same level as Lake Turkana and are up to 80 m deep (Dunkley et al. 1993) but little is known about their physical properties. Winds are strong at Lake Turkana (Ferguson and Harbott 1982; Achiba 2019) but the steep sides of the crater lakes reduce their impact on lake water mixing.

The waters are saline and alkaline and of Na^+–HCO_3^-–CO_3^{2-} composition. Worthington (1932) noted that the salinity and alkalinity of the crater lakes exceeded that of Lake Turkana. In June 2019, Crocodile Lake had a pH of 9.31, and a temperature of 31.3 °C; Flamingo Lake had a pH of 9.53, and a temperature of 37.3 °C; and Tilapia Lake had a pH of 9.76 and a temperature of 29.5 °C. Salinity was highest in Tilapia Lake and lowest in Crocodile Lake. Krienitz (2018) noted that the salinity of Flamingo Lake (15‰) was ten times higher than Lake Turkana.

Beadle (1932a, his Table 15) analysed water from Crocodile Lake and suggested that the latter is "not merely evaporated lake [Turkana] water". The Na/Cl ratio (mg L^{-1}) of Lake Turkana in 1931 was 1.79 whereas the ratio in Crocodile Lake was 1.63. He also reported high concentrations of K^+, Ca^{2+} and Mg^{2+} relative to

Fig. 14.2 Central Island lakes, Lake Turkana. (**a, b**) Crocodile Lake. (**c, d**) Flamingo Lake. H: area of hydrothermal alteration. (**e, f**) Tilapia Lake

contemporary Na^+ and Cl^- in Lake Turkana. Beadle (1932a) also noted the high concentrations of Ca^{2+} and especially Mg^{2+} compared to other rift lakes. The high Mg/Ca ratio (2.77) is exceptional for alkaline Kenya Rift lake waters but high Mg has been reported from other waters in southern Lake Turkana including at Loiyangalani (Dunkley et al. 1993). Beadle (1932a) suggested that the scarcity of phytoplankton maintained a lower pH than Lake Turkana enabling alkaline earths to remain in solution. Dunkley et al. (1993) concluded that the Central Island hot springs (Flamingo Lake) originate from the mixing of Lake Turkana water with deeper geothermal fluids.

The lake waters are almost certainly fed by Lake Turkana in the subsurface, but their long-term behaviour is unclear. Tilapia Lake and Crocodile Lake would flood during a relatively modest rise in the level of Lake Turkana. A fall in the surface height of Lake Turkana would eventually lead to desiccation (Krienitz 2018) although a shrunken Flamingo Lake might survive if hot springs continue to flow—much like Lake Logipi in the Suguta Valley (Chap. 7).

Central Island was established as a National Park in 1983–5, then later as a UNESCO World Heritage Site (Lake Turkana National Parks) based on its crocodile population and natural beauty. Crocodile Lake was home to the largest breeding colony of Nile crocodile although it has been under stress from fishermen who have destroyed their nests (IUCN 2020). The flamingo population is relatively small, but many other birds inhabit the island (Fuchs 1935).

Beadle (1932a, b) examined the plankton in the lakes. He found sparse diatoms (*Cyclotella*, *Rhopalodia*) in Crocodile Lake, but Flamingo Lake and Tilapia Lake had high numbers of *Arthrospira fusiformis* (=*platensis*). Zooplankton in Crocodile Lake included copepods (*Diaptomus*), Cladocera and rotifers. Rotifers were the only zooplankton in the other two alkaline lakes.

14.3 North Barrier Lakes

Small alkaline lakes are present along the northern shore of the Barrier, a rugged Quaternary volcanic complex that separates Lake Turkana from the Suguta Valley to the south. Lake Logipi, a saline alkaline lake partly fed by hot-springs, lies on the southern margin of the Barrier (Chap. 7). The volcanic rocks and geothermal features of the Barrier have been mapped and described by Dodson (1963), Williams et al. (1984), Dunkley et al. (1993) and Auko (2011).

Along the northern shoreline, there are several small, but interesting, lakes. Abili Agituk (Fig. 14.3a, b) is a faulted tuff cone breached on its eastern margin. Its crater lake is sometimes directly linked to Lake Turkana; at other times it is separated by a physical sediment barrier (Fig. 14.3b) produced by longshore drift or falling lake level(s) but is still recharged by seepage from Lake Turkana. When separated from Lake Turkana, the salinity of the crater lake water increases and becomes greener reflecting microbial blooms. To the east, the floor of the neighbouring Nabuyatom cone is usually dry (Fig. 14.3c), but a small wetland rarely forms on the eastern crater floor fed by rainfall, radial drainage inside the crater and seepage. Both Abili Agituk and Nabuyatom show evidence (wave-cut notches) of high-level Holocene palaeoshorelines of Lake Turkana (Chap. 6).

Southeast of the Nabuyatom cone a small lake is present (Nabuyatom Lake; Fig. 14.3a, d). This shallow (<5 m deep?) green alkaline lake lies upon a young mugearite lava flow from Teleki's cone dated at ~1895 AD (Cavendish 1898), and in places is compartmentalised into smaller lakes separated by metre-scale-high ridges (Fig. 14.3d, e). Champion (1935) suggested that those ridges (his Plate 11: Lotarr) were crater rims submerged by rising lake level, but the scale makes this unlikely. They appear to be pressure ridges in the late-nineteenth-century lava flow. Satellite imagery shows that those volcanic rocks have white rims of efflorescent

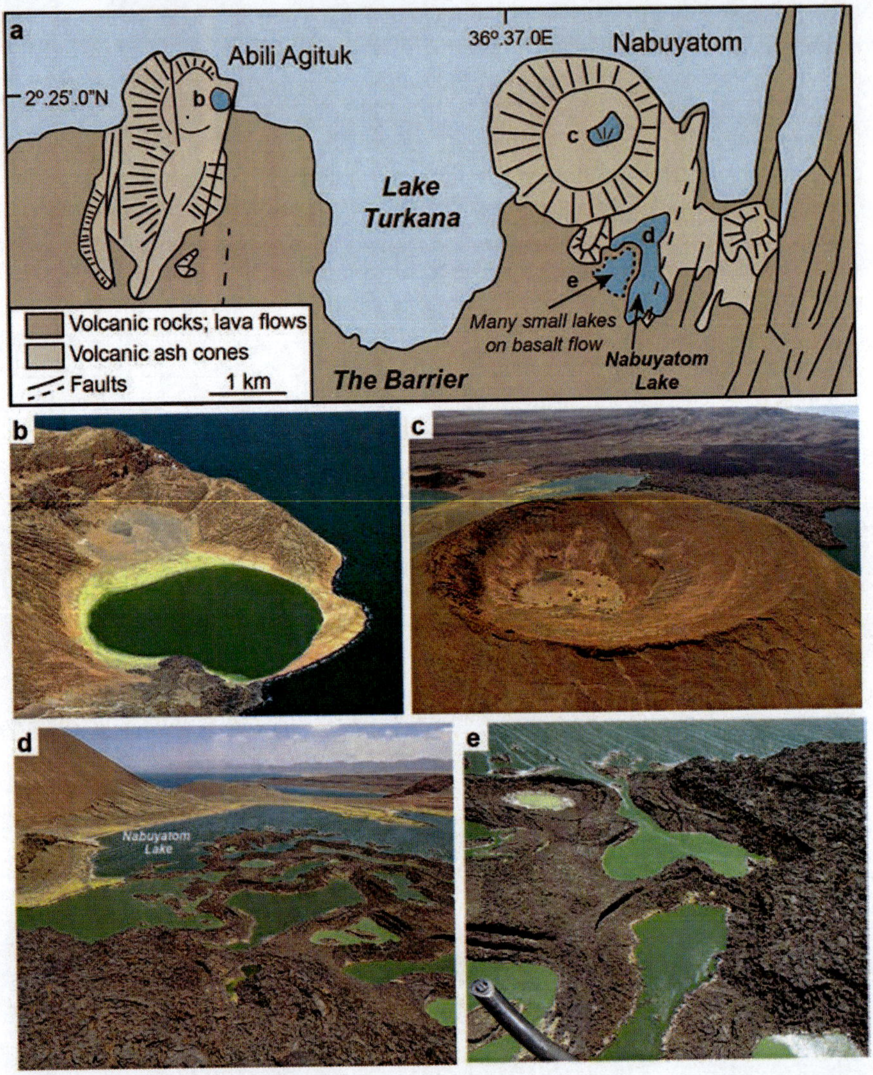

Fig. 14.3 North Barrier lakes. (**a**) Geological setting of the lakes. (**b**) Saline alkaline Abili Agituk lake when cut off from Lake Turkana by a narrow sandy barrier. (**c**) Nabuyatom crater, which only rarely contains a wetland. (**d**) Nabuyatom Lake which lies on a recent mugearite flow. (**e**) Saline alkaline ponds on mugearite flow at southern edge of Nabuyatom Lake

salt crusts (trona or trona-halite) when the level of Lake Turkana falls a few metres. Although of low importance, these small lakes often have a salinity that exceeds modern Lake Turkana. As such, they are possible refugia for saline alkaline species. When relatively fresh, they are a habitat and a breeding site for Nile crocodiles.

14.4 South Suguta (Kangirinyang) Maar Lakes

Three small maar lakes lie at Kangirinyang at the southeastern edge of the Suguta Valley (Fig. 14.4) The maar lakes lie along a NNE-trending fault and fissure zone north of Emuruangogolak volcano (Dunkley et al. 1993). The tuff rings and maars are composed of basaltic tephra mixed with lake sediment and formed when magma moving along the fissure encountered groundwater or wet sediment.

The largest lake is Kageinya Lake (Fig. 14.5a; 0.02 km^2) which is a bluish green saline alkaline lake bounded by crater walls, with alluvial fans along its southern

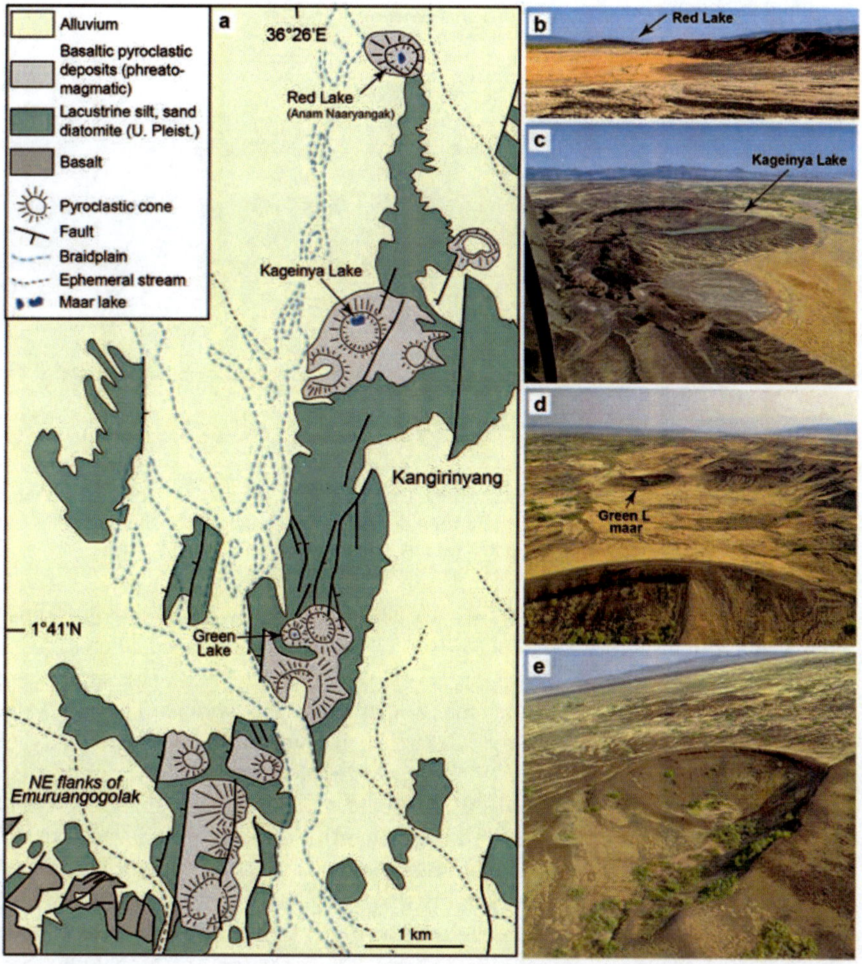

Fig. 14.4 The Kangirinyang maar lakes of the southeastern Lower Suguta. (**a**) Geological map. Based on Dunkley et al. (1993). (**b**) View north showing ridge of maars and location of Red Lake (Anam Naaryangak). (**c**) View south showing Kageinya Lake; Emuruangogolak in distance. (**d**, **e**) Green Lake

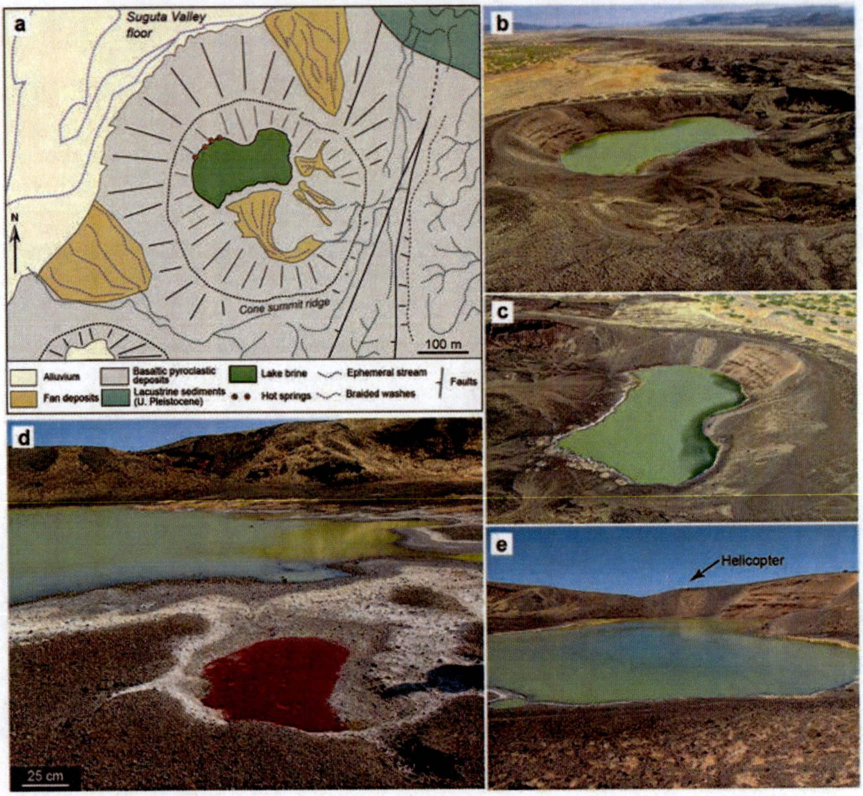

Fig. 14.5 Kageinya Lake. (**a**) Geological setting. (**b**) View north towards Suguta Valley. (**c**) View to southwest, showing dark green zone of spring discharge. (**d**) Southeastern shoreline showing trona efflorescence and red (bacterial) pond. (**e**) Lake from northeastern summit ridge

and southeastern margins (Fig. 14.5a–c, e). The lake is partly fed by springs. Those along the northern shoreline are hot (64–69 °C) and often submerged whereas seeps along the western margins are 43–56 °C (Dunkley et al. 1993). The springs are alkaline and with higher sulfate than many other Kenya Rift springs. The lake water is alkaline (pH 10.13 in February 2018; 9.79 in June 2018) with a salinity of 14–50 g L^{-1} TDS (Dunkley et al. 1993; Grewe 2016, and original measurements). The lake is ~5 m deep (Rohland 2013). Along its southeastern margins several highly alkaline pools are purplish red (Fig. 14.5d). Ponds with similar colour are present along the southern margins of Lake Turkana and eastern South Island, and probably reflect brine with higher salinity than the adjacent water body and with a different microbial community. Little is known about the lake sediments though Grewe (2016) examined aspects of the pore fluid chemistry of a shallow core; Junginger et al. (2019) and Kayalo-Omamo et al. (2023) reported a study using rotifer DNA in shallow sediments to record changes in environmental conditions in the past 200 years. Trona crusts are present in places around the shoreline.

Approximately 2.7 km to the north, Red Lake (Anam Naaryangak) is a small (2140 m^2) maar lake on the western edge of a small crater (Fig. 14.6a–c). The lake water has a pH of 9.23–9.96 and a temperature of 23.8–33.8 °C. Small ambient to warm springs (~30–50 °C; pH 9.45–9.89) discharge along the southwestern and northeastern shoreline (Fig. 14.6d, e). Much of its surface is covered by a pink to red crust of burkeite [$Na_2CO_3(SO_4)_2$] and halite (Dunkley et al. 1993) with local crusts around the northern margin decimetres above lake level. The lake was visited when mostly a dry pan in February 2018 (Fig. 14.6d, f) and a few months later in June 2018 when the crust was submerged by brine but still largely preserved (Fig. 14.6g). Life in the lake is mainly microbial but mammals visit the springs even when the lake is low (e.g. striped hyena). In June 2018 the shoreline was littered with dead amphibians and insects, raising the possibility of asphyxia by toxic gases such as hydrogen sulfide. We smelled sulfurous gases during both our visits, as did Dunkley et al. (1993, p. 76). It is unclear if the gases are geothermal or related to sulfate reduction in the lake. Patches of carbonate pavement are present along the northeastern shoreline (Fig. 14.6h).

Dunkley et al. (1993) mentioned 'Green Lake', a deep crater lake 3.1 km south of Kageinya Lake (Fig. 14.4d, e). Its waters are ambient (24.4 °C) with a pH of 8.5. They reported a waterhole with dilute (pH 6.5) water slightly above the lake, implying little or no direct geothermal influence.

14.5 Spring-Fed Lakes and Wetlands, Emuruangogolak to Silali

Several ambient (cool) springs issue from below the front of a young permeable basalt flow on the northern edge of Emuruangogolak volcano (Figs. 14.7 and 14.8). Dunkley et al. (1993) called them the South Nangarabat springs (SN in Fig. 14.7). Nangarabat, a small (6.6 x 2.9 km) faulted block of basalt with low mesa-like topography, also has warm springs (38 °C; NS in Fig. 14.7) along its southern margin. The South Nangarabat springs discharge into a lush green wetland inhabited by fish (*Tilapia*), crocodiles, molluscs and other aquatic invertebrates, and birds. The largest pools and small lakes have leathery microbial mats along their margins but these are not mineralised (Fig. 14.8). The margins of more distal spring-fed pools locally have trona-encrusted margins. The wetlands extend up to 2 km N-NW from the springs. Flow towards the Suguta River is perennial.

The spring waters are cool and fresh to alkaline (T: 27.9 °C and pH: 9.41 in February 2018; T: 34.3 °C and pH 9.35 in June 2019). Dunkley et al. (1993, their site 197) reported a temperature of 31.1 °C and a pH of 9.15. Using geochemical and stable isotope analyses, they suggested that these springs and those at Kalnang'i (see below) have compositions that imply an origin in groundwater from Lake Baringo (~30%) mixed with water from the rift margins or local volcanic centres.

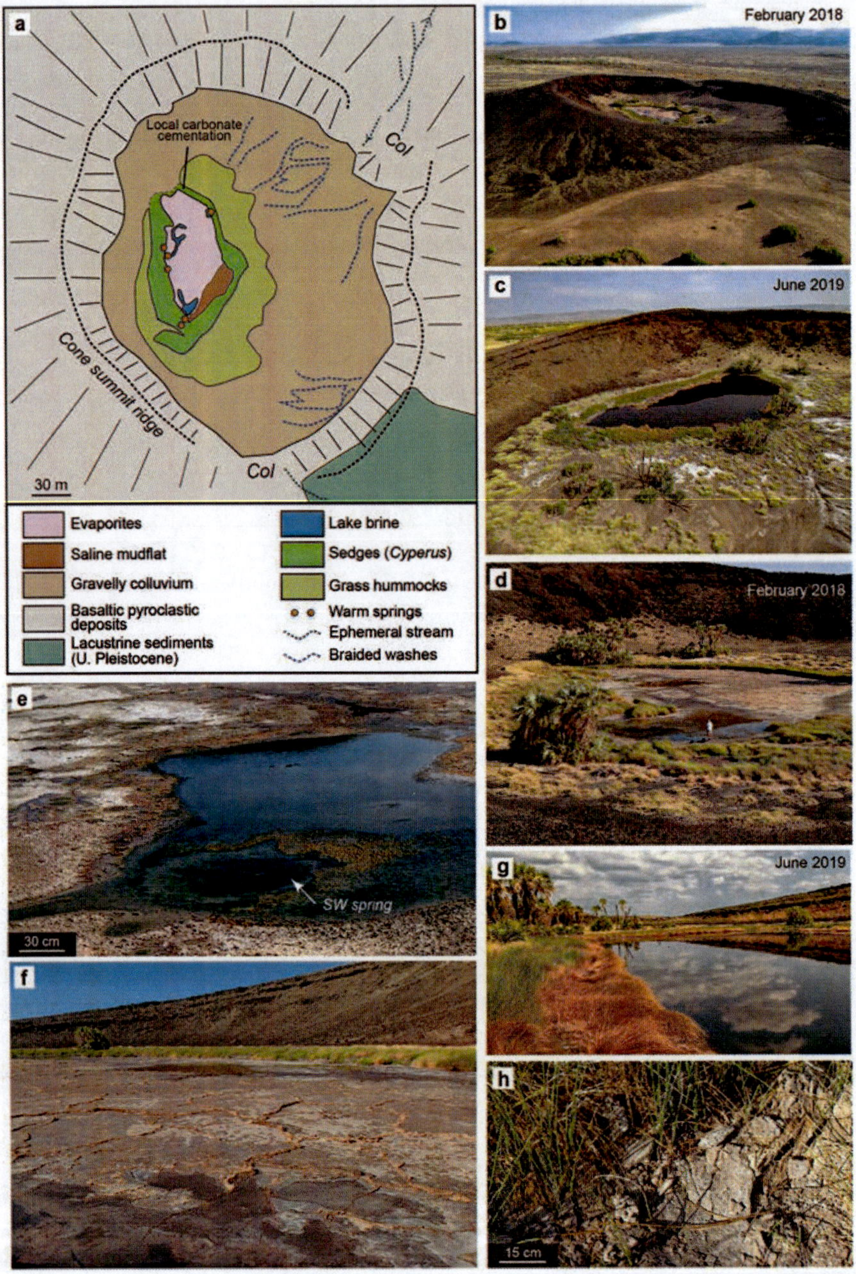

Fig. 14.6 Red Lake (Anam Naaryangak). (**a**) Geological setting. (**b**) Lake with pink evaporite crusts (February 2018). (**c**) Flooded lake in June 2019. (**d**) Lake in February 2018 showing spring discharge and evaporites. (**e**) Spring zone, southwestern shoreline. (**f**) Polygonal pressure ridges in evaporites, eastern shoreline. (**g**) Shallow (<30 cm) brine in June 2019. (**h**) Carbonate pavement along northern shoreline

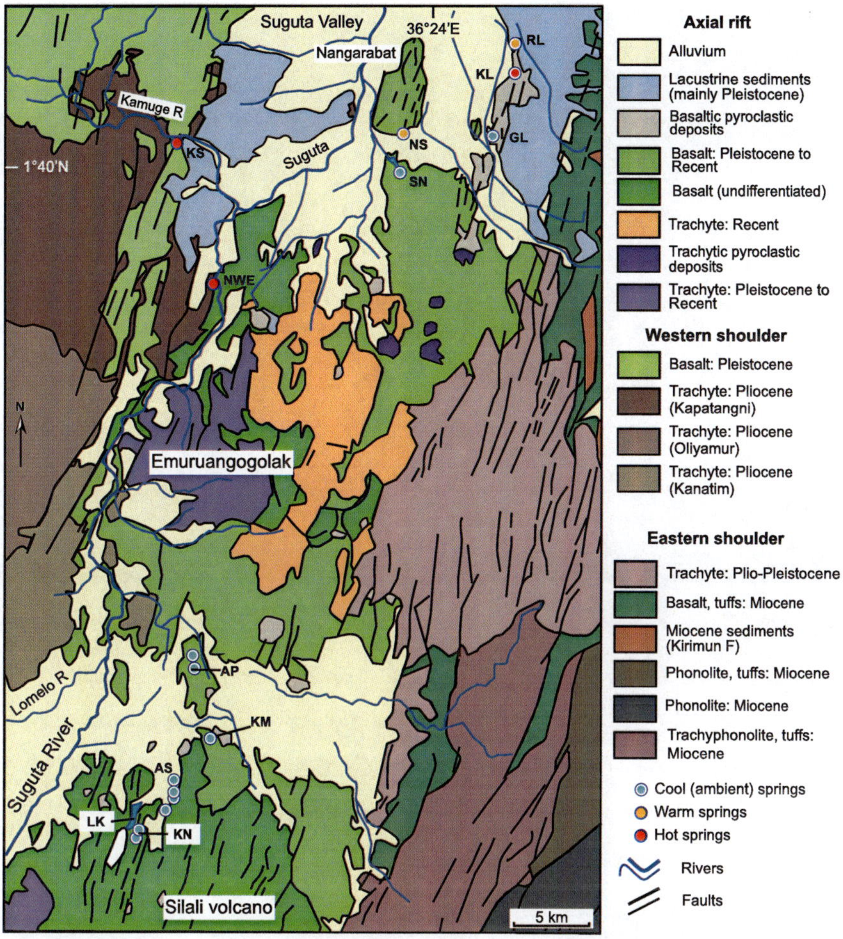

AP Amakat Pools
AS Akiloset springs
ES Kamuge hot springs
GL Green Lake
KL Kageinya Lake
KN Kalnang'i springs
KS Kamuge hot springs
KM Kalomngi spring
LK Lake Kalnang'i
NS Nangarabat springs
NWE NW Emuruangogolak springs
RL Red Lake (Anam Naaryangak)
SN South Nangarabat springs

Fig. 14.7 Geological map of the northern Emuruangogolak-south Suguta region, showing main sites of lakes and spring-fed wetlands. Adapted from maps by Truckle (1977) and Dunkley et al. (1993) with original observations

Several small spring-fed wetlands lie on the southwestern margin of Emuruangogolak and the northwestern margin of Silali volcano (Figs. 14.7 and 14.9). Some springs flow directly into the Suguta River, but others flow into seasonal ephemeral lakes. Several flow towards Loreng and then westwards to the Suguta River. Four areas of wetland are recognised: Amakat Pools, Kalomngi, Akiloset, and Kalnang'i springs and Lake Kalnang'i (Dunkley et al. 1993). The cool and warm springs discharge along faults and fractures in rugged basaltic terrain and are often ponded in small lakes upon former flow surfaces, implying a shallow water table (Fig. 14.9b) where the lavas overlie a less permeable substrate.

Fig. 14.8 Spring-fed wetlands between Nangarabat and northern Emuruangogolak. (**a**, **b**) Fresh springs discharge at the base of a basalt flow overlying older flows along the northern flanks of Emuruangogolak, forming perennial ponds and small lakes. (**c**, **d**) Dense reedbeds in spring outflow

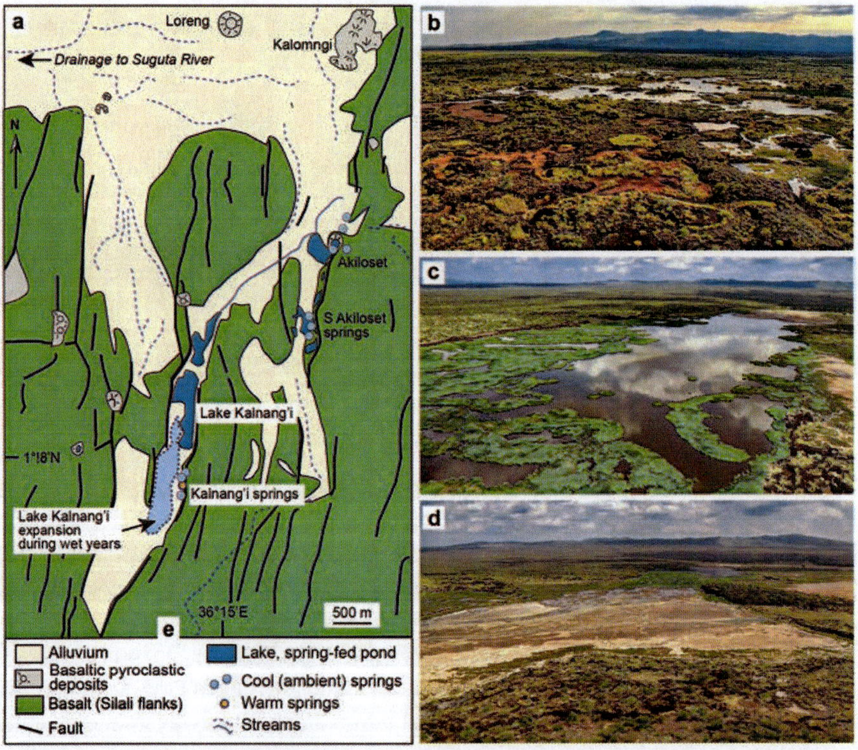

Fig. 14.9 Springs and lakes of the Kalnang'i region. (**a**) Geological setting. (**b**) Akiloset wetlands—shallow lakes on lava flow. (**c**) Lake Kalnang'i. (**d**) Southern extension of Lake Kalnang'i during floods

Lake Kalnang'i and smaller lakes to the north receive inflow from Kalnang'i springs. Lake Kalnang'i occupies a narrow graben and expands seasonally on mudflats to the south (Fig. 14.9c, d). Reedbeds are common and the lake water is commonly green or pink because of microbial growth. Dunkley et al. (1993) report abundant *Tilapia*, crocodiles, flamingos and wading birds. Springs to the northeast at Akiloset range in temperature from ambient to 38.2 °C. Several small local craters have green lakes.

14.6 Spring-Fed Lakes and Wetlands of the Silali-Karosi Region

Hot springs discharge into thermal wetlands from two sites west of Silali caldera: Lorusio and Kapedo (Dunkley et al. 1993; Mutonga 2012; Auko and Kipngok 2021). Lorusio lies 9 km north of Kapedo village (Fig. 14.10a). About 15 hot springs discharge onto a gently sloping alluvial plain about 5 km wide that lies between the rift shoulder and Silali central volcano (Fig. 14.10b). The springs flow into alkaline marshy reedbeds and sedges with *Cyperus laevigatus* and small saline ponds. The water moves eastwards to the Suguta River but much of it evaporates *en route*. The springs discharge Na^{+}–HCO_3^{-}–Cl^{-} waters with a vent pH of 7.5–7.8, increasing in alkalinity downstream (Renaut et al. 1999). Temperatures range from 40 to 82 °C. Travertine, opaline silica and trona precipitate from the spring waters near the vent (Fig. 14.10c, d; Jones and Renaut 1996, 1998).

Kapedo hot springs are the source of the Suguta River (Dunkley et al. 1993). The spring site consists of two groups that extend over 5 km on the eastern side of the Suguta River (Fig. 14.10e, f). Near Kapedo village, four springs with a temperature of ~45–55 °C rise at Adoketabolun and flow westwards before tumbling over a flat scarp forming hot waterfalls. Many small springs discharge in the vicinity locally forming a thermal marsh. Flow rates are very high throughout the year (cumulatively ~1000 L sec^{-1}: Burgess 1986). The waters are of Na^{+}–HCO_3^{-} type, with low Ca^{2+} and Mg^{2+}. Old travertine deposits are locally present.

A remote ephemeral lake, called Lake Tilam or Lake Lokori, lies east of Paka volcano (Dunkley et al. 1993; 0°51′ N and 36°7′ E). We have not visited the lake but from satellite imagery, it appears to be an ephemeral lake (fresh to brackish) in a shallow depression fed by seasonal streams and shallow groundwater in a region of volcanic rocks. The name Lokori is also used for a region in southwestern Turkana (Morgan 1971; Truckle 1977) unrelated to Lake Tilam.

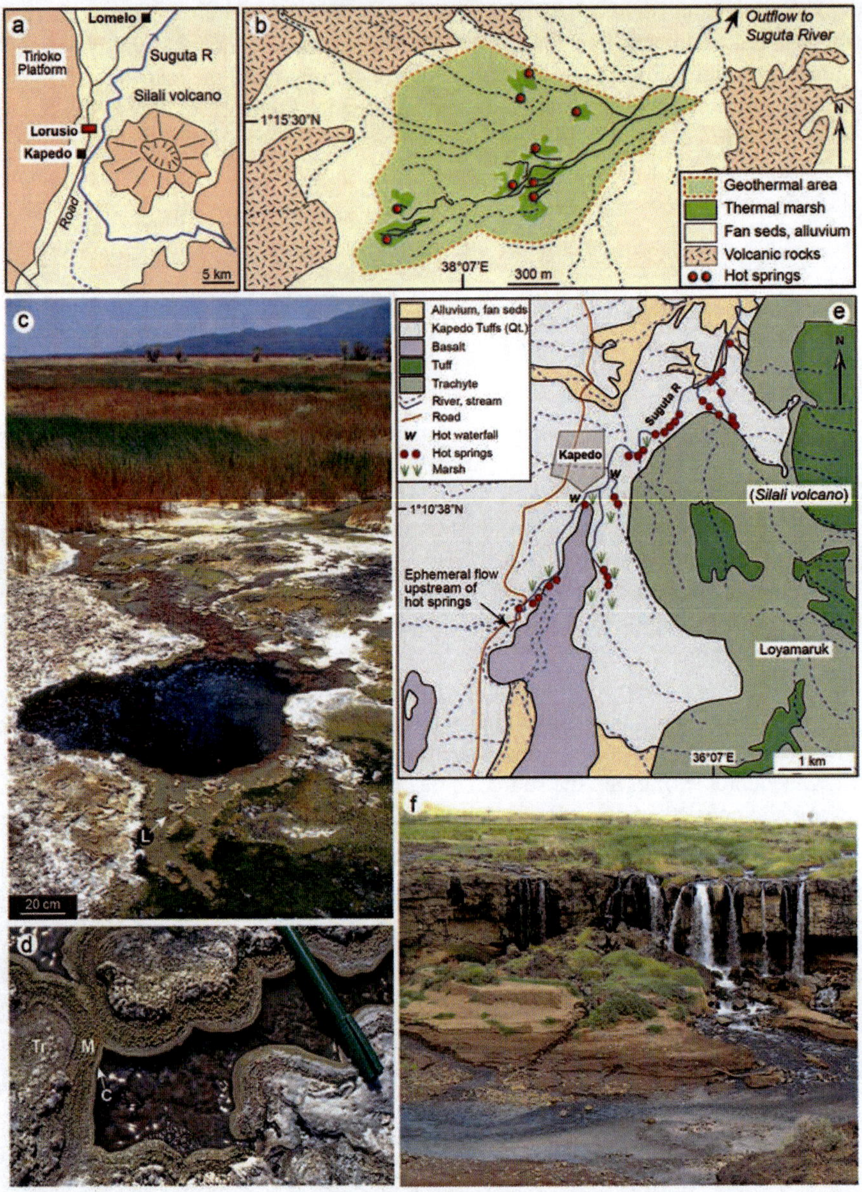

Fig. 14.10 Lorusio and Kapedo geothermal wetlands near Silali. (**a**) Location of the wetlands. (**b**) Plan of the Lorusio wetland locality. (**c**) Saline hot spring emerging and discharging into reedbeds. Deposits around the vent are travertine, opaline silica and trona. L shows the location of carbonate lilypads in (**d**). (**d**) Travertine lilypads forming at the air-water interface. *C*: platy calcite; M: Microstromatolites of opal-A and calcite; Tr: Efflorescent trona (see Renaut et al. 1999 for details). (**e**) Geological setting of the Kapedo hot springs. (**f**) Hot waterfall at Kapedo

14.7 Lake Kichirtit ('Lake 94')

Lake Kichirtit (Kichirtith), informally called 'Lake 94', is a shallow fresh lake ~5 km south of Lake Baringo and is probably the youngest lake in the Kenya Rift to appear on published maps (Fig. 14.11a, b). During the night of 26 May 1994, the Molo River overflowed its banks and changed its course during heavy rains (William Kimosop, pers. comm, 1995). The river, supplemented by runoff from rivers from the eastern rift margin, flooded the Lematacho depression inundating farmland and forming a new lake. Since that avulsion, the lake has frequently expanded and shrunk in both area and volume but remained a lake.

Tarras-Wahlberg et al. (2004) first described the lake, six years after it formed, using data collected in 2000 and 2001. They suggested that the lake was "undisturbed and pristine" but even then the Molo River was draining a basin modified by agriculture and with high soil erosion. The lake flooded an agricultural area (eastern margins of Perkerra Irrigation Scheme) that was not pristine. They provided details of the plankton and compared the trace metal concentrations with those of Lake Baringo and Lake Naivasha but standard hydrochemical data were not given. Water hyacinth has since been reported from the lake and has been flowing downstream into Lake Baringo. When we sampled the lake in June 1995, the water was dilute, slightly alkaline (pH 7.1) and with an ionic composition similar to the adjacent (<200 m) Molo River.

Since its formation, Lake Kichirtit has changed in surface area (Fig. 14.11c–f) but remained hydrologically open and fresh. Several streams flow into the lake including the Molo River which is normally perennial, and that water discharges northwards into Lake Baringo through the migrating channel network of the Molo delta. The residence time of Lake Kichirtit water is unrecorded, but must be brief, especially during the rainy seasons. Its surface level, area and depth respond to seasonal and longer-term changes in rainfall. Recently (2018-) the level of Lake Kichirtit has fluctuated only marginally (<1 m?) above that of Lake Baringo, which is the local base level. Changes in the level of Lake Baringo, therefore, affect Lake Kichirtit. If Lake Baringo continues to rise, as happened in 2011–2021, Lake Kichirtit will be absorbed and submerged, changing the floodplain hydrology. In contrast, a fall in the level of Lake Baringo would modify its upstream drainage and sediment discharge, with channel incision, and lower the water table across the Loboi Plain. Entrenchment, in turn, might affect streams now recharging Lake Kichirtit. With a prolonged arid period, Lake Baringo would fall, and Lake Kichirtit could desiccate or become a muddy temporary swamp like Lake Kamnarok (see below), controlled mainly by fluctuations in the level of the contemporary water table.

Lake Kichirtit is receiving runoff and suspended fine sediments from its inflowing streams – mainly the Molo River. At the same time, Lake Baringo is rising and might incorporate the small lake. Its long-term existence is uncertain. Lake levels

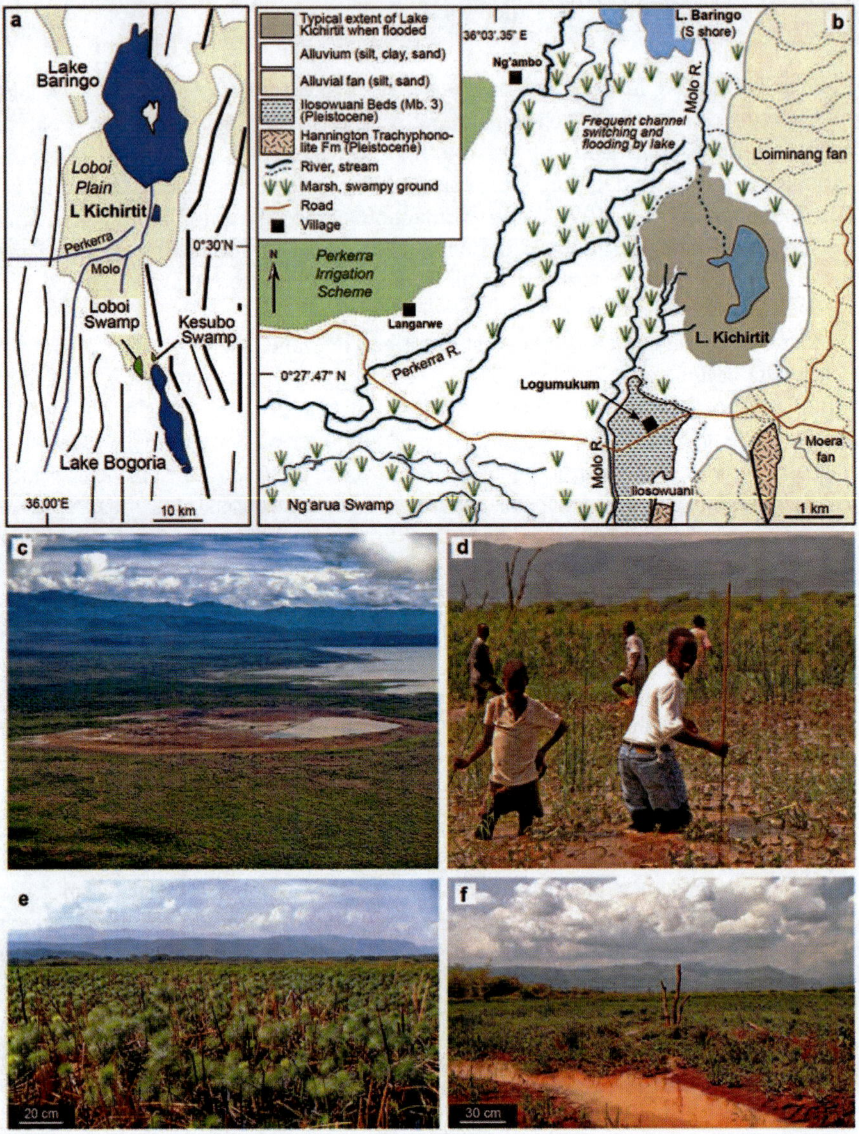

Fig. 14.11 Lake Kichirtit ('Lake 94'). (**a**) Location. (**b**) Geological setting of the lake. (**c**) Lake at relatively low stage during June 2019 when lakes Baringo and Bogoria were both high, possibly indicating that the main river inflow was bypassing the lake. (**d**) Margins of the lake in June 1995, 1 year after it formed. Laikipia Escarpment in the background. (**e**) *Papyrus* swamp near the centre of the lake, June 1995. (**f**) Swamp margins, June 1996

and the impact of climate change will determine its future, but anthropogenic changes in the catchment, including irrigation, and deforestation (in the Mau Highlands and locally) will impact its long-term development. Precursors of Lake Kichirtit might have formed and been lost on the Loboi Plain many times during the last few millennia.

14.8 South Baringo Wetlands

The Loboi Swamp and Kesubo (Kisibor) Marsh lie to the north of Lake Bogoria (Fig. 14.12). Both are threatened by expanding agriculture and competing water usage. The Loboi Swamp formed ~700 years ago. Driese et al. (2004) and Ashley et al. (2004) commented on a possible tectonic origin for the wetland but favoured a climate link with increased rains and spring discharge leading to ponding and the extensive growth of aquatic macrophytes, an environmental change that they correlated with the transition from the Medieval Warm Period (800–1270 AD) to the Little Ice Age (1270–1850 AD). Driese et al. (2004) also noted that the Kesubo Marsh to the east (Fig. 14.12a) may have formed following differential subsidence on the southern part of the Sandai Fan that resulted in the diversion of the Sandai River towards Lake Bogoria (Renaut 1982, 1993). Alternatively, it may reflect an avulsion event, perhaps triggered by climate-related floods, close to the apex of the Sandai Fan that resulted in flow switching from Lake Baringo to Lake Bogoria.

The 1.5 km^2 Loboi Swamp (3 km long; 0.3–0.5 km wide) lies in a westward tilted fault-block, at an elevation of about 1000 m, about 1 km northwest of the Loboi Plain and Lake Bogoria drainage divide (Fig. 14.12a–c) (Ashley et al. 2004). It is bounded to the west and south by Pleistocene Hannington Trachyphonolites, which form the Bogoria Plateau and are cut by sub-parallel N-S-trending faults (Griffiths and Gibson 1980). Pleistocene and Holocene channel and floodplain gravel, sand, silt and clay constitute the low-gradient Loboi Plain to the east and north (Griffiths 1977; Renaut and Owen 1987). Drainage through the swamp eventually reaches Lake Baringo to the north. The local Endoreis community use *Papyrus* from the swamp for cattle fodder, roofing materials, and cooking fuel (Terer et al. 2012).

Mean annual temperatures range between 23 and 25 °C with ~700 mm of precipitation per year and potential evaporation of >2500 mm (LaVigne and Ashley 2001). Rainfall maxima occur in April with a second rainy season in November (Rowntree 1989). Afternoon storms from May to August lead to sheetflow and flashfloods (Ashley et al. 2004). Five- to seven-year El Niño and La Niña rainfall cycles have been reported in the area (LaVigne and Ashley 2001; Johansson and Svensson 2002; Harper et al. 2003).

The Loboi Swamp receives water (pH = 6.4–6.9; Total Dissolved Solids [TDS] = 0.54–0.57 g L^{-1}; Dissolved Oxygen [DO] = 3.5–3.9 mg L^{-1}) from the fault-aligned, Lake Bogoria Hotel and Chelaba warm (35–36 °C) springs to the south, with additional inflow from the smaller Turtle Spring to the west (Ashley et al.

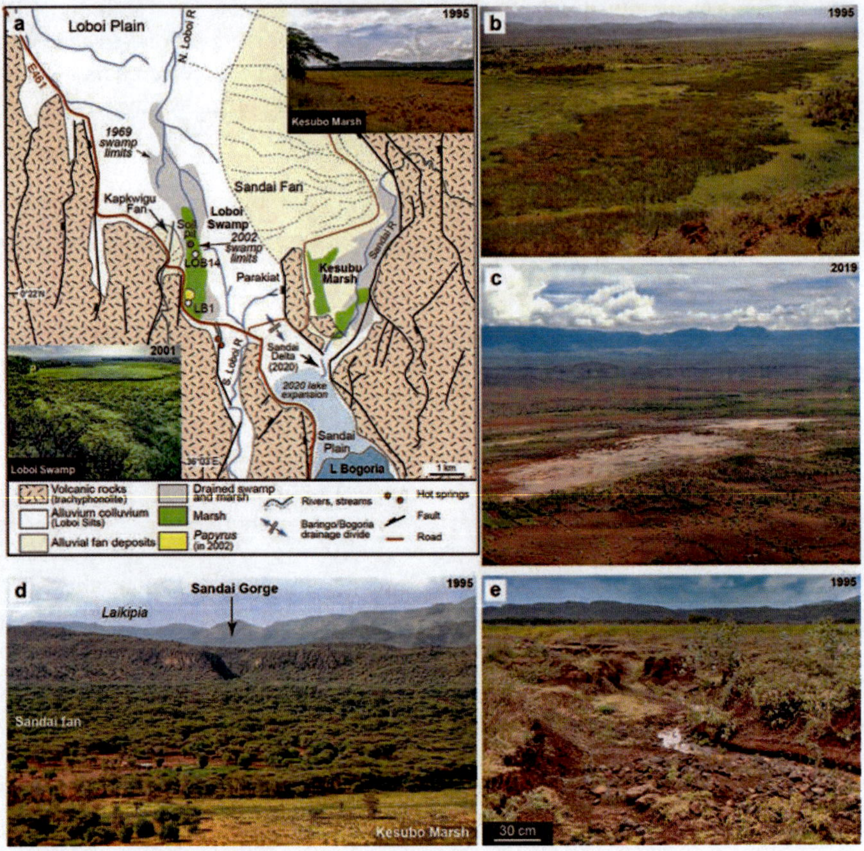

Fig. 14.12 Setting of the Loboi Swamp and Kesubo Marsh. Map modified from Ashley et al. (2004). (**a**) Loboi swamp is located northwest of the Baringo-Bogoria drainage divide at the foot of an east-facing fault-scarp with drainage flowing to Lake Baringo. The 2001 inset shows a view from the adjacent horst block with the central Papyrus zone surrounded by Typha. Kesubo Marsh lies on the Sandai River alluvial fan at a constriction between the Parakiat fault block and faulted lavas to the east. The 1995 inset shows short, waterlogged grasslands. (**b, c**) Show views northwest from Parakiat with agricultural expansion between 1995 and 2019 reducing wetlands in the area. (**d**) View over Kesubo Marsh in 1995 with the wooded Sandai Fan and Gorge beyond. (**e**) Drainage channel on the west side of Kesubo Marsh

2004). These three springs support endemic populations of Nile tilapia (*Oreochromis niloticus*) that are genetically distinct from the same species at Lake Baringo (Nyingi et al. 2009; Ndiwa et al. 2014, 2016). Additional unobserved springs may lie below the swamp itself. Hydrological changes have resulted in a 60% reduction in the swamp area since 1970, reflecting the removal of water for agriculture via a ditch near the Kapkwigu Fan (Fig. 14.12a). The position of the South Loboi River (pH = 8.3; TDS = 0.56 g L^{-1}; DO = 7.3 mg L^{-1}) is also unstable having passed east of the swamp before 2002 and thence to Lake Bogoria. By 2019, the river had switched again and petered out into the plains west of Parakiat (Fig. 14.12a). The poorly aerated (DO = 0.21 mg L^{-1}) fresh waters (pH = 7.4–8.3; TDS = 0.66–1.5 g L^{-1}) of the Loboi Swamp drain northwards via the North Loboi River to Lake Baringo. Ashley et al. (2004) attributed the elevated swamp pH of 8.3, compared with groundwaters (6.4–6.8), to CO_2 degassing and noted that outflows are depleted in DO (4% saturated compared with 60% for spring inflows) due to aerobic decay of organic matter. A modest increase in alkalinity relative to spring inflows was also linked to the input of dissolved CO_2 from organic matter decay with changes in TDS attributed to evapoconcentration.

Von Höhnel (1891) first reported *Papyrus* and reeds (*Typha*?) in the Loboi area in 1888. Today, the central permanently flooded part of the swamp is up to 1.5 m deep and dominated by *Cyperus papyrus* with *Typha domingensis* in adjacent seasonally-flooded areas (Muasya et al. 2004). *Sporobolus spicatus* is also present in a fringe zone around the swamp with *Cyperus laevigatus* around the hot springs (Wetang'ula et al. 2009). Ashley et al. (2004, 2013), Driese et al. (2004), Owen et al. (2004, 2008), Muasya et al. (2004) and Goman et al. (2020) explored various aspects of the surface vegetation and recovered several core and pit samples. Their most detailed stratigraphical record was obtained from a 2-m-long piston core at Site LB1 (Figs. 14.12a and 14.13) which included fluvial silty clay and sandy silt with a weakly developed inceptisol, overlain by wetland organic-rich clay and clayey peat (Fig. 14.13). Organic matter (<5 to 40%), organic carbon (0–20%) and nitrogen (0–2%) all increase upwards whereas magnetic susceptibility decreases, reflecting the proportion of siliciclastics in the sediments.

The LB1 core record indicates two major pollen zones with Poaceae, *Podocarpus*, bushland and herbaceous taxa dominating between 190 and 100 cm (~1540–680 AD) (Fig. 14.13). The upper 100 cm (~680 years ago until present) is associated with increased wetland taxa and fewer montane species as documented by a major decline in *Podocarpus* (Ashley et al. 2004). From about 1850, the full pollen record (Ashley et al. 2004) indicates a decline in *Acacia* (*Vachellia*) and grass, with an increase in herbaceous disturbance indicators, attributed to increased farming activities and associated erosion (Goman et al. 2020). This observation may correlate with increased clay inputs to neighbouring Lake Bogoria and a spike in charcoal input from ~1840 CE (De Cort et al. 2018), which was attributed to regional land clearance. Pollen and non-pollen palynomorphs similarly imply increasing human impacts from about 1850 AD at Lake Baringo (Kiage and Liu 2009).

Diatoms appear at 102 cm depth in LB1 (Fig. 14.13) and continue through to the core top (Owen et al. 2004) with freshwater wetland taxa dominating, including abundant *Diploneis subovalis* and *Rhopalodia gibberula*. Horizons with increased *Nitzschia* probably represent deeper waters than those containing abundant *Rhopalodia–Diploneis*. Gitau and Verschuren (2016) described sediments and aquatic invertebrate remains from a separate core, which included bryozoa, amoebae, acari and dipterans. Subsequently, Goman et al. (2020) reported thirty-five testate amoeba taxa from Loboi surface samples with thirteen genera in three core zones (Fig. 14.13). The most common amoebae were *Arcella* cf. *vulgaris*, *A.* cf. *discoides*, *Centropyxis* cf. *cassis*, *C.* cf. *platystoma* and *Cyclopyxis* cf. *arcelloides*.

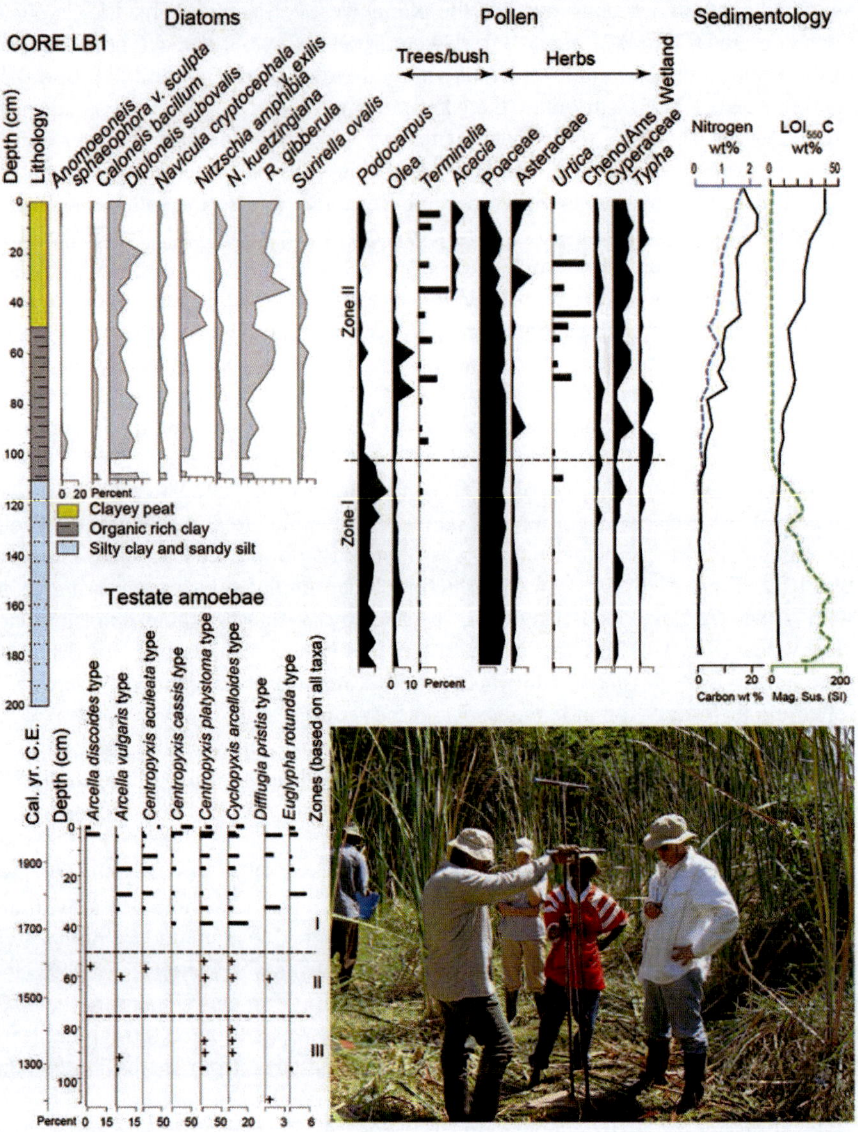

Fig. 14.13 Core LB1 microfossil and sedimentological stratigraphies, Loboi Swamp. Location in Fig. 14.12. Only common diatoms, pollen and testate amoebae are shown. The diatom and testate amoeba records are confined to wetland organic-rich clay and clayey peat in the upper 110 cm. The pollen record extends into the underlying fluvial clayey and sandy silts. Note the increased nitrogen and carbon values in wetland sediments, with magnetic susceptibility increasing in the fluvial deposits. Photograph shows coring at LB1. Compiled from Ashley et al. (2004) and Goman et al. (2020)

In contrast, Kesubo Marsh is a seasonally wet floodplain formed by the slightly alkaline (pH = 8.0; Total alkalinity = 1.85 meq L^{-1}) Sandai River (Fig. 14.12d, e). *Acacia* (*Vachellia*) and C3 scrubland trees and shrubs are present along the river with the western part of the floodplain characterized by open C4 grassland (Driese et al. 2004). The marsh waters are more acidic (pH = 6.5) than those of the river with total alkalinities of 1.03 meq L^{-1} and show lower NO_3^- (0.17 vs. 1.0 mg L^{-1}) and higher SO_4^{2-} (10.85 vs. 5.8 mg L^{-1}) concentrations (Owen et al. 2004). The marshes are also characterised by a diatom flora that is distinct from that of the Loboi Swamp with *Pseudostaurosira brevistriata*, *Eunotia pectinalis*, *Stauroneis anceps*, *Pinnularia acrosphaeria* and *P. maior* present – taxa that are favoured by more acidic conditions.

The importance of environmental information derived from palaeosols is often overlooked in stratigraphical studies. Driese et al. (2004) and Ashley et al. (2013), however, recognised several former soils in the Loboi and Kesubo records (Fig. 14.14). They observed, for example, that average ^{13}C values of −14 to −15‰ PDB for soil organic matter (SOM) in the lower fluvial unit at Loboi Swamp

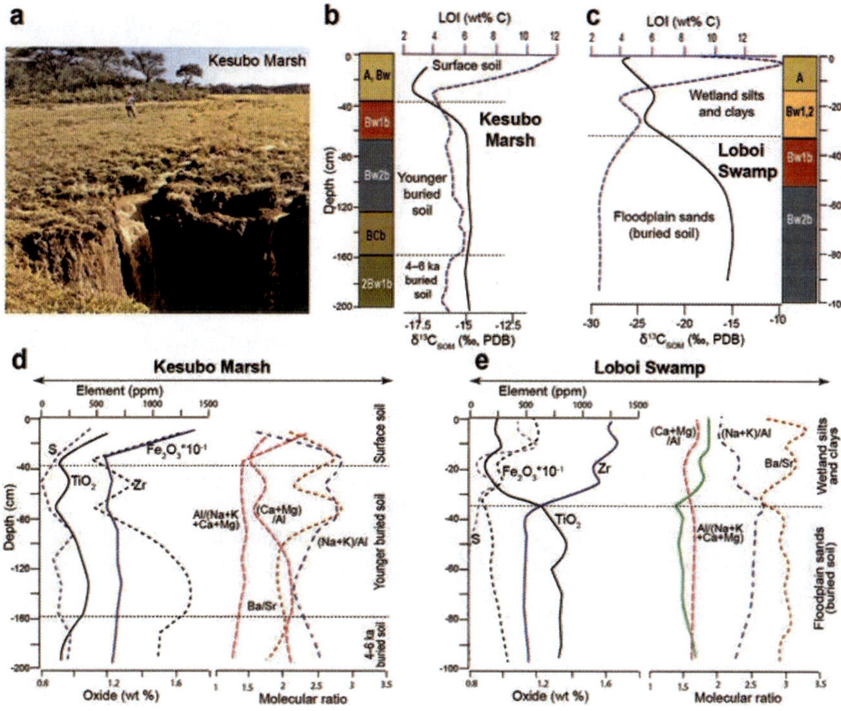

Fig. 14.14 Stable carbon isotopes and element stratigraphies for Loboi Swamp and Kesubo Marsh. Modified after Driese et al. (2004). (**a**) Incised channels into the marsh were used to recover samples. (**b, c**) Soil profiles, LOI (550 °C) and ^{13}C data from Kesubo Marsh and Loboi Swamp. (**d**) Element stratigraphies from Kesubo March. e: Element stratigraphies from Loboi Swamp. $Fe_2O_3*10^{-1}$ and TiO_2 are in wt%; Zr and S are in ppm

indicate that the area was covered by a scrubland mixture of warm-season C4 grasses and C3 shrubs prior to wetland initiation 700 years ago. At Kesubo, they noted only a − 2.5‰ isotope shift from older palaeosols to the surface soil (Fig. 14.14), indicating a slightly greater C3 component. Driese et al. (2004) also provide major and trace element data for Loboi and Kesubo, noting increased Zr in wetland deposits at both sites, which they suggested might be due to surface sheet-flow and (or) aeolian dust. They further reported that high Fe, S and organic C concentrations indicate that Fe and sulfate reduction are occurring in the surface soils at both sites with base element losses, high Ba/Sr ratios and a lack of surface calcification, generally indicating wetter conditions during soil development.

14.9 Lake Kamnarok

Lake Kamnarok is a small freshwater lake in the semi-arid southern Kerio Valley below the Elgeyo Escarpment (Fig. 14.15a–e). The shallow (<2 m deep) lake, which occupies a slight depression on the floodplain east of the Kerio River, is fed mainly by ephemeral streams draining the western dip slope of the Tugen Hills (Kamasia) and flows northwards into the Kerio River. The name Kamnarok, which derives from the Kalenjin word 'Narok', refers to a plant that originally occupied the lake.

The lake had formed before 1961 but major floods from the Kerio River during that exceptionally wet year (Lamb 1966) expanded its area, volume and depth. Although often described as an oxbow lake it lacks classical oxbow morphology and probably formed in a similar manner to Lake Kichirtit. The lake area is from <1 to 2.4 km^2, but it has almost completely dried up in some years. The lake was a swampy wetland before 1900 but then expanded in area following heavy rains and dried up in the 1930s. After it flooded again the lake was colonised by >10,000 Nile crocodiles — the second highest population in Africa after Lake Chad, along with >500 elephants in the neighbouring valley floor and a large bird population, fish (Giddelo et al. 2002; Seegers et al. 2003) and invertebrates. That led to the establishment of the Lake Kamnarok National Reserve in 1983–4. The lake dried up again in 2007–8 and has not recovered. The crocodile population declined substantially and many elephants migrated northwards (Omondi et al. 2002).

The lake has been shrinking rapidly in recent years, related to changes in land use and human activities (Colagiovanni 2002). These include (1) farming and deforestation (charcoal burning and tree felling) in the drainage basin, which have increased soil erosion and lake siltation (Boitt et al. 2020); (2) pastoralists grazing livestock on the former lake bed and the remaining vegetation near the lake; (3) water abstraction, diversion and increased water usage in the catchment (Kimosop 2010); and (4) invasion by water hyacinth (*Eichhornia crassipes*). In addition, short-term climate change has affected the water balance including the flow of the Kerio River. Such factors sometimes have led to serious conflicts among local communities with fresh water a prime concern. Efforts to help the lake to survive have had limited success. These include constructing dykes (gabions) to try to retain lake

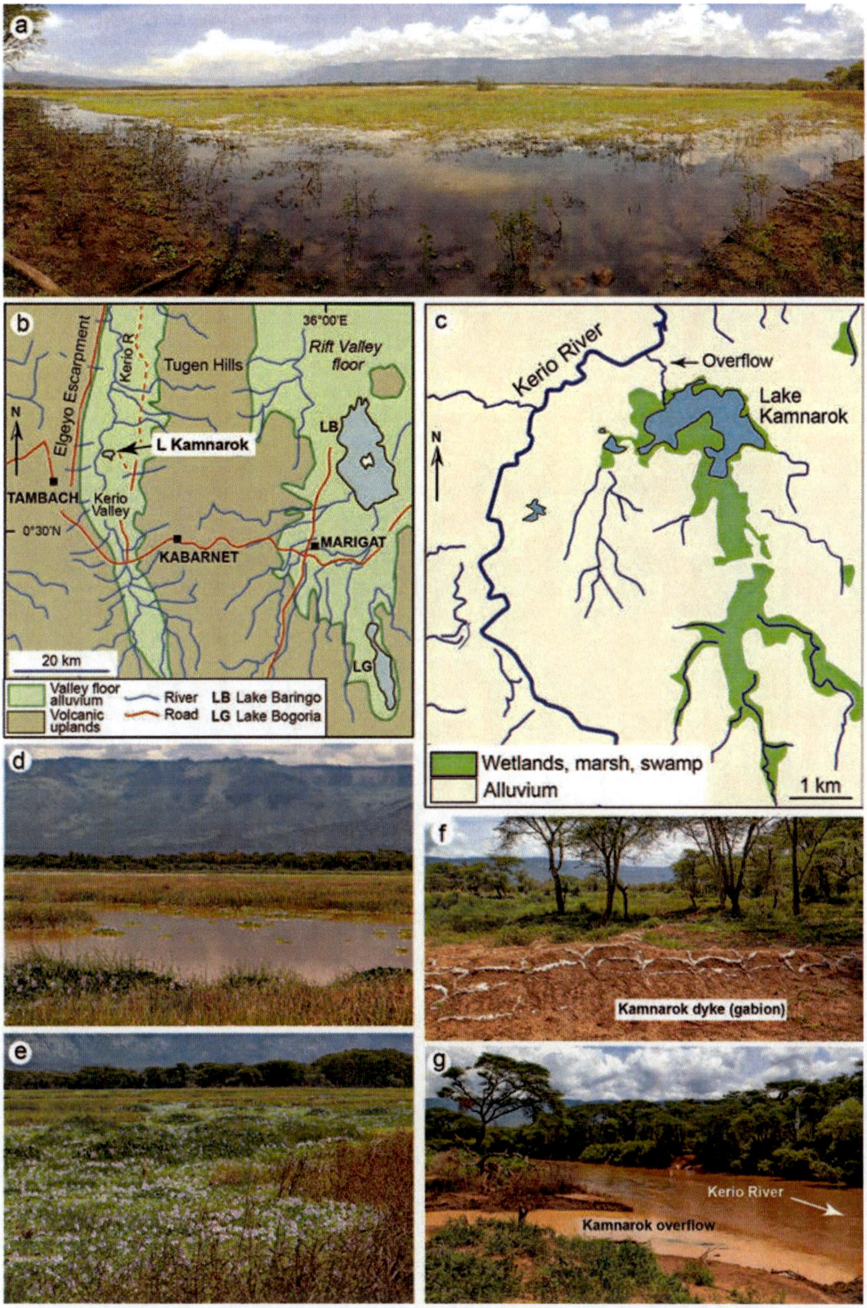

Fig. 14.15 Lake Kamnarok. (**a**) View from eastern shoreline with Elgeyo Escarpment in background (photograph courtesy of Barbara Owen). (**b**) Geological setting of Lake Kamnarok. (**c**) Local setting of the lake. (**d, e**) View across marsh showing patches of open water. (**f**) Dykes constructed at overflow channel. (**g**) Overflow channel to Kerio River

water (Fig. 14.15f) and to control the rate and volume of outflow to the Kerio River (Fig. 14.15g).

14.10 Lake Solai

Lake Solai is a small shallow playa lake in a topographically closed basin ~50 km north of Nakuru and 20 km southeast of Lake Bogoria (Figs. 14.16 and 14.17). The lake occupies an NNE-SSW trending graben at an elevation of ~1512 m. To the east

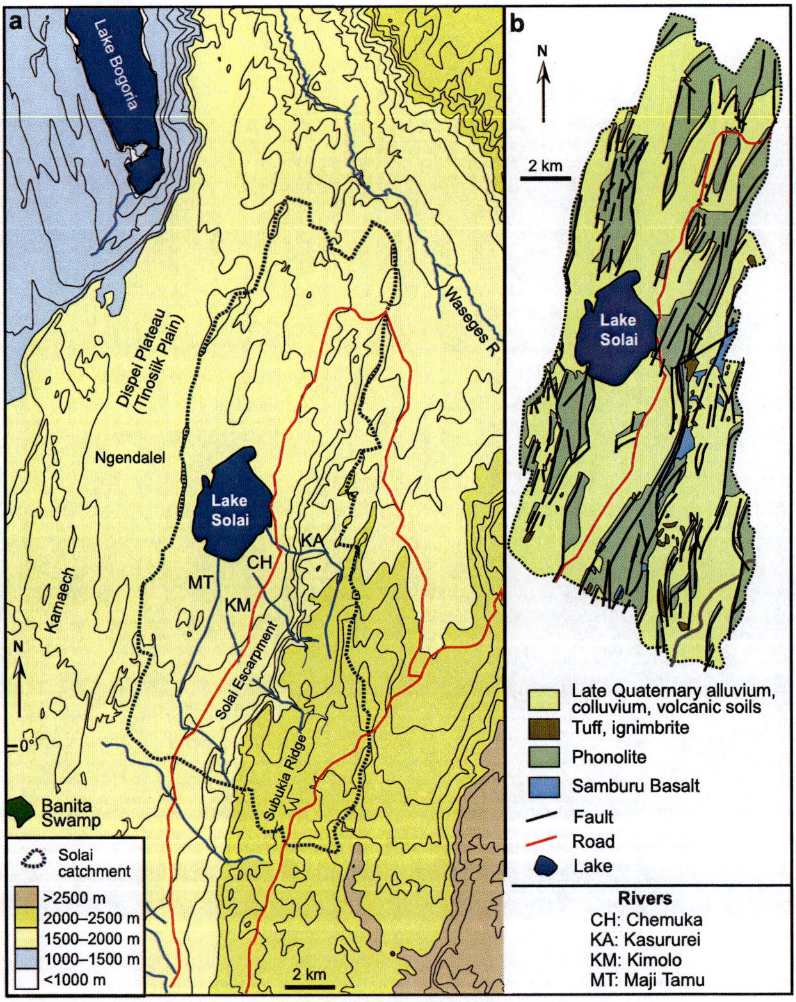

Fig. 14.16 Lake Solai. (**a**) Topography and drainage basin. (**b**) Geology

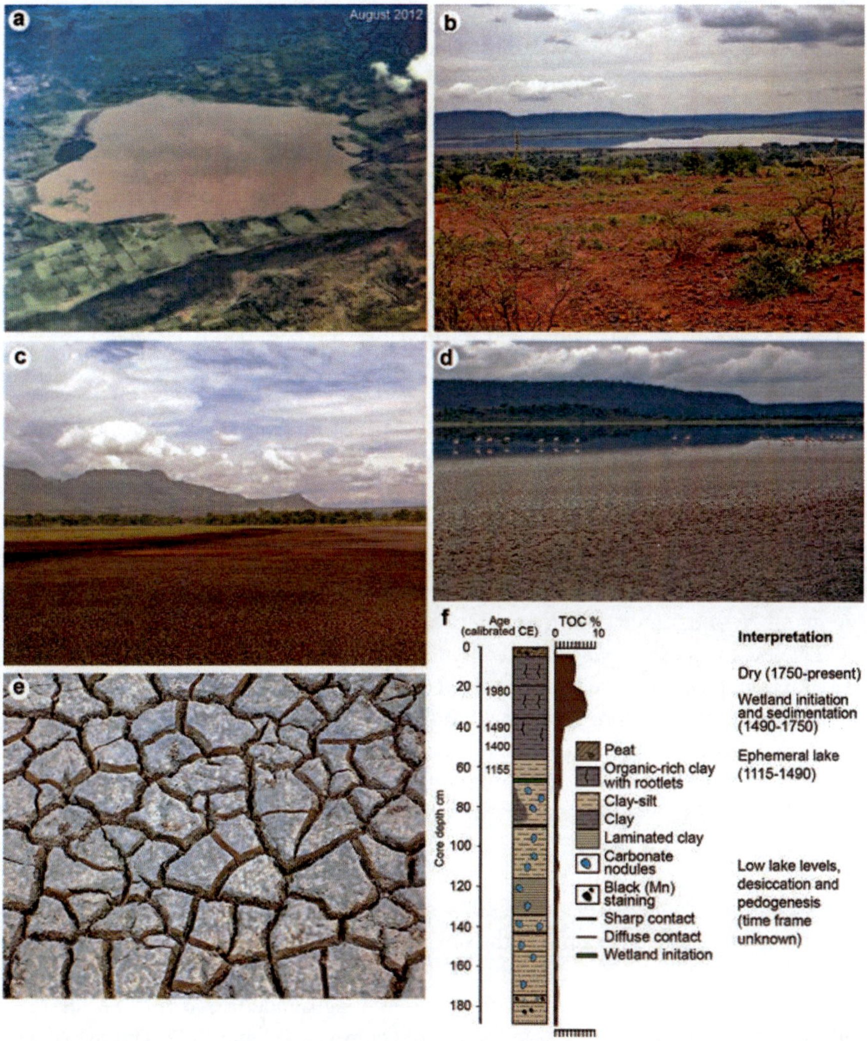

Fig. 14.17 Lake Solai. (**a**) View from west, June 2013; (**b**) View from southeastern margin. (**c**) Mudflats along the southeastern shoreline. (**d**) Mudflat along the eastern shoreline. (**e**) Mud-cracked expandable clays, eastern shoreline. (**f**) Stratigraphy of core from southern shoreline. (Based on Goman et al. (2017))

lies the steep Solai Escarpment and Subukia Ridge which rise ~500 m above the lake floor. To the west are the lower-lying faulted ridges of the Dispei Plateau and Kamaech horst (Griffiths 1977).

The geology of the drainage basin is described by McCall (1967) and Griffiths (1977), and summarised by Nyambok et al. (1993) (Fig. 14.16b). Almost all the bedrock is volcanic. The oldest rocks (Miocene Samburu Basalts), which crop out in the southeastern part of the drainage basin, are overlain by a succession of faulted

Miocene to Pleistocene phonolites, trachytes and tuffs with local intercalations of fluvial and lacustrine deposits of different ages revealed mainly from boreholes.

The drainage basin has an area of ~215 km^2 (Herrnegger et al. 2021). The lake is fed by a stream called Maji Tamu (Chesebet), which though often perennial dries up during extreme drought. Other streams, including the Kasururei, Kimolo and Chemuka, formerly drained into the lake but have been partly diverted for irrigation. Other river water infiltrates the bedrock and contributes to groundwater (Nyambok et al. 1993; De Bock 2004). Hot springs discharge at several sites along the base of the eastern escarpment and to the south (McCall 1967). Much of the lake water comes from direct precipitation and runoff while water loss occurs by evaporation. The watershed receives between 700 and 900 mm of precipitation per year. However, potential evaporation is very high (~1800 mm y^{-1}), resulting in a negative hydrological budget (Goman et al. 2017).

The lake area has varied between ~4 and 11.8 km^2 during the past decade but has been very high since 2011, flooding peripheral mudflats and the road along its eastern margin (GoK 2021). In contrast, McCall's (1967) geological map shows a dry lake with an airstrip where the lake now stands. The lake clearly responds quickly to changes in the hydrological budget or anthropogenic factors. The maximum water depth was typically 1–2 m before 2011, but Herrnegger et al. (2021) report a depth range of 8.5 m based on remotely sensed data.

The local vegetation was mapped by De Bock (2004) and De Bock et al. (2009). They reported stressed Acacia woodland west and south of the lake, with *Cynodon dactylon* along the eastern shoreline. Saline aquatic fringes east and south are dominated by *Sporobolus robustus* near the lake edge with *Cyperus laevigatus* in areas of high evaporation and increasing salinity. Open water wetland occurs on the delta of Maji Tamu where vegetation is represented by *Typha domingensis* and *Cyperus dives* and an aquatic flora (*Pistia stratiotes* and *Ludwigia stolonifera*; De Bock 2004). Land around the lake is used for cattle grazing and maize production (De Bock 2004; Kervyn de Meerendre 2004).

Koskei et al. (2019) examined the physiochemical properties and phytoplankton of the lake between April and August 2018. They reported mean temperature values (23.2 °C), dissolved oxygen (6.35 mg L^{-1}), total suspended solids (220.8 mg L^{-1}), nitrate nitrogen (0.75 mg L^{-1}), and total phosphorus (20.6 mg L^{-1}). The pH ranged from 9.0 to >11.0, peaking during periods of high biomass. Salinity ranged from ~0.7–1.5‰. They identified five groups of plankton, dominated by Cyanobacteria (mainly *Microcystis* and *Ananbaena*), with diatoms (*Nitzschia*, *Navicula*) Chlorophyta, Euglenophyta, and Chrysophyta. Koskei et al. (2019) also noted that high turbidity limits biodiversity.

Shallow cores in Lake Solai have revealed its recent history. Goman et al. (2017) analysed a short (187 cm) core from the southern shoreline. Maharjan (2010) analysed the mineralogy and organic chemistry of that core and four others in the lake and margins. The recent lake sediments are mud (Fig. 14.17e, f) and clay exposed when the lake level is low. Many are organic rich with plant remains; others are broken up by bioturbation or pedoturbation when the lake is low or desiccated. Clays are predominantly smectitic. The lake margins are slightly coarser (silt and fine sand) but the Maji Tamu delta is poorly developed. Satellite imagery at low levels reveals a low-energy broad delta-plain that is densely covered with reeds at high levels, and which merges laterally into mudflats. This pattern is similar to other rift valley lakes, including southern Lake Baringo (Chap. 8) and Lake Nakuru (Chap. 10).

Goman et al. (2017) studied the longest core taken near the southern shoreline (Fig. 14.17f) and analysed the pollen and diatom records. Biological proxies were used to reconstruct environmental change over the last millennium. The diatom record of the lowest part of the core (pre-1115 CE) shows a transition from shallow saline, alkaline water to a deeper fresher lake. Drier conditions (Medieval Climatic Anomaly) then prevailed between 1115 and 1490 CE. Wetter conditions were present during the Little Ice Age with a persistent shallow dilute lake, indicated by planktonic diatoms and rooted and free-floating aquatic plants. After 1750 CE the pollen record and the paucity of diatoms and seeds indicate the onset of drier conditions that have continued until today. Near-surface sediments indicate increasing anthropogenic influence.

McCall (1967) mapped diatomite near Ngendalel (Fig. 14.16) and reported 'diatomite' in boreholes ~100 m deep north of Lake Solai but these appear to be very old. He also reported (p. 74) "several hundred feet of sediments in the Solai basin. . . known from drilling records," and that the grabens west of the lake "contain deposit (sic) of thinly bedded . . . diatomaceous silts". He speculated that these are either evidence of a Quaternary lake that covered the Solai basin or that they represent small local palaeolakes. Little is known about the age(s) and extent of the palaeolakes recorded in the Solai drill cores. The lake might have received overflow from an expanded Lake Nakuru-Elmenteita via Menengai during the African Humid Period (AHP: ~15–5 ka) and might itself have overflowed to feed the palaeodrainage of Lake Bogoria. Evidence is lacking: no high-level Solai palaeoshorelines have been described.

14.11 Lake Ol' Bolossat

Lake Ol' Bolossat, one of the largest and highest ephemeral lakes in the Kenya Rift, lies at ~2330 m elevation just south of the equator (Okumu 2017; Terer et al. 2019). This shallow fresh to brackish alkaline lake, which is normally hydrologically open, drains to the northeast and away from the Kenya Rift unlike most other Kenya Rift lakes (Figs. 14.18 and 14.19). The limnology is poorly known compared to the main rift lakes, but it is often home to many hippopotamuses (Fig. 14.19d) and is important for birds (native species and Eurasian migrants). The Government of Kenya recognised Lake Ol' Bolossat as a 'wetland protected area' in February 2018.

The lake lies on the northern Bahati-Kinangop Plateau, which is tilted eastward towards the base of the Sattima fault escarpment (Fig. 14.18c), which also forms the eastern margin of the lake. The plateau is composed mostly of Pliocene tuffs, trachyte and phonolite that overlie Miocene phonolite. The tuffs and phonolite are

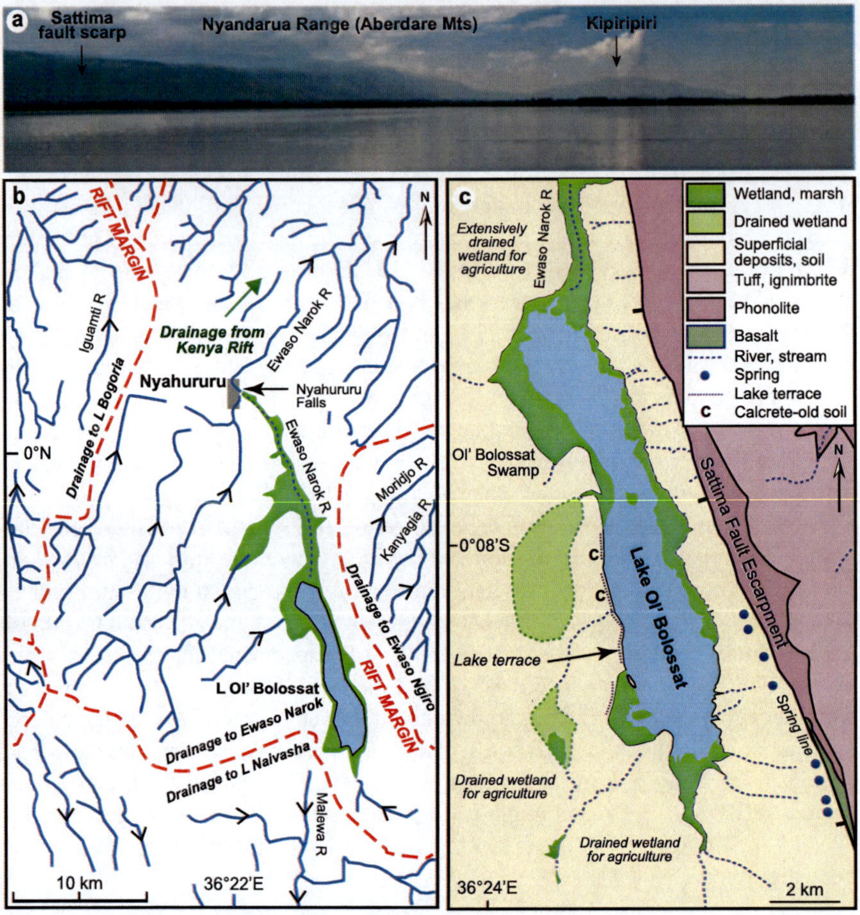

Fig. 14.18 Lake Ol' Bolossat. (**a**) View south from the mid-western shore. (**b**) Drainage pattern at Lake Ol' Bolossat. (**c**) Geological setting of Lake Ol' Bolossat

exposed on the upthrow side of the Sattima fault, with an outcrop of Miocene basalt in the southeast (Fig. 14.18c) (McCall 1967; Conti et al. 2021).

The climate is sub-humid. The mean annual rainfall of the area is ~985 mm and increases southwards and westwards. Rainfall is bimodal, falling mainly between April and June, and between October and November respectively, similar to neighbouring lakes (Solai, Nakuru). Temperatures typically range between about 10 °C and 27 °C.

The drainage basin (4800 km^2: Government of Kenya [GoK] 2020) includes the Sattima Escarpment (2550 m) lying on the NW slopes of the Nyandarua (Aberdare) Mountains and the Dundori Hills-Tumaini Ridge west of the lake (Fig. 14.18b). The lake is fed by short (1–4 km) streams draining the Sattima Escarpment and fault-line springs, mainly along the southeastern lake margins. Some streams drain into the

Fig. 14.19 (**a**) Nyahururu Falls which is fed by Lake Ol' Bolossat and local streams. (**b, c**) View north (**b**) and south (**c**) from the western shore in August 2017. A littoral terrace records a palaeo-shoreline of unknown age. (**d**) Hippopotamus group near mid-eastern shore. (**e, f**) details of sediments forming littoral terrace, mid-western shoreline. 1: greenish grey mud; 2: reddish brown gravelly mud with carbonate-filled mudcracks. 3: White calcrete (medium volcanic gravel cemented by carbonate). 4: Orange gravelly calcrete. 5: Buff structureless sand and silt

subsurface and feed the lake by groundwater discharge. The lake receives water from rivers, a few of which are sometimes perennial, in the north and northwest including the Simba and Nyakariang'a rivers in the Ol' Joro Orok area and the Chamuka and Equator rivers near Nyahururu town (GoK 2020). Small streams flow from the western hills, some spring-fed. In many cases, the natural flow has been dammed or modified for irrigation (NEMA 2012). Mwaura (2006, his Fig. 4) suggested that subsurface outflow from the lake might occur towards the southwest.

The elongate Lake Ol' Bolossat wetland, which is commonly separated into smaller lakes by dense wetlands, is up to 18 km long by 3 km wide. This valley is known as *Ongata Pusi pusi*. This wetland complex is composed of 80% marsh, 15% open water and 5% dry land (GoK 2020). Open water is usually present in the southern part, which is the last area to dry up during desiccation. Northwards the lake passes through dense riparian freshwater marsh to form a headwater of the Ewaso Narok River (Fig. 14.18b) which, in turn, flows over the Nyahururu Falls (Fig. 14.19a) and downstream feeds several wetlands in Laikipia.

The lake has a mean depth of 0.75 to 1 m (McCall 1967; Gichuki and Gichuki 1998). The southern part has an average depth of 0.76 m (range 0.2–1.35 m); the central area has a mean depth of 0.70 m (range 0.3–1.6 m); and the northern part has an average depth of 0.66 m (range 0.3–1.5 m) (GoK 2020). The lake is shallowest on its eastern side. Lake level fluctuates within a 1–2 m range, and the lake occasionally dries up (e.g. 1960, 1984, 1987, 1994, 2000 and 2001) or becomes small, scattered pools of water (Blomqvist 1997; Gichuki and Gichuki 1998) surrounded by rooted and floating marsh. In contrast, it expands in some years, flooding adjacent fields.

Kilham (1971) analysed the lake waters in July 1969 and showed that they were mildly alkaline (pH 7.8) Na^+–Ca^{2+}–HCO_3^-–Cl^- waters like many rivers in the Kenya Rift that have undergone mild to moderate evapoconcentration. According to Terer et al. (2019, cited by GoK 2019) the southern part of the lake has an average pH of 8.47 with a range of 7.64 to 9.03; the central area has pH of 8.54 (range 7.75–9.12); while the northern part has an average pH of 7.85 (range 7.21–8.87). The mean conductivity increases from 81–644 μS cm^{-1} in the north (mean: 335) and up to 1130 μS cm^{-1} (mean: 829) in the south. The southern part of the lake has higher conductivity and salinity than other parts of the lake especially the northern part, which is substantially diluted by the streams entering the lake.

The lake margins are covered by short grasses that support livestock and then pass into swamps with *Cyperus* spp. and *Typha,* with *Papyrus* in the northern marsh (Wamiti et al. 2020). Sedges are abundant, with freshwater reeds near sites of dilute inflow. Saline mudflats are occasionally present in the south when the open water shrinks in the dry season. The lake lies in an agricultural area with maize and vegetables that consume large amounts of spring and river water for irrigation in the dry seasons. The area is also important for livestock grazing. The lake basin hosts abundant hippopotamuses (~800 according to GoK 2020), spotted hyenas, jackals, and occasional predators from Aberdare National Park. Aquatic birds are abundant including Eurasian migrants (Adhola et al. 2009; Okumu 2017). Fish in the lake (catfish, mudfish, carp: East African Wild Life Society [EAWLS] 2017) are augmented by fish farms (Ol Joro Orok). Little information is available on the biological aspects of its limnology. Lind (1968) described the phytoplankton and reported *Microcystis aeruginosa, Melosira (Aulacoseira) ambigua, Microcystis* sp. *Anabaenopsis circularis*, *Chroococcus minor* and *Oscillatoria* sp.. Wafula and Murunga (2020), Kiama et al. (2021) and Kiama (2022) have discussed aspects of the bacterial communities. Rumes et al. (2016) discussed the ostracod fauna; Eggermont et al. (2006) examined the onshore and offshore chironomids.

Little is known about the recent lake sediments, which are most likely to be detrital clay, mud and silt with desiccation surfaces, frequent bioturbation, rootlets and perhaps some carbonate (calcite) in shallowest areas. Shackleton (1945) described Ol' Bolossat as "a dwindling relic of a former lake" that silted up. Leakey (1931), supported by Cole (1950), proposed that the Malewa River, which feeds Lake Naivasha, formerly flowed into Lake Ol' Bolossat and that that river capture expanded the Naivasha drainage basin, contributing to its fresh water. McCall (1967) reported a raised shoreline east of the lake but gave no details. Exposed old lacustrine sediments have not been recognised. McCall (1967) suggested that the most recent (Holocene) palaeolake was never very extensive. A low (1.5–2 m high) littoral terrace along the western shoreline (Fig. 14.19b, c) provides evidence of a former lake highstand(s) overlying a palaeosol with roots, mudcracks and calcrete with carbonate nodules and cement (Fig. 14.19e, f), in turn overlying greenish lacustrine? mud. The sequence implies two lacustrine phases separated by a period of lower lake levels, but the ages of these units are unknown.

In common with many other Kenya Rift lakes, Ol' Bolossat is threatened by increasing population pressures including loss and diversion of inflow for agriculture, siltation, pollution, reduced habitat for the biota, interactions between humans and local biota, and local disputes among residents (Blomqvist 1997; MEMR 2012; NEMA 2012; Mathenge 2013; Zacharia et al. 2013; EAWLS 2017; Okumu 2017; Karuku and Mugo 2019; GoK 2020).

14.12 Lake Sonachi, Naivasha

Three small craters are present along the southern and southwestern shoreline of Lake Naivasha. The crater rims are basaltic ash and agglomerate (Thompson and Dodson 1963; Clarke et al. 1990). Crescent Island crater is breached and fully part of the main lake. Lake Oloidien (Oloiden) is periodically united with Lake Naivasha as a single lake, while the third, Lake Sonachi is a fully separate small lake. Crescent Island and Lake Oloidien are discussed in Chap. 11. Lake Sonachi is considered here.

Lake Sonachi ('barren bull' in Maasai) is a saline, alkaline, meromictic, oval-shaped water body (0.18 km^2) located in a crater ~2.5 km west of Lake Naivasha (Fig. 14.20). It has also been referred to as Naivasha Crater Lake, Crater Lake, and Green Crater Lake (Hecky and Kilham 1973; Peters and Macintyre 1976; Melack 1979a; Macintyre and Melack 1982) and was first examined scientifically by Beadle

Fig. 14.20 Lake Sonachi. (Photographs courtesy of Annett Junginger)

(1932b) and Jenkin (1936). The waters stood at ~1890 m above sea level between 1979 and 1981 with an average depth of 5 m and a maximum depth of 9 m (Macintyre and Melack 1982; Njuguna 1988). The elevation of the lake appears to be directly controlled by rainfall, evaporation and seepage given the lack of major surface streams. The lake surface stands within ~1 m of the level of Lake Naivasha, and both levels often fluctuate synchronously, implying a groundwater connection (Sikes 1936; Njuguna 1982, 1988; Fazi et al. 2021). Verschuren (1999a) and Olaka et al. (2016) suggested that water from Lake Oloidien may flow into Lake Sonachi in the subsurface along faults.

The lake water is of Na^+–HCO_3^-–CO_3^{2-} composition with a pH of 9.5–9.8. Montcoudiol et al. (2019) sampled the lake in 2016–17 and reported a Cl^- concentration of 218–264 mg L^{-1}, 52–58 mg L^{-1} SO_4^{2-}, up 311 mg L^{-1} F^- and 6111 ppm Total Dissolved Solids. The Sonachi crater rim is 30–117 m higher than the lake surface, which reduces the influence of winds, and in turn enhances the potential for stratification with a significant impact on limnological characteristics (Fig. 14.21; Gaudet and Melack 1981; Macintyre and Melack 1982; Njuguna 1982, 1988; Melack and Macintyre 2016). Beadle (1932b), for example, noted higher alkalinity and phosphate in the bottom water and suggested that there was little water mixing. Later studies documented a temperature contrast between the lake floor (18.9–21.6 °C) and surface (19.1–26.7 °C) (Melack 1979a). During 1979–80, conductivities ranged from 5700–7290 µS cm^{-1} in surface water and 8270–14,940 µS cm^{-1} in the bottom water, with a distinct increase in values below 3 m (Njuguna 1988), implying meromixis. Alkalinity data for the same years indicate an increase below 2–3 m from about 40 to >100 meq L^{-1}, with pH remaining steady at around 9–9.6.

In the 1930s, Lake Sonachi was dominated by *Spirulina platensis* (=*Arthrospira fusiformis*) (Beadle 1932a). The lake experienced a dramatic decline of filamentous cyanobacteria from the 1930s to 1970s, which were replaced by small coccoid cyanobacteria such as *Synechococcus bacillaris* (Verschuren et al. 1999). Kilham (1981) recorded high concentrations of pelagic bacteria. Diatoms collected from littoral parts of Lake Sonachi in the 1970s (Melack 1982; Gasse 1986) included *Anomoeoneis sphaerophora*, *Craticula elkab*, *Navicula cryptocephala* var. *veneta*,

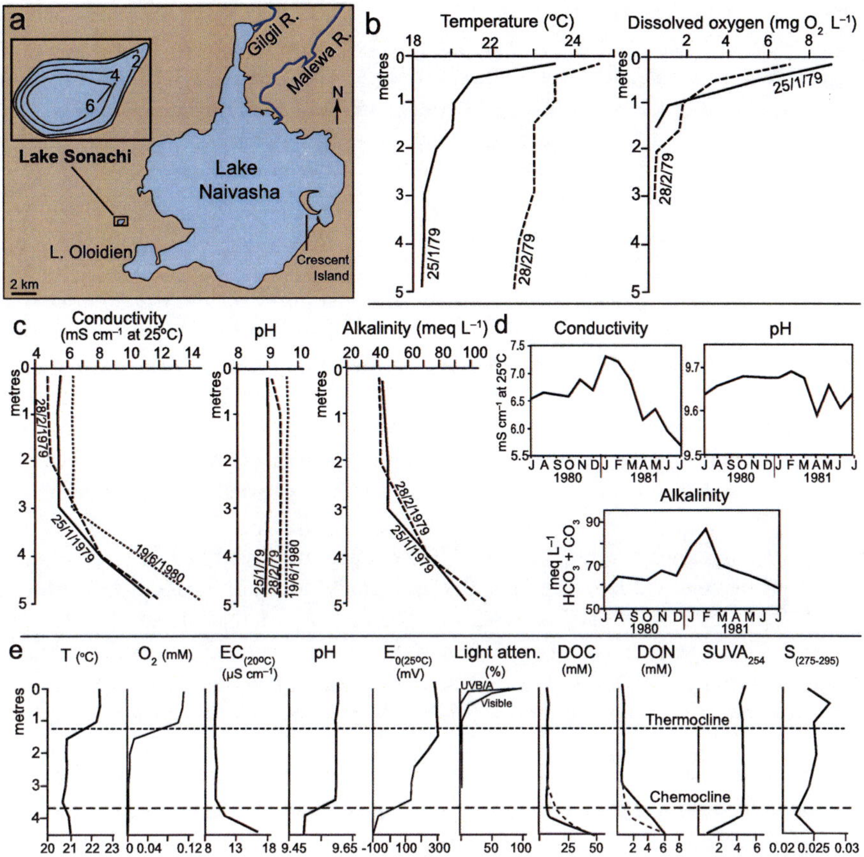

Fig. 14.21 Lake Sonachi. Limnology of Lake Sonachi. ((**a–d**) After Njuguna (1988); (**e**) After Butturini et al. (2020)). (**a**) Location. (**b**) Temperature and dissolved oxygen profiles showing a well-developed thermocline and anoxic bottom waters. (**c**) Chemical profiles showing increased bottom salinity and alkalinity with stable pH at all depths. (**d**) Annual changes in conductivity, pH and alkalinity in surface waters. (**e**) Vertical profiles (June 2017). Temperature, oxygen, electrical conductivity, and pH, plus E0 (redox potential), light attenuation, DOC (Dissolved Organic Carbon), DON (Dissolved Organic Nitrogen, ultraviolet absorbance (SUVA254) and absorbance spectral slope between 275 and 295 nm (S275–295)

Nitzschia frustulum, *Nitzschia* cf. *latens*, *Nitzschia sigma*, and *Rhopalodia gibberula*. In 1980–1981, the Sonachi phytoplankton was dominated by coccoid cyanobacteria and *Synechococcus bacillaris* (Njuguna 1988; Verschuren et al. 1999) together with *Lyngbya limnetica*, *Synechocystis aquatilis*, *Spirulina laxissima*, and *Spirulina platensis*. Planktonic diatoms and green algae were then rare, mainly comprising *Nitzschia* spp. and *Nannochloris* sp. Similar assemblages were present in 1979 together with *Chlorella* sp. and *Oocystis parva* (Kalff and Watson 1986). Studies of epiphytic diatoms in Lake Sonachi indicated the presence of only two taxa and dominance by the cyanobacterium *Arthrospira fusiformis* (Ballot 2004;

Ballot et al. 2005; Kotut et al. 2006; Fazi et al. 2017; Muohi et al. 2018), which contrasted with the 47 diatoms observed in Lake Naivasha (Cocquyt and De Wever 2002). One new fossil taxon, *Staurophora caljonii* (Cocquyt 2004) was reported from sediments in lakes Sonachi and Oloidien (Ballot et al. 2005).

In June 1929 Jenkin (1932) documented the presence of rotifers (*Brachionus pala*) in saline Lake Sonachi with large numbers of *Diaptomus* developing by March 1931. In contrast, the modern saline lake is poor in species. Early research identified *Brachionus didiatus*, *B. quadridentatus* and the calanoid copepod *Paradiaptomus africanus* (De Beauchamp 1932; Beadle 1932a, b; Lowndes 1936; Green 1993). Larval chironomids (*Kiefferulus disparilis*, *Microtendipes* sp. and *Cladotanytarsus pseudomancus*), Hemiptera and Coleoptera spp. have also been reported from littoral areas (Clark et al. 1989; Verschuren et al. 1999). Sediment cores are dominated by chironomids, scarce cladocerans (mainly *Moina micrura* and *Alona rectangula*) with ostracods absent (Verschuren et al. 1999). Genetic analysis of core sediment has also been used to distinguish several haplotypes with similarities to *Brachionus dimidiatus* and *B. plicatilis* (Epp et al. 2010). Ciliates varied considerably during 1998–1999 and included *Cyclidium, Halteria, Cristigera, Vorticella, Linostoma, Spirostomum* and *Frontonia* (Yasindi 2001; Yasindi and Taylor 2016).

Butturini et al. (2020) report a detailed study of Dissolved Organic Matter (DOM) in Lake Sonachi (Fig. 14.21e). They observed that Dissolved Organic Carbon (DOC) was high (~8.1 mM) in surface waters and that meromixis exerted a major control on the DOM pool. Surface DOM contained abundant aliphatic and aromatic molecules, reflecting microbial production and photodegradation. In bottom waters, DOC increased ~5 times with saturated, reduced and non-aromatic DOM indicating strong microbial activity. They also noted that the chemocline was a highly reactive layer for DOM processing. Fazi et al. (2021) examined the biogeochemical and microbial communities and showed an extremely high concentration of biogenic methane in its oxic waters (up to 156 μmol L^{-1}).

The lake attracts many birds including lesser flamingos, which feed mainly on the dense *Arthrospira* blooms that provide the green colour (Barasa et al. 2017). Krienitz et al. (2016) reported that ~1000 lesser flamingos were present at various times between 2002 and 2008, but with a rapid decline early in 2009 when the phytoplankton community crashed and was replaced by "a milky suspension of bacteria." The lesser flamingos soon declined to fewer than 100 birds.

Four stratigraphic units are present in most Naivasha short cores (see Chap. 11 and Verschuren 1999a, b; Verschuren et al. 1999; Grewe 2016). Unit 0, in the Lake Sonachi Core NS93.2-F (Fig. 14.22), for example, represents a drought period (the late 1700s to 1820–30) when oxidation and compaction affected the sediments. The overlying Unit 1 has a low organic matter (OM) and formed during late nineteenth-century lowstands (Åse et al. 1986). OM increases in Unit 2, reflecting higher lake levels (~1890–1940) when lakes Oloidien and Naivasha were joined. Unit 3 has a higher OM content and was laid down during low to intermediate lake levels (post-1940s). Massive clayey mud in Lake Sonachi, and neighbouring water bodies, tended to accumulate during periods of lower to intermediate lake levels, fresher water and holomictic conditions (Verschuren 1999a). In contrast, finely laminated

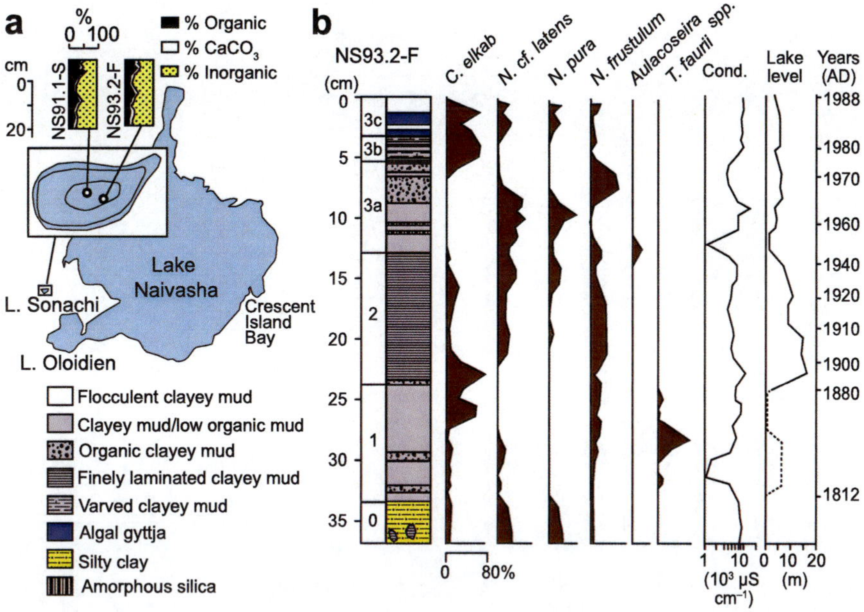

Fig. 14.22 Lake Sonachi sediment and diatom stratigraphy. (**a**) Sediment stratigraphy. After Verschuren (1999a). (**b**) Diatom floras. After Verschuren et al. (1999). Note dominance by saline taxa and high diatom-inferred salinities

clayey mud was deposited during highstands when chemical stratification prevented the development of a bioturbating infauna. Thin horizons of colloidal amorphous silica were precipitated directly from the water column when freshwater inputs caused sudden drops in pH below the chemocline (Verschuren 1999b).

14.13 Small and Ephemeral Lakes of the Southern Kenya Rift

The southern Kenya Rift has much less surface water than central and northern areas with no major water bodies between Lake Naivasha (Chap. 11) and Lake Magadi (Chap. 12). Nonetheless, small ephemeral lakes and swamps are present (Gichuki and Macharia 2003), notably in the region south of Olorgesailie and east of Magadi where the bedrock is mainly Pleistocene Magadi Trachyte but with olivine basalt at Lake Kabongo (Baker 1958, 1963). Three of these lakes are named and periodically hold water: Kwenia, Kabongo and Karamai (Figs. 14.23 and 14.24). Other small temporary lakes (muddy playas) form on graben floors but, except for Lake Loongojit (2°02′07″S; 36°32′14″E; elevation ~1311 m; ~2 km²), about which little is known, they are unnamed on published maps.

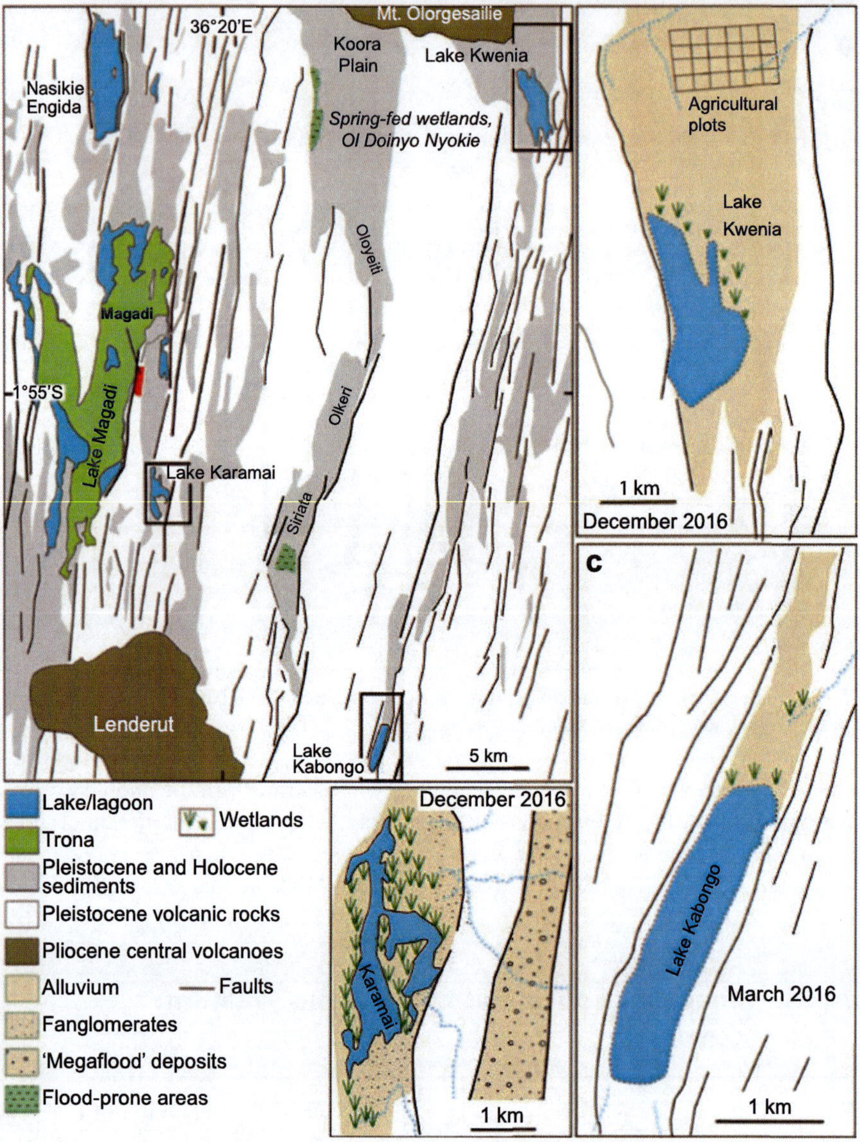

Fig. 14.23 Small, southern Kenya Rift lakes. (**a**) Locations of the main lakes. (**b**) Lake Kwenia. (**c**) Lake Kabongo. (**d**) Lake Karamai

Lake Kwenia (Kwennia) (Fig. 14.24a, b), the best known of this group, has been little studied because it is often a muddy depression without surface water. According to legend, Kwenia means "laughter" . . . local people discovered the lake full of fresh water and were so happy that one laughed until he died. Its fame is mainly

Fig. 14.24 Small, southern Kenya Rift lakes. (**a**) Lake Kwenia viewed from midwest shoreline, July 2018; (**b**) Lake Kwenia viewed from southwest shoreline, July 2018; (**c**) Lake Karamai when dry (June 2014). (**d**) View south over Lake Karamai when dry. (**e**) Lake Karamai when wet (July 2018). (**f**) View across the reedy lake when wet towards eastern fault scarp. Arrow points to one of several dry waterfalls coated in rock varnish that formed during periods of overflow from the Siriata Graben, part of the Koora Basin (see Chap. 19). (**g**) Northern shoreline of Lake Karamai with Magadi Horst in distance

because the cliffs forming the fault scarp east of the lake are a breeding site of the critically endangered Rüppell's griffon vulture (*Gyps rueppellii*) (Virani et al. 2012), which has been considered to be the highest-flying bird in the world, with evidence of flight at least 11,300 m above sea level when one crashed into a plane (Laybourne 1974).

Lake Kwenia is a muddy playa typical of many small grabens on the grid-faulted floor of the Kenya Rift. Most playas are confined by steep young N-S fault scarps and receive axial drainage from one or both (N and S) directions. Only a few receive lateral drainage, commonly linked to local faulting, the influence of oblique (NW-SE) tectonic trends, or related stream piracy. When present, the lake lies at ~1045–1050 m elevation, is fresh and alkaline, and is fed mainly by ephemeral streams that rise in faulted volcanic uplands to the east, one of which feeds the lake along its northeastern margin. Occasionally drainage from the Ol Keju Nyiro River catchment, which drains the landscape north and west of Kwenia, spills over into the northern Kwenia graben (Isaac 1978) and may extend the length of the lake.

During the past three decades, the lake has often been a dry mudflat with desiccation cracks, traversed by animal trails and with patches of rooted vegetation and dead trees near the shoreline. However, during the 'long rains' and periods of extended precipitation (e.g. post-2011), it may be flooded to a depth of at least 30–40 cm and expand in area up to ~11 km^2. The waters were dilute and alkaline in July 2018 when sampled with a pH of 7.9 and a temperature of 25.1 °C. Gichuki et al. (2001) reported a pH of 7.6, a temperature of 29 °C, and a conductivity of 0.94 $\mu S\ cm^{-1}$ in November 1997. Details of the limnology and sedimentology are largely unknown. We know of no records of high salinity. The lake dries up with groundwater at shallow (cm to dm) depth and perhaps very thin (mm-scale) trona crusts that rapidly deflate. Poorly sorted gravel lies along the steep eastern and western shorelines but passes into mud within a few metres.

Lake Kabongo (Fig. 14.23c), an elongate terminal lake that lies in a NNE-SSW graben, is similar in setting to Lake Kwenia. It receives inflow from the Turoka River which rises in volcanic highlands east of the lake. Gichuki et al. (2001) reported a pH of 8.5, a temperature of 31 °C, a conductivity of 1.35 $\mu S\ cm^{-1}$, and high turbidity in November 1997. Baker (1958, 1963) noted that after heavy rains the lake may survive several dry seasons.

Lake Karamai differs in many respects from Lake Kwenia and Lake Kabongo. It lies in a graben south of the East Sub-basin of Lake Magadi (Fig. 14.24c–g). The modern lake is ephemeral. When dry the lakebed is silt and mud covered by yellow grasses and reeds with gravelly margins. When wet, the lake is a green vegetated wetland up to ~30 cm deep with fresh alkaline (pH 9.1) water. Most inflow is from the south, where streams rise on the northern flanks of Lenderut volcano, and from faulted volcanic rocks to the east. The lake is a distal part of a former Quaternary drainage system that included incised channels and waterfalls (Fig. 14.24f) that lies to the east, first recognised by Baker (1958, 1986), Crossley (1976) and Marsden (1979). A large alluvial fan south of the lake was built partly by overflow from waters in the adjacent Koora graben. The surficial sediments are likely underlain by the High Magadi Beds (Chap. 12).

14.14 Lake Simbi and Other Nyanza (Kavirondo) Wetlands

The minor lakes in the Nyanza Rift are Lake Kanyaboli, Lake Namboyo, Lake Sare, and Lake Simbi (Fig. 14.25). The first three lakes are part of the Yala Swamp, the largest (~175 km^2) inland freshwater wetland in Kenya (Aloo 2003; MEMR 2012). The swamp has formed at the delta of the Yala River, is supplemented by lesser inflow from the Nzoia River and minor rivers and is sometimes replenished by an eastward flow of water from Lake Victoria. In contrast, Lake Simbi is a small isolated, saline closed lake that lies south of the Winam (Nyanza) Gulf near the shoreline of Lake Victoria. At the eastern edge of Winam Gulf, the Nyando (Kusa) Swamp lies at the mouth of the Nyando River. The Yala River rises in the Mau Highlands. The average annual rainfall around Yala swamp is ~760 mm, with up to 2000 mm in the highlands to the north and 1600 mm in the lowlands to the south.

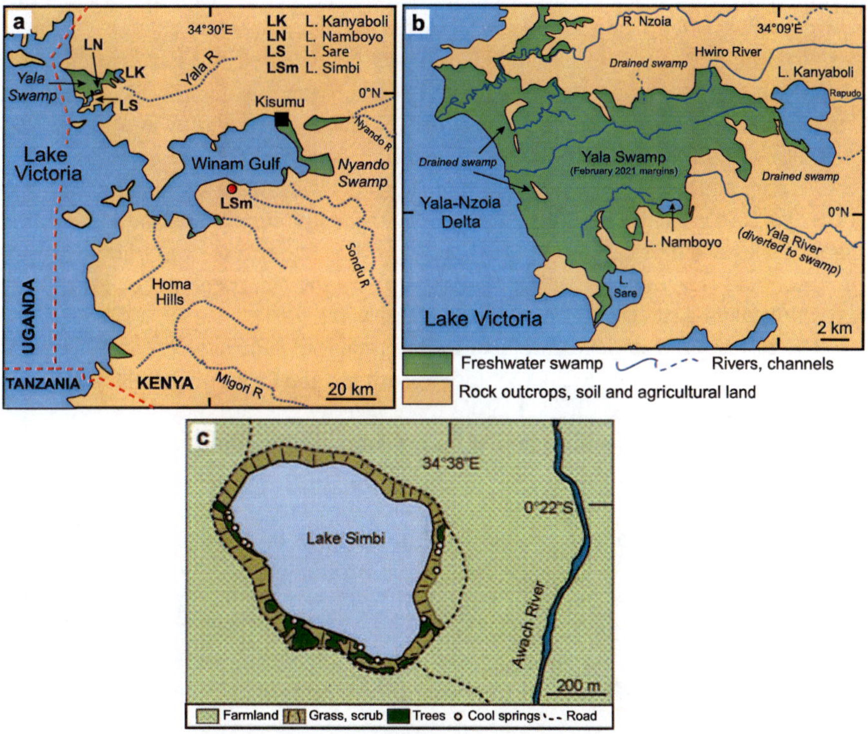

Fig. 14.25 Small lakes of the Nyanza Rift (western Lake Victoria): (**a**) Location of the small lakes. (**b**) Location of the Yala Swamp lakes. (**c**) Location of Lake Simbi

14.14.1 The Yala Swamp Lakes

Lake Kanyaboli (Fig. 14.25b), the largest lake, is ~10 km^2 in extent and 3 m deep. The Yala River used to flow into Lake Kanyaboli and then into the swamp and Lake Victoria. The river was diverted into the main swamp and became cut off from Lake Kanyaboli by a mud dyke. The lake now receives inflow from the Rapudo River and backflow from the swamp. Kondowe et al. (2022) examined seasonal changes in the lake water quality and found that it has fluctuated between eutrophic and hypereutrophic. They noted that anthropogenic nutrient loading and wetland reclamation have impacted water quality in recent years.

Lake Namboyo (4.2 km^2) developed as a patch of open water surrounded by marsh on the southern part of the swamps. In contrast, a barrier erected across a branch of the Yala River cut off a small inlet of Lake Victoria and created Lake Sare by back flooding. Lake Sare lies at 1140 m asl in the southwest part of the Yala swamp. The lake has a surface area of ~5 km^2, is 5 m deep, and forms part of the outlet of the Yala River into L. Victoria. It is surrounded by papyrus swamps that merge with the main Yala Swamp.

Water in the main channels and lakes is well oxygenated, but low oxygen levels develop in stagnant parts of the swamp. The Yala Swamp is dominated by different species of *Cyperus*, including *C. papyrus, C. dives, C. exaltatus* and *C. distans,* together with *Phragmites* and *Typha.* The swamp macrophytes are used for thatching, crafts, and other uses, but are under high pressure from the local population (Thenya 2006; Thenya and Ngecu 2017) with exploitation sometimes exceeding supply. The swamp margins are also shrinking in area because of draining for agricultural land (Kariuki 2011), which also depletes their nutrient status. Population growth is having an impact on the swamp's survival.

The swamp and its lakes are also an important site for haplochromine cichlid fishes that no longer live in Lake Victoria (Kaufman and Ochumba 1993; Abila et al. 2008) having been exterminated in the main lake by the introduced predatory fish *Lates niloticus. Lates niloticus* is present in Lake Sare, which in turn has an impoverished fish fauna compared to Lake Kanyaboli. Other fish are an important food source (Maithya 1998) The swamp is home to many birds and the sitatunga (*Tragelaphus spekii*), a rare swamp-dwelling antelope with long wide hooves that make them adapted to walking through muddy swampland.

14.14.2 Lake Simbi

Lake Simbi (Simbi Nyaima or Simba Nyimbi), which means 'sunken lake' in Luo, is a small (0.3 km^2) crater lake located 2.2 km south of the shoreline of Winam Gulf, eastern Lake Victoria, 2.5 km southwest of the town of Kendu Bay (Fig. 14.25c).

Formation of the volcanic explosion crater may be linked to the Homa Hills volcanic centre, 60 km to the west (Saggerson 1952; Odek et al. 2013) and to the faulting that defines the southern margin of the Nyanza Rift. Unlike Lake Victoria, Simbi is a saline alkaline lake with sodium carbonate water (salinity 10–18‰, conductivity ~15,000–29,000 μS cm^{-1}: Krienitz 2018). The lake surface lies at ~1145 m asl, about 10 m above the level of Lake Victoria. Accordingly, it is unlikely that Lake Victoria seepage is feeding Lake Simbi directly as groundwater. The drainage basin is little bigger than the lake, so it relies on direct precipitation and cool springs (21–26 °C) around the shoreline for recharge (Ochumba and Kibaara 1988). The lake is ~25 m deep.

The lake is both thermally and chemically stratified. Melack (1979b) reported steep temperature gradients (2–6 °C) in the upper 2 m by midday in 1973 and 1974, with less heating (1–1.5 °C) in 1976 when the water was more transparent below the seston. Below the thermocline the temperature fell to ~25 °C, increasing slightly below 10 m depth. Chemoclines were recorded between 6 and 11 m, and 11–15 m depth. Oxygen was strongly supersaturated in surface waters but the water was anoxic below 3 m depth. Ochumba and Kibaara (1988) reported a chemocline at ~4 m depth, with anoxic waters below 2.8 m.

The photosynthetic activity of the plankton is extremely high. Krienitz (2018) reported up to 12,900 mg oxygen per m^3 per hour, which is higher than lakes Bogoria, Elmenteita or Sonachi. The main producers are *Arthrospira fusiformis* accompanied by *Anabaenopsis abijatae*. Ballot et al. (2005) detected the cyanotoxins anatoxin-a and four different microcystins in the phytoplankton.

The lake and its plankton have attracted lesser flamingos and other birds (little grebe, egrets and others) with the area designated a National Sanctuary by the Kenya Wildlife Service.

14.14.3 Nyando (Kusa) Swamp

The Nyando (Kusa) Swamp forms at the mouth of the Nyando River (Fig. 14.25a) and is covered with dense papyrus. Aquatic grass (*Vossia cuspidata*) occurs at the deep-water fringes of the papyrus with reeds (*Phragmites* spp.) dominating the landward margins of the swamp (BirdLife International 2022). Several rare birds inhabit the wetland, which also is an important refuge for ichthyoplankton (Owino and Ryan 2007). Large parts of the swamp and its hinterland have been converted to agriculture. The swamp plays a significant role in filtering sediments and other pollutants from entering the lake, but irrigation upstream and residential and industrial discharge have affected the water quality (MEMR 2012).

14.15 Other Seasonal and Ephemeral Lakes

Nyamweru (1992) and Mavuti (1992) recognised several major classes of wetland, two of which are common in the Kenya Rift: riverine and lacustrine. Each author recognised subdivisions based on size and length of time that surface water is present. Three examples will be highlighted.

The Ewaso Ngiro River (South) is one of the main sources of inflow to Lake Natron. It arises in the equatorial highlands (Mau) and flows southward along the western margin of the Kenya Rift below the Nguruman Escarpment, by-passing Lake Magadi to the east. Its waters might be an important contributor to the subsurface Lake Magadi geothermal recharge (Chap. 12), but they do not directly flow into the lake. They might have fed an expanded Lake Magadi during the Quaternary (Owen et al. 2019). Where the river water approaches Lake Natron, it spreads out to form the Shombole Swamp, then becomes confined again before forming deltaic wetlands near the international border. These often include lakes surrounded by vegetated wetlands (Figs. 12.24c and 14.26a, b).

Lake Lotigipi (Lotikipi) is a seasonal lake that lies on the Lotigipi Plain, located in northwestern Kenya (Fig. 14.26c). The plain is a broad saucer-shaped depression surrounded by, and mainly filled by, volcanic rocks overlain by alluvium and soils (Walsh and Dodson 1969; Morley et al. 1992; Wescott et al. 1995, 1999; Feibel 2011; Rop 2014; Sippel et al. 2017). The remote lake has received little study. It is fed by the Tarach River and local ephemeral streams and experiences an annual rainfall of ~300–500 mm but some areas receive no rainfall in some years. Water is lost by evaporation and infiltration but may overflow to the north and west in wetter years (Walsh and Dodson 1969). The area occupied by the lake is covered by open grassland with scrubland (Olang 1984) when not flooded. Hydrogeological surveys have confirmed the presence of shallow (<80 m) and deeper aquifers below the plain (RTI 2013). The lake might have overflowed during the African Humid Period and contributed to the River Nile system (Owen and Renaut 1986).

In addition, there are many small seasonal and temporary wetlands in the rift that are named and unnamed. Some, such as the freshwater Sawaiti (Barina) swamp (Ol Punyata Swamp of McCall 1967), which lies 19 km southwest of Lake Solai and 9 km southeast of Mogotio town, are small (2 km^2), but are important for resident and migrating waterbirds and other animals (Fig. 14.26d). Geophysical and hydrological investigations imply that the swamp is fed by surface runoff and groundwater (McCall 1967; Nyaberi 2008). The perennial Olobonaita stream, which drains the Bahati uplands and Solai Escarpment to the east, flows into the swamp with a flood spillover into the Rongai-Molo drainage (towards Lake Baringo). Groundwater moves to the swamp from Menengai in the south, and Bahati and Solai to the east, probably in unconfined and semi-confined aquifers. Although the swamp shrinks in area during dry periods, there are no recent indicators of high salinity.

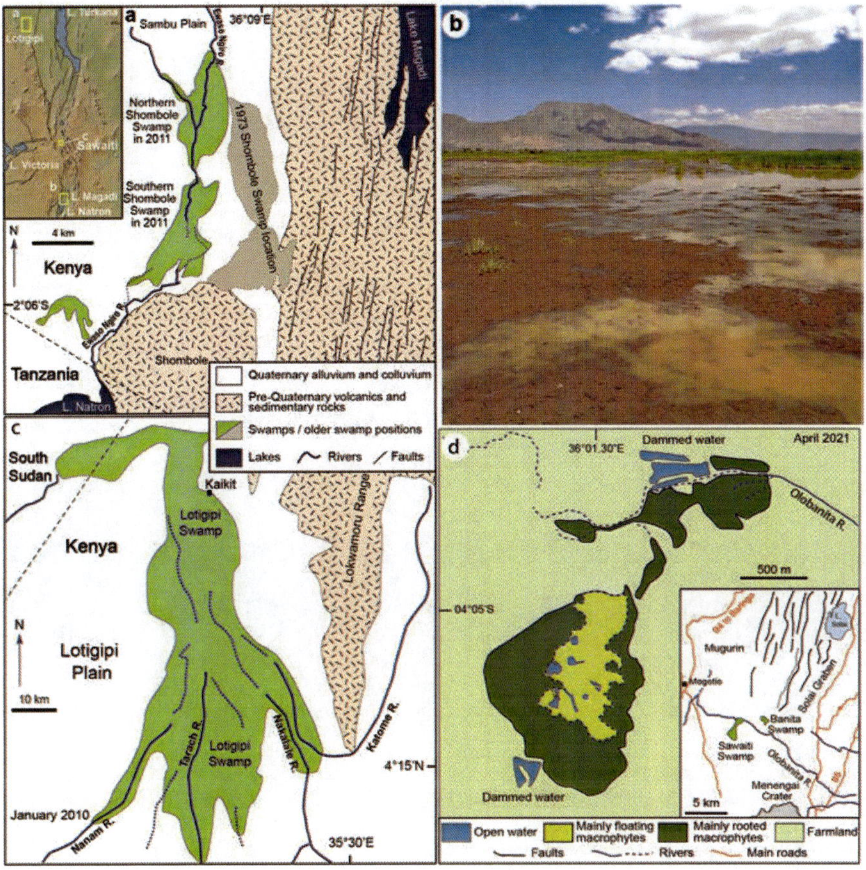

Fig. 14.26 Examples of other wetlands. (**a**) Wetlands of the southern Ewaso Ngiro River (S) showing extent in 1973 and 2011. Adapted from MEMR (2012). (**b**) Shombole Swamp looking southwards. (**c**) Lake Lotigipi, northwestern Turkana. (**d**) Sawaiti Swamp, central Kenya Rift

14.16 Constructed Reservoirs and Ponds

The natural wetlands are supplemented by artificial wetlands resulting from human intervention. These occur on many scales (Fig. 14.27) and are constructed for many reasons, including drainage to avoid frequent flooding, irrigation for crops, water supply for human and non-human consumption, animal waste disposal and hydro-electric power. At the smallest scale, these include widespread small local dams that provide water in ponds, mainly for livestock. At the medium scale, dammed water supplies villages and towns locally or downstream (e.g. the Kirandich dam at

Fig. 14.27 Constructed dams and ponds. (**a**) Chemeron (Yatoi) Dam, completed in 1984, southwest of Lake Baringo. Siltation has built deltas along its western margins. Tugen Hills in background. (**b**) Turkwel Dam, completed in 1991, southwest Turkana, for hydroelectric power generation, water supply and fisheries. (**c, d**) Kirandich Dam, Kabarnet, west of Baringo, constructed mainly for water supply in the Tugen Hills. Phase 1 was completed in 2000. (**c**) Shows dam during construction; (**d**) Shows dam soon after completion. (**e**) Dammed pond, ~3 km SE of Lake Nakuru. (**f**) Small dammed pond for local water supply, ~3 km south of Emening near Mogotio (location of Mogotio is shown in inset for Fig. 14.26d)

Kabarnet: Fig. 14.27c, d). At the largest scale, their outflow can generate hydroelectric power, such as the dam at Turkwel Gorge (Fig. 14.27b) in southwestern Turkana (GoK 2021). Only a few of these artificial water bodies have been studied (e.g. Mwaura 2006; Oketch 2006). Some are important habitats for birds and other fauna (Krienitz 2018). As expected, they are predominantly fresh water, but recent climatic changes have led to salinisation or desiccation of some, or human activities have led to siltation, pollution or loss of inflow.

References

Abila R, Salzburger W, Ndonga MF, Owiti DO, Barluenga M, Meyer A (2008) The role of the Yala swamp lakes in the conservation of Lake Victoria region haplochromine cichlids: evidence from genetic and trophic ecology studies. Lakes Reserv 13:95–104

Achiba GA (2019) Navigating contested winds: development visions and anti-politics of wind energy in northern Kenya. Land 8:7

Adhola T, Ng'weno F, Matiku P, Mulwa R, Ngari A, Barasa F, Madindou I, Wanyiri M, Mwango'mbe J, Musyoki C, Ndonye P, Ogoma M, Machekele J, Musila S (2009) Kenya's important bird areas: status and trends 2008. Nature Kenya, Nairobi

Aloo PA (2003) Biological diversity of the Yala swamp lakes, with special emphasis on fish species composition, in relation to changes in the Lake Victoria Basin (Kenya): threats and conservation measures. Biodivers Conserv 12:905–920

Åse LE, Sernbo K, Syren P (1986) Studies of Lake Naivasha, Kenya and its drainage area. Forkn Naturgeogr Inst Stockholm, p 63

Ashley GM, Maitima Mworia J, Muasya AM, Owen RB, Driese SG, Hover VC, Renaut RW, Goman MF, Mathai S, Blatt SH (2004) Sedimentation and recent history of a freshwater wetland in a semi-arid environment: Loboi Swamp, Kenya, East Africa. Sedimentology 51:1301–1321

Ashley G, Deocampo D, Kahmann-Robinson J, Driese SG (2013) Groundwater-fed wetland sediments and paleosols: it's all about water table. In: Driese SG, Nordt LC (eds) New frontiers in paleopedology and terrestrial paleoclimatology: paleosols and soil surface analog systems, SEPM Special Publication, vol 104. SEPM (Society for Sedimentary Geology), Tulsa, pp 7–61

Auko LO (2011) Evaluation of hydrogeochemical facies of The Barrier volcanic complex geothermal fluids in developing a conceptual model. MSc thesis, University of Nairobi

Auko LO, Kipngok JK (2021) Constraining the fundamental underpinning analogy of the Silali and Emuruangogolak geothermal systems using geochemistry. Geotherm Res Counc Trans 45:925–939

Baker BH (1958) Geology åof the Magadi area. Rep Geol Surv Kenya 42:1–81

Baker BH (1963) Geology of the area south of Magadi area. Rep Geol Surv Kenya 61:1–27

Ballot A (2004) Cyanobacteria in Kenyan Rift Valley lakes – a biological and toxicological study. PhD thesis, Frei Universität, Berlin

Ballot A, Krienitz L, Kotut K, Wiegand C, Pflugmacher S (2005) Cyanobacteria and cyanobacterial toxins in the alkaline crater lakes Sonachi and Simbi, Kenya. Harmful Algae 4:139–150

Barasa M, Mbaria J, Muchemi G, Gakuya F, Kariuki E, Wamiti W (2017) Ecological risk assessment of heavy metals and water bird distribution in Rift Valley lakes, Kenya. J Ecol Natur Environ 9:30–44

Beadle L (1932a) Scientific results of the Cambridge Expedition to the East African Lakes, 1930–31. Part 4. The waters of some East African Lakes in relation to their flora and fauna. J Lin Soc Lond Zool 38:157–211

Beadle L (1932b) Scientific results of the Cambridge Expedition to the East African Lakes, 1930–31. Part 3. Observations on the bionomics of some East African swamps. J Lin Soc Lond Zool 28:135–155

BirdLife International (2022) Important Bird Areas factsheet: Kusa swamp. Downloaded from http://www.birdlife.org on 02/03/2022

Blomqvist T (1997) A hydrological investigation of Lake Ol Bolossat swamp, Nyandarua District, Kenya, vol 94. Earth Sciences Centre Göteborg University, Sweden, Göteborg. (Minor Field Stud 26)

Bloomer SH, Curtis PC, Karson JA (1989) Geochemical variation of Quaternary basaltic volcanics in the Turkana rift, northern Kenya. J Afr Earth Sci 8:511–532

Boitt M, Albright O, Kipkulei H (2020) Assessment of soil erosion and climate variability on Kerio Valley Basin, Kenya. J Geosci Environ Protect 8:97–114

Burgess WG (1986) Report on a visit 17 March to 26 April 1986. Brit Geol Surv Rep WD/OS/86/4

Butturini A, Amalfitano S, Herzsprung P, Lichtenfeld OJ, Venturi S, Olaka LA, Pacino N, Harper DM, Tassi F, Fazi S (2020) Dissolved organic matter in continental hydro-geothermal systems: insights from two hot springs of the East African Rift Valley. Water 12:3512

Cavendish H (1898) Through Somaliland and around and south of Lake Rudolf. Geogr J 11:372–393

Champion AM (1935) Teleki's volcano and the lava fields at the southern end of Lake Rudolf. Geogr J 85:323–341

Clark F, Beeby A, Kirby P (1989) A study of the macro-invertebrates of lakes Naivasha, Oloidien and Sonachi, Kenya. Rev Hydrobiol Tropic 22:21–33

Clarke MCG, Woodhall DG, Allen D, Darling G (1990) Geological, volcanological and hydro-geological controls on the occurrence of geothermal activity in the area surrounding Lake Naivasha, Kenya. Rep. Ministr Energy Gov Kenya, Nairobi

Cocquyt C (2004) *Staurophora caljonii* spec. nov. (Bacillariophyceae, Anomoeoneidaceae), a new halophilic diatom species from sub-recent lake deposits in Kenya. Hydrobiologia 511:37–46

Cocquyt C, De Wever A (2002) Epiphytic diatom communities on herbarium material from Lake Naivasha and Lake Sonachi, Eastern Rift Valley, Kenya. Belg J Bot 135:38–49

Colagiovanni T (2002) Assessing the impact of protected areas on indigenous communities: the case study of Lake Kamnarok NR, Kenya. MSc thesis, University of Guelph, Ontario

Cole SM (1950) An outline of the geology of Kenya. Pitman, London

Conti P, Pistis M, Bernardinetti S, Barbagli A, Zirulia A, Serri L, Colonna T, Guastaldi E, Ghiglieri G (2021) Tectonic setting of the Kenya Rift in the Nakuru Area, based on geophysical prospecting. Geosciences 11:80. https://doi.org/10.3390/geosciences11020080

Crafter SA, Njuguna SG, Howard GW (eds) (1992) Wetlands of Kenya. International Union for Conservation of Nature and Natural Resources, Gland

Crossley R (1976) Structure, stratigraphy and volcanism in the Nguruman Escarpment area of the western side of the Kenya Rift Valley. PhD thesis, University of Lancaster

De Beauchamp P (1932) Reports on the Percy Sladen expedition to some Rift Valley lakes in Kenya in 1929. III. Rotifères des lacs de la Valleé du Rift. Ann Mag Nat Hist Ser 10:158–165

De Bock T (2004) Baseline assessment of the ecohydrology of a seasonal wetland in semi-arid region: Lake Solai (Kenya). MSc thesis, Cranfield University at Silsoe, UK

De Bock T, Kervyn de Meerendre B, Hess T, Gouder de Beauregard A-C (2009) Ecohydrology of a seasonal wetland in the Rift Valley: ecological characterization of Lake Solai. Afr J Ecol 47:289–298

De Cort G, Verschuren D, Ryken E, Wolff C, Renaut RW, Creutz M, Van der Meeren T, Haug G, Olago DO, Mees F (2018) Multi-basin depositional framework for moisture-balance reconstruction during the last 1300 years at Lake Bogoria, Central Kenya Rift Valley. Sedimentology 65:1667–1696

Dodson RG (1963) Geology of the South Horr area. Rep Geol Surv Kenya 60

Driese SG, Ashley GM, Li ZH, Hover VC, Owen RB (2004) Possible late Holocene equatorial palaeoclimate record based upon soils spanning the Medieval Warm Period and Little Ice Age, Loboi Plain, Kenya. Palaeogeogr Palaeoclimatol Palaeoecol 213:231–250

Dugan PJ (ed) (1990) Wetland conservation: a review of current issues and required action. International Union for Conservation of Nature and Natural Resources, Gland

Dunkelman TJ, Rosendahl B, Karson J (1989) Structure and stratigraphy of the Turkana rift from seismic reflection data. J Afr Earth Sci 8:489–510

Dunkley PN, Smith M, Allen DJ, Darling WG (1993) The geothermal activity and geology of the northern sector of the Kenya Rift Valley. Brit Geol Surv Res Rep SC/93/1

East African Wild Life Society (2017) Lake Ol' Bolossat. East African Wild Life Society, Nairobi

Eggermont H, Heiri O, Verschuren D (2006) Fossil Chironomidae (Insecta: Diptera) as quantitative indicators of past salinity in African lakes. Quat Sci Rev 25:1966–1994

Epp LS, Stoof KR, Trauth MH, Tiedemann R (2010) Historical genetics on a sediment core from a Kenyan lake: intraspecific genotype turnover in a tropical rotifer is related to past environmental changes. J Paleolimnol 43:939–954

Fazi S, Butturini A, Tassi F, Amalfitano S, Venturi S, Vazquez E, Clokie M, Wanjal SW, Pacini N, Harper DM (2017) Biogeochemistry and biodiversity in a network of saline-alkaline lakes: implications of ecohydrological connectivity in the Kenyan Rift Valley. Integr Med Res:1–11
Fazi S, Amalfitano S, Venturi S, Pacini N, Vazquez E, Olaka LA, Tassi F, Crognale S, Herzsprung P, Lechtenfeld OJ, Cabassi J, Capecchiacci F, Rossetti S, Yakimov MM, Vaselli O, Harper DM, Butturini A (2021) High concentrations of dissolved biogenic methane associated with cyanobacterial blooms in East African lake surface water. Commun Biol 4:845
Feibel CS (2011) A geological history of the Turkana Basin. Evol Anthropol 20:206–216
Ferguson AJD, Harbott BJ (1982) Geographical, physical, and chemical aspects of Lake Turkana. In: Hopson AJ (ed) Lake Turkana: a report of the findings of the Lake Turkana Project, 1972–1975. Overseas Development Administration, London, pp 1–110
Fuchs VE (1935) The Lake Rudolf Rift Valley expedition, 1934. Geogr J 86:114–137
Gasse F (1986) East African diatoms: taxonomy, ecological distribution (Bibliotheca Diatomologia 11). J Cramer, Berlin
Gaudet JJ, Melack JM (1981) Major ion chemistry in a tropical African lake basin. Freshwat Biol 11:309–333
Gichuki C, Gichuki N (1998) Lake Ol Bolossat. Kenya Past Present 30:25–28
Gichuki NN, Macharia JM (2003) Participation of local communities in the management of wetlands in the Magadi area, Kenya. In: Lemons J, Victor R, Schaffer D (eds) Conserving biodiversity in arid regions. Springer, New York, pp 87–103
Gichuki NN, Oyieke HA, Ndiritu GG (2001) Assessment and monitoring of wetlands for conservation and development in dry lands: a case study of Kajiado District, Kenya. In: Finlayson DM, Davidson NC, Stevenson NJ (eds) Wetland inventory, assessment and monitoring: practical techniques and identification of major issues. Supervising Scientist Report 161, Environment Australia, Darwin, pp 97–111
Giddelo C, Arndt A, Volckaert F (2002) Impact of rifting and hydrography on the genetic structure of *Clarias gariepinus* in eastern Africa. J Fish Biol 60:1252–1266
Gitau PN, Verschuren D (2016) Sedimentological and paleoecological research techniques, with applications to the environmental history of Loboi Swamp, Kenya. J Earth Sci Clim Change 7:1–8
GoK [Government of Kenya] (2020) Lake Ol' Bolossat integrated management plan, 2020–2030. National Environment Management Authority, Nairobi
GoK [Government of Kenya] (2021) Rising water levels in Kenya's Rift Valley lakes, Turkwel Gorge Dam and Lake Victoria: a scoping report. Government of Kenya, Nairobi
Goman M, Ashley GM, Owen RB, Hover VC, Maharjan DK (2017) Late Holocene environmental reconstructions from Lake Solai, Kenya. Prof Geogr 69:434–454
Goman MF, Ashley GM, Owen RB, Driese SG, Muasya AM, Hover VC (2020) A high-resolution climate history of geochemical and biological proxies from a tropical freshwater wetland located in the Kenyan Rift Valley. J Afr Earth Sci 162:103703
Green J (1993) Zooplankton associations in East-African lakes spanning a wide salinity range. Hydrobiologia 267:249–256
Grewe S (2016) Hydro- and biogeochemical investigations of lake sediments in the Kenyan Rift Valley. PhD thesis, Universität Potsdam
Griffiths PS (1977) Geology of the Lake Hannington - Perkerra River area, Kenya Rift Valley. PhD thesis, University of London
Griffiths PS, Gibson I (1980) The geology and petrology of the Hannington Trachyphonolite formation, Kenya Rift Valley. Lithos 13:43–53
Harper DM, Childress RB, Harper MM, Boar RR, Hickley P, Mills SC, Otieno N, Drane T, Vareschi E, Nasirwa O, Mwatha WE, Darlington JPEC, Escuté-Gasulla X (2003) Aquatic biodiversity and saline lakes: Lake Bogoria National Reserve, Kenya. Hydrobiologia 500:259–276
Hecky R, Kilham P (1973) Diatoms in alkaline, saline lakes: ecology and geochemical implications. Limnol Oceanogr 18:53–71

Herrnegger M, Stecher G, Schwatke C, Olang L (2021) Hydroclimatic analysis of rising water levels in the great Rift Valley lakes of Kenya. J Hydrol Reg Stud 36:100857

Isaac GL (1978) The Olorgesailie formation: stratigraphy, tectonics and the palaeogeographic context of the Middle Pleistocene archaeological sites. In: Bishop WW (ed) Geological background to fossil man. Geol Soc Lond Spec Publ 6:173–206

IUCN World Heritage Outlook (2020) Lake Turkana National Parks – 2020 Conservation Outlook Assessment. https://worldheritageoutlook.iucn.org/

Jenkin PM (1932) Reports on the Percy Sladen expedition to some Rift Valley lakes in Kenya 1929. I. Introductory account of the biological survey of five freshwater and alkaline lakes. Ann Mag Nat Hist 9:533–553

Jenkin PM (1936) Reports on the Percy Sladen expedition to some Rift Valley lakes in Kenya 1929. VII. Summary of the ecological results. Ann Mag Nat Hist 18:133–181

Johansson J, Svensson J (2002) Land degradation in the semi-arid catchment of Lake Baringo, Kenya. Earth Sciences Centre, Göteborgs Universitet, Sweden, Projketarabete B3443

Jones B, Renaut RW (1996) Skeletal crystals of calcite and trona from hot-spring deposits in Kenya and New Zealand. J Sediment Res 66:913–927

Jones B, Renaut RW (1998) Origin of platy calcite crystals in hot-spring deposits in the Kenya Rift Valley. J Sediment Res 68:913–927

Junginger A, Kyalo-Omamo M, Krueger J, Epp SE, Stoof-Leichsenring KR, Rohland S, Trauth MH, Tiedmann R (2019) 200 years of environmental impact on the temporal succession of *Brachionus* rotifer haplotypes from sedDNA in two Kenyan crater lakes. Geophys Res Abs 21:EGU2019–6104-1

Kalff J, Watson S (1986) Phytoplankton and its dynamics in two tropical lakes, a tropical and temperate zone comparison. Hydrobiologia 138:161–176

Kariuki P (2011) Kenyan wetlands – an indicator of climate change and human disturbance. SERVIR Africa biodiversity/climate change project phase 1 report. African Conservation Centre, Nairobi

Karson JA, Curtis PC (1994) Quaternary volcanic centres of the Turkana Rift, Kenya. J Afr Earth Sci 18:15–35

Karuku GN, Mugo E (2019) Land use effects on Lake Ol'bolossat watershed conservation, Nyandarua County. JOJ Wildl Divers 1:26–34

Kaufman L, Ochumba PB (1993) Evolutionary and conservation biology of cichlid fishes as revealed by faunal remnants in northern Lake Victoria. Conserv Biol 7:719–730

Kervyn de Meerendre B (2004) Impact assessment of agricultural practices on water resources in the Lake Solai catchment (Kenya). MSc thesis, Cranfield University, Silsoe

Kiage L, Liu K (2009) Paleoenvironmental changes in the Lake Baringo Basin, Kenya, East Africa since AD 1650: evidence from the paleorecord. Prof Geogr 61:438–458

Kiama CW (2022) Microbial biodiversity and community structure within equatorial lakes Olbolosat and Oloiden in Kenya. PhD thesis, Jomo Kenyatta University of Agriculture and Technology, Nairobi

Kiama CW, Njire MM, Kambura AK, Mugweru JN, Matiru VN, Wafula EN (2021) Isolation and characterization of bacteria from lakes Olbolosat and Oloiden, Kenya. Afr J Microbiol Res 15:1–19

Kilham P (1971) Biogeochemistry of African lakes and rivers. PhD thesis, Duke Univ, Durham

Kilham P (1981) Pelagic bacteria: extreme abundances in African saline lakes. Naturwissenschaften 68:380–381

Kimosop P (2010) Comprehensive assessment of hydrological drought and land use change in the Kerio Valley Basin, Kenya. 2010. PhD thesis, University of Cincinnati

Kondowe BN, Masese FO, Raburu PO, Singini W, Riziki J, Walumona RJ (2022) Water quality and ecology of Lake Kanyaboli, Kenya: current status and historical changes. Lakes Reserv 27:e12401

Koskei EC, Kotut K, Nyaga J, Odour SO (2019) Temporal variation in physico-chemical characteristics, phytoplankton composition and biomass in Lake Solai, Kenya. Int J Aquat Sci 10:101–111

Kotut K, Ballot A, Krienitz L (2006) Toxic cyanobacteria and their toxins in standing waters of Kenya: implications for water resource use. J Water Health 4:233–245
Krienitz L (2018) Lesser flamingos: descendants of Phoenix. Springer Nature, Berlin
Krienitz L, Krienitz D, Dadheech PK, Hübener T, Kotut K, Luo W, Teubner K, Versfeld WD (2016) Food algae for Lesser Flamingos: a stocktaking. Hydrobiologia 775:21–50
Kyalo-Omamo M, Junginger A, Krueger J, Epp LS, Stoof-Leichsenring KR, Rohland S, Trauth MH, Tiedemann R (2023) Sedimentary ancient DNA of rotifers reveals responses to 200 years of climate change in two Kenyan crater lakes. Freshw Biol 10.1111/fwb.14093
Lamb HH (1966) Climate in the 1960's. Geogr J 132:183–212
LaVigne M, Ashley GM (2001) Climatology and rainfall patterns: Lake Bogoria National Reserve (1976–2001). Unpublished Report, Dept Geological Sciences, Rutgers University, Piscataway NJ
Laybourne RC (1974) Collision between a vulture and an aircraft at an altitude of 37,000 feet. Wilson Bull 86:461–462
Leakey L (1931) East African lakes. Geogr J 77:497–508
Lind E (1968) Notes on the distribution of phytoplankton in some Kenya waters. Br Phycol Bull 3:481–493
Lowndes AG (1936) Scientific results of the Cambridge Expedition to the East African Lakes 1930–31, 16: the smaller Crustacea. J Linn Soc Zool 40:1–31
MacIntyre S, Melack J (1982) Meromixis in an equatorial African soda lake. Limnol Oceanogr 27:595–609
Maharjan DK (2010) Elemental and isotopic variations in organic matter in wetland and Lake sediment as potential paleoclimate indicators, Lake Solai area, East African Rift Valley, Kenya. MSc thesis, University of Lousiana, Lafayette
Maithya J (1998) A survey of ichthyofauna of Lake Kanyaboli and other small waterbodies in Kenya: alternative refugia for endangered fish species. Naga 21:54–56
Marsden M (1979) Origin and evolution of the Pleistocene Olorgesailie Lake Series, Kenya Rift Valley. PhD thesis. McGill University, Québec
Mathenge MW (2013) Effects of natural resource based conflicts on community livelihoods in Lake Ol'Bolossat catchment area, Nyandarua County, Kenya. MSc thesis, Kenyatta University, Nairobi
Mavuti KM (1992) An account of some important freshwater wetlands of Kenya. In: Crafter SA, Njuguna SG, Howard GW (eds) Wetlands of Kenya. IUCN, Gland, pp 79–84
McCall GJH (1967) Geology of the Nakuru-Thomson's Falls Lake Hannington area. Rep Geol Surv Kenya 78
Melack JM (1979a) Photosynthetic rates in four tropical African fresh waters. Freshw Biol 9:555–571
Melack JM (1979b) Photosynthesis and growth of *Spirulina platensis* (Cyanophyta) in an equatorial Lake (Lake Simbi, Kenya). Limnol Oceanogr 24:753–760
Melack JM (1982) Photosynthetic activity and respiration in an equatorial African soda lake. Freshw Biol 12:381–400
Melack JM, MacIntyre S (2016) Morphometry and physical processes of East African soda lakes. In: Schagerl M (ed) Soda lakes of East Africa. Springer, Cham, pp 61–76
MEMR (2012) Kenya wetlands atlas. Ministry of Environment and Mineral Resources, Nairobi
Montcoudiol N, Burnside NM, Györe D, Mariita N, Mutia T, Boyce A (2019) Surface and groundwater hydrochemistry of the Menengai Caldera geothermal field and surrounding Nakuru County, Kenya. Energies 12:3131
Morgan WTW (1971) The South Turkana expedition: scientific papers IV: land units of the Lokori area. Geogr J 137:14–28
Morley C, Wescott W, Stone D, Harper R, Wigger S, Karanja F (1992) Tectonic evolution of the northern Kenyan Rift. J Geol Soc Lond 149:333–348

Muasya AM, Hover VC, Ashley GM, Owen RB, Goman MF, Kimeli M (2004) Diversity and distribution of macrophytes in a freshwater wetland, Loboi Swamp (Rift Valley) Kenya. J E Afr Nat Hist Soc 93:39–47

Muohi AW, Zauke G-P, Brumsack H-J, Mavuti KM (2018) Bioaccumulation of trace elements in *Arthrospira fusiformis* algal blooms in lakes Bogoria and Sonachi, Kenya: evaluation and verification of toxicokinetic models. Lakes Reserv 78:1–10

Mutonga M (2012) Geothermal geology of Silali volcano in Kenya. Geothermal Res Counc Trans 36:943–950

Mwaura F (2006) Some aspects of water quality characteristics in small shallow tropical man-made reservoirs in Kenya. Afr J Sci Technol 7:82–96

Ndiwa TC, Nyingi DW, Agnèse J-F (2014) An important natural genetic resource of *Oreochromis niloticus* (Linnaeus, 1758) threatened by aquaculture activities in Loboi drainage, Kenya. PLoS One 9(9):e106972

Ndiwa TC, Nyingi DW, Claude J, Agnèse J-F (2016) Morphological variations of wild populations of Nile tilapia (*Oreochromis niloticus*) living in extreme environmental conditions in the Kenyan Rift-Valley. Environ Biol Fish 99:473–485

NEMA (2012) Lake Ol Bolossat integrated management plan 2008–2013. National Environment Management Authority, Nairobi

Njuguna SG (1982) Nutrient productivity relationships in tropical Naivasha basin lakes Kenya. PhD thesis, University of Nairobi

Njuguna SG (1988) Nutrient-phytoplankton relationships in a tropical meromictic soda lake. Hydrobiologia 158:15–28

Nyaberi MD (2008) Geophysical characterization of the lithology and structure of the Olobanita well field, lower Lake Baringo Basin, Kenya rift: implication on groundwater occurrence. MSc thesis, University of Nairobi

Nyambok I, Bhogal P, Bwire-Odhiambo S (1993) Hydrogeology and hydrochemistry of the groundwater resources of the floor of the Rift Valley, Nakuru North, Kenya. Proc 5th conf on geology of Kenya: Geology for sustainable development, Nairobi, pp 92–100

Nyamweru CK (1992) Origins and geomorphology of Kenya's wetlands. In: Crafter SA, Njuguna SG, Howard GW (eds) Wetlands of Kenya. IUCN, Gland, pp 5–12

Nyingi D, De Vos L, Aman R, Agnèse J-F (2009) Genetic characterization of an unknown and endangered native population of the Nile tilapia *Oreochromis niloticus* (Linnaeus, 1758) (Cichlidae; Teleostei) in the Loboi Swamp (Kenya). Aquaculture 297:57–63

Ochumba PBO, Kibaara DI (1988) An instance of thermal instability in Lake Simbi, Kenya. Hydrobiologia 158:247–252

Odek A, Otieno AB, Ambusso WJ, Githiri JG (2013) 2D-Euler deconvolution and forward modelling of gravity data of Homa-Hills geothermal prospect, Kenya. J Agric Sci Technol 15:102–113

Oketch M (2006) The potential role of constructed wetlands in protection and sustainable management of lake catchments in Kenya. In: Odada EO, Olago DO, Ochola W, Ntiba M, Wandiga S, Gichuki N, Oyieke H (eds) Proc 11th world lakes conf, Nairobi, vol 2, pp 41–46

Okumu J (2017) Lake Ol' Bolossat synthesis report. East African Wild Life Society, Nairobi

Olaka LA, Wilke FDH, Olago DO, Odada EO, Mulch A, Musolff A (2016) Groundwater fluoride enrichment in an active rift setting: Central Kenya Rift case study. Sci Total Environ 545–546:641–653

Olang MO (1984) Vegetation cover assessment in Turkana District, Kenya. Proc 3rd workshop on land evaluation for extensive grazing (LEEG), Arusha, Tanzania, pp 183–206

Omondi P, Bitok E, Mayienda R (2002) Decline of elephants and other wildlife in the Nasolot–South Turkana and Kerio Valley-Kamnarok conservation areas, Kenya. Pachyderm 33:69–73

Owen RB, Renaut RW (1986) Sedimentology, stratigraphy and palaeoenvironments of the Holocene Galana Boi Formation, NE Lake Turkana, Kenya. In: Frostick LE, Renaut RW, Reid I (eds) Sedimentation in the African rifts. Geol Soc Spec Publ 25: 311–322

Owen RB, Renaut RW, Hover VC, Ashley GM, Muasya AM (2004) Swamps, springs and diatoms: wetlands of the semi-arid Bogoria-Baringo Rift, Kenya. Hydrobiologia 518:59–78

Owen RB, Renaut RW, Jones B (2008) Geothermal diatoms: a comparative study of floras in hot spring systems of Iceland, New Zealand, and Kenya. Hydrobiologia 610:175–192

Owen RB, Renaut RW, Muiruri VM, Rabideaux NM, Lowenstein TK, McNulty EP, Leet K, Deocampo D, Luo S, Deino AL, Cohen A, Sier MJ, Campisano C, Shen C-C, Billingsley A, Mbuthia A, Stockhecke M (2019) Quaternary history of the Lake Magadi Basin, Southern Kenya Rift: tectonic and climatic controls. Palaeogeogr Palaeoclimatol Palaeoecol 518:97–118

Owino AO, Ryan PG (2007) Recent papyrus swamp habitat loss and conservation implications in western Kenya. Wetl Ecol Manag 15:1–12

Peters R, MacIntyre S (1976) Orthophosphate turnover in East African lakes. Oecologia 25:313–319

Renaut RW (1982) Late Quaternary geology of the Lake Bogoria fault trough, Kenya Rift Valley. PhD thesis, University of London

Renaut RW (1993) Zeolitic diagenesis of late Quaternary fluviolacustrine sediments and associated calcrete formation in the Lake Bogoria Basin, Kenya Rift Valley. Sedimentology 40:271–301

Renaut RW, Owen RB (1987) Les Silts de Loboi, Les Silts de Bogoria, et autres sediments littoraux. In: Tiercelin J-J, Vincens A (eds) Le demi-graben de Baringo-Bogoria, Rift Gregory, Kenya. Bull Centr Rech Explor-Prod Elf-Aquitaine 11:502–508

Renaut RW, Jones B, Le Turdu C (1999) Calcite lilypads and ledges at Lorusio hot springs, Kenya Rift Valley: travertine precipitation at the air-water interface. Can J Earth Sci 36:649–666

Roggeri H (1995) Tropical freshwater wetlands: a guide to current knowledge and sustainable management. Springer, Dordrecht

Rohland S (2013) Sedimentology of the Eight Lake, Kenya. https://www.uni-potsdam.de/fileadmin/projects/umwelt/Studium_Geoeokologie/Geoökologie_BSc_MSc/GÖ_III-Projekte/2013_rohland_web.pdf

Rop BK (2014) Petroleum prospects of the Lotikipi Basin, NW-Kenya: based on gravity and seismic data. Proc Int Conf Sustainable Res Innov, Nairobi 5:95–99

Rouwet D, Németh K, Tamburello G, Calabrese S, ISSA (2021) Volcanic lakes in Africa: the VOLADA_Africa 2.0 database, and implications for volcanic hazard. Front Earth Sci 9:717798

Rowntree KM (1989) Rainfall characteristics, rainfall reliability and the definition of drought: Baringo District, Kenya. S Afr Geogr J 71:74–80

RTI (Radar Technologies International) (2013) Advanced survey of groundwater resources of Northern and Central Turkana County, Kenya. Final Technical Report

Rumes B, Van der Meeren T, Martens K, Verschuren D (2016) Distribution and community structure of Ostracoda (Crustacea) in shallow waterbodies of southern Kenya. Afr J Aquat Sci 41:377–387

Saggerson EP (1952) Geology of the Kisumu District. Rep Geol Surv Kenya 21

Seegers L, de Vos L, Okeyo D (2003) Annotated checklist of the freshwater fishes of Kenya (excluding the lacustrine haplochromines from Lake Victoria). J East Afr Nat Hist 92:11–47

Shackleton RM (1945) Geology of the Nyeri area. Rep Geol Surv Kenya 12:1–26

Sikes HL (1936) Notes on the hydrology of Lake Naivasha. J East Afr Uganda Nat Hist Soc 13:73–89

Sippel J, Meeßen C, Cacace M, Mechie J, Fishwick S, Heine C, Scheck-Wenderoth M, Strecker MR (2017) The Kenya Rift revisited: insights into lithospheric strength through data-driven 3-D gravity and thermal modelling. Solid Earth 8:45–81

Tarras-Wahlberg H, Harper D, Tarras-Wahlberg N (2004) A first limnological description of Lake Kichiritith, Kenya: a possible reference site for the freshwater lakes of the Gregory Rift Valley. S Afr J Sci 99:494–496

Terer T, Muasya AM, Dahdouh-Guebas F, Ndiritu GG, Triest L (2012) Integrating local ecological knowledge and management practices of an isolated semi-arid papyrus swamp (Loboi, Kenya) into a wider conservation framework. J Environ Manag 93:71–84

Terer T, Odeny D, Kirika P, Mbale M, Wamiti W, Njoroge P, Kochey J, Kosgei G, Gathua J, Nyawira M, Yego R, Muchai V, Malonza K (2019) Lake Ol'Bolossat rapid biodiversity assessment results. Birdlife International and E Africa Wildlife Society, Kenya

Thenya T (2006) Analysis of macrophyte biomass productivity, utilization and its impact on various eco-types of Yala Swamp, Lake Victoria Basin, Kenya (Ecology and development series, vol 45). Cuvillier Verlag, Göttingen

Thenya T, Ngecu WM (2017) Indigenous strategies and dynamics of resource utilization in tropical wetland. A case study of Yala swamp, Lake Victoria Basin. Kenya Int J Arts Commerce 6:21–39

Thompson AO, Dodson RG (1963) Geology of the Naivasha area. Rep Geol Surv Kenya 55

Tiner RW (2017) Wetland indicators: a guide to wetland formation, identification, delineation, classification, and mapping, 2nd edn. Taylor and Francis, Boca Raton

Truckle P (1977) The geology of the area to the south of Lokori, South Turkana, Kenya. PhD thesis, University of Leicester

Verschuren D (1999a) Sedimentation controls on the preservation and time resolution of climate-proxy records from shallow fluctuating lakes. Quat Sci Rev 18:821–837

Verschuren D (1999b) Influence of depth and mixing regime on sedimentation in a small, fluctuating tropical soda lake. Limnol Oceanogr 44:1103–1113

Verschuren D, Cocquyt C, Tibby J, Roberts C, Leavitt P (1999) Long-term dynamics of algal and invertebrate communities in a small, fluctuating tropical soda lake. Limnol Oceanogr 44:1216–1231

Virani MZ, Monadjem A, Thomsett S, Kendall C (2012) Seasonal variation in breeding Rüppell's vultures *Gyps rueppellii* at Kwenia, southern Kenya and implications for conservation. Bird Conserv Int 22:260–269

Von Höhnel L (1891) Discovery of lakes Rudolf and Stefanie. Longmans, London

Wafula EN, Murunga SI (2020) Isolation and identification of phosphate solubilizing and nitrogen-fixing bacteria from Lake Ol'Bolossat sediments, Kenya. Mod Appl Sci 14:37

Walsh J, Dodson RG (1969) Geology of Northern Turkana. Rep Geol Surv Kenya 82

Wamiti W, Gichuki N, Sigana D, Njoroge P, Morrison K, Ndung'u G, Rubalcaba J (2020) Water depth, vegetation height and water edge distance are critical in nest-site selection for Grey Crowned Crane at Lake Ol' Bolossat, Kenya. https://doi.org/10.22541/au.158880177.75891321

Wescott WA, Stone DM, Wigger ST (1995) Geological and geophysical reconnaissance of the Lotikipi plain of northwestern Kenya and its relationship to the northern Kenya Rift. J Afr Earth Sci 21:241–251

Wescott WA, Wigger ST, Stone DM, Morley CK (1999) Geology and geophysics of the Lotikipi Plain. In: Morley SK (ed) Geoscience of rift systems–evolution of East Africa. AAPG Stud Geol 44:55–65

Wetang'ula GN, Kubo BM, Were JO (2009) Environmental baseline study for geothermal developments: case study Arus-Bogoria geothermal prospects, Kenya. Short course IV on exploration for geothermal resources, UNU-GTP, KenGen and GDC, Lake Naivasha, November 1–22, 2009

Wilkinson AF (1988) Geology of the Allia Bay area. Rep Mines Geol Dept Kenya 109

Williams LAJ, MacDonald R, Chapman GR (1984) Late Quaternary caldera volcanoes of the Kenya Rift Valley. J Geophys Res 89:8553–8570

Worthington E (1932) Scientific results of the Cambridge Expedition to the East African Lakes, 1930-1. General introduction and station list. J Lin Soc Zool 38:99–119

Yasindi AW (2001) The ecology of planktonic ciliates in tropical lakes of East Africa. PhD thesis, University of Waterloo, Ontario

Yasindi AW, Taylor WD (2016) The protozoa of soda lakes in East Africa. In: Schagerl M (ed) Soda lakes of East Africa. Springer, Cham, pp 179–205

Zacharia M, Elias A, Kiplagat J, Maingi S, Olang LO (2013) Assessment of land cover changes in Lake Olbolosat region of the Central Kenyan Highlands using Landsat satellite imagery aided by indigenous knowledge. J Biodivers Manage Forest 2:2

Part III
The Ancient Kenya Rift Lakes

Chapter 15
The Turkana Basin

15.1 Introduction

Several major lithostratigraphic units have been recognised in the Turkana Basin from outcrop data (Fig. 15.1). These include (1) Precambrian to Lower Palaeozoic basement gneiss and schist; (2) Late Cretaceous to Early Palaeogene immature, coarse-grained arkose, formerly referred to as the Turkana Grits (Arambourg 1933; Murray-Hughes 1933; Fuchs 1939; Dixey 1944; Joubert 1966; Baker et al. 1972; Rhemtulla 1970; Wescott et al. 1993; Agyemang et al. 2019); (3) Late Eocene to Miocene basalt and rhyolite with localised sedimentary intercalations (Bellieni et al. 1987; Brown and McDougall 2011); (4) Oligo-Miocene deposits; and (5) Plio-Pleistocene and Holocene sediments and volcanics (Walsh and Dodson 1969; Baker et al. 1972; Bellieni et al. 1981, 1987; Zanettin et al. 1983; Williamson and Savage 1986).

Controversy developed for many years over whether the Turkana Grits formed a single unit or were deposited during two different time periods. Savage and Williamson (1978) and Williamson and Savage (1986) supported earlier suggestions by Arambourg (1933, 1943) for two distinct cycles of deposition, providing evidence from the Kajong area of southeast Turkana for a Late Mesozoic–Early Palaeogene period of sedimentation with a second phase of deposition during the Miocene. Due to the uncertainties surrounding the term 'Turkana Grits' this designation has since largely been abandoned and local names are used to define each of the sedimentary sequences that are exposed in different locations (Morley et al. 1992).

The older, Late Cretaceous to Lower Palaeogene sediments, formerly placed in the Turkana Grits, include the Lapur (Lappur, Labur) Sandstone Formation (Wescott et al. 1999; Thuo 2009; Tiercelin et al. 2012), the Muruanachok Sandstone Formation (Walsh and Dodson 1969; Wescott et al. 1993; Agyemang et al. 2019) and the

R. W. Renaut, R. B. Owen, *The Kenya Rift Lakes: Modern and Ancient*,
Syntheses in Limnogeology, https://doi.org/10.1007/978-3-642-25055-2_15

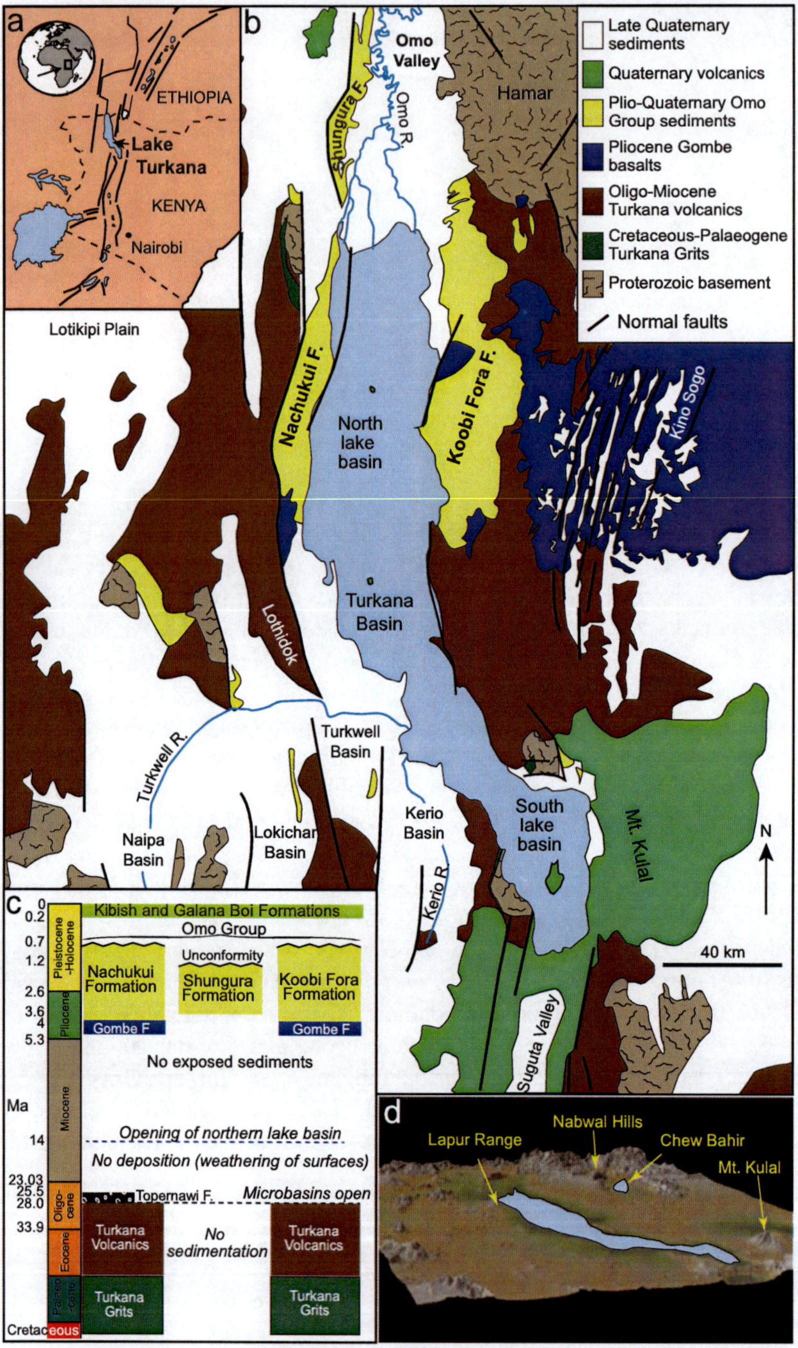

Fig. 15.1 Geology, stratigraphy and topography of the Turkana Basin. Modified after Nutz et al. (2020). (**a**) Regional location. (**b**) Major volcanic and sedimentary units. (**c**) Generalised stratigraphy. (**d**) Topography

sandstones of the Lariu Formation and Sera Iltomia Formation (Fig. 15.2; Morley et al. 1992; Wescott et al. 1993; Tiercelin et al. 2004). Walsh and Dodson (1969) had interpreted these 'Turkana Grits' as being lacustrine (in the broadest sense), based on the presence of current bedding, well-rounded pebbles and the presence of calcareous mudstone, grit (coarse sandstone) and shale. However, the deposits are now recognised as mainly fluvial, including distal alluvial fan and braided stream settings. Shale near the base of the Lapur Sandstone Formation contains burrows and root moulds, and these sediments have been reinterpreted as having formed on a floodplain (Thuo 2009; Tiercelin et al. 2012).

Miocene sediments contrast with older deposits in containing volcanic clasts derived from Oligo-Miocene volcanics (Fig. 15.2). Several depositional sequences, such as the Kajong Formation ('5' in Fig. 15.2a), were previously included in the Turkana Grits and erroneously thought to be of Late Cretaceous-Early Palaeogene age. However, Brown et al. (2016) reported that the formation is capped by a basalt, $^{40}Ar/^{39}Ar$-dated at 19.1 ± 0.1 Ma, which contains volcanic clasts that yielded an age of 20.3 Ma.

15.2 Oligo-Miocene Lacustrine Deposits

15.2.1 The Ekitale Basin

Ragon et al. (2018) described sediments in the Ekitale Basin from the western shoulder of the northern Turkana Depression ('1' in Fig. 15.2a, c) and showed that they are separated from outcrops of the Nachukui Formation by the Murua Rith-Lapur Fault. The basin is up to 5 km wide and includes the 75-m-thick Topernawi Formation (28–25 Ma). Five major units have been described with the lower part including lacustrine fluvial-derived turbidite and fan-delta deposits that give way upwards to fluvial and pyroclastic units.

15.2.2 The Lokichar and North Lokichar Basins

Oligo-Miocene Outcrops

Oligo-Miocene lake sediments in the Turkana region are best described from the Lokichar and North Lokichar basins southwest of Lake Turkana (Macgregor 2015; Boone et al. 2018). Boone et al. (2019) reported denudational cooling of metamorphics that form the bounding basin for rifting at Lokichar and noted that these pre-rifting events took place during the Eocene to Miocene. Subsequently, the Lokichar half-graben developed and was bounded to the west by the Lokichar Border Fault with an opposing flexural margin to the east. The basin contains up to 7 km of fill, with lacustrine sediments attaining their greatest thickness close to the Lokichar Fault (Fig. 15.3a, b, d; Morley 1989, 1999; Morley et al. 1999). Outcrops are

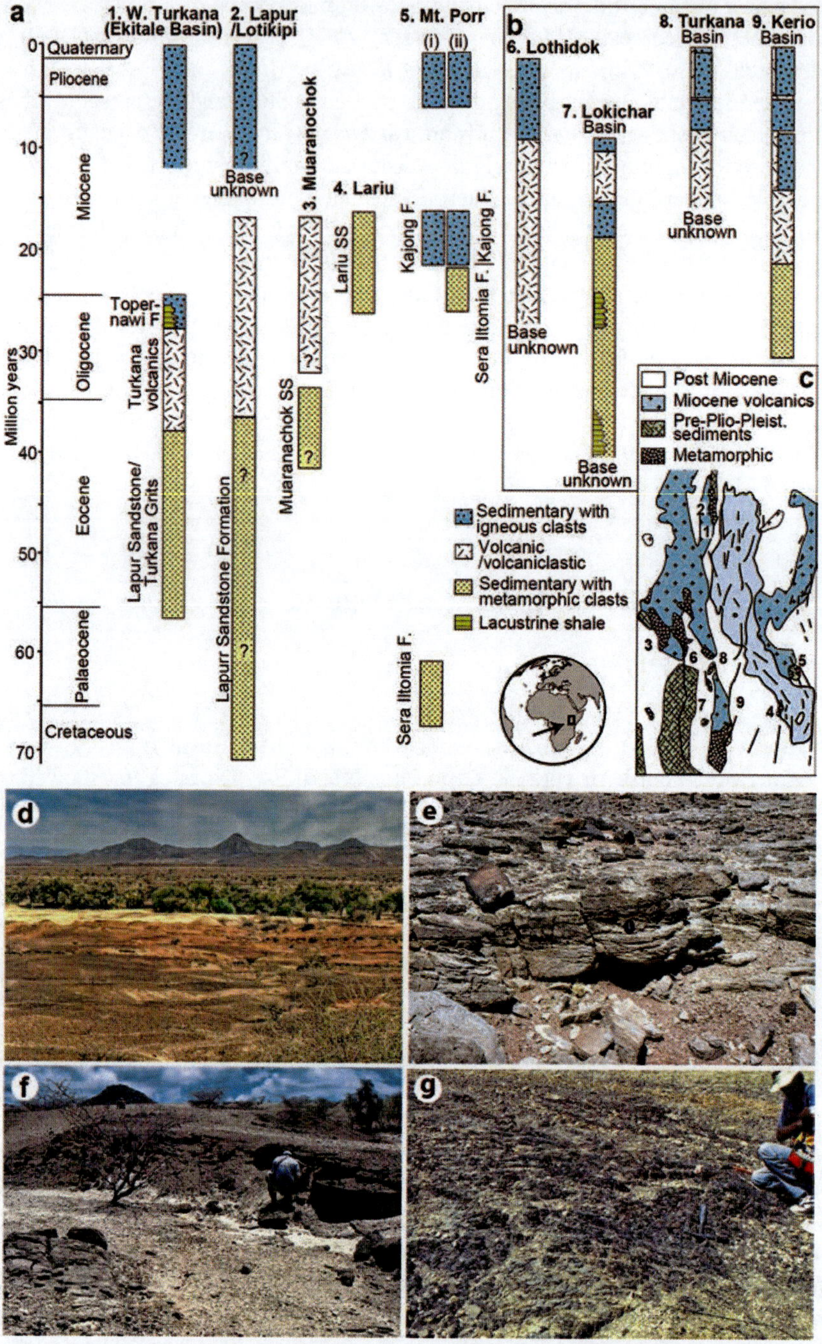

Fig. 15.2 Summary of major sedimentary units in the Turkana region. Based on Morley et al. (1999), Williamson and Savage (1986) and Tiercelin et al. (2012). West Turkana (Ekitale Basin) from Ragon et al. (2018). Sediments derived from the metamorphic basement are mostly fluvial arkose. Younger sediments include volcanic clasts derived from Oligo-Miocene volcanics and

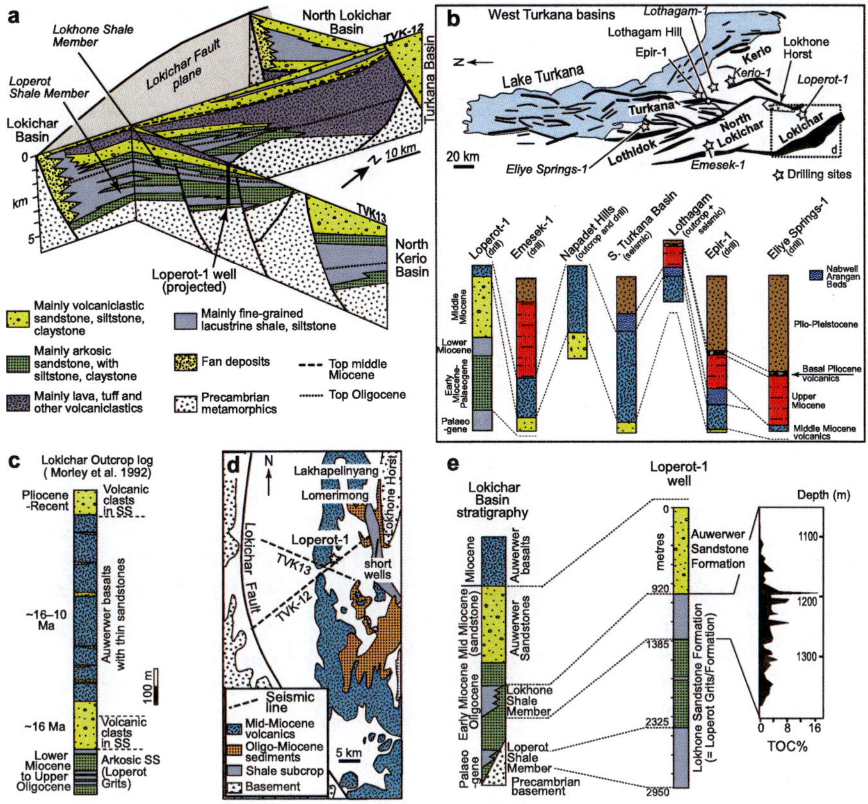

Fig. 15.3 West Turkana Oligo-Miocene basins and sediments. (**a, d**) modified from Talbot et al. (2004); (**b, c**) based on Morley et al. (1999) and Schofield et al. (2021). (**a**) Generalised fence diagram based on seismic sections. (**b**) Location of the main west Turkana basins, major faults (line thickness reflects displacement) and drill cores. Stratigraphic sections based on drill, outcrop and seismic data. (**c**) Lokichar sediment sequence based on outcrops near the Lokhone Horst. Key shown in (**a**). (**d**) Geological map of the Lokichar Basin. Note the location of two seismic profiles shown in (**a**) and the Loperot-1 well. (**e**) Generalised Lokichar Basin stratigraphy and sediment sequence in the Loperot-1 well. The Total Organic Carbon for the Lokhone Shale Member is shown to the right

Fig. 15.2 (continued) were formed in fluvial and lacustrine environments. (**a**) Sections based on outcrops (Lapur section also correlated with seismic data for the Lotikipi Basin). Note the two contrasting models for the Mt. Porr area involving contrasting dates for the Sera Iltomia Formation. Model (i) after Williamson and Savage (1986). Model (ii) after Wescott et al. (1993). (**b**) Sections based on seismic and/or borehole data. Note two prominent lacustrine shales in the Lokichar Basin. (**c**) Location of numbered sections. (**d–g**) Typical outcrops, Lokichar Basin. (**d**) General view of gently dipping bedded sandstones; (**e**) Cross-bedded coarse arkosic sandstone; (**f, g**) Outcrops of organic-rich lacustrine Lokhone Shale exposed on the opposing eastern flexural margin where bone-bearing coarse sandstone with thin gastropod-bearing limestone rests on metamorphic basement of the Lokhone Horst (Morley et al. 1992).

Near Loperot and the Lokhone Horst (Fig. 15.3b), outcrops of basal fluvial arkosic sand and grit (the Loperot Grits/Formation) contain intercalations of lacustrine black and green shale ranging from several to tens of metres in thickness (Figs. 15.2d–g and 15.3c; Morley et al. 1992). These sediments contain freshwater algae (*Pediastrum* and *Botryococcus*) together with spores of *Margocolporites*, *Echitriporites*, *Racemonocolpites*, *Praedapollis protrudentiporatus* and *Retibrevitricoporites* that suggest an Upper Oligocene to Lower Miocene age. The overlying Auwerwer Sandstone Formation (Fig. 15.3c, e) contains volcaniclastic materials, but with only minor fine-grained horizons that could represent ephemeral lake conditions. A fossil beaked-whale has been reported from coarse sand and conglomerate near Loperot (Mead 1975; Wichura et al. 2015) with fossil fish, gastropods, bivalves, crocodiles, testudinids, chelonians and a variety of mammals also present. These accumulated in part of a major river system, which would have been required to explain the presence of a 20 ft (6.1 m) whale. The remaining Loperot sequence is dominated by basalt overlain by Plio-Pleistocene sediments.

Liutkus-Pierce et al. (2019) noted that deposits of similar age to the north of Loperot had been associated with rainforests (Vincens et al. 2006). However, their geochemical proxies and $\delta^{13}C$ data indicated that, at least intermittently, Loperot had shifted from a semiarid climate with riparian forests to a more arid open setting. They also argued against previous suggestions that the early Miocene was characterised by humid closed-canopy forests, inferring a dynamic changeable palaeoenvironment instead.

Sediments resting on metamorphic rocks of the Lokhone Horst, discovered during a search for Oligocene vertebrates (Leakey et al. 2011), are composed mainly of late Early Oligocene to Late Oligocene sandstones (Ducrocq et al. 2010). These have been correlated with the predominantly fluvial Lokhone Sandstone Formation (= Loperot Grits/Formation in outcrop) described from the Loperot-1 drill site (Fig. 15.3d, e; Morley et al. 1999). The deposits include "several decimeter-thick slabs" of fine sandstone with abundant molluscs and a few mudstone horizons with mudcracks. They accumulated along the shallow margins of a large lake that occupied the Lokichar Basin. The sediments also yielded turtle (*Pelomedusoides*) and crocodile fossils, as well as fish teeth and spines. At least five actinopterygian (ray-finned fish) orders have been recorded that collectively imply a freshwater environment. Palaeocurrent data indicate sediment input from the south and thus dominant axial drainage within the Lokichar half-graben.

The Subsurface Sedimentary Record

Evidence for lake environments in the remainder of the Lokichar Basin derives from seismic data and drill logs (Fig. 15.3a). The lacustrine shale observed in outcrop is also present in a series of short wells along seismic line TVK-12, within 3 km of the Lokhone Horst (Fig. 15.3d; Morley et al. 1999), where they are laterally continuous along strike and are up to 100 m thick. Microfossils from the shale include the same freshwater algae found in surface exposures (*Pediastrum* and *Botryococcus*) and a similar Late Oligocene-Early Miocene assemblage of palynomorphs, including *Margocolporites verwijhei*, *Echitriporites kwaense*, *Racemonocolpites hians*, *Praedapollis protrudentiporatus* and *Retibrevitricoporites*. The shale is thermally immature ($R_0 = 0.32–0.37\%$), with Total Organic Carbon (TOC) ranging between 0.05 and 11.0%. The location of the shale near the sand-prone flexural boundary of the basin led to considerable oil exploration interest and stimulated further studies of the sequence closer to the basin centre and the western boundary fault where potential oil source rocks would be more likely.

Shales reported from short cores thicken to 460 m in the Loperot-1 well, across a distance of ~7 km, where they form part of the Late Oligocene to Early Miocene Lokhone Shale Member (LSM) of the Lokhone Sandstone Formation (Fig. 15.3d, e; Morley et al. 1999; Rop 2011, 2015; Rop and Patwardhan 2013). These deposits are separated from the Eocene (?) to Early Oligocene Loperot Shale Member (LoSM) by fluvio-deltaic deposits at core depths of 1385–2325 m (Fig. 15.3e). Further west, the two shales merge and attain a combined thickness of about 1350 m near the Lokichar Fault (Fig. 15.3a). TOC values in the LSM range between 1 and 17% (Fig. 15.3e; Morley et al. 1999) with the LoSM yielding lower values of 0.2–3.3%, but with the greater burial depth resulting in increased maturity with respect to oil ($R_0 = 1.1\%$; Morley et al. 1999).

The LSM in outcrop has yielded TOC values of <1%, with higher values of 0.1–7.15% in the Loperot-1 core (Talbot et al. 2004). Hydrogen Indices (HI) have been reported as from <90 to 611, with petroleum potential S2 values for organic-rich horizons with >5 and occasionally >10 mg HC g^{-1}, which indicates good source-rock potential. RockEval T_{max} values of 438–452 °C for organic-rich shale suggest that parts of the sequence are close to oil-generating conditions. It has also been noted that the underlying LoSM may be close to the change from oil to wet gas generation (Talbot et al. 2004).

The upper part of the LSM contains relatively low TOC and abundant palynomacerals (cuticles, woody fragments), exoskeleton remains and structureless organic matter related to spores and pollen (Talbot et al. 2004). Amorphous Organic Matter (AOM) dominates in the lower part of the member, together with fungal remains, cysts of Prasinophycean algae, *Botryococcus* spp. and *Pediastrum* spp. The latter two taxa indicate freshwater conditions. Pollen indicate a mosaic of semi-deciduous forests and humid woodlands with similarities to the Guineo-Congolian/Zambezia phytogeographical zone (Vincens et al. 2006).

The LoSM is generally associated with lower TOC values and abundant AOM and freshwater coenobial algae, fungal spores, hyphae and a few spores and pollen. Abundant *Pediastrum* and *Botryococcus* in parts of the LoSM indicate eutrophication. Talbot et al. (2004) concluded that the oil-prone parts of the sedimentary sequence formed in a large freshwater lake with permanently anoxic conditions below a chemocline and that the palaeolake might have resembled modern Lakes Tanganyika and Malawi. They also noted that sequences with low organic matter, with no hydrocarbon potential, may have originated in settings subject to greater fluvial inputs of highly degraded organic matter.

Facies Distributions and Sediment Sources

Transgression and regression cycles of the Oligo-Miocene Lokichar palaeolakes are demonstrated in seismic sections by the distribution of the lacustrine shale (Fig. 15.3a). The spatial extent of the palaeolakes was likely determined in part by climate, but also reflects the role of tectonics (Morley and Wescott 1999), especially through the influence of the western boundary fault and related flexural movements of the opposing Lokhone hinge margin. Morley and Wescott (1999, for example, noted that as boundary fault subsidence deepened the western side of the Lokichar basin, a roll back was taking place on the eastern margin, which resulted in transgressive eastward shifts of the Lokichar water body. Initially, two lakes developed with separate western boundary faults (Fig. 15.4a, b). Subsidence rates in the northern basin were probably in balance with sediment inputs resulting in the accumulation of fluvio-deltaic sediments. The southern basin experienced greater subsidence and lacustrine deposition occurred over a wider area. Subsequent fault propagation caused the two boundary faults to merge with time and form a single basin (Fig. 15.4d), but with the area of maximum subsidence shifting northwards and southwards through time, resulting in a series of lake transgressions and regressions. Fluvial progradation into the basin may have also been caused by uplift in source areas, the development of new volcanic centres, or external events such as river capture (Morley and Wescott 1999).

The main sediment sources into the Lokichar Basin were axial and derived from the north or south (Morley et al. 1999). Sediments derived from the east were probably minimal due to the presence of the intervening Oligo-Pliocene Kerio half-graben. With small drainage basins, boundary faults to the west would have provided little sediment. Quartzo-feldspathic arkose constitutes most of the coarse clastics in the Lokhone Formation, suggesting derivation from the granitic basement to the south and southwest (Tiercelin et al. 2004), which is supported by palaeocurrent studies (Wescott et al. 1993). Igneous clasts increase in the Auwerwer Sandstone Formation in the Loperot-1 well and in outcrops at the hinge margin near Lomerimong (Tiercelin et al. 2004) and Lakhapelinyang (Fig. 15.3d; Joubert 1966; Morley et al. 1999). Palaeocurrent data for the lower part of the Lomerimong section, for example, indicate an axial source to the south-southeast of the Lokichar Basin, with increased volcanic clasts linked to the eruption of the Lower Miocene Samburu Basalts Formation. Tiercelin et al. (2004) also noted that extensive Oligo-Miocene volcanism occurred to the north, with much of this material eroding to supply rivers that fed into the Lothidok Basin (Fig. 15.3b).

15.2.3 Other Oligo-Miocene Sediments in the Turkana Region

Schofield et al. (2021) summarised the stratigraphic relationships between the Lokichar, North Lokichar, South Turkana and North Kerio basins (Fig. 15.3b) using outcrop, drill, and seismic data. They noted that the Emesek-1 well, for example, penetrated a 200 m thick sequence of subvolcanic sandstones with minimal volcanic

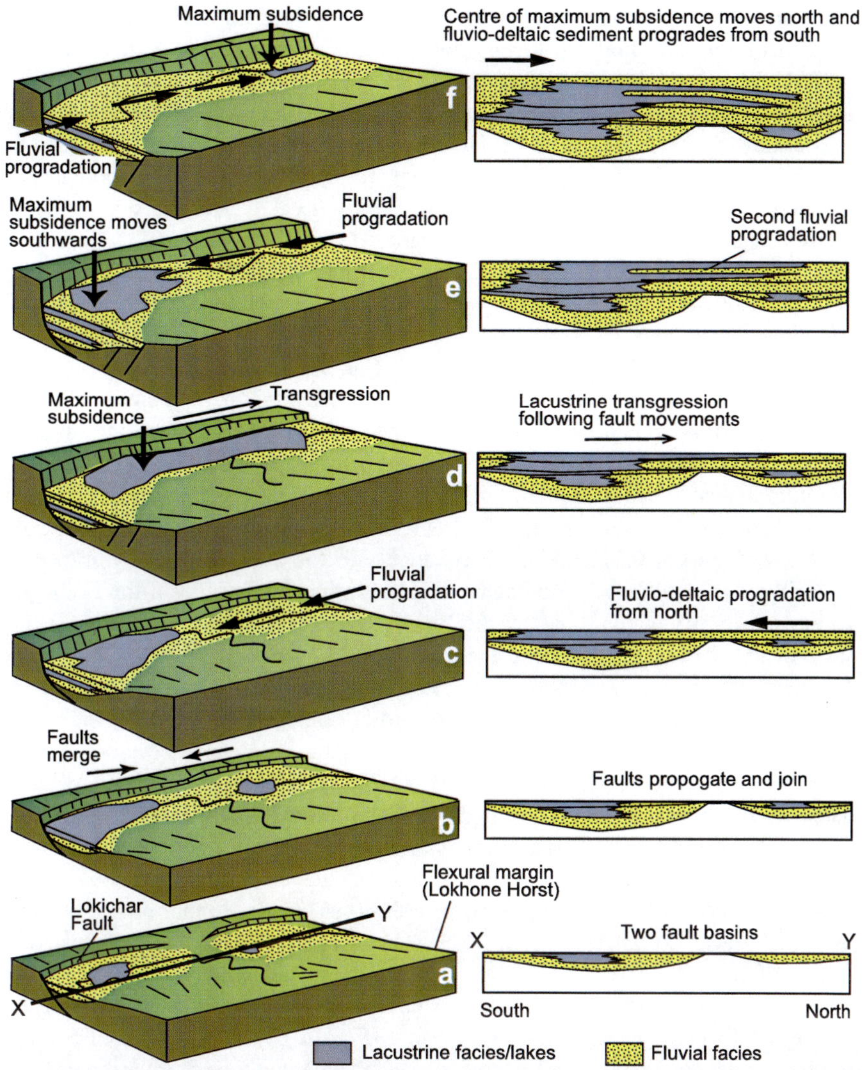

Fig. 15.4 Oligo-Miocene lake evolution in the Lokichar and North Lokichar basins. Modified from Morley and Wescott (1999). (**a**) Two separate basins formed initially. (**b**) Faults propagate towards each other, with a basement high where the faults merge. Subsidence leads to lake expansion in the southern area. (**c**) Greater subsidence to south leads to southward fluvial progradation from the north towards the deepening basin low. (**d**) Fault displacements shift lacustrine and fluvial depocentre towards the boundary fault and cause a northward lake expansion. Maximum subsidence remains in the southern part of the merged basins. (**e**) Southward shift in the zone of maximum subsidence results in a second major phase of fluvial progradation. (**f**) Centre of maximum subsidence moves north, and the lake basin shifts with it. Rivers prograde northwards from the south. Subsequently, the Auwerwer basalts infill the basin followed by further fluvial deposition (not shown)

clasts and pointed out that the Plio–Pleistocene Turkana Fault had developed along an area of former Middle Miocene volcanism. Eruption of the Kalakol Basalts (~28–18 Ma) in the Lothidok Basin occurred at about the same time as lacustrine deposition in the Lokichar Basin (Fig. 15.5; Morley et al. 1999). These basalts are overlain by the fluvial Lower to Middle Miocene Lothidok Formation (~18–13 Ma; Fig. 15.5a), which has yielded mammal fossils, including Hippopotamidae (Boisserie et al. 2017) and important hominoids (Leakey and Leakey 1986a, b; Leakey et al. 1995; Brown and McDougall 2011; Feibel 1983, 2011; Hammond et al. 2019). The lower part of the Lothidok Formation is dominated by reworked volcaniclastic sand, gravel and lahar deposits with several tephra horizons and is overlain by lava flows, lahars and ash, together with fluvial conglomerate, sandstone and mudstone (Boschetto 1988; Boschetto et al. 1992). *Etheria* 'reefs' and biostromes are common in the upper sequence, indicating permanent rivers. The Lothidok Formation is overlain, in turn, by the Loperi Basalt and the Late Miocene to Early Pliocene Turkwell beds, which also crop out at Napadet (Fig. 15.5c).

Oligo-Miocene sediments are present in other basins to the southwest of Lake Turkana, but large parts are buried. Bedrock channels and fluvial sandstone form the lowest sediments of the Kerio Basin sequence and crop out near Lariu (Morley et al. 1992; Wescott et al. 1993). Lothagam Hill lies immediately west of the Lothagam Fault (Fig. 15.3b), which separates the Turkana and Kerio basins and provides exposures of upper Miocene–Pliocene age resting on Miocene alkali basalt (~8–17 Ma) (Powers 1980; Leakey and Harris 2003; Feibel 2003b; Wynn 2003; McDougall and Feibel 2003). The Miocene part of the record at Lothagam (Fig. 15.5a) includes a basal volcaniclastic conglomerate of the Nabwal Arangan beds that were laid down by proximal alluvial fans (Leakey et al. 1996). These sediments are capped by basalt dated at 9.1 Ma, which is overlain by fluvial conglomerate, sandstone and mudstone (Nawata Formation). The Nile oyster, *Etheria elliptica*, is common in channel sandstone of the overlying Lower Nawata Formation, suggesting perennial flow during deposition (Feibel 2011) with aquatic fossils (fish, turtle, hippo, crocodile) also present (Stewart 2003a, b; Storrs 2003; Weston 2003; Wood 2003; Stewart and Rufolo 2018).

Late Miocene to Pliocene lacustrine shale, carbonate and sandstone are present in two short wells – Lothagam-1 and Kerio-1 (Fig. 15.3b) with the former containing freshwater *Pediastrum* and *Botryococcus* in shale and the latter barren of these microfossils (Morley et al. 1999). Both wells were characterised by low organic content. The 2694-metre-long Eliye Springs-1 penetrated Late Miocene to Recent fluvio-lacustrine sandstones and shales (Fig. 15.3b). Microfossil and organic data suggest fluvio-lacustrine, lacustrine and terrestrial (savannah) conditions. Morley et al. (1999) concluded that there was no evidence for a long-lived proto-Lake Turkana. Ongoing petroleum exploration in these basins is providing new core and geophysical data that continue to add knowledge of the Tertiary history of the region.

A Miocene lake of uncertain extent developed at Locherangan, northwest of Lake Turkana (Fig. 15.5c), where 49.1 m of interbedded lacustrine sediment, fluvial sandstone and conglomerate have been dated at ~17.5 Ma (Anyonge 1991). The main facies include red-brown, greenish grey and dark green mudstone,

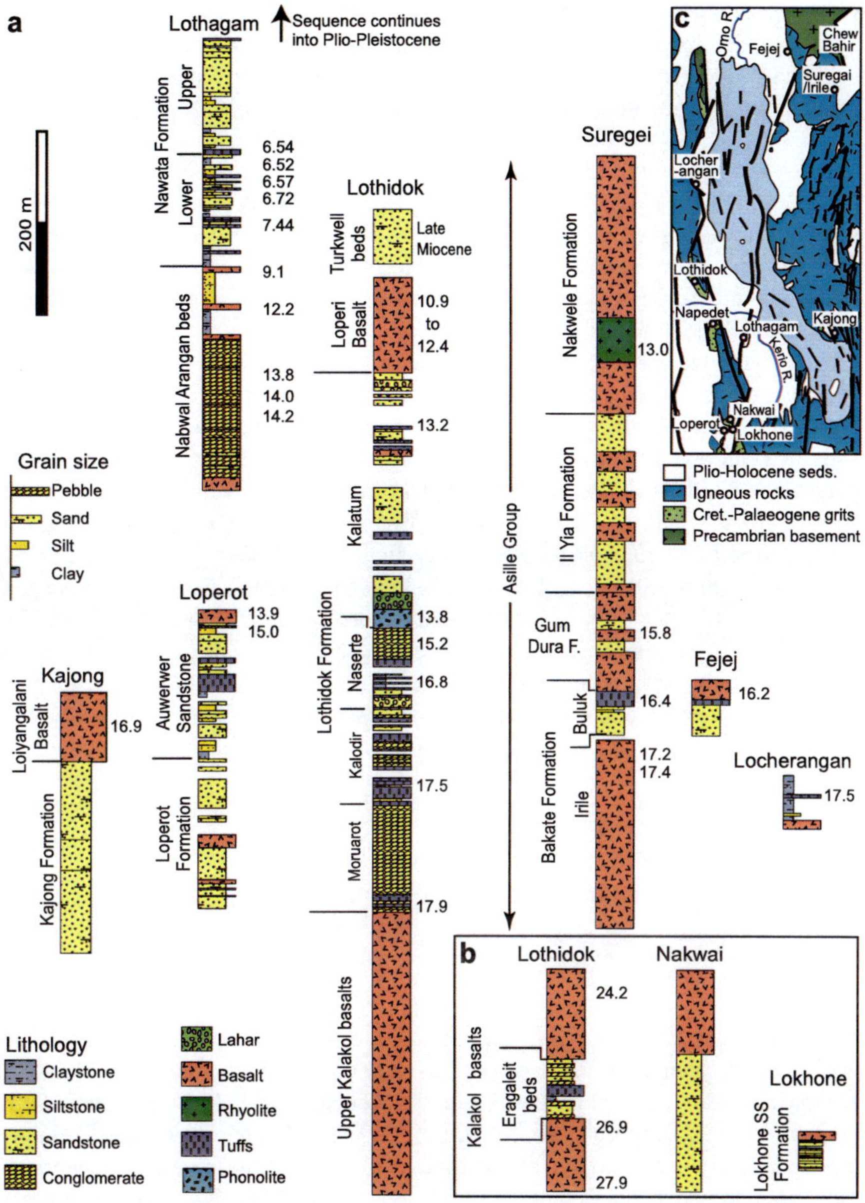

Fig. 15.5 Oligo-Miocene sedimentary formations. Data based on outcrops only and modified from Brown and McDougall (2011). Dates to the right of columns are Ma. (**a**) Miocene sections. Fejej section is schematic. (**b**) Oligocene sections. Eragaleit beds after Boschetto et al. (1992). Nakwai section is schematic. (**c**) Section locations

yellowish-green siltstone, and claystone with fish bone, turtle and crocodile. Olive green to pale brown shale contains abundant fish scales and bone, plus ostracods that locally dominate in thin (<5 cm) fine-grained strata. Fluvial deposits with abundant mammal bones are common at the top of the sequence.

Miocene fluvial sediments have also been reported northeast of the lake at Suregei, forming part of the Asille Group, which includes the Buluk Member of the Bakate Formation (Fig. 15.5a, c; Watkins 1989). These deposits include a basal reddish-brown claystone, originally described as lacustrine (Harris and Watkins 1974). However, this deposit was reinterpreted as forming in a fluvial overbank setting, since it lacks lamination or fossilized fish, and contains carbonate-cemented rootcasts and root nodules (rhizocretions).

Playa lake sediments have been reported southeast of Lake Turkana from the Kajong Formation (Fig. 15.5a, c; Savage and Williamson 1978; Williamson and Savage 1986; Rop 1990), which lacks accurate dating, but is believed to be Miocene and has been divided into four members. The basal Member 1 accumulated in a playa and includes calcrete, halite lenses and stromatolites, with *Etheria* reefs and both freshwater and terrestrial molluscs (*Melanoides*, *Mutela*, *Cerastua*). Fluvial sandstone and calcrete dominate the overlying three members.

Renaut et al. (2002) described a fossil hot-spring fissure ridge and associated spring pools at Ngakoringora, located between the Lothidok Hills and Lodwar. The undated travertine, composed of calcite but which is partly silicified and contains dolomite, could be of Late Miocene age, but because it was emplaced in now denuded, younger strata, could be Pliocene or Pleistocene in age. More study is needed but it provides evidence of hydrothermal discharge into contemporary palaeolakes.

15.3 Plio-Pleistocene Lacustrine Deposits

15.3.1 Historical Background

Plio-Pleistocene sediments crop out in many locations around Lake Turkana. The earliest research on these deposits involved exploratory palaeontological collections in the Omo Valley in 1902 followed by a full expedition in 1930 (Arambourg 1948; Guillemot 1997; Drapeau et al. 2014). From about 1967, the Omo Valley became the focus of intensive geological and palaeontological studies through the work of the 'International Omo Research Expedition' (IORE) leading to many seminal papers (de Heinzelin and Brown 1969; Arambourg et al. 1969; Butzer 1969a, b; Butzer and Thurber 1969; Leakey 1969; Bonnefille et al. 1970; Brown and Lajoie 1970; Coppens 1975; Merrick and Merrick 1976; Peypouquet et al. 1979; de Heinzelin 1980, 1982; Brown 1983; Davidson 1983; Howell and Coppens 1983; Howell et al. 1987; Eck and Jablonski 1987).

IORE and subsequent researchers in the Omo and other parts of the Turkana Basin focussed on fluvio-lacustrine deposits (Harris et al. 2006) in an effort to find and better understand the vertebrate fossils, especially hominids and their evolutionary development (Bobe 2002; Hernández Fernández and Vrba 2006; Bobe et al. 2007, 2022; Joordens 2007; Kingston 2007; Boisserie et al. 2008; Cerling et al. 2011b; Werdelin and Manthi 2012; Geraads et al. 2013; Maurin et al. 2014; Barr 2015; Behrensmeyer 2015; Cerling et al. 2015; Harmand et al. 2015; Negash et al. 2015; Potts and Faith 2015; Rowan et al. 2015; Quinn 2015; Quinn et al. 2020). Butzer and Thurber (1969) first defined the Mursi, Nkalabong and Kibish formations in the Omo Valley and outlined the stratigraphy of the Shungura and Usno formations, which were described by de Heinzelin and Brown (1969), de Heinzelin (1970), de Heinzelin et al. (1976) and Merrick et al. (1973). Butzer (1971, 1976) presented further stratigraphic and sedimentological details for the Plio-Pleistocene sequences. Subsequently, de Heinzelin (1983) included all tilted and faulted Plio-Pleistocene strata, but excluding the Kibish Formation, within the Omo Group, which was later extended to include the Koobi Fora and Nachukui formations along the eastern and western shores of Lake Turkana, respectively.

Correlating the various sedimentary units was a major problem, with scientists using palaeomagnetism (Brock 1974; Brock and Isaac 1976; Brown and Shuey 1976; Brown et al. 1978; Hillhouse et al. 1986; Kidane et al. 2014), fission track, K/Ar and $^{40}Ar/^{39}Ar$ techniques (Brown 1994; McDougall and Feibel 2003). Correlations were made between ash horizons using trace elements in volcanic glass (Cerling 1977; Cerling et al. 1979; Martz 1979; Martz and Brown 1981). Brown (1969) published the first K/Ar results from the Omo region, with Fitch and Miller (1969) reporting $^{40}Ar/^{39}Ar$ data for the Nkalabong Formation. More dates followed (Brown et al. 1970, 1985; Brown and Lajoie 1970; Brown 1972; Brown and Nash 1976; Haileab et al. 2004) and gradually a firmer chronostratigraphic framework was developed for the Turkana Basin (Feibel et al. 1989; Brown and McDougall 2011).

In 1968, Richard Leakey and fellow scientists began work on Plio-Pleistocene sediments northeast of Lake Turkana using many palaeontological, sedimentological, geochemical, palaeoenvironmental and dating methods that continue to be used, and leading to an extensive series of research papers (Behrensmeyer 1970, 1985; Leakey 1970; Maglio 1971; Cerling 1976; Coppens et al. 1976; Frank 1976; Johnson and Raynolds 1976; White 1976; Leakey and Leakey 1978; Leakey et al. 1978; Vondra and Bowen 1978; Cohen 1981, 1982; Brown and Cerling 1982; Cerling and Brown 1982; Mathisen and Vondra 1983; Eisenmann 1985; Brown and Feibel 1985, 1991; Tindall 1985; Hillhouse et al. 1986; Nyamweru 1989; Behrensmeyer et al. 1997; Macho et al. 2003; Abell 2004; van Bocxlaer and van Damme 2009; Rogers 2010; Richmond et al. 2014, 2020; Rowan et al. 2015; Patterson et al. 2017; van Bocxlaer 2020). Early stratigraphic efforts focussed on mapping and correlating tuffs (Vondra et al. 1971, 1978; Bowen and Vondra 1973; Bowen 1974; Vondra and Bowen 1976; Findlater 1976a, b, 1978; Acuff 1976), but this work was fraught with difficulties and miscorrelations that needed fundamental changes in stratigraphic terminology (Brown and Feibel 1986). Current correlations are much improved (Gathogo and Brown 2006a, b; McDougall and Brown 2006, 2008; Brown and McDougall 2011) and, with outcrops in other parts of the basin, have allowed a more complete understanding of the palaeogeographic evolution of the Turkana Basin (Feibel 1988, 2011; Lepre et al. 2007; Quinn et al. 2007; Levin et al. 2011).

Interest in Plio-Pleistocene deposits west of Lake Turkana accelerated rapidly after the discovery of 'Turkana Boy' in 1984 (Nariokotome *Homo erectus* (KNM WT-15000), resulting in many publications on vertebrates, hominin remains and an improved understanding of palaeoenvironments and chronology (Behrensmeyer and Laporte 1981; Harris and Brown 1985; Brown et al. 1985; Walker et al. 1986; Leakey and Leakey 1986a, b; Harris et al. 1988b; Feibel and Brown 1993; Walker and Leakey 1993; Haileab 1995; Leakey et al. 1995, 2001; Feibel 1999, 2003b, 2008; Kibunjia 2002; Stern et al. 2002; Alemseged 2003; Brugal et al. 2003; Wynn 2004; Harris et al. 2006; Brown and Fuller 2008; Feibel et al. 2009; Lepre and Kent 2010; Rogers 2010; Shea and Hildebrand 2010; Tiercelin et al. 2010; Brown and McDougall 2011; Kidane et al. 2014; Lepre and Kent 2015; Robinson et al. 2017; Mana et al. 2019).

After the 1990s, additional research was directed towards the sediments and fauna southwest of Lake Turkana, including the Lothagam and Kanapoi areas (Ward et al. 1999, 2001; Feibel 2003a, b; Harris and Leakey 2003; Leakey and Harris 2003; Stewart 2003a, b; van Bocxlaer 2020; Stewart and Rufolo 2018; Brochu 2020; Bobe et al. 2020; Frost et al. 2020; Manthi et al. 2020) and deposits to the southeast near Loiyangalani (Gathogo et al. 2008).

The resulting revised Plio-Pleistocene stratigraphy has provided an essential framework for investigation of the fossil fauna, which has led to an increased understanding of the environmental context of human evolution and artefact development in the region (Brown and Feibel 1988; Harris et al. 1988a; Suwa et al. 1996; Feibel 1997; Dean et al. 2001; McDougall et al. 2005; Prat et al. 2005; Braun 2006; Behrensmeyer 2006; Fleagle et al. 2008; Bobe and Leakey 2009; Bennett et al. 2009; Braun et al. 2010; Bobe 2011; Wood and Leakey 2011; Aubert et al. 2012; Brown et al. 2012, 2013; Spoor et al. 2007; Ward et al. 2015; Bobe and Carvalho 2019; Hlubik et al. 2019; Cohen et al. 2022a, b). In recent decades, interpretation of the palaeogeography has changed. Initially, the focus was recognising Plio-Pleistocene lakes. Later research showed that lacustrine conditions existed only for a small part of the later Neogene, with fluvial environments dominant for perhaps 85% of the stratigraphic record (Feibel 1988, 2011; Brown and Feibel 1991). Nevertheless, several large and small lakes did develop, some for short periods and others, such as 'Lake Lorenyang', lasting for a few hundred thousand years and extending across much of the Turkana Basin.

15.3.2 Sedimentary Facies

Many Plio-Pleistocene facies and facies associations have been described (Butzer 1971; Bainbridge 1976; Findlater 1976b; Vondra and Bowen 1976; Vondra et al. 1978; Burggraf et al. 1981; Burggraf and Vondra 1982; Abell et al. 1982; Harris et al. 1988a; Feibel 1988, 2003a, b, 2011, 2013). Early studies by Behrensmeyer

Fig. 15.6 Modern analogues for past sedimentary environments. (**a**, **b**) Deep offshore low energy settings, looking north beyond South Island (**a**) and looking east from the Central Island group (**b**). (**c**, **d**) Low-energy coastal lagoons and high-energy beaches, SE Lake Turkana. (**e**) High-energy beach sands, NW Turkana. (**f**) Delta and prodelta settings, Turkwell Delta. (**g**) Meandering river systems, Suguta Valley. (**h**) Alluvial fan with braided to meandering river, NE Turkana. (**i**) Braided distributary channels on a coastal delta, East Turkana. (**j**) Major distributary channel on the Turkwell Delta

(1975) in the Koobi Fora area (NE Lake Turkana) distinguished seven lithofacies that formed in delta margins, delta margins and lagoons, distributary channels and beaches, deltaic mudflats, floodplains and in two types of fluvial channel systems. Subsequent attempts at facies analysis focussed on relating Plio-Pleistocene sediments to deposits forming in and around the modern lake (Fig. 15.6). Findlater (1976a), for example, recognised five sedimentary environments in his scheme:

alluvial plain, alluvial coastal plain, alluvial delta plain, high-energy lake and low-energy lake. Similarly, Bowen (1974) and Burggraf and Vondra (1982) distinguished alluvial fan, fluvial, deltaic and lacustrine settings.

White et al. (1981) provided detailed descriptions of four major lithofacies at Koobi Fora: (1) laminated siltstone (pro-delta and shallow lacustrine), (2) arenaceous bioclastic carbonate (littoral lacustrine, beach, shoreface, lagoon), (3) lenticular fine-grained sandstone and lenticular bedded siltstone (delta plain to distributary channel), and (4) lenticular conglomerate, sandstone and mudstone (fluvial channel and floodplain) (Fig. 15.7a). Subsequently, Brown and Feibel (1991) presented a further modification in which they recognised five major facies associations, including fluvial channel, floodplain, delta, lake margin and lake basin.

Feibel and Brown (1986) and Feibel (1988, 2013) adopted a different approach in which facies were based on their dominant lithology, depositional structures and biological content (Table 15.1a). Gathogo and Brown (2006b) also combined

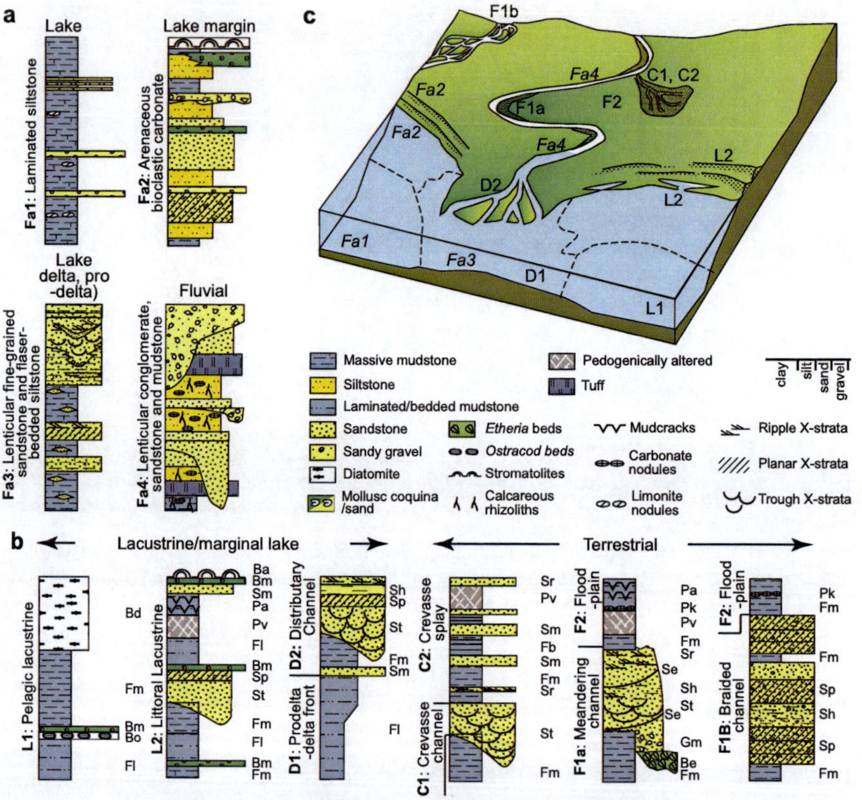

Fig. 15.7 Major lithofacies of the Turkana Basin. Examples of two classification schemes are shown. (**a**) An early example of four major lithofacies recognised by White et al. (1981) reflecting sediment types grouped by environment. (**b**) Facies scheme by Feibel (2013) showing facies codes and associations (see Table 15.1 for details of individual facies). (**c**) General setting of the facies and facies associations shown in (**a**) and (**b**). The facies of White et al. (1981) start with 'Fa' whereas the facies associations of Feibel begin with codes C, D, L or F, as defined in Table 15.1

Table 15.1 Facies classifications and associations for the Plio-Pleistocene Omo Group

(a) Facies classification – Feibel and Brown (1986), Feibel (2013)			
Code	***Lithology***	***Sedimentology, features, fossils***	***Environment***
Gms	Gravel	Matrix-supported massive units.	Debris flow, fluidized channel debris.
Gm	Gravel	Massive or weakly bedded. Horizontal bedding, imbrication. Common at base of fluvial cycles. Rounded pebbles. Reworked pedogenic carbonates, bone debris.	Longitudinal bar, lag deposits.
Gp	Gravel	Thin horizontal beds, planar cross-beds.	Upper surfaces and front of bars.
Sm	Sand	Massive.	Bioturbated deposits.
Se	Sand	Bedded, epsilon cross-strata.	Point bars.
St	Sand	Bedded, trough cross-beds. Small (cm) to large scale (m).	Migrating sand dunes.
Sp	Sand	Bedded, planar cross-beds.	Migrating bars, sand waves.
Sr	Sand	Ripple marks.	Ripples.
Sh	Sand	Horizontally laminated.	Planar bed flow.
Fm	Sand, silt, clay	Massive, bioturbation.	Overbank or lacustrine suspension deposits, biologically reworked.
Fl	Sand, silt, clay	Planar laminated to thinly bedded.	Suspension deposits, possibly pulsed delivery in lakes.
Fb	Sand, silt, clay	Thin to medium bedded.	Proximal flow reduction, levees.
Ba	Cryptalgal carbonate	Mats, oncolites, stromatolites on sand or rock surfaces, often pebble/shell nuclei. Filamentous and coccoid cyanobacteria.	Biogenic (cryptalgal) deposit.
Bd	Diatomite, diatom silt (?)	Massive to weakly laminated, low density. Sometimes recrystallised, *Aulacoseira* dominated, but also *Surirrella* and *Thalassiosira* locally.	Lacustrine.
Bo	Ostracod carbonate	Massive sediments with ostracods.	Lacustrine.
Bm	Molluscan carbonate	Massive or weakly bedded coquinas, bivalves and gastropods.	Mainly shallow lake, life and death accumulations.
Be	*Etheria* carbonate	Bioherm dominated by *Etheria elliptica*, strongly cemented together.	Oyster bioherm/reef often associated with fluvial settings.
Pa	Mud with sand veins	Anastomosing sand veins in fine–grained mud, polygonal fractures in plan view.	Mud-cracks possibly formed below an overlying sand cover.
Pv	Clay	Dish fractures with slickensides.	Compressed vertic sub-soil horizon
Pp	Clay	Prismatic structure.	Structural subsoil horizon.

(continued)

Table 15.1 (continued)

Pk	Carbonate	Pedogenic carbonate, massive or laminated, rhizoliths.		Terrestrial, Bk horizon.
(b) Facies Associations – Gathogo and Brown (2006b)				
Code	***Environment***	***Dominant facies***	***Minor facies***	***Associated features***
F1a	Meandering channel	Se, St, Sp, Sh, Sr	Gm	Basal erosional scour, fining upwards.
F1b	Braided channel	Sp, Sh	Fm	Minor scour, Sp (<30 cm) interbedded with Sh.
F2	Fluvial floodplain	Fm, Pk, Pv, Pa	Pp	Pk lag deposits, overly F1a and F1b associations with both sharp and gradational contacts.
L1	Pelagic lacustrine	Fm, Fl, Bd	Bo	Claystones dominate, mollusks present.
L2	Marginal lacustrine	Fm, Fl, Bm, Ba, Bo, Pa	St, Sr	Dominated by fine-grained clastics, local pedogenic alteration. Thin variant of D1.
D1	Prodeltaic and delta front	Fl, Fm	St, Sr	Thick packages of F1 (10–20 m). Coarsening upwards. Well-developed lamination. Bioturbation.
D2	Distributary channel	Se, St, Sp, Sr	Sh, Be	Often present as lenticular deposits cut into D1 association, otherwise similar to F1.
C1	Crevasse channel	St, Sr	Sm	Sands within erosional scoured channels, merges laterally into C2. Climbing ripples.
C2	Crevasse splay	Sm, Fb, Pv	Sr	Pedogenic overprints on thin, massive sands, silts, clays.

individual facies descriptions with information on bounding surfaces to distinguish associations that reflected sediment origin. In doing so, they recognised three lacustrine and six fluvial facies associations with the former including pelagic lacustrine (L1), littoral lacustrine (L2), and prodelta and delta front (D1) settings (Table 15.1b and Fig. 15.7b).

Feibel (2013) provided descriptions of each association and its component facies. His L1 lacustrine association includes dominant massive (Fm) or finely laminated (Fl) claystone (Table 15.1 and Fig. 15.7b, c), together with diatomite (Bd), ostracod-rich beds (Bo) and molluscs, which reflect deposition from suspension with a low clastic input. In contrast, the littoral lacustrine association (L2) is dominated by massive fine-grained clastics (Fm) with pedogenic alteration (Pv) and includes molluscan carbonate (Bm), cryptalgal biolithite (Ba) and ostracodites (Bo). Thinly laminated siltstone and fine-grained sandstone (Fl) are also present. The third lacustrine association represents prodelta and delta front settings (D1) where clay, silt and fine sand (F1) dominate in relatively thick units (10–20 m), commonly capped by massive fine-grained (Fm) delta-front sand (Sm) (Feibel 2013).

Various other facies classifications have been used for different formations and contrasting sediment types in the Turkana Basin (Tables 15.2 and 15.3). These contrasting schemes have provided the basis for interpreting Lake Turkana Plio-Pleistocene sedimentary sequences (Fig. 15.8a, b; Behrensmeyer 1975; Acuff 1976; Burggraf 1976; Vondra and Bowen 1976; Vondra et al. 1978; Feibel and Brown 1986; Brown and Feibel 1991; Feibel 2003a, b; Lepre et al. 2007; Tiercelin et al. 2010; Bennett et al. 2014; Beck 2015; Nutz et al. 2017, 2020).

Table 15.2 Representative Plio-Pleistocene facies schemes in the Turkana Basin (1974–2007)

Facies	Sedimentology and associated features	Environment
I. Stromatolitic Facies from East Turkana - Johnson (1974)		
A	Polygonal desiccated mats. Abundant gastropods (*Bellamya*). Polygon diameters up to 15 cm. Grades laterally into "Facies B".	Laterally extensive low-energy, shallow water.
B	Concentric, smooth and nearly flat stromatolitic layers or mats. Epifauna scarce, allochthonous clastics.	Low-energy, shallow water. Variable substrates.
C	Discontinuous mats. Varied growth forms: isolated columns, vertically stacked and laterally coalesced columns, hemispheroids.	Restricted shallow littoral environments.
D	Compounded spheroidal and hemsipheroidal stromatolites. Laterally extensive. Mollusc and clastic nuclei.	Distal shallow water environments.
E	Concentrically stacked spheroidal stromatolites (oncolites) coating shell debris, disarticulated vertebrates or clastics. 1–500 mm in diameter.	Higher energy shallow water.
II. East Turkana Facies - Cohen (1982)		
1. Brown silty sands and sandy silts.	Moderately sorted sands and silts, weak pedogenesis locally. Mammal footprints and oriented bones. Mudcracks and mud clasts.	Delta plain.
2. Tuffaceous, laminar to planar crossbeds in sandy silts, sands.	Interdigitating sands and silts. Pumiceous gravel lenses. Fine silt lenses with calcrete. Soft sediment deformation. Climbing ripples, crossbeds and channels. Reworked mollusc lags, burrows, root casts.	Margins of shifting delta and coastal plains, near the littoral zone.
3. Gray and brown, laminated and rippled silts.	Laminated siltstones. Fining and coarsening upward pumiceous gravels and sand. Oolitic-pisolitic beds with stromatolites. Calcareous concretions, fish and mammal fossils, burrows, root casts.	Shallow lake margin to inundated coastal plain.
4. Poorly sorted reddish-brown sandstone-conglomerates.	Fining-up muddy gravels and sand interbedded with moderately sorted silt and sand. Channel scours abundant. Massive beds, discontinuous laterally. Large and common root casts, fossils rare. Tabular calcareous concretions.	Fluvial distributary channels, crevasse splay, overbank.

(continued)

Table 15.2 (continued)

Facies	Sedimentology and associated features	Environment
III. East Turkana Facies - Lepre et al. (2007)		
A	Poorly to moderately sorted medium, coarse and pebbly sand. Massive, laminated, trough to planar cross-beds, fining upwards to mud.	Channels and floodplains, soils and subaerial areas.
B	Poor to moderate sorting of medium, coarse and pebbly sand. Planar crossbeds. Ripples, soft sediment deformation. Broken shells in lenses. Reworked stromatolites.	Beachface and shoreface. Nearshore settings.
C	Poorly to moderately sorted medium, coarse and pebbly sand. Arenaceous bioclastic carbonates. Life and death position stromatolites and molluscs. Calcareous sandstones. Crossbeds, massive, ripple and laminated strata.	Nearshore shallow lake, low clastic inputs. Littoral areas, backbeach settings.
D	Moderately to well-sorted, medium to fine, calcareous mud, massive to laminated, rare life and death mollusc assemblages. Stromatolites absent.	Deep lake, high suspension, pro-delta.

15.3.3 Lacustrine Sediments, Faunas and Floras

Biogenic Carbonate

Biogenic carbonate occurs in a variety of forms (e.g. rhizoliths, ostracods, molluscs) reflecting palaeoenvironmental setting. Calcareous rootcasts, for example, have been reported from Koobi Fora in at least four Pleistocene (post 1.89 Ma) facies that were laid down in delta plain, coastal plain, lake marginal and fluvial settings (Table 15.2; Cohen 1982). The rhizoliths have been grouped into major types (planar root mats; vertical, horizontal and diagonal root casts; root balls) and an additional category (root nodules) observed near Ileret. Horizontal systems (root mats and horizontal root casts) characterised the shallow lacustrine, beach, lagoon and floodplain facies since water was readily available to plants near the palaeosurface. Terrestrial rhizoliths, in contrast, developed longer taproots to reach deeper water sources and consequently were associated with vertical and diagonal root systems (Mount and Cohen 1984).

Ostracods are abundant in both Pliocene and Pleistocene lacustrine deposits east of Lake Turkana and locally constitute ostracodites (Feibel 1988, 2003a, b; Gathogo and Brown 2006a, b; Gravina 2015; Beck 2015), with at least 16 taxa reported (*Cypris*, *Hemicypris*, *Megalocypris*, *Sclerocypris*, *Oncocypris*, *Pseudocypris*, *Stenocypris*, *Plesiocypridopsis*, *Potamocypris*, *Ilyocypris*, *Darwinula*, *Cyprideis*, *Gomphocythere* and *Limnocythere*). Many of the floras are nearly monospecific with a maximum of ten species present. The most common dominants are *Hemicypris*, *Ilyocypris* or *Cyprideis* (Carbonel and Peypouquet 1979; Feibel 1988). The floral compositions are also very uniform through time, including the modern ostracod faunas (Cohen 1981; Cohen et al. 1983; Cohen 1986), suggesting that there

Table 15.3 Plio-Pleistocene facies schemes in the Turkana Basin (2010–2020)

Facies	Sedimentology and associated features	Environment
IV. West Turkana Facies – Tiercelin et al. (2010)		
1	Massive to crudely stratified, matrix or clast supported conglomerate. Horizontal or oblique bedding. Irregular basal erosion surface. Laterally continuous gravel beds.	Ephemeral streams proximal to alluvial fan.
2	Coarse to very coarse sandstones, locally with pebbles. Horizontal and oblique bedding, erosive bases.	Fluvial, distal to alluvial fan.
3	Fine sandstone to siltstone. Massive, faintly laminated, tabular crossbeds, ripple lamination, mud drapes. Rhizoliths, insect traces.	Shallow lake, bioturbated wetland?
4a	Mudstone, massive, rarely fine laminae. Dark brown, smectite. Root marks, burrows, prismatic structure and slickensides. Locally abundant gastropods.	Backshore lagoon.
4b	Mudstone, massive. Finely laminated beige to light brown smectite.	Lacustrine.
5a	Sandy bioclastic. Well-preserved gastropods (up to 98% of sediment), gravel, fish and mammal bone. Passes laterally into molluscan sandstone/ conglomerate.	High-energy, wave-dominated shoreline.
5b	Arenaceous bioclastic. Gastropods and bivalves concentrated in oblique bedding.	High-energy shoreline, storm related.
6	Tephra. Varied sedimentological features including crossbedding, 3D current ripples, laminations, mixing with sand, erosive bases.	Fluvial/lacustrine settings.
V. East Turkana Facies – Bennet et al. (2014)		
1	Sand and gravel sheets with scours, trough cross-beds, normal grading, palaeosols, oolitic stromatolites and carbonate horizons.	Low energy fluvial.
2	Massive silts laminated and rippled sands with scoured bases. Desiccation cracks, pedogenesis. Stromatolites.	Shallow lacustrine, lagoon.
3	Mollusc-rich medium/coarse sands, ripple laminated sands, basal scours, shells inversely graded. Tabular, domal and nodular carbonate.	Transgressive lake /lagoon. Shells.
4	Massive or weakly laminated clay. Mn and Fe staining. Draped basal contacts.	Infilled deep-water channels.
5	Fining upwards silts and fine sands, ripples, soft sediment deformation, rip-up clasts, small scours, trough cross-beds.	Shallow channels, nearshore areas.
VI. West Turkana Facies – Nutz et al. (2017)		
U1:	Black to green siliciclastic muds with mm-scale laminae. Occasional intercalated diatomite and common sand beds with erosive bases. Fish bone common.	Offshore muds in deep lake with turbidites.
U2.1	Alternating well-sorted medium sand and shelly beds. Oblique stratification. Occasional rhizoliths in palaeosols.	Washover fan.

(continued)

Table 15.3 (continued)

Facies	Sedimentology and associated features	Environment
U2.2	Clast-supported rudstones with *Melanoides*, *Bellamya* and *Corbicula*. Oblique stratification.	Beach berms in swash zones.
U2.3	Well-sorted medium to coarse sand alternating with shelly sand. Oblique and planar stratification. Wave and current ripples, vertical burrows.	Shoreface environments.
U3	Well-sorted fine sands grading up into silts. Common clast-supported oncolites (up to 10–50 cm) beds.	Medium/high energy foreshore/shoreface.
VI. West Turkana Facies – Nutz et al. (2020)		
FA1	Black, brown, green fine silt to clay. Horizontal and low-angle laminae, diatomite, fish bones, massive sandstone with erosional bases	Offshore suspension fallout and turbidites
FA2	Mudstone, sandstone topped by ripples/hummocky cross strata and shelly packstone, both with erosional bases.	Transitional depths with suspension deposits, tempestites, gravity flows.
FA3a	Well-sorted medium to coarse sandstone, shelly sandstone, packstone and rudstone. Wave ripples, low to moderate angle planar and cross laminae, heavy mineral concentrations, flat pebbles, burrows.	Shoreface. Oscillatory and traction currents. Surf zone.
FA3b	Sediments as in "3a" with low to moderate-angle planar lamination, burrows.	Beaches/foreshore.
FA3c	Sediments as in "3a" with oblique landward and lakeward dipping cross stratification, abundant shells, ripples, rhizoliths.	Washover fans, backshore.
FA4a	Matrix supported conglomerates. Crude planar lamination. Matrix of poorly sorted sandstone, ripples, shell placers, oncolites.	Debris flows. Coastal alluvial fans.
Fa4b	Well sorted medium tabular and lenticular sandstone. Planar and cross laminae, climbing ripples, load structures.	Turbulent density flows, river mouths.
Fa5a	Poorly sorted mudstone, sandstone, gravel and conglomerate. Massive to horizontally laminated; ripples, termite nests, burrows, rhizoliths, erosive sediment beds in mudstone.	Alluvial plain
Fa5b	Matrix supported fine to coarse conglomerates. Sandy matrix, imbrication, fining upwards, rhizoliths.	Cohesive and hyperconcentrated debris flows and dilute flows on alluvial fans

has been a considerable similarity in ecological conditions in the various palaeo-lakes inhabited by these organisms.

Molluscan packstones (Fig. 15.8c) are common locally and indicate lake marginal conditions. Over 40 cosmopolitan species have been described (Feibel 1988) from a variety of genera that include *Bellamya*, *Bulinus*, *Valvata*, *Pila*, *Gyraulus*, *Burnupia*, *Gabbiella*, *Chambardia*, *Cleopatra*, *Melanoides*, *Pseudobovaria*, *Caelatura*, *Lymnaea*, *Mutela*, *Iridina*, *Pleiodon*, *Corbicula*, *Pisidium* and *Eupera* (Butzer 1972; van Damme and Gautier 1972; Williamson 1981; Feibel 1988; Vonhof

Fig. 15.8 Plio-Pleistocene sediments, Turkana Basin. White bar = 2 cm. (**a**, **b**) Plio-Pleistocene deposits, NE Turkana Basin. (**c**–**h**) Nachukui Formation, NW Turkana. (**a**, **b**) Onlapping Holocene Galana Boi Formation silt (lighter colours) lying disconformably on older Plio-Pleistocene sediments. (**c**) Lacustrine gastropod packstone. (**d**) Lake marginal carbonate oncoids in sandy silt. Lens cap (50 mm) for scale. (**e**) Lacustrine diatomite of the Lorenyang lake. (**f**) Discoidal fish nests (up to ~10 cm across) in shallow-water silt and fine sand. (**g**) Proximal fluvial conglomerate. (**h**) Shoreface, well-bedded silt and carbonate-cemented fine sand, West Turkana

et al. 2013). One common taxon, *Etheria elliptica*, forms dense reefs and biostromes in fluvial or fluvio-lacustrine settings. Van Bocxlair and van Damme (2009) described 11 iridinids including several new forms (*Chambardia feibeli*, *Iridina turkanica*, *I. omoensis*, *I. browni* and *Pleiodon bentoni*) with many not specifically adapted to lakes. Subsequently, van Bocxlair (2020) explored the implications of molluscs in the Pliocene Kanapoi Formation.

Williamson (1979, 1981, 1982) documented 92 Plio-Pleistocene molluscan faunas and argued that they provided evidence for punctuated equilibrium brought about by dry periods that caused lake regression, with small surviving groups subsequently evolving rapidly. However, his model has since been challenged (Cohen and Schwartz 1983; Samadi et al. 2000) because views of the basin history have changed (Brown and Feibel 1991; van Bocxlair et al. 2008). Formerly, it was hypothesised that the basin was filled by an expanding and contracting lake for most of the Plio-Pleistocene. More recently, it has been recognised that the basin was occupied by a large fluvial system for much of the last few million years with lakes representing <15% of the time available (Brown and Feibel 1991; Feibel et al. 1991). Consequently, Williamson's periods of 'punctuated equilibrium' were simply related to brief lacustrine expansions that facilitated episodic molluscan invasion of the basin (van Bocxlair et al. 2008).

Stromatolites ('cryptalgal biolithites') are present in >25 stratigraphic levels in the Pleistocene Koobi Fora Formation northeast of Lake Turkana (Abell et al. 1982; Feibel 1988; Feibel et al. 1991; LaCarrubba 2018) and have been observed in the Koobi Fora Formation at Loiyangalani to the southeast (Hargrave et al. 2014), the Nachukui Formation to the west (Lamond and Tapanila 2003) and the Holocene Galana Boi Formation in many areas around the lake (Fig. 15.8d; Vondra et al. 1971; Johnson 1974; Owen and Renaut 1986; Harris et al. 1988a, b; Lamond and Tapanila 2003). The stromatolites formed in littoral lacustrine settings and are often laterally extensive, which implies low gradients at sites of growth (Ekdale et al. 1989; Feibel et al. 1991). They also tend to be absent in deposits laid down near river mouths and in deltaic sediments, such as in the Omo Valley. Stratigraphically, they are confined to discrete horizons (the 'stromatolite window' of Renaut et al. 2013), which may reflect intermittent periods with raised conductivity and dissolved Ca^{2+}, perhaps driven by greater aridity, decreased inputs of Omo River water, or isolation of stromatolitic areas in lagoons separated from the main lake.

The stromatolites initially grew on pebbles or shell nuclei and were formed where cyanobacteria induced biochemical precipitation of low-Mg calcite (Abell et al. 1982, Hargrave et al. 2014). Several forms have been recognised, including oncolites (oncoids, Fig. 15.8d), domes (hemispheroids) and compound types. Stromatolites with more open, solitary structures have been attributed to offshore locations with coalesced stromatolites having been linked to proximal environments (Johnson 1974). Two kinds of tubular borings (*Sertaterebrites* and *Trypanites*) cut into the stromatolites with a third unnamed and deflected type also recognised (Ekdale et al. 1989; Lamond and Tapanila 2003).

Filamentous cyanobacteria (blue-green algae) are oriented perpendicular to stromatolitic laminae with Chroococcalean spheres also reported (Abell et al. 1982;

Abell and McClory 1986). Small columnar (1–7 mm), columnar-layered and closely-spaced pseudocolumnar fabrics are present with laminae comprised of alternating micrite and radial fibrous calcite. Poorly defined layers of grey peloids (50–90 μm) in microspar also alternate with sparite laminae in some of the stromatolites. Rare carbonate oolites have been reported from a 5–50 cm thick carbonate horizon composed of ooids (2–80%) and biogenic grains (0–10%) in either a micritic low-Mg calcite cement (0–15%) or a blocky/sparry low-Mg calcite cement (15–90%) (Schwartz 1981). These oolitic sediments formed on a coastal plain inundated by a lake transgression.

Siliceous Microfossils

Diatomite and diatomaceous silt constitute important components of several lake sequences (Fig. 15.8e) with two dominant Plio-Pleistocene diatom assemblages reported (Feibel 1988; Feibel et al. 1991; Brindle et al. 2018). The most common taxon, *Aulacoseira* cf. *granulata*, is present together with secondary *Surirella* sp. and suggests freshwater conditions, possibly on the margins of a large lake. Floras characterised by *Thalassiosira* sp. are suggestive of moderate to high alkalinity and pH. Other diatoms reported include *Aulacoseira agassizii*, *Cyclotella meneghiniana* (*Stephanocyclus meneghinianus*) and small numbers of *Rhopalodia* cf. *vermicularis*, *Cocconeis* cf. *placentula*, *Stephanodiscus* sp. and *Cymbella* cf. *tumida* (Cerling 1979; Cohen et al. 1983).

Spongilla lacustris, a freshwater sponge, is present in parts of the Shungura Formation to the north of Lake Turkana (de Heinzelin 1983). Freshwater sponge spicules are also common in diatomites with potamolepids being particularly abundant in Pliocene lake sediments (Feibel 1988). These give way to mixed potamolepid-spongillid assemblages in lacustrine deposits formed between 3.4 and 1.6 Ma, with spongillids dominant in sediments <1 Ma old (Feibel et al. 1991).

Phytoliths have been reported in a long core collected by the Hominin Sites and Paleolakes Drilling Project (HSPDP) in the Nachukui area, northwest of Lake Turkana. This project targeted lacustrine sequences in the rift that shed light on environmental conditions at nearby hominin sites during key stages in hominin evolution (Cohen et al. 2022a, b). The Turkana core sediments were laid down in 'Palaeolake Lorenyang' between 1.87 and 1.38 Ma and contain intermittently abundant phytoliths associated with freshwater marshy conditions (Yost et al. 2014). Yost (2019) reported macrocharcoal and phytolith data to infer past fire regimes. Yost et al. (2021) also noted that phytoliths indicated variable mesic C4 versus xeric C4 grass dominance that varied at precessional periodicities and that other factors such as basin geometry probably also influenced grass composition.

Organic Fossils

Plio-Pleistocene and modern pollen have been described by several researchers (Bonnefille 1976a, b, 1979, 1995, 2010; Vincens 1982a, b). Overall, rainforests appear to have expanded in the drainage basin after about 3.4 Ma, with a trend towards greater aridity from ~2.8–2.5 Ma (Bonnefille 1983; Bonnefille and Vincens 1985). Pollen from the Omo Basin suggest that riverine plants gave way to savannah floras around 2.5–2 Ma ago (Bonnefille 1976b) whereas pollen in lake sediment about 1.5 Ma old tends to be dominated by Gramineae (~43%) and Typhaceae (~10%), with riverine and montane forest taxa more common than in modern lake samples. *Acacia* (*Vachellia*), *Commiphora* and *Salvadora* are also well represented, implying a denser shrub and tree cover than today. The data have been interpreted as indicating cooler and more humid conditions at 1.5 Ma.

Paraffinic hydrocarbons (C_{22} to C_{34}) derived from terrestrial plants and microorganisms have been reported from parts of the Koobi Fora Formation, with algal sources being inferred for hydrocarbons (C_{17} to C_{20}) that are common in lacustrine facies (Margolis 1976). Biomarkers in five lacustrine diatomite-claystone cycles (NE Lake Turkana) deposited in 'Lake Lokochot' at ~3.58–3.4 Ma show distinctive trends (Feakins et al. 2007). The diatomaceous sediment yields lower long-chain n-alkanoic acid abundances than claystone (shifts of up to 10‰) with the data suggesting precession cycles. Lupien et al. (2018) used hydrogen isotopes from terrestrial plant waxes (δD_{wax}) in HSPDP Core WTK13 from west Turkana to infer little change in the long-term average hydroclimate of Lake Lorenyang between 1.9 and 1.4 Ma but noted a strong precession (21 kyr) periodicity that reflected changes in insolation. However, Nutz and Schuster (2018) challenged their assessment of minimal long-term hydroclimate variability, inferring that earlier studies by Nutz et al. (2017) for the 1.95–1.72 Ma interval indicated lake level fluctuations of the order of 40 m that followed 400 ka eccentricity cycles, with superimposed 20 ka precession cycles that exerted only a minimal impact on lake level (<25 m). Subsequently, Lupien (2019) and Lupien et al. (2020) extended their HSPDP core study with a detailed analysis of two time intervals (~1.72 and ~1.60 Ma) with contrasting orbital eccentricities (0.043 and 0.025). They concluded that millennial-scale climate variability was stronger during periods of low summer insolation, which should induce high latitude glaciation and tropical African aridity. Lupien et al. (2022) followed up this work with research on leaf wax biomarkers and reported δD_{precip} data from both the Turkana and Chew Bahir basins that emphasised the importance of 21-kyr precession, modulated by 100-kyr eccentricity, in controlling Pleistocene rainfall.

Research has also been carried out using Loss on Ignition (LOI) techniques for HSPDP Core WTK13 (2–1.4 Ma). The upper 139 m of the core documents a transition from deep to shallow lacustrine environments, reflecting a prograding delta, or a lake regression, or perhaps both (Ortiz et al. 2014, 2015; Sier et al. 2017). Microcharcoal in WTK13 indicates significant precessional (21–19 kyr) variability and correlates with δDwax data, suggesting a link between orbitally forced peaks in precipitation and increased fire episodes (Yost et al. 2021). HSPDP researchers also report that low magnetic susceptibility in the lower core correlated with higher TOC values and indicated deep lake conditions. High Total Inorganic Carbon (TIC) in the lacustrine interval was interpreted as reflecting the presence of shells.

Aquatic Vertebrates

Dish-shaped fish nests (Fig. 15.8f) of the ichnospecies *Piscichnus brownii* are well preserved in parts of the Koobi Fora and Nachukui formations (to the east and west of Lake Turkana) along with trace fossils of *Planolites*, *Pelecypodichnus* and *Piscichnus* (Feibel 1987). Studies by Stewart and Murray (2008) of Plio-Pleistocene

(3.4–1.8 Ma) fossil fish in the Shungura Formation (Omo Valley) included two new species (*Sindacharax omoensis* and *Lates arambourgi*) at the time of their research. Fish have also been reported from Kanapoi to the southwest of Turkana (Stewart and Rufolo 2018). Aquatic resources such as catfish and turtles were in use by hominins at least by 1.95 Ma in the Turkana Basin (Archer et al. 2014) with at least 21 fish taxa reported from the Plio-Pleistocene sediments (Schwartz 1983). All but two of these have been reported in the modern lake or Omo River (Feibel 1988; Brown and Feibel 1991).

Dasyatis africana (a freshwater stingray) has been observed in sediments 1.9–1.3 Ma old and has been used to infer a fluvial connection with the Indian Ocean (Feibel 1993) with lake fish dominated by *Dasyatis*, *Barbus* and *Labeo*, as well as *Sindacharax* and *Hydrocynus*. Late Pleistocene fish from the Kibish Formation in the Omo Valley include catfish (mainly clariids and *Synodontis*), Nile perch (*Lates Niloticus*), a cyprinid (*Barbus*), tigerfish (*Hydrocynus*) and a pufferfish (*Tetraodon*) (Trapani 2008). Nine teleost genera have been discovered compared with 37 genera in the modern lake and Omo River. Interestingly, Broadhurst et al. (1998) suggested that the general availability of fish in East African lakes, with their long-chain polyunsaturated fatty acids, was perhaps an important factor in stimulating brain growth in hominins.

Chelonian fossils are present in both lacustrine and fluvial facies and include *Pelusios sinuatus* and *Trionyx* sp., *Pelusios adansoni* and *Geochelone* sp. (Feibel et al. 1991). Connections with the Nile are suggested by the presence of crocodiles, including *Crocodylus niloticus, C. cataphractus, C. iloidi* and *C. euthecodon brumpti*, which became extinct after the Early Pleistocene. Several species of hippo have also been observed, comprising *Hippopotamus amphibius*, *H. kaisensis* and *H. gorgops*, as well as *Hexaprotodon* with the latter favouring fluvial habitats.

Geochemical Records

Several researchers have attempted to develop geochemical records for Omo Group sediments. Cerling (1979) focussed on exchangeable cations and authigenic minerals in the Koobi Fora area (NE Turkana) and concluded that the Pliocene Kubi Algi Formation (a term now disused) included deposits laid down in a freshwater lake. In contrast, Early Pleistocene lake sediments were characterised as being formed in fresh to slightly brackish waters, with later Pleistocene deposits accumulating under fresh to highly alkaline conditions. Research on Plio-Pleistocene lake carbonates indicated $\delta^{13}C$ values of −15 to +22‰ with $\delta^{18}O$ varying by >10‰, which was attributed to changes in past water budgets, or to variations in the isotopic composition of inflowing water (Cerling et al. 1988). Samples with significantly low $\delta^{13}C$ may also reflect the influence of sulfate-reducing bacteria whereas high values could indicate methanogenesis.

Stable isotopes in Plio-Pleistocene stromatolites suggest a change from cooler and wetter conditions before 1.9 Ma to drier and warmer environments after about 1.4 Ma, with a return to cooler moister conditions at the end of the Pleistocene (Abell et al. 1982). Oxygen isotopes from *Melanoides tuberculata*, *Cleopatra ferruginea* and *Bellamya unicolor*, in the Koobi Fora Formation (~2–0.7 Ma), show a range of 7‰, which has been interpreted in terms of evaporative concentration and alternating open and closed lake basins (Abell 1982, 1985). $\delta^{18}O$ and $\delta^{13}C$ values for bivalves from Lake Lorenyang sediments (~2 Ma) imply wet–dry seasonal changes in water chemistry caused by monsoon rains in the Ethiopian Highlands (Vonhof et al. 2013). In addition, $^{87}Sr/^{86}Sr$ ratios in fish apatite, also from Lake Lorenyang sediments (~150 kyr duration; at ~2 Ma), indicate several precession-forced (~20 ka) wet-dry climate cycles (Joordens et al. 2011, 2013).

Many geochemical studies in the region have focussed on terrestrial sediments rather than lacustrine deposits. Shifting stable isotopes through the Plio-Pleistocene imply a vegetation trend towards wooded grasslands and grasslands (Cerling et al. 2011a; Quinn and Lepre 2021). Soil carbonate isotopes from Omo Group sediments, east and west of Lake Turkana, document a shift from C3 woodland towards C4 grassland in many studies (Cerling et al. 1997, 2003; Schoeninger et al. 2003; Wynn 2000, 2001, 2004; Bonnefille 2010; Feakins et al. 2013) with an abrupt change taking place around 2.7–2.5 Ma. Increases in $\delta^{13}C$ in palaeosol carbonate in parts of the Koobi Fora area (NE Turkana) have similarly been ascribed to grassland expansion between 2.4 and 1.4 Ma (Quinn 2006; Quinn et al. 2007, 2013) with Lepre and Quinn (2022) noting a 100 ka cyclicity and a long-term drying trend after 2.11 Ma. Pedogenic carbonate from the Nachukui and Shungura formations (west and north of Lake Turkana) indicate Plio-Pleistocene soil temperatures at or above those of the present (Passey et al. 2010). More recent work on soil carbonate in the Nachukui, Koobi Fora and Shungura formations has confirmed a steady increase in C4 vegetation through the Plio-Pleistocene, but with C3 plants tending to be relatively more common in the Omo region (Levin et al. 2011).

Oxygen isotopes in tooth enamel reflect differences in C3 and C4 usage by *Homo* versus *Paranthropus boisei* at about 2 Ma, with the data suggesting that *Homo* increased its dependence on C4 plants with time (Cerling et al. 2013). Blumenthal et al. (2017) also used oxygen isotopes in enamel to develop a Plio-Pleistocene aridity index for the Turkana Basin and noted that there was no long-term trend in terms of aridity.

15.3.4 Lake Levels and Sedimentation

The two models in Fig. 15.9 show the impact of lake level change on the Lake Turkana water column and sedimentation, based on comparisons of modern relatively low lake levels and Late Pleistocene to Holocene high-lake levels associated with the African Humid Period (AHP). Lake alkalinity and conductivity increase with basin closure due to evaporative concentration with recessional strandlines and erosional benches standing well above low lake stands (Fig. 15.9a). During low lake levels, most clastic sediments derive from the north leading to a north-south decrease in sedimentation rates, with the South Basin relatively sediment starved. During high lake stands, hypopycnal flows from the Omo River shift northwards as the lake expands across the Omo Valley, reducing inputs to southern Lake Turkana. However,

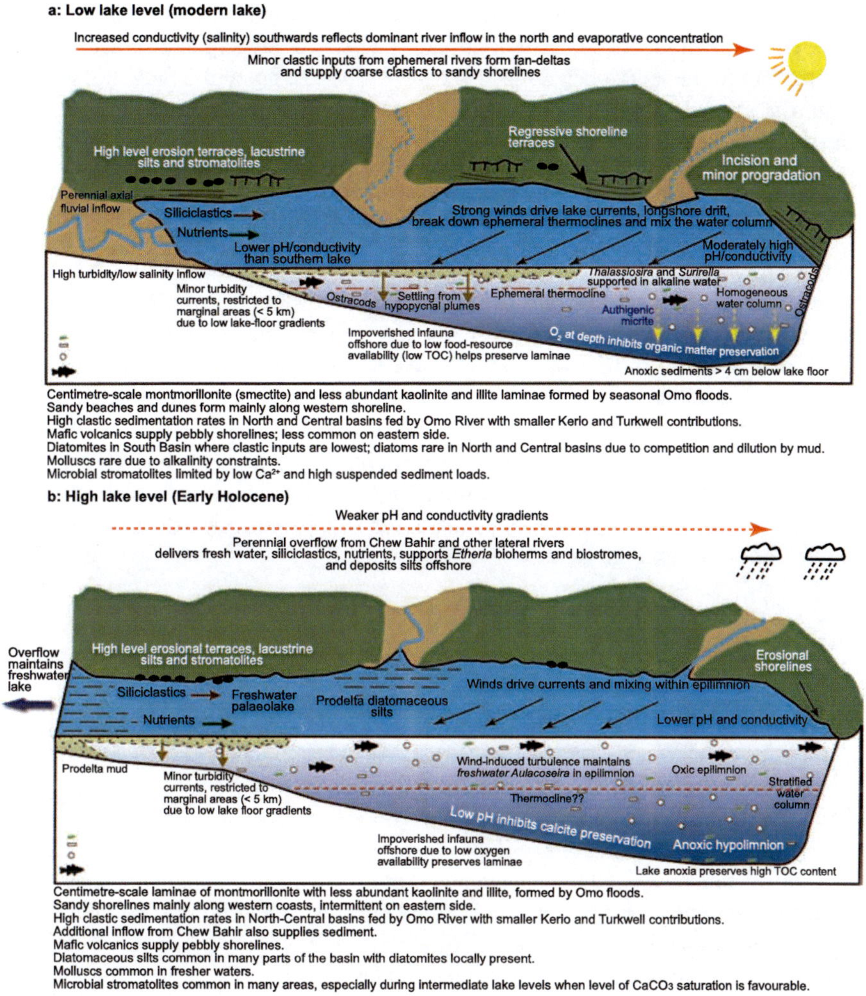

Fig. 15.9 Lake level change, water column characteristics and sedimentation. Major contrasts are indicated for relatively low (modern) and high (Late Pleistocene/Early Holocene) lake levels in the Turkana Basin, **(a)** and **(b)**, respectively. Major changes are noted for organic matter, silica (diatoms) and carbonate (stromatolites, ostracods, micrite) deposition. Water column changes involve the formation or breakdown of stratification and the development of oxic versus anoxic bottom waters. Expanded lakes shift the location of hypopycnal overflow with changes from ephemeral to permanent inflows reflecting increased rainfall and overflow from adjacent basins

additional sediment supplies and water come from lateral basin sources, especially those associated with overflow from the Suguta and Chew Bahir areas, giving rise to thick piles of diatomaceous silt in prodeltaic areas.

Strong winds drive modern lake currents and cause mixing that breaks down ephemeral thermoclines. Furthermore, the relatively shallow waters during low lake level periods allow wind-induced mixing to penetrate to the lake floor producing oxic conditions throughout the water column which, combined with low inputs of terrestrial organics from adjacent desert areas, results in low organic contents (<1%) in bottom sediments (Cohen 1984, 1986, 1989; Morrissey and Scholz 2014). This contrasts strongly with the deeper waters during high stands, when a permanent thermocline and bottom anoxia can develop. This likely explains increased organic matter (up to 6%) in Lake Turkana sediments dating from the AHP (Morrissey and Scholz 2014).

Carbonates are deposited mainly during low stands such as those of today as authigenic micrite and as ostracods in offshore areas (Cohen 1989; Johnson et al. 1991). In contrast, overflow during high lake levels results in lower pH and alkalinity, which reduces the likelihood of carbonate deposition, which explains reduced CaO accumulation during the AHP (Morrissey and Scholz 2014). Renaut et al. (2013), for example, noted that open lakes are dilute and often insufficiently saturated with respect to $CaCO_3$ for stromatolites to develop.

The dominance of *Aulacoseira* diatoms during high stands of the AHP suggests wind-induced turbulence in the epilimnion in order to support their relatively thick and heavy frustules and also implies increased silica, possibly reflecting enhanced fluvial inflow around the lake.

15.3.5 Plio-Pleistocene Stratigraphy, Palaeolakes and Palaeogeography

Pliocene Stratigraphy and Palaeolakes

Early Pliocene volcanism and earth movements created the key features of the modern Turkana Basin (Feibel 2011). Subsequently, the balance between sediment input and subsidence rates, and the development of tectonic and volcanic barriers, determined accommodation space and controlled repeated changes between fluvial- and lacustrine-dominated landscapes.

Sediments laid down in Pliocene lakes and rivers crop out in many parts of the basin and comprise or include parts of the Mursi (>4 Ma), Nkalabong (~4–1.7 Ma), Shungura (3.6–1.0 Ma) and Usno (3.4–3.1 Ma) formations to the north of Lake Turkana (Brown and McDougall 2011; Drapeau et al. 2014) and the lower parts of the Nachukui (west Turkana), Koobi Fora (east Turkana) and Kanapoi (SW Turkana) formations, each of which is subdivided into a series of members (Fig. 15.10) that reflect changing palaeoenvironments. These deposits collectively constitute the lower part of the Plio-Pleistocene Omo Group (de Heinzelin 1983; Harris et al. 1988a, b; Brown and Feibel 1991; McDougall and Feibel 2003; Brown and McDougall 2011; Van Bocxlaer 2017).

Exposures of the Apak Member of the Nachukui Formation at Lothagam document an Early Pliocene (~4.24 Ma and earlier) change from an eroding/pedogenic landscape to a laterally extensive aggrading fluvial system that may also be represented by deposits at Fejej, Mursi and Kanapoi (Fig. 15.10; Wynn 2000; McDougall

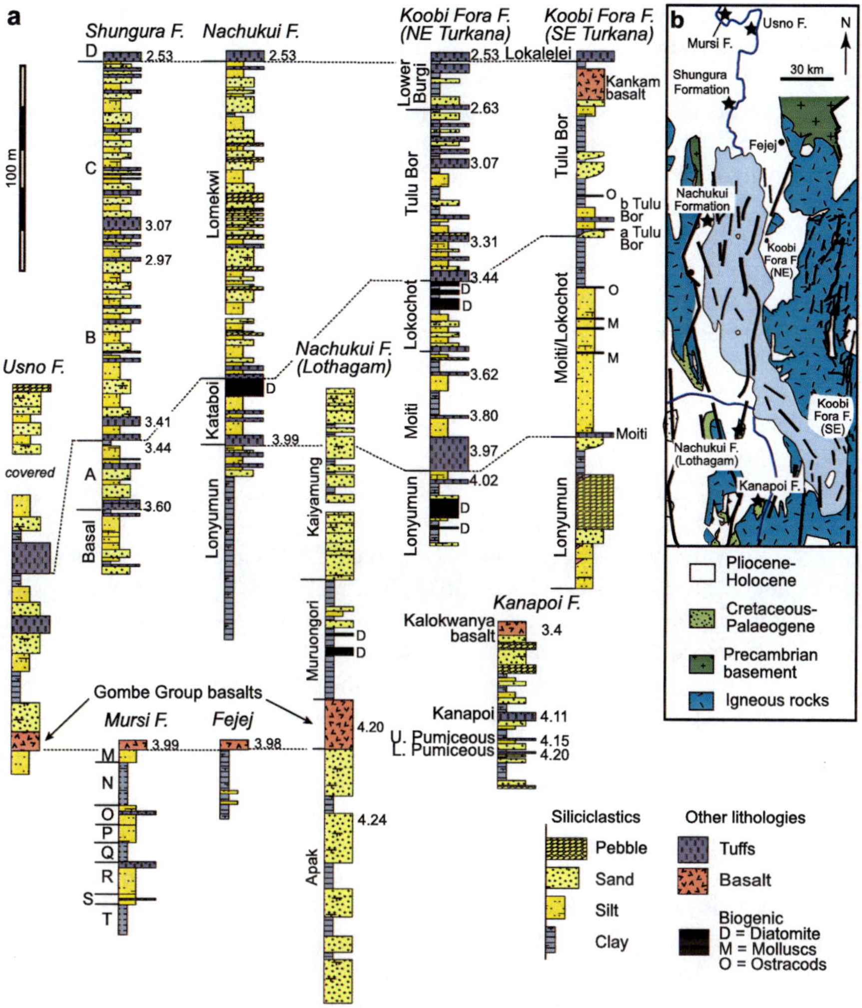

Fig. 15.10 Pliocene sediments and stratigraphy of the Turkana Basin. The Mursi, Usno and Shungura formations, and the Fejej deposits are located north of Lake Turkana. The Nachukui and Kanapoi Formations are present to the west, with the Koobi Fora Formation to the east. Based on Brown and McDougall (2011) with modifications from Harris et al. (1988a, b), Wynn (2000), Feibel (2003a, b), Haileab et al. (2004), Gathogo and Brown (2006a) and Gathogo et al. (2008)

and Feibel 2003; Feibel 2003a, b, 2011; Drapeau et al. 2022), with a heterogeneous mix of habitats at the latter location (Quinn and Lepre 2019). The Apak floodplain likely incorporated water from the precursor Omo, Turkwell and Kerio rivers in a system that drained into the Indian Ocean (Feibel 2011) with isotope data suggesting a mixed C3/C4 vegetation (Uno et al. 2011).

The first and largest of the Omo Group palaeolakes, Lake Lonyumun, appeared by about 4.1 Ma (Feibel 1988; Drapeau et al. 2014) and may have formed in response to normal faulting that dropped the western portion of the contemporary outlet channel so that a natural dam was formed (Fig. 15.11a; Bruhn et al. 2011). As a result, the palaeolake expanded and found a new outlet to the southeast. The Lonyumun lake deposits (up to 120 m thick) are dominated by pale brown to yellowish claystone and *Aulacoseira*-dominated diatomite, with a basal molluscan (*Bellamya*, *Cleopatra*) packstone (Feibel 2011). In the Koobi Fora area, bioherms of *Bellamya*, up to several metres thick, developed near former shorelines with gypsum accumulating in green claystone high in the Lonyumun Member (Feibel 1988). Beach deposits are largely absent due to subsequent erosion but are present with laminated ostracod sand on Sibilot and Jarigole mountains, which may have formed islands in a lake that was at least 36 m deep near Sibilot (Fig. 15.11a; Feibel et al. 1991). Deltaic deposits also accumulated in areas such as Allia Bay. Thin flows of Gombe Group basalt were widely erupted across the Turkana Basin and are intercalated with parts of the Lake Lonyumun sedimentary sequence (Findlater 1976a). The Gombe Group volcanics have been radiometrically dated at about 4.05–3.95 Ma (Watkins 1986; Bruhn et al. 2011) and formed within the period 4.18–4.29 Ma using palaeomagnetic data (Erbello and Kidane 2018).

Lake Lonyumun was largely infilled by about 3.97 Ma and was replaced by the Moiti floodplain (Fig. 15.11b; Feibel 2011; Bruhn et al. 2011), which was crossed by a large meandering river system. Initially, fluvial sediments aggraded, but as subsidence waned the ancestral Omo, Turkwell and Kerio rivers recycled their own alluvium. Deltaic and paludal deposits in the southwest suggest that a body of standing water might have existed in south-central parts of the Turkana region.

Freshwater Lake Lokochot formed by about 3.4 Ma ago and lasted for perhaps only 60,000 years (Fig. 15.11c; Feibel 1988, 2011). The major facies deposited in this relatively shallow lake (~10–15 m deep in the Koobi Fora area) included diatomite (altered to opal-CT) and green clay. East of modern Lake Turkana, the Lokochot water body accumulated brown silty claystone that locally coarsens upwards into fine micaceous sandstone with abundant fossil fish, turtle, crocodile and molluscan moulds. Channel sand bodies may indicate a delta distributary system.

Sediments that comprise the Tulu Bor Member of the Koobi Fora Formation (Fig. 15.10) were laid down on a floodplain that dominated the region between about 3.3 and 2.68 Ma (Fig. 15.11d), with *Etheria* reefs growing in the ancestral Omo and Kerio rivers (Feibel 2011). Two small water bodies, Lakes Waru and Kokiselei, were short-lived and restricted to the northeast of modern Lake Turkana, with the former likely being a floodplain lake in which very fine sand and green clay accumulated together with molluscs (Feibel 1988). Dwarf *Cleopatra* present in a marker bed may indicate stunted growth during at least one interval due to alkaline water. Feibel (1988) also documented abundant crocodile remains in a lenticular catfish packstone up to 1.5 m thick in the upper part of the Tulu Bor Member.

Floodplain conditions continued to dominate until the end of the Pliocene. However, interbedded bioclastic and siliciclastic sediments near the top of the lower Burgi Member (Koobi Fora region; Fig. 15.10) likely represent another episode of lacustrine deposition (Brown and Feibel 1991). Fortelius et al. (2016), using an econometric analysis of mammals, suggest there was a persistent difference between

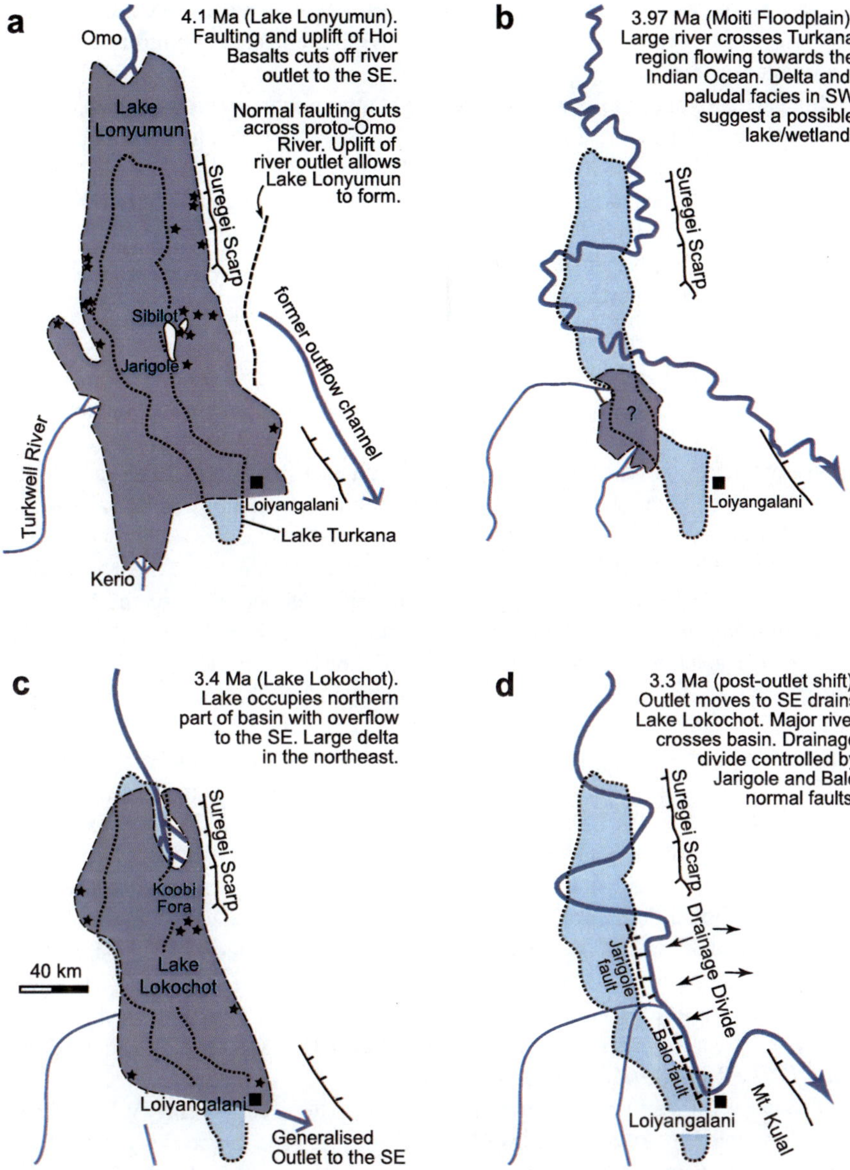

Fig. 15.11 Pliocene lakes and palaeogeography of the Turkana Basin. Black stars represent outcrops of diatomite. (**a**) Early Pliocene Lake Lonyumun represents the largest-known former lake in the basin history. (**b**) The Moiti floodplain with throughflow drainage to the southeast; a shallow paludal lake might have existed in the southern part of the basin. (**c**) Lake Lokochot extended over much of the basin with outflow to the southeast. (**d**) A large meandering river system occupied the basin for much of the Late Pliocene (Tulu Bor Member, Koobi Fora Formation) with outflow to the southeast, but deflected by the rising Mt. Kulal volcanic centre. After Brown and Feibel (1991) and Feibel (2011)

the eastern and western sides of the Turkana Basin through most of the Late Pliocene with wetlands of the shallow eastern side providing enhanced humidity to the terrestrial ecosystems.

Subsequently, Late Pliocene uplift in the Hamar region, to the northeast of Lake Turkana, resulted in erosion and the development of an angular unconformity between the Lower (Pliocene) and Upper (Pleistocene) Burgi Members. While erosion was taking place to the northeast, fluvial sediments accumulated in other parts of the basin (Fig. 15.11d). Late Pliocene volcanism to the southeast of Lake Turkana brought about the growth of Mt. Kulal and, near Loiyangalani, diverted the Late Pliocene river outlet northwards.

The Early Pleistocene Record

Pleistocene fluvial and lacustrine sediments in the Turkana-Omo Basin form the upper part of the Omo Group (Harris et al. 1988a, b; Brown and Feibel 1991; McDougall and Feibel 2003; Kidane et al. 2014; Brugal and Roche 2022). Major exposures occur in the Omo Valley (Shungura and Kibish formations) and to the west (Nachukui Formation) and east (Koobi Fora Formation) of Lake Turkana (Fig. 15.12).

Brown and McDougall (2011) reported that sedimentation was mostly continuous from 2.6 to 1.1 Ma during the deposition of Members D–L of the Shungura Formation (Fig. 15.12; McDougall and Brown 2006) and from 2.6 to 1.2 Ma for the Lokalelei, Kalachoro, Kaitio, Natoo and Nariokotome members of the Nachukui Formation (Fig. 15.12; Harris et al. 1988a; Roche et al. 2003; Brown and McDougall 2011). After an erosional period, deposition to the northeast of Lake Turkana resumed with the accumulation of the Upper Burgi Member (Koobi Fora Formation) at about 2 Ma. Sediments continued to be laid down near Koobi Fora until about 0.75 Ma, producing the KBS, Okote and Chari members with the latter experiencing a period of non-deposition between ~0.7 and 0.2 Ma. Subsequently, Middle to Late Pleistocene sedimentation resumed in the Omo Valley to produce Members I–III of the Kibish Formation (McDougall et al. 2005; Feibel 2008).

The growth of Pleistocene volcanoes had significant impacts on the basin palaeogeography. Eruption centres include Mt Kulal to the southeast (2.4–2.0 Ma; Gathogo et al. 2008), the Barrier Volcanic Complex between Lake Turkana and the Suguta Valley (1.37 Ma to the present; McDougall et al. 2012) and three volcanic islands within Lake Turkana (Karson and Curtis 1994).

The earliest Pleistocene environments in the eastern part of the Turkana Basin were dominated by fluvial settings (Fig. 15.13a) that gave rise to the Tulu Bor and Lower Burgi Members, with lacustrine environments possibly developing in axial areas (Feibel et al. 1991). Plant wax biomarkers from the Nachukui Formation (Uno et al. 2016) imply that C4 plants variously formed 5 to 100% of the vegetation to the west of the modern lake between 2.3 and 1.7 Ma with C3-dominance from 2.3 to 2.2 Ma and a mixed C3–C4 and C4 cover after 2.1 Ma. Uplift in the northeast resulted in erosion or non-deposition, with volcanism and growth of Mt. Kulal in the southeast partially obstructing flow and deflecting the former river outlet northwards (Fig. 15.13b). Mt Kulal may also have acted as a barrier to a freshwater lake

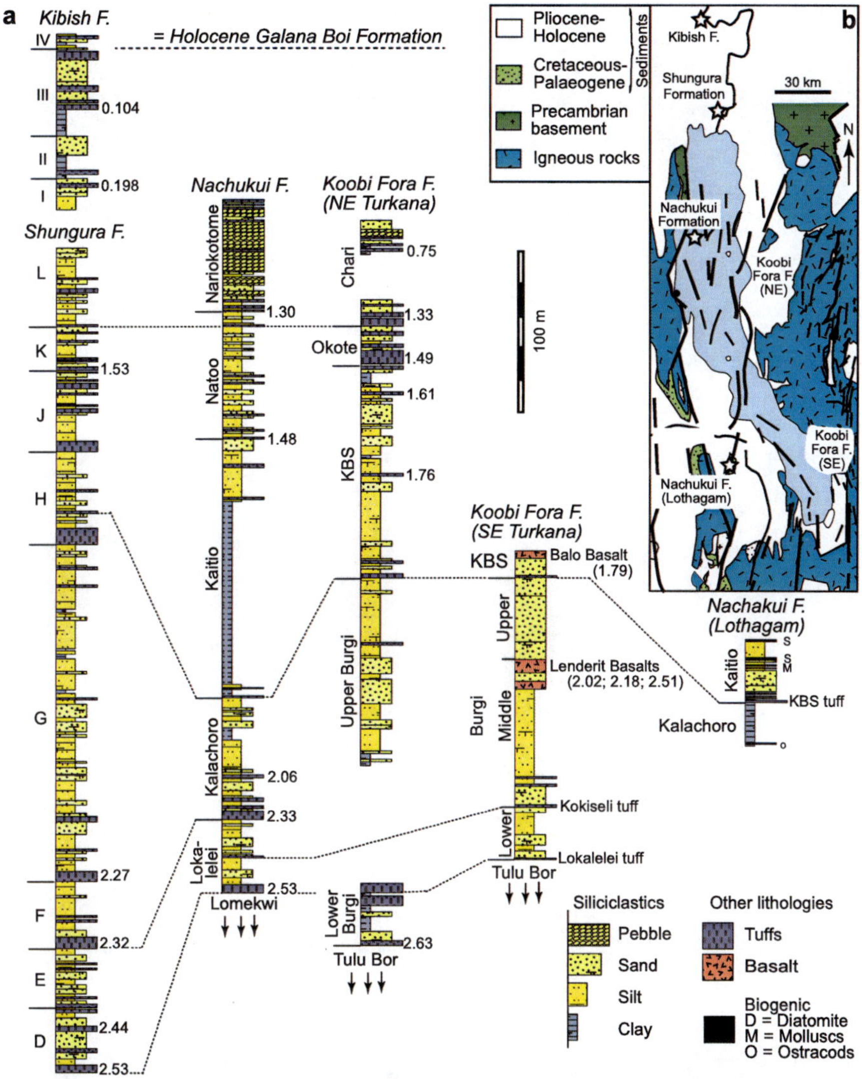

Fig. 15.12 Pleistocene sediment stratigraphy, Turkana Basin. (**a**) Four major formations are recognised. To the north, the Shungura (Members D-L) and Kibish (Members I-III) formations are Pleistocene. Member IV of the Kibish Formation and the Galana Boi Formation (not shown) are Holocene. Five members of the Nachukui Formation (west of Lake Turkana) and four Members of the Koobi Fora Formation (east of Lake Turkana) are also of Pleistocene age. (**b**) Generalised locations of the major formations. After Feibel (2003b) and Brown and McDougall (2011)

near Kargi, to the south of the Chalbi Desert, at about 2.3 Ma (Nyamweru 1989). This lake may have had a connection to the Turkana Basin, given the presence of a Nilotic fauna, with separation related to the eruption of early Pleistocene fissure lavas near the present southeast shoreline of Lake Turkana.

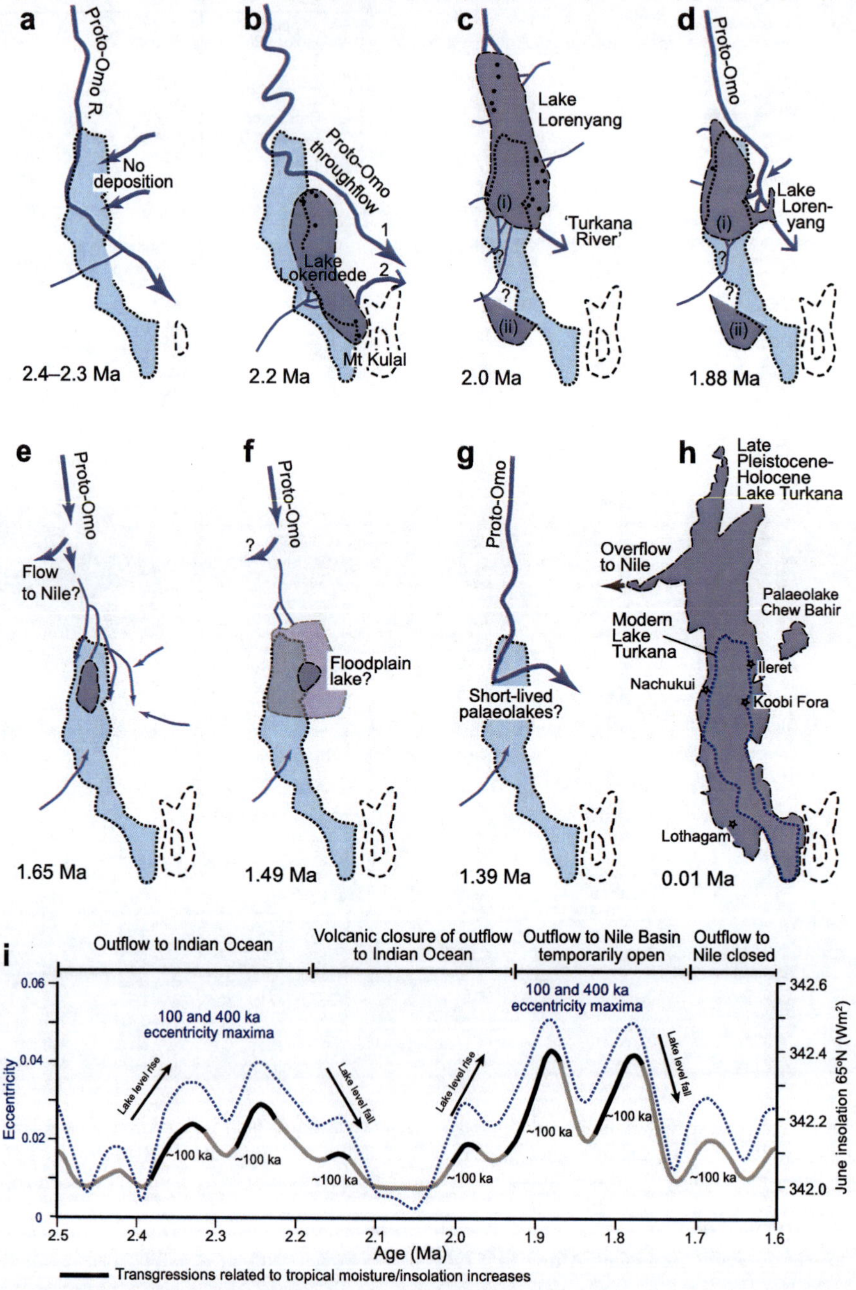

Fig. 15.13 Pleistocene palaeogeography, Turkana Basin. Environments shown for the Burgi (**a**–**c**), KBS (**d**), Okote (**e**, **f**) and Chari (**g**) members of the Koobi Fora Formation. (**a**–**h**) after Feibel (1988, 2011), Brown and Feibel (1988, 1991), and Bruhn et al. (2011). (**a**) A large fluvial system dominated during deposition of the Lower Burgi Member. (**b**) Shallow freshwater Lake Lokeridede. Erosion occurred in NE Turkana associated with the uplift of the Hammar block. Mt. Kulal

Freshwater Lake Lokeridede, dominated by *Aulacoseira* cf. *granulata*, developed by about 2.2 Ma. Monospecific *Hemicypris* ostracods were common in its waters. Williamson (1982) reported ten mollusc taxa belonging to *Bellamya, Gabiella, Cleopatra, Melanoides, Pseudobovaria, Caelatura, Mutela, Corbicula* and *Eupera*, which he argued represented speciation events — a view since challenged (Cohen and Schwartz 1983; van Bocxlaer et al. 2008). Locally, in the Koobi Fora area, *Gymnarchus* and *Hyperopisus* dominated the fish fauna.

Lake Lorenyang formed during Upper Burgi Member times (Figs. 15.12 and 15.13c; Feibel 1988; Feibel et al. 1991; Brown and Feibel 1991) and extended over most of the northern Turkana Basin, perhaps reaching as far south as Lothagam (Feibel 2011). On its eastern side, large deltas entered the water body from the north, shifting position with time (Feibel et al. 2009; Feibel 2011). Lepre (2014) suggested Lake Lorenyang first appeared in the northwest parts of the basin (~2.14 Ma) followed by the Omo area to the north (~2.04 Ma) and extended into NE Turkana a little before 1.95 Ma. The lake therefore formed at about the same time that the Lenderit basalts erupted to the southeast (Gathogo et al. 2008; Bruhn et al. 2011), suggesting possible initiation by damming rather than through climate change (Lepre 2014).

Zawacki et al. (2021, 2022) studied zircons which suggested a variety of Archaean, Cambrian, Proterozoic and Cenozoic bedrock sources for the sediments in the basin. Strontium isotope data from fossil fish indicate precessional variations in Lake Lorenyang water budgets (Joordens et al. 2011). Scott et al. (2017) also reported that ~3–8 m thick sedimentary cycles in an HSPDP core represent alternate flooding and exposure that may have been controlled by orbital cyclicity. Periodically, diatom floras were dominated by *Thalassiosira*, suggesting higher salinity within a broadly freshwater lake. A few horizons, in several localities, contain cryptalgal biolithites (stromatolites) that may also reflect phases of higher salinity (Feibel et al. 1991). Littoral and deep-water ostracods are common as are molluscs (Scholz and Glaubrecht 2010). Molluscan zones 4–9 of Williamson (1982) lie within the Upper Burgi sequence and variously include twenty species from the genera *Bellamya, Valvata, Gabiella, Cleopatra, Melanoides, Bulinus, Pseudobovaria, Caelatura,*

Fig. 15.13 (continued) volcanics forming to SE. Through-drainage (1) suggested by Feibel (1988). Overflow route (2) suggested by Bruhn et al. (2011). (**c**) Freshwater Lake Lorenyang (i) (Feibel 1988; Brown and Feibel 1991). Feibel (2011) notes that lacustrine sediments at Lothagam (ii), to the southwest, may indicate a larger lake that linked (i) and (ii). Outflow to the SE and on towards the Indian Ocean via the Turkana River. (**d**) Lake Lorenyang continued through to 1.88 Ma, but with delta sediments in the northeast. (**e**) Braided river systems flowing from the Ethiopian Highlands deliver pumice to the Turkana region. Possible ponding in the southwest and diversion of the proto-Omo River to the Nile system. (**f**) Lake contraction forms a shallow floodplain (?) lake in the Ileret area. Possible Omo River diversion to the Nile River. (**g**) Large meandering river and floodplain in the northern part of the Turkana Basin. Sedimentation becomes episodic with ephemeral lake deposition. Cessation of sedimentation 0.7–0.2 Ma (except perhaps below Lake Turkana). Subsequently, there were four Pleistocene cycles of lake expansion, plus one Late Pleistocene/Early Holocene high lake stand with overflow to the Nile (**h**). (**i**) Evolution of the basin between 2.5 and 1.6 Ma in relation to volcanism and climate cycles. After Boës et al. (2019)

Mutela, *Pleiodon*, *Corbicula* and *Eupera*. A freshwater stingray (*Dasyatis africana*) made its first appearance at about this time, indicating the opening of a fluvial corridor to the Indian Ocean (Feibel 1993; Feibel et al. 1991). Pollen from lacustrine strata indicate dominance by grasses in the surrounding basin (Vincens 1982b; Bonnefille and Vincens 1985).

Lake Lorenyang continued to exist until ~1.88 Ma when a major delta supplied sediments to the Koobi Fora area (Fig. 15.13d). At Karari, further north, a change to drier conditions during regression of Lake Lorenyang is reflected in a faunal change from water-dependent mammals to taxa favoured by arid settings (O'Brien et al. 2020). *Thalassiosira*-dominated diatoms have been reported from the Nachukui Formation, with laterally extensive stromatolites developing at different times and locations on gently shelving shorelines. Both lines of evidence suggest elevated salinities, at least episodically. Molluscs from this period include *Bellamya*, *Gabiella*, *Cleopatra*, *Melanoides*, *Bulinus*, *Pseudobovaria*, *Caelatura*, *Mutela*, *Corbicula* and *Eupera* (Williamson 1982). Nutz et al. (2017) report that Nachukui Formation sediments (1.95–1.72 Ma) record a long-term regressive-transgressive period with five higher-frequency superimposed cycles. The latter developed between 1.87 and 1.76 Ma and have been attributed to orbital forcing. Boës et al. (2019) used geochemical tracers to explore lake level changes from 2.4 to 1.7 Ma and showed that the lake tended to expand during 100 ka insolation/eccentricity maxima when there were greater monsoonal inputs from the Omo River (Fig. 15.13i). However, they also reported that expansion occurred through an insolation minimum at 2.17–1.95 Ma, noting that this might reflect volcanic closure that likely blocked outflow. Subsequently, the lake rose during a period of insolation/eccentricity maxima (1.9–1.7 Ma) when a new outflow developed to the northwest and the Nile basin.

By 1.65 Ma (Fig. 15.13e) and into Okote Member times (Fig. 15.12), a much-reduced lake occupied the northern Turkana Basin (Feibel et al. 1991). Molluscs were less common, with *Melanoides* dominating in the northeast. Ostracod beds are more common than in the older deposits, with *Iliocypris* dominant. Several stromatolite horizons (LaCarruba 2018) and oolitic beds are present on what would have been gently shelving shores and, with the intermittent occurrence of dolomite, imply more alkaline conditions. The proto-Omo River was more diffuse, possibly braided (Feibel 1988; Feibel et al. 1991) and may have been partially or intermittently diverted to the Nile late in Okote Member times, which caused the palaeolake to be reduced to ephemeral flooded pans (Brown and Feibel 1988). Feibel (1988) notes that a shallow (1–2 m) floodplain lake with abundant fish developed near Ileret, in the northeast, around 1.49 Ma (Fig. 15.13f). Lepre (2017) reported crevasse splay deposits from the Okote Member, which includes hominin and mammalian tracks in muddy deltaic sediment near Ileret at about 1.5 Ma (Roach et al. 2016, 2018; Hatala et al. 2017).

A broad floodplain characterised the northern part of the Turkana Basin by about 1.39 Ma, during early Chari Member times (Figs. 15.12 and 15.13g), with erosion or non-deposition taking place further south. This fluvial system appears to have exited the region to the east. A temporary, poorly-defined, water body, Lake

Nachukui, is represented by beach deposits at Nachukui to the west of the modern lake and by lacustrine strata near Ileret and Silbo as well as parts of the lower Omo Valley (Feibel 2011). A return to floodplain conditions followed with ephemeral lakes developing that contained ostracods, molluscs and fish with the stingray (*Dasyatis africana*) and *Sindacharax* becoming extinct at about this time. Junginger et al. (2023) also report $^{87}Sr/^{86}Sr$ analysis on microfossils in Suguta Valley lake deposits that suggest there may have been two to three palaeolakes in the Turkana-Suguta Basin around 900–700 ka with these water bodies separated by tectono-volcanic barriers. Subsequently, deposition became more intermittent in the Turkana Basin, stopping by about 0.7 Ma, except perhaps in areas below the modern lake.

Nutz et al. (2020) provide a different palaeoenvironmental synthesis based on the Nachukui Formation to the northwest of Lake Turkana (Fig. 15.14). Their reconstructions suggest a return to earlier models (Cerling 1979; Williamson 1981) that inferred greater continuity of lacustrine deposition than those of Brown and Feibel (1988, 1991), who inferred that the basin was river-dominated for 85% of the Plio-Pleistocene. Nutz et al. (2020) suggested that lakes of varying size continuously occupied the North Basin for at least 70% of the time between 4 and 0.75 Ma. They also noted that Brown and Feibel's Lonyumun, Lokochot and Lorenyang paleolakes correlated well with major highstands in their reconstructions (Fig. 15.14) and suggested that the previously recognised Lokeridede lake (~2.5 Ma) represented only a brief transgression during their Lomekwi lowstand. Similarly, they suggest that Brown and Feibel's Silbo (~1 Ma) palaeolake represented a short-term transgression during a broader regressive period that followed their Nachukui Highstand.

The Middle to Late Pleistocene Record

Based on marine and terrestrial core proxy records, Kaboth-Bahr et al. (2021) noted that pan-African climate change drove lake level fluctuations through the last 620 kyr and that rainfall (and lake levels) through this period was related to eccentricity-controlled low-latitude insolation. Unfortunately, the sedimentary record between about 700 and 200 ka is lacking in the Turkana Basin, but recent K and K/Zr geochemical studies at Chew Bahir, to the northeast (Fig. 15.15b), provide an overlapping history for at least the last 600 kyr (Viehberg et al. 2018; Foerster et al. 2020; Duesing et al. 2021a, b; Arnold et al. 2021; Roberts et al. 2021). There, a 280-m-long HSPDP core indicates three environmental phases. Phase I (~600–430 ka) represents a change from humid to arid conditions with increasing climate variability. Phase II (~430–200 ka) was characterised by increased humidity with particularly wet conditions at around ~439–315 and 280–195 ka. Phase III overlaps with the Kibish Formation in the Turkana Basin and was a period of long-term increased aridification and climate variability. Düsing (2020) also noted distinct cyclicities in the HSPDP core record, reporting that drier and more variable climates developed during minima in 400 kyr eccentricity cycles. Trauth et al. (2021) developed this work further, exploring two types of variability in the area. The first involved slow change associated with precession cycles of ~20 kyr. The second reflected more extreme events over short wet or dry intervals (several centuries to millennia) with rapid shifts between wet and dry episodes.

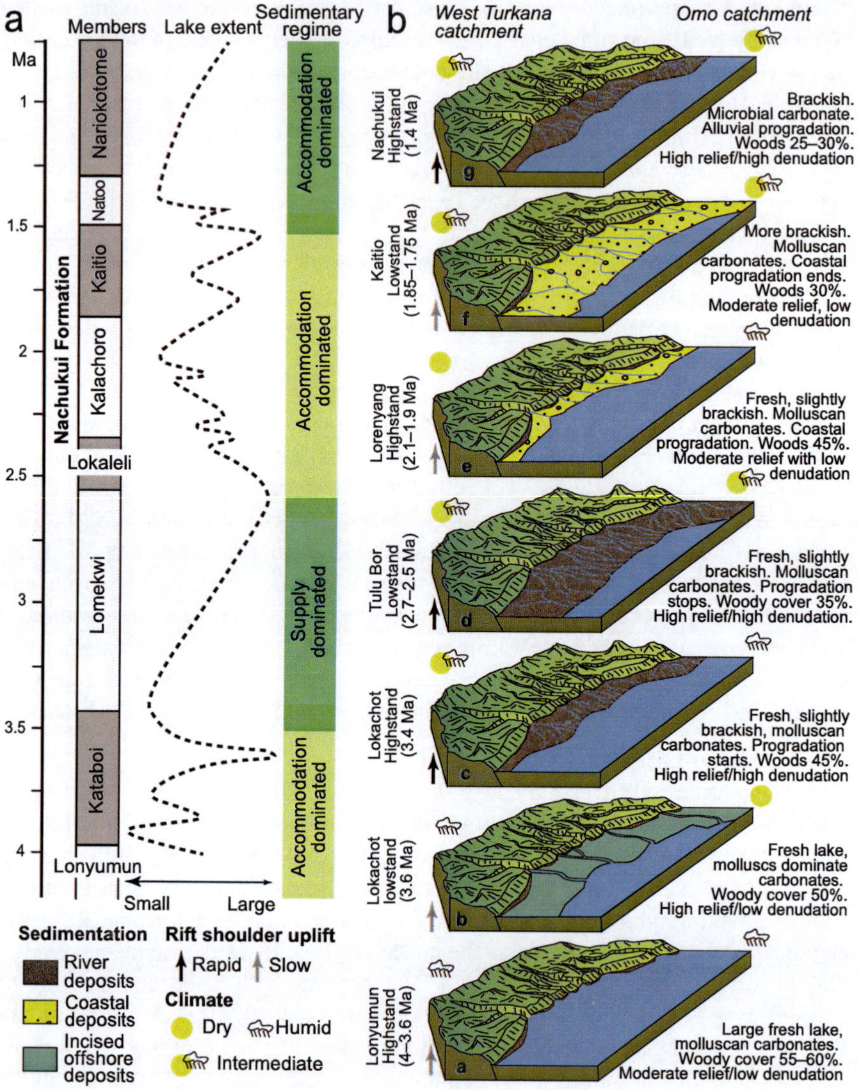

Fig. 15.14 West Turkana palaeolandscapes and Lake Turkana expansion and contraction phases (after Nutz et al. 2020). (**a**) Time scale, Nachukui Formation members and accommodation versus supply-dominated phases. (**b**) Models of West Turkana sedimentation. Note contrasts in climate (see sun vs cloud symbols) between the Omo and West Turkana catchments

Schaebitz et al. (2021) further subdivided Phase III at Chew Bahir into five contrasting climate periods with humid conditions at 200–125 and 14–5 ka. Drier conditions prevailed at 125–60, 60–14 and 5–0 ka with the first of these being also more variable. In addition, they reported that these changes were driven by the effects of

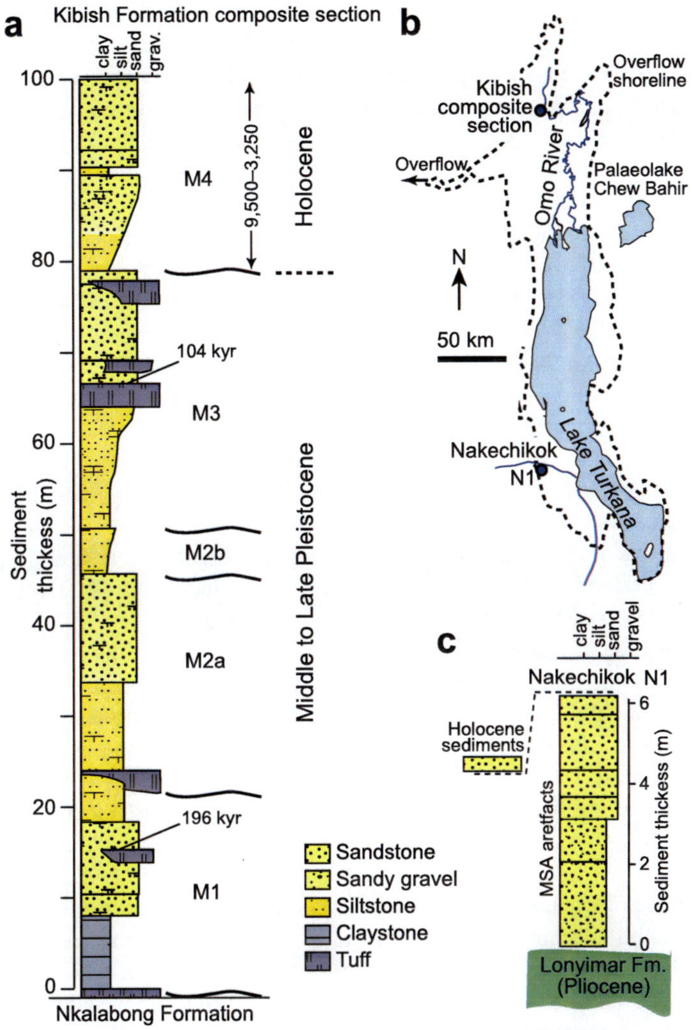

Fig. 15.15 Middle to Late Pleistocene sediments. (**a**) Kibish Formation. After McDougall et al. (2005). Holocene Member 4 is broadly equivalent to the Galana Boi Formation. (**b**) Location of sections in (**a**) and (**c**), and the Late Pleistocene to Early Holocene highstand shoreline associated with overflow to the Nile. (**c**) Sand deposits at Nakechikok containing Late Pleistocene 'Middle Stone Age' artefacts (after Shea and Hildebrand 2010) with a thin Holocene sequence of sand

eccentricity on the amplitude of precession and that "precession minima/maxima, associated with high/low summer insolation, led to a more/less northerly position of the African rainbelt". In turn, these climatic changes influenced episodic hominin migrations from northeast Africa into the Levant and also between mountain refugia and lower levels within the region.

In the Turkana Basin, the Kibish Formation accumulated after a 0.5 Ma gap in deposition, preserving an environmental record for the last 200 kyr (Butzer and Thurber 1969; McDougall et al. 2005, 2008) equivalent to Phase III at Chew Bahir. The Kibish sediments also indicate alternating periods of lake expansion and contraction with the most recent expansion taking place during the Late Pleistocene–Early Holocene (Fig. 15.13h) at which time there was an outflow to the northwest linking Palaeolake Turkana to the Nile.

The older part of the record is preserved by flat-lying deltaic claystone, siltstone, sandstone and tuff laid down in a proto-Omo delta about 100 km north of modern Lake Turkana (Fig. 15.15a). When these sediments were deposited the lake stood at or close to its present overflow height of ~460 m above sea level (Fig.15.15b; Butzer 1970; Garcin et al. 2012). The Kibish sequence was divided into four members using disconformities, which have a relief of up to 30 m (Butzer 1969b; Butzer and Thurber 1969). Subsequently, $^{40}Ar/^{39}Ar$ dating was used to show that Members 1 (M1) and M3 date from about 196–197 and 104 ka, respectively (Fig. 15.15a; McDougall et al. 2005; Vidal et al. 2022) with M1 containing a fossil *H. sapiens* (Omo I) and both members yielding fossil fish and water birds (Louchart 2008). These two sets of dates are also close to the ages of organic-rich sapropels (S7 and S4) in the eastern Mediterranean Sea that were formed by high River Nile discharge. McDougall et al. (2005) speculated that M2a and M2b (Fig. 15.15a), separated by another disconformity, were deposited at the same time as sapropels S5 (119–123 ka) and S6 (172 Ka). However, Vidal et al. (2022) used geochemical fingerprinting of tuffs to infer an older age of 233 ± 22 ka for a tuff in M1, which overlaps in age with sapropel S8. Despite the questions over the precise chronologies, Johnson and Malala (2009) observed that the similarity of the eastern Mediterranean and Lake Turkana records might be linked causally and reflect increased monsoonal rains over the Ethiopian Highlands.

The Kibish Formation has also yielded an important suite of Middle Stone Age (MSA) archaeological sites that include Levallois tools, bifaces and handaxes (Shea 2008). In addition, some of the oldest anatomically modern human fossils have been recovered from M1 of the formation (Butzer 1969b; McDougall et al. 2005; Pearson et al. 2008). Robinson et al. (2016) noted that oxygen and carbon stable isotope data from ungulates suggested that M1 and M3 climates were less humid than those of M4. Elsewhere in the Turkana Basin sediments of equivalent age to M1–3 are scarce, although there are localised outcrops of fluvial to nearshore sediments with MSA artefacts, both to the northeast (Kelly and Harris 1992; Kelly 1996) and west of the lake. Shea and Hildebrand (2010), for example, noted the presence of MSA and Late Stone Age (LSA) artefacts associated with Upper Pleistocene and Holocene terrestrial sediments at Nakechikok (Fig. 15.15c) and suggested that Late Pleistocene sediments reported previously by Whitworth (1960, 1965) are actually of Holocene age.

There is no evidence for any high lake levels between M3 and M4 of the Kibish Formation. However, echo soundings and seismic reflection profiles in Lake Turkana indicate a major erosional unconformity ~60 m below the modern lake surface that probably records a major regression just before ~10,000 y BP (Johnson et al. 1987). Following this period of desiccation, lake levels rose again to the outlet height and gave rise to the Holocene deltaic sediments of M4, which accumulated between ~9500 and 3250 years ago (uncalibrated shell dates; Butzer et al. 1972) and correlate with the Holocene Galana Boi Formation around other parts of Lake Turkana.

References

Abell PI (1982) Paleoclimates at Lake Turkana, Kenya, from oxygen isotope ratios of gastropod shells. Nature 297:321–323

Abell PI (1985) Oxygen isotope ratios in modern African gastropod shells: a data base for paleoclimatology. Chem Geol 58:183–193

Abell PI (2004) Palaeoclimates at Lake Turkana, Kenya, from oxygen isotope ratios of gastropod shells. Nature 297:321–323

Abell PI, Awramik SM, Osborne RH, Tomellini S (1982) Plio-Pleistocene lacustrine stromatolites from Lake Turkana, Kenya: morphology, stratigraphy and stable isotopes. Sediment Geol 32:1–26

Abell PI, McClory JP (1986) Sedimentary carbonates as isotopic marker horizons at Lake Turkana, Kenya. In: Frostick LE, Renaut RW, Reid I, Tiercelin J-J (eds) Sedimentation in the African rifts. Geol Soc Lond Spec Publ 25:153–158

Acuff HN (1976) Late Cenozoic sedimentation in the Allia Bay area, East Rudolf (Turkana) basin, Kenya. PhD thesis, Iowa State University, Ames

Agyemang P, Roberts EM, Downie B, Sertich JJW (2019) Sedimentary provenance and maximum depositional age analysis of the Cretaceous? Lapur and Muruanachok sandstones (Turkana Grits), Turkana Basin, Kenya. Geol Mag 156:1334–1356

Alemseged Z (2003) An integrated approach to taphonomy and faunal change in the Shungura Formation (Ethiopia) and its implication for hominid evolution. J Hum Evol 44:451–478

Anyonge W (1991) Fauna from a new lower Miocene locality west of Lake Turkana, Kenya. J Vertebr Paleontol 11:378–339

Arambourg C (1933) Decouverte d'un gisement de la Mammifères burdigeliens dans le Bassin du lac Rudolphe (Afrique Orientale). C R Acad Sci Paris 14:221–222

Arambourg C (1943) Contribution à l'étude géologique et paleontologique du Bassin du lac Rodolphe et de Basse Vallée de l'Omo. Mem Mus Natl Hist 1(2):157–230

Arambourg C (1948) Contribution à l'etude geologique et paleontologique du basin du lac Rodolphe et de la basse vallee de l'Omo. Deuxieme partie: Paleontologie. Mission Scientifique de l'Omo 1932–1933. Tome I. Fasc. 3 Geologie-Anthropologie. Editions du Museum, Paris, pp 231–562

Arambourg C, Chavaillon J, Coppens Y (1969) Résultats de la novelle Mission de l'Omo (1968). C R Acad Sci Paris D-268:759–776

Archer W, Braun DR, Harris JWK, McCoy JT, Richmond BG (2014) Early Pleistocene aquatic resource use in the Turkana Basin. J Hum Evol 77:74–87

Arnold GE, Foerster V, Trauth MH, Lamb H, Schaebitz F, Asrat A, Szczech C, Günter C (2021) Advanced hyperspectral analysis of sediment core samples from the Chew Bahir Basin, Ethiopian Rift in the spectral range from 0.25 to 17 μm: support for climate proxy interpretation. Front Earth Sci 9:606588

Aubert M, Pike AWG, Stringer C, Bartsiokas A, Kinsley L, Eggins S, Day M, Grün R (2012) Confirmation of a late middle Pleistocene age for the Omo Kibish 1 cranium by direct uranium-series dating. J Hum Evol 63:704–710

Bainbridge RB (1976) Stratigraphy of the Lower Member, Koobi Fora Formation, northern Karari Escarpment, East Turkana Basin, Kenya. MSc thesis, Iowa State University, Ames

Baker BH, Mohr PA, Williams LAJ (1972) Geology of the Eastern Rift System. Spec Pap Geol Soc Am 136

Barr WA (2015) Paleoenvironments of the Shungura Formation (Plio-Pleistocene: Ethiopia) based on ecomorphology of the bovid *Astragalus*. J Hum Evol 88:97–107

Beck CC (2015) The terrestrial climate record from the Turkana Basin, Kenya: a multi-proxy approach. PhD thesis, Rutgers University, Piscataway, New Jersey

Behrensmeyer AK (1970) Preliminary geological interpretation from hominid-bearing sediments. Nature 247:520–523

Behrensmeyer AK (1975) The taphonomy and palaeoecology of Plio-Pleistocene vertebrate assemblages east of Lake Rudolf, Kenya. Bull Mus Comp Zool Harvard Univ 146:473–577

Behrensmeyer AK (1985) Taphonomy and the paleoecologic reconstruction of hominid habitats in the Koobi Fora Formation. In: Coppens Y (ed) L'Environment des Hominides au Plio-Pleistocene. Foundation Singer-Polignac, Paris, pp 309–324

Behrensmeyer AK (2006) Climate change and human evolution. Science 311:476–478

Behrensmeyer AK (2015) Four million years of African herbivory. Proc Natl Acad Sci 112:11428–11429

Behrensmeyer AK, Laporte LF (1981) Footprints of a Pleistocene hominid in northern Kenya. Nature 289:167–169

Behrensmeyer AK, Todd NE, Potts R, McBrinn GE (1997) Late Pliocene faunal turnover in the Turkana basin, Kenya and Ethiopia. Science 278:1589–1594

Bellieni G, Justin-Visentin E, Piccirillo EM, Di Brozolo FR, Rita F (1981) Oligocene transitional tholeiitic magmatism in northern Turkana (Kenya): comparison with the coeval Ethiopian volcanism. Bull Volcanol 44:411–427

Bellieni G, Justin-Visentin E, Piccirillo EM, Zanettin B (1987) Volcanic cycles and magmatic evolution in northern Turkana (Kenya). Tectonophysics 143:161–168

Bennett M, Harris J, Richmond B, Braun D, Mbua E, Kiura P, Olago D, Kibunjia M, Omuombo C, Behrensmeyer AK, Huddart D, Gonzalez S (2009) Early hominin foot morphology based on 1.5-million-year-old footprints from Ileret, Kenya. Science 323:1197–1201

Bennett M, Morse SA, Falkingham PL (2014) Tracks made by swimming Hippopotami: an example from Koobi Fora (Turkana Basin, Kenya). Palaeogeogr Palaeoecol Palaeoclimatol 409:9–23

Blumenthal SA, Levin NE, Brown FH, Brugal J-P, Chritz KL, Harris JM, Jehle GE, Cerling TE (2017) Aridity and hominin environments. Proc Natl Acad Sci 114:7331–7336

Bobe R (2002) Faunal change, environmental variability and late Pliocene hominin evolution. J Hum Evol 42:475–497

Bobe R (2011) Fossil mammals and paleoenvironments in the Omo-Turkana Basin. Evol Anthropol 20:254–263

Bobe R, Behrensmeyer AK, Eck GG, Harris JM (2007) Patterns of abundance and diversity in late Cenozoic bovids from the Turkana and Hadar Basins, Kenya and Ethiopia. In: Bobe R, Alemseged Z, Behrensmeyer AK (eds) Hominin environments in the East African Pliocene: an assessment of the faunal evidence. Springer, Dordrecht, pp 129–157

Bobe R, Carvalho S (2019) Hominin diversity and high environmental variability in the Okote Member, Koobi Fora Formation, Kenya. J Hum Evol 126:91–105

Bobe R, d'Oliveira Coelho J, Carvalho S, Leakey M (2022) Early hominins and paleoecology of the Koobi Fora Formation, Lake Turkana Basin, Kenya. In: Reynolds SC, Bobe R (eds) African paleoecology and human evolution. Cambridge University Press, Cambridge, pp 311–331

Bobe R, Leakey MG (2009) Ecology of Plio-Pleistocene mammals in the Omo-Turkana Basin and the emergence of *Homo*. In: Grine F, Fleagle JG, Leakey RE (eds) The first humans - origin and early evolution of the genus *Homo*. Springer, Dordrecht, pp 173–184

Bobe R, Manthi FK, Ward CV, Plavcan JM, Carvalho S (2020) The ecology of *Australopithecus anamensis* in the early Pliocene of Kanapoi, Kenya. J Hum Evol 140:102717

Boës X, Prat S, Arrighi V, Feibel C, Haileab B, Lewis J, Harmand S (2019) Lake-level changes and hominin occupations in the arid Turkana basin during volcanic closure of the Omo River outflows to the Indian Ocean. Quat Res 91:892–909

Boisserie J-R, Guy F, Delagnes A, Hlusko LJ, Bibi F, Beyene Y, Guillemot C (2008) New palaeoanthropological research in the Plio-Pleistocene Omo Group, Lower Omo Valley, SNNPR (Southern Nations, Nationalities and People Regions) Ethiopia. C R Palevol 7:429–439

Boisserie J-R, Kiarie C, Lihoreau F, Nengo I (2017) Middle Miocene *Kenyapotamus* (Cetartiodactyla, Hippopotamidae) from Napudet, Turkana Basin, Kenya. J Vertebr Pal 37:e1272055–e1272014

Bonnefille R (1976a) Implications of pollen assemblage from the Koobi Fora Formation, East Rudolf, Kenya. Nature 264:403–407

Bonnefille R (1976b) Palynological evidence for an important change in the vegetation of the Omo Basin between 2.5 and 2 million years ago. In: Coppens Y, Howell FC, Isaac GL, Leakey RE (eds) Earliest man and environments in the Lake Rudolf Basin. University of Chicago Press, Chicago, pp 421–431

Bonnefille R (1979) Methode palynologique et reconstitutions paleoclimatiques au Cénozoïque dans le Rift Est Africain. Bull Soc Géol Fr 21:331–342

Bonnefille R (1983) Evidence for a cooler and drier climate in the Ethiopian uplands. 2.5 Myr ago. Nature 303:487–491

Bonnefille R (1995) A reassessment of the Plio-Pleistocene pollen record of East Africa. In: Vrba ES, Denton GH, Partridge TC, Burkle LH (eds) Paleoclimate and evolution. Yale University Press, New Haven, pp 299–310

Bonnefille R (2010) Cenozoic vegetation, climate changes and hominid evolution in tropical Africa. Glob Planet Chang 72:390–411

Bonnefille R, Chavaillon J, Coppens Y (1970) Résultats de la nouvelle Mission de l'Omo (1969). C R Acad Sci Paris D 270:924–927

Bonnefille R, Vincens A (1985) Apport de la palynologie à l'environnement des Hominidés d'Afrique orientale. In: Beden M, Behrensmeyer AK, Boaz NT, Bonnefille R, Brain CK, Cooke R, Coppens Y, Dechamps R, Eisenmann V, Gentry A, Geraads D, Gèze R, Guérin C, Harris J, Koeniguer JC, Letouzey R, Petter G, Vincens A, Vrba E (eds) L'environnement des Hominidés au Plio-Pléistocène. Masson, Paris, pp 237–278

Boone SC, Kohn BP, Gleadow AJW, Morley CK, Seiler C, Foster DA, Chung L (2018) Tectonothermal evolution of a long-lived segment of the East African Rift System: thermochronological insights from the North Lokichar Basin, Turkana, Kenya. Tectonophysics 744:23–46

Boone SC, Kohn BP, Gleadow AJW, Morley CK, Seiler C, Foster DA (2019) Birth of the East African Rift System: nucleation of magmatism and strain in the Turkana Depression. Geology 47:886–890

Boschetto H (1988) Geology of the Lothidok Range, northern Kenya. MSc thesis, University of Utah, Salt Lake City

Boschetto HB, Brown FH, McDougall I (1992) Stratigraphy of the Lothidok Range, northern Kenya, and K/Ar ages of its Miocene primates. J Hum Evol 22:47–71

Bowen B (1974) The geology of the upper Cenozoic sediments in the East Rudolf embayment of the Lake Rudolf basin, Kenya. PhD thesis,. Iowa State University, Ames

Bowen B, Vondra CF (1973) Stratigraphical relationships of the Plio-Pleistocene deposits, East Rudolf, Kenya. Nature 242:391–393

Braun DR (2006) The ecology of Oldowan technology: perspectives from Koobi Fora and Kanjera South (Kenya). PhD thesis, Rutgers University, Piscataway, New Jersey

Braun DR, Harris JW, Levin NE, McCoy JT, Herries AI, Bamford MK, Bishop LC, Richmond BG, Kibunjia M (2010) Early hominin diet included diverse terrestrial and aquatic animals 1.95 Ma in East Turkana, Kenya. Proc Natl Acad Sci 107:10002–10007

Brindle M, Mohan J, Beck C, Stone JR (2018) Three novel species of Bacillariophyta (Diatoms) in the genera *Surirella* and *Thalassiosira* from Pleistocene Paleolake Lorenyang (~2–1.6 Ma) Turkana Basin, Kenya. Phytotaxa 371:230 240

Broadhurst CL, Cunnane SC, Crawford MA (1998) Rift Valley lake fish and shellfish provided brain-specific nutrition for early Homo. Br J Nutr 79:3–21
Brochu CA (2020) Pliocene crocodiles from Kanapoi, Turkana Basin Kenya. J Hum Evol 140:102410
Brock A (1974) Paleomagnetic stratigraphy and chronology of hominid-bearing sediments east of Lake Rudolf, Kenya. Nature 247:344–348
Brock A, Isaac GL (1976) Reversal stratigraphy and its application at East Rudolf. In: Coppens Y, Howell FC, Isaac GL, Leakey REF (eds) Earliest man and environments in the Lake Rudolf Basin. University of Chicago Press, Chicago, pp 148–162
Brown FH (1969) Observations on the stratigraphy and radiometric age of the 'Omo Beds', lower Omo Basin, Southern Ethiopia. Quaternaria 11:1–14
Brown FH (1972) Radiometric dating of sedimentary formations in the lower Omo Valley, Ethiopia. In: Bishop WW, Miller JA (eds) Calibration of hominoid evolution. Scottish Academic Press, Edinburgh, pp 273–287
Brown FH (1983) The Omo Group, chronology. Mus R Afr Centr Tervuren, Belg Ann Sci Geol 85:145–147
Brown FH, Jicha BR, Leakey RE (2016) An age for Kajong, a Miocene fossil site east of Lake Turkana, Kenya. J Afr Earth Sci 114:74–77
Brown FB (1994) Development of Pliocene and Pleistocene chronology of the Turkana Basin, East Africa, and its relation to other studies. In: Corrucini RS, Ciochon RL (eds) Integrative paths to the past: paleoanthropological advances in honor of F. Clark Howell. Prentice Hall, Englewood Cliffs, pp 285–312
Brown FH, Cerling TE (1982) Stratigraphical significance of the Tulu Bor Tuff of the Koobi Fora Formation. Nature 299:631–633
Brown FH, de Heinzelin J, Howell FC (1970) Pliocene/Pleistocene formations in the lower Omo Basin, Southern Ethiopia. Quaternaria 13:247–268
Brown FH, Feibel CS (1985) Stratigraphic notes on the Okote Tuff Complex at Koobi Fora. Nature 316:794–797
Brown FH, Feibel CS (1986) Revision of lithostratigraphic nomenclature in the Koobi Fora region, Kenya. J Geol Soc Lond 143:297–310
Brown FH, Feibel CS (1988) "Robust" hominids and Plio-Pleistocene paleogeography of the Turkana Basin, Kenya and Ethiopia. In: Grine FE (ed) Evolutionary history of the "Robust" australopithecines. Aldine de Gruyter, New York, pp 325–341
Brown FH, Feibel CS (1991) Stratigraphy, depositional environments and palaeogeography of the Koobi Fora Formation. In: Harris JM (ed) Koobi Fora Research Project, The fossil ungulates: geology, fossil artiodactyls, and paleoenvironments, vol 3. Clarendon, Oxford, pp 1–30
Brown FH, Fuller CR (2008) Stratigraphy and tephra of the Kibish Formation, southwestern Ethiopia. J Hum Evol 55:366–403
Brown F, Harris J, Leakey R, Walker A (1985) Early *Homo erectus* skeleton from west Lake Turkana, Kenya. Nature 316:788–792
Brown FH, Lajoie KR (1970) K-Ar ages of the Omo group and fossil localities of the Shungura Formation southwest Ethiopia. Nature 229:483–485
Brown FH, McDougall I (2011) Geochronology of the Turkana Depression of northern Kenya and Southern Ethiopia. Evol Anthropol 20:217–227
Brown FH, McDougall I, Fleagle JG (2012) Correlation of the KHS Tuff of the Kibish Formation to volcanic ash layers at other sites, and the age of early *Homo sapiens* (Omo I and Omo II). J Hum Evol 63:577–585
Brown FH, McDougall I, Gathogo PN (2013) Age ranges of *Australopithecus* species, Kenya, Ethiopia, and Tanzania. In: Reed KE, Fleagle JG, Leakey RE (eds) The paleobiology of *Australopithecus*. Springer, Dordrecht, pp 7–20
Brown FH, Nash WP (1976) Radiometric dating and tuff mineralogy of Omo Group deposits. In: Coppens YY, Howell FC, Isaac GL, Leakey REF (eds) Earliest man and environments in the Lake Rudolf Basin. University of Chicago Press, Chicago, pp 50–63

Brown FH, Shuey RT (1976) Magnetostratigraphy of the Shungura and Usno formations, Lower Omo Valley, Ethiopia. In: Coppens YY, Howell FC, Isaac GL, Leakey REF (eds) Earliest man and environments in the Lake Rudolf Basin. University of Chicago Press, Chicago, pp 67–78

Brown FH, Shuey RT, Croes MK (1978) Magnetostratigraphy of the Shungura and Usno Formations, southwestern Ethiopia: new data and comprehensive reanalysis. Geophys J R Astron Soc 54:519–538

Brugal J-P, Roche H, Kibunjia M (2003) Faunes et paléoenvironnements des principaux sites archéologiques plio-pléistocènes de la formation de Nachukui (Ouest-Turkana, Kenya). C R Palevol 2:675–684

Brugal J, Roche H (2022) Paleoecology and paleoenvironments of Early Quaternary faunal assemblages from the Nachukui Formation in Kenya: insights from the West Turkana Archeological Project. In: Reynolds SC, Bobe R (eds) African paleoecology and human evolution. Cambridge University Press, Cambridge, pp 298–310

Bruhn RL, Brown FH, Gathogo PN, Haileab B (2011) Pliocene volcano-tectonics and paleogeography of the Turkana Basin, Kenya and Ethiopia. J Afr Earth Sci 59:295–312

Burggraf DR (1976) Stratigraphy of the Upper Member, Koobi Fora Formation, Southern Karari Escarpment, East Turkana Basin, Kenya. MSc thesis, Iowa State University, Ames

Burggraf DR, Vondra CF (1982) Rift valley facies and paleoenvironments: an example from the East African Rift System of Kenya and Southern Ethiopia. Z Geomorphol 42:43–73

Burggraf DR, White HJ, Frank HJ, Vondra CF (1981) Hominid habitats in the Rift Valley, part 2. In: Rapp GR, Vondra CF (eds) Hominid sites: their geologic setting (Am Assoc Adv Sci Select Symp 63). Westview Press, Boulder, pp 115–147

Butzer KW (1969a) Early *Homo sapiens* remains from the Omo River region of south-west Ethiopia: geological interpretation of two Pleistocene hominid sites in the Lower Omo Basin. Nature 222:1133–1135

Butzer KW (1969b) Horizontal sediments of the lower Omo Valley: the Kibish Formation. Quaternaria 11:15–29

Butzer KW (1970) Contemporary depositional environments of the Omo delta. Nature 226:425–430

Butzer KW (1971) The lower Omo Basin: geology, fauna and hominids of Plio-Pleistocene formations. Naturwissenschaften 58:7–16

Butzer KW (1972) Molluscan assemblages from the late Cenozoic of the lower Omo Basin, Ethiopia. Quat Res 4:591–592

Butzer KW (1976) The Mursi, Nkalabong and Kibish Formations, Lower Omo Basin, Ethiopia. In: Coppens Y, Howell FC, Isaac GL, Leakey REF (eds) Earliest man and environments in the Lake Rudolf Basin. University of Chicago Press, Chicago, pp 12–23

Butzer KW, Isaac GL, Richardson JL, Washbourn-Kamau C (1972) Radiocarbon dating of east African lake levels. Science 175:1069–1076

Butzer KW, Thurber DL (1969) Some late Cenozoic sedimentary formations of the lower Omo Valley. Nature 222:1138–1143

Carbonel P, Peypouquet JP (1979) Les ostracods des series du bassin de l'Omo. Bull Inst géol Bassin d'Aquit 25:167–199

Cerling TE (1976) Oxygen-isotope studies of the East Rudolf volcaniclastics. In: Coppens YY, Howell FC, Isaac GL, Leakey REF (eds) Earliest man and environments in the Lake Rudolf Basin. University of Chicago Press, Chicago, pp 105–114

Cerling TE (1977) Paleochemistry of Plio-Pleistocene Lake Turkana and diagenesis of its sediments. PhD thesis, University of California, Berkeley

Cerling TE (1979) Paleochemistry of Plio-Pleistocene Lake Turkana, Kenya. Palaeogeogr Palaeoclimatol Palaeoecol 27:247–285

Cerling TE, Brown FH (1982) Tuffaceous marker horizons in the Koobi Fora region and the Lower Omo Valley. Nature 299:216–221

Cerling TE, Andanje SA, Blumenthal SA, Brown FH, Chritz KL, Harris JM, Hart JA, Kirera FM, Kalee P, Leakey LN, Leakey MG, Levin NE, Manthi FK, Passey BH, Uno KT (2015) Dietary changes of large herbivores in the Turkana Basin, Kenya from 4 to 1 Ma. Proc Natl Acad Sci 112:11467–11472

Cerling TE, Bowman JR, O'Neil JR (1988) An isotopic study of a fluvial-lacustrine sequence: the Plio-Pleistocene Koobi Fora sequence, East Africa. Palaeogeogr Palaeoclimatol Palaeoecol 63:335–356

Cerling TE, Brown FH, Cerling BW, Curtis GH, Drake RE (1979) Preliminary correlations between the Koobi Fora and Shungura Formations, East Africa. Nature 279:118–121

Cerling TE, Harris JM, Leakey MG (2003) Isotope paleoecology of the Nawata and Nachukui Formations at Lothagam, Turkana Basin, Kenya. In: Leakey MG, Harris JM (eds) Lothagam: the dawn of humanity in Africa. Columbia University Press, New York, pp 605–624

Cerling TE, Harris JM, MacFadden BJ, Leakey MJ, Quade J, Eisenmann V, Ehleringer JR (1997) Global vegetation change through the Miocene/Pliocene boundary. Nature 389:153–158

Cerling TE, Levin NE, Passey BH (2011a) Stable isotope ecology in the Omo-Turkana Basin. Evol Anthropol 20:228–237

Cerling TE, Wynn JG, Andanje SA, Bird MI, Korir DK, Levin NE, Mace W, Maharia AN, Quade J, Remien CH (2011b) Woody cover and hominin environments in the past 6 million years. Nature 476:51–56

Cerling TE, Manthi FK, Mbua EN, Leakey LN, Leakey MG, Leakey RE, Brown FH, Grine FE, Hart JA, Kaleme P, Roche H, Uno KT, Wood BA (2013) Stable isotope-based diet reconstructions of Turkana Basin hominins. Proc Natl Acad Sci 110:10501–10506

Cohen A (1981) Paleolimnological research at Lake Turkana, Kenya. Palaeoecol Afr 13:61–82

Cohen A (1982) Paleoenvironments of root casts from the Koobi Fora Formation, Kenya. J Sediment Petrol 52:401–414

Cohen A (1984) Effect of zoobenthic standing crop on laminae preservation in tropical lake sediment, Lake Turkana, East Africa. J Paleontol 58:499–510

Cohen A (1986) Distribution and faunal associations of benthic invertebrates at Lake Turkana, Kenya. Hydrobiologia 141:179–197

Cohen AS (1989) Facies relationships and sedimentation in large rift lakes and implications for hydrocarbon exploration: Examples from lakes Turkana and Tanganyika. Palaeogeogr Palaeoclimatol Palaeoecol 70:65–80

Cohen AS, Campisano CJ, Arrowsmith JR, Asrat A, Beck CC, Behrensmeyer AK, Deino AL, Deocampo D, Feibel CS, Foerster V, Kingston JD, Lamb HF, Lowenstein TK, Lupien RL, Muiruri V, Olago DO, Owen RB, Potts R, Russell JM Schaebitz F, Stone JR, Trauth M, Yost CL (2022a) Reconstructing the environmental context of human origins in Eastern Africa through scientific drilling. Annu Rev Earth Planet Sci 50:451–476

Cohen AS, Du A, Rowan J, Yost CL, Billingsley AL, Campisano C, Brown ET, Deino AL, Feibel CS, Grant K, Kingston JD, Lupien R, Muiruri V, Owen RB, Reed KE, Russell J, Stockhecke M (2022b) Plio-Pleistocene African environmental variability and mammalian evolution. Proc Natl Acad Sci 119:e2107393119

Cohen AS, Dussinger R, Richardson J (1983) Lacustrine paleochemical interpretations based on eastern and southern African ostracodes. Palaeogeogr Palaeoclimatol Palaeoecol 43:129–151

Cohen AS, Schwartz HL (1983) Speciation in molluscs from Turkana Basin. Nature 304:659–660

Coppens Y (1975) Évolution des mammifères, de leurs fréquences et de leurs associations, aucours du Plio-Pleistocène dans la basse vallée de l'Omo en Éthiopie. C R Acad Sci D 281:1571–1574

Coppens Y, Howell FC, Isaac GL, Leakey REF (eds) (1976) Earliest man and environments in the Lake Rudolf Basin. University of Chicago Press, Chicago

Davidson A (1983) The Omo River Project. Reconnaissance of geology and geochemistry of parts of Ilubabor, Kefa, Gemu Gofa and Sidamo, Ethiopia. Ministry of Mines and Energy, Ethiopian Institute of Geological Surveys, Addis Ababa, Ethiopia

Dean C, Leakey MG, Reid D, Schrenk F, Schwartz GT, Stringer C, Walker A (2001) Growth processes in teeth distinguish modern humans from *Homo erectus* and earlier hominins. Nature 414:628–631

de Heinzelin J (1970) Op zoek naar fossiele hominiden in Africa. Biol Jaarboek 38:57–63 de Heinzelin J (1980) Cartographie géologique de la Basse-Vallée de l'Omo (Ethiopie) – (Omo Research Expedition). Bull Soc Belge Géol 89:137–140

de Heinzelin J (1982) The Lower Omo and Turkana Basin in Plio-Pleistocene times. Z Geomorphol N F 41:35–41

de Heinzelin J (1983) The Omo Group. Koninklijk Museum voor Midden-Afrika, Tervuren, Geologische Wetenschappen, 85

de Heinzelin J, Brown FH (1969) Some early Pleistocene deposits of the lower Omo valley: the Usno Formation. Quaternaria:31–46

de Heinzelin J, Haesaerts P, Howell FC (1976) Plio-Pleistocene formations of the lower Omo Basin with particular reference to the Shungura formation. In: Coppens YY, Howell FC, Isaac GL, Leakey REF (eds) Earliest man and environments in the Lake Rudolf Basin. University of Chicago Press, Chicago, pp 24–49

Dixey F (1944) Miocene sediments in south Turkana. J Nat Hist Soc 1:13–14

Drapeau MSM, Bobe R, Wynn JG, Campisano CJ, Dumouchel L, Geraads D (2014) The Omo Mursi Formation: a window into the East African Pliocene. J Hum Evol 75:64–79

Drapeau MSM, Wynn JG, Geraads R, Dumouchel L, Campisano CJ, Bobe R (2022) Paleontology and geology of the Mursi Formation. In: Reynolds SC, Bobe R (eds) African paleoecology and human evolution. Cambridge University Press, Cambridge, pp 278–288

Ducrocq S, Boisserie J-R, Tiercelin J-J, Delmer C, Garcia G, Kyalo MF, Leakey MG, Marivaux L, Ostero O, Peigné S, Tassy P, Lithoreau F (2010) New Oligocene vertebrate localities from northern Kenya (Turkana basin). J Vertebr Paleontol 30:293–299

Duesing W, Berner N, Deino AL, Foerster V, Hauke Kraemer K, Marwan N, Trauth MH (2021a) Multiband wavelet age modeling for a ~293 m (~600 kyr) sediment core from Chew Bahir Basin, southern Ethiopian Rift. Front Earth Sci 9:594047

Duesing W, Kaboth-Bahr S, Asrat A, Cohen AS, Foerster V, Lamb HF, Schaebitz F, Trauth MH, Viehberg F (2021b) Changes in the cyclicity and variability of the eastern African paleoclimate over the last 620 kyrs. Quat Sci Rev 273:107219

Düsing W (2020) From changes in the Earth's orbit to African climate variability. Dr rer Natl dissertation, Universität Potsdam

Eck GG, Jablonski NG (1987) The skull of *Theropithecus brumpti* compared with those of other species of the genus *Theropithecus*. In: Coppens Y, Howell FC (eds) Les faunes Plio-Pléistocènes de la basse vallée de l'Omo (Ethiopie), vol 3. Editions CNRS, pp 18–142

Eisenmann V (1985) Family Equidae. In: Harris JM (ed) Koobi Fora Research Project, The fossil ungulates: Proboscidia, Perissodactyla, and Suidae, vol 2. Clarendon, Oxford, pp 156–214

Ekdale AA, Brown FH, Feibel CS (1989) Nonmarine macroborings in Early Pleistocene algal biolithites (stromatolites) of the Turkana Basin, northern Kenya. Palaios 4:389–396

Erbello A, Kidane T (2018) Timing of volcanism and initiation of rifting in the Omo-Turkana depression, southwest Ethiopia: evidence from paleomagnetism. J Afr Earth Sci 139:319–329

Feakins SJ, Eglinton TI, deMenocal PB (2007) A comparison of biomarker records of northeast African vegetation from lacustrine and marine sediments (ca. 3.40 Ma). Org Geochem 38:1607–1624

Feakins SJ, Levin NE, Liddy HM, Sieracki A, Eglinton TI, Bonnefille R (2013) Northeast African vegetation change over 12 m.y. Geology 41:295–298

Feibel CS (1983) Stratigraphy and paleoenvironments of the Koobi Fora Formation along the western Koobi Fora Ridge, East Turkana, Kenya. MSc thesis, Iowa State University, Ames

Feibel CS (1987) Fossil fish nests from the Koobi Fora Formation (Plio-Pleistocene) of northern Kenya. J Paleontol 61:130–134

Feibel CS (1988) Paleoenvironments of the Koobi Fora Formation, Turkana Basin, northern Kenya. PhD thesis, University Utah, Salt Lake City

Feibel CS (1993) Freshwater stingrays from the Plio-Pleistocene of the Turkana Basin, Kenya and Ethiopia. Lethaia 26:359–366

Feibel CS (1997) Debating the environmental factors in hominid evolution. GSA Today 7:1–7

Feibel CS (1999) Tephrostratigraphy and geological context in paleoanthropology. Evol Anthropol 8:87–100

Feibel C (2003a) Stratigraphy and depositional setting of the Pliocene Kanapoi Formation, lower Kerio Valley, Kenya. In: Harris JM, Leakey MG (2003) Geology and vertebrate paleontology of the Early Pliocene site of Kanapoi, northern Kenya. Natl Hist Mus Los Angeles Contrib Sci 498:9–20

Feibel C (2003b) Stratigraphy and depositional history of the Lothagam sequence. In: Leakey MG, Harris JM (eds) Lothagam: the dawn of humanity in Eastern Africa. Columbia University Press, New York, pp 17–30

Feibel CS (2008) Microstratigraphy of the Kibish hominin sites KHS and PHS, Lower Omo Valley, Ethiopia. J Hum Evol 55:404–408

Feibel CS (2011) A geological history of the Turkana Basin. Evol Anthropol 20:206–216

Feibel CS (2013) Facies analysis and Plio-Pleistocene paleoecology. In: Spoonheimer M, Lee-Thorp JA, Reed KE, Ungar PS (eds) Early hominin paleoecology. University of Colorado Press, Boulder, pp 35–58

Feibel CS, Brown FH (1986) Depositional history of the Koobi Fora Formation, northern Kenya. Proc Second Conf on the Geology of Kenya

Feibel CS, Brown FH (1993) Microstratigraphy and paleoenvironments. In: Walker AC, Leakey REF (eds) The Nariokotome *Homo erectus* skeleton. Harvard University Press, Boston, pp 21–39

Feibel CS, Brown FH, McDougall I (1989) Stratigraphic context of fossil hominids from the Omo Group deposits: northern Turkana Basin, Kenya and Ethiopia. Am J Phys Anthropol 78:595–622

Feibel CS, Harris JM, Brown FH (1991) Palaeoenvironmental context for the late Neogene of the Turkana Basin. In: Harris JM (ed) Koobi Fora Research Project, The fossil ungulates: geology, fossil artiodactyls and paleoenvironments, vol 3. Clarendon Press, Oxford, pp 321–370

Feibel CS, Lepre C, Quinn R (2009) Stratigraphy, correlation, and age estimates for fossils from area 123, Koobi Fora. J Hum Evol 57:112–122

Findlater I (1976a) Stratigraphic analysis and palaeoenvironmental interpretation of a Plio/Pleistocene sedimentary basin east of Lake Turkana. PhD thesis, University of London

Findlater I (1976b) Tuffs and the recognition of isochronous mapping units. In: Coppens Y, Howell FC, Isaac GL, Leakey REF (eds) Earliest man and environments in the Lake Rudolf Basin. University of Chicago Press, Chicago, pp 94–104

Findlater I (1978) Isochronous surfaces within the Plio-Pleistocene sediments east of Lake Turkana. In: Bishop WW (ed) Geological background to fossil man. Geol Soc Lond Spec Publ 6:415–420

Fitch FJ, Miller JA (1969) Age determinations on feldspar from the lower Omo Basin. Nature 222:1143

Fleagle JG, Assefa Z, Brown FH, Shea JJ (2008) Paleoanthropology of the Kibish Formation, southern Ethiopia: introduction. J Hum Evol 55:360–365

Foerster VE, Asrat A, Cohen AS, Chapot MS, Deino A, Deocampo DM, Duesing W, Guenter C, Junginger A, Kaboth-Bahr S, Lamb HF, Lane C, Bronk Ramsey C, Roberts HM, Vidal C, Schaebitz F, Trauth MH (2020) Assessing the role of climate change in human evolution and dispersal: a 600,000-year record from Chew Bahir, southern Ethiopia, EGU General Assembly, EGU2020–18292

Fortelius M, Žliobaitė I, Kaya F, Bibi F, Bobe R, Leakey L, Leakey M, Patterson D, Rannikko J, Werdelin L (2016) An ecometric analysis of the fossil mammal record of the Turkana Basin. Phil Trans R Soc Lond Ser B Biol Sci 371:20150232

Frank H (1976) Stratigraphy of the Upper Member, Koobi Fora Formation, northern Karari Escarpment, East Turkana Basin, Kenya. MSc thesis, Iowa State University, Ames

Frost SR, Ward CV, Manthi FK, Plavcan JM (2020) Cercopithecid fossils from Kanapoi, West Turkana, Kenya (2007-2015). J Hum Evol 140:102642

Fuchs VE (1939) The geological history of the Lake Rudolf basin, Kenya Colony. Phil Trans R Soc Lond Ser B 229:219–274

Garcin Y, Melnick D, Strecker MR, Olago D, Tiercelin J-J (2012) East African mid-Holocene wet-dry transition recorded in palaeo-shorelines of Lake Turkana, northern Kenya Rift. Earth Plan Sci Lett 331-332:322–334

Gathogo PN, Brown FH (2006a) Revised stratigraphy of area 123, Koobi Fora, Kenya, and new age estimates of its fossil mammals, including hominins. J Hum Evol 51:471–479

Gathogo PN, Brown FH (2006b) Stratigraphy of the Koobi Fora Formation (Pliocene and Pleistocene) in the Ileret region of northern Kenya. J Afr Earth Sci 45:369–390

Gathogo PN, Brown FH, McDougall I (2008) Stratigraphy of the Koobi Fora Formation (Pliocene and Pleistocene) in the Loiyangalani region of northern Kenya. J Afr Earth Sci 51:277–297

Geraads D, Bobe R, Manthi FK (2013) New ruminants (Mammalia) from the Pliocene of Kanapoi, Kenya, and a revision of previous collections, with a note on the Suidae. J Afr Earth Sci 85:53–61

Gravina A (2015) Paleoecology of ostracods from the Hominin Sites and Paleolakes Drilling Project's West Turkana Kaitio core from West Turkana, Kenya. BSc thesis, Rutgers University, Piscataway, New Jersey

Guillemot C (1997) Recherches paléoanthropologiques dans la basse vallée de l'Omo. Bull Maison Études Éthiop 11:69–78

Haileab B (1995) Geochemistry, geochronology and tephrostratigraphy of tephra from the Turkana Basin, southern Ethiopia and northern Kenya. PhD thesis, University Utah, Salt Lake City, USA

Haileab B, Brown FH, McDougall I, Gathogo PN (2004) Gombe Group basalts and initiation of Pliocene deposition in the Turkana depression, northern Kenya and Southern Ethiopia. Geol Mag 141:41–53

Hammond AS, Foecke KK, Kelley J (2019) Hominoid anterior teeth from the late Oligocene site of Losodok, Kenya. J Hum Evol 128:59–75

Hargrave JE, Hicks MK, Scholz CA (2014) Lacustrine carbonates from Lake Turkana, Kenya: a depositional model of carbonates in an extensional basin. J Sediment Res 84:224–237

Harmand S, Lewis JE, Feibel CS et al (2015) 3.3-million-year-old stone tools from Lomekwi 3, West Turkana, Kenya. Nature 521:310–315

Harris JM, Brown FH (1985) A new hominid locality west of Lake Turkana, Kenya. Natl Geogr Res 1:280–297

Harris JM, Leakey MG (2003) Geology and vertebrate paleontology of the Early Pliocene site of Kanapoi, northern Kenya. Contrib Sci Natur Hist Mus Los Angeles 498:1–134

Harris JM, Brown FH, Leakey MG (1988a) Stratigraphy and paleontology of Pliocene and Pleistocene localities west of Lake Turkana, Kenya. Contrib Sci, Natur Hist Mus Los Angeles 399:1–128

Harris JM, Brown FH, Leakey MG, Walker A, Leakey RE (1988b) Pliocene and Pleistocene hominid-bearing sites from west of Lake Turkana, Kenya. Science 239:27–33

Harris J, Leakey M, Brown FH (2006) A brief history of research at Koobi Fora, northern Kenya. Ethnohistory 53:1–35

Harris JM, Watkins RT (1974) New Early Miocene vertebrate locality near Lake Rudolf, Kenya. Nature 252:576–577

Hatala KG, Roach NT, Ostrofsky KR, Wunderlich RE, Dingwall HL, Villmoare BA, Green DJ, Braun DR, Harris JWK, Behrensmeyer AK, Richmond BG (2017) Hominin track assemblages from Okote Member deposits near Ileret, Kenya, and their implications for understanding fossil hominin paleobiology at 1.5 Ma. J Hum Evol 112:93–104

Hernández Fernández M, Vrba ES (2006) Plio-Pleistocene climatic change in the Turkana Basin (East Africa): evidence from large mammal faunas. J Hum Evol 50:595–626

Hillhouse JW, Cerling TE, Brown FH (1986) Magnetostratigraphy of the Koobi Fora Formation, Lake Turkana, Kenya. J Geophys Res 91:11581–11595

Hlubik S, Cutts R, Braun DR, Berna F, Feibel CS, Harris JWK (2019) Hominin fire use in the Okote Member at Koobi Fora, Kenya: new evidence for the old debate. J Hum Evol 133:214–229

Howell FC, Coppens Y (1983) Introduction. In: de Heinzelin J (ed) The Omo Group. Archives of the International Omo Research Expedition. Mus R l'Afr Centr Tervuren, pp 1–5

Howell FC, Haesaerts P, de Heinzelin J (1987) Depositional environments, archeological occurrences and hominids from Members E and F of the Shungura Formation (Omo basin, Ethiopia). J Hum Evol 16:665–700

Johnson GD (1974) Cainozoic lacustrine stromatolites from hominid-bearing sediments east of Lake Rudolf, Kenya. Nature 247:520–523

Johnson GD, Raynolds R (1976) Late Cenozoic environments of the Koobi Fora Formation: the upper member along the western Koobi Fora Ridge. In: Coppens Y, Howell FC, Isaac GL, Leakey REF (eds) Earliest man and environments in the Lake Rudolf Basin. University of Chicago Press, Chicago, pp 115–122

Johnson TC, Halfman JD, Rosendahl BR, Lister GS (1987) Climatic and tectonic effects on sedimentation in a rift-valley lake: evidence from high-resolution seismic profiles, Lake Turkana, Kenya. Geol Soc Am Bull 98:439–447

Johnson TC, Halfman JD, Showers WJ (1991) Paleoclimate of the last 4000 years at Lake Turkana, Kenya based on the isotopic composition of authigenic calcite. Palaeogeogr Palaeoecol Palaeoclimatol 85:189–198

Johnson TC, Malala JO (2009) Lake Turkana and its link to the Nile. In: Dumont HJ (ed) The Nile: origin, environments, limnology and human use. Springer, Dordrecht, pp 287–304

Joordens JCA (2007) The power of place: hominin evolution and dispersal driven by climate change between ~5 – 2.5 Ma. J Hum Evol 53:443–464

Joordens JCA, Dupont-Nivet G, Feibel CS, Spoor F, Sier MJ, van der Lubbe JHJL, Nielsen TK, Knul MV, Davies GR, Vonhof HB (2013) Improved age control on early *Homo* fossils from the upper Burgi Member at Koobi Fora, Kenya. J Hum Evol 65:731–745

Joordens JCA, Vonhof HB, Feibel CS, Lourens LJ, Dupont-Nivet G, van der Lubbe JHJL, Sier MJ, Davies GR, Kroon D (2011) An astronomically-tuned climate framework for hominins in the Turkana Basin. Earth Planet Sci Lett 307:1–8

Joubert P (1966) Geology of the Loperot area. Geol Surv Kenya Rep 74

Junginger A, Kuebler S, Rosca C, Owen RB, Deino A, Feibel C, Trauth MH, Vonhof H (2023) Mid-Pleistocene volcano-tectonic fragmentation of the Turkana-Suguta megalake, EGU General Assembly 2023, Vienna, Austria, 24–28 Apr 2023, EGU23–17267

Kaboth-Bahr S, Gosling WD, Vogelsang R, Bahr A, Scerri EM, Asrat A, Cohen AS, Düsing W, Foerster V, Lamb HF, Maslin MA, Roberts HM, Schäbitz F, Trauth MH (2021) Paleo-ENSO influence on African environments and early modern humans. Proc Natl Acad Sci 118:e2018277118

Karson JA, Curtis PC (1994) Quaternary volcanic centres of the Turkana Rift, Kenya. J Afr Earth Sci 18:15–35

Kelly AJ (1996) Intra-regional and inter-regional variability in the East Turkana (Kenya) and Kenyan Middle Stone Age. PhD thesis, Rutgers University, Piscataway, New Jersey

Kelly AJ, Harris JWK (1992) Recent findings of Middle Stone Age material from East Turkana. Nyame Akuma 38:29–34

Kibunjia M (2002) Archaeological investigations of Lokalalei 1 (GaJh 5): A late Pliocene site, west of Lake Turkana, Kenya. PhD thesis, Rutgers University, Piscataway, New Jersey

Kidane T, Brown FH, Kidney C (2014) Magnetostratigraphy of the fossil-rich Shungura Formation, southwest Ethiopia. J Afr Earth Sci 97:207–223

Kingston JD (2007) Shifting adaptive landscapes: Progress and challenges in reconstructing early hominid environments. Am J Phys Anthropol 134:20–58

LaCarrubba JM (2018) Palaeoenvironmental interpretations of the A4 stromatolites from the Koobi Fora Formation in Lake Turkana, Kenya. PhD thesis, Rutgers University, Piscataway, New Jersey

Lamond RE, Tapanila L (2003) Embedment cavities in lacustrine stromatolites: evidence of animal interactions from Cenozoic carbonates in USA and Kenya. Palaios 18:445

Leakey MG, Feibel CS, Bernor RL, Cerling TE, Harris JM, Stewart KM, Storrs GW, Walker A, Werdelin L, Winkler AJ (1996) Lothagam: a record of faunal change in the late Miocene of East Africa. J Vertebr Paleontol 16:556–570

Leakey MG, Grossman A, Gutiérrez M, Fleagle JG (2011) Faunal change in the Turkana basin during the late Oligocene and Miocene. Evol Anthropol 20:238–253

Leakey MG, Harris JM (2003) Lothagam: the dawn of humanity in Eastern Africa. Columbia University Press, New York

Leakey MG, Leakey RE (1978) The Koobi Fora Research Project, volume I: the fossil hominids and an introduction to their context 1968–1974. Oxford University Press, Oxford

Leakey MG, Spoor F, Brown FH, Gathogo PN, Kiarie C, Leakey LN, McDougall I (2001) New hominin genus from eastern Africa shows diverse middle Pliocene lineages. Nature 410:433–440

Leakey MG, Ungar PS, Walker A (1995) A new genus of large primate from the Late Oligocene of Lothidok, Turkana District, Kenya. J Hum Evol 28:519–532

Leakey RE, Leakey MG (1986a) A new Miocene hominoid from Kenya. Nature 324:143–146

Leakey RE, Leakey MG (1986b) A second new hominoid from Kenya. Nature 324:146–148

Leakey RE, Leakey MG, Behrensmeyer AK (1978) The hominid catalogue. In: Leakey MG, Leakey RE (eds) Koobi Fora Research project, The fossil hominids and an introduction to their context, 1968–1974, vol 1. Oxford University Press, Oxford

Leakey REF (1969) Early *Homo sapiens* remains from the Omo River region of south-west Ethiopia. Nature 222:1132

Leakey REF (1970) New hominid remains and early artefacts from northern Kenya. Nature 226:223–224

Lepre CJ (2014) Early Pleistocene lake formation and hominin origins in the Turkana-Omo rift. Quat Sci Rev 102:181–191

Lepre CJ (2017) Crevasse-splay and associated depositional environments of the hominin-bearing lower Okote Member, Koobi Fora Formation (Plio-Pleistocene), Kenya. Depos Rec 32:1–26

Lepre CJ, Kent DV (2010) New magnetostratigraphy for the Olduvai Subchron in the Koobi Fora Formation, northwest Kenya, with implications for early Homo. Earth Planet Sci Lett 290:362–374

Lepre CJ, Kent DV (2015) Chronostratigraphy of KNM-ER 3733 and other area 104 hominins from Koobi Fora. J Hum Evol 86:99–111

Lepre CJ, Quinn RL (2022) Aridification and orbital forcing of eastern African climate during the Plio-Pleistocene. Glob Planet Change 208:103684

Lepre CJ, Quinn RL, Joordens JCA, Swisher CC, Feibel CS (2007) Plio-Pleistocene facies environments from the KBS Member, Koobi Fora Formation: implications for climate controls on the development of lake-margin hominin habitats in the northeast Turkana Basin (northwest Kenya). J Hum Evol 53:504–514

Levin NE, Brown FH, Behrensmeyer AK, Bobe R, Cerling TE (2011) Paleosol carbonates from the Omo Group: isotopic records of local and regional environmental change in East Africa. Palaeogeogr Palaeoclimatol Palaeoecol 307:75–89

Liutkus-Pierce CM, Takashita-Bynum KK, Beane LA, Edwards CT, Burns OE, Mana S, Hemming S, Grossman A, Wright JD, Kirera FM (2019) Reconstruction of the Early Miocene critical zone at Loperot, Southwestern Turkana, Kenya. Front Ecol Evol 7:358–320

Lupien RL (2019) The Plio-Pleistocene climate history of East Africa and the role of environmental change on human evolution: studies of leaf wax isotopes from paleolake sediment. PhD thesis, Brown University, Providence

Lupien RL, Russell JM, Feibel C, Beck C, Castañeda I, Deino A, Cohen AS (2018) A leaf wax biomarker record of early Pleistocene hydroclimate from West Turkana, Kenya. Quat Sci Rev 186:225–235

Lupien RL, Russell JM, Grove M, Beck CC, Feibel CS, Cohen AS (2020) Abrupt climate change and its influences on hominin evolution during the early Pleistocene in the Turkana Basin. Kenya Quat Sci Rev 245:106531

Lupien RL, Russell JM, Pearson EJ, Castañeda IS, Asrat A, Foerster V, Lamb HF, Roberts HM, Schäbitz F, Trauth MH, Beck CC, Feibel CS, Cohen AS (2022) Orbital controls on eastern African hydroclimate in the Pleistocene. Sci Rep 12:3170

Louchart A (2008) Fossil birds of the Kibish Formation. J Hum Evol 55:513–520

Macgregor D (2015) History of the development of the East African Rift System: a series of interpreted maps through time. J Afr Earth Sci 101:1–21

Macho GA, Leakey MG, Williamson DK, Jiang Y (2003) Palaeoenvironmental reconstruction: evidence for seasonality at Allia Bay, Kenya, at 3.9 million years. Palaeogeogr Palaeoclimatol Palaeoecol 199:17–30

Maglio VJ (1971) Vertebrate faunas from the Kubi Algi, Koobi Fora, and Ileret areas, East Rudolf, Kenya. Nature 231:240–241

Mana S, Hemming S, Kent DV, Lepre CJ (2019) Temporal and stratigraphic framework for paleoanthropology sites within East-Central Area 130, Koobi Fora, Kenya. Front Earth Sci 7:1584–1513

Manthi FK, Plavcan JM, Ward CV (2020) Introduction to special issue 'Kanapoi: Paleobiology of a Pliocene site in Kenya'. J Hum Evol 140:102718

Margolis MJ (1976) Organic geochemistry of Lake Turkana Basin sediments: normal paraffins of sediments and in situ fossils. PhD thesis, University of Rhode Island, Kingston

Martz AM (1979) Petrology and chemistry of tuffs from the Shungura Formation, southwest Ethiopia. MSc thesis, University of Utah, Salt Lake City

Martz AM, Brown FH (1981) Chemistry and mineralogy of some Plio-Pleistocene tuffs from the Shungura Formation, southwest Ethiopia. Quat Res 16:240–257

Mathisen ME, Vondra CF (1983) Provenance of the Plio-Pleistocene sediments in the East Turkana Basin, northern Kenya. Palaeogeogr Palaeoecol Palaeoclimatol 44:141–168

Maurin T, Delagnes A, Boisserie JR (2014) Spatial behaviours of Early Oldowan toolmakers in the Shungura Formation (Lower Omo Valley, Ethiopia): proposal for an integrated approach. C R Palevol 13:737–746

McDougall I, Brown FH (2006) Precise $^{40}Ar/^{39}Ar$ geochronology for the upper Koobi Fora Formation, Turkana Basin, northern Kenya. J Geol Soc Lond 162:205–220

McDougall I, Brown FH (2008) Geochronology of the pre-KBS Tuff sequence, Omo Group, Turkana Basin. J Geol Soc Lond 165:549–562

McDougall I, Brown FH, Fleagle JG (2005) Stratigraphic placement and age of modern humans from Kibish, Ethiopia. Nature 433:733–736

McDougall I, Brown FH, Fleagle JG (2008) Sapropels and the age of hominins Omo I and Omo II, Kibish, Ethiopia. J Hum Evol 55:409–420

McDougall I, Brown FH, Vasconcelos PM, Cohen BE, Thiede DS, Buchanan MJ (2012) New single crystal $^{40}Ar/^{39}Ar$ ages improve time scale for deposition of the Omo Group, Omo–Turkana Basin, East Africa. J Geol Soc Lond 169:213–226

McDougall I, Feibel CS (2003) Numerical age control for the Miocene-Pliocene succession at Lothagam, a hominoid-bearing sequence in the northern Kenya Rift. In: Leakey MG, Harris JM (eds) Lothagam: the dawn of humanity in Eastern Africa. Columbia University Press, New York, pp 43–64

Mead JG (1975) A fossil beaked whale (Cetacea: Ziphiidae) from the Miocene of Kenya. J Paleontol 49:745–751

Merrick HV, de Heinzelin J, Haesaerts P, Howell FC (1973) Archaeological occurrences of Early Pleistocene age from the Shungura Formation, Lower Omo Valley, Ethiopia. Nature 242:572–575

Merrick HV, Merrick JPS (1976) Archeological occurrences of earlier Pleistocene age from the Shungura Formation. In: Coppens Y, Howell FC, Isaac GL, Leakey REF (eds) Earliest man and environments in the Lake Rudolf Basin. University of Chicago Press, Chicago, pp 574–584

Morley CK (1989) Extension, detachments, and sedimentation in continental rifts (with particular reference to East Africa). Tectonics 8:1175–1192

Morley CK (1999) Boundary fault angle, with particular reference to the Lokichar Fault, Turkana Region, Kenya. In: Morley CK (ed) Geoscience of rift systems – evolution of East Africa. AAPG Stud Geol 44:115–129

Morley CK, Karanja FM, Wescott WA, Stone DM, Harper RM, Wigger ST, Day RA (1999) Geology and geophysics of the Western Turkana Basins, Kenya. In: Morley, CK (ed) Geoscience of rift systems – evolution of East Africa. AAPG Stud Geol 44:19–54

Morley CK, Wescott WA (1999) Sedimentary environments and geometry of sedimentary bodies determined from subsurface studies in East Africa. In: Morley, CK (ed) Geoscience of rift systems – evolution of East Africa. AAPG Stud Geol 44:211–231

Morley C, Wescott W, Stone D, Harper R, Wigger S, Karanja F (1992) Tectonic evolution of the northern Kenyan Rift. J Geol Soc Lond 149:333–348

Morrissey A, Scholz CA (2014) Paleohydrology of Lake Turkana and its influence on the Nile River system. Palaeogr Palaeoclimatol Palaeoecol 403:88–100

Mount JF, Cohen AS (1984) Petrology and geochemistry of rhizoliths from Plio-Pleistocene fluvial and marginal lacustrine deposits, east Lake Turkana, Kenya. J Sediment Res 54:263–275

Murray-Hughes R (1933) Notes on the geological succession, tectonics and economic geology of the western half of the Kenya colony. Rep Geol Surv Kenya 3

Negash EW, Alemseged Z, Wynn JG, Bedaso ZK (2015) Paleodietary reconstruction using stable isotopes and abundance analysis of bovids from the Shungura Formation of South Omo, Ethiopia. J Hum Evol 88:127–136

Nutz A, Schuster M (2018) Comments on: "A leaf wax biomarker record of early Pleistocene hydroclimate from West Turkana, Kenya" by Lupien et al. [Quat. Sci. Rev. 186 (2018), 225–235]. Quat Sci Rev 201:505–507

Nutz A, Schuster M, Boës X, Rubino J-L (2017) Orbitally-driven evolution of Lake Turkana (Turkana Depression, Kenya, EARS) between 1.95 and 1.72 Ma: a sequence stratigraphy perspective. J Afr Earth Sci 125:230–243

Nutz A, Schuster M, Barbon D, Gassier G, Van Bocxlaer B, Robin C, Ragon T, Ghienne JF, Rubino JL (2020) Plio-Pleistocene sedimentation in West Turkana (Turkana Depression, Kenya, East African Rift System): Paleolake fluctuations, paleolandscapes and controlling factors. Earth Sci Rev 211:103415

Nyamweru C (1989) New evidence for the former extent of the Nile drainage system. Geogr J 155:179–188

O'Brien K, Patterson DB, BierNatl MD, Braun DR, Cerling TE, McGrosky A, Faith JT (2020) Ungulate turnover in the Koobi Fora Formation: spatial and temporal variation in the Early Pleistocene. J Afr Earth Sci 161:103658

Ortiz K, Valachovic J, Cohen AS, Feibel C, Kingston J, Deino A, Hill A, Campisano CJ, Beck CC (2015) Paleoclimate and paleoenvironmental forcing on early humans: LOI analysis of three HSPDP drill core sites in Kenya and Ethiopia. Geol Soc Am Abstr Prog 47:368

Ortiz K, Plange K, Cohen AS, Feibel CS, Kingston JD, Deino A, Hill A, Smith P (2014) Preliminary TOC/TIC and geophysical data from two HSPDP drill core sites in Kenya, with implications for Pliocene hominin paleoenvironments. Geol Soc Am Abstr Prog 46:321–323

Owen RB, Renaut RW (1986) Sedimentology, stratigraphy and palaeoenvironments of the Holocene Galana Boi Formation, NE Lake Turkana, Kenya. In: Frostick, LE, Renaut RW, Reid I, Tiercelin-J-J (eds) Sedimentation in the African rifts. Geol Soc Lond Spec Publ 25:311–322

Passey BH, Levin NE, Cerling TE, Brown FH, Eiler JM (2010) High-temperature environments of human evolution in East Africa based on bond ordering in paleosol carbonates. Proc Natl Acad Sci 107:11245–11249

Patterson DB, Braun DR, Behrensmeyer AK, Merritt S, Zliobaite I, Reeves JS, Wood BA, Fortelius M, Bobe R (2017) Ecosystem evolution and hominin paleobiology at East Turkana, northern Kenya between 2.0 and 1.4 Ma. Palaeogeogr Palaeoecol Palaeoclimatol 481:1–13

Pearson OM, Fleagle JG, Grine FE, Royer DF (2008) Further new hominin fossils from the Kibish Formation, southwestern Ethiopia. J Hum Evol 55:444–447

Peypouquet JP, Carbonel P, de Heinzelin J (1979) Les ostracodes: indicateurs de l'evolution des environnements lacustres de la fin du Cenozoique dans la branch orientale du Rift Africain. Bull Inst géol Bassin Aquit 25:201–219

Potts R, Faith JT (2015) Alternating high and low climate variability: the context of natural selection and speciation in Plio-Pleistocene hominin evolution. J Hum Evol 87:5–20

Powers DW (1980) Geology of Mio-Pliocene sediments of the lower Kerio River Valley. PhD thesis, Princeton University New Jersey

Prat S, Brugal JP, Tiercelin J-J, Barrat JA, Bohn M, Delagnes A, Harmand S, Kimeu K, Kibunjia M, Texier PJ, Roche H (2005) First occurrence of early *Homo* in the Nachukui Formation (West Turkana, Kenya) at 2.3–2.4 Myr. J Hum Evol 49:230–240

Quinn R (2006) Stable isotopic evidence for Plio-Pleistocene hominin paleoenvironments of the Koobi Fora Formation, Turkana Basin, northern Kenya. PhD thesis, Rutgers University, Piscataway, New Jersey

Quinn RL (2015) Influence of Plio-Pleistocene basin hydrology on the Turkana hominin enamel carbonate $\delta^{18}O$ values. J Hum Evol 86:13–31

Quinn RL, Lepre CJ (2019) Revisiting the pedogenic carbonate isotopes and paleoenvironmental interpretation of Kanapoi. J Hum Evol 140:102549

Quinn RL, Lepre CJ (2021) Contracting eastern African C4 grasslands during the extinction of *Paranthropus boisei*. Sci Rep 11:7164

Quinn RL, Lepre CJ, Feibel CS, Wright JD, Mortlock RA, Harmand S, Brugal JP, Roche H (2013) Pedogenic carbonate stable isotopic evidence for wooded habitat preference of early Pleistocene tool makers in the Turkana Basin. J Hum Evol 65:65–78

Quinn RL, Lepre CJ, Wright JD, Feibel CS (2007) Paleogeographic variations of pedogenic carbonate $\delta^{13}C$ values from Koobi Fora, Kenya: implications for floral compositions of Plio-Pleistocene hominin environments. J Hum Evol 53:560–573

Quinn RL, Lewis J, Brugal J-P, Lepre CJ, Trifonov A, Harmand S (2020) Influences of dietary niche expansion and Pliocene environmental changes on the origins of stone tool making. Palaeogeogr Palaeoclimatol Palaeoecol 562:11704

Ragon T, Nutz A, Schuster M, Ghienne J-F, Ruffet G, Rubino J-L (2018) Evolution of the northern Turkana Depression (East African Rift System, Kenya) during the Cenozoic rifting: new insights from the Ekitale Basin (28–25.5 Ma). Geol J 119:1–21

Renaut RW, Morley CK, Jones B (2002) Fossil hot-spring travertine in the Turkana Basin, northern Kenya: structure, facies, and genesis. In: Renaut RW, Ashley GM (eds) Sedimentation in continental rifts. SEPM Spec Publ 73:233–246

Renaut RW, Owen RB, Jones B, Tiercelin J-J, Tarits C, Ego JK, Konhauser KO (2013) Impact of lake-level changes on the formation of thermogene travertine in continental rifts: evidence from Lake Bogoria, Kenya Rift Valley. Sedimentology 60:428–468

Rhemtulla S (1970) South Turkana scientific expedition papers III, a geological reconnaissance of South Turkana. Geogr J 136:161–173

Richmond BG, Green DJ, Lague MR, Chirchir H, Behrensmeyer AK, Bobe R, Bamford MK, Griffin NL, Gunz P, Mbua E, Merritt SR, Pobiner B, Kiura P, Kibunjia M, Harris JWK, Braun DR (2020) The upper limb of *Paranthropus boisei* from Ileret, Kenya. J Hum Evol 141:102727

Richmond BG, Hatala KG, Behrensmeyer AK, Bobe R, Braun DR, Green DJ, Kiura P, Ostrofsky K, Roach NT, Villmoare BA, Wunderlich RE, Harris JWK (2014) Hominin ecology and behavior based on 1.5-million-year-old footprint assemblages from Ileret, Kenya. Annu Mtg Europ Soc Stud Hum Evol, Florence, Italy

Roach NT, Du A, Hatala KG, Ostrofsky KR, Reeves JS, Braun DR, Harris JWK, Behrensmeyer AK, Richmond BG (2018) Pleistocene animal communities of a 1.5 million-year-old lake margin grassland and their relationship to *Homo erectus* paleoecology. J Hum Evol 122:70–83

Roach NT, Hatala KG, Ostrofsky KR, Villmoare B, Reeves JS, Du A, Braun DR, Harris JWK, Behrensmeyer AK, Richmond BG (2016) Pleistocene footprints show intensive use of lake margin habitats by *Homo erectus* groups. Sci Rep 6(26374):1–10

Roberts H, Bronk Ramsey C, Chapot M, Deino A, Lane C, Vidal C, Asrat A (2021) Using multiple chronometers to establish a long, directly-dated lacustrine record: constraining >600,000 years of environmental change at Chew Bahir. Ethiopia Quat Sci Rev 266:107025

Robinson JR, Rowan J, Faith JT, Fleagle JG (2016) Paleoenvironmental change in the late Middle Pleistocene–Holocene Kibish Formation, southern Ethiopia: evidence from ungulate isotopic ecology. Palaeogeogr Palaeoclimatol Palaeoecol 450:50–59

Robinson JR, Rowan J, Campisano CJ, Wynn JG, Reed KE (2017) Late Pliocene environmental change during the transition from *Australopithecus* to *Homo*. Nat Ecol Evol 1(0159):1–7

Roche H, Brugal J, Delagnes A, Feibel C, Harmand S, Kibunjia M, Prat S, Texier PJ (2003) Les sites archéologiques Plio-Pléistocènes de la formation de Nachukui, Ouest-Turkana, Kenya: bilan synthétique 1997-2001. C R Palevol 2:663–673

Rogers OM (2010). Stratigraphy of paleontological collecting areas 106, 107 and 109, Koobi Fora region, northern Kenya. MSc thesis, University Utah, Salt Lake City

Rop BK (1990) Stratigraphic and sedimentologic study of Mesozoic-Tertiary strata in Loiyangalani area, Lake Turkana, district, north-west Kenya. MSc thesis, University of Windsor, Ontario

Rop BK (2011) Hydrocarbon characteristics of source rocks in Loperot-1 Well, Northwest Kenya. Proc Kenya Geothermal Conference, Nairobi (Nov 2011) pp 1–6

Rop BK (2015) Hydrocarbon prospects under Lake Turkana Basin, NW-Kenya: based on geophysical and geochemical data investigations. Res J Eng Tech 1:41–56

Rop BK, Patwardhan AM (2013) Hydrocarbon source rocks evaluation study of Lokichar Basin, northwestern Kenya. J Geol Soc Ind 81:557–580

Rowan J, Tyler Faith J, Gebru Y, Fleagle JG (2015) Taxonomy and paleoecology of fossil Bovidae (Mammalia, Artiodactyla) from the Kibish Formation, southern Ethiopia: implications for dietary change, biogeography, and the structure of the living bovid faunas of East Africa. Palaeogeogr Palaeoclimatol Palaeoecol 420:210–222

Samadi S, David P, Jarne P (2000) Variations of shell shape in the clonal snail *Melanoides tuberculata* and its consequences for the interpretation of fossil series. Evolution 54:492–502

Savage RJG, Williamson PG (1978) The early history of the Turkana Depression. In: Bishop WW (ed) Geological background to fossil man. Geol Soc Lond Spec Publ 6:375–394

Schaebitz F, Asrat A, Lamb H, Cohen A, Foerster V (2021) Hydroclimate changes in eastern Africa over the past 200,000 years may have influenced early human dispersal. Commun Earth Environ 2:123

Schoeninger JJ, Reeser H, Hallin K (2003) Paleoenvironment of *Australopithecus anamensis* at Allia Bay, East Turkana, Kenya: evidence from mammalian herbivore enamel stable isotopes. J Anthropol Archaeol 22:2000–2007

Schofield N, Newton R, Thackrey S, Watson D, Jolley D, Morley C (2021) Linking surface and subsurface volcanic stratigraphy in the Turkana Depression of the East African Rift system. J Geol Soc Lond 178:jgs2020–110

Scholz H, Glaubrecht M (2010) A new and open coiled *Valvata* (Gastropoda) from the Pliocene Koobi Fora Formation of the Turkana Basin, Northern Kenya. J Paleontol 84:996–1002

Schwartz HL (1981) Lower Pleistocene quiet-water lacustrine ooliths from Koobi Fora Tuff, East Lake Turkana, north Kenya. Am Ass Petrol Geol Bull 6:989–989

Schwartz HL (1983) Paleoecology of Late Cenozoic fishes from the Turkana Basin, northern Kenya. PhD thesis, University California, Berkeley

Scott JJ, Hannah AJ, Chupik DT (2017) Stratigraphic packaging in Pliocene to recent rift lake successions, Kenya Rift Valley, using lake-margin and terrestrial ichnology. Geol Soc Am Abstr Progr 49(6):309–312

Shea JJ (2008) The Middle Stone Age archaeology of the Lower Omo Valley Kibish Formation: excavations, lithic assemblages, and inferred patterns of early *Homo sapiens* behavior. J Hum Evol 55:448–485

Shea JJ, Hildebrand EA (2010) The Middle Stone Age of West Turkana, Kenya. J Field Archaeol 35:355–364

Sier MJ, Langereis CG, Dupont-Nivet G, Feibel CS, Joordens JCA, van der Lubbe JHJL, Beck CC, Olago D, Cohen A, WTK Science team members (2017) The top of the Olduvai Subchron in a high-resolution magnetostratigraphy from the West Turkana core WTK13, Hominin Sites and Paleolakes Drilling Project (HSPDP). Quat Geochron 42:117–129

Spoor F, Leakey NM, Gathogo PN, Brown FH, Antón SC, McDougall I, Kiarie C, Manthi FK, Leakey LN (2007) Implications of new early *Homo* fossils from Ileret, east of Lake Turkana, Kenya. Nature 448:688–691

Stern N, Porch N, McDougall I (2002) FxJj43: a window into a 1.5-million-year-old palaeo-landscape in the Okote Member of the Koobi Fora Formation, northern Kenya. Geoarchaeology 17:349–392

Stewart KM (2003a) Fossil fish remains from Mio-Pliocene deposits at Lothagam, Kenya. In: Leakey MG, Harris JM (eds) Lothagam: the dawn of humanity in Eastern Africa. Columbia University Press, New York, pp 75–112

Stewart KM (2003b) Fossil fish remains from the Pliocene Kanapoi site, Kenya. Natur Hist Mus Los Angeles Contrib Sci 498:21–39

Stewart KM, Murray AM (2008) Fish remains from the Plio-Pleistocene Shungura Formation, Omo River basin, Ethiopia. Geobios 41:283–295

Stewart KM, Rufolo SJ (2018) Kanapoi revisited: Paleoecological and biogeographical inferences from the fossil fish. J Hum Evol 140:102452

Storrs GW (2003) Late Miocene–Early Pliocene crocodilian fauna of Lothagam, Southwest Turkana Basin, Kenya. In: Leakey MG, Harris JM (eds) Lothagam: the dawn of humanity in Eastern Africa. Columbia University Press, New York, pp 137–160

Suwa G, White TD, Howell FC (1996) Mandibular postcanine dentition from the Shungura Formation, Ethiopia: crown morphology, taxonomic allocations, and Plio-Pleistocene hominid evolution. Am J Phys Anthropol 101:247–282

Talbot MR, Morley C, Tiercelin J-J, Le Hérissé A, Potdevin J, Le Gall B (2004) Hydrocarbon potential of the Meso-Cenozoic Turkana Depression, northern Kenya. II. Source rocks: quality, maturation, depositional environments and structural control. Mar Petrol Geol 21:63–78

Thuo P (2009) Stratigraphic, petrographic and diagenetic evaluation of Cretaceous/Paleogene potential reservoir sandstones of western Turkana, Kenya. Implications on the petroleum potential of northwestern Kenya. PhD thesis, Université de Bretagne Occidentale, Brest

Tiercelin J-J, Potdevin J, Morley CK, Talbot MR, Bellon H, Rio A, Le Gall B, Vétel W (2004) Hydrocarbon potential of the Meso-Cenozoic Turkana Depression, northern Kenya. I. Reservoirs: depositional environments, diagenetic characteristics, and source rock-reservoir relationships. Mar Petr Geol 21:41–62

Tiercelin J-J, Potdevin J-L, Thuo PK, Abdelfettah Y, Schuster M, Bourquin S, Bellon H, Clément J-P, Guillou H, Nalpas T, Ruffet G (2012) Stratigraphy, sedimentology and diagenetic evolution of the Lapur Sandstone in northern Kenya: implications for oil exploration of the Meso-Cenozoic Turkana depression. J Afr Earth Sci 71-72:43–79

Tiercelin J-J, Schuster M, Roche H, Brugal J-P, Thuo P, Prat S, Harmand S, Davtian G, Barrat JA, Bohn M (2010) New considerations on the stratigraphy and environmental context of the oldest (2.34 Ma) Lokalalei archaeological site complex of the Nachukui Formation, West Turkana, northern Kenya Rift. J Afr Earth Sci 58:157–184

Tindall KW (1985) Stratigraphy and sedimentology of the Koobi Fora Formation, Eastern Koobi Fora Ridge, East Turkana, Kenya. MSc thesis, Iowa State Univ, Ames

Trapani J (2008) Quaternary fossil fish from the Kibish Formation, Omo Valley, Ethiopia. J Hum Evol 55:521–530

Trauth MH, Asrat A, Cohen AS, Duesing W, Foerster V, Kaboth-Bahr S, Kraemer KH, Lamb HF, Marwan N, Maslin MA, Schäbitz F (2021) Recurring types of variability and transitions in the ~620 kyr record of climate change from the Chew Bahir basin, Southern Ethiopia. Quat Sci Rev 266:106777

Uno KT, Cerling TE, Harris JM, Kunimatsu Y, Leakey MG, Nakatsukasa M, Nakaya H (2011) Late Miocene to Pliocene carbon isotope record of differential diet change among East African herbivores. Proc Natl Acad Sci 108:6509–6514

Uno KT, Polissar PJ, Kahle E, Feibel C, Harmand S, Roche H, deMenocal PB (2016) A Pleistocene palaeovegetation record from plant wax biomarkers from the Nachukui Formation, West Turkana, Kenya. Phil Trans R Soc Lond Series B Biol Sci 371:20150235

Van Bocxlaer B (2017) Paleoecological insights from fossil freshwater mollusks of the Kanapoi Formation (Omo-Turkana Basin, Kenya). J Hum Evol 140:102341

van Bocxlaer B (2020) Paleoecological insights from fossil freshwater mollusks of the Kanapoi Formation (Omo-Turkana Basin, Kenya). J Hum Evol 140:102341

van Bocxlaer B, van Damme D (2009) Palaeobiology and evolution of the late Cenozoic freshwater molluscs of the Turkana Basin: Iridinidae Swainson, 1840 and Etheriidae Deshayes, 1830 (Bivalvia: Etherioidea). J Syst Palaeontol 7:129–161

van Bocxlaer B, van Damme D, Feibel CS (2008) Gradual versus punctuated equilibrium evolution in the Turkana basin molluscs: evolutionary events or biological invasions? Evolution 62:511–520

van Damme D, Gautier A (1972) Molluscan assemblages from the late Cenozoic of the lower Omo Basin, Ethiopia. Quat Res 2:25–37

Vidal CM, Lane CS, Asrat A, Barfod DN, Mark DF, Tomlinson EL, Tadesse AZ, Yirgu G, Deino A, Hutchison W, Mounier A, Oppenheimer C (2022) Age of the oldest known *Homo sapiens* from eastern Africa. Nature 601:579–583

Viehberg FA, Just J, Dean JR, Wagner B, Franz SO, Klasen N, Kleinen T, Ludwig P, Asrat A, Lamb HF, Leng MJ, Rethemeyer J, Milodowski AE, Claussen M, Schäbitz F (2018) Environmental change during MIS4 and MIS 3 opened corridors in the Horn of Africa for *Homo sapiens* expansion. Quat Sci Rev 202:139–153

Vincens A (1982a) Palynologie, environnements actuels et Plio-Pléistocènes à L'Est du lac Turkana (Kenya). PhD thesis, Université d'Aix-Marseille II

Vincens A (1982b) Recherches palynologiques dans la region du lac Turkana, Kenya. Palaeoecol Afr 14:85–91

Vincens A, Tiercelin J-J, Buchet G (2006) New Oligocene-early Miocene microflora from the southwestern Turkana Basin: Palaeoenvironmental implications in the northern Kenya Rift. Palaeogeogr Palaeoclimatol Palaeoecol 239:470–486

Vondra CF, Bowen BE (1976) Plio-Pleistocene deposits and environments, East Rudolf, Kenya. In: Coppens Y, Howell FC, Isaac GL, Leakey REF (eds) Earliest man and environments in the Lake Rudolf Basin. University of Chicago Press, Chicago, pp 79–93

Vondra CF, Bowen BE (1978) Stratigraphy, sedimentary facies and paleoenvironments, East Lake Turkana, Kenya. In: Bishop WW (ed) Geological background to fossil man. Geol Soc Lond Spec Publ 6:395–414

Vondra CF, Burggraf DR, White HJ (1978) The Plio-Pleistocene: sediments, environments, and geochronology along the Karari Escarpment, East Turkana, Kenya. Trans Nebr Acad Sci 6:19–34

Vondra CF, Johnson GD, Behrensmeyer AK, Bowen BE (1971) Preliminary stratigraphical studies of the East Rudolf basin. Nature 231:245–248

Vonhof HB, Joordens JCA, Noback ML, van der Lubbe JH, Feibel CS, Kroon D (2013) Environmental and climatic control on seasonal stable isotope variation of freshwater molluscan bivalves in the Turkana Basin (Kenya). Palaeogeogr Palaeoclimatol Palaeoecol 383–384:16–26

Walker A, Leakey RE (1993) The Nariokotome *Homo erectus* skeleton. Harvard University Press, Boston

Walker A, Leakey RE, Harris JM, Brown FH (1986) 2.5 Myr *Australopithicus boisei* from west Lake Turkana, Kenya. Nature 322:517–522

Walsh J, Dodson RG (1969) Geology of Northern Turkana. Geol Surv Kenya Rep 82

Ward CV, Feibel CS, Hammond AS, Leakey LN, Moffett EA, Plavcan JM, Skinner MM, Spoor F, Leakey MG (2015) Associated ilium and femur from Koobi Fora, Kenya, and postcranial diversity in early *Homo*. J Hum Evol 81:48–67

Ward CV, Leakey MG, Brown B, Brown F, Harris J, Walker A (1999) South Turkwell: a new Pliocene hominid site in Kenya. J Hum Evol 36:69–95

Ward CV, Leakey MG, Walker A (2001) Morphology of *Australopithecus anamensis* from Kanapoi and Allia Bay, Kenya. J Hum Evol 41:255–368

Watkins RT (1986) Volcano-tectonic control on sedimentation in the Koobi Fora sedimentary basin, Lake Turkana. In: Frostick LE, Renaut RW, Reid I, Tiercelin J-J (eds) Sedimentation in the African rifts. Geol Soc Lond Spec Publ 25:85–95

Watkins RT (1989) The Buluk Member, a fossil hominoid-bearing sedimentary sequence of Miocene age from northern Kenya. J Afr Earth Sci 8:107–112

Werdelin L, Manthi FK (2012) Carnivora from the Kanapoi hominin site, northern Kenya. J Afr Earth Sci 64:1–8

Wescott WA, Morley CK, Karanja FM (1993) Geology of the Turkana Grits in the Lariu Range and Mt Porr areas, southern Lake Turkana, Northwestern Kenya. J Afr Earth Sci 16:425–435

Wescott WA, Wigger ST, Stone DM, Morley CK (1999) Geology and geophysics of the Lotikipi Plain. In: Morley CK (ed) Geoscience of rift systems — evolution of East Africa. AAPG Stud Geol 44:55–65

Weston EM (2003) Fossil Hippopotamidae from Lothagam. In: Leakey MG, Harris JM (eds) Lothagam: the dawn of humanity in Eastern Africa. Columbia University Press, New York, pp 441–484

White HJ (1976) Stratigraphy of the Lower Member, Koobi Fora Formation, southern Karari Escarpment, East Turkana Basin, Kenya. MSc thesis, Iowa State Univ, Ames

White HJ, Burggraf DR, Bainbridge RB, Vondra CF (1981) Hominid habitats in the Rift Valley: part 1. In: Rapp G, Vondra CF (eds) Hominid sites: their geologic settings. Am Assoc Adv Sci Symp 63:57–61

Whitworth T (1960) Fossilized human remains from northern Kenya. Nature 185:947–948

Whitworth T (1965) The Pleistocene lake beds of Kabua, northern Kenya. Durham Univ J 57:88–100

Wichura H, Jacobs LL, Lin A, Polcyn MJ, Manthi FK, Winkler DA, Strecker MR, Clemens M (2015) A 17-my-old whale constrains onset of uplift and climate change in east Africa. Proc Natl Acad Sci 112:3910–3915

Williamson PG (1979) Evolutionary implications of Late Cenozoic mollusc assemblages from the Turkana Basin, north Kenya. PhD thesis, University of Bristol

Williamson PG (1981) Palaeontological documentation of speciation in Cenozoic molluscs from Turkana Basin. Nature 293:437–443

Williamson PG (1982) Molluscan biostratigraphy of the Koobi Fora hominid-bearing deposits. Nature 295:140–142

Williamson PG, Savage RJG (1986) Early rift sedimentation in the Turkana basin, northern Kenya. In: Frostick LE, Renaut RW, Reid I, Tiercelin J-J (eds) Sedimentation in the African rifts. Geol Soc Lond Spec Publ 25:267–283

Wood RC (2003) Fossil turtles from Lothagam. In: Leakey MG, Harris JM (eds) Lothagam: the dawn of humanity in Eastern Africa. Columbia University Press, New York, pp 115–136

Wood B, Leakey M (2011) The Omo-Turkana Basin fossil hominins and their contribution to our understanding of human evolution in Africa. Evol Anthropol 20:264–292

Wynn JG (2000) Paleosols, stable carbon isotopes, and paleoenvironmental interpretation of Kanapoi, Northern Kenya. J Hum Evol 39:411–432

Wynn JG (2001) Paleosols, stable carbon isotopes, and paleoenvironments of hominid evolution in the Neogene Turkana Basin, northern Kenya. PhD thesis, University of Oregon, Eugene

Wynn JG (2003) Miocene and Pliocene palaeosols of Lothagam. In: Leakey MG, Harris JM (eds) Lothagam: the dawn of humanity in Eastern Africa. Columbia University Press, New York, pp 31–42

Wynn JG (2004) Influence of Plio-Pleistocene aridification on human evolution: evidence from paleosols of the Turkana Basin, Kenya. Am J Phys Anthropol 123:106–118

Yost CL (2019) Phytoliths and microcharcoal from East African lake sediments used to reconstruct Plio-Pleistocene vegetation and hydroclimate. PhD thesis, University of Arizona, Tucson

Yost CL, Cohen AS, Feibel CS, Campisano CJ (2014) Preliminary phytolith results from the early Pleistocene West Turkana, Kenya, drill core of the Hominin Sites and Paleolakes Drilling Project (HSPDP) (Paper 292-6). Geol Soc Am Abstr Progr 46(6):711

Yost CL, Lupien RL, Beck C, Feibel CS, Archer SR, Cohen AS (2021) Orbital influence on precipitation, fire, and grass community composition from 1.87 to 1.38 Ma in the Turkana Basin, Kenya. Front Earth Sci 9:56864

Zanettin B, Visentin EJ, Bellieni G, Piccirillo EM, Rita F (1983) The volcanics of North Turkana Basin (Kenya): age, succession and structural evolution. Bull Ctr Rech Explor-Prod Elf-Aquitaine 7:249–255

Zawacki EE (2021) Plio-Pleistocene sediment provenance and erosion rates along the East African Rift System. PhD thesis, Arizona State University, Phoenix

Zawacki EE, van Soest MC, Hodges KV, Scott JJ, Barboni M, Strecker MR, Feibel CS, Campisano CJ, Arrowsmith JR (2022) Sediment provenance and silicic volcano-tectonic evolution of the northern East African Rift System from U/Pb and (U-Th)/He laser ablation double dating of detrital zircons. Earth Planet Sci Lett 580:117375

Chapter 16
The Suguta Basin

16.1 Introduction

The Suguta River rises in the Tugen Hills at heights over 2300 m and flows northwards for about 200 km along the rift valley floor, reaching Lake Logipi at an elevation of ~275 m (Fig. 16.1). The Upper Suguta Valley lies south of the Quaternary volcano Emuruangogolak and consists of an arid lava-floored region punctured by volcanoes (Paka, Silali). Spring-fed wetlands are present in several areas and, locally, deposits such as the Kapedo Tuffs, west of Silali, host Middle Stone Age (135–123 ka) artefacts (Tryon et al. 2008). The Lower Suguta is arid with water delivered mainly by the Suguta River with smaller contributions from the ephemeral Kamuge and Baragoi Rivers (Fig. 16.1) plus occasional rainfall, and rift-marginal streams such as the intermittently flowing Losergoi and Akangbelai rivers to the northeast.

Access is very difficult with only one road entering the Lower Suguta, west of Emuruangogolak where it passes the Suguta Falls. The road crosses the valley and climbs the eastern escarpment but is commonly impassable. The valley is uninhabited, except for a small village in the southwest near the Kamuge River. During the very wet year of 2019, a temporary village was set up on the Baragoi River and a geothermal exploration camp was located to the northeast, on the eastern edge of the Barrier volcanic complex.

Champion (1935) first reported lake deposits in the Lower Suguta with subsequent studies of different parts of the Suguta linked to Geological Survey of Kenya mapping in adjacent areas, including Nyeri (Shackleton 1945), South Horr (Dodson 1963), Baragoi (Baker 1963), Maralal (Key 1987) and Loiyangalani (Ochieng et al. 1988). Reconnaissance studies of the Kerio and Loriu areas also incorporated parts of the Suguta in the late 1960s (Gwynne 1969; Rhemtulla 1970; Caukwell 1971). Basement rocks are present near Mt. Ng'iro, within the Lower Suguta catchment, with Neogene volcanic rocks lapping onto what must have been a contemporary high relief area (Bosworth and Maurin 1993).

R. W. Renaut, R. B. Owen, *The Kenya Rift Lakes: Modern and Ancient*, Syntheses in Limnogeology, https://doi.org/10.1007/978-3-642-25055-2_16

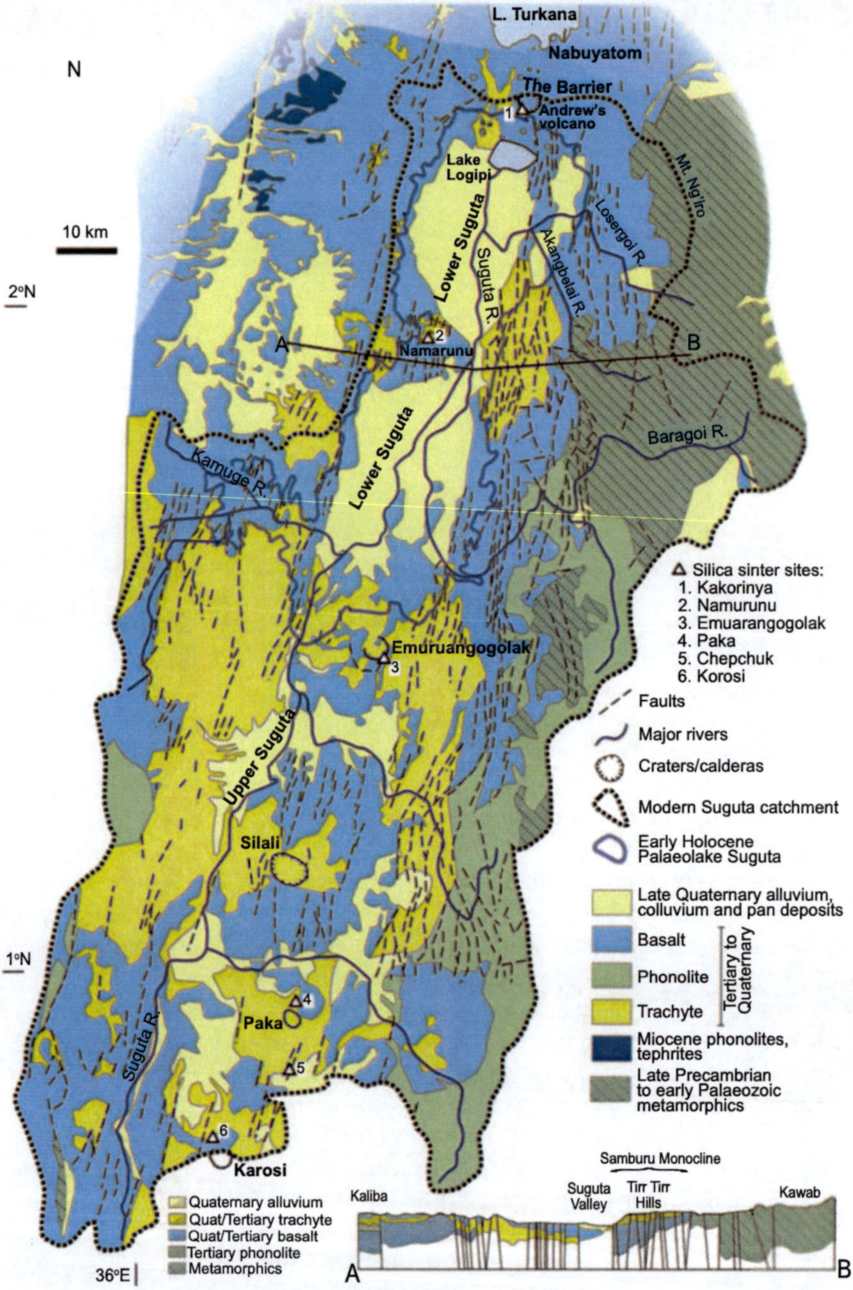

Fig. 16.1 Geological map of the Suguta catchment and adjacent areas. The Upper and Lower Suguta valleys are separated by Emuruangogolak volcano. Simplified from maps by Dunkley et al. (1993); silica sinter sites derived from Sturchio et al. (1993)

Table 16.1 Development of the stratigraphic nomenclature for the volcanic and sedimentary rocks of the Suguta Valley region

	Dodson (1963)	Ochieng et al. (1988)	K-Ar dates (Ma)	Bosworth and Maurin (1993)	
Quaternary	Recent volcanic rocks (Barrier)	Recent volcanic rocks (Barrier)		Not studied	Quaternary
	Quaternary deposits	Quaternary deposits		Quaternary deposits	
	U. Pleistocene sediments	Pleistocene lake beds and limestones		Pleistocene lake beds and travertine	
	Barrier volcanic rocks	Barrier volcanics		Namurinyang and Barrier basalts	
		Suguta Trough basalts/basanites		Upper Suguta Trough basalts	
			2.3 ± 0.4	Lower Suguta Trough basalts	Late Pliocene
	Mt. Kulal lavas	Kulal shield volcanics	2.6 ± 0.1 2.9 ± 0.1 3.0 ± 0.6	Not studied	
	Serima valley lavas	Miocene Plateau Basalts and Longipi shield volcanics		Not studied	
	Murniau sediments	Not discussed		Not studied	
Tertiary	Upper Tirr Basalts	Upper Tirr Basalts		Tirr volcanics	
	Lower Tirr Trachytes	Lower Tirr Trachytes	2.6 ± 0.2 2.7 ± 0.2 2.7 ± 0.2 3.6 ± 0.4 3.8 ± 0.4 3.9 ± 0.4		Early Pliocene
	Iyuk acid volcanic rocks	Iyuk Rhyolites		Not studied	
	Porphyritic basalts	Longipi Shield Volcanics	Overlain by 3.0 ± 0.6 Kulal basalt	Not studied	
	Fine-grained basalts	Parkati volcanics	3.1 ± 1.2 3.8 ± 0.2 4.2 ± 0.3	Losergoi volcanics	
	Tertiary sediments	Sub-volcanic sediments		Not studied	
	Precambrian basement	Ol Doinyo Ng'iro and Kotim gneisses		Not studied	

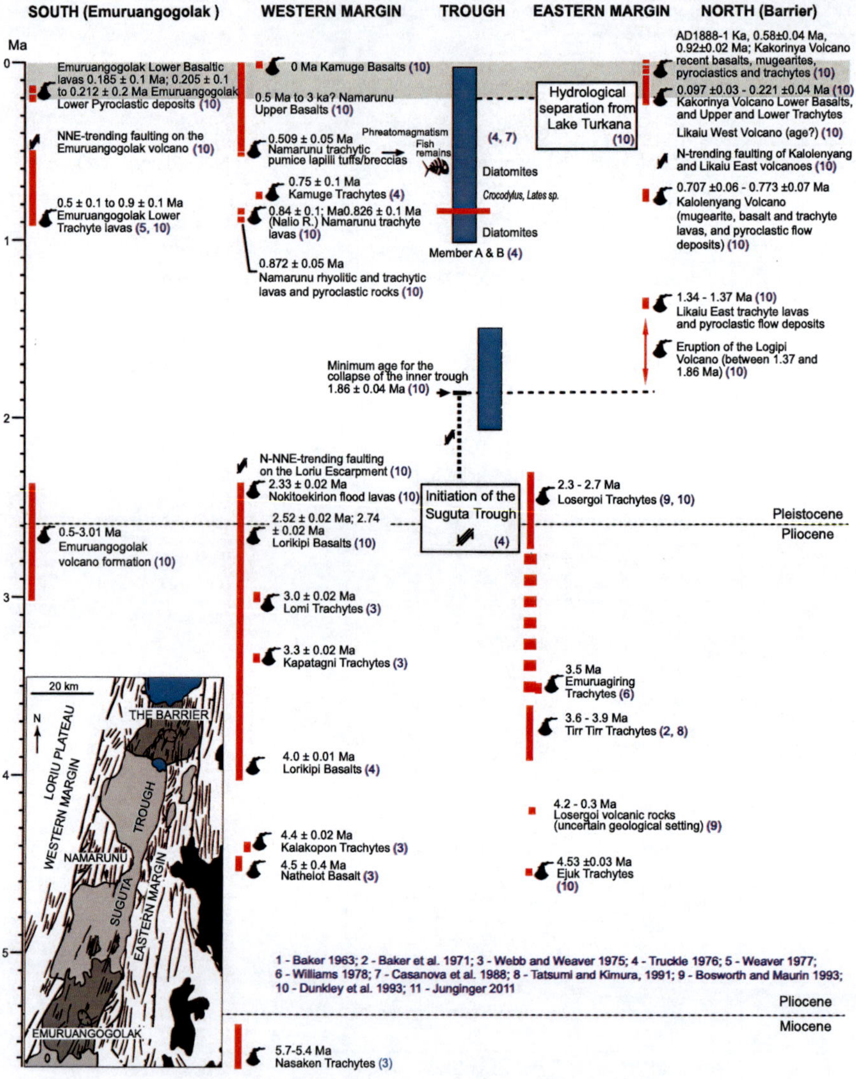

Fig. 16.2 Volcanic stratigraphy of the Suguta Basin. After Junginger (2011)

Several stratigraphical summaries have been presented for the Lower Suguta (Table 16.1). The Pliocene to present volcanic and tectonic history of the Lower Suguta and adjacent areas is also summarised in Fig. 16.2. Junginger (2011) noted that a tectonic depression has existed in the area since at least Miocene times (Matsuda et al. 1984; Williamson and Savage 1986) with faulting spread over a wider region than today prior to ~4 Ma ago (Saggerson and Baker 1965; Cerling and Powers 1977). Axial basin subsidence, monoclinal tilting of the rift margins and the development of antithetic faulting along the eastern margin became prominent at 3–1.8 Ma (Truckle 1976; Dunkley et al. 1993; Uemura 1994; Junginger 2011).

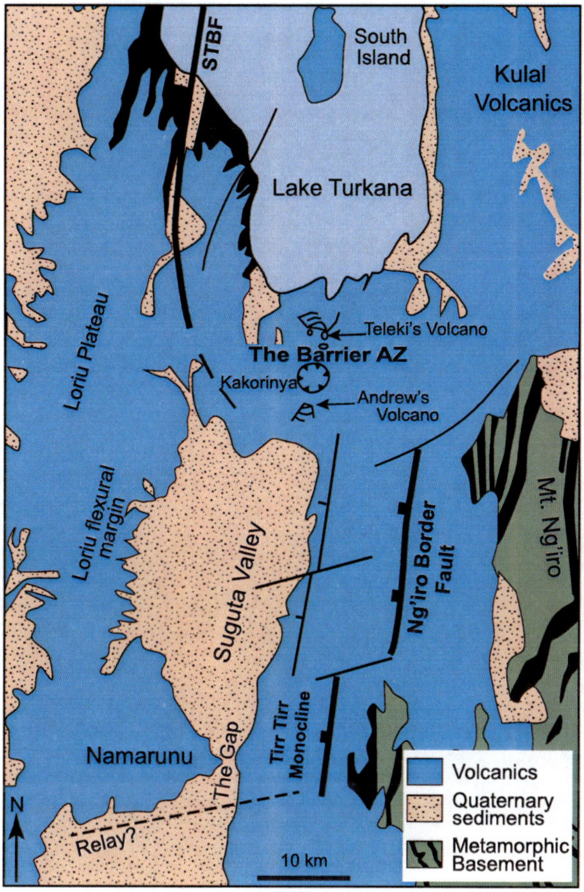

Fig. 16.3 Structural geology of the northern lower Suguta Valley and south Lake Turkana areas. After Bosworth and Maurin (1993). STBF South Turkana Border Fault

Consequently, rift marginal volcanics are tilted towards the rift axis and cut by NNE-trending faults (cross-section in Fig. 16.1; Bosworth and Maurin 1993; Saneyoshi et al. 2006; Acosta et al. 2015). Bosworth and Maurin (1993) noted that Pliocene faulting in the northern Lower Suguta produced an asymmetric basin that thickened to the east and terminated against an inferred border fault at the base of the Ng'iro Range (Fig. 16.3). They also commented that the western rift shoulder of the Lower Suguta experienced less uplift and is characterised by fewer faults that tend to drop to the west, away from the valley.

The Barrier volcanic complex may represent an accommodation zone transferring stresses from the east-facing South Turkana Border Fault to the west-facing Ng'iro Border Fault (Fig. 16.3; Bosworth and Maurin 1993; Lambiase and Bosworth 1995). Gravity profiling suggests that it is underlain by a mafic body in the upper crust (Mariita and Keller 2007). The development of Namarunu and 'The Gap' may also represent a relay zone that divided the Lower Suguta into two parts. The

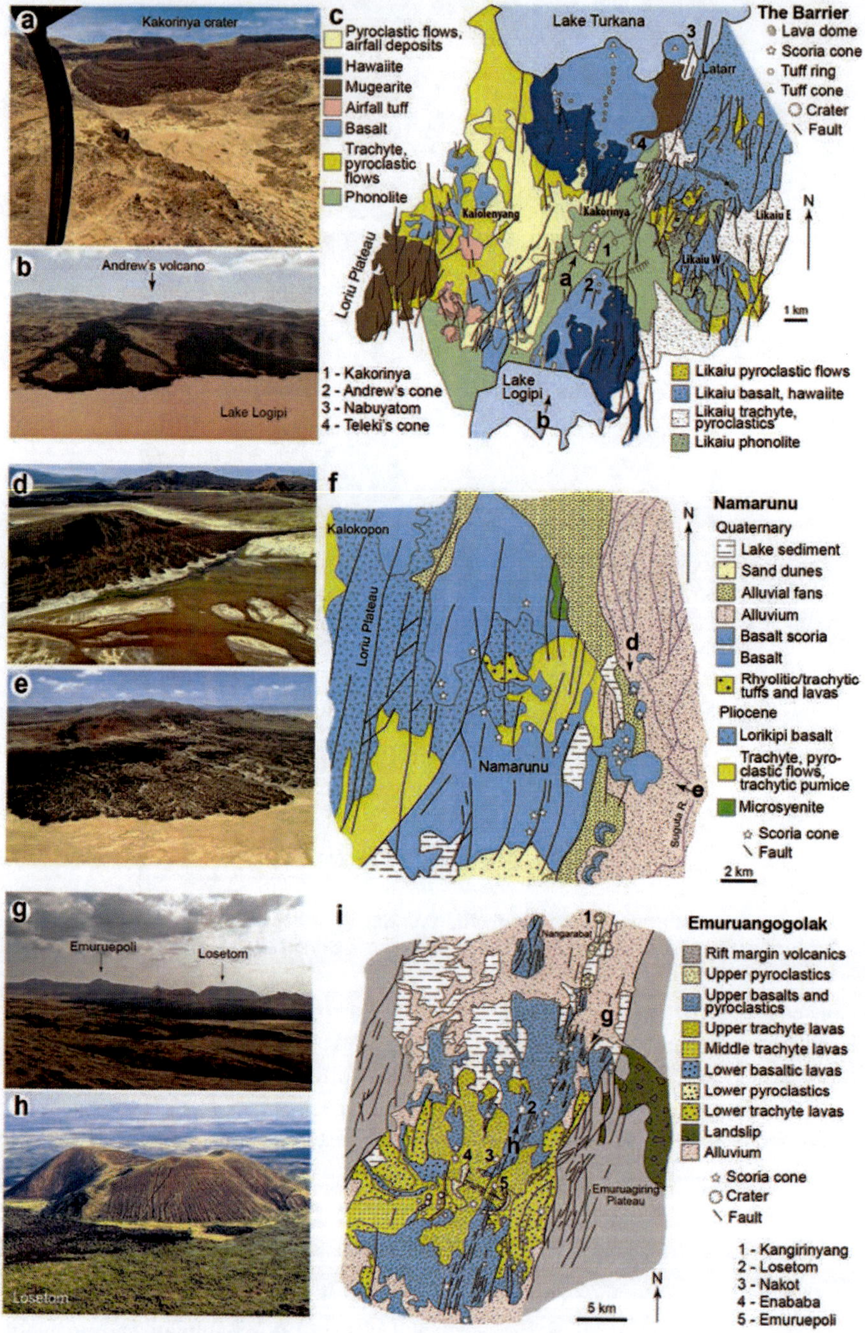

Fig. 16.4 Geology of the Barrier, Namarunu and Emuruangogolak volcanoes. Based on Dunkley et al. (1993). Labelled arrows show the view direction for adjacent photos. Numbered white circles on the Barrier and Emuruangogolak maps indicate prominent cones and craters labelled in relevant keys

trachytic and basaltic Barrier and Emuruangogolak volcanoes (Fig. 16.4) were both initiated around 1.3–1 Ma ago and subsequently defined the northern and southern limits of the Lower Suguta (Champion 1935; Weaver 1977; Junginger 2011). Complete separation of the Lower Suguta from Lake Turkana may have taken place by about 0.2–0.1 Ma ago, with Namarunu (Fig. 16.4) erupting around 0.8 Ma (Dunkley et al. (1993). In the Upper Suguta, the Silali, Paka and Karosi volcanoes (Fig. 16.1) became active at about 460, ~390 and ~500 ka, respectively, with caldera collapse at Silali at ~64 ka, and at Paka at 10 ka (Sceal 1974; Dunkley et al. 1993; Lichoro et al. 2019).

Fluvial and lacustrine sediments first accumulated during the Miocene in what are now the Samburu Hills to the southeast of the Lower Suguta (Dodson 1963; Makinouchi et al. 1984; Pickford 1984; Pickford et al. 1984a, b). More recently, Middle to Late Pleistocene lacustrine deposits have accumulated along the central axial rift (Junginger 2011) as they also did during the African Humid Period. Today, the large palaeolakes in which these deposits accumulated have retreated to form the shallow, saline alkaline Lake Logipi and ephemeral Lake Alablab (Chap. 7).

16.2 Mio-Pliocene Sedimentary Formations

Baker (1963) produced the first geological report (mainly basement studies) of the area between Baragoi and the Suguta. As part of his study, he mapped 1350 ft (~411 m) of "lake beds and interbedded volcanics" (Tm_2 on his map) that included silt, pumice, gravel and diatomaceous sediments near Waitiakwel. He considered them to be of Pleistocene age, but they lie below the Pliocene Tirr Formation and should be stratigraphically older. Detailed research followed with the work of the 1980 Japan-Kenya expedition to the Kirimun District (Ishida et al. 1982). Ishida (1984) questioned some of Baker's earlier correlations and proposed formation names for the major volcanic and sedimentary rocks of the region (Fig. 16.5). In the Nachola area, these included basaltic lavas in the Nanyagaten Formation (Sawada et al. 1987) and basalt, tuffaceous silt, gravel and chert in the Nachola Formation, with sediments yielding fossils of *Kenyapithicus*.

Intensive studies followed in 1982 when a joint expedition of Japanese and Kenya scientists re-examined the Samburu Hills and Nachola area. The geology, dating and sedimentology of the region were described by Koyaguchi (1984), Makinouchi et al. (1984), Matsuda et al. (1984), Itaya and Sawada (1986), Sawada et al. (1987), and Kabeto et al. (2001a, b). Details of the anthropoid, rodent, large mammal and molluscan fossils were documented by Ishida et al. (1984), Kawamura and Nakaya (1984), Nakaya et al. (1984), Pickford (1984), Pickford et al. (1984a, b, 1987), Nakaya (1994), Kunimatsu et al. (2007), and Handa et al. (2018).

Further west, in the Samburu Hills, extensive basalt and minor nearshore and fluvial clastics, siliceous sediments and chert accumulated in the Aka Aiteputh Formation (Fig. 16.6a; Tateishi 1987), which partially overlaps in age with the Upper Nachola Formation (Fig. 16.5a). The Late Miocene fluvio-lacustrine Namarungule

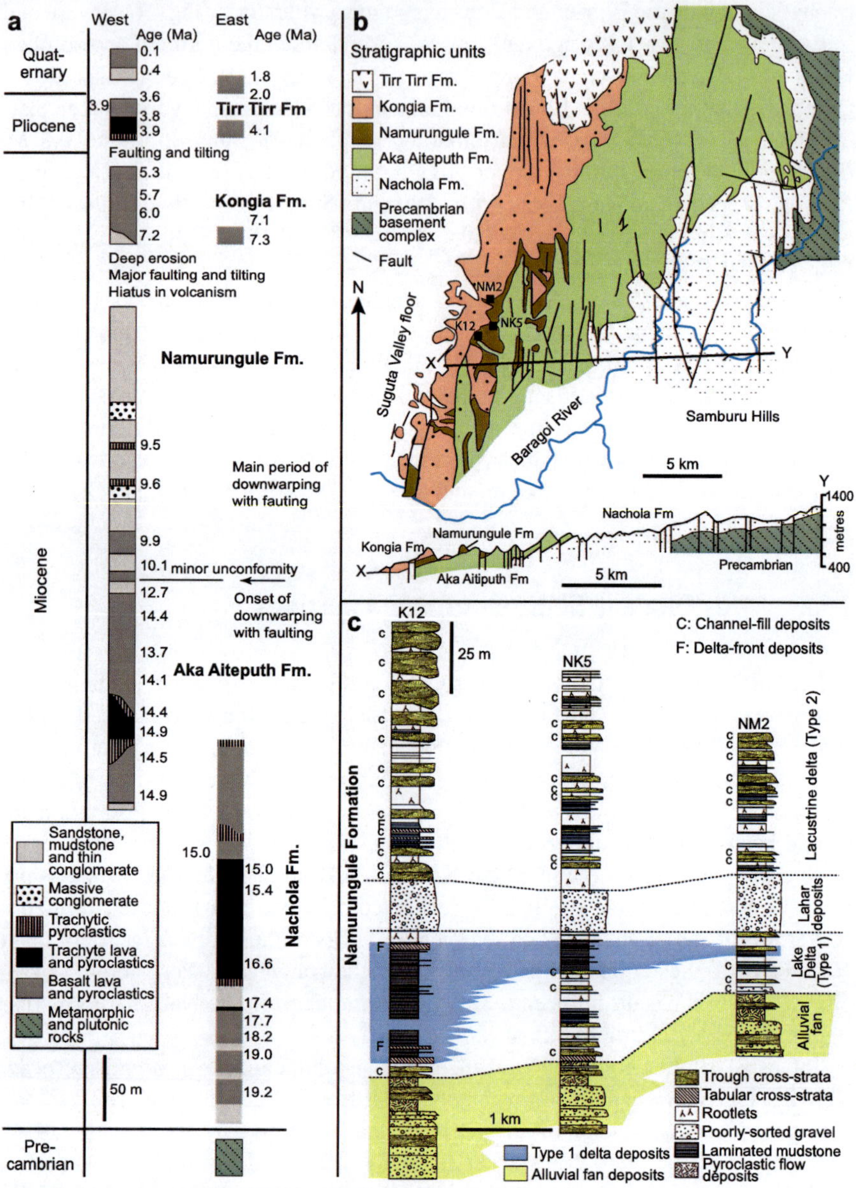

Fig. 16.5 Stratigraphy and geology of the Samburu Hills, SE Suguta catchment. Based on Sakai et al. (2010). (**a**) General stratigraphy. (**b**) Geological map of the Samburu Hills area. (**c**) Detailed succession in the Namurungule Formation. Sections located in '**b**'. C = channel-fill deposit; F = delta-front deposit

Fig. 16.6 Exposed formations at Baragoi Gorge. (**a**) Aka Aiteputh Formation, composed of basalt, red palaeosols and calcrete. (**b**) Namarungule Formation, which includes alluvial fan and deltaic conglomerate, sandstone and mudstone

Formation (Fig. 16.6b) is dominated by sand and mud with minor volcanics. Younger rocks are largely basalt and include the Miocene Kongia (with basal clastic sediments) and Nagubarat formations, and the Pliocene Tirr Tirr Formation, which also includes rhyolite (Makinouchi et al. 1984). Sawada et al. (1987) noted two phases of igneous activity at 19–10 Ma and after 7 Ma that were separated by a quiescent period. The first phase peaked at 15–13 Ma and commenced with eruptions of basalt and trachybasalt that gave way to phonolites and then basalt. The second phase was characterised by basalt with rhyolite erupting around 4 Ma.

The 100–200-m-thick Namarungule Formation, questionably, may correlate with Baker's 'Waitiakwell lake beds' and includes fluvial and lacustrine sediments that rest unconformably on the Aka Aiteputh Formation (Saneyoshi et al. 2006; Sawada et al. 2006). The sediments accumulated in fluvial settings, alluvial fans, deltas and as lahar deposits. Saneyoshi and colleagues recognised two types of freshwater lacustrine delta associations (Fig. 16.5c). Type 1 was dominated by three facies. The first consists of alternating trough and planar cross-laminated and rippled sandstone and burrowed mudstone deposited as fluvial channel fill and floodplain deposits. The two other facies are planar cross-stratified sandstones (delta front) and laminated mudstones with no burrows or rootlets (anoxic lake). In contrast, their type 2 delta includes five facies: (1) gravelly sandstone with trough cross-stratification and ripple lamination (fluvial channel fill), (2) alternating sandstone with current-ripples, rootlets and mudstone (levee or floodplain), (3) massive bioturbated mudstone with rootlets (floodplain), (4) planar cross-stratified sandstone (fluvial), and (5) laminated mudstone (lacustrine).

The continuity of sedimentation in the Namarungule Formation was thought to reflect an underfilled palaeolake (Bohacs et al. 2000) fed by nearby sediment sources, given the lack of metamorphic clasts that more distal rivers would have introduced (Saneyoshi et al. 2006). Retrogradational stacking of sediments in the lower part of the formation was interpreted as being due to the palaeolake surface level rising faster than the sediments were accumulating. The earlier (type 1) deltas contain fewer channel-fill deposits than type 2 deltas, suggesting initial basin immaturity with fewer coarse-grained clastics being introduced. Consequently, particles were dominated by fine sand and silt that could be resuspended by wave action. Increased dissection in the catchment later increased channel and particle sizes with a slower lake level rise promoting greater stability of fluvial channels. Saneyoshi et al. (2006) also compared sedimentation patterns in the Namurungule Formation with those of the Ngorora Formation (Sect. 17.2.5) in the Baringo District to the south. They suggested that (1) early volcanic (ash, lahars) control of rift basins was common; (2) the sedimentation rate tended to decrease with time; and (3) there were distinct relationships between catchment size and sedimentation. They noted, for example, that increased sediment supply caused by catchment expansion was sufficient to overcome the rate of lake-level rise in the Namurungule Basin, but that this was reversed in the larger Ngorora Basin (Tugen Hills, Baringo).

Sakai et al. (2010) reported a major sedimentary shift at 9.6 Ma from red soil-dominated intervals in the upper Aka Aiteputh Formation to fluvio-lacustrine deltaic sediments in the Lower Namurungule Formation, which also contains an open woodland/savanna mammalian fauna. They interpreted this change in terms of a shift from a dry climate to wetter conditions with greater seasonality. Sakai et al. (2010) also noted that this correlated with the intensification of the Indian Ocean summer monsoon, which would have enabled moisture to penetrate further into East Africa.

Sakai et al. (2013) also described Miocene lake sediments of similar age to those near Baragoi from Nakali, east of Silali, in the Upper Suguta Valley (Fig. 16.7). They recognised the sedimentary Nakali and volcanic Nasorut formations with the

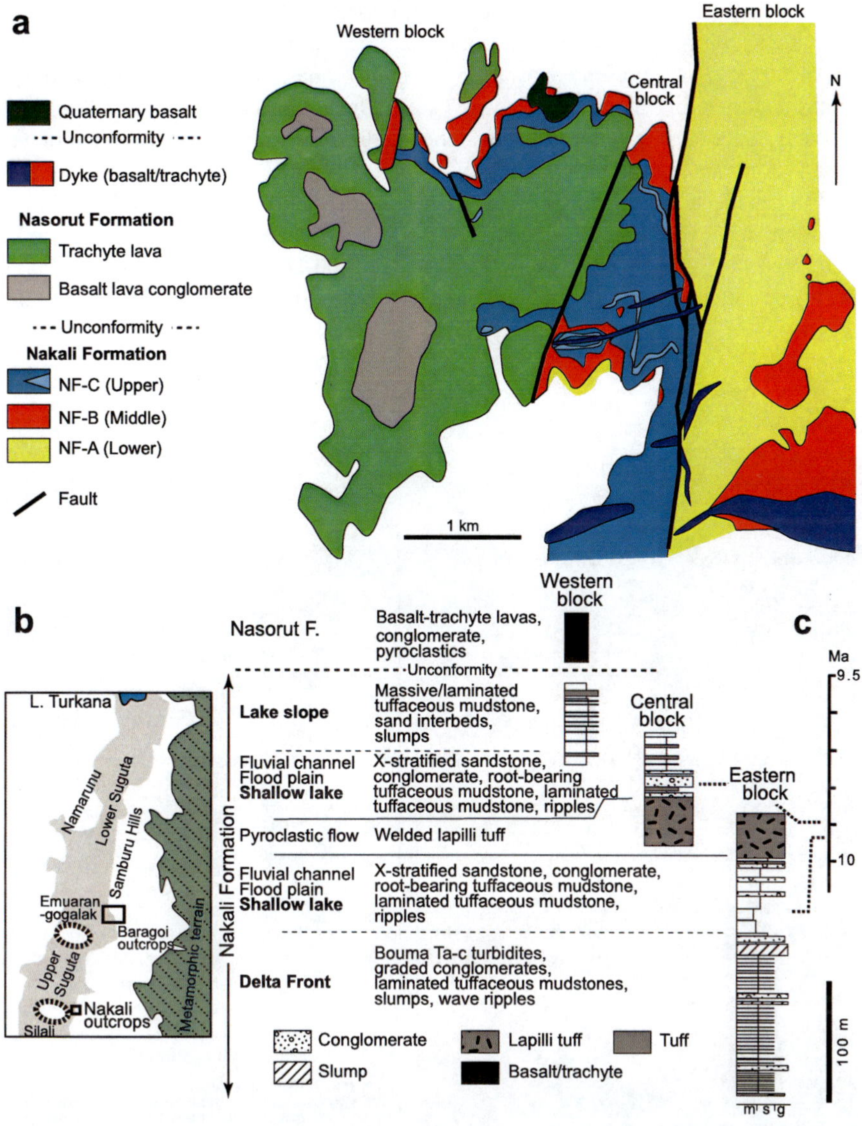

Fig. 16.7 Miocene sedimentary rocks at Nakali. After Sakai et al. (2013). (**a**) Nakali outcrops. (**b**) Location of Nakali and Baragoi Miocene lake sediments. (**c**) Stratigraphy of Nakali Formation and interpreted palaeoenvironments. Lacustrine environments in bold type

former divided into three informal members. The basal part of the lower member of the Nakali Formation includes tuffaceous sandstone and mudstone that accumulated as turbidites and slumps on a delta front. The overlying parts of this member represent a shallowing lake that gave way to a fluvial environment. The middle member is a welded tuff that infilled the basin. A shallow lake was re-established during the deposition of the upper member of the Nakali Formation, with this water body gradually developing into a fluvial environment. In contrast, the uppermost part of the upper member is represented by massive and laminated mudstone with sand and slump features that indicate a sloping lake floor. Variations in sediment thickness also suggest a seesaw pattern of subsidence in the area, reflecting differential movements along bounding faults.

Pliocene lacustrine sediments are scarce and restricted to clastic interbeds within predominantly volcanic sequences on the rift margins. Webb (1971), for example, described the 3–18-m-thick Aterir Beds west of the Upper Suguta (Fig. 16.8a–c), noting that they are dominated by pumice tuffs with diatomite intercalations and are

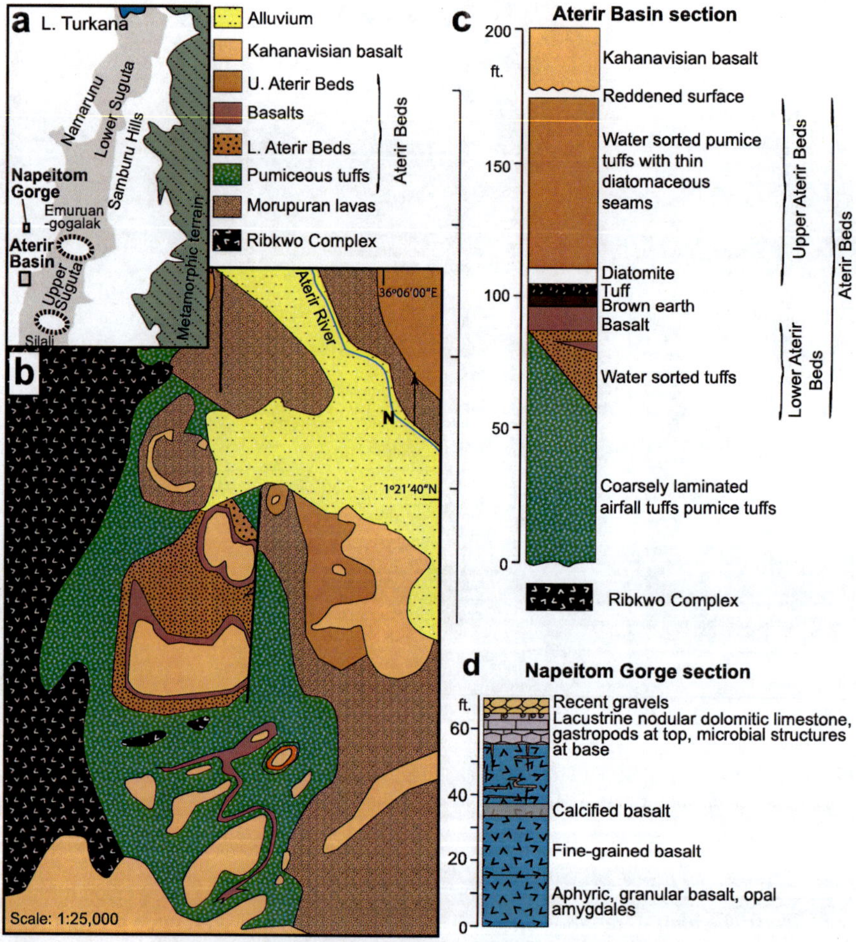

Fig. 16.8 Pliocene deposits west of the Upper Suguta. (**'a–c'** after Webb 1971); (**'d'** after Weaver 1973). (**a**) Location of Aterir Beds and Napeitom Gorge lake sediments. (**b**) Map of Aterir Basin outcrops. (**c**) Section showing Aterir Beds. (**d**) Section showing the Napeitom Gorge sequence

divided into upper and lower sequences by a brown earth. Eruptions of the Morupuran lavas initially defined the basin, with drainage diverted northwards by subsequent eruptions of coarse-grained ash and pumice. This phase created the Aterir aquatic basin with 9 m of mostly fluviatile (?) well-laminated, fine-grained, white tuffs accumulating (Lower Aterir Beds). The Upper Aterir Beds record a basin shift to the east and accumulation of both fluviatile deposits and lacustrine diatomaceous tuffs and diatomites (up to 1.3 m thick). Two laterally extensive diatomite bands (25 cm each) are also present. Weaver (1973) also described Pliocene lacustrine algal stromatolites and nodular to tubular limestones with freshwater (*Viviparus*-like) gastropods from the Napeitom Gorge west of Emuruangogolak (Fig. 16.8a, d). These deposits are up to 5 m thick and crop out over an area of <1 km^2. Weaver suggested a lacustrine origin, although descriptions of the sediments also resemble several spring-fed wetlands with pervasive carbonates in the underlying basalts possibly reflecting such a setting.

16.3 Pleistocene Sediments and Lakes

The axial Lower Suguta had subsided by 1.86 Ma (Fig. 16.2) creating accommodation space north of Emuruangogolak for Quaternary lake sediments to accumulate. Trachytic volcanism began to build the northern Barrier Volcanic Complex between 1.37 and 0.7 Ma (Dunkley et al. 1993) with a similar trachytic shield erupting at Emuruangogolak between about 1 and 0.5 Ma (Weaver 1977), thus defining the margins of a Suguta palaeolake by the Middle Pleistocene.

Truckle (1976, 1978, 1979) presented a composite section for lake sediments in the Lower Suguta (Fig. 16.9a), which he defined as the Suguta Beds. Four members were distinguished. Members A–C are of Early to Middle Pleistocene age with Member D representing Late Pleistocene to Holocene sedimentation. Member A is present near the Kamuge-Suguta river confluence (Fig. 16.9b, c) and includes coarse fluvial clastic sediments that are unconformably overlain by Member B diatomites, both at Kamuge and Rupa (Gwasinyono Arrolio). These members were tilted, faulted and eroded before being covered by a subaerial basalt (Member C) that was also faulted at a later stage. The uppermost Member D includes a lower clastic unit at Gwasinyono Arolio that is overlain by well-bedded diatomites.

Member D is also associated with former shorelines at up to 600 m above sea level. Wave-cut basalts, stromatolites and gastropods are present along the palaeocoastlines with gastropods indicating an age of 9660 ± 210 y BP (Truckle 1976). Casanova (1986) described thick stromatolites that he related to long Middle Pleistocene lacustrine phases. Casanova et al. (1988) documented additional Middle Pleistocene sections that rest on 0.75 ± 0.1 Ma trachytes from deep-water sites near the Kamuge-Suguta confluence and from more littoral locations at Namruy (Fig. 16.9a, c). They also described Late Pleistocene to Holocene sediments and palaeoshorelines associated with stromatolites, noting high lake levels of 460–437 m at ≤31,000–27,000 y BP, and 484 m at 9650 y BP.

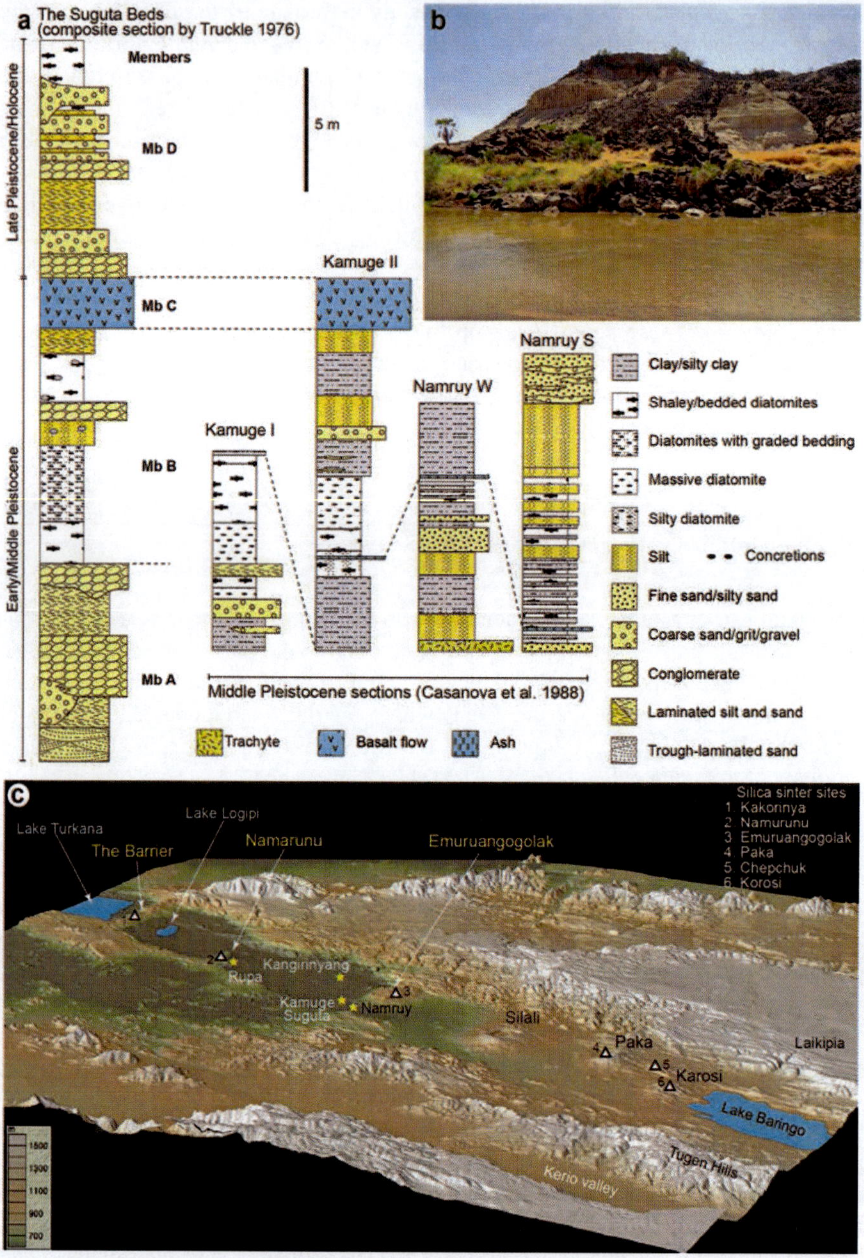

Fig. 16.9 Pleistocene lake sediments in the Suguta Valley. (**a**) Composite Middle and Late Pleistocene/Holocene section by Truckle (1976) to left. Middle Pleistocene sections by Casanova et al. (1988). (**b**) Middle Pleistocene sediments capped by basalt (Member C) near the Suguta and Kamuge confluence. (**c**) Relief model showing main locations mentioned in the text. Black stars are Middle Pleistocene sediment locations. White triangles are silica sinter sites. Major volcanoes and lakes also shown. (Constructed using GeomapApp)

Dunkley et al. (1993) noted up to 17 m of Middle Pleistocene sediments on pillow lavas and hyaloclastites along the Nalio River, southwest of Namarunu. They reported convolute bedding and small clastic dykes in several diatomites that implied loading and dewatering. Mass-flow diatomite breccias, cross-bedded pumice beds, fish-bone beds and a capping algal limestone were also reported. An epiclastic pumice yielded a date of 826 ± 6 ka and thus a minimum age for the sequence.

Examples of Middle Pleistocene sediments are shown in Fig. 16.10. Faulting of Middle Pleistocene deposits is common and has resulted in one diatomite bed being repeated at Kangirinyang (Figs. 16.9c and 16.10a). Sediments at Rupa are shown in Fig. 16.10b–i. Fish bone is common through the Rupa sequence with some beds containing intact fish fossils (Fig. 16.10d). A variety of facies are present including pumice and ash beds (Fig. 16.10e, f), massive and laminated diatomite (Fig. 16.10g), locally with lenses of fish bone (Fig. 16.10h), and thick beds of cross-laminated clayey silt with small 'clay balls' up to 3 mm in diameter (Fig. 16.10i).

Figure 16.11 shows Middle Pleistocene sediment logs at the Kamuge-Suguta confluence and at Rupa (Fig. 16.9c); both are overlain by basalt flows (Truckle 1976). Clay, silt, diatomaceous silt, diatomite, fine to coarse sand, ash and pumice are variously present (Robakiewicz et al. 2023). A logged section (RUPA III) contains diatoms dominated by freshwater planktonic (*Aulacoseira granulata*, *A. granulata* v. *angustissima*, *A. goetzeana*, *A. ambigua*, *Nitzschia spiculum*, *Ulnaria ulna*) and saline planktonic (*Cyclotella meneghiniana* [*Stephanocyclus meneghinianus*], *Thalassiosira faurii*, *T. rudolfii*) taxa, indicating deep waters at the sampling site. Saline taxa dominate at the top of the diatomaceous interval suggesting a more saline lake before its termination. Pollen are commonly degraded in the rift-valley lake sediments. One notable exception of contemporary age is reported from part of a core from hypersaline Lake Magadi (Owen et al. 2018, 2019) where anoxic conditions prevailed in the sediments. The Middle Pleistocene Palaeolake Suguta (at Rupa) did not reach the same salinities, but bottom conditions were likely anoxic given the state of pollen preservation. A diverse assemblage includes Afromontane (e.g. *Juniperus*, *Podocarpus*, *Syzygium*), woodland (e.g. *Ficus*, *Combretum*, *Grewia*) and herbaceous (e.g. *Asteraceae*, *Aloe*, *Myrica*) species in varying percentages, but the details remain unpublished (V Muiruri, personal communication).

Junginger et al. (2008), Junginger and Trauth (2013), and Junginger et al. (2014) note several Middle to Late Pleistocene humid episodes at about 400, 130, 80, 60, 35 ka and 15–5 ky ago. Late Pleistocene lake sediments have been reported from stream sections southwest of Kakorinya on the Barrier complex where Dunkley et al. (1993) described ashy diatomite, silt, clay, pumiceous sandstone and mass-flow deposits at 400–450 m asl. Some of these sediments have been correlated with fossil beaches and stromatolitic limestones exposed further south that were dated by Casanova (1986) and Casanova et al. (1988) at 121 ± 20 ka.

Sturchio et al. (1993) observed that U-series-dated, Middle to Late Pleistocene, hydrothermal sinter accumulated during climatically wetter intervals in the Kenya Rift when elevated water tables would be expected. They also considered the possibility that the oldest silica deposits (350 ± 60, 430 ± 120 ka), located on Chepchuk (Figs. 16.1 and 16.9c), might overlap with an inferred period of high lake levels

Fig. 16.10 Representative Middle Pleistocene sediments at Suguta (locations in Fig. 16.9c). (**a**) Kangirinyang looking north. Basalt lying upon faulted diatomite and diatomaceous silt. One white diatomite bed is repeated at different heights because of faulting. (**b**) General view of Rupa locality. (**c**) Diatomaceous silt, diatomite, clay and tuffs at Rupa. The three marked sections show the sampling sites in following photographs. White arrow marks catfish site in '**d**'. (**d**) *In situ* catfish (?) on bedding plane. (**e**) Pumice in tuffaceous diatomite. (**f**) Tuffaceous bed in silty diatomite. (**g**) Massive diatomite lying upon laminated diatomite. (**h**) Fish bone in laminated diatomite. (**i**) Cross-laminated clayey silt beds

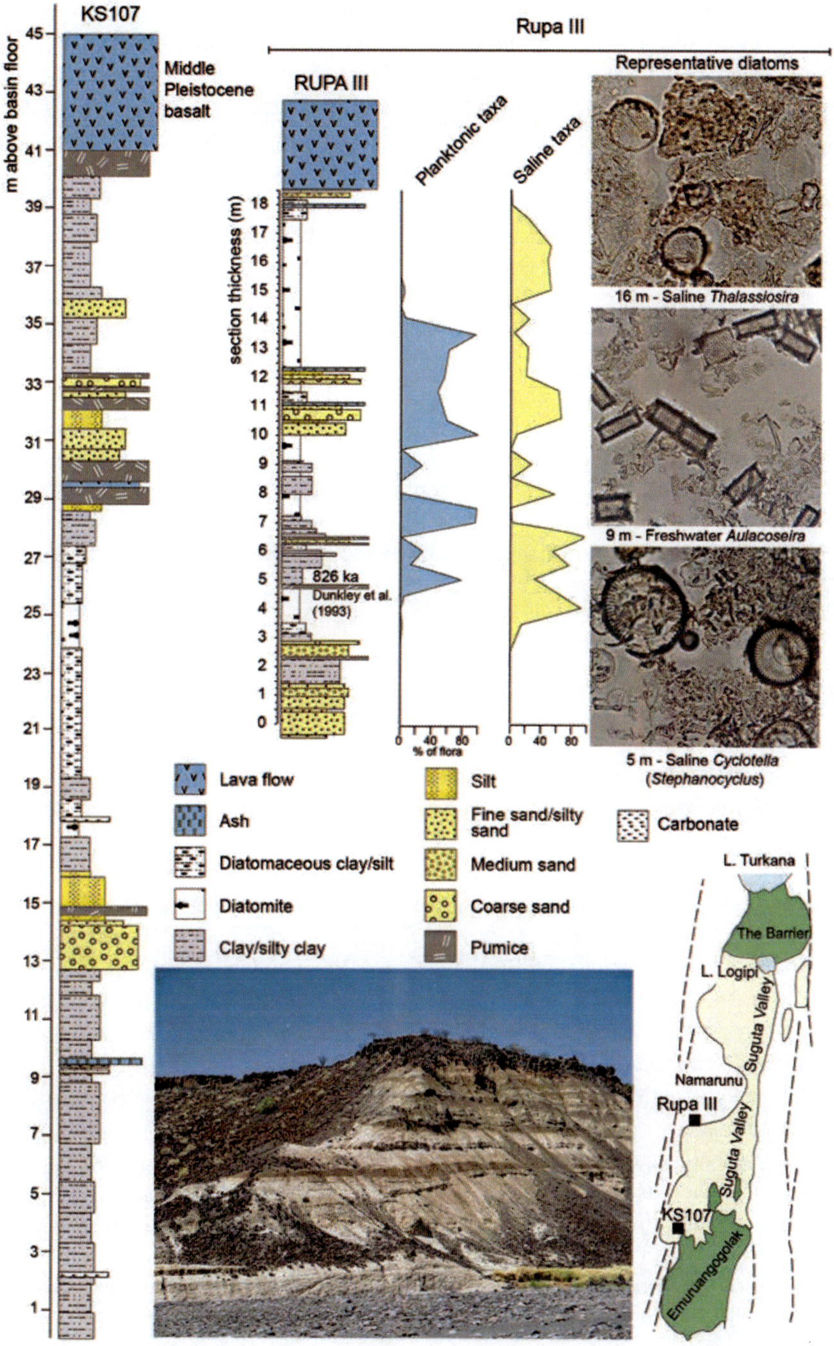

Fig. 16.11 Middle Pleistocene sediment sections and diatom stratigraphy. Sections courtesy of Junginger (2011). Profile KSI07 (Kamuge-Suguta I 2007) is shown in the photograph. RUPA III (at Rupa) shows associated diatom stratigraphy. Diatom photos indicate representative floras with depth from base of section noted

based on stromatolite data from the Magadi-Natron Basin. Better correlations were obtained for silica near the summit of Namarunu, which yielded ages of 133 ± 11 ka that correspond with a period of high lake levels in the Kenya Rift. Locally, stromatolites from high shorelines near Namarunu are dated at 121 ± 20 ka. Younger sinter associated with active fumaroles on Emuruangogolak was dated at 29 ± 3 ka, overlapping with high lake levels between 35 and 25 ka. Furthermore, sinter on the western rim of the Barrier summit caldera was dated at 9.3 ± 0.5 and 6.9 ± 0.3 ka, correlating well with the African Humid Period (AHP) when Palaeolake Suguta was much expanded. Darling and Spiro (2007) noted that $\delta^{18}O$ values from the Barrier sinters were relatively high at the beginning of the Holocene but fell over the following 2.5 ky. Silica veins on western Karosi (Figs. 16.1 and 16.9c) dated at 11.9 ± 0.6 and 12.7 ± 0.6 ka also correlate with this period. In addition, Darling et al. (1993) and Darling (1996) reported oxygen isotope data from sinters (145–125 and 35–25 ka) on Emuruangogolak and Namarunu that suggested that lake water $\delta^{18}O$ values were similar to those of today.

Late Pleistocene to Holocene sediments and shorelines that include the African Humid Period (~15–5 ka) are widespread at many locations in the Lower Suguta (Sect. 7.7) and have been discussed by several authors (e.g. Garcin et al. 2009; Junginger 2011; Junginger and Trauth 2013; Junginger et al. 2014).

16.4 Mio-Pleistocene Palaeogeography

Sakai et al. (2010) commented that the Lower Suguta Basin was probably initiated as a broad downwarp during Miocene times, about 20 million years ago, with the area subsequently characterised by multiple phases of basaltic and trachytic volcanism (Fig. 16.12a, b). Variability in catchment relief was considered by Sakai and colleagues to be minor based on evidence from the Aka Aiteputh and Namarungule formations. Evidence from other parts of the basin is poor and the palaeogeography in Fig. 16.12a, b is speculative for those areas. By ~10 Ma major doming started to tilt the northern Kenya Rift floor downwards to the north with NNE-SSW faulting causing basin fragmentation over a wide area (Sakai et al. 2010). Pollen indicate a shift from forest and dense woodlands to savannah and more open woodland after 10 Ma, which combined with the formation of freshwater lakes suggests a shift to a wetter climate.

The palaeolakes to the southeast of the Lower Suguta had disappeared by about 9 Ma. Over the next 1–2 Ma, the lacustrine sediments and associated lavas were eroded and tilted towards the modern axial trough. Basalt and trachyte volcanism continued through the Plio-Pleistocene (Figs. 16.3 and 16.12c, d). With continued tilting of the rift floor and a lack of volcanic obstructions, it is possible that palaeorivers flowed towards the Turkana Basin which was then occupied by a proto-Omo River (Feibel 2011). Truckle (1976, 1978) observed that the modern Suguta valley began to develop <3 Ma ago with the zone of faulting gradually narrowing towards the modern active rift axis (Fig. 16.12d). Subsequently, much of this

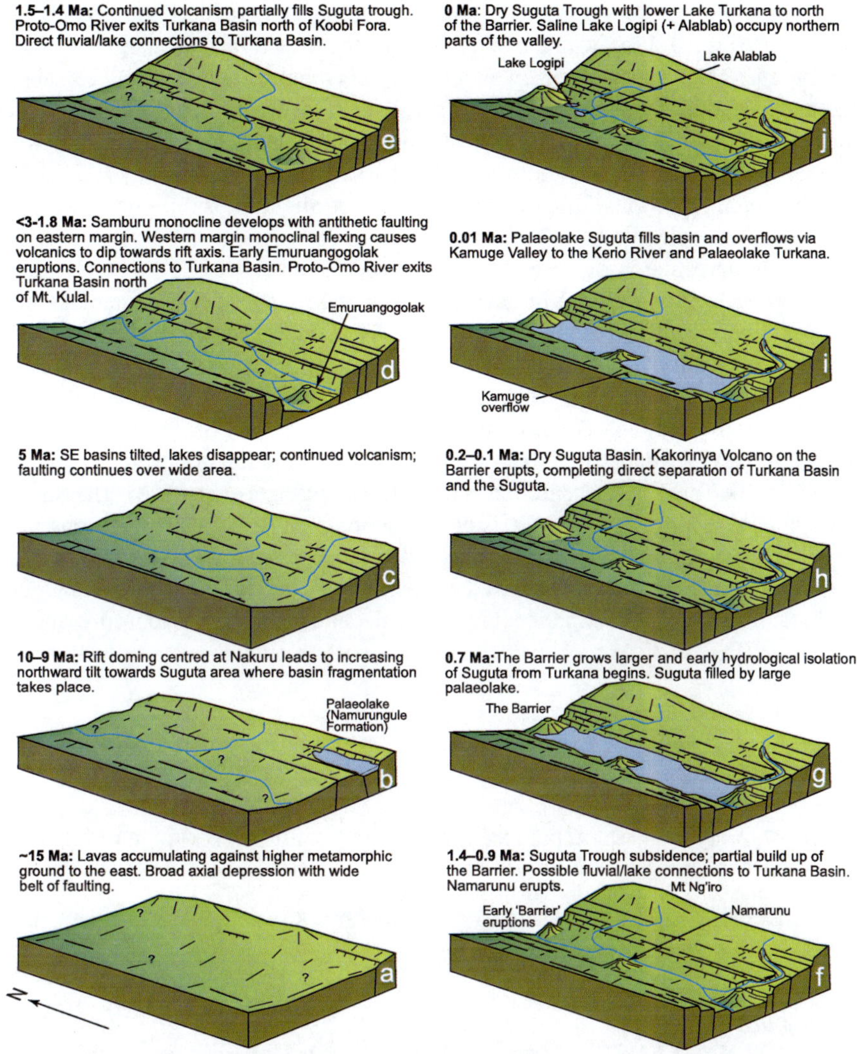

Fig. 16.12 Schematic representation of the evolution of the Suguta Valley. (**a, b**) Miocene downwarping, early faulting over a wide region, and first recognised palaeolakes SE of modern valley. (**c, d**) Initial Pliocene basinward tilting and faulting over wide area followed by development of a narrower axial Suguta Trough. Emuruangogolak begins to erupt. (**e**) Early Pleistocene infilling by volcanics reduces topographic relief. (**f**) Renewed faulting increases topographic relief with the eruption of Barrier volcanoes, Emuruangogolak, and Namarunu, which define the modern Lower Suguta valley. (**g**) Middle Pleistocene lake expansion. Likely separation of Suguta palaeolake from the Turkana Basin by the Barrier Complex. (**h**) Late Pleistocene dry phase. Other periods of lake expansion are poorly understood but were likely of smaller scale than the Middle Pleistocene. (**i**) Lake expansion during the African Humid Period with overflow via the Kamuge Valley. (**j**) Modern Suguta Valley

accommodation space was filled by younger volcanics that reduced the topographic variability (Fig. 16.12e), with the relief between the rift axis and its margins increasing because of further faulting by about 0.75 Ma (Junginger 2011). Between about 1.3 and 0.5 Ma, the modern topography was largely established by the growth of the Barrier Volcanic Complex to the north and Emuruangogolak to the south, with Namarunu constricting the middle part of the basin (Fig. 16.12f; Dunkley et al. 1993).

Champion (1935) had suggested that the Lower Suguta was once occupied by a southward extension of Lake Turkana with volcanic eruptions at the Barrier dividing the former water body. Truckle (1978) noted the presence of Nile Perch (*Lates niloticus*), which was absent in other rift lakes, thus supporting the inference of a direct connection. Research has shown that a large lake occupied the Suguta Basin through parts of the Middle Pleistocene (Fig. 16.12g; Truckle 1976; Casanova et al. 1988; Dunkley et al. 1993). However, the Barrier would have blocked direct connections to the Turkana Basin (Truckle 1978). Studies of $^{87}Sr/^{86}Sr$ isotope ratios from Suguta Valley lacustrine microfossils confirm separation of the Suguta palaeolake from the Turkana Basin around 900–700 ka (Junginger et al. 2023). The Suguta became arid again in the Late Pleistocene, although smaller lake expansions may not be recorded given the current state of knowledge. The basin filled during the African Humid Period and overflowed via the Kamuge Valley (Fig. 16.12i; Junginger and Trauth 2013) before experiencing a rapid decline in lake levels to its present dry condition (Fig. 16.12j).

References

Acosta VT, Bande A, Sobel ER, Parra M, Schildgen TF, Stuart F, Strecker MR (2015) Cenozoic extension in the Kenya Rift from low-temperature thermochronology: links to diachronous spatiotemporal evolution of rifting in East Africa. Tectonics 34:2367–2386

Baker BH (1963) The geology of the Baragoi area. Geol Surv Kenya Rep 53

Bohacs AC, Carroll AR, Neal JE, Mankiewicz PJ (2000) Lake-basin type, source potential, and hydrocarbon character: an integrated sequence-stratigraphic–geochemical framework. In: Gierlowski-Kordesch EH, Kelts KR (eds) Lake basins through space and time. AAPG Stud Geol 46:3–34

Bosworth W, Maurin A (1993) Structure, geochronology and tectonic significance of the northern Suguta Valley (Gregory Rift), Kenya. J Geol Soc Lond 150:751–762

Casanova J (1986) Les stromatolites continenteaux: paléoécologie, paléohydrologie, paléoclimatologie. Application au rift Gregory. PhD thesis, Université d'Aix-Marseille

Casanova J, Hillaire-Marcel C, Pagé N (1988) Stratigraphie et paléohydrologie des épisodes lacustres du Quaternaire récent du rift Suguta (Kenya). C R Acad Sci Paris 307:1251–1258

Caukwell R (1971) The South Turkana Expedition scientific papers VI. Field survey in South Turkana. Geogr J 137:157–164

Cerling TE, Powers DW (1977) Paleorifting between the Gregory and Ethiopian Rifts. Geology 5:441–444

Champion AM (1935) Teleki's volcano and the lava fields at the southern end of Lake Rudolf. Geogr J 85:323–336

Darling WG (1996). The geochemistry of fluid processes in the eastern branch of the East African Rift System. PhD thesis, Open University, UK

Darling WG, Spiro B (2007) Oxygen isotopes in hydrothermal sinters as an indicator of Quaternary lakewater compositions in the Kenya Rift Valley. In: Bullen TD, Wang Y (eds) Water-rock interaction: Proc 12 int symp water-rock interaction, Kunming, China, July–August 2007. CRC Press, London, pp 1475–1480

Darling WG, Allen DJ, Spiro B (1993) Oxygen isotopes in relict sinters as a palaeoindicator of Rift Valley climate. In: Isotope techniques in the study of past and current environmental changes in the hydrosphere and atmosphere. Proc IAEA symp, Vienna, pp 555–559

Dodson RG (1963) Geology of the area south of Lodwar. Geol Surv Kenya Rep 87

Dunkley PN, Smith M, Allen DJ, Darling WG (1993) The geothermal activity and geology of the northern sector of the Kenya Rift Valley. Brit Geol Surv Res Rep SC/93/1

Feibel CS (2011) A geological history of the Turkana Basin. Evol Anthropol 20:206–216

Garcin Y, Junginger A, Melnick D, Olago DO, Strecker MR, Trauth MH (2009) Late Pleistocene–Holocene rise and collapse of Lake Suguta, northern Kenya Rift. Quat Sci Rev 28:911–925

Gwynne M (1969) The South Turkana Expedition scientific papers I. Preliminary report on the 1968 season. Geogr J 135:331–342

Handa N, Nakatsukasa M, Kunimatsu Y, Nakaya H (2018) *Brachypotherium* (Perissodactyla, Rhinocerotidae) from the late Miocene of Samburu Hills, Kenya. Geobios 51:391–399

Ishida H (1984) Outline of 1982 survey in Samburu Hills and Nachola Area, Northern Kenya. Afr Stud Monogr Suppl Iss 2:1–13

Ishida H, Ishida S, Torii M, Matsuda T, Kawamura Y, Koizumi K (1982) Report of field survey in Kirimun, Kenya, 1980. Study of the Tertiary hominoids and their palaeoenvironments in East Africa, vol 1. Osaka University, Osaka

Ishida H, Pickford M, Nakaya H, Nakano Y (1984) Fossil anthropoids from Nachola and Samburu Hills District, Kenya. Afr Stud Monogr 2:73–85

Itaya T, Sawada Y (1986) K-Ar ages of volcanic rocks in the Samburu Hills area, northern Kenya. Afr Stud Monogr Suppl Iss 5:27–45

Junginger A (2011) East African climate variability on different time scales: the Suguta Valley in the African-Asian Monsoon Domain. PhD thesis, Universität Potsdam

Junginger A, Trauth MH (2013) Hydrological constraints of paleo-Lake Suguta in the Northern Kenya Rift during the African Humid Period (15–5 ka BP). Glob Planet Change 111:174–188

Junginger A, Garcin Y, Trauth MH, Maslin M, Melnick D, Strecker MR, Olago D, Graduate School GRK (2008) Multiple Cenozoic lake episodes in the Suguta Valley, northern Kenya Rift. Geophys Res Abstr 10:EGU2008-A-09682

Junginger A, Roller S, Olaka LA, Trauth MH (2014) The effects of solar irradiation changes on the migration of the Congo Air Boundary and water levels of paleo-Lake Suguta, Northern Kenya Rift, during the African Humid Period (15–5 ka BP). Palaeogeogr Palaeoecol Palaeoclimatol 396:1–16

Junginger A, Kuebler S, Rosca C, Owen RB, Deino A, Feibel C, Trauth MH, Vonhof H (2023) Mid-Pleistocene volcano-tectonic fragmentation of the Turkana-Suguta Megalake, EGU General Assembly 2023, Vienna, Austria, 24–28 April 2023, EGU23-17267

Kabeto K, Sawada Y, Iizumi S, Wakatsuki T (2001a) Mantle sources and magma–crust interactions in volcanic rocks from the northern Kenya Rift: geochemical evidence. Lithos 56:111–139

Kabeto K, Sawada Y, Wakatsuki T (2001b) Different evolution trends in alkaline evolved lavas from the northern Kenya Rift. J Afr Earth Sci 32:419–433

Kawamura Y, Nakaya H (1984) Thryonomyid rodent from the Late Miocene Namurungule Formation, Samburu Hills, northern Kenya. Afr Stud Monogr Suppl Iss 2:133–139

Key RM (1987) Geology of the Maralal area. Kenya Mines Geol Dep Rep 105

Koyaguchi T (1984) Volcanic rocks in the Samburu Hills, northern Kenya. Afr Stud Monogr Suppl Iss 2:147–179

Kunimatsu Y, Nakatsukasa M, Sawada Y, Sakai T, Hyodo M, Hyodo H, Itaya T, Nakaya H, Saegusa H, Mazurier A, Saneyoshi M, Tsujikawa H, Yamamoto A, Mbua E (2007) A new Late Miocene great ape from Kenya and its implications for the origins of African great apes and humans. Proc Natl Acad Sci 104:19220–19225

Lambiase JJ, Bosworth W (1995) Structural controls on sedimentation in continental rifts. In: Lambiase JJ (ed) Hydrocarbon habitat in rift basins. Geol Soc Lond Spec Publ 80:117–144

Lichoro CM, Árnason K, Cumming W (2019) Joint interpretation of gravity and resistivity data from the northern Kenya volcanic rift zone: structural and geothermal significance. Geothermics 77:139–150

Makinouchi T, Torii M, Koyaguchi T, Matsuda T (1984) The geology of the Nachola area and the Samburu Hills area, west of Baragoi, northern Kenya. Afr Stud Monogr 2:15–44

Mariita N, Keller G (2007) An integrated geophysical study of the northern Kenya rift. J Afr Earth Sci 48:80–94

Matsuda T, Torii M, Koyaguchi T, Makinouchi T, Mitsushio H, Ishida S (1984) Fission-track, K-Ar age determinations and palaeomagnetic measurements of Miocene volcanic rocks in the western area of Baragoi, northern Kenya: ages of hominoids. Afr Stud Monogr 2:87–131

Nakaya H (1994) Faunal change of Late Miocene Africa and Eurasia: mammalian fauna from the Namurungule Formation, Samburu Hills, Northern Kenya. Afr Stud Monogr 20:1–112

Nakaya H, Pickford M, Nakano Y, Ishida H (1984) The late Miocene large mammal fauna from the Namurungule Formation, Samburu Hills, Northern Kenya. Afr Stud Monogr 2:45–56

Ochieng JO, Wilkinson AF, Kagasi J, Kimiomo S (1988) Geology of the Loiyangalani area. Kenya Mines Geol Dep Rep 107

Owen RB, Muiruri VM, Lowenstein TK, Renaut RW, Rabideaux N, Luo S, Deino AL, Sier MJ, Dupont-Nivet G, McNulty EP, Leet K, Cohen A, Campisano C, Deocampo D, Shen C-C, Billingsley A, Mbuthia A (2018) Progressive aridification in East Africa over the last half million years and implications for human evolution. Proc Natl Acad Sci 115:11174–11179

Owen RB, Renaut RW, Muiruri VM, Rabideaux NM, Lowenstein TK, McNulty EP, Leet K, Deocampo D, Luo S, Deino AL, Cohen A, Sier MJ, Campisano C, Shen C-C, Billingsley A, Mbuthia A, Stockhecke M (2019) Quaternary history of the Lake Magadi Basin, southern Kenya Rift: tectonic and climatic controls. Palaeogeogr Palaeoclimatol Palaeoecol 518:97–118

Pickford M (1984) Fossil Mollusca from the Samburu Hills, Northern Kenya. Afr Stud Monogr 2:141–145

Pickford M, Ishida H, Nakano Y, Nakaya H (1984a) Fossiliferous localities of the Nachola-Samburu Hills area, Northern Kenya. Afr Stud Monogr 2:45–56

Pickford M, Nakaya H, Ishida H, Nakano Y (1984b) The biostratigraphic analysis of the faunas of the Nachola area and Samburu Hills, Northern Kenya. Afr Stud Monogr 2:67–72

Pickford M, Ishida H, Nakaya Y, Yasui K (1987) The Middle Miocene Fauna from the Nachola and Aka Aiteputh Formations, Northern Kenya. Afr Stud Monogr, Suppl Iss 5:141–154

Rhemtulla S (1970) The South Turkana Expedition scientific papers III. A geological reconnaissance of South Turkana. Geogr J 136:61–73

Robakiewicz E, Owen RB, Rosca C, Deino A, Garcin Y, Trauth MH, Kübler S, Junginger, A (2023) Hydroclimate reconstructions in the Suguta Valley, northern Kenya, during the Early-Middle Pleistocene transition. Palaeogeography, Palaeoclimatology, Palaeoecology. https://doi.org/10.1016/j.palaeo.2023.111758

Saggerson EP, Baker BH (1965) Post-Jurassic erosion-surfaces in eastern Kenya and their deformation in relation to rift structure. Q J Geol Soc Lond 121:51–72

Sakai T, Saneyoshi M, Tanaka S, Sawada Y, Nakatsukasa M, Mbua E, Ishida H (2010) Climate shift recorded at around 10 Ma in Miocene succession of Samburu Hills, northern Kenya Rift, and its significance. In: Clift PD, Tada R, Zheng H (eds) Monsoon evolution and tectonics–climate linkage in Asia. Geol Soc Lond Spec Publ 342:109–127

Sakai T, Saneyoshi M, Sawada Y, Nakatsukasa M, Kunimtatsu Y, Mbua E (2013) Early continental rift basin stratigraphy, depositional facies and tectonics in volcaniclastic system: examples from the Miocene successions along the Japan Sea and in the East African Rift Valley (Kenya). In: Itoh Y (ed) Mechanism of sedimentary basin formation-multidisciplinary approach on active plate margins. InTech, Rijeka, pp 83–108

Saneyoshi M, Nakayama K, Sakai T, Sawada Y, Ishida H (2006) Half graben filling processes in the early phase of continental rifting: the Miocene Namurungule Formation of the Kenya Rift. Sediment Geol 186:111–131

Sawada Y, Tateishi M, Ishida S (1987) Geology of the Neogene System in and around the Samburu Hills, Northern Kenya. Afr Stud Monogr Suppl Iss 5:7–26

Sawada Y, Saneyoshi M, Nakayama K, Sakai T, Itaya T, Hyodo M, Mukyokya Y, Pickford M, Senut B, Tanaka S, Chujo T, Ishida H (2006) The ages and geological background of Miocene *Nacholapithecus*, *Samburupithecus* and *Orrorin* from Kenya. In: Ishida H, Tuttle RH, Pickford M, Nakatsukasa M, Ogihara N (eds) Human origins and environmental backgrounds. Springer, New York, pp 71–96

Sceal JSC (1974) The geology of Paka volcano and the country to the east, Baringo District, Kenya. PhD dissertation, University of London

Shackleton RM (1945) Geology of the Nyeri area. Geol Surv Kenya Rep 11

Sturchio N, Dunkley P, Smith M (1993) Climate-driven variations in geothermal activity in the northern Kenya rift valley. Nature 362:233–234

Tateishi M (1987) Sedimentary facies of the Miocene in the Samburu Hills, northern Kenya. Afr Stud Monogr Suppl Iss 5:59–77

Truckle PH (1976) Geology and late Cainozoic lake-sediments of the Suguta Trough, Kenya. Nature 263:380–383

Truckle PH (1978) The geology of the area to the south of Lokori, South Turkana, Kenya. PhD thesis, University of Leicester

Truckle PH (1979) Pahoehoe and pillow lavas from the Suguta Trough, South Turkana, Kenya. Geol Mag 116:139–142

Tryon CA, Roach NT, Logan MAV (2008) The Middle Stone Age of the northern Kenyan Rift: age and context of new archaeological sites from the Kapedo Tuffs. J Hum Evol 55:652–664

Uemura Y (1994) Tectonic landforms and active faulting in the area surrounding Sumburu Hills, northern Gregory Rift Valley, Kenya. J Fac Lett Bukkyo Univ 78:81–98

Weaver SD (1973) The geology of the Nasaken area, South Turkana, Kenya and the petrochemistry of the Pliocene volcanic rocks. PhD thesis, University of London

Weaver SD (1977) The Quaternary caldera volcano Emuruangogolak, Kenya Rift, and the petrology of a bimodal ferrobasalt-pantelleritic trachyte association. Bull Volcanol 40:209–230

Webb PK (1971) The geology of the Tiati Hills, Rift Valley Province, Kenya. PhD thesis, University of London

Williamson PG, Savage RJG (1986) Early rift sedimentation in the Turkana basin, northern Kenya. In: Frostick LE, Renaut RW, Reid I, Tiercelin J-J (eds) Sedimentation in the African rifts. Geol Soc Lond Spec Publ 25:267–283

Chapter 17
The Baringo-Bogoria Basin and Adjacent Parts of the Kenya Rift

17.1 Introduction and Geological Exploration

The Kenya Rift at the latitudes of the Baringo-Bogoria basin is about 90 km wide and is structurally complex (King et al. 1972; King 1978) with numerous faults that expose a series of Miocene and younger lacustrine sediments (Chapman et al. 1978). To the west, the Elgeyo Escarpment stands about 1500 m above the Kerio Valley and marks the position of a major border fault (Fig. 17.1a). In contrast, the Laikipia Escarpment, on the eastern rift margin, consists of a series of parallel faults that step down to the rift floor. The 1300-m-high Tugen Hills are located in the middle of the rift, separating the Kerio and Baringo-Bogoria basins. The hills are formed by a major north-south trending tilt-block that dips westwards from the easterly-facing Kito Pass and Saimo fault scarps (Fig. 17.1a, b; Williams and Chapman 1986). The Kerio River flows northwards along the Kerio Valley to Lake Turkana. In contrast, the Baringo-Bogoria Basin is occupied by freshwater Lake Baringo and saline Lake Bogoria, which today are separated by fluvio-lacustrine sediments below the intervening Loboi Plain.

The Baringo region includes well-exposed outcrops of Neogene sediments that formed in a series of discrete depositional basins. These range in age from Early Miocene to Pleistocene with younger exposed sediments tending to shift eastwards with time (Fig. 17.1c). The basal Sidekh and Saimo Phonolites include thin bands of sediment that form the Chepkuno, Aimeno and Surkai Beds, the last including shales that represent the earliest fully lacustrine strata in the Tugen Hills succession (Pickford et al. 2009). The Scottish geologist Joseph Thomson was the first European to visit the region in 1884, but he erroneously described the Tugen Hills as being metamorphic (Hill et al. 1986; Hill 2002). Gregory (1896) was the next visitor and the first to recognise sediments (Gregory 1921), with Prior (1903) subsequently describing many of Gregory's rock collections and noting the prevalence of phonolite, phonolitic trachyte and basalt. Other early geological explorations included

R. W. Renaut, R. B. Owen, *The Kenya Rift Lakes: Modern and Ancient*,
Syntheses in Limnogeology, https://doi.org/10.1007/978-3-642-25055-2_17

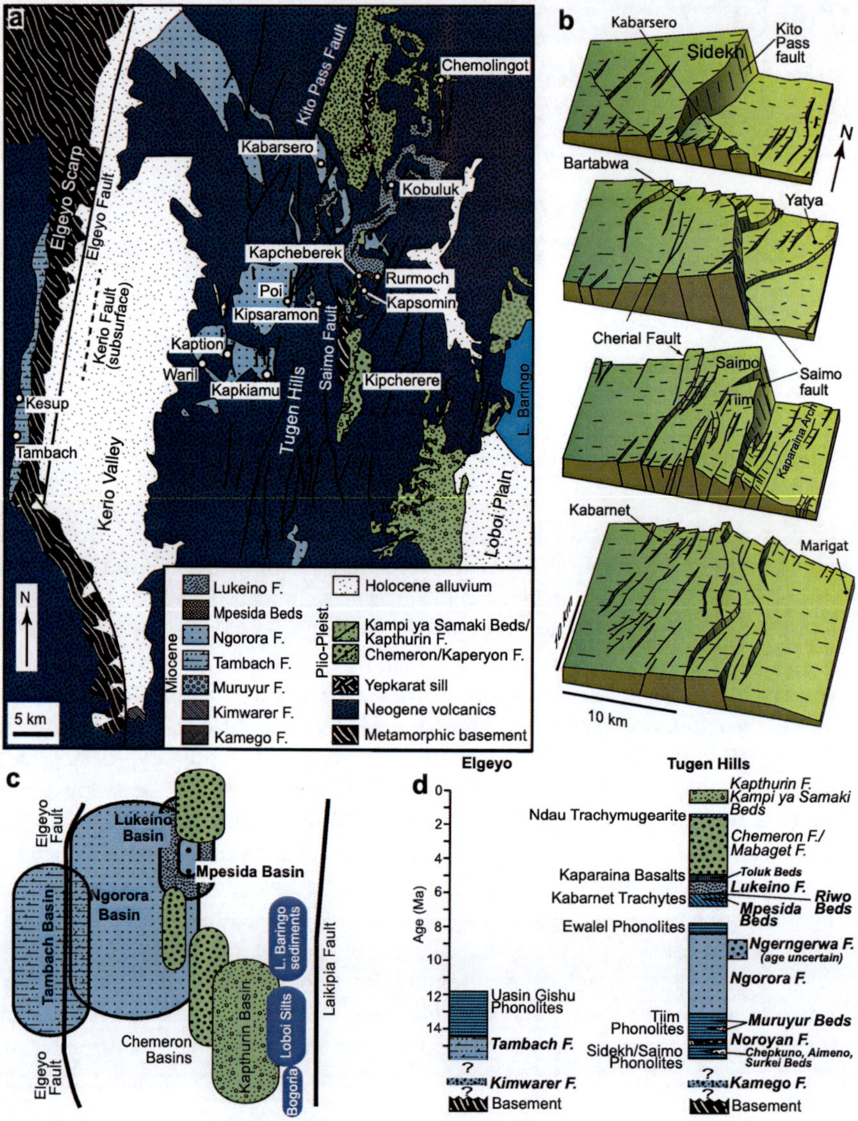

Fig. 17.1 Geology and stratigraphy of the Tugen Hills and Kerio Valley. (**a**) Geology. Based on Chapman et al. (1973). (**b**) Schematic model of the Tugen Hills and regional faulting. Modified from Chapman et al. (1978). (**c**) Major depositional basins. Note the gradual shift eastwards of exposed formations (key shown in (**a**) in stratigraphic order). Miocene basins shown in bold. After Owen and Renaut (2000). (**d**) Regional stratigraphy. After Chapman et al. (1978), Hill (1995) and Renaut et al. 1999). Dating for the Muruyur Beds by Behrensmeyer et al. (2002) implies that they may be older than shown (see text)

studies by Walker (1903), Murray-Hughes (1933), Fuchs (1934), Willis (1936) and Shackleton (1950).

Gregory (1921) described sediments in the Tugen foothills, which he thought to be of Miocene or Oligocene age (under his Nyasan Series) and inferred that they accumulated in an expansive 'Lake Kamasia'. Subsequently, Leakey and Solomon (1929) identified these deposits as Pleistocene and suggested that they were laid down during an 'Eburrian Pluvial', one of several wet intervals that were thought to have affected the region (Kingston and Hill 2005). Leakey (1931) renamed this wet interval the Kamasian Pluvial, after the name adopted by Gregory for the palaeolake that had occupied the basin. Based on stone tools and fossils, Fuchs (1950) supported a Pleistocene age for the Kamasia sediments, with the term Kamasian subsequently being split into the Kamasian (lower Kamasian) and Kanjeran (upper Kamasian) pluvials that were separated by a long dry interpluvial (Leakey 1950, 1955; Bishop 1967). However, Evernden et al. (1965) reported radiometric ages that gradually contributed to a rejection of the pluvial concept and linkages to northern hemisphere glacial–interglacial cycles, with the term Kamasian falling into disuse. Members of the Kenya Geological Survey carried out fieldwork in the 1960s, which led to McCall et al. (1967) and Walsh (1969) dividing the former 'Kamasia' sediments into the Chemeron and Kapthurin beds.

Intensive research followed when Basil King and Bill Bishop initiated and directed the East African Geological Research Unit (EAGRU), which examined the geology of the Tugen Hills in order to improve knowledge of the overall palaeoenvironmental setting in which mammalian and hominid fossils accumulated (e.g. Bishop and Chapman 1970; Chapman 1971; Bishop et al. 1971; Chapman et al. 1973, 1978; Bishop and Pickford 1975; Pickford 1975a, b, 2009a, b; Bishop 1978; Chapman and Brook 1978; Pickford et al. 1983; Hill 1985, 2002; Williams and Chapman 1986; Senut and Pickford 2009). Subsequent work on the palaeontology and geology continued through the efforts of individual researchers and through the Baringo Paleontological Research Group (BPRP), initially led by Andrew Hill (Hill et al. 1985, 1991, 1992; Kingston 1992; McBrearty et al. 1996; Bishop et al. 1999; Hill 1999; Wood 1999; Senut et al. 2001; Tryon and McBrearty 2006). Much effort has also been devoted towards establishing a firm chronostratigraphy for the palaeontological research (Chapman and Brook 1978; Tauxe et al. 1985; Hill et al. 1986; Deino et al. 1990, 2002).

The petroleum industry, including the National Oil Corporation of Kenya (NOCK), developed an interest in the lacustrine succession as possible oil or gas source rocks in the late 1980s (Morley et al. 1992, 1999; Ego 1994; Renaut et al. 1999; Tiercelin et al. 2004, 2012), acquiring gravity and seismic reflection data for the Kerio Basin in 1990 (Tiercelin et al. 2012). A magnetotelluric survey in the Baringo-Bogoria Basin showed that both basins contained about 7–8 km of alternating volcanic rocks and sediments (Mugisha et al. 1997; Hautot et al. 2000; Tiercelin et al. 2012). Broad-based surveys have documented a range of economic resources in the region that include, for example, diatomite, rubies, fluorite, and sites with geotourism potential (Kipseba et al. 2009).

17.2 Miocene Lacustrine Sedimentary Formations

17.2.1 The Kirimun Formation

The Kirimun Formation underlies the Miocene Rumuruti Group phonolite on the Laikipia Plateau, infilling gullies in an irregular sub-Miocene land surface (Shackleton 1946; Hackman 1988). The best exposures lie north of the Palagalagi River and west of Kirimun (~85 km east of Karosi volcano) with fossils (*Dinotherium hobleyi*, *Mastodon* sp., chelonian scutes and crocodile vertebrae) suggesting a Miocene age. The deposits include up to 30 m of sepiolite (altered volcanic ash), white clay, diatomite, marl, shale, chert laminated green siltstone, graded grit (coarse sandstone) and grey calcareous gravel (Hackman 1988). Graded beds have been interpreted as shallow-water deposits, probably fluvial. One thin and relatively pure diatomite suggests localised lacustrine deposition (Shackleton 1946).

17.2.2 The Kimwarer and Kamego Formations

Palaeogene fluvial sandstones constitute the Kimwarer and Kamego formations in the Tugen Hills-Kerio Valley. Sandstones in these deposits contain metamorphic basement-derived quartz and lithic clasts (Figs. 17.2c, g and 17.4; Chapman et al. 1978; Williams and Chapman 1986; Renaut et al. 1999; Tiercelin and Lezzar 2002; Tiercelin et al. 2012). The Kimwarer Formation (Fig. 17.2a–c), exposed at the south end of the Kerio Valley (Fig. 17.1a), is older than 18 Ma (Muia 2015). It comprises conglomerate and sandstone overlain by chert, tuffaceous sandstone, cream or yellow siltstone and white claystone that lack fossil diatoms (Walsh 1969). The formation also includes green, waxy laminated tuffaceous shale that has been interpreted as lacustrine (Fig. 7.2a, b; Lippard 1972; Nyambok and Gaciri 1975; Ego 1994; Renaut et al. 1999), although Muia (2015) reports only alluvial fan, distal fluvial channel and floodplain facies. The Elgeyo Border Fault had minimal or no topographic expression at the time of deposition although the basement surface is believed to have been irregular (Williams and Chapman 1986).

In the Tugen Hills, the easterly dipping Kamego Formation (Figs. 17.1a and 17.2d–g) consists of 'quartzite', sandstone and siltstone deposited on a surface with up to 100 m relief (Williams and Chapman 1986). Muia (2015) obtained $^{40}Ar/^{39}Ar$ dates of 19.9 and 18.0 Ma from a section at Kipsaramon. A pronounced unconformity is present between these deposits and the overlying sediments and volcanics, which generally dip to the west. The river systems that deposited most of the Kimwarer and Kamego formation sediments ultimately flowed northwards, feeding lakes in the Lokichar and North Kerio basins (Chap. 15) of the southwestern Turkana region (Morley and Wescott 1999; Tiercelin et al. 2004, 2012).

Fig. 17.2 Early Miocene sediments. (**a–c**) Kimwarer Formation; (**d–g**) Kamego Formation. (**a**, **b**) Bluish grey sandstones interbedded with lacustrine shales, some with mudcracks. (**c**) Kimwarer sandstone with basement-derived quartz. White scale-bar = 4 mm. (**d**) Kito Pass area where Kamego Formation outcrops are located. Paka volcano in distance. (**e**) Kamego quartz arenite overlying clay-rich palaeosol with rhizoliths (floodplain or shallow lake?). Section is ~40 cm high. (**f**) Quaternary colluvium overlying well-cemented mudstone with rhizoliths, possibly terrestrial. Backpack is 40 cm long. (**g**) Kamego sandstone with basement-derived quartz. White scale bar = 4 mm

17.2.3 The Tambach Formation

The Tambach Basin and Setting

The Tambach Formation is well-exposed between Tambach and Kesup (Fig. 17.1a) along the C51 road and rests on an irregular metamorphic palaeosurface. The top of the sequence is unconformably overlain by the 14.5–12 Ma Uasin Gishu Phonolite Formation (Figs. 17.1d and 17.4; Lippard 1973). These 500-m-thick phonolites cover the Tambach Formation to the west of the rift, where outcrops are lacking. The Tambach sequence includes 200 m of fluvial conglomerate, siltstone and claystone overlain by 200 m of finely laminated lacustrine claystone with fish fossils (Fig. 17.3a–d; Lippard 1973). Phonolitic lava, phonolitic and trachytic tuff, and small basic intrusions are also present in the sequence (Shackleton 1950; Lippard 1972; Ego 1994; Renaut et al. 1999). The sediments and volcanics accumulated in a broad downwarp characterised by colluvial, fluvial and lacustrine environments (Chapman et al. 1978).

During the Early Palaeogene, faulting was focused along the north-south Kerio Fault below the floor of the modern Kerio Valley (Fig. 17.1a; Mugisha et al. 1987).

Fig. 17.3 Middle Miocene Tambach and Muruyur sediments. (**a–d**) Tambach Formation; (**e–f**) Muruyur Beds. (**a**) Inclined bedded deltaic sandstone and conglomerate, lower Tambach Formation, Elgeyo Escarpment. (**b**, **c**) Small Gilbert-type deltaic sandstones with foresets representing progradation into a shallow saline lake. (**d**) Interbedded thin sandstones and shales. (**e**) Member 1 bone bed at Kipsaramon, in clay and silt. (**f**) Antelope footprints on bedding surface, Kipsaramon

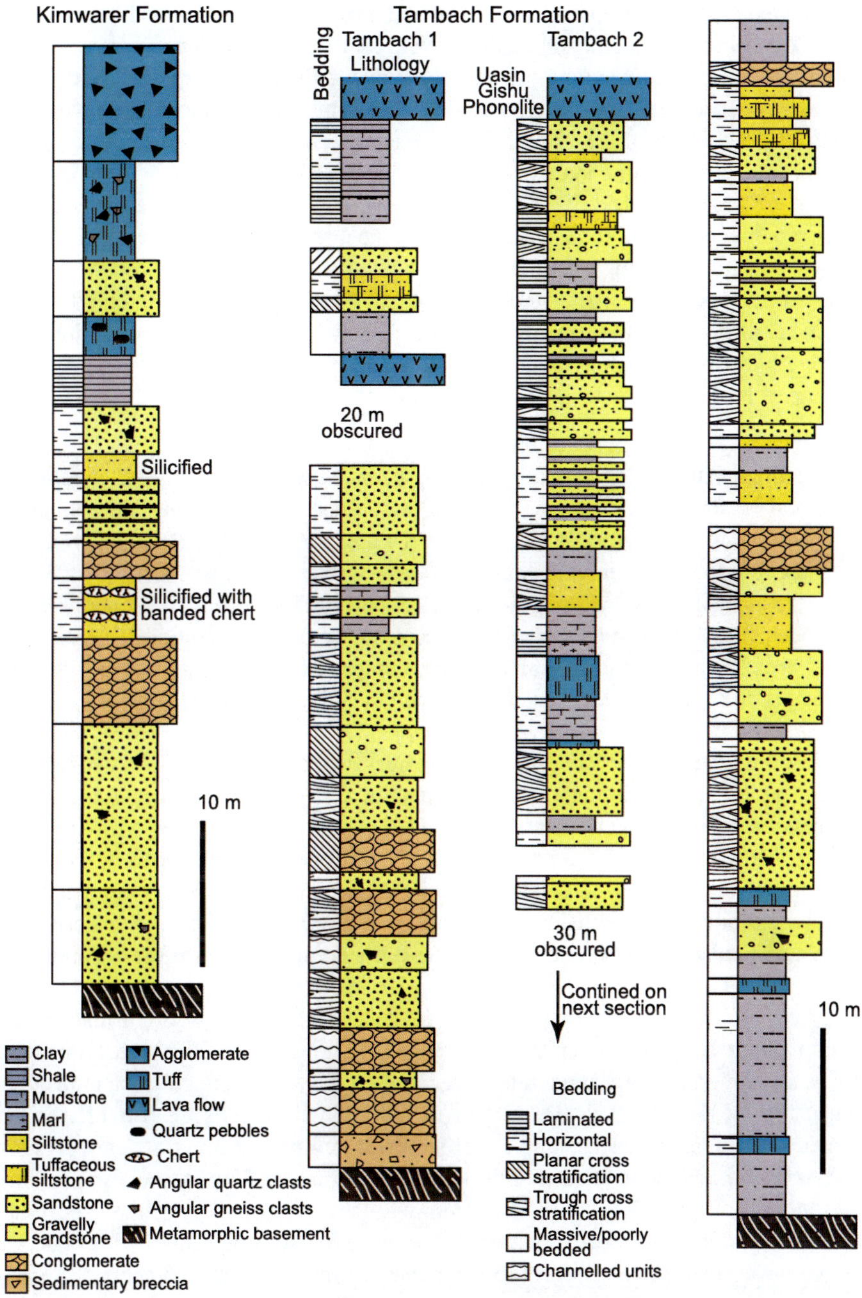

Fig. 17.4 Early to Middle Miocene lacustrine stratigraphic sections. Kimwarer Formation, southern Kerio Valley. After Ego (1994). Tambach Formation, Elgeyo Escarpment. After Renaut et al. (1999). Section 'Tambach 1' is 2–3 km south of Tambach along the main C51 road. Section 'Tambach 2' is from Shackleton (1950)

Subsequently, fault activity diminished, and a broad sag basin formed between 16 and 12 Ma in which Tambach deposits were laid down. The depocentre accumulated >500 m of sediment and lay near the modern Elgeyo Escarpment, which did not exist at that time. The western margin of the Tambach Basin may have extended up to 40–50 km west of the modern Elgeyo Escarpment, where a 6-m-thick sequence of tuff and grit, of similar or older age, rests directly on basement near Eldoret (Sanders 1963; Renaut et al. 1999). Geophysical data indicate that the eastern margin probably lay east of the Tugen Hills, with the northern and southern basin limits poorly constrained (Mugisha et al. 1997).

Sedimentation was terminated by eruption of the Uasin Gishu Phonolites and renewed rifting with movements along the Elgeyo Fault, which downfaulted the Tambach Formation below the Kerio Valley and buried the deposits underneath a younger fill of sediments and lavas (Morley et al. 1992).

Stratigraphy and Sedimentation

The sediment succession generally fines upwards, reflecting a change from alluvial fans and braided stream settings to shallow deltaic and lacustrine environments (Renaut et al. 1999). Airfall and waterlain tuffs are common. Basal polymictic, boulder conglomerate and paraconglomerate (debris flow deposits) rest on an irregular metamorphic palaeotopography. Palaeocurrent indicators for sections near Tambach (Fig. 17.1a) indicate a derivation from W to SW directions (Ego 1994). The lower 200 m of the formation is mainly comprised of discontinuous conglomerate and sandstone interpreted as having accumulated in braided stream and crevasse splay environments.

Basement metamorphic clasts dominate near the base, giving way to nephelinite and phonolite compositions upwards (Renaut et al. 1999). Sandstones in the lower succession are moderately to poorly sorted litharenites and arkoses with volcanic grains increasing upwards. Less commonly, sandstones are well sorted and quartz-rich, indicating basement derivation. Calcite and iron oxide cements dominate with quartz or albite overgrowths in a few sandstones. Authigenic illite and smectite are common. Feldspathic floodplain siltstone is more common towards the top and may be associated with palaeosols, calcareous rhizoliths and thin calcretes.

Lacustrine deposits are present in the lower part of the Tambach Formation but become more frequent higher in the sequence (Fig. 17.3). Lake facies are dominated by waxy pale green, grey and white shale interbedded with waxy greenish mudstone and claystone (Renaut et al. 1999). Several fine-grained deposits are also tuffaceous. Green and white paper shale and claystone, with laminae <2 mm thick, are present in centimetre to decimetre-thick units. Coarser laminae (2–10 mm) are also present, commonly in upward-fining thicker units with siltstone or rippled fine sandstone interbeds and lenses. Locally, deltaic Gilbert-type foresets interfinger with shale.

Massive greenish mudstone and claystone are interbedded with shale and may constitute 1–2 cm granules, with reworking producing intraclast breccia. Mudcracks are also present. Shale often contains ostracods, insects and fragmented fish bone, with intact fish present locally near Kesup. Diatoms are absent. Small leaves, root marks and brown carbonised films, formed by disseminated plant debris, are present locally (Renaut et al. 1999).

Silt and fine sand in the lake facies are dominated by K-feldspar, quartz and sodic plagioclase. Minor calcite is present in coarsely laminated shale. Shale and mudstone commonly contain smectite with subsidiary illite and minor chlorite and kaolinite. Waxy shale and mudstone commonly contain analcime (20–25%) as pore-filling cement with low Si/Al ratios (2.15–2.16) suggesting a clay mineral precursor (Ego 1994). The zeolites imply that the lake periodically became saline and alkaline.

17.2.4 The Muruyur Formation

Martyn (1969) reported the presence of tuffaceous sediments within the Tiim Phonolites Formation, which Chapman (1971) formally described as the Muruyur Beds (Fig. 17.1d) and Pickford et al. (2009) referred to as the Muruyur Formation. These deposits show considerable spatial variation in thickness (3–274 m), which has been attributed to contemporary faulting and differential sedimentation. Fish and crocodile fossils are present in the deposits, which early studies suggested were ~13 to 14 Ma old (Bishop et al. 1971). Subsequently, the faunal list was expanded within a broader stratigraphic context (Lippard 1973; Pickford 1975b; Chapman et al. 1978; Hill et al. 1985, 1986; Hill 1987). Hill and Ward (1988) and Pickford (1988), for example, noted that the presence of *Diamantomys* implied an age closer to 16 Ma, which later received support from Hill et al. (1991) who obtained dates of 15.8 and 15.6 Ma. Behrensmeyer et al. (2002) reported additional ages of 15.63–13.37 Ma for a section at Muruyur and 15.58–15.36 Ma for a section at Kispsaramon (Fig. 17.5).

The Kipsaramon sequence includes three Members (Figs. 17.1a and 17.5; Behrensmeyer et al. 2002). Member 1 consists of lacustrine and subaerial deposits with up to 20 m of laminated green, red, and purple, illite-dominated claystone and siltstone, as well as less common sandstone and oolitic limestone. Palaeosols, pedogenically altered sediment, clay-clast conglomerate and Magadi-type chert are all present. Smectite is a common product of weathering of volcaniclastics in the Kenya Rift, which led to suggestions that the illite resulted from low-grade metamorphic alteration of pre-existing smectite or was formed in response to contemporaneous alteration of smectite in an evaporitic setting (Behrensmeyer et al. 2002). Member 2 is an ash-flow tuff interpreted as having entered a lake. Member 3 consists of >25 m of green claystone with limonitic siltstone and minor sandy or silty claystone. Weak pedogenesis and massive tuff, partially altered to analcime, is also present.

Fish bone, scales and coprolites are present in both Members 1 and 3, with ostracods abundant in Member 1 limestone (Behrensmeyer et al. 2002). Burrowing and bioturbation are minor except in a few limited horizons. Gastropods and bivalves are common near the base of Member 3 with *Pila ovata*, *Melanoides tuberculata*, *Bartoa nilotica* (?) and *Pleiodon* sp. near the top of Member 2. Elsewhere, *Lanistes*

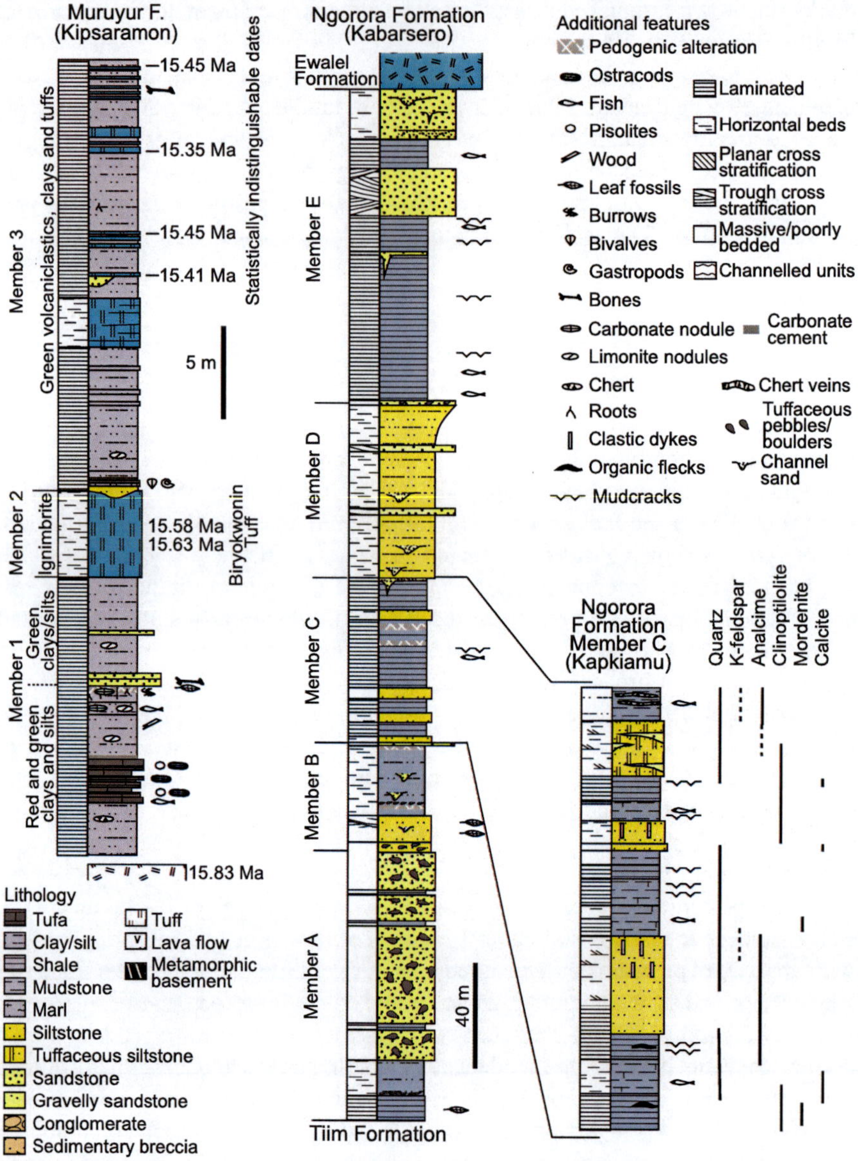

Fig. 17.5 Middle to Late Miocene stratigraphic sections. Muruyur Formation. After Behrensmeyer et al. (2002). Ngorora Formation type section at Kabarsero. After Bishop and Pickford (1975). Member C stratigraphic section (Kapkiamu) shows XRD data for authigenic minerals. After Renaut et al. (1999)

carinatus and *Gabiella* have been observed (Pickford 1988). Crocodile bones and coprolites are common in Member 1. Other fossils reported include silicified wood and poorly preserved leaves and stems, with microbial stromatolites in lake facies

(Pickford 1988). A diverse vertebrate fauna and traces (Fig. 17.3e, f) have been observed, including mammals, reptiles, birds and hominoid remains.

17.2.5 The Ngorora Formation

Geological Background and Stratigraphy

Pickford et al. (2009) dated the Ngorora Formation at ~12–10 Ma using faunal content with Hill (1995) and Hill et al. (2002) suggesting a range of ~13–8.5 Ma. The 450 m of volcaniclastic, colluvial, fluvial and lacustrine sediments rest on the Tiim Phonolite Formation (~13.5 Ma) and lie unconformably below the Ewalel Phonolites Formation (~8 Ma) (Martyn 1969; Bishop and Chapman 1970; Bishop et al. 1971; Chapman 1971; Lippard 1972; Bishop and Pickford 1975; Hill 1995). The rising Elgeyo Escarpment marked the western border of the sedimentary basin (Mugisha et al. 1997) with the eastern side formed by a fractured monocline that would eventually evolve into the Laikipia Escarpment. The sediments thin northwards to the former Tiati volcanic massif with the southern limit controlled by higher elevations on the rift floor (McClenaghan 1971; Pickford 1978a). Kaption volcano was active during early Ngorora times (Pickford 1975a) with downwarping focused along the foot of the Elgeyo Escarpment to the west (Lippard 1972).

The formation has been divided into five members (A–E) at the Kabarsero type section (Figs. 17.1a, 17.5, and 17.6a; Bishop and Chapman 1970). Member A includes basal lake shale and claystone. However, the palaeolake was subsequently filled by fluvial reworked volcaniclastics and lahars, derived mostly from Kaption to the south of the basin (Bishop and Pickford 1975; Pickford 1975a, 1978a). Members B and D are largely fluvial siltstone and sandstone with an abundant mammalian fauna in gritty channel deposits. In contrast, Members C and E are predominantly lacustrine, with sediments that were laid down in multiple complex basins reflecting the gradual rise of the Tugen Hills during deposition.

Member C accumulated in three lakes, with deposits cropping out at Kabarsero, Kapkiamu and near Kaption (Fig. 17.1a). The Kabarsero lake sediments (17.6a) were intermittently fresh and may also have been well-oxygenated, given the presence of potamid crabs (Pickford 1975a, 1978a). Mudcracks and palaeosols indicate periodic regressions. In contrast, the Kapkiamu lake was alkaline throughout most of its existence. Bishop and Pickford (1975) suggested that the Kapkiamu shale (Fig. 17.6b–f) was chemically precipitated and they also noted the presence of stunted *Tilapia grahami*, as well as algal and oolitic limestone with a petroliferous odour. Sediments near Waril formed in an alkaline lake that likely stretched to the Elgeyo Escarpment and included some of the earliest well-preserved cichlid fish (Van Couvering 1982; Rasmussen et al. 2017; Altner et al. 2017, 2020, Altner and Reichenbacher 2021). Diatomite, paper shale, conglomerate and tuff are present in parts of Member E, which developed in an expansive freshwater lake. Diatomite horizons at Kabarsero are dominated by archaic *Aulacoseira granulata, A.*

Fig. 17.6 Middle to Late Miocene sediments. (**a**–**f**) Ngorora Formation; (**g**–**j**) Lukeino Formation. White scale bar = 5 mm; black bars = 10 μm. (**a**) Kabarsero, showing Members C–E. (**b**) Ngorora Formation shale (Member C) with fish impression, Poi, Kapkiamu Basin. (**c**) Silicified sediments at top of a steep road section at Kapkiamu. (**d**) Laminated zeolitic shales, Kapkiamu. (**e**) Westward-dipping bedded shales overlain by colluvial gravel, Poi. Elgeyo Escarpment in distance. (**f**) Rippled (mainly wave) siltstone, Poi, Kapkiamu Basin. (**g**) Lukeino Formation laminated diatomite, north of Yatya. (**h**) Lukeino Formation clay-diatomite laminae, Member B at Kobuluk. (**i**) SEM image of Lukeino Formation diatomite comprised of partially comminuted *Aulacoseira*. (**j**) *Synedra* (*Ulnaria*)-dominated Lukeino Formation diatomite

granulata-agassizii and *A. agassizii* var. *malayensis* suggesting freshwater episodes and a deep lake (Owen 1981; Owen and Renaut 2000).

Palaeoenvironmental investigations include the work of Owen (1981), Kingston et al. (1994) and Renaut et al. (1999). However, most Ngorora research has concentrated on its rich mammalian fauna and rare fragmental hominoid discoveries (e.g. Hill et al. 1985, 2002; Hill and Ward 1988; Hill 1995, 1999; Pickford 2001; Winkler 2002; Morales and Pickford 2005; Leonard et al. 2006) with studies of catchment vegetation based on leaf assemblages (Jacobs and Kabuye 1987, 1989; Jacobs and Winkler 1992; Jacobs 2002) and pollen (Rasmussen et al. 2017). The latter include *Juniperus*, *Podocarpus*, *Acalypha*, *Croton* and herbaceous pollen of Poaceae and Asteraceae that were deposited in sediments at ~13.3–12 Ma, indicating wooded grasslands or grassy woodlands with nearby montane forests at that time. The chronological context for these and other studies is based mainly on $^{40}Ar/^{39}Ar$ dating, K/Ar techniques and palaeomagnetic studies (e.g. Chapman and Brook 1978; Hill et al. 1985; Tauxe et al. 1985; Deino et al. 1990).

Lacustrine Facies in the Kapkiamu Basin

The Ngorora Formation is ~200 m thick in the Kapkiamu Basin, near Kapkiamu and Poi (Fig. 17.1a; Renaut et al. 1999) where Member C lake deposits are ~35 m thick. This lacustrine sequence includes white to pale green, coarsely to finely laminated shale, mudstone, siltstone and fine sandstone (Fig. 17.6d, e), similar to the fine-grained deposits in the Tambach Formation. Wave and current ripples occur in coarser units (Fig. 17.6f). Mudcracks, clay intraclasts and clastic dykes are common. Pseudomorphs after halite and gaylussite (?) and raindrop imprints are also present. The deposits include fish (*Tilapia* and *Clarias* sp.?) (Rassmussen et al. 2017; Altner et al. 2017; Penk et al. 2019; Kevrekidis et al. 2019), abundant bird footprints, carbonaceous plant debris, leaf fossils, rootmarks and clay-rich palaeosols. Sediments in upper Member C have been silicified (Fig. 17.6c) and contain massive porcellaneous chert (10–40 cm thick). Cross-stratified and graded beds of fine to coarse sandstone, conglomerate and tuffaceous sandstone (<50 cm thick) accumulated in shallow ephemeral streams. Petroliferous dark greenish-grey shale (<4.3% Total Organic Carbon: TOC) is present high in Member C at Poi, along with poorly preserved stromatolites, oncoids, marl and limestone (Pickford 1975a).

Member C shale and mudstone include abundant zeolites (clinoptilolite, analcime, minor mordenite) and authigenic clay (smectite and illite) along with calcite, quartz, K-feldspar and opal CT (Fig. 17.6d; Renaut et al. 1999). Locally, the sediments form zeolitite with fluorite present locally in analcime-bearing mudstone. The mineralogy supports earlier interpretations of a saline alkaline palaeolake (Bishop and Chapman 1975; Pickford 1978a) that formed a playa during low stages, allowing gaylussite (?) and halite to develop along with mud intraclasts and mudcracks. Regressions led to palaeosol development with laminated sediments suggesting that climate cyclicity affected the palaeoenvironment. Shale with high TOC may reflect lake stratification and bottom anoxia with modern Lake Bogoria as a possible analogue (Renaut et al. 1999). Calcified stromatolites indicate periods when water was fresh enough to hold dissolved calcium.

Authigenic minerals provide possible palaeoenvironmental clues. Volcanic glass was likely altered to smectite in dilute waters near basin margins with calcite forming cement following near-surface degassing (Renaut et al. 1999). Glass and smectite were initially altered to clinoptilolite in areas with more saline water. Subsequently, further alteration took place as salinity and alkalinity increased, producing analcime and then authigenic K-feldspar. Silica in highly alkaline pore fluids probably contributed to silicification following reduced pH, perhaps related to dilution by fresh water or following organic matter decay (Renaut et al. 1999).

17.2.6 The Ngerngerwa (Ngeringerowa) Formation

These discontinuously exposed deposits crop out among a series of horsts and grabens (Martyn 1969) and occupy an area of about 170 km^2, 20 km north of Kabarnet (Fig. 17.7a, b). Martyn used the name "Ngeringerowa" for the main locality, which Pickford et al. (2009) noted was a misspelling of Ngerngerwa. Pickford also noted that the sediments had been correlated with Members D and E of the upper Ngorora Formation, but he suggested that they were younger because they contained common redbeds with equids and bovids that suggested an age of 10.5–10 Ma. Subsequently, Rossie et al. (2013) radiometrically dated the deposits to 9.5 and 8.8 Ma, overlapping with the upper Ngorora Formation.

Martyn (1969) first described the location with Pickford (1975a) and Benefit and Pickford (1986) subsequently recognising a diverse fauna that included abundant *Hipparion* and a primitive hippopotamid, *Kenyapotamus coryndoni*. The 140-m-thick fluviolacustrine succession is dominated by phonolite clasts with pumice suggesting contemporary volcanism. Lacustrine and fluviatile facies include, for example, coarse conglomerates and shales (Fig. 17.7c). Cross-bedding, ripple marks, cut-and-fill structures, and lensoidal strata are common with beach sediments and abundant *Melanoides tuberculata* present at one locality (Pickford et al. 2009). The fauna includes a wide variety of molluscs, fish, mammals and reptiles.

17.2.7 The Mpesida Beds

Tuffaceous sediments rest on and between lavas of the Kabarnet Trachyte Formation that crop out in several areas between the Tugen Hills and Lake Baringo. They are often difficult to distinguish due to discontinuous exposures and similar lithologies. For example, Martyn (1969) described 47 m of the tuffaceous Riwo Beds, noting that they rested unconformably on the Kabarnet Trachytes. In contrast, the Mpesida Beds are generally described as interbedded with or resting conformably on the same trachyte flows (Fig. 17.1d; Kingston et al. 2002).

The Mpesida Beds were initially dated at ~7.5–6.9 Ma (Baker et al. 1971; Chapman and Brook 1978; Hill et al. 1985; Hill 1987). However, later studies

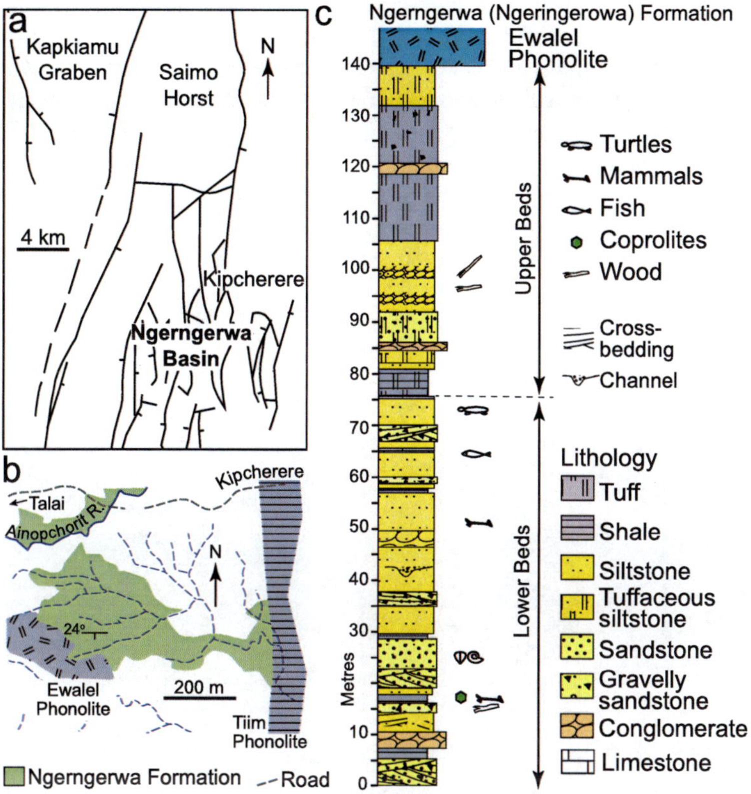

Fig. 17.7 Late Miocene Ngerngerwa (Ngeringerowa) Formation, Tugen Hills. Simplified from Pickford et al. (2009). (**a**) Location and tectonic setting northeast of Kabarnet. (**b**) Geological sketch map of the Ngerngerwa Basin. (**c**) Type section of the Ngerngerwa Formation

indicated ages of 6.36–6.2 Ma for trachyte flows overlying the Mpesida Beds near Yatya (Kingston et al. 2002), as well as 7.2 and 6.7 Ma for a basal trachyte and an overlying reworked tuff near Rurmoch (Fig. 17.1a; Jacobs and Deino 1996). Sediments at the latter location were described as epiclastic and pyroclastic (Chapman 1971) with similar descriptions subsequently applied to all volcanics and sediments within, or conformably overlying, the Kabarnet Trachytes (Kingston et al. 2002). Researchers have also noted geochemical similarities between the Mpesida Beds and the older sedimentary units in the Tugen Hills, with younger sediments reflecting a shift in volcanic geochemistry (Pickford 1975a; Williams and Chapman 1986; Owen et al. 2018).

Kingston et al. (2002) described Mpesida lacustrine facies southwest of Rurmoch (Fig. 17.1a), dated at 6.36 Ma, with an overlying trachyte date of 6.37 Ma. Moderate- to well-bedded pyroclastic and epiclastic ash is present along the eastern basin margin (Section A, Fig. 17.8) and stratigraphically above a massive ash. Pyroclastic flow deposits with silicified wood (De Franceschi et al. 2016) form the base of

Section B (Fig. 17.8), which includes lacustrine massive and well-bedded clay, silt, fine-grained sandstone and diatomite (>2 m thick) laid down in the northwestern Rurmoch Basin. Diatoms indicate a substantial lake system (Edgar, in Kingston et al. 2002). Localised silicification of diatomite has also produced porcellanites. Isotopic studies on carbonate in palaeosols and vertebrate tooth enamel indicate a significant C4 plant community with a minor contribution from C3 plants (Kingston et al. 1994; Kingston 1999). Mpesida lake sediments located outside the Rurmoch Basin contain large *Tilapia* and *Clarias* fish, *Melanoides tuberculata*, varied vertebrates (Pickford 1975a; Hill et al. 1986) and plant fossils (Jacobs and Deino 1996).

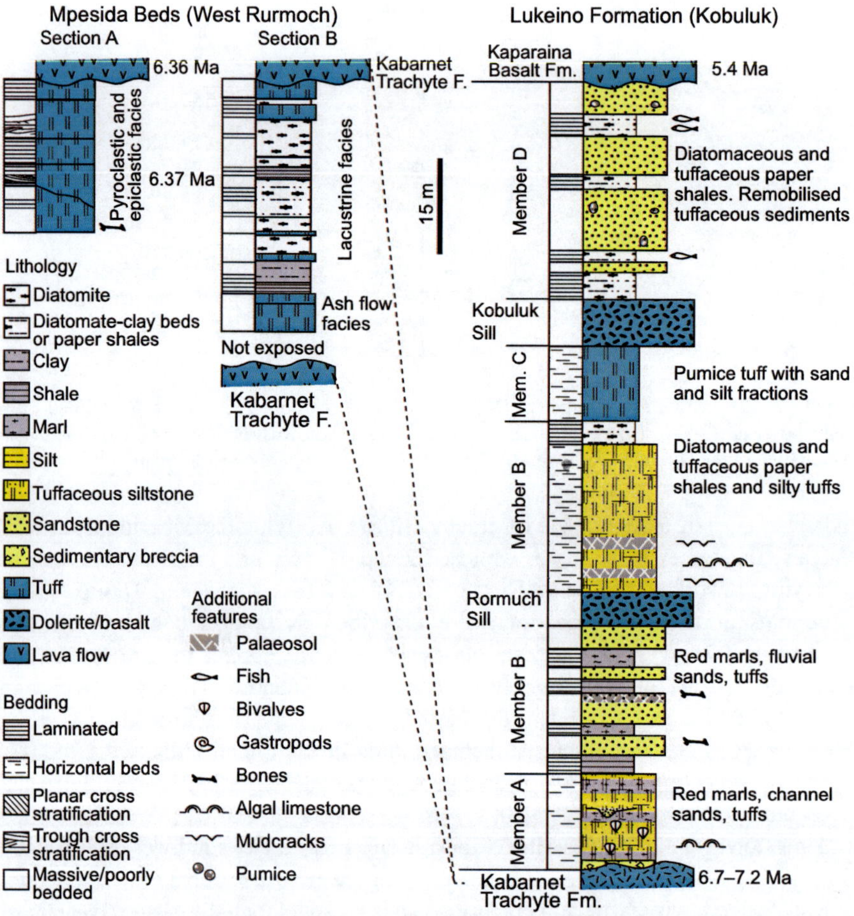

Fig. 17.8 Late Miocene stratigraphic sections. Lithology and stratigraphy of the Mpesida Beds. After Kingston et al. (2002) and the overlying Lukeino Formation. After Pickford (1975a, b)

17.2.8 The Lukeino Formation

Basin Stratigraphy and Facies

Late Miocene sediments at the base of the Kaparaina Basalts Formation were originally assigned to the Lukeino Member (Chapman 1971; Bishop et al. 1971). However, Pickford (1975a, 1978b) raised these deposits to full formation status, noting that they rested unconformably on the Kabarnet Trachyte Formation and below the Kaparaina Basalt Formation. The former has been dated at 7.3, 6.8 and 6.7 Ma with ages for the latter ranging between 8.4 and 4.0 Ma (Bishop et al. 1971; Pickford 1978b; Chapman and Brook 1984; Dagley et al. 1978). Later geochronology narrowed the ages to 5.9–5.1 Ma with a median of 5.6 Ma (Hill et al. 1985). Palaeomagnetic studies further constrained the age of the upper Lukeino Formation, near Kapcherberek, to 5.88–5.72 Ma (Deino et al. 2002) with the same researchers noting an $^{40}Ar/^{39}Ar$ pumice date of 5.73 Ma for the upper part of the sequence and a date of 5.72 Ma for the overlying Kaparaina Basalt Formation. Sawada et al. (2002) suggested ages of 6.0–5.7 Ma and distinguished three members at Kapsomin.

Pickford (1975b) designated a 130-m type section at Kobuluk (Fig. 17.8) that includes primary and reworked sediment intruded by two sills and divided into four members (A–D). Member A includes red marl and pale orange/yellow tuffaceous horizons together with sand-filled, fossiliferous (including a hominid molar) stream channels. The presence of stromatolites and bivalves indicates that palaeostreams entered a lake marginal setting. Member B includes lacustrine diatomaceous and silty tuff, laminated shale and paper shale comprised of diatomite-clay couplets (Fig. 17.6g, h) with minor palaeosols, mudcracks and stromatolites. Member B displays facies change southwards where red clay is present. Member C includes well-bedded lacustrine tuff with freshwater molluscs, giving way southwards to poorly bedded fluvial pumice tuff. Member D is dominated by diatomaceous and silty tuff with minor pumice. Fish impressions are abundant locally.

Dericquebourg et al. (2019) report replacement of bones and formation of carbonate-hydroxyapatite nodules in shallow lake and wetland sediments. Phosphatic horizons in sandy clay at Sunbarua and Kapcheberek have been interpreted as forming during cyclic, and large-scale, inputs of phosphorus to the palaeolake (Dericquebourg et al. 2015). These eutrophication phases were associated with intense rainfall and runoff events that introduced organic matter to the lake and which also stimulated diatom blooms.

Flora and Fauna

The Lukeino Formation includes a diverse fauna (Pickford 1975a, b, 1978b) and has yielded fragments of an early biped named *Orrorin tugenensis* (Pickford et al. 1983; Senut et al. 2001; Pickford and Senut 2001a, b; Mein and Pickford 2006). Reptiles and mammals are mainly present in Members A and B, and include crocodile, Chelonia, bird fossils and varied mammals (e.g. Rodentia, Canidae, Hyenidae, Artiodactyla, Hippopotamidae and Hominidae) mostly associated with lake marginal and stream facies. Lacustrine fossils include bivalves (*Melanoides tuberculata*, two gastropod spp., two bivalve spp.), fish (Cyprinidae, Cichlidae, Clariidae), ostracods and algal limestone. Altner and Reichenbacher (2015) report the presence of cyprinodontiform fossils, including a new species (*Kenyaichthys kipkechi*). They note a scarcity of typical freshwater fish such as cyprinids in their collections and suggest that seasonal aridity might have prevented the widespread establishment of other fish species.

Thirty-eight dicotyledonous leaf morphotypes have been recognized from the upper Lukeino Formation (Bamford et al. 2013). These were interpreted as having grown in a deciduous forest or woodland, with a wet summer. However, $\delta^{13}C$ isotope research on herbivore enamel suggests a significant C4 grass presence and may reflect a more open savanna ecosystem (Roche et al. 2013).

Although scarce through most of the formation, two distinct diatom assemblages have been identified from Members B and D (Fig. 17.6i, j; Owen 1981; Owen and Renaut 2000). Assemblage 1 is dominated by *Aulacoseira granulata* with less common *A. granulata* var. *angustissima*, *A. granulata* var. *muzzanensis*, *Nitzschia frustulum*, *Navicula cryptocephala* and *Synedra* (*Ulnaria*) *ulna*. Assemblage 2 is dominated by *Synedra* (*Ulnaria*) *acus and S. ulna* with common *A. granulata*, *Staurosira construens* and *N. frustulum*. White diatomite laminae are dominated by *A. granulata*, with darker clayey diatomaceous laminae including higher percentages of *Synedra* (*Ulnaria*) suggesting seasonal blooms and a fresh, stratified and deep palaeolake. Changes between these floras may reflect variability in Si/P ratios and (or) turbulence in the palaeolake, and perhaps indicate nutrient consumption following periodic *Aulacoseira* blooms (Owen and Renaut 2000).

17.3 Miocene Lacustrine Palaeogeography and Palaeoenvironments

Shallow-water fluvial settings and minor lakes developed below what is now the Laikipia area, east of Lake Baringo during the Miocene. To the west, in the modern south Kerio Valley, a possible Kimwarer Formation palaeolake was probably small and may have lain on the line of one of several rivers inferred to have been draining north towards the Lokichar and north Kerio Basins (Tiercelin et al. 2012; Chap. 15). In contrast, the Tambach palaeolake was much larger and developed in a thermally induced sag-basin after an initial phase of Palaeogene rift faulting. At its maximum size the palaeolake extended for many kilometres to the west of the modern Elgeyo Escarpment and eastwards beyond the present day Tugen Hills (Fig. 17.9a). Lake levels and water chemistry varied during deposition (Renaut et al. 1999). A thriving fish population may have developed during fresher intervals. However, many of the fish could have been washed in from rivers. Ostracods suggest fresh to slightly brackish water. Proximal settings produced thicker beds with deeper waters accumulating finer-grained laminated sediments. Euhedral analcime indicates saline and highly alkaline phases (Renaut 1993), which likely reflect prolonged periods of evaporative concentration, low lake level and a closed-basin. The resulting highly alkaline lake and pore water would have readily dissolved diatoms. Outflow during fresher intervals was likely towards the north (Tiercelin et al. 2012).

Early research placed the Muruyur Beds (Tugen Hills) within the upper Tiim Phonolites, which were broadly correlated with the Uasin Gishu Phonolite Formation of the Elgeyo Escarpment. This implied that the Muruyur Beds were younger than the Tambach lake sediments (Fig. 17.1d; Chapman and Brook 1978; Hill et al. 1985; Deino et al. 1990). However, Behrensmeyer et al. (2002) indicate an older age (15.98–13.37 Ma), which suggests that the deposits may partially overlap with those of the Tambach Formation. The Muruyur palaeolake had a surface area of about 120 km^2 and was partially blocked by Tiati volcano, 45 km to the north (Fig. 17.9b; Pickford 1988) and formed after a period of erosion of previously

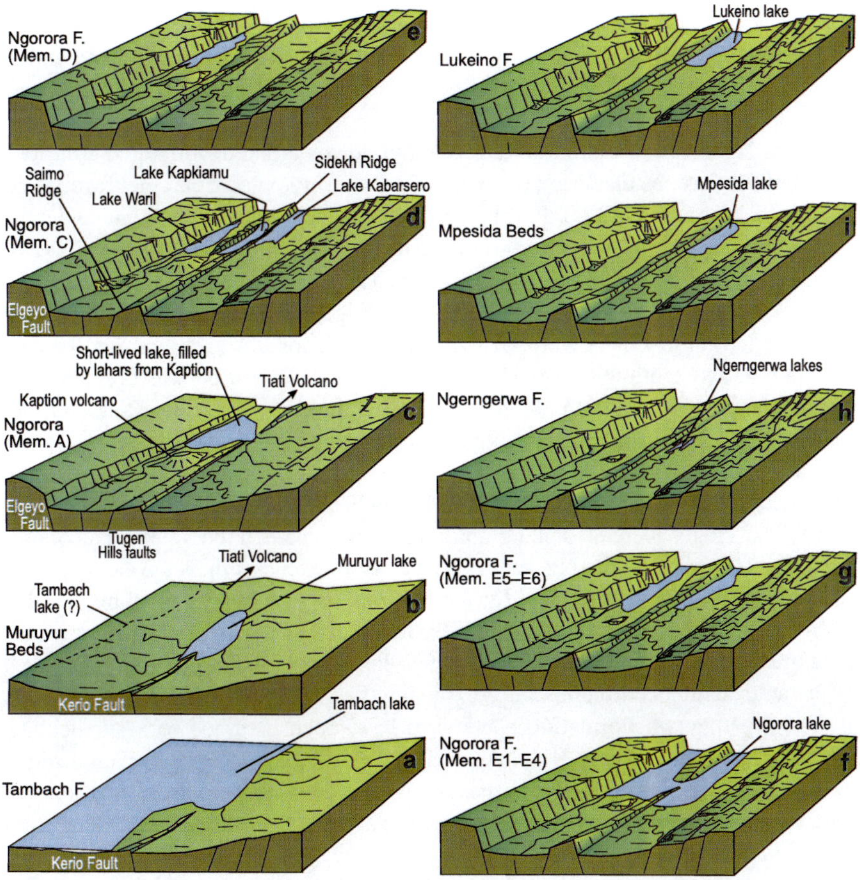

Fig. 17.9 Simplified models for selected stages of major Miocene lakes. Based on data in Bishop and Pickford (1975), Pickford (1975a, b, 1978a, b), Tiercelin and Vincens (1987), Renaut et al. (1999) and Behrensmeyer et al. (2002). (**a**) Tambach Fm.; (**b**) Muruyur Fm.; (**c–g**) Ngorora Fm.; (**h**) Ngerngerwa Fm.; (**i**) = Mpesida Beds; (**j**) Lukeino Fm.. See text for discussion

weathered phonolites (Behrensmeyer et al. 2002). Conditions fluctuated between aquatic (limestone, laminated green clay) and subaerial (red clay, palaeosols) during Member 1 times (Behrensmeyer et al. 2002). Beds with Magadi-type chert indicate very alkaline intervals. Locally, rivers introduced coarser clastics (sand, mud-clast conglomerate, a bonebed) to the palaeolake, which expanded and deepened towards the end of Member 1 times when anoxic bottom conditions restricted benthic fauna. Lacustrine tuff constitutes Member 2. During early Member 3 times gastropods and bivalves thrived in swamps or a shallow and partly oxygenated lake.

The Ngorora Formation sediments were initially laid down on the relatively subdued upper surface of the Tiim Phonolites Formation (Bishop and Pickford 1975; Pickford 1975a, 1978a; Tiercelin and Vincens 1987). The rising Elgeyo Escarpment formed high ground to the west with the northern limits marked by the flanks of Tiati volcano. North-south trending fault-blocks represent the early development of the Tugen Hills. A lake developed early in Member A times in western parts of the basin (Fig. 17.9c) but was filled by lahars from an active Kaption volcano and by fluvial sediments from other parts of the developing rift. Fluvial conditions dominated during deposition of Member B.

Three Ngorora Formation lakes formed in downwarped/downfaulted areas during Member C times that dried out many times to form palaeosols (Pickford 1975a, 1978a; Pickford et al. 2009). Freshwater Lake Kabarsero initially extended beyond the Sidekh Ridge but was later confined to the east of this feature (Fig. 17.9d). Saline alkaline Lake Kapkiamu was confined to a graben in the central rift with Lake Waril located further to the west. The Elgeyo Escarpment, as well as the Sidekh and Saimo ridges were all rising during deposition. Kaption had become extinct and was eroding by Member D times, when deposition was dominated by rivers. Lacustrine conditions were confined to the west with the Kapkiamu and Kabarsero lakes ceasing to exist (Fig. 17.9e).

Member E of the Ngorora Formation was laid down during a period of rapid uplift along the Elgeyo Escarpment (Pickford 1975a, 1978a). Although the Saimo and Sidekh ridges were also being uplifted, a gap remained that allowed a continuous large freshwater Lake Ngorora (Fig. 17.9f) to extend both to the east and west of what now forms the Tugen Hills. An inactive Kaption Volcano had by this time been much reduced in size by erosion. By late Member E times, Lake Ngorora had split into two freshwater bodies (Fig. 17.9g) due to sustained uplift along the Tugen Hills faults, with a continuous ridge extending from Saimo to Sidekh.

The Ngerngerwa Formation appears to have been confined to a smaller area between horst blocks near Kipcherere (Fig. 17.9h). Rapid facies variation reflect accumulation in narrow grabens that may have been isolated from one another. Outcrop distributions suggest a lake or lakes with a north-south length of 17 km that extended east-west for about 10 km (Pickford et al. 2009).

The palaeogeography of the Mpesida Beds is difficult to determine at a regional scale although it probably covered much of the Lukeino Block (Fig. 17.9i). Pickford (1975a) reported that the basin was bounded to the west by the Tugen Hills with a southern limit at about 0°45'N and the northern margin at around 1°10'N. The eastern margin is unknown. Within this area, diatoms, fish and *Melanoides* accumulated in a freshwater lake, or series of mainly fresh lakes, with Pickford (1975a) noting that the local presence of gypsum rosettes might indicate more saline conditions. However, the gypsum could also be a product of sulfide oxidation since it is not normally compatible with the saline alkaline lakes that dominate the region.

The lavas of the Kabarnet Trachyte Formation covered much of the area east of the Tugen Hills, reducing topographic relief in the Lukeino area; the lava surfaces then become weathered (Pickford 1975b, 1978b). However, continued downthrow to the east of the Tugen Hills produced a new basin in which fluvial and lake sediments of Member A of the Lukeino Formation accumulated. Fossil wood suggests the presence of forests. During Member B times a major transgression expanded the freshwater lake to its maximum extent (Fig. 17.9j) with lake marginal and floodplain settings to the south. Member C eruptions introduced ash to the marginal plains and lake.

Subsequently (Member D), diatomaceous sedimentation was re-established in a large freshwater lake. Igneous activity led to intrusion of sills into the Lukeino Formation and the growth of Kobuluk volcano as an island or headland that extended into the palaeolake. Faulting then uplifted and isolated the Lukeino block, which started to erode, but with deposition possibly continuing to the east. Miocene lake sedimentation was finally terminated with the eruption of lavas of the Kaparaina Basalt Formation in the south and the Ribkwo Trachytes to the north.

17.4 Plio-Pleistocene Sediments and Palaeolakes

17.4.1 Plio-Pleistocene Sediments and the Toluk Beds

Several rivers and location names associated with Plio-Pleistocene sequences have changed their names over the years (Namwamba 1993). For example, the Chemeron (or Chemorun) River (source name for the Chemeron Formation) was also referred to as the Nasagum in its upper reaches by McCall et al. (1967) and is locally known as the Losegem, or "place of bees" in Tugen. Similarly, the Kapthurin River (source name for the Kapthurin Formation) has been referred to as the Kathiorin, Katiorin or Kapthiorin.

The Pliocene and Pleistocene deposits (Fig. 17.1d) formed in basins that lay to the east of the proto-Tugen Hills. They include the relatively thin Early Pliocene Toluk Beds and the much thicker Plio-Pleistocene Chemeron (Kaperyon) Formation, which accumulated in several discrete lake basins, as well as the mainly fluvial Pleistocene Kapthurin Formation. The Toluk Beds lie between lavas in the Kaparaina Basalt Formation, which is variously dated at 5.3–4 Ma, with the Toluk sediments likely accumulating during the latter part of this period (Pickford 1975a). They consist of yellow lapilli tuffs with plant remains, fossil wood and mammals. One 3-m-thick section near Pelion (Fig. 17.10a) includes diatoms dominated by *Staurosira construens* and its variety *venter*, together with *Pseudostaurosira*

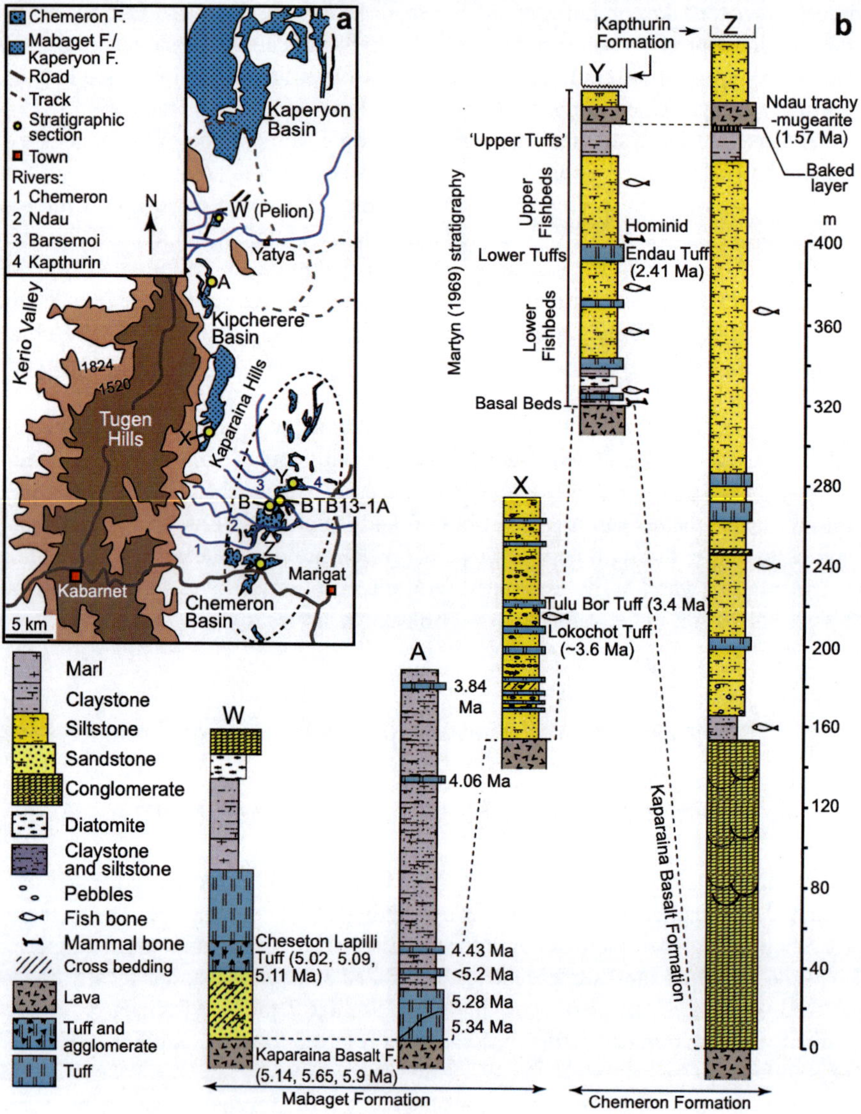

Fig. 17.10 Pliocene sediments of the Tugen Hills. (**a**) Distribution of the Chemeron Formation and regional topography. Sections B and BTB13-1A in (**a**) shown in Fig. 17.13. Map modified from Kingston et al. (2007). (**b**) Representative stratigraphic sections from the three main Pliocene basins. Sections W and Z after Namwamba (1993). Section A after Deino et al. (2002). Lokochot and Tulu Bor tuff dates in Section X modified using later ages reported by Brown and McDougall (2011). Cheseton Lapilli tuff dates from Pickford et al. (1983); Ndau Trachymugearite, Kaparaina Basalt and Endau Tuff dates from Hill et al. (1985)

bevistriata, *Synedra* (*Ulnaria*) *ulna*, *Aulacoseira granulata* and *Cocconeis placentula* (Owen 1981) — an assemblage that suggests shallow fresh water.

17.4.2 The Chemeron and Mabaget Formations

Sediments in the Chemeron, Kipcherere and Kaperyon basins have been variously dated between 5.3 and 1.6 Ma (Kingston et al. 2007; Wilson et al. 2014; Deino et al. 2021). The deposits rest disconformably on the Kaparaina Basalt Formation and are overlain unconformably by the middle Pleistocene Kapthurin Formation (Fig. 17.10b; Pickford 1975a, b; Chapman et al. 1978; Pickford et al. 1983; Hill et al. 1986; Williams and Chapman 1986). Lacustrine sediments form a significant part of the succession, which includes fluvial and lacustrine clay, silt, sand and conglomerate, reworked ash and airfall tuff, with common and locally very pure diatomite. Terrestrial sediments are also present that yield an important mammalian record (Goble 2022).

Unfortunately, the Plio-Pleistocene stratigraphic terminology has changed several times reflecting differing concepts of sedimentation in the area. For example, the Chemeron and Kapthurin beds were raised to formation status based on studies in the Chemeron and Kipcherere basins (Fig. 17.10a). Martyn (1967, 1969) divided the newly-defined Chemeron Formation into five members designated: A (Earths and gravels), B (tuffs silts and diatomites), C (pumiceous tuffs), D (silts and gravels) and E (stratified tuffs). However, only Members A, B and D could be recognized in the Kipcherere Basin.

The Plio-Pleistocene Chemeron Formation was once also thought to include deposits to the west and north of the Kaparaina Hills in the Kipcherere and Kaperyon Basins (Fig. 17.10a; McClenaghan 1971; Chapman 1971; Pickford 1975a, b; Chapman and Brook 1978; Chapman et al. 1978; Pickford et al. 1983, 2009; Hill et al. 1986) where they include fluvial epiclastic tuffs and diatomaceous silts (Fig. 17.11a–d) with lacustrine deposits more common in southwestern outcrops. More recent tuff correlations have since established links between the main Kipcherere Basin outcrops and exposures to the north, near Pelion (Deino et al. 2002). Pickford et al. (2004) and Roche et al. (2013) suggested that earlier Pliocene sediments west and north of the Kaparaina Hills should be renamed the Mabaget Formation with the type area around Mabaget with two good sections located at Pelion (154 m thick) and at Mabaget itself.

The Pelion section (Section 'W' in Fig. 17.10b) accumulated near the base of a major fault scarp (Pickford 1975a, b) with the lowermost sediments comprised of poorly fossiliferous fluvial and lake marginal sand. Early Pliocene volcanism produced the Cheseton Lapilli Tuff (~5.1 Ma), which locally fell into water and incorporated freshwater *Melanoides tuberculata* (Pickford et al. 1983), with reworking of this ash in the catchment generating overlying tuffaceous units. Subsequently, lake marginal grey marl was deposited followed by lacustrine red marl, diatomite and sandy tuff. The diatomites are dominated by *Aulacoseira granulata* together

Fig. 17.11 Representative sediments of the Chemeron Formation, Kaperyon Basin, south of Chemolingot (Fig. 17.1a). (**a**) Pliocene tuffaceous and diatomaceous silt about 2 km south of Chemolingot. The hill is ~10 m high. (**b**) Reworked fluvial tuffaceous silt in Cheptopokwa River, 1 km south of Chemolingot. Section is about 4 m high. (**c, d**) Detailed images of Cheptopokwa deposits. (**c**) Reworked tuffaceous silt with 12 cm cobble. (**d**) Cross-bedded tuffaceous silt with volcanic pebbles

with two less common floras (Owen 1981; Owen and Renaut 2000). The first includes *Stephanodiscus astraea* and *S. minutulus*, with minor *S. hantzschii* var. *pusilla*. The second is more diverse (*A. ambigua*, *Synedra* (*Ulnaria*) *ulna*, *S. ulna* var. *aequalis*, *Synedra rumpens*, *Cocconeis placentula*, *Epithemia argus*, *Staurosirella pinnata* and *Stephanodiscus astraea*), but is also heavily dominated by *A. granulata*. Fish impressions are common with interbedded limestone yielding gastropods and ostracods. Bivalves are common in red marl. Hippo, crocodile and turtle bones are present in the lake sediments with lake marginal facies yielding mammals and reptiles.

Sediments at Tabarin, an important hominid site (Hill 1985; Ward and Hill 1987; Pickford et al. 2022) 10 km south of Pelion (Section A, Fig. 17.10b; Deino et al. 2002) include a basal ignimbrite and an air-fall tuff, possibly equivalent to the Cheseton Lapilli Tuff, which has been dated at 4.96, 5.25 Ma and 5.31 Ma (Hill et al. 1985). Tuffaceous mud, silt and sand with rare limestone and diatomite rest disconformably on the ignimbrite with a time gap of possibly 0.7 Ma (Deino et al. 2002). Tuff horizons within the sequence have been $^{40}Ar/^{39}Ar$ dated between 4.43

and 3.84 Ma. Fossils are abundant and varied, including animals associated with lakes and lake margins, as well as terrestrial taxa (Hill 1985).

The Kipcherere Basin preserves 120 m of alternating clay, silt and gravel conglomerate that are partly cross-bedded (Section X, Fig. 17.10b; Namwamba 1993). Fish bone is present in silt and conglomerate above an ash that has been correlated with the Lokochot Tuff in the Lake Turkana area (Chap. 15). This is succeeded by deltaic silt and gravel above a second ash, the Tulu Bor Tuff, that is also found at Lake Turkana.

The Chemeron Basin (Sections Y and Z, Fig. 17.10b) includes sediments that are recognised by all previous researchers as constituting the Chemeron Formation in the strict sense. They are generally younger (~3.2 to 1.6 Ma) (Hill et al. 1992; Deino and Hill 2002; Kingston et al. 2007; Wilson et al. 2014) with deposits that dip east at 20° and rest on, or are in fault contact with, the Kaparaina Basalt Formation, variously dated at 5.9–5.1 Ma (Chapman and Brook 1978; Hill et al. 1985, 1986; Hill 1999; Deino et al. 2002). The uppermost 3–5 m of the formation, in this area, lies above the 1.57 Ma Ndau Trachymugearite (Figs. 17.10b and 17.12b) with the upper boundary formed by an angular unconformity overlain by the Kapthurin Formation (Fig. 17.12a).

The Chemeron Basin sediments are dominated by fine-grained terrestrial and lacustrine mud, silt and massive to bedded diatomite with interbedded conglomerate and primary or reworked tuff (Fig. 17.12a–c), which may be ripple marked or laminated (Deino et al. 2006; Kingston et al. 2007). Pure diatomite beds (Fig. 17.12d), 3–7 m thick, tend to thicken eastwards, indicating a basin centre in that direction. Separation of *Aulacoseira* and *Stephanodiscus* in fine laminae indicates possible seasonal silica depletion and lake stratification (cf. Owen and Crossley 1992). The diatomites are bracketed above and below by fine sand and fish-bearing silt and clay (Fig. 17.12g) that grade sharply into high-energy fluvial deposits (Fig. 17.12e, f) suggesting rapid transgressions and regressions (Kingston et al. 2007). Palaeosols are common.

The Hominin Sites and Paleolakes Drilling Project (HSPDP) (Cohen et al. 2016, 2022; Campisano et al. 2017) recovered a 230-m-long core through the succession. Core analyses showed that quartz-rich intervals and diatomaceous sequences are distinct from zeolitic zones, indicating variable salinities during deposition (Muriithi 2012; Minkara 2017; Minkara et al. 2016). The interplay between climate, as documented by fluvial-lacustrine transitions, and mammalian faunal shifts have been discussed by Goble (2011). The sediments are dominated by euplanktonic diatoms, mainly *Aulacoseira granulata*, but also *A. agassizii* and *Stephanodiscus spp.*, including *S. hantzschii* and *S. cf. subtransylvanicus* var. *minutula* with less common (<8%) *A. granulata* var. *angustissima*, *A. granulata* f. *curvata*, *Cyclotella meneghiniana* (*Stephanocyclus meneghinianus*) and rare shallow water species (Fig. 17.13a; Westover et al. 2021). The flora is typical of a large freshwater lake (Owen 1981). Long-chained *Aulacoseira* frustules are common, suggesting low-energy deposition. Diatom dissolution results in relative increases in resistant diatom 'ringleistes' compared to *Aulacoseira* frustules. This tends to occur at the top and bottom of diatomites, suggesting increased alkalinity and dissolution during transgressive and

Fig. 17.12 Pliocene deposits in the Chemeron Basin. (**a**) Fluvial erosion-surface cut into Chemeron Formation with overlying Kapthurin Formation conglomerate. Cobbles are up to 20 cm. (**b**) Ndau Trachymugearite overlying Chemeron deposits. (**c**) Bedded silt, diatomaceous silt and diatomite. (**d**) SEM image of partially fragmented *Aulacoseira granulata* in diatomite. Frustules are up to 15 µm long. (**e**) Interbedded diatomite and silt. Hammer is ~30 cm long. (**f**) Ripple laminated fine sand. Scale is in cm. (**g**) Fossilised fish in diatomaceous silty clay. Black bar = 1 cm

regressive phases (Westover et al. 2021). The presence of authigenic analcime in siliciclastic intervals between diatomites indicates greater alkalinities during deposition of the clastic units.

Five main diatomite beds were laid down between ~2.7 and 2.55 Ma (Fig. 17.13b, c; Kingston 2007; Kingston et al. 2007; Ortiz et al. 2015; Wilson 2011; Deino et al. 2021; Westover et al. 2021; Sier et al. 2021), a period when humid phases have been inferred for other East African basins (Trauth et al. 2005, 2007, 2009, 2010; Maslin et al. 2015). Variation in the *Aulacoseira*/*Stephanodiscus* ratio in each diatomite (Fig. 17.13b) suggests the influence of precession forcing with 23 kyr cycles centred

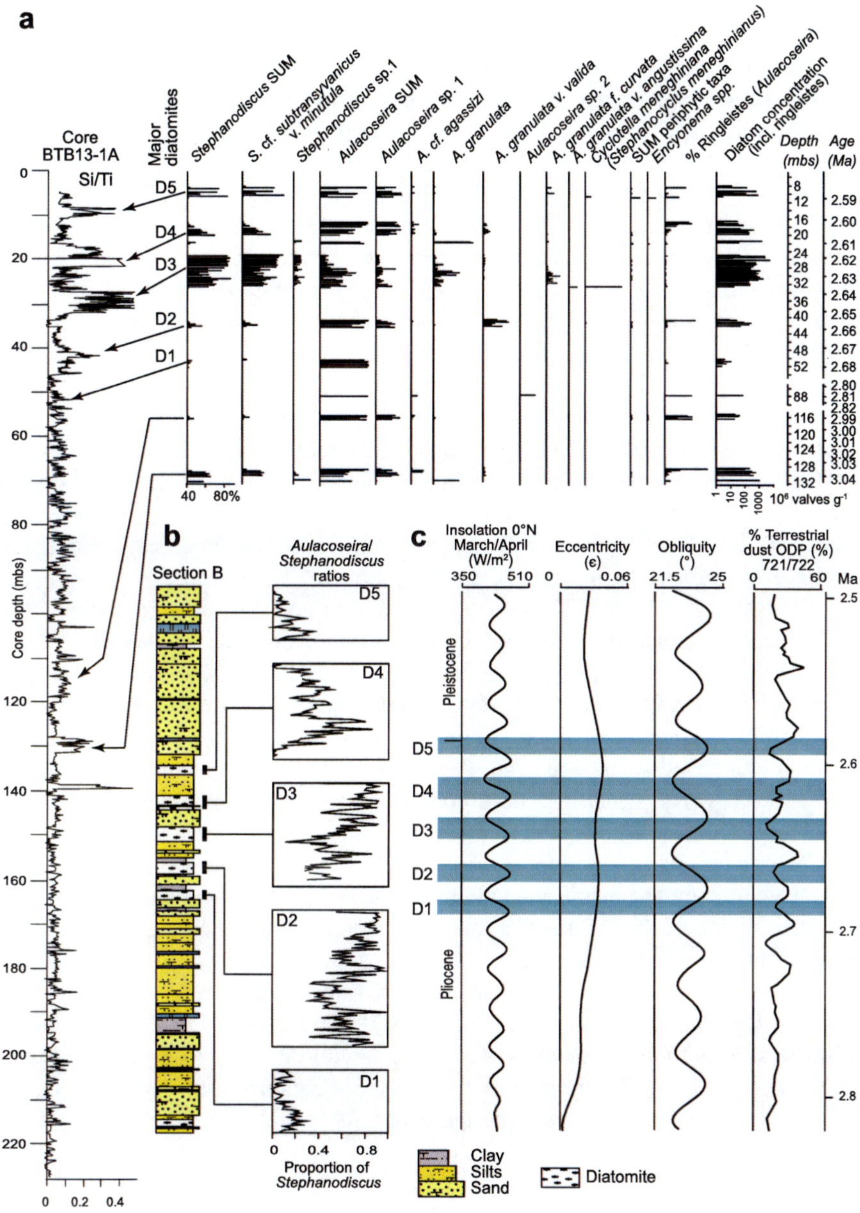

Fig. 17.13 Diatomite beds (D1–D5) and Late Pliocene precession cyclicity in the Chemeron Basin. (**a**) After Westover et al. (2021); (**b**, **c**) After Kingston et al. (2007). Sections B and BTB13-1A are located in Fig. 17.10a. (**a**) Si/Ti ratios, diatom floras and presence of diatom 'ringleistes' in BTB13-1A. Diatoms are scarce or absent, except in diatomites marked by high Si/Ti ratios. (**b**) Outcrop data show variability in the proportion of *Stephanodiscus* in the five major Plio-Pleistocene diatomites (D1–5). Relationship between each diatomite and insolation, eccentricity, obliquity and dust records shown in (**c**). Note how deep-water diatomites correlate well with low dust levels (moist phases) in the correlated Ocean Drilling Project (ODP) core

on a 400 kyr eccentricity maximum (Kingston et al. 2007). Contemporary low aeolian dust inputs in Indian Ocean cores indicate moister periods that correlate well with the individual Chemeron Formation diatomite beds.

Yost (2019) and Yost et al. (2021) noted a lack of pollen but reported charcoal and phytolith data from the HSPDP core. Charcoal peaks were observed about 5000 years before and after insolation peaks. Phytoliths varied in preservation quality and suggest that before 3.2 Ma a woody cover fluctuated with precessional periodicities between savanna (<40%), woodland (40–80%), and forest (>80%) settings. Charcoal abundances have also been used to infer that between 3.26 and 3.01 Ma fire caused a regional change from wooded to open habitats. From ~3.1 Ma, low-elevation forests covered <40% of the landscape with mesic C4 tall-grass peaking in abundance during insolation maxima, and xeric C4 short-grass increasing during insolation minima. Lupien (2019) and Lupien et al. (2021) also noted an opening of habitats using leaf-wax biomarker data ($\delta^{13}C_{wax}$). Between ~3.3 and 3.04 Ma, C3 plants averaged ~53% of the vegetation, which subsequently declined to ~41%, perhaps reflecting either tectonically driven change or part of a larger-scale drying trend. In addition, high amplitude variability in the $\delta^{13}C_{wax}$ record (C3 plants = ~80–0%) was correlated with facies-inferred lake level oscillations and orbital-scale variability in rainfall data derived from $\delta^{2}H_{wax}$ records.

Billingsley (2019) and Billingsley et al. (2021) analysed organic carbon from fish bone, teeth and scales in diatomites in the HSPDP core, which yielded $\delta^{13}C_{VPDB}$ values of −20 to −27‰, suggesting deep-water lake settings. The lack of shallow-water fish likely reflected rapid flooding and drying. In contrast, oxygen isotopes in one diatomite suggested that the transgressions were rapid and were followed by more gradual declines in lake level. It was also noted that there was a cyclicity of ~1400–1700 years, which has been compared with Late Quaternary Dansgaard-Oeschger intervals (Wilson et al. 2014).

Chemeron Formation sedimentation continued into the Lower Pleistocene, after which tectonic movements resulted in moderate tilting and fluvial erosion with termination of deposition before the accumulation of the Kapthurin Formation. Further to the south, at Kilombe near the headwaters of the Molo River (Fig. 8.2), a small caldera lake developed. The sediment sequence ranges between 2.6 and 1.6 Ma and includes Oldowan stone tools (Hoare et al. 2021; Gowlett et al. 2022). Basal lacustrine deposits at the main outcrop site contain phytoliths and diatoms and are dated at about 1.8–1.7 Ma. The neutral to alkaline lake contained macrophytes in its littoral zone and is thought to have existed both prior to and after these dates.

17.4.3 The Kapthurin Formation

The Kapthurin Formation includes 150 km^2 of alluvial fan to littoral plain, colluvial, lacustrine and volcaniclastic sediments (Fuchs 1950; McCall et al. 1967, Martyn 1969; Walsh 1969; Tallon 1976, 1978; Renaut et al. 1999, 2000) that are bounded to the north, south and west by volcanic rocks (Fig. 17.14). To the east, the basin is downfaulted below the Holocene and Late Pleistocene Loboi Silts (Farrand et al. 1976). The Kapthurin sediments are up to 150 m thick and were deposited after about 600–800 ka by rejuvenated ephemeral and perennial rivers, following uplift of the Tugen Hills (Tallon 1976, 1978; Deino and McBrearty 2002).

The deposits rest unconformably on the Plio-Pleistocene Chemeron Formation (Fig. 17.15a) and the whole succession is normally magnetized (Dagley et al. 1978), indicating a maximum age of 775 ka (Johnson and McBrearty 2012). Martyn (1969) subdivided the formation into five units (Fig. 17.16a), including the Lower (K1), Middle (K3) and Upper (K5) Silts and Gravels, which are separated by the Pumice Tuffs (K2) and the Bedded Tuffs (K4), with subsequent modifications added by Tallon (1976, 1978). Dates of 0.61 Ma have been reported for the Lower Kasurein Basalt, 0.54 Ma for the Pumice Tuff, and 0.28 Ma for the Bedded Tuff near the top of the sequence (Deino and McBrearty 2002). Blegen et al. (2018) report additional dates of 465.3 ka from the Lebus Tuff in the Middle Silts and Gravels Member and 395.6, 222.5 and 200.7 ka for the Bedded Tuff.

Feldspathic silt, sand and pebble to boulder conglomerate (Fig. 17.15a, b) were laid down by rivers flowing eastwards towards a Pleistocene palaeolake in the main axial rift, close to modern Lake Baringo and the Loboi Plain (Tallon 1978). Kapthurin palaeosols and calcretes are common. The predominantly fluvial sediments have yielded an important series of Acheulian and Middle Stone Age artefacts and hominin remains (Leakey et al. 1969; Wood and Van Noten 1986;

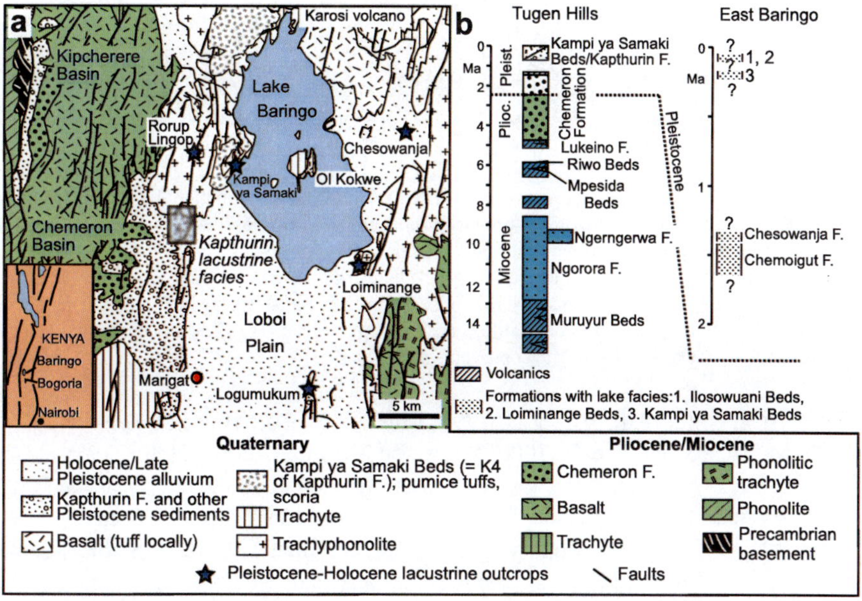

Fig. 17.14 Location and stratigraphy of Pleistocene sediments in the Baringo District. (**a**) General location and geology. Based on Chapman et al. (1973). (**b**) Regional stratigraphy. After Renaut et al. (1999)

Fig. 17.15 Kapthurin Formation sediments. (**a**) Basal Kapthurin Formation silt, sand and conglomerate resting on a fluvial erosion surface entrenched in lacustrine Plio-Pleistocene Chemeron Formation. (**b**) Crudely bedded floodplain silt and channel conglomerate. (**c**) Friable red lacustrine clay lying on weathered basalt. (**d**) Friable black lacustrine clay in section capped by lava. (**e**) Stromatolite (~1 m high) resting on Kasurein Basalt at base of the lacustrine facies of the Kapthurin Formation. (**f**) Lamination in a stromatolite head. Black bar = 5 cm. (**g**) Grey tufa overlain by black lacustrine clay. Backpack (lower right) provides scale

Cornelisson et al. 1990; Hill et al. 1992; McBrearty et al. 1996; McBrearty 1999, 2005; McBrearty and Brooks 2000; Tryon and McBrearty 2002, 2006; Tryon 2002, 2003, 2010; Tryon et al. 2005; Johnson and McBrearty 2010, 2012; Tryon and Faith 2013; Blegen 2017; Shipton 2022) as well as a varied fossil fauna (Leakey et al. 1969; Bishop et al. 1971; Hill et al. 1985, 1986; O'Brien et al. 2021). Stable isotope data from bovid teeth have been used to infer mesic and mildly xeric components in the vegetation between 542 and 509 ka, with riparian woodlands, swamps and grassland ecosystems (Leslie et al. 2016).

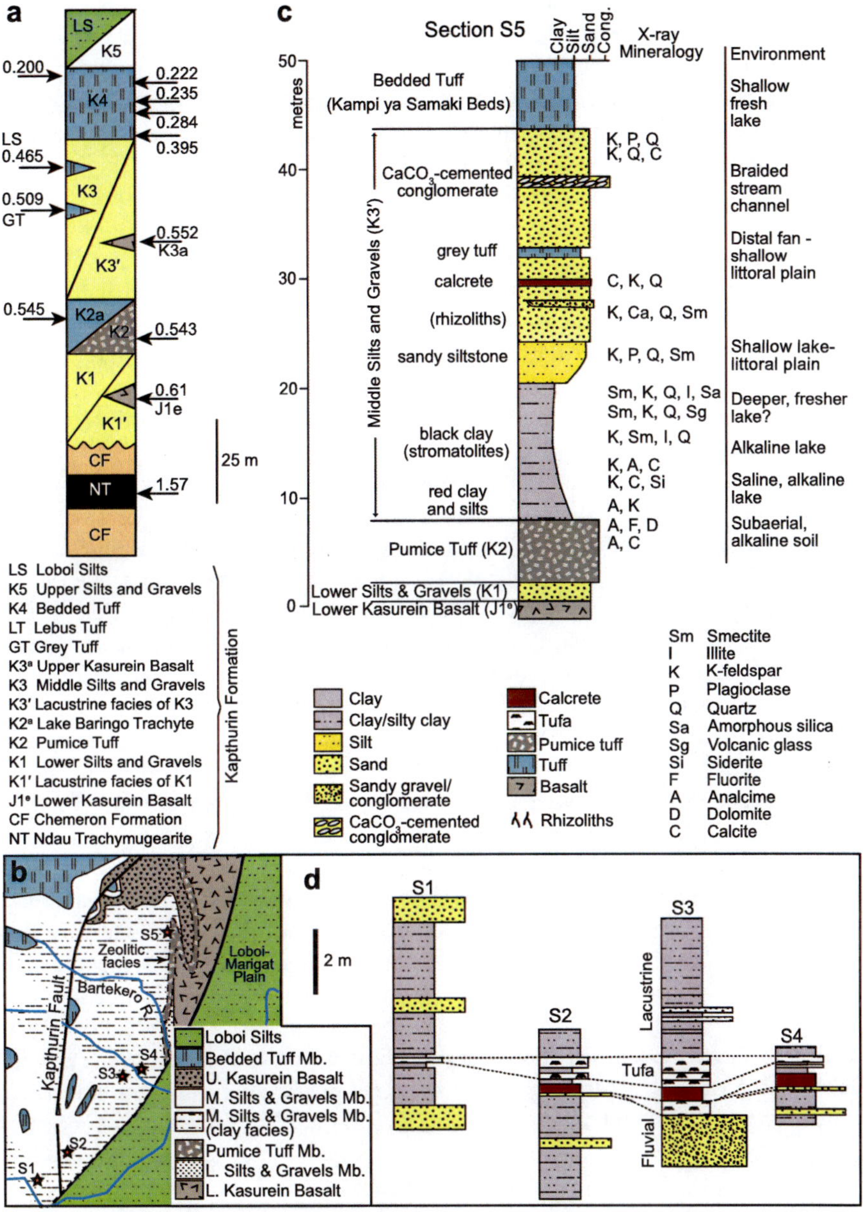

Fig. 17.16 Stratigraphy and lithology of the Kapthurin Formation. (**a**) General stratigraphy for the Kapthurin Formation. After Deino and McBrearty (2002, with supplementary ages from Blegen et al. 2018). (**b**) Geological map of the lacustrine facies and section locations. (**c**) Section S5 with XRD mineralogy. After Tallon (1978 and Renaut et al. (1999). (**d**) Lateral variability of Kapthurin Formation sediments and tufa. After Johnson et al. (2009)

Lacustrine sediments form part of the two Silts and Gravels Members (K1, K3) and the Bedded Tuffs (K4). The lacustrine facies of the Middle Silts and Gravels are well-exposed in northeastern outcrops (Fig. 17.16b–c) and probably represent only part of a much larger area of lake deposits that have been downfaulted below the Loboi Plain. The lake sediments include reddish silty and sandy clay and clayey silt that are locally greenish with scattered gravel and calcite concretions (Fig. 17.15c; Renaut et al. 1999). These deposits pass upwards into dark brown to blackish clay and silt (Fig. 17.15d). Intermittent horizons of sideritic concretions (up to 25 cm long) are present throughout, which have been interpreted as indicating bacterial reduction (Renaut et al. 1999). Large (up to 1 m) columnar stromatolites and oncoids are also present (Fig. 17.15e, f), especially upon the underlying volcanic substrates. Stromatolites are forming today along the Kokwob Murren shoreline at Lake Baringo (Sect. 8.7; Renaut et al. 2013) but are much smaller. The lacustrine clay passes upwards into bedded silt and sand with calcite rhizoliths and thin palaeosols (Renaut et al. 1999).

The reddish clay includes analcime and other zeolites together with authigenic quartz, fluorite, calcite and minor dolomite indicating a saline, alkaline water body (Fig. 17.16c; Renaut et al. 1999). In contrast, the black clay lacks zeolites and is dominated by smectite. The silt and sand in the upper part of the sequence are mainly alkali feldspar, minor plagioclase, lithic grains and quartz. Calcite forms cement, except in the reddish clay. The black clay was laid down in a shallow freshwater body similar to modern Lake Baringo—an inference that is supported by the presence of crocodile, hippo and *Tilapia* bones (McBrearty et al. 1996).

Additional sections have been described to the south where sediments transition into fluvial deposits (Fig. 17.1 6b, d). A spring tufa (Fig. 17.15g) within fluvial and lacustrine sediments accumulated in a lake marginal spring-fed wetland (Johnson et al. 2009). This swampy area was occupied by molluscs, ostracods (*Darwinula stevensoni*, *Cypridopsis ochracea*, *Stenocypris* sp., *Limnocythere africana*), diatoms (*Hantzschia amphioxys*, *Achnanthes exigua (Gogorevia exilis)*, *Pinnularia appendiculata* var. *budensis*, *Staurosira elliptica*) and charophyte oogonia that imply damp to very shallow aquatic environments. Changes in $\delta^{18}O$ in the tufa have been used to infer decreasing influence from fault-derived fluids up section. The presence of a spring system such as this on the margins of a saline-alkaline lake is typical of many modern rift lakes and probably also most of the Neogene palaeolakes of the Baringo District, although sediment preservation (i.e. tufa, travertine, silica sinter) is rare, and many modern springs do not leave a sedimentary record (Sect. 4.9.4).

The Bedded Tuff Member (K4, Fig. 17.16a) was deposited under freshwater (?) lacustrine conditions (Renaut et al. 1999). Several ichnofossils (bird, carnivore, large herbivores) have been observed at Rorup Lingop (Fig. 17.14a; Scott 2005, 2010), which also includes three well-preserved trackways formed by waders and a mongoose. The trace fossils were observed in fluvio-lacustrine/deltaic sediments with low-amplitude symmetrical ripples and microbial mat surface textures. The Bedded Tuff has been dated at 0.235 and 0.284 Ma (Deino and McBrearty 2002) and correlates with the Kampi ya Samaki Beds (Martyn 1969) to the west of Lake

Baringo (Fig. 17.14), which are dated at ~0.23 and 0.24 Ma (Tallon 1976). The lower unit of the Kampi ya Samaki Beds includes lacustrine thinly bedded, silt- and sand-grade tuff with whitened bedding planes whereas the upper unit includes coarser pumice tuff with palaeosols and thin-bedded siltstone (Spooner et al. 1969).

17.4.4 Pleistocene Lake Sediments East of Lake Baringo

Generally, the eastern side of the Baringo Basin experienced greater subsidence than areas to the west and was filled by up to 1.5 km of sediment, possibly laid down on top of middle-Miocene phonolite (Le Turdu et al. 1995, 1999; Hautot et al. 2000). Most of the downfaulted deposits are probably fluvio-lacustrine and lacustrine having accumulated in a former 'central rift lake' (Tallon 1976, 1978; Renaut 1982; Tiercelin and Vincens 1987; Le Turdu et al. 1995) that existed at the same time as the fluvial Kapthurin Formation was being deposited to the west. These sequences are not exposed.

Carney (1972) described outcrops of Pleistocene fluvio-lacustrine sediments from Chesowanja, east of Lake Baringo, which Truckle (1977) divided into the lower and upper 'Chesowanja sediments' on maps of the area. Detailed stratigraphic studies showed that lake deposits are present in parts of the hominid- and artefact-bearing Chemoigut Formation (Fig. 17.17; Bishop and Pickford 1975; Gowlett 1999). The Chemoigut sediments, and basalt in the overlying Chesowanja Formation, were initially dated between 2.46 and 0.71 Ma (Bishop et al. 1978) with a subsequent date of 1.42 Ma reported for a Chesowanja Formation basalt (Hooker and Miller 1979). Trachyte from the overlying Karau Formation (Fig. 17.17b), which includes well-bedded tuffaceous lacustrine facies, has been dated at about 0.54 Ma (Bishop et al. 1978). Floodplain silt and channel conglomerate (Mukutan Beds) complete the sequence.

The 25–30-m-thick Chemoigut Formation includes basal pedogenic red clay that gives way to lacustrine silt and clay with bands of algal? carbonate (Fig. 17.17b; Bishop and Pickford 1975; Bishop et al. 1978). Channel sand and conglomerate, calcrete, abundant root casts, Oldowan stone tools and Australopithecine fossils are also present (Gowlett et al. 1981). A varied fauna has been recovered, including bivalves, gastropods (*Bellamya* sp.), fish, crocodiles and many mammals, including hippopotamus. Earlier interpretations suggested that the deposits formed on the margins of a saline lake with fluctuating water levels (Bishop and Pickford 1975). However, Gowlett et al. (1981) noted that the presence of hippo and crocodile implied periods, or localised settings, with fresher conditions.

The Chemoigut Formation is unconformably overlain by Chesowanja Formation basalt, which in turn is overlain by pumice tuffs of the Karau Formation. The latter are massive in the west but pass eastwards into lacustrine well-bedded tuff with plant impressions, as well as graded and current bedding. Acheulian tools on the top of the Chesowanja Formation may have accumulated at about the same time as the lacustrine tuff (Bishop and Pickford 1975).

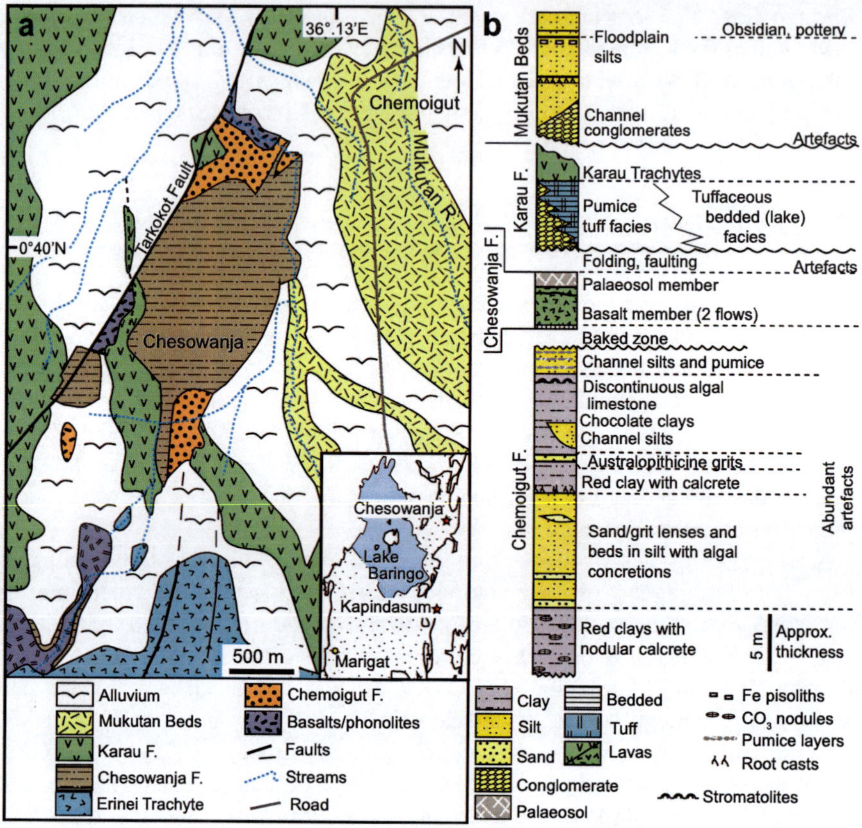

Fig. 17.17 Geology and stratigraphic section for Pleistocene sediments at Chesowanja. After Bishop et al. (1978). (**a**) Outcrop map of the Pleistocene sediments. Stars in the inset show locations of Pleistocene lake sequences. (**b**) General stratigraphy and lithology of the Chemoigut, Chesowanja and Karau formations and the Mukutan Beds

17.4.5 *Late Pleistocene Lake Sediments*

Small (5 km^2) outcrops of the Late Pleistocene (?) Kapindasum Beds (Griffiths, 1977; Hackman 1988) represent lake deposits that accumulated at the present junction of the Ol Arabel and Ngusero rivers (inset map in Fig. 17.17a). They form an isolated sequence of reworked, laminated and cross-bedded, waterlain tuff, up to 15 m thick. The former lake was probably dammed by growth of Paratallo volcano to the north, which also supplied the ash and pumice that constitute the Kapindasum Beds. The Late Pleistocene sediments are of uncertain age but may correlate with the Bedded Tuff Member of the Kapthurin Formation and Kampi ya Samaki Beds located near Loruk. Similar undated sediments are also present at Loiminange, to the southeast where Carney (1972) mapped well-bedded tuffaceous silt and clay (Figs. 17.18a, b and 17.19a). Collectively, the outcrops suggest that a large palaeolake existed at ~250–200 ka across much of the area that is today occupied by Lake Baringo and the Loboi Plain.

A second series of probable lacustrine sediments crop out at Logumukum and have been assigned to the Ilosowuani Beds (Fig. 17.18a, d, e). These deposits are probably Late Pleistocene based on the dipping character (~12°) of the strata, the presence of stone tools (Farrand et al. 1976) and ash, the relationship to overlying Late Pleistocene/Early Holocene Kokwob Formation coquinas in the area and the

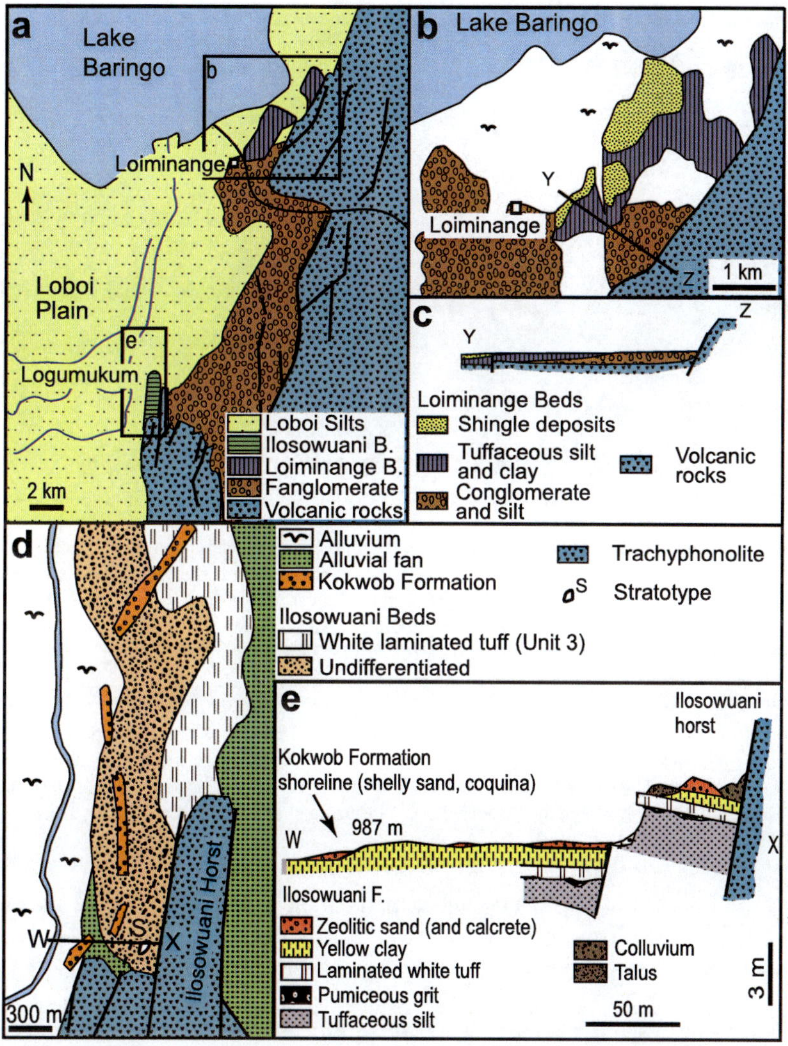

Fig. 17.18 Late Pleistocene sediments. (**b, c**) After Carney (1972); (**d, e**) After Griffiths (1977) and Tiercelin and Vincens (1987). (**a**) Locations of maps and major sites. (**b**) Geological map of the Loiminange Beds. (**c**) Simplified cross-section of the Loiminange Beds. (**d**) Geological map of the Ilosowuani Beds. (**e**) Cross-section of the main outcrop of the Ilosowuani Beds. Location in (**d**)

Fig. 17.19 Late Pleistocene sediments, eastern Baringo Basin. Locations in Fig. 17.14a. (**a**) Well-bedded lacustrine tuff (Loiminange Beds), possibly correlated with the Kampi ya Samaki Beds west of Lake Baringo, Loiminange. (**b**) Logumukum site from the Ilosowuani Horst. Ilosowuani Beds in the foreground forming the elevated mound with end Pleistocene-early Holocene Kokwob Formation shelly strandline (black arrows). (**c**) White laminated tuff forming prominent ledge overlying pumiceous grit and tuffaceous silt, Ilosowuani Beds. (**d**) Reddish analcimic siltstone and sandstone, Ilosowuani Beds. (**e**) Hippo-trampled surface in the uppermost zeolitic sand. (**f**) Loboi Silts. Tuff horizon (black arrows) dips into fluvial channel to the left. Sediments comprised of pedogenically altered silt above and below the tuff. The deposits are incised by channels with abundant pebbles. The latter often include Middle Stone Age tools (inset) reworked from the silt below. (**g**) Termite nest within the Loboi Silts. (**h**) Insect nests in Loboi Silts, possibly bees and (or) wasps (Jenni Scott, pers. comm)

degree of zeolitisation, which resembles parts of the Loboi Silts (Renaut and Owen 1980, 1987). The 3.2-m-thick sequence has been divided into five units (Fig. 17.18e) including a lowermost tuffaceous silt whose base is not exposed, pumice, a laminated white waterlain tuff, yellow clay and overlying zeolitic sand (Renaut 1993) that locally passes into calcrete (Fig. 17.19b–e; Griffiths 1977). Diatoms (*Rhopalodia gibberula* var. *sphaerula*, *R. gibberula* var. *rupestris*, *Anomoeoneis sphaerophora* var. *guntheri*, *Surirella ovalis*, *Thalassiosira rudolfii*, *Cyclotella meneghiniana* [*Stephanocyclus meneghinianus*]) typical of saline, alkaline lakes (up to 80 meq L^{-1}) are present in the yellow clay as well as shallow-water indicators such as *Cocconeis placentula* (Owen 1981). Hippo-trampled surfaces on the yellow clay and zeolitic sand may indicate fresher intervals (contemporary or later) or local freshwater inputs to the palaeolake (Fig. 17.19e; Scott 2005).

Late Pleistocene tectonism resulted in moderate tilting, faulting and erosion of both the Ilosowuani Beds and Kapthurin formations. Deposition continued in the subsiding axial trough probably as lacustrine facies in a 'Lake Loboi' (Farrand et al. 1976; Renaut 1982; Sect. 9.10). However, the rate of sedimentation began to exceed that of subsidence leading to the development of the mostly fluvial Loboi Silts. These lack accurate radiometric dates but contain Middle Stone Age (MSA) tools suggesting that the exposed parts of the sequence are >20 ka old (Farrand et al. 1976; Renaut and Owen 1980). In addition, McCall (1967) mentioned a 15-m-thick section that showed Loboi Silts resting conformably on the Kapthurin Formation in the Molo Valley, which would imply time-transgressive fluvio-lacustrine deposition from roughly 200 ka through to the present day.

The Loboi Silts are comprised of massive to weakly stratified clayey silt, feldspathic sand, lithic gravel and conglomerate (Figs. 17.19f and 17.20a, b) with planar or trough cross-stratification, minor channel incision and common palaeosols (Le

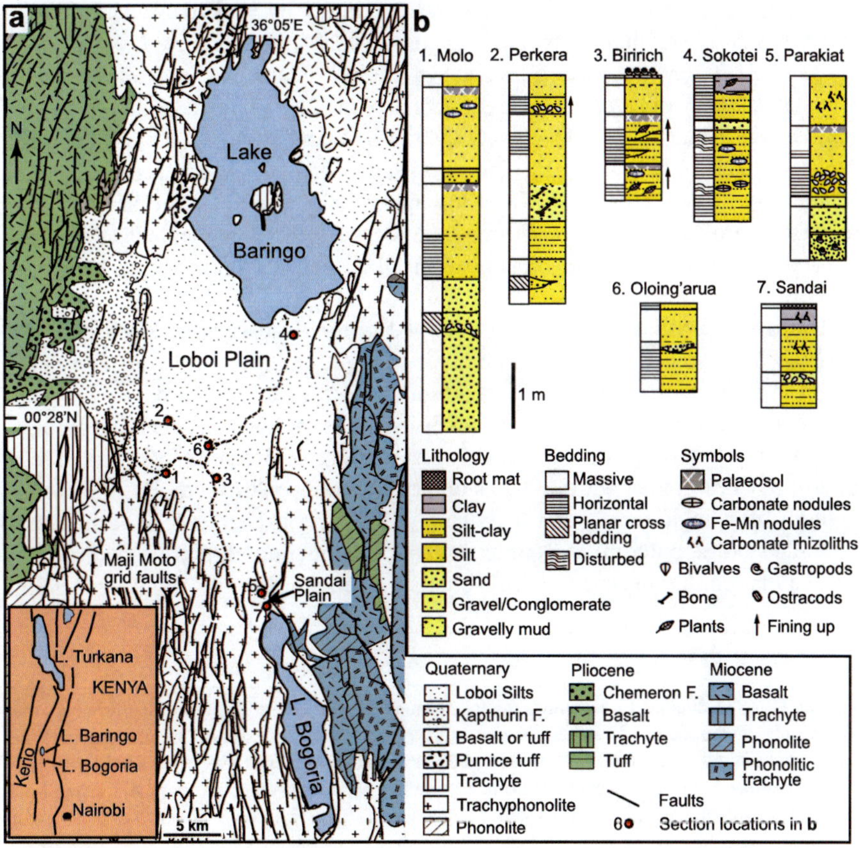

Fig. 17.20 Stratigraphy of the Loboi Silts. (**a**) Geological map of the Baringo area and location of Loboi Silts sections. (**b**) Representative sections in the Loboi Silts. After Renaut (1982)

Turdu et al. 1995). Other common features include Fe-Mn or ferruginous concretions up to 1.5 cm in diameter (Renaut 1993), horizontal and vertical rootmarks, and calcareous rhizoliths. Trace fossils are abundant, including calcified termite and insect nests (Fig. 17.19g, h), invertebrate and terrestrial vertebrate (ground squirrels, warthog and [or] aardvark?) burrows with examples of the latter infilled by tuff near the Nyogonyek badlands (SW Loboi Plain). Calcrete and analcime-cemented Loboi Silts are also present on the Sandai Plain (Fig. 17.20a; Renaut 1993), north of Lake Bogoria, where they have been interpreted as indicating a late Pleistocene 'dry phase' between ~20 and 12 ka (Hamilton 1982). Today, sedimentation on the Loboi Plain continues and is dominated by lacustrine, fluvial overbank, sheetflow and channel deposits that vary according to lake level. These sediments pass northwards into the fluvio-deltaic facies and lacustrine sediments of the Lake Baringo Basin.

17.5 Plio-Pleistocene Palaeogeography

During the earlier Pliocene, there may have been several separate palaeolakes in what are now the Tugen foothills. Lateral continuity between these water bodies is uncertain as are their eastward margins towards what is now Lake Baringo (Fig. 17.21a). By the Late Pliocene to Early Pleistocene, the southernmost Chemeron Basin and areas to the east were occupied by a large lake that fluctuated between fresh and alkaline status, but the northern and eastern extent of this water body is unknown (Fig. 17.21b). The Plio-Pleistocene lake had retreated from the Tugen foothills by about 2.4 Ma following earth movements that caused tilting and erosion of the Chemeron Formation. A lake of unknown extent could have existed towards the Baringo rift axis, but all evidence is now buried below the axial Baringo graben.

Uplift and erosion of the Tugen Hills led to the deposition of Middle to Late Pleistocene fluvial silt, gravel and boulders of the Kapthurin Formation. Fan-deltas in the west, in what are now the Tugen foothills, gave way eastwards to saline lake environments, implying closed basin conditions. The Middle Pleistocene Chemoigut Formation also indicates a saline lake, at least intermittently, on the eastern side of the rift at Chesowanja, but it remains uncertain whether this represents the eastern margins of a single large water body (Fig. 17.21c) or a separate lake basin. Small Late Pleistocene outcrops at Loiminange and Logumukum indicate a shallow lake or lake margin in those areas, with diatoms in the latter location pointing to saline

Fig. 17.21 (continued) to fresh and retreated from areas west of the Kapthurin Fault in response to tectonic movements. (**d**) Lake Loboi was probably fresh and subject to gradual infilling. Separation into two lakes finally occurred close to the Sandai-Waseges River. (**e**) Late Pleistocene-Early Holocene Lakes Baringo and Bogoria were both freshwater bodies. In contrast, modern Lake Baringo is fresh and Lake Bogoria is saline

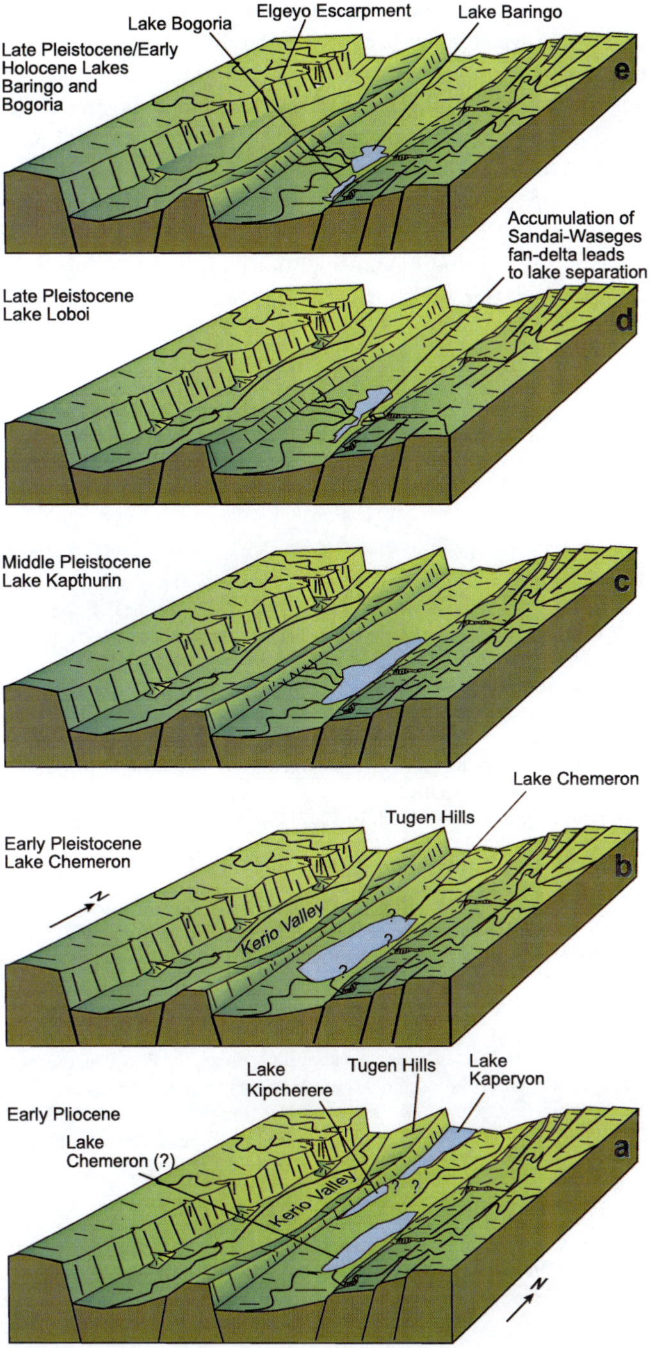

Fig. 17.21 Plio-Pleistocene lake palaeogeography, Baringo basin. (**a**) Late Pliocene palaeolakes – linkages between the water bodies are uncertain as is their eastward extent. (**b**) Early Pleistocene Lake Chemeron ranged from fresh to alkaline. (**c**) Lake Kapthurin ranged from saline and alkaline (continued)

conditions. Black smectite clay and well-bedded tuff of the upper Kapthurin Formation indicate fresher water lake phases that may have extended to the central axial basin.

Late Pleistocene grid faulting resulted in incision and erosion of the Kapthurin Formation sediments as the main axial basin continued to subside. Lake Loboi (Fig. 17.21d) evolved from the earlier Kapthurin Formation water bodies with axial subsidence confining the lake to the area now occupied by Lake Baringo, the Loboi Plain and possibly Lake Bogoria. As sedimentation rates started to exceed subsidence rates fluviolacustrine and fluvial conditions advanced lakewards, splitting the palaeolake into two water bodies that would become modern freshwater Lake Baringo and saline Lake Bogoria (Fig. 17.21e) with these lakes probably remaining separate even during the African Humid Phase when most Kenya lakes experienced a rapid expansion (Sect. 9.10.6).

References

Altner M, Reichenbacher B (2015) *Kenyaichthyidae* fam. nov. and *Kenyaichthys* gen. nov. – first record of a fossil aplocheiloid killifish (Teleostei, Cyprinodontiformes). PLoS One 10:e0123056

Altner M, Ruthensteiner B, Reichenbacher B (2020) New haplochromine cichlid from the upper Miocene (9–10 MYA) of Central Kenya. BMC Evol Biol 20:65

Altner M, Reichenbacher B (2021) A small cichlid species flock from the upper Miocene (9–10 MYA) of Central Kenya. Hydrobiologia 848:3613–3637

Altner M, Schliewen UK, Penk SBR, Reichenbacher B (2017) *Tugenchromis pickfordi*, gen. et sp. nov., from the upper Miocene—a stem-group cichlid of the "East African Radiation". J Vert Paleontol 37:e1297819–e1297820

Baker BH, Williams LAJ, Miller JA, Fitch FJ (1971) Sequence and geochronology of the Kenya Rift volcanics. Tectonophysics 11:191–215

Bamford MK, Senut B, Pickford M (2013) Fossil leaves from Lukeino, a 6-million-year-old formation in the Baringo Basin, Kenya. Geobios 46:253–272

Behrensmeyer AK, Deino AL, Hill A, Kingston JD, Saunders JJ (2002) Geology and geochronology of the middle Miocene Kipsaramon site complex, Muruyur Beds, Tugen Hills, Kenya. J Hum Evol 42:11–38

Benefit B, Pickford MHL (1986) Miocene fossil cercopithecoids from Kenya. Am J Phys Anthrop 69:441–464

Billingsley AL (2019) Pan-African climate variability since the Plio-Pleistocene and possible implications for hominin evolution. PhD thesis, University of Arizona, Tucson

Billingsley AL, Reinthal O, Dettman DL, Kingston JD, Deino AL, Ortiz K, Mohler B, Cohen AS (2021) $\delta^{13}C$ records from fish fossils as paleo-indicators of ecosystem responses to lake levels in the Plio-Pleistocene lakes of Tugen Hills. Kenya Palaeogeogr Palaeoclimatol Palaeoecol 570:109466

Bishop WW (1967) Annotated lexicon of Quaternary stratigraphical nomenclature in East Africa. In: Bishop WW, Clark JD (eds) Background to evolution in Africa. University of Chicago Press, Chicago, pp 375–395

Bishop WW (ed) (1978) Geological background to fossil man. Geological Society London Special Publication, vol 6

Bishop WW, Chapman GR (1970) Early Pliocene sediments and fossils from the northern Kenya Rift Valley. Nature 226:914–918

Bishop WW, Chapman GR, Hill A, Miller JA (1971) Succession of Cainozoic vertebrate assemblages from the northern Kenya Rift Valley. Nature 233:389–394

Bishop L, Hill A, Kingston JD (1999) Palaeoecology of Suidae from the Tugen Hills, Baringo, Kenya. In: Andrews P, Banham P (eds) Late Cenozoic environments and hominid evolution. Geological Society, London, pp 99–111

Bishop W, Hill A, Pickford MHL (1978) Chesowanja: a revised geological interpretation. In: Bishop WW (ed) Geological background to fossil man. Geol Soc Lond Spec Publ 6:309–327

Bishop WW, Pickford MHL (1975) Geology, fauna and palaeoenvironments of the Ngorora Formation, Kenya Rift Valley. Nature 254:185–192

Blegen N (2017) The earliest long-distance obsidian transport: evidence from the ~200 ka Middle Stone Age Sibilo School Road Site, Baringo, Kenya. J Hum Evol 103:1–19

Blegen N, Jicha BR, McBrearty S (2018) A new tephrochronology for early diverse stone tool technologies and long-distance raw material transport in the Middle to Late Pleistocene Kapthurin Formation, East Africa. J Hum Evol 121:75–103

Brown FH, McDougall I (2011) Geochronology of the Turkana depression of northern Kenya and southern Ethiopia. Evol Anthropol 20:217–227

Campisano CJ, Cohen AS, Arrowsmith JR, Asrat A, Behrensmeyer AK, Brown ET, Deino AL, Deocampo DM, Feibel CS, Kingston JD, Lamb HF, Lowenstein TK, Noren A, Olago DO, Owen RB, Pelletier JD, Potts R, Reed KE, Renaut RW, Russell JM, Russell JL, Schäbitz F, Stone JR, Trauth MH, Wynn JG (2017) The Hominin Sites and Paleolakes Drilling Project: acquiring high-resolution paleoclimate records from the East African rift system and their implications for understanding the environmental context of hominin evolution. PaleoAnthropology 2017:1–43

Carney JN (1972) The geology of the eastern part of the rift valley in Baringo District, Kenya. PhD thesis, University of London

Chapman GR (1971) The geological evolution of the northern Kamasia Hills, Baringo District, Kenya. PhD thesis, University of London

Chapman GR, Brook M (1978) Chronostratigraphy of the Baringo Basin, Kenya. In: Bishop WW (ed) Geological background to fossil man. Geol Soc Lond Spec Publ 6:207–223

Chapman GR, Lippard SJ, Martyn JE (1978) The stratigraphy and structure of the Kamasia range, Kenya Rift Valley. J Geol Soc Lond 135:265–281

Chapman GR, Lippard SJ, Martyn JE, Sinclair-Jones N, King BC (1973) Geological map of the northern Tugen Hills. East African Geological Research Unit (Bedford College, University of London) and Mines and Geology Dept, Kenya

Cohen A, Campisano C, Arrowsmith R, Asrat A, Beck C, Behrensmeyer AK, Deino A, Deocampo D, Feibel C, Foerster V, Kingston J, Lamb H, Lowenstein T, Lupien R, Muiruri V, Olago D, Owen RB, Potts R, Russell SF, Stone J, Trauth M, Yost C (2022) Reconstructing the environmental context of human origins in eastern Africa through scientific drilling. Annu Rev Earth Planet Sci 50:451–476

Cohen A, Campisano C, Arrowsmith R, Asrat A, Behrensmeyer AK, Deino A, Feibel C, Hill A, Johnson R, Kingston J, Lamb H, Lowenstein T, Noren A, Olago D, Owen RB, Potts R, Reed K, Renaut R, Schäbitz F, Tiercelin J-J, Trauth M, Wynn J, Ivory S, Brady C, O'Grady R, Rodysill J, Githiri J, Russell J, Foerster V, Dommain R, Rucina S, Deocampo D, Billingsley A, Beck C, Dullo L, Feary D, Garello D, Johnson T, Junginger A, Karanja M, Kimburi E, Mbuthia A, McCartney T, McNulty E, Muiruri V, Nambiro E, Njagi D, Norman J, Rabideaux N, Raub T, Sier M, Smith P, Urban J, Warren M, Wondiymu E, Yost C (2016) The hominin sites and paleolakes drilling project: inferring the environmental context of human evolution from eastern African Rift lake deposits. Sci Drill 21:1–16

Cornelissen E, Boven A, Dabi A, Hus J, Ju Yong K, Keppens E, Langohr R, Moeyersons J, Pasteels P, Pieters M, Uytterschaut H, Van Noten F, Workineh H (1990) The Kapthurin Formation revisited. Afr Archaeol Rev 8:23–76

Dagley P, Mussett AE, Palmer HC (1978) Preliminary observations on the palaeomagnetic stratigraphy of the area west of Lake Baringo, Kenya. In: Bishop WW (ed) Geological background to fossil man. Geol Soc Lond Spec Publ 6:225–235

De Franceschi D, Bamford M, Pickford M, Senut B (2016) Fossil wood from the upper Miocene Mpesida Beds at Cheparain (Baringo District, Kenya): botanical affinities and palaeoenvironmental implications. J Afr Earth Sci 115:271–280

Deino A, Hill A (2002) $^{40}Ar/^{39}Ar$ dating of Chemeron Formation strata encompassing the site of hominid KNM-BC 1, Tugen Hills, Kenya. J Hum Evol 42:141–151

Deino AL, Kingston JD, Glen JM, Edgar RK, Hill A (2006) Precessional forcing of lacustrine sedimentation in the late Cenozoic Chemeron Basin, Central Kenya Rift, and calibration of the Gauss/Matuyama boundary. Earth Planet Sci Lett 247:41–60

Deino AL, McBrearty S (2002) $^{40}Ar/^{39}Ar$ dating of the Kapthurin Formation, Baringo, Kenya. J Hum Evol 42:185–210

Deino AL, Sier MJ, Garello DL, Keller CB, Kingston JD, Scott JJ, Dupont-Nivet G, Cohen AS (2021) Chronostratigraphy of the Baringo-Tugen Hills-Barsemoi (HSPDP-BTB13-1A) core – $^{40}Ar/^{39}Ar$ dating, magnetostratigraphy, tephrostratigraphy, sequence stratigraphy and Bayesian age modeling. Palaeogeogr Palaeoclimatol Palaeoecol 570:109519

Deino A, Tauxe L, Monaghan M, Drake R (1990) $^{40}Ar/^{39}Ar$ age calibration of the Ngorora Formation, Kenya. J Geol 98:567–587

Deino AL, Tauxe L, Monaghan M, Hill A (2002) $^{40}Ar/^{39}Ar$ geochronology and paleomagnetic stratigraphy of the Lukeino and lower Chemeron Formations at Tabarin and Kapcheberek, Tugen Hills, Kenya. J Hum Evol 42:117–140

Dericquebourg P, Person A, Ségalen L, Pickford M, Senut B, Fagel N (2015) Environmental significance of upper Miocene phosphorites at hominid sites in the Lukeino Formation (Tugen Hills, Kenya). Sediment Geol 327:43–54

Dericquebourg P, Person A, Ségalen L, Pickford M, Senut B, Fagel N (2019) Bone diagenesis and origin of calcium phosphate nodules from a hominid site in the Lukeino Formation (Tugen Hills, Kenya). Palaeogeogr Palaeoclimatol Palaeoecol 536:109377

Ego JK (1994) Sedimentology and diagenesis of Neogene sediments in the central Kenya Rift Valley. MSc thesis, University of Saskatchewan, Saskatoon

Evernden JF, Curtis GH, Bishop W, Brace CL, Clark JD, Damon PE, Hay RL, Hopkins DM, Clark Howell F, Knopf A, Kretzoi M, Leakey LSB, Maude HE, Richards JR, Savage DE, Wright HE (1965) The potassium-argon dating of Late Cenozoic Rocks in East Africa and Italy [and Comments and Reply]. Curr Anthropol 6:342–385

Farrand WR, Redding RW, Wolpoff MH, Wright HT (1976) An archaeological investigation on the Loboi Plain, Baringo District, Kenya. Museum of Anthropology, University of Michigan (Ann Arbor) Technical Reports 4

Fuchs V (1934) The geological work of the Cambridge expedition to the East African Lakes, 1930–31. Geol Mag 71:97–112

Fuchs V (1950) Pleistocene events in the Baringo Basin. Geol Mag 87:149–174

Goble ED (2011) Paleontology of the Chemeron Formation, Tugen Hills, Kenya with emphasis on faunal shifts and precessional climatic forcing. PhD thesis, Yale University, New Haven

Goble ED (2022) The southern Chemeron Formation, Tugen Hills, Kenya: a review and a paleoecological analysis of the bovid fauna. In: Reynolds SC, Bobe R (eds) African paleoecology and human evolution. Cambridge University Press, Cambridge, pp 343–359

Gregory JW (1896) The Great Rift Valley. John Murray, London

Gregory JW (1921) The Rift valleys and geology of East Africa. Seeley Service, London

Griffiths P (1977) The geology of the area around Lake Hannington and the Perkerra River, Rift Valley Province, Kenya. PhD thesis, University of London

Gowlett JAJ (1999) Lower and Middle Pleistocene archaeology of the Baringo Basin. In: Andrews P, Banham P (eds) Late Cenozoic environments and hominid evolution: a tribute to Bill Bishop. Geological Society, London, pp 123–141

Gowlett JAJ, Harris J, Walton D, Wood B (1981) Early archaeological sites, hominid remains and traces of fire from Chesowanja, Kenya. Nature 294:125–129

Gowlett JAJ, Stanistreet IG, Albert RM, Blackbird SJ, Herries AIR, Hoare S, Kogai P, Komboh CK, Mark DF, Muriuki RM, Murphy H, Rucina SM, Stollhofen H (2022) New Oldowan localities at high level within Kilombe caldera, Kenya. L'anthropologie 126:102976

Hackman BD (1988) Geology of the Baringo-Laikipia area. Kenya Mines and Geological Department Report, Nairobi, p 104

Hamilton AC (1982) Environmental history of East Africa. Academic, London

Hautot S, Tarits P, Whaler K, Le Gall B, Tiercelin J-J, Le Turdu C (2000) Deep structure of the Baringo Rift Basin (Central Kenya) from three-dimensional magnetotelluric imaging: implications for rift evolution. J Geophys Res 105:23,493–23,518

Hill A (1985) Early hominid from Baringo, Kenya. Nature 315:222–224

Hill A (1987) Causes of perceived faunal change in the later Neogene of East Africa. J Hum Evol 16:583–596

Hill A (1995) Faunal and environmental change in the Neogene of East Africa: evidence from the Tugen Hills sequence, Baringo District, Kenya. In: Vrba ES, Denton GH, Partridge TC, Burckle LH (eds) Paleoclimate and evolution with emphasis on human origins. Yale University Press, New Haven, pp 178–193

Hill A (1999) The Baringo Basin, Kenya: from Bill Bishop to BPRP. In: Andrews P, Banham B (eds) Late Cenozoic environments and hominid evolution: a tribute to Bill Bishop. Geological Society, London, pp 85–97

Hill A (2002) Paleoanthropological research in the Tugen Hills, Kenya. J Hum Evol 42:1–10

Hill A, Behrensmeyer AK, Brown B, Deino A, Rose M, Saunders J, Ward S, Winkler A (1991) Kipsaramon: a lower Miocene hominoid site in the Tugen Hills, Baringo District. J Hum Evol 20:67–75

Hill A, Curtis G, Drake R (1986) Sedimentary stratigraphy of the Tugen Hills. In: Frostick LE, Renaut RW, Reid I, Tiercelin J-J (eds) Sedimentation in the African rifts. Geol Soc London Spec Publ. 25:285–295

Hill A, Drake R, Tauxe L, Monaghan M, Barry JC, Behrensmeyer AK, Curtis G, Jacobs BF, Jacobs L, Johnson N, Pilbeam D (1985) Neogene palaeontology and geochronology of the Baringo Basin, Kenya. J Hum Evol 14:759–773

Hill A, Leakey M, Kingston JD, Ward S (2002) New cercopithecoids and a hominoid from 12·5 Ma in the Tugen Hills succession, Kenya. J Hum Evol 1–2:75–93

Hill A, Ward S (1988) Origin of the Hominidae: the record of African large hominoid evolution between 14 My and 4 My. Am J Phys Anthropol 31:49–83

Hill A, Ward S, Deino A, Garniss C, Drake R (1992) Earliest *Homo*. Nature 355:719–722

Hoare S, Brink JS, Herries AIR, Mark DF, Morgan LE, Onjala I, Rucina SM, Stanistreet IG, Stollhofen H, Gowlett JAJ (2021) Geochronology of a long Pleistocene sequence at Kilombe volcano, Kenya: from the Oldowan to Middle Stone Age. J Archaeol Sci 125:105273

Hooker PJ, Miller JA (1979) K-Ar dating of the Pleistocene fossil hominid site at Chesowanja, North Kenya. Nature 282:710–712

Jacobs BF (2002) Estimation of low-latitude paleoclimates using fossil angiosperm leaves: examples from the Miocene Tugen Hills, Kenya. Paleobiology 28:399–421

Jacobs BF, Deino AL (1996) Test of climate-leaf physiognomy regression models, their application to two Miocene floras from Kenya, and $^{40}Ar/^{39}Ar$ dating of the late Miocene Kapturo site. Palaeogeogr Palaeoclimatol Palaeoecol 123:259–271

Jacobs BF, Kabuye CHS (1987) A middle Miocene (12.2 my old) forest in the East African Rift Valley. Kenya J Hum Evol 16:147–155

Jacobs BF, Kabuye CHS (1989) An extinct species of *Pollia* Thunberg (Commelinaceae) from the Miocene Ngorora Formation, Kenya. Rev Palaeobot Palynol 59:67–76

Jacobs BF, Winkler DA (1992) Taphonomy of a middle Miocene autochthonous forest assemblage, Ngorora Formation, central Kenya. Palaeogeogr Palaoecol Palaeoclimatol 99:31–40

Johnson CR, Ashley GM, de Wet CB, Dvoretsky R, Park L, Hover VC, Owen RB, McBrearty S (2009) Tufa as a record of perennial fresh water in a semi-arid rift basin, Kapthurin Formation, Kenya. Sedimentology 56:1115–1137

Johnson CR, McBrearty S (2010) 500,000 year old blades from the Kapthurin Formation, Kenya. J Hum Evol 58:193–200

Johnson CR, McBrearty S (2012) Archaeology of middle Pleistocene lacustrine and spring paleoenvironments in the Kapthurin Formation, Kenya. J Anthropol Archaeol 31:485–499

Kevrekidis C, Valtl M, Penk SBR, Altner M, Reichenbacher B (2019) *Rebekkachromis* nov. gen. from the middle–upper Miocene (11 MYA) of Central Kenya: the oldest record of a haplotilapiine cichlid fish. Hydrobiologia 832:39–64

King BC (1978) Structural and volcanic evolution of the Gregory Rift Valley. In: Bishop WW (ed) Geological background to fossil man. Geol Soc Lond Spec Publ 6:29–54

King BC, Chapman GR, Robson DA, McConnell RB (1972) Volcanism of the Kenya Rift Valley [and discussion]. Phil Trans Roy Soc Ser A 271:185–208

Kingston JD (1992) Stable isotopic evidence for hominid paleoenvironments in East Africa. PhD thesis, Harvard University, Boston

Kingston JD (1999) Environmental determinants in early hominid evolution: issues and evidence from the Tugen Hills, Kenya. In: Andrews P, Banham P (eds) Late Cenozoic environments and hominid evolution: a tribute to Bill Bishop. Geological Society, London, pp 69–84

Kingston JD (2007) Shifting adaptive landscapes: Progress and challenges in reconstructing early hominid environments. Am J Phys Anthropol 134:20–58

Kingston JD, Deino AL, Edgar RK, Hill A (2007) Astronomically forced climate change in the Kenyan Rift Valley 2.7–2.55 Ma: implications for the evolution of early hominin ecosystems. J Hum Evol 20:487–503

Kingston JD, Hill A (2005) When it rains it pours: legends and realities of the East African pluvials. In: Lieberman DE, Smith RJ, Kelley J (eds) Interpreting the past: essays on human, primate and mammal evolution, American School of Prehistoric Research Monograph Series, vol 5. Brill, Boston, pp 189–205

Kingston JD, Jacobs FJ, Hill A, Deino A (2002) Stratigraphy, age and environments of the late Miocene Mpesida Beds, Tugen Hills, Kenya. J Hum Evol 42:95–116

Kingston JD, Marino BD, Hill A (1994) Isotopic evidence for Neogene hominid paleoenvironments in the Kenya Rift Valley. Science 264:955–959

Kipseba WK, Kotut JJ, Kasitet J (2009) Mineral exploration and assessment of geological materials and geotourism sites in ALRMP project area, Baringo and East Pokot districts. Government of Kenya Report, Nairobi

Leakey LSB (1931) The Stone Age cultures of Kenya Colony. Cambridge University Press, Cambridge

Leakey LSB (1950) The lower limit of the Pleistocene in Africa. Proceedings of the 18th International Geological Congress, pp 62–65

Leakey LSB (1955) The climatic sequence of the Pleistocene in Africa. In: Balout L (ed) Actes de la IIe Session du Congrès Panafricain de Préhistoire, Alger, 1952, pp 293–294

Leakey LSB, Solomon JD (1929) East African archaeology. Nature 124:9

Leakey M, Tobias PV, Martyn JE, Leakey REF (1969) An Acheulian industry with prepared core technique and the discovery of a contemporary hominid mandible at Lake Baringo, Kenya. Proc Prehist Soc 3:48–76

Leonard LM, Dyke GJ, Walker CA (2006) New specimens of a fossil ostrich from the Miocene of Kenya. J Afr Earth Sci 45:391–394

Leslie DE, McBrearty S, Hartman G (2016) A middle Pleistocene intense monsoonal episode from the Kapthurin Formation, Kenya: stable isotopic evidence from bovid teeth and pedogenic carbonates. Palaeogeogr Palaeoclimatol Palaeoecol 449:27–40

Le Turdu C, Coussement C, Tiercelin J-J, Renaut RW, Rolet J, Richert JP, Xavier JP, Coquelet D (1995) Rift basin structure and depositional patterns interpreted using 3D remote sensing approach: the Baringo and Bogoria basins, Central Kenya Rift, East Africa. Bull Ctr Rech Explor-Prod Elf Aquitaine 19:1–37

Le Turdu C, Tiercelin J-J, Richert JP, Rolet J, Xavier JP, Renaut RW, Lezzar KE, Coussement C (1999) Influence of preexisting oblique discontinuities on the geometry and evolution of extensional fault patterns: evidence from the Kenya Rift using SPOT imagery. In: Morley CK (ed) Geoscience of rift systems – evolution of East Africa. AAPG Stud Geol 44:173–191

Lippard SJ (1972) The stratigraphy and structure of the Elgeyo Escarpment, southern Kamasia Hills and adjoining areas, Rift Valley Province, Kenya. PhD thesis, University of London

Lippard SJ (1973) Plateau phonolite lava flows, Kenya. Geol Mag 110:543–549

Lupien RL (2019) The Plio-Pleistocene climate history of East Africa and the role of environmental change on human evolution: studies of leaf wax isotopes from paleolake sediment. PhD thesis, Brown University, Providence

Lupien RL, Russell JM, Yost CL, Kingston JD, Deino AL, Logan J, Schuh A, Cohen AS (2021) Vegetation change in the Baringo Basin, East Africa across the onset of Northern Hemisphere glaciation 3.3–2.6 Ma. Palaeogeogr Palaeoclimatol Palaeoecol 570:109426

Martyn JE (1967) Pleistocene deposits and new fossil localities in Kenya. Nature 215:476–479

Martyn JE (1969) The geological history of the country between Lake Baringo and the Kerio River, Baringo District, Kenya. PhD thesis, University of London

Maslin MA, Shultz S, Trauth MH (2015) A synthesis of the theories and concepts of early human evolution. Phil Trans R Soc Ser B Biol Sci 370:20140064

McBrearty S (1999) Archaeology of the Kapthurin Formation. In: Andrews P, Banham P (eds) Late Cenozoic environments and hominid evolution: a tribute to Bill Bishop. Geological Society, London, pp 143–156

McBrearty S (2005) The Kapthurin Formation: what we know now that we didn't know then. In: Lieberman DE, Smith RJ, Kelley J (eds) Interpreting the past: essays on human primate, and mammal evolution (American School of Prehistoric Research Monograph Series, vol 5). Brill, Boston, pp 263–274

McBrearty S, Bishop L, Kingston J (1996) Variability in traces of Middle Pleistocene hominid behavior in the Kapthurin Formation, Baringo, Kenya. J Hum Evol 30:563–580

McBrearty S, Brooks AS (2000) The revolution that wasn't: a new interpretation of the origin of modern human behavior. J Hum Evol 39:453–563

McCall GJH (1967) Geology of the Nakuru-Thomson's Falls Lake Hannington area. Rep Geol Surv Kenya 78

McCall GJH, Baker BH, Walsh J (1967) Late Tertiary and Quaternary sediments of the Kenya Rift Valley. In: Bishop WW, Clark JD (eds) Background to evolution in Africa. University of Chicago Press, Chicago, pp 191–220

McClenaghan M (1971) Geology of the Ribkwo area, Baringo District. PhD thesis, University of London

Mein P, Pickford M (2006) Late Miocene micromammals from the Lukeino Formation (6.1 to 5.8 Ma), Kenya. Bull Mens Soc Linn Lyon 75:183–223

Minkara K, Rabideaux NM, Deocampo D, Kingston J, Cohen AS (2016) Paleoenvironmental interpretation of drill core from Tugen Hills, Kenya through X ray diffraction. AGU Fall Mtg, San Francisco 12–16 Dec, Abstract PP23A-2313

Minkara K (2017) Zeolite facies and environmental change in the Plio-Pleistocene Baringo Basin, Kenya Rift. MSc thesis, Georgia State University, Atlanta

Morales J, Pickford M (2005) Carnivores from the middle Miocene Ngorora Formation (13–12 Ma), Kenya. Estud Geol 61:271–284

Morley CK, Ngenoh DK, Ego JK (1999) Introduction to the East African Rift System. In: Morley CK (ed) Geoscience of rift systems—evolution of East Africa. AAPG Stud Geol 44:1–18

Morley CK, Wescott WA, Stone DM, Harper RA, Wigger ST, Karanja FM (1992) Tectonic evolution of the northern Kenyan Rift. J Geol Soc Lond 149:333–348

Morley CK, Wescott WA (1999) Sedimentary environments and geometry of sedimentary bodies determined from subsurface studies in East Africa. In: Morley CK (ed) Geoscience of rift systems—evolution of East Africa. AAPG Stud Geol 44:211–231

Mugisha F, Ebinger CJ, Strecker M, Pope D (1997) Two-stage rifting in the Kenya Rift: implications for half-graben models. Tectonophysics 278:63–81

Muia G (2015) The "Turkana Grits": potential hydrocarbon reservoirs of the northern and Central Kenya Basins. PhD thesis, Université de Rennes, France

Muriithi ND (2012) The Plio-Pleistocene history of sedimentation in Lake Baringo Basin, Kenya – a drill core perspective. MSc thesis, University of Nairobi

Murray-Hughes R (1933) Notes on the geological succession, tectonics and economic geology of the western half of the Kenya Colony. Rep Geol Surv Kenya 3

Namwamba FL (1993) Tephrostratigraphy of the Chemeron Formation. MSc thesis, University of Utah, Salt Lake City

Nyambok IO, Gaciri SJ (1975) Geology of the fluorite deposits in Kerio Valley, Kenya. Econ Geol 70:299–307

O'Brien K, Tryon CA, Blegen N, Kimeu B, Rowan J, Faith JT (2021) First appearance of Grévy's zebra (*Equus grevyi*), from the Middle Pleistocene Kapthurin Formation, Kenya, sheds light on the evolution and paleoecology of large zebras. Quat Sci Rev 256:106835

Owen RB (1981) Quaternary diatomaceous sediments and the geological evolution of lakes Turkana, Baringo and Bogoria, Kenya Rift Valley. PhD thesis, University of London

Owen RB, Crossley R (1992) Spatial and temporal distribution of diatoms in sediments of Lake Malawi, Central Africa, and ecological implications. J Paleolimnol 7:55–71

Owen RB, Renaut RW (2000) Miocene and Pliocene diatomaceous lacustrine sediments of the Tugen Hills, Baringo District, central Kenya Rift. In: Gierlowski-Kordesch EH, Kelts KR (eds) Lake basins through space and time. AAPG Stud Geol 46:465–472

Owen RB, Renaut RW, Lowenstein TK (2018) Spatial and temporal geochemical variability in lacustrine sedimentation in the East African Rift System: evidence from the Kenya Rift and regional analyses. Sedimentology 65:1697–1730

Ortiz K, Arrowsmith R, Cohen AS, Feibel CS, Deino A, Hill A, Beck CC, Campisano CJ, Valachovic J, Kingston J (2015) Paleoclimate and paleoenvironmental forcing on early humans: LOI analysis of three HSPDP drill core sites in Kenya and Ethiopia. Geol Soc Am Abstr Prog 47(7):368

Penk SBR, Altner M, Cerwenka A, Schliewen UK, Reichenbacher B (2019) New fossil cichlid from the Middle Miocene of East Africa revealed as oldest known member of the Oreochromini. Sci Rep 9:10198

Pickford M (1975a) Stratigraphy and palaeoecology of five Late Cenozoic formations in the Kenya Rift Valley. PhD thesis, University of London

Pickford MHL (1975b) Late Miocene sediments and fossils from the northern Kenya Rift Valley. Nature 256:279–284

Pickford MHL (1978a) Geology, palaeoenvironments and vertebrate faunas of the mid-Miocene Ngorora Formation. In: Bishop WW (ed) Geological background to fossil man. Geol Soc Lond Spec Publ 6:238–262

Pickford MHL (1978b) Stratigraphy and mammalian palaeontology of the late-Miocene Lukeino Formation, Kenya. In: Bishop WW (ed) Geological background to fossil man. Geol Soc Lond Spec Publ 6:263–278

Pickford M (1988) Geology and fauna of the middle Miocene hominoid site at Muruyur, Baringo district, Kenya. J Hum Evol 3:381–390

Pickford M (2001) Equidae in the Ngorora Formation, Kenya, and the first appearance of the family in East Africa. Rev Esp Paleontol 16:339–345

Pickford M (2009a) Le Rift, le plus grand conservatoire paléontologique d'Afrique. In: Hirsch B, Roussel B (eds) Le Rift Africain. Institut de Recherche pour le Développement, Marseille, pp 95–116

Pickford M (2009b) Le kaleidoscope des paléoenvironments. In: Hirsch B, Roussel B (eds) Le Rift Africain. Institut de Recherche pour le Développement, Marseille, pp 117–133

Pickford M, Johanson DC, Lovejoy CO, White TD, Aronson JL (1983) A hominoid humeral fragment from the Pliocene of Kenya. Am J Phys Anthrop 60:337–346
Pickford M, Senut B (2001a) The geological and faunal context of late Miocene hominid remains from Lukeino, Kenya. C R Acad Sci Paris 332:145–152
Pickford M, Senut B (2001b) "Millenium ancestor", a 6-million-year-old bipedal hominid from Kenya. S Afr J Sci 97:22
Pickford M, Senut B, Cheboi K (2009) The geology and palaeobiology of the Tugen Hills, Kenya: rift tectonics, basin formation, volcanics and sedimentation. Geo-Pal Kenya, vol 1. Egerton University Press, Njoro, pp 4–133
Pickford M, Senut B, Gommery D, Kipkech J (2022) New Pliocene hominid fossils from Baringo County, Kenya. Fossil Imprint 78:451–488
Pickford M, Senut B, Mourer-Chauviré C (2004) Early Pliocene Tragulidae and peafowls in the Rift Valley, Kenya: evidence for rainforest in East Africa. C R Palevol 3:179–189
Prior GT (1903) Contributions to the petrology of British East Africa. Mineral Mag 13:228–263
Rasmussen C, Reichenbacher B, Lenz O, Altner M, Penk SBR, Prieto J, Brüsch D (2017) Middle–late Miocene palaeoenvironments, palynological data and a fossil fish Lagerstätte from the Central Kenya Rift (East Africa). Geol Mag 154:24–56
Renaut RW (1982) Late Quaternary geology of the Lake Bogoria fault-trough, Kenya Rift Valley. PhD thesis, University of London
Renaut RW (1993) Zeolitic diagenesis of late Quaternary fluviolacustrine sediments and associated calcrete formation in the Lake Bogoria Basin, Kenya Rift Valley. Sedimentology 40:271–301
Renaut RW, Ego JK, Tiercelin J-J, Le Turdu C, Owen RB (1999) Saline, alkaline palaeolakes of the Tugen Hills-Kerio Valley Region, Kenya Rift Valley. In: Andrews P, Banham P (eds) Late Cenozoic environments and hominid evolution: a tribute to Bill Bishop. Geological Society, London, pp 41–58
Renaut RW, Owen RB (1980) Late Quaternary fluvio-lacustrine sedimentation and lake levels in the Baringo Basin, northern Kenya Rift Valley. Rech Géol Afr 5:130–133
Renaut RW, Owen RB (1987) Les Silts de Loboi, Les Silts de Bogoria, et autres sediments littoraux. In: Tiercelin J-J, Vincens A (eds) Le demi-graben de Baringo-Bogoria, Rift Gregory, Kenya. Bull Ctr Rech Explor-Prod Elf-Aquitaine 11:502–508
Renaut RW, Owen RB, Jones B, Tiercelin JJ, Tarits C, Ego JK, Konhauser KO (2013) Impact of Lake-level changes on the formation of thermogene travertine in continental rifts: evidence from lake Bogoria, Kenya Rift Valley. Sedimentology 60:428–468
Renaut RW, Tiercelin JJ, Owen RB (2000) Lake Baringo, Kenya Rift Valley, and its Pleistocene precursors. In: Gierlowski-Kordesch EH, Kelts KR (eds) Lake basins through space and time. AAPG Stud Geol 46:561–568
Roche D, Ségalen L, Senut B, Pickford M (2013) Stable isotope analyses of tooth enamel carbonate of large herbivores from the Tugen Hills deposits: Palaeoenvironmental context of the earliest Kenyan hominids. Earth Planet Sci Lett 381:39–51
Rossie JB, Gilbert CC, Hill A (2013) Early cercopithecid monkeys from the Tugen Hills, Kenya. Proc Natl Acad Sci 110:5818–5822
Sanders LD (1963) Geology of the Eldoret area. Geol Surv Kenya Rep 64
Sawada Y, Pickford M, Senut B, Itaya T, Hyodo M, Miura T, Kashine C, Chujo T, Fujii H (2002) The age of *Orrorin tugenensis*, an early hominid from the Tugen Hills, Kenya. C R Palevol 1:293–303
Scott JJ (2005) Taphonomy of modern and ancient vertebrate traces in the marginal sediments of saline, alkaline and freshwater lakes, Baringo-Bogoria Basin, Kenya Rift Valley. MSc thesis, University of Saskatchewan, Saskatoon
Scott JJ (2010) Saline lake ichnology: composition and distribution of Cenozoic traces in the saline, alkaline lakes of the Kenya Rift Valley and Eocene Green River formation. PhD thesis, University of Saskatchewan, Saskatoon

Senut B, Pickford M (2009) Le Rift, un laboratoire pour l'évolution géobologique. In: Hirsch B, Roussel B (eds) Le Rift Africain. Institut de Recherche pour le Développement, Marseille, pp 83–94

Senut B, Pickford M, Gommery D, Mein P, Cheboi K, Coppens Y (2001) First hominid from the Miocene (Lukeino Formation, Kenya). C R Acad Sci Paris 332:137–144

Shackleton RM (1946) Geology of the country between Nanyuki and Maralal. Geol Surv Kenya Rep 11

Shackleton RM (1950) A contribution to the geology of the Kavirondo Rift Valley. Q J Geol Soc Lond 106:345–392

Shipton C (2022) Predetermined refinement: the earliest Levallois of the Kapthurin Formation. J Paleolith Archaeol 5:4

Sier MJ, Dupont-Nivet G, Langereis C, Deino AL, Kingston JD, Cohen AS (2021) Magnetostratigraphy of the Hominin Sites and Paleolakes Drilling Project (HSPDP) Baringo-Tugen Hills-Barsemoi core (Kenya). Palaeogeogr Palaeoclimatol Palaeoecol 570:110190

Spooner E, Buckland P, Bishop WW (1969) The Kampi-ya-Samaki Beds and Kokwob Beds of the Baringo Basin, Kenya. Unpublished report, Bedford College, University of London

Tallon PWJ (1976) The stratigraphy, palaeoenvironments and geomorphology of the Pleistocene Kapthurin Formation, Kenya. PhD thesis, University of London

Tallon PWJ (1978) Geological setting of hominid fossils and Acheulian artefacts from the Kapthurin Formation, Baringo District, Kenya. In: Bishop WW (ed) Geological background to fossil man. Geol Soc Lond Spec Publ 6:361–378

Tauxe L, Monaghan M, Drake R, Curtis G, Staudigel H (1985) Paleomagnetism of Miocene East African rift sediments and the calibration of the geomagnetic reversal time scale. J Geophys Res 90:3639–4646

Tiercelin J-J, Lezzar KE (2002) A 300 million year history of rift lakes in Central and East Africa: an updated broad review. In: Odada EO, Olago DO (eds) The East African Great Lakes: limnology, palaeolimnology biodiversity. Kluwer, Dordrecht, pp 3–60

Tiercelin J-J, Nalpas T, Thuo P, Potdevin JL (2012) Hydrocarbon prospectivity in Mesozoic and early–middle Cenozoic rift basins of central and northern Kenya, Eastern Africa. In: Gao D (ed) Tectonics and sedimentation: implications for petroleum systems. AAPG Mem 100: 179–207

Tiercelin J-J, Potdevin JL, Morley CK, Talbot MR, Bellon H, Rio A, Le Gall B, Vétel W (2004) Hydrocarbon potential of the Meso-Cenozoic Turkana depression, northern Kenya. I. Reservoirs: depositional environments, diagenetic characteristics, and source rock–reservoir relationships. Mar Petrol Geol 21:41–62

Tiercelin J-J, Vincens A (eds) (1987) Le demi-graben de Baringo-Bogoria, Rift Gregory, Kenya. Bull Centr Rech Explor-Prod Elf-Aquitaine 11:249–540

Trauth MH, Maslin MA, Deino AL, Junginger A, Lesoloyia M, Odada EO, Olago DO, Olaka LA, Strecker MR, Tiedemann R (2010) Human evolution in a variable environment: the amplifier lakes of Eastern Africa. Quat Sci Rev 29:2981–2988

Trauth M, Larrasoaña J, Mudelsee M (2009) Trends, rhythms and events in Plio-Pleistocene African climate. Quat Sci Rev 28:399–411

Trauth M, Maslin M, Deino A, Strecker MR (2005) Late Cenozoic moisture history of East Africa. Science 309:2051–2053

Trauth MH, Maslin MA, Deino AL, Strecker MR, Bergner AGN, Dühnforth M (2007) High-and low-latitude forcing of Plio-Pleistocene East African climate and human evolution. J Hum Evol 53:475–486

Truckle PH (1977) Geological map of the Lake Baringo–Laikipia area (scale 1:250,000). Directorate of Overseas Surveys, East African Geological Research Unit, Kenya Rift Valley Project, Bedford College, University of London

Tryon CA (2002) Middle Pleistocene sites from the "southern" Kapthurin Formation of Kenya. Nyame Akuma 57:6–13

Tryon CA (2003) The Acheulian to MSA transition: tephrostratigraphic context for archaeological change in the Kapthurin Formation, Kenya. PhD thesis, University of Connecticut, Storrs

Tryon CA (2010) How the geological record affects our reconstructions of Early Middle Stone Age settlement patterns: the case of an alluvial fan setting for Koimilot (Kapthurin Formation), Kenya. In: Conard J, Delagnes A (eds) Settlement dynamics of the Middle Paleolithic and Middle Stone Age, vol III. Kerns Verlag, Tübingen, pp 39–65

Tryon CA, Faith JT (2013) Variability in the Middle Stone Age of Eastern Africa. Curr Anthropol 54:S234–S254

Tryon CA, McBrearty S (2002) Tephrostratigraphy and the Acheulian to Middle Stone Age transition in the Kapthurin Formation, Kenya. J Hum Evol 42:211–235

Tryon CA, McBrearty S (2006) Tephrostratigraphy of the Bedded Tuff Member (Kapthurin Formation, Kenya) and the nature of archaeological change in the later middle Pleistocene. Quat Res 65:492–507

Tryon CA, McBrearty S, Texier J-P (2005) Levallois lithic technology from the Kapthurin Formation, Kenya: Acheulian origin and Middle Stone Age diversity. Afr Arch Rev 22:199–229

Van Couvering JAH (1982) Fossil cichlid fish of Africa. Spec Pap Palaeontol 29:1–103

Walker EE (1903) Reports on the geology of the East Africa Protectorate. Colonial Reports 11. HMSO, London

Walsh J (1969) Geology of the Eldama Ravine-Kabarnet area. Geol Surv Kenya Rep 83

Ward S, Hill A (1987) Pliocene hominid partial mandible from Tabarin, Baringo, Kenya. Am J Phys Anthropol 72:21–37

Westover KS, Stone JR, Yost CL, Scott JJ, Cohen AS, Rabideaux NM, Stockhecke M, Kingston JD (2021) Diatom paleolimnology of late Pliocene Baringo Basin (Kenya) paleolakes. Palaeogeogr Palaeoclimatol Palaeoecol 570:109382

Williams LAJ, Chapman GR (1986) Relationships between major structures, salic volcanism and sedimentation in the Kenya Rift from the equator northwards to Lake Turkana. In: Frostick LE, Renaut RW, Reid I, Tiercelin J-J (eds) Sedimentation in the African rifts. Geol Soc Lond Spec Publ 25:59–74

Willis B (1936) Studies in comparative seismology; East African plateaus and rift valleys. Carnegie Institution for Science, Washington DC

Wilson KE (2011) Plio-Pleistocene reconstruction of East African and Arabian Sea palaeoclimate. PhD thesis, University of London

Wilson KE, Maslin MA, Leng MJ, Kingston JD, Deino AL, Edgar RK, Mackay AW (2014) East African lake evidence for Pliocene millennial-scale climate variability. Geology 42:955–958

Winkler AJ (2002) Neogene paleobiogeography and East African paleoenvironments: contributions from the Tugen Hills rodents and lagomorphs. J Hum Evol 42:237–256

Wood B (1999) Plio-Pleistocene hominins from the Baringo Region, Kenya. In: Andrews P, Banham P (eds) Late Cenozoic environments and hominid evolution: a tribute to Bill Bishop. Geological Society, London, pp 113–122

Wood B, Van Noten F (1986) Preliminary observations on the BK 8518 mandible from Baringo, Kenya. Am J Phys Anthropol 69:117–127

Yost CL (2019) Phytoliths and microcharcoal from East African lake sediments used to reconstruct Plio-Pleistocene vegetation and hydroclimate. PhD thesis, University of Arizona, Tucson

Yost CL, Ivory SJ, Deino AL, Rabideaux N, Kingston JD, Cohen AS (2021) Phytoliths, pollen, and microcharcoal from the Baringo Basin, Kenya reveal savanna dynamics during the Plio-Pleistocene transition. Palaeogeogr Palaeoclimatol Palaeoecol 570:109779

Chapter 18
The Central Kenya Rift Basins (Nakuru-Elmenteita-Naivasha)

18.1 Volcanic Rocks and Geological Background

The northwest to southeast trending, 80-km-wide, Central Kenya Rift includes three basins, about 70 to 130 km northwest of Nairobi, that contain Lakes Nakuru (1758 m above sea level), Elmenteita (1786 m) and Naivasha (1889 m) (Fig. 18.1). The lakes are located at the crest of the rift valley floor, which drops in height both southwards and northwards. This segment of the rift contains three major Pleistocene volcanoes (Fig. 18.1a). Longonot lies to the south and Menengai to the north, with the volcanic massif of Eburru dividing the rift into two major segments that contain Lake Naivasha (south) and Lakes Elmenteita and Nakuru (north) (Thompson and Dodson 1963; McCall 1967; Onywere 1997). The eastern rift margin is marked by a series of major step faults that rise to the Kinangop Plateau and to heights of >3860 m in the Aberdare (Nyandarua) Mountains (Figs. 18.1b and 18.2a). The western flank is marked by the Mau Escarpment at heights of >3000 m. The area is reached from Nairobi via the A104 highway, with several good tarmac and earth roads providing access to the three drainage basins.

The Central Kenya Rift was initiated when an east-facing half-graben developed about seven million years ago and was filled with Miocene flood phonolite and Mio-Pliocene trachyte (Fig. 18.2b; Leat 1991; Roessner and Strecker 1997; Conti et al. 2021; Richter et al. 2021). Baker et al. (1988) recognised four major volcanic episodes that were separated by four phases of faulting. They suggested that Pliocene ash flows (Mau Tuffs to the west; Kinangop Tuffs to the east) accumulated between 4.5 and 3.4 Ma, after initial development of the eastern border fault (Clarke et al. 1990; Bergner et al. 2003). Major faulting took place along the Sattima Fault during and after eruption of the Kinangop Trachytes between ~3.7 and 3.4 Ma (Fig. 18.2c).

The second phase of faulting (Baker et al. 1988) produced step faults, the west-dipping Naivasha-Kijabe Escarpment, and the Bahati-Kinangop Plateau between 3.4 and <1.9 Ma (Fig. 18.2d; Roessner and Strecker 1997) and which today marks the eastern margin of the modern inner rift (Fig. 18.2e). A thick sequence of

R. W. Renaut, R. B. Owen, *The Kenya Rift Lakes: Modern and Ancient*, Syntheses in Limnogeology, https://doi.org/10.1007/978-3-642-25055-2_18

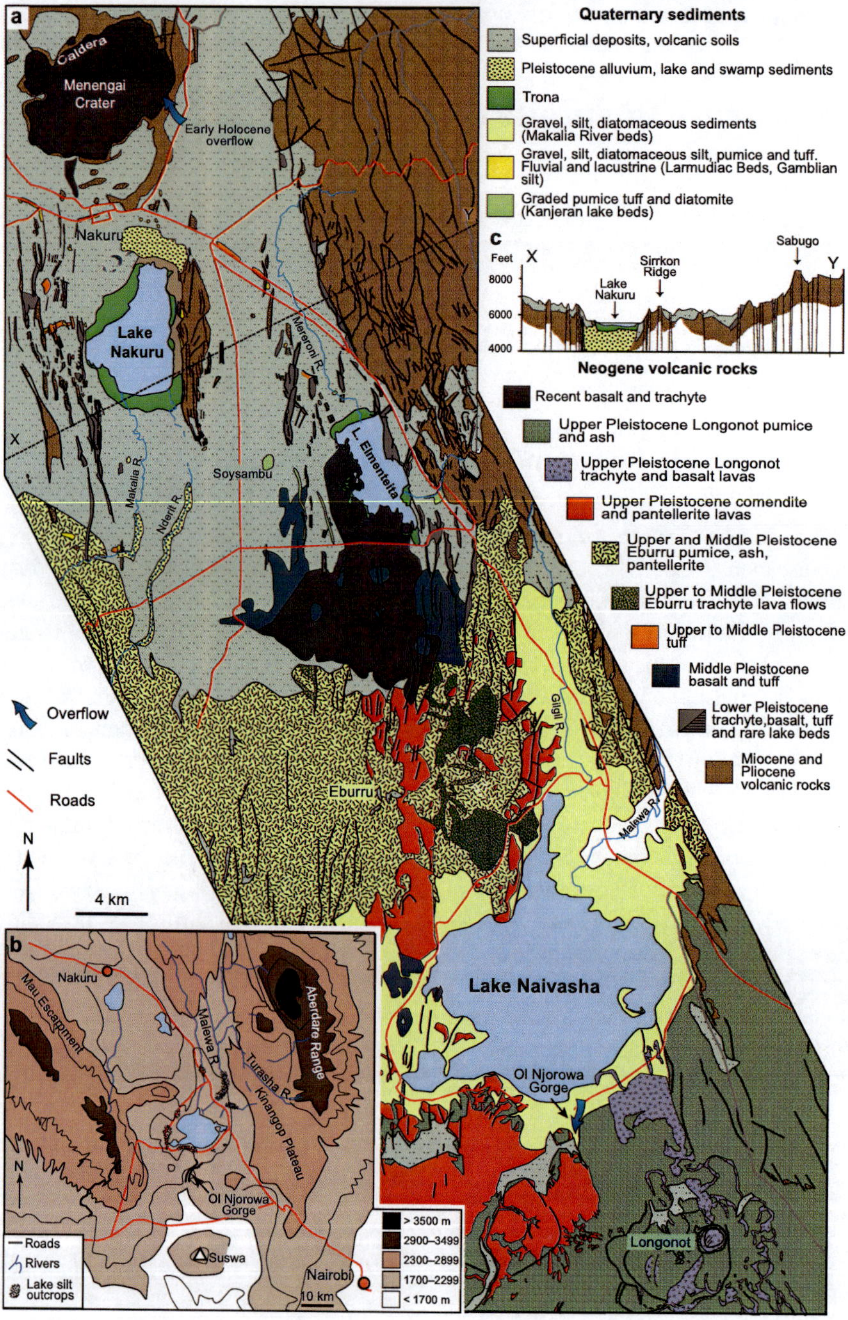

Fig. 18.1 Geological map of the Central Kenya Rift. Simplified after McCall (1967) and Woodhall and Clarke (1988). (**a**) Geology of the rift. (**b**) Topography. Outcrops of high-level lake silt also shown. (**c**) Cross section (X to Y) across the Nakuru Basin

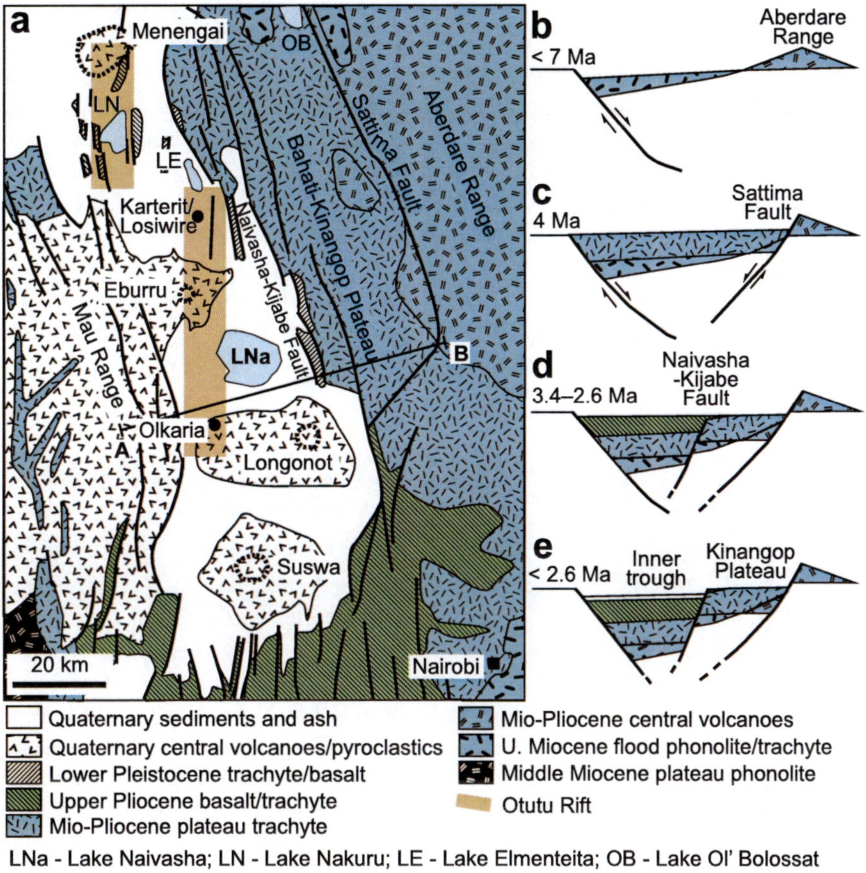

Fig. 18.2 Major tectonic stages in the evolution of the Central Kenya Rift. After Roessner and Strecker (1997). (**a**) Simplified geology. (**b**–**e**) Geological development of the rift valley through the last seven million years

Quaternary flood trachytes (the Limuru Trachytes) was erupted at ~2.64–1.66 Ma with minor basalt prior to 1.7 Ma.

Renewed faulting led to further definition of the modern rift with flood lavas (the Plateau Trachytes Formation) erupting on the rift floor between 1.65 and 0.9 Ma. To the east, lavas of the Kijabe Basalt Formation flowed over pre-existing marginal step-faults. According to Baker et al. (1988), a fourth phase of minor faulting took place between 0.8 and 0.4 Ma, after which Late Quaternary to Recent salic volcanism (trachyte, pumice, comendite and pyroclastics) developed between ~0.4 and 0 Ma. These rocks formed the Longonot, Eburru and Olkaria volcanic centres (Fig. 18.1a; Scott 1980; Clarke et al. 1990; Odongo 1993) and resulted in closure of the Naivasha Basin before ~320 ka (Trauth and Strecker 1996). Strecker et al. (1990) also reported that the direction of extension rotated from east-west to northwest-southeast during the Pleistocene.

More recently, the last of Baker's phases of activity has been related to the one-million-year-old, 30-km-long, N-S trending Otutu Rift (Fig. 18.2a), which forms an offset graben that cuts through the central part of the main rift between Menengai and Olkaria, two locations being developed for geothermal power (Nyawir et al. 2018). The Otutu Rift has also been associated with sub-lacustrine eruptions below an Early to Middle Pleistocene, expanded and combined, Naivasha-Elmenteita-Nakuru lake. The eruptions produced both hyaloclastite rings and ash deposits that interfinger with lacustrine diatomite near the Losiwire and Karterit cones in the Elmenteita Basin (Fig. 18.2a). Eruptions from Menengai (Riedl et al. 2020) and other volcanoes have produced many tuffs that provide chronological control for studies of stone tool assemblages in the region, as well as the largest known Late Pleistocene eruption, which formed the Menengai Tuff at about 36 ka (Blegen et al. 2016).

18.2 Quaternary Shorelines and Pluvial Hypotheses

The Kenya Rift was once thought to have been largely flooded by a very large water body during the Kamasian (Leakey 1934), but this idea was abandoned as further geological evidence accumulated (McCall et al. 1967). Early geological investigations were stimulated by the discovery of stone tools (Wilshaw 2012) and lacustrine sediments located well above the contemporary lakes in the Nakuru-Elmenteita-Naivasha area (Leakey and Solomon 1929; Zeuner 1948). These deposits, and sediments from other parts of the Kenya Rift, were used to infer a series of former high lake levels that were tentatively correlated with glacial advances in northern latitudes under a 'pluvial' hypothesis. Five main 'wet' periods were distinguished: the Kageran (oldest), Kamasian (Eburrian), Gamblian (lower part = Enderrian), Makalian and Nakuran (Leakey 1931a, b, c, 1934; Nilsson 1931, 1935, 1938, 1940; Burkitt 1945; Cole 1950; Kingston and Hill 2005) with these periods formally accepted as climato-stratigraphic units through Resolutions 1–4 at the third Pan African Congress on Pre-history (Clark and Cole 1957).

The pluvial concept evolved through the 1930s to 1960s, leading to several alternative schemes being developed (Table 18.1). For example, the Kamasian was subdivided into the lower and upper Kamasian, which were then renamed the Kamasian and Kanjeran (Leakey 1950, 1955; Leakey and Cole 1952; Cole 1954, 1963). Nilsson (1931) presented evidence for pluvial and interpluvial intervals based on shoreline data (and studies of equatorial glaciers). He recognised seven distinct still-stands, including the modern shoreline, and inferred tilting of the oldest shorelines (Fig. 18.3a, b). Nilsson noted that many of his shorelines might either represent stages in a retreating lake or different stands caused by both rising and falling lakes. However, he also pointed out that lacustrine sediments along the Makalia and Nderit (Enderit) Rivers in the Nakuru Basin (Fig. 18.4a–c) contained disconformities between lacustrine sedimentary packages that supported a pattern of alternating transgressions and regressions as the main explanation for the different lake stands.

Table 18.1 Historical changes in Quaternary climate-related terminology in the Central Kenya and Baringo-Bogoria Rifts

Authors	**Gregory** (1921)	**Leakey** (1931b) **/Fuchs** (1950)	**Nilsson** (1931)		**Leakey** (1950, 1955)	**Cole** (1954)			**Cole** (1963)	**McCall et al.** (1967)		Pluvial terminology abandoned
Location	*Kamasia (Baringo)*	*Kamasia (Baringo)*	*Nakuru-Elmenteita (tentative correlations with European glaciations)*		*Kamasia (Baringo)*	*Kenya Rift*	*Kavirondo*	*Climate phases*	*East Africa*	*Kamasia (Baringo)*	*Nakuru-Elmenteita*	
Holocene		High lake	Nakuran wet phase	Wet period ca. 850 BC		Nakuran shoreline		Wet and dry	Nakuran Drier Makalian Drier Makalian 1?	High lake	Recent lake Nakuran shoreline Makalia/ Kariandus Beds Nderit 'Gamblian' sediments	
			Dry	Climate opt.								
			Makalian wet phase Very dry	Bühl stadial etc. Achen retreat								
Pleistocene		'Kamasian' sediments at Baringo			Lower Kamasian Upper Kamasian	Aeolian sand		Dry	Drier	Upper Kamasian (=Kapthurin Beds)		
			U. Gamblian M. Gamblian L. Gamblian Faulting, volcanism Kamasian pluvial	Würm glacial Riss–Würm Interglacial Riss glacial Mindel–Riss interglacial Gunz–Mindel glaciations/ interglacials		Gamble's Cave deposits/ Malewa Beds		Gamblian Pluvial Dry	Gamblian 3 Drier Gamblian 2 Drier Gamblian 1 Drier		Tuff capping Larmudiac sediments Kariandusi lake beds	
						Olorges-ailie Lake Beds	Kanjera Beds	Kanjeran pluvial	Kanjeran Wetter? Kamasian			
							U. Rawe Beds	Dry				
							L. Rawe Beds	Kamasian pluvial				
						U. Omo Beds		Dry		Lower Kamasian (=Chemeron Beds)		
						L. Omo Beds	Kanam Beds	Kageran pluvial				
Pliocene												

Table 18.1 (continued)

Oligocene/ Miocene (?)	Nyasan System at Baringo: sediments later transferred to the Pleistocene as 'Kamasian' and Plio-Pleistocene as the Lower and Upper Kamasian. Subsequently recognised as forming the Chemeron and Kapthurin Beds (eventually raised to Formation status).	**Correlation of pluvials with faunal stages in** Bishop (1967): Nakuran – Post Olduvai Makalian – Post Olduvai Gamblian – Post Olduvai Kanjeran – Upper Olduvai (Beds III and IV) Kamasian – Lower Olduvai (Beds I and II) Kageran – Omo-Kanam

Climate-related terms were abandoned from the late 1960s. Note dating of boundaries between 'pluvials' was associated with considerable uncertainty

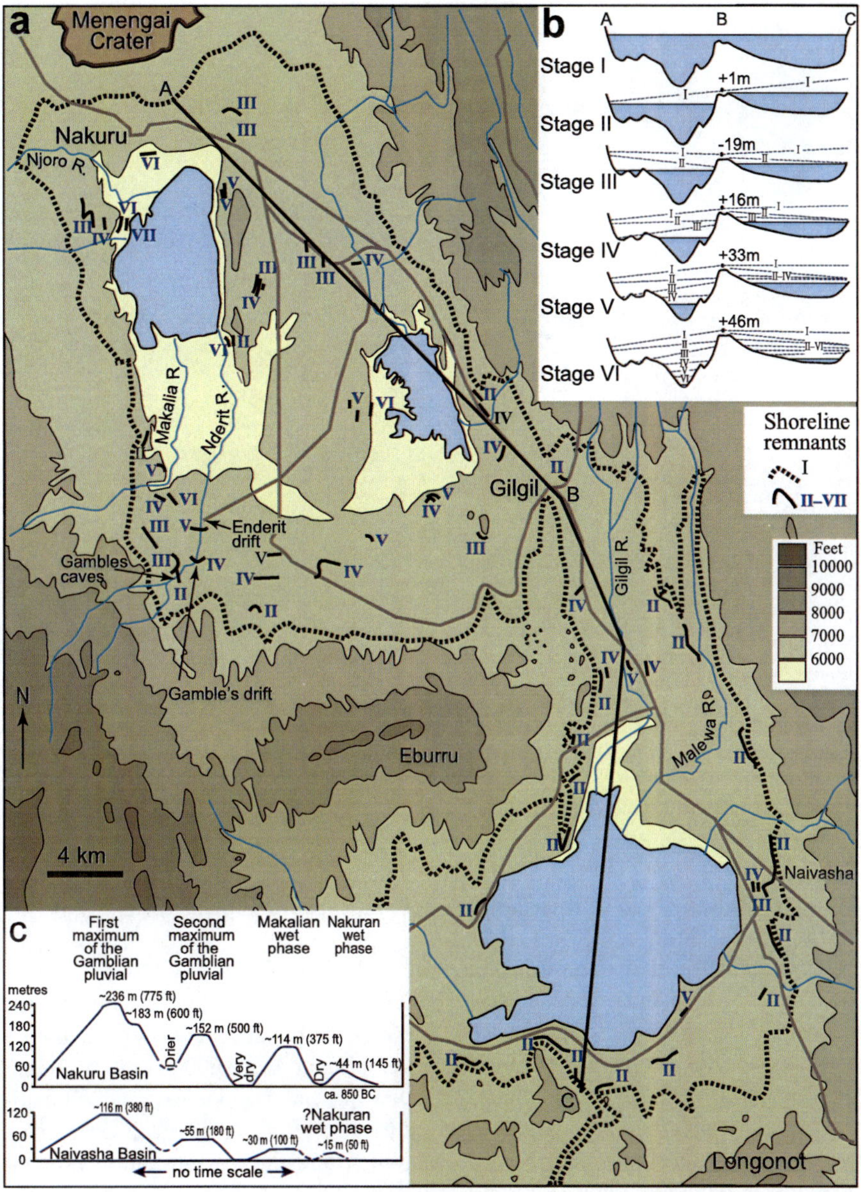

Fig. 18.3 Nilsson's Late Quaternary shorelines in the Nakuru-Elmenteita-Naivasha Basin. Detailed shorelines were mapped and tectonic tilting was inferred, but these data and ideas were subsequently challenged (Butzer et al. 1972) and largely abandoned. (**a**) Reconstructed highest lake shoreline dashed. Lower shorelines indicated by beach deposits and wave-cut cliffs at specific locations (II–VII, from oldest to youngest). After Nilsson (1931). (**b**) Inferred tectonic tilting of the shorelines (I–VII). The deformed shoreline 'I' was believed to be at its highest at 'B' with tilting effects most extreme during the development of Stages I–IV. Younger shorelines formed during Stages V–VII were thought to show less deformation. Filled circle heights are relative to plane A-B-C in Stage I. Length at 1:100,000; Height at 1:20,000. After Nilsson (1931). (**c**) Pluvial episodes and lake levels in the Nakuru and Naivasha basins. After Leakey (1931a). Later research challenged these inferences (Washbourn-Kamau 1971)

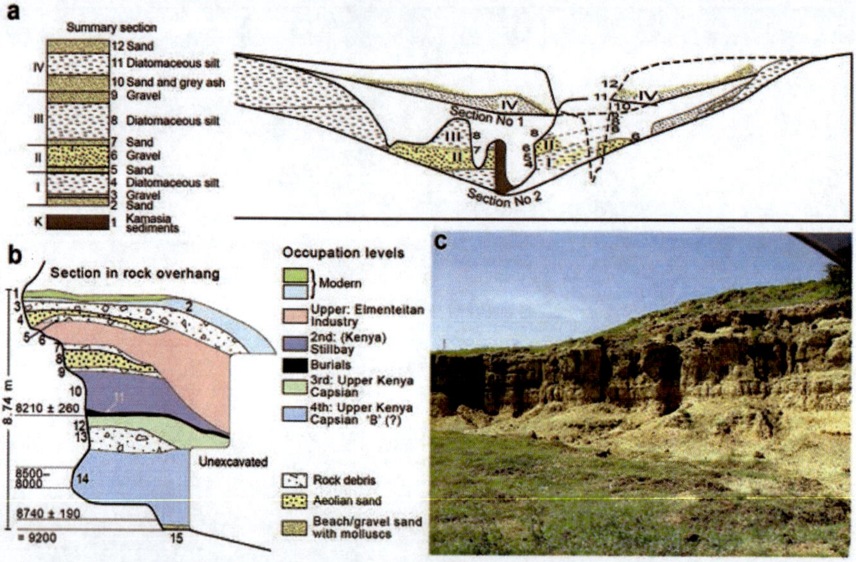

Fig. 18.4 The pluvial hypothesis and evidence for fluctuating lake levels in the Nakuru, Elmenteita and Naivasha basins. (**a**) After Nilsson (1931); (**b**) After Leakey (1931b) with dates added by Protsch (1978). (**a**) Nilsson's cross-section of lake sediments at Nderit Drift. Two sections along the river with a diagrammatic summary to the left, location in Fig. 18.5a. Nilsson suggested that 'Kamasian' sediments were overlain by younger deposits (packages I to IV) and that sediments 4.5 km to the west were associated with a wave-cut terrace that represented a fifth lake stage. Horizontal scale 1:2000; Vertical scale 1:800. (**b**) Multiple Early Holocene occupation levels overlie a beach deposit at the wave-cut Gamble's Caves. Location in Fig. 18.5a. (**c**) 'Makalian' deltaic/lacustrine sediments, Nderit River channel. Location in Fig. 18.5a

Solomon (1931) and Leakey (1931b) placed these fluctuations in a pluvial scheme that they claimed could be recognised in both the Nakuru-Elmenteita and Naivasha basins (Fig. 18.3c).

Although not linked to any absolute time scale, two phases of a 'Gamblian pluvial' were inferred and associated with 775 ft (236 m) and 500 ft (152 m) shorelines in the Nakuru Basin (Fig. 18.3c). These phases were believed to be separated by a dry interval with a temporary strandline developing at about 600 ft (183 m). Over subsequent decades the Gamblian and other pluvial periods became climatostratigraphic units that were widely referred to across East Africa with an inferred correlation, for example, between the Gamblian and the last glaciation (Würm/Wisconsin) at high latitudes (Leakey 1955).

A sequence of thickly bedded, coarse-graded pumice tuffs, with less common silty horizons, thin diatomites and clays, up to at least 16 m thick, constitute the Upper Pleistocene Larmudiac beds (Tables 18.1 and 18.2). These were once correlated with the lower part of the Gamblian, in the Nakuru Basin. However, minor faulting suggests that they are slightly older since inferred Gamblian deposits south of Nakuru are not fractured (McCall 1967).

Table 18.2 Stratigraphy of the Central Kenya Rift

<table>
<tr><td colspan="2">Holocene</td><td colspan="2" rowspan="2">SUPERFICIAL DEPOSITS: Soil, Alluvium etc.
VOLCANIC ROCKS: UPPER MENENGAI VOLCANICS
Trachyte – Lava flows, scoria cones.
SEDIMENTS: Makalia sediments (deltaic?).
------------weak unconformity----------
Gamblian sediments (Nderit and Makalia sections).</td><td rowspan="2">ELMENTEITA BASALTS
Olivine basalt – lava flows, scoria cones, (200 ft?).</td></tr>
<tr><td rowspan="10">Pleistocene</td><td rowspan="4">Upper</td></tr>
<tr><td colspan="3">MINOR FAULTING new fractures and renewals on older fracture lines west of Lake Nakuru.</td></tr>
<tr><td colspan="3">Tuffs/sediments (lacustrine), Larmudiac sediments in Nakuru Basin. Pumice showers from Menengai (possibly older).</td></tr>
<tr><td colspan="3">MINOR FAULTING new fractures and renewals on older fracture lines in Nakuru-Elmenteita Basin.</td></tr>
<tr><td rowspan="6">Lower to Middle</td><td colspan="3">Older basalts of Elmenteita: Olivine basalt flows at Soysambu.</td></tr>
<tr><td colspan="2">Kariandusi lake beds.
'Kanjeran'; Acheulian artifacts.</td><td>Tuff and Agglomerate (Elmenteita cones; Honeymoon Hill cones) – Phreatic?</td></tr>
<tr><td colspan="3">THIRD MAJOR FAULTING possible formation of Menengai Caldera. Complex grid faulting and dropping of rift floor several hundred feet.</td></tr>
<tr><td>EROSION</td><td>Gilgil Trachyte (lava flows).
Enigma Cove sediments (Lacustrine).
Mbaruk Basalt (lava flows). -----
Willan's Farm Trachyte (lava flows).</td><td>Ronda sediments (Lacustrine).
Ronda Phonolite and Trachyte (lava flows).
West Cliff sediments (lacustrine).
————————————
Nakuru Lake Syndicate (<500 ft).</td></tr>
<tr><td colspan="3">Late pumice tuff, welded tuff (ignimbrite) and sediments as unconformable outliers on Kinangop and Bahati tuffs.</td></tr>
<tr><td colspan="3">SECOND MAJOR FAULTING Less complex and less closely spaced than the third; forms tilt blocks dropping rift floor ca. 3000 ft in Nakuru sector.</td></tr>
</table>

(continued)

Table 18.2 (continued)

Pliocene	MAU TUFFS Pumice tuff. (ignimbrite). **Lacustrine sediments.** Intercalated phonolites.	4. BAHATI TUFFS Pumice tuff. **Lacustrine sediments.** 3. **Kisinana sediments (lacustrine).** 2. Lake Hannington phonolites. 1. Goitumet Basalt.	5. KINANGOP TUFFS Pumice tuff, **Lacustrine sediments.** 4. Fragmental Trachytes of Kariandusi. 3. North Kinangop trachytes. **2. Turasha/ Oleolondo Lake Beds.** 1. Turasha Basalt, olivine basalt flow.	LOWER MENEGAI VOLCANIC SERIES and SIRRKON VOLCANIC SERIES
	Total thickness of Pliocene succession >1000 ft			
	FIRST MAJOR FAULTING Less complex and less closely spaced than the third; forms tilt blocks dropping rift floor minor in Nakuru sector.			
??	THOMSON'S FALLS PHONOLITES, OLORONYI BASALT, OL DOINYO OLIEP BASALT, RUMURUTI PHONOLITES, SATTIMA LAVAS. Maximum thickness = 2000 ft.			
Miocene	SAMBURU VOLCANIC SERIES Basalts – numerous flows and feeder dykes. Picrite Basalts, Picrites. **Intercalated tuffs and lacustrine sediments.** No trace of central volcanoes. Thickness 1–2000 ft.		SIMBARA VOLCANIC SERIES Basalts – Thin flows and feeder dykes. Picrite Basalts. Agglomerate and tuff intercalated. Central volcanoes at Kipipiri and Sattima (?). Thickness 1000 ft perhaps more.	
	Subvolcanic sediments containing *Deinotherium hobleyi*, Mastodon sp.			
~20 Ma	FIRST MINOR WARPING OF RIFT VALLEY FORMING BASINS WITH SUBVOLCANIC SEDIMENTS.			
Early Tertiary, Mesozoic, Palaeozoic		Successive erosion cycles and epeirogenic movements.		
Precambrian		Basement System: Precambrian structures partially control Tertiary-Quaternary structures.		

Simplified after McCall (1967)

Lacustrine sequences in bold. Note the base of the Pleistocene had been placed at 1 Ma in original documents, but this has been changed to the currently recognised 2.6 Ma date for this table

Research by Washbourn-Kamau raised many questions about the shorelines identified by Nilsson. She could not find good evidence for the 775 ft shoreline and suggested several identified features might represent breaks of slope related to faulting (Washbourn-Kamau, 1970, 1971, 1975; Kamau 1972). Butzer et al. (1972) reported evidence from the Nakuru Basin that indicated high lake levels prior to

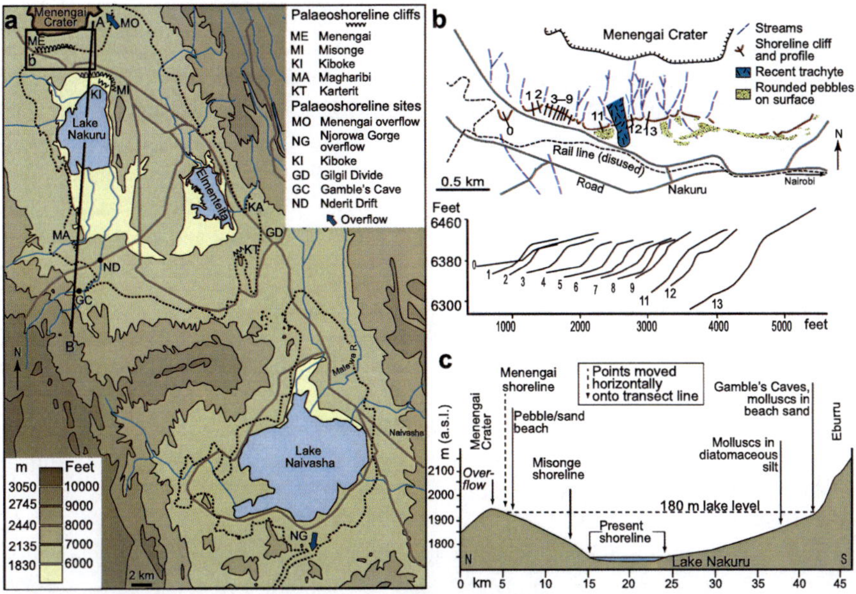

Fig. 18.5 Lake palaeoshorelines. After Washbourn-Kamau (1971). (**a**) Washbourn-Kamau mapped shoreline cliffs in the Nakuru-Elmenteita Basin and palaeoshoreline sites. Dashed line = Maximum Early Holocene ('Makalian') lake shoreline. Note overflows to north and south at Menengai and Njorowa Gorge. (**b**) Transect showing wave-cut cliffs and maps of beach deposits on Menengai Crater. (**c**) Washbourn-Kamau's model, suggesting one major Holocene high lake shoreline (~180 m above Lake Nakuru) that may have been reoccupied several times

about 20 ka followed by a dry period and a return to expanded lakes during the Early Holocene when Lakes Nakuru and Elmenteita formed a single water body at about 180 m (~600 ft level) (Fig. 18.5a–c; Chap. 10). Beach sediments and wave-cut erosion at Gamble's Caves (Fig. 18.4b), an important archaeological site (Leakey 1931b; Protsch 1978), appear to be related to a palaeolake that might have overflowed northwards into Menengai Crater, where Kamau (1972) had recognised near-shore deposits and wave-cut cliffs (Fig. 18.5b). Menengai caldera also appears to have contained a lake with diatomites in its northern and eastern areas dated at 10,440 and 4555 y BP at mean elevations of 1770 m (Riedl et al. 2020). Leat (1984) also reported a caldera lake palaeo-highstand at 1860 m.

Butzer et al. (1972) discussed sediment from the southern Nakuru Basin that had been reported by the University of California Archaeological Research Group in the late 1960s. This work recognised three lacustrine phases based on sedimentary packages (A, B and C) separated by disconformities. Unit A was disconformably overlain by Unit B and yielded a ^{14}C date of 19 ka with the sediments containing Middle Stone Age artefacts and showing two to three lacustrine cycles, depending on location. Unit B was correlated with the Early Holocene Gamble's Cave shoreline, given its height of at least 160 m. Unit C included a diatomaceous silt (at +54 m) that rested on Menengai ash, and which was overlain by silts dated at 3550 y BP. The data implied a correlation with the Misonge shoreline cliff (Fig. 18.5a, c) and an age of about 4–6 ka.

Subsequently, the validity of the pluvial concept was rejected, and its associated terminology was abandoned since it did not fully account for regional tectonic and volcanic influences on sedimentation (Cooke 1958; Flint 1959a, b; Hay 1963). McCall et al. (1967), for example, noted that correlations between high lakes in East Africa and northern hemisphere glaciations were untenable since many of the breaks between different pluvial intervals reflected faulting and (or) volcanism. They also observed that some of the wetter episodes, such as the Gamblian and Makalian pluvials, had been referred to sedimentary packages of different ages. From the late 1960s and early 1970s, research made greater use of geological mapping, detailed logging of sections, correlation with stone tools, palaeontology and radiometric dating to reconstruct environmental histories. These efforts also placed environmental change into a more detailed local/regional context that incorporated a greater recognition of the role of tectonism (Evernden and Curtis 1965; Butzer et al. 1972; Blisniuk et al. 1988; Bower and Nelson 1978; Clarke et al. 1990; Marean et al. 1994; Bergner 2003; Wilshaw 2012; Frahm and Tryon 2018). Furthermore, the advent of lake coring in East Africa significantly improved the overall understanding of Late Quaternary environmental change (Kendall 1969; Richardson 1966, 1972; Richardson and Dussinger 1986).

18.3 Mio-Pleistocene Sediments and Palaeogeography

18.3.1 Mio-Pliocene Sediments and Palaeogeography

Sub-volcanic sediments accumulated during the initial warping of the rift valley some 20 Ma ago (Table 18.2). However, the oldest recognised lacustrine sediments are Miocene well-graded pumice tuffs and diatomites related to ponding by lavas during eruption of the Samburu Volcanic Series. The Pliocene Mau Tuffs, Bahati Tuffs and Kinangop Tuffs also contain intercalations of lacustrine sediments, variously including diatomites, reworked tuffs, silts and gravels. Locally, these have been referred to as the Oleolondo and Turasha beds (McCall 1957, 1967; McCall et al. 1967). For example, a 90-m-thick sequence of diatomite, diatomaceous silt and tuff rests on ~4.6 Ma trachyte in the Turasha River Valley (Fig. 18.1b) and suggests the presence of a freshwater Pliocene lake that may have extended across the Central Kenya Rift (Strecker 1991; Boven 1992; Trauth et al. 2007). In the Oleolondo Valley (Kinangop Plateau), lake sediments contain the remains of monocotyledonous plants (McCall et al. 1967).

18.3.2 Lower to Middle Pleistocene Sediments and Palaeogeography

Three major faulting episodes (Table 18.2) have been used to help stratigraphically distinguish lake sequences where there are few other lines of evidence for dating (McCall 1967). Three such sequences (Enigma Cove, West Cliff and Ronda Hill sediments) lie between the second and third major faulting periods and are thought to be of Lower to Middle Pleistocene age. The Enigma Cove sediments (Table 18.2), near the Gilgil Escarpment, were once correlated with the Kariandusi lake beds. They are comprised of a thin diatomite between two trachyte flows that rapidly thickens northwards and includes pumice tuffs. The sediments lie below the Gilgil Trachyte, confirming that they predate the deposits at Kariandusi. The West Cliff Sediments at Lake Nakuru consist of a few metres of stratified pumice and minor diatomite between the Mbaruk Basalt and Ronda lava flows. At Ronda Hill, the strongly faulted fluvio-lacustrine and lake marginal Ronda Hill sediments rest conformably on Ronda lavas and are comprised of coarse yellowish layered tuff and clay younger than the West Cliff deposits (Table 18.2; Mboya 1993).

Laminated lacustrine silts and sands with ripple marks are also present in a mainly terrestrial volcanic sequence (2.49–0.13 Ma) associated with Oldowan and Acheulian stone tools at Kilombe, an eroded Pleistocene volcano 25 km northwest of Menengai Crater (Gowlett 1978, 1993; Hoare et al. 2021). The Kilombe volcano held a former crater lake in which >100 m of laminated clays were deposited, some selectively cemented, that alternate with tuffs (Jennings 1971; Gowlett et al. 2015). The lake deposits contain phytoliths and diatoms and are dated at 1.8–1.7 Ma (Gowlett et al. 2022). More extensive lacustrine diatomites are present at Kariandusi (SE Nakuru-Elmenteita Basin; Fig. 18.6) where they overlie the Gilgil Trachyte, dated at ~1.05 Ma (Bergner 2003) and 0.99 Ma (Trauth et al. 2005). The Kariandusi lake sediments are overlain by a ~0.95-Ma-old tuff (Evernden and Curtis 1965) – a date that was revised to 0.97 Ma (Durkee and Brown 2014). The Kariandusi deposits have also been correlated with sediments that formed when lakes expanded in the Olorgesailie, Turkana and Afar regions (Trauth and Maslin 2009), although the Olorgesailie palaeolake underwent multiple phases of expansion, contraction, and desiccation during that period (Owen et al. 2008, 2009).

The sediments at Kariandusi (Fig. 18.6c–e) are associated with Acheulian stone tools (Fig. 18.6f; Gowlett and Crompton 1994; Munga 2014). Kübler et al. (2016 suggested that the area was then part of a corridor between edaphically rich wildlife areas. The deposits include up to 45 m of Pleistocene diatomite, diatomaceous silt and volcaniclastic sediments that are well exposed in a former quarry and adjacent areas east of Lake Elmenteita (Figs. 18.6a and 18.7a, b). The diatom floras are dominated by planktonic taxa, including *Stephanodiscus transylvanicus*, *S. niagarae* and *Aulacoseira granulata*, which indicate a fresh deep-water lake (>100 m?) that extended across much of the palaeorift (Bergner 2003). A variety of Fragilariaceae, *Cyclotella meneghiniana* (*Stephanocyclus meneghinianus*) and *Rhopalodia* spp. are present in lower and higher parts of the sequence implying a

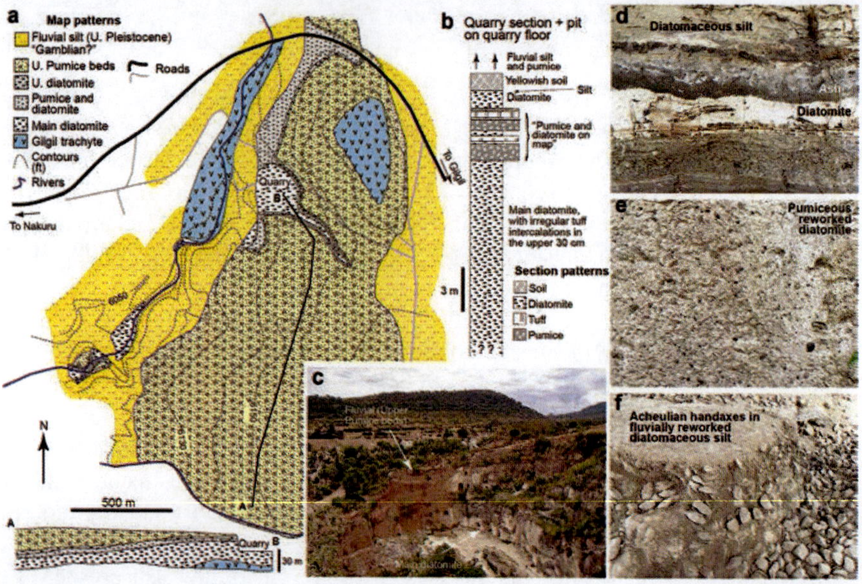

Fig. 18.6 The Kariandusi diatomite quarry and associated sediments. Diatomites, diatomaceous sediments and reworked/airfall tuffs and pumice provide evidence for a large Early to Middle Pleistocene lake that spread across the Naivasha, Nakuru and Elmenteita basins about 1 million years ago. After McCall (1967). (**a**) Map of the Kariandusi area showing extensive diatomite and diatomaceous silt outcrops, marked by caves shown in (**c**). (**b**) Section in the quarry area. (**c**) The Kariandusi quarry. Note white diatomites overlain by trough cross-bedded fluvial pumice. (**d**) Alternating ash, diatomaceous silt and diatomite (5 cm thick). (**e**) Rounded pumice clasts (~0.2–0.4 cm) in reworked diatomite. (**f**) Fluvially concentrated Acheulian handaxes (up to 10 cm long) and cleavers in reworked diatomaceous silt

greater littoral influence (Bergner et al. 2009). Bergner and Trauth (2004) noted that these diatom taxa indicate slightly alkaline water with a pH of ~7–7.5 and low conductivities (<500 μS cm^{-1}). Late Pleistocene fluvial pumice and ash deposits cut deeply into the Kariandusi diatomite (Figs. 18.6c and 18.7a) indicating a subsequent cut-and-fill event of indeterminate age (Bergner 2003).

Bergner et al. (2009) noted a long-term trend in the Central Kenya Rift from deep, large freshwater lakes at about one million years ago towards smaller, more compartmentalised water bodies with increased alkalinities. They attributed this to tectonic, volcanic and climatic factors, noting that reduced basin subsidence, accompanied by higher sediment supply, brought about basin infilling at a time when long-term increased aridity was affecting the region (Trauth et al. 2009). During the Lower Pleistocene, the inner rift was characterised by a broad rift valley with central volcanoes largely absent (Fig. 18.7e). This was filled by a large water body during a wetter interval at about 987 to 977 ka (Fig. 18.7f). Tectonism had developed sufficient accommodation space to allow a large lake to form when climatic conditions were favourable. Trauth et al. (2007, 2009) pointed out the importance of insolation in driving the underlying climate variability that brought about

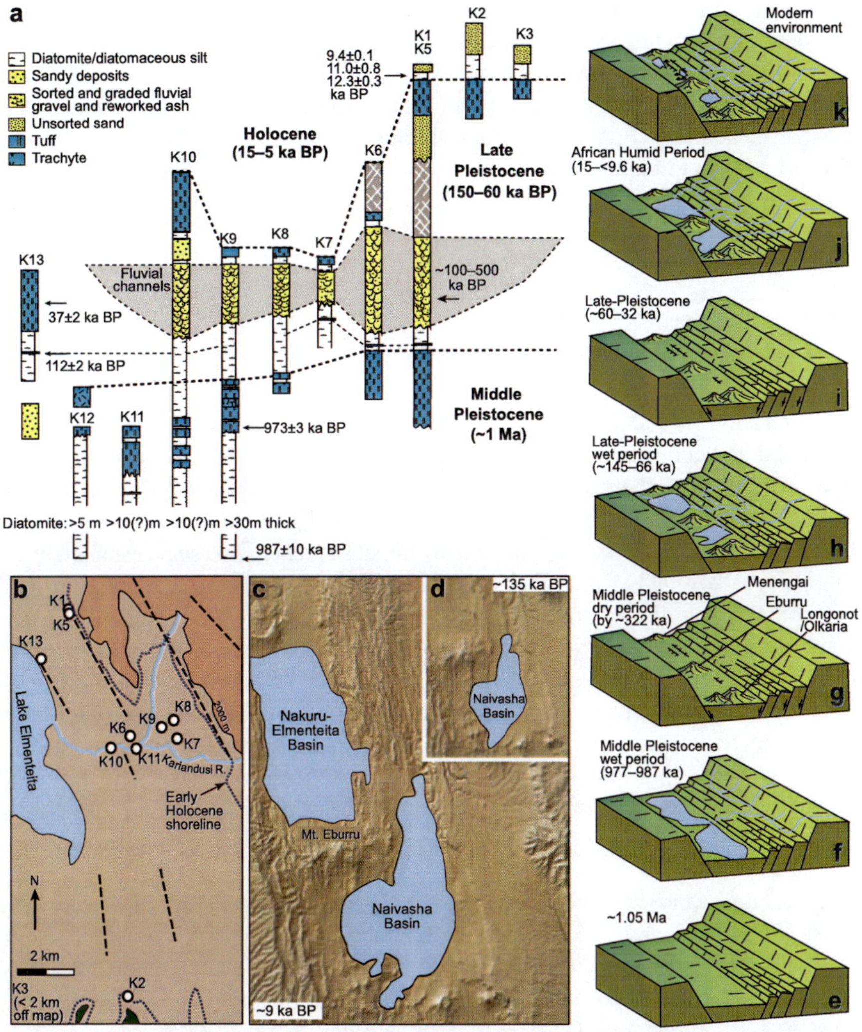

Fig. 18.7 Sediment stratigraphy and evolution of the Central Kenya Rift. Based on Bergner et al. (2003, 2009). **(a)** Sediment stratigraphy and correlation. **(b)** Outcrop locations. **(c, d)** Maps of highstands at 135 ka and 9 ka. The 135-kyr palaeolake stood ~1900 m above sea level and was ~150 m deep with an area of 520 km^2 and volume of 55 km^3. **(e–k)** Models of Pleistocene environmental change. **(e)** Major tectonism, volcanism and eruption of trachytes on rift floor forms large rift basin. **(f)** Single large lake occupies the rift floor. **(g)** Volcanism and tectonism. Closure of Naivasha outlet and formation of two basins with eruption of Eburru at valley centre. **(h)** Two separate lakes fill Nakuru-Elmenteita and Naivasha basins. Overflows at Menengai to north and Ol Njorowa to south. **(i)** Minor tectonism and volcanism. Headward erosion of rivers. Very low lakes due to dry conditions. **(j)** Late Pleistocene–Early Holocene with two separate lakes. **(k)** Present conditions with three separate lakes

lake expansion, partially based on correlations between the Kariandusi wet period and other parts of East Africa between 1.1 and 0.9 Ma, which they observed coincided with a 400-ka eccentricity cyclicity.

During the Middle Pleistocene, the Central Kenya Rift became increasingly compartmentalised (Fig. 18.7g). Eruptions that produced Mt. Eburru closed the Naivasha Basin to the north by about 400 ka (Clarke et al. 1990) with the southern margins being defined by volcanics that erupted at Olkaria around 320 ka (Trauth et al. 2001).

18.3.3 Late Pleistocene Sediments and Palaeogeography

Between about 145 and 66 ka, lakes in the Naivasha and Nakuru-Elmenteita basins expanded and contracted as separate water bodies (Fig. 18.7h). Deposits in the Ol Njorowa Gorge, to the south of Lake Naivasha (Figs. 18.1 and 18.8; Trauth 1992; Trauth and Strecker 1996), document several significant transgression-regression cycles that gave rise to a 60-m-thick depositional sequence. The succession includes intervals with weakly laminated pure-white diatomite, laminated diatomite and massive, greyish clastic diatomite that accumulated under a series of lake highstands (Bergner 2003; Bergner et al. 2003; Bergner and Trauth 2004). Preservation of laminae in some of the diatomites suggests anoxic bottom water during deposition, with burrows in massive units indicating periods of oxic conditions. Clay-rich, dark layers within some of the diatomites have been interpreted as intermittent sediment inputs to the palaeolakes rather than annual or seasonal events. The diatomite beds at Ol Njorowa Gorge are separated by subaerial and lacustrine ash and by zeolitic lacustrine sediments that have been altered by exposure to alkaline brines during periods of lower lake level.

Major highstands at Ol Njorowa have been identified at 145–120, 110, 90, 80 and 66 ka (Figs. 18.7h and 18.8) (Trauth et al. 2001, 2003; Bergner et al. 2003; Trauth 2014). The 145–120 ka lake was probably the largest Late Pleistocene water body and is also represented by diatomites in the Nakuru-Elmenteita Basin at Soysambu, west of Lake Elmenteita (Fig. 18.9; Mboya 1993; Thorpe 2006; Bergner et al. 2009). The younger Ol Njorowa lacustrine transgressions, at 110 and 80 ka, were not reported from Soysambu, but may be equivalent to diatomites near Gilgil and Kariandusi that contain abundant phytoliths, clastic particles and reduced diatom preservation when matched with the earlier large lake phase (Bergner et al. 2009). Compared with the Early Pleistocene lake at Kariandusi, facultative planktonic and epiphytic diatoms are more common, as are *Aulacoseira* spp. that are favoured by shallower water and halophilic taxa such as *Thalassiosira faurii*. Despite similar diatom floras, the Central Kenya Rift probably contained two separate lakes between 145 and 110 ka, given that the Nakuru-Elmenteita Basin diatomites are characterised by higher $\delta^{18}O_{diatom}$ values than those at Ol Njorowa (Fig. 18.9b; Thorpe 2006; Bergner et al. 2009). This indicates higher water input to evaporation ratios (I/E) at Nakuru-Elmenteita than in the Naivasha Basin and suggests that subsurface

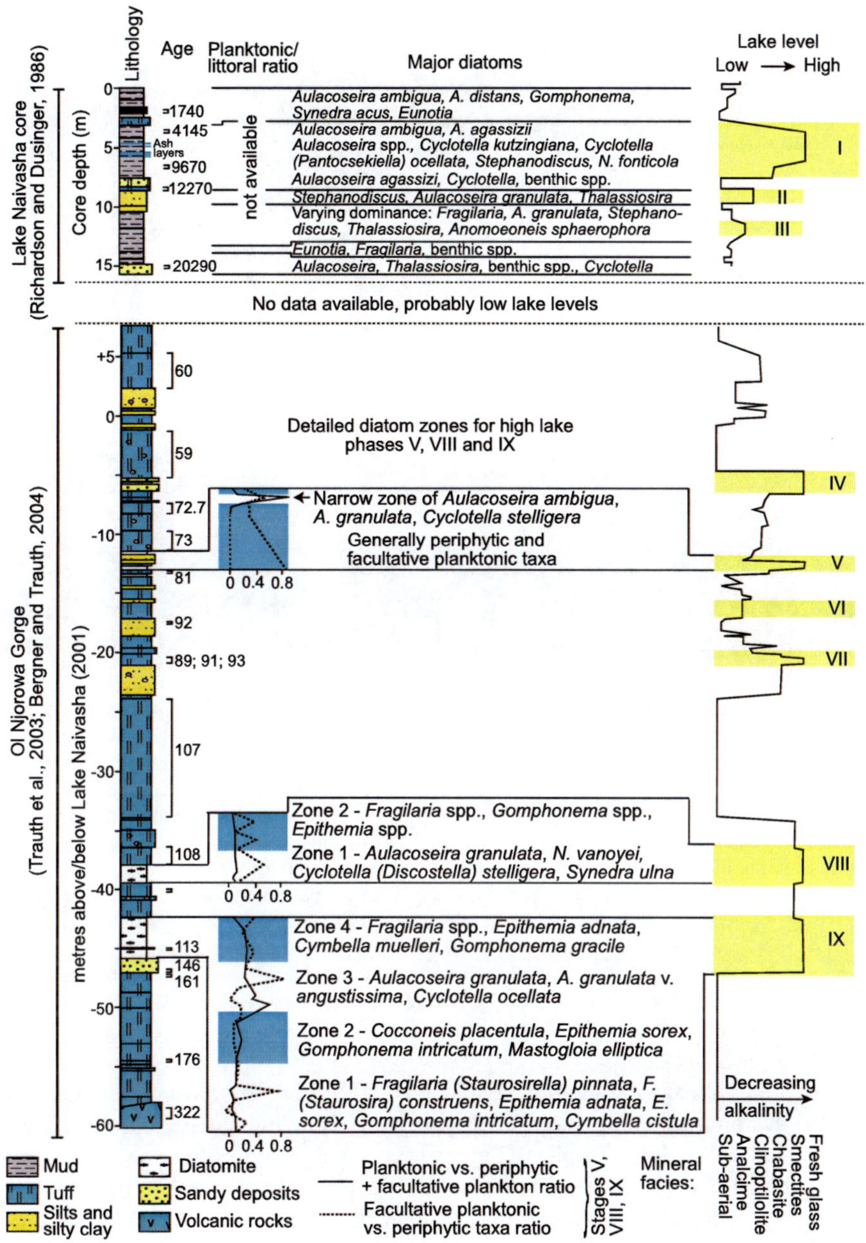

Fig. 18.8 Late Quaternary sediment and diatom stratigraphy of the Naivasha Basin Major lake level highstands (I–IX) noted to the right. Diatom floras for selected horizons shown at centre. Based on Richardson and Dussinger (1986), Trauth et al. (2003) and Bergner and Trauth (2004)

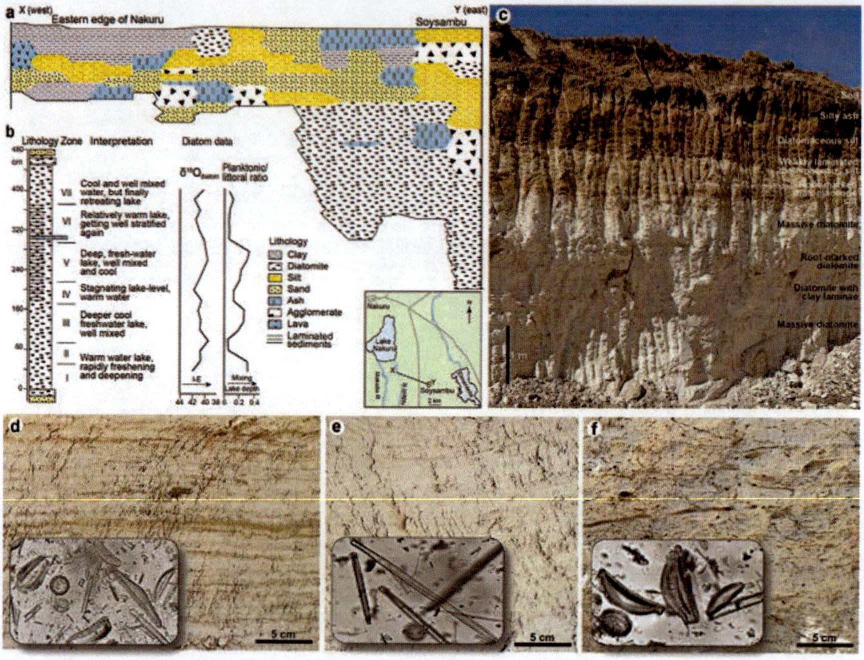

Fig. 18.9 The Soysambu diatomite deposits, Nakuru-Elmenteita Basin. (a) X-Y cross section (location in inset map) of major lithologies between Soysambu and eastern Lake Nakuru. After Mboya (1993). (b) Diatom oxygen isotope data and planktonic/littoral ratios from the Soysambu diatomite mine. Text gives lake status for seven zones. After Bergner et al. (supplementary data, 2009). (a, b) Share the same vertical scale. (c) Soysambu quarry face (different to that shown in (b)). Pure diatomites represent deeper water. (d–f) Show three contrasting diatomaceous facies with differing diatom assemblages. (d) Laminated clay-diatomite couplets. Clay represents seasonal or episodic wetter conditions and clastic inputs to the palaeolake. Inset shows flora dominated by *Rhopalodia gibba*, *Rhopalodia vermicularis*, *Epithemia adnata*, *Encyonema muelleri*, *Pantocsekiella ocellata* and *Ulnaria ulna* fragments. (e) Massive diatomite. Inset showing dominance by *Ulnaria ulna*. (f) Root moulds in diatomite suggest shallow water conditions with macrophytes. Inset includes *Epithemia sorex*, *Encyonema muelleri*, *Aulacoseira granulata* var. *angustissima* and *Pantocsekiella ocellata*

groundwater linkages were minimal (Thorpe 2006). Both Thorpe and Bergner suggested that precession drove the lake level changes due to increased insolation and convection at the Inter Tropical Convergence Zone, resulting in greater regional precipitation. They also inferred that the large magnitude and length of the 145–120 ka highstand might reflect high tropical Indian Ocean sea surface temperatures during the last interglacial and enhanced lateral transport of moisture into the region.

Intermediate and low lake levels prevailed through the Late Pleistocene with lakes possibly drying out (Fig. 18.7i) and with a major lake expansion during the Early Holocene (Fig. 18.7j) (Washbourn-Kamau 1975; Richardson and Richardson 1972). Bergner and Trauth (2004) focused on three of the Ol Njorowa Gorge Pleistocene highstands (IX, VIII and V in Fig. 18.8), noting that their diatom floras indicated deep fresh lakes at ~135 and 80 ka and a shallower, but more expanded water body at ~110 ka, with all three transgressions having spread over areas up to three times larger than the modern lake. Diatoms from Highstand IX indicate fresh conditions (pH ~8; conductivity ~ 400 μS cm^{-1}) with Highstand VIII suggesting slightly more saline water (pH 8–8.5; conductivity 300–1000 μS cm^{-1}) and Highstand V providing evidence for a higher salinity (pH up to 9.5; conductivity 1000–3000 μS cm^{-1}).

Highstands at 145–120 and 10–5.5 ka (Richardson and Dussinger 1986) appear to represent times when the Central Kenya Rift lakes reached their largest extent (Fig. 18.7c, d), expanding at about the same time as several other East African lakes (Trauth et al. 2005, 2007, 2010; Maslin and Trauth 2009; Maslin et al. 2014). Hastenrath and Kutzbach (1983) estimated a minimum increase of 15% in precipitation to support the Early Holocene expansion, assuming no overflow. In contrasting lake-balance models, Richardson and Richardson (1972) and Bergner et al. (2003) inferred that Early Holocene precipitation might have been up to 60% and 30% greater than modern values. Trauth et al. (2001) also reported that highstands at about 135, 110, 90 and 66 ka approximately matched maximum spring insolation periods near the equator (Berger 1978) and suggested that orbitally-induced changes in summer solar radiation accounted for at least part of the hydrological changes. Subsequently, Trauth (2014) and Trauth et al. (2015) used revised age models to infer that the major lake highstands occurred at 155, 127, 113, 108 ka, ~98–93, ~83–80 and ~72–67 ka with drought events at about 91, 88, 83, 78 and 73 ka. They concluded that there was a more complex link between insolation and the hydrological budget of Lake Naivasha than had previously been believed and that high lake levels developed about 2000–5000 years after minimum or maximum precession.

The Lake Naivasha palaeo-outlet at Ol Njorowa Gorge is unusual in having two entrances with the lowest standing at 1943 m ('B' in Fig 18.10b; Kamau 1977). Kamau pointed out that this overflow lies about 50 m above the modern lake, yet the highest shoreline is 100 m above Lake Naivasha (2000 m asl). The Gorge could not have been in its present form when the 2000 m lake was present. Thompson and Dodson (1963), supported by Kamau (1977), suggested an origin that involved faulting and foundation scouring. Kamau noted that downcutting would be unlikely to form two large valleys with near vertical lava cliffs and flat valley floors so close together (Fig. 18.10c, d) and that the form of the valleys is unusual for stream incision. It was suggested that as the lake transgressed to higher levels seepage took place through porous tuffs and agglomerates below less permeable comendite lavas that now form the cliffs. Seepage weakened the underlying volcanics and slowed lake level rise, which may never have reached the overflow height at that time. Seepage emerged to the south carving out the lower gorge along a pre-existing fracture (Fig. 18.10e, f). Ultimately the subsurface washout led to collapse at the northern end of the gorge, perhaps linked to an earthquake. Richardson (in Kamau 1977) suggested that the formation of the gorge through this process must have been complete by 9000 y BP since microfossils in a Lake Naivasha core suggested no sudden

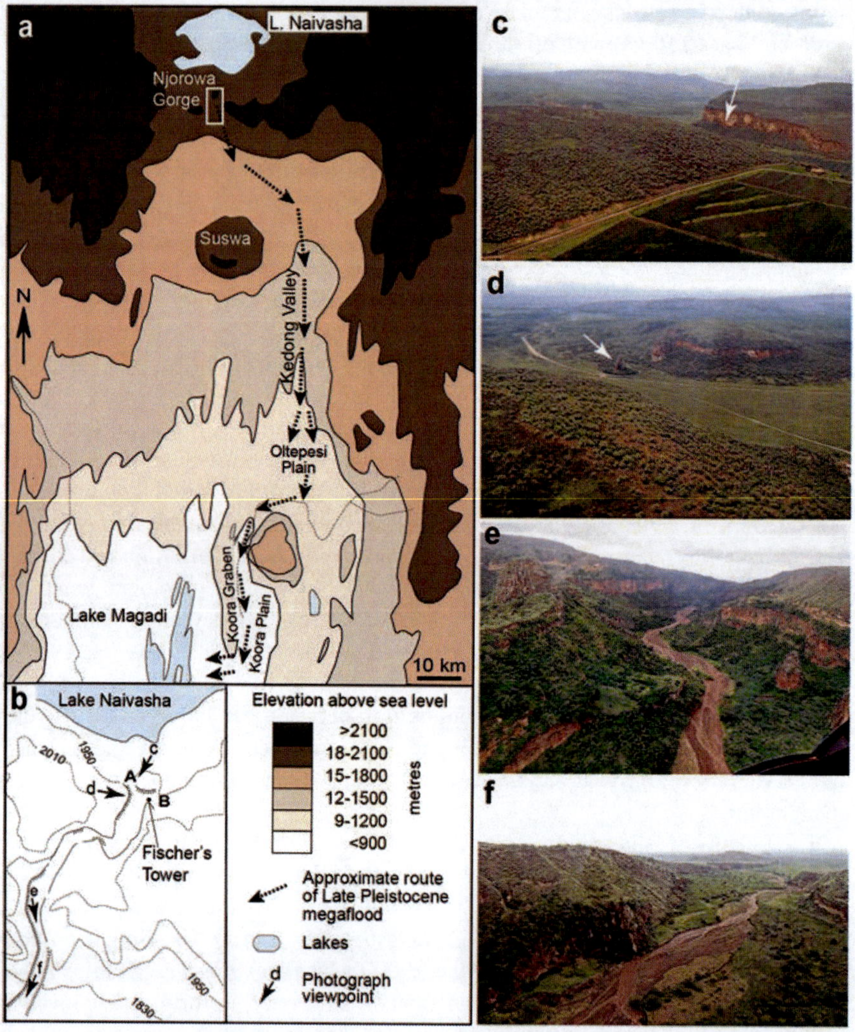

Fig. 18.10 The Ol Njorowa Gorge and megaflood. (**a**) Late Pleistocene megaflood route after descriptions in Baker and Mitchell (1976) and Baker et al. (1988). (**b**) Morphology of the Ol Njorowa Gorge. Map after Kamau (1977). (**c–f**) Aerial images of different sections of the gorge. Location of viewpoints shown in (**b**). Note the double entrances to the gorge (white arrows in (**c**, **d**) with flat floors and steep marginal comendite cliffs. Arrow in (**d**) points to Fischer's Tower plug. (**e**, **f**) Show the narrower lower gorge and its southern termination

decline in lake level between then and 5650 y BP, after which the lake level fell due to climate change.

It appears that instability of the outlet caused a catastrophic lowering of Lake Naivasha at some time during the Late Pleistocene with water escaping from Naivasha via the Ol Njorowa Gorge (Fig. 18.10a; Washbourn-Kamau 1975; Kamau

1977; Baker and Mitchell 1976; Baker et al. 1988). This produced scoured lava surfaces, large boulders (up to 15 m diameter) and polished rock surfaces up to 30 m above valley floors that channelled the escaping megaflood. The waters passed through the Kedong Valley to the south and spread over the Oltepesi Plain, depositing sand and boulders, before flowing onwards through the Koora Graben and Plain, then occupied by a palaeolake, and prior to overflowing westwards across a horst block and into the Magadi Basin where large boulder bars were deposited (Sect. 19.5; Muiruri et al. 2021). Baker et al. (1988) dated the event at about 0.1 Ma. However, Trauth et al. (2003) suggested that a major fall in Lake Naivasha after about 60 ka might be related to this megaflood. Subsequently, there is no evidence for higher lakes in the Naivasha basin until about 21 to 17.5 ka with a major highstand also occurring during the Early Holocene (Richardson and Dussinger 1986; Chap. 10).

References

Baker BH, Mitchell JG (1976) Volcanic stratigraphy and geochronology of the Kedong-Olorgesailie area and the evolution of the South Kenya rift valley. J Geol Soc Lond 132:467–484

Baker BH, Mitchell JG, Williams LAJ (1988) Stratigraphy, geochronology and volcano-tectonic evolution of the Kedong-Naivasha-Kinangop region, Gregory Rift Valley, Kenya. J Geol Soc Lond 145:107–116

Berger A (1978) Long-term variations of daily insolation and Quaternary climatic changes. J Atmos Sci 35:2362–2367

Bergner A (2003) Lake-level fluctuations and Late Quaternary climate change in the Central Kenya Rift. Doctoral thesis, Universität Potsdam

Bergner A, Trauth M (2004) Comparison of the hydrological and hydrochemical evolution of Lake Naivasha (Kenya) during three highstands between 175 and 60 kyr BP. Palaeogeogr Palaeoclimatol Palaeoecol 215:17–36

Bergner A, Trauth M, Bookhagen B (2003) Paleoprecipitation estimates for the Lake Naivasha basin (Kenya) during the last 175 ky using a lake-balance model. Glob Planet Change 36:117–136

Bergner AGN, Strecker MR, Trauth MH, Deino A, Gasse F, Blisniuk P, Dühnforth M (2009) Tectonic and climatic control on evolution of rift lakes in the Central Kenya Rift, East Africa. Quat Sci Rev 28:2804–2816

Bishop WW (1967) Annotated lexicon of Quaternary stratigraphical nomenclature in East Africa. In: Bishop WW, Clark JD (eds) Background to evolution in Africa. University of Chicago Press, Chicago, pp 375–395

Blegen N, Brown FH, Jicha BR, Binetti KM, Faith JT, Ferraro JV, Gathogo PN, Richardson JL, Tryon CA (2016) The Menengai Tuff: A 36 ka widespread tephra and its chronological relevance to Late Pleistocene human evolution in East Africa. Quat Sci Rev 152:152–168

Blisniuk PM, Strecker MR, Eisbacher GH (1988) Sedimentation und tektonik im Lake Elmenteita-Gebiet, zentrales Gregory Rift, Kenya. Z Dtsch Geol Ges 39:62–63

Boven A (1992) The applicability of the K-Ar and $^{40}Ar/^{39}Ar$ techniques to Quaternary volcanics: a fundamental and experimental investigation. PhD thesis, Vrije Universiteit Brussel

Bower JRF, Nelson CM (1978) Early pottery and pastoral cultures of the Central Rift Valley, Kenya. Man 13:554–566

Burkitt MC (1945) Excavations at Hyrax Hill, Nakuru, Kenya Colony. Nature 156:722–723

Butzer KW, Isaac GL, Richardson JL, Washbourn-Kamau C (1972) Radiocarbon dating of East African lake levels. Science 175:1069–1076
Clark JD, Cole S (eds) (1957) Prehistory: Proceedings of the Third Pan-African Congress on Pre-History, Livingstone 1955. Chatto and Windus, London
Clarke MCG, Woodhall DG, Allen D, Darling WG (1990) Geological, volcanological and hydrological controls on the occurrence of geothermal activity in the area surrounding Lake Naivasha, Kenya, Ministry of Energy, Nairobi, Kenya
Cole SM (1950) An outline of the geology of Kenya. Pitman, Nairobi
Cole SM (1954) The prehistory of East Africa. Am Anthropol 56:1026–1050
Cole SM (1963) The prehistory of East Africa. McMillan, London
Conti P, Pistis M, Bernardinetti S, Barbagli A, Zirulia A, Serri L, Colonna T, Guastaldi E, Ghiglieri G (2021) Tectonic setting of the Kenya Rift in the Nakuru area, based on geophysical prospecting. Geosciences 11:80 https://doi.org/10.3390/geosciences11020080
Cooke HBS (1958) Observations relating to Quaternary environment in East and Southern Africa. Geol Soc S Afr Annex 61:1–73
Durkee H, Brown FH (2014) Correlation of volcanic ash layers between the Early Pleistocene Acheulean sites of Isinya, Kariandusi, and Olorgesailie, Kenya. J Archaeol Sci 49:510–517
Evernden JF, Curtis GH (1965) The potassium-argon dating of Late Cenozoic rocks in East Africa and Italy [and Comments and Reply]. Curr Anthropol 6:343–385
Flint RF (1959a) On the basis of Pleistocene correlation in East Africa. Geol Mag 96:265–284
Flint RF (1959b) Pleistocene climates in eastern and southern Africa. Geol Soc Am Bull 70:343–374
Frahm E, Tryon CA (2018) Later Stone Age toolstone acquisition in the Central Rift Valley of Kenya: Portable XRF of Eburran obsidian artifacts from Leakey's excavations at Gamble's Cave II. J Arch Sci Rep 18:475–486
Fuchs V (1950) Pleistocene events in the Baringo Basin. Geol Mag 87:149–174
Gowlett JAJ (1978) Kilombe - an Acheulian site complex in Kenya. In: Bishop WW (ed) Geological background to fossil man. Geol Soc Lond Spec Publ 6:337–360
Gowlett JAJ (1993) Le site Acheuléen de Kilombe: stratigraphie, géochronologie, habitat et industrie lithique. Anthropologie 97:69–84
Gowlett JAJ, Crompton R (1994) Kariandusi: Acheulean morphology and the question of allometry. Afr Archaeol Rev 12:3–42
Gowlett JAJ, Brink JS, Herries AIR, Hoare S, Onjala I, Rucina SM (2015) At the heart of the African Acheulean: the physical, social and cognitive landscapes of Kilombe. In: Coward F, Hosfield R, Wenban-Smith F (eds) Settlement, society and cognition in human evolution: Landscapes in mind. Cambridge University Press, Cambridge, pp 75–93
Gowlett JAJ, Stanistreet IG, Albert RM, Blackbird SJ, Herries AIR, Hoare S, Kogai P, Komboh CK, Mark DF, Muriuki RM, Murphy H, Rucina SM, Stollhofen H (2022) New Oldowan localities at high level within Kilombe Caldera. Kenya. L'anthropologie 126:102976
Gregory JW (1921) The rift valleys and geology of East Africa. Seeley Service, London
Hastenrath S, Kutzbach JE (1983) Paleoclimatic estimates from water and energy budgets of East African lakes. Quat Res 19:141–153
Hay RL (1963) Stratigraphy of Beds I through IV, Olduvai Gorge, Tanganyika. Science 139:829–833
Hoare S, Brink JS, Herries AIR, Mark DF, Morgan LE, Onjala I, Rucina SM, Stanistreet IG, Stollhofen H, Gowlett JAJ (2021) Geochronology of a long Pleistocene sequence at Kilombe volcano, Kenya: from the Oldowan to Middle Stone Age. J Archaeol Sci 125:105273
Jennings DJ (1971) Geology of the Molo area. Rep Geol Surv Kenya 42
Kamau CK (1972) Late Quaternary lake level fluctuations in the Nakuru-Elmenteita Basin, Kenya. In: Reeves CC (ed) Playa Lake Symposium (Oct 1970). Dept Geosciences, International Center for Arid and Semi-arid Land Studies (ICASALS), Lubbock, Texas, Publ 4:185–192
Kamau CK (1977) The Ol Njorowa Gorge, Lake Naivasha basin, Kenya. In: Greer DC (ed) Proc Int Conf Desertic Terminal Lakes, Utah Water Res Lab, Logan, pp 297–307
Kendall R (1969) An ecological history of the Lake Victoria basin. Ecol Monogr 39:121–176

Kingston JD, Hill A (2005) When it rains it pours: legends and realities of the East African pluvials. In: Lieberman DE, Smith RJ, Kelley J (eds) Interpreting the past: Essays on human, primate and mammal evolution. American School of Prehistoric Research Monograph Series, vol 5. Brill, Boston, pp 189–205

Kübler S, Rucina S, Reynolds S, Owenga P, Bailey G, King GCP (2016) Edaphic and topographic constraints on exploitation of the Central Kenya Rift by large mammals and early hominins. Open Quat 2:199–119

Leakey LSB (1931a) East African lakes. Geogr J 77:497–508

Leakey LSB (1931b) The Stone Age cultures of Kenya Colony. Cambridge University Press, Cambridge

Leakey LSB (1931c) Prehistoric man in Kenya. Nature 127:814–815

Leakey LSB (1934) Changes in the physical geography of East Africa in human times. Geogr J 84:296–305

Leakey LSB (1950) The lower limit of the Pleistocene in Africa. Proc Int Geol Congr 18:62–65

Leakey LSB (1955) The climatic sequence of the Pleistocene in Africa. In: Balout L (ed) Actes De La IIe Session Du Congrès Panafricain De Préhistoire, Alger, 1952, pp 293–294

Leakey LSB, Cole SM (eds) (1952) Proceedings Pan-African Congress on Prehistory, 1947. Philosophical Library, New York

Leakey LSB, Solomon J (1929) African pluvial periods and prehistoric man. Man 29:213–214

Leat PT (1984) Geological evolution of the trachytic caldera volcano Menengai, Kenya Rift Valley. J Geol Soc Lond 141:1057–1069

Leat PT (1991) Volcanological development of the Nakuru area of the Kenya Rift Valley. J Afr Earth Sci 13:483–498

Marean C, Mudida N, Reed K (1994) Holocene paleoenvironmental change in the Kenyan Central Rift as indicated by micromammals from Enkapune Ya Muto rockshelter. Quat Res 41:376–389

Maslin MA, Trauth MH (2009) Plio-Pleistocene East African pulsed climate variability and its influence on early human evolution. In: Grine FE, Fleagle JG, Leakey RE (eds) The first humans – Origin and early evolution of the genus *Homo*. Contributions from the third Stony Brook human evolution symposium and workshop, October 3–October 7, 2006, pp 151–158

Maslin MA, Brierley CM, Milner AM, Shultz S, Trauth MH, Wilson KE (2014) East African climate pulses and early human evolution. Quat Sci Rev 101:1–17

Mboya BGO (1993) Lithostratigraphic analysis and palaeoenvironmental interpretation of the Late Tertiary and Quaternary sediments in the Nakuru, Elmentaita and Naivasha Basins. PhD thesis, University of Nairobi

McCall GJH (1957) Geology and groundwater conditions in the Nakuru area. Tech Rep Ministr Works (Hydraul Br) Kenya 3

McCall GJH (1967) Geology of the Nakuru-Thomson's Falls- Lake Hannington area. Geol Surv Kenya Rep 78

McCall GJH, Baker BH, Walsh J (1967) Late Tertiary and Quaternary sediments of the Kenya Rift Valley. In: Bishop WW, Clark JD (eds) Background to evolution in Africa. University of Chicago Press, Chicago, pp 191–220

Muiruri V, Owen RB, Potts R, Deino AL, Behrensmeyer AK, Dommain R, Reidl S, Rabideaux N, Beverly EJ, Renaut RW, Moerman JW, Deocampo D, Faith JT, Noren A, Cohen AS, Shannon KB, Dommain R (2021) Quaternary diatoms and palaeoenvironments of the Koora Plain, southern Kenya Rift. Quat Sci Rev 267:107106

Munga JU (2014) Morphological variability in Acheulean handaxes from Kariandusi and Lewa Downs archaeological sites in Kenya. MA thesis, University of Nairobi

Nilsson E (1931) Quaternary glaciations and pluvial lakes in British East Africa. Geog Annlr 13:249–348

Nilsson E (1935) Traces of ancient changes of climate in East Africa. Geogr Annlr 17:1–21

Nilsson E (1938) Pluvial lakes in East Africa. Geol Fören Stockholm Förh 60:423–433

Nilsson E (1940) Ancient changes of climate in British East Africa and Abyssinia. Geogr Annlr 22:1–79

Nyawir E, Barrat JA, Varet J (2018) Geological, volcanological, petrological and structural studies to evaluate geothermal occurrence within an axial rift segment of the central Kenya Rift Valley. In: Proc 7th Afr Rift Geotherm Conf, Kigali, Rwanda

Odongo MEO (1993) A geological review of Olkaria geothermal reservoir based on structure. In: Proc 15th NZ Geotherm Workshop, pp 169–173

Onywere SM (1997) Structural analysis of the drainage basin of Kenyan rift valley lakes within the Aberdare detachment, using satellite data, GIS and field observations. PhD thesis, University of Nairobi

Owen RB, Potts R, Behrensmeyer A, Ditchfield P (2008) Diatomaceous sediments and environmental change in the Pleistocene Olorgesailie Formation, southern Kenya Rift Valley. Palaeogeogr Palaeoclimatol Palaeoecol 269:17–37

Owen RB, Potts R, Behrensmeyer AK (2009) Reply to the comment on [Diatomaceous sediments and environmental change in the Pleistocene Olorgesailie Formation, southern Kenya Rift Valley. Palaeogeography, Palaeoclimatology, Palaeoecology 269 (2008) 17–37]. Palaeogeogr Palaeoclimatol Palaeoecol 282:147–148

Protsch R (1978) The chronological position of Gamble's Cave II and Bromhead's site (Elmenteita) of the Rift Valley, Kenya. J Hum Evol 7:101–109

Richardson JL (1966) Changes in level of Lake Naivasha, Kenya, during postglacial times. Nature 209:290–291

Richardson JL (1972) Palaeolimnological records from Rift lakes in central Kenya. Palaeoecol Afr 7:131–136

Richardson JL, Dussinger R (1986) Paleolimnology of mid-elevation lakes in the Kenya Rift Valley. Hydrobiologia 143:167–174

Richardson JL, Richardson AE (1972) History of an African rift lake and its climatic implications. Ecol Monogr 42:499–534

Richter MJEA, Brune S, Riedl S, Glerum A, Neuharth D, Strecker MR (2021) Controls on asymmetric rift dynamics: Numerical modeling of strain localization and fault evolution in the Kenya Rift. Tectonics 40:e2020TC006553

Riedl S, Melnick D, Mibei GK, Njue L, Strecker MR (2020) Continental rifting at magmatic centres: structural implications from the Late Quaternary Menengai Caldera, central Kenya Rift. J Geol Soc Lond 177:153–169

Roessner S, Strecker MR (1997) Late Cenozoic tectonics and denudation in the Central Kenya Rift: quantification of long-term denudation rates. Tectonophysics 278:83–94

Scott S (1980) The geology of Longonot volcano, central Kenya: a question of volumes. Phil Trans R Soc Lond Ser A, Math Phys Sci 296:437–465

Solomon JD (1931) The geology of the implementiferous deposits in the Nakuru and Naivasha basins and the surrounding area in Kenya Colony. Appendix A. In: Leakey LSB (ed) The Stone Age cultures of Kenya Colony. Cambridge University Press, Cambridge, pp 245–257

Strecker MR (1991) Das zentrale und südliche Kenia-Rift unter besonderer Berücksichtigung der neotektonischen Entwicklung. PhD thesis, Universität Karlsruhe

Strecker MR, Blisniuk PM, Eisbacher GH (1990) Rotation of extension direction in the central Kenya Rift. Geology 18:299–302

Thompson AC, Dodson RG (1963) Geology of the Naivasha area. Geol Surv Kenya Rep 43

Thorpe JL (2006) Records of late Quaternary climatic change from Tswaing crater lake, South Africa, and the Central Kenya Rift. PhD thesis, University of London

Trauth MH (1992) Tektonik und Sedimentation im Olkaria Gebiet, südlich Lake Naivasha, Kenia. Dissertation, Universität Karlsruhe

Trauth MH (2014) A new probabilistic technique to build an age model for complex stratigraphic sequences. Quat Geochronol 22:65–71

Trauth MH, Maslin MA (2009) Comment on "Diatomaceous sediments and environmental change in the Pleistocene Olorgesailie Formation, southern Kenya Rift". Palaeogeogr Palaeoclimatol Palaeoecol 282:145–146

Trauth MH, Strecker MR (1996) Late Pleistocene lake-level fluctuations in the Naivasha Basin, Kenya. In: Johnson TC, Odada EO (eds) The limnology, climatology and paleoclimatology of the East African Lakes. Gordon and Breach, Amsterdam, pp 549–557

Trauth MH, Deino A, Strecker MR (2001) Response of the East African climate to orbital forcing during the last interglacial (130–117 ka) and the early last glacial (117–60 ka). Geology 29:499–502

Trauth M, Deino A, Bergner A, Strecker M (2003) East African climate change and orbital forcing during the last 175 kyr BP. Earth Planet Sci Lett 206:297–313

Trauth MH, Maslin MA, Deino A, Strecker MR (2005) Late Cenozoic moisture history of East Africa. Science 309:2051–2053

Trauth MH, Maslin MA, Deino AL, Strecker MR, Bergner AGN, Dühnforth M (2007) High and low-latitude forcing of Plio-Pleistocene East African climate and human evolution. J Hum Evol 53:475–486

Trauth MH, Larrasoana JC, Mudelsee M (2009) Trends, rhythms and events in Plio-Pleistocene African climate. Quat Sci Rev 28:399–411

Trauth MH, Maslin MA, Deino AL, Junginger A, Lesoloyia M, Odada EO, Olago DO, Olaka LA, Strecker MR, Tiedemann R (2010) Human evolution in a variable environment: The amplifier lakes of Eastern Africa. Quat Sci Rev 29:2981–2988

Trauth MH, Bergner AGN, Foerster V, Junginger A, Maslin MA, Schaebitz F (2015) Episodes of environmental stability versus instability in Late Cenozoic lake records of Eastern Africa. J Hum Evol 87:21–31

Washbourn-Kamau C (1970) Late Quaternary chronology of the Nakuru–Elmenteita Basin, Kenya. Nature 226:253–254

Washbourn-Kamau C (1971) Late Quaternary lakes in the Nakuru-Elmenteita basin, Kenya. Geogr J 137:522–535

Washbourn-Kamau CK (1975) Late Quaternary shorelines of Lake Naivasha, Kenya. Azania 10:77–92

Wilshaw A (2012) An investigation into the LSA of the Nakuru-Naivasha Basin and surround, Central Rift Valley, Kenya: Technological classifications and population considerations. PhD thesis, University of Cambridge

Woodhall DG, Clarke MCG (1988) Geological map of Longonot Volcano, the Greater Olkaria and Eburru volcanic complexes and adjacent areas. Government of Kenya, Ministry of Energy, Geothermal Section

Zeuner FE (1948) Climate and early man in Kenya. Man 48:13–16

Chapter 19
The South Kenya Rift Basins

19.1 Introduction

The 'South Kenya Rift' lies between the Quaternary volcanoes of Suswa in the north and Shombole to the south and extends from the Nguruman Escarpment in the west to the eastern margins of the Kenya Rift Valley (Fig. 19.1) but excludes the axial Magadi-Natron basin which is treated separately (Chap. 20). The rift floor descends southwards gradually from Lake Naivasha and is cut by dense networks of parallel N-S faults (Fig. 19.1b, c) with grabens that host buried Quaternary sediments such as in the Koora Basin. Deposits in the Pleistocene Olorgesailie Basin are laterally extensive and form a flat plain (Fig. 19.1a). Miocene deposits are exposed by rift marginal faults with well-exposed outcrops at Lemudong'o in the northwest of the region.

This chapter focuses on the lacustrine sediments of the Miocene Lemudong'o Basin, outcrops of Pleistocene sediments at Munya wa Gicheru and Olorgesailie, and subsurface deposits below the Koora Plain. Although larger lakes developed during periods wetter than today at Lake Kwenia, Lake Kabongo and other small ephemeral lakes (Fig. 19.1a), their sediments are poorly exposed and the former extent, age and salinity of those palaeolakes await investigation.

19.2 Miocene Lake Sediments of the Lemudong'o Basin

19.2.1 General Setting of the Miocene Sediments

The western margin of the Kenya Rift lies in the southern Narok District and includes Late Miocene sediments at Lemudong'o that crop out in the headwater tributaries of the southern Ewaso Ngiro ("Uaso Ngiro" according to Wright 1967) and Seyabei rivers. The area lies ~100 km west of Nairobi (Fig. 19.2) with deposits

R. W. Renaut, R. B. Owen, *The Kenya Rift Lakes: Modern and Ancient*, Syntheses in Limnogeology, https://doi.org/10.1007/978-3-642-25055-2_19

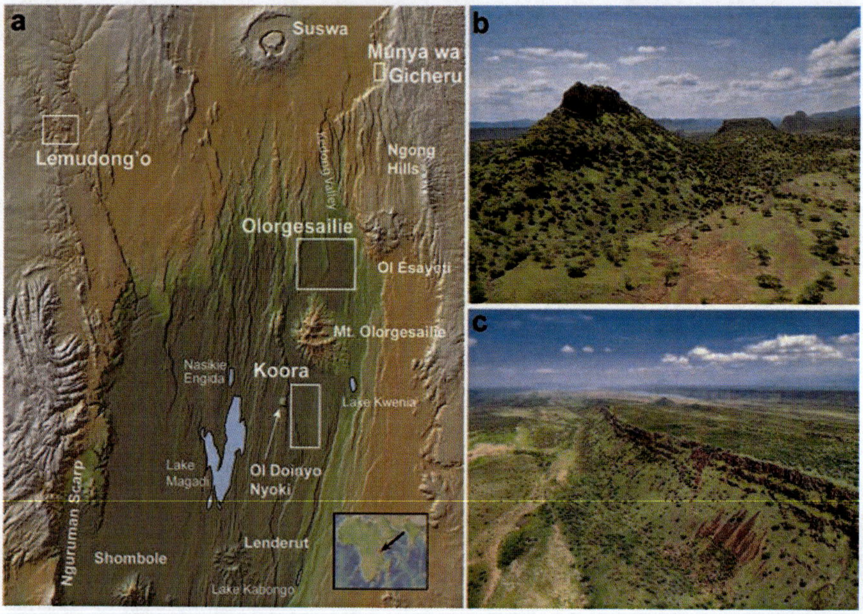

Fig. 19.1 The south Kenya Rift. (**a**) Topography and major basins (white boxes). (**b**) Typical north-south horsts and graben landscape, north of Lake Magadi. (**c**) Axial grid-faulted landscape northeast of Nasikie Engida

that crop out over a N-S distance of ~50 km (Ambrose et al. 2002, 2007a, b; Hlusko et al. 2002; Hlusko and Hailie-Selassie 2007). J.B. Wright mapped the area for the Kenya Geological Survey in 1959 and identified stratified alluvial and lacustrine sediments, including volcanic ash, with many palaeosols that accumulated in three major palaeolakes and two isolated water bodies (Wright 1967).

Wright's oldest lake sediments were recorded as the '2nd (lower elevation) Ewaso Ngiro lake', with the second oldest lake referred to as the '1st Ewaso Ngiro Lake'. Wright identified the youngest lake deposits as forming in a 'Seyabei Lake' and considered the sediments to be of Plio-Pleistocene age. However, the second of his three lakes was later dated to the Late Miocene (Deino and Ambrose 2007), with the sediments assigned to the Lemudong'o Formation and the overall sequence being dated from Late Miocene to Pleistocene in age (Ambrose et al. 2007c).

In an unpublished report, Waibel and McDonough stimulated archaeological interest in 1976 when they documented mammal fossils and phonolite hand axes in deposits that rested on welded tuffs dated to 4.4 Ma (Bower et al. 1977; Ambrose et al. 2007a). An ash stratified within the sequence produced a K/Ar date of 3.0 Ma. After these early investigations, archaeological research identified mammalian and vertebrate fossils in various late Neogene sedimentary units together with Acheulian, Middle to Late Stone Age and possible Oldowan tools (Kyule et al. 1997).

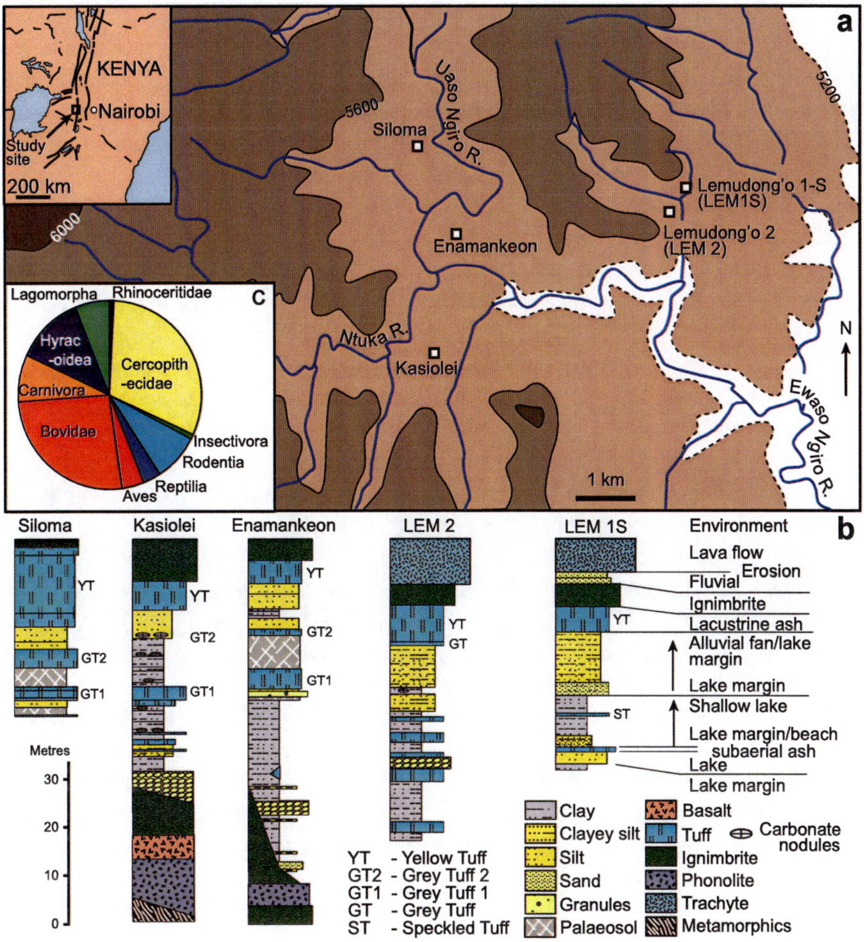

Fig. 19.2 Location and stratigraphy of the Lemudong'o Formation. Modified after Ambrose et al. (2007c). (**a**) Location of major sections and river systems. (**b**) Lithological sections and major tuff units. (**c**) Pie chart showing faunal composition. After Ambrose et al. (2007a)

19.2.2 Sediments, Stratigraphy and Palaeogeography of the Lemudong'o Formation

Wright (1967) estimated that the Miocene palaeolake (his 1st Ewaso Ngiro lake) extended >16 km from north to south by >8 km from east to west. Sediments accumulated in a half-graben with the initial topography undulating up to 65 m and with palaeolake waters lapping onto a small hill, Ol Doinyo Oborosoit, to the south. The sediments that accumulated in this water body are now exposed over an area of 1250 km^2 and have been formally described as the 135-m-thick Lemudong'o Formation (Ambrose et al. 2007c). Three main phases of deposition have been recognised. Phase 1 gave rise to lacustrine mudstone, diatomaceous (?) siltstone, sandstone and fine-grained laminated to massive tuff. Phase 2 is represented by palaeosols in the centre of the basin and by fluvial and subaerial siltstone and sandstone to the east. Lacustrine conditions resumed during Phase 3 with an accumulation of mudstone, siltstone and tuff, and were terminated by an ignimbritic eruption.

Figure 19.2b shows sections through the Lemudong'o Formation. Lake marginal mudstone and siltstone form the basal deposits in Sections LEM 1S and LEM 2, where three tuffs have been dated at 6.09–6.12 Ma (Ambrose et al. 2003; Deino and Ambrose 2007). These sediments gave way to a variety of facies including deeper-water laminated silt, lake marginal and swampy clayey mudstone, beach or deltaic well-sorted sand and sandy gravel, and poorly-sorted or imbricated alluvial/colluvial sand and gravel. Palaeosols, some with carbonate nodules, are also present. A speckled tuff in LEM 1S (Fig. 19.2b) was dated at 6.08 Ma (Deino and Ambrose 2007) and locally contains a micromammal breccia and seeds of *Celtis zenkeri*. Many sedimentary units, especially towards the base, show significant spatial variations in thickness. Dark grey brown/green waxy clay that breaks up into small 'peds' at the modern surface has been interpreted as forming in anoxic alkaline lake-margins (Ambrose et al. 2007c).

Many Late Miocene vertebrates (Fig. 19.2c) have been recovered from the sediments (Bernor 2007; Hlusko and Haile-Selassie 2007; Hlusko et al. 2007; Saegusa and Hlusko 2007), and are dominated by bovids, monkeys (Hlusko 2007a), rodents (Hlusko 2007b; Manthi 2007), lagomorphs (Darwent 2007), carnivores (Howell and Garcia 2007), hyracoidea (Pickford and Hlusko 2007), bird fossils (Stidham 2007) and reptiles, mainly snakes (Head and Bell 2007). Hippopotamus (Boisserie 2007) and crocodile teeth are also present, but the absence of fish implies that the lakes might have been ephemeral (Ambrose et al. 2007c). The dominant land fauna indicates patches of forest within a broader setting of humid grassy woodlands and woodlands (Ambrose et al. 2007c).

19.3 Pleistocene Sediments of the Munya wa Gicheru Basin

19.3.1 Location and History of Research

Small outcrops (~0.5 km^2) of Early Pleistocene lacustrine, swamp and fluvial sediments are present in a north-south trending graben at Munya wa Gicheru (MwG) on the eastern side of the South Kenya Rift (1°11′16″S, 36°32′53″E; Fig. 19.3). The area lies between the Kedong River and the Kikuyu Highlands, west of Nairobi, with Suswa crater located about 21 km to the west (Fig. 19.3a). The deposits are reached by travelling 7 km along the B3 road from Mai Mahiu and then turning south along an unpaved road for 21 km. An alternative 30 km-route via Ngong is possible, but not easy to follow. The deposits are not visible from the road and lie

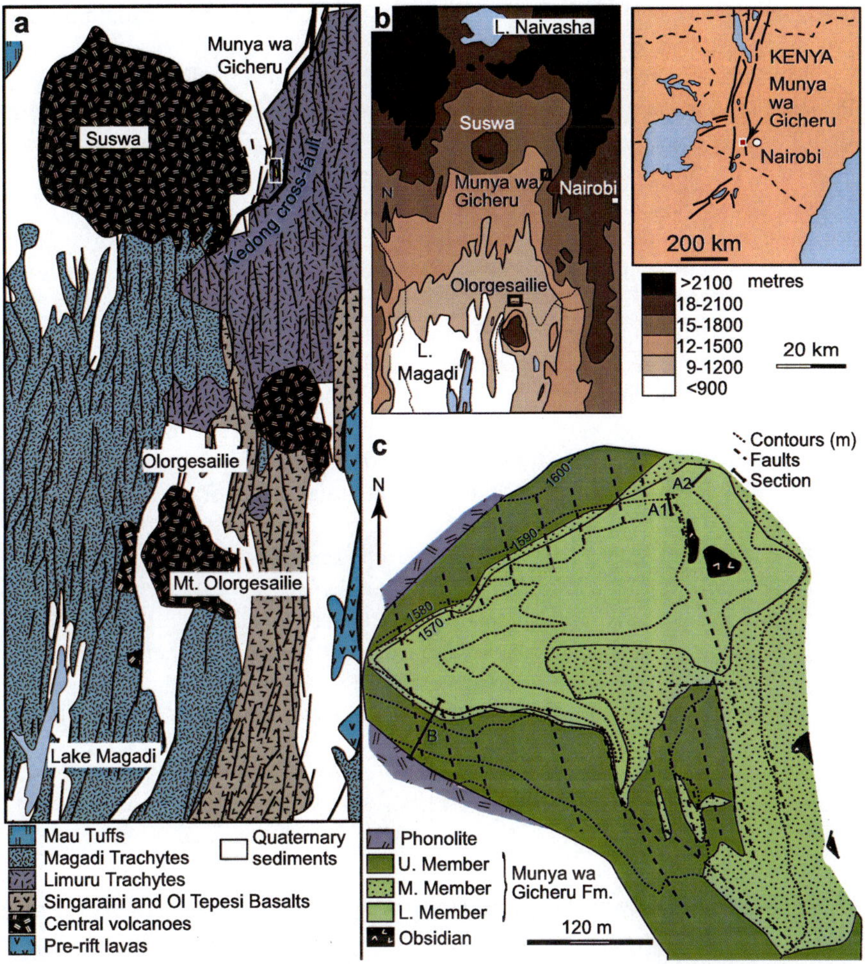

Fig. 19.3 Location and geological map of the Munya wa Gicheru Basin. After Ngecu and Njue (1999) and Owen et al. (2014). (**a**) Regional geology. (**b**) General location and topography. (**c**) Geological map of the main outcrop area

east of a N-S horst block. A rough track provides vehicular access, as does a 0.5 km walk through a river gorge east of the main access road.

While travelling through the Kedong area, Gregory (1896) noted that "strange straight lines along the hill-sides were old lake terraces" and stated, "I at once named the lake that had made them after Prof. Suess". Gregory believed that these terraces represented the eastern shores of a large Pliocene lake (Lake Suess) that extended across the width of the rift and which, he suggested, drained northwards, despite the gradual southward slope of the southern Kenya Rift. Gregory (1921) reassigned the MwG sediments to the Lower Pleistocene. Shackleton (1955) suggested that the palaeolake might have been confined to a narrow trough, which maps presented by Sikes (1926) had shown to be a graben blocked to the south by high ground formed by the Kedong cross-fault (Fig. 19.3a).

Baker and Mitchell (1976) reported 35 m of diatomite associated with a middle Pleistocene fauna and suggested that the deposits correlated with the Legemunge Beds of the Olorgesailie Basin (since disproved). They inferred that the palaeolake post-dated faulting that cuts through the Limuru Trachytes, the upper part of which has been dated at 1.96 Ma (Baker et al. 1988). An ash in the middle of the MwG sequence has been dated at 1.75–1.65 Ma (Fig. 19.4; Strecker 1991; Boven 1992; Trauth et al. 2007), with a pumice lapilli tuff near the top of the succession assigned an age of 1.65 Ma. Phonolite that lies disconformably on the sediments has been dated at 0.724 Ma (Trauth et al. 2007).

The economic geology and potential diatomite reserves were outlined by Saggerson (1991). Njue (1995) and Ngecu and Njue (1999) documented the lithostratigraphy and recognised a lower member comprised of massive diatomite with rhythmic clay laminae; a middle member dominated by pumice and ash with a green sand horizon; and an upper member consisting of reworked and silty diatomite, and diatomite with numerous disconformities. Trauth et al. (2007) described diatomite, silty diatomite, ash, silt and gravel, and reported that diatom assemblages were dominated by *Aulacoseira granulata*, *Stephanodiscus niagarae*, *S. transylvanicus* and *S. carconensis*. Owen et al. (2012, 2014) provided more detailed accounts of diatom floras, geochemistry, facies and lithostratigraphy, together with an environmental reconstruction that reported eight deep-water phases, eight shallow-water episodes and three periods when fluvio-terrestrial conditions prevailed (Fig. 19.4).

19.3.2 Geochemistry

The sediments are very well exposed along two streams and in a disused diatomite quarry (Figs. 19.4 and 19.5a, b). Estimates of overall sedimentation rates range between 0.05 and 0.1 mm y^{-1}, including periods of non-deposition and minor erosion (Owen et al. 2012, 2014). Chemical Index of Alteration (CIA) data range from 60–80, indicating a lower degree of weathering of source rocks compared with the Pleistocene (~1.2–0.5 Ma) Olorgesailie Formation to the south. Rare Earth Element (REE) plots display a decline towards the heavier elements with a distinct dip in Eu values, which reflects patterns displayed by rift trachytes (Owen et al. 2011), implying that these were the dominant source rocks.

The silica in diatom frustules is well preserved with minimal corrosion, which contrasts with diatom silica in the Olorgesailie Formation (Behrensmeyer et al. 2002; Owen et al. 2008a, b, 2011, 2014) where diatoms are also well-preserved, but with generally greater corrosion and with silica overgrowths in parts of the succession. Corrosion becomes greater in the Magadi Basin further south (Barker et al. 1990), which reflects a tendency for pH to increase in surface and groundwater, and with time, southwards in the South Kenya Rift (Owen et al. 2012).

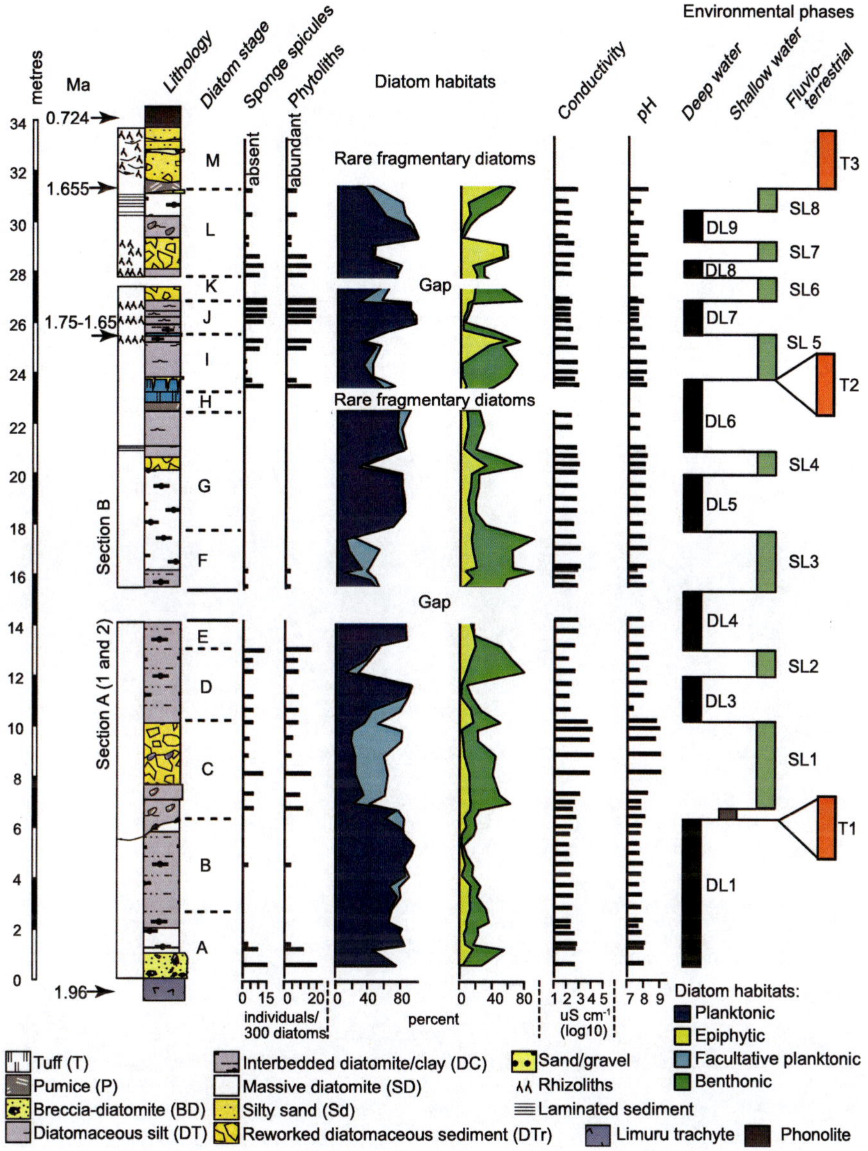

Fig. 19.4 Lithological logs and microfossil stratigraphy of the Munya wa Gicheru sediments. Modified after Owen et al. (2012). Section locations are shown in Fig. 19.3c. Environmental conditions derived from diatom transfer functions and habitat preferences. Environmental phases: DL deep lake, SL shallow lake, T terrestrial

The MwG sediments are also characterised by very low CaO concentrations. The deposits lack calcite rhizoliths and cements, with fossil roots preserved as open moulds lined with carbonaceous material. Molluscs are rare. This again contrasts

Fig. 19.5 General setting and sedimentary facies in the Munya wa Gicheru Basin. White scale bar = 5 cm. Hammer = 30 cm. (**a**) Oblique aerial image of the main outcrops with streams rising to the Ngong Hills. (**b**) General view of Units 9–36. Eastern rift scarp in background. (**c**) Interbedded diatomite and clay facies. (**d**) Massive subdiatomite with organic root moulds. (**e**) Desiccation cracks infilled with speckled silty sand facies. (**f**) Silty sand and gravel facies with angular obsidian fragments in small lensoid channel. (**g**) Tuffaceous deposits of sedimentary units 14–17. Sandy tuff is 4 cm thick

with younger deposits to the south, which variously contain abundant calcite rhizoliths and nodules, calcrete and spring tufa (Lee et al. 2013), and suggests an overfilled MwG basin with a dilute undersaturated fresh lake relatively poor in Ca^{2+}.

19.3.3 Facies, Lithostratigraphy and Microfossils

The Munya wa Gicheru sediments rest directly on obsidian of the Limuru Trachytes, the upper part of which has been dated at 1.96 Ma by Baker et al. (1988). Most sedimentary horizons can be traced laterally several hundred metres but tend to pinch out westwards. Owen et al. (2012) identified ten major facies based on earlier diatomaceous sediment classification schemes (Owen 2002) that were present in 36 sedimentary units of varying thickness (Fig. 19.4 and Table 19.1). The basal contact is sharp with a breccia-diatomite (Facies F10, Table 19.1) comprised of angular obsidian blocks (up to 15 cm) separated by diatomite. These deposits give way to pure white massive diatomite (F3, Table 19.1) and interbedded diatomite and clay (F1, Table 19.1 and Fig. 19.5c). The latter consist of alternations of diatomite and green clay-rich horizons. Thicker (1–2 cm) clay-rich horizons are present at 20–30 cm intervals, with thinner (2–3 mm) clay layers every 2–3 cm. These cyclic variations have been attributed to deposition in a deep-water lake (DL1, Fig. 19.4) with influxes of clay from streams in response to variable rains with an ENSO-style variability (Owen et al. 2012).

Table 19.1 Sedimentary facies of the Munya wa Gicheru deposits

Facies	Lithology	Lithology, structure, and diatom content	Associated features	Environment
F1	Interbedded diatomite and clay (DC).	Alternating greenish diatomaceous clay (C) (3–4 mm) and white diatomite (D) (4–6 cm) (>90% diatoms). Clays (C) thicken at regular intervals of 20–30 cm. >50% diatoms in clay. Planktonic diatoms dominate. Low diatom fragmentation.	Sponge spicules, phytoliths.	Deep-water lake, periodic flood inputs of fine-grained clastics.
F2	Laminated diatomite (Dp).	White to grey, planar (p) laminated diatomite. Diatoms forming >90% of diatomite. Planktonic diatoms dominate. Low diatom fragmentation.	Non-diatom siliceous microfossils rare.	Deep lake, low clastic input.
F3	Massive diatomite (Dmr).	Massive (m), white diatomite (D) with variable clastic and tuffaceous components. Mostly in beds <80 cm thick, sometimes bioturbated. Diatoms form >90% of diatomite. Mixed benthonic and planktonic diatoms.	Phytoliths and sponge spicules vary in abundance. *In situ* root marks (r) present.	Shallow to moderately deep lake.

(continued)

Table 19.1 (continued)

Facies	Lithology	Lithology, structure, and diatom content	Associated features	Environment
F4	Massive subdiatomite (SDmsabr).	White to grey, massive (m) subdiatomite (SD), with siliciclastics (s) and/or volcanic ash (a). Sometimes bioturbated (b). Diatoms form 50–70% of the sediment. Moderate to high diatom fragmentation.	Phytoliths and sponge spicules common. Root marks (r) present. Rare gastropods.	Shallow lake.
F5	Massive diatomaceous silts (DTmsabr).	Brownish-grey, massive (m) diatomaceous sediments (DT), with siliciclastics (s) and/or volcanic ash (a). Sometimes bioturbated (b). Diatoms form <50% of the sediment. Moderate to high diatom fragmentation.	Phytoliths and sponge spicules common. Root marks (r) sometimes common.	Shallow lake/ wetland.
F6	Reworked diatomaceous sediments (DTmibr).	Grey, greenish or greenish brown, massive (m), diatomaceous silts (DT). Angular intraclasts (i) of up to 1 cm may be present. Sometimes bioturbated (b). Diatom percentages very variable. Benthonic diatoms dominate.	Sponge spicules and phytoliths common. Rare gastropods. Root marks (r).	Shallow lake with emergent periods.
F7	Silty sand and gravel (SGleir).	Light brown, greenish, grey, poorly sorted silty sands (S) and gravels (G). Often in lensoid (l) deposits with multiple minor erosion surfaces (e). Intraclasts (i) common.	Phytoliths and broken diatoms. Root marks (r) *in situ* and reworked.	Sheet wash and minor fluvial channels.
F8	Speckled silty sand (Sd).	Greenish-yellow silty sand (S) speckled with small obsidian fragments. 1–5 cm thick infilled desiccation polygons (d).	Rare fragmentary diatoms.	Aeolian sand (?). Arid setting.
F9	Volcaniclastic sands and gravels (T)(P).	Tuff (T) and pumice (P) deposits in beds up to 40 cm thick. Thickens to the east.	Rare, fragmentary diatoms.	Air fall ash.
F10	Breccia-diatomite (BD).	Large (up to 15 cm) poorly sorted, angular clasts (B). Clasts separated by diatomite (D) infilling crack network.	Abundant planktonic and benthonic diatoms with common phytoliths and sponge spicules.	Obsidian lava flow flooded by palaeolake.

The terminology for the diatomaceous sediments follows Owen (2002)

A widespread channelled erosion surface is overlain by scattered obsidian clasts (<8 cm) at about 6 m (Fig. 19.4), with the overlying deposits (up to 10.1 m, Fig. 19.4) pinching out to the west and consisting of massive diatomaceous silt (F5, Table 19.1) with vertical and horizontal root moulds lined by carbonaceous material (Fig. 19.5d) and reworked grey diatomite (F6) with pumice (~1 cm) and small carbonate nodules (<2 cm). The sequence between 10.1 and 22.2 m includes a stratigraphic gap and is highly diatomaceous, consisting of alternating diatomite (up to 10 cm thick) and diatomaceous clay (1–4 cm thick) interbeds of Facies 1 that give way to diatomite (F3), reworked diatomite (F6) and laminated to massive silty subdiatomite (F4).

Diatomaceous, pumiceous, grey sandy tuff and a tuffaceous subdiatomite (F9; Fig. 19.5g) are present between 22.2 and 23.6 m and are capped by a thin speckled, yellow-brown, coarse-grained aeolian sand (F8) that infills large desiccation polygons (~1 m across; Fig. 19.5e). These deposits collectively comprise the middle member as defined by Ngecu and Njue (1999). Overlying facies (23.6–31.3 m) are comprised mainly of diatomaceous silt (F5) and rare laminated diatomite (F2) with increases in organic root traces, burrowing and disconformities that indicate increasingly shallow water.

The uppermost part of the succession (31.3–33.7 m) comprises coarser-grained silt, sand and gravel with reworked blocky volcanic and obsidian clasts, burrows and organic root marks, and multiple shallow (<2 cm) erosion surfaces, some with concentrations of obsidian granules in small channels (Fig. 19.5f). Aquatic microfossils are absent.

The dominant microfossils through the MwG Formation are diatoms with smaller percentages of sponge spicules. Phytoliths form 0–100% of the siliceous flora and fauna, becoming dominant in the upper part of the succession. A total of 107 diatom taxa have been identified and classified into seven major ecological groups (Owen et al. 2012).

Thirteen diatom stages (A–M: Fig. 19.4) are recognised based on the presence of the dominant taxa: *Aulacoseira agassizii, A. ambigua, A. granulata* var. *valida, Cyclotella meneghiniana* (*Stephanocyclus meneghinianus*), *Dicostella pseudostelligera, Pseudostaurosira zeilleri, Staurosirella pinnata, Stephanodiscus niagarae* and *S. transylvanicus*. Diatom preservation ranges from poor to excellent. Corrosion is minor indicating that dissolution is unlikely to have had a significant impact on assemblage compositions. Fragmentation is minimal in the lower sequence but increases upwards and correlates with increases in sponge spicules and phytoliths.

19.3.4 Palaeoenvironments

The Kedong section of the South Kenya Rift had been filled by 300 m of Limuru Trachytes by 1.96 Ma, producing a relatively level land surface (Fig. 19.6a; Baker and Mitchell 1976; Baker 1986; Baker et al. 1988). Subsequently, grid faulting created a series of grabens and horsts. One of these marginal troughs provided accommodation space for the MwG palaeolake. Figure 19.6 shows a summary of lake changes through time. Twenty-five diatom-based zones and subzones have been recognised, reflecting eight deep-lake periods and eight shallow-water phases (Owen et al. 2012, 2014). The lake was possibly stratified during high-level periods,

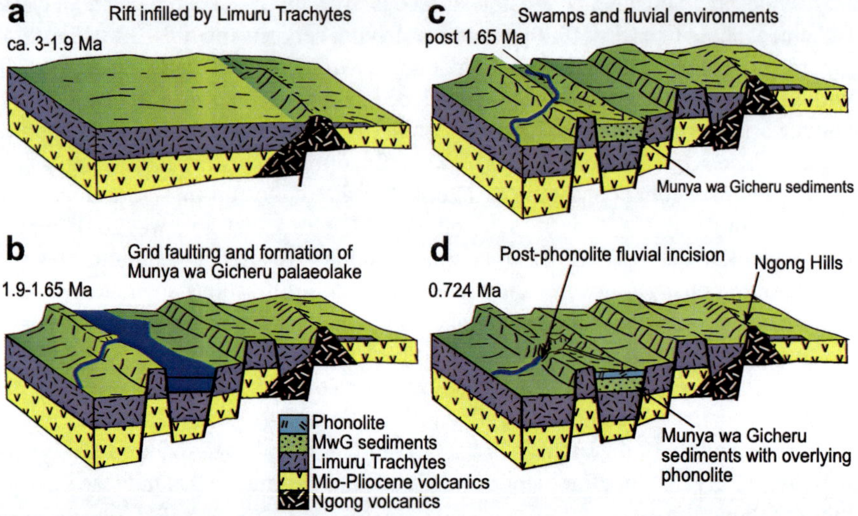

Fig. 19.6 Geological development of the Munya wa Gicheru Basin. After Baker and Mitchell (1976) and Owen et al. (2012). (**a**) Early rift infilling. (**b**) Rift faulting and graben flooding with periodic overflow. (**c**) Lacustrine sedimentation terminated by increasing aridity and outlet downcutting. (**d**) Continued downcutting and eruption of phonolite. See Fig. 19.4 for dating evidence

with pH ranging from 7.2–9 and conductivities between ~70–11,000 μS cm^{-1}, but mostly <1000 μS cm^{-1}.

Diatom Stages A and B reflect deep lakes (DL1, Fig. 19.4) and very fresh water (200–900 μS cm^{-1}; pH 7.5–8) with planktonic *A. granulata* var. *valida* and *A. granulata* var. *angustissima* dominant and with abundant *A. agassizii*, and *Stephanodiscus transylvanicus*. The sharp basal contact implies rapid flooding, which might have been climate-related, but could also indicate tectonic or volcanic damming by lava flows or landslides, or perhaps by river capture. The freshwater diatoms are evidence of an open lake system. An eroded gorge to the west is the likely outlet (Fig. 19.6b). An erosion surface at about 6 m indicates a significant lake retreat and exposure to fluvial conditions (T1, Fig. 19.4).

A mildly saline shallow lake developed during SL1 times (= diatom stage C, Fig. 19.4) when *Discostella pseudostelligera*, *C. meneghiniana*, *Craticula elkab* and *Pseudostaurosira zeilleri* dominated in pools with average conductivities of ~7000–11,000 μS cm^{-1} and pH of ~8.8–9. Subsequently (Stages D–G) the lake expanded and contracted several times giving rise to four deep (DL3–6) and three shallow lake (SL–4) periods. Diatom floras varied significantly, with prominent taxa including planktonic *Aulacoseira goetzeana*, *A. granulata* var. *valida*, *S. transylvanicus*, *S. niagarae* and shallow-water floras with *A. ambigua*, *A. granulata* var. *angustissima* and *A. granulata* var. *angustissima* f. *curvata*, *Discostella stelligera* and *Pseudostaurosira zeilleri*. The waters remained generally fresh with conductivities of 70–1000 μS cm^{-1} and pH of 7.3–8.2.

The high DL6 lake levels were terminated at about the same time as a major influx of ash (~1.71 Ma) and were followed by desiccation and the production of extensive polygonal mudcracks that were infilled by aeolian sand (Ngecu and Njue

1999). Three more high-level phases (DL7–9) followed, but root moulds, increased sponge spicules and phytoliths, and mixed planktonic and benthonic diatoms (*Aulacoseira* spp., *Stephanodiscus* spp., *Cocconeis placentula*, *Cyclotella meneghiniana* (*Stephanocyclus meneghinianus*), *Staurosira pinnata*, *Ulnaria ulna*, pp *zeilleri*, *Gomphonema parvulum*) all indicate shallower waters.

The final depositional phase began about 1.65 Ma (Trauth et al. 2007) and is characterised by rare fragmentary diatoms, many small erosional scours, intraclasts, burrows, and organic root-marks in silt to gravel size sediments that probably formed under subaerial conditions. The end of lacustrine sedimentation could indicate climate change, but more likely reflects loss of accommodation space through infilling of the basin and downcutting of the outlet. Subsequent fluvial incision along the former outlet carved a narrow gorge leading to the modern landscape (Fig. 19.6d).

The MwG sediments have been correlated with deposits at Olduvai (N Tanzania), Turkana (Kenya), and the Ethiopian rift and Afar basins, with the inference that they represent a wetter climate between 1.9 and 1.7 Ma (Trauth et al. 2005, 2007, 2009). Precipitation and evaporation clearly explain some of the cyclic changes in sedimentation within the basin. However, longer-term changes are more difficult to decipher because they may also reflect geomorphological processes, volcanism and local or regional tectonism. Consequently, there has been increasing recognition of the need to be cautious when correlating wet phases between tectonically active regions (Trauth et al. 2005; Owen et al. 2008b; Owen et al. 2009a; Trauth and Maslin 2009).

19.4 Pleistocene Lake Sediments at Olorgesailie

19.4.1 Location and History of Research

The 150 km^2 Olorgesailie Basin (Fig. 19.7) is located at 36°26′E, 1°34′S, 65 km to the southwest of Nairobi and is easily reached via the main Nairobi-Magadi road. North-south trending Late Pleistocene faults (Baker 1986) mark the eastern and western margins of the area with the extinct Pliocene volcanic centre of Mt. Olorgesailie forming a partial barrier to the south (Fig. 19.7a). The Magadi Trachyte Formation comprises much of the rift floor, along with the Ol Tepesi Basalts on the eastern side of the basin (Fig. 19.7b). A trachyte-basalt ridge divides the area into the Legemunge (east) and Oltepesi (west) Plains (Fig. 19.7d). Other nearby volcanic centres include the Pliocene Ol Esakut (Ol Esayeti) to the northeast and the Pleistocene Mt. Suswa to the north. The Koora Graben, to the west of Mt. Olorgesailie, opens into the Koora Plain about 20 km south of Olorgesailie and then narrows into the Siriata Valley further south (Fig. 19.7a).

A variety of facies were laid down in lakes, wetlands/swamps, springs, rivers, and on land surfaces between 1.2 Ma and the present (Fig. 19.7c) that would have

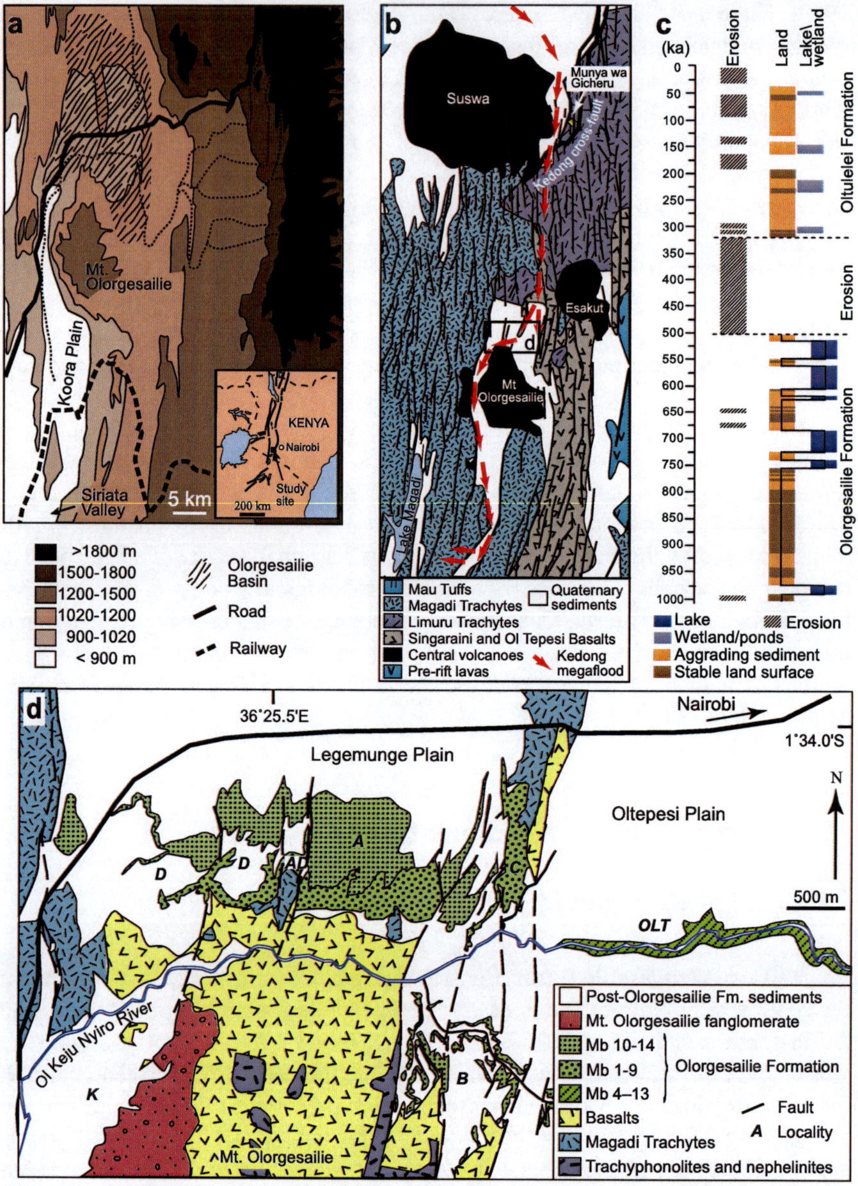

Fig. 19.7 Geology of the Olorgesailie Basin. After Owen et al. (2011). (**a**) General location and topographic setting. (**b**) Regional geology showing dominance of trachyte. (**c**) Environmental record derived from sediments of the Olorgesailie and Oltulelei formations. After Potts et al. (2018). (**d**) Simplified geology of the Olorgesailie Basin and 'locality' designations (A, B, C, D, AD, OLT, K)

influenced hominin distributions (Cuthbert et al. 2017). The oldest deposits, the Olorgesailie Formation, have a composite thickness of about 80 m (Behrensmeyer et al. 2002) and have been the subject of intensive archaeological investigations (Isaac 1968, 1977, 1978; Potts 1989, 1994; Potts et al. 1999, 2004, 2018) since the discovery of Acheulian stone tools at the site (Leakey 1952; Noll 2000; Noll and Petraglia 2003; Tryon and Potts 2011; Walter and Trauth 2013). Middle Stone Age (MSA) tools probably allowed expansions of *Homo sapiens* into a broad range of environments (Timbrell et al. 2022) and are well preserved in Middle to Late Pleistocene terrestrial sediments at Olorgesailie, where they date back to about 320 ka (Haradon 2010; Brooks et al. 2018; Deino et al. 2018). These sediments were previously referred to as the Olkesiteti Formation and the Oltepesi beds (Brooks et al. 2007) or simply as post-Olorgesailie Formation deposits (Owen et al. 2011, 2014), but have now been grouped together as members within the Oltulelei Formation (Behrensmeyer et al. 2018).

Early studies include the work of Parkinson (1914) who described the geology of the South Kenya Rift, referring to Olorgesailie as Lorgosailich. He also noted the presence of lake deposits in the Olorgesailie Basin, which were attributed to a Pliocene 'Lake Suess' that Gregory (1896) believed extended across the rift and which he thought stretched from the Kedong Valley southward to Lake Magadi. Isaac (1978) presented a brief historical review of subsequent research, noting that Gregory camped in the Olorgesailie area in 1919 on his way to Lake Magadi, referring to Mt. Olorgesailie as Ol Gasalik and attributing the origin of the sediments to a Miocene 'Lake Kamasia' (Gregory 1921; Bishop 1978). Gregory also reported the presence of "stone picks", but mistakenly thought they were remnants of Stone Age diatomite mining.

In April 1942, Louis and Mary Leakey (Leakey 1952) rediscovered abundant Acheulian handaxes that had been noted earlier by Gregory (1896, 1921). These lay within diatomite at Olorgesailie, thus indicating that they were of Pleistocene age (Isaac 1978). They returned to excavate between 1943–1945 and invited Dr. R.M. Shackleton to carry out a geological survey between 1944 and 1946 (Shackleton 1944, 1955). Parts of this work, and a detailed 1:600 geological map of the museum site, were incorporated into a Kenya Geological Survey report (Baker 1958) with Shackleton's much-used 1:10,000 map of the Olorgesailie deposits being published some three decades later (Shackleton 1978). Posnansky (1959) described fourteen occupation levels and documented stone tools (Hope Fountain materials) associated with two land surfaces. At the time of these early studies the Olorgesailie sediments were being cited as examples of a pluvial phase within the Middle Pleistocene (Leakey 1955; Cole 1954) although these climate-stratigraphic concepts subsequently fell into disuse (Cooke 1957; Flint 1959a, b; Hay 1965; McCall et al. 1967).

Glynn Isaac carried out archaeological and sedimentological studies of deposits associated with the Acheulian handaxes from 1962 to 1965 as part of his PhD research (Isaac 1968). Baker (1958) had previously referred to the sediments as the 'Olorgesailie Lake Beds' and subsequently as the 'Legemunge Beds' (Baker and Mitchell 1976). Isaac (1978) formalized the terminology, naming the deposits the Olorgesailie Formation. This work stimulated several researchers to examine other aspects of the sedimentology, microfossils and chronology of the deposits (Marsden 1979; Owen 1981; Owen and Renaut 1981, 2000; Koch 1986). Evernden and Curtis (1965) and Miller (1967) reported middle Pleistocene dates for the Olorgesailie Formation with Isaac (1977) later documenting K-Ar ages of 0.42 and 0.48 Ma that he considered compatible with the faunal and archaeological evidence. However, Bye (1984) and Bye et al. (1987) subsequently reported older dates ranging between 0.93–0.6 Ma.

A third phase of research developed from 1985 under the direction of Prof. Rick Potts and through cooperation between the Smithsonian Institution and the National Museums of Kenya. Deino and Potts (1990, 1992) developed a new $^{40}Ar/^{39}Ar$ chronological framework for the Olorgesailie Formation, reporting many dates between 1.2 and 0.493 Ma (Behrensmeyer et al. 2002; Owen et al. 2008a, b). Much of the Olorgesailie work has been focused on stone tools and mammal remains, often associated with land surfaces and palaeosols. However, after many years of study, a ~1-Ma-old *Homo erectus* cranium was discovered in 2003 (Potts et al. 2004). During his studies at Olorgesailie, and at other East African sites, Potts developed ideas on human evolution that related hominins to their contemporary environmental settings, pointing out, in his Variability Selection hypothesis, that environmental instability might be an important driver of physical and cultural change in human ancestors (Potts 1989, 1994, 1996a, b, 1998, 2001, 2002, 2012, 2013). In addition, Potts brought together scientists with a wide variety of expertise in sedimentology, vertebrate palaeontology and palaeoanthropology to reconstruct past landscapes at Olorgesailie (Potts et al. 1996, 1999, 2004; Sikes et al. 1999; Behrensmeyer et al. 2002; Melson and Potts 2002; Owen et al. 2008a, b, 2009a, b, 2011, 2014; Lee et al. 2013; Potts and Faith 2022). Durkee and Brown (2014) have also explored tephrocorrelation between Olorgesailie and other Acheulian tool-bearing sequences at Kariandusi and Isinya.

The Olorgesailie Drilling Project, led by Potts, obtained two long cores from the Koora Plain (Fig. 19.7a) about 22 km south of Olorgesailie, which have provided material for ongoing sedimentological, geochemical and microfossil studies (Pennisi 2013; Rabideaux et al. 2015; Stockhecke et al. 2015; Muiruri 2017; Rabideaux 2018; Potts et al. 2020). Combined with research on two long cores from Magadi, the new data provides an excellent one-million-year record of environmental change for the South Kenya Rift (Owen et al. 2018a, b, 2019).

19.4.2 The Olorgesailie Formation (1.2–0.49 Ma)

Sedimentology

Well-exposed outcrops of the Olorgesailie Formation are present along the southern Legemunge Plain, close to the Ol Keju Nyiro River and on the northeastern slopes of Mt. Olorgesailie (Fig. 19.7d). Shackleton (1944, 1978) divided the formation into 14 layers, which Isaac (1978) later referred to as 'Members' when he defined it as "a series of well-stratified sediments consisting principally of pale, volcanic siltstones, volcanic sandstones, diatomites and brown non-volcanic siltstones" with "lesser quantities of claystones, pyroclastic deposits, and limestones". Isaac also reported a ~45 m thick composite type-section from western Locality D and noted that the sediments locally thickened to ~60 m. Behrensmeyer et al. (2002) subsequently presented a new composite section of ~64 m, with Owen and Renaut (1981) and Owen et al. (2008b, 2009b, 2011) documenting the sedimentological and microfossil stratigraphies of several sections in Localities A, C and D (Fig. 19.7d).

The sediments form laterally extensive beds (Fig. 19.8a) with lake deposits comprising >70% of the total thickness, but only comprising <25% of the time they represent (Fig. 19.7c; Behrensmeyer 2007; Potts et al. 2018). The dominant components include biogenic silica (diatoms, with less common sponge spicules and phytoliths), lithic grains (mainly trachyte, with basalt and rare obsidian), alkali feldspars (mainly anorthoclase and sanidine), as well as subsidiary plagioclase, clinopyroxene, volcanic glass, quartz and other silicates (Owen 1981; Owen and Renaut 1981). Calcite occurs as micrite and sparite associated with rhizoliths, calcrete and cross-cutting calcite veins. Clay is dominated by pedogenically-modified smectite with less abundant kaolinite and minor illite and chlorite. Detrital clay from the volcanic drainage basin has been diagenetically altered through exposure to alkaline, but not highly saline water. The authigenic clay chemistry, especially Mg enrichment/depletion, variably reflects the influences of the original palaeolake waters and post-depositional weathering (Deocampo et al. 2010).

Owen and Renaut (1981) recognized ten lithofacies: (1) Laminated siltstone, (2) Structureless, tuffaceous, diatomaceous siltstone and claystone, (3) Redeposited diatomaceous siltstone and claystone, (4) Massive diatomite, (5) Interbedded volcanic sandstone, siltstone and conglomerate, (6) Pumiceous sandstone and conglomerate, (7) Volcanic ash, (8) Brown siltstone and claystone, (9) Poorly sorted sandy conglomerate, and (10) Calcrete. More recently, Behrensmeyer et al. (2002) distinguished ten major sediment types, which were later modified to more clearly reflect diatom content (Owen 2002), yielding eleven lithofacies (Owen et al. 2008a, b). Subsequently, Owen et al. (2011) reclassified the deposits and recognised fifteen facies and subfacies (F1–F12, Table 19.2).

Geochemical Studies

Kübler et al. (2018, 2021) have explored general links between rock chemistry, soil nutrients and vegetation as possible controls on wildlife movements in and across the Olorgesailie and south Kenya Rift areas. However, more detailed geochemical studies of the sedimentary sequence have been published by others. Principal Components Analysis (PCA) of bulk geochemical data (Fig. 19.9a; Owen et al. 2011) indicate four major facies groups: (1) silica-rich, (2) carbonate-rich, (3) evaporite–carbonate, and (4) lithogenic. The silica-rich facies comprise deposits with abundant diatoms, sponge spicules, phytoliths or opaline silica plant stems (F1–6, Table 19.2 and Fig. 19.8b–k). These facies formed in lakes and wetlands and are associated with relatively high concentrations of U, S and Br that might reflect the presence of phosphate and organics or, partly, volcanic aerosol sources. Facies 6 (Table 19.2), with abundant silica rhizoliths, is largely restricted to M14, and

Fig. 19.8 Olorgesailie Formation sediments. Black bar is 1 cm. Grey bar is 10 µm. Facies (F#) refer to Table 19.2. (**a**) General photograph of the Olorgesailie Formation showing lateral continuity. Note the prominent white diatomite (F1) forming part of Member 12 (M12). Cliffs above are composed of M13–14. Ol Esakut in the background. Foreground surface is formed by a lag of Oltulelei Formation chert and calcrete. (**b**) Clayey diatomaceous sediments (F5) overlain by silty subdiatomite (F3). (**c**) White diatomite (F1) overlain by volcanic ash (F11) and tuffaceous subdiatomite (F3). (**d**) Laminated diatomite (F2). (**e**) SEM of diatomite showing abundant well-preserved *Aulacoseira granulata*. (**f**) Reworked diatomaceous silt with diatomite clasts (F4). (**g**) Burrowed silty subdiatomite (F3). (**h**) Opaline silica rhizoliths (arrow) in diatomaceous silt (F6). (**i**) Large calcite rhizolith (wetted to make visible) in silty subdiatomite (F3). (**j**) Fluid escape structures (arrow) and feeder tubes in evaporite-carbonate marker bed (F7). (**k**) Volcanic sand, gravel (F10) formed by rounded pumice and ash

Table 19.2 Sedimentary facies in the Olorgesailie Formation

Facies	Lithology	Colour	Lithology, structure, & diatom content	Associated features	Environment
F1	Massive diatomite (MD), siliciclastic and tuffaceous diatomite (SD & TD).	White.	Homogeneous, >90% diatoms for MD, >70–90% for SD & TD; benthic taxa common. Mostly in beds <10 cm thick, sometimes bioturbated.	Sponge spicules, phytoliths, ostracods.	Shallow to moderately deep lake.
F1B	Carbonate-cemented diatomite.	White-grey.	Homogeneous, >90% of carbonate free fraction. Mostly in beds <10 cm thick, sometimes bioturbated.	Sponge spicules, phytoliths, ostracods.	Shallow to moderately deep lake.
F2	Laminated diatomite.	White to yellow.	Fine laminations, sometimes contorted, >90% diatoms, mostly planktonic.	Sponge spicules, phytoliths.	Deep lake.
F3	Silty and tuffaceous subdiatomite.	White, light green, yellow.	Massive or contorted bedding, relict lamination. Diatoms generally 50–70%.	Bone, artifacts, gastropods, ostracods, rhizoliths, burrows.	Wetlands
F3B	Carbonate-cemented silty and tuffaceous subdiatomite.	White, light green, yellow.	Massive or contorted bedding, relict lamination. Diatoms generally 50–70%.	Bone, artifacts, gastropods, ostracods.	Wetlands.
F4	Reworked diatomites, subdiatomite and diatomitic silt.	Dirty white, buff-green, greenish brown.	Massive or irregular bedding. Diatom content reflects emergence of a variety of facies. Diatoms often broken, intraclasts may be present.	Rhizoliths, some bones, artifacts.	Lake margin or emergent lakebed.
F5	Clayey diatomite, subdiatomite and diatomitic sediments.	Green, yellow green.	Massive, brecciated diatomaceous sediments. Diatom fragmentation common, intraclasts may be present.	Rhizoliths, some bones, artifacts.	Pedogenically modified, lakebeds.
F6	Cherty-opaline diatomaceous silt.	White, with dark grey chert.	Homogeneous, variable diatom content, sometimes zero.	Cherty and opaline rhizoliths, burrows.	Shallow wetlands, subject to desiccation.

(continued)

Table 19.2 (continued)

Facies	Lithology	Colour	Lithology, structure, & diatom content	Associated features	Environment
F7	Evaporite-carbonate (E-C) silt.	Grey.	Evaporite pseudomorphs, transgressive crack fills., usually <10 cm thick, fluid escape structures.	Diatoms.	Playa evaporative mudflat.
F8	Clay or silty clay.	Dark green, greenish brown.	Ped structures, clay skins, $CaCO_3$ nodules, diatoms absent.	Rhizoliths, some bones, artifacts.	Paleosol.
F9	Silt, sandy silt.	Light brown, buff to greenish brown.	Massive or irregular bedding, $CaCO_3$ nodules.	Rhizoliths, rare silica rhizoliths, artifacts, bones.	Fluvial overbank, paleosol, aeolian (?), hot springs (?) (forming silica rhizoliths).
F9B	Carbonate-cemented silt, sandy silt.	Light brown, buff to greenish brown.	Massive or irregular bedding, $CaCO_3$ nodules.	Rhizoliths, rare silica rhizoliths, artifacts, bones.	Fluvial overbank, pedogenic, aeolian, hot springs.
F10	Volcanic sand, gravel.	Light brown, greenish brown.	Poor sorting, cross stratification, intraclasts, pumice and ash.	Artifacts, bones locally abundant, reworked plants.	Fluvial channels, sheetwash, proximal splays
F11	Silt-grade ash.	Dark to light grey.	Well-sorted, laminated, laterally extensive.	Burrows and rhizoliths, rare silica stems.	Air fall tuffs.
F12	Silty to sandy tuff.	Dark to light grey and grey brown.	Moderate to poor sorting, laminated to massive, laterally variable.	Burrows and rhizoliths, plant impressions, rare bones and artefacts.	Fluvial and aeolian reworked volcanic debris.

The terminology for the diatomaceous sediments follows that suggested by Owen (2002). Facies modified after Behrensmeyer et al. (2002)

probably formed in a spring-fed wetland with silica-enriched water that facilitated permineralization in plants following evapotranspiration (Owen et al. 2009b, 2014).

Group 2 (Fig. 19.9a; F1B, F3B, F9B, Table 19.2) includes sediments with abundant CaO, which is associated with LOI and Sr in PCA analyses (Owen et al. 2011). This represents micritic and microsparitic calcite that precipitated in a variety of syndepositional, pedogenic and early diagenetic settings. The calcite variously occurs in rhizoliths (Fig. 19.8i), carbonate nodules, lenses, discontinuous layers and in vertical crack-filling sparite. Group 3 includes a single distinctive boxwork

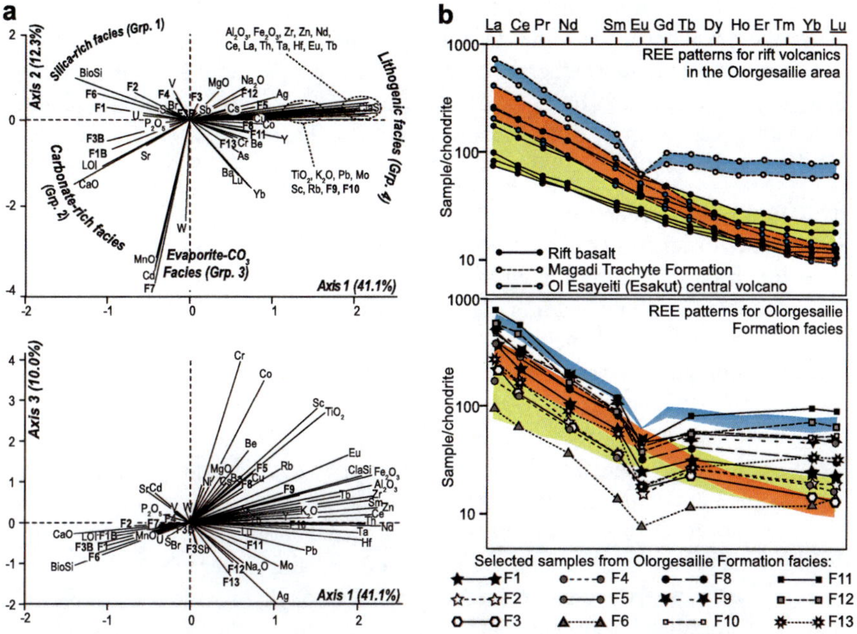

Fig. 19.9 Geochemical data from the Olorgesailie Formation. Modified from Owen et al. (2011). (**a**) Principal Components Analysis of major Olorgesailie Formation facies (see Table 19.2). Four groupings are reflected by the facies and elemental compositions. (**b**) REE data for volcanics and sediments in the Olorgesailie area. Note the similarity in REE patterns between all Olorgesailie sediments and the Magadi Trachyte, which display a prominent Eu-dip and lower HREE values, suggesting that these are the dominant source rocks

evaporite–carbonate marker horizon (F7, Table 19.2 and Fig. 19.8j) near the base of M2, with high Ca, Ba, Cd, W, Cu, Y, Yb and Lu and common MnO dendrites. Calcite pseudomorphs of trona and a cubic mineral also occur. Fluid escape structures are present locally which suggest that the deposit might have formed in response to earthquake triggering, with dewatering of a littoral mudflat. Group 4 (Fig. 19.9a) includes varied sediments with mixed diatom-clastic components (F4, Table 19.2) and non-diatomaceous deposits (F5, F8–12, Table 19.2 and Fig. 19.8k). These represent fluvial and terrestrial deposition, commonly with pedogenesis of pre-existing sediments.

Rare Earth Elements (REE) display similar patterns in all facies with a decline in REE towards the heavier elements and an Eu-dip (Fig. 19.9b; Owen et al. 2011). These characteristics reflect Magadi Trachyte lavas, which dominate in the Olorgesailie Basin. Variable REE and U/Th compositions in fossil bones reflect changes in pore water conditions over short distances (Trueman et al. 2006).

Chemical Index of Alteration (CIA) values range between 54 and 92 and have been used to compare the Early Pleistocene (1.96–1.65 Ma) Munya wa Gicheru sediments (45 km to the north) with the Olorgesailie (1.2–0.49 Ma) and Oltulelei

(0.32–0.05 Ma) formations. The data document a CIA decline through the Quaternary, which reflects either increasing aridity (and reduced basin weathering) or reduced exposure to weathering caused by increasingly rapid erosion and deposition rates (Owen et al. 2011, 2012, 2014). The former suggestion would be compatible with evidence from elsewhere for increasing East African aridity (e.g. Bobe 2006; Fernández and Vrba 2006; Bonnefille 2010).

Ichnofossils and Rhizoliths

Three suites of ichnofossils have been recognized in Member 14 sediments (Owen et al. 2009b). Suite 1 is attributed to insect larvae and/or nematodes and consists of horizontal tunnels along bedding planes that lack linings or walls and suggest saturated substrates. Suite 2 includes vertical open, spherical pellet-filled, and lined, vertical to oblique non-branching burrows <5 mm in diameter formed by beetle and earwig larvae. Larger vertical burrows with a pellet-fill may have been constructed by spiders, beetles, aestivating gastropods or termites and indicate substrates with lower water tables. Suite 3 includes concentrically backfilled and pellet-filled large burrows (4–8 cm diameter) attributed to termites. The latter indicate dry substrates. The coexistence of these three sets of trace fossils, and benthic diatoms, indicate flooding and drying of a wetland, resulting in fluctuating water tables.

Fossilised roots are present in several forms. Mn-oxyhydroxide stains are associated with fossilised root channels and/or clay fills. Micrite and pore-filling microsparite is present as both vertical and subhorizontal rhizoliths with mostly sub-circular cross sections (1–2 cm) and lengths of about 8 cm, but sometimes up to 30 cm (Fig. 19.8i). They commonly possess a central void lined with sparry calcite. The calcite rhizoliths reflect evaporative concentration, evapotranspiration, and (or) CO_2 degassing of Ca-bearing meteoric water in, or adjacent to, former root systems (Owen et al. 2008a; Owen et al. 2009a, b).

Silicified plants occur in Member 14 and the, younger, Oltulelei Formation (Owen et al. 2009b). These are composed of opal-A and generally have a square to triangular cross-section suggestive of a sedge. Typically, they range from 2–8 cm in length and 2–10 mm in width and are straight to mildly undulate with striated surfaces. The stems often occur in clusters that are vertical when *in situ* (Fig. 19.8h) and are composed of opal-A nanospheres that merge to form a homogenous mass of silica that preserves original plant tissues. The build-up of silica in the plant stems may have resulted from an influx of silica-rich hot-spring waters (Owen et al. 2009b, 2014), with silica concentration increasing in the plants due to evapotranspiration (Renaut et al. 2002, their Fig. 10). Chert-like networks of rhizoliths are present near the base of Member 14 (Owen et al. 2008a, b, 2009b, 2014).

Micropalaeontology and Sediment Stratigraphy

Diatoms dominate the siliceous microfossils in the Olorgesailie Formation, with smaller numbers of sponge spicules and phytoliths also present. Seven diatom assemblages have been described representing saline to fresh and deep to shallow palaeolakes, wetlands and swamps (Table 19.3; Owen et al. 2008a, b, 2009b, 2014). Sixteen major diatom zones are recognised through the formation, with several zones further subdivided. In addition to the aquatic sediments, the deposits show

Table 19.3 Dominant diatoms and ecological preferences in the Olorgesailie Formation sediments

Assemblage	*Dominant taxa*	Environment
OF1	*Anomoeoneis sphaerophora, Campylodiscus clypeus, Craticula elkab, Cyclotella atomus.*	Highly saline, benthonic.
OF2	*Thalassiosira rudolfii, Thalassiosira faurii.*	Highly saline, planktonic.
OF3	*Cyclotella meneghiniana* [*Stephanocyclus meneghinianus*], *Cyclotella* [*Pantocsekiella*] *kuetzingiana, Discostella stelligera.*	Mildly to strongly saline, variable water depth.
OF4	*Aulacoseira granulata, A. granulata* var. *valida, A. italica* var. *bacilligera, Stephanodiscus astraea .*	Fresh, deep water.
OF5	*Aulacoseira agassizi, A. ambigua, A. distans* var. *africana, A. granulata* var. *angustissima.*	Fresh, shallow water.
OF6	*Cocconeis placentula, Encyonema muelleri, Epithemia argus, E. adnata, E. sorex.*	Fresh, epiphytic habitats.
OF7	*Sellaphora pupula, Navicula cryptocephala, Caloneis bacillum, Staurosira construens.*	Fresh, benthonic habitats.

After Owen et al. (2014)

evidence for multiple periods of emergence, pedogenesis and erosion (Owen et al. 2011). Major diatom taxa include *Aulacoseira granulata* and its varieties, *A. ambigua*, *A. agassizii*, *Stephanodiscus astraea*, *Cyclotella meneghiniana* (*Stephanocyclus meneghinianus*), *C. atomus*, *Cocconeis placentula*, *Anomoeoneis sphaerophora*, *Encyonema muelleri*, *Epithemia argus*, *E. adnata*, *E. sorex*, *Sellaphora pupula*, *Caloneis bacillum* and many other benthonic taxa.

The diatom record indicates five deep lake phases, as well as eleven shallow lakes and sixteen swamps or wetlands (Fig. 19.10; Owen et al. 2008a, b, 2011). Water conductivities ranged from 200–16,000 µS cm^{-1} with pH between 7.5 and 9.5 (Fig. 19.10). Diatom floras and inferred palaeochemistry vary spatially in several, but not all, sedimentary units (Owen and Renaut 2000). Diatom assemblages in Member 1, for example, indicate fresher conditions (pH ~8.5) in western locations compared with areas further east where pH values reached 9.5. Diatoms in Members 2 and 3 indicate fresher conditions to the west in what was a relatively deep fresh lake. In contrast, a prominent white diatomite in Member 12 is more uniform across the basin with *Cocconeis placentula* dominant in all exposures.

19.4.3 Erosion and the Oltulelei Formation (0.49–0 Ma)

The Olorgesailie Formation sediments ceased to accumulate by about 0.49 Ma with non-deposition and erosion developing between ~0.49 and 0.32 Ma (Fig. 19.7c). Isaac (1978) and Behrensmeyer et al. (2018) related this to faulting and subsidence of the Koora Graben to the south and the consequent lowering of base levels, followed by multiple episodes of erosion (Behrensmeyer 2007). However, variable erosion rates might, in part, also reflect fluctuating water levels of a lake that

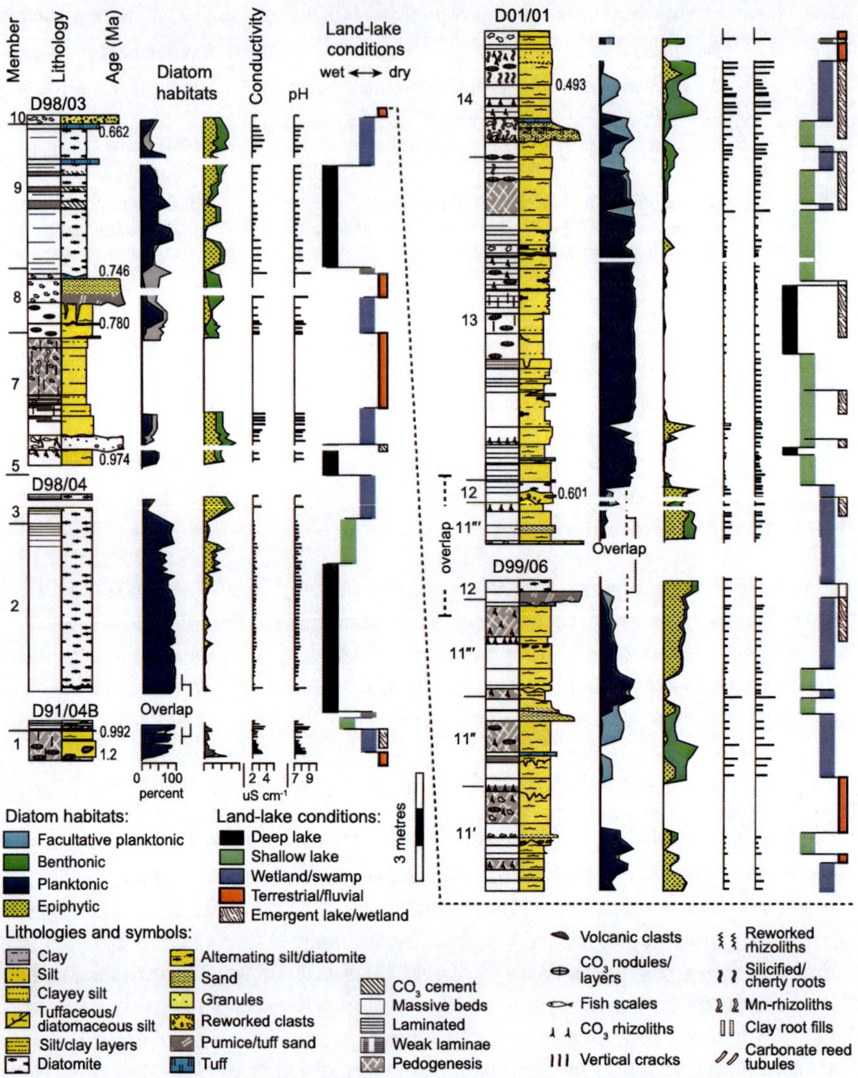

Fig. 19.10 Lithologies and diatom-inferred environmental conditions in the Olorgesailie Formation. Modified from Owen et al. (2014)

occupied the Koora Plain. Kübler et al. (2015, 2018) speculated that Koora subsidence might be related to caldera-style movements around Mt. Olorgesailie.

The younger sediments have been assigned to the Olkesiteti (320–190 ka), Oltepesi (170–50 ka) and Tinga members (~50 ka) of the Oltulelei Formation (Fig. 19.11a, e), with each member separated by periods of fluvial incision of up to 12 m (Behrensmeyer et al. 2018). Oltulelei Formation sediments are associated with Middle Stone Age tools (Brooks et al. 2007, 2018; Haradon 2010; Hatala et al.

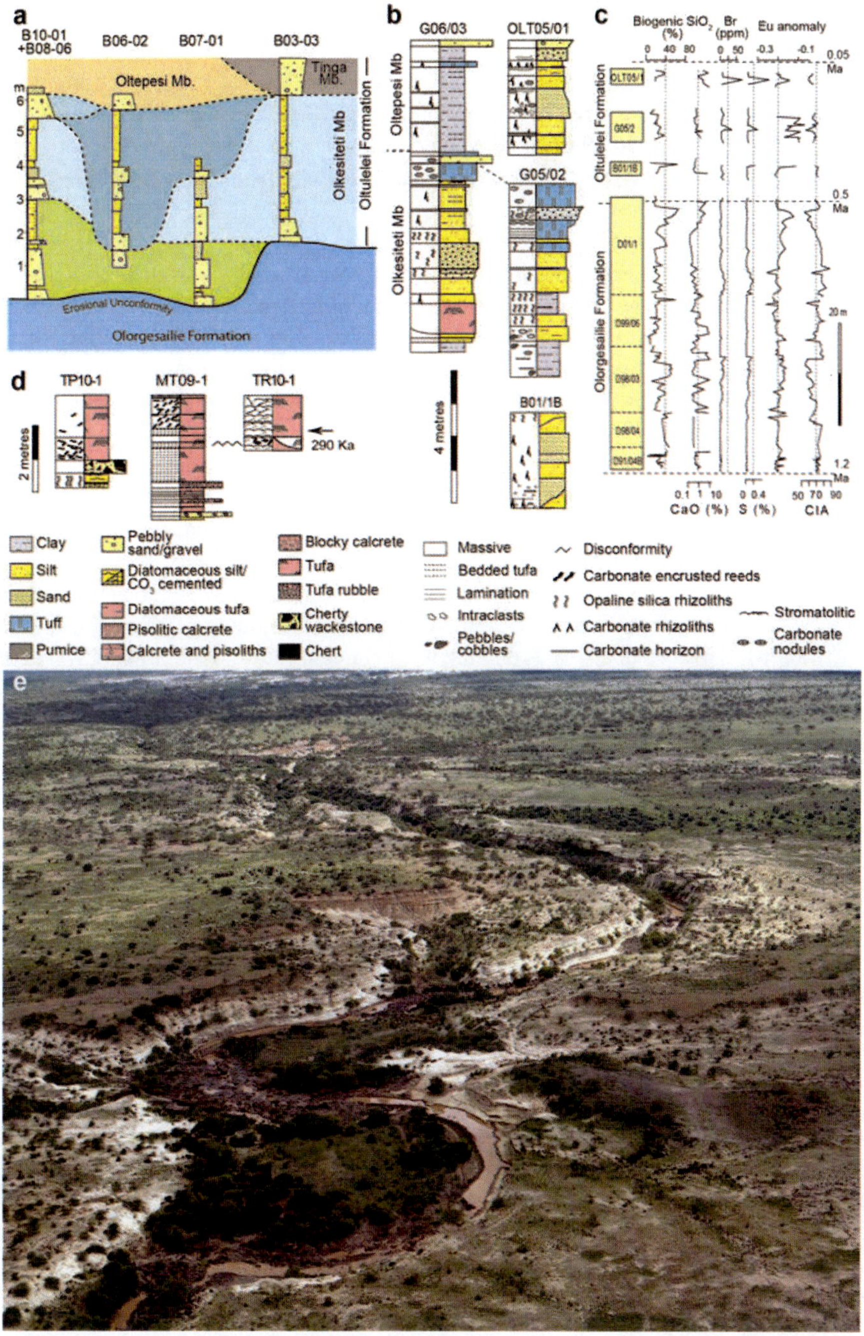

Fig. 19.11 Lithologies and geochemistry of selected Oltulelei Formation sediments. ((**b–d**) after Owen et al. (2014). (**a**) Stratigraphic relationships within the Oltulelei Formation. After Behrensmeyer et al. (2018). (**b**) Selected lithological logs, Oltulelei Formation, showing predominantly siliciclastic/volcanic sequences. (**c**) Geochemistry of selected parameters for the Olorgesailie and Oltululei formations. (**d**) Lithologies of tufa- and calcrete-dominated sequences associated with the Oltulelei Formation. (**e**) The Ol Keju Nyiro River with white/grey diatomites/diatomaceous sediments of the Olorgesailie Formation overlain by darker Oltulelei Formation deposits

2010; Tryon and Faith 2013; Deino et al. 2018) and are located to the north and west of Mt. Olorgesailie in a series of discontinuous outcrops. The deposits have a cumulative thickness of up to 69 m and were laid down in fluvial, volcaniclastic and, less commonly, in shallow pond and wetland settings (Behrensmeyer et al. 2002, 2018; Owen et al. 2014).

The Middle to Late Pleistocene deposits (Figs. 19.11 and 19.12) are dominated by volcaniclastic silt, sand and gravel and rest unconformably on the Olorgesailie Formation. Carbonate rhizoliths and opaline silica plant stems are common (Fig. 19.12a–c). Undated siliceous spring deposits (Fig. 19.12d–f) are present at Eremit, north of the Olorgesailie Basin. Spring waters together with silica in shallow groundwater drawn upwards by evapotranspiration likely contributed to the development of plant silicification.

The Oltulelei Formation also contrasts geochemically with the underlying Olorgesailie Formation (Fig. 19.11c; Owen et al. 2014). Biogenic silica, for example, declines in the younger deposits due to reduced diatom abundance. Spikes in Br and S in the Oltulelei Formation may reflect transport and preservation of organic materials within the basin (Owen et al. 2014) or an influx of volcanic aerosols and ash, with less negative Eu anomalies in sediments likely reflecting changes in source lithologies (Owen et al. 2014). A decline in Chemical Index of Alteration (CIA) data for the younger sediments probably reflects increasing aridity and reduced weathering or more rapid erosion and deposition, which would have minimised exposure times to chemical weathering.

The data for CaO (mainly in calcite) is very variable through the last 1.2 million years (Fig. 19.11c), reflecting the presence or absence of calcareous rhizoliths, nodules and veins. Notably, there appears to be an increase in CaO through the Quaternary across the South Kenya Rift (Owen et al. 2014). CaO, for example, is largely absent, except in siliciclastic minerals, in the older (1.96–1.65 Ma) lacustrine and wetland sediments at Munya wa Gicheru, 45 km north of Olorgesailie. The Olorgesailie Formation and younger sediments both contain abundant calcareous rhizoliths in multiple horizons with calcrete capping parts of the formation (Felske 2016). In contrast, tufa (Figs. 19.11d and 19.12g–l) only appears in deposits younger than 320 ka with one tufa dated at ~290 ka and two other large outcrops likely having formed sometime between 220 and 12 ka (Lee et al. 2013). $\delta^{13}C$ isotopes in the tufa are enriched (0.25–1.96‰) compared with other East African deposits, but are uniform within each set of tufa outcrops, indicating CO_2 that was derived from similar sources for each deposit. The $\delta^{18}O$ values (−7.17 to +5.16‰) indicate a wide range in past evaporation and rainfall and/or water temperature. REE data is variable and indicates spatial and temporal changes in groundwater chemistry during tufa accumulation, which has also been related to faulting and basin breakup (Lee et al. 2013).

Diatoms (Table 19.4) are scarce in the Oltulelei Formation with eight assemblages identified that indicate damp soils, fluvial settings and shallow fresh to mildly saline very shallow lakes and wetlands. No deep-water floras have been observed (Owen et al. 2014). Tufa deposits also contain an additional five assemblages that indicate swamps and small shallow freshwater lakes. Several tufa outcrops also

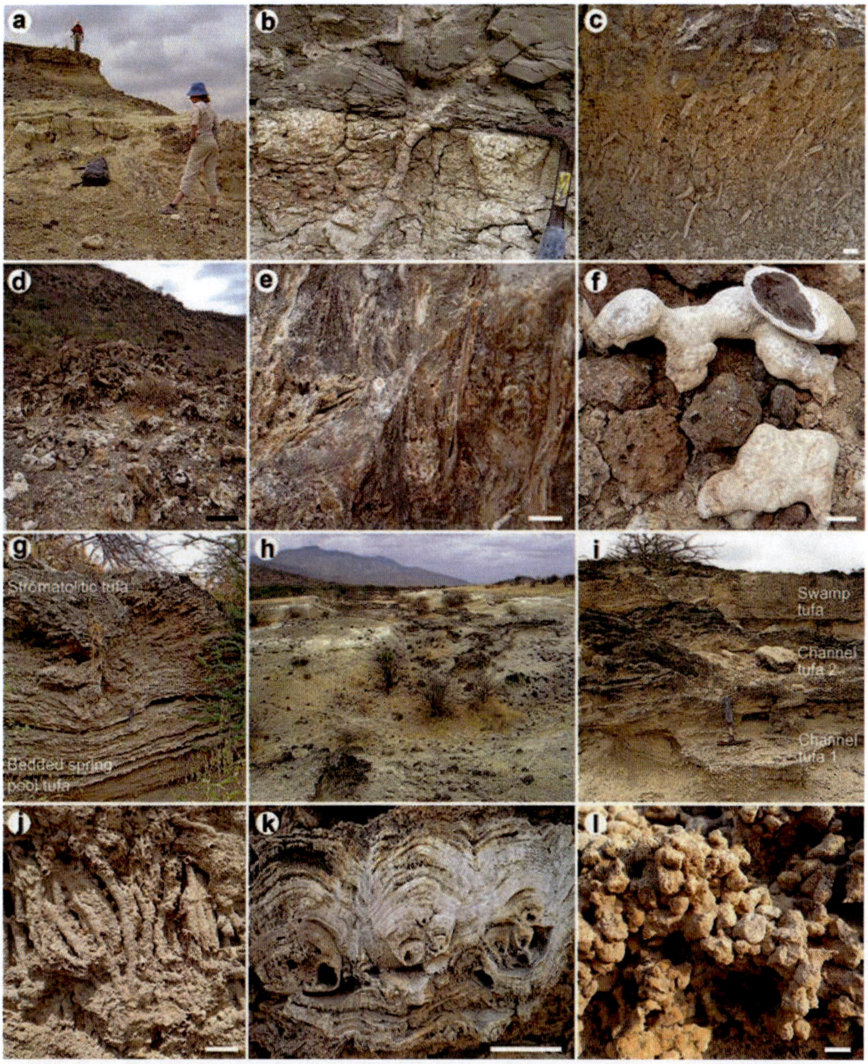

Fig. 19.12 Oltulelei Formation sediments. White bar = 1 cm; black bar = 10 cm. Hammer for scale in (**b**) and (**g**). Meandering channel in (**h**) is about 8–10 m across. (**a**) Section G06/03 also shown in Fig. 19.11b. A basal tufa in G06/03 lies below the ledge to the extreme right in the photograph. (**b**) Volcanic ash resting on moderately diatomaceous silt with large calcareous rhizolith penetrating both deposits. (**c**) White opaline silica plant stems in growth position. (**d**) Fossil spring with massive chert (fossil sinter) forming mound at the spring vent, Eremit. (**e**) Opaline silica rhizoliths (white) in growth position embedded in grey chert located close to spring shown in (**d**), Eremit. (**f**) Flat-lying nodular chert from a laterally extensive subsurface horizon located close to spring (**d**), Eremit. (**g**) Bedded tufa formed in a large spring pool passing up into stromatolitic tufa. (**h**) Meandering fluvial tufa exposed through relief inversion. The former palaeochannel is outlined by white dashes. (**i**) Fluvial tufa in a small channel at base overlain by tufa deposited in a second broader fluvial channel with laterally extensive swamp tufa at top. (**j**) Reeds encrusted by tufa formed in a small spring-fed palaeolake. (**k**) Dense stromatolitic tufa (travertine) encrusting branches deposited in spring pools. (**l**) Pisolitic calcrete

Table 19.4 Dominant diatoms and ecological preferences in the post Oltulelei Formation sediments

	Assemblage	Dominant taxa	Environment
Siliciclastic sediments	OTF 1	*Cyclotella meneghiniana* [*Stephanocyclus meneghinianus*], *Thalassiosira faurii*, *Epithemia argus*	Shallow saline ponds
	OTF 2	*Aulacoseira ambigua*, *A. granulata*, *Encyonema muelleri*, *Epithemia argus*, *E. hyndmanii*	Shallow fresh water, epiphytic habitats
	OTF 3	*Rhopalodia gibberula*, *Diploneis elliptica*, *Nitzschia frustulum*	Very shallow mildly saline water
	OTF 4	*Epithemia argus*, *E. adnata*, *E. hyndmanii*, *Rhopalodia gibberula*, *R. vermicularis*, *Cymbella spp.*	Shallow mildly saline water, epiphytic habitats
	OTF 5	*Caloneis bacillum*, *Cocconeis placentula*, *Epithemia argus*, *Nitzschia frustulum*	Fresh water, epiphytic habitats, flowing water
	OTF 6	*Aulacoseira agassizi*, *A. distans*, *A. granulata*, *A. granulata* var. *angustissima*	Shallow fresh ponds/lakes
	OTF 7	*Mastogloia braunii*, *A. granulata*, *Nitzschia frustulum*	Shallow fresh water bodies
	OTF 8	*Hantzschia amphioxys*, *Navicula* [*Luticola*] *mutica*	Damp soils/swamps
Tufas	OTF 9	*Nitzschia palea*, *N. amphibia*, *Rhopalodia gibberula*	Fresh spring ponds
	OTF 10	*Achnanthidium exiguum* var. *heterovalvum*, *Encyonema muelleri*, *Epithemia sorex*, *Mastogloia braunii*, *Hantzschia amphioxys*	Fresh spring ponds with epiphytic habitats
	OTF 11	*Navicula seminuloides*, *Gomphonema parvulum*, *Aulacoseira granulata*, *A. agassizi*	Shallow fresh spring ponds
	OTF 12	*Nitzschia amphibia*	Moderately deep spring ponds
	OTF 13	*Synedra* [*Ulnaria*] *ulna*, *Rhopalodia gracilis*	Fresh water swamps

After Owen et al. (2014)

contain gastropods and/or cyanobacterial filaments. Based on phytolith studies, Kinyanjui (2012) noted that tall moisture-indicator grasses belonging to C3 Arundinoideae and C4 Panicoideae dominated during the Middle to Late Pleistocene implying a heterogeneous vegetation cover (riverine/riparian, open grasslands, wetlands) between 340 and 220 ka. Subsequent phytolith work by Kinyanjui et al. (2021) also showed that Olorgesailie was relatively wooded between ~670 and 500 ka with riparian habitats also present. In contrast, between 320 and 64 ka there was a shift towards C4 Panicoideae open grasslands with some riparian habitats.

19.5 The Koora Basin

19.5.1 Geomorphology of the Koora Basin

The modern rift floor drops southwards from about 1000 m near Olorgesailie through the Koora Graben and Koora Plain to a low of 776 m about 2 km south of a dry overflow gorge that links the Koora drainage system with the Lake Magadi Basin (O3 in Figs. 19.13 and 19.14a) (Crossley 1976). Outcrops of diatomite and

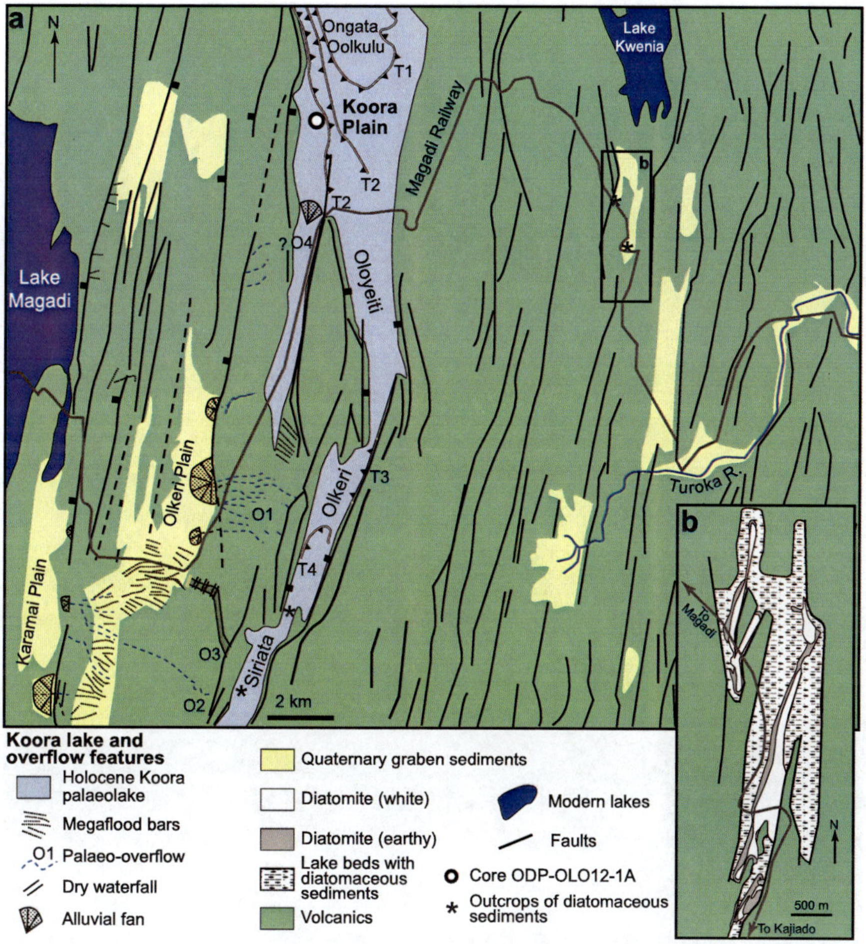

Fig. 19.13 Koora to Magadi palaeodrainage and adjacent areas. Modified from Marsden (1979) with (**b**) based on Matheson (1966). (**a**) Note overflows O1–O4 are located at different elevations, reflecting activity at different times. The O3 outlet is shown in Fig. 19.14b, c. Megaflood bars are formed by alternating, subparallel, cobble to boulder bars and aeolian silt on trachyte (Fig. 19.14d) deposited from a broad overflow front over the western Siriata and Olkeri Grabens. Overflow channels and local drainage terminate at a series of fans on the Olkeri and Karamai Plains. (**b**) Diatomaceous sediments in outcrops south of Lake Kwenia

Fig. 19.14 The Koora Graben and Plain palaeolake and sedimentary facies. Basemap was constructed from GeoMapApp (http://www.geomapapp.org). Cores (**e–i**) are courtesy of the Olorgesailie Drilling Project (ODP) and Project Director, Prof. Rick Potts. (**a**) Approximate size and position of Palaeolake Koora based on the late Pleistocene 30 ft (854 m) strandline. Note

diatomaceous sediments are present at the lowest elevations, which can also be seen on aerial photographs and satellite imagery as terraces and valley floor deposits (Marsden 1979). Although the area has been disturbed by normal faulting (Muiruri et al. 2021c), Baker (1958) reported an undated 30 ft (9.1 m) terrace on his geological map (T1 in Fig. 19.13), which Marsden (1979) reported as consisting of diatomaceous sediment with a variable diatom content and as having a planar upper surface at between 852 and 854 m. Baker also noted a 10–15 ft (3–4.6 m) surface in the northern Koora Plain (T2 in Fig. 19.13), which Marsden suggested was a paired fluvial terrace at 840 m.

A major lake outlet to the southwest (Fig. 19.14a, b) is marked by two hanging valleys at ~830 m (Google Earth data) that merge into a 30-m-deep gorge through the Magadi Trachytes (Baker 1958, 1986; Crossley 1976; Marsden 1979). The channel cuts a straight northerly route for ~2.5 km before changing to a westerly down-dip direction for another 1.5 km (O3 in Fig. 19.13). Breccia at the eastern end of the gorge implies that it might lie along an NW-SE fault. At least three former waterfalls are present along its length (Fig. 19.14c). Other, smaller-scale, parallel, overflow channels terminate on the Olkeri Plain where they have formed a series of alluvial fans (Fig. 19.13). Water likely overflowed at intervals due to enhanced regional rainfall and possibly an expansion of the catchment area related to overflow from Lake Naivasha, which lies 130 km to the north (Roberts and Barker 1993; Dommain et al. 2022).

Terrain west of the main outlet gorge (O3) is marked by a series of subparallel sinuous ridges on Magadi Trachyte that are composed of cobbles and boulders separated by flat areas with thin wind-blown silt (Figs. 19.13 and 19.14d). These bar-like features, which are 500–800 m long (W-E), may have developed following a catastrophic megaflood derived from Lake Naivasha to the north. The waters escaped through the Njorowa Gorge and flowed south along the Kedong Valley before spreading out over the Oltepesi Plain (Behrensmeyer et al. 2018). Floodwaters then followed the Ol Keju Nyiro River to the Koora Graben and Plain (Fig. 19.7b) with the flood eventually overflowing westwards into the Lake Magadi Basin

Fig. 19.14 (continued) overflow channel to the southwest and the position of core ODP-OLO12-1A. (**b**) Koora palaeolake outlet. The Koora palaeolake was located to the right. A second outlet lies 200 m from the viewpoint, hidden behind the central ridge in the photograph. (**c**) Western exit of the Koora overflow channel. Note former waterfall. (**d**) Megaflood deposits west of the overflow in (**a**) and located in Fig. 19.13. These consist of multiple subparallel pebble, cobble and boulder bars (left and right) separated by thin aeolian silt on eroded trachyte (centre). (**e**) Laminated zeolitic silt (highly saline lake). (**f**) Laminated diatomaceous silt and clay (fresh deep lake). Note horizontal burrows (white arrows) and grey ash (black arrow). (**g**) Vaguely bedded basal marl-diatomite with vertical roots and burrows (shallow mildly saline lake). Small erosion channel filled with reworked diatomite, sandy silt and opaline silica plant stems passes upwards into moderately diatomaceous partially bioturbated silt (pedogenically altered shallow, mildly saline, wetland). (**h**) Vaguely bedded diatomite. Post depositional termite burrows and chambers also present (intermediate depth saline lake burrowed after emergence). (**i**) Pedogenically altered diatomaceous clay. Carbonate nodules with Mn dendrites are present near the top (black arrow) with strong bioturbation towards the base (desiccated fresh lake overprinted by pedogenesis)

(Washbourn-Kamau 1975; Baker and Mitchell 1976; Crossley 1976). Baker et al. (1988) suggested the deluge took place about 0.1 million years ago. However, Trauth et al. (2003) proposed that a major fall in the height of Lake Naivasha after about 60 ka might have been related to the loss of water and permanent lowering of the Naivasha outlet following the megaflood. Muiruri et al. (2021c) noted that it may correlate with the end of lacustrine deposition in their core ODP-OLO12-1A from the Koora Plain, which was dated at about 83.5 ka with the flood overtopping a horst block that forms the western margin of the Siriata Valley (Fig. 19.13). In contrast, Behrensmeyer et al. (2018) noted megaflood boulders in the Tinga Member of the Oltulelei Formation, dated at ~50–36 ka.

Areas east of the Koora Basin are characterised by numerous small grabens that host a variety of undated sediments of probable Quaternary age. Several contained palaeolakes with one basin south of Lake Kwenia exposing up to 16 feet (~4.9 m) of diatomites and diatomaceous sediments (Fig. 19.13b). The deposits were discovered by Prof. Simpson and mapped by V.H. Kirkham with his map reproduced in a Kenya Geological Survey Report by Matheson (1966). He noted coarse, green to dark blue-grey sediments with greenish and yellowish mixtures of diatomite and clay. The purest diatomites are white, low density and porous with cylindrical and acicular frustule morphologies that differed from forms described by Creighton (1911) in the Nakuru and Kedong areas to the north.

19.5.2 Sedimentation in the Koora Basin

Aeromagnetic surveys of the south Kenya rift revealed a thick fill of sediments below the Koora Plain (Wohlenberg and Bhatt 1972), which was drilled by the Olorgesailie Drilling Project in September 2012 (Cohen et al. 2016; Campisano et al. 2017; Potts et al. 2020). A basal lava and the upper sediments in OLO12-1A were dated at 1084 ± 4 and 83 ± 3 ka, respectively (Deino et al. 2016; Potts et al. 2020). These Middle to Late Pleistocene sediments correlate with the Olorgesailie and Oltulelei formations and erosional gaps in the Olorgesailie Basin, as well as a one-million-year core record from neighbouring Lake Magadi (Fig. 19.15). The deposits are comprised of 27 m of loose, fluvial and slightly diatomaceous wetland silt (the upper Koora beds) that overlie partially lithified, laminated and massive diatomaceous and zeolitic silt with many pedogenic horizons (Fig. 19.14e–i) (Cappio 2016; Muiruri 2017; Potts et al. 2020). These lower Koora beds document fresh and saline, deep and shallow lakes, wetlands and periods of desiccation (Stockhecke et al. 2015). Major minerals present include analcime and phillipsite from the most saline stages with authigenic calcite forming during lower salinity periods (Rabideaux et al. 2015; Rabideaux, 2018; Potts et al. 2020).

Diatoms in the lower Koora beds of OLO12-1A include, for example, *Aulacoseira granulata* and several of its varieties, as well as *A. ambigua* (Muiruri 2017; Muiruri et al. 2021c). The diatoms are preserved in laminated diatomaceous and clayey diatomite, massive diatomite and diatomaceous silt. Saline phases are documented by the presence of *Thalassiosira rudolfi*, *Anomoeoneis sphaerophora* and other

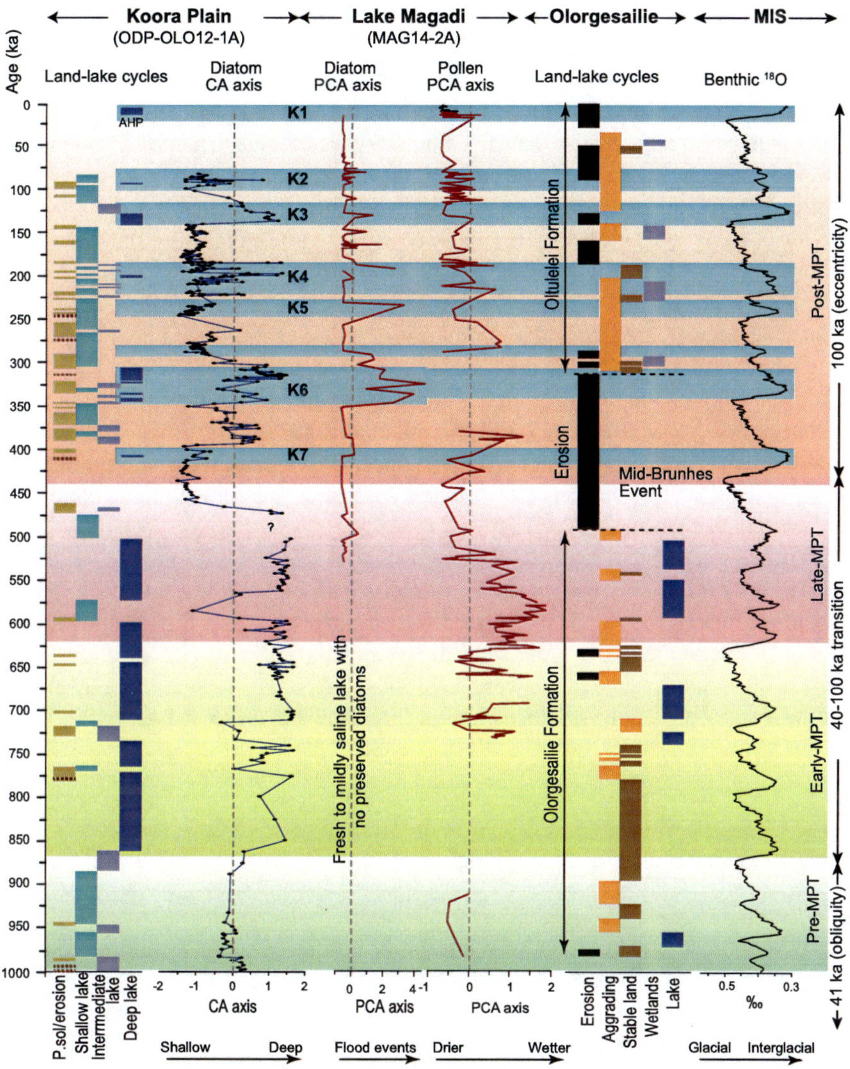

Fig. 19.15 Environmental change across the south Kenya Rift and global correlations. Koora data from Potts et al. (2020) and Muiruri et al. (2021c); Magadi data from Owen et al. (2018a, 2019) and Muiruri et al. (2021a, b). Olorgesailie environments after Potts et al. (2018)). Marine Isotope Stage (MIS) data after Lisieckie and Raymo (2005). ODP-OLO12-1A data show an upward increase in the frequency of palaeosols, implying increasing lake desiccation events. Land-lake cycles based on PCA scores from diatom CA Axis 1 to right with deeper waters indicated by positive CA values. Deep lakes during the post-Mid Pleistocene Transition (MPT) correlate well with Magadi PCA peaks (MAG14-2A) that reflect flooding phases and also correlate with interglacial stages in the MIS global record. Magadi pollen document a long-term reduction in PCA values, reflecting increasing aridity. Note major erosion at Olorgesailie overlaps low lake levels (CA axis) at Koora and pollen evidence for drier conditions at Magadi. Overall, the figure documents drier conditions with wetter intervals after the mid-Brunhes Event

saline taxa (Muiruri 2017). In Fig. 19.15, correspondence analysis (CA) axis 1 correlates with planktonic diatoms with positive values reflecting deeper water (Potts et al. 2020; Muiruri et al. 2021c). The data indicates low lake levels prior to the 'Mid Pleistocene Transition' (MPT) with lakes mostly high during the early and late-MPT. A major change across the MPT boundary was also reported by Luipien et al. (2021) who reported that $\delta^{13}C_{OM}$ (bulk organic matter) in OLO12-1A became more positive after ~790 ka.

Correlations with global marine isotope stages vary, suggesting that lake level changes reflect both climatic and tectonic controls. In contrast, after the mid-Brunhes Event, when 100-ka-eccentricity cycles became dominant, the CA data and palaeosols indicate generally drier conditions that were interrupted by many wetter intervals. Peaks in the CA data (K1–K7) show good correlations with diatom PCA data from Magadi that indicate increased inputs of freshwater benthic diatoms to a saline palaeolake (Muiruri 2017; Owen et al. 2018a, 2019; Potts et al. 2020; Muiruri et al. 2021a). These wetter intervals also appear to correlate well with marine isotope stages indicative of interglacial stages, suggesting a strong climatic control on deposition during post-MPT times.

Termination of the lower Koora beds and lacustrine sedimentation at the OLO12-1A coring site took place soon after 83 ka and might relate to the major megaflood noted above. This traversed the site and overtopped the southwestern Siriata Graben on its way to the Magadi Basin (Muiruri et al. 2021c). Subsequent erosion of an outlet gorge in the western Siriata Graben then lowered lake levels and caused a lake regression and deposition of fluvial silts at OLO12-1A.

19.6 Quaternary Tectonism, Climate and Palaeogeography in the South Kenya Rift

The development of lakes reflects an interplay between climatic, tectonic and volcanic controls. The East African Region, for example, experienced broad uplift through the last 10 Ma that reduced moisture transport from the Indian Ocean, resulting in progressively drier conditions over the region (Sepulchre et al. 2006) Global climate variability also reflects changes in the Earth's orbit around the sun that have altered seasonal insolation (Kaspar et al. 2010). These Milankovitch cycles operate on several different cyclicities, reflecting orbital precession (~23 kyr), obliquity (~41 kyr) and eccentricity (~100 kyr, ~400 kyr) with the interplay between these cycles resulting in complex temporal climatic changes (Kingston et al. 2007).

Lacustrine sediments at Munya wa Gicheru and in the Olorgesailie Formation that date to between 1.9–1.7 and 1.2–0.9 Ma, respectively, have been correlated with sediments in other rift basins in order to infer past high moisture phases when climate also showed greater extremes and climate variability (Trauth et al. 2005, 2007, 2009; Owen et al. 2008b, 2012). These periods of increased climatic fluctuations have, in turn, been related to changes in Earth's eccentricity cycle. However, direct correlation with climate is complicated due to the tectonically active nature of rifted regions. Owen et al. (2009a) and Trauth et al. (2009, 2015) discussed the precise nature of climate variability at Olorgesailie with the former suggesting that the deposits laid down between 1.2 and 0.9 Ma represent a short period of deep lacustrine conditions together with longer episodes characterised by complex environments (shallow waters, land, soil formation) rather than a large, deep and relatively long-lived stable lake. This debate, in turn, has been important for ideas concerning climate-driven human evolution and especially the variability selection hypothesis, which proposes climate instability as a major evolutionary driving force (Potts 2013; Potts and Faith 2015).

Chemical Index of Alteration data for the South Kenya Rift indicate a long-term trend towards aridity through the Quaternary (Owen et al. 2014), which would have favoured more saline water bodies, as observed in both cores and outcrops. Increasing carbonate deposition during the Quaternary (Fig. 19.16) suggests rising $CaCO_3$ saturation of lake and groundwater, which is consistent with progressive desiccation. However, aridity may, at least in part, reflect tectonism with the rise of the rift shoulders and subsidence of the rift floor creating dry conditions through rainshadow effects (Owen et al. 2018b, 2019, Muiruri et al. 2021c). Tectonism is more definitively involved with the progressive southward migration of lake basins that resulted from southward regional tilting and the creation of accommodation space by faulting and volcanic damming. The presence of fossil sinter and geyserite at former spring vents at Eremit (Fig. 19.12), the accumulation of chert in the Magadi Basin (Behr 2002; Owen et al. 2019; Chaps. 12 and 20), and permineralisation of plants by silica after ~0.49 Ma, likely reflect increasing silica in springs and shallow groundwaters, derived from fault-induced tapping of hydrothermal fluids at depth and increasing alkalinity in near-surface fluids. High fluid temperatures and high pH both increase silica solubility.

During the last million years, the topography, lake basins and vegetation of the south Kenya Rift have changed substantially, reflecting the complex interplay of climatic and tectonic controls. Four major phases can be distinguished (Fig. 19.17; Muiruri et al. 2021a, b, c). During Phase I (~1000–750 ka) an initially subdued surface on trachyte flood lavas fractured and started to develop north-south grabens that gradually deepened. Fresh to mildly saline lakes occupied the Magadi and Koora basins with Olorgesailie dominated by emergent land surfaces, except for one lacustrine interval around 970 ka. Conditions became wetter during Phase II (~750–500 ka) when all three basins were occupied, at least intermittently, by large fresh to saline lakes. Other basins such as that at Kwenia (east of Koora) may have also been flooded but the evidence is lacking. Phase III (500–320 ka) was characterised by intense aridity around 470 to 400 ka followed by variable conditions. Saline to freshwater lakes occupied the Magadi and Koora basins with emergent land surfaces at Olorgesailie, the latter being subject to major fluvial incision. Phase IV (320–0 ka) was characterised by highly variable conditions in all basins with many wet-dry cycles. At Olorgesailie, land surfaces were repeatedly eroded by rivers along with intervals of fluvial and wetland deposition. The Koora Basin was subject to many transgressions and regressions along with episodes of lake desiccation.

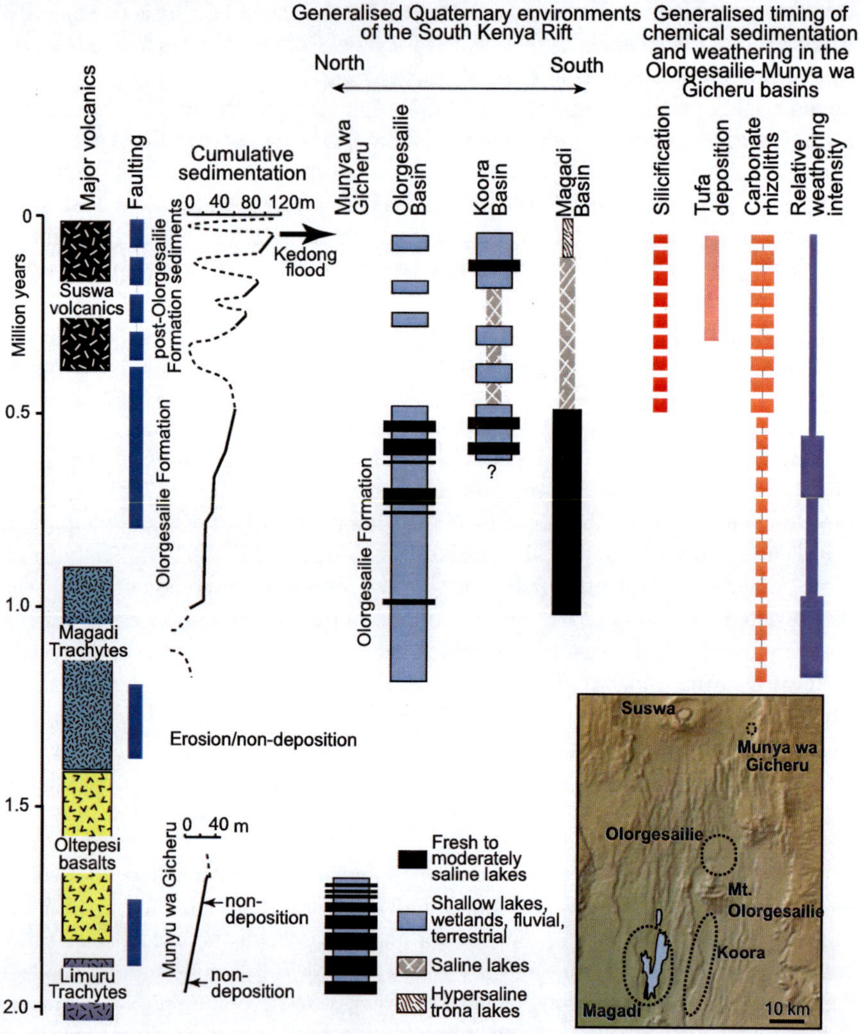

Fig. 19.16 Generalised Quaternary environments and geochemical evolution of the South Kenya Rift. (Map constructed from GeoMapApp (http://www.geomapapp.org). Modified after Owen et al. (2014). Major volcanics and faulting episodes after Baker and Mitchell (1976) and Baker et al. (1988). Cumulative sedimentation after Behrensmeyer et al. (2002) and Owen et al. (2012). Relative weathering summary based on Chemical Index of Alteration data (Owen et al. 2014). Silicification and carbonate trends from Owen et al. (2014) and Lee et al. (2013)

Magadi maintained a saline lake throughout Phase IV, which was episodically subject to freshwater flooding. During the middle to late Holocene, the latter palaeolake evolved into a seasonally flooded trona pan with lakes disappearing from the northern Koora Basin after ~83 ka.

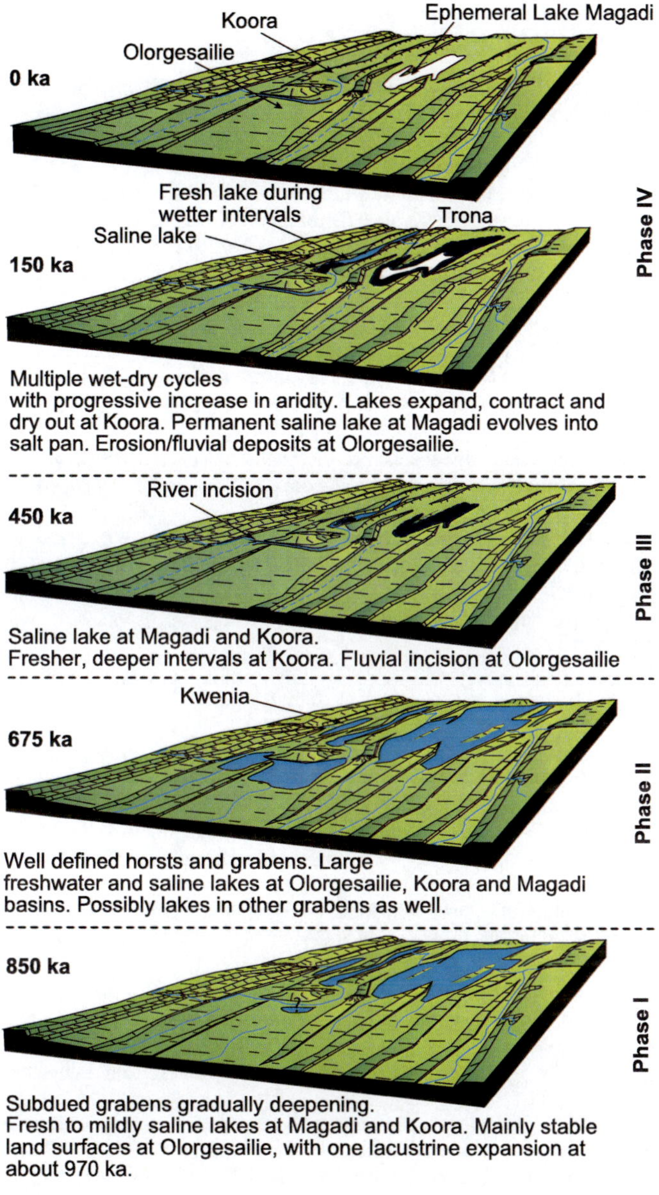

Fig. 19.17 Environmental change across the south Kenya Rift during the last million years. Changes are caused by both tectonic and climate controls

References

Ambrose SH, Hlusko LJ, Kyule MD, Deino A, Williams MAJ (2002) Lemudong'o: a late Miocene fossil site in southern Kenya. Am J Phys Anthropol Suppl 34:37

Ambrose SH, Hlusko LJ, Kyule MD, Deino A, Williams MAJ (2003) Lemudong'o: a new 6 Myr paleontological site in Narok, Kenya. J Hum Evol 44:737–742

Ambrose SH, Bell CJ, Bernor RL, Boisserie JR, Darwent C, Degusta D, Deino A, Garcia N, Haile-Selassie Y, Head JJ, Howell FC, Kyule MD, Manthi FK, Mathu EM, Nyamai CM, Saegusa H, Stidham TA, Williams MAJ, Hlusko LJ (2007a) The paleoecology and paleogeographic context of Lemudong'o Locality 1, a Late Miocene terrestrial fossil site in southern Kenya. Kirtlandia 56:38–52

Ambrose SH, Kyule D, Hlusko L (2007b) History of paleontological research in the Narok District of Kenya. Kirtlandia 56:1–37

Ambrose SH, Nyamai CM, Mathu EM, Williams MAJ (2007c) Geology, geochemistry, and stratigraphy of Lemudongo formation, Kenya Rift Valley. Kirtlandia 56:53–64

Baker BH (1958) Geology of the Magadi area. Geol Surv Kenya Rep 42

Baker BH (1986) Tectonism and volcanism of the Southern Kenya Rift Valley and its influence on sedimentation. In: Frostick LE, Renaut RW, Reid I, Tiercelin J-J (eds) Sedimentation in the African rifts. Geol Soc Lond Spec Publ 25:45–57

Baker BH, Mitchell JG (1976) Volcanic stratigraphy and geochronology of the Kedong-Olorgesailie area and the evolution of the South Kenya rift valley. J Geol Soc Lond 132:467–484

Baker BH, Mitchell JG, Williams LAJ (1988) Stratigraphy, geochronology and volcano-tectonic evolution of the Kedong-Naivasha-Kinangop region, Gregory Rift Valley, Kenya. J Geol Soc Lond 145:107–116

Barker P, Gasse F, Roberts N, Taieb M (1990) Taphonomy and diagenesis in diatom assemblages; a Late Pleistocene palaeoecological study from Lake Magadi, Kenya. Hydrobiologia 214:261–212

Behr HJ (2002) Magadiite and Magadi chert: a critical analysis of the silica sediments in the Lake Magadi Basin. In: Renaut RW, Ashley GM (eds) Sedimentation in continental rifts. SEPM Spec Publ 73:258–273

Behrensmeyer AK (2007) Cycles of erosion and deposition in the Pleistocene Olorgesailie Basin of southern Kenya and their impact on the paleoanthropological record. Geol Soc Am Annu Meet, Denver, Abstr Progr 39:343

Behrensmeyer AK, Potts R, Deino A, Ditchfield P (2002) Olorgesailie, Kenya: a million years in the life of a rift basin In: Renaut RW, Ashley G (eds) Sedimentation in continental rifts. SEPM Spec Publ 73:97–106

Behrensmeyer AK, Potts R, Deino A (2018) The Oltulelei Formation of the southern Kenyan Rift Valley: a chronicle of rapid landscape transformation over the last 500 k.y. Geol Soc Am Bull 130:1474–1492

Bernor RL (2007) The latest Miocene Hipparionine (Equidae) from Lemudong'o, Kenya. Kirtlandia 56:148–151

Bishop WW (ed) (1978) Geological background to fossil man. Geol Soc Lond Spec Publ 6

Bobe R (2006) The evolution of arid ecosystems in eastern Africa. J Arid Environ 66:564–584

Boisserie JR (2007) Late Miocene Hippopotamidae from Lemudong'o, Kenya. Kirtlandia 56:158–162

Bonnefille R (2010) Cenozoic vegetation, climate changes and hominid evolution in tropical Africa. Glob Planet Change 72:390–411

Boven A (1992) The applicability of the K-Ar and $^{40}Ar/^{39}Ar$ techniques to Quaternary volcanics: a fundamental and experimental investigation. PhD thesis, Vrije Universiteit Brussel

Bower JRF, Nelson CM, Waibel AF, Wandibba S (1977) The University of Massachusetts' later Stone Age/Pastoral 'Neolithic' comparative study in central Kenya: an overview. Azania 12:119–146

Brooks A, Behrensmeyer AK, Yellen Y, Deino A, Sharp W, Potts R (2007) After the Acheulean: stratigraphy, dating, and archeology of two new formations in the Olorgesailie basin, southern Kenya Rift. Abstr PaleoAnthropol Soc 2007 Meet, PaleoAnthropology A5

Brooks AS, Yellen JE, Potts R, Behrensmeyer AK, Deino AL, Leslie DE, Ambrose SH, Ferguson JR, d'Errico F, Zipkin AM, Whittaker S, Post J, Veatch EG, Foecke K, Clark JB (2018) Long-distance stone transport and pigment use in the earliest Middle Stone Age. Science 360:90–94

Bye BA (1984) Volcanic stratigraphy and red beds of the Olorgesailie Formation, southern Kenya. MSc thesis, University of Utah, Salt Lake City

Bye BA, Brown FH, Cerling TE, McDougall I (1987) Increased age estimate for the lower Palaeolithic hominid site at Olorgesailie, Kenya. Nature 329:237–239

Campisano CJ, Cohen AS, Arrowsmith JR, Asrat A, Behrensmeyer AK, Brown ET, Deino AL, Deocampo DM, Feibel CS, Kingston JD, Lamb HF, Lowenstein TK, Noren A, Olago DO, Owen RB, Pelletier JD, Potts R, Reed KE, Renaut RW, Russell JM, Russel JL, Schäbitz F, Stone JR, Trauth MH, Wynn JG (2017) The Hominin Sites and Paleolakes Drilling Project: acquiring high-resolution paleoclimate records from the East African Rift System and their implications for understanding the environmental context of hominin evolution. PaleoAnthropology:1–43

Cappio L (2016) Reconstruction of the Pleistocene environment and climate of the Olorgesailie Basin. MSc thesis, University of Minnesota, Minneapolis

Cohen A, Campisano C, Arrowsmith R, Asrat A, Behrensmeyer AK, Deino A, Feibel C, Hill A, Johnson R, Kingston J, Lamb H, Lowenstein T, Noren A, Olago D, Owen RB, Potts R, Reed K, Renaut RW, Schäbitz F, Tiercelin JJ, Trauth MH, Wynn J, Ivory S, Brady K, O'Grady R, Rodysill J, Githiri J, Russell J, Foerster V, Dommain R, Rucina S, Deocampo D, Billingsley A, Beck C, Dorenbeck G, Dullo L, Feary D, Garello D, Gromig R, Johnson T, Junginger A, Karanja M, Kimburi E, Mbuthia A, Mccartney T, McNulty E, Muiruri V, Nambiro E, Negash EW, Njagi D, Wilson JN, Rabideaux N, Raub T, Sier MJ, Smith P, Urban J, Warren M, Yadeta M, Yost C, Zinaye B (2016) The Hominin Sites and Paleolakes Drilling Project: inferring the environmental context of human evolution from eastern African rift lake deposits. Sci Drill 21:1–16

Cole SM (1954) The prehistory of East Africa. Am Anthropol 56:1026–1050

Cooke HBS (1957) Observations relating to Quaternary environment in East and Southern Africa. Trans Geol Soc S Afr 60 Annex:1–73

Creighton JK (1911) The microscopical structure of diatomite. J E Afr Uganda Natur Hist Soc 1:107

Crossley R (1976) Structure, stratigraphy and volcanism in the Nguruman Escarpment area of the western side of the Kenya Rift Valley. PhD thesis, University of Lancaster

Cuthbert MO, Gleeson T, Bennett MR, Newton AC, McCormack CJ, Ashley GM, Reynolds SC (2017) Modelling the role of groundwater hydro-refugia in East African hominin evolution and dispersal. Nature Comm 8:1–11

Darwent CM (2007) Lagomorphs (Mammalia) from late Miocene deposits at Lemudong'o, Southern Kenya. Kirtlandia 56:112–120

Deino AL, Ambrose SH (2007) $^{40}Ar/^{39}Ar$ dating of the Lemudong'o late Miocene fossil assemblages, southern Kenya Rift. Kirtlandia 56:65–71

Deino A, Potts R (1990) Single-crystal $^{40}Ar/^{39}Ar$ dating of the Olorgesailie Formation, southern Kenya Rift. J Geophys Res 95:8453–8470

Deino A, Potts R (1992) Age-probability spectra for examination of single-crystal $^{40}Ar/^{39}Ar$ dating results: examples from Olorgesailie, southern Kenya Rift. Quat Int 13–14:47–53

Deino AL, King J, Heil C, Potts R, Behrensmeyer AK, Dommain R (2016) Chronology of the Olorgesailie Drilling Project Core 1A, Koora Graben, southern Kenya Rift. Geol Soc Am Abstracts Progr 48:7

Deino AL, Behrensmeyer AK, Brooks AS, Yellen JE, Sharp WD, Potts R (2018) Chronology of the Acheulean to Middle Stone Age transition in eastern Africa. Science 360:95–98

Deocampo DM, Behrensmeyer AK, Potts R (2010) Ultrafine clay minerals of the Pleistocene Olorgesailie Formation, southern Kenya Rift: diagenesis and paleoenvironments of early hominins. Clay Miner 58:294–310

Dommain R, Riedl S, Olaka LA, deMenocal P, Deino A, Owen RB, Muiruri V, Müller J, Potts R, Strecker MR (2022) Holocene bidirectional river system along the Kenya Rift and its influence on East African faunal exchange and diversity gradients. Proc Natl Acad Sci 119:e2121388119

Durkee H, Brown FH (2014) Correlation of volcanic ash layers between the Early Pleistocene Acheulean sites of Isinya, Kariandusi, and Olorgesailie, Kenya. J Archaeol Sci 49:510–517

Evernden JF, Curtis CH (1965) Potassium-argon dating of late Cenozoic rocks in East Africa and Italy. Curr Anthropol 6:343–385

Felske GN (2016) Genesis of calcretes and related carbonate rocks in the Southern Kenya Rift. MSc thesis, University of Saskatchewan, Saskatoon

Fernández HM, Vrba E (2006) Plio-Pleistocene climatic change in the Turkana Basin (East Africa): evidence from large mammal faunas. J Hum Evol 50:595–626

Flint RF (1959a) On the basis of Pleistocene correlation in East Africa. Geol Mag 96:265–284

Flint RF (1959b) Pleistocene climates in eastern and southern Africa. Geol Soc Am Bull 70:343–374

Gregory JW (1896) The Great Rift Valley: being the narrative of a journey to Mount Kenya and Lake Baringo. John Murray, London

Gregory JW (1921) The rift valleys and geology of East Africa. Seeley Service, London

Haradon C (2010) The ecological context of the Acheulean to Middle Stone Age transition in Africa. PhD thesis, George Washington University, Washington DC

Hatala KG, Brooks AS, Yellen JE (2010) Variation in the production and use of scrapers during the Middle Stone Age: Evidence from Olorgesailie, Kenya and Aduma, Ethiopia. Paleoanthropol Soc Meet Abstr, St. Louis, MO, 13–14 April

Hay RL (1965) Appendix 1 - Preliminary notes on the stratigraphy of Beds I–IV, Olduvai Gorge, Tanganyika. In: Leakey LSB (author), Olduvai Gorge, vol 1. Cambridge University Press, Cambridge, pp 94–100

Head JJ, Bell CJ (2007) Snakes from Lemudong'o, Kenya Rift Valley. Kirtlandia 56:177–179

Hlusko LJ, Haile-Selassie Y (2007) *Nyanzachoerus syrticus* (Artiodactyla, Suidae) from the late Miocene of Lemudong'o, Kenya. Kirtlandia 56:152–157

Hlusko LJ, Ambrose SH, Bernor R, Deino A, Stidham T (2002) Lemudong'o, a late Miocene mammalian-dominated locality in southern Kenya. J Vertebr Paleontol Suppl 22:65A–66A

Hlusko LJ, Haile-Selassie Y, DeGusta D (2007) Late Miocene Bovidae (Mammalia: Artiodactyla) from Lemudong'o, Narok District, Kenya. Kirtlandia 56:163–172

Howell FC, Garcia N (2007) Carnivora (Mammalia) from Lemudong'o (Late Miocene: Narok District, Kenya). Kirtlandia 56:121–139

Isaac GL (1968) The Acheulian site complex at Olorgesailie, Kenya: a contribution to the interpretation of Middle Pleistocene culture in East Africa. PhD thesis, University of Cambridge

Isaac GL (1977) Olorgesailie, archeological studies of a Mid-Pleistocene lake basin in Kenya. University of Chicago Press, Chicago

Isaac GL (1978) The Olorgesailie Formation: stratigraphy, tectonics and the palaeogeographic context of the Middle Pleistocene archaeological sites. In: Bishop WW (ed) Geological background to fossil man. Geol Soc Lond Spec Publ 6:173–206

Kaspar F, Prömmel K, Cubasch U (2010) Impacts of tectonic and orbital forcing on East African climate: a comparison based on global climate model simulations. Int J Earth Sci 99:1677–1686

Kingston JD, Deino AL, Edgar RK, Hill A (2007) Astronomically forced climate change in the Kenyan Rift Valley 2.7–2.55 Ma: implications for the evolution of early hominin ecosystems. J Hum Evol 20:487–503

Kinyanjui RN (2012) Phytolith analysis as a palaeoecological tool for reconstructing Mid-to Late-Pleistocene environments in the Olorgesailie Basin, Kenya. MSc thesis, University of Cape Town

Kinyanjui RN, Meadows M, Gillson L, Bamford MK, Behrensmeyer AK, Potts R (2021) Reconstructing vegetation history of the Olorgesailie Basin during the Middle to Late Pleistocene using phytolith data. In: Runge J, Gosling WD, Lézine A-M, Scott L (eds) Quaternary vegetation dynamics – the African pollen database. CRC Press, London, pp 85–106

Koch CP (1986) The vertebrate taphonomy and palaeoecology of the Olorgesailie Formation, Middle Pleistocene, Kenya. PhD thesis, University of Toronto

Kübler S, Owenga P, Reynolds SC, Rucina SM, King GCP (2015) Animal movements in the Kenya Rift and evidence for the earliest ambush hunting by hominins. Sci Rep 5:14011

Kübler S, King GCP, Devès MH, Inglis RH, Bailey GN (2018) Tectonic geomorphology and soil edaphics as controls on animal migrations and human dispersal patterns. In: Rasul NMA, Stewart ICF (eds) Geological setting, palaeoenvironment and archaeology of the Red Sea. Springer, Cham, pp 653–673

Kübler S, Rucina S, Aßbichler D, Eckmeier E, King G (2021) Lithological and topographic impact on soil nutrient distributions in tectonic landscapes: implications for Pleistocene human-landscape interactions in the Southern Kenya Rift. Front Earth Sci 9:611687

Kyule MD, Ambrose SH, Noll MP, Atkinson JL (1997) Pliocene and Pleistocene sites in southern Narok District, southwest Kenya. J Hum Evol 32:A9–10

Leakey LSB (1952) The Olorgesailie prehistoric site. In: Leakey LSB, Cole S (eds) Proc Pan-Afr Congr Prehist 1947, p 209

Leakey LSB (1955) The climatic sequence of the Pleistocene in Africa. In: Balout L (ed) Actes de da IIe Session du Congre's Panafricain de Pre'histoire, Alger, 1952, pp 293–294

Lee KL, Owen RB, Renaut RW, Behrensmeyer AK, Potts R, Sharp W (2013) Facies, geochemistry and diatoms of Late Pleistocene Olorgesailie tufas, southern Kenya Rift. Palaeogeogr Palaeoclimatol Palaeoecol 374:196–217

Lisiecki LE, Raymo ME (2005) A Pliocene-Pleistocene stack of 57 globally distributed benthic $\delta^{18}O$ records. Paleoceanography 20:PA1003

Lupien RL, Russell JM, Subramanian A, Kinyanjui R, Beverly EJ, Uno KT, de Menocal P, Dommain R, Potts R (2021) Eastern African environmental variation and its role in the evolution and cultural change of *Homo* over the last 1 million years. J Hum Evol 157:103028

Manthi FK (2007) A preliminary review of the rodent fauna from Lemudong'o, southwestern Kenya, and its implication to the late Miocene paleoenvironments. Kirtlandia 56:92–105

Marsden M (1979) Origin and evolution of the Pleistocene Olorgesailie Lake series. PhD thesis, McGill University, Québec

Matheson FJ (1966) Geology of the Kajiado area. Geol Surv Kenya Rep 70

McCall GJH, Baker BH, Walsh J (1967) Late Tertiary and Quaternary sediments of the Kenya Rift Valley. In: Bishop WW, Clark JD (eds) Background to evolution in Africa. University of Chicago Press, Chicago, pp 191–220

Melson WG, Potts R (2002) Origin of reddened and melted zones in Pleistocene sediments of the Olorgesailie Basin, southern Kenya Rift. J Archaeol Sci 29:307–316

Miller JA (1967) Problems of dating East African Tertiary and Quaternary volcanics by the potassium-argon method. In: Bishop WW, Clark JD (eds) Background to evolution in Africa. University of Chicago Press, Chicago, pp 259–272

Muiruri VM (2017) Late Quaternary diatom and palynomorph stratigraphies and palaeoenvironments of the Koora Graben and Lake Magadi Basin, Kenya Rift Valley. PhD thesis, Hong Kong Baptist University

Muiruri VM, Owen RB, de Cort G, McNulty E, Rabideaux N, Lowenstein T, Renaut RW, Leet K, Deocampo D, Cohen AS, Campisano CJ, Billingsley A, Mbuthia A (2021a) Middle Pleistocene to recent diatom floras and stratigraphy of the Magadi Basin, southern Kenya Rift. J Paleolimnol 65:315–333

Muiruri VM, Owen RB, Lowenstein TK, Renaut RW, Marchant R, Rucina SM, Cohen A, Deino AL, Sier MJ, Luo S, Leet K, Campisano C, Rabideaux N, Deocampo D, Shen CC, Mbuthia A (2021b) A million year vegetation history and palaeoenvironmental record from the Lake Magadi Basin, Kenya Rift Valley. Palaeogeogr Palaeoecol Palaeoclimatol 567:110247

Muiruri V, Owen RB, Potts R, Deino AL, Behrensmeyer AK, Dommain R, Reidl S, Rabideaux N, Beverly EJ, Renaut RW, Moerman JW, Deocampo D, Tyler FJ, Noren A, Cohen AS, Shannon (2021c) Quaternary diatoms and palaeoenvironments of the Koora Plain, southern Kenya Rift. Quat Sci Rev 267:107106

Ngecu W, Njue JN (1999) Lithostratigraphy and distribution of Pleistocene sediments of the Munyu wa Gicheru Formation in South Kenya Rift Valley. J Afr Earth Sci 29:411–421
Njue JN (1995) Stratigraphy and sedimentological investigations of Munya wa Gicheru lake beds in the southern Kenya Rift Valley. MSc thesis, University of Nairobi
Noll M (2000) Components of Acheulean lithic assemblage variability at Olorgesailie, Kenya. PhD thesis, University of Illinois, Urbana-Champaign
Noll M, Petraglia M (2003) Acheulean bifaces and early human behavioral patterns in East Africa and South India. In: Soressi M, Dibble HL (eds) Multiple approaches to the study of bifacial technologies. Univ Penn Mus Archaeol Anthropol, Philadelphia, pp 31–53
Owen RB (1981) Quaternary diatomaceous sediments and the geological evolution of Lakes Turkana, Baringo and Bogoria, Kenya Rift Valley. PhD thesis, University of London
Owen RB (2002) Sedimentological characteristics and origins of diatomaceous deposits in the East African Rift System. In: Renaut RW, Ashley GM (eds) Sedimentation in continental rifts. SEPM Spec Publ 73:233–246
Owen RB, Renaut RW (1981) Paleoenvironments and sedimentology of the Middle Pleistocene Olorgesailie Formation, southern Kenya Rift Valley. Palaeoecol Afr 13:147–174
Owen RB, Renaut RW (2000) Spatial and temporal facies variations in the Pleistocene Olorgesailie Formation, southern Kenya Rift Valley. In: Gierlowski-Kordesch EK, Kelts KR (eds) Lake basins through space and time. Am Assoc Petrol Geol Stud Geol 46:553–560
Owen RA, Owen RB, Renaut R, Scott JJ, Jones B, Ashley GM (2008a) Mineralogy and origin of rhizoliths on the margins of saline, alkaline Lake Bogoria, Kenya Rift Valley. Sediment Geol 203:143–163
Owen RB, Potts R, Behrensmeyer AK, Ditchfield P (2008b) Diatomaceous sediments and environmental change in the Pleistocene Olorgesailie Formation, southern Kenya Rift Valley. Palaeogeogr Palaeoclimatol Palaeoecol 269:17–37
Owen RB, Potts R, Behrensmeyer AK (2009a) Reply to the comment on "Diatomaceous sediments and environmental change in the Pleistocene Olorgesailie Formation, southern Kenya Rift Valley". Palaeogeogr Palaeoclimatol Palaeoecol 282:147–148
Owen RB, Renaut RW, Scott JJ, Potts R, Behrensmeyer AK (2009b) Wetland sedimentation and associated diatoms in the Pleistocene Olorgesailie Basin, southern Kenya Rift Valley. Sediment Geol 222:124–137
Owen RB, Renaut RW, Potts R, Behrensmeyer AK (2011) Geochemical trends through time and lateral variability of diatom floras in the Pleistocene Olorgesailie Formation, southern Kenya Rift Valley. Quat Res 76:167–179
Owen RB, Lee R, Renaut RW (2012) Early Pleistocene lacustrine sedimentation and diatom stratigraphy at Munya wa Gicheru, southern Kenya Rift Valley. Palaeogeogr Palaeoclimatol Palaeoecol 331-332:60–74
Owen RB, Renaut RW, Behrensmeyer AK, Potts R (2014) Quaternary geochemical stratigraphy of the Kedong-Olorgesailie section of the southern Kenya Rift Valley. Palaeogeogr Palaeoecol Palaeoclimatol 396:194–212
Owen RB, Muiruri VM, Lowenstein TK, Renaut RW, Rabideaux N, Luo S, Deino AL, Sier MJ, Dupont-Nivet G, McNulty EP, Leet K, Cohen A, Campisano C, Deocampo D, Shen C-C, Billingsley A, Mbuthia A (2018a) Progressive aridification in East Africa over the last half million years and implications for human evolution. Proc Natl Acad Sci 115:11174–11179
Owen RB, Renaut RW, Lowenstein TK (2018b) Spatial and temporal geochemical variability in lacustrine sedimentation in the East African Rift System: evidence from the Kenya Rift and regional analyses. Sedimentology 65:1697–1730
Owen RB, Renaut RW, Muiruri VM, Rabideaux NM, Lowenstein TK, McNulty EP, Leet K, Deocampo D, Luo S, Deino AL, Cohen A, Sier MJ, Campisano C, Shen C-C, Billingsley A, Mbuthia A, Stockhecke M (2019) Quaternary history of the Lake Magadi Basin, southern Kenya Rift: tectonic and climatic controls. Palaeogeogr Palaeoclimatol Palaeoecol 518:97–118
Parkinson J (1914) The East African trough in the neighbourhood of the soda lakes. Geogr J 44:33–49

Pennisi E (2013) Out of the Kenyan mud, an ancient climate record. Science 341:476–479
Pickford M, Hlusko LJ (2007) Late Miocene procaviid hyracoids (Hyracoidea *Dendrohyrax*) from Lemudong'o, Kenya. Kirtlandia 56:106–111
Posnansky M (1959) A Hope Fountain Site at Olorgesailie, Kenya Colony. South Afr Archaeol Bull 14:83–89
Potts R (1989) Olorgesailie: new excavations and findings in early and middle Pleistocene contexts, southern Kenya Rift Valley. J Hum Evol 18:477–484
Potts R (1994) Variables versus models of early Pleistocene hominid land use. J Hum Evol 27:7–24
Potts R (1996a) Humanities descent: the consequences of ecological instability. William Morrow, New York
Potts R (1996b) Evolution and climate variability. Science 273:922–923
Potts R (1998) Environmental hypotheses of hominin evolution. Am J Phys Anthropol 107:93–136
Potts R (2001) Mid-Pleistocene environmental change and human evolution. In: Barham L, Robson-Brown K (eds) Human roots: Africa and Asia in the Middle Pleistocene. Western Academic Press, Bristol, pp 5–21
Potts R (2002) Complexity and adaptability in human evolution. In: Goodman M, Moffat AS (eds) Probing human origins. American Academy of Arts and Science, Cambridge MA, pp 33–58
Potts R (2012) Environmental and behavioral evidence pertaining to the evolution of early *Homo*. Curr Anthropol 53(S6):S299–S317
Potts R (2013) Hominin evolution in settings of strong environmental variability. Quat Sci Rev 73:1–13
Potts R, Faith JT (2015) Alternating high and low climate variability: the context of natural selection and speciation in Plio-Pleistocene hominin evolution. J Hum Evol 87:5–20
Potts R, Faith JT (2022) Mammalian fauna of the Olorgesailie Basin and southern Kenya Rift. In: Reynolds SC, Bobe R (eds) African paleoecology and human evolution. Cambridge University Press, Cambridge, pp 376–383
Potts R, Jorstad T, Cole D (1996) The role of GIS in the interdisciplinary investigations at Olorgesailie, Kenya, a Pleistocene archaeological locality. In: Aldenderfer M, Maschner H (eds) Anthropology, space and geographic information systems. Oxford University Press, Oxford, pp 202–213
Potts R, Behrensmeyer AK, Ditchfield P (1999) Paleolandscape variation and Early Pleistocene hominid activities: members 1 and 7, Olorgesailie Formation, Kenya. J Hum Evol 37:747–788
Potts R, Behrensmeyer AK, Deino A, Ditchfield P, Clark J (2004) Small mid-Pleistocene hominin associated with East African Acheulian technology. Science 305:75–78
Potts R, Behrensmeyer AK, Faith JT, Tryon CA, Brooks AS, Yellen JE, Deino AL, Kinyanjui R, Clark JB, Haradon C, Levin NE, Meijer HJM, Veatch EG, Owen RB, Renaut RW (2018) Environmental dynamics during the onset of the Middle Stone Age in eastern Africa. Science 360:86–90
Potts R, Dommain R, Moerman JW, Behrensmeyer AK, Deino AL, Beverly EJ, Brown ET, Deocampo D, Kinyanjui R, Lupien R, Owen RB, Rabideaux N, Riedl S, Russell JM, Stockhecke M, deMenocal P, Tyler Faith J, Garcin Y, Noren A, Scott JJ, Western D, Bright J, Clark JB, Cohen AS, Heil CW, Keller CB, King J, Levin NE, Shannon KB, Muiruri V, Renaut RW, Rucina SM, Uno K (2020) Increased ecological resource variability during a critical transition in hominin evolution. Sci Adv 6(43):eabc8075
Rabideaux NM (2018) Late Quaternary East African environmental change based on mineralogical and geochemical analysis of outcrop and core material from the southern Kenya Rift. PhD thesis, Georgia State University, Atlanta
Rabideaux NM, Deocampo DM, Potts R, Behrensmeyer AK, Lowenstein TK, Renaut RW, Owen RB, Chen A (2015) Zeolitic alteration in saline, alkaline paleolake basins in the southern Kenya Rift based on analysis of minerals from Koora Plain (ODP) and Lake Magadi (HSPDP) core samples. Abstr 6 Int Limnogeol Congr, Reno. Open File Rep US Geol Surv 2015–1092:173–174

Renaut RW, Morley CK, Jones B (2002) Fossil hot-spring travertine in the Turkana Basin, northern Kenya: structure, facies, and genesis. In: Renaut RW, Ashley GM (eds) Sedimentation in continental rifts. SEPM Spec Publ 73:233–246

Roberts N, Barker P (1993) Landscape stability and biogeomorphic response to past and future climatic shifts in intertropical Africa. In: Thomas DSG, Allison RJ (eds) Landscape sensitivity. Wiley, Chichester

Saegusa H, Hlusko LJ (2007) New late Miocene elephantoid (Mammalia: Proboscidea) fossils from Lemudong'o, Kenya. Kirtlandia 56:140–147

Saggerson EP (1991) Geology of the Nairobi area. Rep Geol Surv Kenya 98

Sepulchre P, Ramstein G, Fluteau F, Schuster M, Tiercelin J-J, Brunet M (2006) Tectonic uplift and Eastern Africa aridification. Science 313:1419–1423

Shackleton RM (1944) Preliminary report on the Olorgesailie Prehistoric site. Mines and Geological Department, Nairobi, Unpublished report

Shackleton RM (1955) Pleistocene movements in the Gregory Rift Valley. Geol Rundsch 43:257–263

Shackleton RM (1978) A geological map of the Olorgesailie Formation. In Bishop WW (ed) Geological background to fossil man. Geol Soc Lond Spec Publ 6:171–172

Sikes HL (1926) The structure of the eastern flank of the rift valley near Nairobi. Geogr J 68:385–402

Sikes NE, Potts R, Behrensmeyer AK (1999) Early Pleistocene habitat in Member 1 Olorgesailie based on paleosol stable isotopes. J Hum Evol 37:721–746

Stidham TA (2007) Preliminary assessment of the late Miocene avifauna from Lemudong'o, Kenya. Kirtlandia 56:173–176

Stockhecke M, Behrensmeyer AK, Brown E, Cohen A, Deino A, Dommain R, Potts R (2015) Scanning x-ray fluorescence results from a ~1 Ma paleoenvironmental record adjacent to the Olorgesailie archeological sites (Kenya). Abstr 6 Int Limnogeol Congr, Reno. Open File Rep US Geol Surv 2015–1092:210–211

Strecker MR (1991) Das zentrale und südliche Kenia-Rift unter besonderer Berücksichtigung der neotektonischen Entwicklung. PhD thesis, Universität Karlsruhe

Timbrell L, Grove M, Manica A, Rucina S, Blinkhorn J (2022) A spatiotemporally explicit paleoenvironmental framework for the Middle Stone Age of eastern Africa. Sci Rep 12:3689

Trauth MH, Maslin MA (2009) Comment on "Diatomaceous sediments and environmental change in the Pleistocene Olorgesailie Formation, southern Kenya Rift". Palaeogeogr Palaeoclimatol Palaeoecol 282:145–146

Trauth MH, Deino AL, Bergner AGN, Strecker MR (2003) East African climate change and orbital forcing during the last 175 kyr BP. Earth Planet Sci Lett 206:297–313

Trauth MH, Maslin MA, Deino A, Strecker MR (2005) Late Cenozoic moisture history of East Africa. Science 309:2051–2053

Trauth MH, Maslin MA, Deino AL, Strecker MR, Bergner AGN, Dühnforth M (2007) High and low-latitude forcing of Plio-Pleistocene East African climate and human evolution. J Hum Evol 53:475–486

Trauth MH, Larrasoaña JC, Mudelsee M (2009) Trends, rhythms and events in Plio-Pleistocene African climate. Quat Sci Rev 28:399–411

Trauth MH, Bergner AGN, Foerster V, Junginger A, Maslin MA. Schaebitz F (2015) Episodes of environmental stability versus instability in Late Cenozoic lake records of Eastern Africa. J Hum Evol 87:21–31

Trueman CN, Behrensmeyer AK, Potts R, Tuross N (2006) High-resolution records of location and stratigraphic provenance from the rare earth element composition of fossil bones. Geochim Cosmochim Acta 70:4343–4355

Tryon CA, Faith JT (2013) Variability in the Middle Stone Age of Eastern Africa. Curr Anthropol 54:234–S254

Tryon CA, Potts R (2011) Approaches for understanding flake production in the African Acheulean. PaleoAnthropology 2011:376–389

Walter MJ, Trauth MH (2013) A MATLAB based orientation analysis of Acheulean handaxe accumulations in Olorgesailie and Kariandusi, Kenya Rift. J Hum Evol 64:569–581
Washbourn-Kamau CK (1975) Late Quaternary shorelines of Lake Naivasha, Kenya. Azania 10:77–92
Wohlenberg J, Bhatt NV (1972) Report on aeromagnetic surveys of two areas in the Kenya rift valley. Tectonophysics 15:143–149
Wright JB (1967) Geology of the Narok area. Geol Surv Kenya Rep 80

Chapter 20
The Magadi-Natron Basin

20.1 Introduction

The Magadi-Natron Basin lies between the Quaternary volcanoes of Suswa in the north and Oldoinyo Lengai to the south (Fig. 20.1). This part of the Kenya Rift today is dry with no major lakes on the faulted rift floor that descends from Lake Naivasha to Lake Magadi and Lake Natron. To the west lie the Nguruman and Natron border faults, which form steep escarpments that rise 1600 m and 600 m, respectively. The crust below the rift floor is thinner than under the rift flanks and is intruded by dykes and magma pockets that form low-velocity zones of varying shape (Onguso 2019). Below Lake Natron, for example, the crust is ~34 km thick with the adjacent flanks ~37–42 km thick (Plasman et al. 2019). The western border faults are opposed by a more gradual increase in elevation to the east, interrupted by tilt blocks, horsts and grabens. Sedimentary basins are delineated by north-south faults but influenced by northwest-southeast trends that follow the Aswa (Asua) Shear Zone in the underlying basement (Le Turdu et al. 1999; Muirhead and Kattenhorn 2018). These include, for example, the Kordjya Fault lineament, which is comprised of a series of NNW-trending *en echelon* faults and which also marks the southwestern margins of Lake Magadi (Fig. 20.2). Gravimetric surveys by Omondi (2010) suggest that the volcano-sedimentary basin may be up to 8.1 km thick close to the Nguruman Escarpment.

Muirhead et al. (2016) reported that the Magadi Basin started to develop at about 7 Ma with the Natron Basin following at around 3 Ma. In the latter basin, early movements along the Oldoinyo Ogal Border Fault (Fig. 20.1) shifted to the Natron Border Fault at ~1.2–1.0 Ma. They also noted that extensional strain in the Magadi Basin was increasingly accommodated by intra-rift grid faults. In contrast, the strain has remained closely associated with border faults in the younger Natron Basin (Le Gall et al. 2021). Furthermore, magmatic fluids rising along fault systems (Lee et al. 2016) have played an important role in weakening the lithosphere in both basins, contributing to extensional processes. Muirhead et al. (2016) point out that the

R. W. Renaut, R. B. Owen, *The Kenya Rift Lakes: Modern and Ancient*,
Syntheses in Limnogeology, https://doi.org/10.1007/978-3-642-25055-2_20

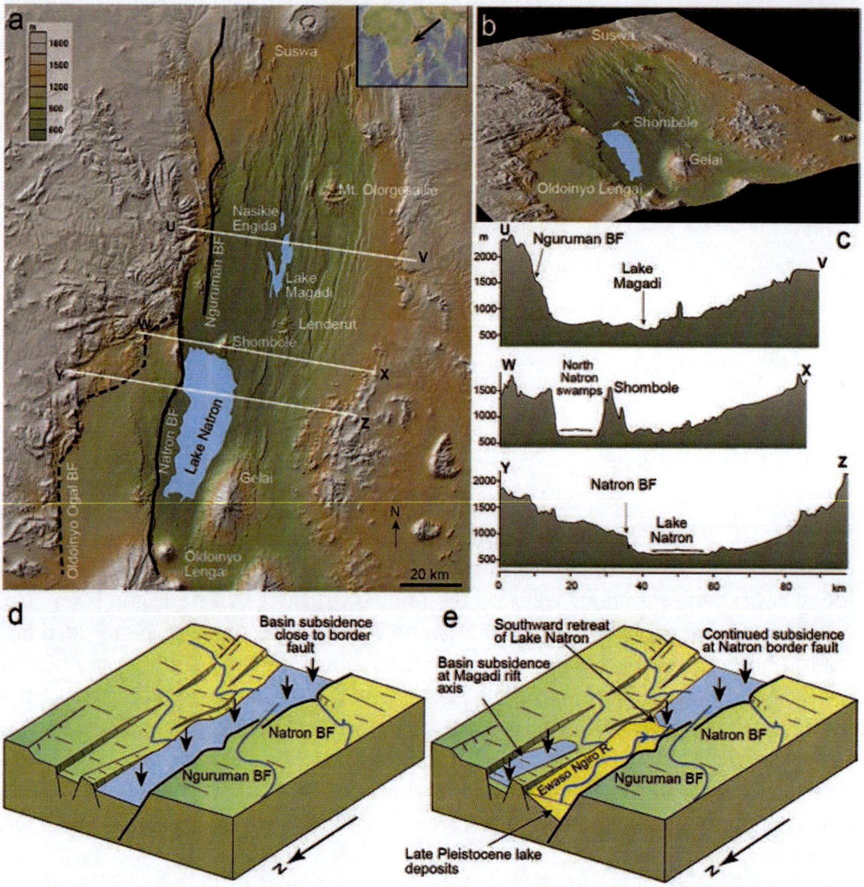

Fig. 20.1 Topography and faulting in the Magadi-Natron Basin. (**a**, **b**) Topography and western border fault locations (Dashed black line represents the inactive Oldoinyo Ogal fault). Border faults and models (**d**) and (**e**) after Muirhead et al. (2016). The topographic grain on the rift floor and eastern margins reflects north-south trending normal faults. (**c**) Cross sections showing the pronounced asymmetry of the rift in this region. (**d–e**) Active border faults with a large adjacent palaeolake (**d**) along the western side of the Magadi and Natron basins. Subsequently, strain shifts eastwards in the Magadi basin (**e**) with lake deposition migrating away from the Nguruman border fault into a grid-faulted area

relocation of the Natron border faulting was also associated with extensive eruptions of Gelai volcano (1.23–0.95 Ma) and Magadi Trachyte lavas (1.4–0.8 Ma) (Baker and Mitchell 1976), which filled the rift and erased earlier topographic features. Subsequently, faulting has dominated with minimal volcanism (Muirhead et al. 2015).

Tectonic models proposed by Muirhead et al. (2016) for the Magadi-Natron Basin suggest that a large Plio-Pleistocene lake developed along the base of the western border faults due to active displacement in those areas (Fig. 20.1d). During

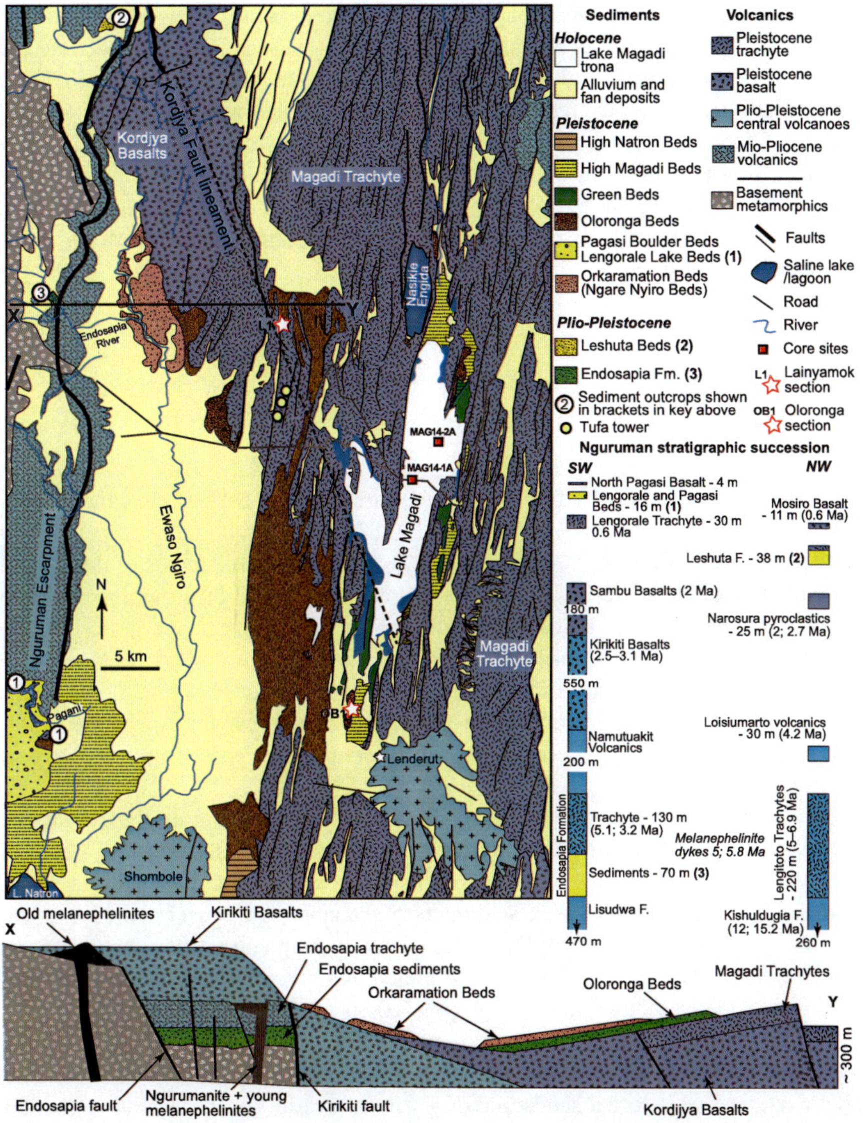

Fig. 20.2 Geological map, stratigraphy and cross section of the Nguruman Escarpment-Lake Magadi area. Circled numbers link locations to the stratigraphic column. Compiled from Baker (1958), Crossley (1979) and Guth (2008)

the later Pleistocene, the area of greatest subsidence in the Magadi Basin migrated eastwards to the central rift where grid faulting was active, giving rise to modern Lake Magadi (Fig. 20.1e). Meanwhile, in the Natron Basin active subsidence remained close to the border fault and led to the formation of modern Lake Natron with the Ewaso Ngiro River (South) valley forming to the north.

The oldest lake sediments are located in the western escarpments and include the Plio-Pleistocene Endosapia and Leshuta formations west of Lake Magadi, and the Pleistocene Hajaro Beds and Peninj Group west of Lake Natron (Table 20.1). Later Quaternary deposits in the region have generally been referred to as 'Beds' (Pagasi Pebble, Oloronga, Orkaramation, Ngare Nyiro, Green and High Magadi Beds) or 'Series' (Chert, Trona Series) (Baker 1958, 1963; Owen et al. 2019) and rest on the Magadi Trachyte Formation in scattered outcrops, especially in the Magadi Basin. The nomenclature has been changed many times with different classifications summarised in chronological order in Table 20.1.

20.2 Miocene to Early Pleistocene Volcanics and Sediments of the Magadi-Natron Basin

Mio-Pliocene fluvial and lacustrine sediments are probably buried below the Magadi Trachytes, which infilled the rift around 1.4–0.8 Ma. The stratigraphy of these hypothetical deposits is unknown (Masinde 2022), but younger sediments of this period crop out in the Nguruman and Natron Escarpments (Table 20.1), providing evidence that water bodies existed at least along the rift margins prior to the eruptions of the Magadi Trachyte lavas. Spink and Stevens (1946) briefly described the Nguruman region. Crossley (1976, 1979, 1980) and Crossley and Knight (1981) outline the volcanic and sedimentary stratigraphy of the Nguruman area with Isaac (1965) providing the first detailed descriptions of ancient lake deposits in the western Natron Basin.

The oldest of these exposed deposits form the lower part of the Endosapia Formation (Figs. 20.2 and 20.3a, b), which postdates the Lisudwa (15.2 Ma) and Kishaldugia (12 Ma) volcanic formations. Exposures in the Endosapia River (Fig. 20.2) reveal a 70-m-thick sedimentary sequence with basal coarse-grained, matrix-supported breccias/conglomerates that include either metamorphic or weathered volcanic clasts (Fig. 20.3d, e) interpreted as debris flows or lahars. Crossley (1976) provided a section in the gorge (Fig. 20.3b) and noted reddening of breccio-conglomerates that suggested desert conditions. Subaqueous laminated, possibly zeolitic, pale green, siltstone and sandstone (Fig. 20.3f) and fluvial quartzose conglomerate and well-sorted sandstone (Fig. 20.3g) overlie basal breccia/conglomerate. Unpublished work by Owen subsequently documented well-preserved diatoms through the lower to middle part of the sequence in a 12-m-long section (Fig. 20.13c). The flora is present in low to moderate abundance and is dominated by benthic taxa (*Rhopalodia*, *Cymbella*, *Epithemia*, *Synedra* (*Ulnaria*), *Nitzschia* spp. and *Aulacoseira granulata* v. *angustissima*) that indicate shallow, fresh to moderately saline conditions. The better sorted deposits suggest high-energy streams that entered the palaeolake. Fluvial, clast-supported, conglomerate (Fig. 20.3h) caps the sequence, which is overlain by Endosapia trachyte dated at 5.1 and 3.2 Ma (Crossley 1976, 1979). The overlying, mostly volcanic, Namutukiat Formation also includes

Table 20.1 Stratigraphic terminology in the Magadi-Natron Basin. Units containing lacustrine deposits shown in bold

Baker (1958)	Baker (1963)	Eugster (1969)	Eugster (1980)	Behr (2002)	Crossley (1976)	Isaac (1965)
MAGADI BASIN					NGURUMAN	NATRON
Evaporite Series (no deformation)	**Evaporite Series** (no deformation)	**Evaporite Series** (slight tilting)	**Evaporite Series** (slight tilting)	**Trona**	**Evaporite Series**	**Trona**
High Magadi Beds (Kunkar limestone) (minor faulting)	**High Magadi Beds** (Kunkar lst.) (grid faulting)	**Upper/Middle High Magadi Beds** (grid faulting)	**Upper and Middle High Magadi Beds** (minor faulting)	**High Magadi Beds** (caliche)	**High Magadi Beds** (grid faulting)	**High Natron Beds**
Chert Series Grid faulting) (Kunkar limestone) / **Ngare Nyiro Beds** (Ewaso Ngiro Basin)	**Ngare Nyiro Beds** (Ewaso Ngiro Basin) **Oloronga Beds** (erosion) Pagasi Pebble Beds **Chert Series**	**Lower High Magadi Beds**	**Lower High Magadi Beds** Trachyte flows (Kunkar limestone)	**Green Beds** (Caliche)	**Orkaramation Beds** (Ewaso Ngiro Basin) **Lengorale Beds** Pagasi Pebble Beds	Lengai and Kerimasi volcanics, aeolian ash. No sediment outcrops reported
Oloronga Beds		**Oloronga Beds**	**Oloronga Beds** (Grid faulting)	**Oloronga Beds**	**Oloronga Beds**	
Magadi Trachyte (1.4–0.7 Ma)						

(continued)

Table 20.1 (continued)

Baker (1958)	Baker (1963)	Eugster (1969)	Eugster (1980)	Behr (2002)	Crossley (1976)	Isaac (1965)	
	Early Pleistocene	**Leshuta F.** Kordija basalts (2.2–1.7 Ma) and Kordija trachytes Ewaso Ngiro trachyte (2.1 Ma) *Faulting and tilting* Narosura pyroclastics (2.7; 2 Ma) (N); Sambu lavas (2.0 Ma) (S)				**Peninj Group** (~1.5–1.15 Ma)	**Moinik Fm.** **Humbu Fm.**
						Hajaro Beds Lavas (2.02–1.7 Ma) and sediments	
	Pliocene	Kirikiti basalts with interbedded fluviatile and waterlain tuffs (3.1–2.5 Ma) *Faulting* Loisiumurto volcanics (4.2 Ma) Namutuakit volcanics **Endosapia Fm. (5.1; 3.2 Ma)**				Sambu lavas (2.5–2 Ma)	
	Miocene	Lengitoto Trachytes (6.9–5 Ma) and *Faulting* Lengitoto Trachytes (6.9–5 Ma) *Faulting* Lisudwa Fm. Kishalduga Fm. (15.2; 12 Ma)					

Fig. 20.3 The Endosapia Formation. (**a**) Composite photographic section showing sandstones, tuffaceous sediments and conglomerates overlain by trachyte. (**b**) Composite section by Crossley (1976). (**c**) Unpublished diatomaceous section from the lower to middle Endosapia Formation. (**d**–**h**) Detailed examples in stratigraphic order from bottom to top. (**d**) Basal matrix-supported breccia with metamorphic clasts (debris flow/lahar). (**e**) Basal matrix-supported breccia with weathered volcanic clasts (debris flow/lahar). (**f**) Laminated, possibly zeolitic, siltstone and sandstone (lacustrine?). (**g**) Quartzose conglomerate on well-sorted sandstone with cross-bedded basal deposits (high energy stream supplying palaeolake). (**h**) Trachyte resting on fluvial orthoconglomerate with volcanic clasts

beds of sandstone and conglomerate indicating repeated episodes of fluvial deposition in the area.

Volcanism continued through the Pliocene giving rise to thick sequences of the Loisiumarto volcanics (~4.2 Ma), Kirikiti basalts (~3.1–2.5 Ma), Narosura pyroclastics (2.7 to 2 Ma) and Sambu basalts (~2 Ma) (Crossley and Knight 1981). Crossley (1979) reports local interbedded fluvial sediments with some tuffaceous waterlain clay near Lisudwa. The next series of lake deposits constitute the 38-m-thick Leshuta Formation (Table 20.1), 25 km north of the Endosapia River (Fig. 20.3). These grey to buff, fine- to medium-grained tuffs contain pumice. Beds are mm- to dm-scale, with cross- and trough-lamination, climbing ripples and parallel lamination (Crossley 1976). Tuff bands can be traced up to 8 m laterally. Crossley (1976) suggested that they accumulated in a wide shallow lake.

Early geological reconnaissance studies northwest of Lake Natron were reported by Uhlig and Jaeger (1942) and Pickering et al. (1962). Guest (1953) identified high lake level 'diatomite'. Subsequently, the discovery of an *Australopithecus robustus* mandible in the area (Leakey and Leakey 1964) stimulated further interest, leading to the first detailed descriptions of Pleistocene sediments by Isaac (1965, 1967), who also reported the presence of Acheulian stone tools (Mturi 1987; de la Torre et al. 2003, 2008; Mora et al. 2003; Domínguez-Rodrigo et al. 2014; Geraads et al. 2022). Isaac noted that Pleistocene lake and alluvial sediments (Fig. 20.4a, b) were interbedded with, and rested on, the Sambu basalts (~2.5–2.0 Ma), forming the Naikuruku and Hajaro Beds and the Peninj Group (Table 20.1; Fig. 20.4d–f). Subsequent palaeomagnetic studies of the Peninj Group suggested that it correlates with the Matuyama Epoch (Thouveny and Taieb 1987; Berggren et al. 1995). Luque et al. (2009a) outlined the overall tectonic and volcanic evolution of the Natron Basin, reporting early volcanism in the southern rift at about 20–16 Ma followed by tectonism between 12 and 7 Ma (Dawson 1992). The Natron Basin was initiated by downfaulting along the Sonjo fault to the west (Fig. 20.4c) at about 7–3.5 Ma. This was followed by eruptions from the Mozonik (Mosinik) and Shombole (Shompole) volcanoes. Faulting was re-activated during the upper Pliocene with many of the volcanic features of the modern basin developing by 2 Ma (Fig. 20.4c).

Isaac's Peninj Group type-section is located at Maritanane, south of the Peninj River (Fig. 20.4a, b, e, f). The basal part included the Hajaro Beds (1.7–2.02 Ma), which are comprised of sediments overlain by lava, which include 10–13 m of reddened sand and clay with a cellular structure that passes laterally into green alluvial, root-marked, clay along the Peninj River. In contrast, areas to the south, near the Moinik River, include lacustrine clay. The overlying Peninj Group is up to 80 m thick (Fig. 20.4d) and accumulated between 1.7 and 1.15 Ma (Curtis 1967; Howell 1972; Isaac and Curtis 1974; MacIntyre et al. 1974; Manega 1993). It includes lacustrine and paludal deposits that accumulated in a shallow and broad proto-rift (Domínguez-Rodrigo et al. 2002; Luque et al. 2009a, b) before the major rise of the Nguruman Escarpment to the north. The Humbu Formation (formerly the Lower Series of the Peninj Group) rests unconformably on lavas and includes about 40 m of analcimic, basaltic tuff, sand and clay (Fig. 20.4e) that mainly accumulated in alluvial fans (Luque et al. 2009a). Fresh to mildly saline, lacustrine, bioturbated

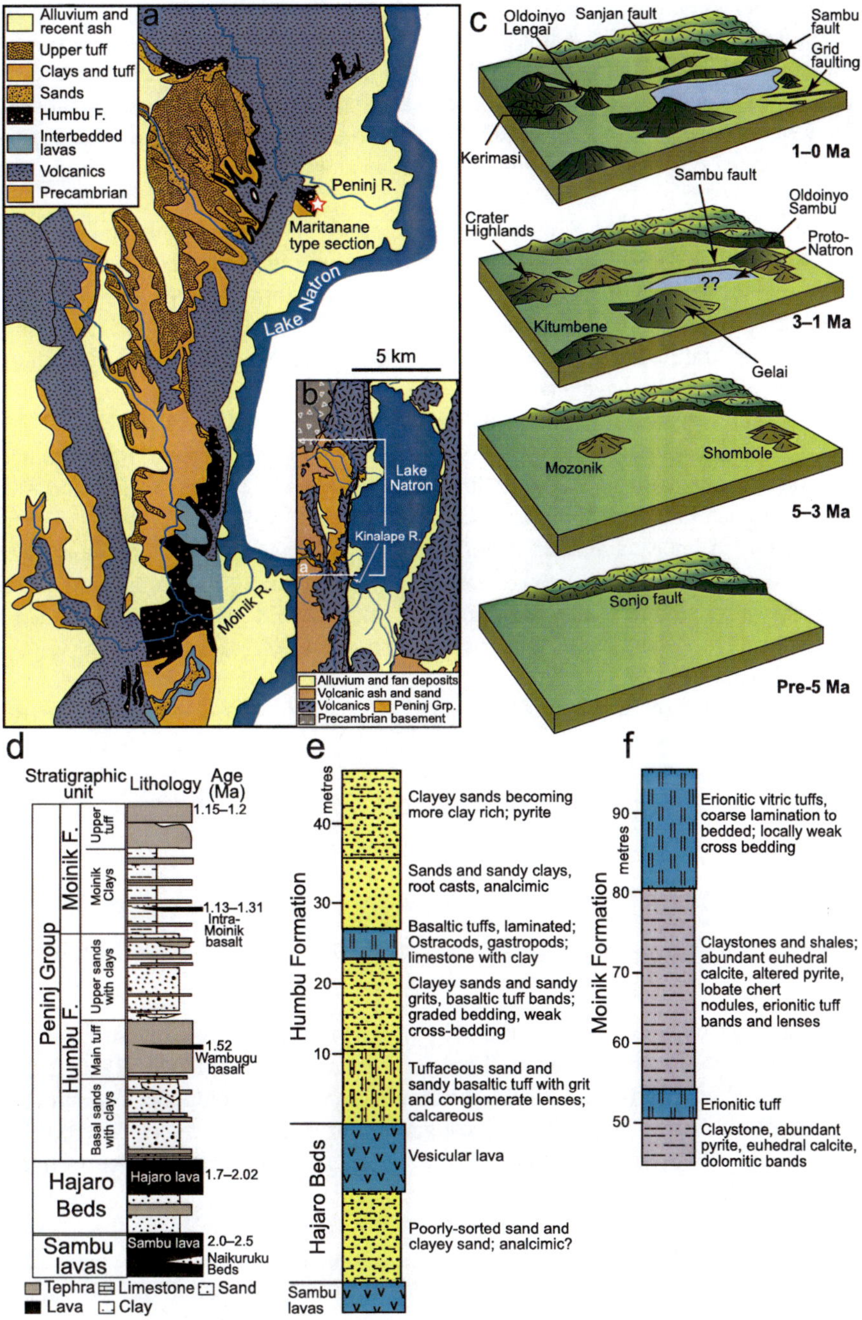

Fig. 20.4 Lacustrine sedimentation in the Natron Basin. ((**a**, **b** and **e**, **f**) after Isaac (1967)). (**a**, **b**) Maps of the Peninj Group sediments. (**c**) Evolution of the Natron Basin, Simplified after Luque et al. (2009a). (**d**) Stratigraphy of the Peninj Group, Hajaro Beds and Naikuruku Beds, NW Lake Natron. (After Domínguez-Rodrigo et al. (2002)). (**e**, **f**) Lithostratigraphy of the Hajaro Beds and Peninj Group

carbonates, dolomitic carbonates and microbial carbonates are also present in parts of the formation. Marshy sediments with carbonate nodules imply fluctuating lake levels (Domínguez-Rodrigo et al. 2002). The fauna includes mammals, gastropods (*Gabbia*), ostracods and fish (Icole et al. 1987) with pollen generally absent. However, Domínguez-Rodrigo et al. (2001) reported a few herbaceous taxa, mostly Graminae (41.6%), with other pollen indicating an open vegetation resembling that of today.

The Moinik Formation (Upper Series) comprises 30–40 m of laminated and thin-bedded claystone, shale and sandstone with trachytic tuff beds and a nephelinite lava (Fig. 20.4f). Calcite and dolomite are present in thin discontinuous horizons. Erionite, phillipsite and clinoptilolite are common with less abundant analcime, indicating a stratified alkaline lake (Icole et al. 1987). Abundant Magadi-type chert, gaylussite and pyrite have also been reported. Icole et al. (2002) recognised three generations of stromatolites in the Natron Basin. At the Kinalape River (Fig. 20.4b), the oldest was correlated with the Moinik Formation and comprises fluorite and fluorite-dolomite-bearing stromatolites interbedded with erionitic clay and silt. These stromatolites typically have a structureless basal bed with Magadi-type chert giving way to finely laminated laterally-linked hemispheroids and columnar stromatolites. Icole et al. (2002) noted that the varying mineralogy was related to elevation above the modern lake and suggested that fluorite and dolomite replaced original calcite when the area was covered by more recent lakes of differing depths and chemistry.

Peninj Group sedimentation terminated with subsidence that gave rise to the modern Natron Basin between about 1.2 and 1.15 Ma, and eruptions of Magadi Trachyte lavas in the Magadi Basin (Luque et al. 2009a).

20.3 The Oloronga Beds

20.3.1 Oloronga Outcrops

Exposed lacustrine sediments in the Magadi Basin postdate eruptions of extensive flood lavas that form the Magadi Trachyte Formation (1.4–0.8 Ma; Baker 1958, 1963) and which overlap in age with the termination of the Moinik Formation in the Natron Basin. Early geological investigations at Magadi were carried out by Parkinson (1914), Gregory (1921), Walter (1922), Coates (cited as Anonymous 1923) and Stevens (1932). There has been confusion over the main stratigraphic units at Magadi, which were initially classified as the Oloronga Beds, Chert Series, High Magadi Beds and the Evaporite Series (Table 20.1). Baker (1958) placed the Oloronga Beds below his 'Chert Series' with an intervening period of faulting, but later placed faulting after the Chert Series. Eugster (1969, 1980, 1986) dropped the term and distinguished older and younger sets of chert, with the former placed in the Oloronga Beds and the latter assigned to the 'lower High Magadi Beds' with both separated by a 'kunkar limestone' (calcrete). Baker (1958) also described the 'Ngare Nyiro' Beds, which were later referred to as the Orkaramation Beds by Crossley (1976) (Fig. 20.2 and Table 20.1). Much of the 'lower High Magadi Beds' of Eugster (1969, 1980) were later termed the 'Green Beds' by Behr and Röhricht (2000).

The Oloronga Beds are the oldest sediments exposed around Lake Magadi where they crop out mainly to the west and southwest of the lake (Figs. 20.2 and 20.5a), but with smaller exposures to the north (Fig. 20.5b, c). They overlie the Magadi

Fig. 20.5 Oloronga Beds in the Magadi Basin. (**a**) Oloronga Beds northwest of hot springs, southwestern Lake Magadi Basin. Calcrete ledge overlies tuffaceous siltstone and sandstone. (**b**) Oloronga Beds and High Magadi Beds northeast of Nasikie Engida (foreground patches left of stream). (**c**) Oloronga Beds resting on trachyte southwest of Nasikie Engida. (**d**) Undulating plates of chert (1–2 cm thick) interbedded with zeolitic silt, southwest of Lake Magadi. (**e**) Granular burrows in Oloronga claystone. (**f**) subhorizontal brown burrows in Oloronga silt and clay. Both (**e**) and (**f**) suggest oxygenated conditions during deposition. (**g**) Branching root-system in greenish black palaeochannel claystone

Trachyte Formation, which formed a broad lava plain with a subdued relief (Baker 1958, 1963; Crossley 1979; Eugster 1980; Behr 2002). This volcanic landscape was subsequently fragmented by north-south grid faulting, producing new accommodation space for lacustrine deposition (Owen et al. 2019). The basal Oloronga sediments in outcrop are dated by a single obsidian radiometric age of 0.78 ± 0.04 Ma from a lava flow northeast of Lake Magadi (Fairhead et al. 1972; Eugster 1980). ^{36}Cl estimates suggest an age of ~0.76 Ma for salt accumulation (Kaufman et al. 1990). Röhricht (1998) and Behr and Röhricht (2000) later reported a U/Th date of 0.3 Ma which suggested the youngest possible age for the 'upper' Oloronga Beds. In contrast, a much-improved Bayesian chronology based on ^{14}C, $^{40}Ar/^{39}Ar$, U-series and palaeomagnetic data was developed for two long cores that reached the trachyte basement (Owen et al. 2018a, b, 2019). These indicate that Oloronga sedimentation began after ~1.08 Ma.

In outcrops near Lake Magadi, the Oloronga Beds are generally <10 m thick and are capped in many areas by a calcrete several dm thick (Eugster 1986; Felske 2016). They consist of indurated, zeolitic, olive-green volcaniclastic sand, silt and clay with wavy plates of chert with a white patina. *Viviparus* limestone (gastropod coquina: grainstone or wackestone), silicified limestone and mud, ostracods and fish have been reported from the lower Oloronga Beds with diatoms noted in chert (Behr 2002) and limestone (Owen et al. 2019). Burrows are locally common in the siliciclastic deposits and include two main types. The first has a cross-section of about 1 cm, cuts across and parallel to bedding, and is filled with granular clay (Fig. 20.5e). The second has a smaller diameter and is sub-horizontal with a brown clay infill (Fig. 20.5f). Branching root systems are well exposed in a greenish-black palaeochannel claystone ~1.6 km south of the Southwest Lagoon (Fig. 20.5g).

Figure 20.6a, b shows a section through the Oloronga Beds at OB1 (Fig. 20.2). The sequence starts with indurated, trough cross-bedded, yellow-brown and pale brown sandstone (Fig. 20.6d) formed by NNE-flowing streams. These are overlain by tuffaceous yellow-green siltstone and sandstone with branching subvertical burrows. The upper part includes tuffaceous silt with chert and silt beds. The 1–2-cm-thick grey cherts have white upper and lower surfaces that show a fine reticulated pattern and form discontinuous, horizontal, wavy plates (Fig. 20.6e, f) with rare desiccation or shrinkage cracks infilled with quartz (Fig. 20.6g). In places, they show high amplitude folds reflecting syndepositional deformation of the soft precursor (magadiite or gel). The cherts contrast with those in the overlying Green Beds by being thinner, more undulating and by their surface reticulations. They likely formed in shallow water following evaporation, precipitation and diagenesis of a magadiite precursor (Eugster 1967; Buatois et al. 2020). The sequence is capped by calcretes, which include massive, bedded and pisolitic types (Fig. 20.6h). Many contain angular quartzose chert clasts, implying early diagenesis of the chert precursor (Fig. 20.6i).

Outcrops west of Magadi in parts of the Oloronga Plateau and areas to the north lack chert and are dominated by crudely bedded beige, yellowish and brownish, tuffaceous, diatomaceous silt and fine sand (Fig. 20.7). White diatomaceous silts with well-developed lamination are also present, with cross-cutting veins of

Fig. 20.6 Representative Oloronga Beds section from southwest Magadi (OB1 in Fig. 20.2). (**a, b**) Lithological log and photograph of Section OB1. (**c–i**) photos from OB1 and adjacent areas. (**c**) Small tufa chimney from the tuffaceous zeolitic silt. Subhorizontal ridges suggest that the small subaqueous vent was episodically exposed. (**d**) Cross-bedded sandstone. (**e**) Thin (<1 cm) undulating chert in lacustrine tuffaceous silt. (**f**) Detail of undulating chert beds. (**g**) Chert surface with desiccation cracks and silica infills. (**h**) Pisolitic calcrete. (**i**) Angular chert clasts in calcrete

carbonate and (or) thin limestone horizons. Several outcrops include boxwork carbonates with cross-cutting calcite veins that form (sub)angular chambers with a white silt infill. Individual chambers range from less than 1 cm to more than 5 cm (Fig. 20.7d, e). Some represent small mudcracks filled by early carbonate cement precipitated from shallow dilute groundwater. The uppermost parts of sections at

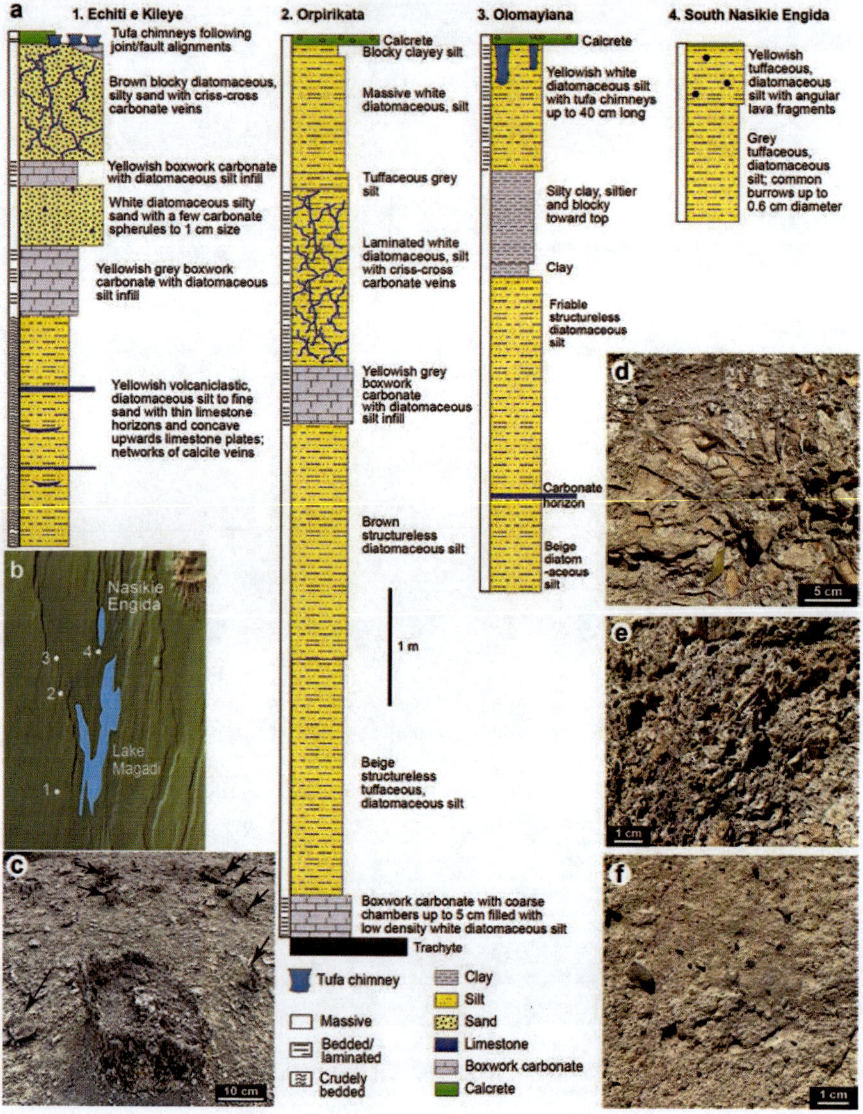

Fig. 20.7 Representative Oloronga Beds sections to the west and north of Lake Magadi. (**a**) Four Oloronga Beds sections dominated by tuffaceous silts, boxwork carbonates, silty clay, silts and tufa chimneys. (**b**) Locations. (**c**) Tufa chimney garden with chimneys commonly aligned parallel to the fault scarp exposing the Oloronga Beds, Echiti e Kileye. (**d, e**) Boxwork carbonates showing the different scales of enclosed chambers filled with silt, Opirikata. (**f**) Tuffaceous silt with angular volcaniclastics, Nasikie Engida

Echiti e Kileye and Olomayiana include small tufa chimneys up to 40 cm high with diameters of 10–15 cm. At the former site, tufa chimneys form 'gardens' (Fig. 20.7c) upon an eroding surface of diatomaceous silt with many chimneys aligned parallel to the fault scarp that exposes the deposits. Calcrete caps most of these sequences. North and south of Nasikie Engida, the Oloronga Beds are dominated by tuffaceous, diatomaceous silts, locally with angular volcanic clasts up to ~8 mm across (Fig. 20.7f). Burrows (up to 1 cm diameter) are present in several outcrops. Diatom abundance varies in most of these western Oloronga Formation sediments, ranging from absent to common. The assemblages are dominated by fresh to mildly saline, epiphytic and benthic (*Epithemia*, *Rhopalodia*, *Synedra* [*Ulnaria*], *Encyonema*, *Cymbella* spp.) taxa indicative of shallow water settings.

A well-studied sequence at Lainyamok, 8 km west of Lake Magadi (L1 in Fig. 20.2; Fig. 20.8a), near Olomayiana, correlates with the Oloronga Beds. The sediments were deposited in a water body that at times may have been linked directly to the main Oloronga palaeolake via Olomayiana. Good exposures have yielded hominid (*Homo*) teeth, mammals and Acheulian tools (Shipman et al. 1983; Faith et al. 2012; Potts and Faith 2022). The deposits thicken northwards implying deeper water in that direction and rest on reversed and normally magnetised trachyte that likely spans the Jaramillo Event at 0.89 Ma. Potts and Deino (1995) reported dates of 890 ± 30 and 920 ± 10 ka for the trachytes, and 392 ± 4 and 330 ± 6 ka for the deposits. The sedimentary sequence (Fig. 20.8b; Potts et al. 1988) includes a basal boxwork carbonate that was interpreted as an evaporative lake-marginal deposit. This is overlain by fluvial, graded, sandy silt and massive white lacustrine silt (Fig. 20.8a) with fish fossils and mudcracks. Diatoms (*Rhopalodia*, *Epithemia*) are present but scarce. A second evaporative lake margin boxwork carbonate (Fig. 20.8c) is present in mid-sequence with an overlying mudflow that consists of zeolitic, poorly-sorted silt with rounded clasts. In places, a dark limestone with tepee structures is exposed, representing carbonate precipitation along a shallow shore zone. A laminated and massive siltstone with fish fossils near the top is capped by calcrete (Fig. 20.8d, e), which is a common basin-wide stratigraphic horizon (Eugster 1980).

Barthelme (1991) and Barthelme et al. (2009) also discuss a 7-m-thick sequence of lacustrine silts, fluvial sediments, and carbonate horizons northwest of Lenderut (white star in Fig. 20.2) that they attributed to lacustrine, shoreline and river settings. There are no age determinations from the site, but Acheulian tools and the faunal context imply an age of ~900–500 ka (Potts and Faith 2022).

Horst blocks across the Magadi Basin generally lack sediments on their surfaces other than thin, wind-blown silt and sand, and recent gravelly soils. However, porous tufa-travertine towers (up to ~8 m high) with cm-scale vertical pipes, subhorizontal plates and carbonate cement follow a broadly N-S fault line across a horst block about 6 km south of Lainyamok (Figs. 20.2 and 20.8f, g). Smaller spring carbonate mounds (up to 2.5 m high) are also present at Olomayiana (Fig. 20.7) along the same alignment. Owen et al. (2019) reported possible eroded rimstone dams on smaller tufa mounds 100–200 m to the south of the highest tower that implied subaerial growth but agreed that the tall towers were subaqueous and formed on a lake floor. The towers lack external drapery that can form when water flows down their sides in subaerial settings.

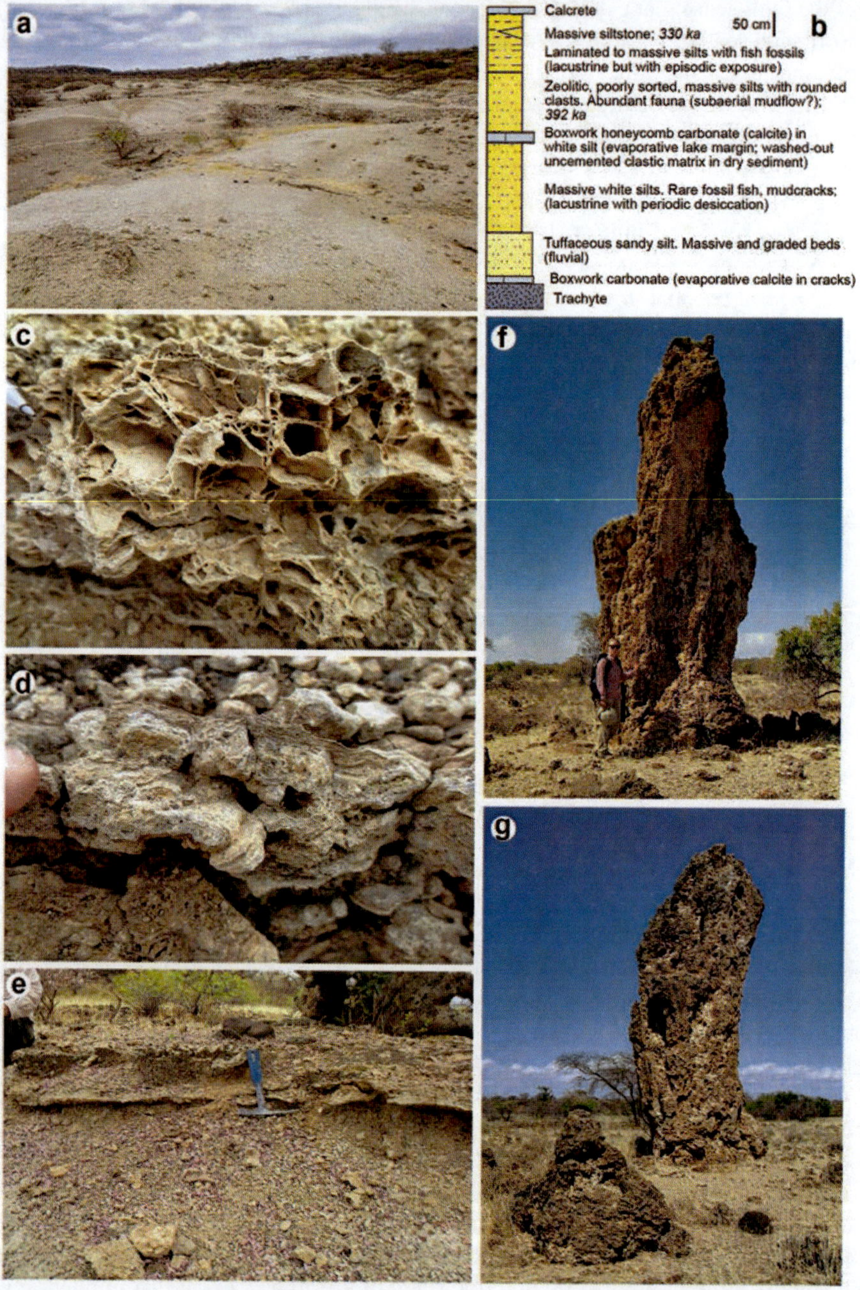

Fig. 20.8 Lainyamok sediments and tufa-travertine towers. (**a**) General view southward showing the Lainyamok graben and white silt overlain by a capping calcrete. (**b**) Lithological section after Potts et al. (1988). (**c**) Honeycomb carbonate with boxwork structure (pen tip for scale). (**d**) Laminated calcrete around silt clasts. (**e**) Capping calcrete. (**f, g**) Tufa-travertine towers (~6 m high) on horst blocks 6 km south of Lainyamok

The towers resemble examples reported from Lake Abhé (Dekov et al. 2014; DeMott et al. 2021), Searles Lake (Guo and Chafetz 2012), and other lakes where they formed under fluctuating lake levels as products of the mixing of sublacustrine hydrothermal springs and lake waters during highstands. Hillaire-Marcel et al. (1986) dated the most prominent tower at 14,380 ± 180 to 10,985 ± 215 ka and inferred that it formed within, or on the margin of, a High Magadi Beds palaeolake just before or during the African Humid Period (AHP). However, the horst surface stands >20 m above adjacent graben floors, and Owen et al. (2019), in common with Baker (1958, 1986) and Behr (2002), noted that there was no other reported evidence for a former terminal Pleistocene-Holocene lake as high as this (>710 m asl). Instead, Owen et al. (2019) suggested that the towers may be older, noting that small tufa chimneys are interbedded with Oloronga Beds sediments along faults elsewhere (Figs. 20.6c and 20.7c) and that core evidence also indicated spring and lacustrine calcite accumulation during Oloronga Beds deposition. Calcite is scarce in geochemical zone G5 in core MAG14-2A (Fig. 20.9; Sect. 20.3.2 below), which Owen et al. (2019) correlated with the Green Beds. However, calcite is present in outcrops of the Green Beds (Sect. 20.5.2 below) but the former extent of the Green Beds palaeolakes is unknown. Although their age remains uncertain, the tufa towers might have been upfaulted to their present level (see discussion in Sect. 12.8.3).

20.3.2 Lake Magadi Cores and Oloronga Deposition

Two long Lake Magadi cores (HSPDP-MAG14-1A and -2A in Fig. 20.2) were collected in 2014 as part of the Hominin Sites and Paleolakes drilling project (HSPDP) with each containing a one-million-year record of deposition (Owen et al. 2018a, b, 2019; Rabideaux 2018; Muiruri et al. 2021a, b; Cohen et al. 2022). Both reached Magadi Trachyte and include a wide variety of facies. Correlation with the Oloronga Beds and other exposed sedimentary units (Green Beds, Evaporite Series, High Magadi Beds) is difficult. However, Owen et al. (2019) noted the presence of calcareous sediments (gastropod limestone, tufa chimneys and tufa-travertine towers) in Oloronga outcrops and suggested that they may be equivalent to their geochemical zones G1–G4 in Core HSPDP-MAG14-2A, because of their relatively high Ca/Na ratios and presence of calcite, Mg-calcite and dolomite (Fig. 20.9). This would imply that the Oloronga Beds broadly correlate with sediments between the HSPDP-MAG14-2A core base and ~102 m core depth, or ~1056 and 380 ka (Fig. 20.9), which is compatible with the limited dating evidence for outcrop sediments. Owen et al. (2019) also noted that dates from the overlying Green Beds suggested that they may be equivalent to parts of their Zone G5 (~102–60 m; ~380–105 ka), which has low Ca concentrations and includes varied zeolites (analcime, erionite, phillipsite, clinoptilolite). Furthermore, high K+Na/Al ratios in Zone G6 that reflect trona imply a correlation with the Evaporite Series. The High Magadi Beds may also be represented by massive and laminated mud at around the 10–20 m depth interval in the HSPDP core.

If these inferred correlations are correct, the deeper parts of Palaeolake Oloronga at the HSPDP-MAG14-2A site initially accumulated carbonate mud that subsequently gave way to laminated and bedded mud with pyrite and chert

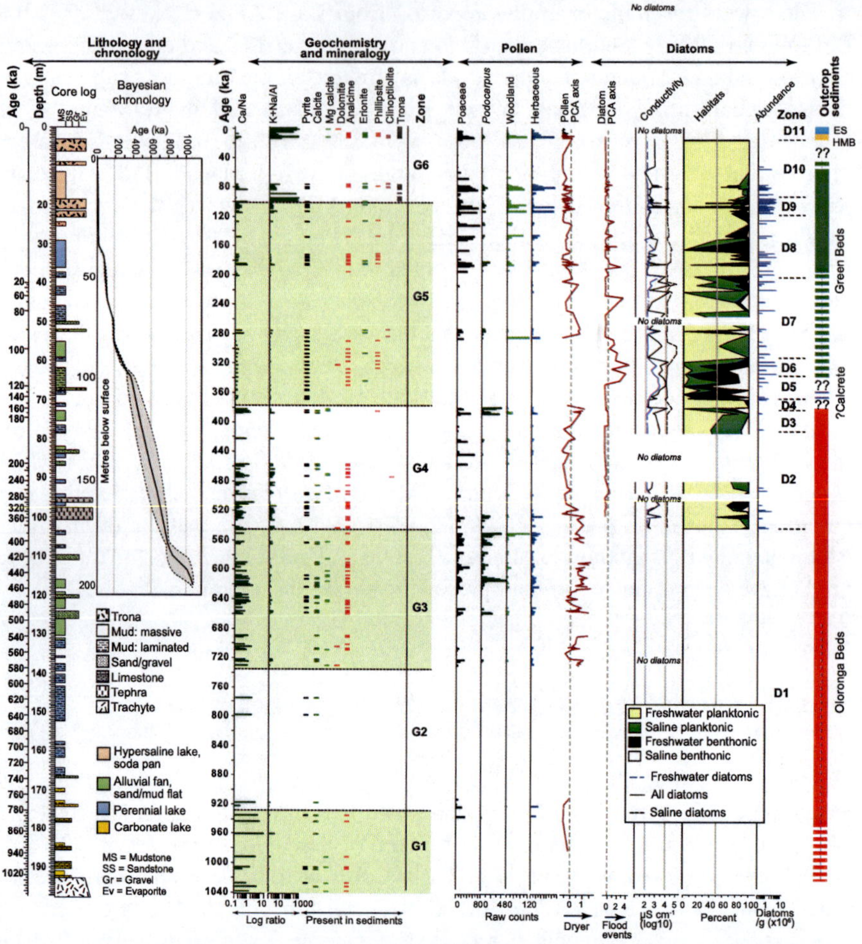

Fig. 20.9 Key datasets from Core HSPDP-MAG14-2A and possible outcrop correlations. Compiled from Owen et al. (2018a, b, 2019). Depth and age scales are on left for the lithological log; other data sets are set against a chronological scale

(Fig. 20.10h–k). At the onset of sedimentation, the nearby HSPDP-MAG14-1A core site stood about 60 m higher and must have been shallower or even emergent at times. The deposits there also include gastropod limestone (Fig. 20.10a), black

Fig. 20.10 (continued) upwards into chert. (**d**) Green to black bedded clay with rhomboidal to cubic crystals and disseminated cubic pyrite. (**e**) Massive grey-green silty clay with disseminated pyrite. (**f**) Well-sorted gravel resting on green laminated erionitic clay. (**g**) Grey to white chert between silicified mudstone and sandstone beds. (**h**) Bedded to laminated brown clay/silt with chert nodule at top. (**i**) Laminated grey and light brown clay with patches of chert. (**j**) Bedded dark grey and brown clay with salt pseudomorphs. (**k**) Bedded grey and white chert fragmented by drilling. (**l**) Laminated grey, green and brown clays with a cherty band near base. (**m**) Greyish green to black clays with white chert horizons and patches. (**n**) Large crystals of trona. (**o**) Massive black silty clay. (**p**) Bedded trona

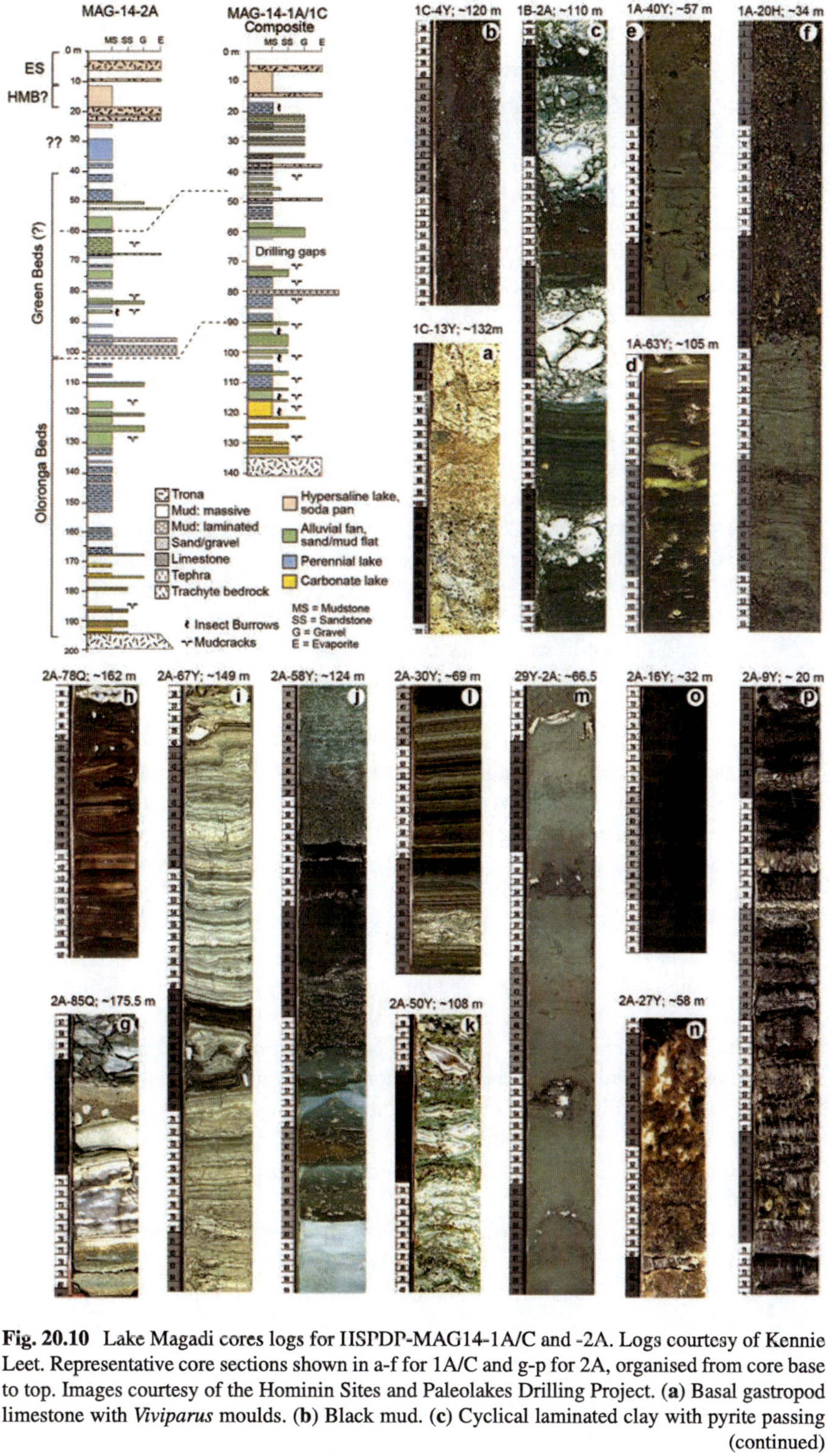

Fig. 20.10 Lake Magadi cores logs for HSPDP-MAG14-1A/C and -2A. Logs courtesy of Kennie Leet. Representative core sections shown in a-f for 1A/C and g-p for 2A, organised from core base to top. Images courtesy of the Hominin Sites and Paleolakes Drilling Project. (**a**) Basal gastropod limestone with *Viviparus* moulds. (**b**) Black mud. (**c**) Cyclical laminated clay with pyrite passing

(continued)

mud (Fig. 20.10b), cyclical laminated clay/chert beds (Fig. 20.10c) and massive grey-green silty clay with pyrite (Fig. 20.10e), sediments that suggest anoxic conditions even in shallower locations. Ferland (2015) discussed the potential of organics to evaluate palaeoclimate. Scott et al. (2021) noted trace fossils (e.g. termites, flies, beetles or earwig burrows) as evidence for shallow water phases in core 1A. HSPDP-MAG14-2A sediments that may correlate with the Green Beds include laminated and massive green, brown and black clay with chert (Fig. 20.10m) with green to black bedded clay and massive grey-green silty clay, both with disseminated pyrite (Fig. 20.10d, e), again implying anoxia. Bedded and massive trona and laminated to massive black and green mud constitute younger deposits (Fig. 20.10f, n–p).

HSPDP-MAG14-2A also yielded detailed pollen and diatom records (Fig. 20.9; Owen et al. 2018a, b, 2019; Muiruri et al. 2021a, b). Pollen data suggest increasing moisture between 1000 and 575 ka followed by an overall drying trend through to the present that was interrupted by many wet–dry cycles. The greatest variability appears to have been associated with periods of high eccentricity-modulated precession. Diatom PCA data start at about 540 ka with planktonic taxa dominating. After the possible termination of Oloronga deposition at ~380 ka, periodic influxes of benthic taxa, as reflected in peaks in diatom PCA data (Fig. 20.9), indicate episodic flooding (Owen et al. 2018a, b; Muiruri et al. 2021a). Pollen PCA data (Fig. 20.9) suggest aridification around 525–400 ka in late Oloronga times, which overlaps with the Mid-Brunhes Event (~430 ka), mammalian extinctions in the South Kenya Rift between 500 and 400 ka, and an erosional period between the last Acheulian tools at Olorgesailie (~500 ka) and their replacement by Middle Stone Age artifacts by at least 320 ka (Potts et al. 2018, 2020).

Muiruri et al. (2021a, b) concluded that four major environmental phases could be distinguished through the last million years. Their Phase I was characterised by fresh to mildly saline lakes at Magadi between 1000 and 740 ka. The palaeolake reached maximum depths during Phase II (~740–500 ka) with conditions becoming gradually drier through to 325 ka (Phase III). In contrast, Phase IV was characterised by fluctuating wet-dry cycles as also reflected in data from the Koora graben to the east (Muiruri et al. 2021c) and the Magadi pollen record. The Magadi diatom record for that time suggests a persistent lake possibly maintained by spring inflow.

The Oloronga palaeolake developed through Phases I–III with contrasting depositional regimes (Fig. 20.11). During Phase I the palaeolake was fresh to moderately saline. Shallow swamps initially favoured the accumulation of gastropod- and ostracod-bearing lime mud. The presence of chert in the modern axial Magadi basin implies elevated alkalinity with sufficient silica to precipitate chert, perhaps linked to spring inflows. However, outcrops to the west lack chert perhaps reflecting regional differences during deposition. Calcite accumulated in deeper water where laminated and bedded mud was laid down. Subaqueous spring inflows during Phse I, and possible later, led to the formation of small tufa chimneys and, locally, tufa-travertine towers. The paucity of pollen in deeper-water sediments implies a well-mixed water column with oxic bottom water. During Phase II (730–540 ka), the lake

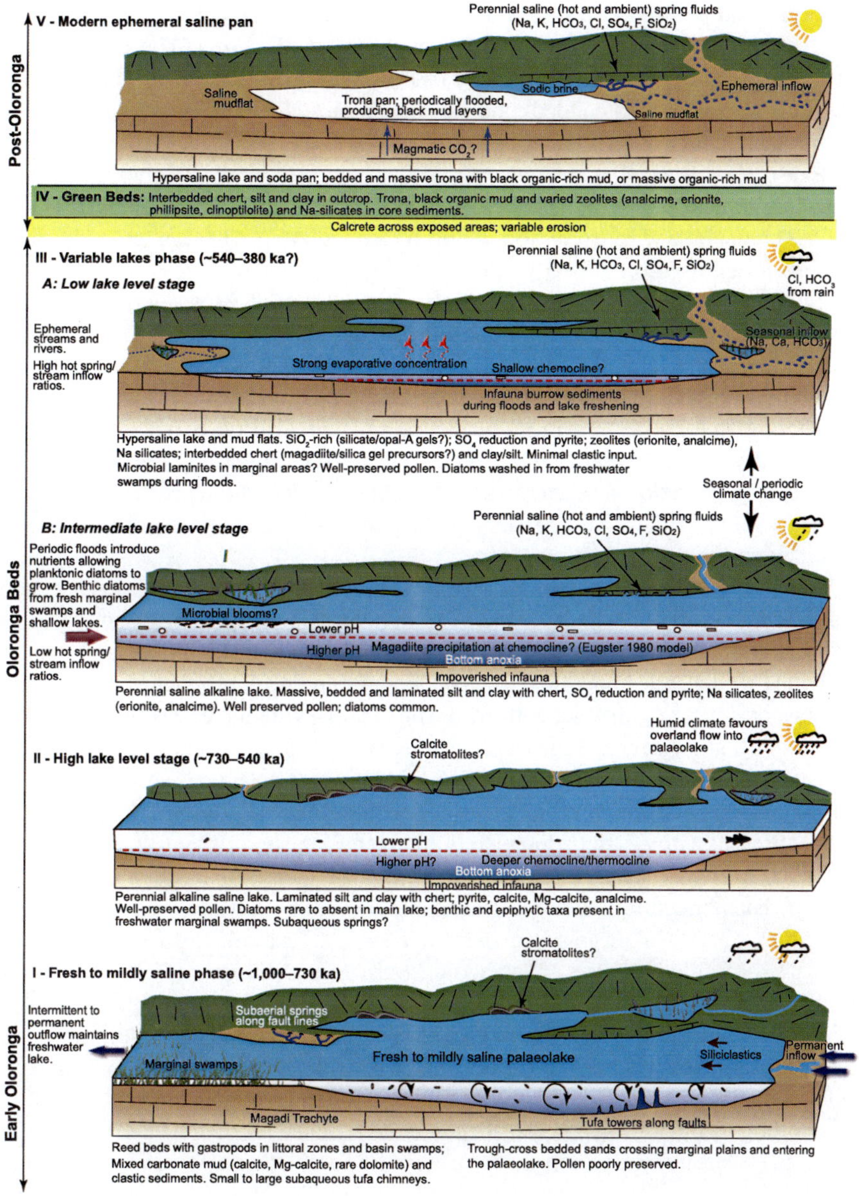

Fig. 20.11 Evolution of lake systems in the Magadi Basin. Phases I to IV follow Owen et al. (2019) and Muiruri et al. (2021a). Phases I to III reflect Oloronga deposition. Phase 1 represents a well-mixed fresh lake that gave way to a stratified water body during Phase II with anoxic and euxinic bottom waters. Fluctuating degrees of anoxia and euxinia continued through to the present. Lake levels varied during Phase III with the lake becoming periodically very shallower. Phase IV was associated with more varied zeolites in lake sediments and cherts of the Green Beds (see Fig. 20.12). Phase V represents the modern seasonally flooded saline pan

deepened and became stratified with euxinic conditions that appear to be linked to intervals of high orbital eccentricity (Deocampo et al. 2021). These conditions favoured sulfate reduction leading to pyrite formation and facilitated the development of well-laminated/bedded mud and pollen preservation. Calcite continued to accumulate in the sediments during these periods.

Phase III (540–380 ka) appears to have been more variable with the lake deepening and shallowing. Freshwater benthic diatoms were introduced to the saline palaeolake during periodic floods with planktonic taxa growing in a freshwater epilimnion during wetter intervals. The palaeolake never completely dried out at the HSPDP-MAG14-2A core site, although a few mudcracks have been observed. In contrast, neighbouring basins such as the Koora Graben dried out frequently, leading to the development of palaeosols. This also may be evidence that the axial Oloronga palaeolake at the 2A coring site might have been maintained by hot (?) springs during periods of aridity. Silica inputs from springs might also have raised silica concentrations sufficiently to preserve diatoms despite the alkaline waters and would have favoured the pervasive replacement of volcanic glass by zeolites—a common early diagenetic process in lake sediments in the Kenya Rift.

Herrick (1972) and Surdam and Eugster (1976) reported the conditions needed for the common zeolites in the Magadi Basin to form (Fig. 20.12). Erionite requires addition of alkaline water to volcanic glass and is favoured by high silica and sodium activities with a lack of alkaline earths. Clinoptilolite evolves when some calcium and magnesium are available, with chabazite forming with increased alkaline earths (often in groundwater) and lower silica activity. Phillipsite develops mainly in fluvial and lake marginal settings with low silica activity. Analcime is common in most deposits as an alteration product of earlier zeolites, particularly erionite. It is commonly present with quartz, which led Surdam and Eugster (1976) to suggest that it develops when silica activity is lowered during the magadiite to chert transformation, perhaps linked to dilute percolating groundwater. They also suggested direct transformation of Na-Al-gels to analcime. Where Ca^{2+} is available, they reported the formation of either fluorite or gaylussite. Surdam and Eugster (1976) noted that calcite, illite, analcime and chabazite are more common in outcrops than in cores, where fluorite and Na-Al-gels were also observed. They attributed this difference to the more varied hydrological conditions in lake marginal settings with concentrated sodium carbonate brines associated mainly with offshore core locations.

Oloronga sediments in outcrop are commonly capped by calcrete across much of the axial basin indicating relatively dry semi-arid conditions around a shrinking palaeolake (Phase IV) (Eugster 1980; Felske 2016). This post-Oloronga phase may correlate with part, or all, of geochemical zone G5 and include the Green Beds in outcrop. Chert and zeolite formation became widespread in both lake marginal and deeper offshore locations. Many cherts in HSPDP cores contain labyrinth patterns that suggest synsedimentary deposition directly from siliceous gels (Leet et al. 2021). Later increases in the evaporation-precipitation ratio and a relative increase

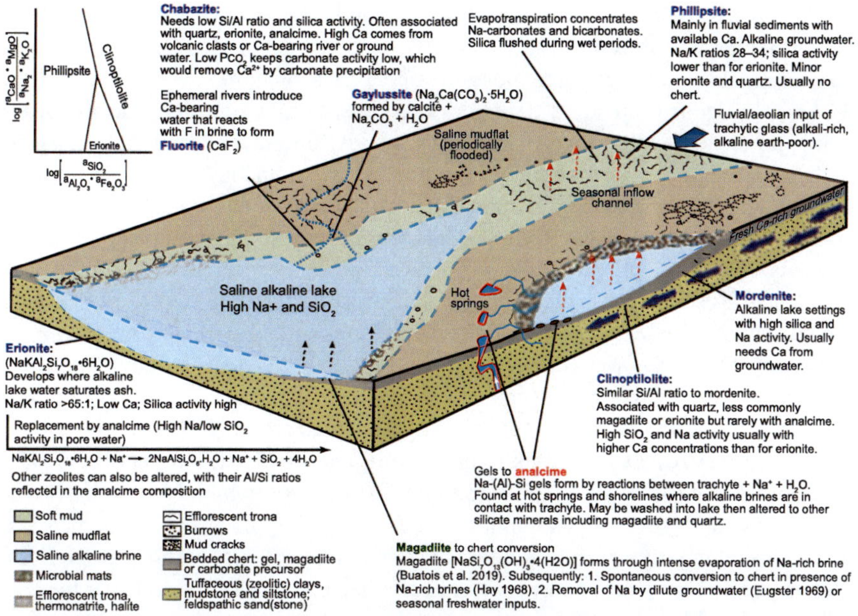

Fig. 20.12 Controlling factors in zeolite alteration of volcanic glass in the Magadi Basin. Based on Surdam and Eugster (1976). Zeolites (erionite, analcime, clinoptilolite, chabazite, mordenite and phillipsite) are marked in bold. The figure also indicates favourable conditions for development of gaylussite and magadiite, which were also present in the alkaline Oloronga and Green Beds palaeolakes

in spring solutes led to trona deposition after about 105 ka in a seasonally or periodically flooded saline lake, gradually leading to the saline pan and evaporite deposits of modern Lake Magadi.

20.4 The Lengorale and Ngare Nyiro (Orkaramation) Lake Beds

The Lengorale lake beds rest on Lengorale Trachyte dated at 0.68–0.69 Ma (Crossley 1976) and are younger than the Oloronga Beds. They are exposed along the Pagasi Gorge and along the Kirikiti fault scarp, up to 1 km south of the Pagasi River (Crossley 1976). The deposits consist of low-density, buff, laminated clay and silt with a blocky texture. Some of the sediments contain mica and fish fossils. They thicken southwards to ~5 m in outcrop but may be up to 20 m thick. They are overlain by up to 10 m of the fluvial Pagasi Pebble Beds, which thicken westwards to ~20 m.

The Ngare Nyiro lake beds, first described by Baker (1958), were referred to as the Orkaramation Beds by Crossley (1976). They rest on eroded Oloronga Beds or volcanic rocks and are exposed along the upper Ewaso Ngiro River (Figs. 20.2 and 20.13). The deposits are up to 10 m thick and consist of low-density, pale brown and white, laminated, diatomaceous clay and silt with thin gravel lenses that gently dip to the west. They too have a blocky structure, but individual layers of up to 1 cm can be traced laterally for tens of metres. Carbonate horizons and scattered nodules are common and may represent drying events. Tufa cuts through and locally coats the deposits. Fish (Fig. 20.13d) and gastropods are present especially in carbonate layers. Trace fossils are also present on lamination surfaces (Fig. 20.13c). Diatoms, which are not abundant, consist mainly of epiphytes and other benthic taxa (*Synedra* [*Ulnaria*], *Cymbella*, *Rhopalodia*, *Epithemia* spp.) and rare planktonic *Aulacoseira granulata*. Rare cross-bedded coarse sands and gravels are present in channels, especially to the south, and record ephemeral streams crossing dry lake flats.

Crossley (1976) also noted fine-grained soils and carbonate stromatolites at altitudes of up to 1400 m along the Nguruman Escarpment, which he correlated with the Orkaramation Beds. This implies about 680 m of uplift of these modified lake sediments along the Kirikiti fault. Given the present distribution of both the Oloronga and Orkaramation Beds, it is likely that a former lake stood near the base of the modern Nguruman Escarpment and that it extended eastwards, at least intermittently, to the modern Lake Magadi area and possibly southwards into the Natron Basin. Crossley (1976) also noted that erosion of the westward-dipping Orkaramation lake beds and a change to alluvial deposition probably reflected major border-fault movements with a switch from west-to-east drainage to the modern north-south-flowing Ewaso Ngiro (S) River, a development also supported by trace element distributions in the Magadi HSPDP core record (Owen et al. 2019).

A lake of similar extent to the Oloronga and Orkaramation water bodies, after this phase of faulting, is recorded by stromatolites with fish remains at heights of up to 700 m asl (Crossley 1976), which Hillaire Marcel et al. (1986) U-Th dated to somewhere between 180–129 ka. However, modern height relationships may be of little value in assessing this possible palaeolake given the Pleistocene tectonic activity in the region.

20.5 The Green Beds

20.5.1 General Characteristics and Age

The Green Beds are lacustrine sediments of mid-Pleistocene age that crop out in the axial depression mainly south and east of Lake Magadi, and in the Northeast Subbasin (Figs. 12.3, 20.2 and 20.14). Small outcrops lie along the faulted midwestern margin of Nasikie Engida and the southwest margin of that lake. The sediments are tuffaceous mud, silt and siltstone, but chert is the defining lithology of the Green

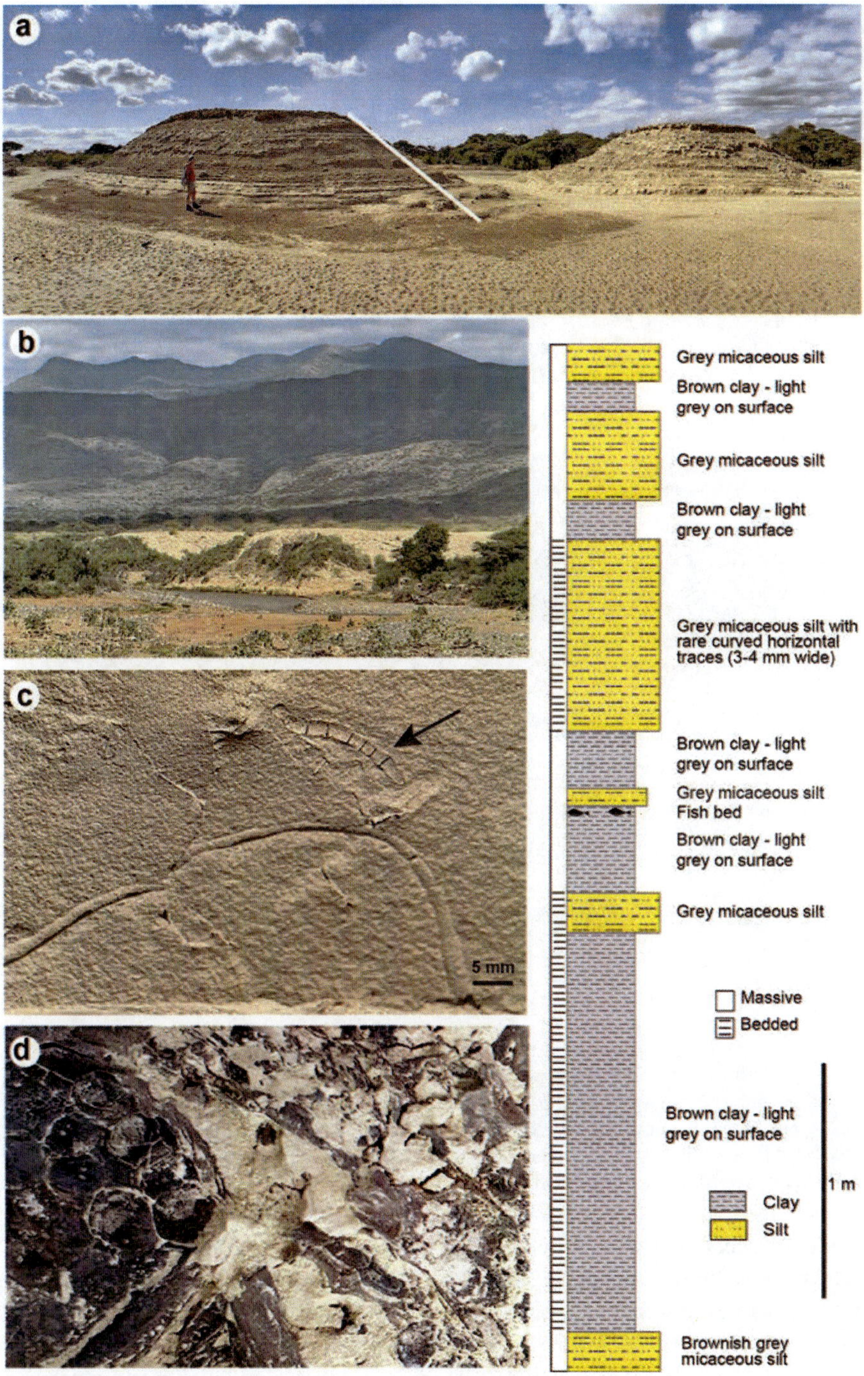

Fig. 20.13 The Orkarimatian Beds. (**a**) View of outliers east of the Ewaso Ngiro. (**b**) View of deposits west of the Ewaso Ngiro River (S) with the Nguruman Escarpment and Loita Hills in the background. (**c**) Insect larvae on bedding plane arrowed with insect larval trail made just below the sediment/water interface (Jenni Scott, pers. comm). (**d**) Fish fossils with scales to left and bones to right. Section through the Orkarimatian Beds to right

Fig. 20.14 (**a**) Southeast peninsula showing location of the section in b, the main area of chert dykes (N-S and W-E), fault ridge cherts, and a large chert mound along the fault line. Platy cherts cover much of the surface in the top half of the photograph, broken in places by E-W dykes and shallow drainage channels. Small spring seepages issue along the faulted margins of the peninsula. (**b**) Green Beds sediments (see details in Fig. 20.15). Location shown in (**a**). (**c**) Chert dyke cutting platy chert, SE peninsula. (**d**) Fault-ridge chert mounds (pillows), eastern SE peninsula

Beds and accounts for an estimated 90% of the chert exposed in the Magadi Basin (Röhricht 1998; Behr and Röhricht 2000). The weathered surfaces of the cherts are dark green in the Northeast Sub-Basin, but mainly red or cream south and east of Magadi town, presumably reflecting the oxidation state of contained iron.

Parkinson (1914) first discussed the Magadi cherts (as "flint") and attributed them to hydrothermal fluids. Temperley (unpublished, cited by Baker 1958) reported a 'Chert Series' that lay above the Oloronga Beds although geological contacts were not recorded (Table 20.1). Baker (1958) described the stratigraphic relationships of the chert-bearing lake deposits but later placed the Chert Series below the Oloronga Beds (Baker 1963). Eugster (1967, 1969, 1980) and Surdam and Eugster (1976) proposed that the cherts originated from the diagenetic alteration of sodium silicate minerals (by loss of Na) and suggested that the Chert Series was part of the Late Pleistocene to Early Holocene High Magadi Beds (HMB: Sect. 12.8.2) despite their clear non-conformable contacts (Fig. 12.17g). U-series dating of chert by Goetz and Hillaire-Marcel (1992), however, showed that some cherts were older (98.5 ± 20 ka and 40.0 ± 6.5 ka). Röhricht (1998), Behr and Röhricht (2000) and Behr (2002) mapped, then reassigned sediments above the Oloronga Beds and below the HMB to the newly defined 'Green Beds' (Table 20.1). The relationships of these rocks are complicated by limited exposure, ambiguous contacts and disconformable relationships, which led to the early stratigraphic confusion. Many Green Beds rocks have been physically emplaced in overlying sediments of similar age because of syndepositional tectonics. Behr (2002) suggested that the Green Beds are up to 30 m thick, but in outcrop, they are typically only a few metres thick. In drill-core, inferred Green Beds are up to ~40 m thick (Owen et al. 2018a, b, 2019). Radiometric ages for Green Beds sediments in outcrop, including intrusive cherts, span the period from 191 to 40 ka.

20.5.2 The Green Beds Sediments in Core

Baker (1958) recorded up to 76 feet (23 m) of porous massive chert with intervening green mud in four early cores (A–D) from Lake Magadi but its age was unknown. Hay (1968) noted chert older than 10.1 ka in 'core B'. In later cores drilled by Magadi Soda Ltd, Surdam and Eugster (1976) reported the distribution of quartz, present mainly as chert, but provided little information on the sediments between the chert beds, possibly because of poor recovery(?).

Leet (2022) and Leet et al. (2016, 2021) studied cherts of different ages preserved in the HSPDP (2014) cores. The cherts spanning the Green Beds time-interval are interbedded with zeolitic mud, silt, tuff, magadiite and evaporites (Fig. 20.9; Owen et al. 2019). Leet et al. (2021) recognised two categories of chert: (1) 'syndepositional' cherts, formed by replacement and crystallisation before burial and compaction; and (2) 'diagenetic cherts', which formed after burial and compaction. Both are quartzose with variable crystal sizes from cryptocrystalline to megaquartz. Syndepositional cherts contain diatoms, permineralised plant and microbial debris, and volcanic glass shards floating in the chert matrix. Some cherts have magadiite or evaporite pseudomorphs preserved without compaction indicating early constant-volume replacement (cf. Schubel and Simonson 1990). Leet et al. (2021) described unusual 'labyrinth structures' in the cherts, which provide evidence for siliceous gel precipitation at or near the surface. Other mudstones in the cores contain cherts that formed in the subsurface following burial. These cherts contain aligned, flattened, rotated, and compacted plant fragments and sediment grains that indicate chert formation after compaction.

20.5.3 The Green Beds Sediments in Outcrop

Chert is widely distributed but other Green Beds sediments are poorly exposed (Kodikara et al. 2012, their Fig. 11a). A 2.6 m section excavated at the southern part of the Southeast Peninsula (Baker 1958, p. 29; Figs. 20.14b and 20.15) revealed massive, burrowed green and orange tuffaceous mud, silt, and siltstone overlain by laminated green and orange mud with white flakes (silicified plant fragments or compressed traces?) on lamination planes, thinly shelled gastropods, and three undulating light and dark grey quartz-chert beds, each <3 cm thick (Fig. 20.15; Owen et al. 2019). Most of the tuffaceous muds and silts have been altered to zeolites (erionite, analcime) with subsidiary opal-A or calcite cement. The detrital sediments are calcareous and contain microbial and plant remains that are commonly silicified (Röhricht 1998; Behr and Röhricht 2000). Scattered thin outcrops of greenish and brown mudstone and siltstone are locally present below protective chert outcrops. Burrows are common in the reddish silts and muds (Buatois et al. 2020; Scott et al. 2021).

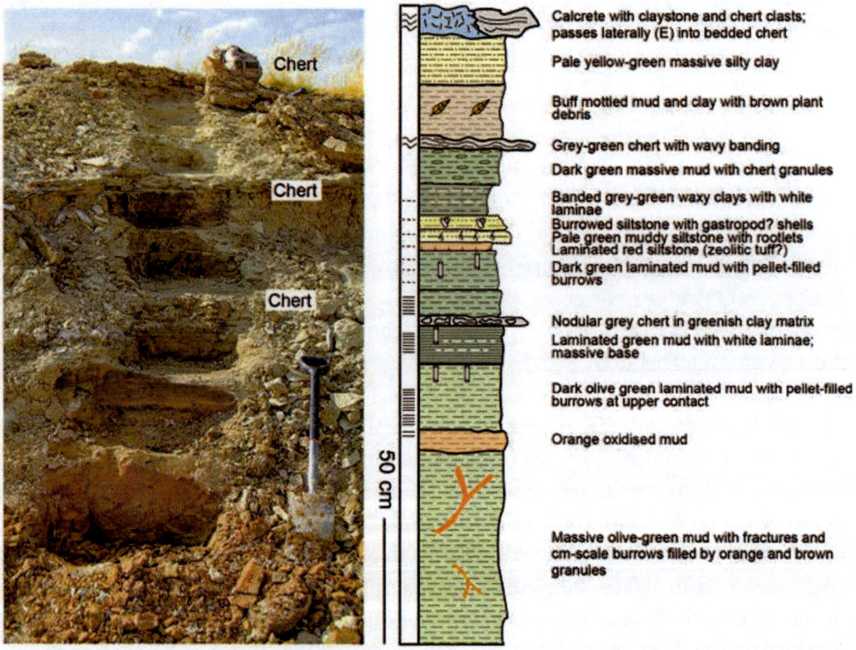

Fig. 20.15 Excavated section through Green Beds, road-cut southeast peninsula. Location shown in Fig. 20.14a. Modified from Owen et al. (2019)

Baker (1958, 1963) made general observations of the cherts, recording their position in several outcrops. Eugster (1969, 1980) provided detailed descriptions and made a case for their origin from the sodium silicate precursors, magadiite and kenyaite. Röhricht (1998) and Behr and Röhricht (2000) recognised nine types of Quaternary chert in the Magadi basin, based on morphology, petrography and age. Two types (1 and 2) are present in the Oloronga Beds (Sect. 20.3; Figs. 20.5d and 20.6e, f). Types 3 to 5 belong to the Green Beds, with the remainder (6–9) in the High Magadi Beds (HMB: Sect. 12.8.2). They proposed that most Green Beds cherts did not have a sodium silicate precursor and that many are microbialites.

Types 3 and 4 cherts, termed massive cherts by Behr and Röhricht (2000), are separated by internal structure, petrography and organic remains. The oldest type (Type 3) consists of stromatolites and bioherms, including thrombolites. The stromatolites include pinnacle and blister mats, oncoids and subspherical laminated bioherms up to 25 cm in diameter (Fig. 20.16). They contain fossil coccoid cyanobacteria, that they tentatively identified on morphology and size as *Pleurocapsa* and *Gloecocapsa*. Filamentous forms are rarely preserved. The bacterial fossils were interpreted to have been originally calcareous, implying relatively dilute lake water, but were replaced by a dense silica phase. The second generation (Type 4) consists of microcrystalline fibrous quartzine and spherulitic chalcedony derived from a silica sol or gel. Type 4 chert commonly cements Type 3 chert.

The youngest generation (Type 5), which extended across much of the axial lake basin, is composed of bedded or platy chert ('plattenchert'of Röhricht 1998). It is most common along the south-central margins of Lake Magadi where angular broken plates of chert litter much of the land surface, locally forming semi-continuous pavements several metres in extent and 1–5 cm thick (Fig. 20.17a, b). Bedded chert is most common on the two (SW, SE) southern peninsulas and up to ~2.5 km south of the southern shoreline. It is also present in the southern part of the Northeast Sub-basin (Fig. 12.3). In outcrop, the chert is cream and red on its weathered upper surface but grey to pale green in section. Bedded chert northeast of the lake is distinctly greener than that to the south, perhaps explaining the name.

The exposed chert surface is flat to gently undulating, but in places preserves current- and wave-ripple bedforms (Fig. 20.17c, d). In cross-section, the chert is banded and wavy on a centimetre scale with diffuse alternating lighter and darker layers, some of which resemble low-amplitude stromatolites or wave ripples. Some layers contain calcified coccoid bacterial cells (Behr and Röhricht 2000). The diffuse chert layers are commonly interbedded with pale green siltstone, which is commonly erionitic and also silicified.

The bedded cherts show many sedimentary structures including wave and current ripples, desiccation cracks on several scales, syneresis cracks, raindrop-impact prints (?), fluid-escape structures and ichnofossils. Common ichnofossils are vertical and subvertical burrows (*Skolithos*), probably produced by tiger beetles, and small-diameter burrows that might have been produced by brine-fly larvae (Röhricht

Fig. 20.16 (**a**) Domal chert mounds (type 4) oriented N-S, Eastern sub-basin, east of Magadi town. Outer chert layers are commonly parallel to external morphology; interiors are brecciated. Dark linear feature below crest of the fault scarp is Magadi-Mombasa railway track. High Magadi Beds (buff) form +13 m terrace at base of the scarp. (**b**) Small chert mound near photograph (**a**) showing external laminae parallel to outer surface. (**c**) Large composite fault-ridge mound, SE peninsula (see Fig. 20.14a for location). (**d**) Stromatolitic mounds adjacent to composite fault-ridge mound in (**c**). Laminae in the laterally linked hemispheroidal mounds are fractured implying early lithification. (**e–i**) Variations in stromatolitic laminae in chert mounds from the Eastern Sub-basin. (**e**) Fault-ridge mound shown in (**c**). Some laminae are selectively replaced by opaque quartz. (**f, g**) Details of lamination. Gas pockets below are commonly filled by white cement. Doublets and triplets of laminae are evident. (**h**) Details of laminae showing internal discontinuities (arrow). (**i**) Preservation of fine continuous laminae

Fig. 20.17 (**a**) Pavement of platy chert, southern Southeast Peninsula. View to SW. (**b**) Fractured surface of platy chert. (**c**) Platy chert in section. Most original detail is masked by recrystallization. (**d**) Bedding plane showing wave ripple, burrows. and crystal pseudomorphs. Absence of mudcracks supports very shallow saline water. (**e**) Bedding plane with large trona? pseudomorphs. (**f**) Crystal pseudomorphs on bedding plane. Lens cap: 50 mm. (**g**, **h**) Wrinkled silicified microbial mats

1998; Scott 2010; Scott et al. 2021). Rare fishbones and peloids are present. Evaporite pseudomorphs (up to 10 cm) are common and have been attributed to trona, gaylussite and calcite (Fig. 20.17e, f; Eugster 1969; Behr and Röhricht 2000; Behr 2002; Brenna 2016). A range of structures attributed to shrinkage and expansion of microbial mats is present (Behr 2002), particularly in the thicker units (Fig. 20.17g, h). These include wrinkles, tepees and other common forms (cf. Schieber et al. 2007).

The Type 5 cherts are preserved essentially *in situ*. The older cherts (Types 3, 4), however, are exposed mainly in intrusive masses penetrating the overlying bedded cherts, and along (mainly) N-S faults and fractures. Behr and Röhricht (2000) mapped their distribution but it is unclear if any formed *in situ* and are undeformed. The older cherts show abundant evidence of deformation (fractures, brecciation, folding), but retain some of the presumed original structure and fabrics (microbialites, lamination) on cm–m scales. They noted that deformation followed new and rejuvenated N-S faulting and movements along E-W structures. The injected older cherts were not fully lithified when mobilised. Pseudomorphs of highly soluble evaporite minerals (e.g. trona) imply crystallisation in a gel or non-lithified precursor.

Behr and Röhricht (2000) recognised several types of intrusive chert. Three types dominate: 'pillow cherts', chert dykes, and fault-ridge cherts. Pillow cherts are ovoid and round domal mounds up to 2 m high and up to ~5 m in diameter that occur alone or coalesce to form groups. More than 200 mounds ('diapirs') lie in the East Sub-basin (Fig. 12.3), east of the Magadi Horst. They also are present along the lake margins (e.g. Bird Rock), the southeastern peninsula and elsewhere (Fig. 20.14d). Internally the pillows are typically brecciated masses of stromatolite and mats cemented by silica. The stromatolites commonly show couplets and triplets of laminae (Fig. 20.16f, g), implying seasonal or repetitive patterns of growth. The outermost surface layers, which are typically bedded parallel to the external surface of the mounds, commonly envelop the brecciated interior evenly, similar to bioherms.

Chert dykes, either as individuals or in clusters, are common on both southern peninsulas, and in the area extending ~2 km south of the lake, and in the East and Northeast Sub-basins (Figs. 12.3, 20.14a, c, and 20.18a, b). Eugster (1969) interpreted them as large-scale desiccation polygons through which magadiite was extruded. The dykes form N-S swarms extending up to 5 km, cut locally by younger E-W dykes. Internally the dykes consist of chaotic breccia (Fig. 20.18c), including microbialite clasts up to 25 cm across, cemented by silica. In a few places, massive silica drapes their external walls.

A third group, present along the faulted margins of Magadi horst near the townsite, the southeast peninsula, and the margins of the Southwest Lagoon, are discontinuous linear ridges of grey and red chert >15 m long and 4–5 m thick in outcrop (Fig. 20.18d–f). They are similar internally to the diapirs but more massive and grey. Some contain volcanic rock fragments, large (>10 cm) trona pseudomorphs and oncoids. Laterally-linked hemispheroidal stromatolites are commonly associated, some in apparent growth position (Fig. 20.16d).

Fig. 20.18 (**a**) Chert dyke penetrating bedded (platy) chert, Southeastern Peninsula. (**b**) Chert dyke cutting chert rubble North-east Sub-basin. Arrows mark contact. (**c**) Chert breccia forming interior of dyke. Coin indicates upper surface. (**d**) Fault-ridge chert, roadside outcrop below Magadi townsite. (**e**) Fault-ridge chert, 140 m west of Magadi hospital. (**f**) Large evaporite (trona) pseudomorphs (arrows) preserved in the chert shown in (**e**). They confirm that the matrix in which the lithified chert intraclasts were entrained was soft, gelatinous and saline when injected

Behr and Röhricht (2000) concluded that most of the intrusive cherts developed from flat-topped calcareous bioherms of coccoid cyanobacteria, thinly bedded filamentous microbial mats, stromatolites, bacterial slimes, and diatoms. Those precursors would therefore underlie the platy Type 5 cherts at a shallow, but unknown depth.

20.5.4 Geochemistry of the Green Beds Sediments

The greenish volcanic silts of the Green Beds are relatively enriched in SiO_2, CaO, and MgO compared to the younger HMB, and have lower contents of Na_2O and Al_2O_3 (Röhricht 1998). O'Neil and Hay (1970) studied cherts from Lake Magadi before the Green Beds were defined. Two samples from Core B had extremely high $\delta^{18}O$ values (>43) attributed to precipitation of the precursor in brine undergoing strong evaporation (see also Leet 2022). Kerrich et al. (2002) and Brenna (2016) reported trace element analyses for the Magadi cherts including the Green Beds. They noted that trace element patterns are similar in all the alkaline lake cherts analysed (including known Magadi-type cherts), but overall concentrations are lower in those of the Green Beds than those in the HMB.

Rock-Eval analyses of Green Beds sediments showed little organic matter (0.08–0.12% Total Organic Carbon), mostly from higher plants and Graminae (J-J. Tiercelin, pers. com. 2013). Reinhardt et al. (2019) examined the diagenesis of organic matter (bitumen, kerogen) in the cherts including samples from the Green Beds. They assessed the impact of hydrothermal processes on the organic matter maturation and molecular markers, and the broader implication of Magadi chert organic matter for Archean cherts.

20.5.5 Depositional Environment of the Green Beds

The Green Beds were laid down in the axial rift depression after a depositional hiatus with semi-arid conditions across much of the basin, commonly recorded by a calcrete that overlies the Oloronga Beds (Eugster 1980; Behr 2002; Felske 2016). During that interval, tectonic movements altered the basin topography and drainage, and the inner graben was faulted to produce several narrow N-S sub-basins. Oloronga Beds were faulted and locally tilted westwards (e.g. as seen 1.4 km north of Nasikie Engida). Subsidence of the graben floor led to compartmentalisation, and modified surface drainage to change a predominant W-E trend to N–S trend (Owen et al. 2018a, b, 2019). Faulting and fracturing of bedrock accompanying the rift extension allowed deep fluids to rise to the surface, some of which mixed with shallower groundwater and (or) discharged lakewards as hot springs. Those deep hot fluids were rich in aqueous silica, fluoride and magmatic (mantle) CO_2. Hydrothermal silica supplemented aqueous silica produced by chemical weathering of the bedrock. Although unequivocal sedimentological evidence of a hydrothermal contribution to chert formation has not yet been found, an exceptional phase of silica precipitation followed in the highly alkaline water.

The Green Beds sediments have indicators of both dilute water and saline alkaline lake water. Dilute water indicators include molluscs, green algae (?), ichnofossils and calcite. Saline alkaline waters are confirmed by evaporite pseudomorphs and bedded cherts (Type 5). The limited outcrop evidence reveals fine sediments that are commonly bioturbated and arranged in cycles of fine sand, silt and mud capped by chert (Fig. 20.15). The lake probably fluctuated between several metres deep and local subaerial exposure when the cherts and evaporite minerals were forming. Dark laminated muds in cores may record periods of deeper stratified water in the axial parts of the contemporary palaeolake.

The inferred setting for the Types 3 and 4 cherts, preserved in intrusive masses, was a fresh to brackish lake bordered by low-angled sloping margins floored at times by microbial mats and small thrombolitic mounds. Siliciclastic sediment was supplied by ephemeral runoff and wind. The setting for the Green Beds when Type 5 platy chert was forming is clear (Figs. 20.19 and 20.20)—a shallow saline alkaline lake with marginal mudflats partly covered by microbial mats and efflorescent salt crusts, similar to today. The lake periodically was inundated by shallow more dilute water in which muddy sediments were deposited, followed by evaporative concentration and shallowing, forming dm-scale saline-upward cycles. The maximum extent of the Green Beds palaeolakes is unknown.

Fig. 20.19 Modern setting of platy chert formation. **(a, b)** Saline mudflats at the southern end of Bird Rock Lagoon. Arrow in b points to large fault ridge mound shown in Fig. 20.14a. **(c, d)** Platy chert from SE Peninsula. **(d)** (arrow) shows rare preservation of microbial laminae that pass laterally into chert, indicating that most of the platy cherts were probably laminites before diagenesis. **(e)** Modern microbial mats from margin of saline mudflat (compare with Fig. 20.17g, h)

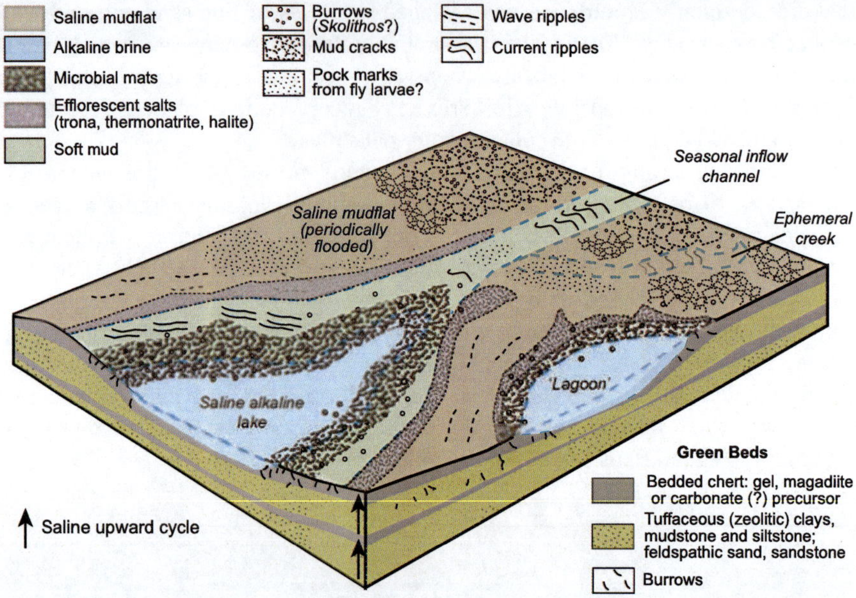

Fig. 20.20 General model for the environment of platy chert formation

20.5.6 *Origin of the Green Beds Cherts*

High silica concentrations in the lake water reflect the silica supplied from chemical weathering of volcanic rocks and ash, the high pH of the alkaline lake water, and silica-rich hydrothermal inflow. Precipitation was caused by a fall in pH, evapoconcentration, and (or) the cooling of thermal spring fluids. Microbes might have played a role, given the abundant microbialites. Behr and Röhricht (2000) proposed a microbially induced decrease in pH, but an influx of geothermal CO_2 might have had a similar effect. The fall in pH lowered silica solubility. Silica then precipitated as a precursor of the cherts (e.g. as opaline silica, a silica gel, or Na-silicate). Eugster (1969, 1980) proposed the formation of the bedded (Type 5) cherts from magadiite precipitated at the chemocline of a stratified lake. Although that is feasible, evapoconcentration is considered a simpler and more likely origin of the cherts (see discussion in Buatois et al. 2020). The Type 5 cherts cap sequences that record rising salinity.

Much of the evidence broadly fits the tectono-sedimentary model proposed by Röhricht (1998). Most silica was emplaced along faults and fractures from below during or following seismic activity. Cherts are rare away from obvious faults. Some silica was already lithified before it was injected while some was soft and probably gelatinous, providing a viscous medium to entrain angular semi-lithified intraclasts. The inferred 'source rocks' for the injected silica (Types 3 and 4) were below the bedded chert (Type 5), which was probably at least semi-lithified during intrusion. The HSPDP cores provide ample evidence for cherts forming within the succession

(Leet et al. 2016) including the common occurrence of present and former silica gels that might have served as a lubricant during the injection.

It is unclear whether the injection of silica occurred subaerially or in a shallow sublacustrine setting, or below the contemporary land surface. Many diapiric mounds resemble bioherms (Fig. 20.16a). The association with N-S faults implies that fluids were being ejected as viscous silica entraining partly lithified clasts. The morphology and size of some mounds resemble those produced at vents of geysers and hot springs at geothermal fields (e.g. Taupo Volcanic Zone in New Zealand, El Tatio in Chile) (Renaut and Jones 2011a). It has been suggested in discussions that the mounds might be sinter and geyserite mounds, which commonly show laminated and bedded structures including stromatolites, fossil microbes, and brecciation. Hot-spring and geyser vents are also commonly aligned along faults. The hydrothermal silica is usually non-crystalline opal-A ($SiO_2.nH_2O$) that alters to quartz during diagenesis (Renaut and Jones 2011b). The domal mounds at Magadi, however, lack clear evidence of subaerial sinter discharge aprons including terraces (meso- and macroscale) and vent depressions. Sublacustrine hydrothermal spring discharge as seeps cannot be ruled out.

Several questions arise from the Behr-Röhricht model. For example, some stromatolites appear to be in growth position (Fig. 20.16d). It is unclear how they preserved this morphology after injection along a relatively narrow fault plane. Many 'pillows' have a chaotic brecciated interior (Type 3) but are externally coated by parallel laminae (Fig. 20.16b) of nearly even thickness (on cm scale). This is presumably the Type 4 chert of Röhricht and Behr (2000). The morphology and external structure (bedding, laminae) of these mounds resemble some subaqueous carbonate bioherms. How did they preserve this form?

As emphasised by Behr (2002), most exposed cherts lack conclusive evidence for a Na-silicate precursor. Many originated from primary siliceous gels (Leet et al. 2021). The evaporite pseudomorphs confirm that the precursor was soft and probably formed in shallow saline water. Modern siliceous gels covered by efflorescent trona form periodically along the northern shoreline of Nasikie Engida (Renaut et al. 2021). Evidence for a Ca-carbonate precursor for some chert is seen locally (Fig. 20.21) in both bedded and laminated chert and in intrusive chert along faulted margins (Fig. 20.21b–d). Rare examples show finely laminated microbial laminae (silica that might have been carbonate originally) replaced laterally by microcrystalline quartz. Oncoids replaced by silica are found in cherts emplaced along fault ramps (Fig. 20.21a) but their original composition is unclear. Plates of Magadi-type chert showing reticulation, locally lie upon older bedded cherts, but are residual deposits from the HMB. Whether original silica or carbonate, the injected sediments underwent early diagenesis to form quartzose chert (Behr 2002; Leet et al. 2021), often by volume-for-volume replacement (Schubel and Simonson 1990).

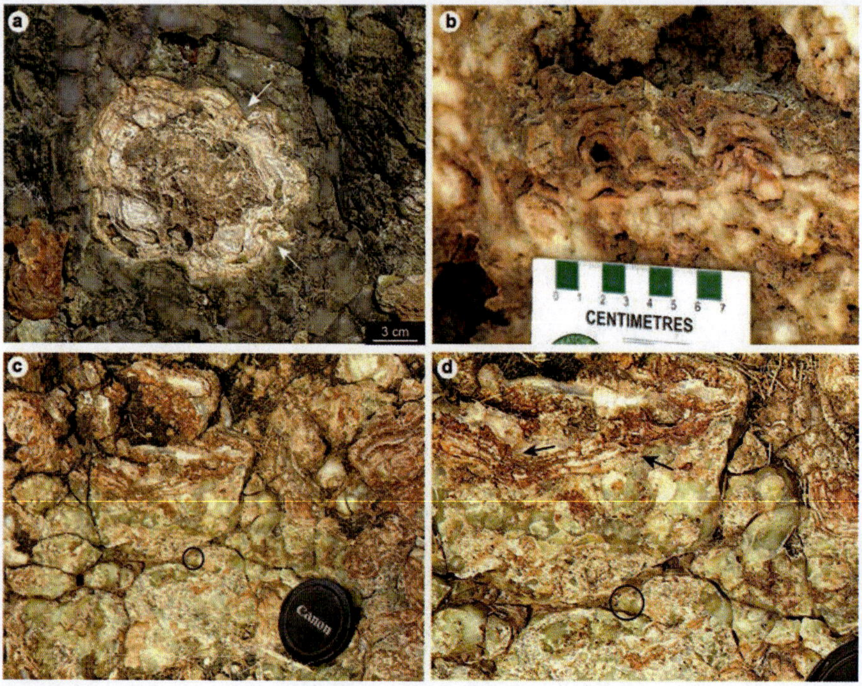

Fig. 20.21 Replacement cherts. (**a**) Large oncoid replaced by silica, massive fault-ridge chert, roadside outcrop 100 m east of Magadi police station. The outer margin of the oncoid shows laminae that appear to have been replaced by massive silica (arrows). (**b**) Microbialites that might have been calcite in massive chert. Outcrop 60 m west of Tata Chemicals administration building, north Magadi townsite. (**c, d**) Stromatolites (originally carbonate?) replaced by silica, fault-ridge chert outcrop on shoreline ~1.3 km southeast of Graham's Lagoon. (**d**) shows replacement of stromatolitic laminae (black circle shows same area in (**c**)). Arrows show sites where pale brown laminae have been replaced by pale green massive silica

References

Anonymous (1923) The soda deposits of Lake Magadi in Kenya. Bull Imp Inst 21:431–444

Baker BH (1958) Geology of the Magadi area: Geol Surv Kenya Rep:42

Baker BH (1963) Geology of the area south of Magadi. Geol Surv Kenya Rep 61

Baker BH (1986) Tectonics and volcanism of the southern Kenya Rift Valley and its influence on rift sedimentation. In: Frostick LE, Renaut RW, Reid I, Tiercelin J-J (eds) Sedimentation in the African rifts. Geol Soc Lond Spec Publ 25:45–57

Baker BH, Mitchell JG (1976) Volcanic stratigraphy and geochronology of the Kedong-Olorgesailie area and the evolution of the South Kenya rift valley. J Geol Soc Lond 132:467–484

Barthelme J (1991) Lenderut: a new Acheulian site in the southern Kenya Rift. Nyame Akuma 35:21–22

Barthelme J, Hunt K, Ngare L, Kipintoi W (2009) Renewed survey and excavations in the Lake Magadi Basin, southern Kenya. Nyame Akuma 71:54–62

Behr HJ (2002) Magadiite and Magadi-chert: a critical analysis of the silica sediments in the Lake Magadi Basin, Kenya. In: Renaut RW, Ashley GM (eds) Sedimentation in continental rifts. SEPM Spec Publ 73:257–273

Behr HJ, Röhricht C (2000) Record of seismotectonic events in siliceous cyanobacterial sediments (Magadi cherts), Lake Magadi, Kenya. Int J Earth Sci 89:268–283

Berggren WA, Kent DV, Swisher CC III, Aubry M-P (1995) A revised Cenozoic geochronology and chronostratigraphy. In: Berggren WA, Kent DV, Aubry M-P, Hardenbol J (eds) Geochronology, time scales, and global stratigraphic correlation. SEPM Spec Publ 54:129–212

Brenna BL (2016) The chemical, physical, and microbial origins of Pleistocene cherts at Lake Magadi, Kenya Rift Valley. MSc thesis, University of Saskatchewan, Saskatoon

Buatois LA, Renaut RW, Owen RB, Behrensmeyer AK, Scott JJ (2020) Animal bioturbation preserved in Pleistocene magadiite at Lake Magadi, Kenya Rift Valley, and its implications for the depositional environment of bedded magadiite. Sci Rep 10:6794

Cohen A, Campisano C, Arrowsmith R, Asrat A, Beck C, Behrensmeyer AK, Deino A, Deocampo D, Feibel C, Foerster V, Kingston J, Lamb H, Lowenstein T, Lupien R, Muiruri V, Olago D, Owen RB, Potts R, Russell SF, Stone J, Trauth M, Yost C (2022) Reconstructing the environmental context of human origins in Eastern Africa through scientific drilling. Annu Rev Earth Planet Sci 50:451–476

Crossley R (1976) Structure, stratigraphy and volcanism in the Nguruman Escarpment area of the western side of the Kenya Rift Valley. PhD thesis, University of Lancaster

Crossley R (1979) The Cenozoic stratigraphy and structure of the western part of the rift valley in southern Kenya. J Geol Soc Lond 136:393–405

Crossley R (1980) Structure and volcanism in the S Kenya Rift. Atti dei Covegni Lincei 47:89–98

Crossley R, Knight R (1981) Volcanism in the western part of the rift valley in southern Kenya. Bull Volcan 44:117–128

Curtis GH (1967) Notes on some Miocene to Pleistocene potassium/argon results. In: Bishop WW, Clark JC (eds) Background to evolution in Africa. University of Chicago Press, Chicago, pp 365–367

Dawson JB (1992) Neogene tectonics and volcanicity in the North Tanzania sector of the Gregory Rift Valley: contrasts with the Kenya sector. Tectonophysics 204:81–92

Dekov VM, Egue NM, Kamenov GD, Bayon G, Lalonde SV, Schmidt M, Liebetrau V, Munnik F, Fouquet Y, Tanimizu M, Awaleh MO, Guirreh I, Le Gall B (2014) Hydrothermal carbonate chimneys from a continental rift (Afar Rift): mineralogy, geochemistry, and mode of formation. Chem Geol 387:87–100

de la Torre I, Mora R, Domínguez-Rodrigo M, de Luque L, Alcalá L (2003) The Oldowan industry of Peninj and its bearing on the reconstruction of the technological skills of Lower Pleistocene hominids. J Hum Evol 44:203–224

de la Torre I, Mora R, Martínez-Moreno J (2008) The early Acheulean in Peninj (Lake Natron, Tanzania). J Anthropol Arch 27:244–264

DeMott L, Scholz CA, Awaleh MO (2021) Lacustrine carbonate towers of Lake Abhe, Djibouti: interplay of hydrological and microbial processes. Sediment Geol 424:10593

Deocampo D, Owen RB, Lowenstein TK, Renaut RW, Rabideaux NM, Billingsley A, Cohen A, Deino AL, Sier MJ, Luo S, Shen CC, Gebregiorgis D, Campisano C, Mbuthia A (2021) Orbital control of Pleistocene euxinia in Lake Magadi, Kenya. Geology 50:42–47

Domínguez-Rodrigo M, de La Torre I, de Luque L, Alcalá L, Mora R, Serrallonga J, Medina V (2002) The ST Site Complex at Peninj, west Lake Natron, Tanzania: implications for early hominid behavioural models. J Archaeol Sci 29:639–665

Domínguez-Rodrigo M, López-Sáez J-A, Vincens A, Alcalá L, Luque L, Serrallonga J (2001) Fossil pollen from the Upper Humbu Formation of Peninj (Tanzania): hominid adaptation to a dry open Plio-Pleistocene savanna environment. J Hum Evol 40:151–157

Domínguez-Rodrigo M, Diez-Martín F, Mabulla A, Baquedano E, Bunn HT, Musiba C (2014) The evolution of hominin behavior during the Oldowan–Acheulean transition: recent evidence from Olduvai Gorge and Peninj (Tanzania). Quat Int 322-323:1–6

Eugster H (1967) Hydrous sodium silicates from Lake Magadi, Kenya: precursors of bedded chert. Science 157:1177–1180

Eugster HP (1969) Inorganic bedded cherts from the Magadi area, Kenya. Contrib Mineral Petrol 22:1–31

Eugster HP (1980) Lake Magadi, Kenya and its precursors. In: Nissenbaum A (ed) Hypersaline brines and evaporites. Elsevier, Amsterdam, pp 195–232

Eugster HP (1986) Lake Magadi, Kenya: a model for rift valley hydrochemistry and sedimentation. In: Frostick LE, Renaut RW, Reid I, Tiercelin J-J (eds) Sedimentation in the African rifts. Geol Soc Lond Spec Publ 25:177–189

Fairhead J, Mitchell J, Williams L (1972) New K/Ar determinations on rift volcanics of S. Kenya and their bearing on age of rift faulting. Nature 238:66–69

Faith JT, Potts R, Plummer TW, Bishop LC, Marean CW, Tryon CA (2012) New perspectives on middle Pleistocene change in the large mammal faunas of East Africa: *Damaliscus hypsodon* sp. nov. (Mammalia, Artiodactyla) from Lainyamok, Kenya. Palaeogeogr Palaeoclimatol Palaeoecol 361-362:84–93

Felske GN (2016) Genesis of calcretes and related carbonate rocks in the southern Kenya Rift. MSc thesis, University of Saskatchewan, Saskatoon

Ferland TM (2015) An evaluation of the organic geochemical potential to reconstruct Mid-Pleistocene palaeoclimate adjacent to an established hominin site: Lake Magadi, Kenya. MSc thesis, University of Pittsburgh

Geraads D, de la Torre I, Denys C (2022) Context and environments of the Lower Pleistocene hominins of Peninj, Tanzania. In: Reynolds SC, Bobe R (eds) African paleoecology and human evolution. Cambridge University Press, Cambridge, pp 384–393

Goetz C, Hillaire-Marcel C (1992) U-series disequilibria in early diagenetic minerals from Lake Magadi sediments: dating potential. Geochim Cosmochim Acta 56:1331–1341

Gregory JW (1921) The Rift Valleys and geology of East Africa. Seeley Service, London

Guest NJ (1953) The geology and petrology of the Engaruka - Oldoinyo Lengai - Lake Natron area of northern Tanganyika Territory. PhD thesis, University of Sheffield

Guo X, Chafetz HS (2012) Large tufa mounds, Searles Lake, California. Sedimentology 59:1509–1535

Guth AL (2008) Evolution of the southern Kenya Rift from Miocene to present with a focus on the Magadi area. MSc thesis, Michigan Technological University, Houghton

Hay RL (1968) Chert and its sodium-silicate precursors in sodium-carbonate lakes of East Africa. Contrib Mineral Petrol 17:255–274

Herrick R (1972) Authigenic minerals in the Pleistocene and recent sediments of Lake Magadi, Kenya. MSc thesis. University of Wyoming, Laramie

Hillaire-Marcel C, Carro O, Casanova J (1986) ^{14}C and Th/U dating of Pleistocene and Holocene stromatolites from East African paleolakes. Quat Res 25:312–329

Howell FC (1972) Pliocene/Pleistocene Hominidae in East Africa: absolute and relative ages. In: Bishop WW, Miller JC (eds) Calibration of hominid evolution. Scottish Academic Press, Edinburgh, pp 331–339

Icole M, Masse J-P, Perinet G, Taieb M (2002) Pleistocene lacustrine stromatolites, composed of calcium carbonate, fluorite, and dolomite, from Lake Natron, Tanzania: depositional and diagenetic processes and their paleoenvironmental significance. Sediment Geol 69:139–155

Icole M, Taieb M, Perinet G, Manega P, Robert C (1987) Minéralogie des sédiments du groupe Peninj (lac Natron, Tanzanie). Reconstitution des paléoenvironnements lacustres. Sci Géol Bull 40:71–82

Isaac GL (1965) The stratigraphy of the Peninj Beds and the provenance of the Natron *Australopithecine* mandible. Quaternaria 7:101–130

Isaac GL (1967) The stratigraphy of the Peninj group – early middle Pleistocene formations west of Lake Natron, Tanzania. In: Bishop WW, Clark JD (eds) Background to evolution in Africa. University of Chicago Press, Chicago, pp 229–257

Isaac GL, Curtis GH (1974) Age of early Acheulian industries from the Peninj Group, Tanzania. Nature 249:624–627

Kaufman A, Margaritz MPM, Hillaire-Marcel C, Hollos G, Boaretto E, Taieb M (1990) The ^{36}Cl ages of the brines in the Magadi-Natron Basin, East Africa. Geochim Cosmochim Acta 54:2827–2833

Kerrich R, Renaut RW, Bonli T (2002) Trace-element composition of cherts from alkaline lakes in the East African Rift: a probe for ancient counterparts. In: Renaut RW, Ashley GM (eds) Sedimentation in continental rifts. SEPM Spec Publ 73:275–294

Kodikara GRL, Woldaib T, van Ruitenbeek FJA, Kuria Z, van der Meer F, Shepherd KD, van Hummel GJ (2012) Hyperspectral remote sensing of evaporate minerals and associated sediments in Lake Magadi area, Kenya. Int J Appl Earth Observ Geoinfo 14:22–32

Le Turdu C, Tiercelin J-J, Richert JP, Rolet J, Xavier JP, Renaut RW, Lezzar KE, Coussement C (1999) Influence of preexisting oblique discontinuities on the geometry and evolution of extensional fault patterns: evidence from the Kenya Rift using SPOT imagery. In: Morley CK (ed) Geoscience of rift systems—evolution of East Africa. AAPG Stud Geol 44:173–191

Leakey LSB, Leakey MD (1964) Recent discoveries of fossil hominids in Tanganyika: at Olduvai and near Lake Natron. Nature 202:5–7

Lee H, Muirhead JD, Fischer TP, Ebinger CJ, Kattenhorn SA, Sharp ZD, Kianji G (2016) Massive and prolonged deep carbon emissions associated with continental rifting. Nat Geosci 9:145–149

Leet K (2022) Origins of chert: Lake Magadi Basin, Kenya. PhD thesis, Binghamton University, New York

Leet K, Lowenstein T, Owen RB, Renaut RW, Deocampo D, Cohen AS, McNulty E, Muiruri V, Rabideaux NM, Billingsley AL, Mbuthia A (2016) Origins of Magadi-type chert: new clues from the HSPDP Lake Magadi drill cores. Geol Soc Am Abstr Progr 48(7):Paper 42–9

Leet K, Lowenstein TK, Renaut RW, Owen RB (2021) Labyrinth patterns in Magadi (Kenya) cherts: evidence for early formation from siliceous gels. Geology 49:1137–1142

Le Gall B, Gama R, Koptev A, Chazot G (2021) The anomalously-propagating South Kenya rift in the context of the North Tanzanian Divergence zone, East Africa. Tectonophysics 824:228968

Luque L, Alcala L, Domínguez-Rodrigo M (2009a) The Peninj Group: tectonics, volcanism, and sedimentary paleoenvironments during the Lower Pleistocene in the Lake Natron Basin (Tanzania). In: Domínguez-Rodrigo M, Alcala L, Luque L (eds) Peninj: a research project on human origins (1995–2005). Oxbow Books, Cambridge, pp 15–48

Luque L, Alcala L, Domínguez-Rodrigo M (2009b) The Peninj Group in type section (Maritanane): an analysis of landscape. In: Domínguez-Rodrigo M, Alcala L, Luque L (eds) Peninj: a research project on human origins (1995–2005). Oxbow Books, Cambridge, pp 49–72

MacIntyre RM, Mitchell JG, Dawson JB (1974) Age of fault movements in Tanzanian sector of East African Rift System. Nature 247:354–356

Manega PC (1993) Geochronology, geochemistry and isotopic study of the Plio-Pleistocene hominid sites and the Ngorongoro Volcanic Highland in northern Tanzania. PhD thesis, University of Colorado, Boulder

Masinde A (2022) Integrated interpretation of sub-basalt structures in the Magadi Basin, an exploration opportunity in the Kenyan Tertiary Rift. Sixth EAGE Eastern Africa Petroleum Geoscience Forum, Cape Town, October 2022, pp. 1–4

Mora R, Domínguez-Rodrigo M, de la Torre I, Luque L, Alcala L (2003) The archaeology of the Peninj "ST Complex" (Lake Natron, Tanzania). In: Martínez-Moreno J, Mora R, de la Torre I (ed) Oldowan: rather more than smashing stones. First hominid technology workshop. Bellaterra, December 2001, vol 9. Treballs d'Arqueologia, Barcelona, pp 77–116

Mturi AA (1987) The archeological sites of Lake Natron (Tanzania). Sci Géol Bull 40:209–215

Muirhead JD, Kattenhorn SA (2018) Activation of preexisting transverse structures in an evolving magmatic rift in East Africa. J Struct Geol 106:1–18

Muirhead JD, Kattenhorn SA, Le Corvec N (2015) Varying styles of magmatic strain accommodation in the East African Rift. Geochem Geophys Geosys 16:2775–2795

Muirhead JD, Kattenhorn SA, Lee H, Mana S, Turrin BD, Fischer TP, Kianji G, Dindi E, Stamps DS (2016) Evolution of upper crustal faulting assisted by magmatic volatile release during early-stage continental rift development in the East African Rift. Geosphere 12:1670 1700

Muiruri VM, Owen RB, de Cort G, McNulty E, Rabideaux N, Lowenstein T, Renaut RW, Leet K, Deocampo D, Cohen AS, Campisano CJ, Billingsley A, Mbuthia A (2021a) Middle Pleistocene to recent diatom floras and stratigraphy of the Magadi Basin, southern Kenya Rift. J Paleolimnol 65:315–333

Muiruri VM, Owen RB, Lowenstein TK, Renaut RW, Marchant R, Rucina SM, Cohen A, Deino AL, Sier MJ, Luo S, Leet K, Campisano C, Rabideaux N, Deocampo D, Mbuthia A (2021b) A million year vegetation history and palaeoenvironmental record from the Lake Magadi Basin, Kenya Rift Valley. Palaeogeogr Palaeoecol Palaeoclimatol 567:110247

Muiruri V, Owen RB, Potts R, Deino AL, Behrensmeyer AK, Dommain R, Reidl S, Rabideaux N, Beverly EJ, Renaut RW, Moerman JW, Deocampo D, Tyler Faith J, Noren A, Cohen AS, Shannon KB (2021c) Quaternary diatoms and palaeoenvironments of the Koora Plain, southern Kenya Rift. Quat Sci Rev 267:107106

O'Neil JR, Hay RL (1973) $^{18}O/^{16}O$ ratios in cherts associated with the saline lake deposits of East Africa. Earth Planet Sci Lett 19:257–266

Omondi OP (2010) Gravimetric investigation of geological structures within Magadi trough in the southern Kenya Rift. MSc thesis, University of Nairobi

Onguso BN (2019) Crustal structure evolution in Magadi area, Kenya, using 3D seismic tomography. MSc thesis, University of Nairobi

Owen RB, Muiruri VM, Lowenstein TK, Renaut RW, Rabideaux N, Luo S, Deino AL, Sier MJ, Dupont-Nivet G, McNulty EP, Leet K, Cohen AS, Campisano C, Deocampo D, Shen C-C, Billingsley A, Mbuthia A (2018a) Progressive aridification in East Africa over the last half million years and implications for human evolution. Proc Natl Acad Sci 115:11174–11179

Owen RB, Renaut RW, Lowenstein TK (2018b) Spatial and temporal geochemical variability in lacustrine sedimentation in the East African Rift System: evidence from the Kenya Rift and regional analyses. Sedimentology 65:1697–1730

Owen RB, Renaut RW, Muiruri VM, Rabideaux NM, Lowenstein TK, McNulty EP, Leet K, Deocampo D, Luo S, Deino AL, Cohen A, Sier MJ, Campisano C, Shen C-C, Billingsley A, Mbuthia A, Stockhecke M (2019) Quaternary history of the Lake Magadi Basin, southern Kenya Rift: tectonic and climatic controls. Palaeogeogr Palaeoclimatol Palaeoecol 518:97–118

Parkinson J (1914) The East African trough in the neighbourhood of the soda lakes. Geogr J 44:33–46

Pickering E, Guest NJ, Dawson JB, James TC (1962) Quarter Degree Sheet 39, Tanganyika Geological Survey

Plasman M, Hautot S, Tarits P, Gautier S, Tiberi C, Le Gall B, Mtelela K, Gama R (2019) Lithospheric structure of a transitional magmatic to amagmatic continental rift system—insights from magnetotelluric and local tomography studies in the North Tanzanian Divergence, East African Rift. Geosciences 9:462–430

Potts R, Behrensmeyer AK, Tyler Faith J, Tryon C, Brooks AS, Yellen J, Deino A, Kinyanjui R, Clark JM, Haradon C, Levin N, Meijer HJM, Veatch EG, Owen RB, Renaut RW (2018) Environmental dynamics during the onset of the Middle Stone Age in eastern Africa. Science 360:86–90

Potts R, Deino A (1995) Mid-Pleistocene change in large mammal faunas of East Africa. Quat Res 43:106–113

Potts R, Dommain R, Moerman JW, Behrensmeyer AK, Deino AL, Beverly EJ, Brown ET, Deocampo D, Kinyanjui R, Lupien R, Owen RB, Rabideaux N, Riedl S, Russell JM, Stockhecke M, deMenocal P, Tyler Faith J, Garcin Y, Noren A, Scott JJ, Western D, Bright J, Clark JB, Cohen AS, Heil CW, Keller CB, King J, Levin NE, Shannon KB, Muiruri V, Renaut R, Rucina SM Uno K (2020) Increased ecological resource variability during a critical transition in hominin evolution. Sci Adv 6(43):eabc8975

Potts R, Faith JT (2022) Mammalian fauna of the Olorgesailie Basin and southern Kenya Rift. In: Reynolds SC, Bobe R (eds) African paleoecology and human evolution. Cambridge University Press, Cambridge, pp 376–383

Potts R, Shipman P, Ingall E (1988) Taphonomy, paleoecology, and hominids of Lainyamok, Kenya. J Hum Evol 17:597–614

Rabideaux N (2018) Late Quaternary East African environmental change based on mineralogical and geochemical analysis of outcrop and core material from the southern Kenya Rift. PhD thesis, Georgia State University, Atlanta

Reinhardt M, Goetz W, Duda J-P, Heim C, Reitner J, Thiel V (2019) Organic signatures in Pleistocene cherts from Lake Magadi (Kenya) -- implications for early earth hydrothermal deposits. Biogeosciences 16:2443–2465

Renaut RW, Jones B (2011a) Hydrothermal environments (Terrestrial). In: Reitner J, Thiel V (eds) Encyclopedia of geobiology. Springer, Berlin, pp 467–479

Renaut RW, Jones B (2011b) Sinter. In: Reitner J, Thiel V (eds) Encyclopedia of geobiology. Springer, Berlin, pp 808–813

Renaut RW, Owen RB, Lowenstein TK, de Cort G, McNulty E, Scott JJ, Mbuthia A (2021) The role of hydrothermal fluids in sedimentation in saline alkaline lakes: evidence from Nasikie Engida, Kenya Rift Valley. Sedimentology 68:108–134

Röhricht C (1998) Lithologie und der Chertserien des Magadi Beckens, Lake Magadi, Kenia. Doctoral thesis, Universität Göttingen

Scott JJ (2010) Saline lake ichnology: composition and distribution of Cenozoic traces in thesaline, alkaline lakes of the Kenya Rift Valley and Eocene Green River Formation, U.S.A. PhDthesis, University of Saskatchewan, Saskatoon

Scott JJ, Renaut RW, Buatois LA, Owen RB, McNulty EP, Stockhecke M, Leet K, Lowenstein T, Mángano GM (2021) Modern and ancient animal traces in the extreme environments of Lake Magadi and Nasikie Engida, Kenya Rift Valley. In: Rosen MR, Finkelstein DB, Park-Boush L, Pueyo SP (eds) Limnogeology: Progress, challenges and opportunities: a tribute to Beth Gierlowski-Kordesch. Springer, Cham, pp 19–66

Schieber J, Bose PK, Eriksson PG, Banerjee S, Sarkar S, Altermann W, Catuneanu O (eds.) (2007) Atlas of microbial mat features preserved within the siliciclastic rock record. Atlases in Geoscience 2. Elsevier, Amsterdam

Schubel KA, Simonson BM (1990) Petrography and diagenesis of cherts from Lake Magadi, Kenya. J Sediment Petrol 60:761–776

Shipman P, Potts R, Pickford M (1983) Lainyamok, a new middle Pleistocene hominid site. Nature 306:365–368

Spink JC, Stevens JA (1946) Notes on the Magadi section of the eastern Rift Valley. Geogr J 10:236–241

Stevens JA (1932) Lake Magadi and its alkaline springs. Unpublished report to Magadi Soda Ltd (summarised by Baker 1958 in Geol Surv Kenya Rep 42)

Surdam RC, Eugster HP (1976) Mineral reactions in the sedimentary deposits of the Lake Magadi region, Kenya. Geol Soc Am Bull 87:1739–1752

Thouveny N, Taieb M (1987) Paleomagnetic study of Plio-Pleistocene formations in the Peninj area (west of Lake Natron, Tanzania). Limits of the magnetostratigraphic interpretation. Sci Géol Bull 40:57–70

Uhlig C, Jaeger F (1942) Die ostafrikanische Bruchstufe und die angrenzenden Gebiete zwischen den Seen Magad und Lawa ja Mweri sowie dem Westfluß des Meru. Part II. Wissenschaftliche Veröffentlichungen des deutschen Instituten für Landeskunde N. S. 10

Walter P (1922) Sodium carbonate minerals of the Mogadi lakes, British East Africa. Am Mineral 7:86–88

Chapter 21
Lake Victoria Basin

21.1 Introduction

Gregory (1921) reported that the Nyanza Rift was formed in response to crustal tension, but since his work several alternative explanations have been proposed. These range from suggestions by early researchers that it was an erosional valley with minimal earth movements (Dixey 1945) to tectonic-based hypotheses involving either crustal tension (Gregory 1921; Shackleton 1950) or compression (Bullard 1936; Kent 1942a, b, 1944). Further research led to proposals that the Nyanza Rift constituted a full graben at approximately right angles to the main Kenya Rift (Saggerson 1952; Jennings 1964). More recently, this rift model was revised with the recognition of the presence of normal, antithetic and oblique faulting styles (Mboya 1983). Jones and Lippard (1979) noted that the junction between the N-S Kenya and E-W Nyanza rifts was a monoclinal structure with nephelinite and phonolite volcanoes erupting between 20 and 7 Ma at Timboroa, followed by alkali basalt-trachyte eruptions towards the axial Kenya Rift.

The Nyanza (Kavirondo) Rift is about 150 km long and up to 50 km wide, fading out about 50 km from the Kenya Rift below extensive piles of volcanics (Jones and Lippard 1979). It extends in an east–west to northeast–southwest direction and includes both the Nyanza Valley and Winam Gulf (Fig. 21.1a, b). Major faults have throws of 500–1000 m with the eastern half of the rift forming a full graben and the western portion defined by antithetically fractured monoclines and a series of normal faults (Pickford 1982). Several large central volcanoes are present that range in age back to the Oligocene (King et al. 1972).

Nyanzian and Kavirondian metamorphic rocks (3–2.6 billion years) underlie areas to the west of the Nandi Fault with Mozambique Belt metamorphics (1.25–0.65 billion years) forming bedrock to the east (Fig. 21.2a). Tectonic downwarping started at around 16 Ma resulting in a broad shallow basin in which Miocene sediments accumulated (Figs 21.1b and 21.2b; Pickford 1986a). This downwarped basin was wider than the modern Nyanza Rift, which subsequently started to

R. W. Renaut, R. B. Owen, *The Kenya Rift Lakes: Modern and Ancient*,
Syntheses in Limnogeology, https://doi.org/10.1007/978-3-642-25055-2_21

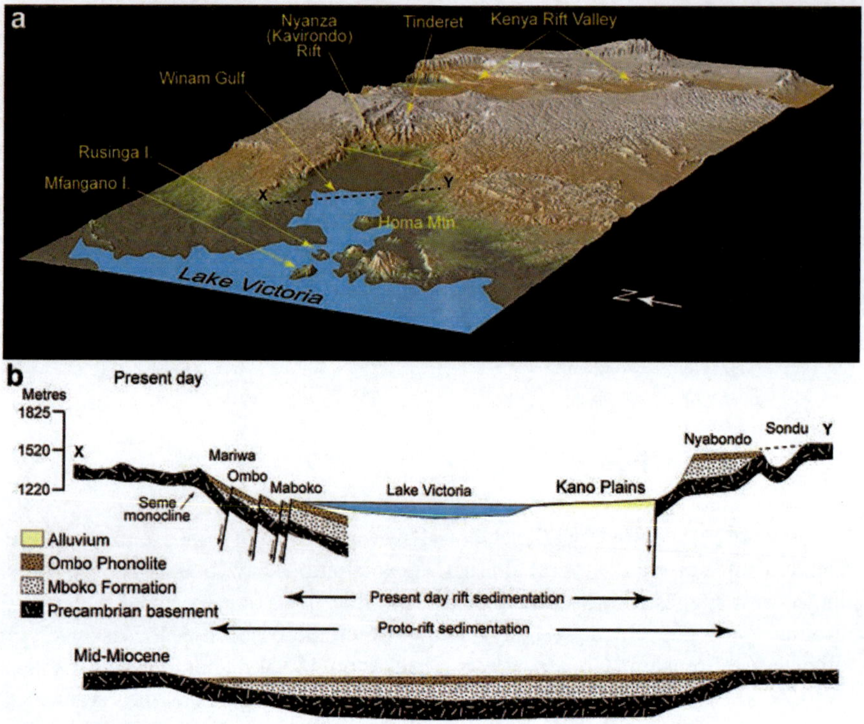

Fig. 21.1 Topography and Miocene warping. (**a**) Topography. Note the contrasting orientations of the Kenya and Nyanza (Kavirondo) rifts. Constructed using GeoMapApp (http://www.geomapapp.org). (**b**) Early downwarping at ~16–17 Ma in the Maboko-Nyakach region affected an area of 60 km (E-W) by 30 km (N-S) with later faulting defining the present rift. Note the smaller lateral extent of the modern rift compared to earlier area of downwarping

accumulate sediments in a series of half grabens that now expose older sediments along the rift margins.

21.2 Miocene Deposits

21.2.1 Miocene Deposits in the Nyanza Rift

Miocene sediments occur in several locations within the Nyanza Rift (Fig. 21.2) (Oswald 1914; Owen 1937; Kent 1941; Bishop and Whyte 1962; Bishop et al. 1969) and have been of particular interest for their rich mammalian and hominoid faunas (Le Gros Clark 1947; Andrews and Van Couvering 1975; Andrews 1981; Shipman et al. 1981; Pickford 1984, 1985; Drake et al. 1988; Walker and Teaford 1988; Retallack et al. 1995, 2002; Wynn and Retallack 2001; Hill et al. 2013; Michel et al. 2014). Early sedimentary and palaeontological research tended to regard most of

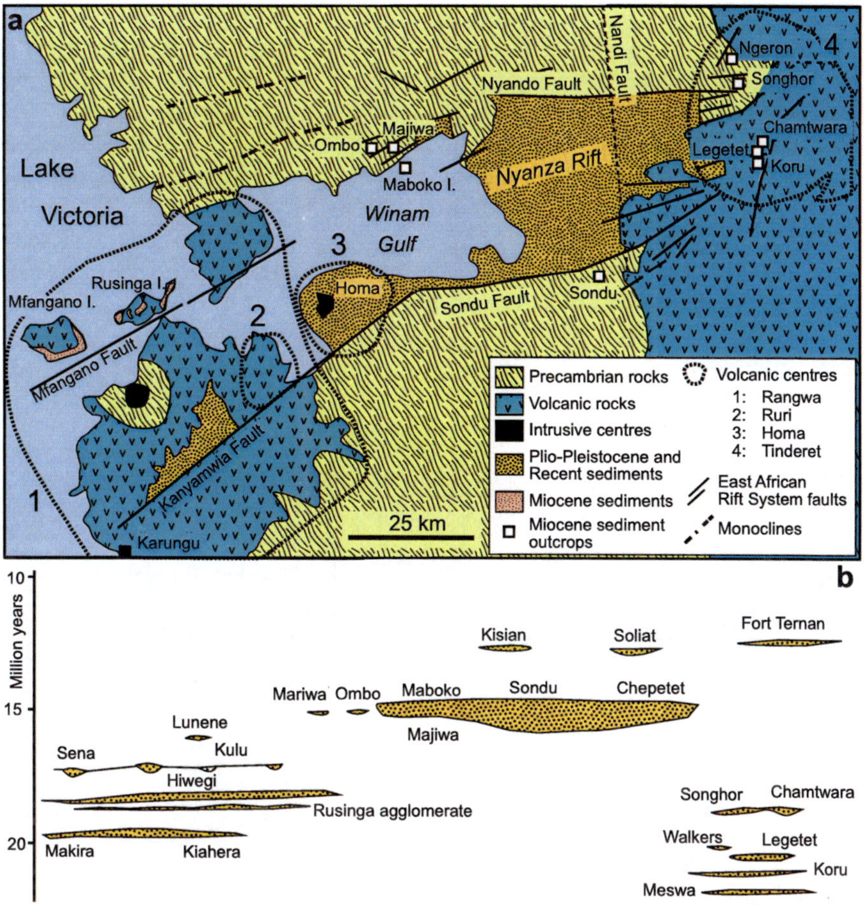

Fig. 21.2 Miocene sediments of the Nyanza Rift. After Pickford (1982). (**a**) Geological setting of the Nyanza Rift and locations of Miocene sediments. (**b**) Stratigraphic and spatial distribution of major Miocene sedimentary units (approximately aligned with locations in map (**a**)). Note greater lateral extent of the older sedimentary units

the layered rocks in these sequences to be lacustrine and consequently many studies identified "lake deposits" (Pickford 1986a). Wayland (1928, 1929a), for example, noted two limestones in the Koru area (Fig. 21.2), one of which he regarded as metamorphic and the other, higher in the sequence and well-bedded, as lacustrine. The former was later reassessed as being a carbonatite tuff (Legetet Carbonatite Formation) and the latter (Walker's Limestone) as a subaerial calcified tuff with a possible paludal influence towards the top of the sequence (Pickford and Andrews 1981; Deans and Roberts 1984; Zaitsev et al. 2013).

Kent (1944) suggested that deposits at Koru, Songhor, Maboko, Ombo, Rusinga, Mfangano and Karungu (Fig. 21.2) formed in a single large Miocene lake that occupied much of the Nyanza Rift (Pickford 1981). Similarly, Shackleton (1950) identified the 'Koru Limestone' at Koru, Songhor and Ngeron and suggested that these deposits also formed in a single large lake (Pickford and Andrews 1981). However, these inferences were later challenged by researchers who had noted that many of the sedimentary sequences were not lacustrine at all and that they were not deposited at the same time (McCall 1958; Binge 1962; Bishop 1964, 1967a, 1968; van Couvering 1972; Andrews et al. 1981; Pickford 1981, 1982, 1983, 1986a; Benefit and McCrossin 1989; Feibel and Brown 1991; Nengo and Rae 1992; Pickford et al. 2006). Nevertheless, some lacustrine sequences are present. The deposits at Karungu, for example, were re-examined by Driese et al. (2016) who reported a thick basal palaeosol (Lukens et al. 2017) and two overlying lacustrine sequences that were separated by fluvial sand, gravel and palaeosols (Fig. 21.3).

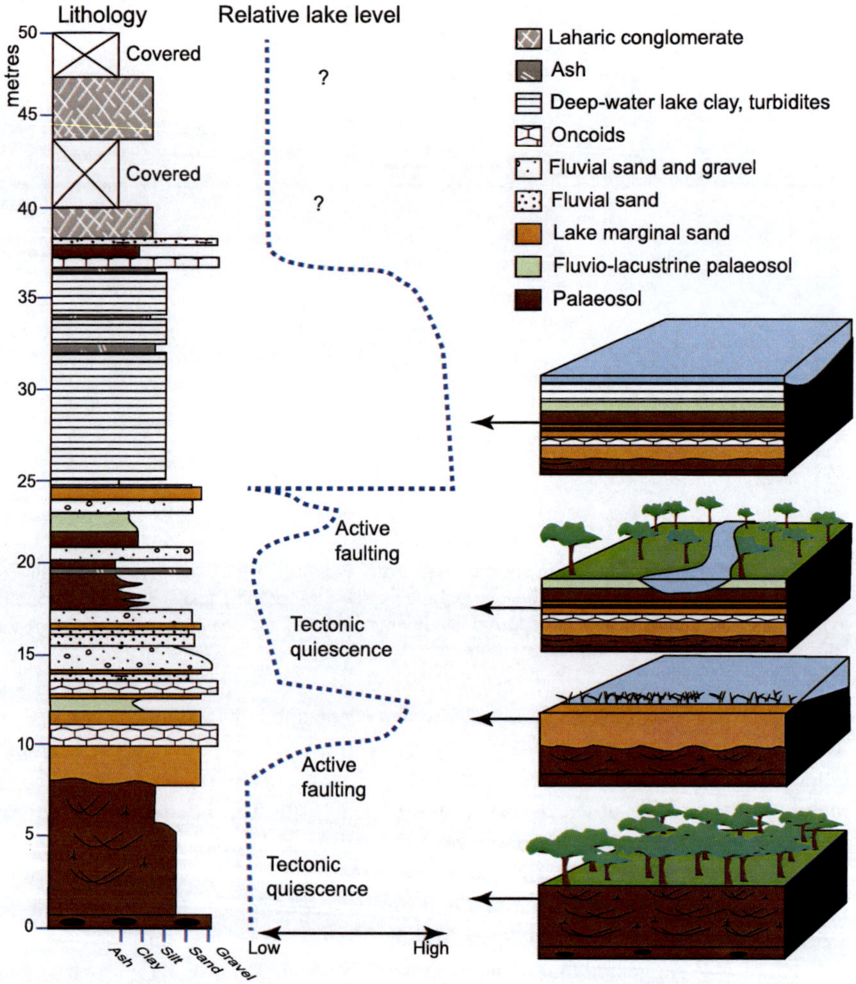

Fig. 21.3 Lacustrine evolution at Karungu. Lithology, lake cyclicity and models. After Driese et al. (2016). Two major lake transgressions are evident. Representative models for specific time-slices shown to the right

Mayr (2014) also reported an avifauna at Maboko that was dominated by aquatic or semi-aquatic birds, suggesting a nearby expanse of water at that location. Arney et al. (2022) subsequently reported 20 fluvio-lacustrine beds from the Maboko Formation with dates of 14.7 (Bed 8) and 13.8 (Bed 20) Ma. Their Bed 3 includes ooids attributed to a low-energy lake or littoral setting with fauna that was buried during transgression events.

Many of the Early Miocene deposits accumulated before faulting of the Nyanza Rift and formed in depressions next to carbonatitic and nephelinitic volcanic centres following volcanic damming and/or isostatic depression rather than through rifting processes (Pickford 1986a). Records of subaerial environments are dominant on the Miocene volcanoes, with lakes and lake margins developing in an 'arena' around their basal margins (Fig. 21.4a). The deposits laid down in these settings include lacustrine shales, silts and sands that interfinger laterally with fluvial volcaniclastics. Whitworth (1953), for example, noted fissile tuffs that may have formed in small ephemeral lakes on floodplains on Rusinga Island. Fossil faunas and floras reflect the varied depositional setting (Fig. 21.4b; Pickford 1981) with lacustrine facies dominated by remains of fish, turtles, crocodiles and other aquatic and amphibious animals. In contrast, Potamidae (freshwater crabs) and molluscs such as *Cleopatra*, *Pila* and the Mutelidae (freshwater mussels) are associated with other subaerial aquatic settings such as large rivers.

21.2.2 Miocene Lacustrine Sediments on Rusinga Island

Rusinga Island includes several outcrops of Miocene sediments (Fig. 21.5), which are divided into the Rusinga (alluvial, lacustrine and pyroclastic deposits with calcareous palaeosols) and Kisingiri (debris flows, nephelinite lavas and tuff breccia) Groups. The sediments accumulated around the periphery of the Early to Mid-Miocene Kisingiri nephelinite-carbonatite central volcano and have yielded a rich mammalian fauna and hominin fossils (Shackleton 1950; Le Gros Clark and Leakey 1951; Whitworth 1953; van Couvering 1969, 1972; Evans et al. 1981; Pickford 1986a; Drake et al. 1988; Bestland and Retallack 1993; Beynon et al. 1998; Peppe et al. 2009; Michel et al. 2014) as well as insects (Wilson and Taylor 1964), terrestrial gastropods (Verdcourt 1972) and plant remains (Chesters 1957; Collinson et al. 2009; Maxbauer et al. 2013; Baumgartner and Peppe 2021).

The Miocene lake sediments were initially described as the 'upper argillaceous series' (Kent 1944) and later as the 'Kulu Series' (and 'Beds') (Shackleton 1950) before being named a formation (van Couvering 1972; Bestland 1991). Their stratigraphic position (Fig. 21.5a) has been controversial with various researchers placing the lacustrine deposits: (1) on top of a thrust fault above other sediments (Kent 1944); (2) above the Hiwegi Formation (Shackleton 1950; Michel et al. 2020); 3) above the Kisingiri Group (van Couvering and Miller 1969); (4) within channels cut into the Hiwegi Formation but below the Kisingiri Group (van Couvering 1972; Drake et al. 1988; Peppe et al. 2009); (5) as lateral facies equivalents of the Hiwegi

a

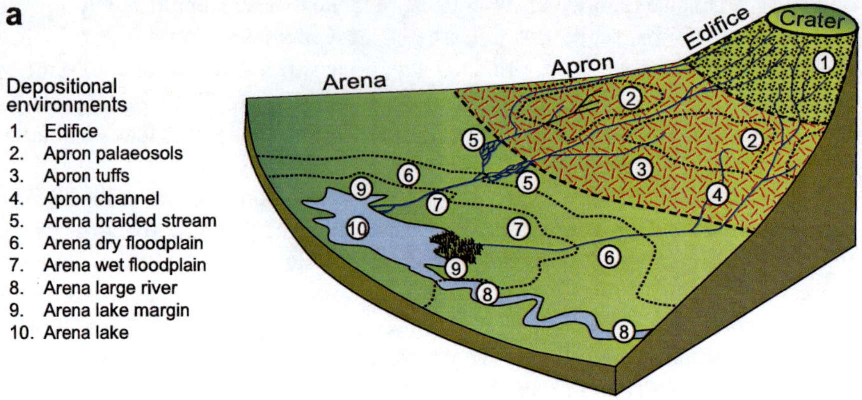

b

	Invertebrate fossils	Braided stream	Dry floodplain	Wet floodplain	Large river	Lake margin	Lake
Aquatic	Mutelidae				⬤		+
	Etheria			+	⬤		
	Lanistes			+	⬤		
	Pila		+	●	⬤	●	
	Potamidae			+	⬤		+
Land	*Cerastua major*	●	⬤	⬤	⬤	●	
	Ligatella	●	●	●	+	+	
	Maizania	●	●	●	+	+	
	Burtoa	●	●	●	+	+	
	Gonaxis protocavalii	+	+	+			
	Marconia			+			
	Krapfiella			+			
	Cerastia miocenica	●	⬤	●			
	Trochozonites		+	+			
	Rachistia		⬤	⬤			
	Achatina			+			
	Halolimnohelix	+	+	+			
	Tayloria	+	+				

	Non-mammalian vertebrates	Braided stream	Dry floodplain	Wet floodplain	Large river	Lake margin	Lake
Aquatic	*Polypterus*			+	⬤		
	Protopterus			+	⬤		
	Gymnarchus				●		
	Cyprinidae						⬤
	Clariidae				⬤		
	Lates				⬤		
	Euthecodon			+	⬤		
	Trionyx		+	⬤	⬤		
	Pelomedusidae		+	⬤	●	●	
	Crocodylus		+	⬤	⬤	⬤	●
Land	Phoenicopteridae	+		●	⬤	⬤	+

Plants

Plants	Braided stream	Dry floodplain	Wet floodplain	Large river	Lake margin	Lake
Tree trunks	+	+				
Wood fragments		+	+			
Leaves		⬤	⬤			
Rootcasts		+	+			
Fruit/seeds	+	⬤	⬤	+		

Trace + Common ● Abundant ⬤

Fig. 21.4 Major depositional environments and associated floras and faunas. Modified from Pickford (1986a). (**a**) Depositional environments around major Miocene volcanic centres. Lacustrine facies are located near the base of the volcano due to damming and (or) isostatic subsidence. (**b**) Diagnostic faunas and floras associated with major depositional 'apron' and 'arena' settings. See Pickford (1986a) for full lists with additional 'apron' and 'edifice' data

Group (Pickford 1984); or (6) as a separate formation equivalent to the Lower Hiwegi Formation (Bestland 1991). The various interpretations imply different ages for the lake sediments. The Bestland interpretation implies an age of about 17.8 Ma whereas that of Peppe et al. (2009) suggests an age of 17 Ma or younger. In contrast,

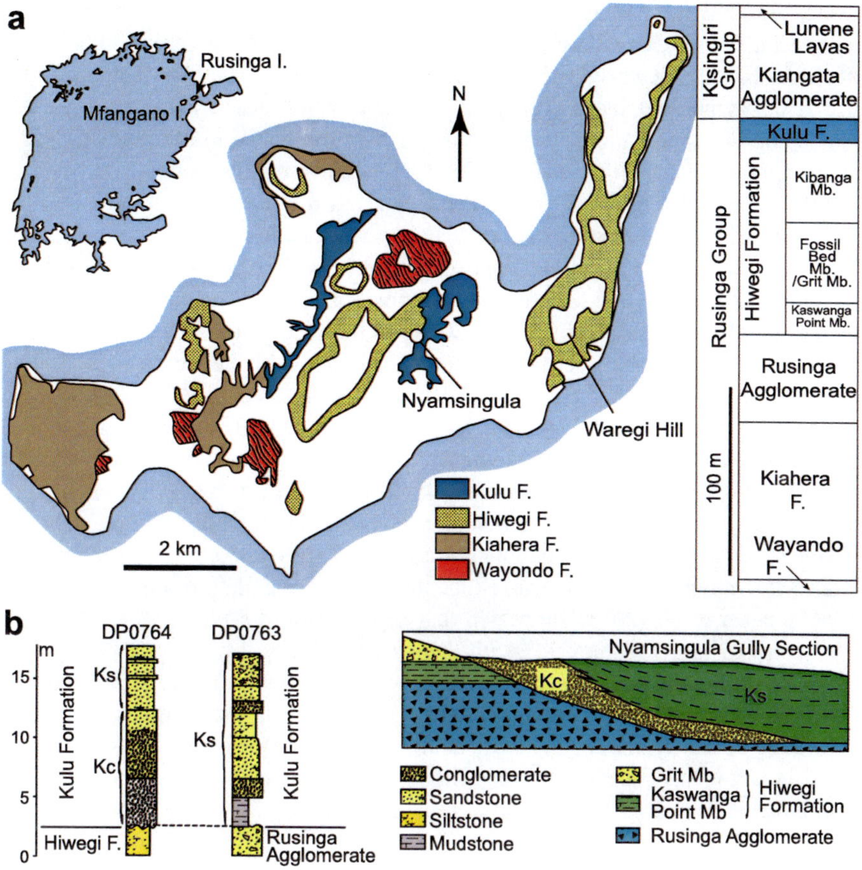

Fig. 21.5 Miocene sediments of Rusinga Island. Modified after Peppe et al. (2009). (**a**) Geological map and general stratigraphy of the Miocene sediments. The lacustrine Kulu Formation is shaded in the stratigraphic column to the right. (**b**) Lithostratigraphic sections and generalised cross-section of the Kulu Formation. Ks = laminated shale, siltstone and sandstone. Kc = matrix supported conglomerate and breccia. Hiwegi Formation subdivisions follow Michel et al. (2020)

palaeontological evidence has been used to infer that the Kulu Formation could be as young as ~15.5–15 Ma. Stratigraphic revisions have complicated earlier attempts to fit depositional patterns in the area into three broad cycles of sedimentation related to the doming and eruption of the Kisingiri stratovolcano (Bestland et al. 1995; Peppe et al. 2009).

The Miocene Kulu Formation (Fig. 21.5a; Peppe et al. 2009) includes both lacustrine and fan-delta facies that were probably laid down in two small, short-lived palaeolakes that developed on either side of an active horst, or perhaps within two arms of a single water body (Bestland 1991; Bestland and Krull 1999). The lake deposits are dominated by laminated shale with cichlid fish fossils first reported by Louis Leakey (Leakey 1931; Greenwood 1951; van Couvering 1982). Fossils of water birds have also been discovered, including the flamingo *Leakeyornis aethiopicus* and the stork *Ciconia minor* (Kent 1944; Dyke and Walker 2008). Other fauna includes tortoises, crocodiles and mammals (Kent 1944). The $\delta^{13}C$ isotopic values for Kulu Formation shales average 25.4‰ but do not discriminate between lake algae and the contemporary C3 land plants that dominated in the drainage basin, although the mean C/N ratios of 4.4 indicate a dominant algal source (Forbes et al. 2004). Palaeosol $\delta^{13}C$ values suggest increased aridity in the basin, although this was not sufficiently severe to cause a major increase in lake salinity that could have decreased algal productivity and led to depletion in the $\delta^{13}C$ values of the sediments.

Gilbert-style fan-deltas of the Kulu Formation are well exposed in gullies at Nyamsingula on Rusinga Island (Fig. 21.5b) with three major lithofacies identified (Bestland 1991). Facies 1 consists of laminated to thinly bedded tuffaceous siltstone (bottomset deposits) that are interbedded with matrix-supported pebbly siltstone with rip-up clasts (debris flows). Facies 2 comprises laterally continuous conglomerate and breccia with reverse grading and weak imbrication. The breccia beds thin distally and represent slumps or debris flows associated with gravity winnowing. Locally, large clasts, up to 5 m in diameter, have been interpreted as colluvial. Facies 3 consists of steeply-dipping pebbly sandstones, matrix-supported cobble conglomerates and fining-upward silty sandstone that formed as delta foresets.

Maxbauer et al. (2013) suggested that rains during the Early Miocene ranged between 400 and 750 mm y^{-1}, with the area dominated by seasonal forests and woodlands that developed near a river. However, they also noted a drier climate when evaporites of the Hiwegi Formation accumulated, with a dry episode also identified at nearby Karungu (Driese et al. 2016). Liutkus-Pierce et al. (2019) suggested that the Miocene forests were later reduced in area due to continental rifting that induced decreased rainfall and increased seasonality.

21.3 Plio-Pleistocene Lakes of the Nyanza Rift

21.3.1 Pliocene to Middle Pleistocene Lacustrine Sediments

Several early researchers set up a complex Quaternary climate stratigraphy for East Africa that related 'pluvial' tropical wet periods to glacial episodes in the northern hemisphere (Leakey 1950, 1960, 1965; Sect. 2.3.2). Five major pluvials were recognised in East Africa: the Kageran (oldest), Kamasian, Gamblian, Makalian and Nakuran (Wayland 1929b; Leakey 1931; Nilsson 1931, 1935, 1938, 1940; Kent 1942a). The Kamasian was subdivided into the lower and upper Kamasian and then redefined as the Kamasian and Kanjeran, respectively (Leakey 1950, 1952, 1955; Cole 1954; Bishop 1967b). The Kanjeran pluvial was named after the Kanjera beds, south of Winam Gulf (Fig. 21.6), which were once thought to have formed in a much-enlarged Lake Victoria. The pluvial concept fell into disuse in the 1960s when

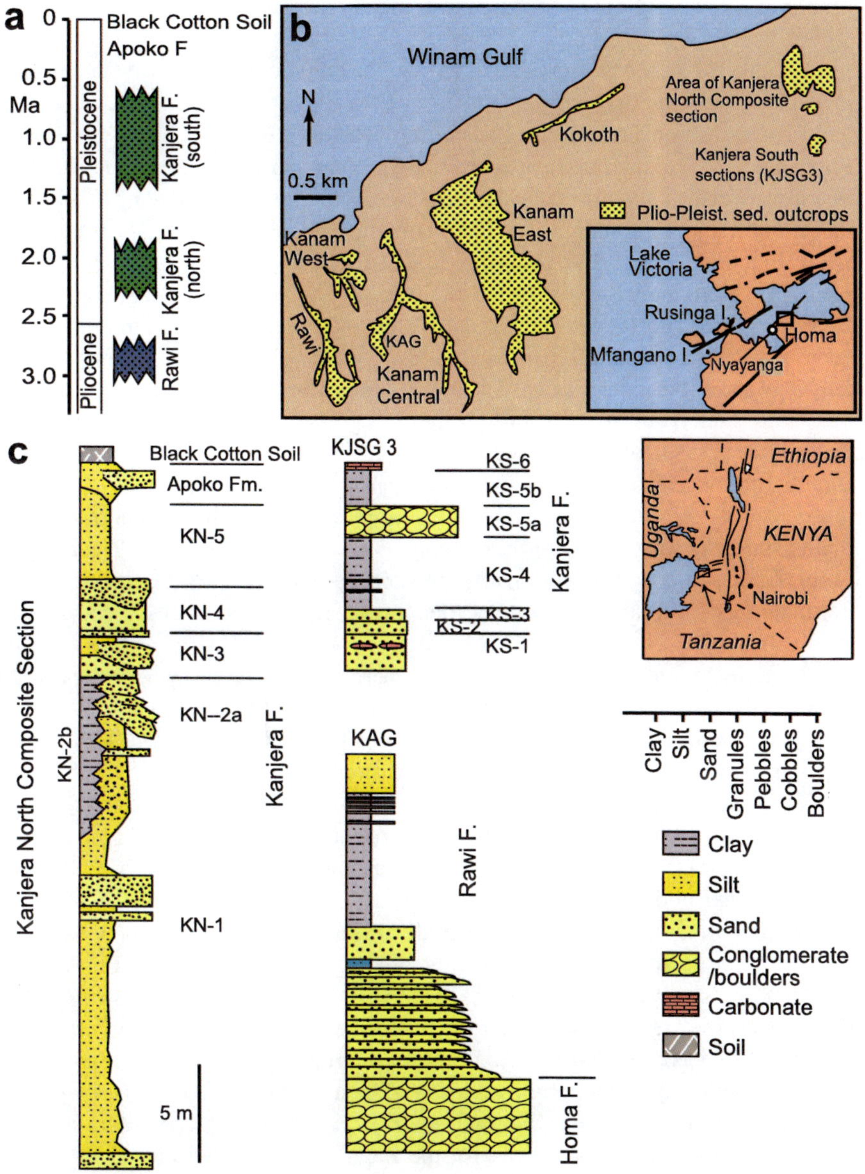

Fig. 21.6 Plio-Pleistocene sediments with lacustrine components from the Homa area, southwestern Nyanza Rift. Modified from Ditchfield et al. (1999). (**a**) Stratigraphic position of the Rawi and Kanjera (north and south) formations. (**b**) Location of Pleistocene sediments in the Homa area. (**c**) Logged sediments in the Kanjera and Rawi formations. Modified Kanjera North Composite Section after Behrensmeyer et al. (1995)

problems arose when applying it in a tectonically active region (Cooke 1957; Flint 1959a, b; Hay 1963).

Oswald (1914), the first researcher to work at Kanjera in the early twentieth century, discovered the first monkey fossils in East Africa. That area includes fluvial and subaerial Miocene and Plio-Pleistocene deposits (Boswell 1935; Kent 1942b; Pilbeam 1974; Plummer and Potts 1989, 1995; Plummer et al. 1999; Vincent 2020) that are overlain unconformably by Pleistocene fluvial sediments and the fluvio-lacustrine Kanjera Formation. The latter contains Oldowan stone tools (Braun et al. 2008; Plummer et al. 2009; Lemorini et al. 2014) and, in northern (KN) outcrops (~1.5–0.5 Ma), consists of five major units numbered KN1–5 (Fig. 21.6c; Behrensmeyer et al. 1995). The basal KN1 deposits were deposited in a transition zone between fluvial and lacustrine settings, possibly part of a prograding alluvial fan. Clay and silt in KN2 may have accumulated offshore in a contemporary palaeo-lake with mudcracks and evaporites suggesting a shallow saline playa, but with waters that occasionally supported fish. The KN3 sediments formed in a swampy lake margin, whereas KN4 deposits represent a mudflat. Homogeneous clay in the lower part of KN5 indicates a stable swamp or lake, with minimal coarse detrital inputs, that gave way to more subaerial conditions in upper KN5 times. The Kanjera Formation is, in turn, unconformably overlain by sediments of the fluvial Apoko Formation (<0.5 Ma) and by late Pleistocene to Holocene black cotton soil (vertisols).

The Kanjera Formation is older in southern (KS) outcrops where deposition began at about 2.3 Ma and terminated around 1.95–1.77 Ma (Ditchfield et al. 1999, 2018). The sediments again indicate nearshore lacustrine facies, but with fluvial deposits being more common. A limestone with plant casts in KS6 may reflect wet conditions near a spring (Fig. 21.6b, c; Behrensmeyer et al. 1995). A saline playa lake lay to the north of Kanjera and occupied an area that is today covered by part of Lake Victoria (Winam Gulf). However, the Kanjera palaeolake was a completely different water body from that of modern Lake Victoria, which dates back only a few hundred thousand years (Johnson et al. 2000).

Lake deposits also occur at Kanam (Kent 1942b) west of Kanjera (Fig. 21.6b). Ditchfield et al. (1999) place the lacustrine (and other) deposits at Kanam in the Rawi Formation, which rests unconformably on sediments of the Homa Formation, dated at ~4–3.5 Ma (Bishop et al. 2022). The latter include lacustrine (stromatolitic) carbonates and clays with correlated lake sediments also reported from Nyayanga, west of the Homa Peninsula (Plummer et al. 2023). Biostratigraphic data imply a potenial age range of 3.36–1.8 Ma for the Rawi Formation (Bishop 1994; White 1995); magnetostratigraphic and other sedimentary evidence favour deposition during the Gauss Chron at ~3.58–2.58 Ma (Ditchfield et al. 1999). The fluvial sandstone and conglomerate that form the lower member of the Rawi Formation are overlain by lacustrine deposits of the upper member comprised of cm-thick couplets of fine sand and silt interbedded with thin tabular fine- to medium-grained sandstone. Laterally-continuous mm-scale laminations are present with fish bone and reworked evaporite (unspecified) pseudomorphs. These sediments were deposited in a shallow palaeolake that periodically dried out and precipitated saline minerals.

Desiccation and increased salinities may also have been responsible for mass fish mortalities represented by well-preserved fish fossils (Kent 1942b; Ditchfield et al. 1999). The sediment couplets likely reflect seasonal climatic cycles with sand sheets laid down from streams that flowed into the palaeolake during increased rainfall. Crocodile fossils and extensive mammalian faunas (hippo, tortoise, pigs, giraffe, antelopes) have also been recovered (Kent 1942b; Ditchfield et al. 1999; Frost et al. 2003). Vertebrate-bearing Middle to Late Pleistocene tufas that formed in springs, pools and streams, have also been reported from Rusinga Island and near Karungu (Beverly et al. 2015a, b) where MSA artefacts are common in terrestrial sediments (Faith et al. 2015).

21.3.2 Middle to Late Pleistocene Origins and Evolution of Lake Victoria

Large Miocene rivers formerly flowed from east to west across the area now occupied by Lake Victoria and discharged into the Atlantic Ocean (Fig. 21.7a; Wayland 1929b, 1930; Groves 1934; Doornkamp and Temple 1966; Ouma 1971; Ebinger 1989; Talbot and Williams 2009). However, by the Pliocene, the uplift of the western branch of the East Africa Rift System (EARS) resulted in interception of these watercourses, with later Middle Pleistocene flow-reversal resulting in ponding of water in an area that is now partially occupied by Lake Victoria (Fig. 21.7c; Talbot and Williams 2009; Van Damme and Van Bocxlaer 2009; Salzburger et al. 2014). Flow reversal, for example, has been documented at Nsongesi in southwest Uganda (Fig. 21.7d), where there is a change from fluvial sediments laid down in a west-flowing Kagera River to sequences dominated by Middle Pleistocene lacustrine sand (Bishop and Trendall 1966; Bishop, 1969; Taylor and Howard 1998). The changes in the drainage eventually resulted in flooding and expansion of Lake Victoria in a broad tectonic sag between the eastern and western branches of the EARS (Ebinger 1989).

Kent (1944) examined sediments in the Nyanza (Kavirondo) Rift and believed that Lake Victoria was up to 1.6 Ma old, although later studies challenged his interpretations (e.g. Pickford 1982, 1986b; Behrensmeyer et al. 1995). Sediments west of Lake Victoria were used to infer that Lake Victoria was significantly younger and probably <0.8 Ma in age (Doornkamp and Temple 1966). More recently, Blegen et al. (2021) used tephrostratigraphy to extend the eastern Lake Victoria Basin chronologies to the rest of the Nyanza Rift as a basis for understanding Middle to Late Pleistocene human origins.

Uplift and tilting gradually resulted in a broad 50-km eastward shift in the Lake Victoria depocentre from its original location and also the exposure of Middle to Late Pleistocene lake sediments up to 100 km to the west of modern Lake Victoria (Bishop and Posnansky 1960; Temple and Doornkamp 1970). Drowned valleys to the north, east and south of the lake provide additional evidence for Pleistocene

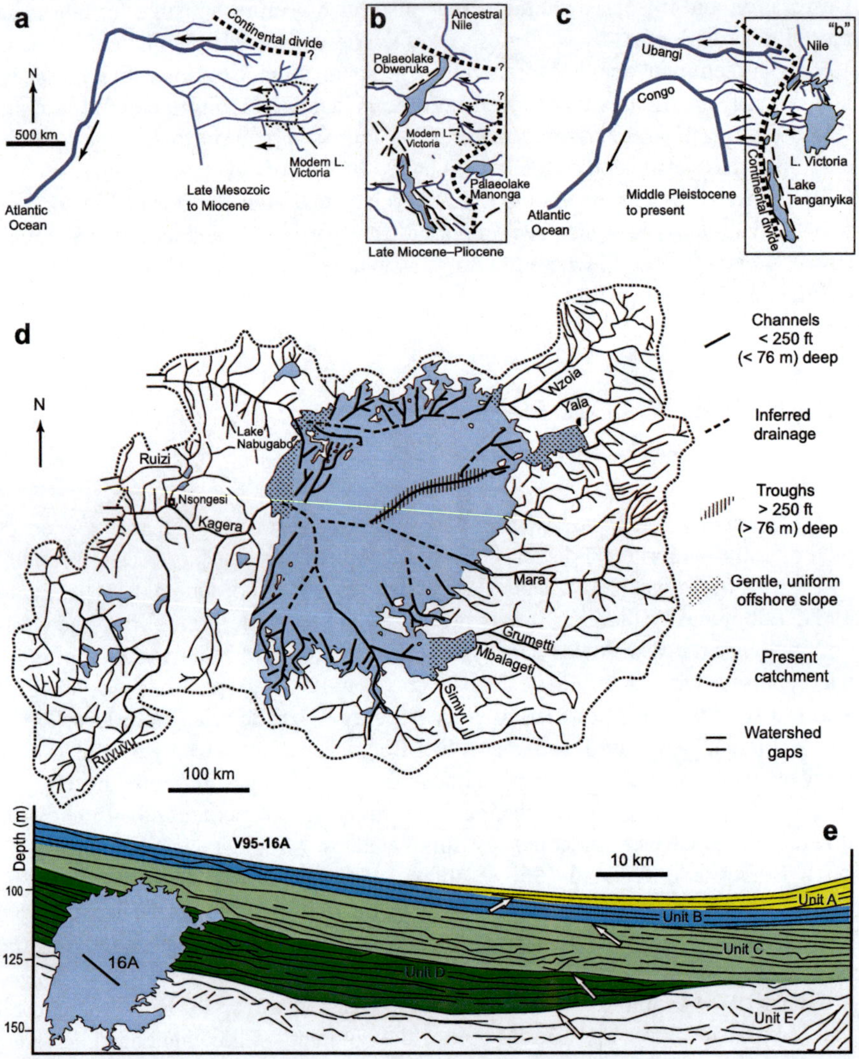

Fig. 21.7 Palaeogeographical evolution of the Victoria Basin and evidence for desiccation. (**a–c**) after Talbot and Williams (2009). (**a**) Late Mesozoic to Miocene with westward flowing rivers. (**b**) Late Miocene–Pliocene times. A large 'Palaeolake Obweruka' intercepts westward-flowing rivers and eventually develops into Lakes Albert and Edward. An east-west continental divide develops to the east and a north-south divide forms in the north. Fish in Palaeolake Manonga have Nilotic affinities. (**c**) Location of the modern drainage divide and palaeolakes. (**d**) Buried river channels below Lake Victoria. Modified from Temple 1966). (**e**) Seismic profile sketches showing five depositional units (A-E) separated by erosional boundaries (arrows). After Johnson et al. (2000)

tilting (Fig. 21.7d; Johnson et al. 1996; Johnson et al. 2000; Stager and Johnson 2008). In addition, two strandlines near Lake Nabugabo, west of Lake Victoria (Fig. 21.7d), lie at heights of about 180 ft (54.9 m) and 80–85 ft (24–26 m) above the pre-1969 average level of Lake Victoria and may reflect the closing phases of regional tilting (Temple 1969). Other high strandlines, at varying altitudes, also reflect the effects of tectonism (Kent 1942b; Bishop 1958; Flint 1959b). Three lower strandlines, at 60 (18.3 m), 45 (13.7 m) and 10 ft (3.05 m), probably formed and were abandoned as the River Nile entrenched its outlet to the modern elevation (Temple 1969).

Seismic reflection data and sediment cores have also been used to reconstruct the Late Pleistocene and Holocene history of Lake Victoria (Fig. 21.7e; Johnson et al. 1996, 2000). The data indicate up to 60 m of sediment that can be divided into at least five discrete depositional sequences, each of which is separated by erosion surfaces that represent major desiccation events. The oldest sequence is thickest in the west with successively younger depositional packages thickening towards the east, supporting other evidence for an eastward shift in the depocentre (Scholz et al. 1998; Johnson et al. 1996, 2000). Estimates of sedimentation rates suggest an age of ~400,000 years (Johnson et al. 1996). However, the erosional hiatuses within the lacustrine sequence could represent a significant span of time and therefore a somewhat older age for the initiation of lake deposition.

Johnson et al. (2000) hypothesised that erosion surfaces in their seismic data represent drier periods that occurred on 100,000-year orbital cycles. Although unproven, data from other East African basins also indicate episodes of desiccation. For example, a megadrought has been reported from basin south of Lake Victoria between about 145 and 60 ka (Fig. 21.8b; Cohen et al. 2007; Scholz et al. 2007, 2011a, b; Stone et al. 2011; Salzburger et al. 2014). Regional aridity during the Last Glacial Maximum (LGM) is evident in several lake basins (Fig. 21.8c; Gasse et al. 1989; Johnson et al. 1996; Beuning et al. 1997; McGlue et al. 2006) and in the Lake Victoria sediments (Johnson et al. 2000; Stager and Johnson 2008), whereas water bodies such as Lake Malawi, to the south, stood at significantly higher levels than during the earlier megadrought phase.

Bishop (1969) suggested that Lake Victoria originally overflowed directly into Lake Albert to the northwest. However, by about 35–25 ka ago, uplift had resulted in an eastward shift of the outlet to its modern northerly location near Jinja and drainage via Lake Kyoga (Figs. 21.7c and 21.8d). The uppermost package of sediments in seismic profiles of Lake Victoria (Unit A; Johnson et al. 1996), of Late Pleistocene to Holocene age, marks a lake expansion and later slight contraction following the very low levels and desiccation of the LGM (Kendall 1969; Johnson et al. 1996; Stager 1984a, b; Stager et al. 2002; Talbot et al. 2000; Talbot and Lærdal 2000; Stager and Johnson 2000, 2008).

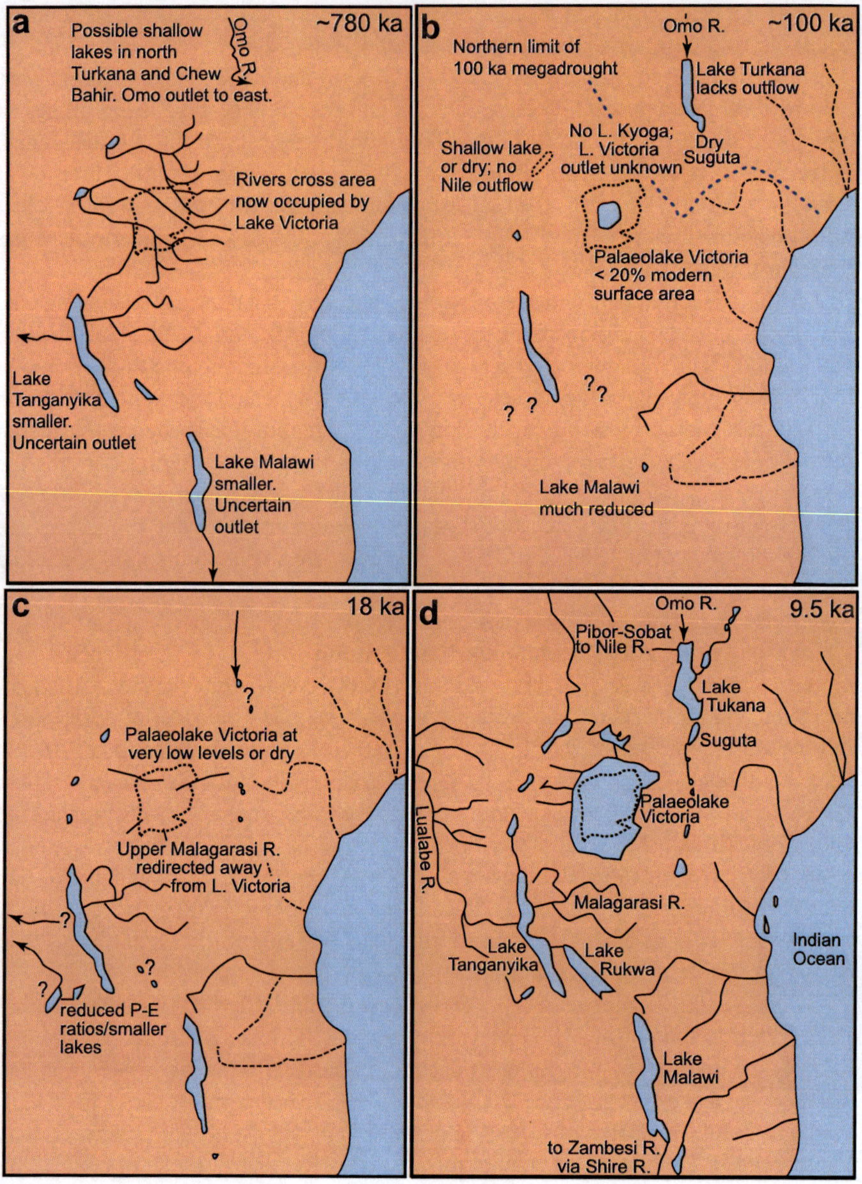

Fig. 21.8 Selected stages in the Pleistocene palaeogeography of lakes in East Africa. Maps centred around Lake Victoria. After Salzburger et al. (2014)

The Wasiriya Beds on Rusinga Island (Fig. 21.1a) were once interpreted as Late Pleistocene raised-beach and lacustrine deposits (Kent 1942b; Van Couvering 1972; Pickford and Thomas 1984; Pickford; 1986b; Van Plantinga 2011). More recent studies of these deposits, and the equivalent or slightly younger sediments (Waware Beds) on Mfangano Island (Fig. 21.1a), have led to their reinterpretation as fluvial and valley-fill deposits with interspersed palaeosols. The Wasirya Beds formed at some time between 33 and 100 ka ago in an arid setting and may have been deposted during a Late Pleistocene low-stand of Lake Victoria (Kendall 1969; Stager et al. 1986; Scholz et al. 1998; Tryon et al. 2010, 2012; Van Plantinga 2011; Blegen et al. 2015, 2017; Garrett et al. 2015). This interpretation is supported by palaeosol evidence (Beverly et al. 2017). Peppe et al. (2011) noted the presence of lake-marginal sediments ~18 m above the modern lake level on Mfangano Island, which they dated at ~38 ka using *Corbicula* shells. Pleistocene tufas indicate former springs, paludal settings and streams that crossed grassy palaeolandscapes (Van Plantinga 2011; Tryon et al. 2010, 2012, 2014; Beverly et al. 2015a, b). Tryon et al. (2012, 2014) and Faith et al. (2012, 2014, 2015) gave details of the Pleistocene palaeontology and Middle Stone Age artifacts in these carbonates. Mant-Melville et al. (2022) outlined the major characteristics of Late Stone Age tools (50–36 ka) that accumulated on grassland palaeolandscapes in the southeastern Nyanza Rift.

21.3.3 Late Quaternary Palaeohydrology and Lake Levels

Beverly et al. (2019) developed hydrological models to account for these major Late Quaternary regressions and transgressions (Fig. 21.9a, b), pointing out that the broad shallow nature of the lake led to large spatial variations for even small depth changes. Their modelling was based on palaeosol-based proxies to derive past precipitation estimates, which were then combined with insolation-based evaluations for evaporation. Their modelling suggested that the lake can switch between modern lake elevations and complete desiccation within a few centuries (Fig. 21.9c), which is rapid enough to explain two previously observed, short-duration, Late Pleistocene, desiccation events between 18 and 14 ka (Fig. 21.9d).

Beverly et al. (2019) also noted that their modelling was compatible with their estimates of palaeoprecipitation and aridity between 94 and 36 ka. They noted that it was unlikely that movements of the Inter Tropical Convergence Zone (ITCZ) could explain this desiccation period since it was unlikely that the ITCZ could have shifted enough to avoid its twice-annual passage over equatorial Lake Victoria. Consequently, they inferred that desiccation may have been related to eccentricity-enhanced precession.

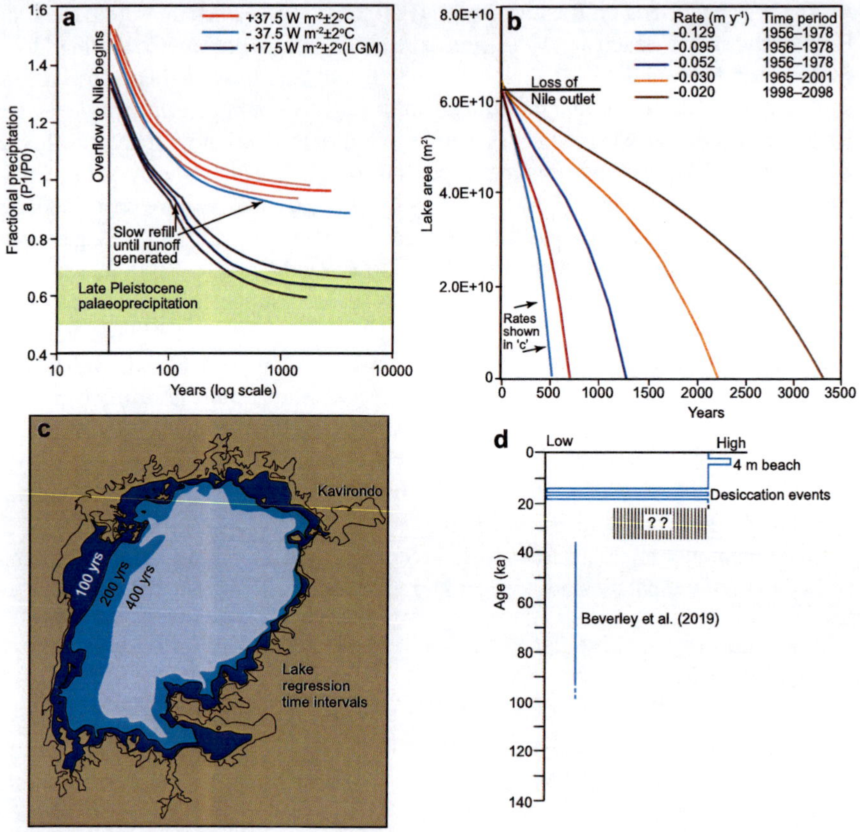

Fig. 21.9 Lake Victoria water-budget models and palaeolake levels. Modified from Beverly et al. (2019). (**a**) Filling rates that indicate how many years are needed to fill the lake. Fractional precipitation = Pleistocene precipitation/modern precipitation. (**b**) Drying rates for modern Lake Victoria. (**c**) Surface area of Lake Victoria for different periods using the fastest regression rate in (**b**). (**d**) Relative lake levels for Lake Victoria. Late Pleistocene data, excluding the African Humid Period, is based on outcrops but lacks the detail available in cored sediments from Lake Victoria

References

Andrews P (1981) A short history of Miocene field palaeontology in western Kenya. J Hum Evol 10:3–9

Andrews P, Meyer GE, Pilbeam DR, Van Couvering JA, Van Couvering JAH (1981) The Miocene fossil beds of Maboko Island, Kenya: geology, age, taphonomy and palaeontology. J Hum Evol 10:35–48

Andrews P, Van Couvering JAH (1975) Palaeoenvironments in the East African Miocene. In: Szalay FS (ed) Approaches to primate paleobiology. Karger, Basel, pp 62–103

Arney I, Benefit BR, McCrossin ML, MacLatchy L, Kingston JD (2022) Herbivore isotopic dietary ecology of the middle Miocene Maboko Formation, Kenya. Palaeogeogr Palaeoclimat Palaeoecol 601:111061

Baumgartner A, Peppe DJ (2021) Paleoenvironmental changes in the Hiwegi Formation (lower Miocene) of Rusinga Island, Lake Victoria, Kenya. Palaeogeogr Palaeoclimat Palaeoecol 574:110458

Behrensmeyer AK, Potts R, Plummer T, Tauxe L, Opdyke N, Jorstad T (1995) The Pleistocene locality of Kanjera, western Kenya: stratigraphy, chronology and paleoenvironments. J Hum Evol 29:247–274

Benefit B, McCrossin M (1989) New primate fossils from the middle Miocene of Maboko Island, Kenya. J Hum Evol 18:493–497

Bestland E (1991) A Miocene Gilbert-type fan-delta from a volcanically influenced lacustrine basin, Rusinga Island, Lake Victoria, Kenya. J Geol Soc Lond 148:1067–1078

Bestland EA, Krull ES (1999) Palaeoenvironments of Early Miocene Kisingiri volcano *Proconsul* sites: evidence from carbon isotopes, palaeosols and hydromagmatic deposits. J Geol Soc Lond 156:965–976

Bestland EA, Retallack GJ (1993) Volcanically influenced calcareous palaeosols from the Miocene Kiahera Formation, Rusinga Island, Kenya. J Geol Soc Lond 150:293–310

Bestland EA, Thackray GD, Retallack GJ (1995) Cycles of doming and eruption of the Miocene Kisingiri volcano, southwest Kenya. J Geol 103:598–607

Beuning KRM, Kelts K, Ito E, Johnson TC (1997) Paleohydrology of Lake Victoria, East Africa, inferred from $^{18}O/^{16}O$ ratios in sediment cellulose. Geology 25:1083–1086

Beverly EJ, Driese SG, Peppe DJ, Johnson CR, Michel LA, Sharp WD (2015a) Recurrent spring-fed rivers in a Late Pleistocene semi-arid grassland: implications for environments of early humans in the Lake Victoria Basin, Kenya. Sedimentology 62:1611–1635

Beverly EJ, Driese SG, Peppe DJ, Arellano LN, Blegen N, Faith JT, Tryon CA (2015b) Reconstruction of a semi-arid late Pleistocene paleocatena from the Lake Victoria region, Kenya. Quat Res 84:368–381

Beverly EJ, Peppe DJ, Driese SG, Blegen N, Faith JT, Tryon CA, Stinchcomb GE (2017) Reconstruction of Late Pleistocene paleoenvironments using bulk geochemistry of paleosols from the Lake Victoria region. Front Earth Sci 5:691–612

Beverly EJ, White JD, Peppe DJ, Faith JT, Blegen N, Tryon CA (2019) Rapid Pleistocene desiccation and the future of Africa's Lake Victoria. Earth Planet Sci Lett 530:115883

Beynon AD, Dean MC, Leakey MG, Reid DJ, Walker A (1998) Comparative dental development and microstructure of *Proconsul* teeth from Rusinga Island, Kenya. J Hum Evol 35:163–209

Binge FW (1962) Geology of the Kericho area. Geol Surv Kenya Rep 50

Bishop LC (1994) Pigs and the ancestors: hominids, suids and environments during the Plio-Pleistocene of East Africa. PhD thesis, Yale University, New Haven

Bishop LC, Plummer T, Braun D, Ditchfield P, Early E, Hertel F, Lemorini C, Oliver JS, Potts R, Vincent T, Whitfield E, Kinyanjui R (2022) Fauna and Paleoenvironments of the Homa Peninsula, Western Kenya. In: Reynolds S, Bobe R (eds) African paleoecology and human evolution. Cambridge University Press, Cambridge, pp 360–375

Bishop WW (1958) A review of the Pleistocene stratigraphy of the Uganda Protectorate. Proc Comm Tech Co-op in Africa meeting, Geology (Leopoldville), pp 91–105

Bishop WW (1964) The later Tertiary and Pleistocene in eastern equatorial Africa. In: Bourlière F, Howell CF (eds) African ecology and human evolution. Methuen, London, pp 246–275

Bishop WW (1967a) The later Tertiary in East Africa—volcanics, sediments and faunal inventory. In: Bishop WW, Clark JD (eds) Background to evolution in Africa. University Chicago Press, Chicago, pp 31–56

Bishop WW (1967b) Annotated lexicon of Quaternary stratigraphical nomenclature in East Africa. In: Bishop WW, Clark JD (eds) Background to evolution in Africa. University of Chicago Press, Chicago, pp 375–395

Bishop WW (1968) The evolution of fossil environments in East Africa. Trans Leics Lit Phil Soc 62:22–44

Bishop WW (1969) Pleistocene stratigraphy in Uganda. Geol Surv Uganda Mem 10:1–128

Bishop WW, Miller JA, Fitch FJ (1969) New potassium-argon age determinations relevant to the Miocene fossil mammal sequences in East Africa. Am J Sci 267:669–699

Bishop WW, Posnanksy M (1960) Pleistocene environments and early man in Uganda. Uganda J 24:44–61

Bishop WW, Trendall AF (1966) Erosion-surfaces, tectonics and volcanic activity in Uganda. Q J Geol Soc Lond 122:385–420

Bishop WW, Whyte F (1962) Tertiary mammalian faunas and sediments in Karamoja and Kavirondo, East Africa. Nature 196:1283–1287

Blegen N, Faith JT, Mant-Melville A, Peppe DJ, Tryon CA (2017) The Middle Stone Age after 50,000 years ago: new evidence from the Late Pleistocene sediments of the eastern Lake Victoria Basin, western Kenya. PaleoAnthropology 2017:1–31

Blegen N, Tryon CA, Faith JT, Peppe DJ, Beverly EJ, Li B, Jacobs Z (2015) Distal tephras of the eastern Lake Victoria basin, equatorial East Africa: correlations, chronology and a context for early modern humans. Quat Sci Rev 122:89–111

Blegen N, Faith JT, Pepe DJ (2021) Tephrostratigraphy of the eastern Lake Victoria Basin including the Nyanza Rift, Kenya: building a stratigraphic and chronological framework for modern human evolution. Quat Sci Rev 256:106823

Boswell PGH (1935) Human remains from Kanam and Kanjera, Kenya Colony. Nature 135:371

Braun DR, Plummer T, Ditchfield P, Ferraro JV, Maina D, Bishop LC, Potts R (2008) Oldowan behavior and raw material transport: perspectives from the Kanjera Formation. J Archaeol Sci 35:2329–2345

Bullard EC (1936) Gravity measurements in East Africa. Phil Trans R Soc Ser A 235:445–531

Chesters KIM (1957) The Miocene flora of Rusinga Island, Lake Victoria, Kenya. Palaeontographica Abt B 101:29–71

Cohen AS, Stone J, Beuning KRM, Park LE, Reinthal PN, Dettman D, Scholz CA, Johnson TC, King JW, Talbot MR, Brown ET, Ivory SJ (2007) Ecological consequences of early Late–Pleistocene megadroughts in tropical Africa. Proc Natl Acad Sci 104:16422–16427

Cole SM (1954) The prehistory of East Africa. Am Anthropol 56:1026–1050

Collinson ME, Andrews P, Bamford MK (2009) Taphonomy of the early Miocene flora, Hiwegi Formation, Rusinga Island, Kenya. J Hum Evol 57:149–162

Cooke HBS (1957) Observations relating to Quaternary environment in East and Southern Africa. Geol Soc S Afr, Alex Du Toit Memorial Lecture 5

Deans T, Roberts B (1984) Carbonatite tuffs and lava clasts of the Tinderet foothills, western Kenya: a study of calcified natrocarbonatites. J Geol Soc Lond 141:563–580

Ditchfield P, Hicks J, Plummer T, Bishop LC, Potts R (1999) Current research on the Late Pliocene and Pleistocene deposits north of Homa Mountain, southwestern Kenya. J Hum Evol 36:123–150

Ditchfield PW, Whitfield E, Vincent T, Plummer T, Braun D, Deino A, Hertel F, Oliver JS, Louys J, Bishop LC (2018) Geochronology and physical context of Oldowan site formation at Kanjera South, Kenya. Geol Mag 156:1190–1200

Dixey F (1945) The relation of the main peneplains of Central Africa to sediments of Lower Miocene Age. Q J Geol Soc Lond 101:243–253

Doornkamp JC, Temple PH (1966) Surface drainage and tectonic instability in part of southern Uganda. Geogr J 132:238–251

Drake RE, Van Couvering JA, Pickford MH, Curtis GH (1988) New chronology for the Early Miocene mammalian faunas of Kisingiri, Western Kenya. J Geol Soc Lond 145:479–491

Driese SG, Peppe DJ, Beverly EJ, DiPietro LM, Arellano LN, Lehmann T (2016) Paleosols and paleoenvironments of the early Miocene deposits near Karungu, Lake Victoria, Kenya. Palaeogeogr Palaeoclimatol Palaeoecol 443:167–182

Dyke GJ, Walker CA (2008) New records of fossil "waterbirds" from the Miocene of Kenya. Am Mus Novitates 3610:1–10

Ebinger CJ (1989) Tectonic development of the western branch of the East African rift system. Geol Soc Am Bull 101:885–903

Evans EM, Van Couvering JA, Andrews P (1981) Palaeoecology of Miocene sites in western Kenya. J Hum Evol 10:99–116

Faith JT, Potts R, Plummer TW, Bishop LC, Marean CW, Tryon CA (2012) New perspectives on middle Pleistocene change in the large mammal faunas of East Africa: *Damaliscus hypsodon* sp. nov. (Mammalia, Artiodactyla) from Lainyamok, Kenya. Palaeogeogr Palaeoclimatol Palaeoecol 361-362:84–93

Faith JT, Tryon CA, Peppe DJ, Beverly EJ, Blegen N (2014) Biogeographic and evolutionary implications of an extinct late Pleistocene impala from the Lake Victoria Basin. J Mamm Evol 21:213–222

Faith JT, Tryon CA, Peppe DJ, Beverly EJ, Blegen N, Blumenthal S, Chritz KL, Driese SG, Patterson D (2015) Paleoenvironmental context of the Middle Stone Age record from Karungu, Lake Victoria Basin, Kenya, and its implications for human and faunal dispersals in East Africa. J Hum Evol 83:28–45

Feibel CS, Brown FH (1991) Age of the primate-bearing deposits on Maboko Island, Kenya. J Hum Evol 21:221–225

Flint RF (1959a) On the basis of Pleistocene correlation in East Africa. Geol Mag 96:265–284

Flint RF (1959b) Pleistocene climates in eastern and southern Africa. Geol Soc Am Bull 70:343–374

Forbes MS, Bestland EA, Krull ES, Dicker DG (2004) Palaeoenvironmental mosaic of *Proconsul* habitats: geochemical and sedimentological interpretation of Kisingiri fossil sites, Western Kenya. J Afr Earth Sci 39:63–79

Frost SR, Plummer T, Bishop LC, Ditchfield P, Ferraro J, Hicks J (2003) Partial cranium of *Cercopithecoides kimeui* Leakey, 1982 from Rawi Gully, southwestern Kenya. Am J Phys Anthropol 122:191–199

Garrett ND, Fox DL, McNulty KP, Faith JT, Peppe DJ, Plantinga AV, Tryon CA (2015) Stable isotope paleoecology of Late Pleistocene Middle Stone Age humans from the Lake Victoria basin, Kenya. J Hum Evol 82:1–14

Gasse F, Lédée V, Massault M, Fontes J-C (1989) Water-level fluctuations of Lake Tanganyika in phase with oceanic changes during the last glaciation and deglaciation. Nature 342:57–59

Greenwood JEG (1951) Fish remains from Miocene deposits of Rusinga Island and Kavirondo Province, Kenya. Ann Mag Nat Hist Ser 12(4):1192–1201

Gregory JW (1921) The Rift valleys and geology of East Africa. Seeley Service, London

Groves AW (1934) The physiography of Uganda: the evolution of the Great Lakes and the Victoria Nile drainage system. J R Afr Soc 33:59–69

Hay RL (1963) Stratigraphy of Beds I through IV, Olduvai Gorge, Tanganyika. Science 139:829–833

Hill A, Nengo IO, Rossie JB (2013) A *Rangwapithecus gordoni* mandible from the early Miocene site of Songhor, Kenya. J Hum Evol 65:490–500

Jennings DJ (1964) Geology of the Kapsabet Plateau area. Geol Surv Kenya Rep 63

Johnson TC, Kelts K, Odada E (2000) The Holocene history of Lake Victoria. Ambio 29:2–11

Johnson TC, Scholz CA, Talbot MR, Kelts K, Ricketts RD, Ngobi G, Beuning K, Ssemanda I, McGill JW (1996) Late Pleistocene desiccation of Lake Victoria and rapid evolution of cichlid fishes. Science 273:1091–1093

Jones WB, Lippard SJ (1979) New age determinations and the geology of the Kenya Rift-Kavirondo Rift junction, W Kenya. J Geol Soc Lond 136:693–704

Kendall RL (1969) An ecological history of the Lake Victoria basin. Ecol Monogr 39:121–176

Kent PE (1941) Miocene deposits in Kenya. Nature 148:169

Kent PE (1942a) Pleistocene climates in Kenya and Abyssinia. Nature 149:736–737

Kent PE (1942b) The country round the Kavirondo Gulf of Victoria Nyanza. Geogr J 100:22–31

Kent PE (1944) The Miocene beds of Kavirondo, Kenya. Q J Geol Soc Lond 100:85–118

King BC, Le Bas MJ, Sutherland DS (1972) The history of the alkaline volcanoes and intrusive complexes of eastern Uganda and western Kenya. J Geol Soc Lond 128:173–205

Leakey LSB (1931) East African Lakes. Geogr J 77:497–508

Leakey LSB (1950) The lower limit of the Pleistocene in Africa. Proc 18 Int Geol Congr, pp 62–65

Leakey LSB (ed) (1952) Proceedings of the Pan-African congress on prehistory, 1947. Philosophical Library, New York
Leakey LSB (1960) Adam's ancestors: the evolution of man and his culture, 4th edn. Harper and Row, New York
Leakey LSB (1955) The climatic sequence of the Pleistocene in Africa. In: Balout L (ed) Actes de la IIe Session du Congrès Panafricain de Préhistoire, Alger, 1952, pp 293–294
Leakey LSB (1965) Olduvai Gorge 1951–1961, vol 1. Cambridge University Press, Cambridge
Le Gros Clark WE (1947) The British-Kenya Miocene expedition, 1947. Nature 160:891–892
Le Gros Clark WE, Leakey LSB (1951) Fossil Mammals of Africa 1: The Miocene Hominoidea of East Africa. British Museum (Natural History), London, pp 1–11
Lemorini C, Plummer TW, Braun DR, Crittenden AN, Ditchfield PW, Bishop LC, Hertel F, Oliver JS, Marlowe FW, Schoeninger MJ, Potts R (2014) Old stones' song: use-wear experiments and analysis of the Oldowan quartz and quartzite assemblage from Kanjera South (Kenya). J Hum Evol 72:10–25
Liutkus-Pierce CM, Takashita-Bynum KK, Beane LA, Edwards CT, Burns OE, Mana S, Hemming S, Grossman A, Wright JD, Kirera FM (2019) Reconstruction of the Early Miocene Critical Zone at Loperot, Southwestern Turkana, Kenya. Front Ecol Evol 7:358–320
Lukens WE, Lehmann T, Peppe DJ, Fox DL, Driese SG, McNulty KP (2017) The Early Miocene Critical Zone at Karungu, western Kenya: an equatorial, open habitat with few primate remains. Front Earth Sci 5:943–922
Mant-Melville A, Blegen N, Tyler Faith JT (2022) Technological diversity in the Late Pleistocene of the Nyanza Rift, Kenya: archaeological excavations at Kapsarok 1 and Anderea's Farm 1. J Archaeol Sci 41:103257
Maxbauer DP, Peppe DJ, Bamford M, McNulty KP, Harcourt-Smith WE, Davis LE (2013) A morphotype catalog and paleoenvironmental interpretations of early Miocene fossil leaves from the Hiwegi Formation, Rusinga Island, Lake Victoria, Kenya. Palaeontol Electronica 16:28A1–19
Mayr G (2014) On the Middle Miocene avifauna of Maboko Island, Kenya. Geobios 47:133–146
Mboya B (1983) The genesis and tectonics of the N.E. Nyanza rift valley, Kenya. J Afr Earth Sci 1:315–320
McCall GJH (1958) Geology of the Gwasi area. Geol Surv Kenya Rep 45
McGlue MM, Scholz CA, Karp T, Lezzar KE, Ongodia B (2006) Facies architecture of flexural margin lowstand delta deposits in Lake Edward, East African Rift: constraints from seismic reflection imaging. J Sediment Res 76:942–958
Michel LA, Lehmann T, McNulty KP, Driese SG, Dunsworth H, Fox DL, Harcourt-Smith WEH, Jenkins K, Peppe DJ (2020) Sedimentological and paleoenvironmental study from Waregi Hill in the Hiwegi Formation (early Miocene) on Rusinga Island, Lake Victoria, Kenya. Sedimentology 67:3567–3594
Michel LA, Peppe DJ, Lutz JA, Driese SG, Dunsworth HM, Harcourt-Smith WEH, Horner WH, Lehmann T, Nightingale S, McNulty KP (2014) Remnants of an ancient forest provide ecological context for early Miocene fossil apes. Nat Commun 5:3236
Nengo IO, Rae TC (1992) New hominoid fossils from the early Miocene site of Songhor, Kenya. J Hum Evol 23:423–429
Nilsson E (1931) Quaternary glaciations and pluvial lakes in British East Africa. Geogr Annlr 13:249–348
Nilsson E (1935) Traces of ancient changes of climate in East Africa. Geogr Annlr 17:1–21
Nilsson E (1938) Pluvial lakes in East Africa. Geol Fören Stockholm Förh 60:423–433
Nilsson E (1940) Ancient changes of climate in British East Africa and Abyssinia. Geogr Annlr 22:1–79
Owen WE (1937) Draft of tentative preliminary report on the July 1937 investigations at Ng'ira Karungu. Unpublished manuscript housed in the archives of the Natural History Museum, London

Ouma JB (1971) Evolution of meander traits in the basin of Lake Victoria. In: Ominde SH (ed) Studies in East African geography and development. University of California Press, Berkeley, pp 29–40
Oswald F (1914) The Miocene Beds of the Victoria Nyanza and the geology of the country between the lake and the Kisii Highlands. Q J Geol Soc Lond 70:128–159
Peppe DJ, Faith JT, Tryon C, Fox DL, Garrett N, Van Plantinga A, Keegan W, Nightingale S, Ogondo J, Zaarur S (2011) Terrestrial evidence for Late Quaternary lake level changes from the Lake Victoria basin, Equatorial Africa. Abstr Progr Geol Soc Am Annu Mtg 43:132–138
Peppe DJ, McNulty KP, Cote SM, Harcourt-Smith WEH, Dunsworth HM, Van Couvering JA (2009) Stratigraphic interpretation of the Kulu Formation (Early Miocene, Rusinga Island, Kenya) and its implications for primate evolution. J Hum Evol 56:447–461
Pickford M (1981) Preliminary Miocene mammalian biostratigraphy for western Kenya. J Hum Evol 10:73–97
Pickford M (1982) The tectonics, volcanics and sediments of the Nyanza Rift Valley, Kenya. Z Geomorphol NF 42:1–33
Pickford M (1983) Sequence and environments of the lower and Middle Miocene hominoids of western Kenya. In: Ciochon RL, Corrucini RS (eds) New interpretations of ape and human ancestry. Plenum, New York, pp 421–439
Pickford M (1984) Kenya Palaeontology Gazetteer, Western Kenya. Nat Mus Kenya Spec Publ 1:1–282
Pickford M (1985) A new look at *Kenyapithecus* based on recent discoveries in Western Kenya. J Hum Evol 14:113–143
Pickford M (1986a) Sedimentation and fossil preservation in the Nyanza Rift system, Kenya. In: Frostick LE, Renaut RW, Reid I, Tiercelin J-J (eds) Sedimentation in the African rifts. Geol Soc Lond Spec Publ 25:345–362
Pickford M (1986b) Cainozoic paleontological sites of Western Kenya. Münch Geowiss Abh 8:1–151
Pickford M, Andrews P (1981) The Tinderet Miocene sequence in Kenya. J Hum Evol 10:11–33
Pickford M, Sawada Y, Tayama R, Matsuda Y-K, Itaya T, Hyodo H, Senut B (2006) Refinement of the age of the Middle Miocene Fort Ternan Beds, Western Kenya, and its implications for Old World biochronology. Compt Rendus Geosci 338:545–555
Pickford M, Thomas H (1984) An aberrant new bovid (Mammalia) in subrecent deposits from Rusinga Island, Kenya. Proc Koninklijke Akad Wetenschappen B87:441–452
Pilbeam D (1974) Hominid-bearing deposits at Kanjera, Nyanza Province, Kenya. Unpublished Report
Plummer T, Bishop LC, Ditchfield P, Hicks J (1999) Research on Late Pliocene Oldowan sites at Kanjera South, Kenya. J Hum Evol 36:151–170
Plummer TW, Ditchfield PW, Bishop LC, Kingston JD, Ferraro JV, Braun DR, Hertal F, Potts R (2009) Oldest evidence of toolmaking hominins in a grassland-dominated ecosystem. PLoS One 4:e7199
Plummer TW, Oliver JS, Finestone EM, Ditchfield PW, Bishop LC, Blumenthal SA, Lemorini C, Caricola I, Bailey SE, Herrie AIR, Parkinson JA, Whitfield E, Hertel F, Kinyanjui RN, Vincent TH, Li Y, Louys J, Frost SR, Braun DR, Reeves JS, Early EDG, Onyango B, Lamela-Lopez R, Forrest FL, He H, Lane TP, Frouin M, Nomade S, Wilson EP, Bartilol SK, Rotich NK, Potts R (2023) Expanded geographic distribution and dietary strategies of the earliest Oldowan hominins and *Paranthropus*. Science 379:561–566
Plummer TW, Potts R (1989) Excavations and new findings at Kanjera, Kenya. J Hum Evol 18:269–276
Plummer TW, Potts R (1995) The hominid fossil sample from Kanjera, Kenya: description, provenance and implications of new and earlier discoveries. Am J Phys Anthropol 96:7–23
Retallack GJ, Bestland EA, Dugas DP (1995) Miocene paleosols and habitats of *Proconsul* on Rusinga Island, Kenya. J Hum Evol 29:53–91

Retallack GJ, Wynn JG, Benefit BR, Mccrossin ML (2002) Paleosols and paleoenvironments of the middle Miocene Maboko formation. J Hum Evol 42:659–703

Saggerson EP (1952) Geology of the Kisumu District. Geol Surv Kenya Rep 21

Salzburger W, Van Bocxlaer B, Cohen AS (2014) Ecology and evolution of the African Great Lakes and their faunas. Annu Rev Ecol Evol Syst 45:519–545

Scholz CA, Cohen AS, Johnson TC (2011a) Southern hemisphere tropical climate over the past 145ka: results of the Lake Malawi Scientific Drilling Project, East Africa. Palaeogeogr Palaeoclimatol Palaeoecol 303:1–2

Scholz CA, Cohen AS, Johnson TC, King J, Talbot MR, Brown ET (2011b) Scientific drilling in the Great Rift Valley: the 2005 Lake Malawi Scientific Drilling Project — an overview of the past 145,000 years of climate variability in Southern Hemisphere East Africa. Palaeogeogr Palaeoclimatol Palaeoecol 303:3–19

Scholz CA, Johnson TC, Cattaneo P, Malinga H, Shana S (1998) Initial results of 1995 IDEAL seismic reflection survey of Lake Victoria, Uganda and Tanzania. In: Lehman JT (ed) Environmental change and response in East African Lakes. Kluwer, Dordrecht, pp 47–57

Scholz CA, Johnson TC, Cohen AS, King JW, Peck J, Overpeck JT, Talbot MR, Brown ET, Kalindekafe L, Amoako PYO, Lyons RP, Shanahan TM, Castañeda IS, Heil CW, Forman SL, McHargue LR, Beuning KR, Gomez J, Pierson J (2007) East African megadroughts between 135–75 kyr ago and bearing on early-modern human origins. Proc Natl Acad Sci 104:16416–16421

Shackleton RM (1950) A contribution to the geology of the Kavirondo Rift Valley. Q J Geol Soc Lond 106:345–392

Shipman P, Walker A, Van Couvering JA, Hooker PJ, Miller JA (1981) The Fort Ternan hominoid site, Kenya: geology, age, taphonomy and paleoecology. J Hum Evol 10:49–72

Stager JC (1984a) Environmental changes in Lake Cheshi, Zambia, since 40,000 years BP with additional information from Lake Victoria, East Africa. PhD thesis, Duke University, Durham

Stager JC (1984b) The diatom record of Lake Victoria (East Africa): the last 17,000 years. In: Mann DG (ed) Proc 7th Int Diatom Symp 1982. Otto Koeltz Science, Philadelphia, pp 455–476

Stager JC, Johnson TC (2000) A 12,400 ^{14}C years offshore diatom record from east central Lake Victoria, East Africa. J Paleolimnol 23:373–383

Stager JC, Johnson TC (2008) The late Pleistocene desiccation of Lake Victoria and the origin of its endemic biota. Hydrobiologia 596:5–16

Stager JC, Mayewski PA, Meeker LD (2002) Cooling cycles, Heinrich event 1, and the desiccation of Lake Victoria. Palaeogeogr Palaeoclimatol Palaeoecol 183:169–178

Stager JC, Reinthal PN, Livingstone DA (1986) A 25,000-year history for Lake Victoria, East Africa, and some comments on its significance for the evolution of cichlid fishes. Freshw Biol 16:15–19

Stone JR, Westover KS, Cohen AS (2011) Late Pleistocene paleohydrography and diatom paleoecology of the central basin of Lake Malawi, Africa. Palaeogeogr Palaeoclimatol Palaeoecol 303:51–70

Talbot MR, Lærdal T (2000) The Late Pleistocene–Holocene palaeolimnology of Lake Victoria, East Africa, based upon elemental and isotopic analyses of sedimentary organic matter. J Paleolimnol 23:141–164

Talbot MR, Williams MAJ (2009) Cenozoic evolution of the Nile Basin. In: Dumont HJ (ed) The Nile: origin, environments, limnology and human use. Springer, Dordrecht, pp 37–60

Talbot MR, Williams MAJ, Adamson DA (2000) Strontium isotopic evidence for late Pleistocene re-establishment of an integrated Nile drainage network. Geology 28:343–346

Taylor RG, Howard K (1998) Post-Palaeozoic evolution of weathered landsurfaces in Uganda by tectonically controlled deep weathering and stripping. Geomorphology 25:173–192

Temple PH (1966) Evidence of changes in the level of Lake Victoria and their significance. PhD Thesis. University of London

Temple PH (1969) Raised strandlines and shoreline evolution in the area of Lake Nabugabo, Masaka District, Uganda. In: Wright HE (ed) Quaternary geology and climate. Proc VII Congr Int Ass Quat Res 16:119–129

Temple PH, Doornkamp JC (1970) Influences controlling lacustrine overlap along the north western margins of Lake Victoria. Z Geomorphol 14:301–317

Tryon CA, Faith JT, Peppe DJ, Fox DL, McNulty KP, Jenkins K, Dunsworth H, Harcourt-Smith W (2010) The Pleistocene archaeology and environments of the Wasiriya Beds, Rusinga Island, Kenya. J Hum Evol 59:657–671

Tryon CA, Peppe DJ, Tyler Faith J, Van Plantinga A, Nightingale S, Ogondo J, Fox DL (2012) Late Pleistocene artefacts and fauna from Rusinga and Mfangano islands, Lake Victoria, Kenya. Azania 47:14–38

Tryon CA, Faith JT, Peppe DJ, Keegan WF, Keegan KN, Jenkins KH, Nightingale S, Patterson D, Van Planting A, Driese S, Johnson CR, Beverly EJ (2014) Sites on the landscape: Paleoenvironmental context of late Pleistocene archaeological sites from the Lake Victoria basin, equatorial East Africa. Quat Int 331:20–30

Van Couvering JA (1969) Miocene stratigraphy and age determinations, Rusinga Island, Kenya. Nature 221:628–632

Van Couvering JA (1972) Geology of Rusinga Island and correlation of the Kenya mid-Tertiary faunas. PhD thesis, University of Cambridge

Van Couvering JA (1982) Fossil cichlid fish of Africa. Spec Pap Palaeontol 29:1–103

Van Couvering JA, Miller JA (1969) Miocene stratigraphy and age determinations, Rusinga Island, Kenya. Nature 221:628–632

Van Damme D, Van Bocxlaer B (2009) Freshwater molluscs of the Nile Basin, past and present. In: Dumont HJ (ed) The Nile: origin, environments, limnology and human use. Springer, Dordrecht, pp 585–629

Van Plantinga A (2011) Geology of the Late Pleistocene artifact-bearing Wasiriya Beds at the Nyamita locality, Rusinga Island, Kenya. MSc thesis, Baylor University, Waco

Verdcourt B (1972) The zoogeography of non-marine Mollusca in East Africa. J Conchol 27:291–348

Vincent TH (2020) A multi-proxy palaeoenvironmental reconstruction of the Homa Peninsula, western Kenya. PhD thesis, Liverpool John Moores University

Walker A, Teaford M (1988) The Kaswanga Primate site: an Early Miocene hominoid site on Rusinga Island, Kenya. J Hum Evol 17:539–544

Wayland EJ (1928) Koru. Annu Rep Geol Surv Uganda for 1927, p 40

Wayland EJ (1929a) Rift valleys and Lake Victoria. C R 15 Int Geol Congr (S Afr 1929) 2:323–353

Wayland EJ (1929b) Notes on the geological history of the Uganda lakes. Geogr J 74:132–134

Wayland EJ (1930) The African Bulge. Geogr J 75:381–383

White TD (1995) African omnivores: global climatic change and Plio-Pleistocene hominids and suids. In: Vrba ES, Denton GH, Partridge TC, Burckle LH (eds) Paleoclimate and evolution with emphasis on human origins. Yale University Press, New Haven, pp 369–384

Whitworth T (1953) A contribution to the geology of Rusinga Island, Kenya. Q J Geol Soc Lond 109:75–92

Wilson EO, Taylor RW (1964) A fossil ant colony: new evidence of social antiquity. Psyche 71:93–103

Wynn JG, Retallack GJ (2001) Paleoenvironmental reconstruction of middle Miocene paleosols bearing *Kenyapithecus* and *Victoriapithecus*, Nyakach Formation, southwestern Kenya. J Hum Evol 40:263–288

Zaitsev AN, Wenzel T, Vennemann T, Markl G (2013) Tinderet volcano, Kenya: an altered natrocarbonatite locality? Mineral Mag 77:213–226

Part IV
Rift Lake Controls, Resources and Environmental Considerations

Chapter 22
Controls on Lacustrine Sedimentation in Rift Settings

22.1 Introduction

The origin of the East African rift lakes and the factors that control their depth, shape, water chemistry and sediments have long been the subject of a debate that has examined the relative roles of tectonism and climate, although other important controlling factors such as pre-rift topography, volcanic activity, bedrock types, weathering, biology and, most recently, human impacts have also been considered (e.g. Tiercelin 1981, 1990; Frostick and Reid 1987, 1989; Frostick 1997; Olago et al. 2009). The published literature on this topic is extensive; we highlight only some of the major factors. Many researchers have focussed on climate as a driver when explaining past and present changes in lacustrine sedimentation (Scholz and Rosendahl 1990; Johnson 1993; Scholz et al. 1998), whereas others have emphasised tectonics and rifting (Tiercelin et al. 1980; Lambiase 1990; Lambiase and Bosworth 1995; Contreras et al. 1997; Lambiase and Morley 1999). The contrasting roles of tectonism and climate operating together are highlighted in many studies of modern and ancient lake sediments (e.g. Olsen 1990; Gierlowski-Kordesch and Kelts 1994; Carroll and Bohacs 1999; Bohacs et al. 2000; Johnson et al. 1987; Bergner et al. 2009; Owen et al. 2019).

Part of the difficulty experienced when assessing tectonic and climatic controls on lake sedimentation in rifts is that they operate at different timescales. Most significant tectonic impacts involve incremental modifications spanning thousands to millions of years. However, some events, such as abrupt fault-controlled changes in drainage and in lakes can operate suddenly over just a few hours or days (e.g. McCall 1967; Ambraseys 1991; Zielke and Strecker 2009). Earthquakes can mix waters in a meromictic lake, or oxygenate dysoxic and anoxic waters, which will have a rapid impact on the contemporary biota, and quickly modify the palaeolimnological record. In other cases, a critical threshold may be exceeded that changes an open lake to a closed lake, or vice versa. Outlet heights (sills) can be physically raised or lowered (usually by tectonics on shorter timescales or erosion on longer

R. W. Renaut, R. B. Owen, *The Kenya Rift Lakes: Modern and Ancient*,
Syntheses in Limnogeology, https://doi.org/10.1007/978-3-642-25055-2_22

timescales), resulting in a lake switching from an open to a closed status, or changing the direction of overflow, which in turn will have a major impact on facies in the basin and (or) the hydrochemistry of the respective lakes.

In contrast, climate generally operates from shorter durations, including daily, seasonal and El Niño scales, to longer Milankovitch cycles of 10^3–10^5 years (e.g. Street and Grove 1976; Tiercelin 1990). Furthermore, interactions between the different Milankovitch cycles result in complex changes in the frequency and amplitude of insolation at the Earth's surface that have in turn led to significant climate variability throughout the Neogene history of the East African Rift. Superimposed on such orbital cycles are secondary controls, including volcanic emissions of CO_2 and non-linear feedback-loops related to variations in albedo, soil moisture or vegetation, which, for example, appear to have been important in causing rapid changes at the start and end of the African Humid Period (AHP) (Shanahan et al. 2015).

In this chapter, we consider some of the important controls on the limnogeology of the Kenya Rift lakes. This brief overview is not comprehensive but highlights some of the factors that have become apparent after studying many of the modern lakes and ancient lacustrine sediments in the Kenya Rift for decades.

22.2 Rifting and Volcano-Tectonic Controls on Kenya Rift Sedimentation

Lakes occupy depressions in the Earth's crust that have multiple origins – from glacial scouring of bedrock to abandoned river-channels to deflation hollows and many other causes (Hutchinson 1975; Wetzel 2001; Cohen 2003; Renaut and Gierlowski-Kordesch 2010). In the Kenya Rift, the main lakes are associated with subsidence linked to faulting or the development of monoclines, damming by central volcanoes or lava flows, and for smaller lakes, flooding of calderas and volcanic craters.

At the largest scale, the Kenya Rift is characterised by a N-S series of segmented subsiding basins, many of which are formed by half grabens that change their polarity (i.e. E-W and *vice versa*) across intervening 'accommodation zones' (Fig. 22.1a, b; Rosendahl et al. 1986, 1992; Dunkelman et al. 1988, 1989). Those half grabens usually comprise a major border-fault with a large (km-scale) downthrow, opposed by a series of smaller grabens, horsts and tilt blocks that control available space, but with the greatest subsidence located near the border faults. Along the rift axis (broadly N-S), the intervening accommodation zones, also called 'transfer zones', form higher ground or oblique ridges, marked by locally intense grid-faulting or by central volcanoes, some of which create lakes by damming axial drainage, while others destroy existing lake basins by filling them with lava or ash. The location of such central volcanoes is controlled by intersections between the N-S Cenozoic Kenya Rift and much older tectonic trends linked to basement lineaments (Le Turdu et al. 1995, 1999).

The importance of volcanism is also reflected in the development of the soda lakes of the Kenya Rift, which owe much of their distinctive chemistry to weathering of alkali volcanic rocks (Eugster 1980, 1986; Deocampo and Renaut 2016; Schagerl and Renaut 2016). Where basins are hydrologically closed, deep enough

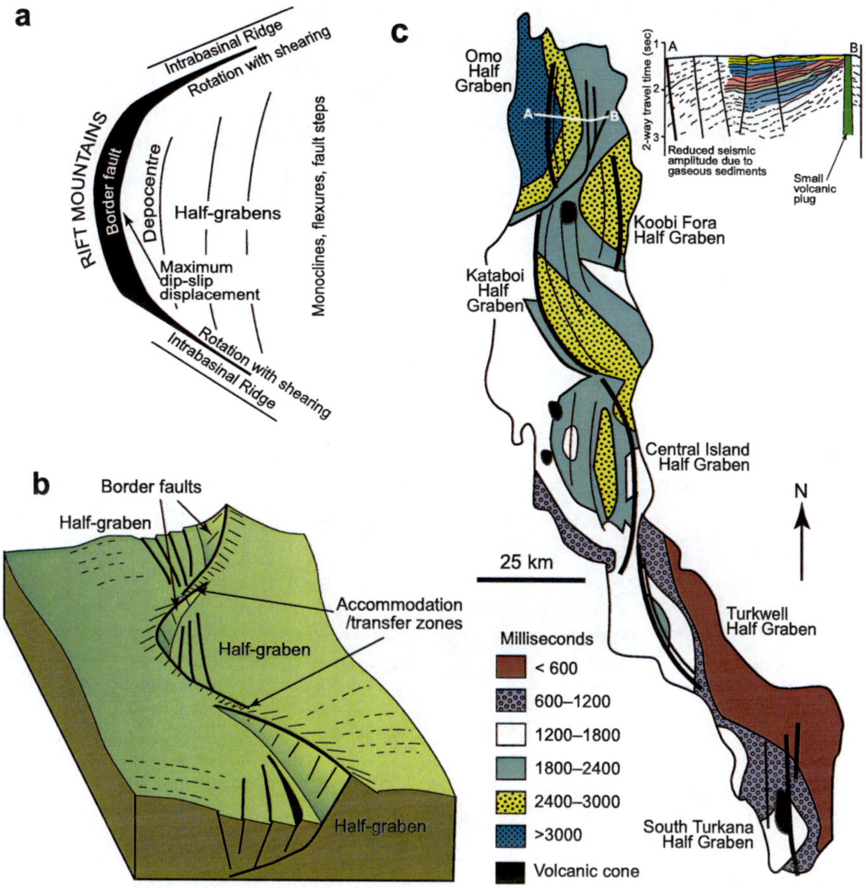

Fig. 22.1 Rift basins and sedimentation. ((**a**, **b**) modified from Rosendahl (1987); (**c**) based on Dunkelman et al. (1989)). (**a**) Schematic border fault and half-graben. (**b**) Linked border faults with opposing polarity show sinusoidal alternating half-grabens. (**c**) Main border faults and depth to basement in Lake Turkana. Note thickening of sediments towards border faults as shown in the A–B cross section

(only metres), and evaporation remains high, the rift lake brines precipitate bedded trona. In contrast, faulting and fracturing of the volcanic bedrock usually increases permeability and, in extensional settings, can increase subsurface drainage, which helps to maintain fresh waters even in lakes lying in topographically closed basins (e.g. Lake Naivasha and Lake Baringo).

As time passes and half-grabens deepen, axial and lateral streams produce a sediment infill with a wedge-shaped cross-section, with the thickest sediment accumulation near the border faults. This is the case at Lake Turkana, which occupies a N-S series of opposing half-grabens, each 25–40 km long (Fig. 22.1c). The most northerly basin, the Omo half-graben, is filled by a typical asymmetrical sedimentary package up to 4 km thick (Dunkelman et al. 1988, 1989). Withjack et al. (2002)

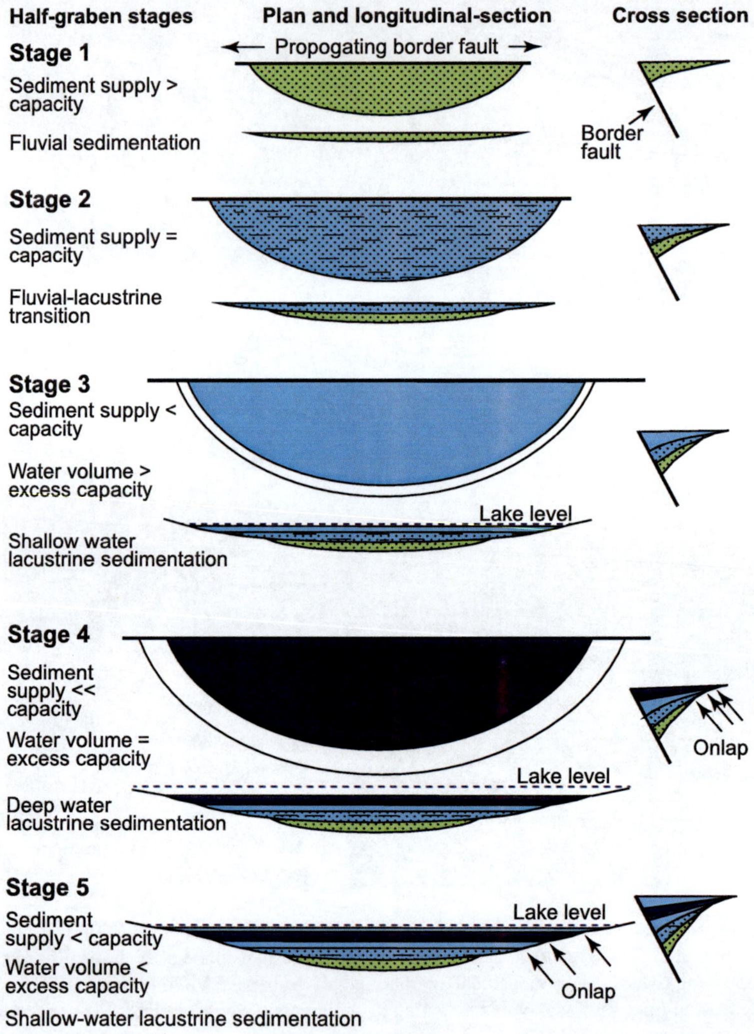

Fig. 22.2 Half-graben filling model in plan, longitudinal and cross-section views. After Withjack et al. (2002). Stages 1 to 5 represent the build-up of a rift basin sequence with conditions during each stage indicated to the left. The interaction of the subsidence rate with sediment and water supply controls whether deposition is fluvial or lacustrine. When water is available and sediment supply exceeds the formation of incremental accommodation space fluvial systems develop. Where sediment supply is less than any incremental increases in accommodation space, and sufficient water is available, a lake will develop. If the water supply exceeds the difference between sediment supply and incremental accommodation space (the excess capacity) then an open lake will form. In contrast, if the excess capacity is greater than the water supply a closed-basin saline lake will develop

noted that in sections parallel to border faults the sediment infill tends to show onlap along the rift axis, reflecting longitudinal growth of border faults with time (Fig. 22.2). In contrast, Morley et al. (1999a, b) suggested that many Kenya Rift basins reached a maximum border-fault length early in their development, resulting in minimal late-stage axial onlap. Individual half-grabens, separated by accommodation-zone highs, tend to develop different stratigraphic records, reflecting contrasts in basin subsidence rates (Rosendahl et al. 1986) with intra-basin highs forming local barriers that further subdivide half-grabens.

Several researchers have discussed infilling of rift basins using earlier ideas (Crossley 1984; Baker 1986) or the later asymmetric border-fault model (e.g. Frostick 1997; Withjack et al. 2002; Zhang et al. 2020). Most have noted that rift morphology (slope angles, relative elevations) and pre-rift factors (pre-rifting lithology and basement structure) strongly influence sedimentation (Owen et al. 2018, 2019). Pre-rift geology and topography, for example, control the location of antecedent drainage, with many modern East African rivers tending to follow Karoo or Cretaceous troughs. Where former fluvial incision equals or exceeds rift flank uplift, rivers deliver large sediment loads to evolving rifts (Frostick and Reid 1989). In contrast, if incision is slower, antecedent rivers may be diverted from the rift or experience flow-reversal, as happened when the shoulders of the western branch of the East African Rift were raised, and the river-flow directions were reversed to form Lake Victoria (Talbot and Williams 2009; Chaps. 13 and 21).

Clastic sediments may be introduced to lakes by several pathways, such as axial drainage along the length of a rift (N-S or S-N in the Kenya Rift), although axial flow may be blocked by topographic highs linked to accommodation zones (Rosendahl et al. 1986; Morley et al. 1990; Lambiase and Bosworth 1995; Wescott et al. 1996). Other paths into rifts include complex river courses that cross monoclines, half-grabens, grabens and tilt blocks on the shoaling side of asymmetric rifts, and short, deeply incised streams that flow down steep border-fault margins, in places involving drainage deflections along marginal fault-ramps and platforms (Scholz 1995). Overall, such areas deliver high sediment loads (Crossley 1984; Scholz et al. 1993). In contrast, uplift along border faults is important in forming short steep catchments that supply smaller volumes of coarse-grained sediments to associated half-grabens (Tiercelin and Vincens 1987; Cohen 1990; Wells et al. 1999; Soreghan et al. 1999).

Basin shape exerts a significant impact on transgressions and regressions, which will in turn impact lacustrine sedimentation. Lakes in steep-sided grabens often respond rapidly to variability in evaporation and precipitation, resulting in major changes in lake level and facies development (Street-Perrott and Harrison 1985; Trauth et al. 2010). Olaka et al. (2010), for example, related Late Pleistocene and Holocene lake-level fluctuations in the Kenya and Ethiopian rifts to an hypsometric integral (HI) and an aridity index (AI), noting that Lake Naivasha (HI = 0.23; AI = 1.20) was a 'sensitive' lake and that the response to lakes in the Suguta Valley (HI = 0.29, AI = 0.43) and the Nakuru-Elmenteita (HI = 0.30, AI = 0.85) basin were the most sensitive in the Kenya Rift to shifts in climate.

Cumulative fault movements also exert important regional controls on climate that, in turn, may affect vegetation and sedimentation (Ring et al. 2018; Hinderer and Einsele 2002; Owen et al. 2018). Tectonic uplift of rift margins, for example, can generate a rain shadow over the adjacent rift floor. Rising rift margins change slope angles and erosion rates, may lead to fluvial incision, and modify the time available for weathering before surface materials are removed (Owen et al. 2018). Changing rift topography has also been linked to major transformations in vegetation. Fer et al. (2017), for example, attributed the fragmentation of a formerly continuous Mio-Pliocene forest across equatorial East Africa to regional uplift. Furthermore, Acosta (2015) noted the impact of dense plant cover in slowing erosion, and that vegetation cover partially reflects the influence of rift subsidence and uplift on regional climate.

22.3 Climate Controls on Kenya Rift Sedimentation

Both tectonic processes and climate interact in diverse ways that control sedimentation. For example, the Kenya Rift is formed on an elevated 'dome' that had a significant impact on regional climates and drainage patterns (Davis and Slack 2002; Burke and Wilkinson 2016). Climate, in turn, is a major control on rift sedimentation. Given the same amount of subsidence, contrasting facies will develop in wetter-versus-drier rift basins. Today, the Kenya Rift is relatively dry because potential evapotranspiration rates exceed rainfall over the N-S rift floor. All the modern lakes lack surface outlets, and most are saline to varying degrees (Beadle 1981; Olaka et al. 2010). Baringo and Naivasha remain exceptions because subsurface outflow, generated by tectonics, inhibits the development of high salinity despite annual evaporation exceeding rainfall across both basins.

Climate variability results in major fluctuations in lake level and depositional regimes over timescales of hundreds to thousands of years but can produce sediment packages that are only 2–10 m thick (Tiercelin and Faure 1978). Many gaps in the sediment record result from non-deposition and erosion, but clastic sediments eroded from the highest parts of lake catchments have few places to go: they either partly fill their regional sumps (e.g. lakes Logipi and Magadi) or they are dispersed unevenly across the landscape in fluvial and alluvial settings. The evidence is often not apparent. Rift subsidence is often implied but not always stated. During the African Humid Period (AHP: ~15–5 ka), for example, most Kenyan lakes reached their outlets and overflowed, but the sediments laid down during that period rarely exceeded 10 m. Clearly local factors, in both tectonics and climate change, control the sediment record. "Broad brush" statements about climate controls that fit expectations based on earlier work are not always correct. Trying to fit new palaeoclimate records to older published records is not the best option.

Shorter timescales can have dramatic impacts on lake levels and sedimentation. Our observations of the Loburu Delta (Lake Bogoria) over four decades have shown that sediments deposited during transgressive phases of higher lake-level were subsequently eroded during regressive phases leading to no obvious change in the sediment record, even though these short climate-induced cycles had a major, but temporary, impact on the conditions in the lake basin. Such brief (years to decadal) climate changes may leave no obvious sedimentary record in outcrop or in offshore core (i.e. mud overlying mud with the same mineralogical and chemical composition), but those environmental windows may also provide opportunities for migration of fauna and flora from the constraints of one ecological window to another.

Rifts in arid settings are associated with shallow saline to hypersaline ephemeral lakes and fluvial settings with high spring/stream inflow ratios (Fig. 22.3a–c). Where at least part of the basin is characterised by wetter climates, typically in upland parts of the catchment, then increased fluvial inflow will contribute to the development of saline and brackish permanent lakes. These also fail to reach their outlets with the basins characterised by lower spring/stream ratios. They are generally deeper, brackish to saline, and may be stratified (Fig. 22.3d–f). In some cases, groundwater losses can result in a freshwater lake evolving despite the lack of surface overflow. Very wet catchments are absent in the modern Kenya Rift, although they are present in the western branch of the East African Rift System (EARS)

Fig. 22.3 Lake levels and sedimentation. (**a–c**) shows examples of sediments in closed hypersaline lake basins. (**a**) Alluvial fan, Suguta Valley. (**b**) Intersecting river and spring-fed streams crossing alluvial plain, Suguta Valley. (**c**) Trona desiccation polygons forming during a drying phase, Lake Magadi. (**d–f**) give examples of closed brackish lake basins with deeper lakes. (**d**) Stratified saline Lake Bogoria is fed mainly by the Sandai River (delta to bottom left) and contains organic-rich mud, trona and littoral gravels. (**e, f**) Brackish Lake Turkana is mainly fed by the Omo River and has no outlet. The upwind coastline (**e**) includes sand bars, lagoons and rocky shorelines. Downwind coastlines include gravel and sandy deposits with locally extensive aeolian dunes (**f**) close to rivers supplying quartz to the basin. (**g–i**) Freshwater open Lake Malawi fully occupies its basin and overflows via the Shire River. Turbidites are common adjacent to border faults (**g**) with fan-deltas also common (**h**). Sublacustrine canyons often develop where streams cut through border faults (**i**), feeding turbidites and sands into deeper axial regions

where higher river inflows and (or) precipitation fill large basins created by high subsidence rates (Fig. 22.3g–i). However, deep freshwater lakes were common in the Kenya Rift as recently as the AHP, just 15,000 to 5000 years ago.

Figure 22.4 shows three generalised models of an asymmetric rift and representative sedimentation patterns that are associated with contrasting climate-induced lake-level states (low, intermediate, high). The rift morphology is kept constant in each case, nevertheless, there are major contrasts in facies that develop. Hypersaline ephemeral to semi-permanent lakes dominate in the most arid basins of the Kenya Rift (e.g. Logipi, Natron-Magadi). Spring and stream waters are depleted in Ca^{2+} and Mg^{2+} as groundwater passes through sediments, soils and spring plumbing (Deocampo and Renaut 2016). Strong evaporation gradually produces saline alkaline water rich in Na^+, HCO_3^-, CO_3^{2-} and Cl^- with salinity depending on the balance between evaporative losses and recharge. Some lakes and playas reach trona saturation and accumulate bedded trona and (or) zeolitic-rich mud with trona and nahcolite (e.g. Magadi, Figs. 22.3c and 22.4d, e; Owen et al. 2019). The evaporites may pass laterally into zeolitic saline mudflats or abut against scarps with saline lagoons near hot and ambient springs. Areas below fault scarps accumulate angular scree with ephemeral rivers feeding sediment into rift marginal alluvial fans and plains (Fig. 22.3b; Tiercelin 1990). Alluvial fans are well developed, for example, in the arid Suguta Valley where they coalesce to form bajadas below the Samburu Mountains (Fig. 22.3a). Paraconglomerate accumulates in proximal locations with mid- to distal-fan orthoconglomerate and coarse sand also present.

Moderately wetter catchments supply greater fluvial inflow and support intermediate lake levels, but with lakes that still fail to reach their outlets in most years. This category includes saline (Bogoria, Fig. 22.3d; Tiercelin 1990) to mildly brackish (Turkana, Fig. 22.3e, f; Owen and Renaut 1986) lakes that fluctuate over a wide range of heights. The saline examples typically have higher spring/stream inflow ratios, are often chemically stratified and accumulate organic-rich (2–6% TOC; Tiercelin 1990) laminated microbial mud, trona and nahcolite. Pebble and granular shorelines dominate in volcanic catchments. Lakeward-dipping granular bedded sediments with distal sublacustrine silt and clay and occasional low-density turbidites develop in fan-deltas. As fluvial inflow increases brackish water bodies such as Lake Turkana may develop (Fig. 22.4b). This large wind-dominated lake lies in a catchment with mixed volcanic and metamorphic rocks and is characterised by sandy and pebbly shorelines with common lagoons (Fig. 22.3e, f). Biogenic facies are present with coquinas and ostracods in littoral and shallower water areas. Stromatolites formed during past higher lake levels. Diatomaceous deposits are associated with areas receiving low clastic inputs. Fan-deltas and deltas are

Fig. 22.4 (continued) silica gel (light colour), Nasikie Engida. (**f**) Lacustrine stromatolitic limestone crust on trachyte, Lake Bogoria. (**g**) Spring tufa, Olorgesailie. ~40 cm thick. (**h**) Fe-Mn nodules (up to 3 mm diameter) from 250 m water depth, Lake Malawi. (**i**) Laminated diatomite-mud couplets from 500 m water depth, core ~10 cm long, Lake Malawi. (**j**) Diatomite from 80 m water depth, core ~9 cm long, Lake Malawi. (**k**) Iron cemented lake floor crust with vivianite $[Fe_3(PO_4)_2{\cdot}8H_2O]$ from 100 m depth, crust is ~8 cm wide, Lake Malawi

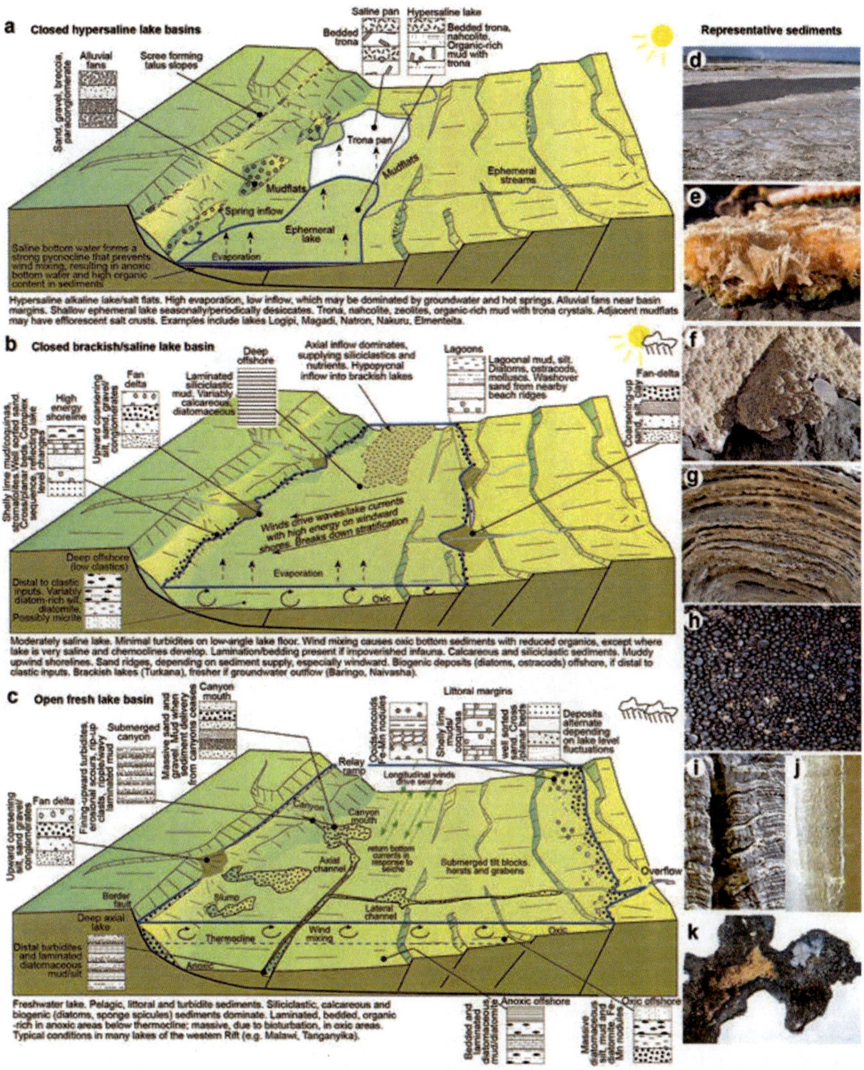

Fig. 22.4 Rift sedimentation, climate and lake levels. The tectonic setting is the same in all three models. (**a**) Closed hypersaline lakes develop during arid phases, forming stratified saline lakes and (or) trona pans. Zeolitic mudflats, alluvial fans and scree deposits are common. (**b**) Closed saline to brackish lakes develop when wetter conditions develop in the catchment, but with insufficient inflow to raise lakes to their outlets. In brackish lakes, laminated and massive muds accumulate offshore with diatomaceous deposits in siliciclastic-deprived areas. Fan-deltas and sandy/gravelly shorelines are common with mud in lagoonal areas. Chemical facies may develop in more highly saline lakes. (**c**) Open basins with overflows are fresher allowing the development of a variety of biogenic deposits (e.g. diatomites, diatomaceous sediments, coquinas, stromatolites) in littoral and offshore areas. Sandy coastlines are common where quartz is available with turbidites and fan-deltas developing near border faults. (**d–k**) Representative facies types. (**d**) General view of polygonal cracks in trona, Lake Magadi. (**e**) Trona resting on nearshore reduced muds with

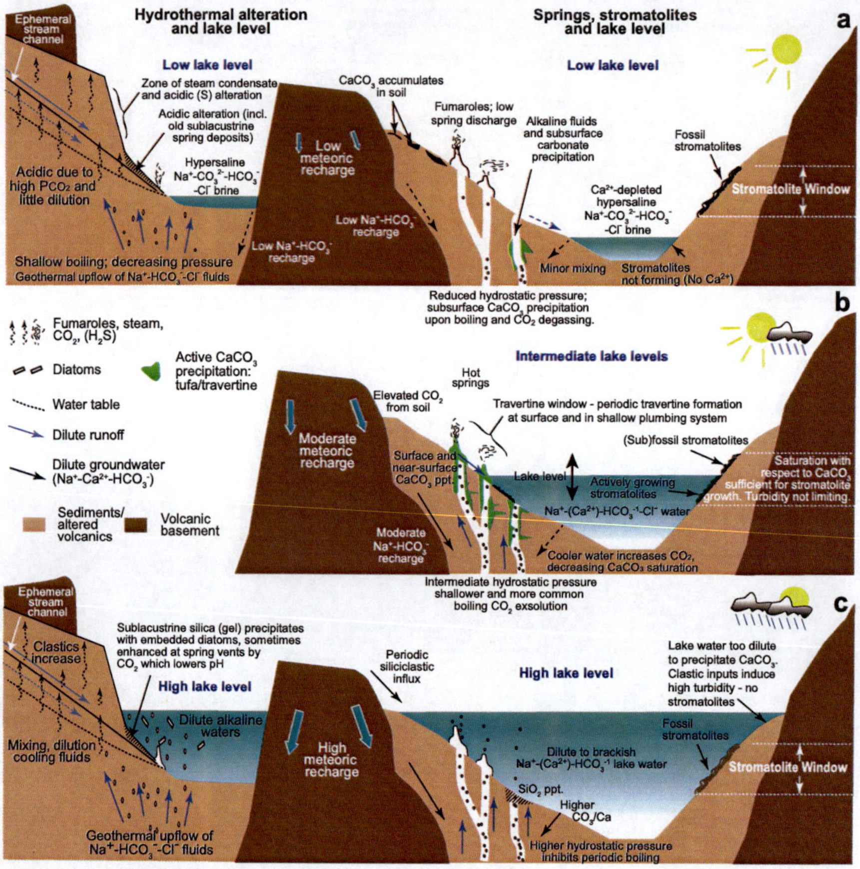

Fig. 22.5 Climate-induced lake level change and impacts on springs and stromatolites (for details see Renaut et al. 2013, 2017). Hydrothermal alteration in left-hand basin. Spring and stromatolite relationships to right. (**a**) Low lake levels (drier climates) are associated with no active stromatolites and subsurface carbonates in springs. Steam condensates and alteration at higher levels above springs shown to left. (**b**) Intermediate lake levels and climates favour active stromatolite growth with surface and near-surface spring carbonates. (**c**) High lake levels (wetter climates) lack stromatolites due to dilution and low Ca^{2+} and higher turbidity with greater clastic inputs. Sublacustrine hydrothermal silica may be present

common with seasonal floods generating hypopycnal overflows in which fine suspended sediments gradually sink and form offshore laminated mud (Reid and Frostick 1985; Cohen 1989; Halfman 1987).

Climate-induced changes in these intermediate lake-level, closed basin lakes may also have major impacts on the growth of nearshore stromatolites and spring tufa/travertine (Figs. 22.4f, g and 22.5). For example, the 'stromatolite window', which varies from basin to basin, represents a narrow range of lake levels within which saturation is suitable for $CaCO_3$ precipitation and low turbidity favours stromatolite growth (Renaut et al. 2013). During dry periods lake levels fall and Ca^{2+} is lost by abiotic and biotic processes in the catchment. Ca-(Mg) carbonates precipitate in sediments and soils near the land surface where they form cements above and below the water table, nodules, concretions and biological nests (e.g. termites). Given sufficient time and aridity, these carbonate cementation mechanisms may produce laterally extensive calcrete. The loss of calcium carbonate in terrestrial settings, in turn, leads to Ca-depleted lake water, preventing calcite precipitation (Fig. 22.5a). Authigenic clay and plankton also reduce the depth of the photic zone, further limiting stromatolite growth. At the other extreme, wetter periods result in raised lake levels and more dilute water that is insufficiently saturated in Ca^{2+} for stromatolite growth, with high clastic inputs causing high turbidity that inhibits photosynthesis.

Hydrothermal springs, which commonly lie along faulted lake shorelines in the Kenya Rift, are also affected by climate-induced fluctuating lake levels with carbonate precipitation restricted to a 'travertine window'. During arid phases minimal mixing of ground-, lake- and thermal waters and reduced hydrostatic pressure on thermal fluids causes $CaCO_3$ precipitation to take place at depth below the ground surface (Fig. 22.5a; Renaut et al. 2013). Moderately wetter periods result in higher lake levels, which allow shallower boiling. Rapid degassing of CO_2 then leads to $CaCO_3$ precipitation at spring vents and near the surface (Fig. 22.5b). During very high lake levels hot springs are submerged and spring waters are rapidly mixed and diluted by lake water (Fig. 22.5c). Combined with cooler fluid temperatures, suppression of the boiling and associated CO_2 degassing, calcite precipitation is inhibited around submerged vents (Fig. 22.5c) but may supply sufficient Ca^{2+} to the lake for lacustrine stromatolites to develop elsewhere. Hydrothermal alteration similarly varies with climate. During low lake levels (Fig. 22.5a) previously alkaline springs may become neutral or weakly acidic fumaroles or steam vents (Renaut and Owen 1988; Renaut et al. 2017) altering pre-existing rocks and older, deeper-water, clastic sediments and spring deposits (Fig. 22.5c), as well as precipitating native sulfur.

Catchments with abundant rainfall have very low spring/stream inflow ratios and fill rift basins to their overflows, resulting in freshwater conditions (Fig. 22.4c). These open lakes are common in the western EARS, but absent in the Kenya Rift today. Sediment-laden inflows into freshwater tend to produce density currents that lay down turbidites, especially close to border faults (Scholz and Rosendahl 1990; Scholz et al. 1993, 1998; Scholz 1995). Sublacustrine canyons and axial channels may transfer coarse clastics to deeper water (Crossley and Owen 1988). Seasonal climates result in laminated diatomite-siliciclastic deposits (Fig. 22.4i; Owen and Crossley 1992). Thermoclines help to preserve lamination and organics (up to ~10%) in deeper anoxic water with massive diatomaceous sediments more common in oxic areas (Fig. 22.4j; Crossley and Owen 1987). Where a thermocline intersects the lake floor and fluctuates in depth, changing redox conditions may help to produce fields of Fe-Mn nodules, which are common in Lake Malawi between depths of 80–160 m (Fig. 22.4h; Owen et al. 1996) and crusts of vivianite (Fig. 22.4k).

22.4 Bedrock, Weathering and Diagenesis

Bedrock is a major control of sedimentation and hydrochemistry in the Kenya Rift lakes. The parent rocks in drainage basins weather to produce detrital sediments with grains of different composition and size, and ions in the lake waters, some of which form new minerals in the lake or its sediment or react with deposited detrital grains to produce new minerals during diagenesis.

Major differences exist between lakes that lie entirely within volcanic catchments and those that are fed by river and spring waters sourced at least in part by weathering of metamorphic basement rocks (Fig. 22.6). Lake Turkana, the most obvious example, has broad sandy beaches (Fig. 22.3e, f), whereas the shorelines of most of the other lakes are typically mud and silt except where bordered by bedrock. The sediment deposited is controlled by hydrodynamics—high energy shorelines, for example, are not characteristic of the smaller rift lakes—but the mineralogical composition of the sediments is also an important factor in sedimentation, as a few examples show.

The modern and ancient Lake Turkana sands are commonly quartzose, reflecting weathering of metamorphic basement rocks in the catchment (Fig. 22.6a, b). In contrast, quartz is uncommon in most of the Kenya Rift volcanic rocks, which are commonly salic and undersaturated or mafic lavas (Fig. 22.6d). Weathering of trachyte, phonolite and basalt releases sand-sized particles, but they are commonly K-feldspar grains that were phenocrysts in the parent lavas or lithic grains (VRFs: volcanic rock fragments) derived from the matrix of porphyritic volcanic rocks or from finely crystalline aphanitic lavas (many basalts and phonolites) (Fig. 22.6e). Quartz is chemically stable and lacks cleavage, so survives physical transport well and commonly forms sand. In contrast, K-feldspars such as sanidine and anorthoclase both have good cleavage and are commonly twinned so are more likely to break during transport, forming silt. They are also more prone to chemical alteration than quartz and alter to clay minerals such as illite.

Sands and sandstones are present in fluvial and lacustrine successions in predominantly volcanic terrains but are much less common than in regions sourced by quartz-bearing bedrock. For example, the Pleistocene Kapthurin Formation west of Lake Baringo is composed of five members, three of which are named 'Silts and Gravels' members (Fig. 22.6f; Tallon 1978; Sect. 17.4.3). In the modern analogous setting, ephemeral stream channels feeding Lake Baringo are floored mainly by gravel and sandy gravel, with silt accumulating in overbank settings near the lake margin. Sand and granules in the modern river channels form small braid bars or the matrix in gravel bars deposited during flow deceleration but thick sand accumulations are uncommon. Where present, modern sands are litharenite or feldspathic arenite in composition. In mixed volcanic-metamorphic terrains, sand and sandstones are more common. The oldest beds of the Miocene Tambach Formation, which crops out in the Elgeyo Escarpment (upper Kerio Valley: Sect. 17.2.3), contain quartzose fluvial sandstone (Fig. 22.6c), whereas younger fluvial deposits are finer grained lithofeldspathic arenites with minor quartz. This partly reflects a

Fig. 22.6 Contrasting weathering and sediments from metamorphic and volcanic terrains. (**a**) Quartz-rich banded gneiss near South Horr, east of Suguta. (**b**) Sandy braided channel of Turkwell River flowing into Lake Turkana. (**c**) Arkosic sandstone, lower Tambach Formation, Elgeyo Escarpment. (**d**) Historic basalt flow, Andrew's volcano, The Barrier, north Suguta. (**e**) Medial sand bar, Parkirichai River, west of Lake Bogoria. VRF volcanic rock fragments (lithic grains). (**f**) Sandy silts and conglomerate, Middle Silts and Gravels Member, Kapthurin Formation, Baringo

gradual transition from erosion of metamorphic basement landscapes to erosion of volcanic landscapes over time (Renaut et al. 1999).

The huge amounts of ash generated by volcanism often mantle the landscape after eruptions and are prone to rapid reworking with particles incorporated in mud, soil or coarser deposits, but distinct ash layers often survive after they settle as bedded and reworked tuffs in low-energy lake sediments and provide food for radiometric dating. Most ash (lithic and vitric) is highly reactive and breaks down rapidly to clay minerals and zeolites in soils and at the land surface, or in lakes (Hay 1970; Mees et al. 2005). Glass and lithic particles may survive in the drier parts of the rift, but most alter rapidly in humid settings. Alteration products vary with the composition of both the tephra and the fluid. Ash from basaltic eruptions, for example, commonly alters to smectite in semi-arid settings.

An unusual feature of the saline alkaline hydrochemical environment is that it can affect products of detrital sedimentation as well as producing new chemical sediments, mainly because silica is soluble in highly alkaline water. Silicates often dissolve in acidic fluids, as in humid tropical weathering, but also at high pH. Particles deposited in relatively fresh water can change not only in mineralogical composition when exposed to alkaline lake and pore fluids, but also in particle size. A sand-size grain of volcanic glass or a reactive silicate mineral (e.g. plagioclase or feldspathoid) deposited by a moderately high energy current can rapidly alter to clay minerals or zeolites when in contact with sodic alkaline fluids. Thus sands can potentially become mud during burial, which can modify the hydrodynamic interpretation of a sediment: particles originally deposited as sand may be preserved as mudstone after the grains interact with pore fluids. Relics of precursor grains (i.e. pseudomorphs) may or may not be preserved, depending on their original composition and degree of chemical interaction with lake or pore fluids. After even moderate compaction such evidence may not always be preserved.

22.5 Geothermal Influences and Controls

As noted throughout this book, hot springs and related geothermal activity are present at almost all the modern lakes or in their drainage basins, and spring deposits are preserved locally in the geological record. The types of geothermal activity and their deposits are summarised in Sect. 4.9 with local details in individual chapters. The aim of this section is to assess the overall role that geothermal processes play in controlling or modifying the sedimentation, limnology and environment of the Kenya Rift lakes.

Geothermal activity is usually ignored or downplayed when discussing the ecology and history of the Kenya Rift lakes. It nonetheless can contribute to the hydrology and chemistry of many lake basins, generally increasing in importance when rainfall and runoff decline. Hot springs contribute to the Kenya Rift lakes and their geological record in many ways. The most important roles of hot springs are (1) in lake water recharge on various timescales, (2) their impact on lake water chemistry, and (3) their contribution to the sedimentary record as spring deposits and in hydrothermal alteration of other sediments and bedrock. In the Kenya Rift, these roles are often not obvious because the geological record of spring activity is not always preserved.

22.5.1 Lake Recharge

Hot springs contribute to lake recharge, but the volume varies from trivial to major and can change over time. In large Kenya Rift lakes such as Lake Turkana, hydrothermal inflow is negligible today. In contrast, thermal-spring fluids often contribute >50% of the annual recharge in Lake Magadi and Nasikie Engida (Chap. 12), and in many small lakes ((e.g. Kageinya, Sect. 14.4). In some small lakes and wetlands, thermal fluids can exceed 80% of the annual recharge.

The volume of geothermal recharge as a proportion of total recharge in each lake is not constant. If the geothermal inflow is maintained at the same volume annually, the proportion of thermal inflow water may increase or decrease from year to year because the volume of non-thermal-water recharge changes with variations in dilute inflow (rain, rivers, unchannelled runoff, groundwater) and evaporation. In other words, in relatively wet years a lake will receive more recharge from runoff, whereas in drier years hot-spring inflow may account for more of the annual recharge (Fig. 22.7a). This applies in the short term (years to decades), and over the longer term, often linked to climatic changes. The impact is greatest in small lakes.

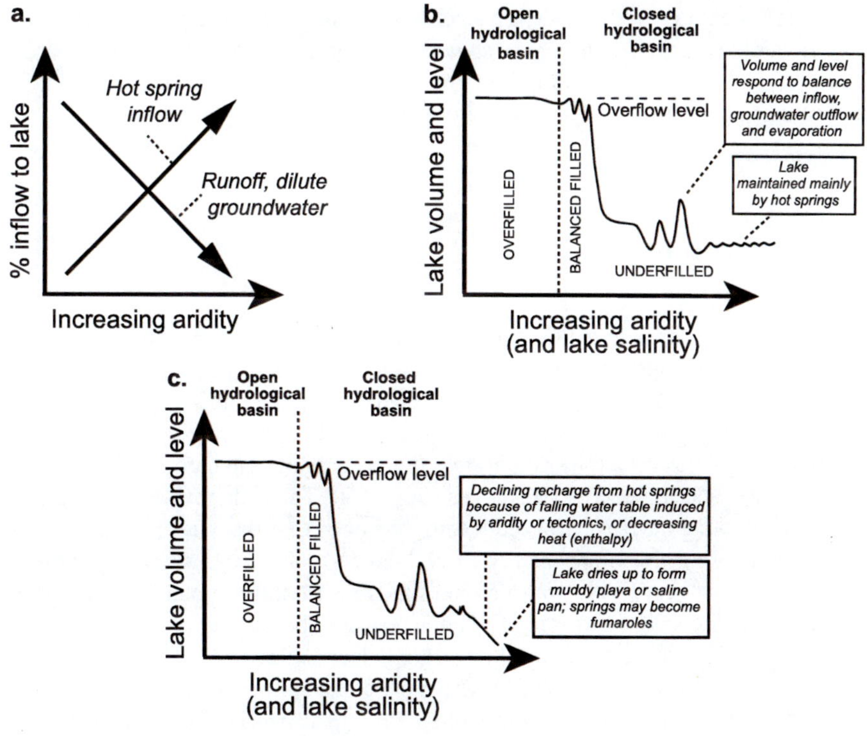

Fig. 22.7 Changing sources of lake inflow with increasing aridity. (**a**) Reciprocal relationship between hot spring inflow and runoff. (**b**) Changes in lake volume and level with increasing aridity but with hydrothermal inflow maintained. (**c**) Changes in lake volume and level with increasing aridity but with decreasing hydrothermal inflow

Figure 22.7b shows what can happen to the volume and level of a lake during a period of increasing aridity when hot spring inflow remains constant. Hydrothermal inflow may be maintained for a long time if the geothermal reservoir is deep and its water has a long residence time or is fed by distant waters in a wetter environment. Under such conditions increasing aridity may not directly impact the spring discharge. The lake initially may be hydrologically open (overfilled) and its level stable, with dilute or slightly brackish water. With increasing aridity lake level may fall below the spillpoint and the lake level will fluctuate with variations in the rainfall/evaporation ratio (balanced filled, becoming underfilled), such as Lake Bogoria. There may come a point when little runoff reaches the lake and hydrothermal fluids provide most of the annual recharge. The lake level may become approximately stable for a period controlled by the balance between hydrothermal inflow and evaporation. The lake water would be saline and might precipitate trona and other evaporite minerals, such as at modern Lake Magadi and Nasikie Engida. If the hydrothermal inflow also responds to increasing aridity (Fig. 22.7c) the lake may desiccate completely and become a playa. With a return to wetter conditions, the lake will rise and the proportion of hydrothermal inflow will decline. This model will vary if there is outflow via the subsurface and if tectonic movements modify the hydrology and hydrogeology.

Spring locations and activity may change in response to lake-level changes. Observations at Lake Bogoria have shown that shoreline spring and geyser activity is closely linked to the prevailing lake level. Falls in lake level may lead to the highest springs ceasing to flow and becoming fumaroles. Rising lake levels may submerge springs or cause changes in their behaviour (Renaut et al. 2008, 2017).

22.5.2 *Impact on Lake Water Chemistry*

Many hot springs have little direct impact on lake water chemistry, which is a reason why they are often ignored. The products of chemical weathering by silicate hydrolysis, where bedrock is altered by carbonic acid induced by rainfall (minor) or CO_2-charged surface and groundwaters (common), are little different to reactions with bedrock that are produced by geothermal fluids charged with carbon dioxide from deep (mantle and lower crustal) sources (Sect. 4.9). Those thermal waters are also fed by waters that have experienced weathering at or near the land surface.

The hot spring fluids are mainly of Na^+–HCO_3^-–Cl^- composition in common with the lake waters. Dilute warm spring fluids commonly contain Ca^{2+} and Mg^{2+} though some are saline and free of alkaline earths. The salinity of the hot springs varies from place to place from dilute (<1 g L^{-1}) to saline (>35 g L^{-1}) at northern Nasikie Engida: Renaut et al. 2021). The most dilute springs are often condensates (e.g. Ol Kokwe, Lake Baringo: Tarits et al. 2006; S. Bogoria: Renaut et al. 2017). The saline hot springs provide 'pre-concentrated' waters to saline lakes enhancing their potential to precipitate evaporites after inflow declines. In rare examples, spring fluids are enriched in other anions, notably sulfate. Maar lakes at Kangirinyang in the southeastern Suguta Valley (Sect. 14.4) precipitate sulfate evaporites (Dunkley et al. 1993). Oxidised S-bearing volcanic gases may be responsible.

Perhaps the most important geothermal fluid discharging into the rift lakes is carbon dioxide. Deeply sourced CO_2 is being released at many sites along the Kenya Rift, often along faults and fractures (Lee et al. 2016, 2017; Muirhead et al. 2016) and discharge is focused at hot springs and fumaroles along fault lines or pervasively through sediments and rocks below lakes (Tarits et al. 2006). This 'excess' CO_2 contributes to alkalinity in the lakes and the precipitation of trona and nahcolite at Lakes Magadi, Nasikie Engida and Lake Bogoria (DeCort et al. 2019; Renaut et al. 2021). Hot springs and associated bacteria can also introduce high concentrations of elements toxic to the higher biota (Krienitz et al. 2003, 2005).

22.5.3 *Hydrothermal Contributions to the Sedimentary Record of the Lake Basins*

As noted (Sect. 4.9.4), spring deposits are a small part of the sedimentary record of the lake basins, but this can belie the importance of hydrothermal recharge. The record of spring deposits is linked to both climate and lake level but not in a direct and simple way (Fig. 22.7). It might be expected that tufa and travertine would be associated with humid phases when dilute to brackish Ca-bearing groundwater was more commonly available. This can be true, but evidence from Lake Bogoria shows that the relationship is less simple. Even Ca-poor fluids (<2 mg L^{-1} Ca^{2+}) precipitate modern spring carbonates by rapid degassing of CO_2 (Renaut et al. 2013). Shoreline hot springs, common at many lakes, are frequently submerged when lake level rises, but many continue to flow from submerged vents. Some produce sublacustrine carbonates as chimneys (Casanova, 1986; Faure et al. 2002; Owen et al. 2019; Sect. 20.3.1) but others mix with the lake water without precipitating carbonate near the vent. When lake level falls, shoreline springs may become fumaroles and carbonate precipitation may cease or, in the case of boiling springs, precipitation may occur below the surface in the shallow plumbing system, as previously discussed.

Silica is precipitated at many modern hot springs but typically only forms thin efflorescent crusts and spicules, often on microbial substrates (Sect. 4.9.4). Subaerial sinter deposits that characterise other geothermal localities are not forming because the aqueous silica concentrations are much lower than those elsewhere and some are controlled by chalcedony solubility rather than quartz (Cioni et al. 1992). Many springs are undersaturated with respect to amorphous silica at the vent. Consequently, they are not precipitating silica (opal-A) unless subject to evaporation in addition to cooling. However, sublacustrine sinter has formed around submerged hot springs at Lake Bogoria (Renaut and Owen, 1988; Renaut et al. 2017) and Lake Baringo (Renaut et al. 2002; Buatois et al. 2017). Incorporated planktonic diatoms indicate formation from a soft gelatinous precursor on the floor of a dilute lake. Hydrothermal alteration is also a minor contributor to lacustrine sedimentation.

22.6 Rift Sedimentation Models

Many have attempted to identify the major controls of facies in continental rift settings (e.g. Cross and Bosence 2008; Withjack et al. 2002; Dorobek 2008; Yang et al. 2018; Mercedes-Martín et al. 2019), though few have reported research undertaken in the Kenya Rift. More generalised classifications have examined lacustrine sedimentation in a variety of settings with considerations of the balance between changes in accommodation space (mostly tectonic) and sediment + water fill (mostly climate) (Carroll and Bohacs 1999). Owen et al. (2018) integrated many of the structural factors that influence sedimentation in the Kenya Rift (Fig. 22.8), distinguishing primary and secondary controls. The former include faulting, subsidence and uplift, magmatic evolution and volcanism, antecedent lithologies and pre-existing landscape geomorphology. They noted that these, in turn, had major impacts on lacustrine sedimentation through secondary controls that included syn-rifting igneous lithologies, rift shoulder uplift and axial subsidence, the development of accommodation space, meteoric and hydrothermal circulation and ascending mantle CO_2.

Rift sedimentation responds to these primary and secondary controls in a variety of ways. Early-stage rifting is associated with a modest rise in the asthenosphere with broad sag basins developing at the surface (Fig. 22.8a). Geothermal gradients are low with minimal development of geothermal fluid reservoirs. Faults begin to form but fail to tap any existing deep hot-water bodies, minimising hot spring development, although shallow meteoric circulation might result in ambient-temperature springs, some perhaps precipitating tufa depending on local geology and climate. Lake catchments would be large and might extend beyond the developing rift margins. Consequently, rivers would likely deliver significant sediment loads. Erosion rates would be relatively slow on the subdued rift margins, which would result in longer exposure to weathering and a higher Chemical Index of Alteration (CIA; Owen et al. 2018) for sediments from these areas.

The subdued topography would tend to be associated with an unfocussed distribution of clouds over the basin, which should allow lower evaporation rates and higher rainfall over the developing rift floor due to a lack of rain shadow effects, which are an important feature of the modern Kenya Rift. Low cumulative subsidence results in the formation of broad, shallow, lake basins with relatively high rift-floor rainfall favouring the generation of overfilled dilute lakes with minimal carbonate accumulation. Silica may be deposited as diatoms with shallow, wind mixed (?) lakes tending to favour low organic matter accumulation.

Mature rifts are associated with changes that impact on the nature and spatial extent of lacustrine sedimentation (Fig. 22.8b). Brittle (faulting) and ductile deformation stretches and thins the crust as the asthenosphere below rises and dykes intrude into higher crustal levels (Fairhead 1976). At the surface, rift shoulder uplift and axial subsidence deepen the valley floor, increasing accommodation space, which trends towards deeper, narrower fault-bounded basins that accumulate diatomaceous sediments. Catchments become smaller as the rising rift flanks intercept

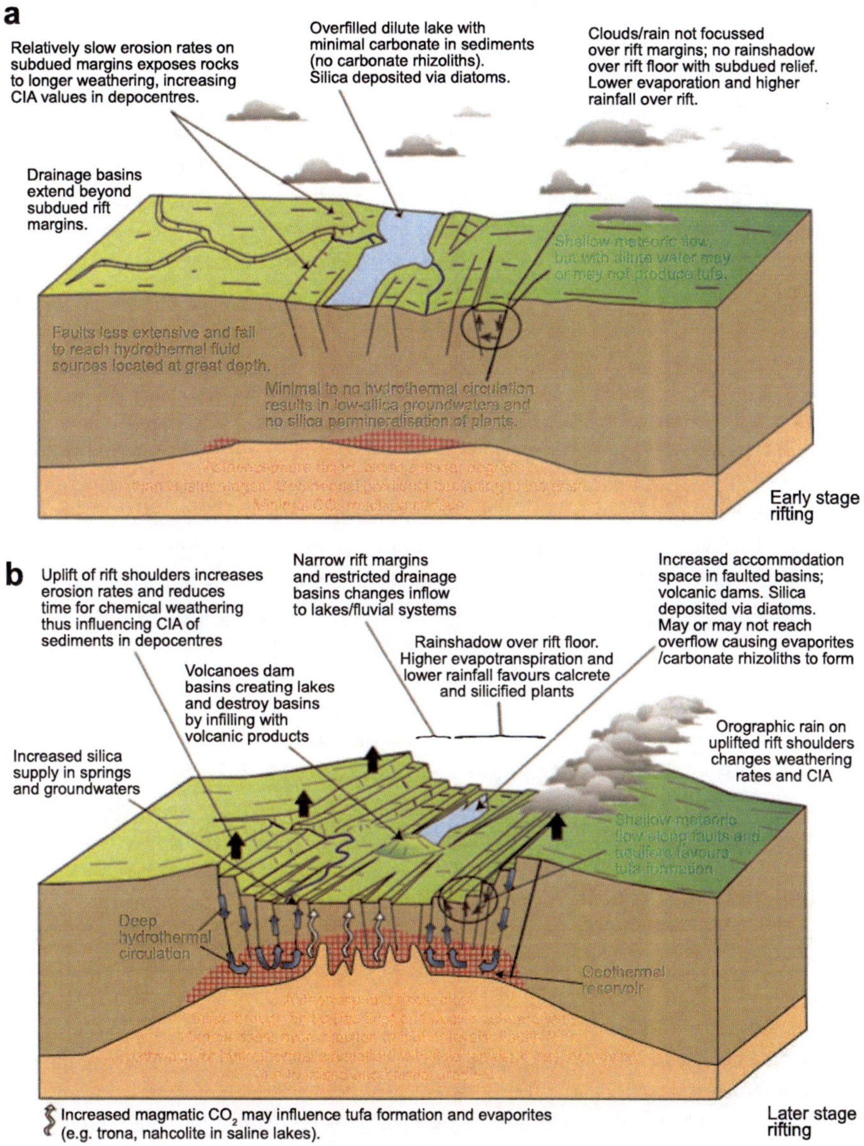

Fig. 22.8 Tectonic influences on spring and lake water in the South Kenya Rift. (**a**) Early-stage rifting is characterised by subdued rift margins. (**b**) Late-stage rifting. Differential subsidence and uplift change accommodation space and drainage basin size. Rifting creates rain shadows, inducing drier conditions over the rift floor, and alters weathering, erosion and deposition rates. Faults control shallow meteoric water, deeper geothermally-sourced springs and the ascent of magmatic CO_2, all of which may have significant impacts on lacustrine sedimentation

and reverse antecedent rivers, which now flow away from the rift. At the same time, slope angles on the rift margins increase, which raises erosion rates and lowers CIA values of sediments. Episodic uplift of the rift flanks may result in temporal changes in the influx of coarse sediment to adjacent lakes, river plains and (or) alluvial fans. Uplift may also increase orographic rains over the rift margins, which complicates erosion patterns. If there is little resulting vegetation change then erosion is likely to increase, but often expansion and compositional changes in vegetation cover and density will stabilise formerly exposed slopes. In contrast, rain shadows over the rift floor raise evaporation/precipitation ratios with drier conditions emerging, which will tend to favour the creation of saline lakes and alluvial fans with calcrete and carbonate rhizoliths present in terrestrial sediments.

Later-stage rifting is also associated with increased geothermal gradients and shallower geothermal reservoirs, with deep fractures allowing chloride-rich waters and gases to reach the surface, as for example at southern Lake Bogoria and Lake Magadi (Renaut et al. 2017; Owen et al. 2019). Combined with rain shadow-induced aridity, hot spring fluids that are rich in silica and poor in calcium proportionally account for more of the annual lake-water recharge. Scarce travertine may form locally. However, $CaCO_3$ precipitation is more common from dilute to moderately saline (<5 g L^{-1} Total Dissolved Solids) meteoric springs with shallower circulation. Silica precipitates as sinter at a few hot spring vents, although evaporation of hot-spring outflow waters may lead to silica saturation (opal-A) downstream and in spring-fed groundwaters, resulting in silica precipitation in microbial crusts and local permineralisation of plants undergoing evapotranspiration. At Lake Magadi, evaporation has led to very high silica concentrations in the highly alkaline lake water. Deeply sourced carbon dioxide may also escape to shallower levels (Darling et al. 1995; Lee et al. 2016) where it can reduce silica solubility in the water column (by lowering pH) and in sediments and thus contribute to the formation of sodium silicates and lacustrine chert (Renaut et al. 2021). Furthermore, magmatic CO_2 is important in causing trona formation in lakes with high evaporation rates (Lake Magadi) and contributing to nahcolite precipitation where CO_2 emissions are exceptionally high (Nasikie Engida) (De Cort et al. 2019; Owen et al. 2019; Renaut et al. 2021).

Volcanic products are scarce in the western EARS but are abundant in the Kenya Rift where Neogene activity has changed temporally and spatially, reflecting magma fractionation processes, and mixing of lithospheric and asthenospheric materials. Volcanoes may block drainage and form new lakes (Baringo) but can also destroy accommodation space when ash and (or) lava infills a basin. Changing trace elements in volcanic rocks also have a significant impact on trace element geochemistry of lake deposits. Owen et al. (2018), for example, noted Neogene geochemical changes in lake sediments from the Baringo area that reflected a transition from alkali basalts and phonolites to weakly alkaline basalts and trachytes (Baker 1987). Evidently, lacustrine sedimentary records are complex in the Kenya Rift and their interpretation requires combined considerations of tectonic, volcanic and climatic controls set against the stage of rifting when the deposits accumulated.

References

Acosta VT (2015) Denudation processes in a tectonically active rift on different time scales: New insights from thermochronology and CRN dating in the Kenya Rift. PhD thesis, Universität Potsdam

Ambraseys NN (1991) Earthquake hazard in the Kenya Rift: the Subukia earthquake 1928. Geophys J Int 105:253–269

Baker BH (1986) Tectonics and volcanism of the southern Kenya Rift Valley and its influence on rift sedimentation. In: Frostick LE, Renaut RW, Reid I, Tiercelin J-J (eds) Sedimentation in the African rifts. Geol Soc Lond Spec Publ 25:45–57

Baker B (1987) Outline of the petrology of the Kenya rift alkaline province. In: Fitton JG, Upton BJG (eds) Alkaline igneous rocks. Geol Soc Lond Spec Publ 30:293–311

Beadle LC (1981) The inland waters of tropical Africa: an introduction to tropical limnology, 2nd edn. Longman, London

Bergner AGN, Strecker MR, Trauth MH, Deino A, Gasse F, Blisniuk P, Duhnforth M (2009) Tectonic and climatic control on evolution of rift lakes in the Central Kenya Rift, East Africa. Quat Sci Rev 28:2804–2816

Bohacs KM, Carroll AR, Neal JE, Mankiewicz PJ (2000) Lake-basin type, source potential, and hydrocarbon character: an integrated sequence-stratigraphic-geochemical framework. In: Gierlowski-Kordesch EH, Kelts KR (eds) Lake basins through space and time. Am Assoc Petrol Geol Stud Geol 46:3–34

Buatois LA, Renaut RW, Scott JJ, Owen RB (2017) An unusual occurrence of the trace fossil *Vagorichnus* preserved in hydrothermal silica at Lake Baringo, Kenya Rift Valley: Taphonomic and paleoenvironmental significance. Palaeogeogr Palaeoclimatol Palaeoecol 485:843–853

Burke K, Wilkinson MJ (2016) Landscape evolution in Africa during the Cenozoic and Quaternary—the legacy and limitations of Lester C. King. Can J Earth Sci 53:1089–1102

Carroll AR, Bohacs KM (1999) Stratigraphic classification of ancient lakes: balancing tectonic and climatic controls. Geology 27:99–102

Casanova J (1986) Les stromatolites continentaux: paléoecologie, paléohydrologie, paléoclimatologie. Application au rift Gregory. Thèse, Université de Marseille II

Cioni R, Fanelli G, Guidi M, Kinyariro JK, Marini L (1992) Lake Bogoria hot springs (Kenya): geochemical features and geothermal implications. J Volcanol Geotherm Res 50:231–246

Cohen AS (1989) Facies relationships and sedimentation in large rift lakes and implications for hydrocarbon exploration: examples from Lakes Turkana and Tanganyika. Palaeogeogr Palaeoecol Palaeoclimatol 70:65–80

Cohen AS (1990) Tectono-stratigraphic model for sedimentation in Lake Tanganyika, Africa. In: Katz BJ (ed) Lacustrine basin exploration: case studies and modern analogs. Am Assoc Petrol Geol Mem 50:137–150

Cohen AS (2003) Paleolimnology: the history and evolution of lake systems. Oxford University Press, Oxford

Contreras J, Scholz CH, King GCP (1997) A model of rift basin evolution constrained by first-order stratigraphic observations. J Geophys Res 102:7673–7690

Cross NE, Bosence D (2008) Tectono-sedimentary models for rift-basin carbonate systems. In: Lukasik J, Simo JA (eds) Controls on carbonate platform and reef development. SEPM Spec Publ 89:83–105

Crossley R (1984) Controls of sedimentation in the Malawi Rift Valley, Central Africa. Sediment Geol 40:33–50

Crossley R, Owen RB (1987) Sedimentation in the Malawi Rift and some implications for hydrocarbon exploration in Africa. In: Matheis G, Schandelmeier H (eds) Current research in African earth sciences. Balkema, Rotterdam, pp 227–230

Crossley R, Owen RB (1988) Sand turbidites and organic-rich diatomaceous muds from Lake Malawi, Central Africa. In: Fleet AJ, Kelts K, Talbot M (eds) Lacustrine petroleum source rocks. Geol Soc Lond Spec Publ 40:369–374

Darling WG, Griesshaber E, Andrews JN, Armannsson H, O'Nions RK (1995) The origin of hydrothermal and other gases in the Kenya Rift Valley. Geochim Cosmochim Acta 59:2501–2512

Davis PM, Slack PD (2002) The uppermost mantle beneath the Kenya dome and relation to melting, rifting and uplift in East Africa. Geophys Res Lett 29/21:1–4

De Cort G, Mees F, Renaut RW, Sinnesael M, Van der Meeren T, Goderis S, Keppens E, Mbuthia A, Verschuren D (2019) Late-Holocene sedimentation and sodium carbonate deposition in hypersaline, alkaline Nasikie Engida, southern Kenya Rift Valley. J Paleolimnol 132:1–22

Deocampo DM, Renaut RW (2016) Geochemistry of African soda lakes. In: Schagerl M (ed) Soda lakes of East Africa. Springer, Cham, pp 77–93

Dorobek SL (2008) Tectonic and depositional controls on syn-rift carbonate platform sedimentation. In: Lukasik J, Simo JA (eds) Controls on carbonate platform and reef development. SEPM Spec Publ 89:57–81

Dunkelman TJ, Karson JA, Rosendahl BR (1988) Structural style of the Turkana rift, Kenya. Geology 16:258–261

Dunkelman TJ, Rosendahl BR, Karson JA (1989) Structure and stratigraphy of the Turkana rift from seismic reflection data. J Afr Earth Sci 8:489–510

Dunkley PN, Smith M, Allen DJ, Darling WG (1993) The geothermal activity and geology of the northern sector of the Kenya Rift Valley. Brit Geol Surv Res Rep SC/93/1

Eugster HP (1980) Lake Magadi, Kenya and its precursors. In: Nissenbaum A (ed) Hypersaline brines and evaporitic environments. Elsevier, Amsterdam, pp 195–232

Eugster HP (1986) Lake Magadi, Kenya: a model for rift valley hydrochemistry and sedimentation. In: Frostick LE, Renaut RW, Reid I, Tiercelin J-J (eds) Sedimentation in the African rifts. Geol Soc Lond Spec Publ 25:177–189

Fairhead JD (1976) The structure of the lithosphere beneath the Eastern Rift, East Africa, deduced from gravity studies. Tectonophysics 30:269–298

Faure H, Walter RC, Grant DR (2002) The coastal oasis: ice age springs on emerged continental shelves. Glob Planet Change 33:47–56

Fer I, Tietjen B, Jeltsch F, Trauth MH (2017) Modelling vegetation change during Late Cenozoic uplift of the East African plateaus. Palaeogeogr Palaeoecol Palaeoclimatol 467:120–130

Frostick LE (1997) The East African rift basins. In: Selley RC (ed) Sedimentary basins of the world volume 3: African basins. Elsevier, Amsterdam, pp 187–209

Frostick LE, Reid I (1987) Tectonic control of desert sediments in rift basins ancient and modern. In: Frostick LE, Reid I (eds) Desert sediments: ancient and modern. Geol Soc Lond Spec Publ 35:53–68

Frostick LE, Reid I (1989) Is structure the main control of river drainage and sedimentation in rifts? J Afr Earth Sci 8:165–182

Gierlowski-Kordesch E, Kelts K (1994) Introduction. In: Gierlowski-Kordesch E, Kelts K (eds) Global geological record of lake basins, vol 1. Cambridge Univ Press, Cambridge, p xvii–xxxiii

Halfman JD (1987) High-resolution sedimentology and paleoclimatology of Lake Turkana, Kenya. PhD thesis, Duke University, Durham, North Carolina

Hay RL (1970) Silicate reactions in three lithofacies of a semi-arid basin, Olduvai Gorge, Tanzania. In: Morgan BA, Jones BF, Kullerud G, Osborn EF, Holser WT (eds) Fiftieth anniversary symposium. Mineral Soc Am Spec Pap 3:237–255

Hinderer M, Einsele G (2002) Rifts as denudation-accumulation systems: concept, models, and examples. In: Renaut RW, Ashley GM (eds) Sedimentation in continental rifts. SEPM Spec Publ 73:83–96

Hutchinson GE (1975) A treatise on limnology, volume 1: geography, physics and chemistry. Wiley, New York

Johnson TC (1993) Sedimentary processes and signals of past climatic change in the large lakes of the East African Rift Valley. In: Odada EO, Johnson TC (eds) The limnology, climatology and paleoclimatology of the East African lakes, Gordon and Breach, Amsterdam, pp 367–412

Johnson TC, Halfman JD, Rosendahl BR, Lister GS (1987) Climatic and tectonic effects on sedimentation in a rift-valley lake: evidence from high-resolution seismic profiles, Lake Turkana, Kenya. Geol Soc Am Bull 98:439–447

Krienitz L, Ballot A, Casper P, Kotut K, Wiegand C, Pflugmacher S (2005) Cyanobacteria in hot springs of East Africa and their potential toxicity. Algol Stud 117:297–306

Krienitz L, Ballot A, Wiegand C, Kotut K, Codd GA, Pflugmacher S (2003) Contribution of hot spring cyanobacteria to the mysterious deaths of lesser flamingos at Lake Bogoria, Kenya. FEMS Microbiol Ecol 43:141–148

Lambiase JJ (1990) A model of tectonic control of lacustrine stratigraphic sequences in continental rift basins. In: Katz BJ (ed) Lacustrine basin exploration—case studies and modern analogs. Am Assoc Petrol Geol Mem 50:265–276

Lambiase JJ, Bosworth W (1995) Structural controls on sedimentation in continental rifts. In: Lambiase JJ (ed) Hydrocarbon habitat in rift basins. Geol Soc Lond Spec Publ 80:117–144

Lambiase JJ, Morley CK (1999) Hydrocarbons in rift basins: the role of stratigraphy. Phil Trans R Soc Lond A 357:877–900

Le Turdu C, Tiercelin J-J, Coussement C, Rolet J, Renaut RW, Richert JP, Xavier J-P, Coquelet D (1995) Basin structure and depositional patterns interpreted using a 3D remote sensing approach: the Baringo–Bogoria basins, central Kenya Rift, East Africa. Bull Cent Rech Explor-Prod Elf-Aquitaine 19:1–37

Le Turdu C, Tiercelin J-J, Richert JP, Rolet J, Xavier JP, Renaut RW, Lezzar KE, Coussement C (1999) Influence of preexisting oblique discontinuities on the geometry and evolution of extensional fault patterns: evidence from the Kenya Rift using SPOT imagery. In: Morley CK (ed) Geoscience of rift systems—evolution of East Africa. AAPG Stud Geol 44:173–191

Lee H, Muirhead JD, Fischer TP, Ebinger CJ, Kattenhorn SA, Sharp ZD, Kianji G (2016) Massive and prolonged deep carbon emissions associated with continental rifting. Nat Geosci 9:145–149

Lee H, Fischer TP, Muirhead JD, Ebinger CJ, Kattenhorn SA, Sharp ZD, Kianji G, Takahata N, Sano Y (2017) Incipient rifting accompanied by the release of subcontinental lithospheric mantle volatiles in the Magadi and Natron basin, East Africa. J Volcanol Geotherm Res 346:118–133

McCall GJH (1967) Geology of the Nakuru-Thomson's Falls Lake Hannington area. Rep Geol Surv Kenya 78

Mees F, Stoops G, Van Ranst E, Paepe R, Elfi Van Overloop E (2005) The nature of zeolite occurrences in deposits of the Olduvai Basin, northern Tanzania. Clays Clay Miner 53:659–673

Mercedes-Martín R, Ayora C, Tritlla J, Sánchez-Román M (2019) The hydrochemical evolution of alkaline volcanic lakes -- a model to understand the South Atlantic Pre-salt mineral assemblages. Earth-Sci Rev 198:102938

Morley CK, Karanja FM, Wescott WA, Stone DM, Harper RM, Wigger ST, Day RA (1999a) Geology and geophysics of the Western Turkana Basins, Kenya. In: Morley CK (ed) Geoscience of rift systems—evolution of East Africa. Am Assoc Petrol Geol Stud Geol 44:19–54

Morley CK, Nelson RA, Patton TL (1990) Transfer zones in the East African rift system and their relevance to hydrocarbon exploration in rifts. Am Assoc Petrol Geol Stud Bull 74:1234–1253

Morley CK, Ngenoh DK, Ego JK (1999b) Introduction to the East African Rift System. In: Morley CK (ed) Geoscience of rift systems–evolution of East Africa. Am Assoc Petrol Geol Stud Geol 44:1–18

Muirhead JD, Kattenhorn SA, Lee H, Mana S, Turrin BD, Fischer TP, Kianji G, Dinde E, Stamps DS (2016) Evolution of upper crustal faulting assisted by magmatic volatile release during early-stage continental rift development in the East African Rift. Geosphere 12:1670–1700

Olaka LA, Trauth MH, Olago DO (2010) The sensitivity of East African rift lakes to climate fluctuations. J Paleolimnol 44:629–644

Olago D, Opere A, Barongo J (2009) Holocene palaeohydrology, groundwater and climate change in the lake basins of the Central Kenya Rift. Hydrol Sci J 54:765–780

Olsen PE (1990) Tectonic, climatic, and biotic modulation of lacustrine ecosystems—examples from the Newark Supergroup of eastern North America. In: Katz BJ (ed) Lacustrine basin exploration—case studies and modern analogs. Am Assoc Petrol Geol Mem 50:209–224

Owen RB, Crossley R (1992) Spatial and temporal distribution of diatoms in sediments of Lake Malawi, Central Africa, and ecological implications. J Paleolimnol 7:55–71

Owen RB, Muiruri VM, Lowenstein TK, Renaut RW, Rabideaux N, Luo S, Deino AL, Sier MJ, McNulty EP, Leet K, Cohen A, Campisano C, Deocampo D, Shen C-C, Billingsley A, Mbuthia A, Stockhecke M (2019) Quaternary history of the Lake Magadi basin, southern Kenya Rift: tectonic and climatic controls. Palaeogeogr Palaeoecol Palaeoclimatol 518:97–118

Owen RB, Renaut RW (1986) Sedimentology, stratigraphy and palaeoenvironments of the Holocene Galana Boi Formation, NE Lake Turkana, Kenya. In: Frostick LE, Renaut RW, Reid I, Tiercelin J-J (eds) Sedimentation in the African rifts. Geol Soc Lond Spec Publ 25:311–322

Owen RB, Renaut RW, Lowenstein TK (2018) Spatial and temporal geochemical variability in lacustrine sedimentation in the East African Rift System: evidence from the Kenya Rift and regional analyses. Sedimentology 65:1697–1730

Owen RB, Renaut RW, Williams TM (1996) Characteristics and origins of ferromanganese nodules in Lake Malawi. In: Johnson TC, Odada E (eds) Limnology climatology and paleoclimatology of the East African lakes. Gordon and Breach, Amsterdam, pp 461–474

Reid I, Frostick LE (1985) Beach orientation, bar morphology and the concentration of metalliferous placer deposits: a case study, Lake Turkana, N Kenya. J Geol Soc Lond 142:837–848

Renaut RW, Ego JK, Tiercelin J-J, Le Turdu C, Owen RB (1999) Saline, alkaline palaeolakes of the Tugen Hills-Kerio Valley Region, Kenya Rift Valley. In: Andrews P, Banham P (eds) Late Cenozoic environments and hominid evolution: a tribute to Bill Bishop. Geological Society, London, pp 41–58

Renaut RW, Jones B, Tiercelin J-J, Tarits C (2002) Sublacustrine precipitation of hydrothermal silica in rift lakes: evidence from Lake Baringo, central Kenya Rift Valley. Sediment Geol 148:235–257

Renaut RW, Gierlowski-Kordesch GK (2010) Lakes. In: James NP, Dalrymple RW (eds) Facies models, 4th edn. Geological Association of Canada, St. John's, pp 541–573

Renaut RW, Owen RB (1988) Opaline cherts associated with sublacustrine hydrothermal springs at Lake Bogoria, Kenya Rift Valley. Geology 16:699–702

Renaut RW, Owen RB, Ego JK (2008) Recent changes in geyser activity at Loburu, Lake Bogoria, Kenya Rift Valley. GOSA Trans 10:4–14

Renaut RW, Owen RB, Ego JK (2017) Geothermal activity and hydrothermal mineral deposits at southern Lake Bogoria, Kenya Rift Valley: impact of lake level changes. J Afr Earth Sci 129:623–646

Renaut RW, Owen RB, Lowenstein T, De Cort G, McNulty E, Scott JJ, Mbuthia A (2021) The role of hydrothermal fluids in sedimentation in saline alkaline lakes: evidence from Nasikie Engida, Kenya Rift Valley. Sedimentology 68:108–134

Renaut RW, Owen RB, Jones B, Tiercelin J-J, Tarits C, Ego JK, Konhauser KO (2013) Impact of lake-level changes on the formation of thermogene travertine in continental rifts: evidence from Lake Bogoria, Kenya Rift Valley. Sedimentology 60:428–468

Ring U, Albrecht C, Schrenk F (2018) The East African Rift System: tectonics, climate and biodiversity. In: Hoorn C, Perrigo A, Antonelli A (eds) Mountains, climate and biodiversity. Wiley-Blackwell, pp 391–412

Rosendahl BR (1987) Architecture of continental rifts with special reference to East Africa. Annu Rev Earth Sci 15:445–503

Rosendahl BR, Kilembe E, Kaczmarick K (1992) Comparison of the Tanganyika, Malawi, Rukwa and Turkana rift zones from analyses of seismic reflection data. Tectonophysics 213:235–256

Rosendahl BR, Reynolds DJ, Lorber P, Scott D, McGill J, Lambiase J, Derksen SJ (1986) Structural expressions of rifting: lessons from Lake Tanganyika. In: Frostick LE, Renaut RW, Reid I, Tiercelin J-J (eds) Sedimentation in the African rifts. Geol Soc Lond Spec Publ 25:29–44

Schagerl M, Renaut RW (2016) Dipping into the soda lakes of East Africa. In: Schagerl M (ed) Soda lakes of East Africa. Springer, Cham, pp 3–24

Scholz CA (1995) Seismic stratigraphy of an accommodation-zone margin rift-lake delta, Lake Malawi, Africa. In: Lambiase JJ (ed) Hydrocarbon habitat in rift basins. Geol Soc Lond Spec Publ 80:183–195

Scholz CA, Johnson TC, McGill JW (1993) Deltaic sedimentation in a rift valley lake; new seismic reflection data from Lake Malawi (Nyasa), East Africa. Geology 21:395–398

Scholz CA, Moore TC, Hutchinson DR, Golmshtok AJ, Klitgord KD, Kurotchkin AG (1998) Comparative sequence stratigraphy of low-latitude versus high-latitude lacustrine rift basins: seismic data examples from the East African and Baikal rifts. Palaeogeogr Palaeoclimatol Palaeoecol 140:401–420

Scholz CA, Rosendahl BR (1990) Coarse-clastic facies and stratigraphic sequence models from Lakes Malawi and Tanganyika, East Africa. In: Katz BJ (ed) Lacustrine basin exploration: case studies and modern analogs. Am Assoc Petrol Geol Mem 50:151–167

Shanahan TM, McKay NP, Hughen KA, Overpeck JT, Otto-Bliesner B, Heil CW, King J, Scholz CA, Peck J (2015) The time-transgressive termination of the African Humid period. Nat Geosci 8:140–144

Soreghan MJ, Scholz CA, Wells JT (1999) Coarse-grained, deep-water sedimentation along a border fault margin of Lake Malawi, Africa: seismic stratigraphic analysis. J Sediment Res 69:832–846

Street FA, Grove AT (1976) Environmental and climatic implications to late Quaternary lake-level fluctuations in Africa. Nature 261:385–390

Street-Perrott FA, Harrison SP (1985) Lake levels and climate reconstruction. In: Hecht AD (ed) Paleoclimate analysis and modelling. Wiley, New York, pp 291–340

Talbot MR, Williams MAJ (2009) Cenozoic evolution of the Nile Basin. In: Dumont HJ (ed) The Nile: origin, environments, limnology and human use. Springer, Dordrecht, pp 37–60

Tallon PWJ (1978) Geological setting of hominid fossils and Acheulian artefacts from the Kapthurin Formation, Baringo District, Kenya. In: Bishop WW (ed) Geological background to fossil man. Spec Publ Geol Soc Lond 6:361–378

Tarits C, Renaut RW, Tiercelin J-J, Le Hérissé A, Cotten J, Cabon J-Y (2006) Geochemical evidence of hydrothermal recharge in Lake Baringo, central Kenya Rift Valley. Hydrol Proc 20:2027–2055

Tiercelin J-J (1981) Rifts continentaux: tectonique, climats, sédiments. Exemples: la sedimentation dans le Nord du Rift Gregory, Kenya, et dans le Rift de l'Afar, Ethiopie, depuis le Miocene. PhD thesis, Université d'Aix Marseille II

Tiercelin J-J (1990) Rift-basin sedimentation: responses to climate, tectonism and volcanism. Examples of the East African Rift. J Afr Earth Sci 10:283–305

Tiercelin J-J, Faure H (1978) Rates of sedimentation and vertical subsidence in neorifts and paleorifts. In: Ramberg IB, Neuman ER (eds) Tectonics and geophysics of continental rifts. D Reidel, Dordrecht, pp 41–47

Tiercelin J-J, Le Fournier J, Herbin JP, Richert JP (1980) Continental rifts: modern sedimentation, tectonic and volcanic control. Example from the Bogoria-Baringo graben, Gregory Rift, Kenya. In: Geodynamic evolution of the Afro Arabic Rift System. Acad Naz dei Lincei, Rome, pp 143–164

Tiercelin J-J, Vincens A (eds) (1987) Le demi–graben de Baringo–Bogoria, Rift Gregory, Kenya: 30,000 ans d'histoire hydrologique et sédimentaire. Bull. Centr Rech Explor-Prod Elf-Aquitaine 11:249–540

Trauth MH, Maslin MA, Deino AL, Junginger A (2010) Human evolution in a variable environment: the amplifier lakes of eastern Africa. Quat Sci 29:2981–2988

Wells JT, Scholz CA, Soreghan MJ (1999) Processes of sedimentation on a lacustrine border-fault margin: interpretation of cores from Lake Malawi, East Africa. J Sediment Res 69:816–831

Wescott WA, Morley CK, Karanja FM (1996) Tectonic controls on the development of rift-basin lakes and their sedimentary character: examples from the East African Rift System. In: Johnson TC, Odada EO (eds) The limnology and paleoclimatology of the East African lakes. Gordon and Breach, Amsterdam, pp 3–21

Wetzel RG (2001) Limnology: lake and river ecosystems, 3rd edn. Academic, London
Withjack MO, Schlische RW, Olsen PE (2002) Rift-basin structure and its influence on sedimentary systems. In: Renaut RW, Ashley GM (eds) Sedimentation in continental rifts. SEPM Spec Publ 73:57–81
Yang T, Cao Y, Liu K, Wang Y, Zavala C, Friis H, Song M, Yuan G, Liang C, Xi K, Wang J (2018) Genesis and depositional model of subaqueous sediment gravity-flow deposits in a lacustrine rift basin as exemplified by the Eocene Shahejie Formation in the Jiyang Depression, Eastern China. Mar Petrol Geol 102:231–257
Zhang C, Scholz CA, Harris AD (2020) Sedimentary fills and sensitivity analysis of deep lacustrine facies in multi-segment rift basins: insights from 3D forward modeling. Sediment Geol 408:105753
Zielke O, Strecker MR (2009) Recurrence of large earthquakes in magmatic continental rifts: insights from a paleoseismic study along the Laikipia–Marmanet Fault, Subukia Valley, Kenya Rift. Bull Seismol Soc Am 99:61–70

Chapter 23
Economic Aspects of the Kenya Rift Lakes and Their Deposits

23.1 Introduction

The 'value' of the Kenya Rift lakes can be measured in many terms. Some lakes, mainly those with freshwater, are a vital water- and food source (mainly fish) for many Kenyans; others are an ecological and biological treasure house with several lakes (Bogoria-Nakuru-Elmenteita and the three Central Island lakes in Lake Turkana) recognised as UNESCO World Heritage Sites and Ramsar Sites, while Lake Baringo and Lake Naivasha are Ramsar Sites (Spawls and Mathews 2012). Nonetheless, several modern lakes and fossil lacustrine mineral deposits also contribute directly to the Kenyan economy (Tiercelin 1991). In this chapter we briefly review the economic value of the lakes and wetlands. Two main groups of resource can be recognised – mineral resources and, in broader terms, human-related resources. Our focus is the former.

23.2 Mineral Resources

In purely economic terms, the most important lacustrine mineral deposits that have been extracted in the past are the soda deposits of Lake Magadi and diatomite in the Nakuru-Elmenteita basin, but other potential resources are present including petroleum (oil shale and petroleum source rocks), which is being developed at SW Turkana, and zeolites. Buried lake sediments also provide aquifers for groundwater (e.g. McCall 1957; Sect. 4.3.1).

R. W. Renaut, R. B. Owen, *The Kenya Rift Lakes: Modern and Ancient*,
Syntheses in Limnogeology, https://doi.org/10.1007/978-3-642-25055-2_23

23.2.1 Soda Deposits of Lake Magadi

The trona deposits of Lake Magadi (Fig. 23.1a, b) are among the largest natural soda deposits in the world. They have been exploited commercially to produce soda ash (Na_2CO_3) since 1911 (Baker 1958; Hill 1964; Fig. 23.1c) and are continuously renewed by evaporation of brine recharged from thermal springs and supplemented by local runoff (Chap. 12). Soda ash is used in glass manufacturing; the production of chemicals, soaps and detergents; the scrubbing of sulfur from smokestack emissions; pulp and paper production; water treatment; oil refining; synthetic rubber production; and producing explosives.

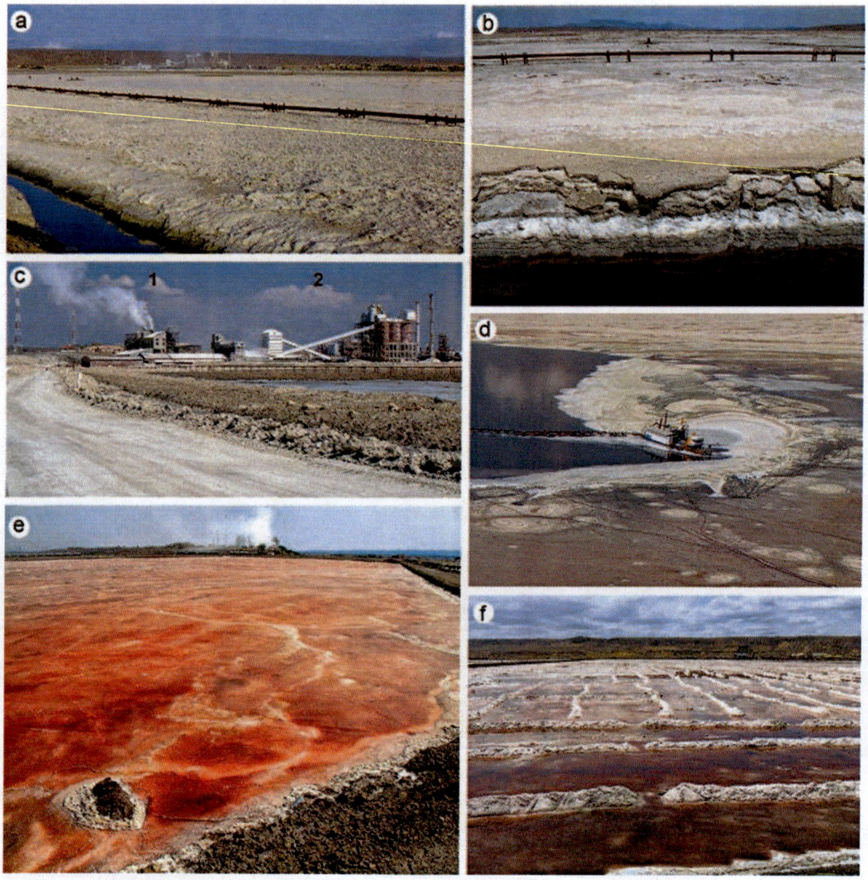

Fig. 23.1 Lake Magadi trona. (**a**) Lake Magadi trona pan with Magadi soda plant in distance. Pipe transporting water from Nguruman Escarpment crosses the trona pan. (**b**) Bedded trona with thin, muddy interbeds. (**c**) Magadi soda plant. 1: the original soda works; 2: the PAM plant. (**d**) 'Dredger' that extracts trona and brine, forming a slurry that is pumped to the soda plant. (**e**) Artificial evaporation pond east of the soda plant. Microbes, mainly Archaea, produce the reddish-purple colour in the brine. (**f**) Halite that is manually extracted from the evaporation ponds

The general extraction process has been summarised by Baker (1958), Tiercelin (1991) and Tata Chemicals (https://www.tatachemicals.com/Operations/Magadi/About-us-Magadi/Lake-Magadi-facility). Using dredgers (Fig. 23.1d), soft bedded trona crystals and interstitial brine are crushed, slurried and pumped to a 'washery plant', where they are cleaned to remove impurities and then blended. The interstitial brine is then displaced by water pumped ~35 km from the Nguruman Escarpment to reduce the NaCl concentration from 11.5% to 0.3% (Fig. 23.1a, b). The 'crushed refined soda' is then transported to rotary calciners – kilns where hot gases chemically decompose sodium sesquicarbonate to soda ash, CO_2 and water vapour. The solid soda ash product is then ground physically and screened before packaging and shipping. Some is used in Kenya (e.g. for glass making) but most is transported to Mombasa by train for export.

Salt (NaCl) is also produced from Magadi brine by fractional crystallization. Brine is pumped to pre-concentration ponds where solid deposition takes place (Fig. 23.1e). Thermonatrite ($Na_2CO_3{\cdot}H_2O$) precipitates in the ponds. When the brine contains equal amounts of sodium chloride and sodium carbonate, it is transferred to ponds to crystallise halite. With more solar evaporation, soda then salt precipitate as solids, with NaCl forming a 1-cm-thick superficial layer that is harvested manually, processed, and shipped for use mainly in Kenya (Fig. 23f; Sutton 1994).

Rising fuel costs and variations in the market value of the product have been recent issues. A new processing plant (Premium Ash Magadi [PAM] plant) built in 2006 (Fig. 23.1c) was mothballed in 2014–15. An increasing problem has been contamination, which lowers the quality of the trona. Although the trona beds are commonly of high purity they contain detrital mud deposited during flooding events (Fig. 23.1b; Chap. 12) and as wind-blown dust. An increase in soil erosion in the northern part of the drainage basin, partly because of deforestation, has led to increased influx of silt in runoff. Attempts are being made to filter the silt before it reaches the saline pan, using earth dams east of northern Nasikie Engida. In addition, some trona (e.g. in the northwestern lake quadrant) contains villiaumite (NaF) which decreases its quality (Nielsen 1999; Nielsen and Dahi 1995). High fluoride in the Kenya Rift reflects weathering of the silicate rocks (Kilham and Hecky 1973) and is often associated with thermal spring discharge and paucity of alkaline earths. Spring fluids at Magadi discharge both subaerially and subaqueously.

Soda ash production is an important contributor to the Kenyan economy. Lake Natron was also targeted for soda production, but plans were halted because of international concerns about damaging the local ecology including the most important breeding grounds of the lesser flamingo in the world.

23.2.2 Diatomite and Diatomaceous Earth

Diatomite is used as a filter, cleaning agent, and a non-toxic abrasive insecticide. Pleistocene diatomite has been quarried in two main locations in the Kenya Rift: Kariandusi east of Lake Elmenteita, and Munya wa Gicheru, southeast of Suswa

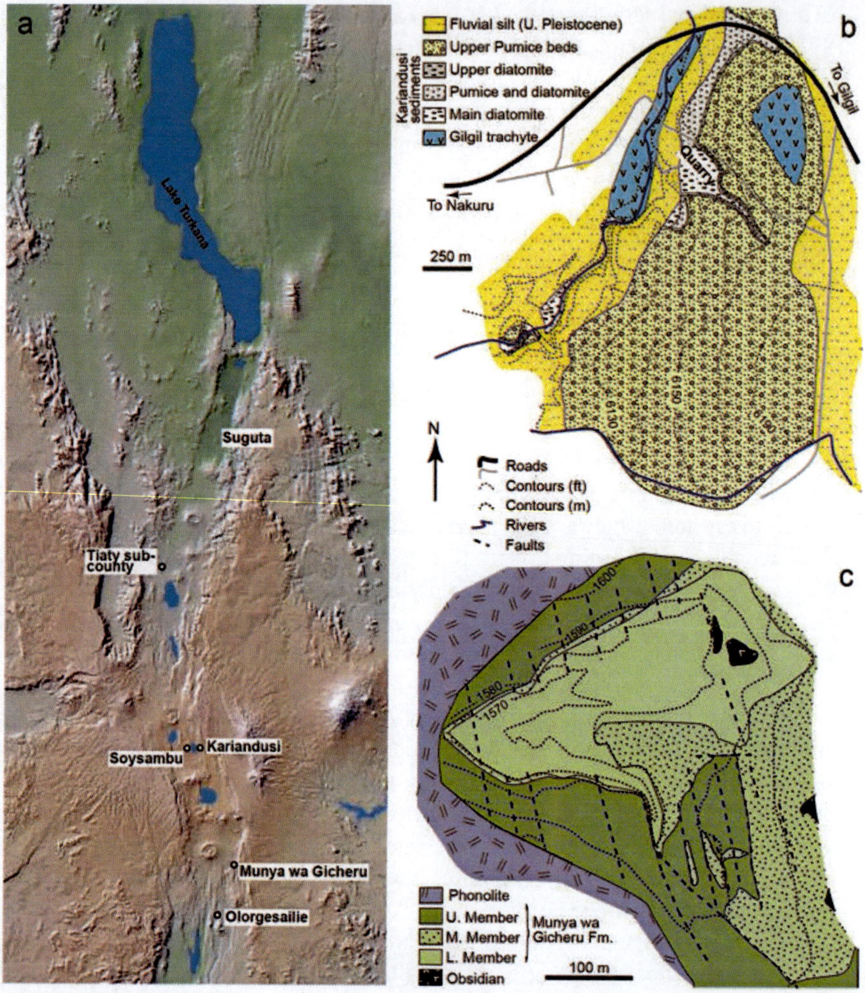

Fig. 23.2 Diatomite resources. (**a**) Distribution of exploited and potential diatomaceous deposits. (**b**) Geological map of the Kariandusi diatomite mine. (Modified from Saggerson 1991). (**c**) Geological map of the Munya wa Gicheru quarry. (After Owen et al. 2012)

volcano (Figs. 23.2 and 23.3). The Kariandusi quarry has been mined since about 1950. The Gicheru deposit is not actively worked but was quarried intermittently a few decades ago (Pulfrey and Walsh 1969; Saggerson 1991; Owen et al. 2012). Diatomite is also produced at Soysambu west of Lake Elmenteita.

A new diatomite quarry and processing plant is being developed at Tiaty (Tiati) in the Tugen Hills where Miocene diatomites (Chap. 17) are planned to be extracted. Quaternary diatomites are present in the Turkana, Suguta and Olorgesailie basins but the economics of production (purity, volume, transport costs) are currently unfavourable.

Fig. 23.3 Exploitation of diatomaceous deposits. (**a**) Kariandusi mine; (**b**) Munya wa Gicheru quarry; (**c**) Aerial view of the Munya wa Gicheru deposits

23.2.3 Petroleum Potential of the Kenya Rift Lakes

The organic deposits of the Kenya Rift lakes had long been suspected of being favourable for producing oil-prone source rocks and this became a focus of research beginning in the 1970s when major petroleum companies invested in geophysical exploration and eventually drilling (Morley 1999; Thuo 2009; Tiercelin et al. 2012a, b). Oil has been discovered in southwest Turkana (South Lokichar) from Miocene deposits and plans for production remain in progress. The source rocks are lacustrine shales; the productive reservoirs are mainly fluvial and deltaic sandstones with provenance in mixed volcanic and basement terrains (Thuo 2009; Tiercelin et al. 2012b; Muia 2015; Fig. 23.4). The basement rocks have provided stable quartz sandstones, unlike the feldspathic sandstones that derive mainly from volcanic rocks (Sect 22.4).

Organic matter in lacustrine petroleum source-rocks in East African rift basins accumulates in two main hydrological settings (Talbot 1988): (1) large, relatively

Fig. 23.4 South Lokichar basin. (**a, b**) Outcrop of the Miocene Lokhone Shale (source rock); (**c, d**) Outcrops of coarse reddish, cross-bedded fluvial sandstone (Auwerwer Sandstone; reservoir facies at depth) formerly known as 'Turkana Grits'

stable, hydrologically open lake-basins with moderate to high rates of planktonic organic matter accumulation – such fresh lakes may be deep, often thermally stratified with little sediment-mixing, and (2) hydrologically closed, chemically stratified, saline alkaline lakes with oxygenated surface waters (mixolimnion) and anoxic bottom waters (monimolimnion) (Fig. 23.5). The soda lakes usually have exceptionally high organic productivity (Melack and Kilham 1974; Melack, 1981; Oduor and Schagerl 2007; Schagerl and Burian 2016) but are typically smaller than fresh lakes and have less stable limnological regimes. The open lakes are typically associated with humid, often well-vegetated environments with high rates of chemical weathering, whereas the latter are common in semi-arid environments with limited vegetation and lower rates of chemical weathering near the lakes.

The source rocks that produce oil in the south Lokichar and north Kerio basins southwest of Lake Turkana (Fig. 23.6) are shales that were deposited in half-graben basins. Talbot et al. (2004) described the context and characteristics of the source rocks. Two main stratigraphic units were recognised, one of which is in the oil window and has the composition to have generated oil. The Lokhone Shale Member is up to 1 km thick and contains Type I, algal-prone organic matter with an average of 2.4% TOC (total organic carbon) but with intervals of up to 17% TOC. The deeper Loperot Shale contains <2% TOC and algal organic matter that was diluted by terrestrial plant material and siliciclastic sediments. The depositional environment for the Lokhone Shale was a large freshwater lake in a humid environment (Tiercelin et al. 2004). High concentrations of algal phytoplankton accumulated in the anoxic

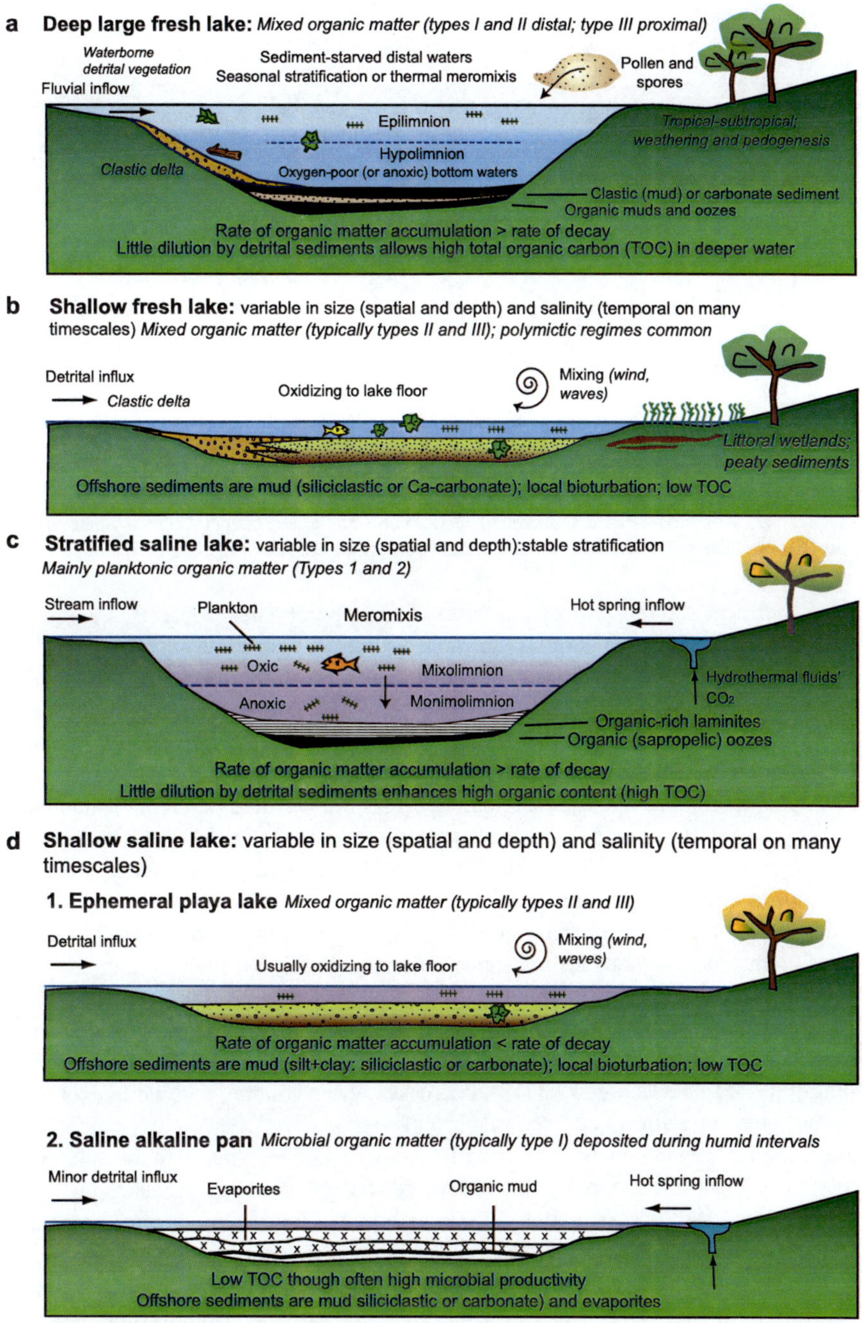

Fig. 23.5 Models for the accumulation of organic matter in the Kenya Rift lakes. (Expanded from a diagram by Renaut and Gierlowski-Kordesch (2010))

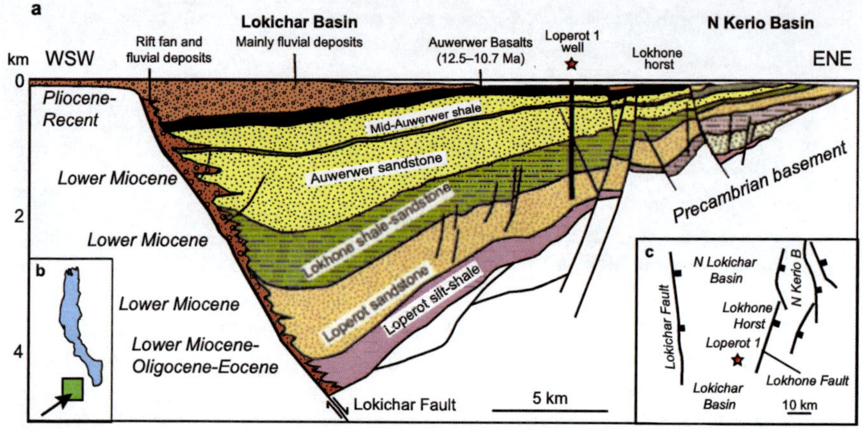

Fig. 23.6 (**a**) Cross section through the south Lokichar Basin, SW Turkana. (After Morley et al. 1999). Loperot 1 well confirmed the presence of oil in the basin in 2012. Since then, several other oil-bearing fields have been discovered. The Lokhone Shale and deeper Loperot Shale are the source rocks. The sandstones are the main reservoirs. (**b**, **c**) Location of Lokichar Basin

deep sediment-starved part of the basin. The Loperot Shale represents an earlier small lake.

Under favourable conditions, saline alkaline rift lakes generate high-quality source rocks. Two factors – evaporative concentration and geothermal fluid inflow – favour the accumulation and preservation of planktonic organic matter in lacustrine offshore muds. The modern, saline alkaline lakes of the Kenya Rift consist of shallow ephemeral lakes (Nakuru, Elmenteita, Logipi), stratified perennial saline lakes (Bogoria, Nasikie Engida), and a hypersaline pan (Magadi). All are variably recharged by hot (~35–98 °C) alkaline Na^{+}–HCO_3^{-} springs and runoff and have exceptionally high microbial productivity. The microbiota in these lakes, dominated today by *Arthrospira fusiformis*, attracts a large avifauna including lesser flamingos, which consume, recycle and introduce nutrients (P, N) that help to maintain high aquatic productivity (Fig. 23.7).

Fluids derived from geothermal sources, combined with evapoconcentration, contribute to the high productivity of these rift lakes. Evaporation increases lake salinity (~20 to >320 mg L^{-1} TDS), decreases water volume, and while excluding many organisms, allows adapted plankton to thrive. The preservation potential of this microbial organic matter varies with lake type. Organic matter in the ephemeral (playa) lakes oxidizes rapidly or is recycled, but perennial saline lakes are more likely to preserve organic matter in their sediments. The stratified (often meromictic) regime of Lake Bogoria has allowed organic matter preservation (1–6% TOC) in the lake-centre muds and oozes below a shallow anoxic monimolimnion (Tiercelin and Vincens 1987). Even Lake Magadi, a saline pan with bedded trona, has cm-scale layers with preserved organic (microbial) remains (bacteria and archaea) that settled during episodes of high productivity when the lake was flooded. Some of

Fig. 23.7 Modern environments at Lake Bogoria show the potential for saline alkaline lakes to generate oil-prone source rocks. (**a**) Green productive lake water at SE Lake Bogoria (Ng'airus) in summer 2005 when the lake was low and saline. Hot springs discharge into the warm lake waters, both onshore and offshore. (**b**) Microbial (mainly *Arthrospira*) scum blown onshore where it degrades as lake level falls. (**c**) Iridescent fluids that rose rapidly to the lake surface during coring to ~20 m depth, with Jean-Jacques Tiercelin (Université de Rennes) during July 1996. These oily substances were not due to contamination. (**d**) Black organic-rich (>7.5% Total Organic Matter: TOC) muds at the base of a core from the southern sub-basin of Lake Bogoria. Lens cap has 50 mm diameter

that microbial organic matter (1–4% TOC) survives in the evaporites, muds and clays in the anoxic subsurface brine.

A common factor in these saline lake basins is the reliable supply of carbon, some from geothermal fluids of Na^+–HCO_3^-–(CO_3^{2-}) composition. Hot springs that provide those fluids discharge along faulted lake-shorelines and on lake floors. They supplement bicarbonate ions in dilute inflow waters (rivers, groundwaters) derived from silicate mineral hydrolysis during weathering of volcanic rocks in their drainage basins. Hypersaline lakes with chloride or sulfate as their dominant anion are often less productive.

Geothermal sources of recharge that are deeply sourced (from >1 km depth) respond more slowly to short- to medium-term climate variations than those that affect surface and near-surface waters. Lakes Bogoria, Nasikie Engida, and Magadi depend strongly on geothermal inflow (Sect. 4.9). About 35–40% of the annual recharge at Lake Bogoria, and perhaps 80–90% of the recharge at Nasikie Engida and Lake Magadi originates from alkaline hot-spring fluids. At Lake Magadi, the modern hot-spring inflow evaporates rapidly and precipitates trona. In contrast, perennial lakes Bogoria and Nasikie Engida shrink and become shallower when the climate becomes drier, but still retain saline, alkaline surface water. Consequently, organic matter in their sediments has a higher preservation potential than that of Lake Nakuru and Lake Elmenteita, which have lower hydrothermal recharge. Geothermal fluids help to preserve organic matter during short periods of climate change (up to thousands of years).

Optimal conditions for generating source rocks in saline, alkaline rift lakes occur in relatively deep (>10 m), steep-sided faulted basins with small catchments that have enough hydrothermal inflow to maintain lake water when the climate is more arid. Planktonic organic matter, favourable for types I and II kerogen, then forms in shallow oxygenated waters during humid periods and is deposited offshore in anoxic muds, often below a chemocline. Small catchments limit coarse sediment supply and macrovegetal organic influx that generates type III kerogen. If climatic conditions become drier after the period when organic-rich muds were deposited, saline hydrothermal fluids may then dominate the lake water recharge. The lake may shrink in surface area and depth but, with steep faulted margins, may maintain surface water that both limits subaerial exposure and reduces oxidation.

Evapoconcentration of alkaline brines after periods of organic mud sedimentation may lead to the precipitation of bedded evaporites or cherts. Both lithologies provide effective caprocks and the trona itself may be a resource. Bedded cherts form because silica solubility decreases when the lake-water temperature falls (cooling of geothermal fluids) or when the pH of the lake brine decreases abruptly (e.g. during magadiite precipitation). High geothermal gradients in active rift basins, in turn, promote the maturation of lacustrine microbial organic matter.

Member C of the Miocene Ngorora Formation (Tugen Hills, W Baringo) is comprised of a series of lacustrine, zeolitic organic-rich shales that become silicified near their top (Renaut et al. 1999; Fig. 23.8). Lake Magadi contains laminated carbonaceous mudstones (High Magadi Beds) capped by trona evaporites, and older organic mudstones interbedded with Pleistocene cherts (Green Beds). A decrease in water temperature and (or) pH both potentially created caprocks above alkaline, lake-mudstone sequences rich in planktonic (Type I) organic matter.

Stratified saline, alkaline lakes fed partly by geothermal fluids provide highly favourable conditions for source-rock generation in continental rifts. Those potentially productive intervals nonetheless are more limited in lateral extent and thickness than those of large, open, freshwater lakes that have lower organic productivity.

23.2.4 *Zeolites*

Zeolitic sediments are common in the modern saline, alkaline lakes (Magadi, Bogoria) and in analogous ancient lake sediments. Zeolites have a crystal structure that enables them to serve as molecular sieves that adsorb gases and liquids. They are widely used in agriculture (animal feed, fertilisers, soil additives), water and air

Fig. 23.8 Zeolitic shales of lacustrine Member C of the Ngorora Formation at Kapkiamu, Tugen Hills. (**a**) trackside outcrop below Kapkiamu village. Section is ~1.5 m high. (**b**) and (**c**) refer to the images on the right. (**b**) Fissile zeolitic shale with silica veining (arrows), linked to seismic shaking and injection from alkaline (or hot) pore water, or precipitation from alkaline groundwater after desiccation. (**c**) Organic-rich laminated shales containing up to 7% TOC. Hammer (arrow) provides scale. This small section (**a**) records a deeper lake that was succeeded by a shallow evaporative saline, alkaline lake

filtration, and as detergent. Some zeolitic deposits are mined for use as dimension stone, aggregate and in cement. Although widespread in the Kenya Rift, few reported deposits are pure enough for exploitation. Some beds in zeolitic tuffs (mainly clinoptilolite and analcime) of Miocene Member C of the Ngorora Formation in the Tugen Hills are of high purity (Ego 1994; Renaut et al. 1999) but have not been evaluated as a resource. Zeolitic lake beds also exist in the Miocene Tambach Formation (Kerio-Elgeyo, west of Baringo) and at several locations in the Turkana and Magadi basins.

23.2.5 Other Lacustrine Mineral Resources

Heavy mineral sands are present along the modern shoreline of Lake Turkana and enriched in some of the sandstones in the region. Reid and Frostick (1985) showed that they are composed mainly of ilmenite and magnetite grains and concentrated in the swash zone of the modern lake (Fig. 23.9a–c). The modern littoral zones of the other rift lakes are of low energy, or lack sources of heavy metals, so metalliferous sands are absent. However, the heavy mineral layers at Lake Turkana are thin and commonly discontinuous. Rare minerals have not been reported so they are unlikely to form an economic resource.

The lacustrine avifauna has produced small local deposits of guano. Deposits are present on islands in Lake Baringo (notably Lesukut) where they coat cliffs

Fig. 23.9 Miscellaneous lacustrine resources. Heavy mineral sands forming in the swash zone of sandy beaches near Ferguson's Gulf (**a**, **b**) and Koobi Fora (**c**), Lake Turkana. (**d**, **e**) Guano accumulation (semi-lithified phosphates) on Lesukut island, Lake Baringo. Cormorants (inset, (**d**)) and fish eagles are responsible

(Fig. 23.9d, e). Similar coatings are present on trachyte fault escarpments west of Lake Baringo. The island guano is produced mainly by cormorants and fish eagles. Similar coatings are found on the islands of Lake Turkana. The richest deposits are commonly linked to major joints where soluble phosphates and nitrates are flushed downwards, and then evaporate forming vertical laminites or miniterraces. Some may be precipitated microbially, either biomediated or using a microbial template. Unlike phosphatic deposits in other rift lakes such as Lake Manyara in Tanzania (fish bones at Minjingu: Schlüter 1987) and Lake Malawi (guano on Boadzulu Island and fish bone is adjacent water: Owen and Renaut 1993), they are too small for more than very local usage as a phosphate fertiliser. Phosphorite in the Miocene Lukeino Formation in the Tugen Hills, Baringo, represents a period of lake eutrophication (Dericquebourg et al. 2015). Small nitrate (nitratine) deposits have been reported at several localities including in diatomaceous sediments at Lake Turkana (Owen and Renaut 1990) and at Lake Magadi (Baker 1958) but none currently merit development.

23.3 Human-Related Resources

The lakes are an obvious resource for humans and their activities, including provision of potable water and food, and providing attractions for tourism. A few examples can be highlighted:

- Fish are a vital food resource for local people in the fresher rift lakes and their affluents: mainly in lakes Victoria, Baringo, Naivasha and Turkana. The main fish caught are *Tilapia, Labeo cylindricus* (a type of carp), catfish (*Clarias*) and lungfish (*P. aethiopicus*) (Sects. 4.7 and 8.6.3; Fig. 8.12). Crocodiles provide food mainly in Lake Turkana.
- The lakes are important water sources for humans and other animals and will likely become more so with increasing aridification providing those lakes maintain fresh waters. Lakes Victoria, Baringo and Naivasha provide drinking water for local communities and for domestic use. Rivers flowing into Lake Naivasha support an important horticultural industry. Small lakes and reservoirs supplement the major lakes. Lake Turkana water is brackish and unpleasant for human consumption but important for local wildlife. The fresh lakes also provide water for local irrigation; some is now pumped for geothermal development.
- Reedbeds along the shoreline on the freshwater lakes and other wetlands are a source of papyrus and sedges that are dried and used for building materials. Water hyacinth has spread across much of East Africa where it clogs the shoreline (e.g. Lake Victoria) and is generally considered a problem. However, when mixed with animal dung and dried it can be used as a biofuel for domestic cooking.
- Half of the Kenya Rift lakes are UNESCO World Heritage sites recognised for their natural beauty, spectacular wildlife (especially birds) and their rich fauna and flora (Melack 1996). The Turkana National Parks are recognised for their importance as a stop-over site for migrating birds and as breeding grounds for the Nile crocodile and hippopotamus. In addition, the sedimentary rocks, fossils and archaeology in northeastern part of the lake near Koobi Fora basin have been a focus of paleoanthropological research for 50 years. Furthermore, important hominin fossil-sites have also been recognised west of the lake.
- The soda lakes host some of the highest bird diversities and densities in the world. With Lake Natron and to a lesser extent Lake Magadi, they encompass the most important habitat and breeding area for lesser flamingos in the world, often

exceeding a million birds. They also are a major breeding and nesting area for other birds including great white pelicans. Lake Bogoria has the bonus of the most spectacular geothermal activity in Africa (at least 20 geysers and almost 200 hot springs) but this has been suppressed since about 2011 by rising lake levels that have drowned many vents (Chap. 9). Lake Baringo and Lake Naivasha are also major destinations for bird enthusiasts. Even the small lakes, such as Kamnarok, Simbi and Ol' Bolossat increasingly attract visitors.

– Ancient lake sediments are commonly associated with archaeology sites either on palaeoshorelines or land surfaces interlayered with lake beds. Examples include the Olorgesailie Prehistoric Site (Chap. 19) and Neolithic sites at Lake Baringo and Lake Turkana. These increase the attraction of many lakes for tourism.

References

Baker BH (1958) Geology of the Magadi area. Geol Surv Kenya Rep 42

Dericquebourg P, Person A, Ségalen L, Pickford M, Senut B, Fagel N (2015) Environmental significance of Upper Miocene phosphorites at hominid sites in the Lukeino Formation (Tugen Hills, Kenya). Sediment Geol 327:43–54

Ego JK (1994) Sedimentology and diagenesis of Neogene sediments in the central Kenya Rift Valley. MSc thesis, University of Saskatchewan, Saskatoon

Hill MF (1964) Magadi; the story of the Magadi Soda Company. Kynoch Press, Birmingham

Kilham P, Hecky RE (1973) Fluoride: geochemical and ecological significance in East African waters and sediments. Limnol Oceanogr 18:932–994

McCall GJH (1957) Geology and groundwater conditions in the Nakuru area. Tech Rep Ministr Works (Hydraul Branch) Kenya 3

Melack JM (1981) Photosynthetic activity of phytoplankton in tropical African soda lakes. Hydrobiologia 81:71–85

Melack JM (1996) Saline and freshwater lakes of the Kenyan rift valley. In: McClanahan TR, Young TP (eds) East African ecosystems and their conservation. Oxford University Press, New York, pp 171–190

Melack JM, Kilham P (1974) Photosynthetic rates of phytoplankton in East African alkaline, saline lakes. Limnol Oceanogr 19:743–755

Morley CL (ed) (1999) Geoscience of rift systems–evolution of East Africa. AAPG Stud Geol 44

Morley CK, Wescott WA, Stone DM, Harper RM, Wigger ST, Day RA (1999) Geology and geophysics of the Western Turkana Basins, Kenya. In: Morley CL (ed) Geoscience of rift systems–evolution of East Africa. AAPG Stud Geol 44:19–54

Muia G (2015) The "Turkana Grits": potential hydrocarbon reservoirs of the Northern and Central Kenya Rifts. PhD Thèse, Université de Rennes 1

Nielsen JMN (1999) East African magadi (trona): fluoride concentration and mineralogical composition. J Afr Earth Sci 29:423–428

Nielsen JMN, Dahi E (1995) The occurrence of fluoride contaminated magadi (trona) in Kenya and Tanzania. In: Dahi E, Bregnhøj B (eds) Proc 1st Int workshop on fluorosis prevention and defluoridation of water, pp 17–22

Oduor SO, Schagerl M (2007) Phytoplankton primary productivity characteristics in response to photosynthetically active radiation in three Kenyan rift valley saline-alkaline lakes. J Plankton Res 29:1041–1050

Owen RB, Lee RKL, Renaut RW (2012) Early Pleistocene lacustrine sedimentation and diatom stratigraphy at Munya wa Gicheru, southern Kenya Rift Valley. Palaeogeogr Palaeoclimatol Palaeoecol 331–332:60–74

Owen RB, Renaut RW (1990) Nitrate deposition in the Turkana basin of northern Kenya. J Afr Earth Sci 10:701–706

Owen RB, Renaut RW (1993) Resource potential of Lake Malawi. International Development Research Centre (Ottawa) Report 3-P-86-1028-02

Pulfrey W, Walsh J (1969) The geology and mineral resources of Kenya (second revision). Geol Surv Kenya Bull 9

Reid I, Frostick LE (1985) Beach orientation, bar morphology and the concentration of metalliferous placer deposits: a case study, Lake Turkana, N Kenya. J Geol Soc Lond 142:837–848

Renaut RW, Ego JK, Tiercelin J-J, Le Turdu C, Owen RB (1999) Saline, alkaline palaeolakes of the Tugen Hills-Kerio Valley region, Kenya Rift Valley. In: Andrews P, Banham P (eds) Late Cenozoic environments and hominid evolution: a tribute to Bill Bishop. Geological Society, London, pp 41–58

Renaut RW, Gierlowski-Kordesch EK (2010) Lakes. In: James NP, Dalrymple RW (eds) Facies models 4. Geological Association of Canada, St John's, pp 541–575

Saggerson EP (1991) Geology of the Nairobi area. Geol Surv Kenya Rep 98

Schagerl M, Burian A (2016) The ecology of African soda lakes: driven by variable and extreme conditions. In: Schagerl M (ed) Soda lakes of East Africa. Springer, Cham, pp 295–320

Schlueter T (1987) Paleoenvironment of lacustrine phosphate deposits at Minjingu, northern Tanzania, as indicated by their fossil record. In: Matheis G, Schandelmeier H (eds) Current research in African earth sciences. Balkema, Rotterdam, pp 223–226

Spawls S, Mathews G (2012) Kenya: a natural history. Bloomsbury, London

Sutton I (1994) Salt in Kenya: a survey of literature. J E Afr Res Develop 24:163–182

Talbot MR (1988) The origins of lacustrine oil source rocks: evidence from the lakes of tropical Africa. In: Fleet AJ, Kelts K, Talbot MR (eds) Lacustrine petroleum source rocks. Geol Soc Lond Spec Publ 40:29–43

Talbot M, Morley CK, Tiercelin J-J, Le Hérissé A, Potdevin J-L, Le Gall B (2004) Hydrocarbon potential of the Meso-Cenozoic Turkana Depression, northern Kenya. II. Source rocks: quality, maturation, depositional environments and structural control. Mar Petrol Geol 21:63–78

Tiercelin J-J (1991) Natural resources in the lacustrine facies of the Cenozoic rift basins of East Africa. In: Anadón P, Cabrera L, Kelts K (eds) Lacustrine facies analysis. IAS Spec Publ 13:3–38

Tiercelin J-J, Potdevin L, Morley CK, Talbot MR, Bellon H, Rio A, Le Gall B, Vétel W (2004) Hydrocarbon potential of the Meso-Cenozoic Turkana Depression, northern Kenya. I. Reservoirs: depositional environments, diagenetic characteristics, and source rock–reservoir relationships. Mar Petrol Geol 21:41–62

Tiercelin J-J, Potdevin J-L, Thuo PK, Abdelfettah Y, Schuster M, Bourquin S, Bellon H, Clément J-P, Guillou H, Nalpas T, Ruffet G (2012b) Stratigraphy, sedimentology and diagenetic evolution of the Lapur Sandstone in northern Kenya: implications for oil exploration of the Meso-Cenozoic Turkana depression. J Afr Earth Sci 71-72:1–37

Tiercelin J-J, Thuo P, Potdevin J, Nalpas T (2012a) Hydrocarbon prospectivity in Mesozoic and Early-Middle Cenozoic Rift Basins of Central and Northern Kenya, Eastern Africa. In: Gao D (ed) Tectonics and sedimentation: implications for petroleum systems. Am Assoc Petrol Geol Mem 100:179–207

Tiercelin J-J, Vincens A (eds) (1987) Le demi–graben de Baringo–Bogoria, Rift Gregory, Kenya: 30,000 ans d'histoire hydrologique et sédimentaire. Bull. Centr Rech Explor-Prod Elf-Aquitaine 11:249–540

Thuo P (2009) Stratigraphic, petrographic and diagenetic evaluation of Cretaceous/ Paleogene potential reservoir sandstones of western Turkana, Kenya. Implications on the petroleum potential of Northwestern Kenya. PhD thesis, Université de Bretagne Occidentale, Brest

Chapter 24
Afterthoughts and Perspectives

24.1 Drivers of Change in the Contemporary Rift Valley Lakes

The modern alkaline lakes of the Kenya Rift are successors to a long series of former lakes that have existed since the Miocene. They are globally important given their diverse and unusual alkaline chemistry (Schagerl and Renaut 2016), distinctive mineralogy (Hay 1970; Surdam and Eugster 1976), and often unique macro- and micro-floras and faunas (Schagerl 2016). They lie in an imposing, but beautiful tectonic landscape formed by large border-fault escarpments, smaller fault blocks and regularly spaced volcanoes. Volcanic ash and lavas are present along the rift with hot springs, geysers and fumaroles locally recording present and former thermal activity. The altitudinal variability of the rift floor and its lakes (~275–1890 m) contributes to the diverse climates of the region, ranging from relatively moist and cool around Lakes Nakuru, Elmenteita and Naivasha to hot and more arid conditions at Lakes Magadi, Logipi and Turkana. Rainshadow effects amplify a negative precipitation-evaporation balance over the rift floor, which is replenished with water via rivers from the steep rift margins and from rift-floor hot springs.

The Kenya Rift lakes are ecologically important because they support many unusual organisms including endemic species, and locally abundant wildlife, including ~1.5 million lesser flamingos. The lake waters also provide important feeding stations for migratory birds (Harper et al. 2016). However, a combination of rapid human population growth, economic activity, and the effects of planetary global warming have had deleterious impacts that threaten current ecosystems, biodiversity and the existence of several Kenyan lakes. In this chapter, we comment on only some issues related to the rift lakes. There is a large and growing literature on the environmental problems faced by Kenya in the next few decades.

R. W. Renaut, R. B. Owen, *The Kenya Rift Lakes: Modern and Ancient*,
Syntheses in Limnogeology, https://doi.org/10.1007/978-3-642-25055-2_24

24.1.1 *Pristine Lakes and Catchments?*

The Kenya Rift lakes (Fig. 24.1) have changed dramatically since the first arrival of Europeans more than 120 years ago. At that time, the lakes and their drainage basins appeared to have changed little from ancient times with their abundant wildlife and

Fig. 24.1 Kenya Rift lakes. (**a**) View northwest from volcanic South Island with moderately saline Lake Turkana extending beyond the horizon. Major threats are related to the construction of dams and potential agricultural developments along the Omo River, its dominant source of water. (**b**) Saline Lake Logipi in the Suguta provides a home for flamingos when other lakes become too fresh to support their major food sources. Geothermal development could impact Logipi's spring inputs, which maintain the lake during dry years (**c**) Freshwater Lake Baringo expands and contracts significantly and has an ecology that is drastically different from that of the nineteenth century, with different fish species and recent (post-2016) growth of water hyacinth, which invaded from Lake Kichirtit. (**d**) Saline Lake Bogoria lies at the centre of a national reserve and has become fresher in recent years due to increased water inputs. These resulted in flooding across much of the Sandai Plain at its northern end (foreground). Grazing pressure in adjacent areas causes problems, with cattle sometimes using reserve lands. (**e**) Saline Lake Nakuru is surrounded by a game park but is under a pollution threat from the expansion of nearby Nakuru city and its industries. (**f**) Saline Lake Elmenteita is a relatively small lake in a catchment dominated by farms and a nature conservancy. (**g**) Freshwater Lake Naivasha competes for water resources with a thriving horticulture industry and a growing population. (**h**) Saline Nasikie Engida is maintained by hot spring inflows with CO_2 gases resulting in unusual evaporite minerals. Relative remoteness has helped protect the area to date, but ecotourism may result in future changes to its catchment where eroding roads have exposed silt to deflation. (**i**) Saline Lake Magadi is impacted by a trona mining industry and increasing demands for ecotourism. (**j**) Saline Lake Natron, in Tanzania, is the major home for nesting flamingos, but is of interest for its trona reserves

flourishing vegetation. However, even then they were not pristine lakes unaffected by humans. Marchant et al. (2018), for example, reported that the first large-scale human influences across East Africa involved the introduction of domesticated livestock from ~4000 y BP with extensive forest clearances taking place at ~2500 y BP. They suggested that during the last 1300 years, the major drivers of change were population growth, regional movements of people, and expansion of trade routes. For instance, at Lake Bogoria, van der Plas et al. (2019) found evidence of a decline in forests after about 1430 AD that they attributed to human activity, with an increase in siliciclastic influx to the lake, implying catchment erosion after ~1800 AD.

During the late nineteenth century, European exploration was carried out mainly by adventurers, hunters and missionaries, who produced often crude maps of the areas that they had visited together with stories of their "new discoveries". There were no baseline scientific investigations to document the geology, hydrology or life in the wide-ranging habitats through which they briefly travelled with others. The first serious scientific research expeditions that examined the geology (i.e. looked at the rocks) and biology (i.e. recorded the flora and fauna) of the lakes took place in the 1920s and 1930s, although Gregory (1896) and others had visited the area in the 1890s. Although the volcanic bedrock has not changed since then, sedimentation in the lakes has been altered, with modifications in their biology, hydrology and chemistry. There are no pristine water bodies. Even the lakes first examined by Worthington (1932), Beadle (1932), Rich (1933), Worthington and Ricardo (1936a, b), Jenkin (1936) and others had already been modified by human activities when studied and analysed. Tarras-Walberg et al. (2003) claimed that their study of Lake Kichirtit (Baringo) was an investigation of a "pristine" new lake that had formed in 1997, but it had formed in 1994 and by then was likely receiving effluent from the Perkerra River irrigation scheme and other agricultural sources (cf. Ochuka et al. 2019). Many environmental pressures have increased since the early pioneering investigations and more alterations will occur in the future.

24.1.2 Global Warming and the Rift Lakes: Past, Present and Future

The water levels and volumes of the rift valley lakes are very sensitive to climate variability as demonstrated by the 300 m fall in Palaeolake Suguta over the last few thousand years of natural climate change. Most of the lakes have been at very high levels through much of the last decade, flooding formerly riparian habitats and causing the loss of agricultural land and many homes (Government of Kenya and UNDP 2021). Lake Baringo, for example, flooded large areas to its east and south. Lake Bogoria expanded across the Sandai Plain to its north. Many game park roads around Lake Nakuru were flooded in 2015–2023. Lake Naivasha was also relatively high, but hydrological estimates suggest it would have been even higher were it not for water abstraction for the local flower industry (Becht and Harper 2002). The causeway across Lake Magadi was flooded and impassable in 2018–2019. These high lakes were considered a disastrous anomaly, yet historical and palaeoenvironmental records show similar and higher levels during the last 200 years.

Looking forward, global warming is likely to have significant impacts on rainfall, seasonality and storm events, with major implications for the rift lakes (Githumbi et al. 2019). Much of the palaeoclimatic variability reconstructed from the Kenya Rift has been previously interpreted in terms of changes in sea surface temperatures related to developments in the El Nino Southern Oscillation and the Indian Ocean Dipole, along with their consequent impacts on pressure gradients and the position of the Congo Air Boundary (CAB) and the Inter Tropical Convergence Zone (ITCZ) (Junginger and Trauth 2013; Junginger et al. 2014). Predictive computer models show considerable variability in these parameters (Rowell and Chadwick 2018). In some studies, increases in precipitation are predicted for the later part of the twenty-first century, after a period of hydrological stress with a warmer and variable climate through the next few decades (Shongwe et al. 2011; Seimon et al. 2013). Seimon and Picton Phillips (2012) suggest that rainfall increases will develop due to lengthening of the short rainy season. In contrast, others have inferred increasingly dry conditions into the distant future (Funk et al. 2008). In addition to uncertainties in climate models, individual lakes may respond quite differently to changing evaporation and precipitation rates, reflecting contrasts in basin morphology and elevation, with seasonality varying according to the geographic location relative to the positions of the CAB and ITCZ.

Given historical variability and future global warming scenarios, it is probable that recent high lake levels will be reversed over the short to medium term and it is even plausible for Lakes Nakuru, Elmenteita and Natron to dry up as they have done in the past. Should such events take place they would have significant impacts on wildlife and particularly flamingos. Indeed, palaeoenvironmental research has documented past desiccation in several basins during recent centuries. Lake Bogoria, for example, was reduced to shallow spring-fed brine pools between 690 and 950 AD and was very low with two separate lakes from 1350 to 1800 AD (de Cort et al. 2018) with Lake Baringo also dry around 230 years ago (Bessems et al. 2008). Such changes are dramatic and often fast.

24.1.3 Human Impacts on Future Lake Levels

There is an increasing need for better management of water resources in order to avoid worse-case scenarios, including the possibility of several lakes completely drying out. For example, increasing water abstraction from Lake Naivasha for horticulture could pose a risk of desiccation should there be a return to drier climates, which can be expected given the known long-term wet-dry cyclicity of these water bodies. Planning also needs to consider the potential for even greater changes related to the effects of global warming on the regional East African climate.

International planning with neighbouring countries is becoming increasingly important given that hydrological modelling of even the largest of all the Kenya lakes, Lake Turkana. For example, water levels dropped during reservoir filling at the Gibe dams along the Omo River in Ethiopia. The Omo supplies about 90% of the inflow to Lake Turkana (Avery and Tebbs 2018) and any reduction in its contribution could have disastrous effects. Large-scale sugar irrigation schemes are also being developed/proposed in the Omo Valley which are an even greater threat given the increased evaporation rates that can be expected. Estimates suggest a possible Omo water loss of up to 50%, which could reduce lake level by as much as 20 m (Avery 2014). Should the lake continue to be reduced in volume there would also be a consequent increase in dissolved solutes that would increase its salinity, change the phytoplankton composition and alter primary productivity. These impacts would likely have a deleterious effect on the Turkana fishery, which, although unstable, helps to provide protein to local peoples.

Irrigation has had relatively minor impacts on other lakes in the rift system to date. However, future agricultural development is likely to increase abstraction from river inflows to the lakes. Boreholes may also have an increasing impact on groundwater supplies as these are drilled to supplement river water for agricultural use and to supply drinking water to local communities.

24.1.4 Faunal and Floral Stability-Instability of the Lakes and their Catchments

The floras and faunas of the rift lakes are highly dependent on natural lake level variability and climate, which change considerably with time. Lower lake level periods are associated with higher salinities and, under favourable conditions, lead to *Arthrospira fusiformis* growth, which in turn promotes thriving lesser flamingo populations (Vareschi and Jacobs 1985; Harper et al. 2016). In contrast, during wetter intervals, lake levels rise, and waters become fresher with *A. fusiformis* replaced by other Cyanobacteria and (or) diatoms less suitable for lesser flamingos. Hence there are natural variations for each lake with flamingos compensating by migrating between different water bodies. Such lake level and ecological variability and consequent avian changes can be expected to continue albeit with global warming possibly imposing a different long-term climatic pattern.

Major ecological changes have also been brought about by deliberate and accidental introductions of fauna and flora. Fish populations in the fresh waters of Lakes Baringo, Naivasha and Victoria have already changed markedly following deliberate introductions of new fish species such as Nile perch as well as plants such as water hyacinth (Mwanuzi et al. 2005; Sitoki et al. 2010). Fisheries have, for example, intermittently crashed at lakes Turkana, Baringo and Victoria following ecological change, poor management policies and minimal catch enforcement (Kolding 1993; Britton and Harper 2008; Hickley et al. 2008). Lake Victoria exemplifies a recurrent problem in a dramatic way. Increased populations and demand resulted in overfishing and a drift to catches of smaller species (Chapman et al. 2008). Large Nile perch were introduced to compensate for this trend, in the belief that they would make use of the smaller fish. This seemed to work for a number of years, but eventually the fishery collapsed, perhaps partly also due to increased eutrophication.

24.1.5 Protection Measures and Drivers of Environmental Change

Tourism and ecotourism have been a driver for the protection of the Kenya Rift lakes and parts of their catchments. Lake Turkana is a designated UNESCO World Heritage Site in recognition of its many archaeological sites with Sibiloi (NE Turkana), Central and South Islands designated as parks to protect wildlife. Lakes Nakuru, Elmenteita and Bogoria also have UNESCO status and, along with Lake Baringo and Lake Naivasha, protection under the 1972 Ramsar Convention on Wetlands. Lake Nakuru is designated a national park and Lake Bogoria a national reserve. Much of Lake Elmenteita lies within the Soysambu Conservancy with other local conservancies such as Ruko and Kaptuya at Lake Baringo, Chuine and Irong at Lake Bogoria, and the Lake Magadi Conservancy helping to increase environmental awareness in recent years, albeit with questions over land rights causing some concerns (Greiner 2012). Landscapes and wildlife form the focus at Longonot and Hell's Gate national parks, south of Lake Naivasha.

Despite the recognition of these varied reserves, environmental degradation remains an important issue with rapid population growth being a major driver of problems. Given population trends, it is probable that environmental issues will increase in the future. Oduor and Kotut (2016) and Harper et al. (2016) have both pointed out the importance of flamingos for tourism development in the area and that these birds require multiple lakes to survive in large numbers. Consequently, it is important for the future to think of Kenya lakes as an integrated system rather than as individual water bodies.

24.1.6 Development Issues and their Impacts

In the mid-1970s tarred roads stopped at Nakuru with Lakes Baringo, Bogoria, Logipi and Turkana all being protected by their relatively remote locations. The paved road network was subsequently expanded to Baringo by the 1980s. More recently, economic developments such as the discovery of oil southwest of Lake Turkana and in South Sudan have led to an expansion of tarred roads north of Lake Baringo. A similar road from Nanyuki, outside the rift, now reaches Loiyangalani on the shores of southeast Lake Turkana, stimulated by the development of a 160 km^2 wind-power station. Geothermal power exploration has established new or improved earth roads into the northeast Suguta Valley, near Lake Logipi, and on the Quaternary volcanoes north of Lake Baringo. These developments provide arterial routes for power projects but also open areas for other forms of much-needed economic development that may have secondary environmental implications. Water, for example, is being extracted from NW Lake Baringo for use in geothermal exploration with the possibility of water also being distributed to local people along the pipeline. The new roads north of Baringo will inevitably increase tourism to the Suguta Valley and Lake Logipi-- one of the most inaccessible parts of the Kenya Rift. The spectacular scenery, in an area still largely unspoilt, will lead to more visitors, infrastructure and development. Planning for this expansion is essential to protect the natural environment.

In the 1970s, the Nakuru area had a relatively low population density, but this has grown significantly as has the city of Nakuru. Similarly, increased populations and hotel developments at Elmenteita have changed the environmental pressures at that lake (Oduor and Kotut 2016) with overgrazing and clearance of original vegetation leading to increased soil erosion and sediment load delivery to the central Kenya Rift lakes. Siltation is an increasing problem at the northern end of Lake Magadi thanks to soil erosion upstream following deforestation as far as Suswa and Narok. Trucks have been used to move some of the silt and barriers have been constructed to trap sediment, but better land management is required in the problematic source areas. Tata Chemicals Magadi (formerly the Magadi Soda Company), which mines trona at Lake Magadi, reports reduced trona purity from 99% to 95% due to siltation (Harper et al. 2016). Small lakes are particularly prone to catchment change with most already shrinking or damaged. For example, Peter Kahare, writing for the Inter Press Service (Kenya: Key Lakes Succumb to Human Activities, 19 November 2019), reported that tree felling, charcoal production, grazing, farming activities and sand extraction contributed to siltation and contraction of Lake Kamnarok (Kerio Valley), with crocodile numbers reduced from 30,000 to <5000. Similar changes are happening at other lakes.

Eutrophication has been reported from several lakes and has been a particular problem at Lake Victoria, where it is associated with deforestation, fertiliser use and erosion of surrounding soils (Mwirigi et al. 2005; Olago et al. 2021). Industrial pollution is comparatively limited as yet, but population and economic growth may lead to increasing pressures. Lake Magadi has been mined for trona for a century with mining also proposed for Lake Natron, which would have threatened the major breeding ground for lesser flamingos. However, development plans were temporarily dropped there in response to opposition from local Maasai (Harper et al. 2016). The growth of industry in Nakuru and inferred pollution release to Lake Nakuru has been proposed to explain flamingo mortality there, but the causes are still controversial (Harper et al. 2016; Krienitz 2018). Active geothermal exploration may lead to new power plants that could negatively alter the plumbing systems of springs and geysers, which play an important role in maintaining several lakes and which have important influences on their water chemistry. Geothermal development may also increase water abstraction. The hot springs at Lake Bogoria, for example, are a major tourist attraction and have sustained the lake during periods of drought. A full impact assessment would be essential before any development that might damage this World Heritage Site.

All of these developments reflect the key driver of population expansion driven by natural growth and immigration to the lakes region (Marchant 2021). Such impacts are also motivated by a natural desire to improve lifestyles. The future of the rift lakes requires enhanced environmental awareness. Improved continuous monitoring of lake levels, ecology and habitats is essential in order to assist with good management of the lakes and their catchments, particularly with regard to their water resources. Furthermore, several of the environmental issues require international co-operation given cross-border river systems, migratory wildlife and global warming threats. These requirements are being recognised locally with institutional frameworks being developed to enhance cooperation and scientific research through organisations such as the Centre for Aquatic Research and Education (ACARE) (Obiero et al. 2020).

24.2 Non-Anthropogenic Drivers and the Importance of Time

24.2.1 Tectonic, Volcanic and Sedimentological Events and Future Lakes

There are also a few non-anthropogenic events that could prove important for the evolution of the rift lakes over the near future. Most volcanoes in the rift valley appear to be dormant or extinct, but the extensive spreads of lava and ash across the rift testify to past large-scale activity. Such large-scale events are unlikely to repeat on a human time scale, but there have been a few recent eruptions (Wadge et al. 2016). Maasai oral records indicate Longonot erupted on its northern slopes sometime around the 1840s (Thompson and Dodson 1963) with Teleki's cone, on The Barrier, active in 1888 and Emuruangogolak lavas, south of the Lower Suguta, having been dated to 1910. Lava flows have the potential to divert and (or) block streams, leading to sudden changes in lake levels or the creation of new water bodies. Changes in geothermal activity can also be rapid (Renaut et al. 2008) and, although on a small scale compared to volcanism, can nonetheless impact the environment including the lakes. The hot springs at the south end of Lake Magadi are quiet seepages today, but geysers and boiling springs and mud were active there in 1958 (Crane 1981). Earthquakes can modify their activity, discharge and temperature.

Small earthquakes are a common feature of the rift with current peak activity taking place north of Nasikie Engida (Albaric et al. 2009). Faults can also lead to drainage diversion and changes in hot spring reservoirs at depth, both of which could have important influences on surface lakes in the region. Drainage changes may also be induced by heavy rainfall events that can cause rivers to avulse or suddenly change direction, especially as they enter an area of gentler gradients. In some cases, rivers may escape their channels through crevasse splays and form new floodplain lakes, as happened with the creation of Lake Kichiritit in 1994 (Tarras-Wahlberg et al. 2003).

These are all relatively rare events on a human time scale but are important over geological periods and cannot be excluded as future impacts. Improved understanding of these geological processes and their impacts on lake sedimentation has driven a considerable amount of research in the rift and has led to new understandings and conceptual models of relevance to environmental planners.

24.2.2 The Climate Versus Tectonic Debate and Change over Geological Time

Arguments concerning the relative importance of climate vs. tectonics as controls on lacustrine sedimentation are more complex than previously thought and also involve a consideration of time. One fruitful area of research on the rift lakes has involved attempts to understand lake change on historical scales and through the last few thousand years, often with the subsidiary aim of understanding what might happen in the near- and mid-term future. When considering thousands and hundreds of thousands of years, climate change is very important as a driver for lake expansion and contraction with climate variability reflecting the interaction of different Earth-Sun orbital Milankovitch cycles (eccentricity, precession, obliquity). Over shorter time scales, and particularly through the last few decades, humans have also proved to be increasingly important drivers of climate change, leading to suggestions of a new geological period called the 'Anthropocene' – one that is clearly applicable to the rift lakes (Waters et al. 2014).

However, over longer time scales geological agencies become more important, particularly in terms of creating and destroying lake basins. For example, many of the palaeolakes that formed and disappeared through the last 15 million years in the Baringo Basin were responses to both faulting and volcanism, which created subsiding basins, changed lake outlet heights, dammed lake basins with lavas and also destroyed lakes by filling them with ash and lava before a renewed cycle of lake formation developed.

24.2.3 The Development of New Rift Models

Recent research in the Kenya Rift has made a major contribution to the understanding of contemporary and Quaternary lake deposition, which can be applied to the interpretation of lake sediments in ancient rifts elsewhere. Early models considered faulting and volcanism as important in creating space for lakes but focussed more on climate when interpreting lacustrine histories (e.g. Bergner et al. 2009; Trauth et al. 2010). A few studies have combined climatic and tectonic controls of lake sedimentation. Bohacs et al. (2000), for example, related sediment and water infilling rates (reflecting the precipitation-evaporation balance) against the rate of development of accommodation space (a tectonic control), leading to a threefold classification of overfilled, balanced-filled and underfilled lakes. Several other researchers have emphasised the role of rift faulting in lake sedimentation (Hinderer and Einsele 2002).

New research from the Kenya Rift has outlined how other tectonic and volcanic controls exert an important influence on lacustrine sedimentation. Owen et al. (2018b, 2019) point out, for example, the importance of both faulting and volcanism in creating and destroying space for lake deposits and controlling drainage inflow to lakes. They also emphasised the value of volcanic rock geochemistry in interpreting lake deposits and noted that hydrology and hydrogeology exert a critical role in determining the water chemistry of rift lakes, observing that hot springs play a much more important role than previously appreciated, especially in underfilled basins. Recent studies from Kenya have also demonstrated that magmatic gases and hydrothermal fluids rising along faults can play a key role in developing some of the distinctive minerals (e.g. trona, nahcolite, chert) associated with the Kenya Rift lakes. Interpretations of ancient rift systems around the world need to consider some of the lessons from the rift lakes of Kenya.

24.2.4 *The Rift Valley Lakes and Hominin-Environment Interactions*

On one of our first evenings in Kenya, while sitting on the shore of Lake Baringo being bitten by malarial mosquitos, our supervisor, Bill Bishop, told us that the rift valley was "a great place to die". He went on to explain that rift valley lake basins provide an excellent depositional environment for fossil preservation. Although the lake waters are calcium poor, much of the Ca^{2+} from weathering of volcanic rocks is locked up as calcite in the surrounding soils, where it plays an important role in preserving bone. Furthermore, the rift is a sump for sediments and water and can allow rapid burial of bone before it decays at the land surface.

The lake basins have played a long-term role in supporting human populations, and their sediments have yielded many important hominin fossils and stone tools, leading to the global significance of this region for palaeoanthropology. Few bones and artefacts are found in the lake sediments themselves except near shorelines, but potable water and food (e.g. fish, crocodile, hippo and others) were, and remain, a critical resource. Throughout our ancestors' existence in the rift, and other parts of Africa, they relied on local environments for sustenance. They surely must have ventured beyond the rift margins and onto the adjacent plains, but these were often sites of erosion and unfavourable for fossil preservation, so in the absence of evidence, interpretations about this setting are often speculative.

During particularly severe periods of change, or greater environmental variability, hominins were subject to increased evolutionary pressure (Potts 2013). One example of this, reported from the Magadi Basin, was a phase of intense aridification at 525–400 ka, which overlaps with mammalian extinctions and the last Acheulian tools in the region (Owen et al. 2018a; Potts et al. 2020). Middle Stone Age artefacts appeared by about 320 ka with *Homo sapiens* emerging in Africa at around 315 ka. The data needs comparisons with other locations and time periods but suggest that climate might have played a role in our physical and cultural evolution. In contrast, modern humans are modifying the environment at a rapid rate and perhaps unsustainably. We should remember that although we can change the environment, the environment also affects us in significant and sometimes unpredictable ways.

24.3 A Personal Perspective

We first visited the Kenya Rift Valley in the summer of 1976 as young, enthusiastic PhD students at Queen Mary College (University of London) (Fig. 24.2) and have continued to do research in East Africa for more than 40 years. The impact of that new experience was immense, involving meeting new peoples and exploring new lands that differed tremendously from anything we had known previously. During our first trip to Nairobi National Park, just outside the city, we saw thousands of wildebeest, zebra and other animals of which we had little or no knowledge. Our first research trip to Lake Baringo involved flash floods, dirt roads, overturned vehicles, multiple insect bites, and camping to the sound of hippos. That world still exists but has changed and continues to do so. The Uhuru Highway through Nairobi could be driven in minutes whereas travelling the same distance today can take hours (the new [2022] Nairobi Expressway should help!). Wildlife in the Nairobi National Park has been reduced in numbers with urban expansion disrupting migration routes. The road to Baringo is now paved all the way, and beyond.

In many ways, we were in the transition from an old way of scientific field research (monochrome aerial photos, plane table surveying, typewriters) to a newer one involving easier access to study areas and the availability of more advanced research techniques, all of which invigorated scientific study in the rift. During our first visit, research permits to work in Suguta were revoked due to security problems and we diverted to study areas at NE Turkana, Bogoria and Olorgesailie. Our supervisor, Prof. Bill Bishop, passed away before the second field season and we continued under Bill French, an igneous petrologist with no East African experience. Consequently, we had to do much of the logistical organisation for our second trip ourselves, supporting each other, which led to a life-long friendship. Work then took us in different directions: Robin to Canada, Bernie to Malawi. Initially, we continued with joint studies of Lake Malawi and research in Kenya, but after 8 years in

Fig. 24.2 The effects of decades of research. (**a**) Bernie Owen at Olorgesailie in 1976. (**b**) Robin Renaut at Olorgesailie in 1976. (**c**) Bernie and Robin, in 2018, somewhat more aged and standing on dune fields in the Suguta Valley, which was the target of their first field season. At that time, security problems resulted in the revocation of their research permits. Both are smiling because they finally made it, forty-three years later

Malawi Bernie left for Hong Kong and drifted away from African research, studying marine environments. Robin studied saline lakes in British Columbia, and hot spring deposits in New Zealand, Iceland and South America. In the 1990s he joined lake projects in the Kenya Rift with Jean-Jacques Tiercelin (CNRS) and colleagues, in collaboration with John Ego and others at NOCK (National Oil Corporation of Kenya). However, after several years, with the occasional holiday in Africa, the urge to return became too much and Bernie joined up with Robin again. Except for the Covid-19 pandemic of 2020/21, one or both of us have been in Africa at least once or twice a year since 2000, driven by the desire to advance limnogeological and palaeoclimatological research in the Kenya Rift.

We have written more than 250 research papers over the years, mostly on Africa, and many together. We have interacted with many of the 'big names' of African

Fig. 24.3 Robin and Bernie, in 2017, consulting a geology map before a day's fieldwork at Ol' Bolossat – a high-altitude lake in the central Kenya Rift

research often sitting around campfires or camp dining tables, as well as eating from tin cans of baked beans (and sometimes custard) during smaller-scale trips. We began in leaky tents but in recent years have been able to take advantage of the better-quality accommodation (Fig. 24.3) that is springing up around Kenya. Oranges remain our favourite break in the field. Typically, we work through the lunch hour and finish a little earlier than other field researchers. There have been many moments when we have looked around in silence and wondered how we have been so lucky to work in Africa for our careers doing what was really our hobby. We hope that some of our research has been helpful to others and will be read by new young scientists, but we mostly hope they discover the East African lakes, landscapes, and rocks for themselves.

Kenya has changed dramatically since we first arrived when the population stood at about 14 million. In 2022, this figure was about 56 million, quadruple the 1976 total. The impact can be seen almost everywhere through deforestation, changed lake systems, soil erosion and lake siltation and pollution. At least plastic shopping bags have been banned. Wildlife has been reduced, although protected in some areas. The people are better off now with mobile phones providing a 'leapfrogging' technology that has stimulated local economies for the better and improved economic growth. Nevertheless, the environment is changing rapidly, sometimes reflecting planetary pressures rather than local problems. Back in 1976, the July-August glaciers and snow on Kilimanjaro covered the entire top of the mountain. Now, viewed from Amboseli, just a few sad patches of white remain, all in less than 50 years.

The population of Kenya is young and dynamic. Its people are also becoming more environmentally aware, but poverty still causes many problems. The country is rich in resources and talent, but sometimes a little overly bureaucratic. Obtaining research permits and visiting research sites, for example, is much more complicated than ever before, even for local scientists. On the other hand, we know and have known many Kenyans who care deeply about their home and who work hard to protect it and enhance the lives of its people. They still have a chance to avoid many of the losses of natural habitat that have taken place in Europe and other regions and can readily move in more environmentally friendly directions. This is already happening in areas such as wind power development and geothermal energy, but even these technologies must be implemented carefully to avoid irreparable damage to the environment. In common with many parts of the world, however, the key resource is water. Over the next few decades, the decisions made concerning its use and availability will be of critical importance for the human population in East Africa, but also for the Kenya Rift lakes and their flora and fauna.

References

Albaric J, Déverchère J, Petit C, Perrot J, Le Gall B (2009) Crustal rheology and depth distribution of earthquakes: insights from the central and southern East African Rift System. Tectonophysics 468:28–41

Avery ST (2014) What future for Lake Turkana and its wildlife. SWARA Magazine (E Afr Wild Life Soc), January–March edition, pp 24–30

Avery ST, Tebbs EJ (2018) Lake Turkana, major Omo River developments, associated hydrological cycle change and consequent lake physical and ecological change. J Great Lakes Res 44:1164–1182

Beadle L (1932) The waters of some East African Lakes in relation to their fauna and flora. J Linn Soc Lond, Zool 38:157–211

Becht R, Harper D (2002) Towards an understanding of human impact upon the hydrology of Lake Naivasha, Kenya. Hydrobiologia 488:1–11

Bergner AGN, Strecker MR, Trauth MH, Deino A, Gasse F, Blisniuk P, Dühnforth M (2009) Tectonic and climatic control on evolution of rift lakes in the Central Kenya Rift, East Africa. Quat Sci Rev 28:2804–2816

Bessems I, Verschuren D, Russell JM, Hus J, Mees F, Cumming BF (2008) Palaeolimnological evidence for widespread late 18th century drought across equatorial East Africa. Palaeogeogr Palaeoclimatol Palaeoecol 259:107–120

Bohacs KM, Carroll AR, Neal JE, Mankiewicz PJ (2000) Lake-basin type, source potential, and hydrocarbon character: an integrated-sequence-stratigraphic-geochemical framework. In: Gierlowski-Kordesch EH, Kelts KR (eds) Lake basins through space and time. Am Assoc Petrol Geol Stud Geol 46:3–34

Britton JR, Harper DM (2008) Juvenile growth of two tilapia species in lakes Naivasha and Baringo, Kenya. Ecol Freshw Fish 17:481–488

Chapman LJ, Chapman CA, Kaufman L, Witte F, Balirwa J (2008) Biodiversity conservation in African inland waters: lessons of the Lake Victoria region. Verh Int Verein Limnol 30:16–34

Crane K (1981) Thermal variations in the Gregory Rift Valley of southern Kenya (?). Tectonophysics 74:239–262

De Cort G, Verschuren D, Ryken E, Wolff C, Renaut RW, Creutz M, Van der Meeren T, Haug G, Olago DO, Mees F (2018) Multi-basin depositional framework for moisture-balance reconstruction during the last 1300 years at Lake Bogoria, central Kenya Rift Valley. Sedimentology 10:458–445

Funk C, Dettinger MD, Michaelsen JC, Verdin JP, Brown ME, Barlow M, Hoell A (2008) Warming of the Indian Ocean threatens eastern and southern African food security but could be mitigated by agricultural development. Proc Natl Acad Sci 105:11081–11086

Githumbi E, Marchant R, Olago D (2019) Using the past to inform a sustainable future: Palaeoecological insights from East Africa. In: Ramutsindela M, Mickler D (eds) Africa and the sustainable development goals. Springer, Cham, pp 187–196

Government of Kenya and UNDP (2021) Rising water levels in Kenya's Rift Valley lakes, Turkwel Gorge Dam and Lake Victoria: a scoping report. Government of Kenya, Nairobi

Greiner C (2012) Unexpected consequences: wildlife conservation and territorial conflict in northern Kenya. Hum Ecol 40:415–425

Gregory JW (1896) The Great Rift Valley. John Murray, London

Harper DM, Tebbs E, Bell O, Robinson VJ (2016) Conservation and management of East Africa's soda lakes. In: Schagerl M (ed) Soda lakes of East Africa. Springer, Cham, pp 345–364

Hay RL (1970) Silicate reactions in three lithofacies of a semi-arid basin, Olduvai Gorge, Tanzania. In: Morgan BA, Jones BF, Kullerud G, Osborne HWT (eds) Fiftieth anniversary symposia. Mineral Soc Am Spec Pap 3:237–255

Hickley P, Muchiri M, Britton JR, Boar R (2008) Economic gain versus ecological damage from the introduction of non-native freshwater fish: case studies from Kenya. Open Fish Sci J 1:36–46

Hinderer M, Einsele G (2002) Rifts as denudation-accumulation systems: concept, models and examples. In: Renaut RW, Ashley GM (eds) Sedimentation in continental rifts. SEPM Spec Publ 73:83–96

Jenkin PM (1936) Reports on the Percy Sladen expedition to some Rift Valley lakes in Kenya in 1929. VII. Summary of the ecological results, with special reference to the alkaline lakes. Ann Mag Nat Hist Ser 10 18:133–181

Junginger A, Roller S, Olaka LA, Trauth MH (2014) The effects of solar irradiation changes on the migration of the Congo Air Boundary and water levels of paleo-Lake Suguta, Northern Kenya Rift, during the African Humid Period (15–5 ka BP). Palaeogeogr Palaeoecol Palaeoclimatol 396:1–16

Junginger A, Trauth MH (2013) Hydrological constraints of paleo-Lake Suguta in the Northern Kenya Rift during the African Humid Period (15–5 ka BP). Glob Planet Change 111:174–188

Kolding J (1993) Trophic interrelationships and community structure at two different periods of Lake Turkana, Kenya: a comparison using the ECOPATH II box model. In: Chistensen V, Pauly D (eds) Trophic models of aquatic ecosystems. ICLARM Conf Proc 6 Manila, pp 116–123

Kreinitz L (2018) Lesser Flamingos: descendants of Phoenix. Springer, Cham

Marchant R (2021) East Africa's human environment interactions: historical perspectives for a sustainable future. Palgrave Macmillan, Cham

Marchant R, Richer S, Boles O, Capitani C, Courtney-Mustaphi CJ, Lane P, Prendergast ME, Stump D, De Cort G, Kaplan JO, Phelps L, Kay A, Olago D, Petek N, Platts PJ, Punwong P, Widgren M, Wynne-Jones S, Ferro-Vázquez C, Benard J, Boivin N, Crowther A, Cuní-Sanchez A, Deere NJ, Ekblom A, Farmer J, Finch J, Fuller D, Gaillard-Lemdahl M-J, Gillson L, Githumbi E, Kabora T, Kariuki R, Kinyanjui R, Kyazike E, Lang C, Lejju J, Morrison KD, Muiruri V, Mumbi C, Muthoni R, Muzuka A, Ndiema E, Nzabandora CK, Onjala I, Schrijver AP, Rucina S, Shoemaker A, Thornton-Barnett S, van der Plas G, Watson EE, Williamson D, Wright D (2018) Drivers and trajectories of land cover change in East Africa: human and environmental interactions from 6000 years ago to present. Earth-Sci Rev 178:322–378

Mwanuzi FL, Abuodha JOX, Muyodi FJ, Hecky RE (eds) (2005) Lake Victoria Environment Management Project (LVEMP) water quality and ecosystem status. South Eastern Kenya University, Kitui

Mwirigi PM, Rutagemwa DK, Gikuma-Njuru P, Matovu A, Waya RK, Mwebaza-Ndawula L, Ssenfuma-Nsubuga M, Kinobe J, Abuodha JOZ, Hecky RE (2005) Lake Victoria: the changing lake. In: Mwanuzi FL, Abuodha JOX, Muyodi FJ, Hecky RE (eds) Lake Victoria Environment Management Project (LVEMP) water quality and ecosystem status. South Eastern Kenya University, Kitui, pp 62–80

Obiero K, Lawrence T, Ives J, Smith S, Njaya F, Kayanda R, Waidbacher H, Olago D, Miriti E, Hecky RE (2020) Advancing Africa's Great Lakes research and academic potential: answering the call for harmonized, long-term, collaborative networks and partnerships. J Great Lakes Res 46:1240–1250

Ochuka M, Ikporukpo CO, Ogendi GM, Mijinyawa Y (2019) Spatial variability in physico-chemical parameters of water in Lake Baringo Basin, Kenya. Curr World Environ 14:334–457

Oduor SO, Kotut K (2016) Soda lakes of the East African Rift System: the past, the present and the future. In: Schagerl M (ed) Soda lakes of East Africa. Springer, Cham, pp 365–374

Olago DO, Raini J, Omuombo C, Ogonda G, Muli J, Okello C, Memo W, Ong'ang'a O (2021) Lentic-lotic water system response to anthropogenic and climatic factors in Kenya and their sustainable management. In: Diop S, Scheren P, Niang A (eds) Climate change and water resources in Africa. Springer-Nature, Cham, pp 193–218

Owen RB, Muiruri VM, Lowenstein TK, Renaut RW, Rabideaux N, Luo S, Deino AL, Sier MJ, Dupont-Nivet G, McNulty EP, Leet K, Cohen A, Campisano C, Deocampo D, Shen C-C, Billingsley A, Mbuthia A (2018a) Progressive aridification in East Africa over the last half million years and implications for human evolution. Proc Natl Acad Sci 115(44):11174–11179

Owen RB, Renaut RW, Lowenstein TK (2018b) Spatial and temporal geochemical variability in lacustrine sedimentation in the East African Rift System: evidence from the Kenya Rift and regional analyses. Sedimentology 65:1697–1730

Owen RB, Renaut RW, Muiruri VM, Rabideaux NM, Lowenstein TK, McNulty EP, Leet K, Deocampo D, Luo S, Deino AL, Cohen A, Sier MJ, Campisano C, Shen C-C, Billingsley A, Mbuthia A, Stockhecke M (2019) Quaternary history of the Lake Magadi Basin, southern Kenya Rift: tectonic and climatic controls. Palaeogeogr Palaeoclimatol Palaeoecol 518:97–118

Potts R (2013) Hominin evolution in settings of strong environmental variability. Quat Sci Rev 73:1–13

Potts R, Dommain R, Moerman JW, Behrensmeyer AK, Deino AL, Riedl S, Beverly EJ, Brown ET, Deocampo D, Kinyanjui R, Lupien R, Owen RB, Rabideaux N, Russell JM, Stockhecke M, deMenocal P, Faith JT, Garcin Y, Noren A, Scott JJ, Western D, Bright J, Clark JB, Cohen AS, Keller CB, King J, Levin NE, Shannon KB, Muiruri V, Renaut RW, Rucina SM, Uno K (2020) Increased ecological resource variability during a critical transition in hominin evolution. Sci Adv 6:eabc8975

Renaut RW, Owen RB, Ego JK (2008) Recent changes in geyser activity at Loburu, Lake Bogoria, Kenya Rift Valley. GOSA Trans 10:4–14

Rich F (1933) Reports on the Percy Sladen Expedition to some Rift Valley lakes in Kenya in 1929. IV. Phytoplankton from the Rift Valley lakes in Kenya. Ann Mag Nat Hist 10:233–262

Rowell DP, Chadwick R (2018) Causes of the uncertainty in projections of tropical terrestrial rainfall change: East Africa. J Clim 31:5977–5995

Seimon A, Ingram JC, Watson JEM (2013) Climatology of the East African Great Lakes region and potential impacts of climate change on its biodiversity and ecosystem services. In: Gordon I (ed) Conservation strategy in the Great Lakes region. MacArthur Foundation, Chicago, pp 103–134

Seimon A, Picton Phillips G (2012) Regional climatology of the Albertine Rift. In: Plumptre A (ed) Long-term changes in Africa's Rift Valley. Nova Science Publishers, Hauppauge, pp 10–30

Schagerl M (ed) (2016) Soda lakes of East Africa. Springer, Cham

Schagerl M, Renaut RW (2016) Dipping into the soda lakes of East Africa. In: Schagerl M (ed) Soda lakes of East Africa. Springer, Cham, pp 3–23

Sitoki L, Gichuki J, Ezekiel C, Wanda F, Mkumbo OC, Marshall BE (2010) The environment of Lake Victoria (East Africa): current status and historical changes. Int Rev Hydrobiol 95:209–223

Shongwe ME, van Oldenborgh GJ, van den Hurk B, van Aalst M (2011) Projected changes in mean and extreme precipitation in Africa under global warming. Part II: East Africa J Clim 24:3718–3733

Surdam RC, Eugster HP (1976) Mineral reactions in the sedimentary deposits of the Lake Magadi region, Kenya. Geol Soc Am Bull 87:1739–1752

Tarras-Wahlberg H, Harper DM, Tarras-Wahlberg N (2003) A first limnological description of Lake Kichiritith, Kenya: a possible reference site for the freshwater lakes of the Gregory Rift Valley. S Afr J Sci 99:494–496

Thompson AO, Dodson RG (1963) Geology of the Naivasha area. Geol Surv Kenya Rep 55

Trauth MH, Maslin MA, Deino AL, Junginger A, Lesoloyia M, Odada EO, Olago DO, Olaka LA, Strecker MR, Tiedemann R (2010) Human evolution in a variable environment: the amplifier lakes of Eastern Africa. Quat Sci 19:2981–2988

van der Plas GW, De Cort G, Petek-Sargeant N, Wuytack T, Colombaroli D, Lane PJ, Verschuren D (2019) Distinct phases of natural landscape dynamics and intensifying human activity in the central Kenya Rift Valley during the past 1300 years. Quat Sci Rev 218:91–106

Vareschi E, Jacobs J (1985) The ecology of Lake Nakuru VI. Synopsis of production and energy flow. Oecologia 65:412–424

Wadge G, Biggs J, Lloyd R, Kendall JM (2016) Historical volcanism and the state of stress in the East African Rift System. Front Earth Sci 4:1808–1824

Waters CN, Zalasiewicz JA, Williams M, Ellis M, Snelling AM (eds) (2014) A stratigraphical basis for the Anthropocene. Geol Soc Lond Spec Publ 395

Worthington EB (1932) Scientific results of the Cambridge Expedition to the East African Lakes, 1930-1. Fishes other than Cichlidae. J Linn Soc Lond Zool 38:121–134

Worthington E, Ricardo C (1936a) Scientific results of the Cambridge Expedition to the East African Lakes 1931. 17. The vertical distribution and movements of the plankton in Lakes Rudolf, Naivasha. Edward and Bunyoni. J Lin Soc Lond Zool 40:33–69

Worthington EB, Ricardo CK (1936b) Scientific results of the Cambridge Expedition to the East African Lakes, 1930-1. No. 15. The fish of Lake Rudolf and Lake Baringo. J Linn Soc Lond Zool 39:353–389

Index

A
Accommodation zones, 34, 40–42, 47, 140, 173, 697, 910, 913
Aeolian, 101–103, 154, 193, 364, 393, 395, 485, 489, 514, 598, 744, 802–804, 812, 821–823, 843, 915
African Humid Period, 130
Air boundaries, 79, 80, 192, 252, 533, 557, 560, 954
Alablab, 221, 232, 233, 238, 243, 578, 699
Amplifier lakes, 133, 245, 445
Anthropogenic impacts, 290
Archaeology, 8, 17, 18, 21, 168, 200, 348, 420, 947, 948
Aterir Beds, 704, 705

B
Background information, 77
Baringo, 129
Basalts, 37, 39, 46, 48, 49, 54, 56–59, 62, 63, 97, 137, 140, 164–165, 169, 170, 173, 223, 226, 228, 237, 260, 263, 306, 308, 366–369, 395, 396, 421, 422, 467, 578, 585, 588, 601, 604, 611, 631, 633, 636, 638–640, 662, 667, 695, 699, 701, 705–708, 710, 717, 733, 737–739, 741, 745, 746, 749, 769, 779, 805, 809, 844, 846, 920, 921, 928
Bedrock types, 909
Biological discoveries, 163, 612, 657, 890, 892, 953, 963
Biology, x, 7, 15, 129, 168, 186–190, 238–242, 278–282, 306, 331–333, 387–391, 433–440, 482, 515, 544–554, 909, 953
Book layout, 3–8
Border faults, 40–42, 44, 47, 50, 52, 140, 141, 173, 226, 258, 269, 334, 421, 467, 468, 633, 697, 717, 720, 767, 839–841, 862, 910–913, 915–917, 919, 951
Brine, 94, 139, 140, 153, 154, 326, 327, 341, 350, 351, 381, 395, 463, 464, 469, 476, 478–482, 484, 485, 489, 490, 493–495, 499, 507, 513, 514, 584–586, 782, 860, 872, 911, 936, 937, 943, 944, 954

C
$CaCO_3$ precipitation, 114, 117, 121, 918, 928
Carbonates, 94–96, 98, 100, 115–119, 121, 129, 132, 136–139, 142–144, 149, 154, 155, 185, 193, 201, 202, 238, 242, 243, 246, 247, 276, 280, 283, 306, 317, 327, 329, 344, 345, 350, 353, 364, 384, 393, 395, 440, 441, 463, 468, 476, 479–481, 489, 492, 494, 500, 506, 507, 514, 557, 558, 585, 586, 590, 605, 607, 617, 640, 646–648, 650, 651, 653, 655, 657–660, 705, 732, 749, 796, 803, 809, 811–813, 818, 822–823, 827, 828, 848, 851–855, 860, 862, 875, 876, 892, 897, 918, 919, 925, 926, 928, 937
Central Island, 164, 180, 183, 184, 188, 578–581, 645, 935

R. W. Renaut, R. B. Owen, *The Kenya Rift Lakes: Modern and Ancient*, Syntheses in Limnogeology, https://doi.org/10.1007/978-3-642-25055-2

Chemical divides, 138, 139, 479
Chert, 93, 129, 151, 154, 155, 322, 384, 463, 464, 468, 485, 489, 494, 495, 497, 499–501, 503, 505–507, 699, 720, 725, 729, 736, 810, 811, 819, 827, 842, 843, 848–851, 855–860, 862, 864–876, 928, 944, 960
Clastic shorelines, 131, 140
Climate, 7, 17, 18, 24–26, 77–83, 88, 90, 93, 98, 100, 103, 104, 108, 117, 119, 129, 132, 133, 152, 174–177, 205, 206, 226, 229, 245, 250, 251, 257, 264–267, 310, 341, 343, 351, 363, 364, 370–372, 381, 397, 398, 404, 405, 423–425, 443, 445, 447, 468–469, 507, 532–535, 553, 557, 559, 560, 593, 598, 604, 636, 638, 656, 658, 666–667, 669, 670, 672, 702, 710, 729, 741, 780, 786, 805, 826–830, 890, 909, 910, 913–919, 925, 926, 943, 944, 951, 953–955, 959–961
Core stratigraphy, 401
Cores, x, 7, 18, 25–27, 87, 192, 193, 198–205, 242, 243, 245, 251, 267, 272, 283–288, 290, 323, 333, 334, 339–342, 350–351, 391, 393–395, 399, 401, 402, 422, 423, 438, 439, 441, 442, 445–447, 465, 484, 489, 492, 495, 496, 498, 499, 503, 507, 513, 514, 546, 547, 555, 557–560, 584, 595, 596, 601–603, 610, 635, 637, 640, 655, 656, 667, 669, 707, 741, 743, 744, 785, 808, 822–824, 827, 850, 855–858, 860, 862, 865–866, 872–874, 895, 915–917, 943
Correlations, 18, 26, 81, 82, 169, 199–201, 247, 251, 278, 282, 350, 403, 423, 557, 643, 699, 710, 739, 774, 778, 781, 782, 825–827, 855, 856
Cyanobacteria, 117, 119, 121, 149, 152, 156, 238, 283, 305, 323, 333, 341, 388–391, 435, 436, 482, 484, 515, 547, 549, 550, 555, 559, 602, 608, 609, 647, 654, 867, 871, 955
Cyclic sedimentation, 805

D

Dams, 6, 120, 143, 177, 178, 190, 317, 534, 619, 620, 662, 853, 937, 952, 955
Deflation, 103, 154, 363, 365, 393, 425, 910, 952
Deforestation, 268, 271, 376, 392, 485, 536, 545, 557, 593, 598, 937, 957, 963
Delta, 7, 33, 51, 92, 94, 101, 131, 142, 144–146, 154, 174, 180, 184, 187–193, 231, 233, 272, 282, 283, 304, 311, 313, 317, 318, 324, 334–340, 348–350, 366, 393, 395, 396, 419, 440, 463, 485, 487, 488, 503, 512, 516, 578, 591, 602, 603, 615, 620, 645, 646, 648–650, 656, 662, 666–668, 672, 702, 703, 890, 914–916
Diatomaceous sediments, 373, 387, 403, 445, 656, 699, 780, 801, 802, 810–812, 817, 821, 823, 824, 916–917, 919, 926, 947
Diatomite, 149–151, 195, 196, 198, 245, 368, 370, 398, 603, 647, 648, 651–653, 655, 662, 663, 704, 705, 707, 708, 719, 720, 727, 728, 732, 734, 739–744, 770, 774, 777–780, 782, 784, 798, 800–803, 807, 809–811, 815, 817, 821–824, 846, 916–917, 935, 937–938
Diatoms, xxvi, 23, 26, 90, 121, 130, 142, 148–152, 156, 185, 186, 193, 199–202, 241, 276, 277, 279, 283, 286, 287, 289, 320, 323, 331, 332, 350, 388–391, 394, 399, 401, 402, 430, 435, 436, 442–446, 480, 484, 495, 496, 499, 500, 515, 544, 546–549, 555–559, 581, 595–597, 602, 603, 608–611, 647, 655, 659, 660, 667, 668, 707, 709, 720, 724, 732–737, 741, 743, 744, 748, 752, 754, 779, 780, 782–785, 798, 799, 801–805, 809, 811, 812, 814, 815, 818, 820, 823–826, 842, 850, 853, 858, 860, 862, 865, 871, 925, 926, 955
Drainage, 7, 14, 26, 63, 83–88, 90, 93, 94, 100, 101, 108, 115, 129–132, 142, 163–165, 169–170, 173, 174, 224, 226, 227, 229, 232, 234, 238, 242, 243, 246, 257, 259, 263–264, 268, 272, 277, 282, 283, 290, 305, 308, 310, 311, 322, 334, 338, 339, 352, 363, 367, 370, 373–375, 378, 384, 387, 392, 417, 418, 468–474, 503, 527, 529, 531, 532, 535, 536, 555, 577, 578, 581, 591, 593, 594, 598, 600–602, 604, 607, 614, 617–619, 636, 638, 655, 663, 705, 767, 809, 821, 862, 864, 872, 890, 893–895,

909–911, 913, 914, 920, 922, 927, 928, 937, 943, 952, 958, 960
Drilling, xii, 23, 25–28, 151, 204, 308, 403, 465, 499, 603, 655, 741, 743, 808, 822–824, 855–857, 939
Drowned valleys, 893

E
Early explorers, 12, 16, 166, 225
East African Geological Research Unit (EAGRU), xxv, 19, 20, 22, 225, 257, 261, 719
Elmenteita, 129
Emuruangogolak, 47, 48, 63, 116, 221, 223, 226, 228–230, 578, 583, 585–589, 693, 694, 698, 699, 705, 710–712, 958
Endosapia Formation, 842, 845
Environmental changes, 6, 267, 306, 351, 402, 530, 593, 603, 778, 781, 808, 825, 829, 956
Environmental history, 24, 402, 778
Environmental models, 6
Environments, xii, 12, 24, 26, 77, 87, 94, 98, 109, 140, 141, 185, 188, 221, 225, 309, 336, 350, 388, 392, 396, 402, 403, 420, 440–443, 464, 496, 513, 515, 577, 634–636, 644–649, 651, 652, 654, 656, 657, 664, 666–667, 703, 704, 722, 724, 748, 754, 807, 815, 820, 825, 827, 828, 872–874, 887, 888, 922, 924, 940, 943, 957, 958, 960–964
Erosion, 34, 36, 57, 59, 98–101, 103, 113, 130, 131, 149, 197, 263, 271, 272, 282, 290, 333, 338, 339, 350, 351, 367, 376, 379, 381, 392, 393, 397, 402, 403, 440, 485, 499, 531, 536, 545, 556, 591, 595, 598, 651, 662, 664, 666–668, 735, 736, 744, 746, 753, 754, 756, 777, 781, 798, 802–804, 814–823, 825, 826, 843, 862, 895, 909, 914, 921, 926–928, 937, 953, 957, 961, 963
Eutrophication, 434, 435, 439, 445, 530, 536, 546, 548, 550, 553, 554, 637, 733, 947, 956, 957
Evaporites, 4, 93, 98, 129, 131, 132, 136, 140, 152–156, 243, 314, 329, 339–342, 350, 384, 387, 393, 395, 398, 463, 464, 469, 471, 476, 482, 489–494, 496, 499, 510, 513, 586, 809, 812, 813, 843, 848, 855, 861, 865, 870–873, 875, 890, 892, 916, 924, 925, 943, 944, 952
Exploration, xi, 7, 11, 12, 15, 23, 24, 85, 167, 168, 528, 637, 693, 939, 953, 957

F
Facies, 7, 8, 117, 120, 129–131, 140–156, 337, 440–443, 554, 638, 640, 644–652, 656, 657, 662, 702, 707, 720, 724–726, 729–731, 733, 736, 740, 746–749, 753, 754, 784, 796, 798, 800–803, 805, 809–813, 822–823, 855, 887–890, 892, 910, 913, 914, 916–917, 926, 940
Faults, 14, 36, 91, 132, 170, 226, 260, 306, 367, 417, 465, 527, 577, 633, 697, 717, 767, 793, 839, 883, 909, 951
Fish, 4, 90, 138, 167, 224, 258, 333, 389, 426, 463, 535, 585, 636, 707, 722, 796, 848, 887, 955
Fish introductions, 439, 554
Flamingos, 4, 106, 107, 164, 180, 188, 235, 237, 238, 241, 305, 324, 325, 332, 333, 336, 363–366, 372, 376, 381, 387–391, 436, 482–484, 515, 579–581, 589, 610, 617, 890, 937, 942, 947, 951, 952, 954–957
Flow reversal, 893, 913
Fluvial deposits, 18, 22, 403, 596, 642, 741, 748, 892, 920
Fossils, 15–19, 100, 112, 115, 118–121, 141, 155, 163, 168, 169, 200, 243, 277, 317, 319, 320, 322, 325, 344, 348, 349, 420, 436, 438, 439, 484, 495–497, 499, 515, 547, 610, 636, 640, 642–644, 647, 649, 655–657, 662, 667, 672, 699, 707, 719, 720, 722, 725, 726, 729, 732–734, 737, 741, 746, 748, 749, 754, 794, 796, 799, 813, 814, 819, 827, 853, 858, 861–863, 867, 875, 887, 889, 892, 893, 935, 947, 960, 961

G
Geochemistry, 193, 731, 798–800, 817, 872, 928, 960
Geological controls, 38
Geological evolution, 19, 57–63
Geological explorations, 717–719
Geological mapping, 18–21, 46, 168, 778

Geological setting, 222, 226–228, 307, 321, 367–370, 465, 582, 584, 586, 588, 590, 592, 599, 604, 885
Geology, x, xxvi, 3, 7, 11, 12, 14, 16, 18, 19, 23, 33–63, 83, 88, 93, 138, 164–165, 168, 225, 235, 257, 284, 306, 364, 368, 384, 398, 420, 421, 423, 464–466, 507, 531, 600, 601, 632, 697–700, 718, 719, 745, 750, 768, 769, 797, 798, 806, 807, 913, 926, 953, 963
Geothermal, 7, 23, 77, 134, 221, 282, 305, 373, 419, 474, 580, 693, 874, 922, 957
Geothermal power, 49, 109, 370, 439, 770, 957
Geothermal studies, 108
Global warming, 77, 951, 953–955, 958
Green Beds, 468, 495–497, 843, 849, 850, 855, 858–876, 944
Groundwater, 23, 84, 129, 174, 229, 264, 312, 372, 420, 469, 577, 784, 798, 851, 915, 955
Groundwater recharge, 233, 375, 376, 391
Guano, 98, 945, 946

H

Half-graben basins, 940
Halite, 94, 138, 139, 243, 329, 341, 393, 479, 489, 494, 514, 585, 642, 729, 936, 937
Heavy minerals, 141, 191, 652, 945, 946
High Magadi Beds, 145, 485, 487, 495–501, 504–507, 512, 614, 842, 843, 848, 849, 855, 865, 867, 868, 944
Hiwegi Formation, 887–890
Holocene, 4, 47, 82, 142, 163, 226, 260, 320, 363, 422, 495, 552, 578, 631, 705, 745, 777, 828, 855, 892, 913
Holocene stratigraphy, 558
Hominin Sites and Paleolakes Drilling Project (HSPDP), 151
Hot springs, xxv, xxvi, 18, 51, 52, 95, 108–116, 118–121, 138, 143, 144, 154, 188, 232–240, 242, 243, 260, 264, 277–278, 290, 303, 305, 309, 310, 312, 314–324, 327, 329, 331, 333, 336, 340, 341, 349, 350, 370, 374, 375, 384, 388, 391, 395, 420, 425, 463, 468, 469, 471–476, 479–487, 489, 494, 495, 506, 510, 511, 513, 514, 578–581, 589, 590, 595, 602, 642, 812, 814, 849, 872, 875, 919, 922–926, 928, 943, 948, 951, 952, 958, 960, 962
Humbu Formation, 846
Hydrology, 4, 7, 77, 83–93, 174–177, 229–238, 251, 263–267, 310–314, 354, 367, 372–375, 381, 420, 423–425, 469–474, 510–513, 532–535, 591, 922, 924, 953, 960
Hydrolysis, 96, 137, 138, 384, 482, 924, 943

K

Kamnarok, 129
Kangirinyang, 129
Kanjera Formation, 892
Kenya Rift, ix–xi, xxv, xxvi, 3–8, 11–26, 33–63, 77–121, 129–134, 136, 139, 140, 142–144, 149–152, 156, 170, 171, 174, 205, 206, 221, 225, 226, 228, 229, 243–245, 247, 248, 257, 260, 262, 263, 288, 303, 305, 306, 317, 325, 327, 331, 363, 364, 367, 368, 370, 377, 381, 387, 392, 397, 398, 402, 417, 418, 421, 422, 463, 465, 467, 468, 476, 507, 527, 530, 577–620, 707, 710, 717–756, 767–787, 793–830, 839, 858, 860, 883, 910–920, 922, 923, 925–928, 935–948, 951, 952, 954, 956, 957, 959–964
Kichirtit, 82, 129, 264, 578, 591–593, 598, 952, 953
Koora cores, 151
Kulu Formation, 889, 890

L

Lacustrine models, 6
Lake Baringo, ix–xii, 12–14, 19–21, 23–25, 39, 47–50, 56, 63, 82, 84–86, 88, 104–107, 114, 115, 121, 132, 144, 167, 205, 221, 226, 229, 233, 257–290, 306–308, 312, 314, 327, 335, 345, 352–354, 398, 585, 591, 593–595, 603, 618, 620, 717, 730, 734, 745, 748–749, 751, 752, 754–756, 911, 920, 924, 925, 935, 945, 946, 948, 952–954, 956, 957, 960, 961
Lake basins, 3, 8, 16, 24, 34, 87, 94, 101, 114–121, 131, 180, 203, 370, 372, 376, 464, 606, 639, 646, 658, 737,

754, 827, 867, 895, 910, 914, 915, 922, 925, 926, 940, 943, 959, 960
Lake Bogoria, v, xi, xii, xxv, 4, 14, 22, 24, 25, 42, 50–52, 56, 63, 82, 86, 87, 90, 95, 101–103, 106, 107, 111, 112, 114, 115, 118–119, 121, 136, 141–147, 152, 205, 246, 260, 263, 268, 271, 278, 290, 303–354, 363, 390, 403, 422, 495, 578, 593–595, 600, 603, 717, 729, 754–756, 914–917, 921, 924, 925, 928, 942, 943, 948, 952–954, 956, 958
Lake dimensions, 945
Lake Elmenteita, 15, 25, 51, 84, 85, 106, 107, 314, 363–405, 422, 779, 782, 937, 938, 944, 952, 956
Lake Kamnarok, 88, 591, 598–600, 957
Lake Kwenia, 578, 612–614, 793, 821, 824
Lake levels, 6, 16, 18, 82, 86, 93, 132, 133, 140, 142, 144, 145, 152, 174–178, 180, 186, 190, 192, 194, 196–198, 200, 201, 203–205, 235, 238, 242, 243, 245, 247, 248, 250–252, 257, 259, 265, 267, 272, 277, 282, 283, 285, 286, 289, 290, 306, 310, 314, 316, 317, 320, 322, 325, 326, 332–337, 339, 341, 343, 345, 348–350, 352, 353, 363, 364, 367, 377–379, 381, 389, 391–393, 395, 398, 399, 402, 403, 405, 424–426, 428, 430, 436, 437, 439–443, 445, 447, 471, 473, 496, 500, 502, 503, 509, 513, 515, 533–535, 549, 557–559, 578, 581, 585, 591, 602, 606, 607, 610, 633–642, 656, 658–660, 668, 669, 672, 673, 702, 705, 707, 710, 712, 734, 744, 754, 770, 773, 774, 776, 782–786, 804, 825, 826, 846, 848, 855, 859, 897, 898, 913–919, 924, 925, 943, 948, 954–955, 958
Lake Magadi, x, 3, 4, 14, 26, 27, 42, 46, 52–54, 63, 84, 85, 101, 102, 106, 108, 112, 114, 115, 118–119, 145, 152, 154, 155, 238, 243, 306, 325, 341, 350, 447, 463–516, 611, 614, 618, 707, 794, 807, 821, 823, 824, 839, 841, 842, 849, 850, 852, 853, 855–862, 865, 867, 872, 915–917, 923, 924, 928, 935–937, 942–944, 947, 952, 953, 956–958
Lake management, 530
Lake Naivasha, 17, 19, 23, 25, 44, 49, 56, 57, 62, 84, 85, 106, 145, 257, 283, 288, 312–314, 367, 373, 375, 397, 399, 417–447, 482, 591, 607, 608, 610, 611, 767, 782, 785–787, 793, 823, 824, 839, 911, 913, 935, 947, 948, 952–954, 956
Lake Nakuru, 18, 85, 100, 106, 107, 245, 248, 250, 314, 354, 363–405, 603, 620, 777, 779, 784, 944, 952, 953, 956, 957
Lake Natron, x, 3, 4, 7, 18, 52, 54, 84, 88, 105, 108, 116, 143, 325, 463, 465, 467, 469, 470, 482, 484, 494, 495, 500, 502, 507–516, 618, 839, 841, 842, 846, 847, 937, 947, 952, 957
Lake Oloidien, 417, 427, 429–431, 434–436, 438, 439, 441–443, 607, 608
Lake Simbi, 615–617
Lake Solai, 4, 50, 51, 578, 600–603, 618
Lake Sonachi, 417, 431, 607–611
Lake Turkana, x, 3, 4, 14, 16, 19–24, 35, 39, 42, 44, 46–49, 60, 62, 63, 80, 84–86, 89, 92, 94, 100, 102, 103, 105, 106, 108, 109, 129, 134, 140–145, 147, 149, 163–206, 221, 225, 226, 228, 229, 233, 243, 244, 248, 250, 258, 420, 578–582, 584, 633, 640, 642–645, 649, 650, 654–656, 658, 660, 662, 664–667, 669, 670, 672, 673, 697, 699, 712, 717, 741, 911, 915, 916, 920, 921, 923, 935, 940, 945–948, 952, 955–957
Lake Victoria, 3, 7, 8, 11–15, 25, 84, 88, 100, 105, 134, 282, 527–560, 615–617, 883–898, 913, 947, 955, 957
Lakes, 3, 11, 33, 77, 129, 163, 221, 257, 303, 363, 417, 463, 527, 633, 693, 719, 767, 793, 839, 885, 909, 935, 951
Lamination, 135, 149, 192, 201, 337, 344, 350, 498, 642, 648, 651, 652, 702, 746, 811, 846, 850, 862, 866, 868, 870, 892, 919
Late Quaternary, 8, 18, 78, 121, 194–206, 225, 231, 244–252, 286–290, 306, 325, 333, 341–354, 368–370, 395–405, 421, 464, 515–516, 556–560, 744, 769, 773, 778, 783, 897
Lemudong'o Formation, 794–796
Limnogeological explorers, 962
Limnology, x, 7, 15, 77, 87, 132, 168, 180–185, 236, 238, 270–278, 306, 314, 325–331, 377–387, 428–433, 536–544, 547, 603, 607, 609, 614, 922

Loboi, 25, 39, 40, 89, 102, 260, 263, 264, 274, 278, 282, 288, 290, 304, 306, 308, 312, 322, 325, 338, 339, 345, 347, 348, 351, 353, 354, 591, 593–598, 717, 745, 748, 751–756
Logipi, 4, 5, 19, 23, 24, 48, 49, 84, 85, 89, 90, 103, 106, 115, 129, 134, 138, 142, 144, 155, 156, 205, 206, 221–252, 285, 578, 581, 693, 699, 914, 916, 942, 951, 952, 956, 957

M
Magadi, 129
Magadi cores, 27, 151, 479, 855–861
Magadiite, 154, 155, 331, 341, 494, 495, 499–501, 506–507, 850, 860, 861, 865, 867, 870, 874, 944
Magadi-Natron, 8, 23, 84, 92, 94, 465, 468, 469, 500, 710, 793, 839–875
Magmatic CO_2, 96, 111, 116, 139, 154, 156, 480, 872, 927, 928
Meromictic, 134–136, 149, 152, 326, 350, 607, 909, 942
Microfossils, 142, 148, 149, 201–203, 401, 596, 636, 640, 655, 669, 712, 785, 799, 801–803, 808, 809, 814
Mineral resources, 935–947
Miocene, 3, 35, 149, 168, 228, 260, 306, 367, 420, 601, 631, 696, 717, 767, 793, 883, 920, 951
Miocene palaeolakes, 795
Mio-Pleistocene, 8, 249, 710–712
Mio-Pliocene sediments, 778
Modern lakes, x, 4, 6–8, 19, 24, 60, 62, 114, 151, 152, 243, 244, 257, 306, 308, 326, 350, 394, 420, 430, 438, 445, 467, 468, 494, 495, 502, 514, 515, 532, 535, 546, 550, 582, 614, 637, 645, 656, 657, 659, 662, 664, 669, 672, 673, 729, 745, 748, 754–755, 785, 841, 848, 861, 862, 892, 893, 897, 898, 910, 914, 922, 924, 935, 945
Moinik Formation, 848
Munya wa Gicheru, 151
Munya wa Gicheru Beds, 61, 63, 150, 151, 793, 796–805, 813, 818, 937–939

N
Nahcolite, 155
Naivasha, 129
Nakali Formation, 703, 704
Nakuru, 129, 154
Nakuru-Elmenteita, 364, 368, 370, 395–405, 603, 774, 777, 779, 781, 782, 784, 913, 935
Namarungule Formation, 701, 702, 710
Namarunu, 63, 89, 223, 228, 231, 232, 234, 236, 239, 697–699, 707, 710–712
Nasikie Engida, 4, 6, 7, 25, 53, 95, 101, 108, 115, 129, 136, 138, 144, 145, 154, 155, 350, 463–516, 794, 849, 852, 853, 862, 872, 875, 916–917, 923–925, 928, 937, 942–944, 952, 958
Natron, 144
Nitratine, 198, 947
Nutrients, 23, 100, 129, 134–136, 151, 185, 276, 381, 399, 433–437, 445, 515, 536, 544–546, 549, 550, 616, 734, 809, 942
Nyanza Rift, x, 5, 24, 52, 59, 60, 85, 263, 367, 527, 530, 615, 617, 883–887, 890–897

O
Ol' Bolossat, 4, 5, 7, 51, 84, 105, 353, 578, 603–607, 948, 963
Oligo-Miocene, 46, 169, 260, 631, 633–642
Ol Njorowa Gorge, 419, 426, 782, 784–786
Olorgesailie Formation, 101, 798, 807–818, 826
Oloronga Beds, 27, 468, 473, 480, 495–498, 500–501, 503, 505, 843, 848–862, 865, 867, 872
Oltulelei Formations, 467, 807, 810, 814–821
Omo Group, 163–165, 169, 170, 643, 657, 658, 660, 661, 664
Organic matter (OM), 18, 24, 117, 130, 131, 135, 138, 152, 156, 241, 242, 283, 286, 339, 341, 350, 393, 402, 425, 440, 441, 479, 492, 499, 507, 513, 514, 545, 546, 555, 559, 595, 597, 610, 637, 659, 660, 730, 733, 826, 872, 926, 939–944
Organic sediments, 147
Orkaramation Beds, 843, 849, 862

P
Palaeoanthropological research, 21–22
Palaeoclimate studies, 24–26
Palaeoenvironments, 403, 557, 636, 644, 660, 703, 729, 734–737, 803–805

Palaeogeography, 8, 204, 644, 660–673, 710–712, 734–737, 754–756, 778–787, 795–796, 826–830, 896
Palaeolakes, 4, 6, 8, 24, 118–119, 130, 136, 149–152, 203, 204, 221, 225, 228, 244–246, 248, 250, 251, 263, 364, 402, 403, 445–447, 465, 473, 484, 495, 496, 499–506, 515, 532, 603, 607, 637, 638, 642, 652, 655, 660–673, 699, 702, 705, 707, 710–712, 719, 727, 729, 733–756, 777, 779, 781, 782, 784, 787, 793–795, 797, 798, 802, 803, 809, 814, 819, 822–824, 826, 828, 840, 842, 845, 853, 855, 858, 860, 862, 873, 889, 892–894, 898, 953, 959
Pelagic deposits, 134
Personal perspectives, 961–964
Petroleum exploration, 23–24, 640
Petroleum potential, 24, 637, 939–944
Phonolites, 36, 49, 50, 54, 57, 60, 62, 95, 137, 140, 169, 260, 263, 306, 317, 319, 367, 422, 602, 603, 701, 717, 720, 721, 724, 725, 727, 730, 735, 736, 749, 767, 794, 798, 804, 883, 920, 928
Phytoplankton, 4, 129, 148, 149, 154, 186–188, 278–279, 332, 387, 389, 390, 435–436, 530, 534, 544–550, 559, 580, 602, 607, 609, 610, 617, 940, 955
Playa lakes, 14, 87, 372, 378, 392, 578, 600, 642, 892
Pleistocene, 8, 16, 52, 144, 168, 221, 260, 308, 369, 420, 463, 532, 578, 642, 699, 717, 767, 793, 841, 892, 913
Plio-Pleistocene, 16, 36, 46, 52, 54, 61, 62, 142, 163–165, 169, 464, 631, 633–673, 710, 737–756, 778–787, 794, 840, 842, 890–897
Plio-Pleistocene palaeolakes, 644, 890–898
Pluvials, 16–18, 395–398, 403, 719, 770–778, 807, 890
Population growth, 6, 424, 439, 530, 616, 951, 953, 956
Potable water, 947, 960
Pre-rift topography, 909
Protection, 221, 437, 956

Q

Quaternary palaeohydrology, 897
Quaternary shorelines, 770–778
Quaternary stratigraphy, 397, 401

R

Research expedition, 642, 953
Research history, 7, 11–27, 221–226, 796–798, 805–808
Reservoirs, 7, 23, 24, 120, 177, 178, 225, 246, 247, 277, 278, 350, 373, 425, 474, 480–482, 513, 514, 619–621, 924, 926, 928, 939, 940, 942, 947, 955, 958
Rhizoliths, 100, 101, 138, 142, 144, 149, 345, 348, 349, 648, 650–652, 721, 724, 748, 754, 799, 800, 809–812, 814, 818, 819, 928
Rivers, 7, 16, 34, 82, 131, 163, 229, 264, 306, 372, 417, 463, 527, 577, 636, 693, 717, 770, 793, 841, 887, 910, 951
Rusinga Island, 887–890, 893, 897

S

Saline alkaline lakes, 22, 24, 106, 145, 163, 221, 232, 305, 363, 469, 581, 583, 617, 699, 736, 737, 748, 872, 873, 940, 942, 943
Saline mudflats, 154
Sedimentary basins, 7, 33, 39, 57, 63, 260, 727, 839
Sedimentation, xi, xxv, 7, 8, 19, 24, 26, 27, 77, 93, 94, 103, 108–121, 129–156, 170, 185, 190–193, 203–206, 226, 229, 242–243, 249, 257, 260, 282–286, 308, 333–354, 363, 367, 375, 391–395, 397, 435, 440–445, 468, 485–507, 513–516, 554–556, 631, 658–660, 664, 666–667, 670, 702, 705, 724–725, 737, 739, 744, 753, 754, 756, 778, 798, 804, 805, 824–826, 828, 847, 848, 850, 856, 889, 895, 909–928, 944, 953, 959, 960
Sedimentation controls, 8, 141, 243, 909–928
Sediments, 6, 14, 46, 85, 129, 163, 225, 257, 306, 363, 420, 463, 530, 578, 631, 699, 717, 770, 793, 842, 883, 909, 957
Seismic reflection, 578, 672, 719, 895
Siliciclastic, 129, 131, 140–142, 144, 149, 152, 192, 193, 243, 257, 306, 314, 333, 339, 348, 350, 368, 393, 443, 485, 499, 595, 651, 662, 742, 802, 811, 817, 818, 820, 850, 873, 940, 953
Simbi, 129

Small lakes, 117, 577, 581, 582, 585, 587, 588, 591, 607, 615, 644, 923, 942, 947, 948, 952, 957
Soils, 7, 16, 77, 84, 85, 96–100, 103, 105, 116, 138, 139, 149, 154, 180, 198, 242, 245, 263, 271, 272, 329, 331, 333, 339, 351, 367, 376, 377, 379, 399, 420, 467, 479, 485, 530, 536, 545, 546, 591, 597, 598, 618, 650, 658, 809, 818, 820, 827, 853, 862, 892, 910, 916, 919, 921, 922, 937, 944, 957, 960, 963
Sonachi, 129
Source rocks, 637, 719, 798, 813, 874, 935, 939, 940, 942–944
Spring/stream ratios, 915
Stratification, 7, 101, 129, 132, 134–136, 142, 180, 185, 236, 271, 326, 377, 380, 402, 428, 513, 536–540, 544, 547, 549, 553, 554, 608, 611, 651, 652, 659, 729, 741, 812
Stratigraphic sections, 498, 635, 723, 726, 732, 738, 750
Stratigraphy, 18, 59, 169, 225, 286, 336, 397, 441, 464, 557, 643, 724, 799, 842, 890
Stromatolites, 131, 143, 144, 195, 196, 238, 245, 246, 283–285, 290, 320, 344, 345, 354, 393, 500, 502, 503, 509, 516, 642, 647, 649–651, 654, 655, 657, 659, 660, 667, 668, 705, 710, 726, 729, 733, 746, 748, 848, 862, 867, 870, 871, 875, 876, 916–919
Structures, 7, 14, 24, 38, 40, 52, 100, 117, 131, 190, 225, 228, 260, 262, 325, 388, 392, 420, 464, 499, 531, 578, 646, 647, 651, 652, 654, 730, 801, 810–813, 846, 853, 854, 862, 865, 867, 870, 875, 883, 913, 944
Suguta, 4, 19, 47, 85, 133, 152, 163, 221, 285, 581, 659, 693, 913, 915, 953
Suguta Beds, 245, 248, 705
Swamps, 3, 21, 25, 105, 190, 266, 268, 278, 282, 288, 306, 311, 338, 339, 351, 375, 417, 418, 426, 430, 435–437, 467, 469, 512, 513, 536, 549, 559, 578, 591–597, 606, 611, 615–619, 736, 746, 796, 805, 814, 815, 818–820, 858, 892

T

Tectonic models, 45, 840
Tectonics, xii, 3, 4, 18, 19, 26, 33–36, 40, 42–45, 47, 52, 57, 63, 77, 85, 92, 93, 108, 129, 130, 140, 170–173, 206, 243, 247, 252, 257, 260–263, 290, 306, 308, 352, 353, 367, 372, 381, 393, 403, 464, 473, 502, 530, 593, 614, 638, 660, 696, 731, 744, 754–755, 769, 773, 778, 780, 804, 826, 827, 829, 846, 862, 865, 872, 883, 893, 909, 910, 913, 914, 916–917, 924, 926–928, 951, 958–960
Tectonism, 7, 34, 753, 778, 780, 781, 805, 826–830, 846, 895, 909
Tourism, 221, 364, 947, 948, 956, 957
Trachytes, 27, 39, 48, 49, 52–54, 56, 57, 60–63, 93, 95, 97, 112, 137, 140, 226, 228, 260, 263, 266, 268, 288, 320, 367–370, 422, 425, 465, 467, 468, 471, 480, 489, 495, 497, 500–501, 505, 506, 602, 603, 611, 695, 705, 710, 717, 730, 731, 733, 737, 749, 767, 769, 778, 779, 781, 798, 801, 803, 805, 806, 809, 813, 821–823, 827, 840, 842–845, 848–850, 853, 855, 861, 916–917, 920, 928, 946
Travertine, 112, 116–121, 138, 143, 144, 149, 243, 314, 317–320, 331, 341, 348, 349, 393, 473, 489, 589, 590, 642, 695, 748, 819, 918, 919, 925, 928
Trona, 4, 27, 95, 102, 139, 145, 151–156, 231, 234, 237, 243, 322, 329, 336, 340, 341, 351, 363, 365, 384, 393, 395, 396, 463, 464, 469, 472, 476, 479, 480, 482, 485, 489–496, 500, 509, 513–516, 582, 584, 589, 590, 614, 813, 828, 842, 843, 855–858, 861, 869–871, 875, 911, 915–917, 924, 925, 928, 936, 937, 942–944, 952, 957, 960
Tufa, 118–119, 121, 143, 144, 243, 322, 331, 393, 489, 502, 748, 800, 818, 819, 852, 853, 862, 893, 916–918, 925, 926
Tufa chimneys, 851, 852, 855, 858

V

Vegetation, 7, 77, 78, 90, 96, 103–105, 111, 117, 177–180, 242, 243, 245, 267–270, 272, 277, 324–325, 336, 349, 371, 375–376, 402, 405, 418, 425–428, 447, 468–469, 507, 535–536, 546, 559, 595, 598, 602, 614, 658, 661, 664, 729, 744, 746, 809, 820, 827, 848, 910, 913, 914, 928, 940, 953, 957

Volcanic, 7, 18, 33, 85, 129, 163, 221, 260, 306, 363, 417, 463, 530, 578, 631, 693, 719, 767, 794, 842, 883, 909, 951

Volcanism, 34, 36, 37, 39, 54–59, 61–63, 111, 116, 171, 260, 368, 467, 468, 514, 638, 640, 660, 664, 666–667, 705, 710, 730, 739, 769, 778, 781, 805, 840, 846, 910, 921, 926, 958–960

W

Water abstraction, 282, 378, 379, 424, 425, 439, 598, 953, 954, 958

Water chemistry, 6, 15, 114, 129, 132, 200, 272, 276, 420, 431–432, 476, 515, 534, 540, 544, 658, 734, 909, 922, 924–925, 957, 960

Weathering, 7, 27, 57, 77, 85, 93–98, 138, 142, 242, 384, 430, 440, 476, 482, 546, 725, 798, 809, 814, 818, 828, 872, 874, 909, 910, 914, 920–922, 924, 926, 927, 937, 940, 943, 960

Wetlands, x, xi, 24, 25, 27, 108, 111, 117, 120, 180, 221, 223, 228, 236, 311, 318, 323, 325, 353, 364, 369, 395, 437, 469, 530, 534, 546, 549, 555, 577–620, 651, 664, 693, 705, 733, 748, 802, 805, 809, 811, 812, 814, 815, 818, 820, 822–824, 827, 923, 935, 947, 956

Wildlife, xi, 7, 12, 14, 77, 106, 107, 177–180, 258, 267–270, 324–325, 333, 363, 375–376, 419, 426–428, 468–469, 507, 535–537, 577, 617, 779, 809, 947, 951, 952, 954, 956, 958, 961, 963

Y

Yala Swamp, 615, 616

Z

Zeolites, 26, 93, 116, 138, 140, 151, 152, 154, 242, 331, 339, 341, 348, 350, 384, 463, 479, 495, 499, 500, 514, 725, 729, 748, 855, 859–861, 866, 922, 935, 944–945

Zooplankton, 4, 148, 149, 152, 187–188, 279–282, 333, 389, 390, 437–438, 484, 547, 550–551, 581